Student's Solutions Manual

Algebra and Trigonometry

Keedy/Bittinger

5th Edition

Judith A. Penna

ADDISON-WESLEY PUBLISHING COMPANY
Reading, Massachusetts • Menlo Park, California • New York
Don Mills, Ontario • Wokingham, England • Amsterdam • Bonn
Sydney • Singapore • Tokyo • Madrid • San Juan

Reproduced by Addison-Wesley from camera-ready copy supplied by the author.

ISBN 0-201-14997-4
ABCDEFGHIJ-BA-943210

TABLE OF CONTENTS

Special thanks are extended to Patsy Hammond and Julie Stephenson for their typing and to Pam Smith for her proofreading. Their efficiency, patience, and good humor made the authors' work much easier.

Exercise Set 1.1

1. - 6. Consider the numbers

$-6, 0, 3, -\frac{1}{2}, \sqrt{3}, -2, -\sqrt{7}, \sqrt[3]{2}, \frac{5}{8}, 14, -\frac{9}{4},$

$8.53, 9\frac{1}{2}$.

1. Natural numbers: 3, 14

2. Whole numbers: 0, 3, 14

3. Irrational numbers: $\sqrt{3}, -\sqrt{7}, \sqrt[3]{2}$

4. Rational numbers: $-6, 0, 3, -\frac{1}{2}, -2, \frac{5}{8}, 14, -\frac{9}{4},$ $8.53, 9\frac{1}{2}$

5. Integers: -6, 0, 3, -2, 14

6. Real numbers: All of them

7. $-\frac{6}{5}$ can be expressed as $\frac{-6}{5}$, a quotient of integers. Thus, $-\frac{6}{5}$ is rational.

8. Rational

9. -9.032 is an ending decimal. Thus, -9.032 is rational.

10. Rational

11. $4.\overline{516}$ is a repeating decimal. Thus, $4.\overline{516}$ is rational.

12. Rational

13. 4.303003000300003... has a pattern but not a pattern formed by a repeating block of digits. The numeral is an unending nonrepeating decimal and thus irrational.

14. Irrational

15. $\sqrt{6}$ is irrational because 6 is not a perfect square.

16. Irrational

17. $-\sqrt{14}$ is irrational because 14 is not a perfect square.

18. Irrational

19. $\sqrt{49}$ is rational because 49 is a perfect square.

$\sqrt{49} = \sqrt{7^2} = 7 = \frac{7}{1}$ (Quotient of integers)

20. Rational

21. $\sqrt[3]{5}$ is irrational because 5 is not a perfect cube.

22. Irrational

23. "-x" means "find the additive inverse of x." If x = -7, then we are finding the additive inverse of -7.

$-(-7) = 7$ (The additive inverse of -7 is 7.)

"$-1\cdot x$" means "find negative one times x." If x = -7, then we are finding -1 times -7.

$-1\cdot(-7) = 7$ (The product of two negative numbers is positive.)

24. $\frac{10}{3}, \frac{10}{3}$

25. "-x" means "find the additive inverse of x." If x = 57, then we are finding the additive inverse of 57.

$-(57) = -57$ (The additive inverse of 57 is -57.)

"$-1\cdot x$" means "find negative one times x." If x = 57, then we are finding -1 times 57.

$-1\cdot 57 = -57$ (The product of a negative number and a positive number is negative.)

26. $-\frac{13}{14}, -\frac{13}{14}$

27. $-(x^2 - 5x + 3)$

$= -(12^2 - 5\cdot 12 + 3)$ (Substituting 12 for x)

$= -(144 - 60 + 3)$

$= -87$

28. -107

29. $-(7 - y)$

$= -[7 - (-9)]$ (Substituting -9 for y)

$= -[7 + 9]$

$= -16$

30. 12

31. $-3.1 + (-7.2)$

The sum of two negative real numbers is a negative real number.

$-3.1 + (-7.2) = -10.3$

32. -416

33. $\frac{9}{2} + \left(-\frac{3}{5}\right)$

The positive number $\left(\frac{9}{2}\right)$ is farther from 0 on a number line, so the answer is positive.

$\frac{9}{2} + \left(-\frac{3}{5}\right) = \frac{39}{10}$

34. -20

35. $-7(-4) = 28$
The product of two negative numbers is positive.

36. 12

37. $(-8.2) \times 6 = -49.2$
The product of a negative number and a positive number is negative.

38. -48

39. $-7(-2)(-3)(-5)$
$= 14 \cdot 15$ (The product of two negative numbers is positive.)
$= 210$

40. 16.33

41. $-\frac{14}{3}\left(-\frac{17}{5}\right)\left(-\frac{21}{2}\right)$
$= -\frac{17}{5}\left[\left(-\frac{14}{3}\right)\left(-\frac{21}{2}\right)\right]$ (Using associativity and commutativity)
$= -\frac{17}{5}\left[\left(-\frac{14}{2}\right)\left(-\frac{21}{3}\right)\right]$
$= -\frac{17}{5}[(-7)(-7)]$
$= -\frac{17}{5} \cdot 49$ (The product of two negative numbers is positive.)
$= -\frac{833}{5}$ (The product of a negative number and a positive number is negative.)

42. $\frac{598}{5}$

43. $\frac{-20}{-4} = 5$ (The quotient of two negative numbers is positive.)

44. -7

45. $\frac{-10}{70}$
$= -\frac{10}{70}$ (The quotient of a negative number and a positive number is negative.)
$= -\frac{1}{7}$ (Simplifying)

46. -5

47. $\frac{2}{7} \div \left(-\frac{14}{3}\right)$
$= \frac{2}{7} \cdot \left(-\frac{3}{14}\right)$ (Multiplying by the reciprocal)
$= \frac{2}{7} \cdot \left(-\frac{3}{2 \cdot 7}\right)$
$= \frac{2}{2} \cdot \left(-\frac{3}{7 \cdot 7}\right)$
$= -\frac{3}{49}$

48. $\frac{7}{10}$

49. $-\frac{10}{3} \div \left(-\frac{2}{15}\right)$
$= -\frac{10}{3} \cdot \left(-\frac{15}{2}\right)$ (Multiplying by the reciprocal)
$= \frac{150}{6}$
$= 25$

50. 8

51. $11 - 15$
$= 11 + (-15)$ $[a - b = a + (-b)]$
$= -4$

52. -29

53. $12 - (-6)$
$= 12 + 6$ $[a - b = a + (-b)]$
$= 18$

54. -9

55. $15.8 - 27.4$
$= 15.8 + (-27.4)$ $[a - b = a + (-b)]$
$= -11.6$

56. -34.8

57. $-\frac{21}{4} - \left(-\frac{7}{8}\right)$
$= -\frac{21}{4} + \frac{7}{8}$ $[a - b = a + (-b)]$
$= -\frac{42}{8} + \frac{7}{8}$
$= -\frac{35}{8}$

58. $\frac{59}{12}$

59. a) $(1.4)^2 = 1.96$
$(1.41)^2 = 1.9881$
$(1.414)^2 = 1.999396$
$(1.4142)^2 = 1.999962$
$(1.41421)^2 = 1.999990$
b) $\sqrt{2}$

60. a) 9.261000
9.938375
9.993948
9.999517
9.999935
b) $\sqrt[3]{10}$

61. $k + 0 = k$ Additive Identity Property

62. Commutativity of Multiplication

63. $-1(x + y) = (-1x) + (-1y)$
Distributivity of Multiplication over Addition

64. Associativity of Addition

65. $c + d = d + c$ Commutativity of Addition

66. Multiplicative Identity Property

67. $4(xy) = (4x)y$ Associativity of Multiplication

68. Distributivity of Multiplication over Addition

69. $y\left(\frac{1}{y}\right) = 1,\ y \neq 0$ Multiplicative Inverse Property

70. Additive Inverse Property

71. $a + (b + c) = a + (c + b)$
Commutativity of Addition

72. Commutativity of Multiplication

73. $a(b + c) = a(c + b)$
Commutativity of Addition

74. Distributivity of Multiplication over Addition

75. $7 - 5 \neq 5 - 7$
$7 - 5 = 2$
$5 - 7 = -2$
$2 \neq -2$

76. $16 \div 8 \neq 8 \div 16$
$16 \div 8 = 2$
$8 \div 16 = \frac{1}{2}$

77. $16 \div (4 \div 2) \neq (16 \div 4) \div 2$
$16 \div (4 \div 2) = 16 \div 2 = 8$
$(16 \div 4) \div 2 = 4 \div 2 = 2$
$8 \neq 2$

78. $7 - (5 - 2) \neq (7 - 5) - 2$
$7 - (5 - 2) = 4$
$(7 - 5) - 2 = 0$

79. $\pi = 3.1415926535...$
$3.14 = 3.14$
$\frac{22}{7} = 3.\overline{142857}$
$\frac{3927}{1250} = 3.1416$
Of these three, $\frac{3927}{1250}$ is the best approximation to π.

80. Third

81. Let $n = 0.\overline{9} = 0.999...$
Then $10n = 9.999...$
$n = 0.999$
$9n = 9$ Subtracting
$n = 1.$

82. $\frac{371}{99}$

83. Let $n = 18.3\overline{245} = 18.3245245...$
Then $10{,}000n = 183{,}245.245...$
$10n = 183.245...$
$9990n = 183{,}062$
$n = \frac{183{,}062}{9990}.$

84. $\frac{12{,}346{,}418}{999{,}900}$

85. Prove: $(b + c)a = ba + ca$
Proof: $(b + c)a = a(b + c)$, by commutativity ($\times$)
$= ab + ac$, by the distributive property
$= ba + ca$, by commutativity ($\times$).

86. Let a be a positive number. Then by the definition of subtraction,
$a - 0 = c$ if and only if $0 + c = a$.
Now $0 + c = c$ (identity property of addition), so $c = a$ and $a - 0 = a$ where a is a positive number. Then $a - 0$ is positive, so
$0 < a$ by the definition of <
and $a > 0$ by the definition of >.

Exercise Set 1.2

1. $2^3 \cdot 2^{-4} = 2^{3+(-4)} = 2^{-1} = \frac{1}{2}$

2. $\frac{1}{3}$

3. $b^2 \cdot b^{-2} = b^{2+(-2)} = b^0 = 1$

4. 1

5. $4^2 \cdot 4^{-5} \cdot 4^6 = 4^{2+(-5)+6} = 4^3$

6. 5^3

7. $2x^3 \cdot 3x^2 = 2 \cdot 3 \cdot x^{3+2} = 6x^5$

8. $12y^7$

9. $(5a^2b)(3a^{-3}b^4) = 5 \cdot 3 \cdot a^{2+(-3)} \cdot b^{1+4} = 15a^{-1}b^5$
$= \frac{15b^5}{a}$

10. $\dfrac{12y^7}{x^3}$

11. $(2x)^3(3x)^2 = 2^3x^3\cdot 3^2x^2 = 8\cdot 9\cdot x^{3+2} = 72x^5$

12. $432y^5$

13. $(6x^5y^{-2}z^3)(-3x^2y^3z^{-2})$
$= 6\cdot(-3)\cdot x^{5+2}y^{-2+3}z^{3+(-2)}$
$= -18x^7yz$

14. $-10x^6yz$

15. $\dfrac{b^{40}}{b^{37}} = b^{40-37} = b^3$

16. a^7

17. $\dfrac{x^2y^{-2}}{x^{-1}y} = x^{2-(-1)}y^{-2-1} = x^3y^{-3} = \dfrac{x^3}{y^3}$

18. $\dfrac{x^4}{y^5}$

19. $\dfrac{9a^2}{(-3a)^2} = \dfrac{9a^2}{9a^2} = 1$

20. 1

21. $\dfrac{24a^5b^3}{8a^4b} = 3a^{5-4}b^{3-1} = 3ab^2$

22. $6x^3y^2$

23. $\dfrac{12x^2y^3z^{-2}}{21xy^2z^3} = \dfrac{12}{21}x^{2-1}y^{3-2}z^{-2-3} = \dfrac{4}{7}xyz^{-5} = \dfrac{4xy}{7z^5}$

24. $\dfrac{x^2y^3}{3z^8}$

25. $(2ab^2)^3 = 2^3a^3b^{2\cdot 3} = 8a^3b^6$

26. $16x^2y^6$

27. $(-2x^3)^4 = (-2)^4x^{3\cdot 4} = 16x^{12}$

28. $81x^8$

29. $-(2x^3)^4 = -2^4x^{3\cdot 4} = -16x^{12}$

30. $-81x^8$

31. $(6a^2b^3c)^2 = 6^2a^{2\cdot 2}b^{3\cdot 2}c^2 = 36a^4b^6c^2$

32. $25x^6y^4z^2$

33. $(-5c^{-1}d^{-2})^{-2} = (-5)^{-2}c^{-1(-2)}d^{-2(-2)} = \dfrac{1}{25}c^2d^4$

$$\left[(-5)^{-2} = \frac{1}{(-5)^2} = \frac{1}{25}\right]$$

34. $\dfrac{1}{16}x^2z^4$

35. $\dfrac{4^{-2} + 2^{-4}}{8^{-1}} = \dfrac{\frac{1}{4^2} + \frac{1}{2^4}}{\frac{1}{8}} = \dfrac{\frac{1}{16} + \frac{1}{16}}{\frac{1}{8}} = \dfrac{\frac{2}{16}}{\frac{1}{8}} = \dfrac{\frac{1}{8}}{\frac{1}{8}}$
$= 1$

36. $\dfrac{119}{72}$

37. $\dfrac{(-2)^4 + (-4)^2}{(-1)^8} = \dfrac{16 + 16}{1} = \dfrac{32}{1} = 32$

38. 25

39. $\dfrac{(3a^2b^{-2}c^4)^3}{(2a^{-1}b^2c^{-3})^2}$
$= \dfrac{3^3a^6b^{-6}c^{12}}{2^2a^{-2}b^4c^{-6}}$
$= \dfrac{27}{4}a^{6-(-2)}b^{-6-4}c^{12-(-6)}$
$= \dfrac{27}{4}a^8b^{-10}c^{18}$
$= \dfrac{27a^8c^{18}}{4b^{10}}$

40. $\dfrac{8a^{11}c^{19}}{9b^3}$

41. $\dfrac{6^{-2}x^{-3}y^2}{3^{-3}x^{-4}y} = \dfrac{3^3}{6^2}x^{-3-(-4)}y^{2-1} = \dfrac{27}{36}xy = \dfrac{3}{4}xy$

42. $\dfrac{8}{25}xy^2$

43. $\left[\dfrac{24a^{10}b^{-8}c^7}{3a^6b^{-3}c^5}\right]^5 = \left[\dfrac{8a^4c^2}{b^5}\right]^5 = \dfrac{8^5a^{20}c^{10}}{b^{25}} =$
$\dfrac{32{,}768a^{20}c^{10}}{b^{25}}$

44. $\dfrac{q^{80}r^{148}}{625p^{16}}$

45. When $x = 5$, $-x^2 = -5^2 = -25$
and $(-x)^2 = (-5)^2 = 25$

46. -49, 49

47. When $x = -1.08$, $-x^2 = -(-1.08)^2 = -1.1664$
and $(-x)^2 = [-(-1.08)]^2 = (1.08)^2 = 1.1664$

48. -3, 3

49. $58{,}000{,}000 = 58{,}000{,}000 \times 10^{-7} \times 10^7$
$= 5.8 \times 10^7$

50. 2.7×10^4

51. $365{,}000 = 365{,}000 \times 10^{-5} \times 10^5$
$= 3.65 \times 10^5$

52. 3.645×10^3

53. $0.0000027 = 0.0000027 \times 10^6 \times 10^{-6}$
$= 2.7 \times 10^{-6}$

54. 6.58×10^{-5}

55. $0.027 = 0.027 \times 10^2 \times 10^{-2}$
$= 2.7 \times 10^{-2}$

56. 3.8×10^{-3}

57. 0.000000000000000000000000000911

We convert mentally. To position the decimal point between the 9 and the first 1 we must move it 28 places. Since the number is small, the exponent is negative. Thus,

$0.000000000000000000000000000911 = 9.11 \times 10^{-28}$

58. 9.3×10^7

59. 3,664,000,000

We convert mentally. To position the decimal point between the 3 and the first 6 we must move it 9 places. Since the number is large the exponent is positive. Thus,

$3{,}664{,}000{,}000 = 3.664 \times 10^9$

60. 10^{-9}

61. $4 \times 10^5 = 400{,}000$

62. 0.0005

63. $6.2 \times 10^{-3} = 0.0062$

64. 7,800,000

65. $7.69 \times 10^{12} = 7{,}690{,}000{,}000{,}000$

66. 0.000000854

67. 5.67 E - 7 = 0.000000567

68. 1,314,000,000,000

69. $9.46 \times 10^{12} = 9{,}460{,}000{,}000{,}000$

70. 0.000066

71. 7.69 E - 8 = 0.0000000769

72. 0.0000000008603

73. 2.567 E 8 = 256,700,000

74. 111,300,000,000

75. The amount of water discharged in one hour is the amount discharged per second times the number of seconds in one hour.

$4{,}200{,}000 \times 3600$
$= (4.2 \times 10^6) \times (3.6 \times 10^3)$ Using scientific notation
$= (4.2 \times 3.6) \times (10^6 \times 10^3)$
$= 15.12 \times 10^9$ Multiplying
$= (1.512 \times 10) \times 10^9$
$= 1.512 \times 10^{10}$ Scientific notation

In one hour the discharge of water is 1.512×10^{10} cubic feet.

76. 1.095×10^9 gallons

77. We begin by expressing one billionth as a fraction.

$\frac{1}{1{,}000{,}000{,}000} = \frac{1}{10^9}$ Using scientific notation
$= 10^{-9}$

Scientific notation for 1 nanosecond is 10^{-9} sec.

78. 5.8404×10^8 mi

79. The distance light travels in 13 weeks is the portion of a year in 13 weeks times the number of miles that light travels in one year. (See Example 23.) We will use 52 weeks for 1 year.

$0.25 \times (5.88 \times 10^{12})$
$= (2.5 \times 10^{-1}) \times (5.88 \times 10^{12})$ Using scientific notation
$= (2.5 \times 5.88) \times (10^{-1} \times 10^{12})$
$= 14.7 \times 10^{11}$
$= (1.47 \times 10) \times 10^{11}$
$= 1.47 \times 10^{12}$

Light travels 1.47×10^{12} miles in 13 weeks.

80. 2.4×10^{-2} m^3, or 2.4×10^7 mm^3

81. $3\cdot 2 + 4\cdot 2^2 - 6(3 - 1)$
$= 3\cdot 2 + 4\cdot 2^2 - 6\cdot 2$ Working inside parentheses
$= 3\cdot 2 + 4\cdot 4 - 6\cdot 2$ Evaluating 2^2
$= 6 + 16 - 12$ Multiplying
$= 22 - 12$ Adding in order
$= 10$ from left to right

82. 18

83. $\frac{4(8 - 6)^2 + 4\cdot3 - 2\cdot8}{3^1 + 19^0}$

$= \frac{4\cdot2^2 + 4\cdot3 - 2\cdot8}{3 + 1}$ Calculating in the numerator and in the denominator

$= \frac{4\cdot4 + 4\cdot3 - 2\cdot8}{4}$

$= \frac{16 + 12 - 16}{4}$

$= \frac{28 - 16}{4}$

$= \frac{12}{4}$

$= 3$

84. -5

85. $16 \div 4\cdot4 \div 2\cdot256$

$= 4\cdot4 \div 2\cdot256$ Multiplying and dividing in order from left to right

$= 16 \div 2\cdot256$

$= 8\cdot256$

$= 2048$

86. 2

87. $\left[\frac{5^2}{8} + 5(5) - \frac{5^3}{12}\right] - \left[\frac{(-4)^2}{8} + 5(-4) - \frac{(-4)^3}{12}\right]$

$= \left[\frac{25}{8} + 5(5) - \frac{125}{12}\right] - \left[\frac{16}{8} + 5(-4) - \frac{-64}{12}\right]$

$= \left[\frac{25}{8} + 25 - \frac{125}{12}\right] - \left[2 - 20 + \frac{16}{3}\right]$

$= \frac{425}{24} - \left[-\frac{38}{3}\right]$

$= \frac{729}{24}$

$= \frac{243}{8}$

88. $\frac{9}{8}$

89. The distance of 12 from 0 is 12, so $|12| = 12$.

90. 2.56

91. The distance of -47 from 0 is 47, so $|-47| = 47$.

92. 0

93. $|-7a| = |-7|\cdot|a| = 7|a|$

94. $10|m||n|$

95. $|-8x^6| = |-8|\cdot|x^6| = 8x^6$

96. $5x^4y^8$

97. $|9xy| = |9|\cdot|xy| = 9|xy|$ or $9|x||y|$

98. y^4

99. $|3a^2b| = |3|\cdot|a^2|\cdot|b| = 3a^2|b|$

100. $\frac{4|a|}{b^2}$

101. $(x^t\cdot x^{3t})^2 = (x^{4t})^2 = x^{4t\cdot2} = x^{8t}$

102. 1

103. $(t^{a+x}\cdot t^{x-a})^4 = (t^{2x})^4 = t^{2x\cdot4} = t^{8x}$

104. m^{2yt}

105. $(x^ay^b\cdot x^by^a)^c = (x^{a+b}y^{a+b})^c = x^{ac+bc}y^{ac+bc}$
$= (xy)^{ac+bc}$

106. $(mn)^{x^2}$

107. $\left[\frac{(3x^ay^b)^3}{(-3x^ay^b)^2}\right]^2 = \left[\frac{27x^{3a}y^{3b}}{9x^{2a}y^{2b}}\right]^2$

$= [3x^ay^b]^2$

$= 9x^{2a}y^{2b}$

108. $\frac{x^{6r}}{y^{18t}}$

109. $M = P\left[\frac{\frac{i}{12}\left(1 + \frac{i}{12}\right)^n}{\left(1 + \frac{i}{12}\right)^n - 1}\right]$

$M = 78,000\left[\frac{\frac{0.1075}{12}\left(1 + \frac{0.1075}{12}\right)^{300}}{\left(1 + \frac{0.1075}{12}\right)^{300} - 1}\right]$ (Substituting)

$= 78,000 \cdot \frac{0.008958333(14.52009644)}{14.52009644 - 1}$

$= 78,000(0.009620927)$

$= \$750.43$

110. \$791.88, \$728.12

111. $x^4(x^3)^2 = x^4\cdot x^5 = x^9$

Exponents were added instead of multiplied.

$x^4(x^3)^2 = x^4\cdot x^6 = x^{10}$ (Correct)

112. Exponents were added instead of subtracted; $\frac{x^6}{y^{12}}$

113. $(2x^{-4}\ y^6\ z^3)^3 = 6x^{-1}\ y^3\ z^6$
① ② ③ ④ ① ② ③ ④

Four errors were made.

① $2\cdot 3$ instead of 2^3

② Exponents were added instead of multiplied.

③ Exponents were subtracted instead of multiplied.

④ Exponents were added instead of multiplied.

$(2x^{-4}y^6z^3)^3 = 2^3x^{-4\cdot 3}y^{6\cdot 3}z^{3\cdot 3} = 8x^{-12}y^{18}z^9$

(Correct)

Exercise Set 1.3

1. $-11x^4 - x^3 + x^2 + 3x - 9$

Term	Degree	
$-11x^4$	4	
$-x^3$	3	
x^2	2	
$3x$	1	$(3x = 3x^1)$
-9	0	$(-9 = -9x^0)$

The degree of a polynomial is the same as that of its term of highest degree.

The term of highest degree is $-11x^4$. Thus, the degree of the polynomial is 4.

2. 3, 2, 1, 0; 3

3. $y^3 + 2y^6 + x^2y^4 - 8$

Term	Degree	
y^3	3	
$2y^6$	6	
x^2y^4	6	$(2 + 4 = 6)$
-8	0	$(-8 = -8x^0)$

The terms of highest degree are $2y^6$ and x^2y^4. Thus, the degree of the polynomial is 6.

4. 2, 5, 7, 0; 7

5. $a^5 + 4a^2b^4 + 6ab + 4a - 3$

Term	Degree	
a^5	5	
$4a^2b^4$	6	$(2 + 4 = 6)$
$6ab$	2	$(1 + 1 = 2)$
$4a$	1	$(4a = 4a^1)$
-3	0	$(-3 = -3a^0)$

The term of highest degree is $4a^2b^4$. Thus, the degree of the polynomial is 6.

6. 6, 8, 4, 2, 0; 8

7. $(5x^2y - 2xy^2 + 3xy - 5)+(-2x^2y - 3xy^2 + 4xy + 7)$
$= (5 - 2)x^2y + (-2 - 3)xy^2 + (3 + 4)xy + (-5 + 7)$
$= 3x^2y - 5xy^2 + 7xy + 2$

8. $2x^2y - 7xy^2 + 8xy + 5$

9. $(-3pq^2 - 5p^2q + 4pq + 3)+(-7pq^2 + 3pq - 4p + 2q)$
$= (-3 - 7)pq^2 - 5p^2q + (4 + 3)pq - 4p + 2q + 3$
$= -10pq^2 - 5p^2q + 7pq - 4p + 2q + 3$

10. $-9pq^2 - 3p^2q + 11pq - 6p + 4q + 5$

11. $(2x+3y+z-7) + (4x-2y-z+8) + (-3x+y-2z-4)$
$= (2+4-3)x + (3-2+1)y + (1-1-2)z + (-7+8-4)$
$= 3x + 2y - 2z - 3$

12. $7x^2 + 12xy - 2x - y - 9$

13. $\left[7x\sqrt{y} - 3y\sqrt{x} + \frac{1}{5}\right] + \left[-2x\sqrt{y} - y\sqrt{x} - \frac{3}{5}\right]$
$= (7 - 2)x\sqrt{y} + (-3 - 1)y\sqrt{x} + \left[\frac{1}{5} - \frac{3}{5}\right]$
$= 5x\sqrt{y} - 4y\sqrt{x} - \frac{2}{5}$

14. $7x\sqrt{y} - 5y\sqrt{x} + 1$

15. The additive inverse of $5x^3 - 7x^2 + 3x - 6$ can be symbolized as
$-(5x^3 - 7x^2 + 3x - 6)$ (With parentheses)
or
$-5x^3 + 7x^2 - 3x + 6$ (Without parentheses)

16. $-(-4y^4 + 7y^2 - 2y - 1)$, $4y^4 - 7y^2 + 2y + 1$

17. $(3x^2 - 2x - x^3 + 2) - (5x^2 - 8x - x^3 + 4)$
$= (3x^2 - 2x - x^3 + 2) + (-5x^2 + 8x + x^3 - 4)$
$= (3 - 5)x^2 + (-2 + 8)x + (-1 + 1)x^3 + (2 - 4)$
$= -2x^2 + 6x - 2$

18. $-4x^2 + 8xy - 5y^2 + 3$

19. $(4a - 2b - c + 3d) - (-2a + 3b + c - d)$
$= (4a - 2b - c + 3d) + (2a - 3b - c + d)$
$= (4 + 2)a + (-2 - 3)b + (-1 - 1)c + (3 + 1)d$
$= 6a - 5b - 2c + 4d$

20. $8a - 8b - 2c + 6d$

21. $(x^4 - 3x^2 + 4x) - (3x^3 + x^2 - 5x + 3)$
$= (x^4 - 3x^2 + 4x) + (-3x^3 - x^2 + 5x - 3)$
$= x^4 - 3x^3 + (-3 - 1)x^2 + (4 + 5)x - 3$
$= x^4 - 3x^3 - 4x^2 + 9x - 3$

22. $2x^4 - 5x^3 - 7x^2 + 10x - 5$

23. $(7x\sqrt{y} - 4y\sqrt{x} + 7.5) - (-2x\sqrt{y} - y\sqrt{x} - 1.6)$
$= (7x\sqrt{y} - 4y\sqrt{x} + 7.5) + (2x\sqrt{y} + y\sqrt{x} + 1.6)$
$= (7 + 2)x\sqrt{y} + (-4 + 1)y\sqrt{x} + (7.5 + 1.6)$
$= 9x\sqrt{y} - 3y\sqrt{x} + 9.1$

24. $13x\sqrt{y} - 5y\sqrt{x} + \frac{5}{3}$

25. $(0.565p^2q - 2.167pq^2 + 16.02pq - 17.1)$
$+ (-1.612p^2q - 0.312pq^2 - 7.141pq - 87.044)$
$= (0.565 - 1.612)p^2q + (-2.167 - 0.312)pq^2 +$
$(16.02 - 7.141)pq + (-17.1 - 87.044)$
$= -1.047p^2q - 2.479pq^2 + 8.879pq - 104.144$

26. $2859.6xy^{-2} - 6153.8xy + 7243.4\sqrt{xy} - 10{,}259.12$

Exercise Set 1.4

1.
$$\begin{array}{rl} 2x^2 + 4x + 16 & \\ 3x - 4 & \\ \hline -8x^2 - 16x - 64 & \text{(Multiplying by -4)} \\ 6x^3 + 12x^2 + 48x \quad & \text{(Multiplying by 3x)} \\ \hline 6x^3 + 4x^2 + 32x - 64 & \text{(Adding)} \end{array}$$

2. $6y^3 + 3y^2 + 9y + 27$

3.
$$\begin{array}{r} 4a^2b - 2ab + 3b^2 \\ ab - 2b + 1 \\ \hline 4a^2b - 2ab + 3b^2 \\ -8a^2b^2 + 4ab^2 - 6b^3 \qquad\qquad \\ 4a^3b^2 - 2a^2b^2 \qquad\qquad\qquad\qquad\qquad + 3ab^2 \\ \hline 4a^3b^2 - 10a^2b^2 + 4ab^2 - 6b^3 + 4a^2b - 2ab + 3b^2 + 3ab^3 \end{array}$$

4. $2x^4 - x^2y^2 - 4x^3y - 2y^4 + 3xy^3$

5.
$$\begin{array}{l} a^2 + ab + b^2 \\ \qquad a - b \\ \hline a^3 + a^2b + ab^2 \\ \quad - a^2b - ab^2 - b^3 \\ \hline a^3 \qquad\qquad\quad - b^3 \end{array}$$

The product is $a^3 - b^3$.

6. $t^3 + 1$

7. $(2x + 3y)(2x + y)$
$= 4x^2 + 2xy + 6xy + 3y^2$ (FOIL)
$= 4x^2 + 8xy + 3y^2$

8. $4a^2 - 8ab + 3b^2$

9. $\left(4x^2 - \frac{1}{2}y\right)\left(3x + \frac{1}{4}y\right)$
$= 12x^3 + x^2y - \frac{3}{2}xy - \frac{1}{8}y^2$ (FOIL)

10. $6y^4 - \frac{1}{2}xy^3 + \frac{3}{5}xy - \frac{1}{20}x^2$

11. $(2p^2q^3 - r^2)(5pq - 2r)$
$= 10p^3q^4 - 4p^2q^3r - 5pqr^2 + 2r^3$ (FOIL)

12. $9y^3 - 3xy^2 - 6y + 2x$

13. $(2x + 3y)^2$
$= (2x)^2 + 2(2x)(3y) + (3y)^2$
$[(A + B)^2 = A^2 + 2AB + B^2]$
$= 4x^2 + 12xy + 9y^2$

14. $25x^2 + 20xy + 4y^2$

15. $(2x^2 - 3y)^2$
$= (2x^2)^2 - 2(2x^2)(3y) + (3y)^2$
$[(A - B)^2 = A^2 - 2AB + B^2]$
$= 4x^4 - 12x^2y + 9y^2$

16. $16x^4 - 40x^2y + 25y^2$

17. $(2x^3 + 3y^2)^2$
$= (2x^3)^2 + 2(2x^3)(3y^2) + (3y^2)^2$
$[(A + B)^2 = A^2 + 2AB + B^2]$
$= 4x^6 + 12x^3y^2 + 9y^4$

18. $25x^6 + 20x^3y^2 + 4y^4$

19. $\left(\frac{1}{2}x^2 - \frac{3}{5}y\right)^2$
$= \left(\frac{1}{2}x^2\right)^2 - 2\left(\frac{1}{2}x^2\right)\left(\frac{3}{5}y\right) + \left(\frac{3}{5}y\right)^2$
$[(A - B)^2 = A^2 - 2AB + B^2]$
$= \frac{1}{4}x^4 - \frac{3}{5}x^2y + \frac{9}{25}y^2$

20. $\frac{1}{16}x^4 - \frac{1}{3}x^2y + \frac{4}{9}y^2$

21. $(0.5x + 0.7y^2)^2$
$= (0.5x)^2 + 2(0.5x)(0.7y^2) + (0.7y^2)^2$
$[(A + B)^2 = A^2 + 2AB + B^2]$
$= 0.25x^2 + 0.7xy^2 + 0.49y^4$

22. $0.09x^2 + 0.48xy^2 + 0.64y^4$

23. $(3x - 2y)(3x + 2y)$
$= (3x)^2 - (2y)^2$ $[(A - B)(A + B) = A^2 - B^2]$
$= 9x^2 - 4y^2$

24. $9x^2 - 25y^2$

25. $(x^2 + yz)(x^2 - yz)$
$= (x^2)^2 - (yz)^2$ $[(A + B)(A - B) = A^2 - B^2]$
$= x^4 - y^2z^2$

26. $4x^4 - 25x^2y^2$

27. $(3x^2 - \sqrt{2})(3x^2 + \sqrt{2})$
$= (3x^2)^2 - (\sqrt{2})^2$ $[(A - B)(A + B) = A^2 - B^2]$
$= 9x^4 - 2$

28. $25x^4 - 3$

29. $(2x + 3y + 4)(2x + 3y - 4)$
$= (2x + 3y)^2 - 4^2$ $[(A + B)(A - B) = A^2 - B^2]$
$= 4x^2 + 12xy + 9y^2 - 16$

30. $25x^2 + 20xy + 4y^2 - 9$

31. $(x^2 + 3y + y^2)(x^2 + 3y - y^2)$
$= (x^2 + 3y)^2 - (y^2)^2$ $[(A + B)(A - B) = A^2 - B^2]$
$= x^4 + 6x^2y + 9y^2 - y^4$

32. $4x^4 + 4x^2y + y^2 - y^4$

33. $(x + 1)(x - 1)(x^2 + 1)$
$= (x^2 - 1)(x^2 + 1)$
$= x^4 - 1$

34. $y^4 - 16$

35. $(2x + y)(2x - y)(4x^2 + y^2)$
$= (4x^2 - y^2)(4x^2 + y^2)$
$= 16x^4 - y^4$

36. $625x^4 - y^4$

37. $(0.051x + 0.04y)^2$
$= (0.051x)^2 + 2(0.051x)(0.04y) + (0.04y)^2$
$= 0.002601x^2 + 0.00408xy + 0.0016y^2$

38. $1.065024x^2 - 5.184768xy + 6.310144y^2$

39. $(37.86x + 1.42)(65.03x - 27.4)$
$= 2462.0358x^2 - 1037.364x + 92.3426x - 38.908$
$= 2462.0358x^2 - 945.0214x - 38.908$

40. $169.625105x^2 - 711.87827x - 546.525$

41. $(y + 5)^3$
$= y^3 + 3\cdot y^2\cdot 5 + 3\cdot y\cdot 5^2 + 5^3$
$= y^3 + 15y^2 + 75y + 125$

42. $t^3 - 21t^2 + 147t - 343$

43. $(m^2 - 2n)^3$
$= [m^2 + (-2n)]^3$
$= (m^2)^3 + 3(m^2)^2(-2n) + 3(m^2)(-2n)^2 + (-2n)^3$
$= m^6 - 6m^4n + 12m^2n^2 - 8n^3$

44. $27t^6 + 108t^4 + 144t^2 + 64$

45. $(\sqrt{2}\,x^2 - y^2)(\sqrt{2}\,x - 2y)$
$= 2x^3 - 2\sqrt{2}\,x^2y - \sqrt{2}\,xy^2 + 2y^3$ (FOIL)

46. $3y^3 - \sqrt{3}\,xy^2 - 2\sqrt{3}\,y + 2x$

47. $(a^n + b^n)(a^n - b^n)$
$= (a^n)^2 - (b^n)^2$
$= a^{2n} - b^{2n}$

48. $t^{2a} - 3t^a - 28$

49. $(x^m - t^n)^3$
$= [x^m + (-t^n)]^3$
$= (x^m)^3 + 3(x^m)^2(-t^n) + 3(x^m)(-t^n)^2 + (-t^n)^3$
$= x^{3m} - 3x^{2m}t^n + 3x^mt^{2n} - t^{3n}$

50. $y^{3n+3}z^{n+3} - 4y^4z^{3n}$

51. $(x - 1)(x^2 + x + 1)(x^3 + 1)$
$= (x^3 - 1)(x^3 + 1)$
$= (x^3)^2 - 1^2$
$= x^6 - 1$

52. $a^{2n} + 2a^nb^n + b^{2n}$

53. $[(2x - 1)^2 - 1]^2$
$= [4x^2 - 4x + 1 - 1]^2$
$= [4x^2 - 4x]^2$
$= (4x^2)^2 - 2(4x^2)(4x) + (4x)^2$
$= 16x^4 - 32x^3 + 16x^2$

54. $-a^4 - 2a^3b + 25a^2 + 2ab^3 + b^4 - 25b^2$

55. $(x^{a-b})^{a+b}$
$= x^{(a-b)(a+b)}$
$= x^{a^2-b^2}$

56. $t^{2m^2+2n^2}$

57. $(a + b + c)^2$
$= (a + b + c)(a + b + c)$
$= a^2 + ab + ac + ba + b^2 + bc + ca + cb + c^2$
$= a^2 + b^2 + c^2 + 2ab + 2ac + 2bc$

58. $a^3 + 3a^2b + 3ab^2 + b^3 + 3a^2c + 6abc + 3b^2c + 3ac^2 + 3bc^2 + c^3$

59. $(a + b)^4$
$= (a + b)(a + b)^3$
$= (a + b)(a^3 + 3a^2b + 3ab^2 + b^3)$
$= a^4 + 3a^3b + 3a^2b^2 + ab^3 + a^3b + 3a^2b^2 + 3ab^3 + b^4$
$= a^4 + 4a^3b + 6a^2b^2 + 4ab^3 + b^4$

60. $x^5 - y^5$

61. $(m + t)(m^4 - m^3t + m^2t^2 - mt^3 + t^4)$
$= m^5 - m^4t + m^3t^2 - m^2t^3 + mt^4 +$
$m^4t - m^3t^2 + m^2t^3 - mt^4 + t^5$
$= m^5 + t^5$

62. $a^8 - b^8$

63. $(3a + b)^2 = 3a^2 + b^2$

The first term is $9a^2$, not $3a^2$.
The middle term, $2\cdot 3a\cdot b$, is missing.
$(3a + b)^2 = (3a)^2 + 2\cdot 3a\cdot b + b^2$
$= 9a^2 + 6ab + b^2$ (Correct)

64. The middle term, $-12xy$, is missing; $-9y^2$ should be $+9y^2$; correct answer is $4x^2 - 12xy + 9y^2$.

65. $2x(x + 3) + 4(x^2 - 3)$
$= 2x^2 + 3x + 4x^2 - 3$ (1)
The 3x should be 6x; the 2 and 3 were not multiplied. Similarly, -3 should be -12; the 4 and -3 were not multiplied.
$= 6x^2 + x$
$3x - 3$ is not x since they are not like terms.
$2x(x + 3) + 4(x^2 - 3)$
$= 2x^2 + 6x + 4x^2 - 12$
$= 6x^2 + 6x - 12$ (Correct)

66. In (1) the middle term, $-5ab$, is missing; in going from (1) to (2) the factor 6 has been omitted; correct answer is $6a^2 - 5ab - 6b^2$.

67. $x^{n-1} + x^{n-2}y + x^{n-3}y^2 + \cdots + x^2y^{n-3} + xy^{n-2} + y^{n-1}$
$\underline{\qquad\qquad\qquad\qquad\qquad x - y}$
$-x^{n-1}y - x^{n-2}y^2 - \cdots - x^2y^{n-2} - xy^{n-1} - y^n$
$\underline{x^n + x^{n-1}y + x^{n-2}y^2 + \cdots + x^2y^{n-2} + xy^{n-1}}$
$x^n \qquad\qquad\qquad\qquad -y^n$

The product is $x^n - y^n$.

Exercise Set 1.5

1. $p^2 + 6p + 8$
We look for factors of 8 whose sum is 6. By trial we determine the factors to be 4 and 2.
$p^2 + 6p + 8 = (p + 4)(p + 2)$

2. $(w - 5)(w - 2)$

3. $n^2 + n - 56$
We look for factors of -56 whose sum is 1. By trial we determine the factors to be 8 and -7.
$n^2 + n - 56 = (n + 8)(n - 7)$

4. $(y + 7)(y - 2)$

5. $y^4 - 4y^2 - 21$
Think of this polynomial as $u^2 - 4u - 21$, where we have mentally substituted u for y^2. Then we look for factors of -21 whose sum is -4. By trial we determine the factors to be -7 and 3, so
$u^2 - 4u - 21 = (u - 7)(u + 3).$
Then, substituting y^2 for u, we obtain the factorization of the original trinomial.
$y^4 - 4y^2 - 21 = (y^2 - 7)(y^2 + 3)$

6. $(m^2 - 10)(m^2 + 9)$

7. $18a^2b - 15ab^2$
$= 3ab\cdot 6a - 3ab\cdot 5b$
$= 3ab(6a - 5b)$

8. $4xy(x + 3y)$

9. $a(b - 2) + c(b - 2)$
$= (a + c)(b - 2)$

10. $(a - 2)(x^2 - 3)$

11. $x^3 + 3x^2 + 6x + 18$
$= x^2(x + 3) + 6(x + 3)$
$= (x^2 + 6)(x + 3)$

12. $(x^2 - 6)(3x + 1)$

13. $y^3 - 3y^2 - 4y + 12$
$= y^2(y - 3) - 4(y - 3)$
$= (y^2 - 4)(y - 3)$
$= (y + 2)(y - 2)(y - 3)$

14. $(p + 3)(p - 3)(p - 2)$

15. $9x^2 - 25$
$= (3x)^2 - 5^2$
$= (3x + 5)(3x - 5)$

16. $(4x - 3)(4x + 3)$

17. $4xy^4 - 4xz^2$
$= 4x(y^4 - z^2)$
$= 4x[(y^2)^2 - z^2]$
$= 4x(y^2 + z)(y^2 - z)$

18. $5x(y^2 + z^2)(y + z)(y - z)$

19. $y^2 - 6y + 9$
$= y^2 - 2\cdot y\cdot 3 + 3^2$
$= (y - 3)^2$

20. $(x + 4)^2$

21. $1 - 8x + 16x^2$
$= 1^2 - 2 \cdot 1 \cdot 4x + (4x)^2$
$= (1 - 4x)^2$

22. $(1 + 5x)^2$

23. $4x^2 - 5$
$= (2x)^2 - (\sqrt{5})^2$
$= (2x + \sqrt{5})(2x - \sqrt{5})$

24. $(4x - \sqrt{7})(4x + \sqrt{7})$

25. $x^2y^2 - 14xy + 49$
$= (xy)^2 - 2 \cdot xy \cdot 7 + 7^2$
$= (xy - 7)^2$

26. $(xy - 8)^2$

27. $4ax^2 + 20ax - 56a$
$= 4a(x^2 + 5x - 14)$
$= 4a(x + 7)(x - 2)$

28. $y(7x - 4)(3x + 2)$

29. $a^2 + 2ab + b^2 - c^2$
$= (a + b)^2 - c^2$
$= [(a + b) + c][(a + b) - c]$
$= (a + b + c)(a + b - c)$

30. $(x - y + z)(x - y - z)$

31. $x^2 + 2xy + y^2 - a^2 - 2ab - b^2$
$= (x^2 + 2xy + y^2) - (a^2 + 2ab + b^2)$
$= (x + y)^2 - (a + b)^2$
$= [(x + y) + (a + b)][(x + y) - (a + b)]$
$= (x + y + a + b)(x + y - a - b)$

32. $(r + s + t - v)(r + s - t + v)$

33. $5y^4 - 80x^4$
$= 5(y^4 - 16x^4)$
$= 5(y^2 + 4x^2)(y^2 - 4x^2)$
$= 5(y^2 + 4x^2)(y + 2x)(y - 2x)$

34. $6(y^2 + 4x^2)(y - 2x)(y + 2x)$

35. $x^3 + 8$
$= x^3 + 2^3$
$= (x + 2)(x^2 - 2x + 4)$

36. $(y - 4)(y^2 + 4y + 16)$

37. $3x^3 - \frac{3}{8}$
$= 3\left(x^3 - \frac{1}{8}\right)$
$= 3\left(x - \frac{1}{2}\right)\left(x^2 + \frac{1}{2}x + \frac{1}{4}\right)$

38. $5\left(y + \frac{1}{3}\right)\left(y^2 - \frac{1}{3}y + \frac{1}{9}\right)$

39. $x^3 + 0.001$
$= x^3 + (0.1)^3$
$= (x + 0.1)(x^2 - 0.1x + 0.01)$

40. $(y - 0.5)(y^2 + 0.5y + 0.25)$

41. $3z^3 - 24$
$= 3(z^3 - 8)$
$= 3(z^3 - 2^3)$
$= 3(z - 2)(z^2 + 2z + 4)$

42. $4(t + 3)(t^2 - 3t + 9)$

43. $a^6 - t^6$
$= (a^3)^2 - (t^3)^2$
$= (a^3 + t^3)(a^3 - t^3)$
$= (a + t)(a^2 - at + t^2)(a - t)(a^2 + at + t^2)$

44. $(4m^2 + y^2)(16m^4 - 4m^2y^2 + y^4)$

45. $16a^7b + 54ab^7$
$= 2ab(8a^6 + 27b^6)$
$= 2ab[(2a^2)^3 + (3b^2)^3]$
$= 2ab(2a^2 + 3b^2)(4a^4 - 6a^2b^2 + 9b^4)$

46. $3a^2x(2x - 5a^2)(4x^2 + 10a^2x + 25a^4)$

47. $x^2 - 17.6$
$= x^2 - (\sqrt{17.6})^2$
$= (x + \sqrt{17.6})(x - \sqrt{17.6})$
$= (x + 4.19524)(x - 4.19524)$

48. $(x + 2.8337)(x - 2.8337)$

49. $37x^2 - 14.5y^2$
$= 37(x^2 - 0.391892y^2)$
$= 37[x^2 - (0.626y)^2]$
$= 37(x^2 + 0.626y)(x^2 - 0.626y)$

50. $1.96(x + 2.980y)(x - 2.980y)$

51. $(x + h)^3 - x^3$
$= [(x + h) - x][(x + h)^2 + x(x + h) + x^2]$
$= (x + h - x)(x^2 + 2xh + h^2 + x^2 + xh + x^2)$
$= h(3x^2 + 3xh + h^2)$

52. $0.02(x + 0.005)$

53. $y^4 - 84 + 5y^2$
$= y^4 + 5y^2 - 84$
$= (y^2)^2 + 5y^2 - 84$
$= (y^2 + 12)(y^2 - 7)$

54. $(x^2 + 16)(x^2 - 5)$, or $(x^2 + 16)(x + \sqrt{5})(x - \sqrt{5})$

55. $y^2 - \frac{8}{49} + \frac{2}{7}y$
$= y^2 + \frac{2}{7}y - \frac{8}{49}$
$= \left(y + \frac{4}{7}\right)\left(y - \frac{2}{7}\right)$

56. $\left(x + \frac{4}{5}\right)\left(x - \frac{1}{5}\right)$, or $\frac{1}{25}(5x + 4)(5x - 1)$

57. $t^2 - 0.27 + 0.6t$
$= t^2 + 0.6t - 0.27$
$= (t + 0.9)(t - 0.3)$

58. $(m - 0.1)(m + 0.5)$

59. $x^{2n} + 5x^n - 24$
$= (x^n)^2 + 5x^n - 24$
$= (x^n + 8)(x^n - 3)$

60. $(2x^n - 3)(2x^n + 1)$

61. $x^2 + ax + bx + ab$
$= x(x + a) + b(x + a)$
$= (x + b)(x + a)$

62. $(by + a)(dy + c)$

63. $\frac{1}{4}t^2 - \frac{2}{5}t + \frac{4}{25}$
$= \left(\frac{1}{2}t\right)^2 - 2 \cdot \frac{1}{2}t \cdot \frac{2}{5} + \left(\frac{2}{5}\right)^2$
$= \left(\frac{1}{2}t - \frac{2}{5}\right)^2$

64. $\frac{1}{3}\left(\frac{2}{3}r + \frac{1}{2}s\right)^2$, or $\frac{1}{108}(4r + 3s)^2$,
or $\left(\frac{1}{9}r + \frac{1}{12}s\right)\left(\frac{4}{3}r + s\right)$

65. $25y^{2m} - (x^{2n} - 2x^n + 1)$
$= (5y^m)^2 - (x^n - 1)^2$
$= [5y^m + (x^n - 1)][5y^m - (x^n - 1)]$
$= (5y^m + x^n - 1)(5y^m - x^n + 1)$

66. $2(x^{2a} + 3)(2x^{2a} + 5)$

67. $3x^{3n} - 24y^{3m}$
$= 3(x^{3n} - 8y^{3m})$
$= 3[(x^n)^3 - (2y^m)^3]$
$= 3(x^n - 2y^m)(x^{2n} + 2x^ny^m + 4y^{2m})$

68. $(x^{2a} - t^b)(x^{4a} + x^{2a}t^b + t^{2b})$

69. $(y - 1)^4 - (y - 1)^2$
$= (y - 1)^2[(y - 1)^2 - 1]$
$= (y - 1)^2[y^2 - 2y + 1 - 1]$
$= (y - 1)^2(y^2 - 2y)$
$= y(y - 1)^2(y - 2)$

70. $(x + 1)(x^2 + 1)(x - 1)^3$

71. $5x - 9 = 5 \cdot x - 5 \cdot \frac{9}{5}$
$= 5\left(x - \frac{9}{5}\right)$

72. $\frac{2}{3}\left(x - \frac{21}{2}\right)$

73. a)

$$\begin{array}{r} x^2 - x + 1 \\ x^3 + x^2 - 1 \\ \hline -x^2 + x - 1 \\ x^4 - x^3 + x^2 \\ x^5 - x^4 + x^3 \\ \hline x^5 \qquad\qquad + x - 1 \end{array}$$

The product is $x^5 + x - 1$.

b) We use the result in part (a).
$x^5 + x - 1 = (x^2 - x + 1)(x^3 + x^2 - 1)$

Exercise Set 1.6

1. $\frac{3x - 3}{x(x - 1)}$

Since division by zero is not defined, any number that makes the denominator zero is not a meaningful replacement. When x is replaced by 0 or 1, the denominator is 0. Thus, all real numbers except 0 and 1 are meaningful replacements.

2. All real numbers except -2, 1, and -1

3. We first factor the denominator completely.

$$\frac{7x^2 - 28x + 28}{(x^2 - 4)(x^2 + 3x - 10)} = \frac{7x^2 - 28x + 28}{(x+2)(x-2)(x+5)(x-2)}$$

We see that $x + 2 = 0$ when $x = -2$, $x - 2 = 0$ when $x = 2$, and $x + 5 = 0$ when $x = -5$. Thus, all real numbers except -2, 2, and -5 are meaningful replacements.

4. All real numbers except 0, 3, and -2

5. $\frac{3x - 3}{x(x - 1)} = \frac{3(x - 1)}{x(x - 1)}$
$= \frac{3}{x} \cdot \frac{x - 1}{x - 1}$
$= \frac{3}{x}$

All real numbers except 0 are meaningful replacements in the simplified expression.

6. $\frac{x - 2}{x - 1}$; all real numbers except 1

7. $\frac{7x^2 - 28x + 28}{(x^2 - 4)(x^2 + 3x - 10)} = \frac{7(x^2 - 4x + 4)}{(x+2)(x-2)(x+5)(x-2)}$

$= \frac{7(x - 2)^2}{(x+2)(x+5)(x-2)^2}$

$= \frac{7}{(x + 2)(x + 5)} \cdot \frac{(x - 2)^2}{(x - 2)^2}$

$= \frac{7}{(x + 2)(x + 5)}$

All real numbers except -2 and -5 are meaningful replacements in the simplified expression.

8. $\frac{7x - 3}{x(x - 3)}$; all real numbers except 0 and 3

9. $\frac{25x^2y^2}{10xy^2} = \frac{5xy^2 \cdot 5x}{5xy^2 \cdot 2}$

$= \frac{5xy^2}{5xy^2} \cdot \frac{5x}{2}$

$= \frac{5x}{2}$

All real numbers are meaningful replacements in the simplified expression.

10. $\frac{x - 2}{x + 3}$; all real numbers except -3

11. $\frac{x^2 - 3x + 2}{x^2 + x - 2} = \frac{(x - 2)(x - 1)}{(x + 2)(x - 1)}$

$= \frac{x - 2}{x + 2} \cdot \frac{x - 1}{x - 1}$

$= \frac{x - 2}{x + 2}$

All real numbers except -2 are meaningful replacements in the simplified expression.

12. $\frac{1}{(a - 2)(a + 5)}$; all real numbers except 2 and -5

13. $\frac{x^2 - y^2}{(x - y)^2} \cdot \frac{1}{x + y}$

$= \frac{(x + y)(x - y) \cdot 1}{(x - y)(x - y)(x + y)}$

$= \frac{1}{x - y}$

14. 1

15. $\frac{x^2 - 2x - 35}{2x^3 - 3x^2} \cdot \frac{4x^3 - 9x}{7x - 49}$

$= \frac{(x - 7)(x + 5)(x)(2x + 3)(2x - 3)}{x^2(2x - 3)(7)(x - 7)}$

$= \frac{(x + 5)(2x + 3)}{7x}$

16. $\frac{(x - 5)(3x + 2)}{7x}$

17. $\frac{a^2 - a - 6}{a^2 - 7a + 12} \cdot \frac{a^2 - 2a - 8}{a^2 - 3a - 10}$

$= \frac{(a - 3)(a + 2)(a - 4)(a + 2)}{(a - 4)(a - 3)(a - 5)(a + 2)}$

$= \frac{a + 2}{a - 5}$

18. $\frac{(a + 3)^2}{(a - 6)(a + 4)}$

19. $\frac{m^2 - n^2}{r + s} \div \frac{m - n}{r + s}$

$= \frac{m^2 - n^2}{r + s} \cdot \frac{r + s}{m - n}$

$= \frac{(m + n)(m - n)(r + s)}{(r + s)(m - n)}$

$= m + n$

20. a - b

21. $\frac{3x + 12}{2x - 8} \div \frac{(x + 4)^2}{(x - 4)^2}$

$= \frac{3x + 12}{2x - 8} \cdot \frac{(x - 4)^2}{(x + 4)^2}$

$= \frac{3(x + 4)(x - 4)(x - 4)}{2(x - 4)(x + 4)(x + 4)}$

$= \frac{3(x - 4)}{2(x + 4)}$

22. $\frac{a + 1}{a - 3}$

23. $\frac{x^2 - y^2}{x^3 - y^3} \cdot \frac{x^2 + xy + y^2}{x^2 + 2xy + y^2}$

$= \frac{(x + y)(x - y)(x^2 + xy + y^2)}{(x - y)(x^2 + xy + y^2)(x + y)(x + y)}$

$= \frac{1}{x + y}$

24. c - 2

25. $\frac{(x - y)^2 - z^2}{(x + y)^2 - z^2} \div \frac{x - y + z}{x + y - z}$

$= \frac{(x - y)^2 - z^2}{(x + y)^2 - z^2} \cdot \frac{x + y - z}{x - y + z}$

$= \frac{(x - y + z)(x - y - z)(x + y - z)}{(x + y + z)(x + y - z)(x - y + z)}$

$= \frac{x - y - z}{x + y + z}$

26. $\frac{a + b - 3}{a - b + 3}$

27. $\frac{3}{2a + 3} + \frac{2a}{2a + 3}$

$= \frac{3 + 2a}{2a + 3}$

$= 1$

28. 2

29. $\dfrac{y}{y-1}+\dfrac{2}{1-y}$

$=\dfrac{y}{y-1}+\dfrac{-1}{-1}\cdot\dfrac{2}{1-y}$

$=\dfrac{y}{y-1}+\dfrac{-2}{y-1}$

$=\dfrac{y-2}{y-1}$

30. 1

31. $\dfrac{x}{2x-3y}-\dfrac{y}{3y-2x}$

$=\dfrac{x}{2x-3y}-\dfrac{-1}{-1}\cdot\dfrac{y}{3y-2x}$

$=\dfrac{x}{2x-3y}-\dfrac{-y}{2x-3y}$

$=\dfrac{x+y}{2x-3y}$ $\quad [x-(-y)=x+y]$

32. $\dfrac{5a}{3a-2b}$

33. $\dfrac{3}{x+2}+\dfrac{2}{x^2-4}$

$=\dfrac{3}{x+2}+\dfrac{2}{(x+2)(x-2)}$, LCM = $(x+2)(x-2)$

$=\dfrac{3}{x+2}\cdot\dfrac{x-2}{x-2}+\dfrac{2}{(x+2)(x-2)}$

$=\dfrac{3x-6}{(x+2)(x-2)}+\dfrac{2}{(x+2)(x-2)}$

$=\dfrac{3x-4}{(x+2)(x-2)}$

34. $\dfrac{5a+13}{(a+3)(a-3)}$

35. $\dfrac{y}{y^2-y-20}+\dfrac{2}{y+4}$

$=\dfrac{y}{(y+4)(y-5)}+\dfrac{2}{y+4}$, LCM = $(y+4)(y-5)$

$=\dfrac{y}{(y+4)(y-5)}+\dfrac{2}{y+4}\cdot\dfrac{y-5}{y-5}$

$=\dfrac{y}{(y+4)(y-5)}+\dfrac{2y-10}{(y+4)(y-5)}$

$=\dfrac{3y-10}{(y+4)(y-5)}$

36. $\dfrac{-5y-9}{(y+3)^2}$

37. $\dfrac{3}{x+y}+\dfrac{x-5y}{x^2-y^2}$

$=\dfrac{3}{x+y}+\dfrac{x-5y}{(x+y)(x-y)}$, LCM = $(x+y)(x-y)$

$=\dfrac{3}{x+y}\cdot\dfrac{x-y}{x-y}+\dfrac{x-5y}{(x+y)(x-y)}$

$=\dfrac{3x-3y}{(x+y)(x-y)}+\dfrac{x-5y}{(x+y)(x-y)}$

$=\dfrac{4x-8y}{(x+y)(x-y)}$

38. $\dfrac{2a}{(a+1)(a-1)}$

39. $\dfrac{9x+2}{3x^2-2x-8}+\dfrac{7}{3x^2+x-4}$

$=\dfrac{9x+2}{(3x+4)(x-2)}+\dfrac{7}{(3x+4)(x-1)}$,

LCM = $(3x+4)(x-2)(x-1)$

$=\dfrac{9x+2}{(3x+4)(x-2)}\cdot\dfrac{x-1}{x-1}+\dfrac{7}{(3x+4)(x-1)}\cdot\dfrac{x-2}{x-2}$

$=\dfrac{9x^2-7x-2}{(3x+4)(x-2)(x-1)}+\dfrac{7x-14}{(3x+4)(x-1)(x-2)}$

$=\dfrac{9x^2-16}{(3x+4)(x-2)(x-1)}$

$=\dfrac{(3x+4)(3x-4)}{(3x+4)(x-2)(x-1)}$

$=\dfrac{3x-4}{(x-2)(x-1)}$

40. $\dfrac{y}{(y-2)(y-3)}$

41. $\dfrac{5a}{a-b}+\dfrac{ab}{a^2-b^2}+\dfrac{4b}{a+b}$

$=\dfrac{5a}{a-b}+\dfrac{ab}{(a+b)(a-b)}+\dfrac{4b}{a+b}$,

LCM = $(a+b)(a-b)$

$=\dfrac{5a}{a-b}\cdot\dfrac{a+b}{a+b}+\dfrac{ab}{(a+b)(a-b)}+\dfrac{4b}{a+b}\cdot\dfrac{a-b}{a-b}$

$=\dfrac{5a^2+5ab}{(a+b)(a-b)}+\dfrac{ab}{(a+b)(a-b)}+\dfrac{4ab-4b^2}{(a+b)(a-b)}$

$=\dfrac{5a^2+10ab-4b^2}{(a+b)(a-b)}$

42. $\dfrac{6a^2+9ab+3b^2+5}{(a+b)(a-b)}$

43. $\dfrac{7}{x+2}-\dfrac{x+8}{4-x^2}+\dfrac{3x-2}{4-4x+x^2}$

$=\dfrac{7}{x+2}-\dfrac{x+8}{(2+x)(2-x)}+\dfrac{3x-2}{(2-x)^2}$

LCM = $(2+x)(2-x)^2$

$=\dfrac{7}{2+x}\cdot\dfrac{(2-x)^2}{(2-x)^2}-\dfrac{x+8}{(2+x)(2-x)}\cdot\dfrac{2-x}{2-x}+$

$\dfrac{3x-2}{(2-x)^2}\cdot\dfrac{2+x}{2+x}$

$=\dfrac{28-28x+7x^2-(16-6x-x^2)+3x^2+4x-4}{(2+x)(2-x)^2}$

$=\dfrac{28-28x+7x^2-16+6x+x^2+3x^2+4x-4}{(2+x)(2-x)^2}$

$=\dfrac{11x^2-18x+8}{(2+x)(2-x)^2}$

44. $\dfrac{33-32x+9x^2}{(3+x)(3-x)^2}$

45. $\dfrac{1}{x + 1} - \dfrac{x}{x - 2} + \dfrac{x^2 + 2}{x^2 - x - 2}$

$= \dfrac{1}{x + 1} - \dfrac{x}{x - 2} + \dfrac{x^2 + 2}{(x + 1)(x - 2)}$, LCM = $(x + 1)(x - 2)$

$= \dfrac{1}{x + 1} \cdot \dfrac{x - 2}{x - 2} - \dfrac{x}{x - 2} \cdot \dfrac{x + 1}{x + 1} + \dfrac{x^2 + 2}{(x + 1)(x - 2)}$

$= \dfrac{x - 2}{(x + 1)(x - 2)} - \dfrac{x^2 + x}{(x + 1)(x - 2)} + \dfrac{x^2 + 2}{(x + 1)(x - 2)}$

$= \dfrac{x - 2 - x^2 - x + x^2 + 2}{(x + 1)(x - 2)}$

$= \dfrac{0}{(x + 1)(x - 2)}$

$= 0$

46. $\dfrac{3}{x + 2}$

47. $\dfrac{\frac{x^2 - y^2}{xy}}{\frac{x - y}{y}} = \dfrac{x^2 - y^2}{xy} \cdot \dfrac{y}{x - y}$

$= \dfrac{(x + y)(x - y)y}{xy(x - y)}$

$= \dfrac{x + y}{x}$

48. $\dfrac{a}{a + b}$

49. $\dfrac{a - a^{-1}}{a + a^{-1}} = \dfrac{a - \frac{1}{a}}{a + \frac{1}{a}} = \dfrac{a \cdot \frac{a}{a} - \frac{1}{a}}{a \cdot \frac{a}{a} + \frac{1}{a}}$

$= \dfrac{\frac{a^2 - 1}{a}}{\frac{a^2 + 1}{a}}$

$= \dfrac{a^2 - 1}{a} \cdot \dfrac{a}{a^2 + 1}$

$= \dfrac{a^2 - 1}{a^2 + 1}$

50. $\dfrac{a^2(b - 1)}{b^2(a - 1)}$

51. $\dfrac{c + \frac{8}{c^2}}{1 + \frac{2}{c}} = \dfrac{c \cdot \frac{c^2}{c^2} + \frac{8}{c^2}}{1 \cdot \frac{c}{c} + \frac{2}{c}}$

$= \dfrac{\frac{c^3 + 8}{c^2}}{\frac{c + 2}{c}}$

$= \dfrac{c^3 + 8}{c^2} \cdot \dfrac{c}{c + 2}$

$= \dfrac{(c + 2)(c^2 - 2c + 4)c}{c^2(c + 2)}$

$= \dfrac{c^2 - 2c + 4}{c}$

52. $\dfrac{x^2y^2}{x^2 - xy + y^2}$

53. $\dfrac{x^2 + xy + y^2}{\frac{x^2}{y} - \frac{y^2}{x}} = \dfrac{x^2 + xy + y^2}{\frac{x^2}{y} \cdot \frac{x}{x} - \frac{y^2}{x} \cdot \frac{y}{y}}$

$= \dfrac{x^2 + xy + y^2}{\frac{x^3 - y^3}{xy}}$

$= (x^2 + xy + y^2) \cdot \dfrac{xy}{x^3 - y^3}$

$= \dfrac{(x^2 + xy + y^2)xy}{(x - y)(x^2 + xy + y^2)}$

$= \dfrac{xy}{x - y}$

54. $\dfrac{a + b}{ab}$

55. $\dfrac{\frac{x}{y} - \frac{y}{x}}{\frac{1}{y} + \frac{1}{x}} = \dfrac{\frac{x^2 - y^2}{xy}}{\frac{x + y}{xy}}$

$= \dfrac{x^2 - y^2}{xy} \cdot \dfrac{xy}{x + y}$

$= \dfrac{(x + y)(x - y)xy}{xy(x + y)}$

$= x - y$

56. $-a - b$

57. $\dfrac{x^2y^{-2} - y^2x^{-2}}{xy^{-1} + yx^{-1}} = \dfrac{\frac{x^2}{y^2} - \frac{y^2}{x^2}}{\frac{x}{y} + \frac{y}{x}}$

$= \dfrac{\frac{x^4 - y^4}{x^2y^2}}{\frac{x^2 + y^2}{xy}}$

$= \dfrac{x^4 - y^4}{x^2y^2} \cdot \dfrac{xy}{x^2 + y^2}$

$= \dfrac{(x^2 + y^2)(x + y)(x - y)xy}{x^2y^2(x^2 + y^2)}$

$= \dfrac{(x + y)(x - y)}{xy}$, or $\dfrac{x^2 - y^2}{xy}$

58. $\dfrac{a^2 + b^2}{ab}$

59. $\dfrac{\frac{a}{1 - a} + \frac{1 + a}{a}}{\frac{1 - a}{a} + \frac{a}{1 + a}} = \dfrac{\frac{a}{1 - a} \cdot \frac{a}{a} + \frac{1 + a}{a} \cdot \frac{1 - a}{1 - a}}{\frac{1 - a}{a} \cdot \frac{1 + a}{1 + a} + \frac{a}{1 + a} \cdot \frac{a}{a}}$

$= \dfrac{\frac{a^2 + (1 - a^2)}{a(1 - a)}}{\frac{(1 - a^2) + a^2}{a(1 + a)}}$

$= \dfrac{1}{a(1 - a)} \cdot \dfrac{a(1 + a)}{1}$

$= \dfrac{1 + a}{1 - a}$

60. $\dfrac{1 - x}{1 + x}$

61. $$\frac{\frac{1}{a^2}+\frac{2}{ab}+\frac{1}{b^2}}{\frac{1}{a^2}-\frac{1}{b^2}} = \frac{\frac{1}{a^2}\cdot\frac{b^2}{b^2}+\frac{2}{ab}\cdot\frac{ab}{ab}+\frac{1}{b^2}\cdot\frac{a^2}{a^2}}{\frac{1}{a^2}\cdot\frac{b^2}{b^2}-\frac{1}{b^2}\cdot\frac{a^2}{a^2}}$$
$$= \frac{\frac{b^2+2ab+a^2}{a^2b^2}}{\frac{b^2-a^2}{a^2b^2}}$$
$$= \frac{(b+a)(b+a)}{a^2b^2}\cdot\frac{a^2b^2}{(b+a)(b-a)}$$
$$= \frac{b+a}{b-a}$$

62. $\frac{y+x}{y-x}$

63. $$\frac{(x+h)^2-x^2}{h} = \frac{x^2+2xh+h^2-x^2}{h}$$
$$= \frac{2xh+h^2}{h}$$
$$= \frac{h(2x+h)}{h}$$
$$= 2x+h$$

64. $\frac{-1}{x(x+h)}$

65. $$\frac{(x+h)^3-x^3}{h} = \frac{x^3+3x^2h+3xh^2+h^3-x^3}{h}$$
$$= \frac{3x^2h+3xh^2+h^3}{h}$$
$$= \frac{h(3x^2+3xh+h^2)}{h}$$
$$= 3x^2+3xh+h^2$$

66. $\frac{-2x-h}{x^2(x+h)^2}$

67. $$\left[\frac{\frac{x+1}{x-1}+1}{\frac{x+1}{x-1}-1}\right]^5 = \left[\frac{\frac{(x+1)+(x-1)}{x-1}}{\frac{(x+1)-(x-1)}{x-1}}\right]^5$$
$$= \left[\frac{2x}{x-1}\cdot\frac{x-1}{2}\right]^5$$
$$= x^5$$

68. $\frac{5x+3}{3x+2}$

69. $\frac{a}{b} \div \left(\frac{a}{3}+\frac{b}{4}\right)$

$= \frac{a}{b}\cdot\left(\frac{3}{a}+\frac{4}{b}\right)$ (1) Step (1) uses the wrong reciprocal. The reciprocal of a sum is not the sum of the reciprocals.

$= \frac{a}{b}\cdot\left(\frac{3b+4a}{ab}\right)$ (2) Step (2) would be correct if step (1) had been correct.

$= \frac{a(3b+4a)}{ab^2}$ (3) Step (3) would be correct if steps (1) and (2) had been correct.

$= \frac{4a+3b}{b}$ (4) Step (4) has an improper simplification of the b in the denominator.

69. (continued)

$$\frac{a}{b} \div \left(\frac{a}{3}+\frac{b}{4}\right) = \frac{a}{b} \div \left(\frac{4a+3b}{12}\right)$$
$$= \frac{a}{b}\cdot\frac{12}{4a+3b}$$
$$= \frac{12a}{b(4a+3b)} \quad \text{(Correct answer)}$$

70. In step (1) the $\frac{5x^2}{4xy}$ should be $\frac{10x^2}{4xy}$.

Step (2) would be correct if step (1) had been.

In going from step (2) to step (3) the quantity x is not a factor of both terms in the numerator of (2).

In going from step (3) to step (4) the quantity y is not a factor of both terms in the numerator of (3).

The correct answer is $\frac{10x^2+3y^2}{4xy}$.

71. $$\frac{n(n+1)(n+2)}{2\cdot 3}+\frac{(n+1)(n+2)}{2}$$
$$= \frac{n(n+1)(n+2)}{2\cdot 3}+\frac{(n+1)(n+2)}{2}\cdot\frac{3}{3}, \text{ LCM} = 2\cdot 3$$
$$= \frac{n(n+1)(n+2)+3(n+1)(n+2)}{2\cdot 3}$$
$$= \frac{(n+1)(n+2)(n+3)}{2\cdot 3} \quad \text{(Factoring the numerator by grouping)}$$

72. $\frac{(n+1)(n+2)(n-3)(n+4)}{2\cdot 3\cdot 4}$

Exercise Set 1.7

1. $\sqrt{x-3}$

   We substitute -2 for x in x - 3: -2 - 3 = -5. Since the radicand is negative, -2 is not a meaningful replacement.

   We substitute 5 for x in x - 3: 5 - 3 = 2. Since the radicand is not negative, 5 is a meaningful replacement.

2. Yes, No

3. $\sqrt{3-4x}$

   We substitute -1 for x in 3 - 4x: 3 - 4(-1) = 7. Since the radicand is not negative, -1 is a meaningful replacement.

   We substitute 1 for x in 3 - 4x: 3 - 4·1 = -1. Since the radicand is negative, 1 is not a meaningful replacement.

4. Yes, Yes

5. $\sqrt{1 - x^2}$

We substitute 1 for x in $1 - x^2$: $1 - 1^2 = 0$. Since the radicand is not negative, 1 is a meaningful replacement.

We substitute 3 for x in $1 - x^2$: $1 - 3^2 = -8$. Since the radicand is negative, 3 is not a meaningful replacement.

6. Yes, Yes

7. $\sqrt[3]{2x + 7}$

We substitute -4 for x in 2x + 7: 2(-4) + 7 = -1. Since every real number, positive, negative, or zero has a cube root, -4 is a meaningful replacement.

We substitute 5 for x in 2x + 7: 2(5) + 7 = 17. Since every real number, positive, negative, or zero has a cube root, 5 is a meaningful replacement.

8. No, No

9. $\sqrt{(-11)^2} = |-11| = 11$

10. 1

11. $\sqrt{16x^2} = \sqrt{(4x)^2} = |4x| = 4|x|$

12. $6|t|$

13. $\sqrt{(b + 1)^2} = |b + 1|$

14. $|2c - 3|$

15. $\sqrt[3]{-27x^3} = \sqrt[3]{(-3x)^3} = -3x$

16. -2y

17. $\sqrt{x^2 - 4x + 4} = \sqrt{(x - 2)^2} = |x - 2|$

18. $|y + 8|$

19. $\sqrt[5]{32} = \sqrt[5]{2^5} = 2$

20. -2

21. $\sqrt{180} = \sqrt{36\cdot 5} = \sqrt{36} \cdot \sqrt{5} = 6\sqrt{5}$

22. $4\sqrt{3}$

23. $\sqrt[3]{54} = \sqrt[3]{27\cdot 2} = \sqrt[3]{27} \cdot \sqrt[3]{2} = 3\sqrt[3]{2}$

24. $3\sqrt[3]{5}$

25. $\sqrt{128c^2d^4} = \sqrt{64c^2d^4\cdot 2} = |8cd^2|\sqrt{2} = 8|c|d^2\sqrt{2}$

26. $9c^2|d^3|\sqrt{2}$

27. $\sqrt{3}\sqrt{6} = \sqrt{18} = \sqrt{9\cdot 2} = 3\sqrt{2}$

28. $4\sqrt{3}$

29. $\sqrt{2x^3y}\sqrt{12xy} = \sqrt{24x^4y^2} = \sqrt{4x^4y^2\cdot 6} = 2x^2y\sqrt{6}$

30. $2y^2z\sqrt{15}$

31. $\sqrt[3]{3x^2y}\ \sqrt[3]{36x} = \sqrt[3]{108x^3y} = \sqrt[3]{27x^3\cdot 4y} = 3x\sqrt[3]{4y}$

32. $2xy\sqrt[5]{x^2}$

33. $\sqrt[3]{2(x + 4)}\ \sqrt[3]{4(x + 4)^4} = \sqrt[3]{8(x + 4)^5}$

$= \sqrt[3]{8(x + 4)^3 \cdot (x + 4)^2}$

$= 2(x + 4)\sqrt[3]{(x + 4)^2}$

34. $2(x + 1)\sqrt[3]{9(x + 1)}$

35. $\dfrac{\sqrt{21ab^2}}{\sqrt{3ab}} = \sqrt{\dfrac{21ab^2}{3ab}} = \sqrt{7b}$

36. $\dfrac{2\sqrt{2ab}}{a}$

37. $\dfrac{\sqrt[3]{40m}}{\sqrt[3]{5m}} = \sqrt[3]{\dfrac{40m}{5m}} = \sqrt[3]{8} = 2$

38. $\sqrt{5y}$

39. $\dfrac{\sqrt[3]{3x^2}}{\sqrt[3]{24x^5}} = \sqrt[3]{\dfrac{3x^2}{24x^5}} = \sqrt[3]{\dfrac{1}{8x^3}} = \dfrac{1}{2x}$

40. $y\sqrt[3]{5}$

41. $\dfrac{\sqrt{a^2 - b^2}}{\sqrt{a - b}} = \sqrt{\dfrac{a^2 - b^2}{a - b}} = \sqrt{\dfrac{(a + b)(a - b)}{a - b}}$

$= \sqrt{a + b}$

42. $\sqrt{x^2 + xy + y^2}$

43. $\sqrt{\dfrac{9a^2}{8b}} = \sqrt{\dfrac{9a^2}{8b} \cdot \dfrac{2b}{2b}} = \sqrt{\dfrac{9a^2}{16b^2} \cdot 2b} = \dfrac{3a\sqrt{2b}}{4b}$

44. $\dfrac{b\sqrt{15a}}{6a}$

45. $\sqrt[3]{\dfrac{2x^22y^3}{25z^4}} = \sqrt[3]{\dfrac{2x^22y^3}{25z^4} \cdot \dfrac{5z^2}{5z^2}} = \sqrt[3]{\dfrac{y^3}{125z^6} \cdot 20x^2z^2}$

$= \dfrac{y\sqrt[3]{20x^2z^2}}{5z^2}$

46. $\dfrac{2x}{y}$

47. $\frac{(\sqrt[3]{32x^4y})^2}{(\sqrt[3]{xy})^2} = \left(\sqrt[3]{\frac{32x^4y}{xy}}\right)^2 = (\sqrt[3]{32x^3})^2$

$= (\sqrt[3]{2^5x^3})^2$

$= \sqrt[3]{2^{10}x^6}$

$= \sqrt[3]{2^9x^6\cdot 2}$

$= 2^3x^2\sqrt[3]{2}$

$= 8x^2\sqrt[3]{2}$

48. $4\sqrt[3]{4x^2}$

49. $\frac{3\sqrt{a^2b^2}\sqrt{4xy}}{2\sqrt{a^{-1}b^{-2}}\sqrt{9x^{-3}y^{-1}}} = \frac{3}{2}\sqrt{\frac{a^2b^2}{a^{-1}b^{-2}}}\sqrt{\frac{4xy}{9x^{-3}y^{-1}}}$

$= \frac{3}{2}\sqrt{a^3b^4}\sqrt{\frac{4}{9}x^4y^2}$

$= \frac{3}{2}\cdot ab^2\sqrt{a}\cdot\frac{2}{3}x^2y$

$= ab^2x^2y\sqrt{a}$

50. xy^2a^3b

51. $9\sqrt{50} + 6\sqrt{2} = 9\sqrt{25\cdot 2} + 6\sqrt{2}$

$= 9\cdot 5\sqrt{2} + 6\sqrt{2}$

$= 45\sqrt{2} + 6\sqrt{2}$

$= (45 + 6)\sqrt{2}$

$= 51\sqrt{2}$

52. $29\sqrt{3}$

53. $8\sqrt{2} - 6\sqrt{20} - 5\sqrt{8}$

$= 8\sqrt{2} - 6\sqrt{4\cdot 5} - 5\sqrt{4\cdot 2}$

$= 8\sqrt{2} - 6\cdot 2\sqrt{5} - 5\cdot 2\sqrt{2}$

$= 8\sqrt{2} - 12\sqrt{5} - 10\sqrt{2}$

$= -2\sqrt{2} - 12\sqrt{5}$

54. $4\sqrt{3}$

55. $2\sqrt[3]{8x^2} + 5\sqrt[3]{27x^2} - 3\sqrt[3]{x^3}$

$= 2\cdot 2\sqrt[3]{x^2} + 5\cdot 3\sqrt[3]{x^2} - 3x$

$= 4\sqrt[3]{x^2} + 15\sqrt[3]{x^2} - 3x$

$= 19\sqrt[3]{x^2} - 3x$

56. $(5a^2 - 3b^2)\sqrt{a + b}$

57. $3\sqrt{3y^2} - \frac{y\sqrt{48}}{\sqrt{2}} + \sqrt{\frac{12}{4y^{-2}}}$

$= 3\sqrt{3y^2} - y\sqrt{24} + \sqrt{3y^2}$

$= 3y\sqrt{3} - 2y\sqrt{6} + y\sqrt{3}$

$= 4y\sqrt{3} - 2y\sqrt{6}$

58. $(3x - 2)\sqrt[3]{x^2}$

59. $(\sqrt{3} - \sqrt{2})(\sqrt{3} + \sqrt{2})$

$= (\sqrt{3})^2 - (\sqrt{2})^2$

$= 3 - 2$

$= 1$

60. -12

61. $(1 + \sqrt{3})^2 = 1^2 + 2\cdot 1\cdot\sqrt{3} + (\sqrt{3})^2$

$= 1 + 2\sqrt{3} + 3$

$= 4 + 2\sqrt{3}$

62. $27 - 10\sqrt{2}$

63. $(\sqrt{t} - x)^2$

$= (\sqrt{t})^2 - 2\cdot\sqrt{t}\cdot x + x^2$

$= t - 2x\sqrt{t} + x^2$

64. $a + 2 + a^{-1}$, or $\frac{(a + 1)^2}{a}$

65. $5\sqrt{7} + \frac{35}{\sqrt{7}}$

$= 5\sqrt{7} + \frac{35}{\sqrt{7}}\cdot\frac{\sqrt{7}}{\sqrt{7}}$

$= 5\sqrt{7} + \frac{35\sqrt{7}}{7}$

$= 5\sqrt{7} + 5\sqrt{7}$

$= 10\sqrt{7}$

66. $2a^2b + 5a\sqrt{by} - 3y$

67. $(\sqrt{x + 3} - \sqrt{3})(\sqrt{x + 3} + \sqrt{3})$

$= (\sqrt{x + 3})^2 - (\sqrt{3})^2$

$= (x + 3) - 3$

$= x$

68. h

69. We substitute 90 for L in the formula.

$r = 2\sqrt{5L}$

$= 2\sqrt{5\cdot 90}$

$= 2\sqrt{450}$

$\approx 2(21.213)$ Using a calculator

≈ 42.43

The speed of the car was about 42.43 mph.

70. About 46.90 mph

71. $T = 2\pi\sqrt{\frac{L}{32}}$

$T = 2(3.14)\sqrt{\frac{2}{32}}$

$= 6.28 \cdot \frac{1}{4}$

$= 1.57$ sec

$T = 2(3.14)\sqrt{\frac{8}{32}}$

$= 6.28 \cdot \frac{1}{2}$

$= 3.14$ sec

$T = 2(3.14)\sqrt{\frac{64}{32}}$

$= 2(3.14)\sqrt{2}$

$\approx 6.28(1.414)$

≈ 8.88 sec

$T = 2(3.14)\sqrt{\frac{100}{32}}$

$= 2(3.14)\sqrt{3.125}$

$\approx 6.28(1.768)$

≈ 11.10 sec

72. $65\sqrt{2} \approx 91.92$ ft

73. We use the Pythagorean theorem to find b, the airplane's horizontal distance from the airport. We have a = 3700 and c = 14,200.

$$c^2 = a^2 + b^2$$
$$14{,}200^2 = 3700^2 + b^2$$
$$201{,}640{,}000 = 13{,}690{,}000 + b^2$$
$$187{,}950{,}000 = b^2$$
$$13{,}709.5 \approx b \quad \text{Using a calculator}$$

The airplane is about 13,709.5 ft horizontally from the airport.

74. About 102.8 ft

75. $h^2 + \left(\frac{a}{2}\right)^2 = a^2$ (Pythagorean theorem)

$$h^2 + \frac{a^2}{4} = a^2$$
$$h^2 = \frac{3a^2}{4}$$
$$h = \sqrt{\frac{3a^2}{4}}$$
$$h = \frac{a}{2}\sqrt{3}$$

76. $A = \frac{a^2}{4}\sqrt{3}$

77.

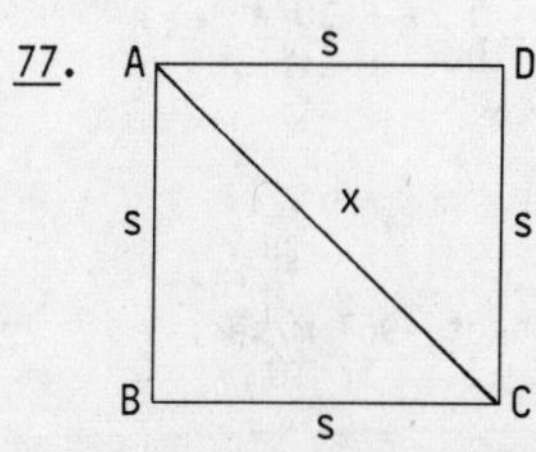

Let x represent the length of $\overline{AC}$. Then using the Pythagorean theorem we have

$$s^2 + s^2 = x^2$$
$$2s^2 = x^2$$
$$\sqrt{2s^2} = x$$
$$s\sqrt{2} = x$$

The length of $\overline{AC}$ is $\sqrt{2}\,s$.

78. $\sqrt{2}\,s$

79.

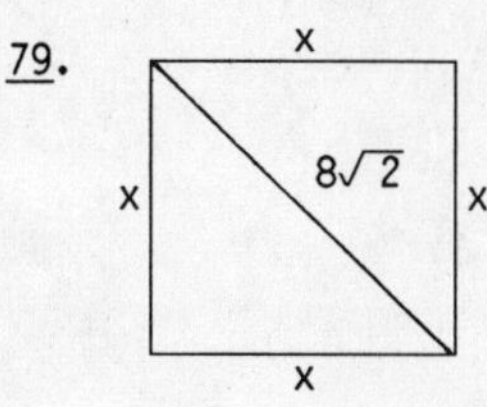

$x^2 + x^2 = (8\sqrt{2})^2$ (Pythagorean theorem)

$$2x^2 = 128$$
$$x^2 = 64$$
$$x = 8$$

80. 50 ft^2

81. $\frac{6}{3 + \sqrt{5}} = \frac{6}{3 + \sqrt{5}} \cdot \frac{3 - \sqrt{5}}{3 - \sqrt{5}}$

$$= \frac{6(3 - \sqrt{5})}{9 - 5}$$
$$= \frac{6(3 - \sqrt{5})}{4}$$
$$= \frac{3(3 - \sqrt{5})}{2}$$

82. $\sqrt{3} + 1$

83. $\sqrt[3]{\frac{16}{9}} = \sqrt[3]{\frac{16}{9} \cdot \frac{3}{3}} = \sqrt[3]{\frac{48}{27}} = \sqrt[3]{\frac{8}{27} \cdot 6} = \frac{2\sqrt[3]{6}}{3}$

84. $\frac{\sqrt[3]{4}}{2}$

85. $\frac{4\sqrt{x} - 3\sqrt{xy}}{2\sqrt{x} + 5\sqrt{y}} = \frac{4\sqrt{x} - 3\sqrt{xy}}{2\sqrt{x} + 5\sqrt{y}} \cdot \frac{2\sqrt{x} - 5\sqrt{y}}{2\sqrt{x} - 5\sqrt{y}}$

$$= \frac{8x - 20\sqrt{xy} - 6x\sqrt{y} + 15y\sqrt{x}}{4x - 25y}$$

86. $\frac{15x + 10\sqrt{xy} + 6x\sqrt{y} + 4y\sqrt{x}}{9x - 4y}$

87. $\frac{\sqrt{2} + \sqrt{5a}}{6} = \frac{\sqrt{2} + \sqrt{5a}}{6} \cdot \frac{\sqrt{2} - \sqrt{5a}}{\sqrt{2} - \sqrt{5a}}$

$$= \frac{2 - 5a}{6(\sqrt{2} - \sqrt{5a})}$$

88. $\frac{3 - 5y}{4(\sqrt{3} - \sqrt{5y})}$

89. $\frac{\sqrt{x + 1} + 1}{\sqrt{x + 1} - 1} = \frac{\sqrt{x + 1} + 1}{\sqrt{x + 1} - 1} \cdot \frac{\sqrt{x + 1} - 1}{\sqrt{x + 1} - 1}$

$$= \frac{(x + 1) - 1}{(x + 1) - 2\sqrt{x + 1} + 1}$$
$$= \frac{x}{x - 2\sqrt{x + 1} + 2}$$

90. $\frac{x}{x + 8 + 4\sqrt{x + 4}}$

91. $\dfrac{\sqrt{a+3}-\sqrt{3}}{3} = \dfrac{\sqrt{a+3}-\sqrt{3}}{3} \cdot \dfrac{\sqrt{a+3}+\sqrt{3}}{\sqrt{a+3}+\sqrt{3}}$

$= \dfrac{(a+3)-3}{3(\sqrt{a+3}+\sqrt{3})}$

$= \dfrac{a}{3(\sqrt{a+3}+\sqrt{3})}$

92. $\dfrac{1}{\sqrt{a+h}+\sqrt{a}}$

93. $\sqrt{8.2x^3y}\,\sqrt{12.5xy} = \sqrt{102.5x^4y^2} \approx 10.124x^2y$

94. $0.1251y^2z$

95. $\sqrt{\dfrac{6.03a^2}{17.13b}} \approx \sqrt{0.352014\,\dfrac{a^2}{b}\cdot\dfrac{b}{b}} \approx \dfrac{0.5933a\sqrt{b}}{b}$

96. $\dfrac{0.1974b\sqrt{a}}{a}$

97. $\sqrt{1+x^2} + \dfrac{1}{\sqrt{1+x^2}}$

$= \sqrt{1+x^2}\cdot\dfrac{1+x^2}{1+x^2} + \dfrac{1}{\sqrt{1+x^2}}\quad\dfrac{\sqrt{1+x^2}}{\sqrt{1+x^2}}$

$= \dfrac{(1+x^2)\sqrt{1+x^2}}{1+x^2} + \dfrac{\sqrt{1+x^2}}{1+x^2}$

$= \dfrac{(2+x^2)\left(\sqrt{1+x^2}\right)}{1+x^2}$

98. $\dfrac{(2-3x^2)\sqrt{1-x^2}}{2(1-x^2)}$

99. Let a = 16 and b = 9.

Then $\sqrt{a+b} = \sqrt{16+9} = \sqrt{25} = 5$

and $\sqrt{a} + \sqrt{b} = \sqrt{16} + \sqrt{9} = 4 + 3 = 7$.

.00. $\left(\sqrt{5+\sqrt{24}}\right)^2 = 5 + \sqrt{24} = 5 + 2\sqrt{6}$

$\left(\sqrt{2}+\sqrt{3}\right)^2 = \left(\sqrt{2}\right)^2 + 2\sqrt{2}\,\sqrt{3} + \left(\sqrt{3}\right)^2$

$= 2 + 2\sqrt{6} + 3$

$= 5 + 2\sqrt{6}$

101. We substitute 0.1 for h, 0.03 for d, and 0.6 for v_0 in the formula.

$w = d\sqrt{\dfrac{v_0}{\sqrt{v_0^2 + 19.6h}}}$

$= 0.03\sqrt{\dfrac{0.6}{\sqrt{(0.6)^2 + 19.6(0.1)}}}$

$= 0.03\sqrt{\dfrac{0.6}{\sqrt{0.36 + 1.96}}}$

$= 0.03\sqrt{\dfrac{0.6}{\sqrt{2.32}}}$

$\approx 0.03\sqrt{\dfrac{0.6}{1.523}}$

$\approx 0.03\sqrt{0.394}$

$\approx 0.03(0.628)$

≈ 0.0188

The width of the stream is about 0.0188 m.

102. $51 - 10\sqrt{26}$

103. a) We substitute 6.672×10^{-11} for G, 5.97×10^{24} for M, 6.37×10^6 for R in the formula.

$V_0 = \sqrt{\dfrac{2GM}{R}}$

$= \sqrt{\dfrac{2 \times 6.672 \times 10^{-11} \times 5.97 \times 10^{24}}{6.37 \times 10^6}}$

$\approx \sqrt{12.5061 \times 10^7}$

$\approx 11{,}183$

The escape velocity is about 11,183 m/sec.

b) We substitute 6.672×10^{-11} for G, 6.27×10^{23} for M, and 3.3917×10^6 for R in the formula.

$V_0 = \sqrt{\dfrac{2GM}{R}}$

$= \sqrt{\dfrac{2 \times 6.672 \times 10^{-11} \times 6.27 \times 10^{23}}{3.3917 \times 10^6}}$

$\approx \sqrt{24.6681 \times 10^6}$

≈ 4967

The escape velocity is about 4967 m/sec.

Exercise Set 1.8

1. $x^{3/4} = \sqrt[4]{x^3}$

2. $\sqrt[5]{y^2}$

3. $16^{3/4} = (16^{1/4})^3 = \left(\sqrt[4]{16}\right)^3 = 2^3 = 8$

4. 128

5. $125^{-1/3} = \frac{1}{125^{1/3}} = \frac{1}{\sqrt[3]{125}} = \frac{1}{5}$

6. $\frac{1}{16}$

7. $a^{5/4}\,b^{-3/4} = \frac{a^{5/4}}{b^{3/4}} = \frac{\sqrt[4]{a^5}}{\sqrt[4]{b^3}} = \frac{a\sqrt[4]{a}}{\sqrt[4]{b^3}}$, or $\frac{a\sqrt[4]{ab}}{b}$

8. $\sqrt[5]{x^2y^{-1}}$

9. $\sqrt[3]{20^2} = 20^{2/3}$

10. $17^{3/5}$

11. $\left(\sqrt[4]{13}\right)^5 = \sqrt[4]{13^5} = 13^{5/4}$, or $\sqrt[4]{13^5} = 13\sqrt[4]{13}$

12. $12^{4/5}$

13. $\sqrt[3]{\sqrt{11}} = \left(\sqrt{11}\right)^{1/3} = (11^{1/2})^{1/3} = 11^{1/6}$

14. $7^{1/12}$

15. $\sqrt{5}\,\sqrt[3]{5} = 5^{1/2} \cdot 5^{1/3} = 5^{1/2 + 1/3} = 5^{5/6}$

16. $2^{5/6}$

17. $\sqrt[5]{32^2} = 32^{2/5} = (32^{1/5})^2 = 2^2 = 4$

18. $\frac{1}{16}$

19. $\sqrt[3]{8y^6} = (8y^6)^{1/3} = (2^3y^6)^{1/3} = 2^{3/3}y^{6/3} = 2y^2$

20. $2c^2d^3$

21. $\sqrt[3]{a^2 + b^2} = (a^2 + b^2)^{1/3}$

22. $(a^3 - b^3)^{1/4}$

23. $\sqrt[3]{27a^3b^9} = (3^3a^3b^9)^{1/3} = 3^{3/3}a^{3/3}b^{9/3} = 3ab^3$

24. $3x^2y^2$

25. $\sqrt[6]{\frac{m^{12}n^{24}}{64}} = \left[\frac{m^{12}n^{24}}{2^6}\right]^{1/6} = \frac{m^{12/6}n^{24/6}}{2^{6/6}} = \frac{m^2n^4}{2}$

26. $\frac{m^2n^3}{2}$

27. $(2a^{3/2})(4a^{1/2}) = 8a^{3/2 + 1/2} = 8a^2$

28. $24a\sqrt{a}$

29. $\left[\frac{x^6}{9b^{-4}}\right]^{-1/2} = \left[\frac{x^6}{3^2b^{-4}}\right]^{-1/2} = \frac{x^{-3}}{3^{-1}b^2} = \frac{3}{x^3b^2}$

30. $\frac{2\sqrt[3]{x^2}}{xy}$

31. $\frac{x^{2/3}y^{5/6}}{x^{-1/3}y^{1/2}} = x^{2/3-(-1/3)}y^{5/6-1/2} = xy^{1/3} = x\sqrt[3]{y}$

32. $\sqrt[4]{ab}$

33. $\sqrt[3]{6}\,\sqrt{2} = 6^{1/3}2^{1/2} = 6^{2/6}2^{3/6}$
$= (6^22^3)^{1/6}$
$= \sqrt[6]{36 \cdot 8}$
$= \sqrt[6]{288}$

34. $2\sqrt[4]{2}$

35. $\sqrt[4]{xy}\,\sqrt[3]{x^2y} = (xy)^{1/4}(x^2y)^{1/3} = (xy)^{3/12}(x^2y)^{4/12}$
$= \left[(xy)^3(x^2y)^4\right]^{1/12}$
$= [x^3y^3x^8y^4]^{1/12}$
$= \sqrt[12]{x^{11}y^7}$

36. $b\sqrt[6]{a^5b}$

37. $\sqrt[3]{a^4\sqrt{a^3}} = \left(a^4\sqrt{a^3}\right)^{1/3} = (a^4a^{3/2})^{1/3}$
$= (a^{11/2})^{1/3}$
$= a^{11/6}$
$= \sqrt[6]{a^{11}}$
$= a\sqrt[6]{a^5}$

38. $a\sqrt[6]{a^5}$

39. $\frac{\sqrt{(a + x)^3}\,\sqrt[3]{(a + x)^2}}{\sqrt[4]{a + x}} = \frac{(a + x)^{3/2}(a + x)^{2/3}}{(a + x)^{1/4}}$
$= \frac{(a + x)^{26/12}}{(a + x)^{3/12}}$
$= (a + x)^{23/12}$
$= \sqrt[12]{(a + x)^{23}}$
$= (a + x)\sqrt[12]{(a + x)^{11}}$

40. $\frac{\sqrt[3]{x + y}}{x + y}$

41. $\left(\sqrt[4]{13}\right)^5 = 13^{5/4} = 13^{1.25} \approx 24.685$

42. 8.372

43. $12.3^{3/2} = 12.3^{1.5} \approx 43.138$

44. 2.098

45. $105.6^{3/4} = 105.6^{0.75} \approx 32.942$

46. 11.671

47. $L = \frac{0.000169d^{2.27}}{h}$
$= \frac{0.000169(180)^{2.27}}{4}$
≈ 5.56 ft

48. 1.46 ft

49. $L = \frac{0.000169d^{2.27}}{h}$

$L = \frac{0.000169(200)^{2.27}}{4}$

≈ 7.07 ft

50. 17.74 ft

51. $T = 34\ x^{-0.41}$

When x = 1:
$T = 34(1)^{-0.41}$
$= 34$ hr

When x = 6:
$T = 34(6)^{-0.41}$
≈ 16.3 hr

When x = 8:
$T = 34(8)^{-0.41}$
≈ 14.5 hr

When x = 10:
$T = 34(10)^{-0.41}$
≈ 13.2 hr

When x = 32:
$T = 34(32)^{-0.41}$
≈ 8.2 hr

When x = 64:
$T = 34(64)^{-0.41}$
≈ 6.2 hr

52. 25.6 hr, 21.7 hr, 17.6 hr, 13.8 hr, 12.7 hr, 10.2 hr, 5.1 hr

53. $a^{-2}b^5 - a^3b^{-5}$

a) The variables common to both terms involve powers of a and b.

b) a^{-2} has a smaller exponent than a^3.
b^{-5} has a smaller exponent than b^5.

c) $a^{-2}b^{-5}$ factored out of $a^{-2}b^5$ leaves
$a^{-2-(-2)}b^{5-(-5)} = a^0b^{10} = b^{10}$.
$a^{-2}b^{-5}$ factored out of a^3b^{-5} leaves
$a^{3-(-2)}b^{-5-(-5)} = a^5b^0 = a^5$.

d) Thus,

$$a^{-2}b^5 - a^3b^{-5} = a^{-2}b^{-5}(b^{10} - a^5) = \frac{b^{10} - a^5}{a^2b^5}.$$

54. $\frac{p^{13} + q^6}{q^2p^5}$

55. $5a^{2/3}b^{-1/2} + 2a^{-1/3}b^{1/2}$

a) The variables common to both terms involve powers of a and b.

b) $a^{-1/3}$ has a smaller exponent than $a^{2/3}$.
$b^{-1/2}$ has a smaller exponent than $b^{1/2}$.

c) $a^{-1/3}b^{-1/2}$ factored out of $5a^{2/3}b^{-1/2}$ leaves
$5a^{2/3-(-1/3)}b^{-1/2-(-1/2)} = 5a^1b^0 = 5a$.
$a^{-1/3}b^{-1/2}$ factored out of $2a^{-1/3}b^{1/2}$ leaves
$2a^{-1/3-(-1/3)}b^{1/2-(-1/2)} = 2a^0b^1 = 2b$.

d) Thus,
$5a^{2/3}b^{-1/2} + 2a^{-1/3}b^{1/2} =$

$$a^{-1/3}b^{-1/2}(5a + 2b) = \frac{5a + 2b}{a^{1/3}b^{1/2}}.$$

56. $\frac{p + 2q^4}{p^{1/5}q^2}$

57. $x^{-1/3}y^{3/4} - x^{2/3}y^{-1/4}$

a) The variables common to both terms involve powers of x and y.

b) $x^{-1/3}$ has a smaller exponent than $x^{2/3}$.
$y^{-1/4}$ has a smaller exponent than $y^{3/4}$.

c) $x^{-1/3}y^{-1/4}$ factored out of $x^{-1/3}y^{3/4}$ leaves
$x^{-1/3-(-1/3)}y^{3/4-(-1/4)} = x^0y^1 = y$.
$x^{-1/3}y^{-1/4}$ factored out of $x^{2/3}y^{-1/4}$ leaves
$x^{2/3-(-1/3)}y^{-1/4-(-1/4)} = x^1y^0 = x$.

d) Thus,
$x^{-1/3}y^{3/4} - x^{2/3}y^{-1/4} = x^{-1/3}y^{-1/4}(y - x) =$

$$\frac{y - x}{x^{1/3}y^{1/4}}.$$

58. $\frac{2(2a - 3b)}{a^{1/2}b^{3/4}}$

59. $(2x - 3)^{-3}(x + 1)^{5/4} + (2x - 3)^{-2}(x + 1)^{1/4}$

$= (2x - 3)^{-3}(x + 1)^{1/4}[(x + 1) + (2x - 3)]$ Factoring

$= (2x - 3)^{-3}(x + 1)^{1/4}(3x - 2)$ Simplifying

$= \frac{(x + 1)^{1/4}(3x - 2)}{(2x - 3)^3}$

60. $\frac{x(13x + 6)}{(5x + 3)^{1/3}}$

61. $2(x+1)^{1/2}(3x+4)^{-3/4} - 10(x+1)^{-1/2}(3x+4)^{1/4}$

$= 2(x + 1)^{-1/2}(3x + 4)^{-3/4}[(x + 1) - 5(3x + 4)]$ Factoring

$= 2(x + 1)^{-1/2}(3x + 4)^{-3/4}(x + 1 - 15x - 20)$ Simplifying

$= 2(x + 1)^{-1/2}(3x + 4)^{-3/4}(-14x - 19)$

$= \frac{2(-14x - 19)}{(x + 1)^{1/2}(3x + 4)^{3/4}}$, or $\frac{-2(14x + 19)}{(x + 1)^{1/2}(3x + 4)^{3/4}}$

62. $\dfrac{-4(-x+11)}{(2x-5)^3(3x+1)^{2/3}}$, or $\dfrac{4(x-11)}{(2x-5)^3(3x+1)^{2/3}}$

63. $3(x^2+1)^3 + 3(3x-5)(x^2+1)^2(2x)$
$= 3(x^2+1)^2[(x^2+1) + (3x-5)(2x)]$ Factoring
$= 3(x^2+1)^2(x^2+1+6x^2-10x)$ Simplifying
$= 3(x^2+1)^2(7x^2-10x+1)$

64. $x^2(x-1)^3(7x-3)$

65. $\dfrac{x^3(2x) - (x^2+1)(3x^2)}{x^6}$
$= \dfrac{2x^4 - 3x^4 - 3x^2}{x^6}$
$= \dfrac{-x^4 - 3x^2}{x^6} = \dfrac{x^2(-x^2-3)}{x^2\cdot x^4}$
$= \dfrac{-x^2-3}{x^4}$

66. $\dfrac{7x^3-4}{2x^{1/2}}$

67. $\dfrac{x^2(x^2+1)^{-1/2}(x) - (2x)(x^2+1)^{1/2}}{x^4}$
$= \dfrac{x(x^2+1)^{-1/2}[x^2 - 2(x^2+1)]}{x^4}$
$= \dfrac{x(x^2+1)^{-1/2}(x^2-2x^2-2)}{x^4}$
$= \dfrac{x(x^2+1)^{-1/2}(-x^2-2)}{x\cdot x^3}$
$= \dfrac{-x^2-2}{x^3(x^2+1)^{1/2}}$

68. $\dfrac{x-3}{2(x-1)^{3/2}}$

69. $\left[\sqrt{a\sqrt{a}}\right]^{\sqrt{a}} = \left[a\sqrt{a}/2\right]^{\sqrt{a}} = a^{a/2}$

70. $48\cdot 2^{1/3}\cdot a^{34/3}\cdot b^{47/9}\cdot c^{34/35}$

Exercise Set 1.9

1. $36 \text{ ft} \cdot \dfrac{1 \text{ yd}}{3 \text{ ft}}$
$= \dfrac{36}{3}\cdot\dfrac{\text{ft}}{\text{ft}}\cdot \text{yd}$
$= 12$ yd

2. 96 oz

3. $6 \text{ kg} \cdot 8\,\dfrac{\text{hr}}{\text{kg}}$
$= 6\cdot 8\cdot \dfrac{\text{kg}}{\text{kg}}\cdot \text{hr}$
$= 48$ hr

4. 27 km

5. $3 \text{ cm}\cdot\dfrac{2\text{g}}{2 \text{ cm}}$
$= \dfrac{3\cdot 2}{2}\cdot\dfrac{\text{cm}}{\text{cm}}\cdot \text{g}$
$= 3$ g

6. 18 km

7. $6\text{m} + 2\text{m}$
$= (6+2)\text{m}$
$= 8$ m

8. 16 tons

9. $5 \text{ ft}^3 + 7 \text{ ft}^3$
$= (5+7) \text{ ft}^3$
$= 12 \text{ ft}^3$

10. 27 yd^3

11. $\dfrac{3 \text{ kg}}{5\text{m}}\cdot\dfrac{7 \text{ kg}}{6\text{m}}$
$= \dfrac{3\cdot 7}{5\cdot 6}\cdot\dfrac{\text{kg}}{\text{m}}\cdot\dfrac{\text{kg}}{\text{m}}$
$= \dfrac{7 \text{ kg}^2}{10 \text{ m}^2}$

12. 180

13. $\dfrac{2000 \text{ lb}\cdot(6 \text{ mi/hr})^2}{100 \text{ ft}}$
$= 2000 \text{ lb}\cdot\dfrac{36 \text{ mi}^2}{\text{hr}^2}\cdot\dfrac{1}{100 \text{ ft}}$
$= \dfrac{2000\cdot 36}{100}\cdot \text{lb}\cdot\dfrac{\text{mi}^2}{\text{hr}^2}\cdot\dfrac{1}{\text{ft}}$
$= 720\,\dfrac{\text{lb-mi}^2}{\text{hr}^2\text{-ft}}$

14. $14\,\dfrac{\text{m-kg}}{\text{sec}^2}$

15. $\dfrac{6 \text{ cm}^2\cdot 5 \text{ cm/sec}}{2 \text{ sec}^2/\text{cm}^2\cdot 2\,\dfrac{1}{\text{kg}}}$
$= 6 \text{ cm}^2\cdot\dfrac{5 \text{ cm}}{\text{sec}}\cdot\dfrac{\text{cm}^2}{2 \text{ sec}^2}\cdot\dfrac{\text{kg}}{2}$
$= \dfrac{6\cdot 5}{2\cdot 2}\cdot\dfrac{\text{cm}^2\cdot\text{cm}\cdot\text{cm}^2\cdot\text{kg}}{\text{sec}\cdot\text{sec}^2}$
$= \dfrac{15}{2}\,\dfrac{\text{cm}^5\text{-kg}}{\text{sec}^3}$

16. 125 ft-lb

17. $72 \text{ in.} = 72 \text{ in.}\cdot\dfrac{1 \text{ ft}}{12 \text{ in.}}$
$= \dfrac{72}{12}\cdot\dfrac{\text{in.}}{\text{in.}}\cdot \text{ft}$
$= 6$ ft

18. 1020 min

19. $2 \text{ days} = 2 \text{ days} \cdot \frac{24 \text{ hr}}{1 \text{ day}} \cdot \frac{60 \text{ min}}{1 \text{ hr}} \cdot \frac{60 \text{ sec}}{1 \text{ min}}$

$= 2 \cdot 24 \cdot 60 \cdot 60 \cdot \frac{\text{day}}{\text{day}} \cdot \frac{\text{hr}}{\text{hr}} \cdot \frac{\text{min}}{\text{min}} \cdot \text{sec}$

$= 172{,}800 \text{ sec}$

20. 0.1 hr

21. $60 \frac{\text{kg}}{\text{m}} = 60 \frac{\text{kg}}{\text{m}} \cdot \frac{1000 \text{ g}}{1 \text{ kg}} \cdot \frac{1 \text{ m}}{100 \text{ cm}}$

$= \frac{60 \cdot 1000}{100} \cdot \frac{\text{kg}}{\text{kg}} \cdot \frac{\text{m}}{\text{m}} \cdot \frac{\text{g}}{\text{cm}}$

$= 600 \frac{\text{g}}{\text{cm}}$

22. $30 \frac{\text{mi}}{\text{hr}}$

23. $216 \text{ m}^2 = 216 \cdot \text{m} \cdot \text{m}$

$= 216 \cdot 100 \text{ cm} \cdot 100 \text{ cm}$

$= 2{,}160{,}000 \text{ cm}^2$

24. $0.81 \frac{\text{ton}}{\text{yd}^3}$

25. $\frac{\$36}{\text{day}} = \frac{\$36}{\text{day}} \cdot \frac{100¢}{\$1} \cdot \frac{1 \text{ day}}{24 \text{ hr}}$

$= \frac{36 \cdot 100}{24} \cdot \frac{\$}{\$} \cdot \frac{\text{day}}{\text{day}} \cdot \frac{¢}{\text{hr}}$

$= 150 \frac{¢}{\text{hr}}$

26. 60 man-days

27. $1.73 \frac{\text{mL}}{\text{sec}} = 1.73 \frac{\text{mL}}{\text{sec}} \cdot \frac{1 \text{ L}}{1000 \text{ mL}} \cdot \frac{60 \text{ sec}}{1 \text{ min}} \cdot \frac{60 \text{ min}}{1 \text{ hr}}$

$= \frac{1.73 \cdot 60 \cdot 60}{1000} \cdot \frac{\text{mL}}{\text{mL}} \cdot \frac{\text{sec}}{\text{sec}} \cdot \frac{\text{min}}{\text{min}} \cdot \frac{\text{L}}{\text{hr}}$

$= 6.228 \frac{\text{L}}{\text{hr}}$

28. $180 \frac{\text{cg}}{\text{mL}}$

29. $186{,}000 \frac{\text{mi}}{\text{sec}}$

$= 186{,}000 \frac{\text{mi}}{\text{sec}} \cdot \frac{60 \text{ sec}}{1 \text{ min}} \cdot \frac{60 \text{ min}}{1 \text{ hr}} \cdot \frac{24 \text{ hr}}{1 \text{ day}} \cdot \frac{365 \text{ days}}{1 \text{ yr}}$

$= 186{,}000 \cdot 60 \cdot 60 \cdot 24 \cdot 365 \frac{\text{sec}}{\text{sec}} \cdot \frac{\text{min}}{\text{min}} \cdot \frac{\text{hr}}{\text{hr}} \cdot \frac{\text{day}}{\text{day}} \cdot \frac{\text{mi}}{\text{yr}}$

$= 5{,}865{,}696{,}000{,}000 \frac{\text{mi}}{\text{yr}}$

30. $6{,}570{,}000 \frac{\text{mi}}{\text{yr}}$

31. $89.2 \frac{\text{ft}}{\text{sec}} = 89.2 \frac{\text{ft}}{\text{sec}} \cdot \frac{1 \text{ m}}{3.3 \text{ ft}} \cdot \frac{60 \text{ sec}}{1 \text{ min}}$

$= \frac{89.2(60)}{3.3} \cdot \frac{\text{ft}}{\text{ft}} \cdot \frac{\text{sec}}{\text{sec}} \cdot \frac{\text{m}}{\text{min}}$

$\approx 1621.8 \frac{\text{m}}{\text{min}}$

32. 774.5 m^3

33. $640 \text{ mi}^2 = 640 \cdot \text{mi} \cdot \text{mi}$

$= 640 \cdot 1.6 \text{ km} \cdot 1.6 \text{ km}$

$= 1638.4 \text{ km}^2$

34. $208.13 \frac{\text{lb}}{\text{ft}}$

35. $\frac{3}{2} = \frac{w}{5}$ (Using a proportion)

$5 \cdot 3 = 2 \cdot w$

$15 = 2w$

$7.5 = w$

A steel rod 3 cm long weighs 7.5 g.

$\frac{500}{2} = \frac{w}{5}$ (Using a proportion)
(5 m = 500 cm)

$5 \cdot 500 = 2 \cdot w$

$2500 = 2w$

$1250 = w$

A steel rod 5 m long weighs 1250 g.

36. a) 1.72 cm

b) 0.08 cm

37. 1 mole of oxygen is 32 grams.

$\frac{50}{1} = \frac{x}{32}$ (Using a proportion)

$50 \cdot 32 = x$

$1600 = x$

50 moles of oxygen is 1600 grams of oxygen.

38. 888.8 g

39. 1 mole of neon is 20.2 grams

$\frac{303}{20.2} = \frac{x}{1}$ (Using a proportion)

$15 = x$

303 grams of neon is 15 moles of neon.

40. 11.8 moles

41. $E = mc^2$

Substitute 5000 for m and 2.9979×10^8 for c.

$E = 5000(2.9979 \times 10^8)^2$

$\approx 44{,}937 \times 10^{16} = 4.4937 \times 10^4 \times 10^{16}$

$= 4.4937 \times 10^{20} \text{ g m}^2/\text{sec}^2$

42. 15.48 g

43. From Exercise 41 we know that the speed of light is 2.9979×10^8 m/sec. We convert m/sec to m/nanosecond.

$$2.9979 \times 10^8 \frac{\text{m}}{\text{sec}} \cdot \frac{10^{-9}\text{ sec}}{1\text{ nanosecond}} =$$

$$2.9979 \times 10^{-1} \frac{\text{m}}{\text{nanosecond}},\text{ or } 0.29979 \frac{\text{m}}{\text{nanosecond}}$$

Light travels 0.29979 m in one nanosecond.

Exercise Set 2.1

1. $4x + 12 = 60$
$4x + 12 - 12 = 60 - 12$
$4x = 48$
$\frac{1}{4} \cdot 4x = \frac{1}{4} \cdot 48$
$x = 12$
The solution set is {12}.

2. {24}

3. $4 + \frac{1}{2}x = 1$
$4 + \frac{1}{2}x - 4 = 1 - 4$
$\frac{1}{2}x = -3$
$2 \cdot \frac{1}{2}x = 2(-3)$
$x = -6$
The solution set is {-6}.

4. {14}

5. $y + 1 = 2y - 7$ or $y + 1 = 2y - 7$
$1 + 7 = 2y - y$ $\quad$ $y - 2y = -7 - 1$
$8 = y$ $\quad$ $-y = -8$
$y = 8$
The solution set is {8}.

6. $\left\{\frac{18}{5}\right\}$

7. $5x - 2 + 3x = 2x + 6 - 4x$
$8x - 2 = 6 - 2x$
$8x + 2x = 6 + 2$
$10x = 8$
$x = \frac{8}{10}$
$x = \frac{4}{5}$
The solution set is $\left\{\frac{4}{5}\right\}$.

8. {-8}

9. $1.9x - 7.8 + 5.3x = 3.0 + 1.8x$
$7.2x - 7.8 = 3.0 + 1.8x$
$7.2x - 1.8x = 3.0 + 7.8$
$5.4x = 10.8$
$x = \frac{10.8}{5.4}$
$x = 2$
The solution set is {2}.

10. {-3}

11. $7(3x + 6) = 11 - (x + 2)$
$21x + 42 = 11 - x - 2$
$21x + 42 = 9 - x$
$21x + x = 9 - 42$
$22x = -33$
$x = -\frac{33}{22}$
$x = -\frac{3}{2}$
The solution set is $\left\{-\frac{3}{2}\right\}$.

12. $\left\{-\frac{27}{14}\right\}$

13. $2x - (5 + 7x) = 4 - [x - (2x + 3)]$
$2x - 5 - 7x = 4 - x + 2x + 3$
$-5x - 5 = 7 + x$
$-5 - 7 = x + 5x$
$-12 = 6x$
$-\frac{12}{6} = x$
$-2 = x$
The solution set is {-2}.

14. ∅

15. $(2x - 3)(3x - 2) = 0$
$2x - 3 = 0$ or $3x - 2 = 0$
$2x = 3$ or $3x = 2$
$x = \frac{3}{2}$ or $x = \frac{2}{3}$
The solution set is $\left\{\frac{3}{2}, \frac{2}{3}\right\}$.

16. $\left\{\frac{2}{5}, -\frac{3}{2}\right\}$

17. $x(x - 1)(x + 2) = 0$
$x = 0$ or $x - 1 = 0$ or $x + 2 = 0$
$x = 0$ or $x = 1$ or $x = -2$
The solution set is {0, 1, -2}.

18. {0, -2, 3}

19. $3x^2 + x - 2 = 0$
$(3x - 2)(x + 1) = 0$
$3x - 2 = 0$ or $x + 1 = 0$
$x = \frac{2}{3}$ or $x = -1$
The solution set is $\left\{\frac{2}{3}, -1\right\}$.

20. $\left\{\frac{3}{5}, 1\right\}$

21. $(x - 1)(x + 1) = 5(x - 1)$
$x^2 - 1 = 5x - 5$
$x^2 - 5x + 4 = 0$
$(x - 4)(x - 1) = 0$
$x - 4 = 0$ or $x - 1 = 0$
$x = 4$ or $x = 1$
The solution set is {4, 1}.

22. {8, 3}

23. $x[4(x - 2) - 5(x - 1)] = 2$
$x(4x - 8 - 5x + 5) = 2$
$x(-x - 3) = 2$
$-x^2 - 3x = 2$
$0 = x^2 + 3x + 2$
$0 = (x + 2)(x + 1)$
$x + 2 = 0$ or $x + 1 = 0$
$x = -2$ or $x = -1$
The solution set is {-2, -1}.

24. {5, 10}

25. $(3x^2 - 7x - 20)(2x - 5) = 0$
$(3x + 5)(x - 4)(2x - 5) = 0$
$3x + 5 = 0$ or $x - 4 = 0$ or $2x - 5 = 0$
$x = -\frac{5}{3}$ or $x = 4$ or $x = \frac{5}{2}$
The solution set is $\left\{-\frac{5}{3}, 4, \frac{5}{2}\right\}$.

26. $\left\{-\frac{11}{8}, -\frac{1}{4}, \frac{2}{3}\right\}$

27. $16x^3 = x$
$16x^3 - x = 0$
$x(16x^2 - 1) = 0$
$x(4x + 1)(4x - 1) = 0$
$x = 0$ or $4x + 1 = 0$ or $4x - 1 = 0$
$x = 0$ or $x = -\frac{1}{4}$ or $x = \frac{1}{4}$
The solution set is $\left\{0, -\frac{1}{4}, \frac{1}{4}\right\}$.

28. $\left\{0, \frac{1}{3}, -\frac{1}{3}\right\}$

29. $2x^2 = 6x$
$2x^2 - 6x = 0$
$2x(x - 3) = 0$
$2x = 0$ or $x - 3 = 0$
$x = 0$ or $x = 3$
The solution set is {0, 3}.

30. {0, -2}

31. $3y^3 - 5y^2 - 2y = 0$
$y(3y^2 - 5y - 2) = 0$
$y(3y + 1)(y - 2) = 0$
$y = 0$ or $3y + 1 = 0$ or $y - 2 = 0$
$y = 0$ $y = -\frac{1}{3}$ or $y = 2$
The solution set is $\left\{0, -\frac{1}{3}, 2\right\}$

32. $\left\{0, 1, \frac{2}{3}\right\}$

33. $(2x - 3)(3x + 2)(x - 1) = 0$
$2x - 3 = 0$ or $3x + 2 = 0$ or $x - 1 = 0$
$x = \frac{3}{2}$ or $x = -\frac{2}{3}$ or $x = 1$
The solution set is $\left\{\frac{3}{2}, -\frac{2}{3}, 1\right\}$.

34. $\left\{4, -3, -\frac{1}{2}\right\}$

35. $(2 - 4y)(y^2 + 3y) = 0$
$2(1 - 2y)y(y + 3) = 0$
$1 - 2y = 0$ or $y = 0$ or $y + 3 = 0$
$y = \frac{1}{2}$ or $y = 0$ or $y = -3$
The solution set is $\left\{\frac{1}{2}, 0, -3\right\}$.

36. {3, -3, 6, -6}

37. $x + 4 = 8 + x$
$-x + x + 4 = -x + 8 + x$
$4 = 8$
We get a false equation. There are no solutions, so the solution set is Ø.

38. Ø

39. $7x^3 + x^2 - 7x - 1 = 0$
$x^2(7x + 1) - (7x + 1) = 0$
$(x^2 - 1)(7x + 1) = 0$
$(x + 1)(x - 1)(7x + 1) = 0$
$x + 1 = 0$ or $x - 1 = 0$ or $7x + 1 = 0$
$x = -1$ or $x = 1$ or $x = -\frac{1}{7}$
The solution set is $\left\{-1, 1, -\frac{1}{7}\right\}$.

40. $\left\{2, -2, -\frac{1}{3}\right\}$

41. $y^3 + 2y^2 - y - 2 = 0$
$y^2(y + 2) - (y + 2) = 0$
$(y^2 - 1)(y + 2) = 0$
$(y + 1)(y - 1)(y + 2) = 0$

$y + 1 = 0$ or $y - 1 = 0$ or $y + 2 = 0$
$y = -1$ or $y = 1$ or $y = -2$

The solution set is $\{-1, 1, -2\}$.

42. $\{-1, 5, -5\}$

43. $11 + x = x + 11$
$-x + 11 + x = -x + x + 11$
$11 = 11$

We get a true equation. The solution set is the set of all real numbers.

44. All real numbers

45. $x + 6 < 5x - 6$ or $x + 6 < 5x - 6$
$6 + 6 < 5x - x$ | $x - 5x < -6 - 6$
$12 < 4x$ | $-4x < -12$
$\frac{12}{4} < x$ | $x > \frac{12}{4}$
$3 < x$ | $x > 3$

The solution set is $\{x|x > 3\}$.

46. $\left\{x|x > -\frac{4}{5}\right\}$

47. $3x - 3 + 2x \geqslant 1 - 7x - 9$
$5x - 3 \geqslant -7x - 8$
$5x + 7x \geqslant -8 + 3$
$12x \geqslant -5$
$x \geqslant -\frac{5}{12}$

The solution set is $\left\{x|x \geqslant -\frac{5}{12}\right\}$.

48. $\left\{y|y \leqslant -\frac{1}{12}\right\}$

49. $14 - 5y \leqslant 8y - 8$ or $14 - 5y \leqslant 8y - 8$
$14 + 8 \leqslant 8y + 5y$ | $-5y - 8y \leqslant -8 - 14$
$22 \leqslant 13y$ | $-13y \leqslant -22$
$\frac{22}{13} \leqslant y$ | $y \geqslant \frac{22}{13}$

The solution set is $\left\{y|y \geqslant \frac{22}{13}\right\}$.

50. $\{x|x < 5\}$

51. $-\frac{3}{4}x \geqslant -\frac{5}{8} + \frac{2}{3}x$
$\frac{5}{8} \geqslant \frac{3}{4}x + \frac{2}{3}x$
$\frac{5}{8} \geqslant \frac{9}{12}x + \frac{8}{12}x$
$\frac{5}{8} \geqslant \frac{17}{12}x$
$\frac{12}{17} \cdot \frac{5}{8} \geqslant \frac{12}{17} \cdot \frac{17}{12}x$
$\frac{15}{34} \geqslant x$

The solution set is $\left\{x|x \leqslant \frac{15}{34}\right\}$.

52. $\left\{x|x \geqslant -\frac{3}{14}\right\}$

53. $4x(x - 2) < 2(2x - 1)(x - 3)$
$4x(x - 2) < 2(2x^2 - 7x + 3)$
$4x^2 - 8x < 4x^2 - 14x + 6$
$-8x < -14x + 6$
$-8x + 14x < 6$
$6x < 6$
$x < \frac{6}{6}$
$x < 1$

The solution set is $\{x|x < 1\}$.

54. $\{x|x > -1\}$

55. $\{x|x > 2.5\}$

56. $\{y|y \leqslant -7\}$

57. $\{t|t^2 = 5\}$

58. $\{m|m^3 + 3 = m^2 - 2\}$

59. $\sqrt{x - 3}$

The radicand must be nonnegative. We set $x - 3 \geqslant 0$ and solve for x.

$x - 3 \geqslant 0$
$x \geqslant 3$

The meaningful replacements are $\{x|x \geqslant 3\}$.

60. $\left\{x|x \geqslant \frac{5}{2}\right\}$

61. $\sqrt{3 - 4x}$

The radicand must be nonnegative. We set $3 - 4x \geqslant 0$ and solve for x.

$3 - 4x \geqslant 0$
$-4x \geqslant -3$
$x \leqslant \frac{3}{4}$

The meaningful replacements are $\left\{x|x \leqslant \frac{3}{4}\right\}$.

62. $\{x \mid x \text{ is a real number}\}$

63. $$\begin{aligned} 2.905x - 3.214 + 6.789x &= 3.012 + 1.805x \\ 9.694x - 3.214 &= 3.012 + 1.805x \\ 9.694x - 1.805x &= 3.012 + 3.214 \\ 7.889x &= 6.226 \\ x &= \frac{6.226}{7.889} \\ x &\approx 0.7892 \end{aligned}$$

The solution set is $\{0.7892\}$.

64. $\{-1.3053, 1.9892\}$

65. $3.12x^2 - 6.715x = 0$

$x(3.12x - 6.715) = 0$

$x = 0$ or $3.12x - 6.715 = 0$

$x = 0$ or $3.12x = 6.715$

$x = 0$ or $x \approx 2.1522$

The solution set is $\{0, 2.1522\}$.

66. $\{0, -1.9492\}$

67. $$\begin{aligned} 1.52(6.51x + 7.3) &< 11.2 - (7.2x + 13.52) \\ 9.8952x + 11.096 &< 11.2 - 7.2x - 13.52 \\ 9.8952x + 11.096 &< -7.2x - 2.32 \\ 9.8952x + 7.2x &< -2.32 - 11.096 \\ 17.0952x &< -13.416 \\ x &< -\frac{13.416}{17.0952} \\ x &< -0.7848 \end{aligned}$$

The solution set is $\{x \mid x < -0.7848\}$.

68. $\{y \mid y \geqslant -2.2353\}$

69. $$\begin{aligned} (x + 1)^3 &= (x - 1)^3 + 26 \\ x^3 + 3x^2 + 3x + 1 &= x^3 - 3x^2 + 3x - 1 + 26 \\ x^3 + 3x^2 + 3x + 1 &= x^3 - 3x^2 + 3x + 25 \\ 6x^2 - 24 &= 0 \\ 6(x^2 - 4) &= 0 \\ 6(x + 2)(x - 2) &= 0 \end{aligned}$$

$x + 2 = 0$ or $x - 2 = 0$

$x = -2$ or $x = 2$

The solution set is $\{-2, 2\}$.

70. $\{1\}$

71. $(x^2 - x - 20)(x^2 - 25) = 0$

$(x - 5)(x + 4)(x + 5)(x - 5) = 0$

$x - 5 = 0$ or $x + 4 = 0$ or $x + 5 = 0$ or $x - 5 = 0$

$x = 5$ or $x = -4$ or $x = -5$ or $x = 5$

The solution set is $\{5, -4, -5\}$.

72. $\{4, -4, 3, -3, -1\}$

Exercise Set 2.2

1. $3x + 5 = 12$

$3x = 7$

$x = \frac{7}{3}$

The solution set is $\left\{\frac{7}{3}\right\}$.

$3x = 7$

$x = \frac{7}{3}$

The solution set is $\left\{\frac{7}{3}\right\}$.

The equations are equivalent.

2. No

3. $x = 3$

The solution set is $\{3\}$.

$x^2 = 9$

$x = \pm 3$

The solution set is $\{-3, 3\}$.

The equations are not equivalent.

4. Yes

5. $\frac{(x - 3)(x + 9)}{(x - 3)} = x + 9$, Note: $x \neq 3$

$x + 9 = x + 9$

$9 = 9$

The solution set is the set of all numbers except 3.

$x + 9 = x + 9$

$9 = 9$

The solution set is the set of all numbers.

Thus, the equations are not equivalent.

6. No

7. $\frac{1}{4} + \frac{1}{5} = \frac{1}{t}$, LCM = 20t

$$\begin{aligned} 20t\left(\frac{1}{4} + \frac{1}{5}\right) &= 20t \cdot \frac{1}{t} \\ 20t \cdot \frac{1}{4} + 20t \cdot \frac{1}{5} &= 20t \cdot \frac{1}{t} \\ 5t + 4t &= 20 \\ 9t &= 20 \\ t &= \frac{20}{9} \end{aligned}$$

The solution set is $\left\{\frac{20}{9}\right\}$.

Check:

$\frac{1}{4} + \frac{1}{5}$	$= \frac{1}{t}$
$\frac{1}{4} + \frac{1}{5}$	$\frac{1}{\frac{20}{9}}$
$\frac{5}{20} + \frac{4}{20}$	$1 \cdot \frac{9}{20}$
$\frac{9}{20}$	$\frac{9}{20}$

8. $\{-2\}$

9. $\frac{3}{x-8} = \frac{x-5}{x-8}$, LCM = x - 8

$(x-8)\cdot\frac{3}{x-8} = (x-8)\cdot\frac{x-5}{x-8}$

$3 = x - 5$

$8 = x$

The solution set is ∅.

Check:

$\frac{3}{x-8} = \frac{x-5}{x-8}$	
$\frac{3}{8-8}$	$\frac{8-5}{8-8}$
$\frac{3}{0}$	$\frac{3}{0}$

Division by zero is undefined.

10. ∅

11. $\frac{x+2}{4} - \frac{x-1}{5} = 15$, LCM = 20

$20\left[\frac{x+2}{4} - \frac{x-1}{5}\right] = 20\cdot 15$

$5(x+2) - 4(x-1) = 300$

$5x + 10 - 4x + 4 = 300$

$x + 14 = 300$

$x = 286$

The solution set is {286}.

12. {-1}

13. $x + \frac{6}{x} = 5$, LCM = x

$x\left[x + \frac{6}{x}\right] = x\cdot 5$

$x^2 + 6 = 5x$

$x^2 - 5x + 6 = 0$

$(x-3)(x-2) = 0$

$x - 3 = 0$ or $x - 2 = 0$

$x = 3$ or $x = 2$

The solution set is {3, 2}.

Check:

For x = 3:

$x + \frac{6}{x} = 5$	
$3 + \frac{6}{3}$	5
$3 + 2$	
5	

For x = 2:

$x + \frac{6}{x} = 5$	
$2 + \frac{6}{2}$	5
$2 + 3$	
5	

14. {4,-3}

15. $\frac{x+2}{2} + \frac{3x+1}{5} = \frac{x-2}{4}$, LCM = 20

$20\left[\frac{x+2}{2} + \frac{3x+1}{5}\right] = 20\cdot\frac{x-2}{4}$

$10(x+2) + 4(3x+1) = 5(x-2)$

$10x + 20 + 12x + 4 = 5x - 10$

$22x + 24 = 5x - 10$

$22x - 5x = -10 - 24$

$17x = -34$

$x = -2$

The solution set is {-2}.

16. {2}

17. $\frac{1}{2} + \frac{2}{x} = \frac{1}{3} + \frac{3}{x}$, LCM = 6x

$6x\left[\frac{1}{2} + \frac{2}{x}\right] = 6x\left[\frac{1}{3} + \frac{3}{x}\right]$

$3x + 12 = 2x + 18$

$3x - 2x = 18 - 12$

$x = 6$

The solution set is {6}.

Check:

$\frac{1}{2} + \frac{2}{x} = \frac{1}{3} + \frac{3}{x}$	
$\frac{1}{2} + \frac{2}{6}$	$\frac{1}{3} + \frac{3}{6}$
$\frac{1}{2} + \frac{1}{3}$	$\frac{1}{3} + \frac{1}{2}$

18. $\left\{\frac{11}{30}\right\}$

19. $\frac{4}{x^2-1} - \frac{2}{x-1} = \frac{3}{x+1}$,

LCM = (x + 1)(x - 1)

$(x+1)(x-1)\left[\frac{4}{(x+1)(x-1)} - \frac{2}{x-1}\right] = (x+1)(x-1)\cdot\frac{3}{x+1}$

$4 - 2(x+1) = 3(x-1)$

$4 - 2x - 2 = 3x - 3$

$2 - 2x = 3x - 3$

$2 + 3 = 3x + 2x$

$5 = 5x$

$1 = x$

Check:

$\frac{4}{x^2-1} - \frac{2}{x-1} = \frac{3}{x+1}$	
$\frac{4}{1^2-1} - \frac{2}{1-1}$	$\frac{3}{1+1}$
$\frac{4}{0} - \frac{2}{0}$	$\frac{3}{2}$

Division by zero is undefined.

The solution set is ∅.

20. ∅

21. $\frac{1}{2t} - \frac{2}{5t} = \frac{1}{10t} - 3$, LCM = 10t

$10t\left[\frac{1}{2t} - \frac{2}{5t}\right] = 10t\left[\frac{1}{10t} - 3\right]$

$5\cdot 1 - 2\cdot 2 = 1\cdot 1 - 10t\cdot 3$

$5 - 4 = 1 - 30t$

$1 = 1 - 30t$

$30t = 0$

$t = 0$

Division by zero is undefined.

The solution set is ∅.

22. ∅

23. $1 - \frac{3}{x} = \frac{40}{x^2}$, LCM = x^2

$$x^2\left[1 - \frac{3}{x}\right] = x^2 \cdot \frac{40}{x^2}$$
$$x^2 - 3x = 40$$
$$x^2 - 3x - 40 = 0$$
$$(x - 8)(x + 5) = 0$$
$$x - 8 = 0 \text{ or } x + 5 = 0$$
$$x = 8 \text{ or } \quad x = -5$$

The solution set is {8,-5}.

Check:

For 8:

$1 - \frac{3}{x} = \frac{40}{x^2}$	
$1 - \frac{3}{8}$	$\frac{40}{8^2}$
$\frac{5}{8}$	$\frac{40}{64}$
	$\frac{5}{8}$

For -5:

$1 - \frac{3}{x} = \frac{40}{x^2}$	
$1 - \frac{3}{-5}$	$\frac{40}{(-5)^2}$
$\frac{5}{5} + \frac{3}{5}$	$\frac{40}{25}$
$\frac{8}{5}$	$\frac{8}{5}$

24. {5,-3}

25. $\frac{11 - t^2}{3t^2 - 5t + 2} = \frac{2t + 3}{3t - 2} - \frac{t - 3}{t - 1}$,

LCM = $(3t - 2)(t - 1)$

$$(3t-2)(t-1) \cdot \frac{11 - t^2}{(3t-2)(t-1)} = (3t-2)(t-1)\left[\frac{2t+3}{3t-2} - \frac{t-3}{t-1}\right]$$
$$11 - t^2 = (t-1)(2t+3)-(3t-2)(t-3)$$
$$11 - t^2 = (2t^2+t-3) - (3t^2-11t+6)$$
$$11 - t^2 = 2t^2+t-3-3t^2+11t-6$$
$$11 - t^2 = -t^2 + 12t - 9$$
$$11 = 12t - 9$$
$$20 = 12t$$
$$\frac{20}{12} = t$$
$$\frac{5}{3} = t$$

The value checks. The solution set is $\left\{\frac{5}{3}\right\}$.

26. {1}

27. $\frac{7x}{x - 3} - \frac{21}{x} + 11 = \frac{63}{x^2 - 3x}$, LCM = $x(x - 3)$

We note at the outset that 0 and 3 are not meaningful replacements.

$$x(x - 3)\left[\frac{7x}{x - 3} - \frac{21}{x} + 11\right] = x(x - 3) \cdot \frac{63}{x(x - 3)}$$
$$7x^2 - 21(x - 3) + 11x(x - 3) = 63$$
$$7x^2 - 21x + 63 + 11x^2 - 33x = 63$$
$$18x^2 - 54x = 0$$
$$18x(x - 3) = 0$$
$$18x = 0 \text{ or } x - 3 = 0$$
$$x = 0 \text{ or } \quad x = 3$$

Neither number checks. The solution set is ∅.

28. ∅

29. $\frac{2.315}{y} - \frac{12.6}{17.4} = \frac{6.71}{7} + 0.763$,

LCM = 7(17.4)y

$$7(17.4)y\left[\frac{2.315}{y} - \frac{12.6}{17.4}\right] = 7(17.4)y\left[\frac{6.71}{7} + 0.763\right]$$
$$281.967 - 88.2y = 116.754y + 92.9334y$$
$$281.967 - 88.2y = 209.6874y$$
$$281.967 = 297.8874y$$
$$0.94656 \approx y$$

The value checks. The solution set is {0.94656}.

30. {0.0855}

31. $\frac{2x^2}{x - 3} + \frac{4x - 6}{x + 3} = \frac{108}{x^2 - 9}$, LCM = $(x + 3)(x - 3)$

We note at the outset that 3 and -3 are not meaningful replacements.

$$(x+3)(x-3)\left[\frac{2x^2}{x-3} + \frac{4x-6}{x+3}\right] = (x+3)(x-3) \cdot \frac{108}{(x+3)(x-3)}$$
$$2x^2(x+3) + (x-3)(4x-6) = 108$$
$$2x^3+6x^2+4x^2-18x+18 = 108$$
$$2x^3 + 10x^2 - 18x + 18 = 108$$
$$2x^3 + 10x^2 - 18x - 90 = 0$$
$$2x^2(x + 5) - 18(x + 5) = 0$$
$$(2x^2 - 18)(x + 5) = 0$$
$$2(x + 3)(x - 3)(x + 5) = 0$$
$$x + 3 = 0 \text{ or } x - 3 = 0 \text{ or } x + 5 = 0$$
$$x = -3 \text{ or } \quad x = 3 \text{ or } \quad x = -5$$

The number -5 checks, but -3 and 3 do not. The solution set is {-5}.

32. $\left\{\frac{8}{3}\right\}$

33. $\frac{24}{x^2 - 2x + 4} = \frac{3x}{x + 2} + \frac{72}{x^3 + 8}$

LCM = $(x + 2)(x^2 - 2x + 4)$

We note at the outset that -2 is not a meaningful replacement.

$$(x+2)(x^2-2x+4) \cdot \frac{24}{x^2-2x+4} =$$
$$(x+2)(x^2-2x+4)\left[\frac{3x}{x+2} + \frac{72}{(x+2)(x^2-2x+4)}\right]$$
$$24(x + 2) = 3x(x^2 - 2x + 4) + 72$$
$$24x + 48 = 3x^3 - 6x^2 + 12x + 72$$
$$0 = 3x^3 - 6x^2 - 12x + 24$$
$$0 = 3x^2(x - 2) - 12(x - 2)$$
$$0 = (3x^2 - 12)(x - 2)$$
$$0 = 3(x + 2)(x - 2)(x - 2)$$
$$x + 2 = 0 \text{ or } x - 2 = 0 \text{ or } x - 2 = 0$$
$$x = -2 \text{ or } \quad x = 2 \text{ or } \quad x = 2$$

The number 2 checks, but -2 does not. The solution set is {2}.

34. {3}

35. $\frac{5}{x - 1} + \frac{9}{x^2 + x + 1} = \frac{15}{x^3 - 1}$

LCM = $(x - 1)(x^2 + x + 1)$

We note at the outset that 1 is not a meaningful replacement.

$$(x-1)(x^2+x+1)\left[\frac{5}{x-1} + \frac{9}{x^2+x+1}\right] = (x-1)(x^2+x+1) \cdot \frac{15}{(x-1)(x^2+x+1)}$$

$$5(x^2 + x + 1) + 9(x - 1) = 15$$
$$5x^2 + 5x + 5 + 9x - 9 = 15$$
$$5x^2 + 14x - 4 = 15$$
$$5x^2 + 14x - 19 = 0$$
$$(5x + 19)(x - 1) = 0$$

$5x + 19 = 0$ or $x - 1 = 0$

$5x = -19$ or $x = 1$

$x = -\frac{19}{5}$ or $x = 1$

The number $-\frac{19}{5}$ checks, but 1 does not. The solution set is $\left\{-\frac{19}{5}\right\}$.

36. $\left\{\frac{23}{7}\right\}$

37. $\frac{7}{x - 9} - \frac{7}{x} = \frac{63}{x^2 - 9x}$, LCM = $x(x - 9)$

We note at the outset that 9 and 0 are not meaningful replacements.

$$x(x - 9)\left[\frac{7}{x - 9} - \frac{7}{x}\right] = x(x - 9) \cdot \frac{63}{x(x - 9)}$$
$$7x - 7(x - 9) = 63$$
$$7x - 7x + 63 = 63$$
$$63 = 63$$

We get a true equation. The solution set is the set of all real numbers.

38. All real numbers

39. $\frac{(x - 3)^2}{x - 3} = x - 3$, LCM = $x - 3$

Note: $x \neq 3$
Division by 0 is undefined.

$$(x - 3) \cdot \frac{(x - 3)^2}{x - 3} = (x - 3)(x - 3)$$
$$(x - 3)^2 = (x - 3)^2$$

The solution set is $\{x|x \neq 3\}$.

40. $\{x|x \neq 2\}$

41. $\frac{x^3 + 8}{x + 2} = x^2 - 2x + 4$, LCM = $x + 2$

Note: $x \neq -2$
Division by 0 is undefined.

$$(x + 2) \cdot \frac{x^3 + 8}{x + 2} = (x + 2)(x^2 - 2x + 4)$$
$$x^3 + 8 = x^3 + 8$$

The solution set is $\{x|x \neq -2\}$.

42. $\{x|x \neq 2\}$

43. Find the solution set of each equation.

Equation (1): $x^2 - x - 20 = x^2 - 25$

$-x = -5$

$x = 5$

The solution set is $\{5\}$.

Equation (2): $(x - 5)(x + 4) = (x - 5)(x + 5)$

$x^2 - x - 20 = x^2 - 25$

Multiplying, we get equation (1), so the solution set of equation (2) is also $\{5\}$.

Equation (3): $x + 4 = x + 5$

$4 = 5$ Adding $-x$

We get a false equation, so the solution set is $\emptyset$.

Equation (4): $4 = 5$

This is a false equation, so the solution set is $\emptyset$.

(1) and (2) have the same solution set, so they are equivalent; (2) and (3) do not have the same solution set, so they are not equivalent; (3) and (4) have the same solution set, so they are equivalent.

44. Identity

45. $x + 4 = 4 + x$

All real numbers are meaningful replacements. Adding $-x$ to both sides, we get $4 = 4$, a true equation. Thus, the equation is an identity.

46. Identity

47. $\frac{x^3 + 8}{x^2 - 4} = \frac{x^2 - 2x + 4}{x - 2}$

$\frac{x^3 + 8}{(x + 2)(x - 2)} = \frac{x^2 - 2x + 4}{x - 2}$

All real numbers except -2 and 2 are meaningful replacements.

$(x+2)(x-2) \cdot \frac{x^3 + 8}{(x+2)(x-2)} = (x+2)(x-2) \cdot \frac{x^2-2x+4}{x - 2}$

$x^3 + 8 = (x + 2)(x^2 - 2x + 4)$

$x^3 + 8 = x^3 + 8$

$8 = 8$

We get an equation that is true for all meaningful replacements of the variables. Thus, the equation is an identity.

48. Not an identity

49. $\sqrt{x^2 - 16} = x - 4$

The set of meaningful replacements is $\{x | x \leqslant -4 \text{ or } x \geqslant 4\}$. Substitute 5, a meaningful replacement, for x in the equation.

$\sqrt{5^2 - 16} = 5 - 4$

$\sqrt{9} = 1$

$3 = 1$

We get a false equation. Thus, the equation is not true for all meaningful replacements. It is not an identity.

50. $\left\{-\frac{7}{2}\right\}$

Exercise Set 2.3

1. $P = 2\ell + 2w$

$P - 2\ell = 2w$

$\frac{P - 2\ell}{2} = w$

2. $r = \frac{C}{2\pi}$

3. $A = \frac{1}{2}bh$

$2A = bh$

$\frac{2A}{h} = b$

4. $\pi = \frac{r^2}{A}$

5. $d = rt$

$\frac{d}{t} = r$

6. $a = \frac{F}{m}$

7. $E = IR$

$\frac{E}{R} = I$

8. $m_2 = \frac{Fd^2}{km_1}$

9. $\frac{P_1V_1}{T_1} = \frac{P_2V_2}{T_2}$

$T_1T_2 \cdot \frac{P_1V_1}{T_1} = T_1T_2 \cdot \frac{P_2V_2}{T_2}$

$T_2P_1V_1 = T_1P_2V_2$

$\frac{T_2P_1V_1}{P_2V_2} = T_1$

10. $V_2 = \frac{T_2P_1V_1}{T_1P_2}$

11. $S = \frac{H}{m(V_1 - V_2)}$ or $S = \frac{H}{m(V_1 - V_2)}$

$m(V_1 - V_2)S = H$ | $m(V_1 - V_2)S = H$

$mSV_1 - mSV_2 = H$ | $V_1 - V_2 = \frac{H}{mS}$

$mSV_1 = H + mSV_2$ | $V_1 = \frac{H}{mS} + V_2$

$V_1 = \frac{H + mSV_2}{mS}$

12. $V_2 = V_1 - \frac{H}{Sm}$

13. $\frac{1}{F} = \frac{1}{m} + \frac{1}{p}$

$Fmp \cdot \frac{1}{F} = Fmp\left(\frac{1}{m} + \frac{1}{p}\right)$

$mp = Fp + Fm$

$mp - Fp = Fm$

$p(m - F) = Fm$

$p = \frac{Fm}{m - F}$

14. $F = \frac{mp}{p + m}$

15. $(x + a)(x - b) = x^2 + 5$

$x^2 - bx + ax - ab = x^2 + 5$

$ax - bx = 5 + ab$

$x(a - b) = 5 + ab$

$x = \frac{5 + ab}{a - b}$

16. $x = \frac{c}{2}$

17. $10(a + x) = 8(a - x)$

$10a + 10x = 8a - 8x$

$10x + 8x = 8a - 10a$

$18x = -2a$

$x = -\frac{2a}{18}$

$x = -\frac{a}{9}$

18. $x = a - 7b$

19. Familiarize: We let x = the percent. The percent symbol, %, will have to be added to the answer.

Translate:

79.2 is what percent of 180?

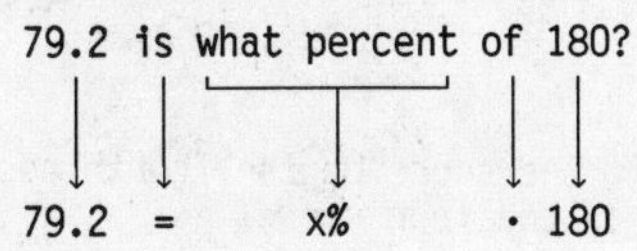

$$79.2 = x\% \cdot 180$$

Carry out:

$79.2 = x\% \cdot 180$

$79.2 = x \cdot 0.01 \cdot 180$

$79.2 = x \cdot 1.8$

$\frac{79.2}{1.8} = x$

$44 = x$

Check:

$44\% \cdot 180 = 0.44 \cdot 180 = 79.2$

State:

The answer is 44%.

20. 8000

21. Familiarize: We let x = the percent. The percent symbol, %, will have to be added to the answer.

Translate:

What percent of 28 is 1.68?

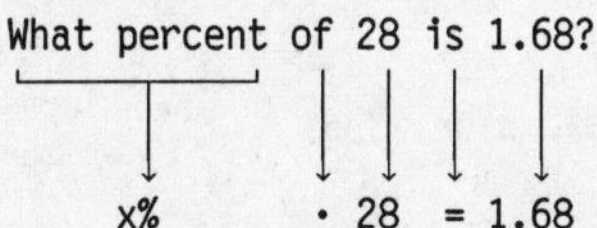

$$x\% \cdot 28 = 1.68$$

Carry out:

$x\% \cdot 28 = 1.68$

$x \cdot 0.01 \cdot 28 = 1.68$

$x \cdot 0.28 = 1.68$

$x = \frac{1.68}{0.28}$

$x = 6$

Check:

$6\% \cdot 28 = 0.06 \cdot 28 = 1.68$

State:

The answer is 6%.

22. 3.171

23. Familiarize: We let x = the old salary. Then the new salary is $x + 1595$.

Translate:

11% of the old salary is $1595.

$$11\% \cdot x = 1595$$

Carry out:

$11\% \cdot x = 1595$

$0.11x = 1595$

$x = \frac{1595}{0.11}$

$x = 14{,}500$

Check:

$11\% \cdot 14{,}500 = 0.11 \cdot 14{,}500 = 1595$

State:

The old salary is $14,500. The new salary is $14,500 + $1595, or $16,095.

24. $21,000, $23,520

25. Familiarize: We restate the situation.

The amount invested plus the interest is $702. We let x = the amount originally invested.

Translate:

Invested amount plus 8% of invested amount is $702.

$$x + 8\% \cdot x = 702$$

Carry out:

$x + 8\%x = 702$

$x + 0.08x = 702$

$(1 + 0.08)x = 702$

$1.08x = 702$

$x = 650$

Check:

$650 + 8% of $650 = $650 + $52 = $702

State:

The amount originally invested was $650.

26. $850

27. Familiarize: We make a drawing.

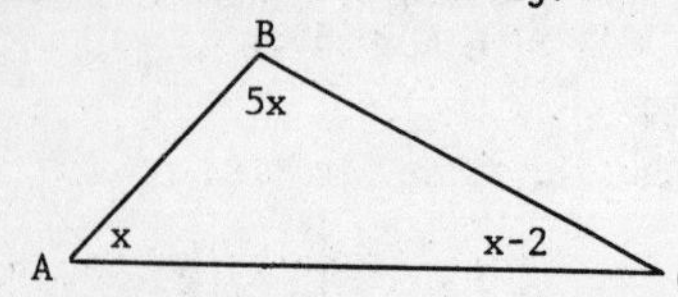

We let x represent the measure of angle A. Then $5x$ represents angle B, and $x - 2$ represents angle C. The sum of the angle measures is 180°.

Translate:

Measure of angle A + Measure of angle B + Measure of angle C = 180

$x + 5x + x - 2 = 180$

Carry out:

$$x + 5x + x - 2 = 180$$
$$7x - 2 = 180$$
$$7x = 182$$
$$x = 26$$

If $x = 26$, then $5x = 5 \cdot 26$, or 130, and $x - 2 = 26 - 2$, or 24.

Check:

The measure of angle B, 130°, is five times the measure of angle A, 26°. The measure of angle C, 24°, is 2° less than the measure of angle A, 26°. The sum of the angle measures is 26° + 130° + 24°, or 180°.

State:

Angle A measures 26°. Angle B measures 130°, and angle C measures 24°.

28. 40°, 80°, 60°

29. Familiarize: We make a drawing.

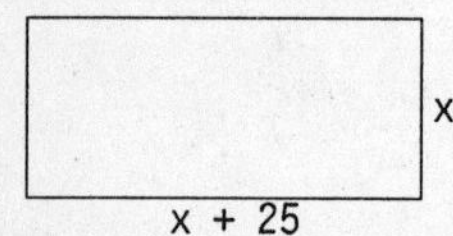

We let x represent the width. Then $x + 25$ represents the length.

Translate:

Perimeter = 2 × length + 2 × width

$322 = 2(x + 25) + 2(x)$

Carry out:

$$322 = 2x + 50 + 2x$$
$$322 = 4x + 50$$
$$272 = 4x$$
$$68 = x$$

If $x = 68$, then $x + 25 = 68 + 25$, or 93.

Check:

The length is 25 m more than the width: $93 = 68 + 25$. The perimeter is $2 \cdot 93 + 2 \cdot 68$, or $186 + 136$, or 322 m.

29. (continued)

State:

The length is 93 m; the width is 68 m.

30. 13 m, 6.5 m

31. Familiarize:

We let x represent the score on the fourth test. Then the average of the four scores is

$$\frac{87 + 64 + 78 + x}{4}.$$

Translate:

The average must be 80.

$$\frac{87 + 64 + 78 + x}{4} = 80$$

Carry out:

$$87 + 64 + 78 + x = 4 \cdot 80$$
$$229 + x = 320$$
$$x = 91$$

Check:

The average of 87, 64, 78, and 91 is $\frac{87 + 64 + 78 + 91}{4}$, or $\frac{320}{4}$, or 80.

State:

The score on the fourth test must be 91%.

32. 83%

33. Familiarize: We make a drawing.

x x x 10-2x x x 20-2x x x x

We let x represent the length of a side of the square in each corner. Then the length and width of the resulting base are represented by $20 - 2x$ and $10 - 2x$.

Translate:

The area of the base is 96 cm².

$$(20 - 2x)(10 - 2x) = 96$$

Carry out:

$$200 - 60x + 4x^2 = 96$$
$$4x^2 - 60x + 104 = 0$$
$$x^2 - 15x + 26 = 0$$
$$(x - 13)(x - 2) = 0$$

$$x - 13 = 0 \quad \text{or} \quad x - 2 = 0$$
$$x = 13 \quad \text{or} \quad x = 2$$

Since the length and width of the base cannot be negative, we only consider $x = 2$.

Check:

If $x = 2$, then $20 - 2x = 16$ and $10 - 2x = 6$. The area of the base is $16 \cdot 6$, or 96 cm².

33. (continued)

State:

The length of the sides of the squares is 2 cm.

34. 8 cm

35. Translate:

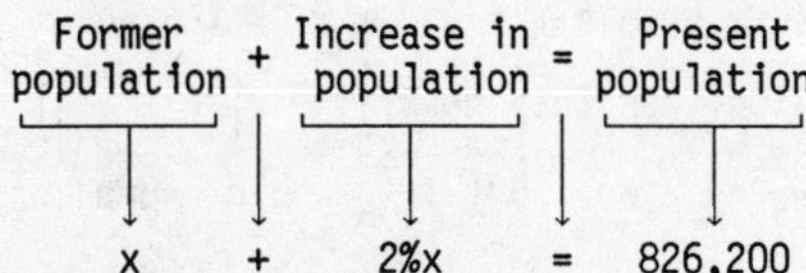

Carry out:

$$x + 0.02x = 826,200$$
$$1.02x = 826,200$$
$$x = \frac{826,200}{1.02}$$
$$x = 810,000$$

Check:

$2\% \cdot 810,000 = 0.02 \cdot 810,000 = 16,200$. The present population is 810,000 + 16,200, or 826,200.

State:

The former population was 810,000.

36. 720,000

37. Familiarize:

We first make a drawing. We let r represent the speed of the boat in still water. Then the speed downstream is r + 3, and the speed upstream is r - 3. The time to go downstream is the same as the time to go upstream. We call it t.

t hours — Downstream →
50 km — r + 3 km/h

t hours — ← Upstream
30 km — r - 3 km/h

We organize the information in a table.

	Distance	Speed	Time
Downstream	50	r + 3	t
Upstream	30	r - 3	t

Translate:

Using t = d/r we can get two expressions for t from the table.

$$t = \frac{50}{r+3} \quad \text{and} \quad t = \frac{30}{r-3}$$

Thus

$$\frac{50}{r+3} = \frac{30}{r-3}$$

37. (continued)

Carry out:

$$(r + 3)(r - 3) \cdot \frac{50}{r+3} = (r + 3)(r - 3) \cdot \frac{30}{r-3}$$

(Multiplying by the LCM)

$$50(r - 3) = 30(r + 3)$$
$$50r - 150 = 30r + 90$$
$$50r - 30r = 90 + 150$$
$$20r = 240$$
$$r = 12$$

Check:

If r = 12, the speed downstream is 12 + 3, or 15 km/h, and thus the time downstream is 50/15, or $3.\overline{3}$ hr. If r = 12, the speed upstream is 12 - 3, or 9 km/h, and thus the time upstream is 30/9 or $3.\overline{3}$ hr. The times are equal; the value checks.

State:

The speed of the boat in still water is 12 km/h.

38. 4 km/h

39. Familiarize:

We organize the information in a table. We let r represent the speed of train B. Then the speed of train A is r - 12. The time, t, is the same for each train.

	Distance	Speed	Time
Train A	230	r - 12	t
Train B	290	r	t

Translate:

Using t = d/r we get two expressions for t from the table.

$$t = \frac{230}{r-12} \quad \text{and} \quad t = \frac{290}{r}$$

Thus

$$\frac{230}{r-12} = \frac{290}{r}$$

Carry out:

$$r(r - 12) \cdot \frac{230}{r-12} = r(r - 12) \cdot \frac{290}{r}$$
$$230r = 290(r - 12)$$
$$230r = 290r - 3480$$
$$3480 = 60r$$
$$58 = r$$

Check:

If r = 58, then the speed of train A is 58 - 12, or 46 mph. Thus the time for train A is 230/46, or 5 hr. If the speed of train B is 58 mph, the time for train B is 290/58, or 5 hr. The times are the same; the value checks.

State:

The speed of train A is 46 mph; the speed of train B is 58 mph.

40. Passenger: 80 mph, freight: 66 mph

41. Familiarize:

We first make a drawing.

Chicago → 475 mph, t hours, d_1 miles

Cleveland ← 500 mph, $t - \frac{1}{3}$ hours, d_2 miles

← 350 miles →

We organize the information in a table.

	Distance	Speed	Time
From Chicago	d_1	475	t
From Cleveland	d_2	500	$t - \frac{1}{3}$

Translate:

Using d = rt we get two equations from the table.

$d_1 = 475t$ $\qquad$ $d_2 = 500\left(t - \frac{1}{3}\right)$

We also know that $d_1 + d_2 = 350$. Thus

$475t + 500\left(t - \frac{1}{3}\right) = 350$

Carry out:

$$475t + 500t - \frac{500}{3} = 350$$

$$975t = \frac{1050}{3} + \frac{500}{3}$$

$$975t = \frac{1550}{3}$$

$$2925t = 1550$$

$$t = \frac{1550}{2925}$$

$$t = \frac{62}{117}$$

Substitute $\frac{62}{117}$ for t and solve for d_2.

$$d_2 = 500\left(t - \frac{1}{3}\right)$$

$$d_2 = 500\left(\frac{62}{117} - \frac{1}{3}\right) = 500 \cdot \frac{23}{117} \approx 98.3$$

Check:

If $t = \frac{62}{117}$, then $d_1 = 475 \cdot \frac{62}{117}$, or ≈ 251.7. The sum of the distances is 98.3 + 251.7, or 350 miles. The value checks.

State:

When the trains meet, they are 98.3 miles from Cleveland.

42. 450 km

43. Familiarize:

A does $\frac{1}{3}$ of the job in 1 hr.

B does $\frac{1}{5}$ of the job in 1 hr.

C does $\frac{1}{7}$ of the job in 1 hr.

Together they do $\frac{1}{3} + \frac{1}{5} + \frac{1}{7} = \frac{71}{105}$ of the job in 1 hr. Together they do $2\left(\frac{1}{3}\right) + 2\left(\frac{1}{5}\right) + 2\left(\frac{1}{7}\right) = \frac{142}{105}$ or $1\frac{37}{105}$ in 2 hr. But $1\frac{37}{105}$ would represent more than 1 job.

We let t = the number of hours required for A, B, and C, working together, to do the job.

Translate:

$t\left(\frac{1}{3}\right) + t\left(\frac{1}{5}\right) + t\left(\frac{1}{7}\right) = 1$, or

$$\frac{t}{3} + \frac{t}{5} + \frac{t}{7} = 1$$

Carry out:

$$\frac{t}{3} + \frac{t}{5} + \frac{t}{7} = 1, \text{ LCM} = 3 \cdot 5 \cdot 7, \text{ or } 105$$

$$105\left(\frac{t}{3} + \frac{t}{5} + \frac{t}{7}\right) = 105 \cdot 1$$

$$35t + 21t + 15t = 105$$

$$71t = 105$$

$$t = \frac{105}{71}$$

$$t = 1\frac{34}{71}$$

Check:

In $1\frac{34}{71}$ hr, A will do $\left(\frac{1}{3}\right)\left(\frac{105}{71}\right)$, or $\frac{35}{71}$ of the job. Then B will do $\frac{1}{5}\left(\frac{105}{71}\right)$, or $\frac{21}{71}$ of the job and C will do $\frac{1}{7}\left(\frac{105}{71}\right)$, or $\frac{15}{71}$, or the job. Together they will $\frac{35}{71} + \frac{21}{71} + \frac{15}{71}$, or 1 complete job.

We have another partial check in noting from the familiarization step that the entire job can be done between 1 hr and 2 hr.

State: It will take $1\frac{34}{71}$ hr for A, B, and C to do the job together.

44. $\frac{30}{13}$ hr

45. Familiarize: Let t = the amount of time it takes B to do the job, working alone. In 1 hr A can do $\frac{1}{3.15}$ of the job and B can do $\frac{1}{t}$ of the job. Then in 2.09 hr A can do $2.09\left(\frac{1}{3.15}\right)$ of the job and B can do $2.09\left(\frac{1}{t}\right)$ of the job. If we add these fractional parts we get the entire job, represented by 1.

Translate: We add as described in the previous step and set the result equal to 1.

$$2.09\left(\frac{1}{3.15}\right) + 2.09\left(\frac{1}{t}\right) = 1, \text{ or } \frac{2.09}{3.15} + \frac{2.09}{t} = 1$$

Carry out: We first multiply by the LCM, 3.15t.

$$3.15t\left(\frac{2.09}{3.15} + \frac{2.09}{t}\right) = 3.15t(1)$$

$$2.09t + 6.5835 = 3.15t$$

$$6.5835 = 1.06t$$

$$6.21 \approx t$$

Check: In 2.09 hr A will do $\frac{2.09}{3.15}$ of the job and B will do $\frac{2.09}{6.21}$ of the job. Together they will do $\frac{2.09}{3.15} + \frac{2.09}{6.21} \approx 0.66 + 0.34$, or 1 complete job. The answer checks.

State: It would take B about 6.21 hr to do the job, working alone.

46. A: 23.95 hr, B: 51.02 hr

47. a) $A = P(1 + i)^t$

$A = 1000(1 + 0.1375)^1$

$= \$1137.50$

b) $A = P\left(1 + \frac{i}{2}\right)^{2t}$

$A = 1000\left(1 + \frac{0.1375}{2}\right)^2$

$= \$1142.23$

c) $A = P\left(1 + \frac{i}{4}\right)^{4t}$

$A = 1000\left(1 + \frac{0.1375}{4}\right)^4$

$= \$1144.75$

d) $A = P\left(1 + \frac{i}{365}\right)^{365t}$

$A = 1000\left(1 + \frac{0.1375}{365}\right)^{365}$

$= \$1147.37$

e) $A = P\left(1 + \frac{i}{8760}\right)^{8760t}$

$A = 1000\left(1 + \frac{0.1375}{8760}\right)^{8760}$

$= \$1147.40$

48. a) \$2055.46

b) \$2109.47

c) \$2139.05

d) \$2170.24

e) \$2170.58

49. Familiarize:
We organize the information in a table.

	Distance	Speed	Time
Slow trip	144	r	t
Fast trip	144	r + 4	$t - \frac{1}{2}$

We let r represent the speed for the slow trip and t the time for the slow trip. Then r + 4 and $t - \frac{1}{2}$ represent the speed and time respectively for the fast trip.

Translate:
Using t = d/r we get two equations from the table.

$$t = \frac{144}{r} \quad \text{and} \quad t - \frac{1}{2} = \frac{144}{r + 4}$$

$$\text{or } t = \frac{144}{r + 4} + \frac{1}{2}$$

This gives us the following equation:

$$\frac{144}{r} = \frac{144}{r + 4} + \frac{1}{2}$$

Carry out:

$$2r(r + 4) \cdot \frac{144}{r} = 2r(r + 4)\cdot\left(\frac{144}{r + 4} + \frac{1}{2}\right)$$

(Multiplying by the LCM)

$$288(r + 4) = 288r + r(r + 4)$$

$$288r + 1152 = 288r + r^2 + 4r$$

$$0 = r^2 + 4r - 1152$$

$$0 = (r + 36)(r - 32)$$

$r + 36 = 0$ or $r - 32 = 0$

$r = -36$ or $r = 32$

Check:
We only consider r = 32 since the speed in this problem cannot be negative. If r = 32, then the time for the slow trip is $\frac{144}{32}$, or 4.5 hr. If r = 32, then r + 4 = 36. Thus, the time for the fast trip is $\frac{144}{36}$, or 4 hr. The time for the fast trip is $\frac{1}{2}$ hr less than the time for the slow trip. The value checks.

State:
The car's speed is 32 mph.

50. 35 mph

51. Familiarize:

Freeway: $D = rt$

$D = 55 \cdot 3$, or 165

City: $t = \frac{d}{r}$

$t = \frac{10}{35}$, or $\frac{2}{7}$

We organize the information in a table.

	Distance	Speed	Time
Freeway	165	55	3
City	10	35	$\frac{10}{35}$, or $\frac{2}{7}$
Totals	175		$3\frac{2}{7}$ or $\frac{23}{7}$

Translate:

$$\text{Average speed} = \frac{\text{Total distance}}{\text{Total time}}$$

$$\text{Average speed} = \frac{175}{\frac{23}{7}}$$

Carry out: We simplify.

$$\text{Average speed} = \frac{175}{\frac{23}{7}}$$

$$= 175 \cdot \frac{7}{23}$$

$$= \frac{1225}{23}, \text{ or } 53\frac{6}{23}$$

Check: At $53\frac{6}{23}$ mph, the distance traveled in $3\frac{2}{7}$ hr is $\left[53\frac{6}{23}\right] \cdot \left[3\frac{2}{7}\right] = \frac{1225}{23} \cdot \frac{23}{7} = 175$ mi.

This is the total distance the student traveled, so the answer checks.

State: The average speed was $53\frac{6}{23}$ mph.

52. 48 km/h

53. Familiarize:

Making a drawing is helpful.

40 mph $\frac{d}{40}$ hours r mph $\frac{d}{r}$ hours

d miles d miles

2d miles

We let d represent the distance of the first half of the trip. Then 2d represents the total distance. We also let r represent the speed for the second half of the trip. The times are represented by $\frac{d}{40}$ and $\frac{d}{r}$. The total time is $\frac{d}{40} + \frac{d}{r}$.

53. (continued)

Translate:

$$\text{Average speed} = \frac{\text{Total distance}}{\text{Total time}}$$

Substituting we get

$$45 = \frac{2d}{\frac{d}{40} + \frac{d}{r}}.$$

Carry out:

$$45 = \frac{2d}{\frac{d}{40} + \frac{d}{r}}$$

$$45\left[\frac{d}{40} + \frac{d}{r}\right] = 2d$$

$$45\left[\frac{1}{40} + \frac{1}{r}\right]d = 2d$$

$$\frac{45}{40} + \frac{45}{r} = 2$$

$$\frac{45}{r} = 2 - \frac{45}{40}$$

$$\frac{45}{r} = \frac{35}{40}$$

$$40 \cdot 45 = r \cdot 35$$

$$1800 = 35r$$

$$\frac{1800}{35} = r$$

$$51\frac{3}{7} = r$$

Check: This value checks.

State: The speed for the second half of the trip would have to be $51\frac{3}{7}$ mph.

54. $4{:}21\frac{9}{11}$

55. Familiarize:

It is helpful to make drawings.

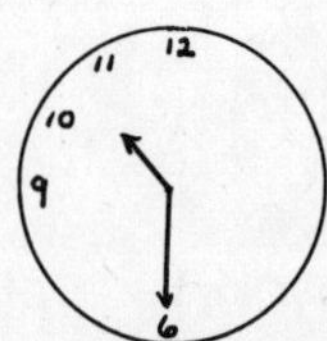

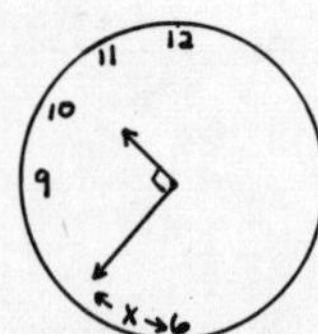

At 10:30 the minute hand is 30 units after the 12. The hour hand is $52\frac{1}{2}$ units after the 12. Let x represent the number of units the minute hand moves before the hands are perpendicular for the first time. When the minute hand moves x units, the hour hand moves $\frac{1}{12}$ x units. Then the minute hand is 30 + x units after the 12, and the hour hand is $52\frac{1}{2} + \frac{1}{12}x$ after the 12. When the hands are perpendicular, they must be 15 units apart.

55. (continued)

Translate: We subtract to find the number of units between the hands. This difference must be 15 units.

$$\left[52\tfrac{1}{2} + \tfrac{1}{12}x\right] - (30 + x) = 15$$

Carry out:

$$\left[52\tfrac{1}{2} + \tfrac{1}{12}x\right] - (30 + x) = 15$$

$$52\tfrac{1}{2} + \tfrac{1}{12}x - 30 - x = 15$$

$$-\tfrac{11}{12}x + 22\tfrac{1}{2} = 15$$

$$\tfrac{15}{2} = \tfrac{11}{12}x$$

$$\tfrac{12}{11} \cdot \tfrac{15}{2} = x$$

$$8\tfrac{2}{11} = \tfrac{90}{11} = x$$

Check: This answer checks:

State: After 10:30, the hands will first be perpendicular in $8\frac{2}{11}$ minutes, or at 10:38 $\frac{2}{11}$.

56. $\frac{2}{3}$ hr

57. Familiarize: Let t = the taxable income. Then t - 29,751 represents the amount of taxable income that exceeds $29,751.

Translate:

\$4462	plus	28%	of	taxable income that exceeds \$29,751	is	\$8711.
↓	↓	↓	↓	↓	↓	↓
4462	+	28%	·	(t - 29,751)	=	8711

Carry out:

$$4462 + 28\%(t - 29{,}751) = 8711$$

$$4462 + 0.28(t - 29{,}751) = 8711$$

$$4462 + 0.28t - 8330.28 = 8711$$

$$0.28t - 3868.28 = 8711$$

$$0.28t = 12{,}579.28$$

$$t = 44{,}926$$

Check: $44,926 - $29,751 = $15,175; 0.28($15,175) = 4249; and $4462 + $4249 = $8711. The answer checks.

State: The taxable income was $44,926.

58. $35,526

59. a) Familiarize: We organize the information in a table. We let d = the distance from Los Angeles at which it takes the same amount of time to return to Los Angeles as it does to go on to Honolulu. We let t = the time for either part of the trip.

	Distance	Speed	Time
Return to L.A.	d	750 - 50, or 700	t
Go on to Honolulu	2574 - d	750 + 50, or 800	t

Translate: Using the formula t = d/r and the rows of the table we get two equations:

$$t = \frac{d}{700} \quad \text{and} \quad t = \frac{2574 - d}{800}$$

Thus,

$$\frac{d}{700} = \frac{2574 - d}{800}.$$

Carry out:

$$\frac{d}{700} = \frac{2574 - d}{800}$$

$$700(800) \cdot \frac{d}{700} = 700(800) \cdot \frac{2574 - d}{800}$$

$$800d = 700(2574 - d)$$

$$800d = 1{,}801{,}800 - 700d$$

$$1500d = 1{,}801{,}800$$

$$d = 1201.2$$

Check: This answer checks.

State: At a distance of 1201.2 mi from Los Angeles it takes the same amount of time to return to Los Angeles as it does to go on to Honolulu.

b) Since 1187 mi < 1201.2 mi, less time is required to return to Los Angeles.

60. 12 mi

61. b = 1.2a, c = 1.25b, d = (1 - 0.01k)c.

Since a = d, we have

a = (1 - 0.01k)c

a = (1 - 0.01k)(1.25b) Substituting for c

a = (1 - 0.01k)(1.25)(1.2a) Substituting for b

Solving for k, we get $k = \frac{100}{3}$.

62. 11.4621%

63. Familiarize: Let m = the amount of money in your father's pocket at the outset. After giving your mother half of his money, your father has $\frac{1}{2} \cdot m$, or $\frac{m}{2}$ left. After giving your sister one-fourth of $\frac{m}{2}$, he has $\frac{3}{4} \cdot \frac{m}{2}$, or $\frac{3m}{8}$, left. After giving your brother one-third of $\frac{3m}{8}$, he has $\frac{2}{3} \cdot \frac{3m}{8}$, or $\frac{m}{4}$, left. After giving you one-half of $\frac{m}{4}$, he has $\frac{1}{2} \cdot \frac{m}{4}$, or $\frac{m}{8}$, left. This final amount is $2.

Translate:

Final amount is $2.

$$\frac{m}{8} = 2$$

Carry out:

$$\frac{m}{8} = 2$$

$$m = 16$$

Check: This value checks.

State: Your father had $16 at the outset.

64. 15

Exercise Set 2.4

1. $\sqrt{-15} = \sqrt{-1 \cdot 15} = \sqrt{-1}\,\sqrt{15} = i\sqrt{15}$

2. $i\sqrt{17}$

3. $\sqrt{-81} = \sqrt{-1 \cdot 81} = \sqrt{-1}\,\sqrt{81} = 9i$

4. $5i$

5. $-\sqrt{-12} = -\sqrt{-1 \cdot 4 \cdot 3} = -\sqrt{-1}\,\sqrt{4}\,\sqrt{3} = -2i\sqrt{3}$

6. $-2i\sqrt{5}$

7. $\sqrt{-16} + \sqrt{-25} = i\sqrt{16} + i\sqrt{25} = 4i + 5i = 9i$

8. $4i$

9. $\sqrt{-7} - \sqrt{-10} = i\sqrt{7} - i\sqrt{10} = (\sqrt{7} - \sqrt{10})i$

10. $(\sqrt{5} + \sqrt{7})i$

11. $\sqrt{-5}\,\sqrt{-11} = i\sqrt{5} \cdot i\sqrt{11} = i^2\,\sqrt{5}\,\sqrt{11} = -\sqrt{55}$

12. $-2\sqrt{14}$

13. $-\sqrt{-4}\,\sqrt{-5} = -(i\sqrt{4} \cdot i\sqrt{5}) = -(i^2 \cdot 2 \cdot \sqrt{5})$
$= -(-1)2\sqrt{5}$
$= 2\sqrt{5}$

14. $3\sqrt{7}$

15. $\frac{-\sqrt{5}}{\sqrt{-2}} = \frac{-\sqrt{5}}{i\sqrt{2}} = \frac{-\sqrt{5}}{i\sqrt{2}} \cdot \frac{i}{i} = \frac{-i\sqrt{5}}{i^2\sqrt{2}}$
$= \frac{-i\sqrt{5}}{-1\sqrt{2}}$
$= \sqrt{\frac{5}{2}}\,i$

16. $-\sqrt{\frac{7}{5}}\,i$

17. $\frac{\sqrt{-9}}{-\sqrt{4}} = \frac{3i}{-2} = -\frac{3}{2}i$

18. $\frac{5}{4}i$

19. $\frac{-\sqrt{-36}}{\sqrt{-9}} = \frac{-i\sqrt{36}}{i\sqrt{9}} = -\frac{6}{3} = -2$

20. $-\frac{5}{4}$

21. $i^{18} = (i^2)^9 = (-1)^9 = -1$

22. -1

23. $i^{15} = (i^2)^7 \cdot i = (-1)^7 \cdot i = -1 \cdot i = -i$

24. 1

25. $i^{39} = (i^2)^{19} \cdot i = (-1)^{19} \cdot i = -1 \cdot i = -i$

26. 1

27. $i^{46} = (i^2)^{23} = (-1)^{23} = -1$

28. 1

29. $(2 + 3i) + (4 + 2i) = 2 + 4 + 3i + 2i = 6 + 5i$

30. $11 + i$

31. $(4 + 3i) + (4 - 3i) = 4 + 4 + 3i - 3i = 8$

32. 0

33. $(8 + 11i) - (6 + 7i) = 8 + 11i - 6 - 7i$
$= 8 - 6 + 11i - 7i$
$= 2 + 4i$

34. $5 - 7i$

35. $2i - (4 + 3i) = 2i - 4 - 3i = -4 - i$

36. $-5 + i$

37. $(1 + 2i)(1 + 3i) = 1 + 3i + 2i + 6i^2$
$= 1 + 5i - 6 \qquad (i^2 = -1)$
$= -5 + 5i$

38. $13 + i$

39. $(1 + 2i)(1 - 3i) = 1 - 3i + 2i - 6i^2$
$= 1 - i + 6 \qquad (i^2 = -1)$
$= 7 - i$

40. 13

41. $3i(4 + 2i) = 12i + 6i^2 = 12i - 6$, or $-6 + 12i$

42. $20 + 15i$

43. $(2 + 3i)^2 = 4 + 12i + 9i^2$
$= 4 + 12i - 9 \qquad (i^2 = -1)$
$= -5 + 12i$

44. $5 - 12i$

45. $4x^2 + 25y^2 = (2x + 5yi)(2x - 5yi)$

Check by multiplying:
$(2x + 5yi)(2x - 5yi) = 4x^2 - 25y^2i^2 = 4x^2 + 25y^2$

46. $(4a + 7bi)(4a - 7bi)$

47.

$x^2 - 2x + 5 = 0$	
$(1 + 2i)^2 - 2(1 + 2i) + 5$	0
$1 + 4i + 4i^2 - 2 - 4i + 5$	
$4i^2 + 4$	
$-4 + 4$	
0	

The number $1 + 2i$ is a solution.

48. Yes

49. $4x + 7i = -6 + yi$

We equate the real parts and solve for x.
$4x = -6$
$x = -\frac{3}{2}$

We equate the imaginary parts and solve for y.
$7i = yi$
$7 = y$

50. $x = 3,\ y = 2$

51. $\frac{4 + 3i}{1 - i} = \frac{4 + 3i}{1 - i} \cdot \frac{1 + i}{1 + i}$
$= \frac{4 + 7i + 3i^2}{1 - i^2} = \frac{1 + 7i}{2} = \frac{1}{2} + \frac{7}{2}i$

52. $\frac{22}{41} - \frac{7}{41}i$

53. $\frac{\sqrt{2} + i}{\sqrt{2} - i} = \frac{\sqrt{2} + i}{\sqrt{2} - i} \cdot \frac{\sqrt{2} + i}{\sqrt{2} + i}$
$= \frac{2 + 2\sqrt{2}\,i + i^2}{2 - i^2} = \frac{1 + 2\sqrt{2}\,i}{3} = \frac{1}{3} + \frac{2\sqrt{2}}{3}i$

54. $\frac{1}{2} + \frac{\sqrt{3}}{2}i$

55. $\frac{3 + 2i}{i} = \frac{3 + 2i}{i} \cdot \frac{-i}{-i}$
$= \frac{-3i - 2i^2}{-i^2} = \frac{2 - 3i}{1} = 2 - 3i$

56. $3 - 2i$

57. $\frac{i}{2 + i} = \frac{i}{2 + i} \cdot \frac{2 - i}{2 - i}$
$= \frac{2i - i^2}{4 - i^2} = \frac{1 + 2i}{5} = \frac{1}{5} + \frac{2}{5}i$

58. $\frac{15}{146} + \frac{33}{146}i$

59. $\frac{1 - i}{(1 + i)^2} = \frac{1 - i}{1 + 2i + i^2} = \frac{1 - i}{2i}$
$= \frac{1 - i}{2i} \cdot \frac{-2i}{-2i} = \frac{-2i + 2i^2}{-4i^2}$
$= \frac{-2 - 2i}{4} = -\frac{1}{2} - \frac{1}{2}i$

60. $-\frac{1}{2} + \frac{1}{2}i$

61. $\frac{3 - 4i}{(2 + i)(3 - 2i)} = \frac{3 - 4i}{6 - i - 2i^2}$
$= \frac{3 - 4i}{8 - i} \cdot \frac{8 + i}{8 + i}$
$= \frac{24 - 29i - 4i^2}{64 - i^2}$
$= \frac{28 - 29i}{65} = \frac{28}{65} - \frac{29}{65}i$

62. $\frac{719}{3233} + \frac{955}{3233}i$

63. $\frac{1 + i}{1 - i} \cdot \frac{2 - i}{1 - i} = \frac{2 + i - i^2}{1 - 2i + i^2}$
$= \frac{3 + i}{-2i} \cdot \frac{2i}{2i}$
$= \frac{6i + 2i^2}{-4i^2}$
$= \frac{-2 + 6i}{4} = -\frac{1}{2} + \frac{3}{2}i$

64. $-\frac{1}{2} - \frac{3}{2}i$

65. $\frac{3 + 2i}{1 - i} + \frac{6 + 2i}{1 - i} = \frac{3 + 6 + 2i + 2i}{1 - i}$
$= \frac{9 + 4i}{1 - i} \cdot \frac{1 + i}{1 + i}$
$= \frac{9 + 13i + 4i^2}{1 - i^2}$
$= \frac{5 + 13i}{2} = \frac{5}{2} + \frac{13}{2}i$

66. $-\frac{1}{2} - \frac{13}{2}i$

67. The reciprocal of $4 + 3i$ is $\frac{1}{4+3i}$, or
$\frac{1}{4+3i} \cdot \frac{4-3i}{4-3i} = \frac{4-3i}{16-9i^2} = \frac{4-3i}{25} = \frac{4}{25} - \frac{3}{25}i.$

68. $\frac{4}{25} + \frac{3}{25}i$

69. The reciprocal of $5 - 2i$ is $\frac{1}{5-2i}$, or
$\frac{1}{5-2i} \cdot \frac{5+2i}{5+2i} = \frac{5+2i}{25-4i^2} = \frac{5+2i}{29} = \frac{5}{29} + \frac{2}{29}i.$

70. $\frac{2}{29} - \frac{5}{29}i$

71. The reciprocal of i is $\frac{1}{i}$, or
$\frac{1}{i} \cdot \frac{-i}{-i} = \frac{-i}{-i^2} = \frac{-i}{1} = -i.$

72. i

73. The reciprocal of $-4i$ is $\frac{1}{-4i}$, or
$\frac{1}{-4i} \cdot \frac{4i}{4i} = \frac{4i}{-16i^2} = \frac{4i}{16} = \frac{i}{4}.$

74. $-\frac{i}{5}$

75.
$$(3 + i)x + i = 5i$$
$$(3 + i)x = 4i$$
$$x = \frac{4i}{3+i}$$
$$x = \frac{4i}{3+i} \cdot \frac{3-i}{3-i}$$
$$x = \frac{12i - 4i^2}{9 - i^2} = \frac{4 + 12i}{10}$$
$$x = \frac{2}{5} + \frac{6}{5}i$$

76. $\frac{12}{5} - \frac{1}{5}i$

77.
$$2ix + 5 - 4i = (2 + 3i)x - 2i$$
$$5 - 4i + 2i = (2 + 3i)x - 2ix$$
$$5 - 2i = (2 + i)x$$
$$\frac{5-2i}{2+i} = x$$
$$\frac{2-i}{2-i} \cdot \frac{5-2i}{2+i} = x$$
$$\frac{8-9i}{5} = \frac{10 - 9i + 2i^2}{4 - i^2} = x$$
$$\frac{8}{5} - \frac{9}{5}i = x$$

78. $\frac{8}{29} + \frac{9}{29}i$

79.
$$(1 + 2i)x + 3 - 2i = 4 - 5i + 3ix$$
$$(1 + 2i)x - 3ix = 4 - 5i - 3 + 2i$$
$$(1 - i)x = 1 - 3i$$
$$x = \frac{1-3i}{1-i}$$
$$x = \frac{1-3i}{1-i} \cdot \frac{1+i}{1+i}$$
$$x = \frac{1 - 2i - 3i^2}{1 - i^2} = \frac{4 - 2i}{2}$$
$$x = 2 - i$$

80. $-\frac{1}{5} + \frac{7}{5}i$

81.
$$(5 + i)x + 1 - 3i = (2 - 3i)x + 2 - i$$
$$(5 + i)x - (2 - 3i)x = 2 - i - 1 + 3i$$
$$(3 + 4i)x = 1 + 2i$$
$$x = \frac{1+2i}{3+4i}$$
$$x = \frac{1+2i}{3+4i} \cdot \frac{3-4i}{3-4i}$$
$$x = \frac{3 + 2i - 8i^2}{9 - 16i^2} = \frac{11 + 2i}{25}$$
$$x = \frac{11}{25} + \frac{2}{25}i$$

82. $\frac{4}{5} + \frac{3}{5}i$

83. For example, $\sqrt{-1}\sqrt{-1} = i^2 = -1$ but $\sqrt{(-1)(-1)} = \sqrt{1} = 1$.

84. For example, $\sqrt{\frac{4}{-1}} = \sqrt{-4} = 2i$, but $\frac{\sqrt{4}}{\sqrt{-1}} = \frac{2}{i} = -2i$

85. Let $z = a + bi$. Then $z \cdot \bar{z} = (a + bi)(a - bi) = a^2 - b^2i^2 = a^2 + b^2$. Since a and b are real numbers, so is $a^2 + b^2$. Thus $z \cdot \bar{z}$ is real.

86. Let $z = a + bi$. Then $z + \bar{z} = (a + bi) + (a - bi) = 2a$. Since a is a real number, 2a is real. Thus $z + \bar{z}$ is real.

87. Let $z = a + bi$ and $w = c + di$. Then $\overline{z + w} = \overline{(a + bi) + (c + di)} = \overline{(a + c) + (b + d)i}$, by adding.

We now take the conjugate and obtain $(a + c) - (b + d)i$. Now $\bar{z} + \bar{w} = \overline{(a + bi)} + \overline{(c + di)} = (a - bi) + (c - di)$, taking the conjugate. We will now add to obtain $(a + c) - (b + d)i$, the same result as before. Thus $\overline{z + w} = \bar{z} + \bar{w}$.

88. Let $z = a + bi$ and $w = c + di$. Then $\overline{z \cdot w} =$ $\overline{(a + bi)(c + di)} = \overline{(ac - bd) + (ad + bc)i} =$ $(ac + bd) - (ad + bc)i$, taking the conjugate. Now $\overline{z} \cdot \overline{w} = \overline{(a + bi)} \cdot \overline{(c + di)} = (a - bi)(c - di) =$ $(ac + bd) - (ad + bc)i$, the same result as before. Thus $\overline{z \cdot w} = \overline{z} \cdot \overline{w}$.

89. By definition of exponents the conjugate of z^n is the conjugate of the product of n factors of z. Using the result of Exercise 88, the conjugate of n factors of z is the product of n factors of $\overline{z}$. Thus $\overline{z^n} = \overline{z}^n$.

90. If z is a real number, then $z = a + 0i = a$ and $\overline{z} = a - 0i = a$. Thus $\overline{z} = z$.

91. $\overline{3z^5 - 4z^2 + 3z - 5}$

$= \overline{3z^5} - \overline{4z^2} + \overline{3z} - \overline{5}$ By Exercise 87

$= \overline{3}\ \overline{z^5} - \overline{4}\ \overline{z^2} + \overline{3}\ \overline{z} - \overline{5}$ By Exercise 88

$= 3\ \overline{z^5} - 4\ \overline{z^2} + 3\ \overline{z} - 5$ By Exercise 90

$= 3\overline{z}^5 - 4\overline{z}^2 + 3\overline{z} - 5$ By Exercise 89

92. 1

93. Solve $5z - 4\overline{z} = 7 + 8i$ for z.

Let $z = a + bi$. Then $\overline{z} = a - bi$.

$$5z - 4\overline{z} = 7 + 8i$$
$$5(a + bi) - 4(a - bi) = 7 + 8i$$
$$5a + 5bi - 4a + 4bi = 7 + 8i$$
$$a + 9bi = 7 + 8i$$

Equate the real parts.

$$a = 7$$

Equate the imaginary parts.

$$9b = 8$$
$$b = \frac{8}{9}$$

Then $z = a + bi$

$$z = 7 + \frac{8}{9}i$$

94. a

95. Let $z = a + bi$. Then $\overline{z} = a - bi$.

$$\frac{1}{2}(\overline{z} - z) = \frac{1}{2}[(a - bi) - (a + bi)]$$
$$= \frac{1}{2}(-2bi) = -bi$$

96. $1 - i,\ -1 + i$

97. $\frac{(a + bi)^2}{4} = i$ or $(a^2 - b^2) + (2ab)i = 0 + 4i$.

Then $\left\{\begin{array}{l} a^2 - b^2 = 0 \\ 2ab = 4 \end{array}\right\}$ gives $a = \sqrt{2}$, $b = \sqrt{2}$ and and $a = -\sqrt{2}$, $b = -\sqrt{2}$. Hence the two values for z are $\sqrt{2} + i\sqrt{2}$ and $-\sqrt{2} - i\sqrt{2}$.

98. $\frac{a}{a^2 + b^2} + \frac{-b}{a^2 + b^2}i$

99. $\frac{w}{z} = \frac{c + di}{a + bi} \cdot \frac{a - bi}{a - bi} = \frac{ac + bd}{a^2 + b^2} + \frac{ad - bc}{a^2 + b^2}i$

Exercise Set 2.5

1. $3x^2 = 27$

$x^2 = 9$ Multiplying by $\frac{1}{3}$

$x = 3$ or $x = -3$ Taking square roots

The solution set is $\{3, -3\}$, or $\{\pm 3\}$.

2. $\{\pm 4\}$

3. $x^2 = -1$

$x = \sqrt{-1}$ or $x = -\sqrt{-1}$ Taking square roots

$x = i$ or $x = -i$ Simplifying

The solution set is $\{i, -i\}$, or $\{\pm i\}$.

4. $\{\pm 2i\}$

5. $4x^2 = 20$

$x^2 = 5$

$x = \sqrt{5}$ or $x = -\sqrt{5}$

The solution set is $\{\pm \sqrt{5}\}$.

6. $\{\pm \sqrt{7}\}$

7. $10x^2 = 0$

$x^2 = 0$

$x = 0$

The solution set is $\{0\}$.

8. $\{0\}$

9. $2x^2 - 3 = 0$

$2x^2 = 3$

$x^2 = \frac{3}{2}$

$x = \sqrt{\frac{3}{2}}$ or $x = -\sqrt{\frac{3}{2}}$

$x = \sqrt{\frac{3}{2} \cdot \frac{2}{2}}$ or $x = -\sqrt{\frac{3}{2} \cdot \frac{2}{2}}$

$x = \frac{\sqrt{6}}{2}$ or $x = -\frac{\sqrt{6}}{2}$

The solution set is $\left\{\pm \frac{\sqrt{6}}{2}\right\}$.

10. $\left\{\pm \frac{\sqrt{21}}{3}\right\}$

11. $2x^2 + 14 = 0$

$2x^2 = -14$

$x^2 = -7$

$x = \sqrt{-7}$ or $x = -\sqrt{-7}$

$x = i\sqrt{7}$ or $x = -i\sqrt{7}$

The solution set is $\{\pm i\sqrt{7}\}$.

12. $\{\pm i\sqrt{5}\}$

13. $ax^2 = b$

$x^2 = \frac{b}{a}$

$x = \pm\sqrt{\frac{b}{a}}$, or $\pm \frac{\sqrt{ba}}{a}$

The solution set is $\left\{\pm\sqrt{\frac{b}{a}}\right\}$.

14. $\left\{\pm\sqrt{\frac{k}{\pi}}\right\}$

15. $(x - 7)^2 = 5$

$x - 7 = \pm\sqrt{5}$

$x = 7 \pm \sqrt{5}$

The solution set is $\{7 \pm \sqrt{5}\}$.

16. $\{-3 \pm \sqrt{2}\}$

17. $\frac{4}{9}x^2 - 1 = 0$

$\frac{4}{9}x^2 = 1$

$x^2 = \frac{9}{4}$

$x = \sqrt{\frac{9}{4}}$ or $x = -\sqrt{\frac{9}{4}}$

$x = \frac{3}{2}$ or $x = -\frac{3}{2}$

The solution set is $\left\{\pm \frac{3}{2}\right\}$.

18. $\left\{\pm \frac{5}{4}\right\}$

19. $(x - h)^2 - 1 = a$

$(x - h)^2 = a + 1$

$x - h = \pm\sqrt{a + 1}$

$x = h \pm \sqrt{a + 1}$

The solution set is $\{h \pm \sqrt{a + 1}\}$.

20. $\left\{h \pm \sqrt{\frac{y - k}{a}}\right\}$

21. $x^2 + 6x + 4 = 0$

$x^2 + 6x + 9 - 9 + 4 = 0$ $\left[\frac{1}{2} \cdot 6 = 3,\ 3^2 = 9;\right.$ thus we add $9 - 9.\left.\right]$

$(x + 3)^2 - 5 = 0$

$(x + 3)^2 = 5$

$x + 3 = \pm\sqrt{5}$

$x = -3 \pm \sqrt{5}$

The solution set is $\{-3 \pm \sqrt{5}\}$.

22. $\{3 \pm \sqrt{13}\}$

23. $y^2 + 7y - 30 = 0$

$y^2 + 7y + \frac{49}{4} - \frac{49}{4} - 30 = 0$ $\left[\frac{1}{2} \cdot 7 = \frac{7}{2},\ \left(\frac{7}{2}\right)^2 = \frac{49}{4};\right.$ thus we add $\left.\frac{49}{4} - \frac{49}{4}\right]$

$\left(y + \frac{7}{2}\right)^2 - \frac{169}{4} = 0$

$\left(y + \frac{7}{2}\right)^2 = \frac{169}{4}$

$y + \frac{7}{2} = \pm\sqrt{\frac{169}{4}} = \pm\frac{13}{2}$

$y = -\frac{7}{2} \pm \frac{13}{2} = \frac{-7 \pm 13}{2}$

$y = \frac{-7 + 13}{2}$ or $y = \frac{-7 - 13}{2}$

$y = 3$ or $y = -10$

The solution set is $\{3, -10\}$.

24. $\{10, -3\}$

25.
$$5x^2 - 4x - 2 = 0$$
$$x^2 - \frac{4}{5}x - \frac{2}{5} = 0 \quad \left[\text{Multiplying by } \frac{1}{5}\right]$$
$$x^2 - \frac{4}{5}x + \frac{4}{25} - \frac{4}{25} - \frac{2}{5} = 0 \quad \left[\text{Adding } \frac{4}{25} - \frac{4}{25}\right]$$
$$\left(x - \frac{2}{5}\right)^2 - \frac{14}{25} = 0$$
$$\left(x - \frac{2}{5}\right)^2 = \frac{14}{25}$$
$$x - \frac{2}{5} = \pm\sqrt{\frac{14}{25}} = \pm\frac{\sqrt{14}}{5}$$
$$x = \frac{2}{5} \pm \frac{\sqrt{14}}{5} = \frac{2 \pm \sqrt{14}}{5}$$

The solution set is $\left\{\frac{2 \pm \sqrt{14}}{5}\right\}$.

26. $\left\{\frac{7 \pm \sqrt{13}}{12}\right\}$

27.
$$2x^2 + 7x - 15 = 0$$
$$x^2 + \frac{7}{2}x - \frac{15}{2} = 0 \quad \left[\text{Multiplying by } \frac{1}{2}\right]$$
$$x^2 + \frac{7}{2}x + \frac{49}{16} - \frac{49}{16} - \frac{15}{2} = 0 \quad \left[\text{Adding } \frac{49}{16} - \frac{49}{16}\right]$$
$$\left(x + \frac{7}{4}\right)^2 - \frac{169}{16} = 0$$
$$\left(x + \frac{7}{4}\right)^2 = \frac{169}{16}$$
$$x + \frac{7}{4} = \pm\sqrt{\frac{169}{16}} = \pm\frac{13}{4}$$
$$x = -\frac{7}{4} \pm \frac{13}{4} = \frac{-7 \pm 13}{4}$$

$x = \frac{-7 + 13}{4}$ or $x = \frac{-7 - 13}{4}$

$x = \frac{3}{2}$ or $x = -5$

The solution set is $\left\{\frac{3}{2}, -5\right\}$.

28. $\left\{\frac{5}{3}\right\}$

29.
$$x^2 + 4x = 5$$
$$x^2 + 4x - 5 = 0$$

$a = 1, \quad b = 4, \quad c = -5$

$$x = \frac{-b \pm \sqrt{b^2 - 4ac}}{2a}$$
$$x = \frac{-4 \pm \sqrt{4^2 - 4(1)(-5)}}{2(1)} = \frac{-4 \pm \sqrt{16 + 20}}{2}$$
$$x = \frac{-4 \pm \sqrt{36}}{2} = \frac{-4 \pm 6}{2}$$

$x = \frac{-4 + 6}{2}$ or $x = \frac{-4 - 6}{2}$

$x = 1$ or $x = -5$

The solution set is $\{1, -5\}$.

30. $\{-3, 5\}$

31. $2y^2 - 3y - 2 = 0$

$a = 2, \quad b = -3, \quad c = -2$

$$y = \frac{-b \pm \sqrt{b^2 - 4ac}}{2a}$$
$$y = \frac{-(-3) \pm \sqrt{(-3)^2 - 4(2)(-2)}}{2(2)} = \frac{3 \pm \sqrt{9 + 16}}{4}$$
$$y = \frac{3 \pm \sqrt{25}}{4} = \frac{3 \pm 5}{4}$$

$y = \frac{3 + 5}{4}$ or $y = \frac{3 - 5}{4}$

$y = 2$ or $y = -\frac{1}{2}$

The solution set is $\left\{2, -\frac{1}{2}\right\}$.

32. $\left\{-1, \frac{2}{5}\right\}$

33. $3t^2 + 8t + 3 = 0$

$a = 3, \quad b = 8, \quad c = 3$

$$t = \frac{-b \pm \sqrt{b^2 - 4ac}}{2a}$$
$$t = \frac{-8 \pm \sqrt{8^2 - 4\cdot3\cdot3}}{2\cdot3} = \frac{-8 \pm \sqrt{64 - 36}}{6}$$
$$t = \frac{-8 \pm \sqrt{28}}{6} = \frac{-8 \pm 2\sqrt{7}}{6} = \frac{2(-4 \pm \sqrt{7})}{6}$$
$$t = \frac{-4 \pm \sqrt{7}}{3}$$

The solution set is $\left\{\frac{-4 \pm \sqrt{7}}{3}\right\}$.

34. $\{3 \pm \sqrt{7}\}$

35.
$$3 + u^2 = 12u$$
$$u^2 - 12u + 3 = 0$$

$a = 1, \quad b = -12, \quad c = 3$

$$u = \frac{-b \pm \sqrt{b^2 - 4ac}}{2a}$$
$$u = \frac{-(-12) \pm \sqrt{(-12)^2 - 4\cdot1\cdot3}}{2\cdot1} = \frac{12 \pm \sqrt{144 - 12}}{2}$$
$$u = \frac{12 \pm \sqrt{132}}{2} = \frac{12 \pm 2\sqrt{33}}{2}$$
$$u = \frac{2(6 \pm \sqrt{33})}{2} = 6 \pm \sqrt{33}$$

The solution set is $\{6 \pm \sqrt{33}\}$.

36. $\{-2, -4\}$

37. $x^2 - x + 1 = 0$

$a = 1, \quad b = -1, \quad c = 1$

$$x = \frac{-b \pm \sqrt{b^2 - 4ac}}{2a}$$

$$x = \frac{-(-1) \pm \sqrt{(-1)^2 - 4\cdot1\cdot1}}{2\cdot1}$$

$$x = \frac{1 \pm \sqrt{1 - 4}}{2} = \frac{1 \pm \sqrt{-3}}{2}$$

$$x = \frac{1 \pm i\sqrt{3}}{2}$$

The solution set is $\left\{\frac{1 \pm i\sqrt{3}}{2}\right\}$.

38. $\left\{\frac{-1 \pm i\sqrt{7}}{2}\right\}$

39. $x^2 + 13 = 4x$

$x^2 - 4x + 13 = 0$

$a = 1, \quad b = -4, \quad c = 13$

$$x = \frac{-b \pm \sqrt{b^2 - 4ac}}{2a}$$

$$x = \frac{-(-4) \pm \sqrt{(-4)^2 - 4\cdot1\cdot13}}{2\cdot1}$$

$$x = \frac{4 \pm \sqrt{16 - 52}}{2} = \frac{4 \pm \sqrt{-36}}{2}$$

$$x = \frac{4 \pm 6i}{2} = \frac{2(2 \pm 3i)}{2} = 2 \pm 3i$$

The solution set is $\{2 \pm 3i\}$.

40. $\left\{\frac{-1 \pm 2i}{5}\right\}$

41. $5x^2 = 13x + 17$

$5x^2 - 13x - 17 = 0$ Standard form

$a = 5, b = -13, c = -17$

$$x = \frac{-b \pm \sqrt{b^2 - 4ac}}{2a}$$

$$x = \frac{-(-13) \pm \sqrt{(-13)^2 - 4(5)(-17)}}{2\cdot5}$$

$$x = \frac{13 \pm \sqrt{169 + 340}}{10}$$

$$x = \frac{13 \pm \sqrt{509}}{10}$$

The solution set is $\left\{\frac{13 \pm \sqrt{509}}{10}\right\}$.

42. $\left\{\frac{5 \pm \sqrt{73}}{6}\right\}$

43. $0.03 + 0.08v = v^2$

$0 = v^2 - 0.08v - 0.03$ Standard form

$a = 1, b = -0.08, c = -0.03$

$$v = \frac{-b \pm \sqrt{b^2 - 4ac}}{2a}$$

$$v = \frac{-(-0.08) \pm \sqrt{(-0.08)^2 - 4(1)(-0.03)}}{2\cdot1}$$

$$v = \frac{0.08 \pm \sqrt{0.0064 + 0.12}}{2} = \frac{0.08 \pm \sqrt{0.1264}}{2}$$

$$v = \frac{0.08 \pm \sqrt{0.0016\cdot79}}{2} = \frac{0.08 \pm 0.04\sqrt{79}}{2}$$

$$v = \frac{2(0.04 \pm 0.02\sqrt{79})}{2} = 0.04 \pm 0.02\sqrt{79}$$

The solution set is $\{0.04 \pm 0.02\sqrt{79}\}$.

44. $\left\{\frac{3 \pm \sqrt{153}}{9}\right\}$

45. $\frac{1}{x} + \frac{1}{x + 3} = 7$, LCM is $x(x + 3)$

$$x(x + 3)\left[\frac{1}{x} + \frac{1}{x + 3}\right] = x(x + 3)\cdot7$$

$$x + 3 + x = 7x^2 + 21x$$

$$2x + 3 = 7x^2 + 21x$$

$0 = 7x^2 + 19x - 3$ Standard form

$a = 7, b = 19, c = -3$

$$x = \frac{-b \pm \sqrt{b^2 - 4ac}}{2a}$$

$$x = \frac{-19 \pm \sqrt{19^2 - 4(7)(-3)}}{2\cdot7}$$

$$x = \frac{-19 \pm \sqrt{361 + 84}}{14} = \frac{-19 \pm \sqrt{445}}{14}$$

The solution set is $\left\{\frac{-19 \pm \sqrt{445}}{14}\right\}$.

46. $\left\{\pm \frac{2\sqrt{30}}{5}\right\}$

47. $$1 + \frac{x + 5}{(x + 1)^2} - \frac{2}{x + 1} = 0, \text{ LCM is } (x + 1)^2$$

$$(x + 1)^2\left[1 + \frac{x + 5}{(x + 1)^2} - \frac{2}{x + 1}\right] = (x + 1)^2 \cdot 0$$

$$(x + 1)^2 + x + 5 - 2(x + 1) = 0$$

$$x^2 + 2x + 1 + x + 5 - 2x - 2 = 0$$

$$x^2 + x + 4 = 0$$

$a = 1$, $b = 1$, $c = 4$

$$x = \frac{-b \pm \sqrt{b^2 - 4ac}}{2a}$$

$$x = \frac{-1 \pm \sqrt{1^2 - 4\cdot 1\cdot 4}}{2\cdot 1} = \frac{-1 \pm \sqrt{1 - 16}}{2}$$

$$x = \frac{-1 \pm \sqrt{-15}}{2} = \frac{-1 \pm \sqrt{-1\cdot 15}}{2}$$

$$x = \frac{-1 \pm i\sqrt{15}}{2}$$

The solution set is $\left\{\frac{-1 \pm i\sqrt{15}}{2}\right\}$.

48. $\left\{\frac{-1 \pm i\sqrt{43}}{2}\right\}$

49. $x^2 - 6x + 9 = 0$

$a = 1$, $b = -6$, $c = 9$

We compute the discriminant.

$$b^2 - 4ac = (-6)^2 - 4\cdot 1\cdot 9 = 36 - 36 = 0$$

Since $b^2 - 4ac = 0$, there is just one solution, and it is a real number.

50. One real solution

51. $x^2 + 7 = 0$

$a = 1$, $b = 0$, $c = 7$

We compute the discriminant.

$$b^2 - 4ac = 0^2 - 4\cdot 1\cdot 7 = -28$$

Since $b^2 - 4ac < 0$, there are two nonreal solutions.

52. Two nonreal solutions

53. $x^2 - 2 = 0$

$a = 1$, $b = 0$, $c = -2$

We compute the discriminant.

$$b^2 - 4ac = 0^2 - 4\cdot 1\cdot (-2) = 8$$

Since $b^2 - 4ac > 0$, there are two real solutions.

54. Two real solutions

55. $4x^2 - 12x + 9 = 0$

$a = 4$, $b = -12$, $c = 9$

We compute the discriminant.

$$b^2 - 4ac = (-12)^2 - 4\cdot 4\cdot 9 = 144 - 144 = 0$$

Since $b^2 - 4ac = 0$, there is just one solution, and it is a real number.

56. Two real solutions

57. $x^2 - 2x + 4 = 0$

$a = 1$, $b = -2$, $c = 4$

We compute the discriminant.

$$b^2 - 4ac = (-2)^2 - 4\cdot 1\cdot 4 = 4 - 16 = -12$$

Since $b^2 - 4ac < 0$, ther are two nonreal solutions.

58. Two nonreal solutions

59. $9t^2 - 3t = 0$

$a = 9$, $b = -3$, $c = 0$

We compute the discriminant.

$$b^2 - 4ac = (-3)^2 - 4\cdot 9\cdot 0 = 9 - 0 = 9$$

Since $b^2 - 4ac > 0$, there are two real solutions.

60. Two real solutions

61. $y^2 = \frac{1}{2}y + \frac{3}{5}$

$y^2 - \frac{1}{2}y - \frac{3}{5} = 0$ (Standard form)

$a = 1$, $b = -\frac{1}{2}$, $c = -\frac{3}{5}$

We compute the discriminant.

$$b^2 - 4ac = \left(-\frac{1}{2}\right)^2 - 4\cdot 1\cdot\left(-\frac{3}{5}\right) = \frac{1}{4} + \frac{12}{5} = \frac{53}{20}$$

Since $b^2 - 4ac > 0$, there are two real solutions.

62. Two real solutions

63. $4x^2 - 4\sqrt{3}\,x + 3 = 0$

$a = 4, \quad b = -4\sqrt{3}, \quad c = 3$

We compute the discriminant.

$$b^2 - 4ac = (-4\sqrt{3})^2 - 4\cdot 4\cdot 3 = 48 - 48 = 0$$

Since $b^2 - 4ac = 0$, there is just one solution, and it is a real number.

64. Two real solutions

65. The solutions are -11 and 9.

$x = -11$ or $x = 9$

$x + 11 = 0$ or $x - 9 = 0$

$(x + 11)(x - 9) = 0$

$x^2 + 2x - 99 = 0$

66. $x^2 - 16 = 0$

67. The solutions are both 7.

$x = 7$ or $x = 7$

$x - 7 = 0$ or $x - 7 = 0$

$(x - 7)(x - 7) = 0$

$x^2 - 14x + 49 = 0$

68. $x^2 + \frac{4}{3}x + \frac{4}{9} = 0$, or $9x^2 + 12x + 4 = 0$

69. The solutions are $-\frac{2}{5}$ and $\frac{6}{5}$.

$x = -\frac{2}{5}$ or $x = \frac{6}{5}$

$x + \frac{2}{5} = 0$ or $x - \frac{6}{5} = 0$

$\left(x + \frac{2}{5}\right)\left(x - \frac{6}{5}\right) = 0$

$x^2 - \frac{4}{5}x - \frac{12}{25} = 0$

or

$25x^2 - 20x - 12 = 0$

70. $x^2 + \frac{3}{4}x + \frac{1}{8} = 0$, or $8x^2 + 6x + 1 = 0$

71. The solutions are $\frac{c}{2}$ and $\frac{d}{2}$.

$x = \frac{c}{2}$ or $x = \frac{d}{2}$

$x - \frac{c}{2} = 0$ or $x - \frac{d}{2} = 0$

$\left(x - \frac{c}{2}\right)\left(x - \frac{d}{2}\right) = 0$

$x^2 - \left(\frac{c + d}{2}\right)x + \frac{cd}{4} = 0$

or

$4x^2 - 2(c + d)x + cd = 0$

72. $x^2 - \left(\frac{k}{3} + \frac{m}{4}\right)x + \frac{km}{12} = 0$, or

$12x^2 - (4k + 3m)x + km = 0$

73. The solutions are $\sqrt{2}$ and $3\sqrt{2}$.

$x = \sqrt{2}$ or $x = 3\sqrt{2}$

$x - \sqrt{2} = 0$ or $x - 3\sqrt{2} = 0$

$(x - \sqrt{2})(x - 3\sqrt{2}) = 0$

$x^2 - 4\sqrt{2}\,x + 6 = 0$

74. $x^2 - \sqrt{3}\,x - 6 = 0$

75. The solutions are $3i$ and $-3i$.

$x = 3i$ or $x = -3i$

$x - 3i = 0$ or $x + 3i = 0$

$(x - 3i)(x + 3i) = 0$

$x^2 - 9i^2 = 0$

$x^2 + 9 = 0$

76. $x^2 + 16 = 0$

77. $x^2 - 0.75x - 0.5 = 0$

$a = 1, \quad b = -0.75, \quad c = -0.5$

$$x = \frac{-(-0.75) \pm \sqrt{(-0.75)^2 - 4(1)(-0.5)}}{2(1)}$$

$$x = \frac{0.75 \pm \sqrt{0.5625 + 2}}{2} = \frac{0.75 \pm \sqrt{2.5625}}{2}$$

$$x = \frac{0.75 + \sqrt{2.5625}}{2} \text{ or } x = \frac{0.75 - \sqrt{2.5625}}{2}$$

$x \approx 1.1754$ or $x \approx -0.4254$

The solution set is {1.1754, -0.4254}.

78. {1.8693, -0.3252}

79. $x + \frac{1}{x} = \frac{13}{6}$, LCM = 6x

$6x\left(x + \frac{1}{x}\right) = 6x \cdot \frac{13}{6}$

$6x^2 + 6 = 13x$

$6x^2 - 13x + 6 = 0$

$(3x - 2)(2x - 3) = 0$

$3x - 2 = 0$ or $2x - 3 = 0$

$x = \frac{2}{3}$ or $x = \frac{3}{2}$

The solution set is $\left\{\frac{2}{3}, \frac{3}{2}\right\}$.

80. $\left\{6, \frac{3}{2}\right\}$

81. $t^2 + 0.2t - 0.3 = 0$

$a = 1, \quad b = 0.2, \quad c = -0.3$

$$t = \frac{-0.2 \pm \sqrt{(0.2)^2 - 4(1)(-0.3)}}{2\cdot 1}$$

$$t = \frac{-0.2 \pm \sqrt{0.04 + 1.2}}{2} = \frac{-0.2 \pm \sqrt{1.24}}{2}$$

$$t = \frac{-0.2 \pm 2\sqrt{0.31}}{2} = \frac{2(-0.1 \pm \sqrt{0.31})}{2}$$

$$t = -0.1 \pm \sqrt{0.31}$$

The solution set is $\{-0.1 \pm \sqrt{0.31}\}$.

82. $\left\{\frac{-0.3 \pm \sqrt{0.89}}{2}\right\}$, or $\{-0.15 \pm 0.05\sqrt{89}\}$

83. $x^2 + x - \sqrt{2} = 0$

$a = 1, \quad b = 1, \quad c = -\sqrt{2}$

$$x = \frac{-1 \pm \sqrt{1^2 - 4(1)(-\sqrt{2})}}{2\cdot 1} = \frac{-1 \pm \sqrt{1 + 4\sqrt{2}}}{2}$$

The solution set is $\left\{\frac{-1 \pm \sqrt{1 + 4\sqrt{2}}}{2}\right\}$.

84. $\left\{\frac{1 \pm \sqrt{1 + 4\sqrt{3}}}{2}\right\}$

85. $x^2 + \sqrt{5}\,x - \sqrt{3} = 0$

$a = 1, \quad b = \sqrt{5}, \quad c = -\sqrt{3}$

$$x = \frac{-\sqrt{5} \pm \sqrt{(\sqrt{5})^2 - 4(1)(-\sqrt{3})}}{2\cdot 1}$$

$$x = \frac{-\sqrt{5} \pm \sqrt{5 + 4\sqrt{3}}}{2}$$

The solution set is $\left\{\frac{-\sqrt{5} \pm \sqrt{5 + 4\sqrt{3}}}{2}\right\}$.

86. $\frac{-\sqrt{3} \pm \sqrt{3 + 8\pi}}{4}$

87. $\sqrt{2}\,x^2 - \sqrt{3}\,x - \sqrt{5} = 0$

$a = \sqrt{2}, \quad b = -\sqrt{3}, \quad c = -\sqrt{5}$

$$x = \frac{-(-\sqrt{3}) \pm \sqrt{(-\sqrt{3})^2 - 4(\sqrt{2})(-\sqrt{5})}}{2\cdot\sqrt{2}}$$

$$x = \frac{\sqrt{3} \pm \sqrt{3 + 4\sqrt{10}}}{2\sqrt{2}} \cdot \frac{\sqrt{2}}{\sqrt{2}}$$ (Rationalizing the denominator)

$$x = \frac{\sqrt{6} \pm \sqrt{6 + 8\sqrt{10}}}{4}$$

The solution set is $\left\{\frac{\sqrt{6} \pm \sqrt{6 + 8\sqrt{10}}}{4}\right\}$.

88. $\left\{\frac{-5\sqrt{2} \pm \sqrt{34}}{4}\right\}$

89. $(2t - 3)^2 + 17t = 15$

$4t^2 - 12t + 9 + 17t = 15$

$4t^2 + 5t - 6 = 0$

$(4t - 3)(t + 2) = 0$

$4t - 3 = 0$ or $t + 2 = 0$

$t = \frac{3}{4}$ or $t = -2$

The solution set is $\left\{\frac{3}{4}, -2\right\}$.

90. $\{2, -3\}$

91. $(x + 3)(x - 2) = 2(x + 11)$

$x^2 + x - 6 = 2x + 22$

$x^2 - x - 28 = 0$

$a = 1, \quad b = -1, \quad c = -28$

$$x = \frac{-(-1) \pm \sqrt{(-1)^2 - 4(1)(-28)}}{2\cdot 1}$$

$$x = \frac{1 \pm \sqrt{1 + 112}}{2} = \frac{1 \pm \sqrt{113}}{2}$$

The solution set is $\left\{\frac{1 \pm \sqrt{113}}{2}\right\}$.

92. $\{-4, 3\}$

93. $2x^2 + (x - 4)^2 = 5x(x - 4) + 24$

$2x^2 + x^2 - 8x + 16 = 5x^2 - 20x + 24$

$0 = 2x^2 - 12x + 8$

$0 = x^2 - 6x + 4$

$a = 1, \quad b = -6, \quad c = 4$

$$x = \frac{-(-6) \pm \sqrt{(-6)^2 - 4\cdot 1\cdot 4}}{2\cdot 1} = \frac{6 \pm \sqrt{36 - 16}}{2}$$

$$x = \frac{6 \pm \sqrt{20}}{2} = \frac{6 \pm 2\sqrt{5}}{2}$$

$$x = \frac{2(3 \pm \sqrt{5})}{2} = 3 \pm \sqrt{5}$$

The solution set is $\{3 \pm \sqrt{5}\}$.

94. $\{-12, 4\}$

95. a) $\frac{-b + \sqrt{b^2 - 4ac}}{2a} + \frac{-b - \sqrt{b^2 - 4ac}}{2a} = \frac{-2b}{2a} = -\frac{b}{a}$

b) $\frac{-b + \sqrt{b^2 - 4ac}}{2a} \cdot \frac{-b - \sqrt{b^2 - 4ac}}{2a}$

$= \frac{b^2 - (b^2 - 4ac)}{4a^2}$

$= \frac{4ac}{4a^2}$

$= \frac{c}{a}$

96. a) $k = 2$

b) $\frac{11}{2}$

97. a) $k(-3)^2 - 2(-3) + k = 0$ gives $k = -\frac{3}{5}$.

b) Then $-\frac{3}{5}x^2 - 2x - \frac{3}{5} = 0$ gives the other solution, $-\frac{1}{3}$.

98. a) $k = 2$

b) $1 - i$

99. a) $(3)^2 - (6 + 3i)(3) + k = 0$ gives $k = 9 + 9i$.

b) Then $x^2 - (6 + 3i)x + (9 + 9i) = 0$ gives the other solution, $3 + 3i$.

100. $k = -1$

101. Write $ax^2 + bx + c = 0$ as $x^2 - \left(-\frac{b}{a}\right)x + \frac{c}{a} = 0$ or $x^2 - Sx + P = 0$ where $S = -\frac{b}{a} = x_1 + x_2$ and $P = \frac{c}{a} = x_1 \cdot x_2$. Here $S = \sqrt{3}$ and $P = 8$. Then the required equation is $x^2 - \sqrt{3}\,x + 8 = 0$.

102. a) $x^2 - 2x + \frac{1}{4} = 0$, or $4x^2 - 8x + 1 = 0$

b) $x^2 - \frac{g^2 - h^2}{gh}x - 1 = 0$, or

$ghx^2 - (g^2 - h^2)x - gh = 0$

c) $x^2 - 4x + 29 = 0$

103. $3x^2 - hx + 4k = 0$ has $x_1 + x_2 = -\frac{b}{a} = -\frac{-h}{3} = -12$ and $x_1 \cdot x_2 = \frac{c}{a} = \frac{4k}{3} = 20$. Then $h = -36$ and $k = 15$.

104. 1

105. Let $1 \cdot x^2 + 2kx - 5 = 0$ have solutions r and s. Then $r + s = \frac{-2k}{1}$ and $rs = -5$. Adding $2rs = -10$ and $r^2 + s^2 = 26$ we obtain $r^2 + 2rs + s^2 = 16$ so that $r + s = \pm 4$. Then $r + s = -2k = \pm 4$ or $|k| = 2$.

106. $ax^2 + bx + c = 0$ has solution set

$$S_1 = \left\{\frac{-b + \sqrt{b^2 - 4ac}}{2a}, \frac{-b - \sqrt{b^2 - 4ac}}{2a}\right\}.$$

$cy^2 + by + a = 0$ has $y = \frac{-b \pm \sqrt{b^2 - 4ac}}{2c}$.

$$\text{Then } \frac{1}{y} = \frac{2c}{-b \pm \sqrt{b^2 - 4ac}} \cdot \frac{-b \mp \sqrt{b^2 - 4ac}}{-b \mp \sqrt{b^2 - 4ac}}$$

$= \frac{-b \mp \sqrt{b^2 - 4ac}}{2a}$, so that the reciprocals of the solutions of the second equation form the set

$S_2 = \left\{\frac{-b - \sqrt{b^2 - 4ac}}{2a}, \frac{-b + \sqrt{b^2 - 4ac}}{2a}\right\}$ which is the same as S_1.

Exercise Set 2.6

1.
$$F = \frac{kM_1M_2}{d^2}$$
$$Fd^2 = kM_1M_2$$
$$d^2 = \frac{kM_1M_2}{F}$$
$$d = \sqrt{\frac{kM_1M_2}{F}}$$

2. $c = \sqrt{\frac{E}{m}}$

3.
$$S = \frac{1}{2}at^2$$
$$2S = at^2$$
$$\frac{2S}{a} = t^2$$
$$\sqrt{\frac{2S}{a}} = t$$

4. $r = \sqrt{\frac{V}{4\pi}}$ or $\frac{1}{2}\sqrt{\frac{V}{\pi}}$

5.
$$s = -16t^2 + v_0t$$
$$0 = -16t^2 + v_0t - s$$
$$a = -16, \quad b = v_0, \quad c = -s$$
$$t = \frac{-v_0 \pm \sqrt{v_0^2 - 4(-16)(-s)}}{2(-16)}$$
$$t = \frac{-v_0 \pm \sqrt{v_0^2 - 64s}}{-32}$$
or
$$t = \frac{v_0 \pm \sqrt{v_0^2 - 64s}}{32}$$

6. $r = \frac{-3\pi h + \sqrt{9\pi^2h^2 + 8A\pi}}{4\pi}$

7. $d = \frac{n^2 - 3n}{2}$

$2d = n^2 - 3n$

$0 = n^2 - 3n - 2d$

$a = 1, \quad b = -3, \quad c = -2d$

$n = \frac{-(-3) \pm \sqrt{(-3)^2 - 4(1)(-2d)}}{2\cdot 1}$

$n = \frac{3 \pm \sqrt{9 + 8d}}{2}$, or just

$n = \frac{3 + \sqrt{9 + 8d}}{2}$ since the negative square root would result in a negative solution.

8. $t = \frac{\pi \pm \sqrt{\pi^2 - 12k\sqrt{2}}}{2\sqrt{2}}$

9. $A = P(1 + i)^2$

$\frac{A}{P} = (1 + i)^2$

$\pm\sqrt{\frac{A}{P}} = 1 + i$

$-1 \pm \sqrt{\frac{A}{P}} = i$, or just

$-1 + \sqrt{\frac{A}{P}} = i$ since the negative square root would result in a negative solution.

10. $i = 2\left[-1 + \sqrt{\frac{A}{P}}\right]$

11. $A = P(1 + i)^t$

$7220 = 5120(1 + i)^2$ (Substituting)

$\frac{7220}{5120} = (1 + i)^2$

$\pm\sqrt{\frac{722}{512}} = 1 + i$

$\pm\sqrt{\frac{361}{256}} = 1 + i$

$\pm\frac{19}{16} = 1 + i$

$-1 \pm \frac{19}{16} = i$

$-1 + \frac{19}{16} = i$ or $-1 - \frac{19}{16} = i$

$\frac{3}{16} = i$ or $-\frac{35}{16} = i$

Since the interest rate cannot be negative, $i = \frac{3}{16} = 0.1875 = 18.75\%$.

12. 10%

13. $A = P(1 + i)^t$

$9856.80 = 8000(1 + i)^2$ (Substituting)

$\frac{9856.80}{8000} = (1 + i)^2$

$\pm\sqrt{\frac{9856.80}{8000}} = 1 + i$

$-1 \pm 1.11 = i$

$-1 + 1.11 = i$ or $-1 - 1.11 = i$

$0.11 = i$ or $-2.11 = i$

Since the interest rate cannot be negative, $i = 0.11 = 11\%$.

14. 12.75%

15. $d = \frac{n^2 - 3n}{2}$

$27 = \frac{n^2 - 3n}{2}$ (Substituting 27 for d)

$54 = n^2 - 3n$

$0 = n^2 - 3n - 54$

$0 = (n - 9)(n + 6)$

$n - 9 = 0$ or $n + 6 = 0$

$n = 9$ or $n = -6$

The number of sides must be positive. Thus the number of sides is 9.

16. 11

17. Familiarize: We make a drawing and label it with the known and unknown data.

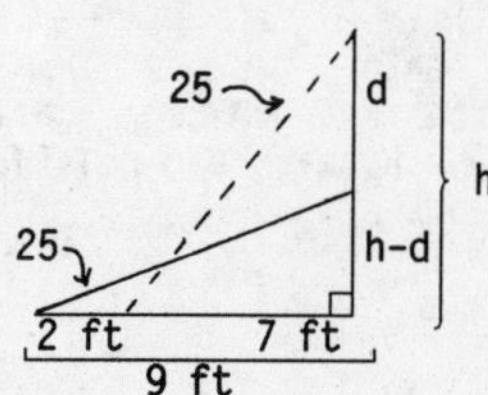

Translate: We use the Pythagorean theorem. From the taller triangle we get

$h^2 + 7^2 = 25^2$. (1)

From the other triangle we get

$(h - d)^2 + 9^2 = 25^2$. (2)

Carry out: We solve equation (1) for h and get h = 24. Then we substitute 24 for h in equation (2).

$(24 - d)^2 + 9^2 = 25^2$, or

$d^2 - 48d + 32 = 0$

Using the quadratic formula, we get

$d = \frac{48 \pm \sqrt{2176}}{2} = \frac{48 \pm 8\sqrt{34}}{2} = 24 \pm 4\sqrt{34}$.

Check and State: The length $24 + 4\sqrt{34}$ is not a solution since it exceeds the original length. The number $24 - 4\sqrt{34} \approx 24 - 4(5.831) = 0.676$ checks and is the solution. Therefore, the top of the ladder moves down the wall about 0.676 ft when the bottom is moved 2 ft.

18. $\sqrt{209} - \sqrt{176} \approx 1.190$ ft

19. Familiarize and Translate:

Using the Pythagorean theorem we determine the length of the other leg.

$10^2 = 6^2 + a^2$

$100 = 36 + a^2$

$64 = a^2$

$8 = a$

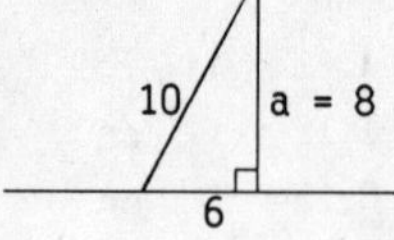

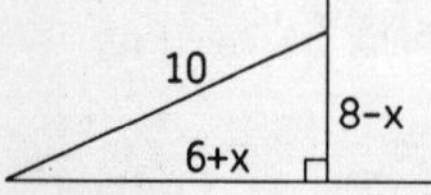

We let x represent the amount the ladder is pulled away and pulled down. The legs then become 6 + x and 8 - x. Then we again use the Pythagorean theorem.

$10^2 = (6 + x)^2 + (8 - x)^2$

Carry out:

$100 = 36 + 12x + x^2 + 64 - 16x + x^2$

$100 = 2x^2 - 4x + 100$

$0 = 2x^2 - 4x$

$0 = x^2 - 2x$

$0 = x(x - 2)$

$x = 0$ or $x - 2 = 0$

$x = 0$ or $x = 2$

We only consider x = 2 since x ≠ 0 if the ladder is pulled away.

Check and State:

When x = 2, the legs of the triangle become 6 + 2, or 8, and 8 - 2, or 6. The Pythagorean relationship still holds.

$8^2 + 6^2 = 10^2$

$64 + 36 = 100$

The ladder is pulled away 2 ft and pulled down 2 ft.

20. 7 ft

21. Familiarize:

We make a drawing. We let h represent the height and h + 3 represent the base.

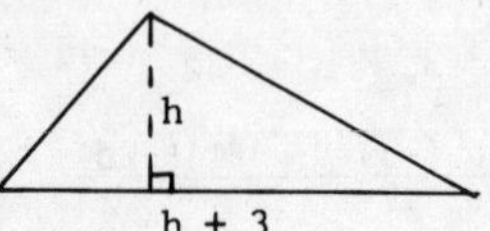

Translate:

$A = \frac{1}{2} \cdot \text{base} \cdot \text{height}$ (Area of triangle)

$18 = \frac{1}{2}(h + 3)h$ (Substituting)

Carry out:

$36 = h^2 + 3h$

$0 = h^2 + 3h - 36$

$a = 1, \quad b = 3, \quad c = -36$

$h = \frac{-3 \pm \sqrt{3^2 - 4(1)(-36)}}{2 \cdot 1}$

$= \frac{-3 \pm \sqrt{9 + 144}}{2} = \frac{-3 \pm \sqrt{153}}{2}$

$h = \frac{-3 + \sqrt{153}}{2}$ (h cannot be negative)

≈ 4.685

Check and State:

When h = 4.685, h + 3 = 7.685 and the area of the triangle is $\frac{1}{2}(7.685)(4.685) \approx 18$.

The height is 4.685 cm.

22. $90\sqrt{2} = 127.28$ ft

23. Familiarize: We make a drawing.

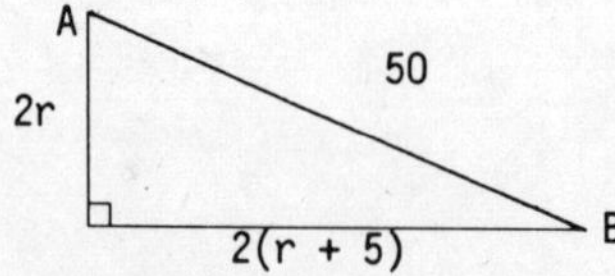

We let r represent the speed of train A. Then r + 5 represents the speed of train B. The distances they travel in 2 hours are 2r and 2(r + 5). After 2 hours they are 50 miles apart.

Translate:

Using the Pythagorean theorem we have an equation.

$(2r)^2 + [2(r + 5)]^2 = 50^2$

23. (continued)

Carry out:

$4r^2 + 4(r + 5)^2 = 2500$

$r^2 + (r + 5)^2 = 625$

$r^2 + r^2 + 10r + 25 = 625$

$2r^2 + 10r - 600 = 0$

$r^2 + 5r - 300 = 0$

$(r + 20)(r - 15) = 0$

$r + 20 = 0$ or $r - 15 = 0$

$r = -20$ or $r = 15$

Check and State:

Since speed in this problem must be positive, we only check 15. If $r = 15$, then $r + 5 = 20$. In 2 hours train A will travel $2 \cdot 15$, or 30 miles, and train B will travel $2 \cdot 20$, or 40 miles.

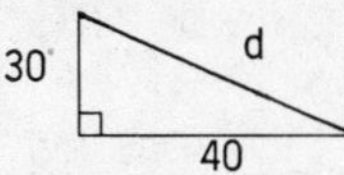

We now calculate how far apart trains A and B will be after 2 hours.

$30^2 + 40^2 = d^2$

$900 + 1600 = d^2$

$2500 = d^2$

$50 = d$

The value checks.

The speed of train A is 15 mph; the speed of train B is 20 mph.

24. A: 24 km/h, B: 10 km/h

25. a) $s = 4.9t^2 + v_0t$

$75 = 4.9t^2 + 0 \cdot t$ (Substituting 75 for s and 0 for v_0)

$75 = 4.9t^2$

$\frac{75}{4.9} = t^2$

$\sqrt{\frac{75}{4.9}} = t$

$3.91 \approx t$

It takes 3.91 sec to reach the ground.

25. (continued)

b) $s = 4.9t^2 + v_0t$

$75 = 4.9t^2 + 30t$ (Substituting 75 for s and 30 for v_0)

$0 = 4.9t^2 + 30t - 75$

$a = 4.9, \quad b = 30, \quad c = -75$

$t = \frac{-30 \pm \sqrt{30^2 - 4(4.9)(-75)}}{2(4.9)}$

$= \frac{-30 \pm \sqrt{900 + 1470}}{9.8}$

$t = \frac{-30 + \sqrt{2370}}{9.8}$ (t must be positive)

≈ 1.906

It takes 1.906 sec to reach the ground.

c) $s = 4.9t^2 + v_0t$

$s = 4.9(2)^2 + 30(2)$ (Substituting 2 for t and 30 for v_0)

$s = 19.6 + 60$

$s = 79.6$

The object will fall 79.6 m.

26. a) 10.1 sec
b) 7.49 sec
c) 272.5 m

27. Familiarize and Translate:

It helps to make a drawing.

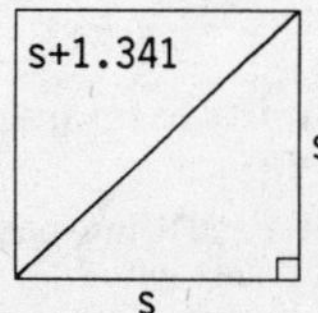

We let s represent the length of the side. Then s + 1.341 represents the diagonal. Using the Pythagorean theorem we have an equation.

$$s^2 + s^2 = (s + 1.341)^2$$

Carry out:

$2s^2 = s^2 + 2.682s + 1.798281$

$s^2 - 2.682s - 1.798281 = 0$

$s = \frac{-(-2.682) \pm \sqrt{(-2.682)^2 - 4(1)(-1.798281)}}{2 \cdot 1}$

$= \frac{2.682 \pm \sqrt{7.193124 + 7.193124}}{2}$

$s = \frac{2.682 + \sqrt{14.386248}}{2}$

≈ 3.237

Check and State:

The value checks. The length of the side is 3.237 cm.

28. 2.2199 cm, 8.0101 cm

29. Familiarize and Translate: From the drawing in the text we see that the area of the original garden is $80 \cdot 60$ and the area of the new garden is $(80 - 2x)(60 - 2x)$, where x is the width of the sidewalk. We know the area of the new garden is 2/3 of the original area, so we can write an equation.

$$(80 - 2x)(60 - 2x) = \frac{2}{3} \cdot 80 \cdot 60, \text{ or}$$

$$x^2 - 70x + 400 = 0$$

Carry out: Using the quadratic formula, we get

$$x = \frac{70 \pm \sqrt{3300}}{2} = \frac{70 \pm 10\sqrt{33}}{2} = 35 \pm 5\sqrt{33}.$$

Check and State: The number $35 + 5\sqrt{33}$ is not a solution since twice that number exceeds both the original length and width of the garden. The number $35 - 5\sqrt{33} \approx 35 - 5(5.745) \approx 6.28$ checks and is the solution. Therefore, the sidewalk is about 6.28 ft wide.

30. $\sqrt{304} \approx 17.4$ mph

31. Familiarize: We make a drawing, letting x represent the length of a side of the square cut from each corner.

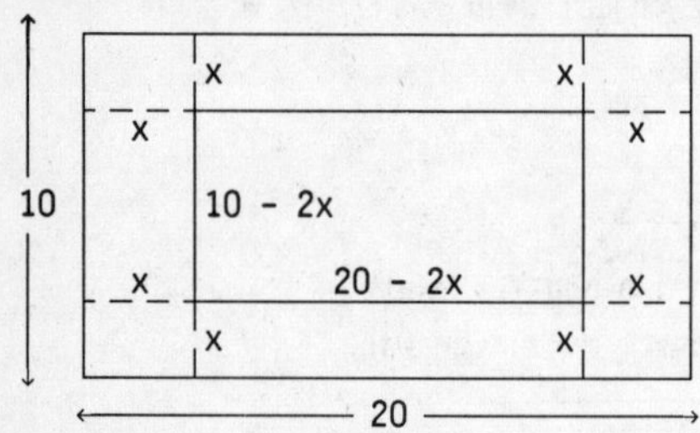

The dimensions of the resulting base are $10 - 2x$ and $20 - 2x$.

Translate: The area of the resulting base is 90 cm^2, so we can write an equation.

$$(20 - 2x)(10 - 2x) = 90, \text{ or}$$

$$2x^2 - 30x + 55 = 0$$

Carry out: Using the quadratic formula, we get

$$x = \frac{30 \pm \sqrt{460}}{4} = \frac{30 \pm 2\sqrt{115}}{4} = \frac{15 \pm \sqrt{115}}{2}.$$

Check and State: The number $\frac{15 + \sqrt{115}}{2}$ is not a solution since twice that number exceeds the width of the piece of tin. The number $\frac{15 - \sqrt{115}}{2} \approx 2.14$ checks and is the solution. Therefore, the length of the sides of the squares is about 2.14 cm.

32. $15 - \sqrt{51} \approx 7.86$ cm

33. Familiarize: We can use the formula $d = rt$ in the form $t = \frac{d}{r}$. Let x represent the speed on the first part of the trip. Then $x - 10$ represents the speed on the second part of the trip. The time for the first part of the trip is $\frac{50}{x}$, and the time for the second part is $\frac{80}{x - 10}$.

Translate: We know that the total time for the trip is 2 hr, so we can write an equation.

$$\frac{50}{x} + \frac{80}{x - 10} = 2, \text{ or}$$

$$x^2 - 75x + 250 = 0 \quad \text{Clearing fractions and simplifying}$$

Carry out: Using the quadratic formula, we get

$$x = \frac{75 \pm \sqrt{4625}}{2} = \frac{75 \pm 5\sqrt{185}}{2}.$$

Check and State: The number $x = \frac{75 - 5\sqrt{185}}{2}$ is not a solution since $x - 10 < 0$. The number $x = \frac{75 + 5\sqrt{185}}{2} \approx 71.5$ mph checks and is the solution. This is the speed on the first part of the trip. The speed on the second part is approximately $71.5 - 10$, or 61.5 mph.

34. 20 mph

35. Familiarize: We will use the formula $t = \frac{d}{r}$ as we did in Exercise 33. Let x represent the speed of the boat in still water. Then the time to travel upstream is $\frac{12}{x - 3}$ and the time to travel downstream is $\frac{12}{x + 3}$.

Translate: We know the total time for the round trip is 2 hr, so we can write an equation.

$$\frac{12}{x - 3} + \frac{12}{x + 3} = 2, \text{ or}$$

$$x^2 - 12x - 9 = 0 \quad \text{Clearing fractions and simplifying}$$

Carry out: Using the quadratic formula, we get

$$x = \frac{12 \pm \sqrt{180}}{2} = \frac{12 \pm 6\sqrt{5}}{2} = 6 \pm 3\sqrt{5}$$

Check and State: The number $6 - 3\sqrt{5}$ is not a solution, because the speed cannot be negative. The number $6 + 3\sqrt{5} \approx 12.7$ checks and is the solution. The speed of the boat in still water is $6 + 3\sqrt{5} \approx 12.7$ mph.

36. $\frac{3 + \sqrt{89}}{2} \approx 6.2$ hr

37. Familiarize and Translate:

We let x represent the number of students in the group at the outset. Then x - 3 represents the number of students at the last minute.

$$\text{Total cost} = (\text{Cost per person})\cdot(\text{Number of persons})$$

$$\frac{\text{Total cost}}{\text{Number of persons}} = \text{Cost per person}$$

Using this formula we get the following:

$$\text{Cost per person at the outset} = \frac{140}{x}$$

$$\text{Cost per person at the last minute} = \frac{140}{x - 3}$$

The cost at the last minute was $15 more than the cost at the outset. This gives us an equation.

$$\frac{140}{x - 3} = 15 + \frac{140}{x}$$

Carry out:

$$x(x - 3)\cdot\frac{140}{x - 3} = x(x - 3)\left[15 + \frac{140}{x}\right]$$

(Multiplying by the LCM)

$$140x = 15x(x - 3) + 140(x - 3)$$

$$140x = 15x^2 - 45x + 140x - 420$$

$$0 = 15x^2 - 45x - 420$$

$$0 = x^2 - 3x - 28$$

$$0 = (x - 7)(x + 4)$$

$x - 7 = 0$ or $x + 4 = 0$

$x = 7$ or $x = -4$

Check and State:

Since the number of people cannot be negative, we only check x = 7. When x = 7, the cost at the outset is 140/7, or $20. When x = 7, the cost at the last minute is 140/(7 - 3), or 140/4, or $35 and $35 is $15 more than $20. The value checks.

Thus, 7 students were in the group at the outset.

38. x = 12

39. Familiarize and Translate:

Let x represent the number of shares bought. Then $\frac{720}{x}$ represents the cost of each share.

$$\text{Total cost} = (\text{Cost per item})\cdot(\text{Number of items})$$

$$720 = \left[\frac{720}{x}\right]\cdot(x)$$

When the cost per share is reduced by $15, the new cost is $\frac{720}{x} - 15$. When the number of shares purchased is increased by 4, the new number of shares is x + 4. The total cost remains the same.

$$720 = \left[\frac{720}{x} - 15\right]\cdot(x + 4)$$

39. (continued)

Carry out:

$$720 = 720 + \frac{2880}{x} - 15x - 60$$

$$60 = \frac{2880}{x} - 15x$$

$$60x = 2880 - 15x^2$$

$$15x^2 + 60x - 2880 = 0$$

$$x^2 + 4x - 192 = 0$$

$$(x + 16)(x - 12) = 0$$

$x + 16 = 0$ or $x - 12 = 0$

$x = -16$ or $x = 12$

Check and State:

The number of shares must be positive. We only check x = 12. The cost per share when x = 12 is 720/12, or $60. The cost per share when the number is increased by 4 is 720/16, or $45. Thus, $45 is $15 less than $60, and the value checks.

Thus, 12 shares of stock were bought.

40. $8

41. $kx^2 + (3 - 2k)x - 6 = 0$

$a = k,\quad b = 3 - 2k,\quad c = -6$

$$x = \frac{-(3 - 2k) \pm \sqrt{(3 - 2k)^2 - 4(k)(-6)}}{2k}$$

$$= \frac{-3 + 2k \pm \sqrt{4k^2 + 12k + 9}}{2k}$$

$$= \frac{-3 + 2k \pm \sqrt{(2k + 3)^2}}{2k}$$

$$= \frac{-3 + 2k \pm (2k + 3)}{2k}$$

$$x = \frac{-3 + 2k + 2k + 3}{2k} \quad\text{or}\quad x = \frac{-3 + 2k - 2k - 3}{2k}$$

$$x = \frac{4k}{2k} \quad\text{or}\quad x = \frac{-6}{2k}$$

$$x = 2 \quad\text{or}\quad x = -\frac{3}{k}$$

The solution set is $\left\{2, -\frac{3}{k}\right\}$.

42. $\left\{1, \frac{1}{1 - k}\right\}$

43. $(m + n)^2x^2 + (m + n)x = 2$

$(m + n)^2x^2 + (m + n)x - 2 = 0$

$a = (m + n)^2, \quad b = (m + n), \quad c = -2$

$$x = \frac{-(m + n) \pm \sqrt{(m + n)^2 - 4(m + n)^2(-2)}}{2(m + n)^2}$$

$$= \frac{-(m + n) \pm \sqrt{9(m + n)^2}}{2(m + n)^2}$$

$$= \frac{-(m + n) \pm 3(m + n)}{2(m + n)^2}$$

$$x = \frac{2(m + n)}{2(m + n)^2} \quad \text{or} \quad x = \frac{-4(m + n)}{2(m + n)^2}$$

$$x = \frac{1}{m + n} \quad \text{or} \quad x = \frac{-2}{m + n}$$

The solution set is $\left\{\frac{1}{m + n}, -\frac{2}{m + n}\right\}$.

44. a) $\{4y, -y\}$

b) $\left\{-x, \frac{x}{4}\right\}$

45. $A = P(1 + i)^t$

$13{,}704 = 9826(1 + i)^3$ (Substituting)

$$\frac{13{,}704}{9826} = (1 + i)^3$$

$$\sqrt[3]{\frac{13{,}704}{9826}} = 1 + i$$

$$-1 + \sqrt[3]{\frac{13{,}704}{9826}} = i$$

$0.117 \approx i$

The interest rate is 11.7%.

46. $\frac{27}{4}\sqrt{3}$ or 11.69

47. Familiarize:

We first make a drawing.

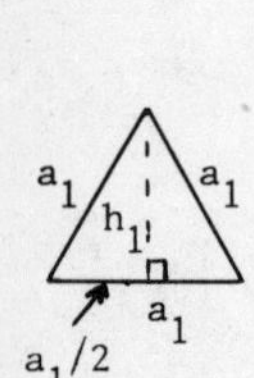

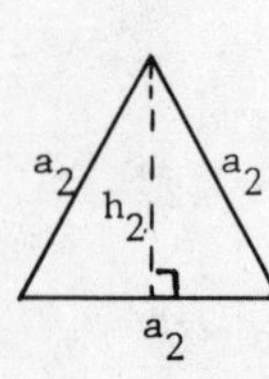

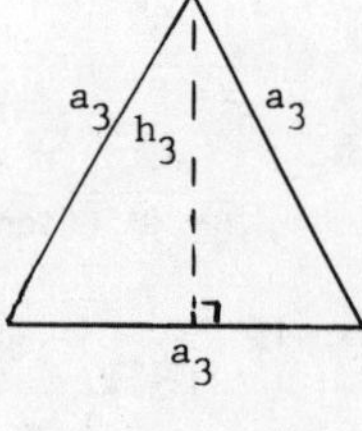

We find the height of the first triangle.

$$\left(\frac{a_1}{2}\right)^2 + (h_1)^2 = a_1^2$$

$$(h_1)^2 = a_1^2 - \frac{a_1^2}{4} = \frac{3a_1^2}{4}$$

$$h_1 = \frac{a_1\sqrt{3}}{4}$$

Next we find the area of the first triangle.

$$A_1 = \frac{1}{2} \cdot a_1 \cdot \frac{a_1\sqrt{3}}{4} = \frac{a_1^2\sqrt{3}}{8}$$

47. (continued)

Similarly we find the areas of the other two triangles:

$$A_2 = \frac{a_2^2\sqrt{3}}{8} \quad \text{and} \quad A_3 = \frac{a_3^2\sqrt{3}}{8}.$$

Translate:

The sum of the areas of the first two triangles equals the area of the third triangle.

$$\frac{a_1^2\sqrt{3}}{8} + \frac{a_2^2\sqrt{3}}{8} = \frac{a_3^2\sqrt{3}}{8}$$

Carry out:

$$\frac{\sqrt{3}}{8}(a_1^2 + a_2^2) = \frac{\sqrt{3}}{8} \cdot a_3^2$$

$$a_1^2 + a_2^2 = a_3^2$$

$\sqrt{a_1^2 + a_2^2} = a_3$ Taking the positive square root

Check and State: The answer checks.

Thus the length of a side of the third triangle is $\sqrt{a_1^2 + a_2^2}$.

48. They have the same area: 300 ft^2.

49.

$\frac{6 - y}{x} = \frac{6}{8}$ and $xy = 12$ give $x = 4$ cm, $y = 3$ cm.

50. 3.3 sec

Exercise Set 2.7

1. $\sqrt{3x - 4} = 1$

$(\sqrt{3x - 4})^2 = 1^2$

$3x - 4 = 1$

$3x = 5$

$x = \frac{5}{3}$

Check:

$\sqrt{3x - 4}$	$= 1$
$\sqrt{3 \cdot \frac{5}{3} - 4}$	1
$\sqrt{5 - 4}$	
$\sqrt{1}$	
1	

The solution set is $\left\{\frac{5}{3}\right\}$.

2. $\{-63\}$

3. $\sqrt[4]{x^2 - 1} = 1$

$(\sqrt[4]{x^2 - 1})^4 = 1^4$

$x^2 - 1 = 1$

$x^2 = 2$

$x = \pm\sqrt{2}$

Check:

$\sqrt[4]{x^2 - 1} = 1$	
$\sqrt[4]{(\pm\sqrt{2})^2 - 1}$	1
$\sqrt[4]{2 - 1}$	
$\sqrt[4]{1}$	
1	

The solution set is $\{\pm\sqrt{2}\}$.

4. {168}

5. $\sqrt{y - 1} + 4 = 0$

$\sqrt{y - 1} = -4$

Note:

The principal root is never negative. Thus, there is no solution.

If we do not observe the above fact, we could continue and reach the same answer.

$(\sqrt{y - 1})^2 = (-4)^2$

$y - 1 = 16$

$y = 17$

Check:

$\sqrt{y - 1} + 4 = 0$	
$\sqrt{17 - 1} + 4$	0
$\sqrt{16} + 4$	
$4 + 4$	
8	

Since 17 does not check, there is no solution. The solution set is Ø.

6. Ø

7. $\sqrt{x - 3} + \sqrt{x + 5} = 4$

$\sqrt{x + 5} = 4 - \sqrt{x - 3}$

$(\sqrt{x + 5})^2 = (4 - \sqrt{x - 3})^2$

$x + 5 = 16 - 8\sqrt{x - 3} + (x - 3)$

$x + 5 = 13 - 8\sqrt{x - 3} + x$

$8\sqrt{x - 3} = 8$

$\sqrt{x - 3} = 1$

$(\sqrt{x - 3})^2 = 1^2$

$x - 3 = 1$

$x = 4$

Check:

$\sqrt{x - 3} + \sqrt{x + 5} = 4$	
$\sqrt{4 - 3} + \sqrt{4 + 5}$	4
$\sqrt{1} + \sqrt{9}$	
$1 + 3$	
4	

The solution set is {4}.

8. {9}

9. $\sqrt{3x - 5} + \sqrt{2x + 3} + 1 = 0$

$\sqrt{3x - 5} + \sqrt{2x + 3} = -1$

Note:

The principal root is never negative. Thus the sum of two principal roots cannot equal -1. There is no solution. The solution set is Ø.

10. {2}

11. $\sqrt[3]{6x + 9} + 8 = 5$

$\sqrt[3]{6x + 9} = -3$

$(\sqrt[3]{6x + 9})^3 = (-3)^3$

$6x + 9 = -27$

$6x = -36$

$x = -6$

Check:

$\sqrt[3]{6x + 9} + 8 = 5$	
$\sqrt[3]{6(-6) + 9} + 8$	5
$\sqrt[3]{-27} + 8$	
$-3 + 8$	
5	

The solution set is {-6}.

12. $\left\{\frac{28}{3}\right\}$

13. $\sqrt{6x + 7} = x + 2$

$(\sqrt{6x + 7})^2 = (x + 2)^2$

$6x + 7 = x^2 + 4x + 4$

$0 = x^2 - 2x - 3$

$0 = (x - 3)(x + 1)$

$x - 3 = 0$ or $x + 1 = 0$

$x = 3$ or $x = -1$

Both values check. The solution set is {3, -1}.

14. $\left\{\frac{1}{3}, -1\right\}$

15. $\sqrt{20 - x} = \sqrt{9 - x} + 3$

$(\sqrt{20 - x})^2 = (\sqrt{9 - x} + 3)^2$

$20 - x = (9 - x) + 6\sqrt{9 - x} + 9$

$20 - x = 18 - x + 6\sqrt{9 - x}$

$2 = 6\sqrt{9 - x}$

$1 = 3\sqrt{9 - x}$

$1^2 = (3\sqrt{9 - x})^2$

$1 = 9(9 - x)$

$1 = 81 - 9x$

$9x = 80$

$x = \frac{80}{9}$

The value checks. The solution set is $\left\{\frac{80}{9}\right\}$.

16. {-1}

17. $\sqrt{x} - \sqrt{3x - 3} = 1$

$$\sqrt{x} = \sqrt{3x - 3} + 1$$
$$(\sqrt{x})^2 = (\sqrt{3x - 3} + 1)^2$$
$$x = (3x - 3) + 2\sqrt{3x - 3} + 1$$
$$2 - 2x = 2\sqrt{3x - 3}$$
$$1 - x = \sqrt{3x - 3}$$
$$(1 - x)^2 = (\sqrt{3x - 3})^2$$
$$1 - 2x + x^2 = 3x - 3$$
$$x^2 - 5x + 4 = 0$$
$$(x - 4)(x - 1) = 0$$

$x = 4$ or $x = 1$

The number 4 does not check, but 1 does. The solution set is {1}.

18. {0, 4}

19. $\sqrt{2y - 5} - \sqrt{y - 3} = 1$

$$\sqrt{2y - 5} = \sqrt{y - 3} + 1$$
$$(\sqrt{2y - 5})^2 = (\sqrt{y - 3} + 1)^2$$
$$2y - 5 = (y - 3) + 2\sqrt{y - 3} + 1$$
$$y - 3 = 2\sqrt{y - 3}$$
$$(y - 3)^2 = (2\sqrt{y - 3})^2$$
$$y^2 - 6y + 9 = 4(y - 3)$$
$$y^2 - 6y + 9 = 4y - 12$$
$$y^2 - 10y + 21 = 0$$
$$(y - 7)(y - 3) = 0$$

$y = 7$ or $y = 3$

Both numbers check. The solution set is {7, 3}.

20. $\{6 + 2\sqrt{3}\}$

21. $\sqrt{7.35x + 8.051} = 0.345x + 0.067$

$$(\sqrt{7.35x + 8.051})^2 = (0.345x + 0.067)^2$$
$$7.35x + 8.051 = 0.119025x^2 + 0.04623x + 0.004489$$
$$0 = 0.119025x^2 - 7.30377x - 8.046511$$
$$x = \frac{-(-7.30377) \pm \sqrt{(-7.30377)^2 - 4(0.119025)(-8.046511)}}{2(0.119025)}$$
$$\approx \frac{7.30377 \pm \sqrt{57.176}}{0.23805}$$

$x \approx 62.4459$ or $x \approx -1.0826$

Since -1.0826 does not check and 62.4459 does check, the solution set is {62.4459}.

22. {0.1444}

23. $x^{1/3} = -2$

$$(x^{1/3})^3 = (-2)^3 \qquad (x^{1/3} = \sqrt[3]{x})$$
$$x = -8$$

The value checks. The solution set is {-8}.

24. {32}

25. $t^{1/4} = 3$

$$(t^{1/4})^4 = 3^4 \qquad (t^{1/4} = \sqrt[4]{t})$$
$$t = 81$$

The value checks. The solution set is {81}.

26. $\emptyset$

27. $8 = \frac{1}{\sqrt{x}}$

$$8\sqrt{x} = 1$$
$$\sqrt{x} = \frac{1}{8}$$
$$(\sqrt{x})^2 = \left(\frac{1}{8}\right)^2$$
$$x = \frac{1}{64}$$

The value checks. The solution set is $\left\{\frac{1}{64}\right\}$.

28. $\left\{\frac{1}{9}\right\}$

29. $\sqrt[3]{m} = -5$

$$(\sqrt[3]{m})^3 = (-5)^3$$
$$m = -125$$

The value checks. The solution set is {-125}.

30. $\emptyset$

31. For L:

$$T = 2\pi\sqrt{\frac{L}{g}}$$
$$\frac{T}{2\pi} = \sqrt{\frac{L}{g}}$$
$$\left(\frac{T}{2\pi}\right)^2 = \left(\sqrt{\frac{L}{g}}\right)^2$$
$$\frac{T^2}{4\pi^2} = \frac{L}{g}$$
$$\frac{gT^2}{4\pi^2} = L$$

For g:

$$T = 2\pi\sqrt{\frac{L}{g}}$$
$$\frac{T}{2\pi} = \sqrt{\frac{L}{g}}$$
$$\left(\frac{T}{2\pi}\right)^2 = \left(\sqrt{\frac{L}{g}}\right)^2$$
$$\frac{T^2}{4\pi^2} = \frac{L}{g}$$
$$gT^2 = 4\pi^2 L$$
$$g = \frac{4\pi^2 L}{T^2}$$

32. $c = \sqrt{H^2 - d^2}$

33. $V = 1.2\sqrt{h}$

$$V = 1.2\sqrt{30,000} \qquad \text{(Substituting)}$$
$$\approx 1.2(173.205)$$
$$\approx 208$$

You can see about 208 miles to the horizon.

34. 10 mi

35. $V = 1.2\sqrt{h}$

$144 = 1.2\sqrt{h}$ (Substituting)

$\frac{144}{1.2} = \sqrt{h}$

$120 = \sqrt{h}$

$14{,}400 = h$

The airplane is 14,400 ft high.

36. 84 ft

37. $(x - 5)^{2/3} = 2$

$\sqrt[3]{(x - 5)^2} = 2$

$(\sqrt[3]{(x - 5)^2})^3 = 2^3$

$(x - 5)^2 = 2^3$

$x^2 - 10x + 25 = 8$

$x^2 - 10x + 17 = 0$

$a = 1, \quad b = -10, \quad c = 17$

$x = \frac{-(-10) \pm \sqrt{(-10)^2 - 4\cdot1\cdot17}}{2\cdot1}$

$= \frac{10 \pm \sqrt{100 - 68}}{2} = \frac{10 \pm \sqrt{32}}{2}$

$= \frac{10 \pm 4\sqrt{2}}{2} = 5 \pm 2\sqrt{2}$

The solution set is $\{5 \pm 2\sqrt{2}\}$.

38. $\{3 \pm 2\sqrt{2}\}$

39. $\frac{x + \sqrt{x + 1}}{x - \sqrt{x + 1}} = \frac{5}{11}$

$11(x + \sqrt{x + 1}) = 5(x - \sqrt{x + 1})$

$11x + 11\sqrt{x + 1} = 5x - 5\sqrt{x + 1}$

$6x = -16\sqrt{x + 1}$

$(6x)^2 = (-16\sqrt{x + 1})^2$

$36x^2 = 256(x + 1)$

$36x^2 - 256x - 256 = 0$

$9x^2 - 64x - 64 = 0$

$(9x + 8)(x - 8) = 0$

$9x + 8 = 0$ or $x - 8 = 0$

$x = -\frac{8}{9}$ or $x = 8$

Only $-\frac{8}{9}$ checks. The solution set is $\left\{-\frac{8}{9}\right\}$.

40. $\emptyset$

41. $\sqrt{x + 2} - \sqrt{x - 2} = \sqrt{2x}$

$\sqrt{x + 2} = \sqrt{2x} + \sqrt{x - 2}$

$(\sqrt{x + 2})^2 = (\sqrt{2x} + \sqrt{x - 2})^2$

$x + 2 = 2x + 2\sqrt{2x(x - 2)} + (x - 2)$

$x + 2 = 3x + 2\sqrt{2x(x - 2)} - 2$

$4 - 2x = 2\sqrt{2x(x - 2)}$

$(2 - x)^2 = (\sqrt{2x^2 - 4x})^2$

$4 - 4x + x^2 = 2x^2 - 4x$

$0 = x^2 - 4$

$0 = (x + 2)(x - 2)$

$x + 2 = 0$ or $x - 2 = 0$

$x = -2$ or $x = 2$

Only 2 checks. The solution set is {2}.

42. {1}

43. $\sqrt[4]{x + 2} = \sqrt{3x + 1}$

$(\sqrt[4]{x + 2})^4 = (\sqrt{3x + 1})^4$

$x + 2 = (3x + 1)^2$

$x + 2 = 9x^2 + 6x + 1$

$0 = 9x^2 + 5x - 1$

$a = 9, \quad b = 5, \quad c = -1$

$x = \frac{-5 \pm \sqrt{5^2 - 4(9)(-1)}}{2\cdot9} = \frac{-5 \pm \sqrt{25 + 36}}{18}$

$= \frac{-5 \pm \sqrt{61}}{18}$

Only $\frac{-5 + \sqrt{61}}{18}$ checks. The solution set is $\left\{\frac{-5 + \sqrt{61}}{18}\right\}$.

44. $\left\{\frac{5}{4}\right\}$

45. $\frac{14}{3 + \sqrt{7 + x}} - \frac{\sqrt{7 + x}}{2} = 0$, LCM is $2(3 + \sqrt{7 + x})$

$2(3 + \sqrt{7 + x})\left[\frac{14}{3 + \sqrt{7 + x}} - \frac{\sqrt{7 + x}}{2}\right] = 2(3 + \sqrt{7 + x})\cdot0$

$28 - 3\sqrt{7 + x} - (7 + x) = 0$

$21 - x = 3\sqrt{7 + x}$

$(21 - x)^2 = (3\sqrt{7 + x})^2$

$441 - 42x + x^2 = 9(7 + x)$

$441 - 42x + x^2 = 63 + 9x$

$x^2 - 51x + 378 = 0$

$(x - 9)(x - 42) = 0$

$x = 9$ or $x = 42$

Only 9 checks. The solution set is {9}.

46. {6}

47. $\sqrt{3x + 1} - \sqrt{2x} = \dfrac{5}{\sqrt{3x + 1}}$,

LCM is $\sqrt{3x + 1}$

$\sqrt{3x + 1}(\sqrt{3x + 1} - \sqrt{2x}) = \sqrt{3x + 1} \cdot \dfrac{5}{\sqrt{3x + 1}}$

$(3x + 1) - \sqrt{6x^2 + 2x} = 5$

$3x - 4 = \sqrt{6x^2 + 2x}$

$(3x - 4)^2 = (\sqrt{6x^2 + 2x})^2$

$9x^2 - 24x + 16 = 6x^2 + 2x$

$3x^2 - 26x + 16 = 0$

$(3x - 2)(x - 8) = 0$

$x = \frac{2}{3}$ or $x = 8$

Only 8 checks. The solution set is {8}.

48. $\left\{\frac{25}{13}\right\}$

49. $\sqrt{15 + \sqrt{2x + 80}} = 5$

$\left[\sqrt{15 + \sqrt{2x + 80}}\right]^2 = 5^2$

$15 + \sqrt{2x + 80} = 25$

$\sqrt{2x + 80} = 10$

$(\sqrt{2x + 80})^2 = 10^2$

$2x + 80 = 100$

$2x = 20$

$x = 10$

This number checks. The solution set is {10}.

50. {-1}

Exercise Set 2.8

1. $x - 10\sqrt{x} + 9 = 0$

We substitute u for $\sqrt{x}$.

$u^2 - 10u + 9 = 0$

$(u - 9)(u - 1) = 0$

$u - 9 = 0$ or $u - 1 = 0$

$u = 9$ or $u = 1$

Now we substitute $\sqrt{x}$ for u and solve for x.

$\sqrt{x} = 9$ or $\sqrt{x} = 1$

$x = 81$ or $x = 1$

1. (continued)

Check:

For 81:

$x - 10\sqrt{x} + 9 = 0$	
$81 - 10\sqrt{81} + 9$	0
$81 - 10\cdot 9 + 9$	
$81 - 90 + 9$	
0	

For 1:

$x - 10\sqrt{x} + 9 = 0$	
$1 - 10\sqrt{1} + 9$	0
$1 - 10 + 9$	
0	

The solutions are 81 and 1. The solution set is {81, 1}.

2. $\left\{16, \frac{1}{4}\right\}$

3. $x^4 - 10x^2 + 25 = 0$

We substitute u for x^2.

$u^2 - 10u + 25 = 0$

$(u - 5)(u - 5) = 0$

$u - 5 = 0$ or $u - 5 = 0$

$u = 5$ or $u = 5$

Now we substitute x^2 for u and solve for x.

$x^2 = 5$ or $x^2 = 5$

$x = \pm\sqrt{5}$ or $x = \pm\sqrt{5}$

The solutions are $\pm\sqrt{5}$. The solution set is $\{\pm\sqrt{5}\}$.

4. $\{1, -1, \pm\sqrt{2}\}$

5. $t^{2/3} + t^{1/3} - 6 = 0$

We substitute u for $t^{1/3}$.

$u^2 + u - 6 = 0$

$(u + 3)(u - 2) = 0$

$u + 3 = 0$ or $u - 2 = 0$

$u = -3$ or $u = 2$

We now substitute $t^{1/3}$ for u and solve for t.

$t^{1/3} = -3$ or $t^{1/3} = 2$

$t = (-3)^3$ or $t = 2^3$

$t = -27$ or $t = 8$

The solutions are -27 and 8. The solution set is {-27, 8}.

6. {64, -8}

7. $z^{1/2} = z^{1/4} + 2$

$z^{1/2} - z^{1/4} - 2 = 0$

We substitute u for $z^{1/4}$.

$u^2 - u - 2 = 0$

$(u - 2)(u + 1) = 0$

$u - 2 = 0$ or $u + 1 = 0$

$u = 2$ or $u = -1$

Next we substitute $z^{1/4}$ for u and solve for z.

$z^{1/4} = 2$ or $z^{1/4} = -1$

$z = 2^4$ or $z = (-1)^4$

$z = 16$ or $z = 1$

The number 16 checks, but 1 does not. The solution set is {16}.

8. {729}

9. $(x^2 - 6x)^2 - 2(x^2 - 6x) - 35 = 0$

We substitute u for $x^2 - 6x$.

$u^2 - 2u - 35 = 0$

$(u - 7)(u + 5) = 0$

$u - 7 = 0$ or $u + 5 = 0$

$u = 7$ or $u = -5$

Next we substitute $x^2 - 6x$ for u and solve for x.

$x^2 - 6x = 7$ or $x^2 - 6x = -5$

$x^2 - 6x - 7 = 0$ or $x^2 - 6x + 5 = 0$

$(x - 7)(x + 1) = 0$ or $(x - 5)(x - 1) = 0$

$x - 7 = 0$ or $x + 1 = 0$ or $x - 5 = 0$ or $x - 1 = 0$

$x = 7$ or $x = -1$ or $x = 5$ or $x = 1$

The solutions are 7, -1, 5, and 1. The solution set is {7, -1, 5, 1}.

10. {1}

11. $(y^2 - 5y)^2 + (y^2 - 5y) - 12 = 0$

We substitute u for $y^2 - 5y$.

$u^2 + u - 12 = 0$

$(u + 4)(u - 3) = 0$

$u + 4 = 0$ or $u - 3 = 0$

$u = -4$ or $u = 3$

Next we substitute $y^2 - 5y$ for u and solve for y.

$y^2 - 5y = -4$ or $y^2 - 5y = 3$

$y^2 - 5y + 4 = 0$ or $y^2 - 5y - 3 = 0$

$(y - 4)(y - 1) = 0$ or $y = \frac{-(-5) \pm\sqrt{(-5)^2-4(1)(-3)}}{2\cdot 1}$

$y - 4 = 0$ or $y - 1 = 0$ or $y = \frac{5 \pm \sqrt{25 + 12}}{2}$

$y = 4$ or $y = 1$ or $y = \frac{5 \pm \sqrt{37}}{2}$

11. (continued)

The solutions are 4, 1, and $\frac{5 \pm \sqrt{37}}{2}$. The solution set is $\left\{4, 1, \frac{5 \pm \sqrt{37}}{2}\right\}$.

12. $\left\{1, -1, \frac{1}{2}, -\frac{3}{2}\right\}$

13. $w^4 - 4w^2 - 2 = 0$

First we substitute u for w^2.

$u^2 - 4u - 2 = 0$

$u = \frac{-(-4) \pm \sqrt{(-4)^2 - 4(1)(-2)}}{2\cdot 1}$

$= \frac{4 \pm \sqrt{16 + 8}}{2} = \frac{4 \pm \sqrt{24}}{2}$

$= \frac{4 \pm 2\sqrt{6}}{2} = 2 \pm \sqrt{6}$

Next we substitute w^2 for u and solve for w.

$w^2 = 2 + \sqrt{6}$ or $w^2 = 2 - \sqrt{6}$

$w = \pm\sqrt{2 + \sqrt{6}}$ or $w = \pm\sqrt{2 - \sqrt{6}}$

The solutions are $\pm\sqrt{2 + \sqrt{6}}$ and $\pm\sqrt{2 - \sqrt{6}}$.

The solution set is $\left\{\pm\sqrt{2 + \sqrt{6}}, \pm\sqrt{2 - \sqrt{6}}\right\}$.

14. $\left\{\pm\sqrt{\frac{5 + \sqrt{5}}{2}}, \pm\sqrt{\frac{5 - \sqrt{5}}{2}}\right\}$

15. $x^{-2} - x^{-1} - 6 = 0$

We substitute u for x^{-1}.

$u^2 - u - 6 = 0$

$(u - 3)(u + 2) = 0$

$u - 3 = 0$ or $u + 2 = 0$

$u = 3$ or $u = -2$

Then we substitute x^{-1} for u and solve for x.

$x^{-1} = 3$ or $x^{-1} = -2$

$\frac{1}{x} = 3$ or $\frac{1}{x} = -2$

$x = \frac{1}{3}$ or $x = -\frac{1}{2}$

The solutions are $\frac{1}{3}$ and $-\frac{1}{2}$. The solution set is $\left\{\frac{1}{3}, -\frac{1}{2}\right\}$.

16. $\left\{-1, \frac{4}{5}\right\}$

17. $2x^{-2} + x^{-1} = 1$

$2x^{-2} + x^{-1} - 1 = 0$

We substitute u for x^{-1}.

$2u^2 + u - 1 = 0$

$(2u - 1)(u + 1) = 0$

$2u - 1 = 0$ or $u + 1 = 0$

$u = \frac{1}{2}$ or $u = -1$

Next we substitute x^{-1} for u and solve for x.

$x^{-1} = \frac{1}{2}$ or $x^{-1} = -1$

$\frac{1}{x} = \frac{1}{2}$ or $\frac{1}{x} = -1$

$x = 2$ or $x = -1$

The solutions are 2 and -1. The solution set is {2, -1}.

18. $\left\{-\frac{1}{10}, 1\right\}$

19. $x^4 - 24x^2 - 25 = 0$

We substitute u for x^2.

$u^2 - 24u - 25 = 0$

$(u - 25)(u + 1) = 0$

$u - 25 = 0$ or $u + 1 = 0$

$u = 25$ or $u = -1$

Then we substitute x^2 for u and solve for x.

$x^2 = 25$ or $x^2 = -1$

$x = \pm 5$ or $x = \pm i$

The solutions are ± 5 and ± i. The solution set is {± 5, ± i}.

20. {± 3, ± 2i}

21. $\left[\frac{x^2 - 2}{x}\right]^2 - 7\left[\frac{x^2 - 2}{x}\right] - 18 = 0$

We first substitute u for $\frac{x^2 - 2}{x}$.

$u^2 - 7u - 18 = 0$

$(u - 9)(u + 2) = 0$

$u - 9 = 0$ or $u + 2 = 0$

$u = 9$ or $u = -2$

Then we substitute $\frac{x^2 - 2}{x}$ for u and solve for x.

$\frac{x^2 - 2}{x} = 9$ or $\frac{x^2 - 2}{x} = -2$

$x^2 - 2 = 9x$ or $x^2 - 2 = -2x$

$x^2 - 9x - 2 = 0$ or $x^2 + 2x - 2 = 0$

$x = \frac{-(-9)\pm\sqrt{(-9)^2-4(1)(-2)}}{2\cdot 1}$ or $x = \frac{-2\pm\sqrt{2^2-4(1)(-2)}}{2\cdot 1}$

$x = \frac{9 \pm \sqrt{89}}{2}$ or $x = \frac{-2 \pm 2\sqrt{3}}{2}$

or $x = -1 \pm \sqrt{3}$

The solutions are $\frac{9 \pm \sqrt{89}}{2}$ and $-1 \pm \sqrt{3}$. The solution set is $\left\{\frac{9 \pm \sqrt{89}}{2}, -1 \pm \sqrt{3}\right\}$.

22. $\left\{\frac{5 \pm \sqrt{21}}{2}, \frac{3 \pm \sqrt{5}}{2}\right\}$

23. $\frac{x}{x - 1} - 6\sqrt{\frac{x}{x - 1}} - 40 = 0$

We substitute u for $\sqrt{\frac{x}{x - 1}}$.

$u^2 - 6u - 40 = 0$

$(u - 10)(u + 4) = 0$

$u - 10 = 0$ or $u + 4 = 0$

$u = 10$ or $u = -4$

We then substitute $\sqrt{\frac{x}{x - 1}}$ for u and solve for x.

$\sqrt{\frac{x}{x - 1}} = 10$ or $\sqrt{\frac{x}{x - 1}} = -4$

$\frac{x}{x - 1} = 100$ No solution

$x = 100x - 100$

$100 = 99x$

$\frac{100}{99} = x$

The solution is $\frac{100}{99}$. The solution set is $\left\{\frac{100}{99}\right\}$.

24. ∅

25. $$5\left[\frac{x+2}{x-2}\right]^2 = 3\left[\frac{x+2}{x-2}\right] + 2$$

$$5\left[\frac{x+2}{x-2}\right]^2 - 3\left[\frac{x+2}{x-2}\right] - 2 = 0$$

Substitute u for $\frac{x+2}{x-2}$.

$$5u^2 - 3u - 2 = 0$$
$$(5u + 2)(u - 1) = 0$$

$5u + 2 = 0$ or $u - 1 = 0$

$u = -\frac{2}{5}$ or $u = 1$

Then substitute $\frac{x+2}{x-2}$ for u and solve for x.

$\frac{x+2}{x-2} = -\frac{2}{5}$ or $\frac{x+2}{x-2} = 1$

$5(x + 2) = -2(x - 2)$ or $x + 2 = x - 2$

$5x + 10 = -2x + 4$ or $2 = -2$

$7x = -6$ No solution

$x = -\frac{6}{7}$

The solution is $-\frac{6}{7}$. The solution set is $\left\{-\frac{6}{7}\right\}$.

26. $\left\{-\frac{5}{2}, -5\right\}$

27. $$\left[\frac{x^2-1}{x}\right]^2 - \left[\frac{x^2-1}{x}\right] - 2 = 0$$

Substitute u for $\frac{x^2-1}{x}$.

$$u^2 - u - 2 = 0$$
$$(u - 2)(u + 1) = 0$$

$u - 2 = 0$ or $u + 1 = 0$

$u = 2$ or $u = -1$

Then substitute $\frac{x^2-1}{x}$ for u and solve for x.

$\frac{x^2-1}{x} = 2$ or $\frac{x^2-1}{x} = -1$

$x^2 - 1 = 2x$ or $x^2 - 1 = -x$

$x^2 - 2x - 1 = 0$ or $x^2 + x - 1 = 0$

$x = \frac{-(-2)\pm\sqrt{(-2)^2-4(1)(-1)}}{2\cdot1}$ or $x = \frac{-1\pm\sqrt{1^2-4(1)(-1)}}{2\cdot1}$

$x = \frac{2 \pm \sqrt{8}}{2}$ $\qquad x = \frac{-1 \pm \sqrt{5}}{2}$

$x = \frac{2 \pm 2\sqrt{2}}{2}$

$x = 1 \pm \sqrt{2}$

The solutions are $1 \pm \sqrt{2}$ and $\frac{-1 \pm \sqrt{5}}{2}$. The solution set is $\left\{1 \pm \sqrt{2}, \frac{-1 \pm \sqrt{5}}{2}\right\}$.

28. $\left\{0, \frac{56}{5}\right\}$

29. Familiarize and Translate:

We let t_1 represent the time it takes for the object to fall to the ground and t_2 represent the time it takes the sound to get back. The total amount of time, 3 sec, is the sum of t_1 and t_2. This gives us an equation.

$t_1 + t_2 = 3$ (1)

We use the formula $s = 16t^2 + v_0t$ to find an expression for t_1. Since the stone is dropped, v_0 is 0.

$$s = 16t_1^2$$
$$\frac{s}{16} = t_1^2$$
$$\sqrt{\frac{s}{16}} = t_1$$
$$\frac{\sqrt{s}}{4} = t_1 \qquad (2)$$

We use $d = rt$ (here we use $s = rt$) to find an expression for t_2.

$s = 1100\cdot t_2$ (Substituting 1100 for r)

$\frac{s}{1100} = t_2$ (3)

We substitute (2) and (3) in (1) and obtain

$$\frac{\sqrt{s}}{4} + \frac{s}{1100} = 3$$

Carry out:

$$275\sqrt{s} + s = 3300$$
$$s + 275\sqrt{s} - 3300 = 0$$

Substitute u for $\sqrt{s}$.

$$u^2 + 275u - 3300 = 0$$

Using the quadratic formula, we get $u \approx 11.5176$.

Then $u = \sqrt{s} \approx 11.5176$ and $s \approx 132.66$.

Check and State:

The value checks. The cliff is 132.66 ft high.

30. 229.94 ft

31. Familiarize: We will use the formula $A = P(1 + i)^t$. At the beginning of the third year the \$2000 investment has grown to $2000(1 + i)^2$ and the \$1200 investment has grown to $1200(1 + i)$.

Translate: We know the total in both accounts is \$3573.80, so we can write an equation.

$$2000(1 + i)^2 + 1200(1 + i) = 3573.80, \text{ or}$$
$$2000(1 + i)^2 + 1200(1 + i) - 3573.80 = 0$$

Carry out: This equation is reducible to quadratic with $u = 1 + i$. Substituting, we get

$$2000u^2 + 1200u - 3573.80 = 0.$$

Using the quadratic formula and taking the positive solution, we get $u = 1.07$. Substituting $1 + i$ for u, we get

$$1 + i = 1.07$$
$$i = 0.07.$$

31. (continued)

Check: We check by substituting 0.07 for i back through the related equations.

State: The interest rate is 0.07, or 7%.

32. 9%

33. $$6.75x = \sqrt{35x} + 5.36$$

$6.75x - \sqrt{35}\sqrt{x} - 5.36 = 0 \qquad (\sqrt{35x} = \sqrt{35}\sqrt{x})$

Substitute u for $\sqrt{x}$.

$6.75u^2 - \sqrt{35}u - 5.36 = 0$

$$u = \frac{-(-\sqrt{35}) \pm \sqrt{(-\sqrt{35})^2 - 4(6.75)(-5.36)}}{2(6.75)}$$

$$= \frac{\sqrt{35} \pm \sqrt{35 + 144.72}}{13.5}$$

$$= \frac{\sqrt{35} \pm \sqrt{179.72}}{13.5}$$

$$\approx \frac{5.9160798 \pm 13.4059688}{13.5}$$

$u \approx 1.4312629$ or $u \approx -0.5548066$

Substitute $\sqrt{x}$ for u and solve for x.

$\sqrt{x} \approx 1.4312629$ or $\sqrt{x} \approx -0.5548066$

$x \approx 2.0485$ No solution

The solution set is {2.0485}.

34. $\{\pm 1.9863, \pm 0.8966i\}$

35. $9x^{3/2} - 8 = x^3$

$0 = x^3 - 9x^{3/2} + 8$

$0 = ((\sqrt{x})^2)^3 - 9(\sqrt{x})^3 + 8$

$0 = ((\sqrt{x})^3)^2 - 9(\sqrt{x})^3 + 8$

Substitute u for $(\sqrt{x})^3$.

$0 = u^2 - 9u + 8$

$0 = (u - 8)(u - 1)$

$u - 8 = 0$ or $u - 1 = 0$

$u = 8$ or $u = 1$

Substitute $(\sqrt{x})^3$ for u and solve for x.

$(\sqrt{x})^3 = 8$ or $(\sqrt{x})^3 = 1$

$\sqrt{x} = 2$ or $\sqrt{x} = 1$

$x = 4$ or $x = 1$

Both values check. The solution set is {4, 1}.

36. $\left\{-\frac{3}{2}, -1\right\}$

37. $\sqrt{x - 3} - \sqrt[4]{x - 3} = 2$

Substitute u for $\sqrt[4]{x - 3}$.

$u^2 - u - 2 = 0$

$(u - 2)(u + 1) = 0$

$u - 2 = 0$ or $u + 1 = 0$

$u = 2$ or $u = -1$

Substitute $\sqrt[4]{x - 3}$ for u and solve for x.

$\sqrt[4]{x - 3} = 2$ or $\sqrt[4]{x - 3} = -1$

$x - 3 = 16$ No solution

$x = 19$

The solution set is {19}.

38. {9}

39. $x^6 - 28x^3 + 27 = 0$

Substitute u for x^3.

$u^2 - 28u + 27 = 0$

$(u - 27)(u - 1) = 0$

$u = 27$ or $u = 1$

Substitute x^3 for u and solve for x.

$x^3 = 27$ or $x^3 = 1$

$x^3 - 27 = 0$ or $x^3 - 1 = 0$

$(x-3)(x^2 + 3x + 9) = 0$ or $(x-1)(x^2 + x + 1) = 0$

Using the principle of zero products and, where necessary, the quadratic formula, we find that the solution set is

$$\left\{3, \frac{-3 \pm 3i\sqrt{3}}{2}, 1, \frac{-1 \pm i\sqrt{3}}{2}\right\}.$$

40. {-2, 1}

41. $(x^2 - 5x - 2)^2 - 5(x^2 - 5x - 2) + 4 = 0$

Substitute u for $x^2 - 5x - 2$.

$u^2 - 5u + 4 = 0$

$(u - 4)(u - 1) = 0$

$u = 4$ or $u = 1$

Substitute $x^2 - 5x - 2$ for u and solve for x.

$x^2 - 5x - 2 = 4$ or $x^2 - 5x - 2 = 1$

$x^2 - 5x - 6 = 0$ or $x^2 - 5x - 3 = 0$

$(x + 1)(x - 6) = 0$ or $x = \dfrac{-(-5) \pm \sqrt{(-5)^2-4(1)(-3)}}{2\cdot 1}$

$x + 1 = 0$ or $x - 6 = 0$ or $x = \dfrac{5 \pm \sqrt{25 + 12}}{2}$

$x = -1$ or $x = 6$ or $x = \dfrac{5 \pm \sqrt{37}}{2}$

The solution set is $\left\{-1, 6, \frac{5 \pm \sqrt{37}}{2}\right\}$.

42. $\left\{\frac{4}{9}, \frac{9}{4}\right\}$

43. $\left[y + \frac{2}{y}\right]^2 + 3y + \frac{6}{y} = 4$

$\left[y + \frac{2}{y}\right]^2 + 3\left[y + \frac{2}{y}\right] - 4 = 0$

Substitute u for $y + \frac{2}{y}$.

$u^2 + 3u - 4 = 0$

$(u + 4)(u - 1) = 0$

$u = -4$ or $u = 1$

Substitute $y + \frac{2}{y}$ for u and solve for y.

$y + \frac{2}{y} = -4$ or $y + \frac{2}{y} = 1$

$y^2 + 2 = -4y$ or $y^2 + 2 = y$

$y^2 + 4y + 2 = 0$ or $y^2 - y + 2 = 0$

$y = \frac{-4 \pm \sqrt{4^2 - 4\cdot 1\cdot 2}}{2\cdot 1}$ or $y = \frac{-(-1) \pm \sqrt{(-1)^2 - 4\cdot 1\cdot 2}}{2\cdot 1}$

$y = \frac{-4 \pm \sqrt{8}}{2}$ or $y = \frac{1 \pm \sqrt{-7}}{2}$

$y = \frac{-4 \pm 2\sqrt{2}}{2}$ or $y = \frac{1 \pm i\sqrt{7}}{2}$

$y = -2 \pm \sqrt{2}$ or $y = \frac{1 \pm i\sqrt{7}}{2}$

The solution set is $\left\{-2 \pm \sqrt{2}, \frac{1 \pm i\sqrt{7}}{2}\right\}$.

44. $\left\{\frac{-3 \pm \sqrt{39 + 2\sqrt{33}}}{2}\right\}$

45. $\frac{2x + 1}{x} = 3 + 7\sqrt{\frac{2x + 1}{x}}$

$\frac{2x + 1}{x} - 7\sqrt{\frac{2x + 1}{x}} - 3 = 0$

Substitute u for $\sqrt{\frac{2x + 1}{x}}$.

$u^2 - 7u - 3 = 0$

$u = \frac{-(-7) \pm \sqrt{(-7)^2 - 4(1)(-3)}}{2\cdot 1}$

$u = \frac{7 \pm \sqrt{61}}{2}$

Substitute $\sqrt{\frac{2x + 1}{x}}$ for u and solve for x.

$\sqrt{\frac{2x + 1}{x}} = \frac{7 + \sqrt{61}}{2}$ or $\sqrt{\frac{2x + 1}{x}} = \frac{7 - \sqrt{61}}{2}$

$\left[\sqrt{\frac{2x + 1}{x}}\right]^2 = \left[\frac{7 + \sqrt{61}}{2}\right]^2$ No solution $\left[\frac{7 - \sqrt{61}}{2} < 0\right]$

$\frac{2x + 1}{x} = \frac{49 + 14\sqrt{61} + 61}{4}$

$8x + 4 = 110x + 14x\sqrt{61}$

$4 = 102x + 14x\sqrt{61}$

$4 = (102 + 14\sqrt{61})x$

$\frac{4}{102 + 14\sqrt{61}} = x$

45. (continued)

Rationalizing the denominator, we get $x = \frac{-51 + 7\sqrt{61}}{194}$. The solution set is $\left\{\frac{-51 + 7\sqrt{61}}{194}\right\}$.

46. 8%

Exercise Set 2.9

1. $y = kx$

$0.6 = k(0.4)$ (Substituting)

$\frac{0.6}{0.4} = k$

$\frac{3}{2} = k$

The equation of variation is $y = \frac{3}{2}x$.

2. $y = \frac{125}{32}x$

3. $y = \frac{k}{x}$

$125 = \frac{k}{32}$ (Substituting)

$4000 = k$

The equation of variation is $y = \frac{4000}{x}$.

4. $y = \frac{0.32}{x}$

5. $y = kx$

$8.6 = k(1.6)$ (Substituting)

$\frac{8.6}{1.6} = k$

$5.375 = k$

The equation of variation is $y = 5.375x$.

6. $y = \frac{20.52}{x}$

7. $y = \frac{k}{x^2}$

$0.15 = \frac{k}{(0.1)^2}$ (Substituting)

$0.15 = \frac{k}{0.01}$

$0.01(0.15) = k$

$0.0015 = k$

The equation of variation is $y = \frac{0.0015}{x^2}$.

8. $y = 0.8xz$

9. $y = k \cdot \frac{xz}{w}$

$\frac{3}{2} = k \cdot \frac{2 \cdot 3}{4}$ (Substituting)

$\frac{3}{2} = k \cdot \frac{6}{4}$

$\frac{4}{6} \cdot \frac{3}{2} = k$

$1 = k$

The equation of variation is $y = \frac{xz}{w}$.

10. $y = 0.3xz^2$

11. $y = k \cdot \frac{xz}{w^2}$

$\frac{12}{5} = k \cdot \frac{16 \cdot 3}{5^2}$ (Substituting)

$\frac{12}{5} = k \cdot \frac{48}{25}$

$\frac{25}{48} \cdot \frac{12}{5} = k$

$\frac{5}{4} = k$

The equation of variation is $y = \frac{5}{4} \cdot \frac{xz}{w^2}$.

12. $y = \frac{1}{5} \cdot \frac{xz}{wp}$

13. $y = kx$

We double x.

$y = k(2x)$

$y = 2 \cdot kx$

Thus, y is doubled.

14. y is multiplied by $\frac{1}{3}$.

15. $y = \frac{k}{x^2}$

We multiply x by n.

$y = \frac{k}{(nx)^2}$

$y = \frac{k}{n^2x^2}$

$y = \frac{1}{n^2} \cdot \frac{k}{x^2}$

Thus, y is mulitplied by $\frac{1}{n^2}$.

16. y is multiplied by n^2.

17. $A = kN$

$42{,}600 = k \cdot 60{,}000$ (Substituting)

$0.71 = k$

The equation of variation is $A = 0.71N$.

$A = 0.71N$

$A = 0.71(750{,}000)$ (Substituting)

$A = 532{,}500$

Thus, 532,500 tons will enter the atmosphere.

18. 220 in^3

19. $L = \frac{kwh^2}{\ell}$

The width and height are doubled, and the length is halved.

$L = \frac{k \cdot 2w \cdot (2h)^2}{\frac{\ell}{2}}$

$L = k \cdot 2w \cdot 4h^2 \cdot \frac{2}{\ell}$

$L = 16 \cdot \frac{kwh^2}{\ell}$

Thus, L is multiplied by 16.

20. 256

21. $d = k\sqrt{h}$

$28.97 = k\sqrt{19.5}$ (Substituting)

$\frac{28.97}{\sqrt{19.5}} = k$

$d = \frac{28.97}{\sqrt{19.5}} \sqrt{h}$ (Equation of variation)

$54.32 = \frac{28.97}{\sqrt{19.5}} \sqrt{h}$ (Substituting)

$\frac{54.32\sqrt{19.5}}{28.97} = \sqrt{h}$

$\left[\frac{54.32\sqrt{19.5}}{28.97}\right]^2 = (\sqrt{h})^2$

$68.6 \approx h$

One must be about 68.6 m above sea level.

22. 1.263 ohms

23. $A = ks^2$

$168.54 = k(5.3)^2$ (Substituting)

$168.54 = 28.09k$

$6 = k$

$A = 6s^2$ (Equation of variation)

$A = 6(10.2)^2$ (Substituting)

$A = 6(104.04)$

$A = 624.24$

The area is 624.24 m^2.

24. 22.5 W/m^2

25. $A = \frac{kR}{I}$

$2.92 = \frac{k \cdot 85}{262}$ (Substituting)

$2.92 \cdot \frac{262}{85} = k$

$9 \approx k$

$A = \frac{9R}{I}$ (Equation of variation)

$2.92 = \frac{9R}{300}$ (Substituting)

$\frac{2.92(300)}{9} = k$

$97 \approx k$

He would have given up 97 runs.

26. 220 cm^3

27. If p varies directly as q, then $p = kq$.

Thus, $q = \frac{1}{k}p$, so q varies directly as p.

28. $u = \frac{k}{v}$ gives $v = \frac{k}{u}$; and $u = \frac{k}{v}$ gives $\frac{1}{u} = \left(\frac{1}{k}\right)v$ or $\frac{1}{u} = (k_2)v$, where $k_2 = \frac{1}{k}$.

29. $A = kd^2$

$\pi r^2 = kd^2$

$\pi\left(\frac{d}{2}\right)^2 = kd^2$

$\frac{\pi}{4} \cdot d^2 = kd^2$

$\frac{\pi}{4} = k$

30. t varies directly as $\sqrt{P}$.

31. a) $N = \frac{kP_1P_2}{d^2}$

$11,153 = \frac{k(744,624)(452,524)}{(174)^2}$

$\frac{(11,153)(174)^2}{(744,624)(452,524)} = k$

$0.001 \approx k$

The equation of variation is $N = \frac{0.001P_1P_2}{d^2}$.

b) $N = \frac{0.001(744,624)(1,511,482)}{(446)^2}$

$N \approx 5658$

The average number of daily phone calls is about 5658.

31. (continued)

c) $4270 = \frac{0.001(744,624)(7,895,563)}{d^2}$

$d^2 = \frac{0.001(744,624)(7,895,563)}{4270}$

$d \approx 1173$

The distance is about 1173 km.

d) Division by 0 is undefined.

Exercise Set 3.1

1. We take the first member of each ordered pair from the tail, or left end, of an arrow and the second member from the tip, or right end, and gather the pairs into a set.

   {(New York, Mets), (New York, Giants), (Atlanta, Braves), (Atlanta, Falcons), (Houston, Astros), (Houston, Oilers), (San Diego, Padres), (San Diego, Chargers)}

2. {(1959, 209), (1969, 231), (1979, 255), (1989, 282), (1999, 312)}

3. {(-3,3), (3,3), (-2,2), (2,2), (-1,1), (1,1) (0,0)}

4. {(-16, -0.0625), (3/4, 4/3), (-2, -0.5), (1, 1), (0.875, 8/7), (10, 0.1), (5^3, 5^{-3})}

5. A = {0,2,4,5}, B = {a,b,c}

   First members of A × B come from set A and second members come from set B.

   A × B = {(0,a), (0,b), (0,c), (2,a), (2,b), (2,c), (4,a), (4,b), (4,c), (5,a), (5,b), (5,c)}

6. {(h,1), (h,4), (h,7), (q,1), (q,4), (q,7), (t,1), (t,4), (t,7), (w,1), (w,4), (w,7)}

7. B = {x,y,z}, C = {1,2}

   First members of B × C come from set B and second members come from set C.

   B × C = {(x,1), (x,2), (y,1), (y,2), (z,1), (z,2)}

8. {(5,a), (5,z), (7,a), (7,z), (10,a), (10,z)}

9. D = {5,6,7,8}

   D × D = {(5,5), (5,6), (5,7), (5,8), (6,5), (6,6), (6,7), (6,8), (7,5), (7,6), (7,7), (7,8), (8,5), (8,6), (8,7), (8,8)}

10. {(-2,-2), (-2,0), (-2,2), (-2,4), (0,-2), (0,0), (0,2), (0,4), (2,-2), (2,0), (2,2), (2,4), (4,-2), (4,0), (4,2), (4,4)}

11. A = {0,2}, B = {a,b,c}

    a) A × B = {(0,a), (0,b), (0,c), (2,a), (2,b), (2,c)}

    b) B × A = {(a,0), (a,2), (b,0), (b,2), (c,0), (c,2)}

    c) A × A = {(0,0), (0,2), (2,0), (2,2)}

    d) B × B = {(a,a), (a,b), (a,c), (b,a), (b,b), (b,c), (c,a), (c,b), (c,c)}

12. a) {(1,d), (1,e), (1,f), (3,d), (3,e), (3,f), (5,d), (5,e), (5,f), (9,d), (9,e), (9,f)}

    b) {(d,1), (d,3), (d,5), (d,9), (e,1), (e,3), (e,5), (e,9), (f,1), (f,3), (f,5), (f,9)}

    c) {(1,1), (1,3), (1,5), (1,9), (3,1), (3,3), (3,5), (3,9), (5,1), (5,3), (5,5), (5,9), (9,1), (9,3), (9,5), (9,9)}

    d) {(d,d), (d,e), (d,f), (e,d), (e,e), (e,f), (f,d), (f,e), (f,f)}

In Exercises 13-18, consider E × E given below:

E × E = {(-7,-7), (-7,-3), (-7,1), (-7,2), (-7,5), (-3,-7), (-3,-3), (-3,1), (-3,2), (-3,5), (1,-7), (1,-3), (1,1), (1,2), (1,5), (2,-7), (2,-3), (2,1), (2,2), (2,5), (5,-7), (5,-3), (5,1), (5,2), (5,5)}

13. We list the set of all ordered pairs in E × E in which the first member is less than the second member.

    {(-7,-3), (-7,1), (-7,2), (-7,5), (-3,1), (-3,2), (-3,5), (1,2), (1,5), (2,5)}

14. {(-3,-7), (1,-7), (1,-3), (2,-7), (2,-3), (2,1), (5,-7), (5,-3), (5,1), (5,2)}

15. We list the set of all ordered pairs in E × E in which the first member is less than or equal to the second member.

    {(-7,-7), (-7,-3), (-7,1), (-7,2), (-7,5), (-3,-3), (-3,1), (-3,2), (-3,5), (1,1), (1,2), (1,5), (2,2), (2,5), (5,5)}

16. {(-7,-7), (-3,-7), (-3,-3), (1,-7), (1,-3), (1,1), (2,-7), (2,-3), (2,1), (2,2), (5,-7), (5,-3), (5,1), (5,2), (5,5)}

17. We list the set of all ordered pairs in E × E in which the first member is the same as the second member.

    {(-7,-7), (-3,-3), (1,1), (2,2), (5,5)}

18. {(-7,-3), (-7,1), (-7,2), (-7,5), (-3,-7), (-3,1), (-3,2), (-3,5), (1,-7), (1,-3), (1,2), (1,5), (2,-7), (2,-3), (2,1), (2,5), (5,-7), (5,-3), (5,1), (5,2)}

19. Answers will vary. One example is given.

Doubling	
Number, x	2x
-2 →	-4
-1 →	-2
0 →	0
1 →	2
2 →	4

20. Answers will vary.

21. {(5,2), (6,4), (8,6)}

 The domain is the set of all first members of ordered pairs in the relation, and the range is the set of all second members.

 Domain: {5,6,8}; range: {2,4,6}

22. Domain: {7,8,9}; range: {1,2,5}

23. {(6,0), (7,5), (8,5), (-4,-7)}

 Domain: {6,7,8,-4}; range: {0,5,-7}

24. Domain: {8,10,6,-2}; range: {2,-10,3,5}

25. {(8,1), (-8,1), (5,1), (-3,1)}

 Domain: {8,-8,5,-3}; range: {1}

26. Domain: {6,-5,0,-3}; range: {2}

27. {(5,-6)}

 Domain: {5}; range: {-6}

28. Domain: {-7}; range: {4}

29. The relation "Sports Teams" is given in Exercise 1.

 Domain: {New York, Atlanta, Houston, San Diego};

 range: {Mets, Giants, Braves, Falcons, Astros, Oilers, Padres, Chargers}

30. Domain: {1959, 1969, 1979, 1989, 1999};

 range: {209, 231, 255, 282, 312}

31. The relation "Absolute Value" is given in Exercise 3.

 Domain: {-3,3,-2,2,-1,1,0}; range: {3,2,1,0}

32. Domain: {-16, 3/4, -2, 1, 0.875, 10, 5^3}

 range: {-0.0625, 4/3, -0.5, 1, 8/7, 0.1, 5^{-3}}

33. The relation "less than" is given in Exercise 13.

 Domain: {-7,-3,1,2}; range: {-3,1,2,5}

34. Domain: {-3,1,2,5}; range: {-7,-3,1,2}

35. The relation "less than or equal to" is given in Exercise 15.

 Domain: {-7,-3,1,2,5}; range: {-7,-3,1,2,5}

36. Domain: {-7,-3,1,2,5}; range: {-7,3,1,2,5}

37. C = {-1,0,1,2}

 a),b) C × C = {(-1,-1), (-1,0), (-1,1), (-1,2),
 (0,-1), (0,0), (0,1), (0,2),
 (1,-1), (1,0), (1,1), (1,2),
 (2,-1), (2,0), (2,1), (2,2)}

 c) Domain: {0,1}; range: {0,1,2}

38. a),b) D × D = {(-1,-1), (-1,1), (-1,3), (-1,5),
 (1,-1), (1,1), (1,3), (1,5),
 (3,-1), (3,1), (3,3), (3,5),
 (5,-1), (5,1), (5,3), (5,5)}

 c) Domain: {-1,1}; range: {1,3}

39. Q = {2,3,4,5}

 Q × Q = {(2,2), (2,3), (2,4), (2,5),
 (3,2), (3,3), (3,4), (3,5),
 (4,2), (4,3), (4,4), (4,5),
 (5,2), (5,3), (5,4), (5,5)}

 {(x,y) | y > x + 1} is {(2,4), (2,5), (3,5)}

40. {(-1,-1), (-1,1), (0,0), (1,-1), (1,1), (2,2)}

Exercise Set 3.2

---

1. The origin has coordinates (0,0). The x-coordinate of each ordered pair tells us how far to move to the left or right of the y-axis. The y-coordinate tells us how far to move up or down from the x-axis.

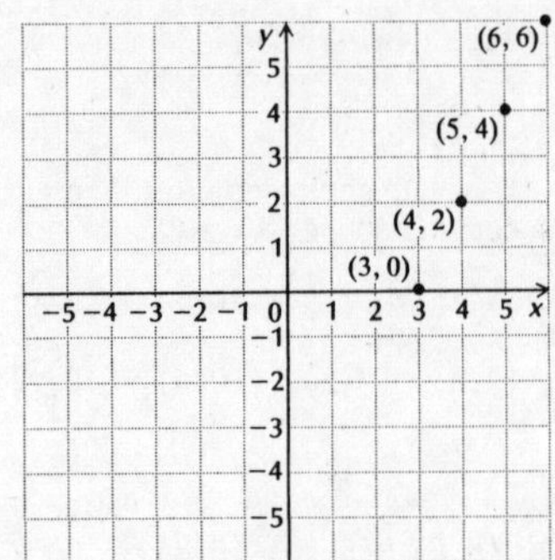

2.

(4, 7)
(3, 5)
(2, 3)
(1, 1)

3.

(3, 0)
(3, −1)
(3, −2)
(3, −3)
(3, −4)

4. 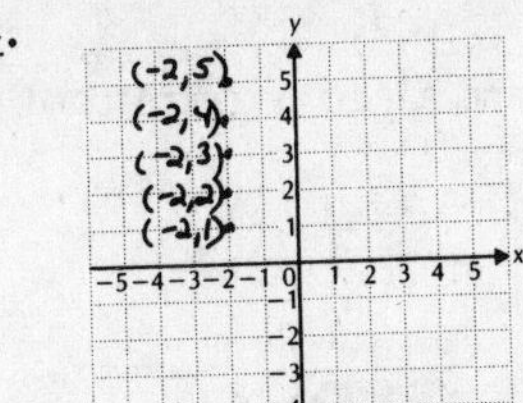

5. 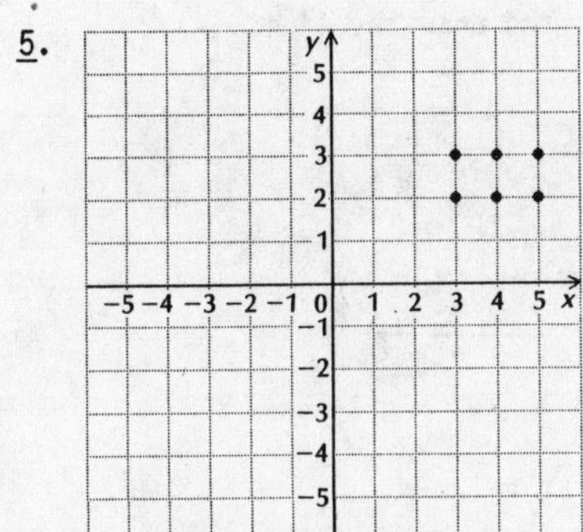

6. 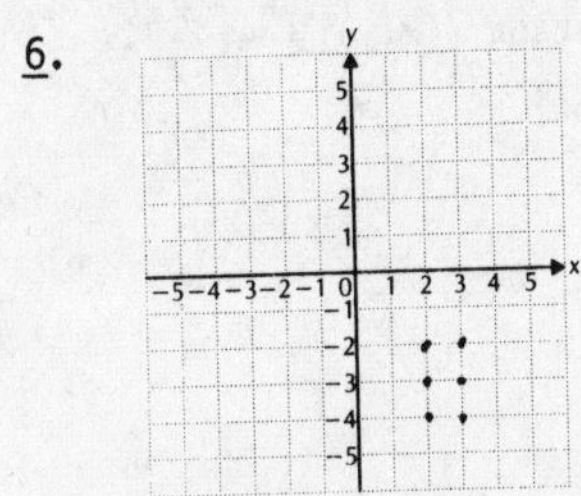

7. 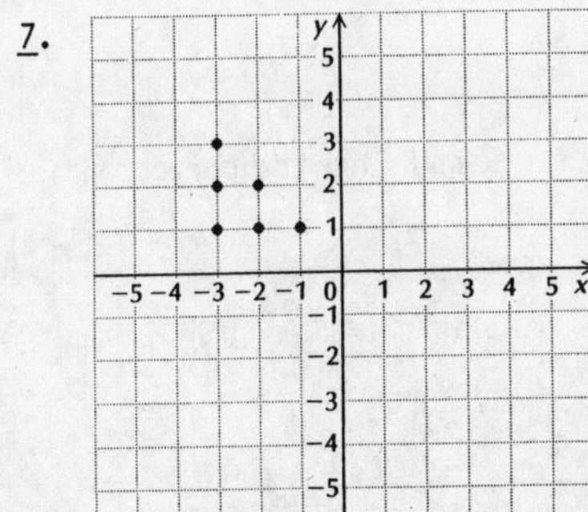

8.

9.

$y = 2x - 3$	
-1	$2 \cdot 1 - 3$
	$2 - 3$
	-1

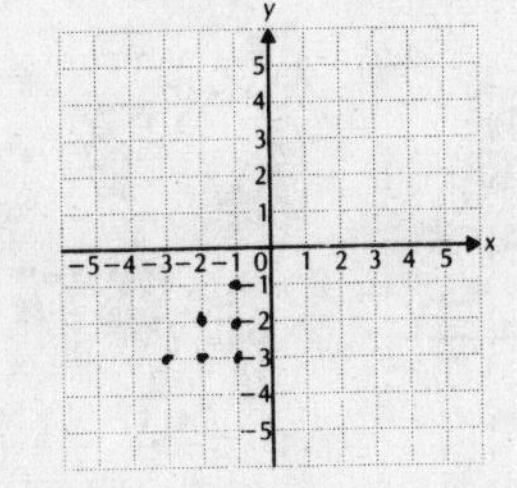

We substitute 1 for x and -1 for y (alphabetical order of variables).

The equation -1 = -1 is true, so (1,-1) is a solution.

$y = 2x - 3$	
3	$2 \cdot 0 - 3$
	$0 - 3$
	-3

The equation 3 = -3 is false, so (0,3) is not a solution.

10. Yes, yes

11.

$3s + t = 4$	
$3 \cdot 3 + 4$	4
$9 + 4$	
13	

We substitute 3 for s and 4 for t (alphabetical order of variables).

The equation 13 = 4 is false, so (3,4) is not a solution.

$3s + t = 4$	
$3(-3) + 5$	4
$-9 + 5$	
-4	

The equation -4 = 4 is false, so (-3,5) is not a solution.

12. No, no

13.

$4x - y = 7$	
$4 \cdot 3 - 5$	7
$12 - 5$	
7	

(3,5) is a solution.

$4x - y = 7$	
$4(-2) - (-15)$	7
$-8 + 15$	
7	

(-2,-15) is a solution.

14. Yes, no

15.

$2a + 5b = 3$	
$2 \cdot 0 + 5 \cdot \frac{3}{5}$	3
$0 + 3$	
3	

$\left(0,\frac{3}{5}\right)$ is a solution.

$2a + 5b = 3$	
$2\left(-\frac{1}{2}\right) + 5\left(-\frac{4}{5}\right)$	3
$-1 - 4$	
-5	

$\left(-\frac{1}{2}, -\frac{4}{5}\right)$ is not a solution.

16. Yes, yes

17.

$4r + 3s = 5$	
$4\cdot2 + 3(-1)$	5
$8 - 3$	
5	

(2,-1) is a solution.

$4r + 3s = 5$	
$4(-0.75) + 3(2.75)$	5
$-3 + 8.25$	
5.25	

(-0.75, 2.75) is not a solution.

18. Yes, yes

19.

$-3x + 2y = -4$	
$-3\cdot3 + 2\cdot2$	4
$-9 + 4$	
-5	

(3,2) is not a solution.

$-3x + 2y = -4$	
$-3\cdot22 + 2\cdot31$	-4
$-66 + 62$	
-4	

(22,31) is a solution.

20. Yes, no

21. $y = x$

We select numbers for x and find the corresponding values for y.

x	y	(x,y)
-4	-4	(-4,-4)
-2	-2	(-2,-2)
0	0	(0,0)
1	1	(1,1)
3	3	(3,3)

Plot these points and draw the line.

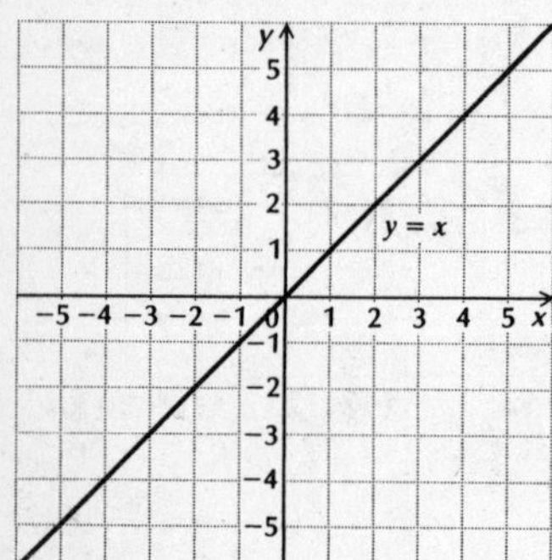

The graph shows the relation $\{(x,y) \mid y = x\}$.

22.

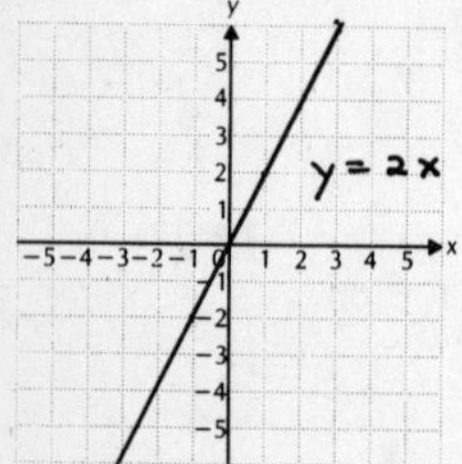

23. $y = -2x$

We select numbers for x and find the corresponding values for y.

x	y	(x,y)
-2	4	(-2,4)
0	0	(0,0)
1	-2	(1,-2)
3	-6	(3,-6)

Plot these points and draw the line.

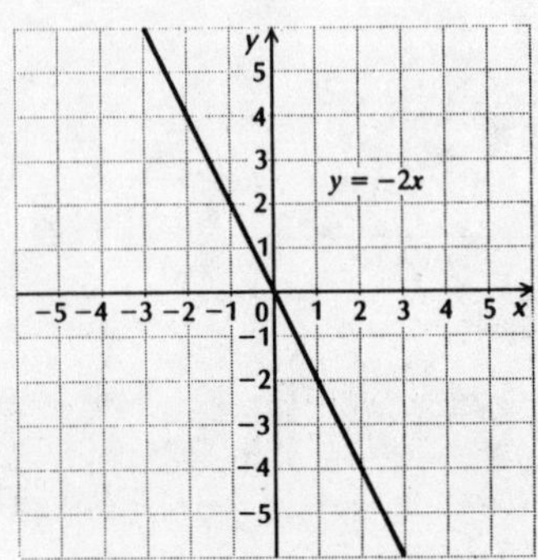

The graph shows the relation $\{(x,y) \mid y = -2x\}$.

24.

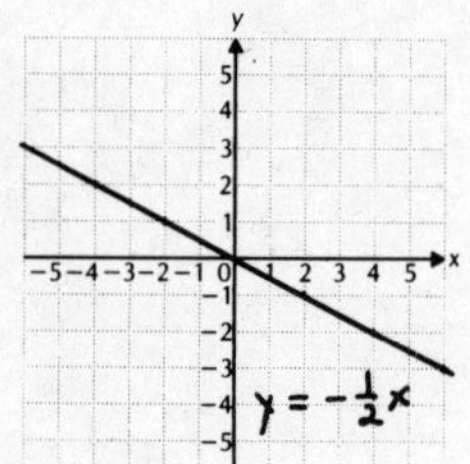

25. $y = x + 3$

We select numbers for x and find the corresponding values for y.

x	y	(x,y)
0	3	(0,3)
2	5	(2,5)
-3	0	(-3,0)
-5	-2	(-5,-2)

Plot these solutions and draw the line.

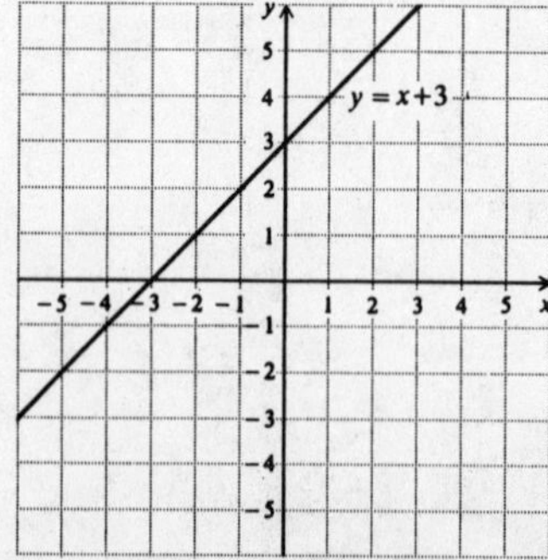

The graph shows the relation $\{(x,y) \mid y = x + 3\}$.

26.

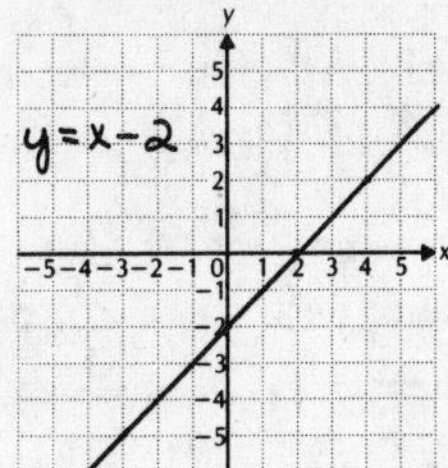

27. $y = 3x - 2$

We select numbers for x and find the corresponding values for y.

x	y	(x,y)
2	4	(2,4)
0	-2	(0,-2)
-1	-5	(-1,-5)

Plot these solutions and draw the line.

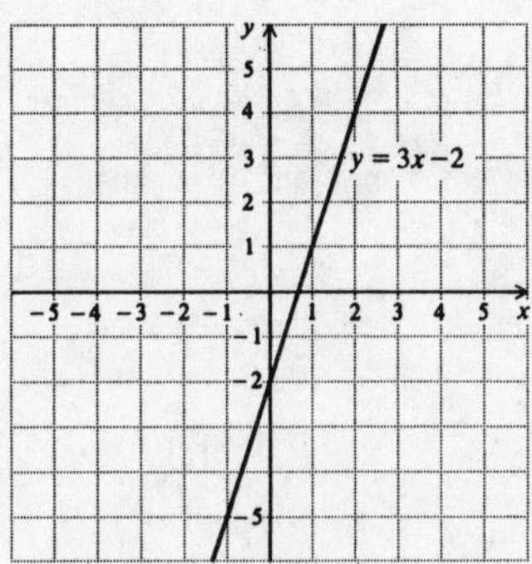

The graph shows the relation $\{(x,y)|y = 3x - 2\}$.

28.

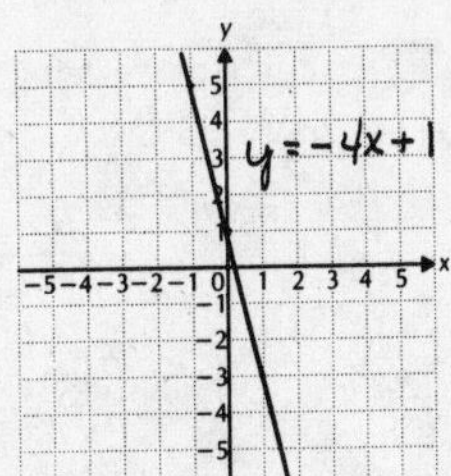

29. $y = x^2$

We select numbers for x and find the corresponding values for y.

x	y	(x,y)
-2	4	(-2,4)
-1	1	(-1,1)
0	0	(0,0)
1	1	(1,1)
2	4	(2,4)

Plot these ordered pairs and connect the points with a smooth curve.

29. (continued)

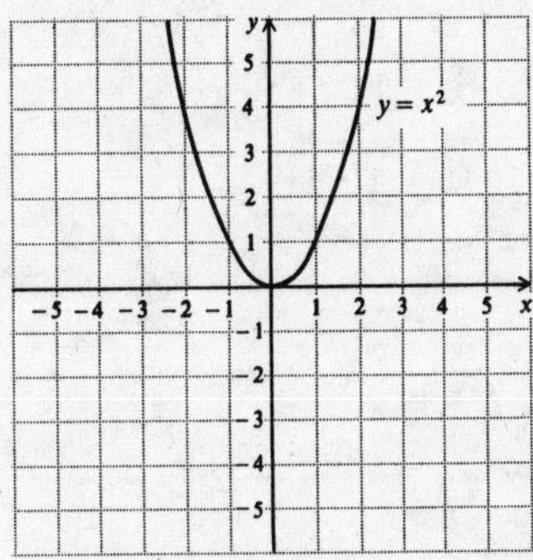

The graph shows the relation $\{(x,y)|y = x^2\}$.

30.

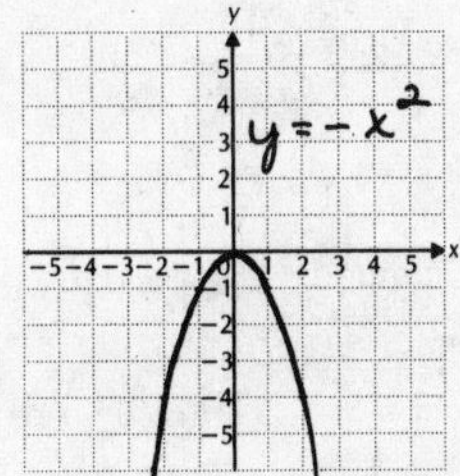

31. $y = x^2 + 2$

We select numbers for x and find the corresponding values for y.

x	y
-2	6
-1	3
0	2
1	3
2	6

We plot these ordered pairs and connect the points with a smooth curve.

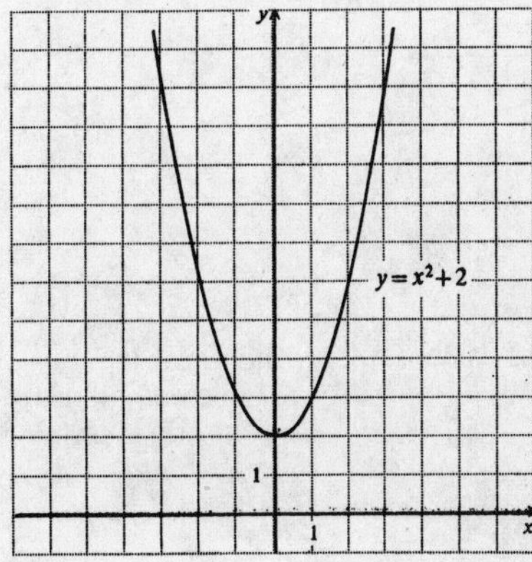

The graph shows the relation $\{(x,y)|y = x^2 + 2\}$.

32.

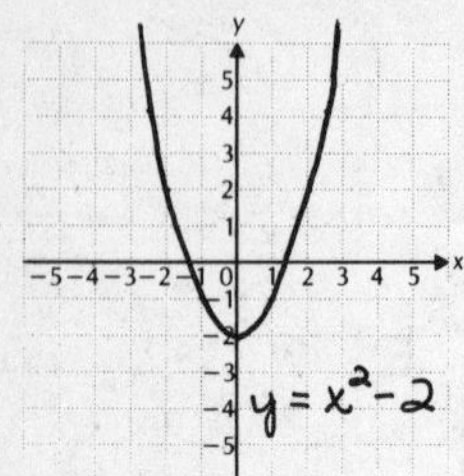

33. $x = y^2 + 2$

Here we select numbers for y and find the corresponding values for x.

x	y
6	-2
3	-1
2	0
3	1
6	2

We plot these points and connect them with a smooth curve.

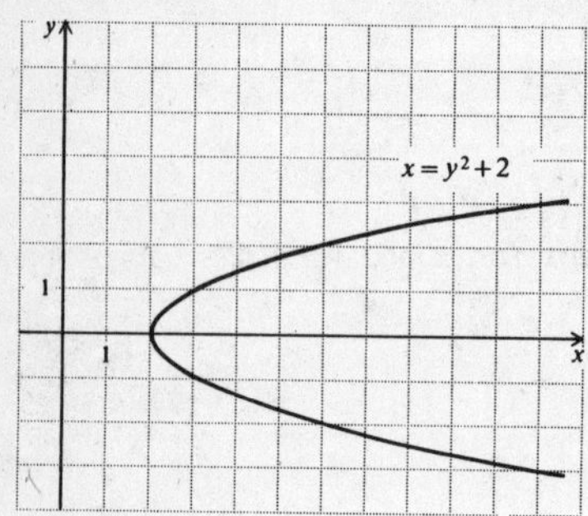

The graph shows the relation $\{(x,y) | x = y^2 + 2\}$.

34.

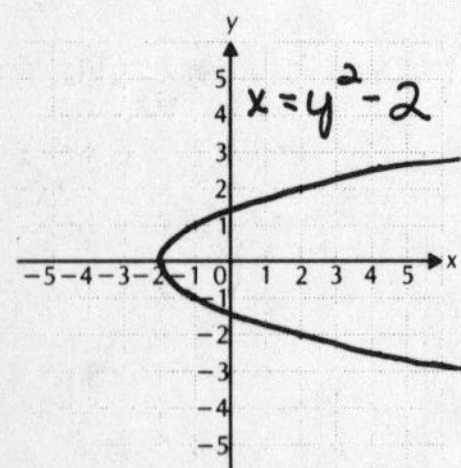

35. $y = |x + 1|$

We find numbers that satisfy the equation.

x	y
-5	4
-3	2
-1	0
2	3
4	5

We plot these points and connect them.

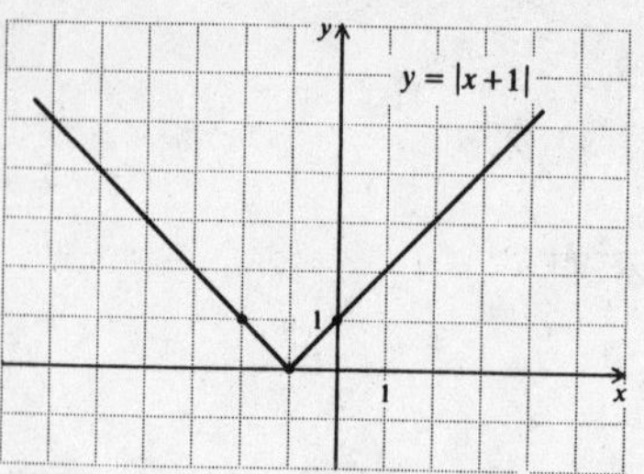

The graph shows the relation $\{(x,y) | y = |x + 1|\}$.

36.

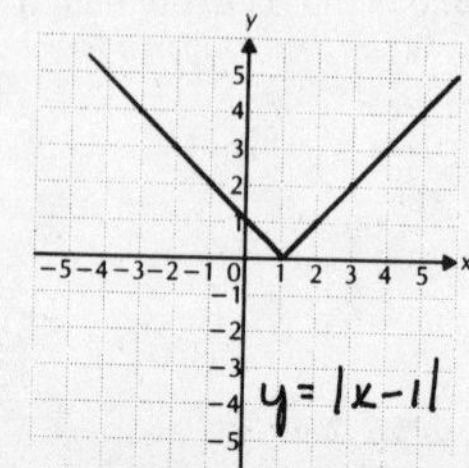

37. $x = |y + 1|$

Here we choose numbers for y and find the corresponding values for x.

x	y
4	-5
1	-2
0	-1
1	0
4	3
5	4

We plot these points and connect them.

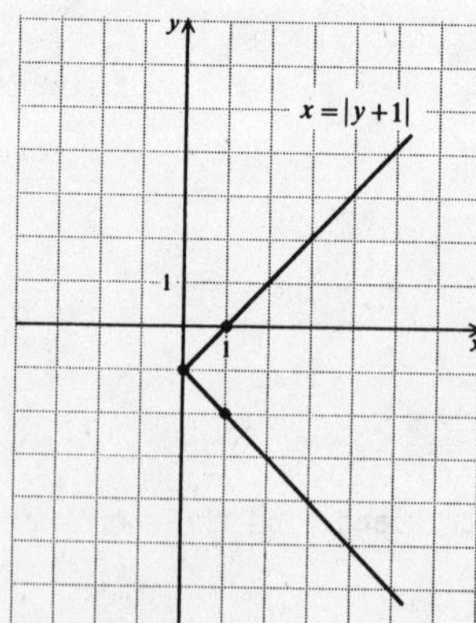

The graph shows the relation $\{(x,y) | x = |y + 1|\}$.

38.

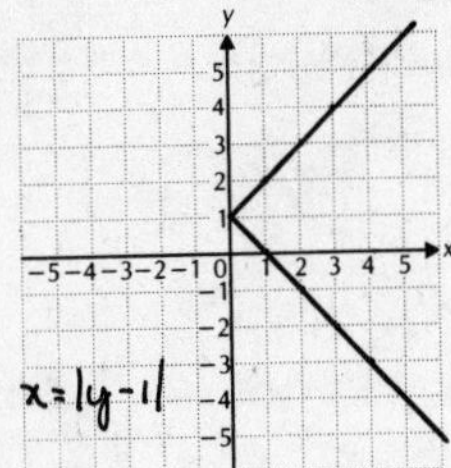

39. $xy = 10$

We find numbers that satisfy the equation. We then plot these points and connect them. Note that neither x nor y can be 0.

x	y
-6	$-\frac{5}{3}$, or $-1\frac{2}{3}$
-5	-2
-3	$-\frac{10}{3}$, or $-3\frac{1}{3}$
-2	-5

x	y
-1	-10
1	10
2	5
3	$3\frac{1}{3}$
5	2
6	$\frac{5}{3}$

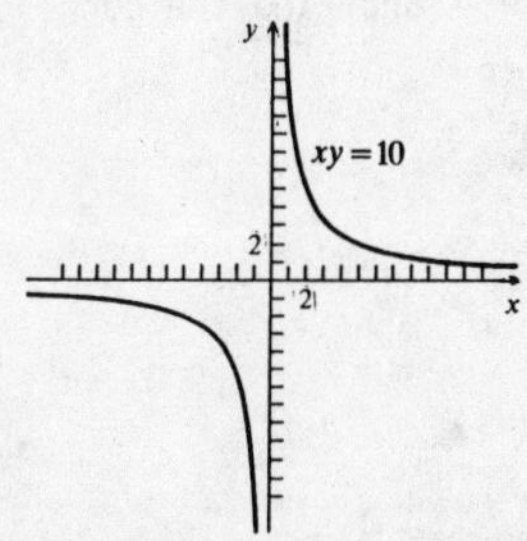

The graph shows the relation $\{(x, y) | xy = 10\}$.

40.

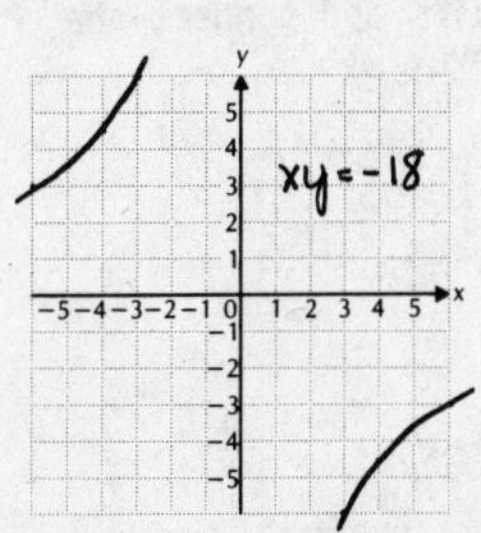

41. $y = \frac{1}{x}$

We find numbers that satisfy the equation. We then plot these points and connect them. Note that neither x nor y can be 0.

x	y
-5	$-\frac{1}{5}$
-3	$-\frac{1}{3}$
-2	$-\frac{1}{2}$
-1	-1
$-\frac{1}{2}$	-2
$-\frac{1}{4}$	-4

x	y
$\frac{1}{4}$	4
$\frac{1}{2}$	2
1	1
2	$\frac{1}{2}$
3	$\frac{1}{3}$
5	$\frac{1}{5}$

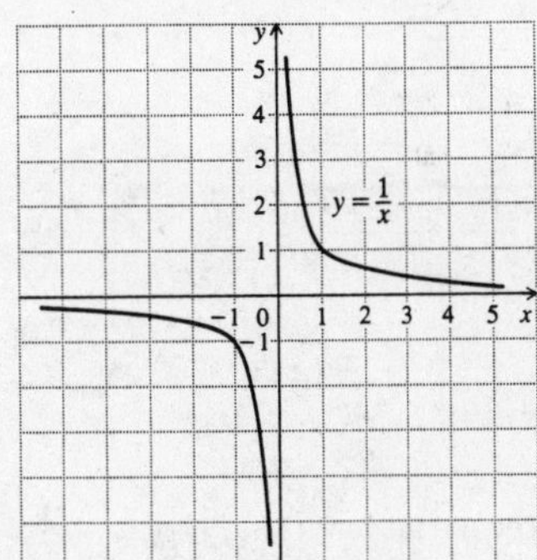

The graph shows the relation $\left\{(x, y) \middle| y = \frac{1}{x}\right\}$.

42.

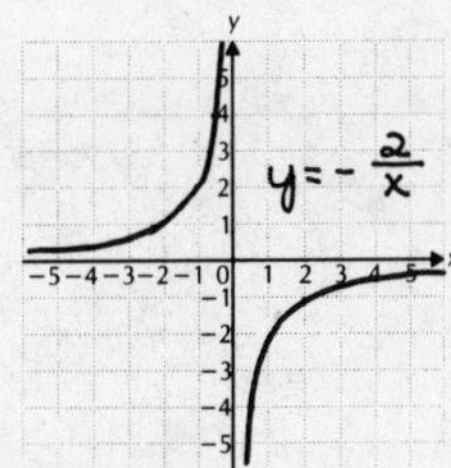

43. $y = \frac{1}{x^2}$

We find numbers that satisfy the equation. We plot these points and connect them. Note that neither x nor y can be 0.

x	y	x	y
-3	$\frac{1}{9}$	3	$\frac{1}{9}$
-2	$\frac{1}{4}$	2	$\frac{1}{4}$
-1	1	1	1
$-\frac{1}{2}$	4	$\frac{1}{2}$	4
$-\frac{1}{4}$	16	$\frac{1}{4}$	16

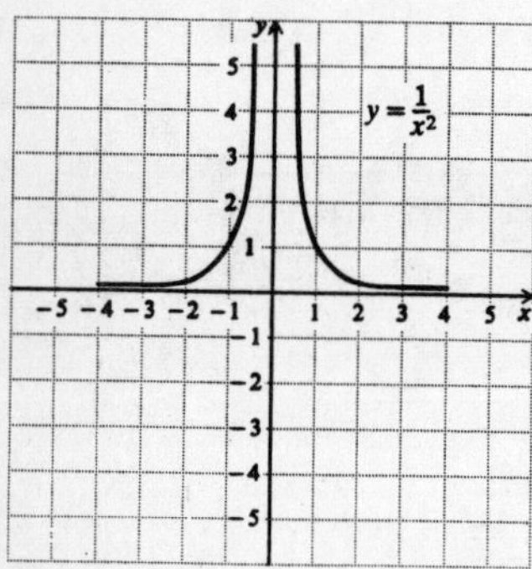

The graph shows the relation $\{(x, y)|y = \frac{1}{x^2}\}$.

44.

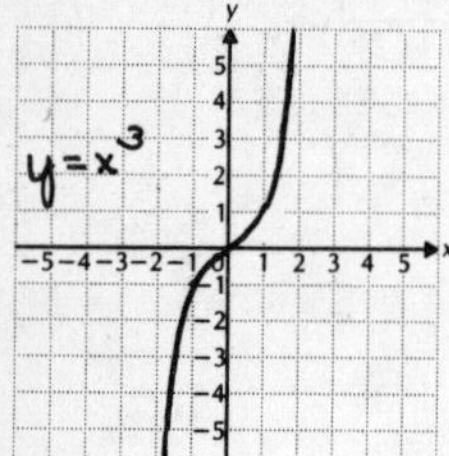

45. $y = \sqrt{x}$

The meaningful replacements for x are $x \geqslant 0$. We choose meaningful replacements for x and find corresponding values of y.

x	y
0	0
2	1.414
4	2
6	2.449
9	3

We plot these points and connect them.

45. (continued)

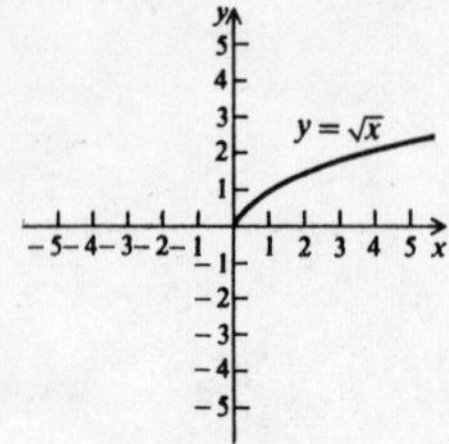

The graph shows the relation $\{(x, y)|y = \sqrt{x}\}$.

46.

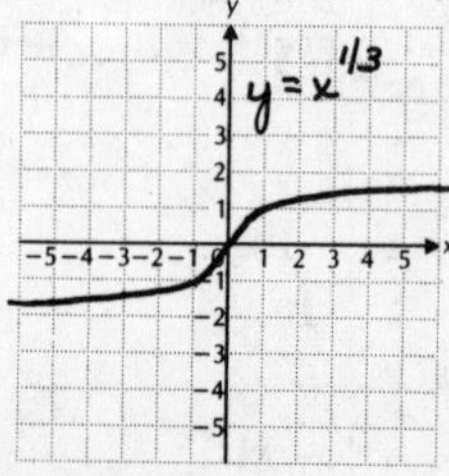

47. $y = 8 - x^2$

We choose numbers for x and find the corresponding values of y.

x	y
-3	-1
-2	4
-1	7
0	8
1	7
2	4
3	-1

We plot these ordered pairs and connect the points with a smooth curve.

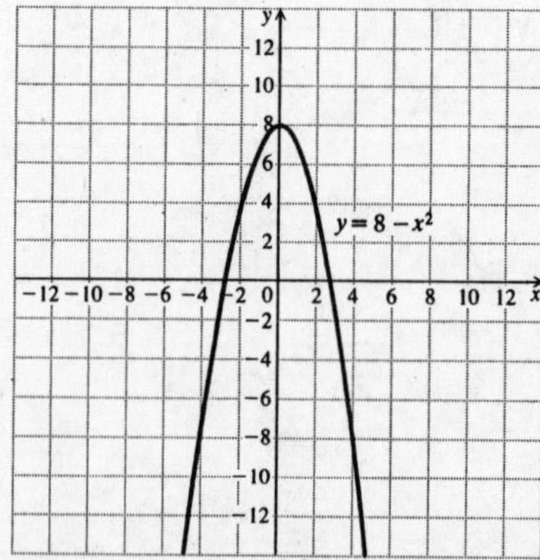

The graph shows the relation $\{(x, y)|y = 8 - x^2\}$.

48.

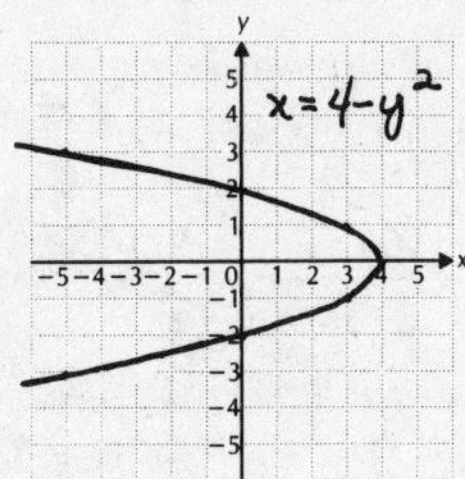

49. $y = x^2 + 1$

x	y
0	1
1	2
-1	2
2	5
-2	5

$y = (-x)^2 + 1$

x	y
0	1
1	2
-1	2
2	5
-2	5

The graphs are the same. The equations are equivalent.

50. $y = x^2 - 2$ $\qquad$ $y = 2 - x^2$

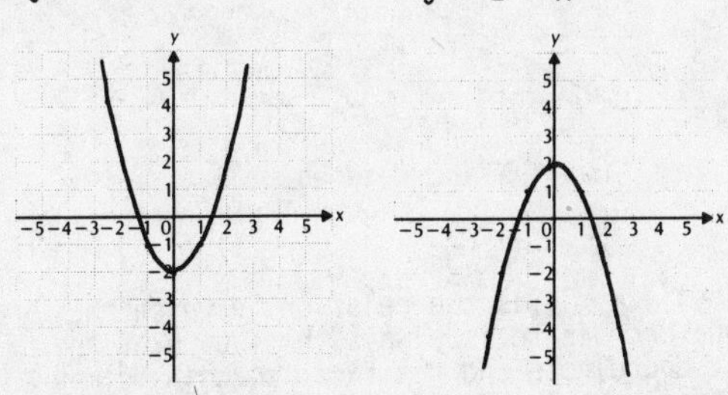

The graphs are not the same. The equations are not equivalent.

51. a)

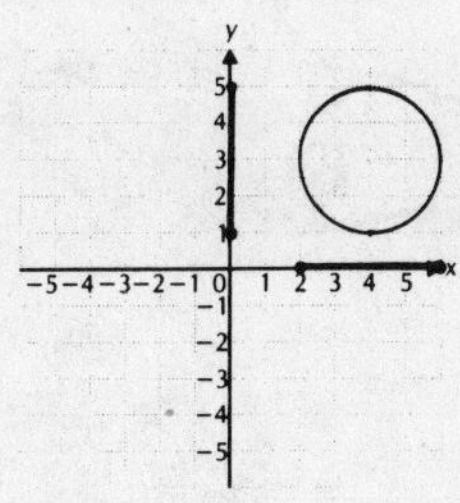

51. (continued)

b) Domain: $\{x \mid 2 \leqslant x \leqslant 6\}$

c) Range: $\{y \mid 1 \leqslant y \leqslant 5\}$

52. a)

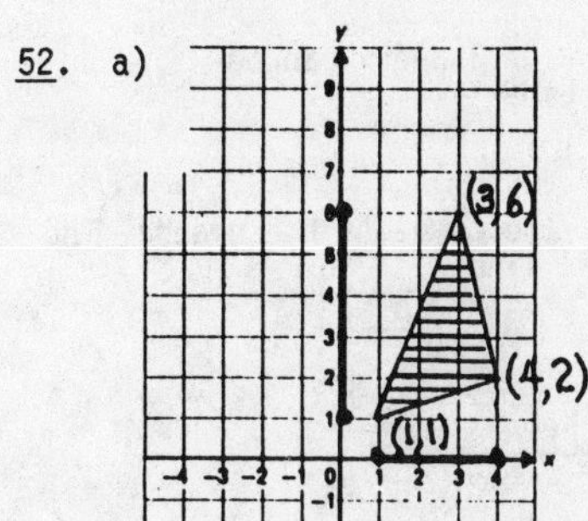

b) Domain: $\{x \mid 1 \leqslant x \leqslant 4\}$

c) Range: $\{y \mid 1 \leqslant y \leqslant 6\}$

53. $y = -x^2$

Domain: The set of real numbers

Range: $\{y \mid y \leqslant 0\}$

54. Domain: The set of real numbers

Range: The set of real numbers

55. $x = |y + 1|$

Domain: $\{x \mid x \geqslant 0\}$

Range: The set of real numbers

56. Domain: $\{x \mid x \neq 0\}$

Range: $\{y \mid y \neq 0\}$

57. $y = \sqrt{x}$

Domain: $\{x \mid x \geqslant 0\}$

Range: $\{y \mid y \geqslant 0\}$

58. Domain: The set of real numbers

Range: The set of real numbers

59. $y = 8 - x^2$

Domain: The set of real numbers

Range: $\{y \mid y \leqslant 8\}$

60. Domain: $\{x \mid x \leqslant 4\}$

Range: The set of real numbers

61. $\{(x, y) | y = 3\}$ is the relation in which the second coordinate is always 3 and the first coordinate may be any real number. Thus, all ordered pairs (x, 3) are solutions. For example,

x	y
-4	3
0	3
3	3

(y must be 3 and x can be any number)

Plot these solutions and complete the graph. The graph is a horizontal line.

$y = 3$

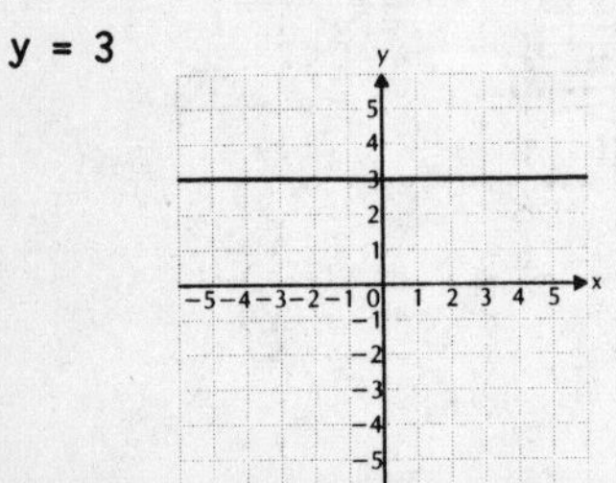

62. $x = -3$

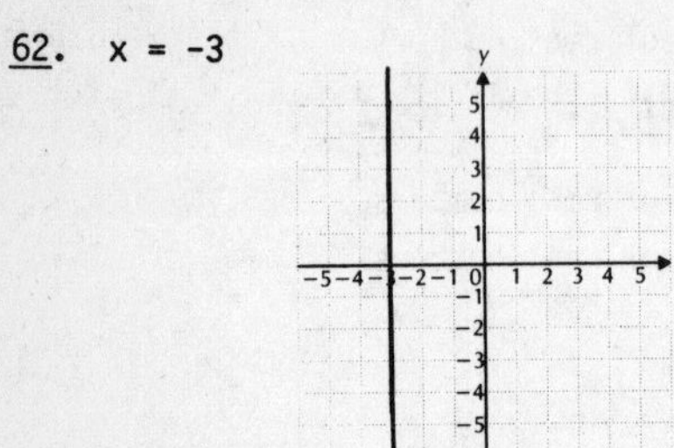

63. $\{(x, y) | y = x + 1\}$ is the relation in which the second coordinate is always 1 more than the first coordinate and the first coordinate may be any real number.

x	y
-5	-4
-2	-1
0	1
1	2
4	5

We plot these solutions and draw the graph.

$y = x + 1$

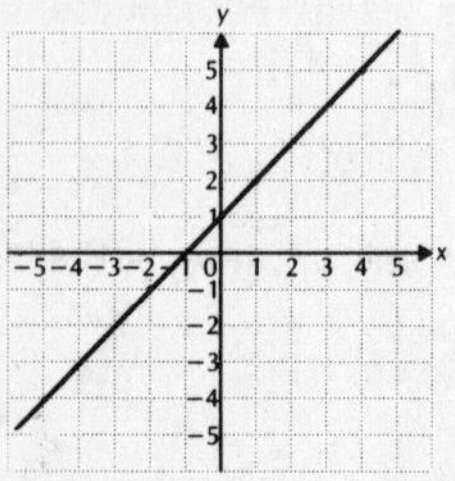

64. $y = x - 1$

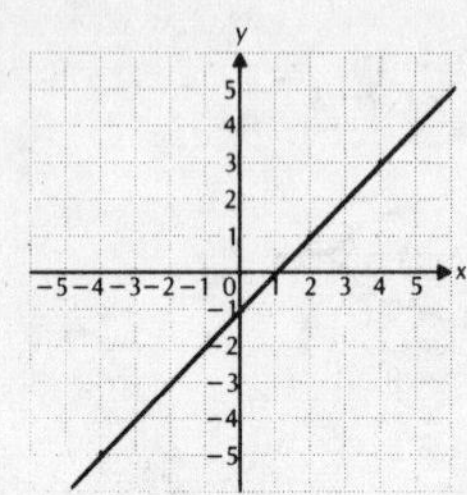

65. $\{(x, y) | y = 2x\}$ is the relation in which the second coordinate is always twice the first coordinate and the first coordinate may be any real number.

x	y
-3	-6
0	0
2	4

We plot these solutions and draw the graph.

$y = 2x$

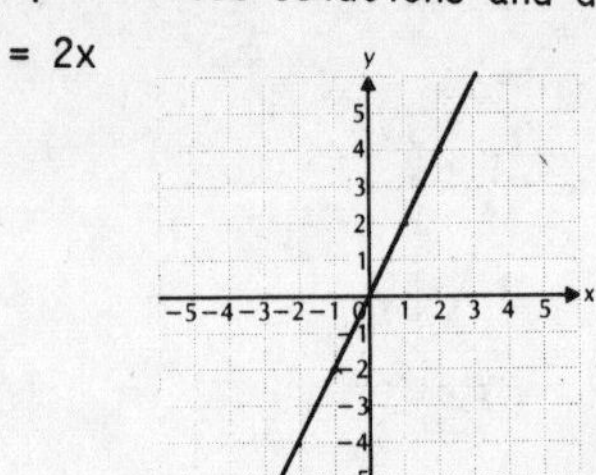

66. $y = \frac{x}{2}$

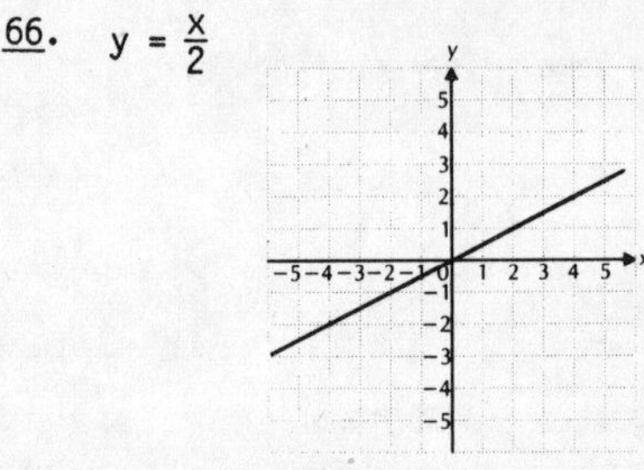

67. $\{(x, y) | y = x^2\}$ is the relation in which the second coordinate is always the square of the first coordinate and the first coordinate may be any real number. See Exercise 29.

68. $x = y^2$

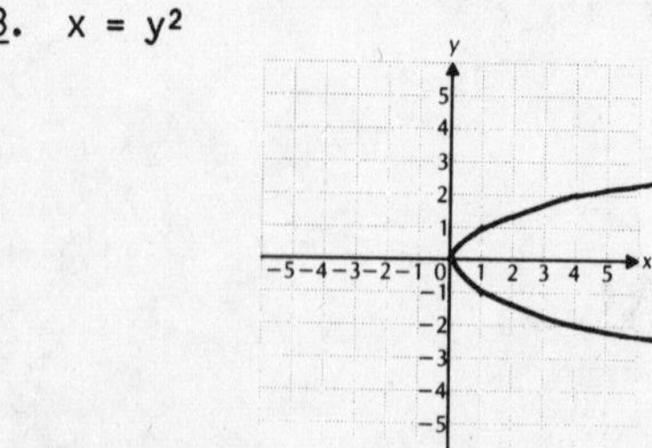

69. $y = |x| + x$

x	y
-5	0
-3	0
-1	0
0	0
1	2
3	6
5	10

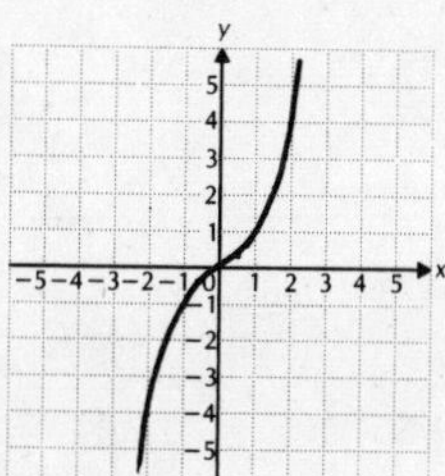

70. $y = x|x|$

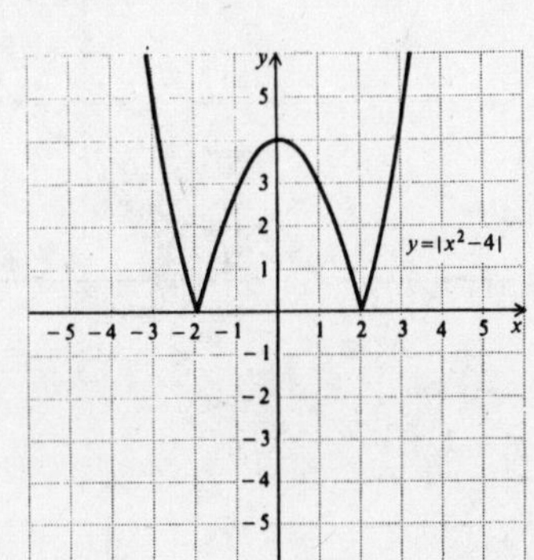

71. $y = |x^2 - 4|$

x	y
3	5
-3	5
2	0
-2	0
1	3
-1	3
0	4

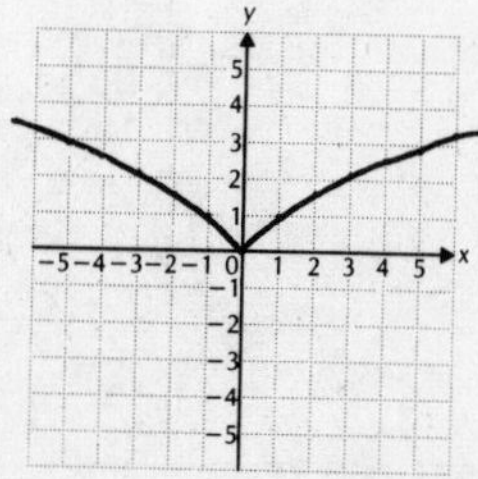

72. $y = x^{\frac{2}{3}}$

73. $|y| = x + 1$, or $x = |y| - 1$

Here we choose values for y and find corresponding values of x.

x	y
3	4
3	-4
2	3
2	-3
1	2
1	-2
0	1
0	-1
-1	0

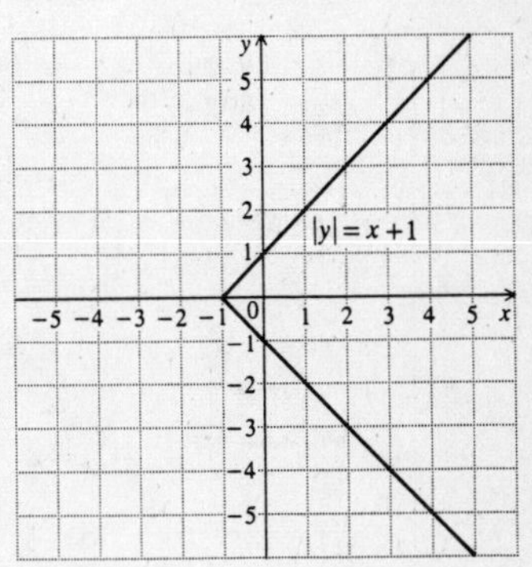

74. $y = |x^3|$

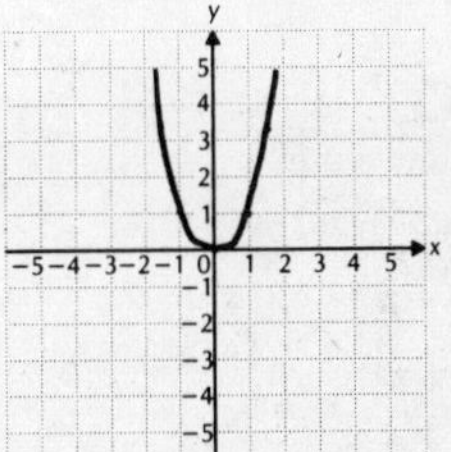

75. $|y| = |x|$

x	y
0	0
1	1
1	-1
-1	1
-1	-1
4	4
4	-4
-4	4
-4	-4

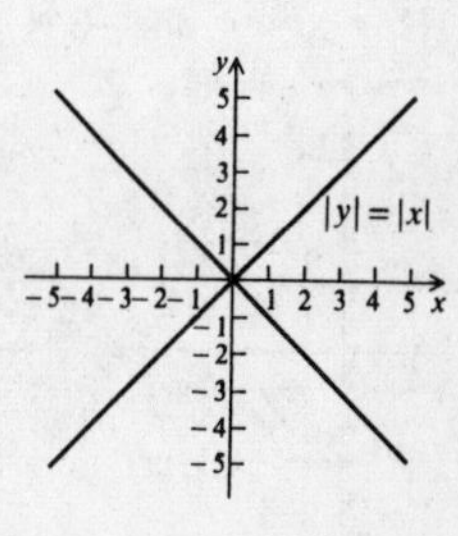

76. Graph consists only of the origin (0, 0).

77. $|xy| = 1$

$xy = 1$ or $xy = -1$

x	y
4	$\frac{1}{4}$
2	$\frac{1}{2}$
1	1
$\frac{1}{2}$	2
$\frac{1}{4}$	4
-4	$-\frac{1}{4}$
-2	$-\frac{1}{2}$
-1	-1
$-\frac{1}{2}$	-2
$-\frac{1}{4}$	-4

x	y
4	$-\frac{1}{4}$
2	$-\frac{1}{2}$
1	-1
$\frac{1}{2}$	-2
$\frac{1}{4}$	-4
-4	$\frac{1}{4}$
-2	$\frac{1}{2}$
-1	1
$-\frac{1}{2}$	2
$-\frac{1}{4}$	4

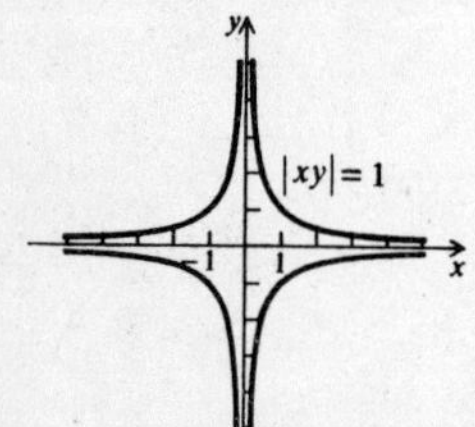

78. $\{(x, y) | 1 < x < 4 \text{ and } -3 < y < -1\}$

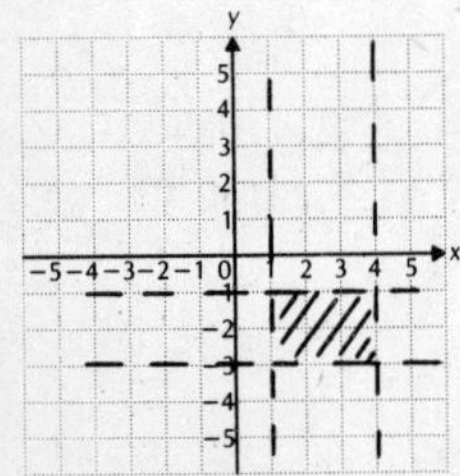

79.

x	y
-3	$-\frac{16}{3} = -5.3$
-2	0
-1.5	1.04
-1	$\frac{4}{3} = 1.3$
-0.5	1.12
0	$\frac{2}{3} = 0.67$
0.5	0.21
1	0
1.5	0.29
2	$\frac{4}{3} = 1.3$
3	$\frac{20}{3} = 6.67$

$y = \frac{1}{3}x^3 - x + \frac{2}{3}$

79. (continued)

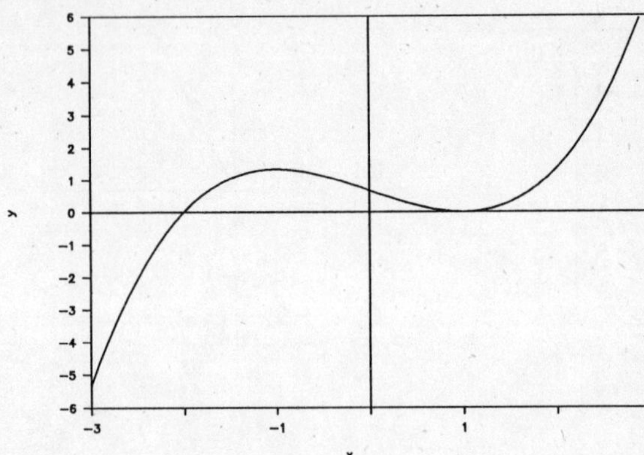

80. $y = \frac{1}{3}x^3 - \frac{1}{2}x^2 - 2x + 1$

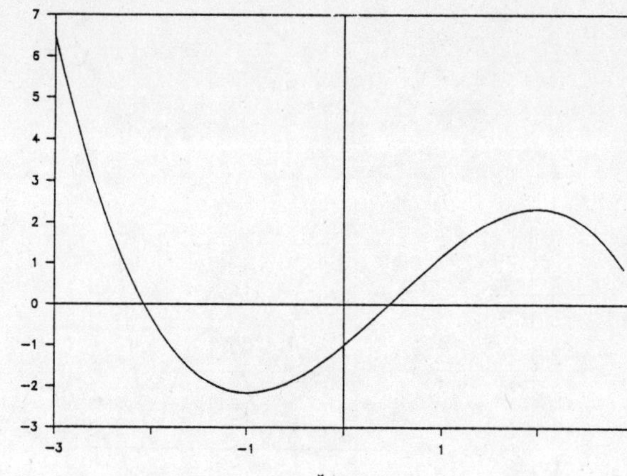

81. $y = 1 + \sqrt{2 - x}$

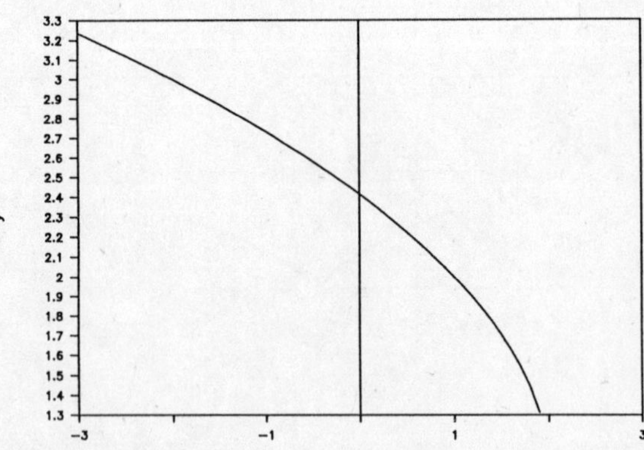

82. $y = 1 + \frac{2}{x + 1}$

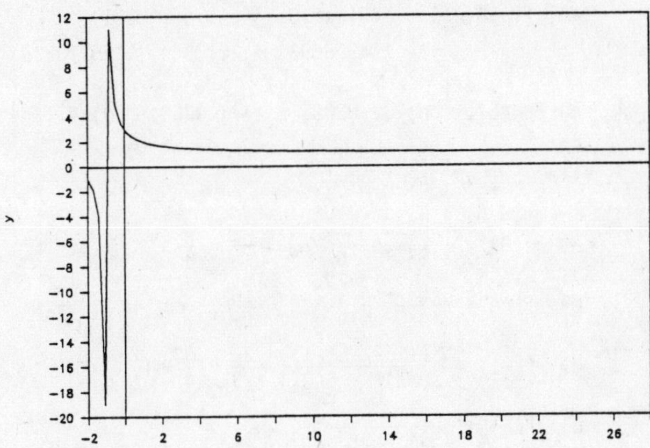

83.

x	y
-1	0
$-\frac{1}{2}$	$\pm\frac{1}{2}$
0	± 1
$\frac{1}{2}$	$\pm\frac{1}{2}$
1	0

$|x| + |y| = 1$

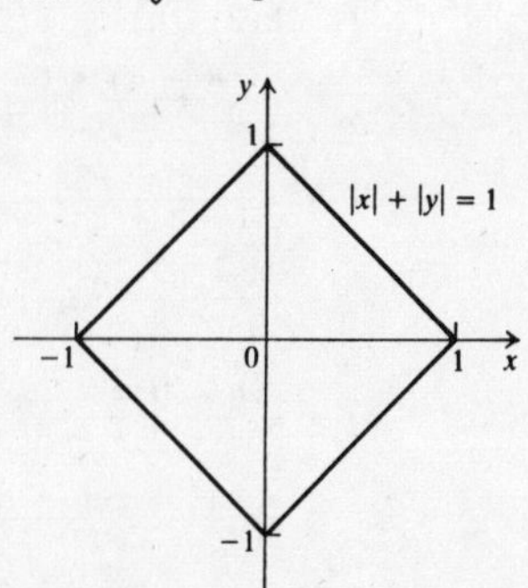

84. $\{(x, y) \mid |x| \leqslant 1 \text{ and } |y| \leqslant 2\}$

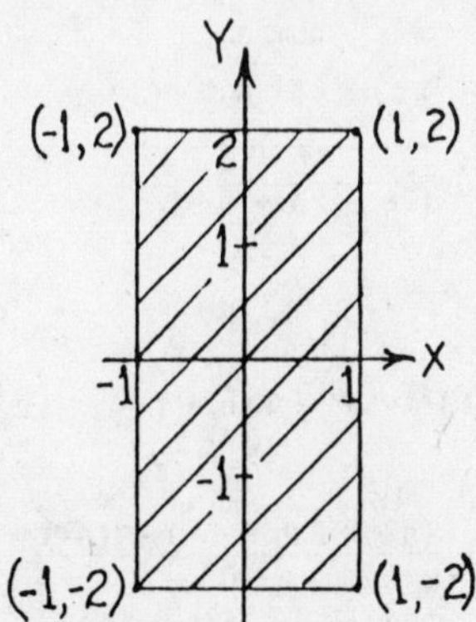

Exercise Set 3.3

1. This is not a function. There are two members of the range, Mets and Giants, that correspond to New York. There are other instances that also show that this is not a function, but only one case is necessary.

2. Yes

3. This is a function. Each member of the domain corresponds to exactly one member of the range.

4. Yes

5. This is a function, because no two ordered pairs have the same first coordinate and different second coordinates.

6. No

7. This is not a function, because the ordered pairs (-4,-4) and (-4,4) have the same first coordinate and different second coordinates.

8. No

9. This is a function, because no two ordered pairs have the same first coordinate and different second coordinates.

10. No

11. This is not the graph of a function, because it fails the vertical line test. We can find a vertical line which intersects the graph at more than one point.

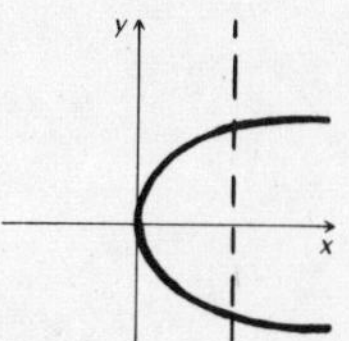

12. Yes

13. This is the graph of a function. There is no vertical line which intersects the graph at more than one point.

14. Yes

15. This is not the graph of a function. We can find a vertical line which intersects the graph in more than one point.

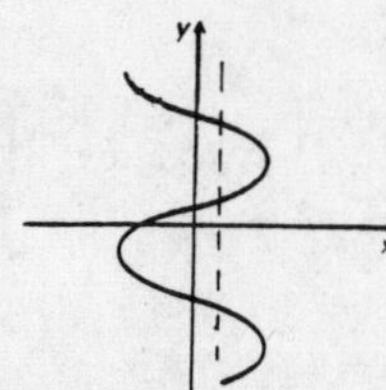

16. Yes

17. This is the graph of a function. There is no vertical line which intersects the graph in more than one point.

18. Yes

19. $f(x) = 5x^2 + 4x$

a) $f(0) = 5\cdot 0^2 + 4\cdot 0 = 0 + 0 = 0$

b) $f(-1) = 5(-1)^2 + 4(-1) = 5 - 4 = 1$

c) $f(3) = 5\cdot 3^2 + 4\cdot 3 = 45 + 12 = 57$

d) $f(t) = 5t^2 + 4t$

e) $f(t - 1) = 5(t - 1)^2 + 4(t - 1)$
$= 5(t^2 - 2t + 1) + 4(t - 1)$
$= 5t^2 - 10t + 5 + 4t - 4$
$= 5t^2 - 6t + 1$

f) $f(a + h) = 5(a + h)^2 + 4(a + h)$
$= 5(a^2 + 2ah + h^2) + 4(a + h)$
$= 5a^2 + 10ah + 5h^2 + 4a + 4h$

$f(a) = 5a^2 + 4a$

$$\frac{f(a + h) - f(a)}{h} = \frac{(5a^2+10ah+5h^2+4a+4h)-(5a^2+4a)}{h} = \frac{10ah + 5h^2 + 4h}{h} = \frac{h(10a + 5h + 4)}{h} = 10a + 5h + 4$$

20. a) 1, b) 6, c) 22, d) $3t^2 - 2t + 1$, e) $3a^2 + 6ah + 3h^2 - 2a - 2h + 1$, f) $6a + 3h - 2$

21. $f(x) = 2|x| + 3x$

a) $f(1) = 2|1| + 3\cdot 1 = 2 + 3 = 5$

b) $f(-2) = 2|-2| + 3(-2) = 4 - 6 = -2$

c) $f(-4) = 2|-4| + 3(-4) = 8 - 12 = -4$

d) $f(2y) = 2|2y| + 3(2y) = 4|y| + 6y$

e) $f(a + h) = 2|a + h| + 3(a + h)$
$= 2|a + h| + 3a + 3h$

f) $$\frac{f(a + h) - f(a)}{h} = \frac{(2|a + h| + 3a + 3h) - (2|a| + 3a)}{h} = \frac{2|a + h| - 2|a| + 3h}{h}$$

22. a) -1, b) -4, c) -56, d) $27y^3 - 6y$, e) $h^3 + 6h^2 + 10h + 4$, f) $h^2 + 6h + 10$

23. $f(x) = 4.3x^2 - 1.4x$

a) $f(1.034) = 4.3(1.034)^2 - 1.4(1.034)$
$= 4.5973708 - 1.4476 = 3.14977$

b) $f(-3.441) = 4.3(-3.441)^2 - 1.4(-3.441)$
$= 50.9140683 + 4.8174 = 55.73147$

c) $f(27.35) = 4.3(27.35)^2 - 1.4(27.35)$
$= 3216.49675 - 38.29 = 3178.20675$

23. (continued)

d) $f(-16.31) = 4.3(-16.31)^2 - 1.4(-16.31)$
$= 1143.86923 + 22.834 = 1166.70323$

24. a) 6.4467, b) 12.0308, c) 1.9259, d) 4.1744

25. $f(x) = \dfrac{x^2 - x - 2}{2x^2 - 5x - 3}$

a) $f(0) = \dfrac{0^2 - 0 - 2}{2\cdot 0^2 - 5\cdot 0 - 3} = \dfrac{-2}{-3} = \dfrac{2}{3}$

b) $f(4) = \dfrac{4^2 - 4 - 2}{2\cdot 4^2 - 5\cdot 4 - 3} = \dfrac{10}{9}$

c) $f(-1) = \dfrac{(-1)^2 - (-1) - 2}{2(-1)^2 - 5(-1) - 3} = \dfrac{0}{4} = 0$

d) $f(3) = \dfrac{3^2 - 3 - 2}{2\cdot 3^2 - 5\cdot 3 - 3} = \dfrac{4}{0}$

We cannot divide by 0. This function value does not exist.

e) $$f(2 - h) = \frac{(2 - h)^2 - (2 - h) - 2}{2(2 - h)^2 - 5(2 - h) - 3} = \frac{4 - 4h + h^2 - 2 + h - 2}{2(4 - 4h + h^2) - 10 + 5h - 3} = \frac{-3h + h^2}{8 - 8h + 2h^2 - 10 + 5h - 3} = \frac{-3h + h^2}{-5 - 3h + 2h^2}, \text{ or } \frac{h^2 - 3h}{2h^2 - 3h - 5}$$

f) $$f(a + b) = \frac{(a + b)^2 - (a + b) - 2}{2(a + b)^2 - 5(a + b) - 3} = \frac{a^2 + 2ab + b^2 - a - b - 2}{2(a^2 + 2ab + b^2) - 5a - 5b - 3} = \frac{a^2 + 2ab + b^2 - a - b - 2}{2a^2 + 4ab + 2b^2 - 5a - 5b - 3}$$

26. a) $\dfrac{\sqrt{26}}{5}$, b) $\dfrac{\sqrt{2}}{3}$, c) Does not exist as a real number, d) Does not exist as a real number, e) $\sqrt{\dfrac{5 + 3h}{11 + 2h}}$, f) $\sqrt{\dfrac{3a - 3b - 4}{2a - 2b + 5}}$

27. $f(x) = x^2$

$f(a + h) = (a + h)^2 = a^2 + 2ah + h^2$

$f(a) = a^2$

$$\frac{f(a + h) - f(a)}{h} = \frac{(a^2 + 2ah + h^2) - (a^2)}{h} = \frac{2ah + h^2}{h} = \frac{h(2a + h)}{h} = 2a + h$$

28. $3a^2 + 3ah + h^2$

29. $f(x) = x + \sqrt{x^2 - 1}$

$f(0) = 0 + \sqrt{0^2 - 1} = 0 + \sqrt{-1}$

Since $\sqrt{-1}$ is not a real number, f(0) does not exist as a real number.

$f(2) = 2 + \sqrt{2^2 - 1} = 2 + \sqrt{4 - 1} = 2 + \sqrt{3}$

$f(10) = 2 + \sqrt{10^2 - 1} = 2 + \sqrt{100 - 1} = 2 + \sqrt{99} =$

$2 + \sqrt{9 \cdot 11} = 2 + 3\sqrt{11}$

30. g(0) = 0, g(3) does not exist as a real number, $g\left(\frac{1}{2}\right) = \frac{\sqrt{3}}{3}$

31. Since (-1,2) is in the function, f(-1) = 2.
Since (7,9) is in the function, f(7) = 9.
Since (5,-6) is in the function, f(5) = -6.
Since (-3,4) is in the function, f(-3) = 4.

The domain is {-1,-3,5,7}.
The range is {2,4,-6,9}.

32. g(8) = -7, g(6) = -7, g(-4) = 5, g(2) = -3;
domain: {2,-4,6,8}; range: {-3,5,-7}

33. We locate x on the x-axis and then find f(x) on the y-axis. From the graph, we see that g(-2) = 0, g(-3) = -3, g(0) = 3, and g(2) = 2.

34. h(-2) = 0, h(0) = -2, h(3) = 0, h(-3) = 3

35. $f(x) = 7x + 4$

There are no restrictions on the numbers we can substitute into this formula. Thus, the domain is the entire set of real numbers.

36. All real numbers

37. $f(x) = 4 - \frac{2}{x}$

Division by 0 is undefined. Thus, $x \neq 0$. The domain is all real numbers except 0, or $\{x|x \neq 0\}$.

38. $\{x|x \geqslant 3\}$

39. $f(x) = \sqrt{7x + 4}$

Since this formula is meaningful only if the radicand is nonnegative, we want the replacements for x which make the following inequality true.

$7x + 4 \geqslant 0$

$7x \geqslant -4$

$x \geqslant -\frac{4}{7}$

The domain is $\left\{x \middle| x \geqslant -\frac{4}{7}\right\}$.

40. $\{x|x \neq 3 \text{ and } x \neq -3\}$

41. $f(x) = \frac{1}{x^2 - 4}$

This formula is meaningful as long as a replacement for x does not make the denominator 0. To find those replacements which do make the denominator 0, we solve $x^2 - 4 = 0$.

$x^2 - 4 = 0$

$(x + 2)(x - 2) = 0$

$x = -2$ or $x = 2$

Thus the domain consists of all real numbers except -2 and 2. This set can be named $\{x|x \neq -2 \text{ and } x \neq 2\}$.

42. $\{x|x \neq 0 \text{ and } x \neq 2 \text{ and } x \neq -2\}$

43. $f(x) = \frac{4x^3 + 4}{4x^2 - 5x - 6}$

This formula is meaningful as long as a replacement for x does not make the denominator 0. To find those replacements which do make the denominator 0, we solve $4x^2 - 5x - 6 = 0$.

$4x^2 - 5x - 6 = 0$

$(4x + 3)(x - 2) = 0$

$4x + 3 = 0$ or $x - 2 = 0$

$x = -\frac{3}{4}$ or $x = 2$

Thus the domain consists of all real numbers except $-\frac{3}{4}$ and 2. This set can be named $\left\{x \middle| x \neq -\frac{3}{4} \text{ and } x \neq 2\right\}$.

44. $\{x|x \neq 2 \text{ and } x \neq -2\}$

45. $f(x) = \frac{4x^3 + 4}{x(x + 2)(x - 1)}$

This formula is meaningful as long as a replacement for x does not make the denominator 0. To find those replacements which do make the denominator 0, we solve $x(x + 2)(x - 1) = 0$.

$x(x + 2)(x - 1) = 0$

$x = 0$ or $x = -2$ or $x = 1$

Thus the domain consists of all real numbers except 0, -2, and 1. This set can be named $\{x|x \neq 0 \text{ and } x \neq -2 \text{ and } x \neq 1\}$.

46. All real numbers

47. $f(z) = z^2 - 4z + i$

$$\begin{aligned} f(3 + i) &= (3 + i)^2 - 4(3 + i) + i \\ &= 9 + 6i + i^2 - 12 - 4i + i \\ &= i^2 + 3i - 3 \\ &= -1 + 3i - 3 \\ &= -4 + 3i \end{aligned}$$

48. 8 - 11i

49. $f(x) = \frac{1}{x}$ $f(x + h) = \frac{1}{x + h}$

$$\frac{f(x + h) - f(x)}{h} = \frac{\frac{1}{x + h} - \frac{1}{x}}{h}$$

$$= \frac{\frac{1}{x + h} \cdot \frac{x}{x} - \frac{1}{x} \cdot \frac{x + h}{x + h}}{h}$$

$$= \frac{\frac{x - (x + h)}{x(x + h)}}{h}$$

$$= \frac{-h}{x(x + h)} \cdot \frac{1}{h}$$

$$= \frac{-1}{x(x + h)}$$

50. $\frac{-2x - h}{x^2(x + h)^2}$

51. $f(x) = \sqrt{x}$ $f(x + h) = \sqrt{x + h}$

$$\frac{f(x + h) - f(x)}{h} = \frac{\sqrt{x + h} - \sqrt{x}}{h}$$

$$= \frac{\sqrt{x + h} - \sqrt{x}}{h} \cdot \frac{\sqrt{x + h} + \sqrt{x}}{\sqrt{x + h} + \sqrt{x}}$$

$$= \frac{(x + h) - x}{h(\sqrt{x + h} + \sqrt{x})}$$

$$= \frac{h}{h(\sqrt{x + h}) + \sqrt{x})}$$

$$= \frac{1}{\sqrt{x + h} + \sqrt{x}}$$

52. $\left\{x \middle| x \neq \frac{5}{2} \text{ and } x \geqslant 0\right\}$

53. $f(x) = \frac{\sqrt{x + 3}}{x^2 - x - 2}$

The replacements for x cannot make the denominator 0. To find those replacements which do make the denominator 0, we solve $x^2 - x - 2 = 0$.

$x^2 - x - 2 = 0$

$(x - 2)(x + 1) = 0$

$x - 2 = 0$ or $x + 1 = 0$

$x = 2$ or $x = -1$

Thus the domain cannot include 2 and -1.

The radicand in the numerator must be nonnegative, thus we want the replacements for x which make the following inequality true.

$x + 3 \geqslant 0$

$x \geqslant -3$

The domain for f(x) is the set of all real numbers greater than or equal to -3 except 2 and -1. This set can be named $\{x | x \neq 2 \text{ and } x \neq -1 \text{ and } x \geqslant -3\}$.

54. $\{x | x > 0\}$

55. $f(x) = \sqrt{x^2 + 1}$

Since $x^2 + 1 \geqslant 0$ for all x, the domain is the set of all real numbers.

56. No

Exercise Set 3.4

1. $f(x) = |x| + 2$

We compute some function values.

$f(-3) = |-3| + 2 = 3 + 2 = 5$

$f(-1) = |-1| + 2 = 1 + 2 = 3$

$f(0) = |0| + 2 = 0 + 2 = 2$

$f(2) = |2| + 2 = 2 + 2 = 4$

$f(4) = |4| + 2 = 4 + 2 = 6$

x	f(x)	(x, f(x))
-3	5	(-3,5)
-1	3	(-1,3)
0	2	(0,2)
2	4	(2,4)
4	6	(4,6)

We plot these points, look for a pattern, and sketch the graph.

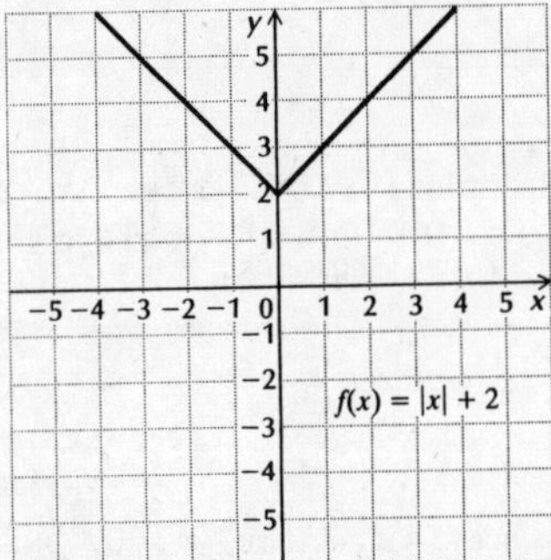

2.

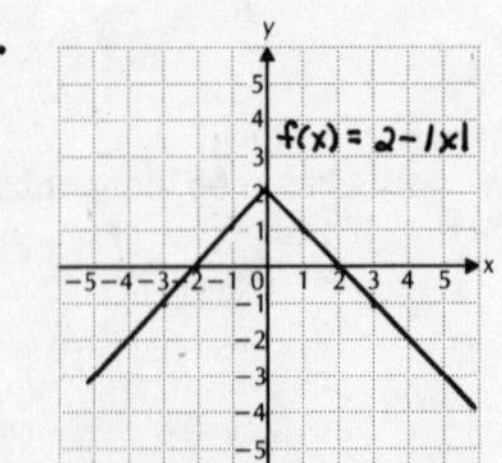

3. $g(x) = 4 - x^2$

We compute some function values.

$g(-3) = 4 - (-3)^2 = 4 - 9 = -5$

$g(-2) = 4 - (-2)^2 = 4 - 4 = 0$

$g(0) = 4 - 0^2 = 4 - 0 = 4$

$g(1) = 4 - 1^2 = 4 - 1 = 3$

$g(2) = 4 - 2^2 = 4 - 4 = 0$

x	g(x)	(x, g(x))
-3	-5	(-3,-5)
-2	0	(-2,0)
0	4	(0,4)
1	3	(1,3)
2	0	(2,0)

We plot these points, look for a pattern, and sketch the graph.

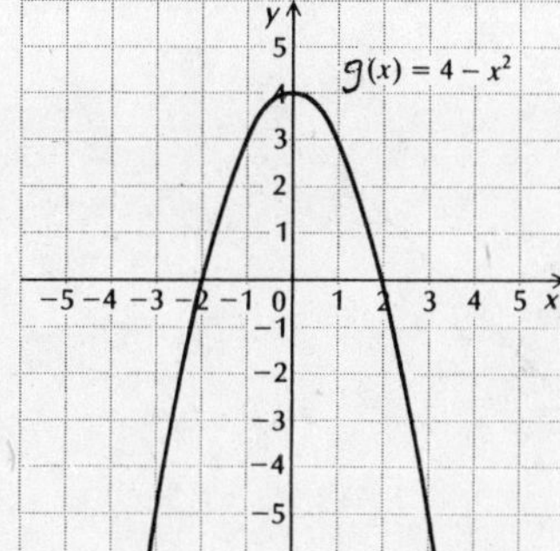

4.

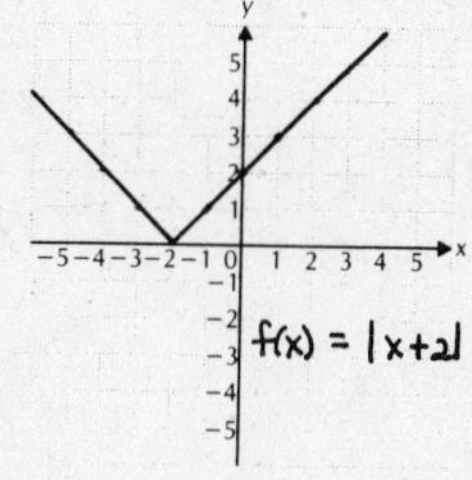

5. $f(x) = \frac{2}{x}$

We compute some function values. Note that x cannot be 0.

$f(-4) = \frac{2}{-4} = -\frac{1}{2}$ $\quad f\left(\frac{1}{2}\right) = \frac{2}{\frac{1}{2}} = 4$

$f(-2) = \frac{2}{-2} = -1$ $\quad f(2) = \frac{2}{2} = 1$

$f\left(-\frac{1}{2}\right) = \frac{2}{-\frac{1}{2}} = -4$ $\quad f(4) = \frac{2}{4} = \frac{1}{2}$

x	f(x)	(x, f(x))
-4	$-\frac{1}{2}$	$\left(-4,-\frac{1}{2}\right)$
-2	-1	(-2,-1)
$-\frac{1}{2}$	-4	$\left(-\frac{1}{2},-4\right)$
$\frac{1}{2}$	4	$\left(\frac{1}{2},4\right)$
2	1	(2,1)
4	$\frac{1}{2}$	$\left(4,\frac{1}{2}\right)$

We plot these points, look for a pattern, and sketch the graph.

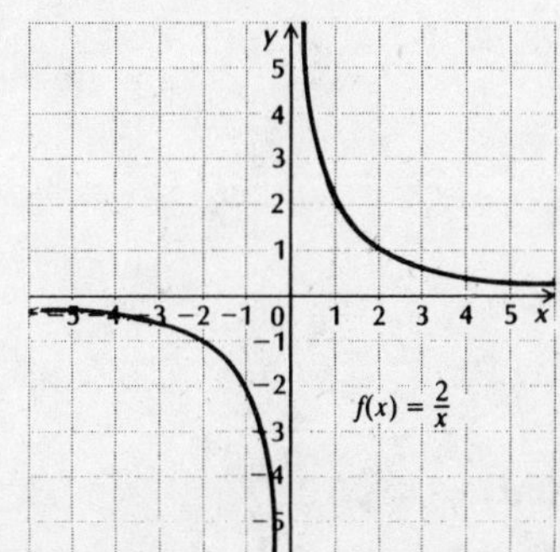

6.

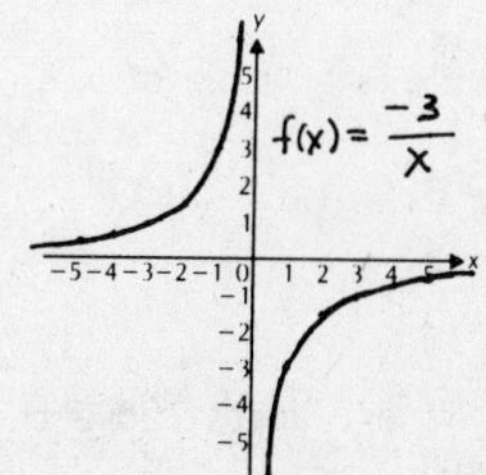

7. $f(x) = \frac{1}{2}|x|$

We compute some function values.

$f(-4) = \frac{1}{2}|-4| = \frac{1}{2} \cdot 4 = 2$

$f(-2) = \frac{1}{2}|-2| = \frac{1}{2} \cdot 2 = 1$

$f(0) = \frac{1}{2}|0| = \frac{1}{2} \cdot 0 = 0$

$f(1) = \frac{1}{2}|1| = \frac{1}{2} \cdot 1 = \frac{1}{2}$

$f(3) = \frac{1}{2}|3| = \frac{1}{2} \cdot 3 = \frac{3}{2}$

We plot these points, look for a pattern, and sketch the graph.

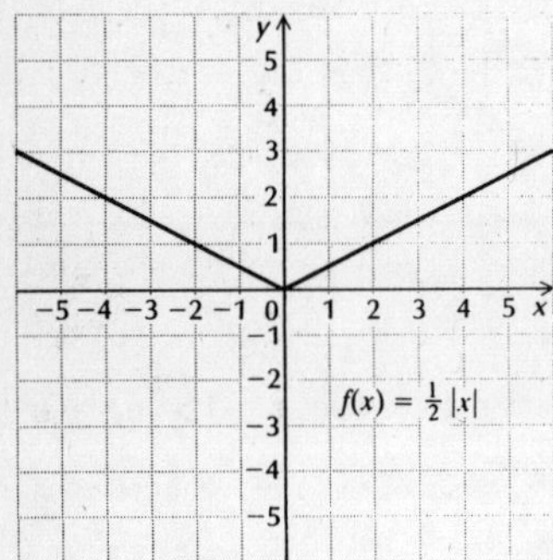

8.

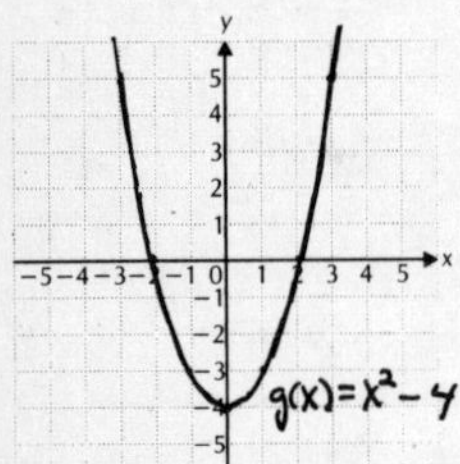

9. $g(x) = 3$

The function value is 3 for every value of x. For example,

$g(-5) = 3,$ $g(1) = 3,$

$g(-3) = 3,$ $g(2) = 3,$

$g(0) = 3,$ $g(4) = 3.$

The graph is a horizontal line 3 units above the x-axis. We sketch the graph.

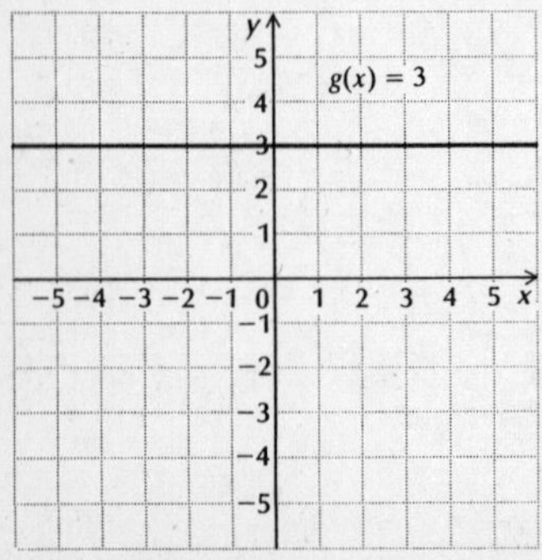

10.

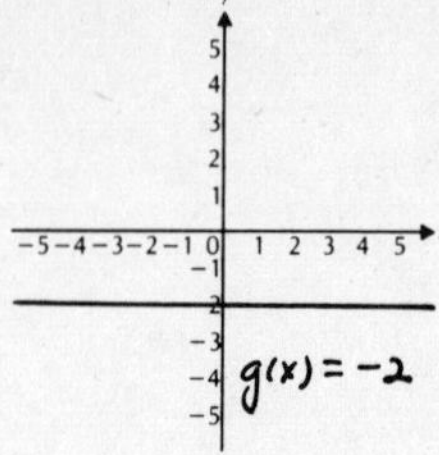

11. $f(x) = |x - 1|$

We compute some function values.

$f(-4) = |-4 - 1| = |-5| = 5$

$f(-2) = |-2 - 1| = |-3| = 3$

$f(0) = |0 - 1| = |-1| = 1$

$f(1) = |1 - 1| = |0| = 0$

$f(3) = |3 - 1| = |2| = 2$

$f(5) = |5 - 1| = |4| = 4$

We plot these points, look for a pattern, and sketch the graph.

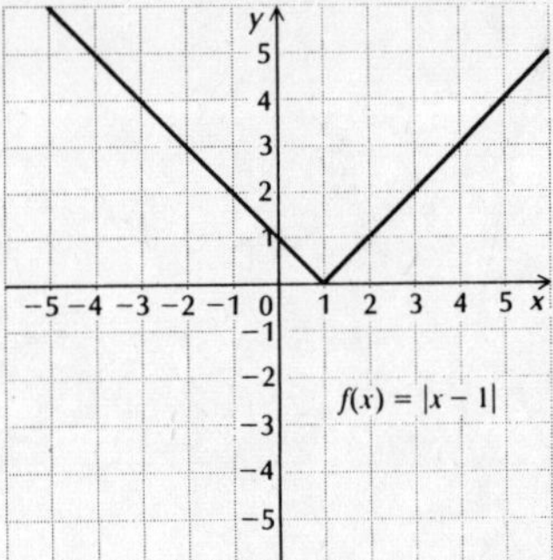

12.

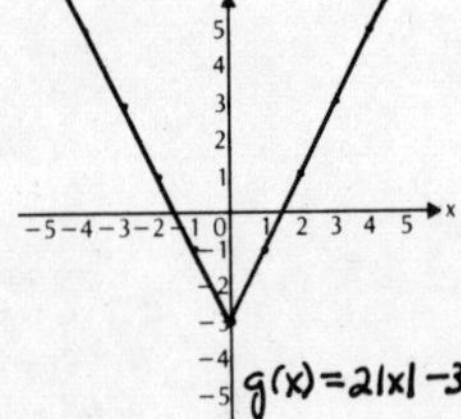

13. $f(x) = |x| + x$

We compute some function values.

$f(-4) = |-4| + (-4) = 4 - 4 = 0$

$f(-2) = |-2| + (-2) = 2 - 2 = 0$

$f(0) = |0| + 0 = 0 + 0 = 0$

$f(1) = |1| + 1 = 1 + 1 = 2$

$f(3) = |3| + 3 = 3 + 3 = 6$

We plot these points, look for a pattern, and sketch the graph.

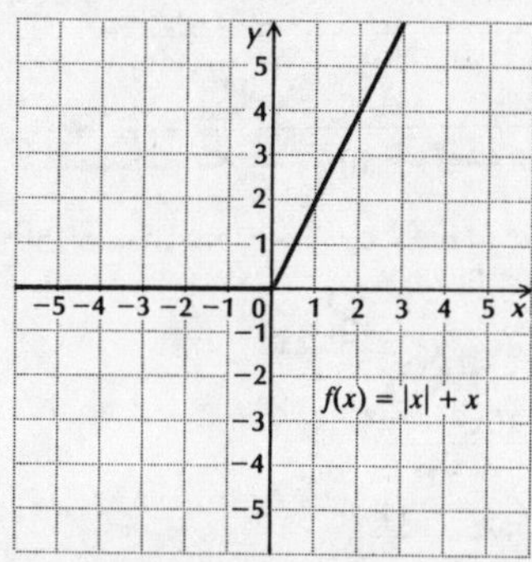

14.

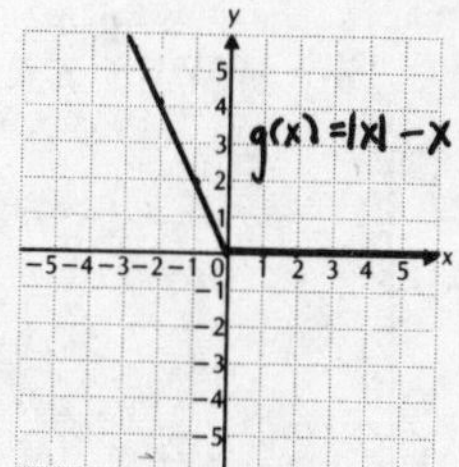

15. $f(x) = \sqrt{x + 3}$

We compute some function values. Note that the meaningful replacements for x are $x \geqslant -3$.

$f(-3) = \sqrt{-3 + 3} = \sqrt{0} = 0$

$f(-2) = \sqrt{-2 + 3} = \sqrt{1} = 1$

$f(1) = \sqrt{1 + 3} = \sqrt{4} = 2$

$f(6) = \sqrt{6 + 3} = \sqrt{9} = 3$

We plot these points, look for a pattern, and sketch the graph.

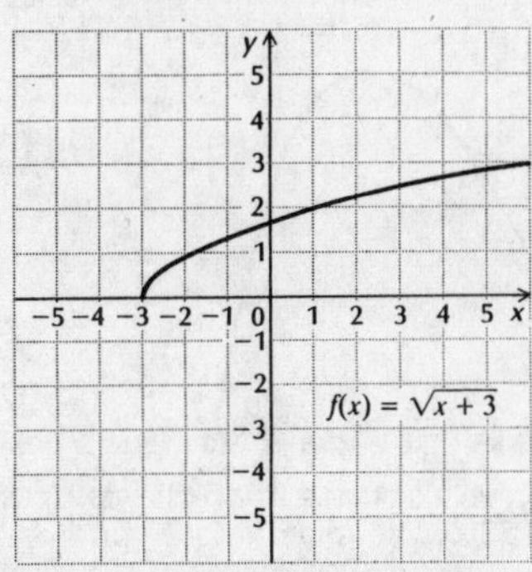

16.

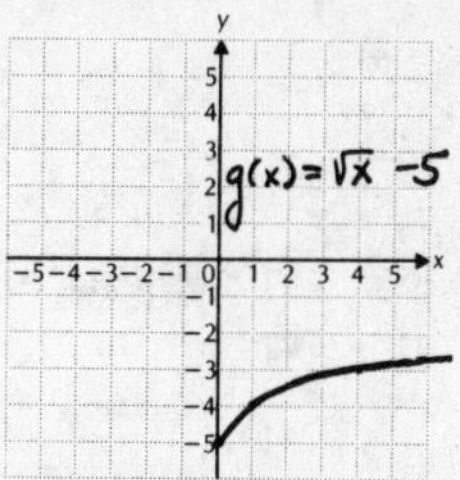

17. $A(t) = 0.08t + 19.7$

a) $A(0) = 0.08(0) + 19.7 = 0 + 19.7 = 19.7$

$A(1) = 0.08(1) + 19.7 = 0.08 + 19.7 = 19.78$

$A(10) = 0.08(10) + 19.7 = 0.8 + 19.7 = 20.5$

$A(30) = 0.08(30) + 19.7 = 2.4 + 19.7 = 22.1$

$A(40) = 0.08(40) + 19.7 = 3.2 + 19.7 = 22.9$

b) 1996 is 46 years after 1950, so we find A(46).

$A(46) = 0.08(46) + 19.7 = 3.68 + 19.7 = 23.38$

The median age of women at first marriage in 1996 will be 23.38.

18. a) \$4.44, \$5.05, \$5.81

b) \$2001

19. $D(n) = \dfrac{n^2 - 3n}{2}$

$D(5) = \dfrac{5^2 - 3\cdot 5}{2} = \dfrac{25 - 15}{2} = \dfrac{10}{2} = 5$

A pentagon has 5 diagonals.

$D(8) = \dfrac{8^2 - 3\cdot 8}{2} = \dfrac{64 - 24}{2} = \dfrac{40}{2} = 20$

An octagon has 20 diagonals.

$D(10) = \dfrac{10^2 - 3\cdot 10}{2} = \dfrac{100 - 30}{2} = \dfrac{70}{2} = 35$

A decagon has 35 diagonals.

$D(12) = \dfrac{12^2 - 3\cdot 12}{2} = \dfrac{144 - 36}{2} = \dfrac{108}{2} = 54$

A dodecagon has 54 diagonals.

20. a) 30

b) 90

21. $E(T) = 1000(100 - T) + 580(100 - T)^2$

a) $E(99.5) = 1000(100 - 99.5) + 580(100 - 99.5)^2$

$= 1000(0.5) + 580(0.5)^2$

$= 500 + 580(0.25) = 500 + 145$

$= 645$ m above sea level

b) $E(100) = 1000(100 - 100) + 580(100 - 100)^2$

$= 1000\cdot 0 + 580(0)^2 = 0 + 0$

$= 0$ m above sea level, or at sea level

22. 0.4 acres, 20.4 acres, 50.6 acres, 125.5 acres, 416.9 acres, 1033.6 acres

23. a) $L = W + 4$, so $W = L - 4$.
Area = length × width, so
$A(L) = L(L - 4)$, or $L^2 - 4L$.

b) $L = W + 4$, and area = length × width, so
$A(W) = (W + 4)W$, or $W^2 + 4W$.

24. a) $A(h) = \frac{1}{2}(2h - 3)h$, or $\frac{1}{2}h(2h - 3)$

b) $A(b) = \frac{1}{2}b\left[\frac{b + 3}{2}\right]$

25. First express y in terms of x.

$$2x + 2y = 34$$
$$2y = 34 - 2x$$
$$y = \frac{1}{2}(34 - 2x)$$
$$y = 17 - x$$

The area is given by $A = xy$, so
$A(x) = x(17 - x)$, or $17x - x^2$.

26. $A(x) = x(24 - x)$, or $24x - x^2$.

27. We make a drawing.

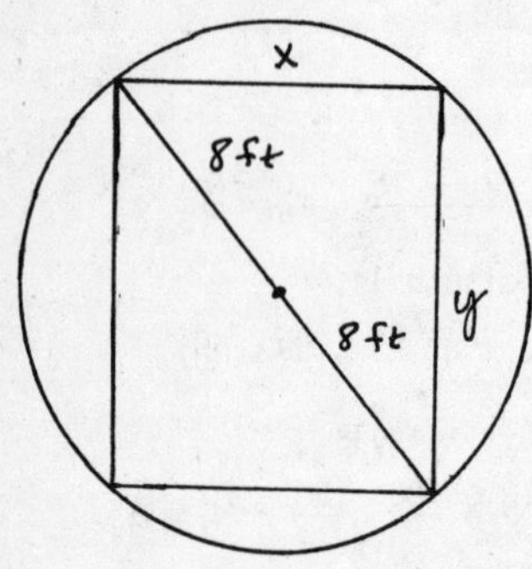

The diameter of the circle is 2·8, or 16 ft. We can use the Pythagorean theorem to express y in terms of x.

$$x^2 + y^2 = 16^2$$
$$x^2 + y^2 = 256$$
$$y^2 = 256 - x^2$$
$$y = \sqrt{256 - x^2} \quad \text{Taking the positive square root}$$

The area is given by $A = xy$, so
$A(x) = x\sqrt{256 - x^2}$.

28. $A(x) = x\sqrt{400 - x^2}$.

29. We will use the formula $A = P(1 + i)^t$ with $P = 1000$ and $t = 4$.

$A(i) = 1000(1 + i)^4$

30. $A(t) = 1000(1.08)^t$

31. a) We make a drawing.

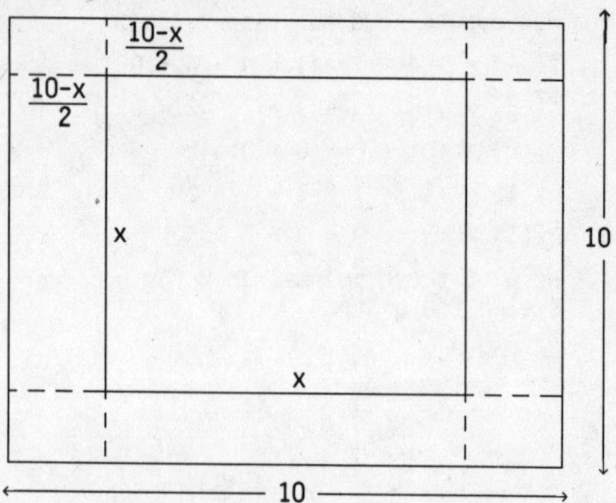

The volume is given by V = length × width × height, so we have

$$V(x) = x \cdot x \cdot \frac{10 - x}{2}$$
$$V(x) = \frac{10x^2 - x^3}{2}$$
$$V(x) = 5x^2 - \frac{x^3}{2}.$$

b) We complete a table of function values. Note that the problem restricts the domain of the function to values of x such that $0 \leqslant x \leqslant 10$.

x	$V(x) = 5x^2 - \frac{x^3}{2}$
0	0
1	4.5
2	16
3	31.5
4	48
5	62.5
6	72
7	73.5
8	64
9	40.5
10	0

We sketch the graph.

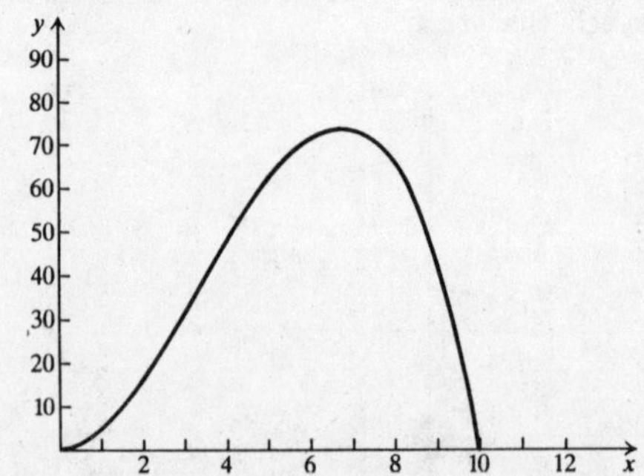

Using the graph, we estimate that the maximum volume is about 73.5 in^3 when x is 7 in.

32. a) $A(x) = x(30 - x)$, or $30x - x^2$

b)

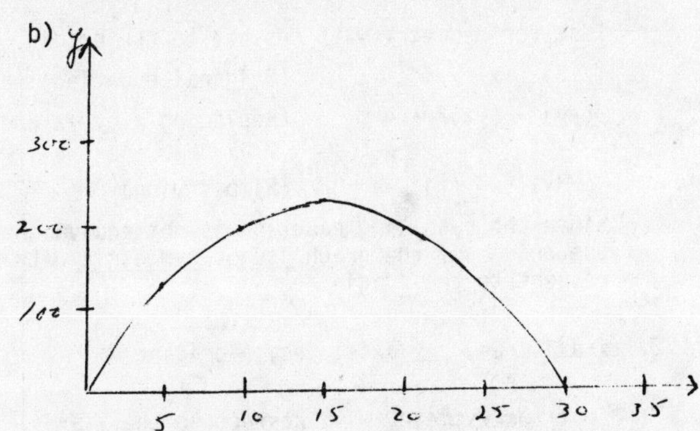

We estimate the maximum area is 225 ft^2 when the dimensions are 15 ft by 15 ft.

33. First we use the volume to express y in terms of x.

$$x \cdot x \cdot y = 108$$

$$y = \frac{108}{x^2}$$

Then we write an expression for the surface area.

$$SA(x) = x^2 + 4 \cdot x \cdot \frac{108}{x^2}$$

$$SA(x) = x^2 + \frac{432}{x}$$

34. $C(x) = 2.5x^2 + \frac{3200}{x}$

35. We will use similar triangles, expressing all distances in feet. $\left[6 \text{ in.} = \frac{1}{2} \text{ ft}, \ s \text{ in.} = \frac{s}{12} \text{ ft}, \text{ and } d \text{ yd} = 3d \text{ ft}\right]$ We have

$$\frac{3d}{7} = \frac{\frac{1}{2}}{\frac{s}{12}}$$

$$\frac{s}{12} \cdot 3d = 7 \cdot \frac{1}{2}$$

$$\frac{sd}{4} = \frac{7}{2}$$

$$d = \frac{4}{s} \cdot \frac{7}{2}, \text{ so}$$

$$d(s) = \frac{14}{s}.$$

36. $A(x) = 24x - \left[\frac{\pi}{2} + 2\right]x^2$

37. We will use the Pythagorean theorem.

$$h^2 + 3700^2 = d^2$$

$$h^2 = d^2 - 3700^2$$

$$h^2 = d^2 - 13{,}690{,}000$$

$$h(d) = \sqrt{d^2 - 13{,}690{,}000}$$ Taking the positive square root

38. a) $V(r) = \frac{4}{3}\pi r^3 + 6\pi r^2$

b) $S(r) = 4\pi r^2 + 12\pi r$

39. We make a drawing.

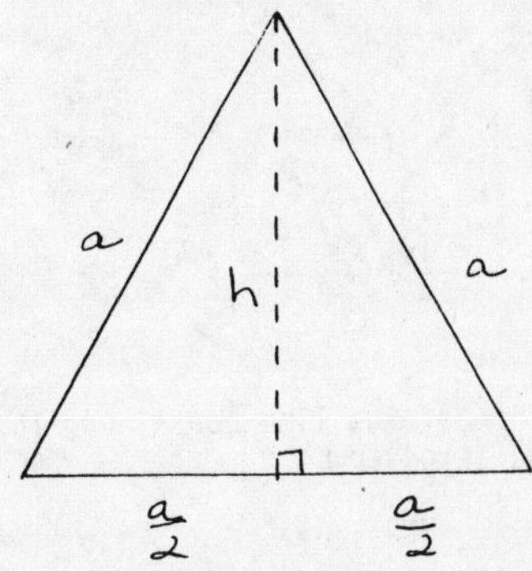

We first use the Pythagorean theorem to express the height h in terms of a.

$$h^2 + \left(\frac{a}{2}\right)^2 = a^2$$

$$h^2 + \frac{a^2}{4} = a^2$$

$$h^2 = \frac{3a^2}{4}$$

$$h = \frac{a\sqrt{3}}{2}$$

Then we write an expression for the area in terms of a.

$$A(a) = \frac{1}{2} \cdot a \cdot \frac{a\sqrt{3}}{2}$$

$$A(a) = \frac{a^2\sqrt{3}}{4}$$

40. $D(x) = \sqrt{x^2 + 4225}$

41. First we find the radius r of a circle with circumference x:

$$2\pi r = x$$

$$r = \frac{x}{2\pi}$$

Then we find the length s of a side of a square with perimeter 24 - x:

$$4s = 24 - x$$

$$s = \frac{24 - x}{4}$$

Finally, we express the sum of the areas as a function of x:

S = area of circle + area of square

$$S = \pi r^2 + s^2$$

$$S(x) = \pi\left[\frac{x}{2\pi}\right]^2 + \left[\frac{24 - x}{4}\right]^2$$

$$S(x) = \frac{x^2}{4\pi} + \left[\frac{24 - x}{4}\right]^2$$

42. $C(x) = 3000(4 - x) + 5000(\sqrt{1 + x^2})$.

43. a) We position a coordinate system as shown below.

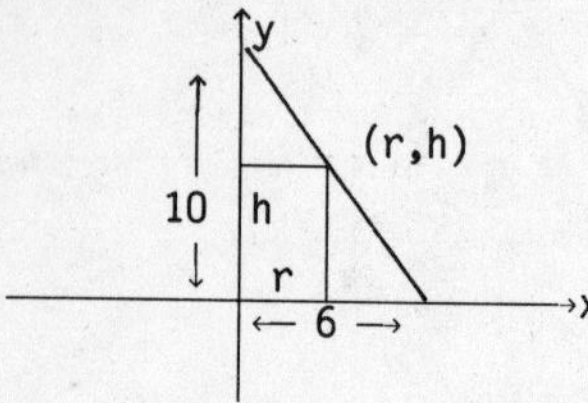

The equation of the line containing (6,0) and (0,10) is given by

$y - 10 = \frac{10 - 0}{0 - 6}(x - 0)$, or $y = -\frac{5}{3}x + 10$.

Now (r,h) is a point on the line so,

$h = -\frac{5}{3}r + 10$, or

$h(r) = \frac{30 - 5r}{3}$.

b) $V = \pi r^2 h$

$V(r) = \pi r^2\left[\frac{30 - 5r}{3}\right]$ Substituting for h

c) We first express h in terms of r.

$h = \frac{30 - 5r}{3}$

$3h = 30 - 5r$

$5r = 30 - 3h$

$r = \frac{30 - 3h}{5}$

$V = \pi r^2 h$

$V(h) = \pi\left[\frac{30 - 3h}{5}\right]^2 h$ Substituting for r

Exercise Set 3.5

1. Test for symmetry with respect to the x-axis.

$3y = x^2 + 4$ (Original equation)

$3(-y) = x^2 + 4$ (Replacing y by -y)

$-3y = x^2 + 4$ (Simplifying)

Since the resulting equation is not equivalent to the original, the graph is not symmetric with respect to the x-axis.

Test for symmetry with respect to the y-axis.

$3y = x^2 + 4$ (Original equation)

$3y = (-x)^2 + 4$ (Replacing x by -x)

$3y = x^2 + 4$ (Simplifying)

Since the resulting equation is equivalent to the original, the graph is symmetric with respect to the y-axis.

1. (continued)

Test for symmetry with respect to the origin.

$3y = x^2 + 4$ (Original equation)

$3(-y) = (-x)^2 + 4$ (Replacing x by -x and y by -y)

$-3y = x^2 + 4$ (Simplifying)

Since the resulting equation is not equivalent to the original, the graph is not symmetric with respect to the origin.

2. x-axis: No, y-axis: Yes, origin: No

3. Test for symmetry with respect to the x-axis.

$y^3 = 2x^2$ (Original equation)

$(-y)^3 = 2x^2$ (Replacing y by -y)

$-y^3 = 2x^2$ (Simplifying)

Since the resulting equation is not equivalent to the original, the graph is not symmetric with respect to the x-axis.

Test for symmetry with respect to the y-axis.

$y^3 = 2x^2$ (Original equation)

$y^3 = 2(-x)^2$ (Replacing x by -x)

$y^3 = 2x^2$ (Simplifying)

Since the resulting equation is equivalent to the original, the graph is symmetric with respect to the y-axis.

Test for symmetry with respect to the origin.

$y^3 = 2x^2$ (Original equation)

$(-y)^3 = 2(-x)^2$ (Replacing x by -x and y by -y)

$-y^3 = 2x^2$ (Simplifying)

Since the resulting equation is not equivalent to the original, the graph is not symmetric with respect to the origin.

4. x-axis: No, y-axis: Yes, origin: No

5. Test for symmetry with respect to the x-axis.

$2x^4 + 3 = y^2$ (Original equation)

$2x^4 + 3 = (-y)^2$ (Replacing y by -y)

$2x^4 + 3 = y^2$ (Simplifying)

Since the resulting equation is equivalent to the original, the graph is symmetric with respect to the x-axis.

Test for symmetry with respect to the y-axis.

$2x^4 + 3 = y^2$ (Original equation)

$2(-x)^4 + 3 = y^2$ (Replacing x by -x)

$2x^4 + 3 = y^2$ (Simplifying)

Since the resulting equation is equivalent to the original, the graph is symmetric with respect to the y-axis.

5. (continued)

Test for symmetry with respect to the origin.

$2x^4 + 3 = y^2$ (Original equation)

$2(-x)^4 + 3 = (-y)^2$ (Replacing x by -x and y by -y)

$2x^4 + 3 = y^2$ (Simplifying)

Since the resulting equation is equivalent to the original, the graph is symmetric with respect to the origin.

6. x-axis: Yes, y-axis: Yes, origin: Yes

7. Test for symmetry with respect to the x-axis.

$2y^2 = 5x^2 + 12$ (Original equation)

$2(-y)^2 = 5x^2 + 12$ (Replacing y by -y)

$2y^2 = 5x^2 + 12$ (Simplifying)

Since the resulting equation is equivalent to the original, the graph is symmetric with respect to the x-axis.

Test for symmetry with respect to the y-axis.

$2y^2 = 5x^2 + 12$ (Original equation)

$2y^2 = 5(-x)^2 + 12$ (Replacing x by -x)

$2y^2 = 5x^2 + 12$ (Simplifying)

Since the resulting equation is equivalent to the original, the graph is symmetric with respect to the y-axis.

Test for symmetry with respect to the origin.

$2y^2 = 5x^2 + 12$ (Original equation)

$2(-y)^2 = 5(-x)^2 + 12$ (Replacing x by -x and y by -y)

$2y^2 = 5x^2 + 12$ (Simplifying)

Since the resulting equation is equivalent to the original, the graph is symmetric with respect to the origin.

8. x-axis: Yes, y-axis: Yes, origin: Yes

9. Test for symmetry with respect to the x-axis.

$2x - 5 = 3y$ (Original equation)

$2x - 5 = 3(-y)$ (Replacing y by -y)

$2x - 5 = -3y$ (Simplifying)

Since the resulting equation is not equivalent to the original, the graph is not symmetric with respect to the x-axis.

Test for symmetry with respect to the y-axis.

$2x - 5 = 3y$ (Original equation)

$2(-x) - 5 = 3y$ (Replacing x by -x)

$-2x - 5 = 3y$ (Simplifying)

Since the resulting equation is not equivalent to the original, the graph is not symmetric with respect to the y-axis.

9. (continued)

Test for symmetry with respect to the origin.

$2x - 5 = 3y$ (Original equation)

$2(-x) - 5 = 3(-y)$ (Replacing x by -x and y by -y)

$-2x - 5 = -3y$ (Simplifying)

or

$2x + 5 = 3y$ (Multiplying by -1)

Since the resulting equation is not equivalent to the original, the graph is not symmetric with respect to the origin.

10. x-axis: No, y-axis: No, origin: No

11. Test for symmetry with respect to the a-axis.

$3b^3 = 4a^3 + 2$ (Original equation)

$3(-b)^3 = 4a^3 + 2$ (Replacing b by -b)

$-3b^3 = 4a^3 + 2$ (Simplifying)

Since the resulting equation is not equivalent to the original, the graph is not symmetric with respect to the a-axis.

Test for symmetry with respect to the b-axis.

$3b^3 = 4a^3 + 2$ (Original equation)

$3b^3 = 4(-a)^3 + 2$ (Replacing a by -a)

$3b^3 = -4a^3 + 2$ (Simplifying)

Since the resulting equation is not equivalent to the original, the graph is not symmetric with respect to the b-axis.

Test for symmetry with respect to the origin.

$3b^3 = 4a^3 + 2$ (Original equation)

$3(-b)^3 = 4(-a)^3 + 2$ (Replacing a by -a and b by -b)

$-3b^3 = -4a^3 + 2$ (Simplifying)

or

$3b^3 = 4a^3 - 2$ (Multiplying by -1)

Since the resulting equation is not equivalent to the original, the graph is not symmetric with respect to the origin.

12. p-axis: No, q-axis: No, origin: No

13. $3x^2 - 2y^2 = 3$ (Original equation)

$3(-x)^2 - 2(-y)^2 = 3$ (Replacing x by -x and y by -y)

$3x^2 - 2y^2 = 3$ (Simplifying)

Since the resulting equation is equivalent to the original equation, the graph is symmetric with respect to the origin.

14. Yes

15.
$$5x - 5y = 0 \quad \text{(Original equation)}$$
$$5(-x) - 5(-y) = 0 \quad \text{(Replacing x by -x and y by -y)}$$
$$-5x + 5y = 0 \quad \text{(Simplifying)}$$
$$5x - 5y = 0 \quad \text{(Multiplying by -1)}$$

Since the resulting equation is equivalent to the original equation, the graph is symmetric with respect to the origin.

16. Yes

17.
$$3x + 3y = 0 \quad \text{(Original equation)}$$
$$3(-x) + 3(-y) = 0 \quad \text{(Replacing x by -x and y by -y)}$$
$$-3x - 3y = 0 \quad \text{(Simplifying)}$$
$$3x + 3y = 0 \quad \text{(Multiplying by -1)}$$

Since the resulting equation is equivalent to the original equation, the graph is symmetric with respect to the origin.

18. Yes

19.
$$3x = \frac{5}{y} \quad \text{(Original equation)}$$
$$3(-x) = \frac{5}{-y} \quad \text{(Replacing x by -x and y by -y)}$$
$$-3x = -\frac{5}{y} \quad \text{(Simplifying)}$$
$$3x = \frac{5}{y} \quad \text{(Multiplying by -1)}$$

Since the resulting equation is equivalent to the original, the graph is symmetric with respect to the origin.

20. Yes

21.
$$y = |2x| \quad \text{(Original equation)}$$
$$-y = |2(-x)| \quad \text{(Replacing x by -x and y by -y)}$$
$$-y = |-2x|$$
$$-y = |2x|$$

Since the resulting equation is not equivalent to the original, the graph is not symmetric with respect to the origin.

22. No

23.
$$3a^2 + 4a = 2b \quad \text{(Original equation)}$$
$$3(-a)^2 + 4(-a) = 2(-b) \quad \text{(Replacing a by -a and b by -b)}$$
$$3a^2 - 4a = -2b$$
or
$$-3a^2 + 4a = 2b$$

Since the resulting equation is not equivalent to the original, the graph is not symmetric with respect to the origin.

24. No

25.
$$3x = 4y \quad \text{(Original equation)}$$
$$3(-x) = 4(-y) \quad \text{(Replacing x by -x and y by -y)}$$
$$-3x = -4y \quad \text{(Simplifying)}$$
$$3x = 4y \quad \text{(Multiplying by -1)}$$

Since the resulting equation is equivalent to the original, the graph is symmetric with respect to the origin.

26. Yes

27.
$$xy = 12 \quad \text{(Original equation)}$$
$$(-x)(-y) = 12 \quad \text{(Replacing x by -x and y by -y)}$$
$$xy = 12 \quad \text{(Simplifying)}$$

Since the resulting equation is equivalent to the original, the graph is symmetric with respect to the origin.

28. Yes

29. a) The graph is symmetric with respect to the y-axis. Thus the function is even.

Reflect the graph across the origin. Are the graphs the same? No.

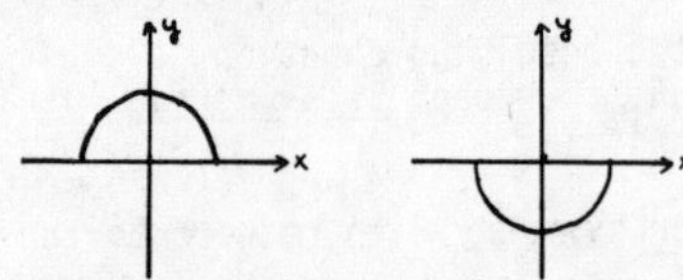

The graph is not symmetric with respect to the origin. Thus the function is not odd.

b) The graph is symmetric with respect to the y-axis. Thus the function is even.

Reflect the graph across the origin. Are the graphs the same? No.

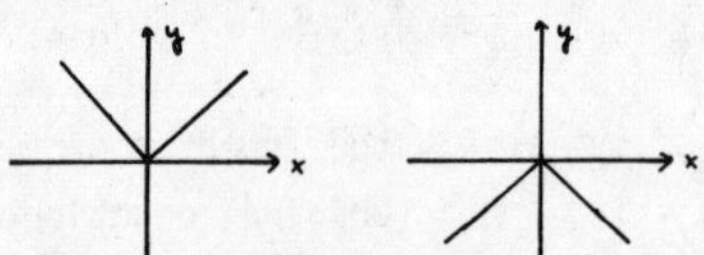

The graph is not symmetric with respect to the origin. Thus the function is not odd.

c) Reflect the graph across the y-axis. Are the graphs the same? No.

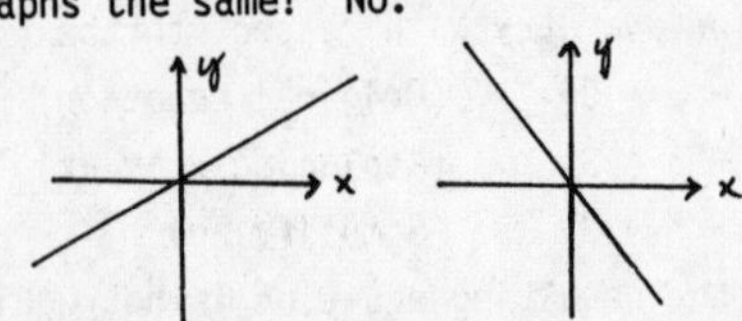

The graph is not symmetric with respect to the y-axis. Thus the function is not even.

29. (continued)

Reflect the graph across the origin. Are the graphs the same? Yes.

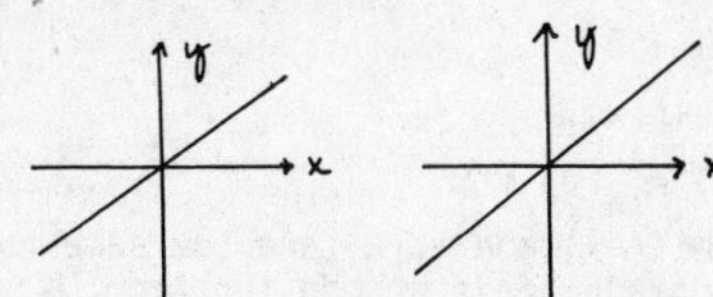

The graph is symmetric with respect to the origin. Thus the function is odd.

d) Reflect the graph across the y-axis. Are the graphs the same? No.

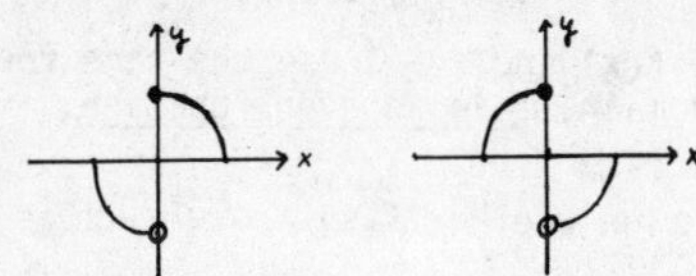

The graph is not symmetric with respect to the y-axis. Thus the function is not even.

Reflect the graph across the origin. Are the graphs the same? No.

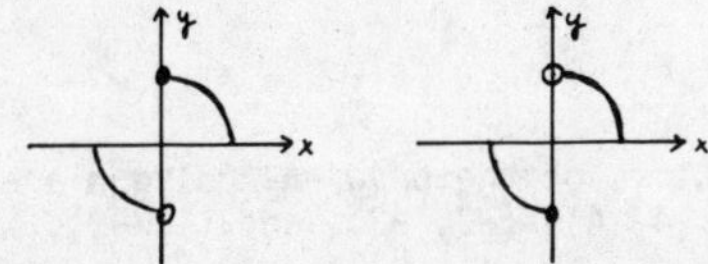

The graph is not symmetric with respect to the origin. Thus the function is not odd.

Therefore the function is neither even nor odd.

30. a) Odd, b) Even, c) Neither, d) Neither

31. Determine whether $f(x) = 2x^2 + 4x$ is even.

Find $f(-x)$.

$f(-x) = 2(-x)^2 + 4(-x) = 2x^2 - 4x$

Since $f(x)$ and $f(-x)$ are not the same for all x in the domain, f is not an even function.

Determine whether $f(x) = 2x^2 + 4x$ is odd.

Find $f(-x)$.

$f(-x) = 2(-x)^2 + 4(-x) = 2x^2 - 4x$

Find $-f(x)$.

$-f(x) = -(2x^2 + 4x) = -2x^2 - 4x$

Since $f(-x)$ and $-f(x)$ are not the same for all x in the domain, f is not an odd function.

Thus, $f(x) = 2x^2 + 4x$ is neither even nor odd.

32. Odd

33. Determine whether $f(x) = 3x^4 - 4x^2$ is even.

Find $f(-x)$.

$f(-x) = 3(-x)^4 - 4(-x)^2 = 3x^4 - 4x^2$

Since $f(x)$ and $f(-x)$ are the same for all x in the domain, f is an even function.

Determine whether $f(x) = 3x^4 - 4x^2$ is odd.

Find $f(-x)$.

$f(-x) = 3(-x)^4 - 4(-x)^2 = 3x^4 - 4x^2$

Find $-f(x)$.

$-f(x) = -(3x^4 - 4x^2) = -3x^4 + 4x^2$

Since $f(-x)$ and $-f(x)$ are not the same for all x in the domain, f is not an odd function.

34. Even

35. Determine whether $f(x) = 7x^3 + 4x - 2$ is even.

Find $f(-x)$.

$f(-x) = 7(-x)^3 + 4(-x) - 2 = -7x^3 - 4x - 2$

Since $f(x)$ and $f(-x)$ are not the same for all x in the domain, f is not an even function.

Determine whether $f(x) = 7x^3 + 4x - 2$ is odd.

Find $f(-x)$.

$f(-x) = 7(-x)^3 + 4(-x) - 2 = -7x^3 - 4x - 2$

Find $-f(x)$.

$-f(x) = -(7x^3 + 4x - 2) = -7x^3 - 4x + 2$

Since $f(-x)$ and $-f(x)$ are not the same for all x in the domain, f is not an odd function.

Thus, $f(x) = 7x^3 + 4x - 2$ is neither even nor odd.

36. Odd

37. Determine whether $f(x) = |3x|$ is even.

Find $f(-x)$.

$f(-x) = |3(-x)| = |-3x| = |3x|$

Since $f(x)$ and $f(-x)$ are the same for all x in the domain, f is an even function.

Determine whether $f(x) = |3x|$ is odd.

Find $f(-x)$.

$f(-x) = |3(-x)| = |-3x| = |3x|$

Find $-f(x)$.

$-f(x) = -|3x|$

Since $f(-x)$ and $-f(x)$ are not the same for all x in the domain, f is not an odd function.

38. Even

39. Determine whether $f(x) = x^{17}$ is even.

Find $f(-x)$.

$f(-x) = (-x)^{17} = -x^{17}$

Find $f(x)$ and $f(-x)$ are not the same for all x in the domain, f is not an even function.

Determine whether $f(x) = x^{17}$ is odd.

Find $f(-x)$.

$f(-x) = (-x)^{17} = -x^{17}$

Find $-f(x)$.

$-f(x) = -(x^{17}) = -x^{17}$

Since $f(-x)$ and $-f(x)$ are the same for all x in the domain, f is an odd function.

40. Odd

41. Determine whether $f(x) = x - |x|$ is even.

Find $f(-x)$.

$f(-x) = (-x) - |(-x)| = -x - |x|$

Since $f(x)$ and $f(-x)$ are not the same for all x in the domain, f is not an even function.

Determine whether $f(x) = x - |x|$ is odd.

Find $f(-x)$.

$f(-x) = (-x) - |(-x)| = -x - |x|$

Find $-f(x)$.

$-f(x) = -(x - |x|) = -x + |x|$

Since $f(-x)$ and $-f(x)$ are not the same for all x in the domain, f is not an odd function.

Therefore, $f(x) = x - |x|$ is neither odd nor even.

42. Neither

43. Determine whether $f(x) = \sqrt[3]{x}$ is even.

Find $f(-x)$.

$f(-x) = \sqrt[3]{-x} = -\sqrt[3]{x}$

Since $f(x)$ and $f(-x)$ are not the same for all x in the domain, f is not an even function.

Determine whether $f(x) = \sqrt[3]{x}$ is odd.

Find $f(-x)$.

$f(-x) = \sqrt[3]{-x} = -\sqrt[3]{x}$

Find $-f(x)$.

$-f(x) = -(\sqrt[3]{x}) = -\sqrt[3]{x}$

Since $f(-x)$ and $-f(x)$ are the same for all x in the domain, f is an odd function.

44. Even

45. Determine whether $f(x) = 0$ is even.

Find $f(-x)$.

$f(-x) = 0$

Since $f(x)$ and $f(-x)$ are the same for all x in the domain, f is an even function.

45. (continued)

Determine whether $f(x) = 0$ is odd.

Find $f(-x)$.

$f(-x) = 0$

Find $-f(x)$.

$-f(x) = -(0) = 0$

Since $f(-x)$ and $-f(x)$ are the same for all x in the domain, f is an odd function.

46. Neither

47. Determine whether $f(x) = \sqrt{x^2 + 1}$ is even.

Find $f(-x)$.

$f(-x) = \sqrt{(-x)^2 + 1} = \sqrt{x^2 + 1}$

Since $f(x)$ and $f(-x)$ are the same for all x in the domain, f is an even function.

Determine whether $f(x) = \sqrt{x^2 + 1}$ is odd.

Find $f(-x)$.

$f(-x) = \sqrt{(-x)^2 + 1} = \sqrt{x^2 + 1}$

Find $-f(x)$.

$-f(x) = -\sqrt{x^2 + 1}$

Since $f(-x)$ and $-f(x)$ are not the same for all x in the domain, f is not an odd function.

48. Odd

49. The vertices of the original polygon are (0, 4), (4, 4), (-2, -2), and (1, -2).

When the graph is reflected across the x-axis the vertices will be (0, -4), (4, -4), (-2, 2), and (1, 2).

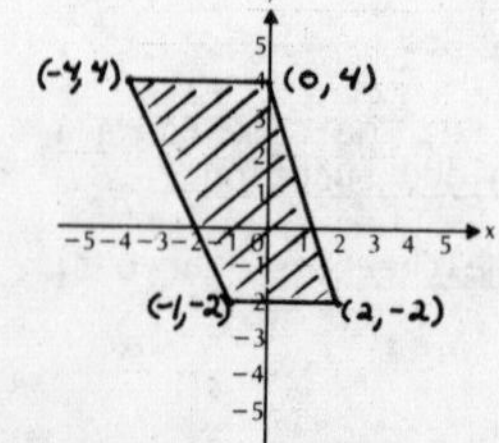

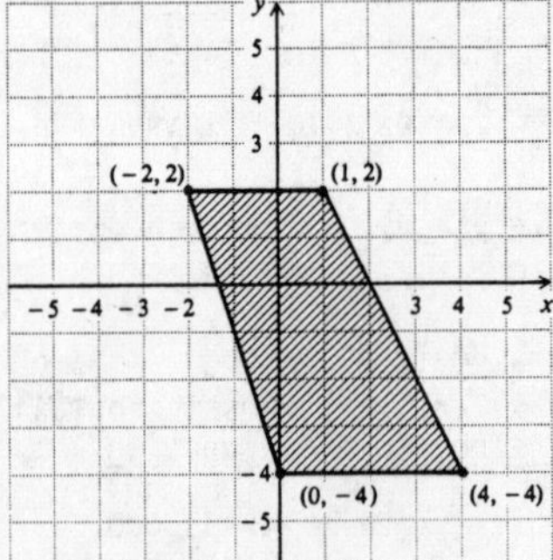

50.

(-4,4) (0,4)

(-1,-2) (2,-2)

51. The vertices of the original polygon are (0, 4), (4, 4), (1, -2), and (-2, -2).

    When the graph is reflected across the line y = x, the vertices will be (4, 0), (4, 4), (-2, 1), and (-2, -2).

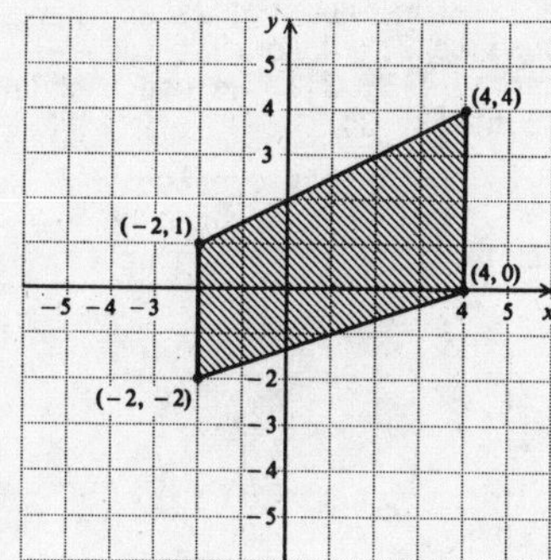

52.

53. See the answer section in the text.

Exercise Set 3.6

---

1. The endpoints 0 and 5 are not included in the interval, so we use parentheses. Interval notation is (0,5).

2. [-1,2]

3. The endpoint -9 is included in the interval, so we use a bracket before the -9. The endpoint -4 is not included, so we use a parenthesis after the -4. Interval notation is [-9,4).

4. (-9,-5]

5. Both endpoints are included in the interval, so we use brackets. Interval notation is [x, x + h].

6. (x, x + h]

7. The endpoint p is not included in the interval, so we use a parenthesis before the p. The interval is of unlimited extent in the positive direction, so we use the infinity symbol ∞. Interval notation is (p,∞).

8. (-∞,q]

9. This is a closed interval, so we use brackets. Interval notation is [-3,3].

10. (-4,4)

11. This is a half-open interval. We use a bracket on the left and a parenthesis on the right. Interval notation is [-14,-11).

12. (6,20]

13. This interval is of unlimited extent in the negative direction, and the endpoint -4 is included. Interval notation is (-∞,-4].

14. (-5,∞)

15. This interval is of unlimited extent in the negative direction, and the endpoint 3.8 is not included. Interval notation is (-∞, 3.8).

16. $[\sqrt{3},\infty)$

17. a) No, b) Yes, c) Yes, d) No

18. a) Yes, b) No, c) No, d) Yes

19. The length of the shortest recurring interval is 4. Thus, the period is 4.

20. 2

21. a) Yes, b) Yes, c) No, d) Yes, e) Yes

22. a) Yes, b) Yes, c) No, d) No, e) Yes

23. The function has discontinuities where x = -3 and x = 2.

24. x = 0

25. a) Increasing, b) Neither
    c) Decreasing, d) Neither

26. a) Neither, b) Decreasing
    c) Increasing, d) Neither

27. We set D(x) = S(x) and solve.

$$-2x + 8 = x + 2$$
$$6 = 3x$$
$$2 = x$$

Thus, $x_E = 2$ (units). To find p_E, we substitute x_E into either D(x) or S(x). We use S(x).

$$p_E = S(x_E) = 2 + 2 = \$4$$

The equilibrium point is (2, $4).

28. (6, $5)

29. We set D(x) = S(x) and solve.

$$(x - 3)^2 = x^2 + 2x + 1$$
$$x^2 - 6x + 9 = x^2 + 2x + 1$$
$$8 = 8x$$
$$1 = x$$

Thus, $x_E = 1$ (unit). To find p_E, we substitute x_E into either D(x) or S(x). We use D(x).

$$p_E = D(x_E) = (1 - 3)^2 = (-2)^2 = \$4$$

The equilibrium point is (1, $4).

30. (1, $9)

31. We set D(x) = S(x) and solve.

$$(x - 4)^2 = x^2$$
$$x^2 - 8x + 16 = x^2$$
$$16 = 8x$$
$$2 = x$$

Thus $x_E = 2$ (units). To find p_E, we substitute x_E into either D(x) or S(x). We use S(x).

$$p_E = S(x_E) = 2^2 = \$4$$

The equilibrium point is (2, $4).

32. (3, $9)

33. Graph $f(x) = \begin{cases} 1 & \text{for } x < 0 \\ -1 & \text{for } x \geqslant 0. \end{cases}$

We graph f(x) = 1 for inputs less than 0. Note that f(x) = 1 only for numbers less than 0. We use an open circle at the point (0, 1).

We graph f(x) = -1 for inputs greater than or equal to 0. Note that f(x) = -1 only for numbers greater than or equal to 0. We use a solid circle at the point (0, -1).

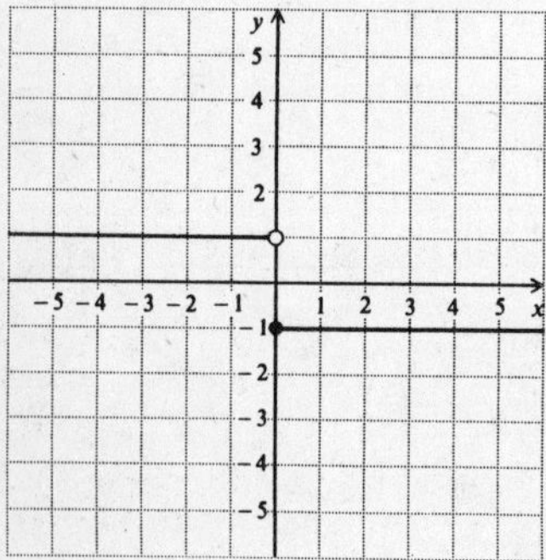

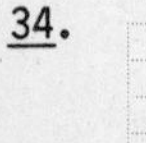

34.

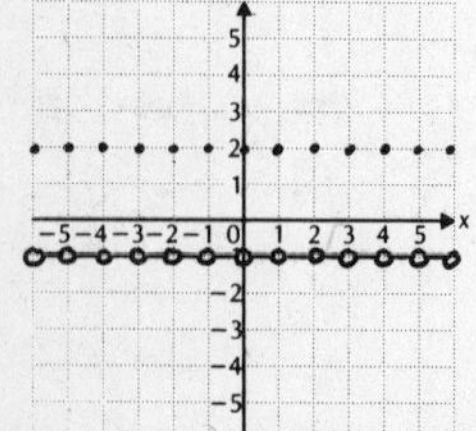

35. $f(x) = \begin{cases} 3 & \text{for } x \leqslant -3 \\ |x| & \text{for } -3 < x \leqslant 3 \\ -3 & \text{for } x > 3 \end{cases}$

We graph f(x) = 3 for inputs less than or equal to -3. Note that f(x) = 3 only for numbers less than or equal to -3.

We graph f(x) = |x| for inputs greater than -3 and less than or equal to 3. Note that f(x) = |x| only on the interval (-3, 3]. We use a solid circle at the point (3, 3).

We graph f(x) = -3 for inputs greater than 3. Note that f(x) = -3 only for numbers greater than 3. We use an open circle at the point (3, -3).

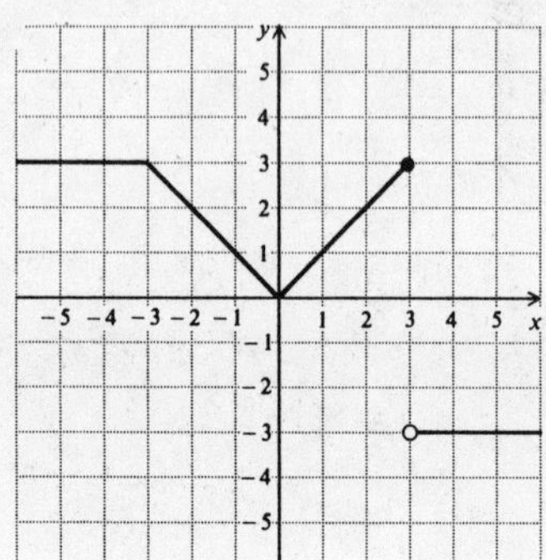

36.

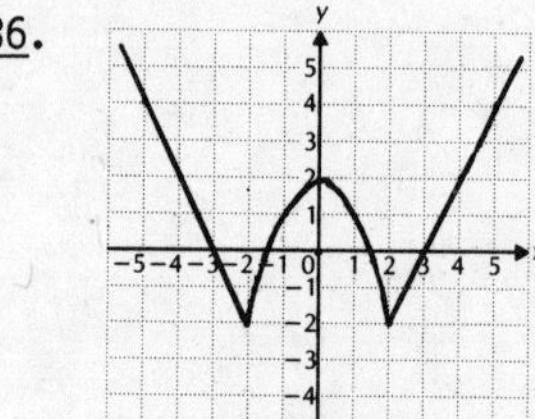

37. $f(x) = \begin{cases} \dfrac{x^2 - 1}{x - 1}, & \text{for } x \neq 1 \\ -2, & \text{for } x = 1 \end{cases}$

When $x \neq 1$, the denominator of $\frac{x^2 - 1}{x - 1}$ is nonzero, so we can simplify.

$$\frac{x^2 - 1}{x - 1} = \frac{(x + 1)(x - 1)}{x - 1} = x + 1$$

Thus, f(x) = x + 1 for $x \neq 1$. The graph of this part of the function consists of a line with a hole at the point (1,2). At x = 1, f(1) = -2, so the point (1,-2) is plotted below (1,2).

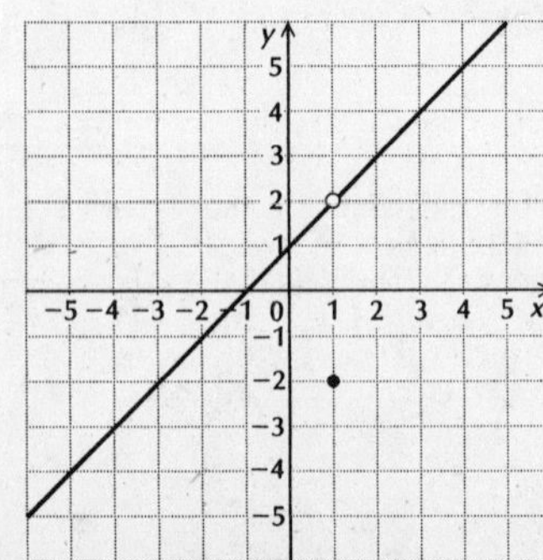

38.

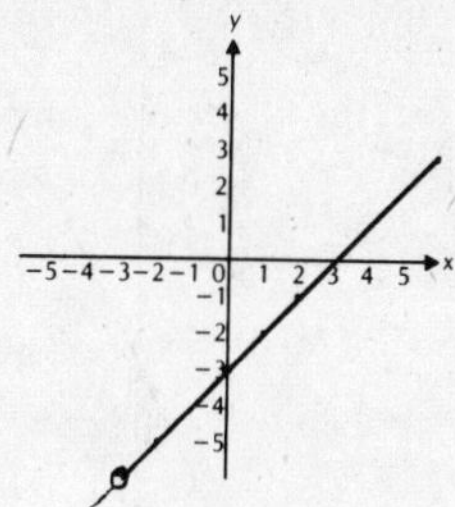

39.

$$p(x) = \begin{cases} 25¢ \text{ if } 0 < x \leqslant 1 \\ 45¢ \text{ if } 1 < x \leqslant 2 \\ 65¢ \text{ if } 2 < x \leqslant 3 \\ 85¢ \text{ if } 3 < x \leqslant 4 \\ \$1.05 \text{ if } 4 < x \leqslant 5 \\ \$1.25 \text{ if } 5 < x \leqslant 6 \\ \$1.45 \text{ if } 6 < x \leqslant 7 \\ \$1.65 \text{ if } 7 < x \leqslant 8 \\ \$1.85 \text{ if } 8 < x \leqslant 9 \\ \$2.05 \text{ if } 9 < x \leqslant 10 \\ \$2.25 \text{ if } 10 < x \leqslant 11 \\ \$2.45 \text{ if } 11 < x \leqslant 12 \end{cases}$$

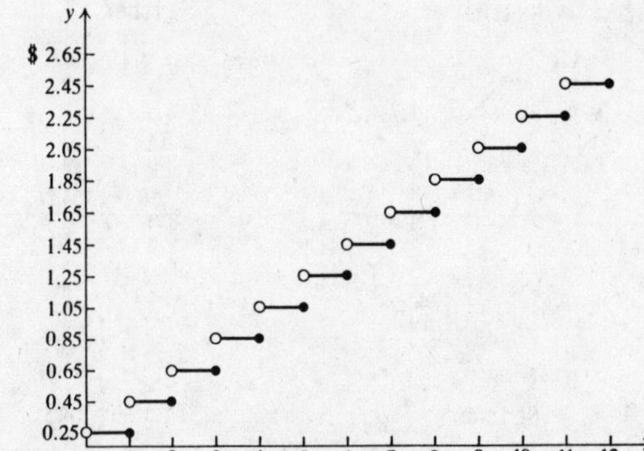

40.

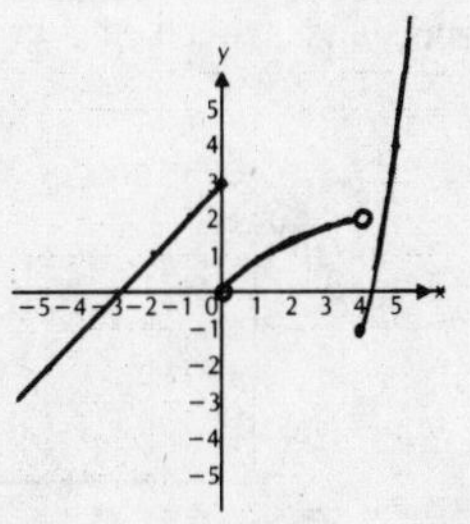

41. f(x) = INT(x - 2)

We can define this function by a piecewise function with an infinite number of statements.

$$f(x) = INT(x - 2) = \begin{cases} \cdot \\ \cdot \\ \cdot \\ -4 \text{ if } -2 \leqslant x < -1 \\ -3 \text{ if } -1 \leqslant x < 0 \\ -2 \text{ if } 0 \leqslant x < 1 \\ -1 \text{ if } 1 \leqslant x < 2 \\ 0 \text{ if } 2 \leqslant x < 3 \\ 1 \text{ if } 3 \leqslant x < 4 \\ \cdot \\ \cdot \\ \cdot \end{cases}$$

We graph the function.

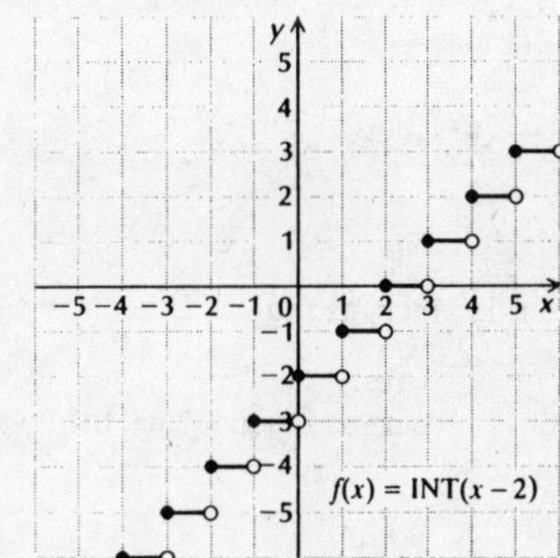

42.

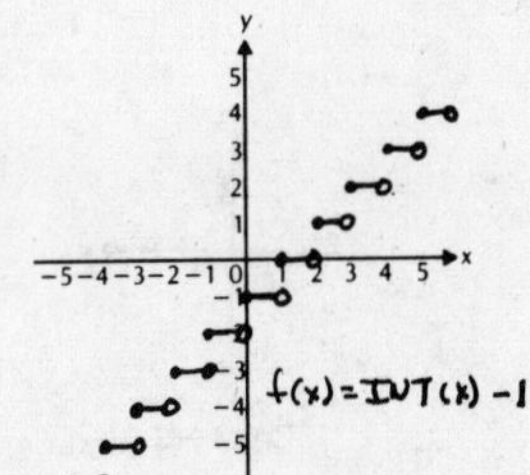

43. f(x) = INT(x) + 2

We can define the function by a piecewise function with an infinite number of statements.

$$f(x) = INT(x) + 2 = \begin{cases} \cdot \\ \cdot \\ \cdot \\ -1 \text{ if } -3 \leqslant x < -2 \\ 0 \text{ if } -2 \leqslant x < -1 \\ 1 \text{ if } -1 \leqslant x < 0 \\ 2 \text{ if } 0 \leqslant x < 1 \\ \cdot \\ \cdot \\ \cdot \end{cases}$$

43. (continued)

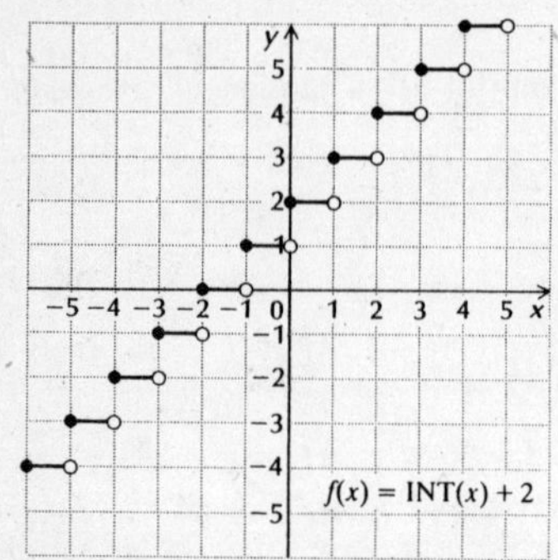

44.

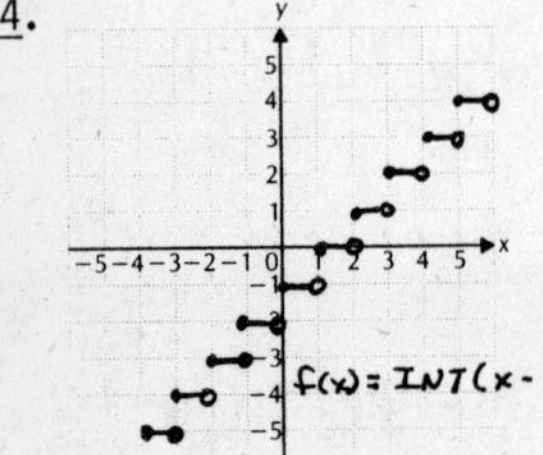

45. a) The function is increasing on the interval [0, 1].

b) The function is decreasing on the interval [-1, 0].

46. a) Increasing: [1, 3]

b) Decreasing: [-3, 1]

47. a)

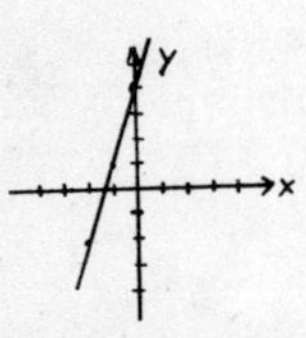
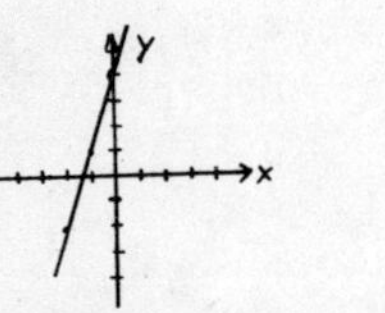
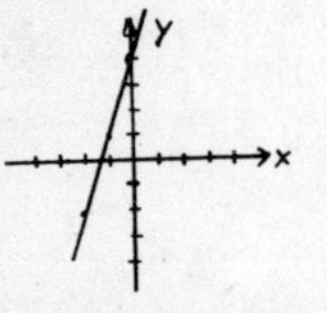

Increasing

b)

Decreasing

c)

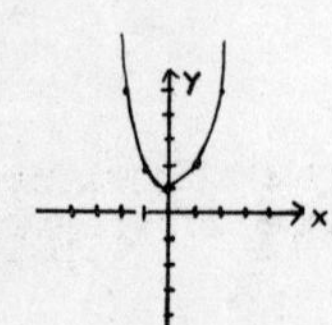

Neither

d)

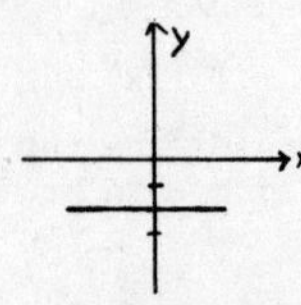

Neither

e)

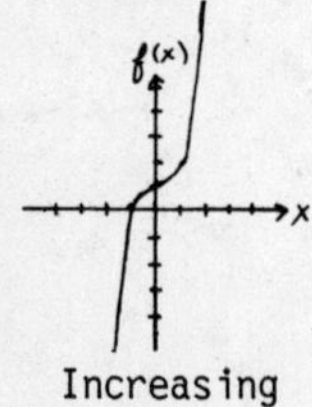

Increasing

f)

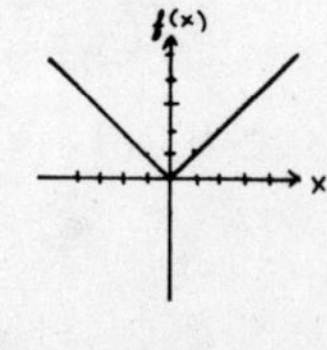

Neither

48. a)

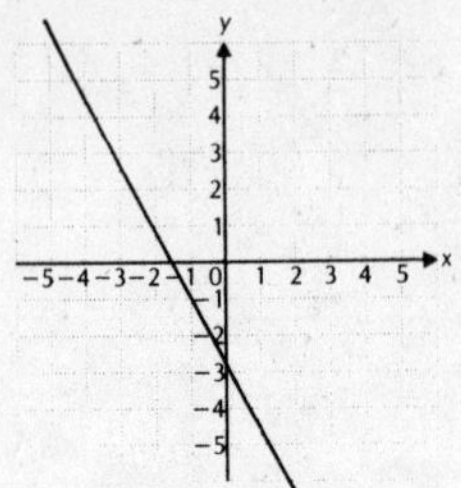

Decreasing

b)

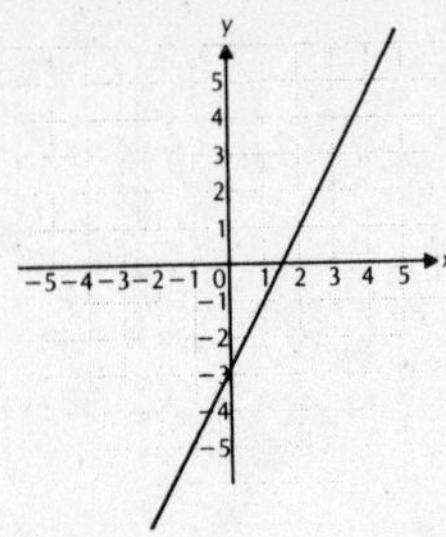

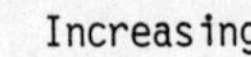

Increasing

c)

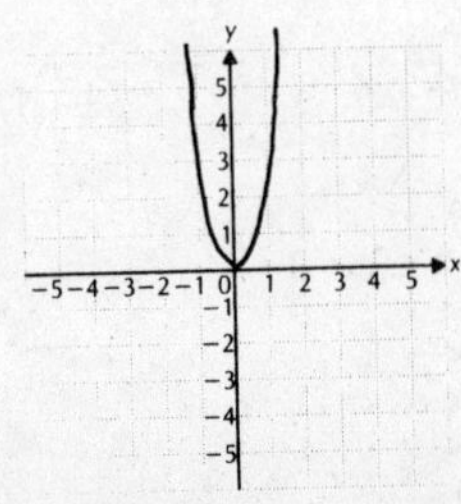

Neither

d)

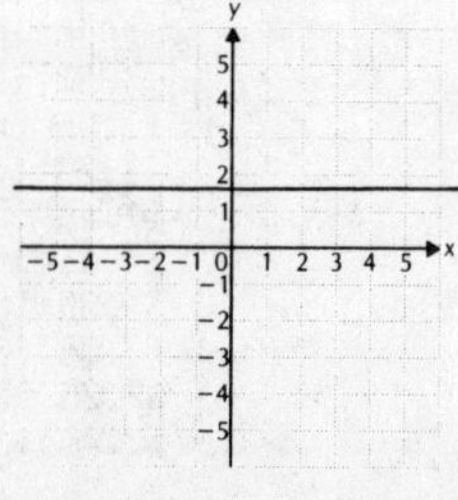

Neither

e)

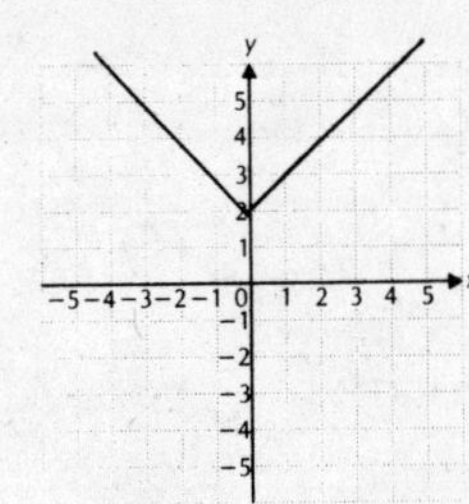

Neither

f)

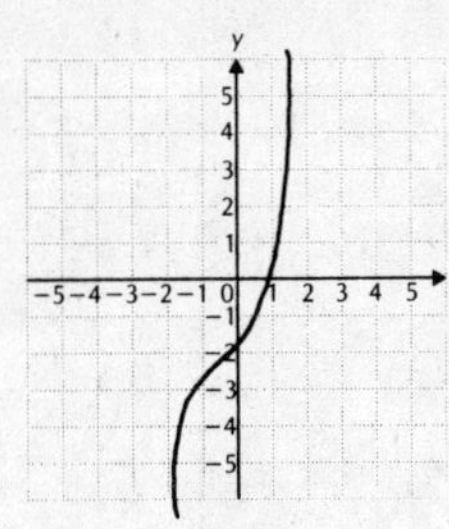

Increasing

49. $f(x) = \frac{|x|}{x}$

We compute some function values, plot these points, look for a pattern, and sketch the graph. Note that x cannot be zero.

x	f(x)
-3	-1
-2	-1
-1	-1
1	1
2	1
3	1

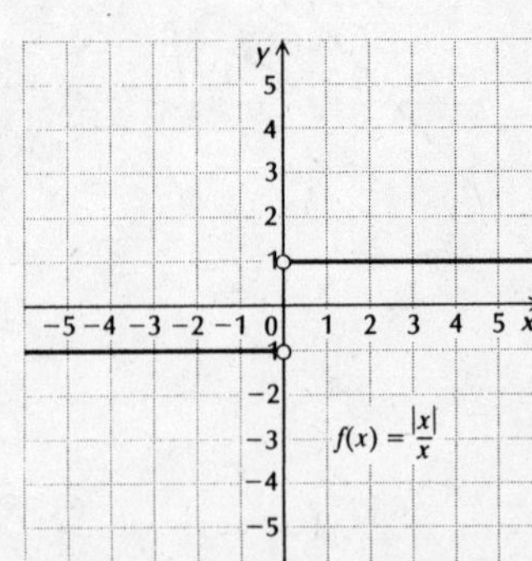

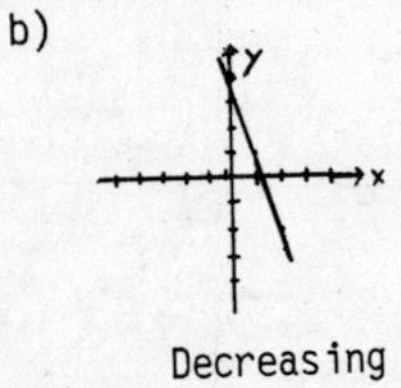

50.

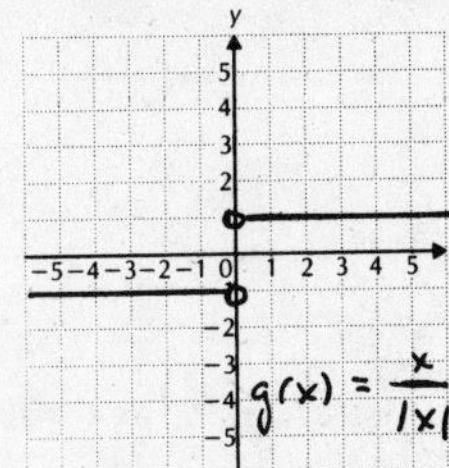

51. INT(x) = 4 for $4 \leqslant x < 5$, so the possible inputs are $\{x \mid 4 \leqslant x < 5\}$.

52. $\{x \mid -3 \leqslant x < 2\}$

53. a) $\text{INT}\left[\frac{547}{3}\right] = \text{INT}\left[182\frac{1}{3}\right] = 182$

The bowling average is 182.

b) $\text{INT}\left[\frac{4621}{27}\right] = \text{INT}\left[171\frac{4}{27}\right] = 171$

The bowling average is 171.

54. a) a) 0, 1, 3, 6

b) Domain: The set of all real numbers

Range: {0,1,2,3,4,5,6,7,8,9}

c) 22

55.

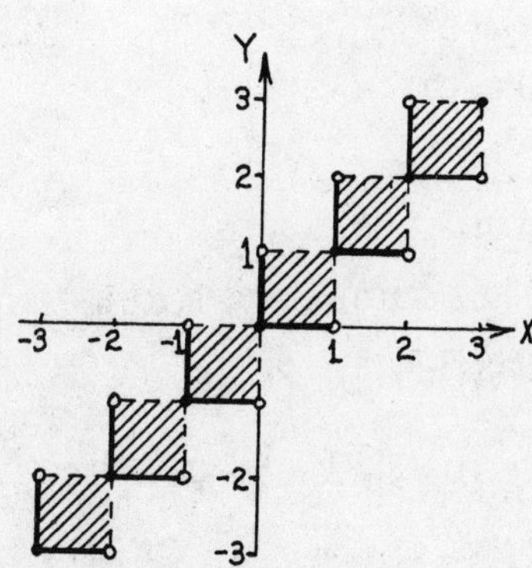

This is not the graph of a function, because it fails the vertical line test.

Exercise Set 3.7

1. $f(x) = x - 3$, $g(x) = x + 4$

a) $(f + g)(x) = f(x) + g(x) = (x - 3) + (x + 4) = 2x + 1$

$(f - g)(x) = f(x) - g(x) = (x - 3) - (x + 4) = -7$

$fg(x) = f(x)g(x) = (x - 3)(x + 4) = x^2 + x - 12$

$ff(x) = [f(x)]^2 = (x - 3)^2 = x^2 - 6x + 9$

$(f/g)(x) = \frac{f(x)}{g(x)} = \frac{x - 3}{x + 4}$

$(g/f)(x) = \frac{g(x)}{f(x)} = \frac{x + 4}{x - 3}$

$f \circ g(x) = f(g(x)) = f(x + 4) = (x + 4) - 3 = x + 1$

$g \circ f(x) = g(f(x)) = g(x - 3) = (x - 3) + 4 = x + 1$

b) The domain of f is the set of all real numbers. The domain of g is the set of all real numbers. Then the domain of f + g, f - g, fg, and ff is the set of all real numbers. Since $g(-4) = 0$, the domain of f/g is $\{x \mid x \text{ is a real number and } x \neq -4\}$. Since $f(3) = 0$, the domain of g/f is $\{x \mid x \text{ is a real number and } x \neq 3\}$. The domain of $f \circ g$ and of $g \circ f$ is the set of all real numbers.

2. a) $(f + g)(x) = x^2 + 2x + 4$

$(f - g)(x) = x^2 - 2x - 6$

$fg(x) = 2x^3 + 5x^2 - 2x - 5$

$ff(x) = x^4 - 2x^2 + 1$

$(f/g)(x) = \frac{x^2 - 1}{2x + 5}$

$(g/f)(x) = \frac{2x + 5}{x^2 - 1}$

$f \circ g = 4x^2 + 20x + 24$

$g \circ f(x) = 2x^2 + 3$

b) All reals, all reals, all reals, all reals, all reals, all reals, all reals except $-\frac{5}{2}$, all reals except 1 and -1, all reals, all reals

3. $f(x) = x^3$, $g(x) = 2x^2 + 9x - 3$

a) $(f + g)(x) = f(x) + g(x) = x^3 + 2x^2 + 9x - 3$
$(f - g)(x) = f(x) - g(x) = x^3 - 2x^2 - 9x + 3$
$fg(x) = f(x)g(x) = x^3(2x^2 + 9x - 3) =$
$2x^5 + 9x^4 - 3x^3$
$ff(x) = [f(x)]^2 = (x^3)^2 = x^6$
$(f/g)(x) = \frac{f(x)}{g(x)} = \frac{x^3}{2x^2 + 9x - 3}$
$(g/f)(x) = \frac{2x^2 + 9x - 3}{x^3}$
$f \circ g(x) = f(g(x)) = f(2x^2 + 9x - 3) =$
$(2x^2 + 9x - 3)^3$
$g \circ f(x) = g(f(x)) = g(x^3) = 2x^6 + 9x^3 - 3$

b) The domain of f is the set of all real numbers. The domain of g is the set of all real numbers. Then the domain of f + g, f - g, fg, and ff is the set of all real numbers. Since g(x) = 0 when $x = \frac{-9 \pm \sqrt{105}}{4}$, the domain of f/g is $\left\{x \mid x \text{ is a real number and } x \neq \frac{-9 \pm \sqrt{105}}{4}\right\}$. Since f(0) = 0, the domain of g/f is $\{x \mid x$ is a real number and $x \neq 0\}$. The domain of $f \circ g$ and of $g \circ f$ is the set of all real numbers.

4. a) $(f + g)(x) = x^2 + \sqrt{x}$, $(f - g)(x) = x^2 - \sqrt{x}$, $fg(x) = x^2\sqrt{x}$, $ff(x) = x^4$, $(f/g)(x) = x\sqrt{x}$, $(g/f)(x) = \frac{\sqrt{x}}{x^2}$, $f \circ g(x) = x$, $g \circ f(x) = |x|$

b) All reals, $\{x \mid x \geq 0\}$, $\{x \mid x \geq 0\}$, $\{x \mid x \geq 0\}$, $\{x \mid x \geq 0\}$, all reals, $\{x \mid x > 0\}$, $\{x \mid x > 0\}$, $\{x \mid x \geq 0\}$, all reals

In Exercises 5 - 20, $f(x) = x^2 - 4$ and $g(x) = 2x + 5$.

5. $(f - g)(3) = f(3) - g(3) = (3^2 - 4) - (2\cdot3 + 5) =$
$5 - 11 = -6$

6. 0

7. $(f - g)(x) = f(x) - g(x) = (x^2 - 4) - (2x + 5) =$
$x^2 - 2x - 9$

8. $x^2 + 2x + 1$

9. $fg(3) = f(3)g(3) = (3^2 - 4)(2\cdot3 + 5) = 5\cdot11 = 55$

10. -1

11. $(g/f)(-2) = \frac{g(-2)}{f(-2)} = \frac{2(-2) + 5}{(-2)^2 - 4} = \frac{-4 + 5}{4 - 4} = \frac{1}{0}$, which does not exist

12. Does not exist

13. $fg(x) = f(x)g(x) = (x^2 - 4)(2x + 5) =$
$2x^3 + 5x^2 - 8x - 20$

14. $\frac{x^2 - 4}{2x + 5}$

15. $(g/f)(x) = \frac{g(x)}{f(x)} = \frac{2x + 5}{x^2 - 4}$

16. $x^4 - 8x^2 + 16$

17. $f \circ g(x) = f(g(x)) = f(2x + 5) = (2x + 5)^2 - 4 =$
$4x^2 + 20x + 25 - 4 = 4x^2 + 20x + 21$

18. $x^4 - 8x + 12$

19. $g \circ g(x) = g(g(x)) = g(2x + 5) = 2(2x + 5) + 5 =$
$4x + 10 + 5 = 4x + 15$

20. $2x^2 - 3$

21. $R(x) = 60x - 0.4x^2$, $C(x) = 3x + 13$

a) $P(x) = R(x) - C(x)$
$= (60x - 0.4x^2) - (3x + 13)$
$= -0.4x^2 + 57x - 13$

b) $R(20) = 60(20) - 0.4(20)^2$
$= 1200 - 0.4(400) = 1200 - 160$
$= 1040$

$C(20) = 3(20) + 13 = 60 + 13$
$= 73$

$P(20) = R(20) - C(20)$
$= 1040 - 73$
$= 967$

22. a) $P(x) = -0.001x^2 + 13.8x - 60$

b) $R(100) = 1500$, $C(100) = 190$, $P(100) = 1310$

23. $f \circ g(x) = f(g(x)) = f\left(\frac{5}{4}x\right) = \frac{4}{5} \cdot \frac{5}{4}x = x$

$g \circ f(x) = g(f(x)) = g\left(\frac{4}{5}x\right) = \frac{5}{4} \cdot \frac{4}{5}x = x$

24. $f \circ g(x) = x$, $g \circ f(x) = x$

25. $f \circ g(x) = f(g(x)) = f\left(\frac{x + 7}{3}\right) = 3\left(\frac{x + 7}{3}\right) - 7 =$
$x + 7 - 7 = x$

$g \circ f(x) = g(f(x)) = g(3x - 7) = \frac{(3x - 7) + 7}{3} =$
$\frac{3x}{3} = x$

26. $f \circ g(x) = x$, $g \circ f(x) = x$

27. $f \circ g(x) = f(g(x)) = f(\sqrt[3]{x + 1}) = (\sqrt[3]{x + 1})^3 - 1 =$
$x + 1 - 1 = x$

$g \circ f(x) = g(f(x)) = g(x^3 - 1) = \sqrt[3]{(x^3 - 1) + 1} =$
$\sqrt[3]{x^3} = x$

28. $f \circ g(x) = x$, $g \circ f(x) = x$

29. $f \circ g(x) = f(g(x)) = f(x^2 - 5) = \sqrt{x^2 - 5 + 5} = \sqrt{x^2} = |x|$

$g \circ f(x) = g(f(x)) = g(\sqrt{x + 5}) = (\sqrt{x + 5})^2 - 5 = x + 5 - 5 = x$

30. $f \circ g(x) = x$, $g \circ f(x) = |x|$

31. $f \circ g(x) = f(g(x)) = f\left(\frac{1}{1 + x}\right) = \frac{1 - \left(\frac{1}{1 + x}\right)}{\frac{1}{1 + x}} =$

$\frac{\frac{1 + x - 1}{1 + x}}{\frac{1}{1 + x}} = \frac{x}{1 + x} \cdot \frac{1 + x}{1} = x$

$g \circ f(x) = g(f(x)) = g\left(\frac{1 - x}{x}\right) = \frac{1}{1 + \left(\frac{1 - x}{x}\right)} =$

$\frac{1}{\frac{x + 1 - x}{x}} = \frac{1}{\frac{1}{x}} = 1 \cdot \frac{x}{1} = x$

32. $f \circ g(x) = \frac{-8x^2 - 2x + 6}{17x^2 - 22x + 10}$, $g \circ f(x) = \frac{-x^2 - 7}{3x^2 - 7}$

33. $f \circ g(x) = f(g(x)) = f(12) = -6$

$g \circ f(x) = g(f(x)) = g(-6) = 12$

34. $f \circ g(x) = 20$, $g \circ f(x) = 5$

35. $h(x) = (4 - 3x)^5$

This is 4 - 3x to the 5th power. The most obvious answer is $f(x) = x^5$ and $g(x) = 4 - 3x$.

36. $f(x) = \sqrt[3]{x}$, $g(x) = x^2 - 8$

37. $h(x) = \frac{1}{(x - 1)^4}$

This is 1 divided by (x - 1) to the 4th power. One obvious answer is $f(x) = \frac{1}{x^4}$ and $g(x) = (x - 1)$.

38. $f(x) = \frac{1}{\sqrt{x}}$, $g(x) = 3x + 7$

39. $f(x) = \frac{x - 1}{x + 1}$, $g(x) = x^3$

40. $f(x) = |x|$, $g(x) = 9x^2 - 4$

41. $f(x) = x^6$, $g(x) = \frac{2 + x^3}{2 - x^3}$

42. $f(x) = x^4$, $g(x) = \sqrt{x} - 3$

43. $f(x) = \sqrt{x}$, $g(x) = \frac{x - 5}{x + 2}$

44. $f(x) = \sqrt{1 + x}$, $g(x) = \sqrt{1 + x}$

45. $f(x) = x^5 + x^4 + x^3$, $g(x) = x + 3$

46. $f(x) = 4x^{2/3} + 5$, $g(x) = x - 1$

47. a) Recall that distance = rate × time. Then

$a(t) = 250t$.

b) The plane's distance from the control tower is the sum of 300 ft and the distance a the plane has traveled down the runway. Thus,

$P(a) = 300 + a$.

c) $(P \circ a)(t) = P(a(t)) = 300 + 250t$

This function gives the plane's distance from the control tower in terms of the time t the plane travels.

48. a) $r(t) = 3t$

b) $A(r) = \pi r^2$

c) $(A \circ r)(t) = \pi(3t)^2$, or $9\pi t^2$

49. First we graph both functions on the same axes.

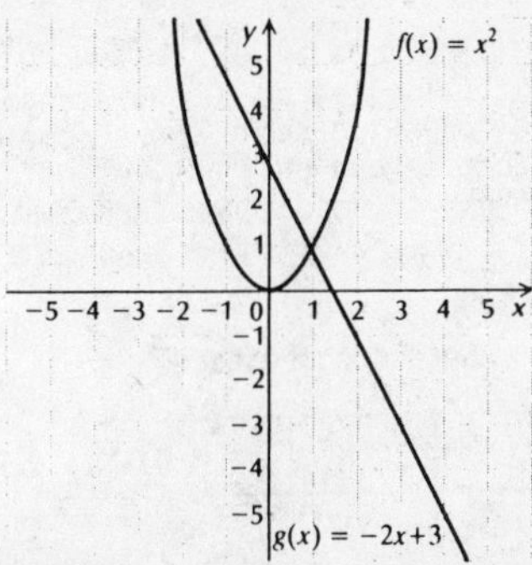

Then we graph $y = (f + g)(x)$ by adding second coordinates. The following table shows some examples.

First coordinate	Second coordinates		
x	f(x)	g(x)	(f + g)(x)
-1	1	5	6
0	0	3	3
1	1	1	2
2	4	-1	3
3	9	-3	6

We sketch the graph of $y = (f + g)(x)$.

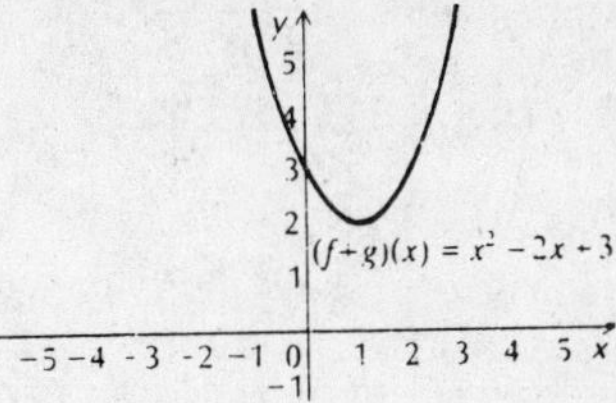

50.

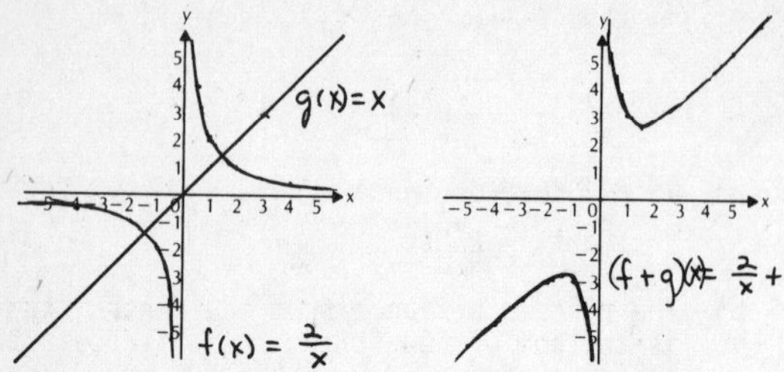

52.

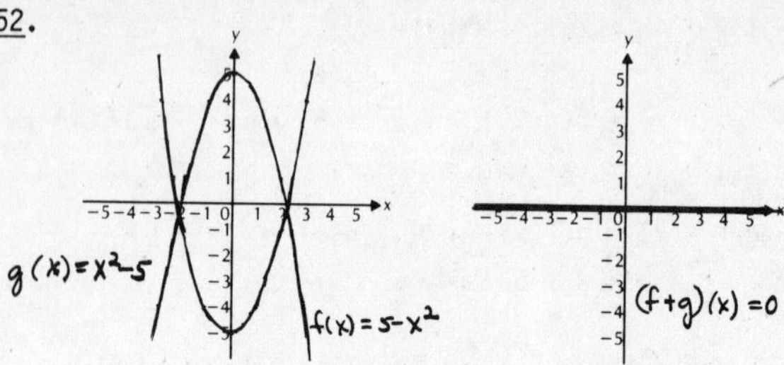

51. First we graph both functions on the same axes.

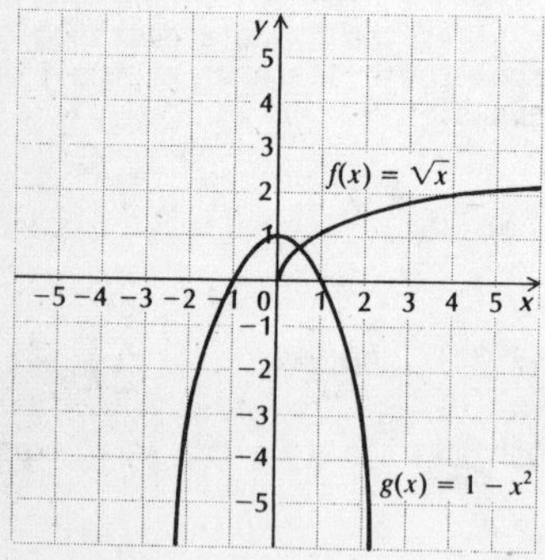

Then we graph y = (f + g)(x) by adding second coordinates. The following table shows some examples. Note that we can only add second coordinates for first coordinates x such that $x \geqslant 0$.

First coordinate	Second coordinates		
x	f(x)	g(x)	(f + g)(x)
0	0	1	1
0.5	0.707	0.75	1.457
1	1	0	1
1.5	1.225	-1.25	-0.025
2	1.414	-3	-1.586
3	1.732	-8	-6.268

We sketch the graph of y = (f + g)(x).

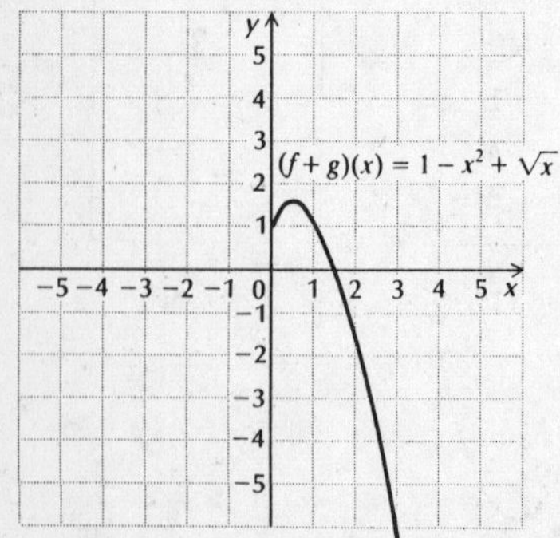

53. $f \circ g(x) = f(g(x)) = f(2x - 1) = 3(2x - 1) + b =$
$6x - 3 + b$

$g \circ f(x) = g(f(x)) = g(3x + b) = 2(3x + b) - 1 =$
$6x + 2b - 1$

Now $f \circ g(x) = g \circ f(x)$ means

$6x - 3 + b = 6x + 2b - 1$, or

$-2 = b$

54. $b = \frac{4}{3}$, $m = \frac{1}{3}$

55. $f \circ f(x) = f(f(x)) = \dfrac{1}{1 - \dfrac{1}{1 - x}} = \dfrac{1}{\dfrac{1 - x - 1}{1 - x}} =$

$\dfrac{1}{\dfrac{-x}{1 - x}} = \dfrac{1 - x}{-x}$, or $\dfrac{x - 1}{x}$

$f \circ f \circ f(x) = f[f(f(x))] = \dfrac{1}{1 - \dfrac{x - 1}{x}} =$

$\dfrac{1}{\dfrac{x - (x - 1)}{x}} = \dfrac{1}{\dfrac{1}{x}} = x$

56. Let f(x) and g(x) be even functions. Then by definition f(x) = f(-x) and g(x) = g(-x). Now $f \circ g(x) = f(g(x)) = f(g(-x)) = f \circ g(-x)$ and $g \circ f(x) = g(f(x)) = g(f(-x)) = g \circ f(-x)$. Thus the composition of two even functions is even.

57. See the answer section in the text.

58. Let f(x) and g(x) be odd functions. By definition f(-x) = -f(x), or f(x) = -f(-x), and g(-x) = -g(x), or g(x) = -g(-x). Then $fg(x) = f(x)\cdot g(x) = [-f(-x)]\cdot[-g(-x)] = f(-x)\cdot g(-x) = fg(-x)$, so fg is even.

59. See the answer section in the text.

60. Let f(x) be an even function, and let g(x) be an odd function. By definition f(x) = f(-x) and g(-x) = -g(x), or g(x) = -g(-x). Then $fg(x) = f(x)\cdot g(x) = f(-x)\cdot[-g(-x)] = -f(-x)\cdot g(-x) = -fg(-x)$, and fg is odd.

61. See the answer section in the text.

62. $E(-x) = \frac{f(-x) + f(-(-x))}{2} = \frac{f(-x) + f(x)}{2} = E(x)$.

Thus E(x) is even.

63. See the answer section in the text.

64. $E(x) + O(x) = \frac{f(x) + f(-x)}{2} + \frac{f(x) - f(-x)}{2} =$

$\frac{2f(x)}{2} = f(x)$

65. $f(x) = |x| + \text{INT}(x)$

This function can also be expressed as a piecewise function with an infinite number of statements:

$$f(x) = |x| + \text{INT}(x) = \begin{cases} \vdots \\ -x - 3 & \text{if } -3 \leqslant x < -2 \\ -x - 2 & \text{if } -2 \leqslant x < -1 \\ -x - 1 & \text{if } -1 \leqslant x < 0 \\ x & \text{if } 0 \leqslant x < 1 \\ x + 1 & \text{if } 1 \leqslant x < 2 \\ x + 2 & \text{if } 2 \leqslant x < 3 \\ \vdots \end{cases}$$

We sketch a graph of the function.

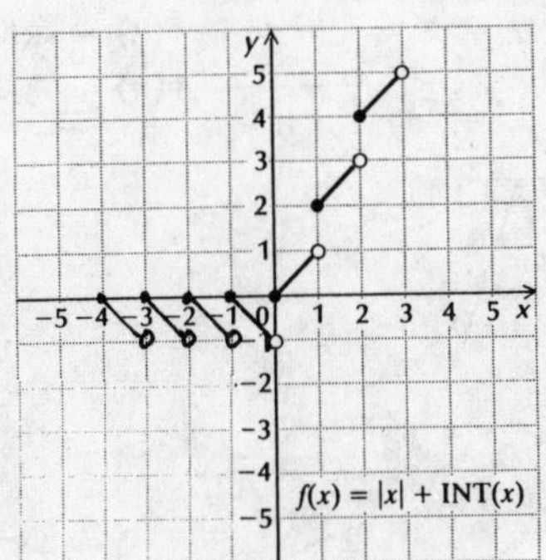

Exercise Set 3.8

1. $f(x) = |x| - 3$

By Theorem 1F, the graph of $f(x) = |x| - 3$ is a translation of the graph of $f(x) = |x|$ downward 3 units.

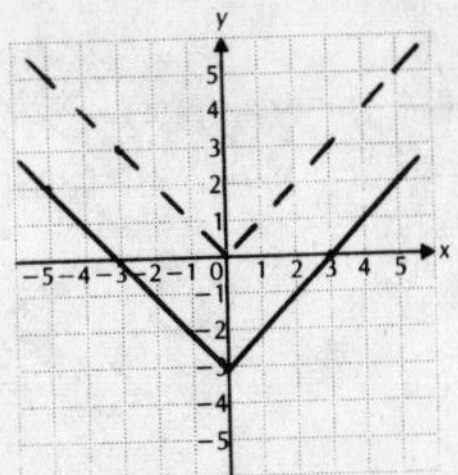

----- $f(x) = |x|$
—— $f(x) = |x| - 3$

2. $f(x) = 2 + |x|$

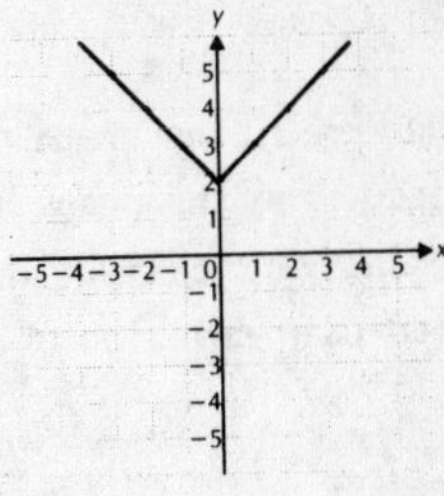

3. $f(x) = |x - 1|$

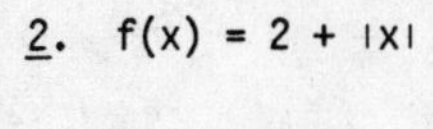

The equation $f(x) = |x - 1|$ is obtained from the equation $f(x) = |x|$ is obtained by replacing x by x - 1. By Theorem 2F, the graph of $f(x) = |x - 1|$ is a translation of the graph of $f(x) = |x|$ 1 unit to the right.

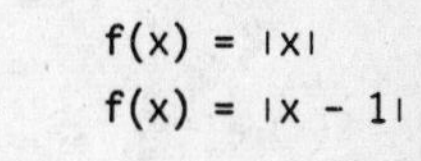

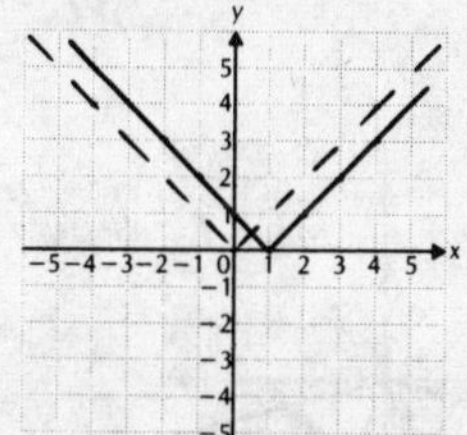

----- $f(x) = |x|$
—— $f(x) = |x - 1|$

4. $f(x) = |x + 2|$

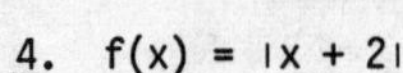

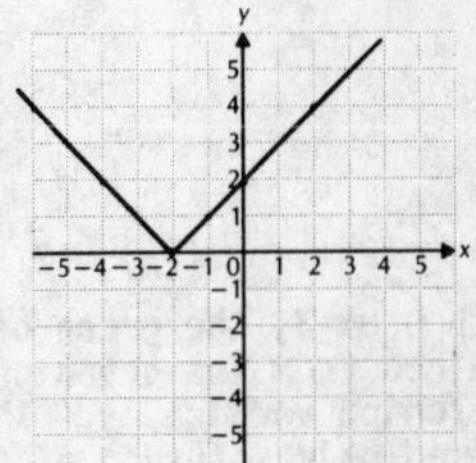

5. $f(x) = -4|x|$

By Theorem 3, the graph of $f(x) = -4|x|$ can be obtained from the graph of $f(x) = |x|$ by stretching vertically and reflecting across the x-axis. The y-coordinate of each ordered-pair solution of $f(x) = |x|$ is multiplied by -4.

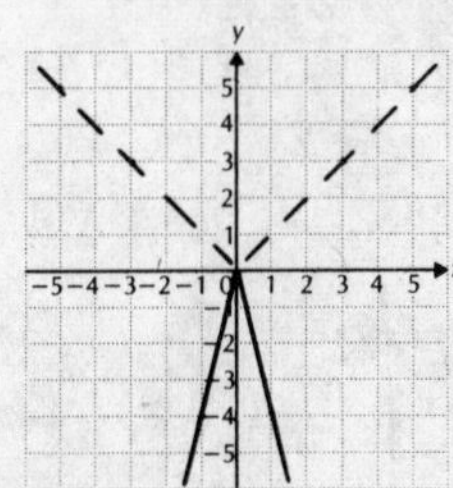

----- $f(x) = |x|$
—— $f(x) = -4|x|$

6. $f(x) = 3|x|$

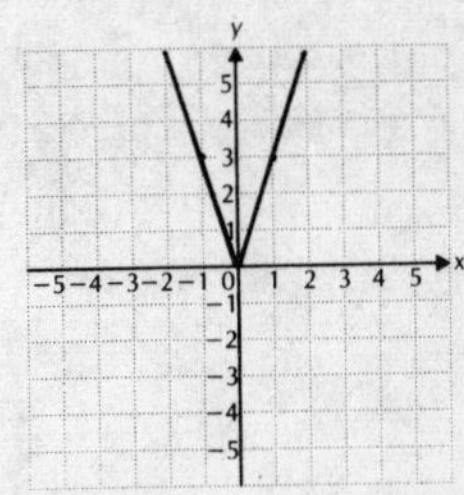

7. $f(x) = \frac{1}{3}|x|$

By Theorem 3, the graph of $f(x) = \frac{1}{3}|x|$ can be obtained from the graph of $f(x) = |x|$ by shrinking vertically. The y-coordinate of each ordered-pair solution of $f(x) = |x|$ is multiplied by $\frac{1}{3}$.

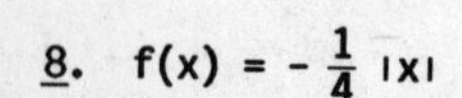

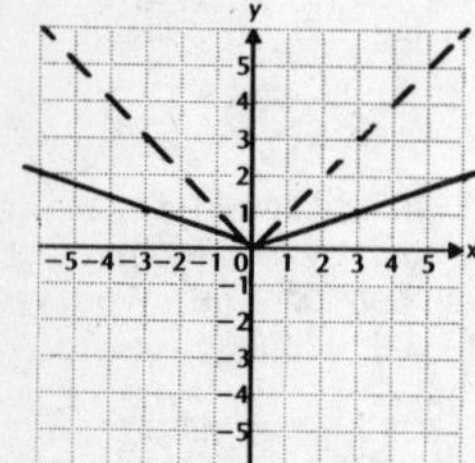

----- $f(x) = |x|$

—— $f(x) = \frac{1}{3}|x|$

8. $f(x) = -\frac{1}{4}|x|$

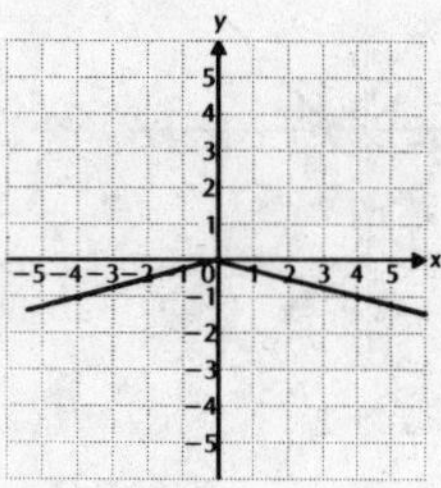

9. $f(x) = |2x|$, or $f(x) = 2|x|$

By Theorem 3, the graph of $f(x) = |2x|$ can be obtained from the graph of $f(x) = |x|$ by stretching vertically. The y-coordinate of each ordered-pair solution of $f(x) = |x|$ is multiplied by 2. (Equivalently, we could use Theorem 4 and shrink the graph of $f(x) = |x|$ horizontally. The result is the same in either case.)

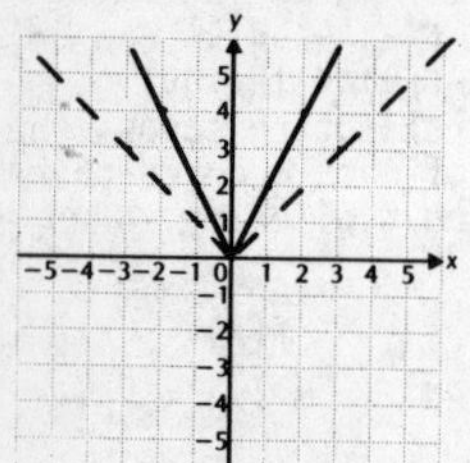

----- $f(x) = |x|$

—— $f(x) = |2x|$

10. $f(x) = \left|\frac{x}{3}\right|$

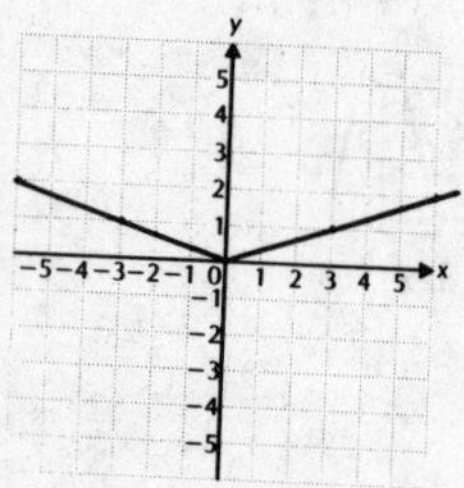

11. $f(x) = |x - 2| + 3$

We translate the graph of $f(x) = |x|$ 2 units to the right (Theorem 2F) and upward 3 units (Theorem 1F).

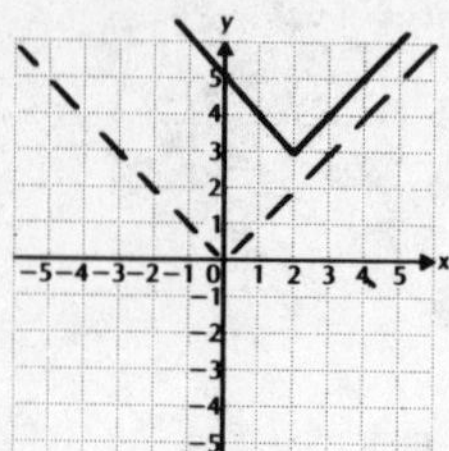

----- $f(x) = |x|$

—— $f(x) = |x - 2| + 3$

12. $f(x) = 2|x + 1| - 3$

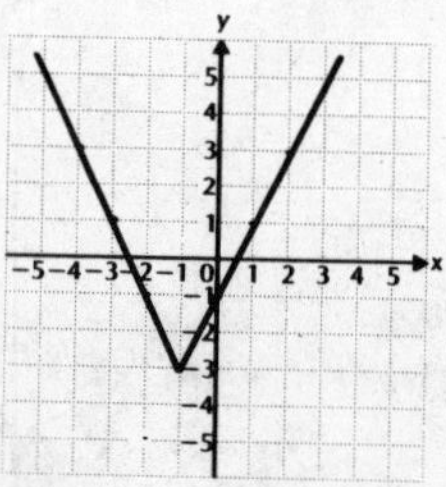

13. $f(x) = -3|x - 2|$

We translate the graph of $f(x) = |x|$ to the right 2 units (Theorem 2F), stretch it vertically, and reflect it across the x-axis (Theorem 3).

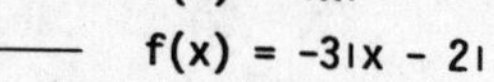

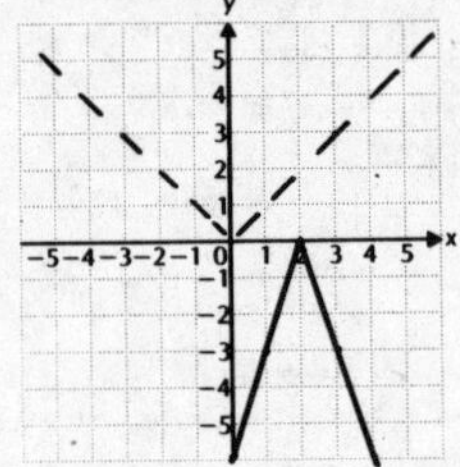

----- $f(x) = |x|$

—— $f(x) = -3|x - 2|$

14. $f(x) = \frac{1}{3}|x + 2| + 1$

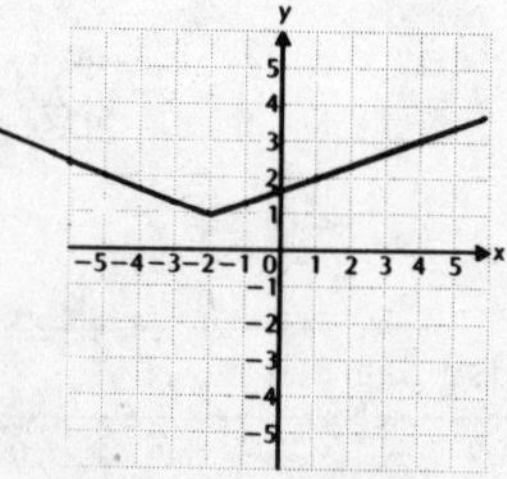

15. $y = 2 + f(x)$

We translate the graph of $y = f(x)$ upward 2 units (Theorem 1F).

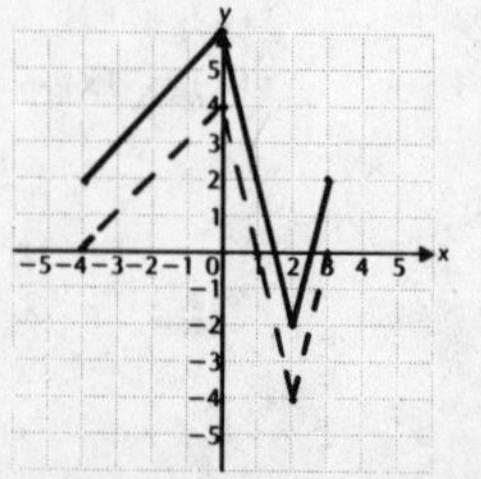

----- $y = f(x)$

—— $y = 2 + f(x)$

16. $y + 1 = f(x)$

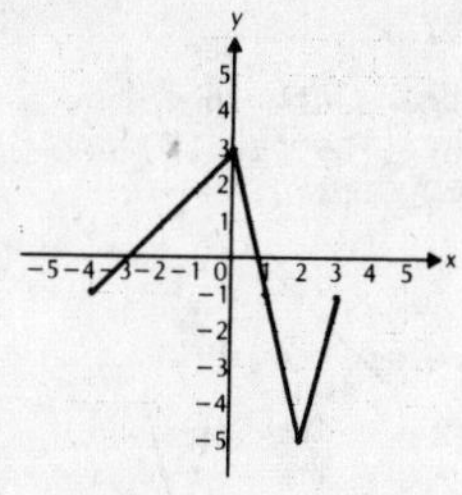

17. $y = f(x - 1)$

We translate the graph of $y = f(x)$ 1 unit to the right (Theorem 2F).

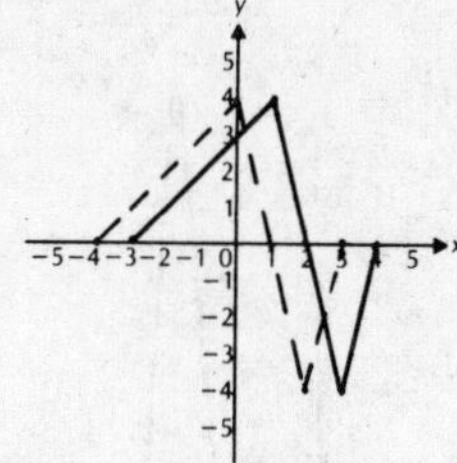

18. $y = f(x + 2)$

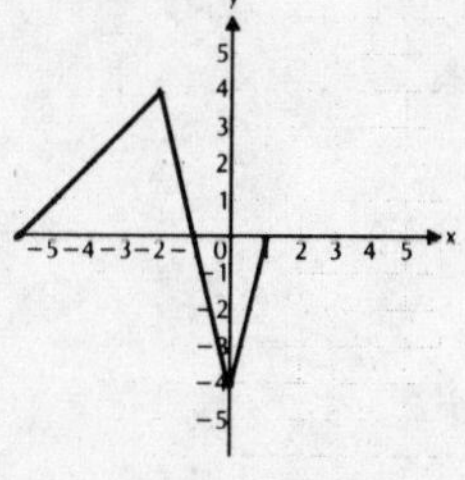

19. $\frac{y}{-2} = f(x)$, or $y = -2f(x)$

We stretch the graph of $y = f(x)$ vertically and reflect it across the x-axis (Theorem 3). The y-coordinate of each ordered-pair solution of $y = f(x)$ is multiplied by -2.

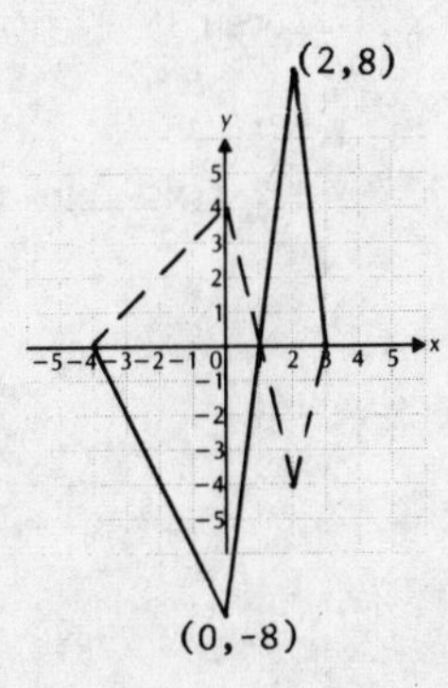

20. $y = 3f(x)$

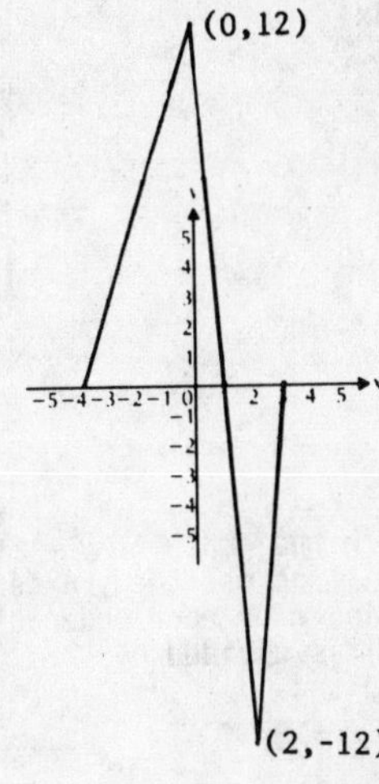

21. $y = \frac{1}{3}f(x)$

We shrink the graph of $y = f(x)$ vertically (Theorem 3). The y-coordinate of each ordered-pair solution of $y = f(x)$ is multiplied by $\frac{1}{3}$.

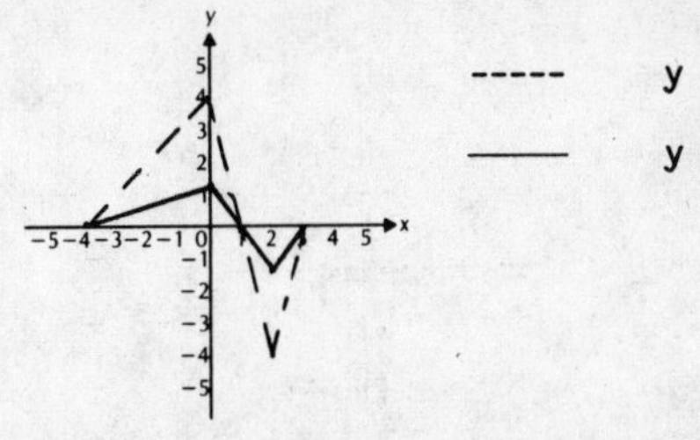

22. $y = -\frac{1}{2}f(x)$

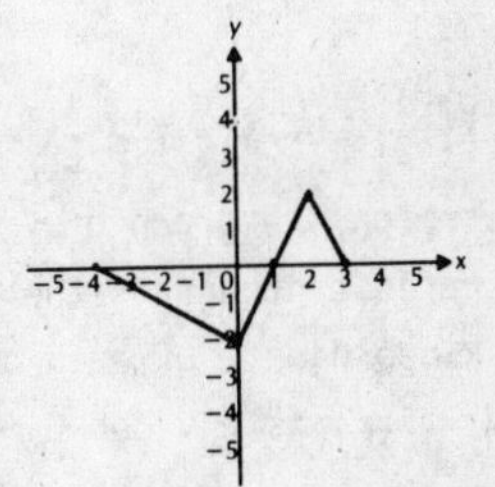

23. $y = f(2x)$

We shrink the graph of $y = f(x)$ horizontally (Theorem 4). The x-coordinate of each ordered-pair solution of $y = f(x)$ is divided by 2.

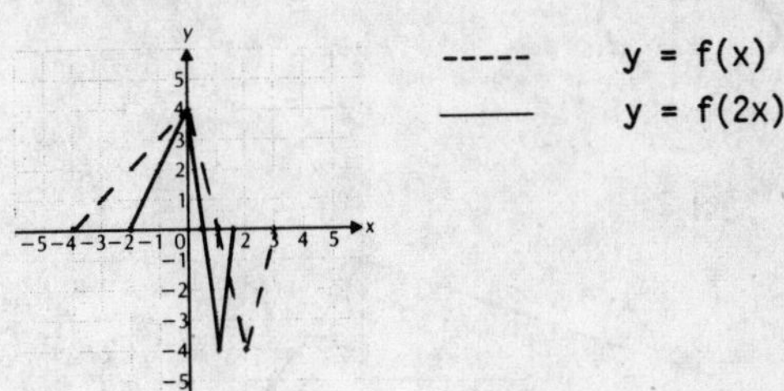

24. $y = f(3x)$

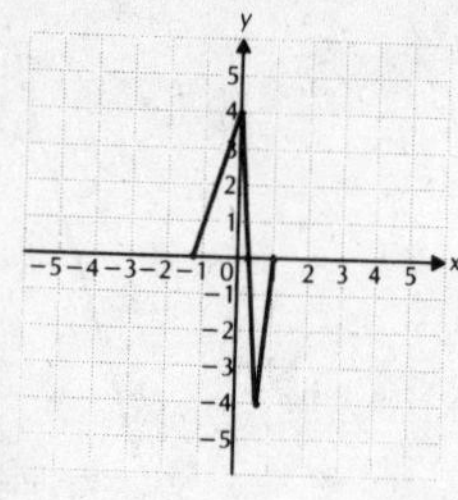

25. $y = f(-2x)$

We shrink the graph of $y = f(x)$ horizontally and reflect it across the y-axis (Theorem 4). The x-coordinate of each ordered-pair solution of $y = f(x)$ is divided by -2.

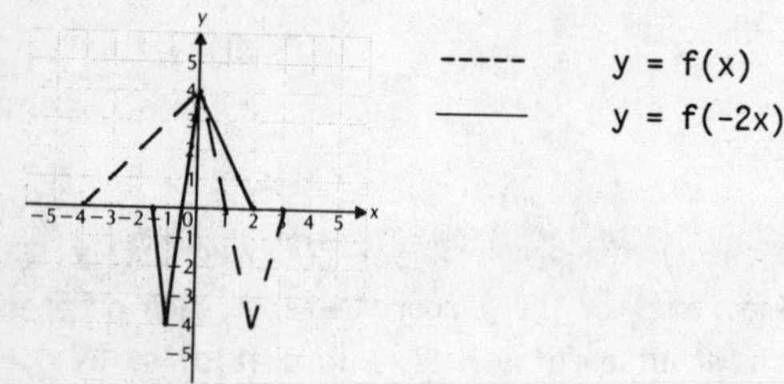

26. $y = f(-3x)$

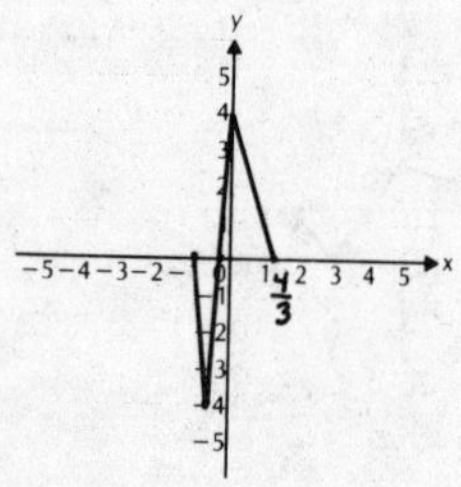

27. $y = f\left(\frac{x}{-2}\right)$, or $y = f\left(-\frac{1}{2}x\right)$

We stretch the graph of $y = f(x)$ horizontally and reflect it across the y-axis (Theorem 4). The x-coordinate of each ordered-pair solution of $y = f(x)$ is divided by $-\frac{1}{2}$.

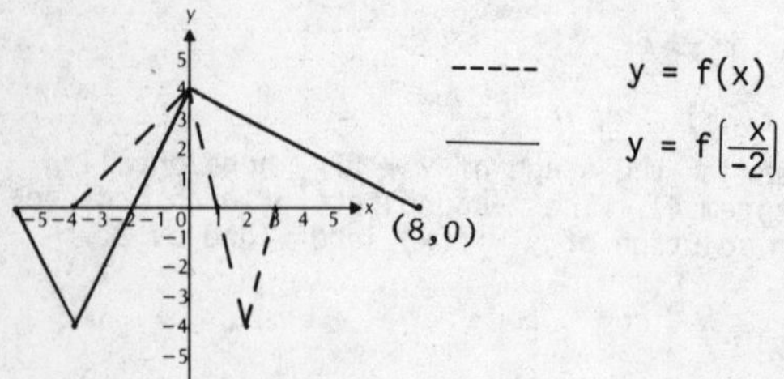

28. $y = f\left(\frac{1}{3}x\right)$

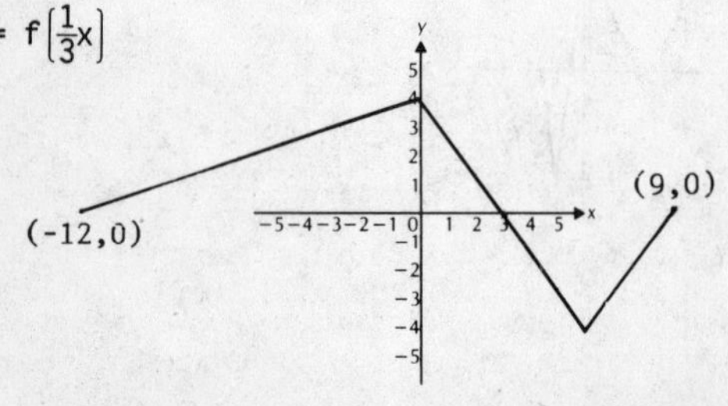

29. $y = f(x - 2) + 3$

We translate the graph of $y = f(x)$ 2 units to the right (Theorem 2F) and upward 3 units (Theorem 1F).

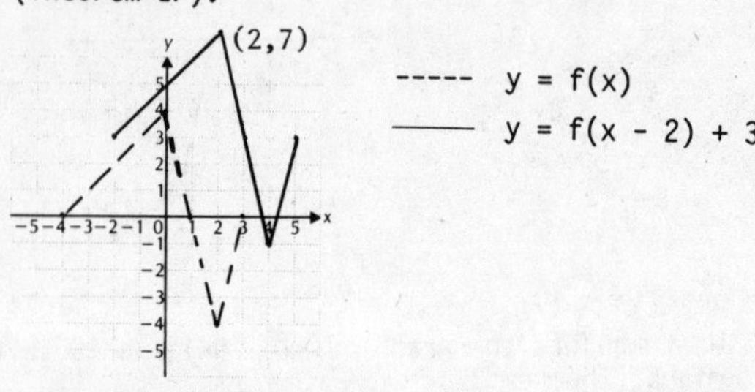

30. $y = -3f(x - 2)$

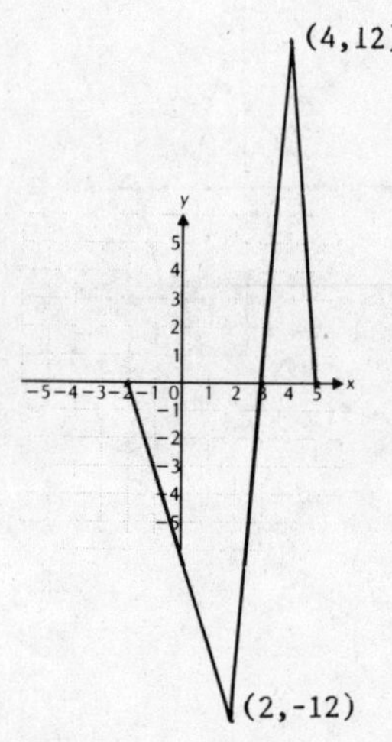

31. $y = 2\cdot f(x + 1) - 2$

We translate the graph of $y = f(x)$ 1 unit to the left to obtain the graph of $y = f(x + 1)$. (Theorem 2F)

Then we stretch the graph of $y = f(x + 1)$ vertically, multiplying each y-coordinate by 2, to obtain the graph of $y = 2\cdot f(x + 1)$. (Theorem 3) Finally we translate the graph of $y = 2\cdot f(x + 1)$ downward 2 units to obtain the graph of $y = 2\cdot f(x + 1) - 2$. (Theorem 1F)

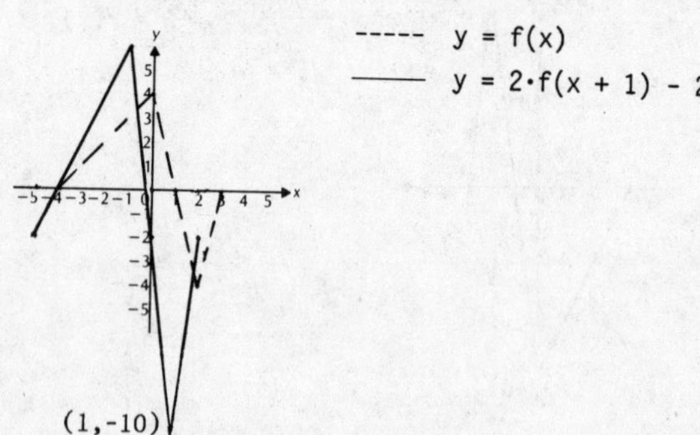

32. $y = \frac{1}{2}f(x + 2) - 1$

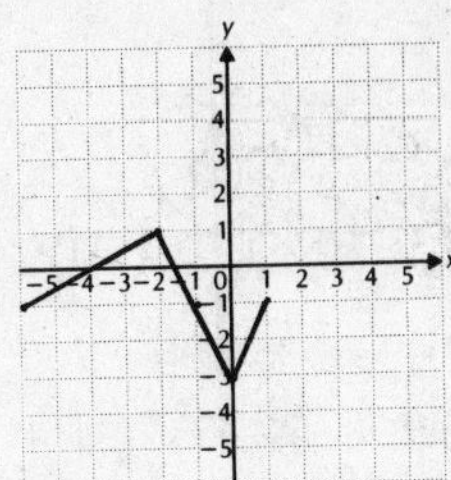

33. $y = -\frac{1}{2}f(x - 3) + 2$

We translate the graph of $y = f(x)$ 3 units to the right to obtain the graph of $y = f(x - 3)$. (Theorem 2F)

Then we shrink the graph of $y = f(x - 3)$ vertically and reflect it across the x-axis, multiplying each y-coordinate by $-\frac{1}{2}$, to obtain the graph of $y = -\frac{1}{2}f(x - 3)$. (Theorem 3)

Finally we translate the graph of $y = -\frac{1}{2}f(x - 3)$ upward 2 units to obtain the graph of $y = -\frac{1}{2}f(x - 3) + 2$. (Theorem 1F)

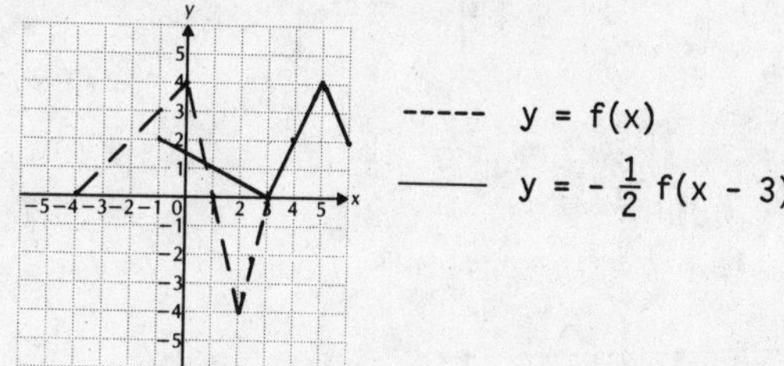

34. $y = -3f(x - 1) + 4$

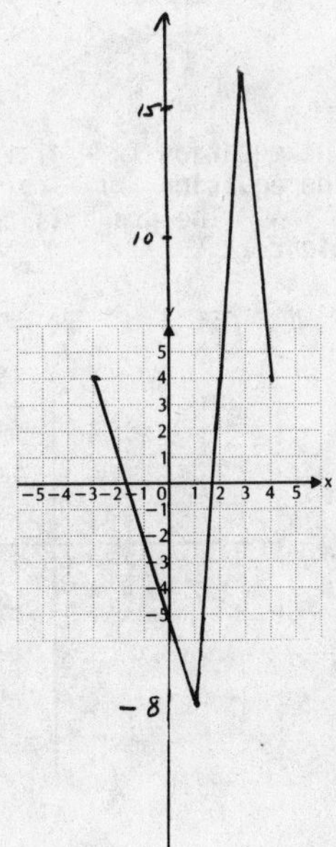

35. $y = -2f(x + 1) - 1$

We translate the graph of $y = f(x)$ 1 unit to the left to obtain the graph of $y = f(x + 1)$. (Theorem 2F)

Then we stretch the graph of $y = f(x + 1)$ vertically and reflect it across the x-axis, multiplying each y-coordinate by -2, to obtain the graph of $y = -2f(x + 1)$. (Theorem 3) Finally we translate the graph of $y = -2f(x + 1)$ downward 1 unit to obtain the graph of $y = -2f(x + 1) - 1$. (Theorem 1F)

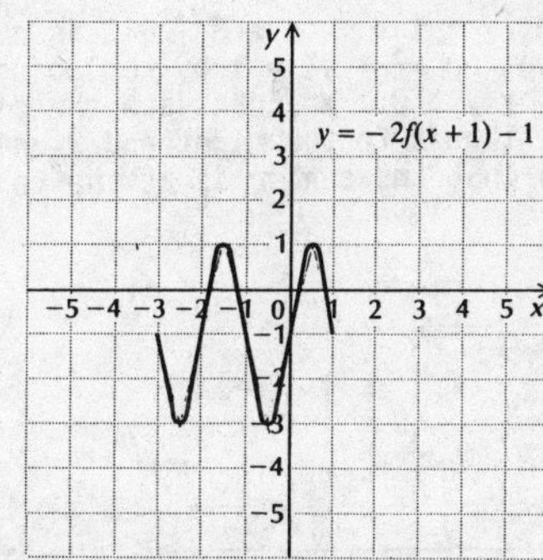

36.

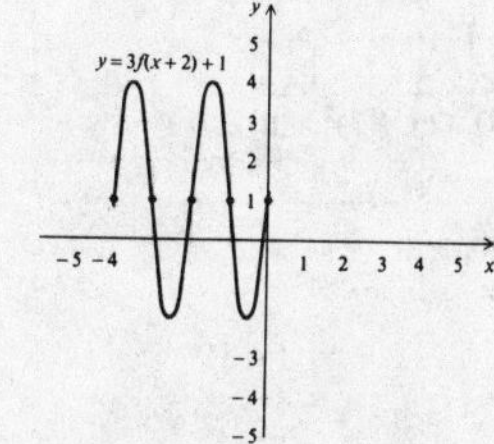

37. $y = \frac{5}{2}f(x - 3) - 2$

We translate the graph of $y = f(x)$ 3 units to the right to obtain the graph of $y = f(x - 3)$. (Theorem 2F)

Then we stretch the graph of $y = f(x - 3)$ vertically, multiplying each y-coordinate by $\frac{5}{2}$, to obtain the graph of $y = \frac{5}{2}f(x - 3)$. (Theorem 3)

Finally we translate the graph of $y = \frac{5}{2}f(x - 3)$ downward 2 units to obtain the graph of $y = \frac{5}{2}f(x - 3) - 2$. (Theorem 1F)

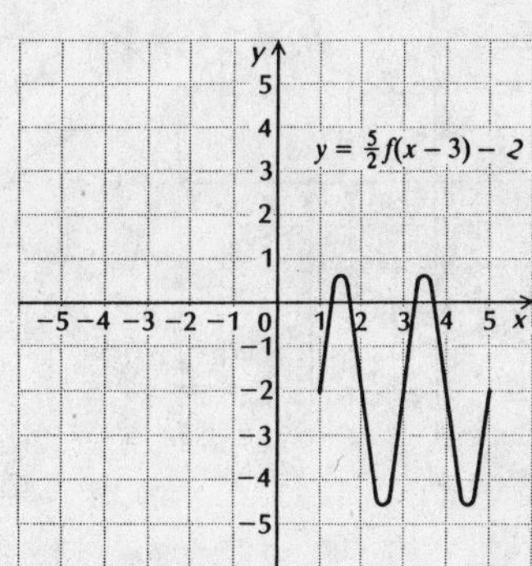

38. $y = -\frac{2}{3}f(x - 4) + 3$

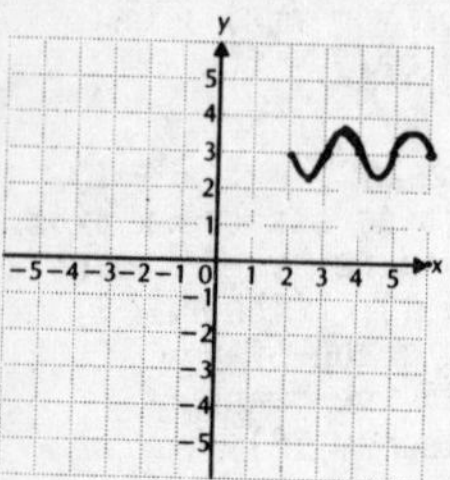

39. The equation $(x - 1)^2 + (y + 3)^2 = 1$ is obtained from the equation $x^2 + y^2 = 1$ by replacing x by x - 1 and y by y + 3. We translate the circle $x^2 + y^2 = 1$, 1 unit to the right and 3 units downward, so that the center is at the point (1,-3).

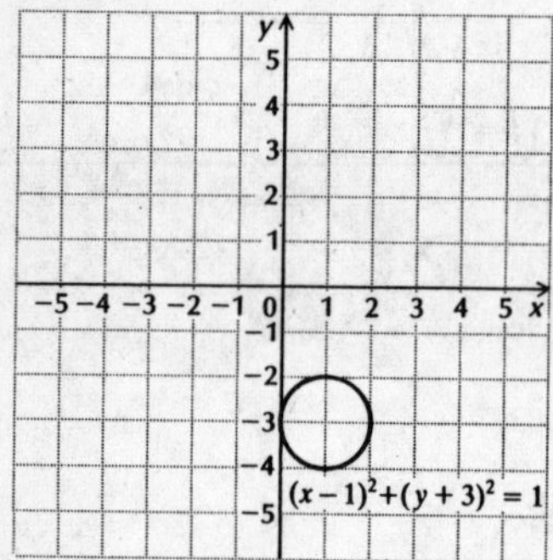

40.

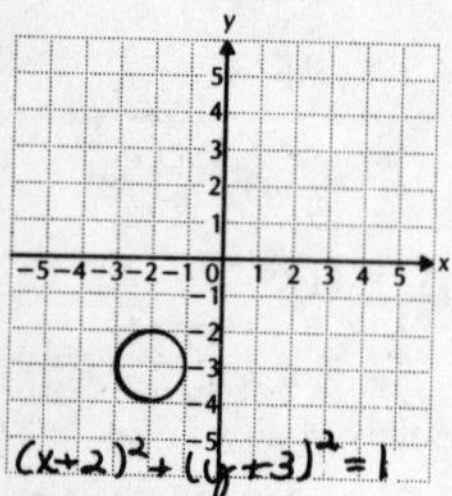

41. The equation $x^2 + (y - 2)^2 = 1$ is obtained from the equation $x^2 + y^2 = 1$ by replacing y by y - 2. We translate the circle $x^2 + y^2 = 1$ upward 2 units, so that the center is at (0,2).

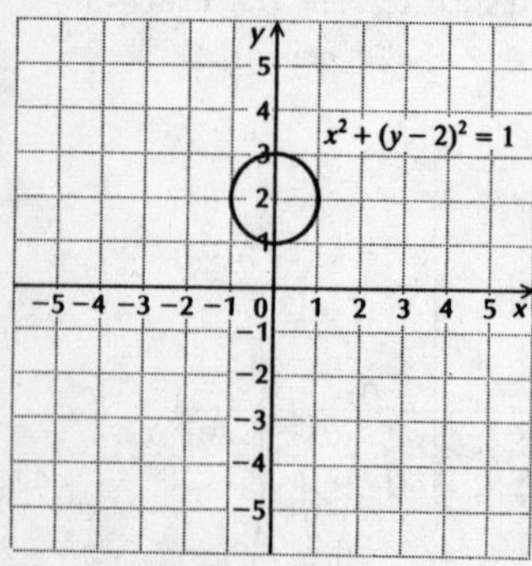

42.

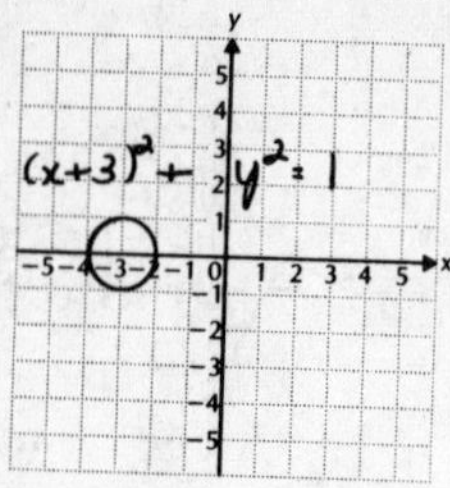

43. The equation $|x| + |y + 3| = 1$ is obtained from the equation $|x| + |y| = 1$ by replacing y by y + 3. The graph is translated downward 3 units.

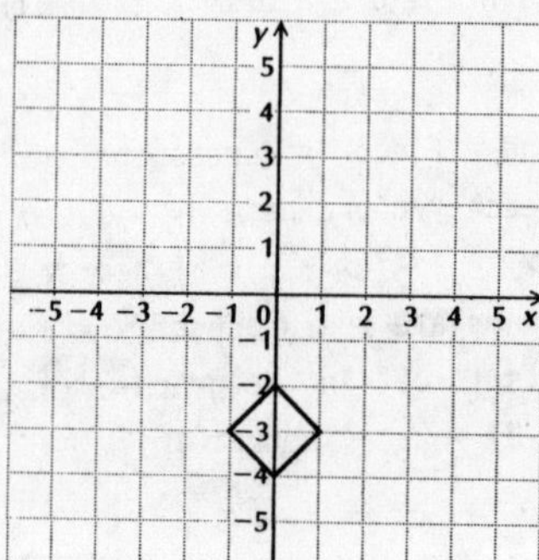

44.

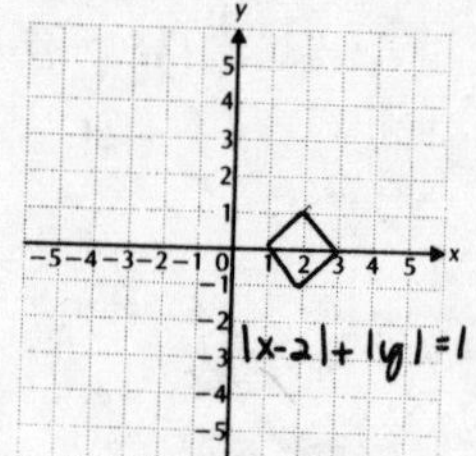

45. The equation $|x - 4| + |y| = 1$ is obtained from the equation $|x| + |y| = 1$ by replacing x by x - 4. The graph is translated 4 units to the right.

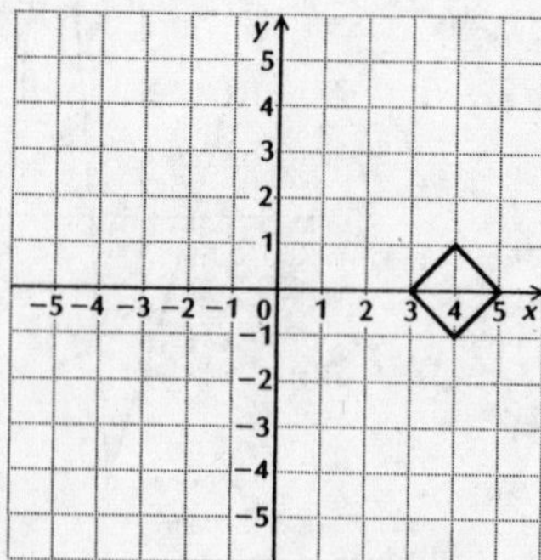

46.

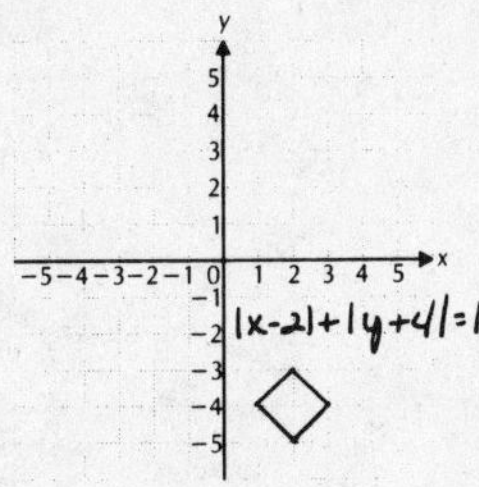

47. $y = -\sqrt{2}\, f(x + 1.8)$

The graph of $y = f(x)$ is translated 1.8 units to the left to obtain the graph of $y = f(x + 1.8)$. Then the graph of $y = f(x + 1.8)$ is stretched vertically and reflected across the x-axis, multiplying each y-coordinate by $-\sqrt{2}$.

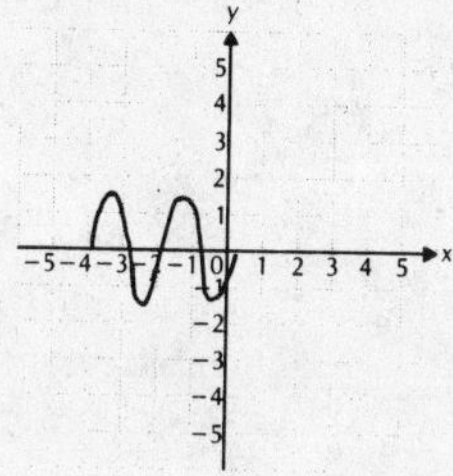

48. $y = \frac{\sqrt{3}}{2} \cdot f(x - 2.5) - 5.3$

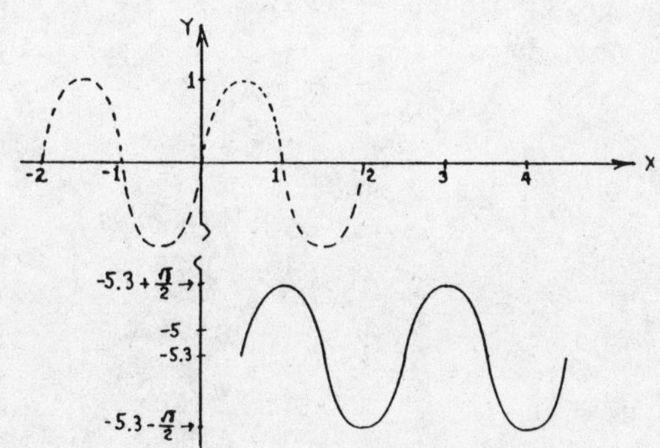

49. ---------- (1) $y = f(x)$ is the given function

— — (2) $\frac{y}{3} = f(2x)$ is the result of shrinking (1) horizontally by a factor of $\frac{1}{2}$ and stretching vertically by a factor of 3

——— (3) $\frac{y}{3} = f\left[2\left(x + \frac{1}{4}\right)\right]$, the required graph, is a translation of (2) to the left through $\frac{1}{4}$ units.

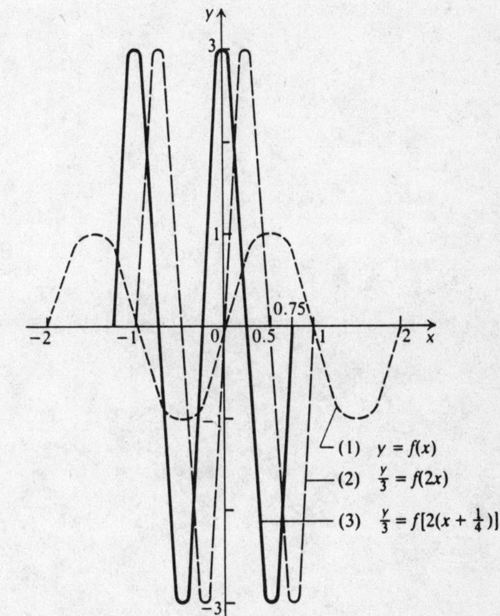

50. $y = -4 \cdot f(5x + 10)$

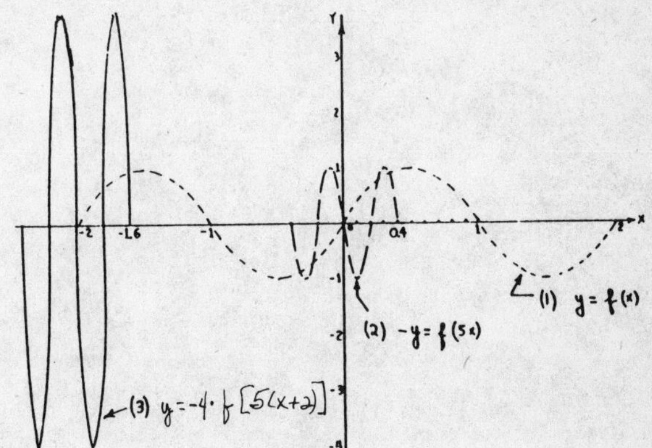

Exercise Set 4.1

1. a) The equation $3y = 2x - 5$ is linear because it is equivalent to $2x - 3y = 5$. Here $A = 2$, $B = -3$, and $C = 5$.

   b) The equation $5x + 3 = 4y$ is linear because it is equivalent to $5x - 4y = -3$. Here $A = 5$, $B = -4$, and $C = -3$.

   c) The equation $3y = x^2 + 2$ is not linear because the x is squared.

   d) The equation $y = 3$ is linear because it is equivalent to $0x + y = 3$. Here $A = 0$, $B = 1$, and $C = 3$.

   e) The equation $xy = 5$ is not linear because the product xy occurs.

   f) The equation $3x^2 + 2y = 4$ is not linear because the x is squared.

   g) The equation $3x + \frac{1}{y} = 4$ is not linear because y is in a denominator.

   h) The equation $5x - 2 = 4y$ is linear because it is equivalent to $5x - 4y = 2$.

2. a, b, g, h

3. Graph: $8x - 3y = 24$

First find two of its points. Here the easiest points are the intercepts.

y-intercept: Set $x = 0$ and find y.

$$8x - 3y = 24$$
$$8\cdot0 - 3y = 24$$
$$-3y = 24$$
$$y = -8$$

The y-intercept is (0, -8).

x-intercept: Set $y = 0$ and find x.

$$8x - 3y = 24$$
$$8x - 3\cdot0 = 24$$
$$8x = 24$$
$$x = 3$$

The x-intercept is (3, 0).

Plot these two points and draw a line through them.

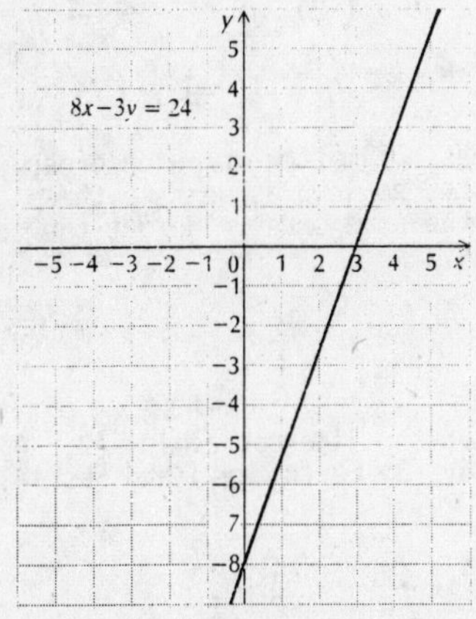

3. (continued)

A third point should be used as a check.

Set $x = 2$ and find y.

$$8x - 3y = 24$$
$$8\cdot2 - 3y = 24$$
$$16 - 3y = 24$$
$$-3y = 8$$
$$y = -\frac{8}{3}$$

The ordered pair $\left(2, -\frac{8}{3}\right)$ is also a point on the line.

4. $5x - 10y = 50$

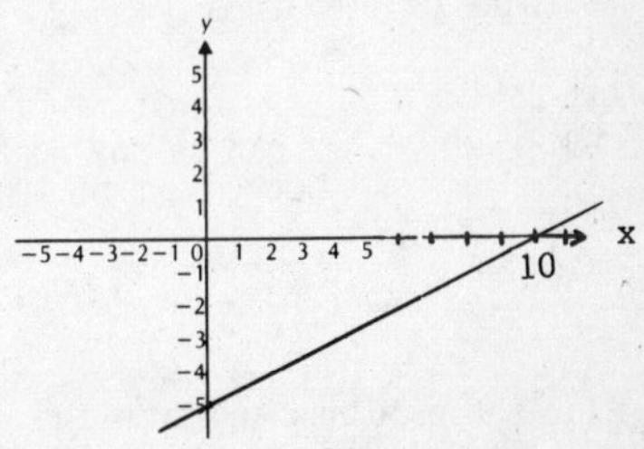

5. Graph: $3x + 12 = 4y$

First find two of its points. Here the easiest points to find are the intercepts.

y-intercept: Set $x = 0$ and find y.

$$3x + 12 = 4y$$
$$3\cdot0 + 12 = 4y$$
$$12 = 4y$$
$$3 = y$$

The y-intercept is (0, 3).

x-intercept: Set $y = 0$ and find x.

$$3x + 12 = 4y$$
$$3x + 12 = 4\cdot0$$
$$3x + 12 = 0$$
$$3x = -12$$
$$x = -4$$

The x-intercept is (-4, 0).

Plot these points and draw a line through them. A third point should be used as a check.

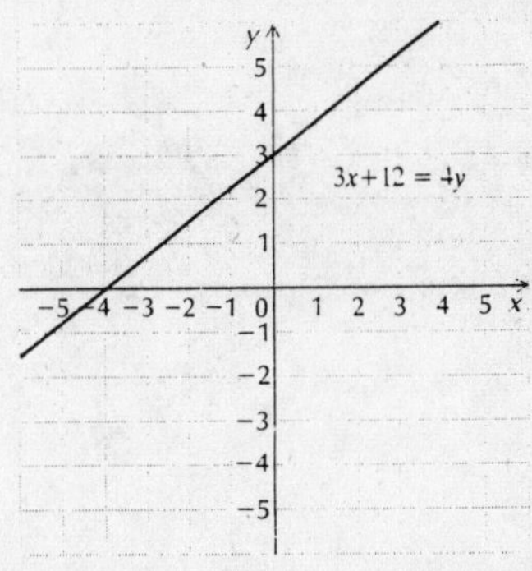

6. $4x - 20 = 5y$

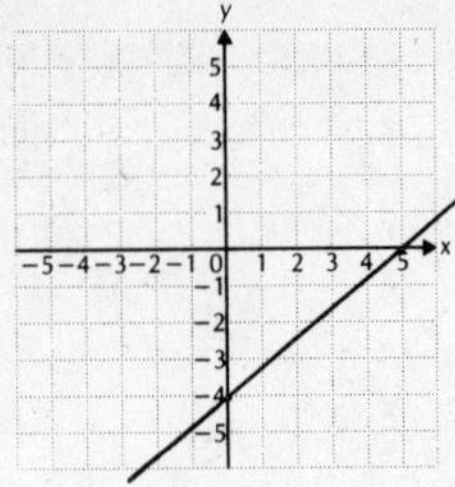

7. Graph: $y = -2$

The equation $y = -2$ says that the second coordinate of every ordered pair of the graph is -2. Below is a table of a few ordered pairs that are solutions of $y = -2$. (It might help to think of $y = -2$ as $0 \cdot x + y = -2$.)

x	y
-3	-2
0	-2
1	-2
4	-2

(x can be any number, but y must be -2)

Plot these points and draw the line through them.

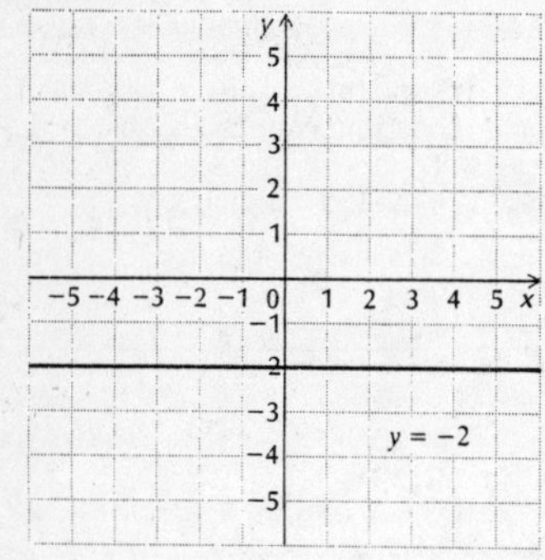

8. $2y - 3 = 9$

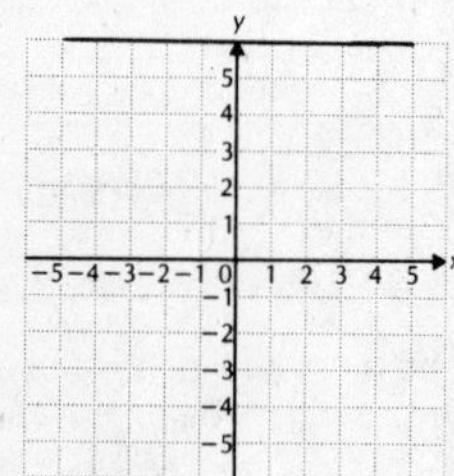

9. Graph: $5x + 2 = 17$

$$5x = 15$$

$$x = 3$$

The equation $5x + 2 = 17$ is equivalent to $x = 3$. The equation $x = 3$ says that the first coordinate of every ordered pair of the graph is 3. Below is a table of a few ordered pairs that are solutions of $x = 3$. (It might help to think of $x = 3$ as $x + 0 \cdot y = 3$.)

x	y
3	0
3	-2
3	4
3	-1

(y can be any number, but x must be 3)

Plot these point and draw a line through them.

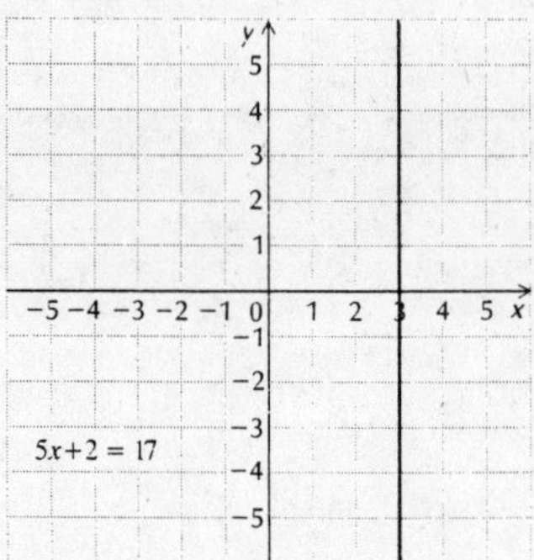

10. $19 = 5 - 2x$

11. Find the slope of the line containing (6, 2) and (-2, 1).

$$m = \frac{\text{change in } y}{\text{change in } x} = \frac{2 - 1}{6 - (-2)} = \frac{1}{8}$$

$$\text{or} = \frac{1 - 2}{-2 - 6} = \frac{-1}{-8} = \frac{1}{8}$$

Note that it does not matter in which order we choose the points, so long as we take the differences in the same order. We get the same slope either way.

12. $\frac{3}{2}$

13. Find the slope of the line containing (2, -4) and (4, -3). Let (4, -3) be (x_1, y_1) and (2, -4) be (x_2, y_2).

$$m = \frac{y_2 - y_1}{x_2 - x_1} = \frac{-4 - (-3)}{2 - 4} = \frac{-1}{-2} = \frac{1}{2}$$

14. $-\frac{11}{10}$

15. Find the slope of the line containing $(\pi, 5)$ and $(\pi, 4)$. Let $(\pi, 4)$ be (x_1, y_1) and $(\pi, 5)$ be (x_2, y_2).

$$m = \frac{y_2 - y_1}{x_2 - x_1} = \frac{5 - 4}{\pi - \pi} = \frac{1}{0}$$

The slope is not defined, because we cannot divide by 0.

16. 0

17. Find the slope of the line containing $(\sqrt{2}, 13)$ and $(\pi, 13)$. Let $(\sqrt{2}, 13)$ be (x_1, y_1) and $(\pi, 13)$ be (x_2, y_2).

$$m = \frac{y_2 - y_1}{x_2 - x_1} = \frac{13 - 13}{\pi - \sqrt{2}} = \frac{0}{\pi - \sqrt{2}} = 0$$

18. Not defined

19. $m = \frac{920.58}{13,740} = 0.067$

The road grade is 6.7%.

We can think of an equation giving the height y as a function of the horizontal distance x as an equation of variation where y varies directly as x and the variation constant is 0.067. The equation is $y = 0.067x$.

20. 4%; $y = 0.04x$

21. The road grade will be negative since the road is dropping.

$$m = -\frac{158.4}{5280} = -0.03$$

The road grade is -3%.

22. 0.045

23. We express the slope of the treadmill, 8%, as a decimal quantity. Let h = the height of the vertical end of the treadmill.

$$0.08 = \frac{h}{5}$$
$$0.4 = h$$

The end of the tread is 0.4 ft high.

24. 30 ft

25. Find an equation of the line through (3, 2) with m = 4.

$(y - y_1) = m(x - x_1)$ (Point-slope equation)

$(y - 2) = 4(x - 3)$ (Substituting)

$y - 2 = 4x - 12$

$y = 4x - 10$

26. $y = -2x + 15$

27. Find an equation of the line with y-intercept -5 and m = 2.

$y = mx + b$ (Slope-intercept equation)

$y = 2x + (-5)$ (Substituting)

$y = 2x - 5$

28. $y = \frac{1}{4}x + \pi$

29. Find an equation of the line through (-4, 7) with $m = -\frac{2}{3}$.

$(y - y_1) = m(x - x_1)$ (Point-slope equation)

$(y - 7) = -\frac{2}{3}[x - (-4)]$ (Substituting)

$y - 7 = -\frac{2}{3}(x + 4)$

$y - 7 = -\frac{2}{3}x - \frac{8}{3}$

$y = -\frac{2}{3}x + \frac{13}{3}$

30. $y = \frac{3}{4}x - \frac{11}{4}$

31. Find an equation of the line through (5, -8) with m = 0

$(y - y_1) = m(x - x_1)$ (Point-slope equation)

$[y - (-8)] = 0(x - 5)$

$y + 8 = 0$

$y = -8$

32. $x = 5$

33. Find an equation of the line containing (1, 4) and (5, 6). Let $(x_1, y_1) = (1, 4)$ and $(x_2, y_2) = (5, 6)$.

$y - y_1 = \frac{y_2 - y_1}{x_2 - x_1}(x - x_1)$ (Two-point equation)

$y - 4 = \frac{6 - 4}{5 - 1}(x - 1)$ (Substituting)

$y - 4 = \frac{2}{4}(x - 1)$

$y - 4 = \frac{1}{2}x - \frac{1}{2}$

$y = \frac{1}{2}x + \frac{7}{2}$

34. $y = \frac{3}{4}x + \frac{3}{2}$

35. Find an equation of the line containing (-2, 5) and (-4, -7). Let $(x_1, y_1) = (-2, 5)$ and $(x_2, y_2) = (-4, -7)$.

$y - y_1 = \frac{y_2 - y_1}{x_2 - x_1}(x - x_1)$ (Two-point equation)

$y - 5 = \frac{-7 - 5}{-4 - (-2)}[x - (-2)]$ (Substituting)

$y - 5 = \frac{-12}{-2}(x + 2)$

$y - 5 = 6x + 12$

$y = 6x + 17$

36. $y = -\frac{54}{7}x + \frac{152}{35}$

37. Find an equation of the line containing (3, 6) and (-2, 6). Let (x_1, y_1) be (3, 6) and (x_2, y_2) be (-2, 6).

$y - y_1 = \frac{y_2 - y_1}{x_2 - x_1}(x - x_1)$ (Two-point equation)

$y - 6 = \frac{6 - 6}{-2 - 3}(x - 3)$ (Substituting)

$y - 6 = \frac{0}{-5}(x - 3)$

$y - 6 = 0$

$y = 6$

38. $x = -\frac{3}{8}$

39. $y = 2x + 3$

$\uparrow \quad \uparrow$

$y = mx + b$ (Slope-intercept equation)

The slope is 2, and the y-intercept is 3.

40. $m = -1$, $b = 6$

41. $2y = -6x + 10$

$y = -3x + 5$

$\uparrow \quad \uparrow$

$y = mx + b$ (Slope-intercept equation)

The slope is -3, and the y-intercept is 5.

42. $m = 4$, $b = -3$

43. $3x - 4y = 12$

$-4y = -3x + 12$

$y = \frac{3}{4}x - 3$

$\uparrow \quad \uparrow$

$y = mx + b$ (Slope-intercept equation)

The slope is $\frac{3}{4}$, and the y-intercept is -3.

44. $m = -\frac{5}{2}$, $b = -\frac{7}{2}$

45. $3y + 10 = 0$

$3y = -10$

$y = -\frac{10}{3}$

or

$y = 0 \cdot x + \left(-\frac{10}{3}\right)$

$\uparrow \quad \uparrow$

$y = mx + b$ (Slope-intercept equation)

The slope is 0, and the y-intercept is $-\frac{10}{3}$.

46. $m = 0$, $b = 7$

47. Graph $y = -\frac{3}{2}x$.

First we plot the y-intercept (0, 0). We think of the slope as $\frac{-3}{2}$. Starting at the y-intercept and using the slope, we find another point by moving 3 units down and 2 units right. We get to a new point (2, -3). By thinking of the slope as $\frac{3}{-2}$ we can start again at the y-intercept and find another point by moving 3 units up and 2 units left. We get to another point on the line, (-2, 3). We draw the line through these points.

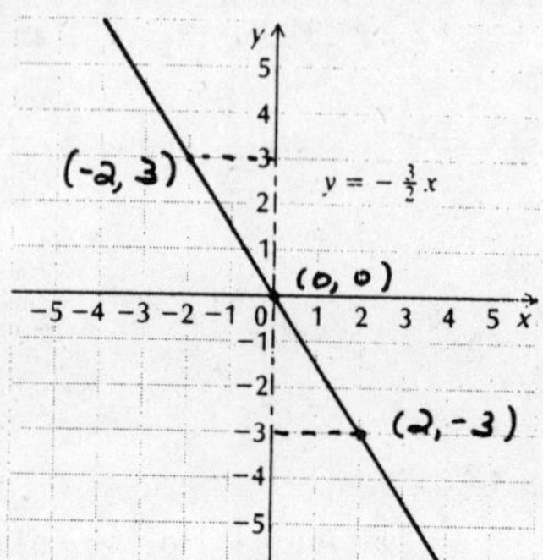

48.

49. Graph $y = -\frac{5}{2}x - 2$.

First we plot the y-intercept (0, -2). We think of the slope as $\frac{-5}{2}$. Starting at the y-intercept and using the slope, we find another point by moving 5 units down and 2 units right. We get to a new point, (2, -7). By thinking of the slope as $\frac{5}{-2}$ we can start again at the y-intercept and find another point by moving 5 units up and 2 units left. We get to another point on the line, (-2, 3). We draw the line through these points.

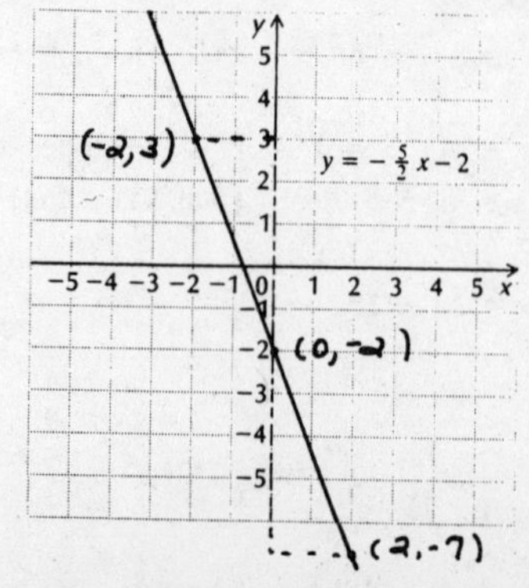

50.

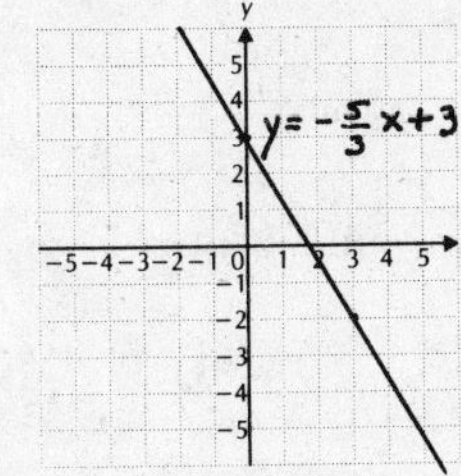

51. Graph $y = \frac{1}{2}x + 1$.

First we plot the y-intercept (0, 1). We consider the slope $\frac{1}{2}$. Starting at the y-intercept and using the slope, we find another point by moving 1 unit up and 2 units right. We get to a new point, (2, 2). By thinking of the slope as $\frac{-1}{-2}$ we can start again at the y-intercept and find another point by moving 1 unit down and 2 units left. We get to another point on the line, (-2, 0). We draw the line through these points.

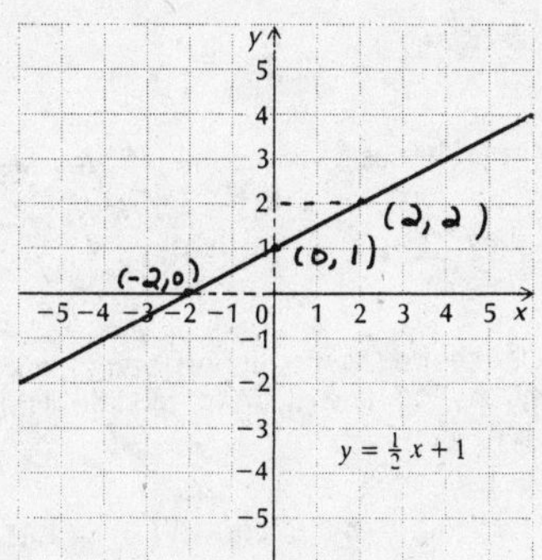

52.

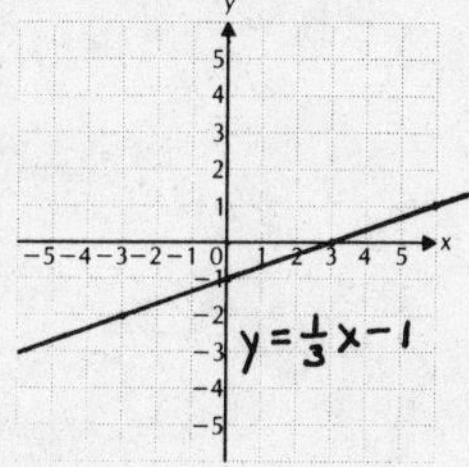

53. Graph $y = \frac{4}{3} - \frac{1}{3}x$, or $y = -\frac{1}{3}x + \frac{4}{3}$.

First we plot the y-intercept $\left(0, \frac{4}{3}\right)$. We think of the slope as $\frac{-1}{3}$. Starting at the y-intercept and using the slope, we find another point by moving 1 unit down and 3 units right. We get to a new point, $\left(3, \frac{1}{3}\right)$. By thinking of the slope as $\frac{1}{-3}$ we can start again at the y-intercept and find another pont by moving 1 unit up and 3 units left. We get to another point on the line, $\left(-3, \frac{7}{3}\right)$. We draw the line through these points.

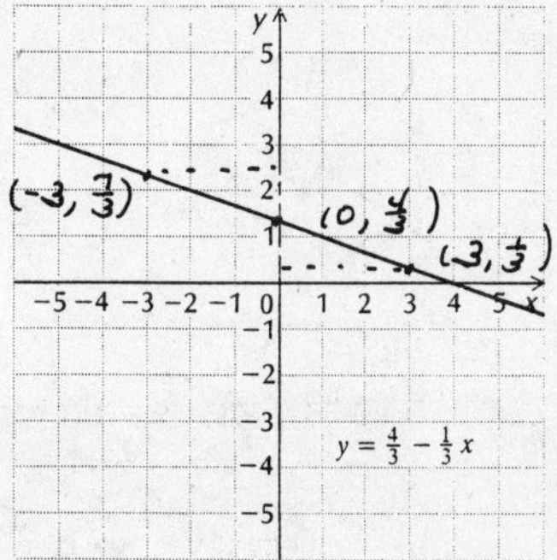

54.

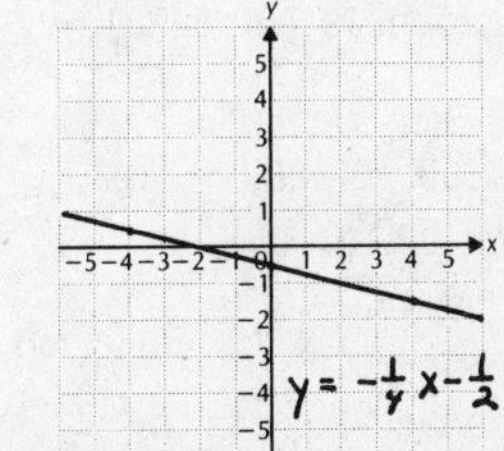

55.
$$y - y_1 = m(x - x_1) \quad \text{(Point-slope equation)}$$
$$y - (-2.563) = 3.516(x - 3.014) \quad \text{(Substituting)}$$
$$y + 2.563 = 3.516x - 10.597224$$
$$y = 3.516x - 13.1602$$

56. $y = -0.00014x - 17.624276$

57.
$$y - y_1 = \frac{y_2 - y_1}{x_2 - x_1}(x - x_1) \quad \text{(Two-point equation)}$$
$$y - 2.443 = \frac{11.012 - 2.443}{8.114 - 1.103}(x - 1.103) \quad \text{(Substituting)}$$
$$y - 2.443 = \frac{8.569}{7.011}(x - 1.103)$$
$$y - 2.443 = 1.222(x - 1.103)$$
$$y - 2.443 = 1.222x - 1.348$$
$$y = 1.2222x + 1.0949$$

58. $y = -1.4213x + 1583.5835$

59. T(d) = 10d + 20

a) T(5 km) = 10·5 + 20 = 50 + 20 = 70°C
T(20 km) = 10·20 + 20 = 200 + 20 = 220°C
T(1000 km) = 10·1000 + 20 = 10,000 + 20 = 10,020°C

b) This is a linear function. We plot the points (5, 70), (20, 220), and (1000, 10,020) and draw the graph.

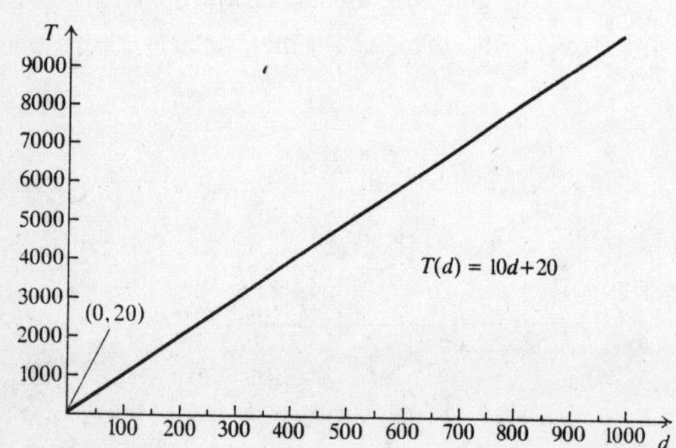

c) The depth must be nonnegative, so d ⩾ 0. If we go more than 5600 km into the earth, we begin to emerge on the other side. Thus d ⩽ 5600.

The domain of the function is [0, 5600].

60. a) C(0) = 0.32, C(10) = 0.59, C(100) = 3.02, C(153) = 4.451

b)

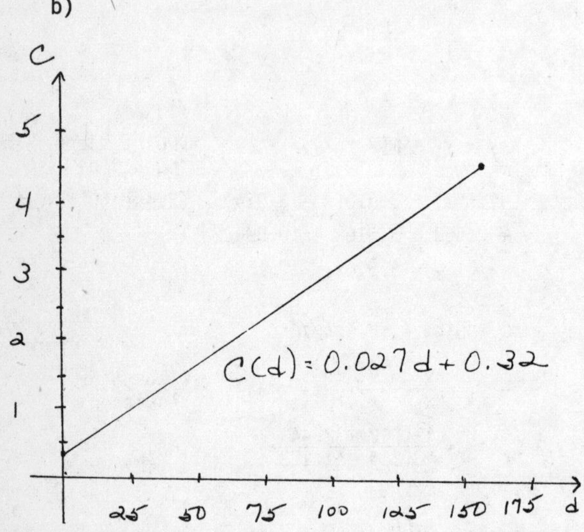

c) [0, 153]

61. T(L) = 0.143L - 1.18

a) T(0 mm) = 0.143(0) - 1.18 = 0 - 1.18 = -1.18 mm
T(50 mm) = 0.143(50) - 1.18 = 7.15 - 1.18 = 5.97 mm
T(80 mm) = 0.143(80) - 1.18 = 11.44 - 1.18 = 10.26 mm

b) This is a linear function. We plot the points (50, 5.97) and (80, 10.26) and draw the graph.

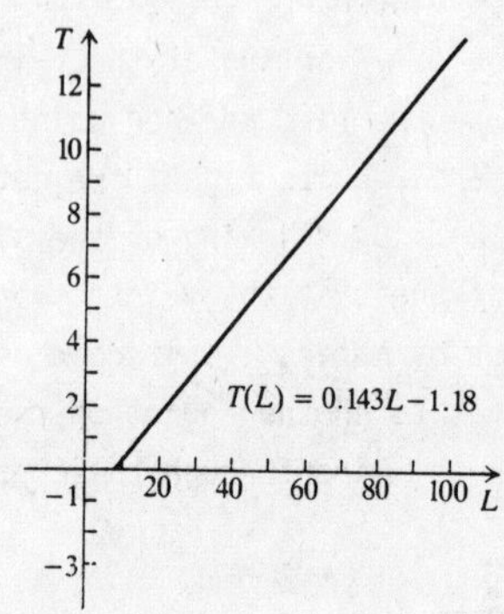

c) Since tail length must be nonnegative, we solve:

0.143L - 1.18 ⩾ 0
L ⩾ 8.25

We can assume there is no upper bound on the tail length, so the domain of the function is [8.25, ∞).

62. a) V(0) = $5200, V(1) = $4687.50, V(2) = $4175, V(3) = $3662.50, V(8) = $1100

b)

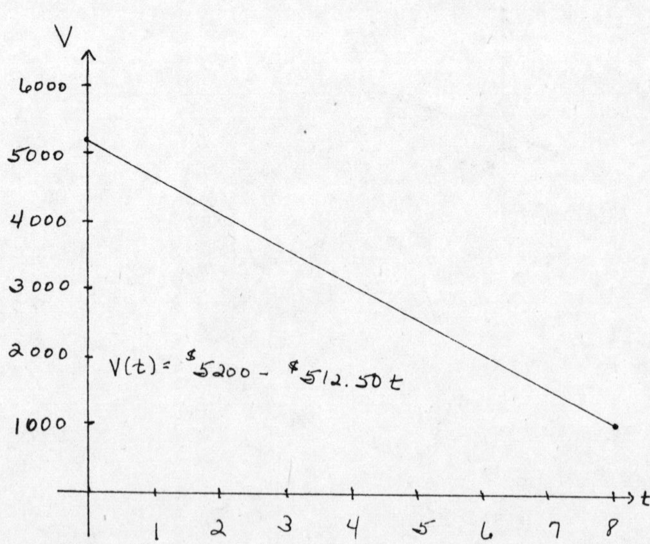

63. $A(t) = 1.1t + 2$

a) $A(0) = 1.1(0) + 2 = 0 + 2 = 2\ mi^2$
$A(1) = 1.1(1) + 2 = 1.1 + 2 = 3.1\ mi^2$
$A(4) = 1.1(4) + 2 = 4.4 + 2 = 6.4\ mi^2$
$A(10) = 1.1(10) + 2 = 11 + 2 = 13\ mi^2$

b) This is a linear function. We plot the points (0, 2), (1, 3.1), (4, 6.4), and (10, 13) and draw the graph.

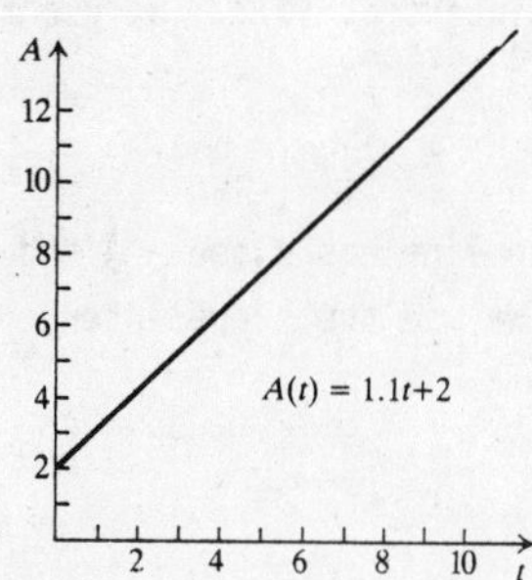

c) The time must be nonnegative, so $t \geqslant 0$. We can assume there is no upper bound on the length of time the organism spreads, so the domain is $[0, \infty)$.

64. a) 200.69 cm

b) 195.23 cm

c) $[0, \infty)$

65.
$f(x) = mx + b$ (Linear function)
$f(3x) = 3f(x)$ (Given)
$m(3x) + b = 3(mx + b)$
$3mx + b = 3mx + 3b$
$b = 3b$
$0 = 2b$
$0 = b$

Thus, if $f(3x) = 3f(x)$, then
$f(x) = mx + 0$
$f(x) = mx$

66. $f(x) = mx$

67.
$f(x) = mx + b$ (Linear function)
$f(x + 2) = f(x) + 2$ (Given)
$m(x + 2) + b = mx + b + 2$
$mx + 2m + b = mx + b + 2$
$2m = 2$
$m = 1$

Thus, if $f(x + 2) = f(x) + 2$, then
$f(x) = 1 \cdot x + b$
$f(x) = x + b$

68. $f(x) = b$

69. $f(x) = mx + b$ (Linear function)
$f(c + d) = m(c + d) + b = mc + md + b$
$f(c) + f(d) = (mc + b) + (md + b) = mc + md + 2b$

$mc + md + b \neq mc + md + 2b$

Thus, $f(c + d) = f(c) + f(d)$ is false.

70. False

71. $f(x) = mx + b$ (Linear function)
$f(kx) = m(kx) + b = mkx + b$
$kf(x) = k(mx + b) = mkx + kb$

$mkx + b \neq mkx + kb$

Thus, $f(kx) = kf(x)$ is false.

72. False

73. For $m > 0$, if $c < d$, then $mc < md$ and $mc + b < md + b$. Thus it is true that f is increasing.

74. True

75. If A(9, 4), B(-1, 2), and C(4, 3) are on the same line, then the slope of the segment $\overline{AB}$ must be the same as the slope of the segment $\overline{BC}$. (Note that B is a point of each segment.)

Slope of $\overline{AB} = \frac{4 - 2}{9 - (-1)} = \frac{2}{10} = \frac{1}{5}$

Slope of $\overline{BC} = \frac{2 - 3}{-1 - 4} = \frac{-1}{-5} = \frac{1}{5}$

Since the slopes are the same, and B is on both lines, A, B, and C are on the same line.

76. No

77.

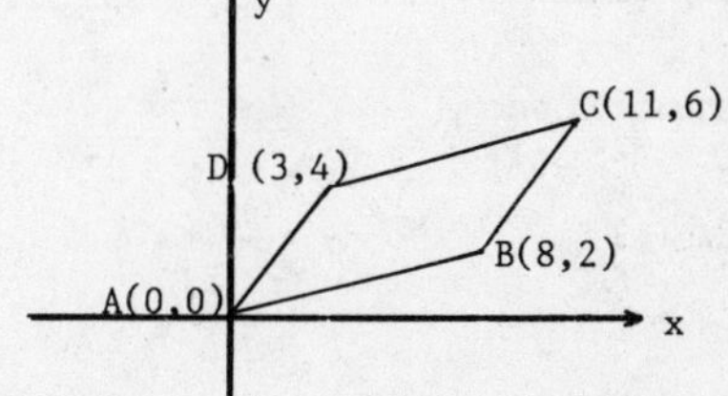

Slope of $\overline{AB} = \frac{2 - 0}{8 - 0} = \frac{2}{8} = \frac{1}{4}$

Slope of $\overline{CD} = \frac{6 - 4}{11 - 3} = \frac{2}{8} = \frac{1}{4}$

Slope of $\overline{BC} = \frac{6 - 2}{11 - 8} = \frac{4}{3}$

Slope of $\overline{DA} = \frac{4 - 0}{3 - 0} = \frac{4}{3}$

$\overline{AB}$ and $\overline{CD}$ have the same slope.
$\overline{BC}$ and $\overline{DA}$ have the same slope.

78. $m_{EG} = \frac{1}{3}$ and $m_{FH} = -3$.

Hence $m_{EG} = -\frac{1}{m_{FH}}$

79. Consider a line containing the ordered pairs (0, 32) and (100, 212).

The slope of the line is $\frac{212 - 32}{100 - 0}$, or $\frac{9}{5}$.

$F(C) = mC + b$ (Linear function)

$F(C) = \frac{9}{5}C + b$ [Substituting $\frac{9}{5}$ for m]

We now determine b.

$32 = \frac{9}{5} \cdot 0 + b$ (Substituting 0 for C and 32 for F)

$32 = b$

The linear function is

$F(C) = \frac{9}{5}C + 32$.

80. $C = \frac{5}{9}[F - 32]$

81. $P = mQ + b,\quad m \neq 0$ (Linear function)

We solve for Q.

$P - b = mQ$

$\frac{P - b}{m} = Q$

$\frac{P}{m} - \frac{b}{m} = Q$ (Linear function)

82. By definition, $y = kx$. Thus, y is a linear function.

Exercise Set 4.2

1. Solve each equation for y.

$2x - 5y = -3$ | $2x + 5y = 4$

$-5y = -2x - 3$ | $5y = -2x + 4$

$y = \frac{2}{5}x + \frac{3}{5}$ | $y = -\frac{2}{5}x + \frac{4}{5}$

The slopes are $\frac{2}{5}$ and $-\frac{2}{5}$. The slopes are not equal. The product of the slopes is not -1. Thus, the lines are neither parallel nor perpendicular.

2. Parallel

3. Solve each equation for y.

$y = 4x - 5$ | $4y = 8 - x$

$y = -\frac{1}{4}x + 2$

The slopes are 4 and $-\frac{1}{4}$. Their product is -1, so the lines are perpendicular.

4. Perpendicular

5. We first solve for y.

$3x - y = 7$

$-y = -3x + 7$

$y = 3x - 7$

The slope of the given line is 3. The line parallel to the given line will have slope 3. Since the line contains the point (0, 3), we know that the y-intercept is 3. We use the slope-intercept equation.

$y = mx + b$

$y = 3x + 3$

The perpendicular line has slope $-\frac{1}{3}$ and y-intercept 3. We use the slope-intercept equation again.

$y = mx + b$

$y = -\frac{1}{3}x + 3$

6. $y = -2x - 13$, $y = \frac{1}{2}x - 3$

7. The line x = 2 is vertical, so any line parallel to it must be vertical. The line we seek has one x-coordinate which is 3, so all x-coordinates on the line must be 3. The equation is x = 3.

The line perpendicular to x = 4 must be horizontal and has one y-coordinate which is 8, so all y-coordinates must be 8. The equation is y = 8.

8. x = 3, y = -3

9. The line y = 4 is horizontal, so any line parallel to it must be horizontal. The line we seek has one y-coordinate which is -3, so all y-coordinates on the line must be -3. The equation is y = -3.

The line perpendicular to y = 4 must be vertical and has one x-coordinate which is -2, so all x-coordinates must be -2. The equation is x = -2.

10. y = 2, x = -3

11. We first solve for y.

$5x - 2y = 4$

$-2y = -5x + 4$

$y = \frac{5}{2}x - 2$

The slope of the given line is $\frac{5}{2}$. The line parallel to the given line will have slope $\frac{5}{2}$. We use the point-slope equation to write an equation with slope $\frac{5}{2}$ and containing the point (-3, -5).

$y - y_1 = m(x - x_1)$

$y - (-5) = \frac{5}{2}[x - (-3)]$

$y + 5 = \frac{5}{2}x + \frac{15}{2}$

$y = \frac{5}{2}x + \frac{5}{2}$

11. (continued)

The slope of the perpendicular line is $-\frac{2}{5}$.

$y - y_1 = m(x - x_1)$ (Point-slope equation)

$y - (-5) = -\frac{2}{5}[x - (-3)]$ [Substituting -3 for x_1, -5 for y_1, and $-\frac{2}{5}$ for m]

$y + 5 = -\frac{2}{5}(x + 3)$

$y + 5 = -\frac{2}{5}x - \frac{6}{5}$

$y = -\frac{2}{5}x - \frac{31}{5}$

12. $y = -\frac{3}{4}x + \frac{1}{4}$, $y = \frac{4}{3}x - 6$

13. The line x = 1 is vertical, so any line parallel to it must be vertical. The line we seek has one x-coordinate which is 0, so all x-coordinates on the line must be 0. The equation is x = 0.

The line perpendicular to x = 1 must be horizontal and has one y-coordinate which is 3, so all y-coordinates must be 3. The equation is y = 3.

14. x = -2, y = -2

15. The line y = 2 is horizontal, so any line parallel to it must be horizontal. The line we seek has one y-coordinate which is -7, so all y-coordinates on the line must be -7. The equation is y = -7.

The line perpendicular to y = 2 must be vertical and has one x-coordinate which is -3, so all x-coordinates must be -3. The equation is x = -3.

16. y = -5, x = 4

17. We first solve for y.

$4.323x - 7.071y = 16.61$

$-7.071y = -4.323x + 16.61$

$y = 0.611x - 2.349$

The slope of the given line is 0.611. The slope of the line parallel to y = 0.611x - 2.349 and containing (-2.603, 1.818) must also be 0.611.

$y - y_1 = m(x - x_1)$ (Point-slope equation)

$y - 1.818 = 0.611[x - (-2.603)]$

(Substituting -2.603 for x_1, 1.818 for y_1, and 0.611 for m)

$y - 1.818 = 0.611x + 1.590$

$y = 0.611x + 3.408$

18. y = 0.642x - 4.930

19. Let $(x_1, y_1) = (-3, -2)$ and $(x_2, y_2) = (1, 1)$.

Then use the distance formula.

$d = \sqrt{(x_1 - x_2)^2 + (y_1 - y_2)^2}$

$d = \sqrt{(-3 - 1)^2 + (-2 - 1)^2} = \sqrt{(-4)^2 + (-3)^2}$

$= \sqrt{16 + 9} = \sqrt{25} = 5$

20. $3\sqrt{5}$

21. Let $(x_1, y_1) = (0, -7)$ and $(x_2, y_2) = (3, -4)$.

Then use the distance formula.

$d = \sqrt{(x_1 - x_2)^2 + (y_1 - y_2)^2}$

$d = \sqrt{(0 - 3)^2 + [-7 - (-4)]^2} = \sqrt{(-3)^2 + (-3)^2}$

$= \sqrt{9 + 9} = \sqrt{18} = 3\sqrt{2}$

22. $4\sqrt{2}$

23. Let $(x_1, y_1) = (a, -3)$ and $(x_2, y_2) = (2a, 5)$.

$d = \sqrt{(x_1 - x_2)^2 + (y_1 - y_2)^2}$

$d = \sqrt{(a - 2a)^2 + (-3 - 5)^2}$

$= \sqrt{(-a)^2 + (-8)^2} = \sqrt{a^2 + 64}$

24. $\sqrt{64 + k^2}$

25. Let $(x_1, y_1) = (0, 0)$ and $(x_2, y_2) = (a, b)$.

$d = \sqrt{(x_1 - x_2)^2 + (y_1 - y_2)^2}$

$d = \sqrt{(0 - a)^2 + (0 - b)^2}$

$= \sqrt{(-a)^2 + (-b)^2} = \sqrt{a^2 + b^2}$

26. $\sqrt{5}$

27. Let $(x_1, y_1) = (\sqrt{a}, \sqrt{b})$ and $(x_2, y_2) = (-\sqrt{a}, \sqrt{b})$.

$d = \sqrt{(x_1 - x_2)^2 + (y_1 - y_2)^2}$

$d = \sqrt{[\sqrt{a} - (-\sqrt{a})]^2 + (\sqrt{b} - \sqrt{b})^2}$

$= \sqrt{(2\sqrt{a})^2 + 0^2} = \sqrt{4a} = 2\sqrt{a}$

28. $2\sqrt{c^2 + d^2}$

29. Let $(x_1, y_1) = (7.3482, -3.0991)$ and $(x_2, y_2) = (18.9431, -17.9054)$.

$d = \sqrt{(x_1 - x_2)^2 + (y_1 - y_2)^2}$

$d = \sqrt{(7.3482 - 18.9431)^2 + [-3.0991-(-17.9054)]^2}$

$= \sqrt{(-11.5949)^2 + (14.8063)^2}$

$= \sqrt{134.4417 + 219.2265}$

$= \sqrt{353.6682} = 18.8061$

30. 404.4729

31. First we find the square of the distances between the points:

Between (9, 6) and (-1, 2)

$d^2 = [9 - (-1)]^2 + (6 - 2)^2 = 100 + 16 = 116$

Between (-1, 2) and (1, -3)

$d^2 = (-1 - 1)^2 + [2 - (-3)]^2 = 4 + 25 = 29$

Between (1, -3) and (9, 6)

$d^2 = (1 - 9)^2 + (-3 - 6)^2 = 64 + 81 = 145$

Since the sum of 116 and 29 is 145, it follows from the Pythagorean theorem that the points are vertices of a right triangle.

32. Yes

33. Let $(x_1, y_1) = (-4, 7)$ and $(x_2, y_2) = (3, -9)$.

$\left(\frac{x_1 + x_2}{2}, \frac{y_1 + y_2}{2}\right)$ (Midpoint formula)

$\left(\frac{-4 + 3}{2}, \frac{7 + (-9)}{2}\right)$ (Substituting)

$\left(\frac{-1}{2}, \frac{-2}{2}\right)$

The midpoint is $\left(-\frac{1}{2}, -1\right)$.

34. (5, -1)

35. Let $(x_1, y_1) = (a, b)$ and $(x_2, y_2) = (a, -b)$.

$\left(\frac{x_1 + x_2}{2}, \frac{y_1 + y_2}{2}\right)$ (Midpoint formula)

$\left(\frac{a + a}{2}, \frac{b + (-b)}{2}\right)$ (Substituting)

$\left(\frac{2a}{2}, \frac{0}{2}\right)$

The midpoint is (a, 0).

36. (0, d)

37. Let $(x_1, y_1) = (-3.895, 8.1212)$ and $(x_2, y_2) = (2.998, -8.6677)$.

$\left(\frac{x_1 + x_2}{2}, \frac{y_1 + y_2}{2}\right)$ (Midpoint formula)

$\left(\frac{-3.895 + 2.998}{2}, \frac{8.1212 + (-8.6677)}{2}\right)$ (Substituting)

$\left(\frac{-0.897}{2}, \frac{-0.5465}{2}\right)$

The midpoint is (-0.4485, 0.27325).

38. (4.652, 0.03775)

39. We write standard form.

$(x - 0)^2 + (y - 0)^2 = 6^2$

The center is (0, 0), and the radius is 6.

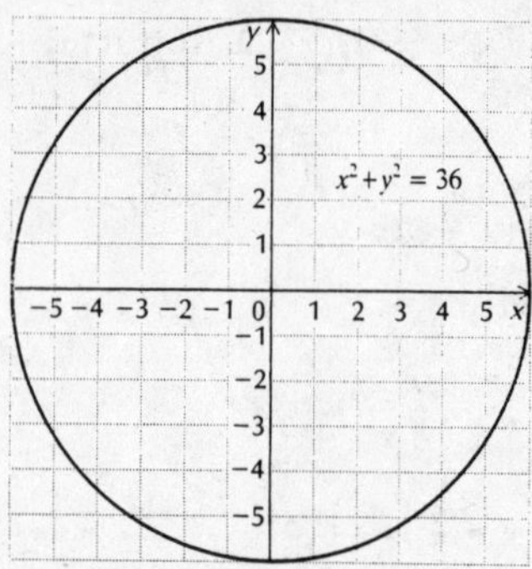

40. Center: (0, 0); radius: 5

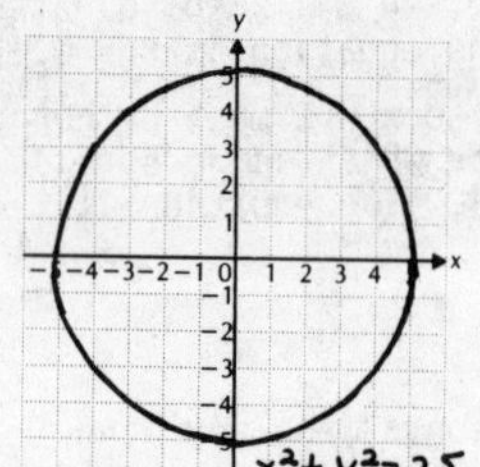

41. We write standard form.

$(x - 0)^2 + (y - 0)^2 = (\sqrt{3})^2$

The center is (0, 0), and the radius is $\sqrt{3}$.

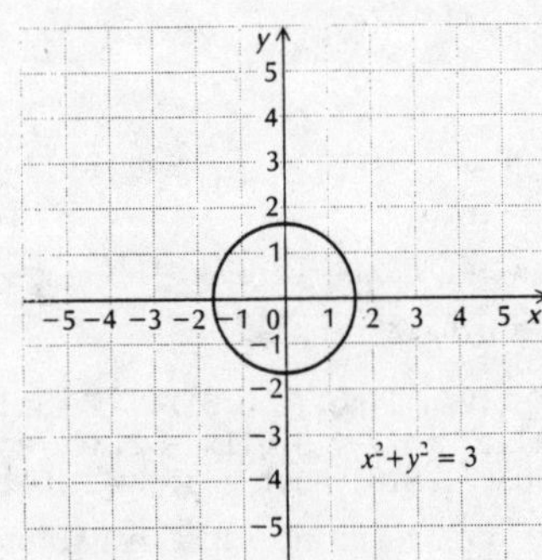

42. Center: (0, 0); radius: $\sqrt{2}$

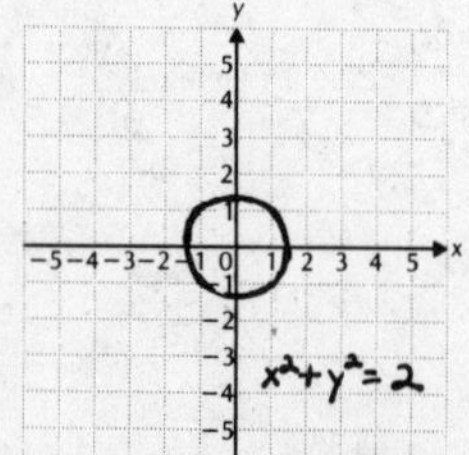

43. We write standard form.

$[x - (-1)^2] + [y - (-3)]^2 = 2^2$

The center is (-1, -3), and the radius is 2.

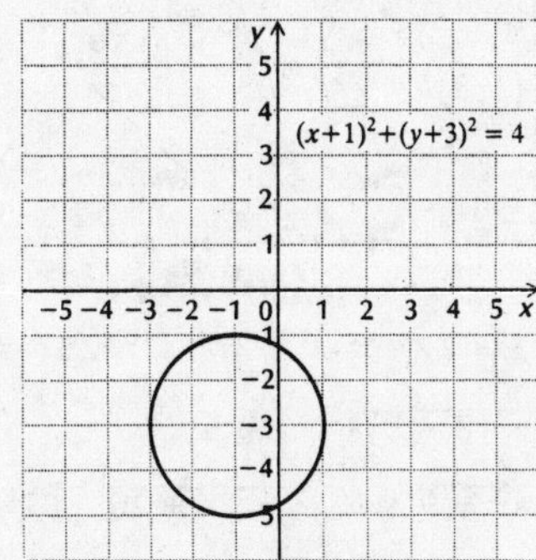

44. Center: (2, -3); radius: 1

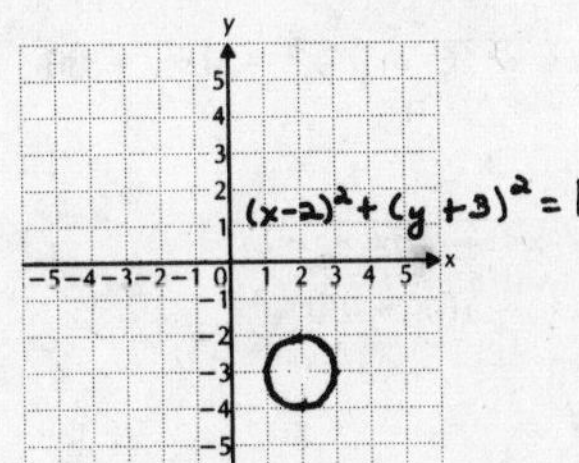

45. $(x - 8)^2 + (y + 3)^2 = 40$

$(x - 8)^2 + [y - (-3)]^2 = (\sqrt{40})^2 = (2\sqrt{10})^2$

Center: (8, -3), Radius: $2\sqrt{10}$

46. Center: (-5, 1), Radius: $5\sqrt{3}$

47. $(x - 3)^2 + y^2 = \frac{1}{25}$

$(x - 3)^2 + (y - 0)^2 = \left(\frac{1}{5}\right)^2$

Center: (3, 0), Radius: $\frac{1}{5}$

48. Center: (0, 1), Radius: $\frac{1}{2}$

49. $x^2 + y^2 + 8x - 6y - 15 = 0$

We complete the square twice to get standard form:

$(x^2 + 8x \qquad) + (y^2 - 6y \qquad) = 15$

We take half the coefficient of the x-term and square it, obtaining 16. We add 16 - 16 in the first parentheses. Similarly, we add 9 - 9 in the second parentheses.

$(x^2 + 8x + 16 - 16) + (y^2 - 6y + 9 - 9) = 15$

Next we rearrange and factor:

$(x^2 + 8x + 16) + (y^2 - 6y + 9) - 16 - 9 = 15$

$(x + 4)^2 + (y - 3)^2 = 40$

$[x - (-4)]^2 + (y - 3)^2 = (2\sqrt{10})^2$

Center: (-4, 3), Radius: $2\sqrt{10}$

50. Center: $\left(-\frac{25}{2}, -5\right)$, Radius: $\frac{\sqrt{677}}{2}$

51. $x^2 + y^2 + 6x = 0$

$(x^2 + 6x \qquad) + y^2 = 0$

$(x^2 + 6x + 9 - 9) + y^2 = 0$ Completing the square once

$(x^2 + 6x + 9) + y^2 - 9 = 0$

$(x + 3)^2 + y^2 = 9$

$[x - (-3)]^2 + (y - 0)^2 = 3^2$

Center: (-3, 0), Radius: 3

52. Center: (2, 0), Radius: 2

53. $x^2 + y^2 + 8x = 84$

$(x^2 + 8x \qquad) + y^2 = 84$

$(x^2 + 8x + 16 - 16) + y^2 = 84$ Completing the square once

$(x^2 + 8x + 16) + y^2 - 16 = 84$

$(x + 4)^2 + y^2 = 100$

$[x - (-4)]^2 + (y - 0)^2 = 10^2$

Center: (-4, 0), Radius: 10

54. Center: (0, 5), Radius: 10

55. $x^2 + y^2 + 21x + 33y + 17 = 0$

$(x^2+21x \qquad)+(y^2+33y \qquad) = -17$

$\left(x^2+21x + \frac{441}{4} - \frac{441}{4}\right)+\left(y^2+33y + \frac{1089}{4} - \frac{1089}{4}\right) = -17$

Completing the square twice

$\left(x^2+21x + \frac{441}{4}\right)+\left(y^2+33y + \frac{1089}{4}\right) - \frac{441}{4} - \frac{1089}{4} = -17$

$\left(x + \frac{21}{2}\right)^2 + \left(y + \frac{33}{2}\right)^2 = \frac{1462}{4}$

$\left[x - \left(-\frac{21}{2}\right)\right]^2 + \left[y - \left(-\frac{33}{2}\right)\right]^2 = \left(\frac{\sqrt{1462}}{2}\right)^2$

Center: $\left(-\frac{21}{2}, -\frac{33}{2}\right)$, Radius: $\frac{\sqrt{1462}}{2}$

56. Center: $\left(\frac{7}{2}, -\frac{3}{2}\right)$, Radius: $\frac{7\sqrt{2}}{2}$

57. $x^2 + y^2 + 8.246x - 6.348y - 74.35 = 0$

$(x^2 + 8.246x \qquad) + (y^2 - 6.348y \qquad) = 74.35$

$(x^2 + 8.246x + 16.999 - 16.999) + (y^2 - 6.348y + 10.074 - 10.074) = 74.35$

$(x^2 + 8.246x + 16.999) + (y^2 - 6.348y + 10.074) - 16.999 - 10.074 = 74.35$

$(x + 4.123)^2 + (y - 3.174)^2 = 101.423$

$[x - (-4.123)]^2 + (y - 3.174)^2 = (10.071)^2$

Center: (-4.123, 3.174), Radius: 10.071

58. Center: (-12.537, -5.002), Radius: 13.044

59. $9x^2 + 9y^2 = 1$

$x^2 + y^2 = \frac{1}{9}$ Multiplying by $\frac{1}{9}$

$(x - 0)^2 + (y - 0)^2 = \left(\frac{1}{3}\right)^2$

Center: (0, 0), Radius: $\frac{1}{3}$

60. Center: (0, 0), Radius: $\frac{1}{4}$

61. Since the center is (0, 0), we have
$(x - 0)^2 + (y - 0)^2 = r^2$ or $x^2 + y^2 = r^2$

The circle passes through (-3, 4). We find r^2 by substituting -3 for x and 4 for y.

$(-3)^2 + 4^2 = r^2$

$9 + 16 = r^2$

$25 = r^2$

Then $x^2 + y^2 = 25$ is an equation of the circle.

62. $(x - 3)^2 + (y + 2)^2 = 64$

63. Since the center is (-4, 1), we have
$[x - (-4)]^2 + (y - 1)^2 = r^2$, or
$(x + 4)^2 + (y - 1)^2 = r^2$.

The circle passes through (-2, 5). We find r^2 by substituting -2 for x and 5 for y.

$(-2 + 4)^2 + (5 - 1)^2 = r^2$

$4 + 16 = r^2$

$20 = r^2$

Then $(x + 4)^2 + (y - 1)^2 = 20$ is an equation of the circle.

64. $(x + 3)^2 + (y + 3)^2 = 54.4$

65. The slope of the line containing (-1, 4) and (2, -3) is $\frac{4 - (-3)}{-1 - 2}$, or $-\frac{7}{3}$. The slope of the line parallel to this line and containing (4, -2) is also $-\frac{7}{3}$.

$y - y_1 = m(x - x_1)$ (Point-slope equation)

$y - (-2) = -\frac{7}{3}(x - 4)$ [Substituting 4 for x_1, -2 for y_1, and $-\frac{7}{3}$ for m]

$y + 2 = -\frac{7}{3}x + \frac{28}{3}$

$y = -\frac{7}{3}x + \frac{22}{3}$

The slope of the line perpendicular to the given line is $\frac{3}{7}$.

$y - y_1 = m(x - x_1)$ (Point-slope equation)

$y - (-2) = \frac{3}{7}(x - 4)$ (Substituting)

$y + 2 = \frac{3}{7}x - \frac{12}{7}$

$y = \frac{3}{7}x - \frac{26}{7}$

66. $y = \frac{2}{5}x + \frac{17}{5}$, $y = -\frac{5}{2}x + \frac{1}{2}$

67.

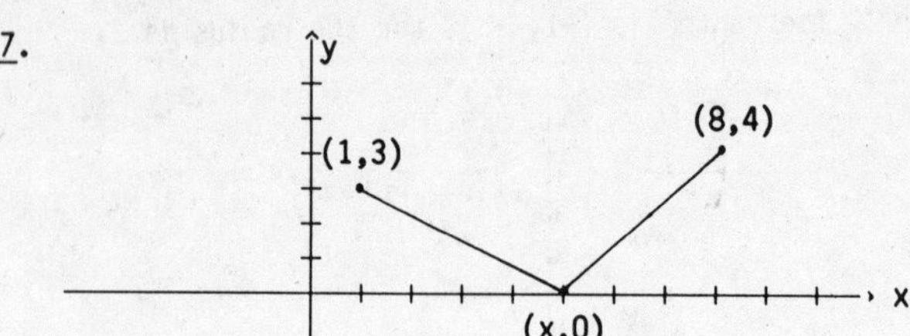

Between (1, 3) and (x, 0):

$d = \sqrt{(1 - x)^2 + (3 - 0)^2}$

$= \sqrt{1 - 2x + x^2 + 9} = \sqrt{x^2 - 2x + 10}$

Between (8, 4) and (x, 0):

$d = \sqrt{(8 - x)^2 + (4 - 0)^2}$

$= \sqrt{64 - 16x + x^2 + 16} = \sqrt{x^2 - 16x + 80}$

Thus,

$\sqrt{x^2 - 2x + 10} = \sqrt{x^2 - 16x + 80}$

$x^2 - 2x + 10 = x^2 - 16x + 80$

$14x = 70$

$x = 5$

The point on the x-axis equidistant from the points (1, 3) and (8, 4) is (5, 0).

68. (0, 4)

69. The slope of the line containing (-3, k) and (4, 8) is

$\frac{8 - k}{4 - (-3)} = \frac{8 - k}{7}$.

The slope of the line containing (6, 4) and (2, -5) is

$\frac{-5 - 4}{2 - 6} = \frac{-9}{-4} = \frac{9}{4}$.

The lines are parallel when their slopes are equal:

$\frac{8 - k}{7} = \frac{9}{4}$

$32 - 4k = 63$

$-4k = 31$

$k = -\frac{31}{4}$

70. $k = \frac{100}{9}$

71. The slope of the line segment with endpoints $(-1, 3)$ and $(-6, 7)$ is $\frac{3 - 7}{-1 - (-6)} = -\frac{4}{5}$.

The midpoint of the given line segement is $\left(\frac{-1 + (-6)}{2}, \frac{3 + 7}{2}\right)$, or $\left(-\frac{7}{2}, 5\right)$.

The perpendicular bisector of the given line segment has slope $\frac{5}{4}$ and contains the point $\left(-\frac{7}{2}, 5\right)$.

$$y - y_1 = m(x - x_1)$$
$$y - 5 = \frac{5}{4}\left[x - \left(-\frac{7}{2}\right)\right]$$
$$y - 5 = \frac{5}{4}x + \frac{35}{8}$$
$$y = \frac{5}{4}x + \frac{75}{8}$$

72. $y = -\frac{2}{3}x - 8$

73. If the circle with center $(2, 4)$ is tangent to the x-axis, the radius is 4. We substitute 2 for h, 4 for k, and 4 for r.

$$(x - h)^2 + (y - k)^2 = r^2$$
$$(x - 2)^2 + (y - 4)^2 = 4^2 = 16$$

74. $r = 3$, so $(x + 3)^2 + (y + 2)^2 = 9$

75.

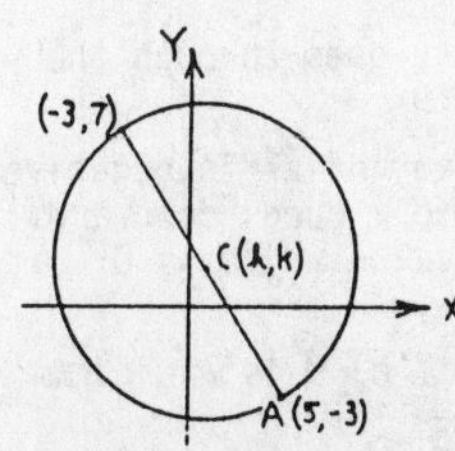

The center of the circle is the midpoint of the line segment with endpoints $(5, -3)$ and $(-3, 7)$:

$$\left(\frac{5 + (-3)}{2}, \frac{-3 + 7}{2}\right), \text{ or } (1, 2)$$

The radius is the distance from $(1, 2)$ to $(5, -3)$ (or to $(-3, 7)$):

$$r = \sqrt{(1 - 5)^2 + [2 - (-3)]^2} = \sqrt{16 + 25} = \sqrt{41}$$

The equation of the circle is

$$(x - 1)^2 + (y - 2)^2 = (\sqrt{41})^2, \text{ or}$$
$$(x - 1)^2 + (y - 2)^2 = 41.$$

76. $\left(x - \frac{5}{2}\right)^2 + \left(y - \frac{5}{2}\right)^2 = \frac{145}{2}$

77.
$$C = 2\pi r$$
$$10\pi = 2\pi r$$
$$5 = r$$

Then $[x - (-8)]^2 + (y - 5)^2 = 5^2$, or $(x + 8)^2 + (y - 5)^2 = 25$.

78. $(x + 3)^2 + (y + 8)^2 = 36$

79.

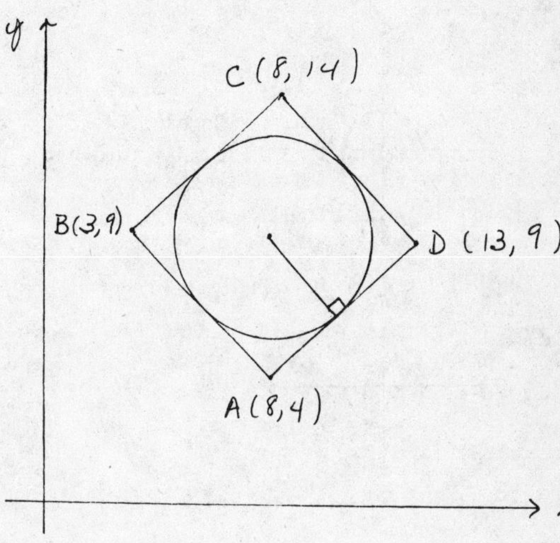

The center is the midpoint of AC or BD. We use AC:

$$\left(\frac{8 + 8}{2}, \frac{4 + 14}{2}\right), \text{ or } (8, 9)$$

The radius is the distance from the center to the midpoint of a side of the square. We will use side AD. The midpoint of AD is

$$\left(\frac{8 + 13}{2}, \frac{4 + 9}{2}\right), \text{ or } \left(\frac{21}{2}, \frac{13}{2}\right). \text{ Then}$$

$$r = \sqrt{\left(8 - \frac{21}{2}\right)^2 + \left(9 - \frac{13}{2}\right)^2} = \sqrt{\frac{25}{2}} = \frac{5\sqrt{2}}{2}.$$

80. $x^2 + (y - 30.8)^2 = 640.09$

81. The center is $(0, 0)$, and the radius is 8.4. Thus the equation is $(x - 0)^2 + (y - 0)^2 = (8.4)^2$, or $x^2 + y^2 = 70.56$.

82. $a_1 \approx 2.68$ ft, $a_2 \approx 37.32$ ft

ɔn and

83.
$$(x^2 + y^2) - 2(ky + hx) + (h^2 + k^2) = r^2$$
$$x^2 - 2hx + h^2 + y^2 - 2yk + k^2 = r^2$$
$$(x - h)^2 + (y - k)^2 = r^2$$

84. a)

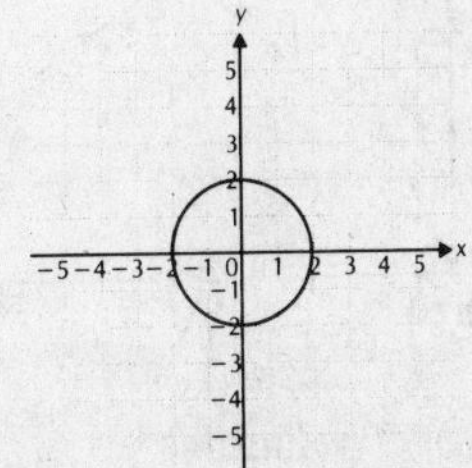

No

b) $y = \pm\sqrt{4 - x^2}$

84. (continued)

c) 

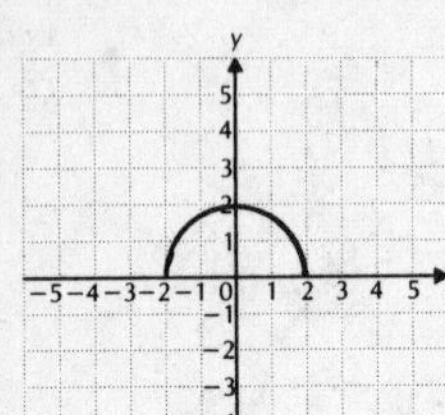

Yes

Domain: $\{x \mid -2 \leqslant x \leqslant 2\}$
Range: $\{y \mid 0 \leqslant y \leqslant 2\}$

d)

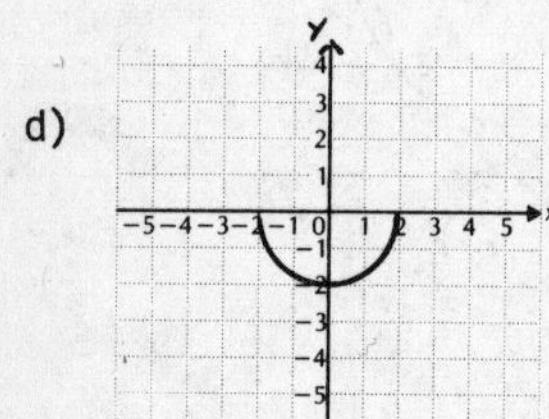

Yes

85.

$$\begin{array}{c|c} x^2 + y^2 & 1 \\ \hline (-1)^2 + 0^2 & 1 \\ 1 & \end{array}$$

(-1, 0) lies on the unit circle.

86. Yes

87.

$$\begin{array}{r|c} x^2 + y^2 & 1 \\ \hline (0.838670568)^2 + (-0.544639035)^2 & 1 \\ 0.703368321 + 0.296631678 & \\ 1 & \end{array}$$

(0.838670568, -0.544639035) lies on the unit circle.

88. No

89. See the answer section in the text.

90. The midpoints are

$X\left(\frac{a_1}{2}, \frac{a_2}{2}\right)$

$Y\left(\frac{a_1 + b_1}{2}, \frac{a_2 + b_2}{2}\right)$

$Z\left(\frac{b_1 + c_1}{2}, \frac{b_2 + c_2}{2}\right)$

$W\left(\frac{c_1}{2}, \frac{c_2}{2}\right)$

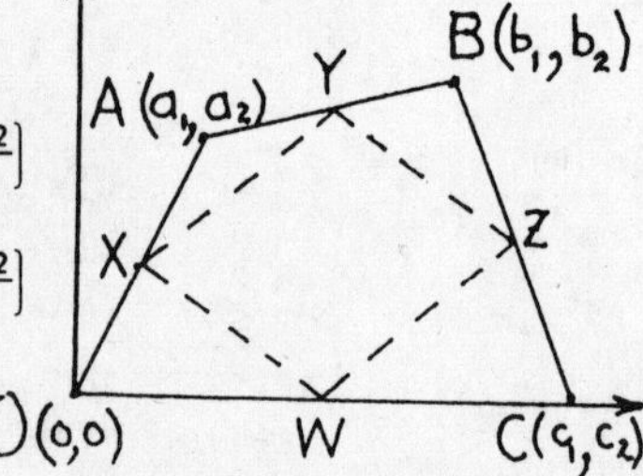

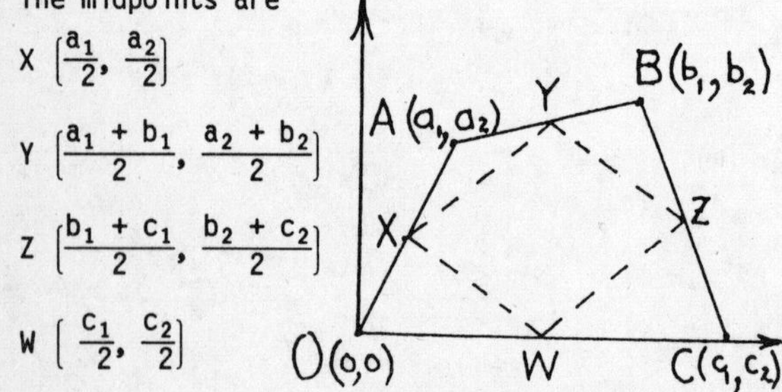

Then $m_{XY} = \frac{b_2}{b_1} = m_{ZW}$ and $\overline{XY} \parallel \overline{ZW}$.

Also $m_{XW} = \frac{a_2 - c_2}{a_1 - c_1} = m_{YZ}$ and $\overline{XW} \parallel \overline{YZ}$.

Hence XYZW is a parallelogram.

91. See the answer section in the text.

92. $x^2 + y^2 = a^2$ gives $b^2 + c^2 = a^2$ or $c^2 = -(b^2 - a^2)$. Also

$$m_{AB} \cdot m_{BC} = \frac{c - 0}{b - (-a)} \cdot \frac{c - 0}{b - a}$$

$$= \frac{c^2}{b^2 - a^2} = \frac{-(b^2 - a^2)}{b^2 - a^2} = -1.$$

$\therefore AB \perp BC$

Exercise Set 4.3

1. $f(x) = x^2$

$f(x) = (x - 0)^2 + 0$ [In the form $f(x) = a(x - h)^2 + k$]

a) The vertex is (0, 0).

b) The line of symmetry goes through the vertex; it is the line x = 0.

c) Since the coefficient of x^2 is positive ($1 > 0$), the parabola opens upward. Thus we have a minimum value which is 0.

2. a) (0, 0), b) x = 0, c) 0 is a maximum

3. $f(x) = -2(x - 9)^2$

$f(x) = -2(x - 9)^2 + 0$ [In the form $f(x) = a(x - h)^2 + k$]

a) The vertex is (9, 0).

b) The line of symmetry goes through the vertex; it is the line x = 9.

c) Since the coefficient of x^2 is negative ($-2 < 0$), the parabola opens downward. Thus we have a maximum value which is 0.

4. a) (7, 0), b) x = 7, c) 0 is a minimum

5. $f(x) = 2(x - 1)^2 - 4$

$f(x) = 2(x - 1)^2 + (-4)$ [In the form $f(x) = a(x - h)^2 + k$]

a) The vertex is (1, -4)

b) The line of symmetry goes through the vertex; it is the line x = 1.

c) Since the coefficient of x^2 is positive ($2 > 0$), the parabola opens upward. Thus we have a minimum value which is -4.

6. a) (-4, -3), b) x = -4, c) -3 is a maximum

7. a) $f(x) = -x^2 + 2x + 3$
$= -(x^2 - 2x) + 3$
$= -(x^2 - 2x + 1 - 1) + 3$
(Adding 1 - 1 inside parentheses)
$= -(x^2 - 2x + 1) + 1 + 3$
$= -(x - 1)^2 + 4$ [In the form $f(x) = a(x - h)^2 + k$]

b) The vertex is (1, 4).

c) Since the coefficient of x^2 is negative $(-1 < 0)$, the parabola opens downward. Thus we have a maximum value which is 4.

8. a) $f(x) = -(x - 4)^2 + 9$

b) (4, 9)

c) 9 is a maximum

9. a) $f(x) = x^2 + 3x$
$= x^2 + 3x + \frac{9}{4} - \frac{9}{4}$ $\left(\text{Adding } \frac{9}{4} - \frac{9}{4}\right)$
$= \left(x + \frac{3}{2}\right)^2 - \frac{9}{4}$
$= \left[x - \left(-\frac{3}{2}\right)\right]^2 + \left(-\frac{9}{4}\right)$
[In the form $f(x) = a(x - h)^2 + k$]

b) The vertex is $\left(-\frac{3}{2}, -\frac{9}{4}\right)$.

c) Since the coefficient of x^2 is positive $(1 > 0)$, the parabola opens upward. Thus we have a minimum value which is $-\frac{9}{4}$.

10. a) $f(x) = \left(x - \frac{9}{2}\right)^2 - \frac{81}{4}$

b) $\left(\frac{9}{2}, -\frac{81}{4}\right)$

c) $-\frac{81}{4}$ is a minimum

11. a) $f(x) = -\frac{3}{4}x^2 + 6x$
$= -\frac{3}{4}(x^2 - 8x)$
$= -\frac{3}{4}(x^2 - 8x + 16 - 16)$
(Adding 16 - 16 inside parentheses)
$= -\frac{3}{4}(x^2 - 8x + 16) + 12$
$= -\frac{3}{4}(x - 4)^2 + 12$
[In the form $f(x) = a(x - h)^2 + k$]

b) The vertex is (4, 12).

c) Since the coefficient of x^2 is negative $\left(-\frac{3}{4} < 0\right)$, the parabola opens downward. Thus we have a maximum value which is 12.

12. a) $f(x) = \frac{3}{2}(x + 1)^2 - \frac{3}{2}$

b) $\left(-1, -\frac{3}{2}\right)$

c) $-\frac{3}{2}$ is a minimum

13. a) $f(x) = 3x^2 + x - 4$
$= 3\left(x^2 + \frac{1}{3}x\right) - 4$
$= 3\left(x^2 + \frac{1}{3}x + \frac{1}{36} - \frac{1}{36}\right) - 4$
$\left(\text{Adding } \frac{1}{36} - \frac{1}{36} \text{ inside parentheses}\right)$
$= 3\left(x^2 + \frac{1}{3}x + \frac{1}{36}\right) - \frac{1}{12} - 4$
$= 3\left(x + \frac{1}{6}\right)^2 - \frac{49}{12}$
$= 3\left[x - \left(-\frac{1}{6}\right)\right]^2 + \left(-\frac{49}{12}\right)$
[In the form $f(x) = a(x - h)^2 + k$]

b) The vertex is $\left(-\frac{1}{6}, -\frac{49}{12}\right)$.

c) Since the coefficient of x^2 is positive $(3 > 0)$, the parabola opens upward. Thus we have a minimum value which is $-\frac{49}{12}$.

14. a) $f(x) = -2\left(x - \frac{1}{4}\right)^2 - \frac{7}{8}$

b) $\left(\frac{1}{4}, -\frac{7}{8}\right)$

c) $-\frac{7}{8}$ is a maximum

15. $f(x) = -5(x + 2)^2$
$f(x) = -5[x - (-2)]^2 + 0$
The value 0 is a maximum since the coefficient -5 is negative.

16. Minimum: 0

17. $f(x) = 8(x - 1)^2 + 5$
The value 5 is a minimum since the coefficient 8 is positive.

18. Maximum: -11

19. $f(x) = -4x^2 + x - 13$
$a = -4$, $b = 1$, $c = -13$
Since we want only the maximum or minimum value of the function, we find the second coordinate of the vertex:
$$-\frac{b^2 - 4ac}{4a} = -\frac{1^2 - 4(-4)(-13)}{4(-4)}$$
$$= -\frac{207}{16}$$
The value $-\frac{207}{16}$ is a maximum since the coefficient of x^2 is negative.

20. Minimum: 3.89625

21. $f(x) = \frac{1}{2}x^2 - \frac{2}{5}x - \frac{67}{100}$

$a = \frac{1}{2},\ b = -\frac{2}{5},\ c = -\frac{67}{100}$

The second coordinate of the vertex is

$$-\frac{b^2 - 4ac}{4a} = -\frac{\left(-\frac{2}{5}\right)^2 - 4\left(\frac{1}{2}\right)\left(-\frac{67}{100}\right)}{4\left(\frac{1}{2}\right)}$$

$$= -\frac{3}{4}.$$

The value $-\frac{3}{4}$ is a minimum since the coefficient of x^2 is positive.

22. Maximum: -16.0106

23. $q(x) = -\$120{,}000x^2 + \$430{,}000x - \$240{,}000$
$a = -\$120{,}000,\ b = \$430{,}000,\ c = -\$240{,}000$

The second coordinate of the vertex is

$$-\frac{b^2 - 4ac}{4a} =$$

$$-\frac{(\$430{,}000)^2 - 4(-\$120{,}000)(-\$240{,}000)}{4(-\$120{,}000)} =$$

$$\$\frac{435{,}625}{3}$$

The value $\$\frac{435{,}625}{3}$ is a maximum since the coefficient of x^2 is negative.

24. Minimum: $\frac{\pi^2 - 8\sqrt{10}}{8\sqrt{2}} \approx 1.3637$

25. $f(x) = -x^2 + 2x + 3$
$= -(x^2 - 2x) + 3$
$= -(x^2 - 2x + 1 - 1) + 3$ (Adding 1 - 1)
$= -(x^2 - 2x + 1) + 1 + 3$
$= -(x - 1)^2 + 4$

It is not necessary to rewrite the function in the form $f(x) = a(x - h)^2 + k$, but it can be helpful. Since the coefficient of x^2 is negative, the parabola opens downward. The vertex is (1, 4). The axis of symmetry is $x = 1$. We calculate several input-output pairs and plot these points.

x	f(x)
1	4
0	3
2	3
-1	0
3	0

$f(x) = -x^2 + 2x + 3$

26. $f(x) = x^2 - 3x - 4$

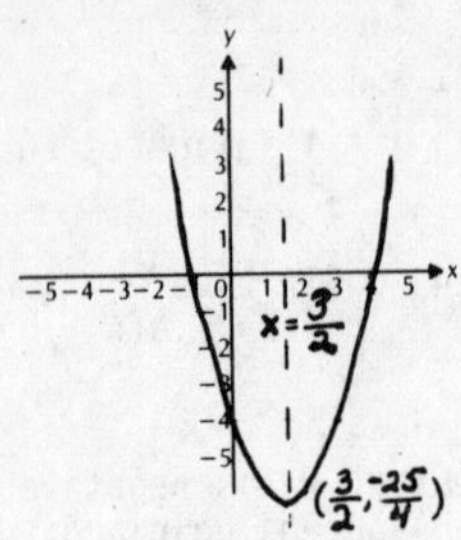

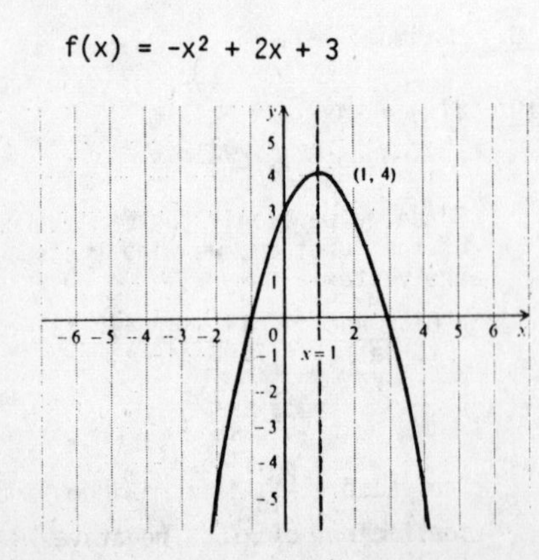

27. $f(x) = x^2 - 8x + 19$
$= (x^2 - 8x + 16 - 16) + 19$ (Adding 16 - 16)
$= (x^2 - 8x + 16) - 16 + 19$
$= (x - 4)^2 + 3$

Since the coefficient of x^2 is positive, the parabola opens upward. The vertex is (4, 3). The axis of symmetry is $x = 4$. We calculate several input-output pairs and plot these points.

x	f(x)
4	3
3	4
5	4
2	7
6	7

$f(x) = x^2 - 8x + 19$

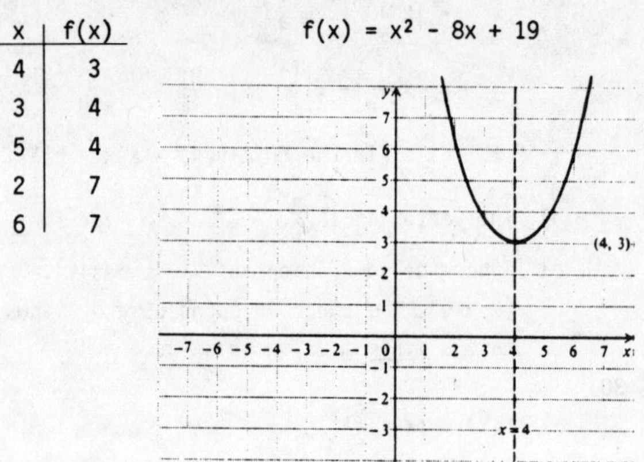

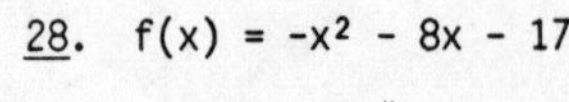

28. $f(x) = -x^2 - 8x - 17$

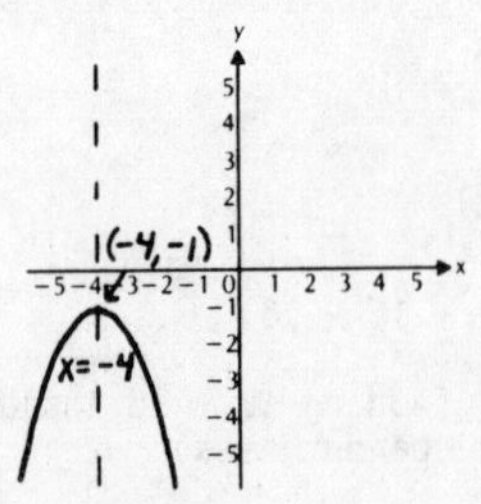

29. $f(x) = -\frac{1}{2}x^2 - 3x + \frac{1}{2}$

$= -\frac{1}{2}(x^2 + 6x) + \frac{1}{2}$

$= -\frac{1}{2}(x^2 + 6x + 9 - 9) + \frac{1}{2}$ (Adding 9 - 9)

$= -\frac{1}{2}(x^2 + 6x + 9) + \frac{9}{2} + \frac{1}{2}$

$= -\frac{1}{2}(x + 3)^2 + 5$

$= -\frac{1}{2}[x - (-3)]^2 + 5$

Since the coefficient of x^2 is negative, the parabola opens downward. The vertex is (-3, 5). The axis of symmetry is x = -3. We calculate several input-output pairs and plot these points.

x	f(x)
-3	5
-4	$\frac{9}{2}$
-2	$\frac{9}{2}$
-5	3
-1	3

$f(x) = -\frac{1}{2}x^2 - 3x + \frac{1}{2}$

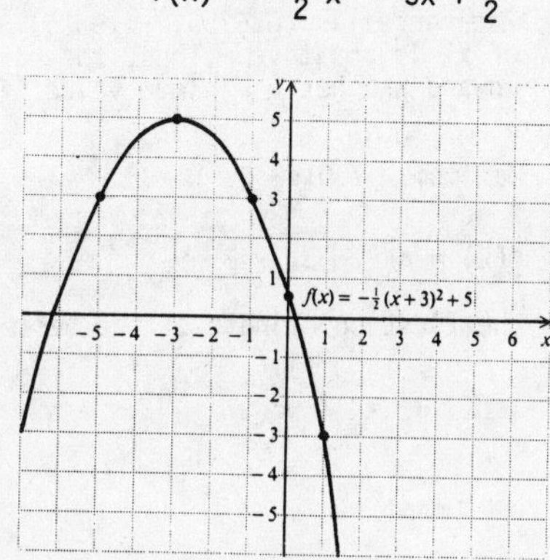

30. $f(x) = 2x^2 - 4x - 2$

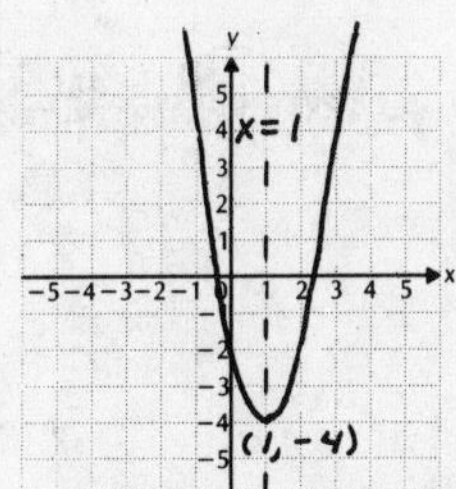

31. $f(x) = 3x^2 - 24x + 50$

$= 3(x^2 - 8x) + 50$

$= 3(x^2 - 8x + 16 - 16) + 50$ (Adding 16 - 16)

$= 3(x^2 - 8x + 16) - 48 + 50$

$= 3(x - 4)^2 + 2$

Since the coefficient of x^2 is positive, the parabola opens upward. The vertex is (4, 2). The axis of symmetry is x = 4. We calculate several input-output pairs and plot these points.

31. (continued)

x	f(x)
4	2
3	5
5	5
2	14
6	14

$f(x) = 3x^2 - 24x + 50$

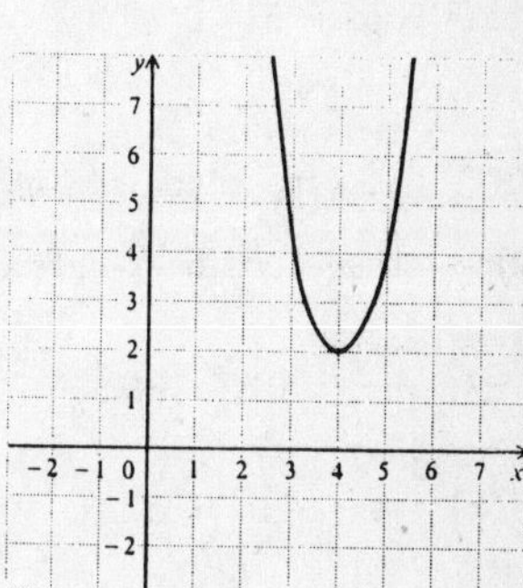

32. $f(x) = -2x^2 + 2x + 1$

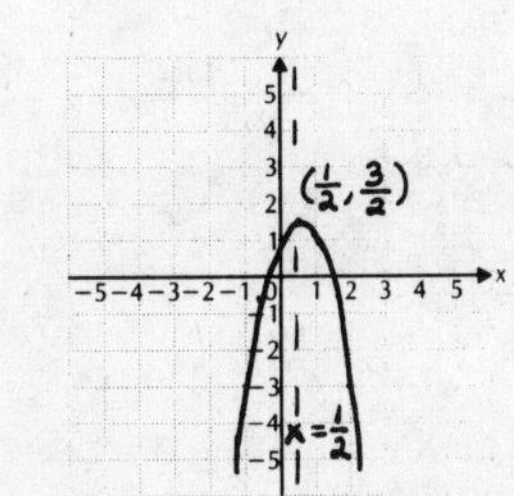

33. $f(x) = -x^2 + 2x + 3$

We solve the equation

$-x^2 + 2x + 3 = 0.$

$x^2 - 2x - 3 = 0$ (Muliplying by -1)

$(x + 1)(x - 3) = 0$

$x + 1 = 0$ or $x - 3 = 0$

$x = -1$ or $x = 3$

Thus the x-intercepts are (-1, 0) and (3, 0).

34. (-1, 0), (4, 0)

35. $f(x) = x^2 - 8x + 5$

We solve the equation

$x^2 - 8x + 5 = 0.$

a = 1, b = -8, c = 5

$$x = \frac{-(-8) \pm \sqrt{(-8)^2 - 4\cdot1\cdot5}}{2\cdot1} = \frac{8 \pm \sqrt{64 - 20}}{2}$$

$$= \frac{8 \pm \sqrt{44}}{2} = \frac{8 \pm 2\sqrt{11}}{2} = 4 \pm \sqrt{11}$$

Thus the x-intercepts are $(4 - \sqrt{11}, 0)$ and $(4 + \sqrt{11}, 0)$.

36. None

37. $f(x) = -5x^2 + 6x - 5$

We solve the equation

$-5x^2 + 6x - 5 = 0.$

$5x^2 - 6x + 5 = 0$

$a = 5, \quad b = -6, \quad c = 5$

$b^2 - 4ac$ (the discriminant) $= -64 < 0$

Thus, there are no x-intercepts.

38. $\left(\frac{-1 - \sqrt{41}}{4}, 0\right), \quad \left(\frac{-1 + \sqrt{41}}{4}, 0\right)$

39. $f(x) = ax^2 + bx + c$

$$= a\left(x^2 + \frac{b}{a}x\right) + c$$

$$= a\left(x^2 + \frac{b}{a}x + \frac{b^2}{4a^2} - \frac{b^2}{4a^2}\right) + c$$

$$= a\left(x^2 + \frac{b}{a}x + \frac{b^2}{4a^2}\right) - \frac{b^2}{4a} + c$$

$$= a\left(x + \frac{b}{2a}\right)^2 + \frac{-b^2 + 4ac}{4a}$$

$$= a\left[x - \left(-\frac{b}{2a}\right)\right]^2 + \frac{4ac - b^2}{4a}$$

40. $f(x) = 3\left(x + \frac{m}{6}\right)^2 + \frac{11m^2}{12}$

41. $f(x) = |x^2 - 1|$

First graph $f(x) = x^2 - 1$.

x	f(x)
0	-1
-1	0
1	0
2	3
-2	3

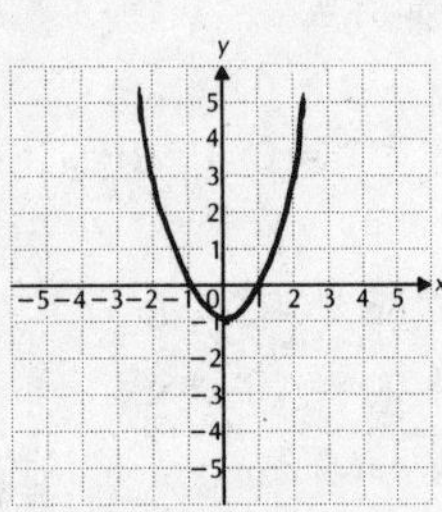

Then reflect the negative values across the x-axis. The point (0, -1) is now (0, 1).

$f(x) = |x^2 - 1|$

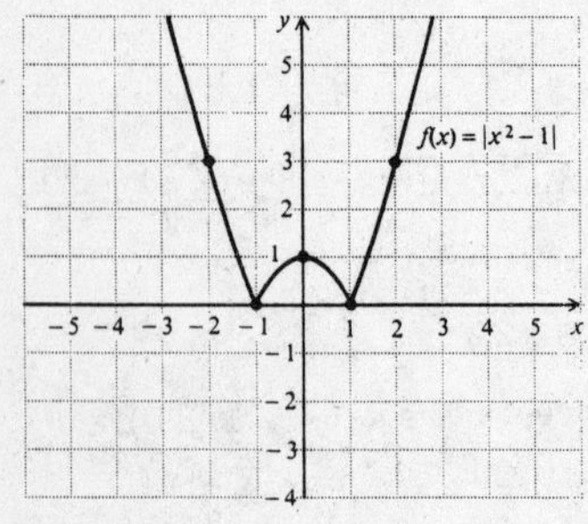

42. $f(x) = |3 - 2x - x^2|$

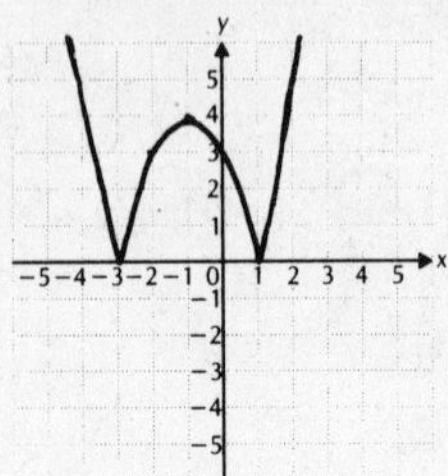

43. $f(x) = 2.31x^2 - 3.135x - 5.89$

$$= 2.31(x^2 - 1.3571x) - 5.89$$

$$= 2.31(x^2 - 1.3571x + 0.4604 - 0.4604) - 5.89$$

$$= 2.31(x^2 - 1.3571x + 0.4604) - 1.0635 - 5.89$$

$$= 2.31(x - 0.6785)^2 - 6.9535$$

The vertex is (0.6785, -6.9535). The coefficient of x^2 is positive. Thus, the parabola opens upward and has a minimum value of -6.9535.

44. Maximum: 7.014

45. $f(x) = ax^2 + 3x - 8$

The first coordinate of the vertex, $-\frac{b}{2a}$, is -2:

$$-\frac{3}{2a} = -2$$

$$\frac{3}{4} = a$$

46. $b = \pm 4\sqrt{17}$

47. $f(x) = -0.2x^2 - 3x + c$

The second coordinate of the vertex, $-\frac{b^2 - 4ac}{4a}$, is -225:

$$-\frac{(-3)^2 - 4(-0.2)(c)}{4(-0.2)} = -225$$

$$\frac{9 + 0.8c}{0.8} = -225$$

$$9 + 0.8c = -180$$

$$0.8c = -189$$

$$c = -236.25, \text{ or } -\frac{945}{4}$$

48. $f(x) = \frac{6}{49}x^2 - \frac{48}{49}x - \frac{149}{49}$

49. $f(x) = qx^2 - 2x + q$

The first coordinate of the vertex is

$$-\frac{b}{2a} = -\frac{-2}{2q} = \frac{1}{q}.$$

We substitute to find the second coordinate of the vertex:

$$f\left(\frac{1}{q}\right) = q\left(\frac{1}{q^2}\right) - 2\left(\frac{1}{q}\right) + q$$

$$= \frac{1}{q} - \frac{2}{q} + q$$

$$= \frac{-1 + q^2}{q}, \text{ or } \frac{q^2 - 1}{q}$$

The vertex is $\left(\frac{1}{q}, \frac{q^2 - 1}{q}\right)$.

50. $x = \frac{p}{6}$

51. $f(x) = x^2 - 6x + c$

The second coordinate of the vertex, $-\frac{b^2 - 4ac}{4a}$, is 0.

$$-\frac{(-6)^2 - 4\cdot 1\cdot c}{4\cdot 1} = 0$$

$$36 - 4c = 0$$

$$36 = 4c$$

$$9 = c$$

52. $b = 0$

Exercise Set 4.4

1. a) Familiarize: The function is of the form $E = mt + b$, where t is the number of years since 1950.

Translate: We use the data points (0, 72) and (20, 75) and the two-point equation.

$$E - E_1 = \frac{E_2 - E_1}{t_2 - t_1}(t - t_1)$$

$$E - 72 = \frac{75 - 72}{20 - 0}(t - 0) \quad \text{(Substituting)}$$

Carry out: We simplify to the following function:

$$E = \frac{3}{20}t + 72, \text{ or } E = 0.15t + 72$$

Check: We can go over our work. We can also substitute $t = 20$. We do get $E = 75$.

State: A linear function that fits the two data points is $E = 0.15t + 72$.

b) In 1993, $t = 1993 - 1950$, or 43. We find E when $t = 43$:

$$E = 0.15(43) + 72 = 78.45$$

In 1997, $t = 1997 - 1950$, or 47. We find E when $t = 47$:

$$E = 0.15(47) + 72 = 79.05$$

2. a) $E = 0.15t + 65$

b) 71.45, 72.05

3. a) Familiarize: The function is of the form $R = mt + b$, where t is the number of years since 1920.

Translate: We use the data points (0, 10.43) and (50, 9.93) and the two-point equation.

$$R - R_1 = \frac{R_2 - R_1}{t_2 - t_1}(t - t_1)$$

$$R - 10.43 = \frac{9.93 - 10.43}{50 - 0}(t - 0)$$

Carry out: We simplify to the following function:

$$R = -0.01t + 10.43$$

Check: We can go over our work. We can also substitute $t = 50$. We do get $R = 9.93$.

State: A linear function that fits the two data points is $R = -0.01t + 10.43$.

b) In 1996, $t = 1996 - 1920$, or 76. We find R when $t = 76$:

$$R = -0.01(76) + 10.43 = 9.67 \text{ sec}$$

In 2000, $t = 2000 - 1920$, or 80. We find R when $t = 80$:

$$R = -0.01(80) + 10.43 = 9.63 \text{ sec}$$

c) We find t when $R = 9.0$:

$$9.0 = -0.01t + 10.43$$

$$-1.43 = -0.01t$$

$$143 = t$$

The record will be 9.0 sec 143 years after 1920, or in 2063.

4. a) $D = 0.2t + 19$.

b) 28 quadrillion BTU, 29 quadrillion BTU

5. a) Familiarize: The function is of the form $P = mt + b$, where t is the number of years since 1971.

Translate: We will express percents in decimal notation. We use the data points (0, 0.016) and (16, 0.102) and the two-point equation.

$$P - P_1 = \frac{P_2 - P_1}{t_2 - t_1}(t - t_1)$$

$$P - 0.016 = \frac{0.102 - 0.016}{16 - 0}(t - 0)$$

Carry out: We simplify to the following function:

$$P = 0.005375t + 0.016$$

Check: We can go over our work. We can also substitute $t = 16$. We do get $P = 0.102$, or 10.2%.

State: A linear function that fits the data points is $P = 0.005375t + 0.016$.

b) In 1997, $t = 1997 - 1971$, or 26. We find P when $t = 26$:

$$P = 0.005375(26) + 0.016 = 0.15575, \text{ or } 15.575\%$$

In 2010, $t = 2010 - 1971$, or 39. We find P when $t = 39$:

$$P = 0.005375(39) + 0.016 = 0.225625, \text{ or } 22.5625\%$$

6. a) $N = 68{,}750t + 260{,}000$

b) 1,291,250; 2,803,750

7. a) $C(x)$ = Variable costs + Fixed costs

$C(x) = 40x + 22{,}500$

b) $R(x) = 85x$

c) $P(x) = R(x) - C(x)$

$P(x) = 85x - (22{,}500 + 40x)$

$= 45x - 22{,}500$

d) $P(x) = 45x - 22{,}500$

$P(3000) = 45(3000) - 22{,}500$

$= 135{,}000 - 22{,}500$

$= 112{,}500$

A profit of \$112,500 will be realized if 3000 pairs of skis are sold.

e) To find the break-even value, we solve $R(x) = C(x)$:

$R(x) = C(x)$

$85x = 40x + 22{,}500$

$45x = 22{,}500$

$x = 500$

The break-even value is 500.

8. a) $C(x) = 20x + 100{,}000$

b) $R(x) = 45x$

c) $P(x) = 25x - 100{,}000$

d) \$3,650,000 profit

e) 4000

9. Familiarize and Translate: We have $x + y = 50$, so $y = 50 - x$. Then Q is given by

$Q = xy$

$Q(x) = x(50 - x) = 50x - x^2$. (Substituting $50 - x$ for y)

Carry out: We want to maximize this function. We first find the vertex of the function. The first coordinate of the vertex is

$$-\frac{b}{2a} = -\frac{50}{2(-1)} = 25.$$

We find the second coordinate by substituting

$Q(25) = 50(25) - (25)^2 = 1250 - 625 = 625$

The vertex is (25, 625). Since the leading coefficient of $Q(x) = 50x - x^2$, -1, is negative, we know that the function has a maximum value. The maximum value is 625 when $x = 25$. Note that $y = 50 - 25$, or 25, when $x = 25$.

Check: We go over our work. We could also examine the graph of the function and estimate the maximum value.

State: The two numbers that have the maximum product are 25 and 25. The maximum product is 625.

10. -8, 8

11. Familiarize and Translate: We first draw a picture. Let h = the height and b = the base.

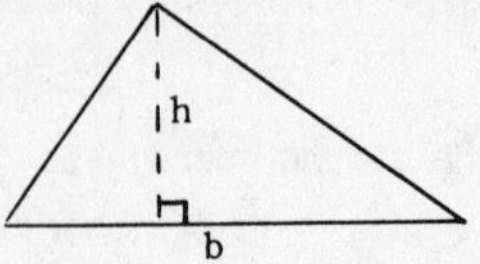

Since the sum of the base and height is 20 cm, we have

$b + h = 20$ and $b = 20 - h$.

Then the area is given by

$$A = \frac{1}{2}bh$$

$$A(h) = \frac{1}{2}(20 - h)h = 10h - \frac{1}{2}h^2.$$

Carry out: We find the first coordinate of the vertex:

$$-\frac{b}{2a} = -\frac{10}{2\left(-\frac{1}{2}\right)} = 10$$

Since the leading coefficient of the function, $-\frac{1}{2}$, is negative, we know that the function has a maximum value. The maximum value occurs when $h = 10$. Note that $b = 20 - 10$, or 10, when $h = 10$.

Check: We go over our work. We could also examine the graph of the function and estimate the maximum value.

State: The area is a maximum when the base is 10 cm and the height is 10 cm.

12. 69 yd by 69 yd

13. Familiarize and Translate: Using the drawing in the text, we see that the area is given by $A(x) = (120 - 2x)x$, or $A(x) = 120x - 2x^2$.

Carry out: We find the vertex of the function. The first coordinate of the vertex is

$$-\frac{b}{2a} = -\frac{120}{2(-2)} = 30.$$

We find the second coordinate by substituting:

$A(30) = 120(30) - 2(30)^2 = 3600 - 1800 = 1800$

The vertex is (30, 1800). Since the leading coefficient of $A(x)$, -2, is negative, we know that the function has a maximum value. The maximum value is 1800 when $x = 30$. Note that $120 - 2x = 120 - 2(30) = 60$ when $x = 30$.

Check: We go over our work. We could also examine the graph of the function and estimate the maximum value.

State: Dimensions of 30 yd by 60 yd will maximize the area. The maximum area is 1800 yd^2.

14. 40 yd by 40 yd, 4800 yd^2

15. Familiarize and Translate: We make a drawing. Let ℓ = the length and w = the width of the room.

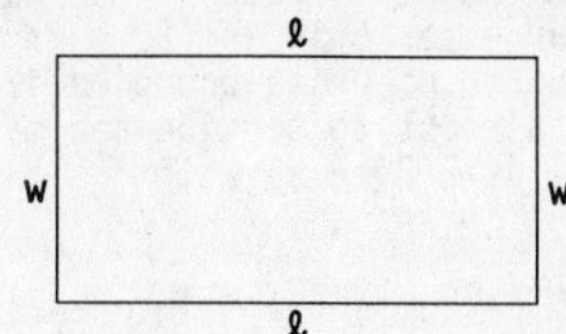

Since the perimeter is 54 ft, we have

$2\ell + 2w = 54$ and $\ell = 27 - w$.

Then the area is given by

$A = \ell w$

$A(w) = (27 - w)w = 27w - w^2$.

Carry out: We find the vertex of the function. The first coordinate of the vertex is given by

$$-\frac{b}{2a} = -\frac{27}{2(-1)} = 13.5.$$

We find the second coordinate by substituting:

$A(13.5) = 27(13.5) - (13.5)^2 = 364.5 - 182.25 = 182.25$

The vertex is (13.5, 182.25). Since the leading coefficient of A(w), -1, is negative, we know the function has a maximum value. The maximum value is 182.25 when w = 13.5. Note that $\ell = 27 - 13.5 = 13.5$ when w = 13.5.

Check: We go over our work. We could also examine the graph of the function and estimate the maximum value.

State: The dimensions are 13.5 ft by 13.5 ft. The area is 182.25 ft^2.

16. 8.5 ft by 8.5 ft, 72.25 ft^2

17. $P(x) = R(x) - C(x)$

$P(x) = (50x - 0.5x^2) - (4x + 10)$

$P(x) = -0.5x^2 + 46x - 10$

We find the vertex of P(x). The first coordinate is

$$-\frac{b}{2a} = -\frac{46}{2(-0.5)} = 46.$$

We substitute to find the second coordinate:

$$\begin{aligned} P(46) &= -0.5(46)^2 + 46(46) - 10 \\ &= -1058 + 2116 - 10 \\ &= 1048 \end{aligned}$$

The vertex is (46, 1048). Since the leading coefficient of P(x), -0.5, is negative, we know that the function has a maximum value. Thus, the maximum profit is $1048, and it occurs when 46 units are produced and sold.

18. 40 units, $797

19. $P(x) = R(x) - C(x)$

$P(x) = 2x - (0.01x^2 + 0.6x + 30)$

$P(x) = -0.01x^2 + 1.4x - 30$

We find the vertex of P(x). The first coordinate is

$$-\frac{b}{2a} = -\frac{1.4}{2(-0.01)} = 70.$$

We find the second coordinate by substituting:

$$\begin{aligned} P(70) &= -0.01(70)^2 + 1.4(70) - 30 \\ &= -49 + 98 - 30 \\ &= 19 \end{aligned}$$

The vertex is (70, 19). Since the leading coefficient of P(x), -0.01, is negative, we know that the function has a maximum value. Thus, the maximum profit is $19, and it occurs when 70 units are produced and sold.

20. 1900 units, $3550

21. a) $R(x) = xD(x)$

$R(x) = x(150 - 0.5x)$

$R(x) = 150x - 0.5x^2$

b) $P(x) = R(x) - C(x)$

$P(x) = (150x - 0.5x^2) - (4000 + 0.25x^2)$

$P(x) = -0.75x^2 + 150x - 4000$

c) We find the first coordinate of the vertex of P(x):

$$-\frac{b}{2a} = -\frac{150}{2(-0.75)} = 100$$

Since the leading coefficient of P(x), -0.75, is negative, we know the function has a maximum value. Thus, 100 suits must be produced and sold in order to maximize profit.

d) We find P(100):

$$\begin{aligned} P(100) &= -0.75(100)^2 + 150(100) - 4000 \\ &= -7500 + 15{,}000 - 4000 \\ &= 3500 \end{aligned}$$

The maximum profit is $3500.

e) We find p when x = 100:

$p = D(100) = 150 - 0.5(100) = 100$

The price per suit must be $100.

22. a) $R(x) = 280x - 0.4x^2$

b) $P(x) = -x^2 + 280x - 5000$

c) 140

d) $14,600

e) $224

23. $s(t) = -4.9t^2 + v_0t + h$

$s(t) = -4.9t^2 + 147t + 560$ (Substituting)

a) We find the vertex of s(t). The first coordinate is

$$-\frac{b}{2a} = -\frac{147}{2(-4.9)} = 15.$$

We find the second coordinate by substituting:

$s(15) = -4.9(15)^2 + 147(15) + 560 = 1662.5$

The vertex is (15, 1662.5). Since the leading coefficient of s(t), -4.9, is negative, we know the function has a maximum value. Thus, the rocket has a maximum height of 1662.5 m, and it is attained 15 sec after the end of the burn.

b) We set s(t) = 0 and solve for t. We use the quadratic formula.

$-4.9t^2 + 147t + 560 = 0$

$a = -4.9,\ b = 147,\ c = 560$

$$t = \frac{-b \pm \sqrt{b^2 - 4ac}}{2a}$$

$$t = \frac{-147 - \sqrt{147^2 - 4(-4.9)(560)}}{2(-4.9)}$$ Taking the negative square root (t > 0)

$t \approx 33.4$

The rocket will reach the ground about 33.4 sec after the end of the burn.

24. $s(t) = -4.9t^2 + 245t + 1240$

a) 4302.5 m attained 25 sec after the end of the burn

b) after 54.6 sec

25. Familiarize and Translate: The maximum amount of light will enter when the area of the window is maximized. In Exercise 36, Exercise Set 3.4, we expressed y as $12 - \frac{\pi x}{2} - x$ and we expressed the area of the window as a function of x as $A(x) = 24x - \left(\frac{\pi}{2} + 2\right)x^2$.

Carry out: We find the first coordinate of the vertex of A(x):

$$-\frac{b}{2a} = -\frac{24}{2\left[-\left(\frac{\pi}{2} + 2\right)\right]} = \frac{24}{\pi + 4}$$

The leading coefficient of A(x), $-\left(\frac{\pi}{2} + 2\right)$, is negative, so we know the function has a maximum value. It occurs when $x = \frac{24}{\pi + 4}$. Note that

$$y = 12 - \frac{\pi x}{2} - x = 12 - \frac{\pi}{2}\left(\frac{24}{\pi + 4}\right) - \frac{24}{\pi + 4} = \frac{24}{\pi + 4}$$

when $x = \frac{24}{\pi + 4}$.

Check: We go over our work.

State: The maximum amount of light will enter when the dimensions of the rectangular part of the window are $\frac{24}{\pi + 4}$ ft by $\frac{24}{\pi + 4}$ ft.

26. \$5.75; 72,500

27. Familiarize: When the daily rate is increased x dollars, the charge per unit is (20 + x) dollars and the number of units occupied is (30 - x). The total cost to service and maintain the occupied units is 2(30 - x).

Translate:

$$\text{Profit} = \begin{pmatrix}\text{Charge}\\ \text{per unit}\end{pmatrix} \cdot \begin{pmatrix}\text{Number}\\ \text{of units}\\ \text{occupied}\end{pmatrix} - \begin{pmatrix}\text{Maintenance}\\ \text{and service}\\ \text{cost}\end{pmatrix}$$

$P(x) = (20 + x) \cdot (30 - x) - 2(30 - x)$

$P(x) = 600 + 10x - x^2 - 60 + 2x$

$P(x) = -x^2 + 12x + 540$

Carry out: We find the value of x for which P(x) is a maximum. The first coordinate of the vertex of P(x) is

$$-\frac{b}{2a} = -\frac{12}{2(-1)} = 6.$$

Since the leading coefficient of P(x), -1, is negative, we know the function has a maximum value. It occurs when x = 6. Note that when x = 6, the charge per unit is 20 + 6, or \$26.

Check: We go over our work. We could also examine the graph of the function and estimate the first coordinate of the vertex.

State: Profit is maximized when the charge per unit is \$26.

28. 25

29. Familiarize: Let x = the number of \$0.10 increases in the admission price. Then the admission price is \$3.00 + \$0.10x, or \$(3 + 0.1x), and the average attendance is 100 - x.

Translate:

$$\text{Revenue} = \begin{pmatrix}\text{Price per}\\ \text{ticket}\end{pmatrix} \cdot \begin{pmatrix}\text{Number}\\ \text{attending}\end{pmatrix}$$

$R(x) = (3 + 0.1x) \cdot (100 - x)$

$R(x) = 300 + 7x - 0.1x^2$

Carry out: We find the value of x for which R(x) is a maximum. The first coordinate of the vertex of R(x) is

$$-\frac{b}{2a} = -\frac{7}{2(-0.1)} = 35.$$

Since the leading coefficient of R(x), -0.1, is negative, we know the function has a maximum value. It occurs when x = 35. Note that, when x = 35, the admission price is 3 + 0.1(35), or \$6.50.

Check: We go over our work. We could also examine the graph of the function and estimate the first coordinate of the vertex.

State: Revenue is maximized when the admission price is \$6.50.

30. The minimum occurs when $x = \frac{24\pi}{\pi + 4}$.

31. By similar triangles, $\frac{9}{12} = \frac{y}{12 - x}$ or $y = 9 - \frac{3}{4}x$.
The area function is $A = xy$ or $A = x\left[9 - \frac{3}{4}x\right]$ or
$A = -\frac{3}{4}(x - 6)^2 + 27$.

Thus the maximum area is 27 cm^2 and this occurs for x = 6. Hence the dimensions x, y are 6 cm and 4.5 cm.

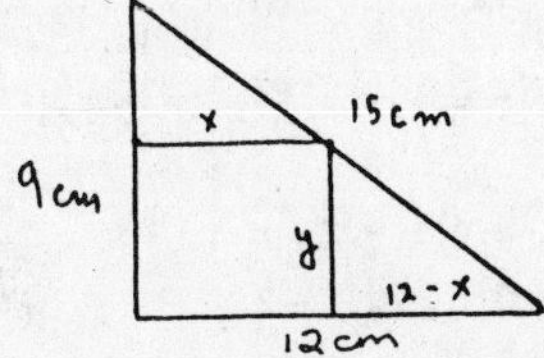

Exercise Set 4.5

1. The intersection of two sets consists of those element common to the sets.
{3, 4, 5, 8, 10} ∩ {1, 2, 3, 4, 5, 6, 7}
= {3, 4, 5}

Only the elements 3, 4, and 5 are in both sets.

2. {1, 2, 3, 4, 5, 6, 7, 8, 10}

3. The union of two sets consists of the elements that are in one or both of the sets.
{0, 2, 4, 6, 8} ∪ {4, 6, 9} = {0, 2, 4, 6, 8, 9}

Note that the elements 4 and 6 are in both sets, but each is listed only once in the union.

4. {4, 6}

5. {a, b, c} ∩ {c, d} = {c}

Since the element c is the only element which is in both sets, the intersection only contains c.

6. {a, b, c, d}

7. Graph: $\{x|7 \leqslant x\} \cup \{x|x < 9\}$
Graph the solution sets separately, and then find the union.

Graph: $\{x|7 \leqslant x\}$ $(7 \leqslant x$ means $x \geqslant 7)$

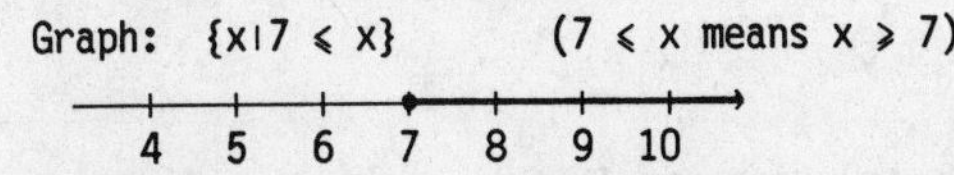

The solid circle at 7 indicates that 7 is in the solution set.

Graph: $\{x|x < 9\}$

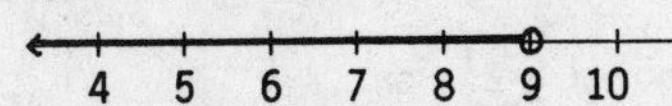

The open circle at 9 indicates that 9 is not in the solution set.

Now graph the union:
$\{x|7 \leqslant x\} \cup \{x|x < 9\}$ = The set of all real numbers, or $(-\infty, \infty)$.

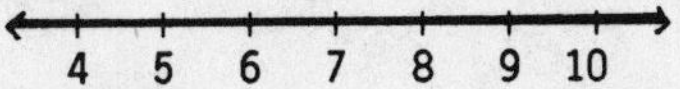

8.

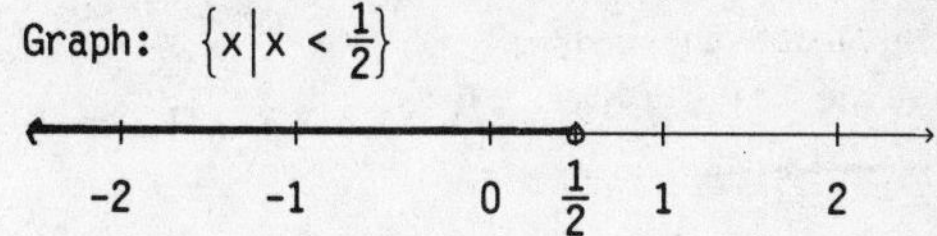

9. Graph: $\left\{x\middle|-\frac{1}{2} \leqslant x\right\} \cap \left\{x\middle|x < \frac{1}{2}\right\}$

Graph the solution sets separately, and then find the intersection.

Graph: $\left\{x\middle|-\frac{1}{2} \leqslant x\right\}$ $\left[-\frac{1}{2} \leqslant x \text{ means } x \geqslant -\frac{1}{2}\right]$

-2 -1 -1/2 0 1 2

The solid circle indicates that $-\frac{1}{2}$ is in the solution set.

Graph: $\left\{x\middle|x < \frac{1}{2}\right\}$

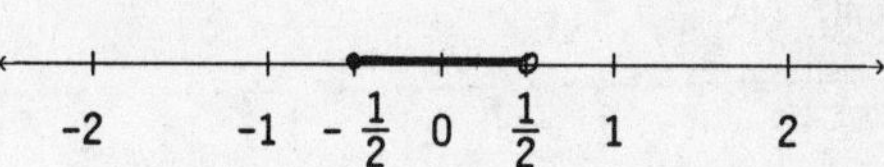

The open circle at $\frac{1}{2}$ indicates that $\frac{1}{2}$ is not in the solution set.

Now graph the intersection:

$\left\{x\middle|-\frac{1}{2} \leqslant x\right\} \cap \left\{x\middle|x < \frac{1}{2}\right\} = \left\{x\middle|-\frac{1}{2} \leqslant x < \frac{1}{2}\right\}$, or $\left[-\frac{1}{2}, \frac{1}{2}\right)$

-2 -1 -1/2 0 1/2 1 2

10.

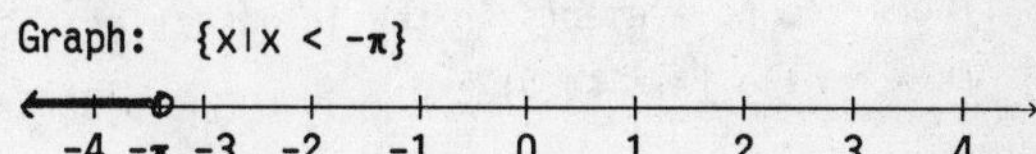

11. Graph: $\{x|x < -\pi\} \cup \{x|x > \pi\}$
Graph the solution sets separately, and then find the union.

Graph: $\{x|x < -\pi\}$

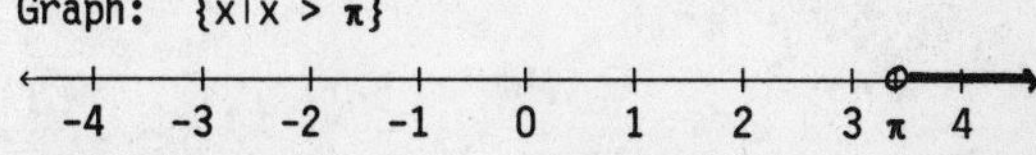

The open circle at $-\pi$ indicates that $-\pi$ is not in the solution set.

Graph: $\{x|x > \pi\}$

-4 -3 -2 -1 0 1 2 3 π 4

The open circle at π indicates that π is not in the solution set.

Now graph the union:
$\{x|x < -\pi\} \cup \{x|x > \pi\} = \{x|x < -\pi \text{ or } x > \pi\}$, or $(-\infty, -\pi) \cup (\pi, \infty)$

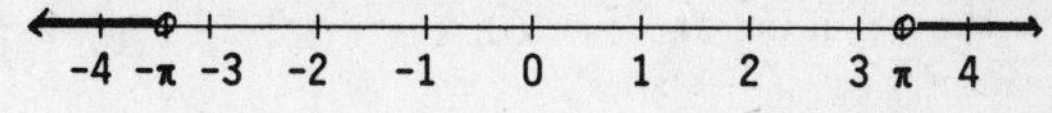

12.

13. Graph: $\{x|x < -7\} \cup \{x|x = -7\}$

Graph the solution sets separately, and then find the union.

Graph: $\{x|x < -7\}$

The open circle at -7 indicates that -7 is not in the solution set.

Graph: $\{x|x = -7\}$

The solid circle at -7 indicates that -7 is in the solution set.

Now graph the union:

$\{x|x < -7\} \cup \{x|x = -7\} = \{x|x \leqslant -7\}$, or $(-\infty, -7]$

14.

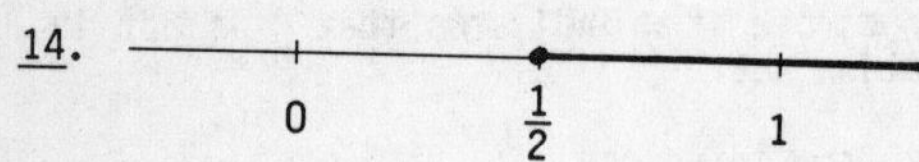

15. Graph: $\{x|x \geqslant 5\} \cap \{x|x \leqslant -3\}$

Graph the solution sets separately, and then find the intersection.

Graph: $\{x|x \geqslant 5\}$

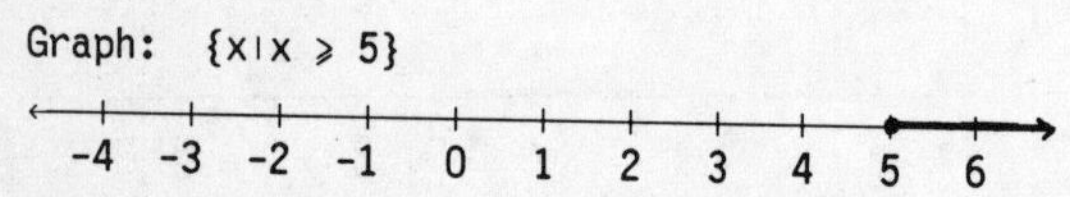

The solid circle at 5 indicates that 5 is in the solution set.

Graph: $\{x|x \leqslant -3\}$

The solid circle at -3 indicates that -3 is in the solution set.

There are no elements in the intersection.

$\{x|x \geqslant 5\} \cap \{x|x \leqslant -3\} = \emptyset$

16.

17. $-2 \leqslant x + 1 < 4$

$-3 \leqslant x < 3$ (Adding -1)

The solution set is $\{x|-3 \leqslant x < 3\}$, or $[-3, 3)$.

18. $\{x|-5 < x \leqslant 3\}$, or $(-5, 3]$

19. $5 \leqslant x - 3 \leqslant 7$

$8 \leqslant x \leqslant 10$ (Adding 3)

The solution set is $\{x|8 \leqslant x \leqslant 10\}$, or $[8, 10]$.

20. $\{x|3 < x < 11\}$, or $(3, 11)$

21. $-3 \leqslant x + 4 \leqslant -3$

$-7 \leqslant x \leqslant -7$

The solution set is $\{-7\}$.

22. $\emptyset$

23. $-2 < 2x + 1 < 5$

$-3 < 2x < 4$ (Adding -1)

$-\frac{3}{2} < x < 2$ $\left[\text{Multiplying by } \frac{1}{2}\right]$

The solution set is $\left\{x\middle|-\frac{3}{2} < x < 2\right\}$, or $\left(-\frac{3}{2}, 2\right)$.

24. $\left\{x\middle|-\frac{4}{5} \leqslant x \leqslant \frac{2}{5}\right\}$, or $\left[-\frac{4}{5}, \frac{2}{5}\right]$

25. $-4 \leqslant 6 - 2x < 4$

$-10 \leqslant -2x < -2$ (Adding -6)

$5 \geqslant x > 1$ $\left[\text{Multiplying by } -\frac{1}{2}\right]$

or $1 < x \leqslant 5$

The solution set is $\{x|1 < x \leqslant 5\}$, or $(1, 5]$.

26. $\{x|-1 \leqslant x < 2\}$, or $[-1, 2)$

27. $-5 < \frac{1}{2}(3x + 1) \leqslant 7$

$-10 < 3x + 1 \leqslant 14$ (Multiplying by 2)

$-11 < 3x \leqslant 13$ (Adding -1)

$-\frac{11}{3} < x \leqslant \frac{13}{3}$ $\left[\text{Multiplying by } \frac{1}{3}\right]$

The solution set is $\left\{x\middle|-\frac{11}{3} < x \leqslant \frac{13}{3}\right\}$, or $\left(-\frac{11}{3}, \frac{13}{3}\right]$.

28. $\left\{x\middle|\frac{7}{4} < x \leqslant \frac{13}{6}\right\}$, or $\left(\frac{7}{4}, \frac{13}{6}\right]$

29. $3x \leqslant -6$ or $x - 1 > 0$

$x \leqslant -2$ or $x > 1$

The solution set is $\{x|x \leqslant -2 \text{ or } x > 1\}$, or $(-\infty, -2] \cup (1, \infty)$.

30. The set of all real numbers, or $(-\infty, \infty)$

31. $2x + 3 \leqslant -4$ or $2x + 3 \geqslant 4$

$2x \leqslant -7$ or $2x \geqslant 1$

$x \leqslant -\frac{7}{2}$ or $x \geqslant \frac{1}{2}$

The solution set is $\left\{x\middle|x \leqslant -\frac{7}{2} \text{ or } x \geqslant \frac{1}{2}\right\}$, or $\left(-\infty, -\frac{7}{2}\right] \cup \left[\frac{1}{2}, \infty\right)$.

32. $\left\{x\middle|x < -\frac{4}{3} \text{ or } x > 2\right\}$, or $\left(-\infty, -\frac{4}{3}\right) \cup (2, \infty)$

33. $2x - 20 < -0.8$ or $2x - 20 > 0.8$

$2x < 19.2$ or $2x > 20.8$

$x < 9.6$ or $x > 10.4$

The solution set is $\{x \mid x < 9.6 \text{ or } x > 10.4\}$, or $(-\infty, 9.6) \cup (10.4, \infty)$.

34. $\left\{x \middle| x \leqslant -3 \text{ or } x \geqslant -\frac{7}{5}\right\}$, or

$(-\infty, -3] \cup \left[-\frac{7}{5}, \infty\right)$

35. $x + 14 \leqslant -\frac{1}{4}$ or $x + 14 \geqslant \frac{1}{4}$

$x \leqslant -\frac{57}{4}$ or $x \geqslant -\frac{55}{4}$

The solution set is $\left\{x \middle| x \leqslant -\frac{57}{4} \text{ or } x \geqslant -\frac{55}{4}\right\}$, or

$\left(-\infty, -\frac{57}{4}\right] \cup \left[-\frac{55}{4}, \infty\right)$.

36. $\left\{x \middle| x < \frac{17}{2} \text{ or } x > \frac{19}{2}\right\}$, or

$\left(-\infty, \frac{17}{2}\right) \cup \left(\frac{19}{2}, \infty\right)$

37. We first make a drawing.

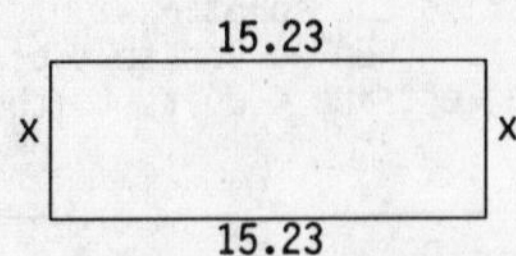

Let x represent the width. Then $2x + 2(15.23)$, or $2x + 30.46$, represents the perimeter. The perimeter is greater than 40.23 cm and less than 137.8 cm. We now have the following inequality.

$40.23 < 2x + 30.46 < 137.8$

$9.77 < 2x < 107.34$ (Adding -30.46)

$4.885 < x < 53.67$ $\left[\text{Multiplying by } \frac{1}{2}\right]$

Widths greater than 4.885 cm and less than 53.67 cm will give a perimeter greater than 40.23 cm and less than 137.8 cm. We can express this as $\{x \mid 4.885 \text{ cm} < x < 53.67 \text{ cm}\}$.

38. $\{b \mid 0 \text{ m} < b \leqslant 40.72 \text{ m}\}$

39. Let x represent the score on the fourth test. The average of the four scores is $\frac{83 + 87 + 93 + x}{4}$. This average must be greater than or equal to 90. We solve the following inequality.

$\frac{83 + 87 + 93 + x}{4} \geqslant 90$

$83 + 87 + 93 + x \geqslant 360$

$263 + x \geqslant 360$

$x \geqslant 97$

The score on the fourth test must be greater than or equal to 97. It will of course be less than or equal to 100%. We can express this as $\{x \mid 97\% \leqslant x \leqslant 100\%\}$. Yes, an A is possible.

40. $\{x \mid 132\% \leqslant x \leqslant 172\%\}$; no

41. We substitute $\frac{5}{9}(F - 32)$ for C in the inequality.

$1083° \leqslant \frac{5}{9}(F - 32) \leqslant 2580°$

$1949.4° \leqslant F - 32 \leqslant 4644°$ $\left[\text{Multiplying by } \frac{9}{5}\right]$

$1981.4° \leqslant F \leqslant 4676°$

42. $\{d \mid 33 \text{ ft} \leqslant d \leqslant 297 \text{ ft}\}$

43. $2 < 0.027d + 0.32 < 4$

$2000 < 27d + 320 < 4000$ (Clearing decimals)

$1680 < 27d < 3680$

$\frac{1680}{27} < d < \frac{3680}{27}$

$62\frac{2}{9} < d < 136\frac{8}{27}$

The solution set is $\left\{d \middle| 62\frac{2}{9} \text{ mi} < d < 136\frac{8}{27} \text{ mi}\right\}$.

44. $\left\{t \middle| 2\frac{14}{41} \text{ yr} < t < 6\frac{10}{41} \text{ yr}\right\}$

45. $x \leqslant 3x - 2 \leqslant 2 - x$

$x \leqslant 3x - 2$ and $3x - 2 \leqslant 2 - x$

$-2x \leqslant -2$ and $4x \leqslant 4$

$x \geqslant 1$ and $x \leqslant 1$

The word "and" corresponds to set intersection.

The solution set is

$\{x \mid x \geqslant 1\} \cap \{x \mid x \leqslant 1\}$, or $\{1\}$.

46. $\left\{x \middle| -\frac{1}{4} < x \leqslant \frac{5}{9}\right\}$, or $\left(-\frac{1}{4}, \frac{5}{9}\right]$

47. $(x + 1)^2 > x(x - 3)$

$x^2 + 2x + 1 > x^2 - 3x$

$2x + 1 > -3x$

$5x > -1$

$x > -\frac{1}{5}$

The solution set is $\left\{x \middle| x > -\frac{1}{5}\right\}$, or $\left(-\frac{1}{5}, \infty\right)$.

48. $\left\{x \middle| x > \frac{13}{5}\right\}$, or $\left(\frac{13}{5}, \infty\right)$

49. $(x + 1)^2 \leqslant (x + 2)^2 \leqslant (x + 3)^2$

$x^2 + 2x + 1 \leqslant x^2 + 4x + 4 \leqslant x^2 + 6x + 9$

$2x + 1 \leqslant 4x + 4 \leqslant 6x + 9$

$2x + 1 \leqslant 4x + 4$ and $4x + 4 \leqslant 6x + 9$

$-2x \leqslant 3$ and $-2x \leqslant 5$

$x \geqslant -\frac{3}{2}$ and $x \geqslant -\frac{5}{2}$

The word "and" corresponds to set intersection.

The solution set is

$\left\{x \middle| x \geqslant -\frac{3}{2}\right\} \cap \left\{x \middle| x \geqslant -\frac{5}{2}\right\}$, or $\left\{x \middle| x \geqslant -\frac{3}{2}\right\}$, or $\left[-\frac{3}{2}, \infty\right]$.

50. $\{x|-1 < x \leqslant 1\}$, or $(-1, 1]$

51. $f(x) = \frac{\sqrt{x + 2}}{\sqrt{x - 2}}$

The radicand in the denominator must be greater than 0. Thus, $x - 2 > 0$, or $x > 2$. Also the radicand in the numerator must be greater than or equal to 0. Thus, $x + 2 \geqslant 0$, or $x \geqslant -2$. The domain of $f(x)$ is the intersection of the two solution sets.

$\{x|x > 2\} \cap \{x|x \geqslant -2\} = \{x|x > 2\}$.

The domain of $f(x)$ is $\{x|x > 2\}$, or $(2, \infty)$.

52. $\{x|-5 < x \leqslant 3\}$, or $(-5, 3]$

Exercise Set 4.6

1. $|x| = 7$

To solve we look for all numbers x whose distance from 0 is 7. There are two of them, so there are two solutions, -7 and 7. The graph is as follows.

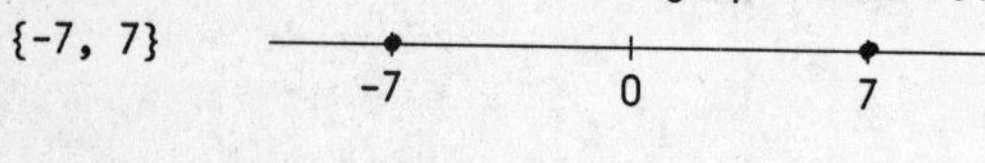

$\{-7, 7\}$

2. $\{-\pi, \pi\}$

3. $|x| < 7$

To solve we look for all numbers x whose distance from 0 is less than 7. These are the numbers between -7 and 7. The solution set and its graph are as follows.

$\{x|-7 < x < 7\}$, or $(-7, 7)$

4. $\{x|-\pi \leqslant x \leqslant \pi\}$, or $[-\pi, \pi]$

5. $|x| \geqslant \pi$

To solve we look for all numbers x whose distance from 0 is greater than or equal to π. The solution set and its graph are as follows.

$\{x|x \leqslant -\pi$ or $x \geqslant \pi\}$, or $(-\infty, -\pi] \cup [\pi, \infty)$

6. $\{x|x < -7$ or $x > 7\}$, or $(-\infty, -7) \cup (7, \infty)$

7. $|x - 1| = 4$

Method 1:

We translate the graph of $|x| = 4$ to the right 1 unit.

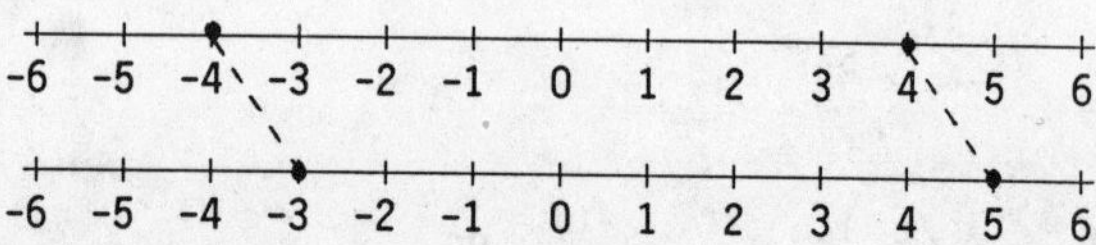

The solution set is { -3, 5}

Method 2:

The solutions are those numbers x whose distance from 1 is 4. Thus to find the solutions graphically we locate 1. Then we locate those numbers 4 units to the left and 4 units to the right.

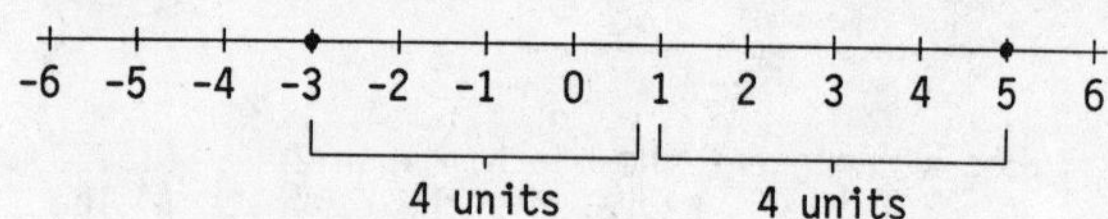

The solution set is {-3, 5}.

Method 3:

$|x - 1| = 4$

$x - 1 = -4$ or $x - 1 = 4$ (Property i)

$x = -3$ or $x = 5$ (Adding 1)

The solution set is {-3, 5}.

8. {2, 12}

9. $|x + 8| < 9$, or $|x - (-8)| < 9$

Method 1:

We translate the graph of $|x| < 9$ to the left 8 units.

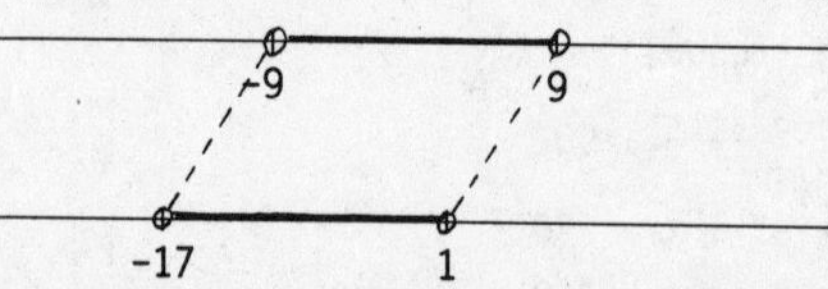

The solution set is $\{x|-17 < x < 1\}$, or $(-17, 1)$.

9. (continued)

Method 2:

The solutions are those numbers x whose distance from -8 is less than 9. Thus to find the solutions graphically we locate -8. Then we locate those numbers that are less than 9 units to the left and less than 9 units to the right.

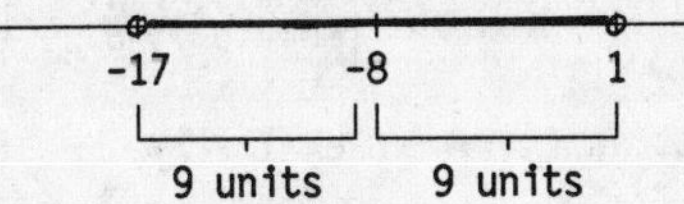

The solution set is $\{x \mid -17 < x < 1\}$, or $(-17, 1)$.

Method 3:

$|x + 8| < 9$

$-9 < x + 8 < 9$ (Property ii)

$-17 < x < 1$ (Adding -8)

The solution set is $\{x \mid -17 < x < 1\}$, or $(-17, 1)$.

10. $\{x \mid -16 \leqslant x \leqslant 4\}$, or $[-16, 4]$

11. $|x + 8| \geqslant 9$, or $|x - (-8)| \geqslant 9$

Method 1:

We translate the graph of $|x| \geqslant 9$ to the left 8 units.

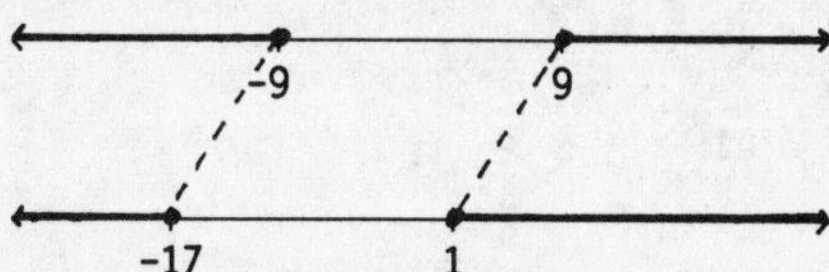

The solution set is $\{x \mid x \leqslant -17 \text{ or } x \geqslant 1\}$, or $(-\infty, -17] \cup [1, \infty)$.

Method 2:

The solutions are those numbers x whose distance from -8 is greater than or equal to 9. Thus to find the solutions graphically we locate -8. Then we locate those numbers that are greater than or equal to 9 units to the left and greater than or equal to 9 units to the right.

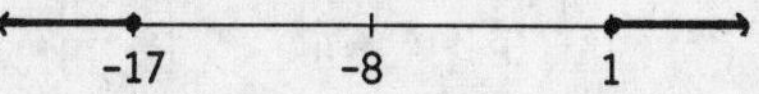

The solution set is $\{x \mid x \leqslant -17 \text{ or } x \geqslant 1\}$, or $(-\infty, -17] \cup [1, \infty)$.

Method 3:

$|x + 8| \geqslant 9$

$x + 8 \leqslant -9$ or $x + 8 \geqslant 9$ (Property iii)

$x \leqslant -17$ or $x \geqslant 1$ (Adding -8)

The solution set is $\{x \mid x \leqslant -17 \text{ or } x \geqslant 1\}$, or $(-\infty, -17] \cup [1, \infty)$.

12. $\{x \mid x < -16 \text{ or } x > 4\}$, or $(-\infty, -16) \cup (4, \infty)$

13. $\left|x - \frac{1}{4}\right| < \frac{1}{2}$

Method 1:

We translate the graph of $|x| < \frac{1}{2}$ to the right $\frac{1}{4}$ unit.

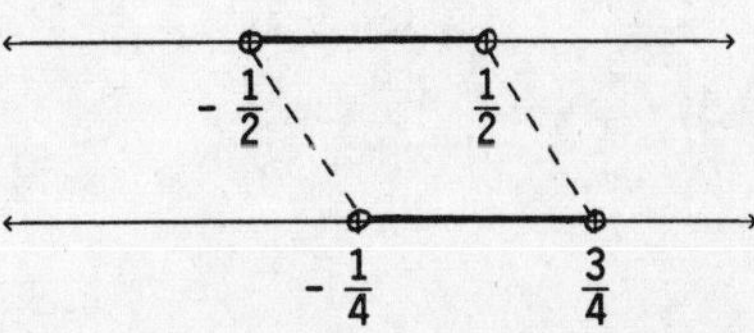

The solution set is $\left\{x \middle| -\frac{1}{4} < x < \frac{3}{4}\right\}$, or $\left(-\frac{1}{4}, \frac{3}{4}\right)$.

Method 2:

The solutions are those numbers x whose distance from $\frac{1}{4}$ is less than $\frac{1}{2}$. Thus to find the solutions graphically we locate $\frac{1}{4}$. Then we locate those numbers that are less than $\frac{1}{2}$ unit to the left and less than $\frac{1}{2}$ unit to the right.

$-\frac{1}{4}$ $\frac{1}{4}$ $\frac{3}{4}$

The solution set is $\left\{x \middle| -\frac{1}{4} < x < \frac{3}{4}\right\}$, or $\left(-\frac{1}{4}, \frac{3}{4}\right)$.

Method 3:

$\left|x - \frac{1}{4}\right| < \frac{1}{2}$

$-\frac{1}{2} < x - \frac{1}{4} < \frac{1}{2}$ (Property ii)

$-\frac{1}{4} < x < \frac{3}{4}$ $\left(\text{Adding } \frac{1}{4}\right)$

The solution set is $\left\{x \middle| -\frac{1}{4} < x < \frac{3}{4}\right\}$, or $\left(-\frac{1}{4}, \frac{3}{4}\right)$.

14. $\{x \mid 0.3 \leqslant x \leqslant 0.7\}$, or $[0.3, 0.7]$

15. $|3x| = 1$

$3x = -1$ or $3x = 1$ (Property i)

$x = -\frac{1}{3}$ or $x = \frac{1}{3}$ $\left(\text{Multiplying by } \frac{1}{3}\right)$

The solution set is $\left\{-\frac{1}{3}, \frac{1}{3}\right\}$.

16. $\left\{-\frac{4}{5}, \frac{4}{5}\right\}$

17. $|3x + 2| = 1$

$3x + 2 = -1$ or $3x + 2 = 1$ (Property i)

$3x = -3$ or $3x = -1$ (Adding -2)

$x = -1$ or $x = -\frac{1}{3}$ $\left(\text{Multiplying by } \frac{1}{3}\right)$

The solution set is $\left\{-1, -\frac{1}{3}\right\}$.

18. $\left\{-\frac{4}{7}, \frac{12}{7}\right\}$

19. $|3x| < 1$

$-1 < 3x < 1$ (Property ii)

$-\frac{1}{3} < x < \frac{1}{3}$ $\left[\text{Multiplying by } \frac{1}{3}\right]$

The solution set is $\left\{x \middle| -\frac{1}{3} < x < \frac{1}{3}\right\}$, or $\left(-\frac{1}{3}, \frac{1}{3}\right)$.

20. $\left\{x \middle| -\frac{4}{5} \leqslant x \leqslant \frac{4}{5}\right\}$, or $\left[-\frac{4}{5}, \frac{4}{5}\right]$

21. $|2x + 3| \leqslant 9$

$-9 \leqslant 2x + 3 \leqslant 9$ (Property ii)

$-12 \leqslant 2x \leqslant 6$ (Adding -3)

$-6 \leqslant x \leqslant 3$ $\left[\text{Multiplying by } \frac{1}{2}\right]$

The solution set is $\{x|-6 \leqslant x \leqslant 3\}$, or $[-6, 3]$.

22. $\{x|-8 < x < 5\}$, or $(-8, 5)$

23. $|x - 5| > 0.1$

$x - 5 < -0.1$ or $x - 5 > 0.1$ (Property iii)

$x < 4.9$ or $x > 5.1$ (Adding 5)

The solution set is $\{x|x < 4.9 \text{ or } x > 5.1\}$, or $(-\infty, 4.9) \cup (5.1, \infty)$.

24. $\{x|x \leqslant 6.6 \text{ or } x \geqslant 7.4\}$, or $(-\infty, 6.6] \cup [7.4, \infty)$

25. $\left|x + \frac{2}{3}\right| \leqslant \frac{5}{3}$

$-\frac{5}{3} \leqslant x + \frac{2}{3} \leqslant \frac{5}{3}$ (Property ii)

$-\frac{7}{3} \leqslant x \leqslant \frac{3}{3}$ $\left[\text{Adding } -\frac{2}{3}\right]$

The solution set is $\left\{x \middle| -\frac{7}{3} \leqslant x \leqslant 1\right\}$, or $\left[-\frac{7}{3}, 1\right]$.

26. $\left\{x \middle| -1 < x < -\frac{1}{2}\right\}$, or $\left(-1, -\frac{1}{2}\right)$

27. $|6 - 4x| \leqslant 8$

$-8 \leqslant 6 - 4x \leqslant 8$ (Property ii)

$-14 \leqslant -4x \leqslant 2$ (Adding -6)

$\frac{14}{4} \geqslant x \geqslant -\frac{2}{4}$ $\left[\text{Multiplying by } -\frac{1}{4}\right]$

$\frac{7}{2} \geqslant x \geqslant -\frac{1}{2}$

The solution set is $\left\{x \middle| -\frac{1}{2} \leqslant x \leqslant \frac{7}{2}\right\}$, or $\left[-\frac{1}{2}, \frac{7}{2}\right]$.

28. $\left\{x \middle| x > \frac{15}{2} \text{ or } x < -\frac{5}{2}\right\}$, or $\left(-\infty, -\frac{5}{2}\right) \cup \left(\frac{15}{2}, \infty\right)$

29. $\left|\frac{2x + 1}{3}\right| > 5$

$\frac{2x + 1}{3} < -5$ or $\frac{2x + 1}{3} > 5$ (Property iii)

$2x + 1 < -15$ or $2x + 1 > 15$ (Multiplying by 3)

$2x < -16$ or $2x > 14$ (Adding -1)

$x < -8$ or $x > 7$ $\left[\text{Multiplying by } \frac{1}{2}\right]$

The solution set is $\{x|x < -8 \text{ or } x > 7\}$, or $(-\infty, -8) \cup (7, \infty)$.

30. $\left\{x \middle| -\frac{22}{3} \leqslant x \leqslant 6\right\}$, or $\left[-\frac{22}{3}, 6\right]$

31. $\left|\frac{13}{4} + 2x\right| > \frac{1}{4}$

$\frac{13}{4} + 2x < -\frac{1}{4}$ or $\frac{13}{4} + 2x > \frac{1}{4}$ (Property iii)

$2x < -\frac{14}{4}$ or $2x > -\frac{12}{4}$

$2x < -\frac{7}{2}$ or $2x > -3$

$x < -\frac{7}{4}$ or $x > -\frac{3}{2}$

The solution set is $\left\{x \middle| x < -\frac{7}{4} \text{ or } x > -\frac{3}{2}\right\}$, or $\left(-\infty, -\frac{7}{4}\right) \cup \left(-\frac{3}{2}, \infty\right)$.

32. $\left\{x \middle| -\frac{2}{3} < x < \frac{1}{9}\right\}$, or $\left(-\frac{2}{3}, \frac{1}{9}\right)$

33. $\left|\frac{3 - 4x}{2}\right| \leqslant \frac{3}{4}$

$-\frac{3}{4} \leqslant \frac{3 - 4x}{2} \leqslant \frac{3}{4}$ (Property ii)

$-3 \leqslant 6 - 8x \leqslant 3$ (Multiplying by 4)

$-9 \leqslant -8x \leqslant -3$ (Adding -6)

$\frac{9}{8} \geqslant x \geqslant \frac{3}{8}$ $\left[\text{Multiplying by } -\frac{1}{8}\right]$

The solution set is $\left\{x \middle| \frac{3}{8} \leqslant x \leqslant \frac{9}{8}\right\}$, or $\left[\frac{3}{8}, \frac{9}{8}\right]$.

34. $\left\{x \middle| x \leqslant -\frac{3}{4} \text{ or } x \geqslant \frac{7}{4}\right\}$, or $\left(-\infty, -\frac{3}{4}\right] \cup \left[\frac{7}{4}, \infty\right)$

35. $|x| = -3$

Since $|x| \geqslant 0$ for all x, there is no x such that $|x|$ would be negative. There is no solution. The solution set is $\emptyset$.

36. $\emptyset$

37. $|2x - 4| < -5$

Since $|2x - 4| \geqslant 0$ for all x, there is no x such that $|2x - 4|$ would be less than -5. There is no solution. The solution set is $\emptyset$.

38. $\emptyset$

39. $|x - 17.217| > 5.0012$

$x + 17.217 < -5.0012$ or $x + 17.217 > 5.0012$ (Property iii)

$x < -22.2182$ or $x > -12.2158$ (Adding -17.217)

The solution set is $\{x|x < -22.2182$ or $x > -12.2158\}$, or $(-\infty, -22.2182) \cup (-12.2158, \infty)$.

40. $\{x|1.9234 < x < 2.1256\}$, or $(1.9234, 2.1256)$

41. $|-2.1437x + 7.8814| \geqslant 9.1132$

$-2.1437x + 7.8814 \leqslant -9.1132$

$-2.1437x \leqslant -16.9946$

$x \geqslant 7.9277$

or

$-2.1437x + 7.8814 \geqslant 9.1132$

$-2.1437x \geqslant 1.2318$

$x \leqslant -0.5746$

The solution set is $\{x|x \leqslant -0.5746$ or $x \geqslant 7.9277\}$, or $(-\infty, -0.5746] \cup [7.9277, \infty)$.

42. $\{x|0.9841 \leqslant x \leqslant 4.9808\}$, or $[0.9841, 4.9808]$

43. $|2x - 8| = |x + 3|$

$2x - 8 = x + 3$ or $2x - 8 = -(x + 3)$

$x = 11$ or $2x - 8 = -x - 3$

$x = 11$ or $3x = 5$

$x = 11$ or $x = \frac{5}{3}$

The solution set is $\left\{\frac{5}{3}, 11\right\}$.

44. $\left\{-\frac{11}{2}, \frac{3}{4}\right\}$

45. $\left|\frac{2x + 3}{6}\right| = \left|\frac{4 - 5x}{8}\right|$

$\frac{2x + 3}{6} = \frac{4 - 5x}{8}$ or $\frac{2x + 3}{6} = -\left(\frac{4 - 5x}{8}\right)$

$4(2x + 3) = 3(4 - 5x)$ or $\frac{2x + 3}{6} = \frac{5x - 4}{8}$

$8x + 12 = 12 - 15x$ or $4(2x + 3) = 3(5x - 4)$

$23x = 0$ or $8x + 12 = 15x - 12$

$x = 0$ or $24 = 7x$

$x = 0$ or $\frac{24}{7} = x$

The solution set is $\left\{0, \frac{24}{7}\right\}$.

46. $\left\{-\frac{5}{86}, \frac{85}{14}\right\}$

47. $|x - 2| = x - 2$

By the definition of absolute value,

$|x - 2| = x - 2$ for $x - 2 \geqslant 0$, or $x \geqslant 2$.

The solution set is $\{x|x \geqslant 2\}$, or $[2, \infty)$.

48. $\left\{\frac{4}{5}, 2\right\}$

49. $|7x - 2| = x + 5$

$7x - 2 = x + 5$ or $7x - 2 = -(x + 5)$

$6x = 7$ or $7x - 2 = -x - 5$

$x = \frac{7}{6}$ or $8x = -3$

$x = \frac{7}{6}$ or $x = -\frac{3}{8}$

The solution set is $\left\{-\frac{3}{8}, \frac{7}{6}\right\}$.

50. $\left\{\frac{5}{2}, \frac{11}{4}\right\}$

51. $|3x - 4| \leqslant 2x + 1$

$-(2x + 1) \leqslant 3x - 4 \leqslant 2x + 1$

$-2x - 1 \leqslant 3x - 4 \leqslant 2x + 1$

$-1 \leqslant 5x - 4 \leqslant 4x + 1$

We write the conjunction, solve each inequality separately, and then find the intersection of the solution sets.

$-1 \leqslant 5x - 4$ and $5x - 4 \leqslant 4x + 1$

$3 \leqslant 5x$ and $x \leqslant 5$

$\frac{3}{5} \leqslant x$ and $x \leqslant 5$

The solution set is $\left[\frac{3}{5}, \infty\right] \cap \left[-\infty, 5\right] = \left[\frac{3}{5}, 5\right]$.

52. $\left[-6, -\frac{1}{2}\right]$

53. $|3x - 4| \geqslant 2x + 1$

$3x - 4 \leqslant -(2x + 1)$ or $3x - 4 \geqslant 2x + 1$

$3x - 4 \leqslant -2x - 1$ or $x \geqslant 5$

$5x \leqslant 3$ or $x \geqslant 5$

$x \leqslant \frac{3}{5}$ or $x \geqslant 5$

The solution set is $\left[-\infty, \frac{3}{5}\right] \cup [5, \infty)$.

54. $\left[-\infty, -6\right] \cup \left[-\frac{1}{2}, \infty\right]$

55. $|4x - 5| = |x| + 1$

If $x \geq 0$, then $|x| = x$ and we solve:

$4x - 5 = x + 1$ or $4x - 5 = -(x + 1)$

$3x = 6$ or $4x - 5 = -x - 1$

$x = 2$ or $5x = 4$

$x = 2$ or $x = \frac{4}{5}$

If $x < 0$, then $|x| = -x$ and we solve:

$4x - 5 = -x + 1$ or $4x - 5 = -(-x + 1)$

$5x = 6$ or $4x - 5 = x - 1$

$x = \frac{6}{5}$ or $3x = 4$

$x = \frac{6}{5}$ or $x = \frac{4}{3}$

Since $\frac{6}{5} > 0$ and $\frac{4}{3} > 0$, this case yields no solution.

The solution set is $\left\{\frac{4}{5}, 2\right\}$.

56. $\{5, -11\}$

57. $\big||x| - 1\big| = 3$

$|x| - 1 = -3$ or $|x| - 1 = 3$ (Property i)

$|x| = -2$ or $|x| = 4$

The solution set of $|x| = -2$ is $\emptyset$.

The solution set of $|x| = 4$ is $\{-4, 4\}$.

$\emptyset \cup \{-4, 4\} = \{-4, 4\}$

The solution set of $\big||x| - 1\big| = 3$ is $\{-4, 4\}$.

58. $(-\infty, \infty)$

59. $|x + 2| \leq |x - 5|$

Divide the set of reals into three intervals:

$x \geq 5$

$-2 \leq x < 5$

$x < -2$

Find the solution set of $|x + 2| \leq |x - 5|$ for each interval. Then take the union of the three solution sets.

If $x \geq 5$, then $|x + 2| = x + 2$ and $|x - 5| = x - 5$.

Solve: $x + 2 \leq x - 5$

$2 \leq -5$ (False)

The solution set for this interval is $\emptyset$.

If $-2 \leq x < 5$, $|x + 2| = x + 2$ and $|x - 5| = -(x - 5)$.

Solve: $x + 2 \leq -(x - 5)$

$x + 2 \leq -x + 5$

$2x \leq 3$

$x \leq \frac{3}{2}$

The solution set for this interval is $\left\{x \middle| x \leq \frac{3}{2}\right\}$.

59. (continued)

If $x < -2$, then $|x + 2| = -(x + 2)$ and $|x - 5| = -(x - 5)$.

Solve: $-(x + 2) \leq -(x - 5)$

$-x - 2 \leq -x + 5$

$-2 \leq 5$ (True for any x such that $x < -2$)

The solution set for this interval is $\{x|x < -2\}$.

The union of the above three solution sets is $\left\{x \middle| x \leq \frac{3}{2}\right\}$, or $\left[-\infty, \frac{3}{2}\right]$. This set is the solution of $|x + 2| \leq |x - 5|$.

60. $\left\{x \middle| x < \frac{1}{2}\right\}$, or $\left[-\infty, \frac{1}{2}\right]$

61. $|x| + |x - 1| < 10$

Divide the set of reals into three intervals:

$x \geq 1$

$0 \leq x < 1$

$x < 0$

Find the solution set of $|x| + |x - 1| < 10$ for each interval. Then take the union of the three solution sets.

If $x \geq 1$, then $|x| = x$ and $|x - 1| = x - 1$.

Solve: $x + (x - 1) < 10$

$2x - 1 < 10$

$2x < 11$

$x < \frac{11}{2}$

The solution set for this interval is $\left\{x \middle| 1 \leq x < \frac{11}{2}\right\}$.

If $0 \leq x < 1$, then $|x| = x$ and $|x - 1| = -(x - 1)$.

Solve: $x + [-(x - 1)] < 10$

$x - x + 1 < 10$

$1 < 10$ (True for any x such that $0 \leq x < 1$)

The solution set for this interval is $\{x|0 \leq x < 1\}$.

If $x < 0$, then $|x| = -x$ and $|x - 1| = -(x - 1)$.

Solve: $-x + [-(x - 1)] < 10$

$-x - x + 1 < 10$

$-2x + 1 < 10$

$-2x < 9$

$x > -\frac{9}{2}$

The solution set for this interval is $\left\{x \middle| -\frac{9}{2} < x < 0\right\}$.

The union of the above three solution sets is $\left\{x \middle| -\frac{9}{2} < x < \frac{11}{2}\right\}$, or $\left(-\frac{9}{2}, \frac{11}{2}\right)$. This set is the solution set of $|x| + |x - 1| < 10$.

62. $(-\infty, \infty)$

63. $|x - 3| + |2x + 5| > 6$

Divide the set of reals into three intervals:

$x \geqslant 3$

$-\frac{5}{2} \leqslant x < 3$

$x < -\frac{5}{2}$

Find the solution set of $|x - 3| + |2x + 5| > 6$ for each interval. Then take the union of the three solution sets.

If $x \geqslant 3$, then $|x - 3| = x - 3$ and $|2x + 5| = 2x + 5$.

Solve: $x - 3 + 2x + 5 > 6$

$3x > 4$

$x > \frac{4}{3}$

The solution set for this interval is $[3, \infty)$.

If $-\frac{5}{2} \leqslant x < 3$, then $|x - 3| = -(x - 3)$ and $|2x + 5| = 2x + 5$.

Solve: $-(x - 3) + 2x + 5 > 6$

$-x + 3 + 2x + 5 > 6$

$x > -2$

The solution set for this interval is $(-2, 3)$.

If $x < -\frac{5}{2}$, then $|x - 3| = -(x - 3)$ and $|2x + 5| = -(2x + 5)$.

Solve: $-(x - 3) + [-(2x + 5)] > 6$

$-x + 3 - 2x - 5 > 6$

$-3x > 8$

$x < -\frac{8}{3}$

The solution set for this interval is $\left[-\infty, -\frac{8}{3}\right]$.

The union of the above solution sets is $\left[-\infty, -\frac{8}{3}\right] \cup (-2, \infty)$. This is the solution set of $|x - 3| + |2x + 5| > 6$.

64. $\emptyset$

65. $|x - 3| + |2x + 5| + |3x - 1| = 12$ (1)

If $x \leqslant 0$, we have $-(x - 3) + |2x + 5| - (3x - 1) = 12$ or $|2x + 5| = 4x + 8$; then $x = -\frac{3}{2}$. (The other value, $-\frac{13}{6}$, is extraneous). [We now notice that x must be less than, say, 3; otherwise the left member of Eq. (1) would be greater than 12].

If $x > 0$ (and $x < 3$), we have from (1), $-(x - 3) + (2x + 5) + |3x - 1| = 12$ or $|3x - 1| = 4 - x$; then $x = \frac{5}{4}$ $\left[\text{and } -\frac{3}{2}, \text{ again}\right]$.

The solution set is $\left\{-\frac{3}{2}, \frac{5}{4}\right\}$.

66. $(x_0 - \delta, x_0 + \delta)$

67. PROVE: $-|a| \leqslant a \leqslant |a|$.

If $a = 0$, then $-|0| \leqslant 0 \leqslant |0|$.

If $a < 0$, then $-|a| = a \leqslant a \leqslant |a|$ or $-|a| \leqslant a \leqslant |a|$.

If $a > 0$, then $-|a| = -a \leqslant a \leqslant |a|$ or $-|a| \leqslant a \leqslant |a|$.

68. PROVE: $|a + b| \leqslant |a| + |b|$

From Exercise 67 we have $-|a| \leqslant a \leqslant |a|$
and $-|b| \leqslant b \leqslant |b|$.

Adding, we have $-[|a| + |b|] \leqslant (a + b) \leqslant |a| + |b|$ or $|a + b| \leqslant |a| + |b|$.

69. SHOW: If $|a| < \frac{e}{2}$ and $|b| < \frac{e}{2}$, then $|a + b| < e$.

By the triangle inequality, $|a + b| < |a| + |b| < \frac{e}{2} + \frac{e}{2}$; hence $|a + b| < e$.

70. a) $-\frac{b - a}{2} < x - \frac{a + b}{2} < \frac{b - a}{2}$ gives $a < x < b$.

b) $|x| < 5$

c) $|x| < 6$

d) $|x - 3| < 4$

e) $|x + 2| < 3$

71. 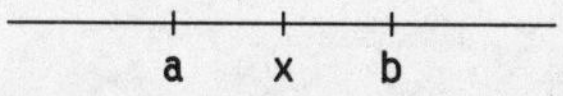

Assume: $a \neq b$.

Since x is the midpoint, $|x - a| = |x - b|$ or $x - a = \pm (x - b)$. The + sign gives $a = b$, not admissable; the - sign gives $x = \frac{a + b}{2}$.

72. $(L - \varepsilon, L + \varepsilon)$

Exercise Set 4.7

1. $(x + 5)(x - 3) > 0$

The solutions of $(x + 5)(x - 3) = 0$ are -5 and 3. They are not solutions of the inequality, but they divide the real-number line in a natural way. The product $(x + 5)(x - 3)$ is positive or negative, for values other than -5 and 3, depending on the signs of the factors $x + 5$ and $x - 3$.

Sign of $x + 5$: - - - - | + + + + | + + + +

Sign of $x - 3$: - - - - | - - - - | + + + +

Sign of product: + + + + | - - - - | + + + +

-5 3

$x + 5 > 0$ when $x > -5$
$x + 5 < 0$ when $x < -5$

$x - 3 > 0$ when $x > 3$
$x - 3 < 0$ when $x < 3$

For the product $(x + 5)(x - 3)$ to be positive, both factors must be positive or both factors must be negative. We see from the diagram that numbers satisfying $x < -5$ or $x > 3$ are solutions. The solution set of the inequality is $\{x | x < -5 \text{ or } x > 3\}$, or $(-\infty,-5) \cup (3,\infty)$.

2. $\{x | x < -4 \text{ or } x > 1\}$, or $(-\infty,-4) \cup (1,\infty)$

3. $(x - 1)(x + 2) \leqslant 0$

The solutions of $(x - 1)(x + 2) = 0$ are 1 and -2. They divide the number line into three intervals as shown:

A B C

-2 1

We try test numbers in each interval.

A: Test -3, $f(-3) = (-3 - 1)(-3 + 2) = 4$;

B: Test 0, $f(0) = (0 - 1)(0 + 2) = -2$;

C: Test 3, $f(3) = (3 - 1)(3 + 2) = 10$

Since $f(0)$ is negative, the function value will be negative for all numbers in the interval containing 0. The inequality symbol is $\leqslant$, so we need to include the intercepts.

The solution set is $\{x | -2 \leqslant x \leqslant 1\}$, or $[-2,1]$.

4. $\{x | -5 \leqslant x \leqslant 3\}$, or $[-5,3]$

5. $x^2 + x - 2 < 0$

$(x - 1)(x + 2) < 0$ Factoring

See the diagram and test numbers in Exercise 3. The solution set is $\{x | -2 < x < 1\}$, or $(-2,1)$.

6. $\{x | -1 < x < 2\}$, or $(-1,2)$

7. $x^2 \geqslant 1$, or $x^2 - 1 \geqslant 0$

Consider the function $f(x) = x^2 - 1$. The inputs that produce outputs that are greater than or equal to 0 are the solutions of the inequality.

Set $f(x) = 0$ and factor to find the x-intercpets.

$x^2 - 1 = 0$

$(x + 1)(x - 1) = 0$

$x + 1 = 0$ or $x - 1 = 0$

$x = -1$ or $x = 1$

The x-intercepts are $(-1, 0)$ and $(1, 0)$.

The table below lists a few ordered pairs that are also solutions of the function. Plot these points and draw the graph.

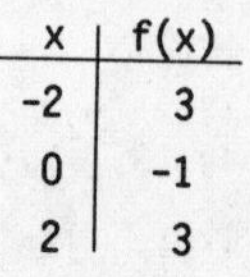

x	f(x)
-2	3
0	-1
2	3

$f(x) = x^2 - 1$

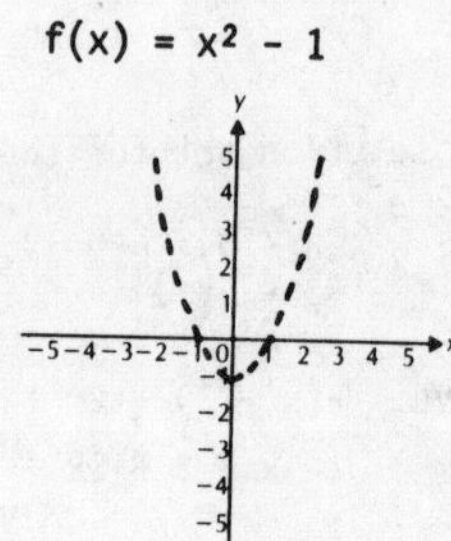

From the graph we can easily see that the solution set is $\{x | x \leqslant -1 \text{ or } x \geqslant 1\}$, or $(-\infty, -1] \cup [1, \infty)$.

8. $\{x | -5 < x < 5\}$, or $(-5, 5)$

9. $9 - x^2 \leqslant 0$

$(3 - x)(3 + x) \leqslant 0$

The solutions of $(3 - x)(3 + x) = 0$ are 3 and -3. They divide the real-number line in a natural way. The product $(3 - x)(3 + x)$ is positive or negative, for values other than 3 and -3, depending on the signs of the factors $3 - x$ and $3 + x$.

Sign of $3 - x$: + + + + | + + + + | - - - -

Sign of $3 + x$: - - - - | + + + + | + + + +

Sign of product: - - - - | + + + + | - - - -

-3 3

$3 - x > 0$ when $x < 3$
$3 - x < 0$ when $x > 3$

$3 + x > 0$ when $x > -3$
$3 + x < 0$ when $x < -3$

For the product $(3 - x)(3 + x)$ to be negative, one factor must be positive and the other negative. We see from the diagram that numbers satisfying $x < -3$ or $x > 3$ are solutions. The intercepts are also solutions. The solution set of the inequality is $\{x | x \leqslant -3 \text{ or } x \geqslant 3\}$, or $(-\infty,-3] \cup [3,\infty)$.

10. $\{x | -2 \leqslant x \leqslant 2\}$, or $[-2,2]$

11. $x^2 - 2x + 1 \geqslant 0$

$(x - 1)^2 \geqslant 0$

The solution of $(x - 1)^2 = 0$ is 1. For all real-number values of x except 1, $(x - 1)^2$ will be positive. Thus the solution set is $\{x|x \text{ is a real number}\}$, or $(-\infty,\infty)$.

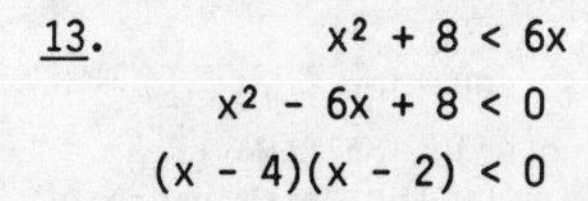

12. $\emptyset$

13. $x^2 + 8 < 6x$

$x^2 - 6x + 8 < 0$

$(x - 4)(x - 2) < 0$

The solutions of $(x - 4)(x - 2) = 0$ are 4 and 2. They are not solutions of the inequality, but they divide the real-number line in a natural way. The product $(x - 4)(x - 2)$ is positive or negative, for values other than 4 and 2, depending on the signs of the factors $x - 4$ and $x - 2$.

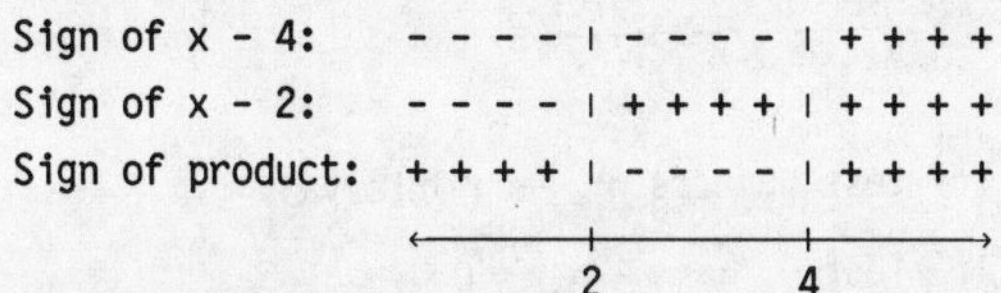

$x - 4 > 0$ when $x > 4$
$x - 4 < 0$ when $x < 4$

$x - 2 > 0$ when $x > 2$
$x - 2 < 0$ when $x < 2$

For the product $(x - 4)(x - 2)$ to be negative one factor must be positive and the other negative. The only situation in the table for which this happens is when $2 < x < 4$. The solution set of the inequality is $\{x|2 < x < 4\}$, or $(2,4)$.

14. $\{x|x < -2 \text{ or } x > 6\}$, or $(-\infty,-2) \cup (6,\infty)$

15. $4x^2 + 7x < 15$

$4x^2 + 7x - 15 < 0$

$(4x - 5)(x + 3) < 0$

The solutions of $(4x - 5)(x + 3) = 0$ are $\frac{5}{4}$ and -3. They divide the number line into three intervals as shown:

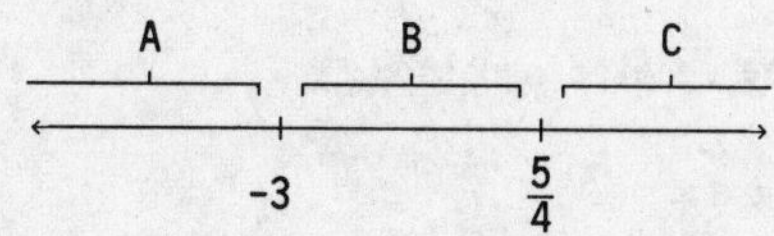

We try test numbers in each interval.

A: Test -4, $f(-4) = [4(-4) - 5][-4 + 3] = 21$

B: Test 0, $f(0) = (4 \cdot 0 - 5)(0 + 3) = -15$

C: Test 2, $f(2) = (4 \cdot 2 - 5)(2 + 3) = 15$

Since $f(0)$ is negative, the function value will be negative for all numbers in the interval containing 0. The solution set is $\left\{x\middle|-3 < x < \frac{5}{4}\right\}$, or $\left(-3, \frac{5}{4}\right)$.

16. $\left\{x\middle|x \leqslant -3 \text{ or } x \geqslant \frac{5}{4}\right\}$, or $(-\infty, -3] \cup \left[\frac{5}{4}, \infty\right)$

17. $2x^2 + x > 5$

$2x^2 + x - 5 > 0$

The solutions are $2x^2 + x - 5 = 0$ are $\frac{-1 \pm \sqrt{41}}{2}$. They divide the number line into three intervals as shown:

A B C

$\frac{-1 - \sqrt{41}}{2}$ $\frac{-1 + \sqrt{41}}{2}$

We try test numbers in each interval. Note that $\frac{-1 - \sqrt{41}}{2} \approx -3.7$ and $\frac{-1 + \sqrt{41}}{2} \approx 2.7$.

A: Test -4, $f(-4) = 2(-4)^2 + (-4) - 5 = 23$

B: Test 0, $f(0) = 2(0)^2 + 0 - 5 = -5$

C: Test 3, $f(3) = 2(3)^2 + 3 - 5 = 16$

Since $f(-4)$ and $f(3)$ are positive, the function value will be positive for all numbers in the intervals containing -4 and 3. The solution set is $\left\{x|x < \frac{-1 - \sqrt{41}}{2} \text{ or } x > \frac{-1 + \sqrt{41}}{2}\right\}$, or $\left(-\infty, \frac{-1 - \sqrt{41}}{2}\right) \cup \left(\frac{-1 + \sqrt{41}}{2}, \infty\right)$.

18. $\left\{x\middle|\frac{-1 - \sqrt{17}}{4} \leqslant x \leqslant \frac{-1 + \sqrt{17}}{4}\right\}$, or $\left[\frac{-1 - \sqrt{17}}{4}, \frac{-1 + \sqrt{17}}{4}\right]$

19. $3x(x + 2)(x - 2) < 0$

The solutions of $3x(x + 2)(x - 2) = 0$ are 0, -2, and 2. They divide the number line into four intervals as shown.

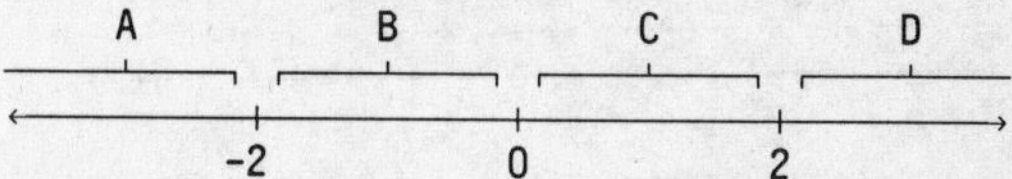

We try test numbers in each interval.

A: Test -3, $f(-3) = 3(-3)(-3 + 2)(-3 - 2) = -45$

B: Test -1, $f(-1) = 3(-1)(-1 + 2)(-1 - 2) = 9$

C: Test 1, $f(1) = 3(1)(1 + 2)(1 - 2) = -9$

D: Test 3, $f(3) = 3(3)(3 + 2)(3 - 2) = 45$

Since $f(-3)$ and $f(1)$ are negative, the function value will be negative for all numbers in the intervals containing -3 and 1. The solution set is $\{x|x < -2 \text{ or } 0 < x < 2\}$, or $(-\infty,-2) \cup (0,2)$.

20. $\{x|-1 < x < 0 \text{ or } x > 1\}$, or $(-1,0) \cup (1,\infty)$

21. $(x + 3)(x - 2)(x + 1) > 0$

The solutions of $(x + 3)(x - 2)(x + 1) = 0$ are -3, 2, and -1. They are not solutions of the inequality, but they divide the real number line in a natural way. The product $(x + 3)(x - 2)(x + 1)$ is positive or negative, for values other than -3, 2, and -1, depending on the signs of the factors $x + 3$, $x - 2$, and $x + 1$.

Sign of x + 3: - - - -|+ + + +|+ + + +|+ + + +

Sign of x - 2: - - - -|- - - -|- - - -|+ + + +

Sign of x + 1: - - - -|- - - -|+ + + +|+ + + +

Sign of product: - - - -|+ + + +|- - - -|+ + + +

-3 -1 2

$x + 3 > 0$ when $x > -3$
$x + 3 < 0$ when $x < -3$

$x - 2 > 0$ when $x > 2$
$x - 2 < 0$ when $x < 2$

$x + 1 > 0$ when $x > -1$
$x + 1 < 0$ when $x < -1$

The product of three numbers is positive when all three are positive or two are negative and one is positive. We see from the diagram that numbers satisfying $-3 < x < -1$ or $x > 2$ are solutions. The solution set of the inequality is $\{x|-3 < x < -1 \text{ or } x > 2\}$, or $(-3,-1) \cup (2,\infty)$.

22. $\{x|x < -2 \text{ or } 1 < x < 4\}$, or $(-\infty,-2) \cup (1,4)$

23. $(x + 3)(x + 2)(x - 1) < 0$

The solutions of $(x + 3)(x + 2)(x - 1) = 0$ are -3, -2, and 1. They divide the number line into four intervals as shown:

A B C D

-3 -2 1

We try test numbers in each interval.

A: Test -4, $f(-4) = (-4 + 3)(-4 + 2)(-4 - 1) = -10$

B: Test $-\frac{5}{2}$, $f\left(-\frac{5}{2}\right) = \left(-\frac{5}{2} + 3\right)\left(-\frac{5}{2} + 2\right)\left(-\frac{5}{2} - 1\right) = \frac{7}{8}$

C: Test 0, $f(0) = (0 + 3)(0 + 2)(0 - 1) = -6$

D: Test 2, $f(2) = (2 + 3)(2 + 2)(2 - 1) = 20$

The function value will be negative for all numbers in intervals A and C. The solution set is $\{x|x < -3 \text{ or } -2 < x < 1\}$, or $(-\infty,-3) \cup (-2,1)$.

24. $\{x|x < -1 \text{ or } 2 < x < 3\}$, or $(-\infty,-1) \cup (2,3)$

25. $\frac{1}{4 - x} < 0$

We write the related equation by changing the < symbol to =:

$$\frac{1}{4 - x} = 0$$

Then we solve the related equation:

$$(4 - x) \cdot \frac{1}{4 - x} = (4 - x)\cdot 0$$

$$1 = 0 \quad \text{False equation}$$

The related equation has no solution. Next we find the replacements that make the denominator 0:

$$4 - x = 0$$
$$4 = x$$

The number 4 divides the number line as shown:

A B

4

Try test numbers in each interval:

A: Test 0, $\frac{1}{4 - x} < 0$; $\frac{1}{4 - 0}$ | 0; $\frac{1}{4}$

The number 0 is not a solution of the inequality, so the interval A is not part of the solution set.

B: Test 5, $\frac{1}{4 - x} < 0$; $\frac{1}{4 - 5}$ | 0; -1

The number 5 is a solution of the inequality, so the interval B is part of the solution set.

The solution set is $\{x|x > 4\}$, or $(4, \infty)$.

26. $\left\{x\middle|x < -\frac{5}{2}\right\}$, or $\left(-\infty, -\frac{5}{2}\right)$

27. $3 < \frac{1}{x}$

Solve the related equation.

$$3 = \frac{1}{x}$$

$$x = \frac{1}{3}$$

Find the replacements that are not meaningful.

$$x = 0$$

Use the numbers $\frac{1}{3}$ and 0 to divide the number line into intervals as shown.

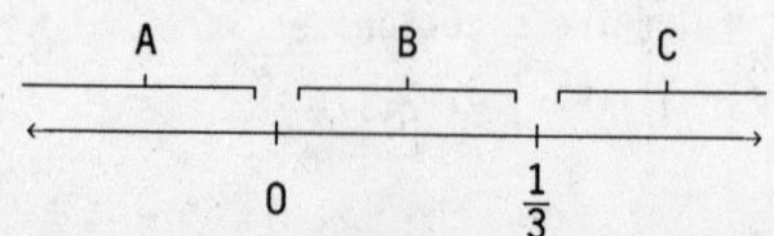

27. (continued)

Try test numbers in each interval.

A: Test -1, $3 < \frac{1}{x}$

$$3 \;\Big|\; \frac{1}{-1}$$
$$\;\Big|\; -1$$

The number -1 is not a solution of the inequality, so the interval A is not in the solution set.

B: Test $\frac{1}{4}$, $3 < \frac{1}{x}$

$$3 \;\Big|\; \frac{1}{\frac{1}{4}}$$
$$\;\Big|\; 4$$

The number 4 is a solution of the inequality, so the interval B is in the solution set.

C: Test 1, $3 < \frac{1}{x}$

$$3 \;\Big|\; \frac{1}{1}$$
$$\;\Big|\; 1$$

The number 1 is not a solution of the inequality, so the interval C is not in the solution set.

The solution set is $\left\{x \middle| 0 < x < \frac{1}{3}\right\}$, or $\left(0, \frac{1}{3}\right)$.

28. $\left\{x \middle| x < 0 \text{ or } x \geqslant \frac{1}{5}\right\}$, or $(-\infty, 0) \cup \left[\frac{1}{5}, \infty\right)$

29. $\frac{3x + 2}{x - 3} > 0$

Solve the related equation.

$$\frac{3x + 2}{x - 3} = 0$$
$$3x + 2 = 0$$
$$3x = -2$$
$$x = -\frac{2}{3}$$

Find replacements that are not meaningful.

$$x - 3 = 0$$
$$x = 3$$

Use the numbers $-\frac{2}{3}$ and 3 to divide the number line into intervals as shown.

A B C

$-\frac{2}{3}$ 3

Try test numbers in each interval.

29. (continued)

A: Test -1, $\frac{3x + 2}{x - 3} > 0$

$$\frac{3(-1) + 2}{-1 - 3} \;\Big|\; 0$$
$$\frac{-1}{-4} \;\Big|$$
$$\frac{1}{4} \;\Big|$$

The number -1 is a solution of the inequality, so the interval A is part of the solution set.

B: Test 0, $\frac{3x + 2}{x - 3} > 0$

$$\frac{3\cdot 0 + 2}{0 - 3} \;\Big|\; 0$$
$$\frac{2}{-3} \;\Big|$$
$$-\frac{2}{3} \;\Big|$$

The number 0 is not a solution of the inequality, so the interval B is not part of the solution set.

C: Test 4, $\frac{3x + 2}{x - 3} > 0$

$$\frac{3\cdot 4 + 2}{4 - 3} \;\Big|\; 0$$
$$14 \;\Big|$$

The number 4 is a solution of the inequality, so the interval C is part of the solution set.

$\left\{x \middle| x < -\frac{2}{3} \text{ or } x > 3\right\}$, or $\left(-\infty, -\frac{2}{3}\right) \cup (3, \infty)$.

30. $\left\{x \middle| x < -\frac{3}{4} \text{ or } x > \frac{5}{2}\right\}$, or $\left(-\infty, -\frac{3}{4}\right] \cup \left[\frac{5}{2}, \infty\right)$

31. $\frac{x + 2}{x} \leqslant 0$

Solve the related equation.

$$\frac{x + 2}{x} = 0$$
$$x + 2 = 0$$
$$x = -2$$

Find the replacements that are not meaningful.

$$x = 0$$

The numbers -2 and 0 divide the number line into intervals as shown.

A B C

-2 0

Try test numbers in each interval.

A: Test -3, $\frac{x + 2}{x} \leqslant 0$

$$\frac{-3 + 2}{-3} \;\Big|\; 0$$
$$\frac{1}{3} \;\Big|$$

The number -3 is not a solution of the inequality, so the interval A is not in the solution set.

31. (continued)

B: Test -1,

$\frac{x+2}{x} \leqslant 0$	
$\frac{-1+2}{-1}$	0
-1	

The number -1 is a solution of the inequality, so the interval B is in the solution set.

C: Test 1,

$\frac{x+2}{x} \leqslant 0$	
$\frac{1+2}{1}$	0
3	

The number 1 is not a solution of the inequality, so the interval C is not in the solution set.

The solution set includes the interval B. The number -2 is also included since the inequality symbol is $\leqslant$ 1 and -2 is a solution of the related equation. The number 0 is not included, since it is not a meaningful replacement. The solution set is $\{x \mid -2 \leqslant x < 0\}$, or $[-2, 0)$.

32. $\{x \mid x \leqslant 0 \text{ or } x > 3\}$, or $(-\infty, 0] \cup (3, \infty)$

33. $\frac{x+1}{2x-3} \geqslant 1$

Solve the related equation.

$$\frac{x+1}{2x-3} = 1$$

$$x + 1 = 2x - 3$$

$$4 = x$$

Find replacements that are not meaningful.

$$2x - 3 = 0$$

$$x = \frac{3}{2}$$

Use the numbers 4 and $\frac{3}{2}$ to divide the number line as shown.

A | B | C, with divisions at $\frac{3}{2}$ and 4

Try test numbers in each interval.

A: Test 0,

$\frac{x+1}{2x-3} \geqslant 1$	
$\frac{0+1}{2\cdot0-3}$	1
$-\frac{1}{3}$	

The interval A is not in the solution set.

B: Test 2,

$\frac{x+1}{2x-3} \geqslant 1$	
$\frac{2+1}{2\cdot2-3}$	1
3	

The interval B is in the solution set.

33. (continued)

C: Test 5,

$\frac{x+1}{2x-3} \geqslant 1$	
$\frac{5+1}{2\cdot5-3}$	1
$\frac{6}{7}$	

The interval C is not in the solution set. The solution set includes the interval B. The number 4 is also included, since the inequality is $\geqslant$ 0 and 4 is a solution of the related equation. The number $\frac{3}{2}$ is not included, since it is not a meaningful replacement. The solution set is $\left\{x \middle| \frac{3}{2} < x \leqslant 4\right\}$, or $\left(\frac{3}{2}, 4\right]$.

34. $\left\{x \middle| 2 < x \leqslant \frac{5}{2}\right\}$, or $\left(2, \frac{5}{2}\right]$

35. $\frac{x+1}{x+2} \leqslant 3$

Solve the related equation.

$$\frac{x+1}{x+2} = 3$$

$$x + 1 = 3(x + 2)$$

$$x + 1 = 3x + 6$$

$$-5 = 2x$$

$$-\frac{5}{2} = x$$

Find the replacements that are not meaningful.

$$x + 2 = 0$$

$$x = -2$$

Use the numbers $-\frac{5}{2}$ and -2 to divide the number line as shown.

A | B | C, with divisions at $-\frac{5}{2}$ and -2

Try test numbers in each interval.

A: Test -3,

$\frac{x+1}{x+2} \leqslant 3$	
$\frac{-3+1}{-3+2}$	3
2	

The interval A is in the solution set.

35. (continued)

B: Test $-\frac{9}{4}$,

$\frac{x+1}{x+2} \leqslant 3$	
$\frac{-\frac{9}{4}+1}{-\frac{9}{4}+2}$	3
$\frac{-\frac{5}{4}}{-\frac{1}{4}}$	
5	

The interval B is not in the solution set.

C: Test 0,

$\frac{x+1}{x+2} \leqslant 3$	
$\frac{0+1}{0+2}$	3
$\frac{1}{2}$	

The interval C is in the solution set.

The number $-\frac{5}{2}$ is in the solution set since the inequality is $\leqslant$, but -2 is not included since it is not a meaningful replacement. The solution set is $\left\{x\middle|x \leqslant -\frac{5}{2} \text{ or } x > -2\right\}$, or $\left[-\infty, -\frac{5}{2}\right] \cup (-2, \infty)$.

36. $\left\{x\middle|x < \frac{3}{2} \text{ or } x \geqslant 4\right\}$, or $\left[-\infty, \frac{3}{2}\right] \cup [4, \infty)$

37. $\frac{x-6}{x} > 1$

Solve the related equation.

$$\frac{x-6}{x} = 1$$
$$x - 6 = x$$
$$-6 = 0$$

The related equation has no solution. We see that 0 is not a meaningful replacement. Use the number 0 to divide the number line as shown.

A | B, divided at 0

Try test numbers in each interval.

A: Test -1,

$\frac{x-6}{x} > 1$	
$\frac{-1-6}{-1}$	1
7	

The interval A is in the solution set.

B: Test 1,

$\frac{x-6}{x} > 1$	
$\frac{1-6}{1}$	1
-5	

The interval B is not in the solution set.
The solution set is $\{x|x < 0\}$, or $(-\infty, 0)$.

38. $\left\{x\middle|x < -3 \text{ or } x > -\frac{3}{2}\right\}$, or $(-\infty, -3) \cup \left[-\frac{3}{2}, \infty\right]$

39. $(x + 1)(x - 2) > (x + 3)^2$

Solve the related equation.

$$(x + 1)(x - 2) = (x + 3)^2$$
$$x^2 - x - 2 = x^2 + 6x + 9$$
$$-11 = 7x$$
$$-\frac{11}{7} = x$$

All real numbers are meaningful replacements. Use the number $-\frac{11}{7}$ to divide the number line as shown.

A | B, divided at $-\frac{11}{7}$

Try test numbers in each interval.

A: Test -2,

$(x + 1)(x - 2)$ >	$(x + 3)^2$
$(-2 + 1)(-2 - 2)$	$(-2 + 3)^2$
4	1

The interval A is in the solution set.

B: Test 0,

$(x + 1)(x - 2)$ >	$(x + 3)^2$
$(0 + 1)(0 - 2)$	$(0 + 3)^2$
-2	9

The interval B is not in the solution set.
The solution set is $\left\{x\middle|x < -\frac{11}{7}\right\}$, or $\left[-\infty, -\frac{11}{7}\right]$.

40. $\{x|x > 13\}$, or $(13, \infty)$

41. $x^3 - x^2 > 0$
$x^2(x - 1) > 0$
The solutions of $x^2(x - 1) = 0$ are 0 and 1. They divide the number line into intervals as shown.

A | B | C, divided at 0 and 1

Try test numbers in each interval.

A: Test -1, $f(-1) = (-1)^3 - (-1)^2 = -2$

B: Test $\frac{1}{2}$, $f\left(\frac{1}{2}\right) = \left(\frac{1}{2}\right)^3 - \left(\frac{1}{2}\right)^2 = -\frac{1}{8}$

C: Test 2, $f(2) = 2^3 - 2^2 = 4$

Since f(2) is positive, the interval C is the solution set. That is, the solution set is $\{x|x > 1\}$, or $(1, \infty)$.

42. $\{x|-2 < x < 0 \text{ or } x > 2\}$, or $(-2, 0) \cup (2, \infty)$

43. $x + \frac{4}{x} > 4$

Solve the related equation.

$$x + \frac{4}{x} = 4$$
$$x^2 + 4 = 4x$$
$$x^2 - 4x + 4 = 0$$
$$(x - 2)(x - 2) = 0$$
$$x = 2$$

We see that 0 is not a meaningful replacement. Use the numbers 2 and 0 to divide the number line as shown.

A B C

0 2

Try test numbers in each interval.

A: Test -1,

$x + \frac{4}{x} > 4$	
$-1 + \frac{4}{-1}$	4
-5	

The interval A is not in the solution set.

B: Test 1,

$x + \frac{4}{x} > 4$	
$1 + \frac{4}{1}$	4
5	

The interval B is in the solution set.

C: Test 4,

$x + \frac{4}{x} > 4$	
$4 + \frac{4}{4}$	4
5	

The interval C is in the solution set. The solution set is $\{x \mid 0 < x < 2 \text{ or } x > 2\}$, or $(0, 2) \cup (2, \infty)$.

44. $\{x \mid 0 < x \leqslant 1\}$, or $(0, 1]$

45. $\frac{1}{x^3} \leqslant \frac{1}{x^2}$

Solve the related equation.

$$\frac{1}{x^3} = \frac{1}{x^2}$$
$$x^3 \cdot \frac{1}{x^3} = x^3 \cdot \frac{1}{x^2}$$
$$1 = x$$

We see that 0 is not a meaningful replacement. Use the numbers 1 and 0 to divide the number line as shown.

A B C

0 1

Try test numbers in each interval.

45. (continued)

A: Test -1,

$\frac{1}{x^3} \leqslant \frac{1}{x^2}$	
$\frac{1}{(-1)^3}$	$\frac{1}{(-1)^2}$
-1	1

The interval A is in the solution set.

B: Test $\frac{1}{2}$,

$\frac{1}{x^3} \leqslant \frac{1}{x^2}$	
$\frac{1}{\left(\frac{1}{2}\right)^3}$	$\frac{1}{\left(\frac{1}{2}\right)^2}$
8	4

The interval B is not in the solution set.

C: Test 2,

$\frac{1}{x^3} \leqslant \frac{1}{x^2}$	
$\frac{1}{2^3}$	$\frac{1}{2^2}$
$\frac{1}{8}$	$\frac{1}{4}$

The interval C is in the solution set. The number 1 is also in the solution set since the inequality is $\leqslant$, but 0 is not included since it is not a meaningful replacement. The solution set is $\{x \mid x < 0 \text{ or } x \geqslant 1\}$, or $(-\infty, 0) \cup [1, \infty)$.

46. $\{x \mid 0 < x < 1 \text{ or } x > 1\}$, or $(0, 1) \cup (1, \infty)$

47. $\frac{2 + x - x^2}{x^2 + 5x + 6} < 0$

Solve the related equation.

$$\frac{2 + x - x^2}{x^2 + 5x - 6} = 0$$
$$2 + x - x^2 = 0$$
$$(2 - x)(1 + x) = 0$$
$$x = 2 \quad \text{or} \quad x = -1$$

Find the replacements that are not meaningful.

$$x^2 + 5x + 6 = 0$$
$$(x + 3)(x + 2) = 0$$
$$x = -3 \quad \text{or} \quad x = -2$$

Use the numbers -3, -2, -1, and 2 to divide the number line into intervals as shown.

A B C D E

-3 -2 -1 2

Try test numbers in each interval.

47. (continued)

A: Test -4

$$\begin{array}{c|c} \frac{2 + x - x^2}{x^2 + 5x + 6} & < 0 \\ \hline \frac{2 + (-4) - (-4)^2}{(-4)^2 + 5(-4) + 6} & 0 \\ \frac{-18}{4} & \\ -\frac{9}{2} & \end{array}$$

The interval A is in the solution set.

B: Test $-\frac{5}{2}$

$$\begin{array}{c|c} \frac{2 + x - x^2}{x^2 + 5x + 6} & < 0 \\ \hline \frac{2 + \left(-\frac{5}{2}\right) - \left(-\frac{5}{2}\right)^2}{\left(-\frac{5}{2}\right)^2 + 5\left(-\frac{5}{2}\right) + 6} & 0 \\ \frac{-\frac{27}{4}}{-\frac{1}{4}} & \\ 27 & \end{array}$$

The interval B is not in the solution set.

C: Test $-\frac{3}{2}$

$$\begin{array}{c|c} \frac{2 + x - x^2}{x^2 + 5x + 6} & < 0 \\ \hline \frac{2 + \left(-\frac{3}{2}\right) - \left(-\frac{3}{2}\right)^2}{\left(-\frac{3}{2}\right)^2 + 5\left(-\frac{3}{2}\right) + 6} & 0 \\ \frac{-\frac{7}{4}}{\frac{3}{4}} & \\ -\frac{7}{3} & \end{array}$$

The interval C is in the solution set.

D: Test 0

$$\begin{array}{c|c} \frac{2 + x - x^2}{x^2 + 5x + 6} & < 0 \\ \hline \frac{2 + 0 - 0^2}{0^2 + 5\cdot 0 + 6} & 0 \\ \frac{2}{6} & \\ \frac{1}{3} & \end{array}$$

The interval D is not in the solution set.

E: Test 3

$$\begin{array}{c|c} \frac{2 + x - x^2}{x^2 + 5x + 6} & < 0 \\ \hline \frac{2 + 3 - 3^2}{3^2 + 5\cdot 3 + 6} & 0 \\ -\frac{4}{30} & \\ -\frac{2}{15} & \end{array}$$

The interval E is in the solution set.

The solution set is $\{x|x < -3 \text{ or } -2 < x < -1 \text{ or } x > 2\}$, or $(-\infty, -3) \cup (-2, -1) \cup (2, \infty)$.

48. $\{x|-2 < x < 0 \text{ or } 0 < x < 2\}$, or $(-2, 0) \cup (0, 2)$

49. $x^4 - 2x^2 \leqslant 0$

$x^2(x^2 - 2) \leqslant 0$

The solutions of $x^2(x^2 - 2) = 0$ are 0, $-\sqrt{2}$, and $\sqrt{2}$. They divide the number line as shown.

A B C D

$-\sqrt{2}$ 0 $\sqrt{2}$

Try test numbers in each interval. Note that $-\sqrt{2} \approx -1.4$ and $\sqrt{2} \approx 1.4$.

A: Test -2, $f(-2) = (-2)^4 - 2(-2)^2 = 8$

B: Test -1, $f(-1) = (-1)^4 - 2(-1)^2 = -1$

C: Test 1, $f(1) = 1^4 - 2(1)^2 = -1$

D: Test 2, $f(2) = 2^4 - 2(2)^2 = 8$

The solution set is $\{x|-\sqrt{2} \leqslant x \leqslant \sqrt{2}\}$, or $[-\sqrt{2}, \sqrt{2}]$.

50. $\{x|x < -\sqrt{3} \text{ or } x > \sqrt{3}\}$, or $(-\infty, -\sqrt{3}) \cup (\sqrt{3}, \infty)$

51. $\left|\frac{x + 3}{x - 4}\right| < 2$

$-2 < \frac{x + 3}{x - 4} < 2$

$-2 < \frac{x + 3}{x - 4}$ and $\frac{x + 3}{x - 4} < 2$

First solve $-2 < \frac{x + 3}{x - 4}$.

We solve the related equation.

$$-2 = \frac{x + 3}{x - 4}$$
$$-2(x - 4) = x + 3$$
$$-2x + 8 = x + 3$$
$$5 = 3x$$
$$\frac{5}{3} = x$$

Find the replacements that are not meaningful.

$x - 4 = 0$

$x = 4$

Use the number $\frac{5}{3}$ and 4 to divide the number line as shown.

A B C

$\frac{5}{3}$ 4

Try test numbers in each interval.

A: Test 0,

$$\begin{array}{c|c} -2 & < \frac{x + 3}{x - 4} \\ \hline -2 & \frac{0 + 3}{0 - 4} \\ & -\frac{3}{4} \end{array}$$

The interval A is in the solution set.

51. (continued)

B: Test 3,

$$\begin{array}{c|c} \multicolumn{2}{c}{-2 < \frac{x+3}{x-4}} \\ \hline -2 & \frac{3+3}{3-4} \\ & -6 \end{array}$$

The interval B is not in the solution set.

C: Test 5,

$$\begin{array}{c|c} \multicolumn{2}{c}{-2 < \frac{x+3}{x-4}} \\ \hline -2 & \frac{5+3}{5-4} \\ & 8 \end{array}$$

The interval C is in the solution set.

The solution set for this portion of the conjunction is $\left\{x \middle| x < \frac{5}{3} \text{ or } x > 4\right\}$, or $\left[-\infty, \frac{5}{3}\right] \cup (4, \infty)$.

Next solve $\frac{x+3}{x-4} < 2$.

We solve the related equation.

$$\frac{x+3}{x-4} = 2$$
$$x + 3 = 2(x - 4)$$
$$x + 3 = 2x - 8$$
$$11 = x$$

As before, 4 is not a meaningful replacement. Use the numbers 4 and 11 to divide the number line as shown.

A B C

4 11

Try test numbers in each interval.

A: Test 0,

$$\begin{array}{c|c} \multicolumn{2}{c}{\frac{x+3}{x-4} < 2} \\ \hline \frac{0+3}{0-4} & 2 \\ -\frac{3}{4} & \end{array}$$

The interval A is in the solution set.

B: Test 5,

$$\begin{array}{c|c} \multicolumn{2}{c}{\frac{x+3}{x-4} < 2} \\ \hline \frac{5+3}{5-4} & 2 \\ 8 & \end{array}$$

The interval B is not in the solution set.

C: Test 12,

$$\begin{array}{c|c} \multicolumn{2}{c}{\frac{x+3}{x-4} < 2} \\ \hline \frac{12+3}{12-4} & 2 \\ \frac{15}{8} & \end{array}$$

The interval C is in the solution set. The solution set for this portion of the conjunction is $\{x \mid x < 4 \text{ or } x > 11\}$, or $(-\infty, 4) \cup (11, \infty)$.

51. (continued)

The solution set of the original inequality is the intersection of the solution sets for the two portions of the conjunction.

$$\left[\left[-\infty, \frac{5}{3}\right] \cup (4, \infty)\right] \cap [(-\infty, 4) \cup (11, \infty)] =$$

$$\left[-\infty, \frac{5}{3}\right] \cup (11, \infty)$$

52. $\{x \mid -\sqrt{5} \leqslant x \leqslant \sqrt{5}\}$, or $[-\sqrt{5}, \sqrt{5}]$

53. $(7 - x)^{-2} < 0$

$$\frac{1}{(7 - x)^2} < 0$$

Since $(7 - x)^2 \geqslant 0$ for all values of x, $\frac{1}{(7 - x)^2} > 0$ for all meaningful replacements. The solution set is $\emptyset$.

54. $\{x \mid x < 1\}$, or $(-\infty, 1)$

55. $\left|1 + \frac{1}{x}\right| < 3$

$$-3 < 1 + \frac{1}{x} < 3$$

$$-3 < 1 + \frac{1}{x} \text{ and } 1 + \frac{1}{x} < 3$$

First solve $-3 < 1 + \frac{1}{x}$

We solve the related equation.

$$-3 = 1 + \frac{1}{x}$$
$$-4 = \frac{1}{x}$$
$$-4x = 1$$
$$x = -\frac{1}{4}$$

We see that 0 is not a meaningful replacement. Use the numbers $-\frac{1}{4}$ and 0 to divide the number line as shown.

A B C

$-\frac{1}{4}$ 0

Try test numbers in each interval.

A: Test -1,

$$\begin{array}{c|c} \multicolumn{2}{c}{-3 < 1 + \frac{1}{x}} \\ \hline -3 & 1 + \frac{1}{-1} \\ & 0 \end{array}$$

The interval A is in the solution set.

55. (continued)

B: Test $-\frac{1}{8}$,

$-3 < 1 + \frac{1}{x}$	
-3	$1 + \frac{1}{-\frac{1}{8}}$
	$1 - 8$
	-7

The interval B is not in the solution set.

C: Test 1,

$-3 < 1 + \frac{1}{x}$	
-3	$1 + \frac{1}{1}$
	2

The interval C is in the solution set.

The solution set for this portion of the conjunction is $\left\{x \middle| x < -\frac{1}{4} \text{ or } x > 0\right\}$, or $\left(-\infty, -\frac{1}{4}\right) \cup (0, \infty)$.

Next we solve $1 + \frac{1}{x} < 3$.

We solve the related equation.

$$1 + \frac{1}{x} = 3$$
$$\frac{1}{x} = 2$$
$$1 = 2x$$
$$\frac{1}{2} = x$$

Again, 0 is not a meaningful replacement. We use the numbers $\frac{1}{2}$ and 0 to divide the number line as shown.

A B C

0 $\frac{1}{2}$

Try test numbers in each interval.

A: Test -1,

$1 + \frac{1}{x} < 3$	
$1 + \frac{1}{-1}$	3
0	

The interval A is in the solution set.

B: Test $\frac{1}{4}$,

$1 + \frac{1}{x} < 3$	
$1 + \frac{1}{\frac{1}{4}}$	3
$1 + 4$	
5	

The interval B is not in the solution set.

55. (continued)

C: Test 1,

$1 + \frac{1}{x} < 3$	
$1 + \frac{1}{1}$	3
2	

The interval C is in the solution set.

The solution set for this portion of the conjunction is $\left\{x \middle| x < 0 \text{ or } x > \frac{1}{2}\right\}$, or $(-\infty, 0) \cup \left(\frac{1}{2}, \infty\right)$.

The solution set of the original inequality is the intersection of the solution sets for the two portions of the conjunction.

$$\left[\left(-\infty, -\frac{1}{4}\right) \cup (0, \infty)\right] \cap \left[(-\infty, 0) \cup \left(\frac{1}{2}, \infty\right)\right] =$$
$$\left(-\infty, -\frac{1}{4}\right) \cup \left(\frac{1}{2}, \infty\right)$$

56. $\{x | x \neq -5\}$, or $(-\infty, -5) \cup (-5, \infty)$

57.
$$|x|^2 - 4|x| + 4 \geqslant 9$$
$$|x|^2 - 4|x| - 5 \geqslant 0$$
$$(|x| - 5)(|x| + 1) \geqslant 0$$

The solutions of $(|x| - 5)(|x| + 1) = 0$ are -5 and 5. Then divide the number line as shown.

A B C

-5 5

Try test numbers in each interval.

A: Test -6, $f(-6) = |-6|^2 - 4|-6| - 5 = 7$

B: Test 0, $f(0) = |0|^2 - 4|0| - 5 = -5$

C: Test 6, $f(6) = |6|^2 - 4|6| - 5 = 7$

The solution set is $\{x | x \leqslant -5 \text{ or } x \geqslant 5\}$, or $(-\infty, -5] \cup [5, \infty)$.

58. $\left\{x \middle| -\frac{3}{2} \leqslant x \leqslant \frac{3}{2}\right\}$, or $\left[-\frac{3}{2}, \frac{3}{2}\right]$

59. $\left|2 - \frac{1}{x}\right| \leqslant 2 + \left|\frac{1}{x}\right|$

Note that $\frac{1}{x}$ is not defined when $x = 0$. Thus $x \neq 0$.

Divide the set of reals into three intervals:

$x < 0$

$0 < x < \frac{1}{2}$

$x \geqslant \frac{1}{2}$

Find the solution set of $\left|2 - \frac{1}{x}\right| \leqslant 2 + \left|\frac{1}{x}\right|$ for each interval. Then take the union of the three solution sets.

59. (continued)

If $x < 0$, then $\left|2 - \frac{1}{x}\right| = 2 - \frac{1}{x}$ and $\left|\frac{1}{x}\right| = -\frac{1}{x}$.

Solve: $2 - \frac{1}{x} \leqslant 2 - \frac{1}{x}$

$2 \leqslant 2$

True for any $x < 0$. The solution set for this interval is $\{x|x < 0\}$.

If $0 < x < \frac{1}{2}$, then $\left|2 - \frac{1}{x}\right| = -\left(2 - \frac{1}{x}\right)$ and $\left|\frac{1}{x}\right| = \frac{1}{x}$.

Solve: $-\left(2 - \frac{1}{x}\right) \leqslant 2 + \frac{1}{x}$

$-2 + \frac{1}{x} \leqslant 2 + \frac{1}{x}$

$-2 \leqslant 2$

True for all x such that $0 < x < \frac{1}{2}$. The solution set for this interval is $\left\{x\middle|0 < x < \frac{1}{2}\right\}$.

If $x \geqslant \frac{1}{2}$, then $\left|2 - \frac{1}{x}\right| = 2 - \frac{1}{x}$ and $\left|\frac{1}{x}\right| = \frac{1}{x}$.

Solve: $2 - \frac{1}{x} \leqslant 2 + \frac{1}{x}$

$-\frac{1}{x} \leqslant \frac{1}{x}$

$-1 \leqslant 1$

True for all x such that $x \geqslant \frac{1}{2}$. The solution set for this interval is $\left\{x\middle|x \geqslant \frac{1}{2}\right\}$.

The union of the above three solution sets is the set of all reals except 0, or $(-\infty, 0) \cup (0, \infty)$.

60. $\{x|-2 < x \leqslant -1 \text{ or } 3 \leqslant x < 4 \text{ or } x = 2\}$, or $(-2, -1] \cup [3, 4) \cup [2, 2]$

61. $|x^2 + 3x - 1| < 3$

$-3 < x^2 + 3x - 1 < 3$

$-3 < x^2 + 3x - 1$ and $x^2 + 3x - 1 < 3$, or

$0 < x^2 + 3x + 2$ and $x^2 + 3x - 4 < 0$

First we solve $0 < x^2 + 3x + 2$. We solve the related equation.

$0 = x^2 + 3x + 2$

$0 = (x + 2)(x + 1)$

$x = -2$ or $x = -1$

Use the numbers -2 and -1 to divide the number line as shown.

A B C

-2 -1

Try test numbers in each interval.

61. (continued)

A: Test -3, $f(-3) = (-3)^2 + 3(-3) + 2 = 2$

B: Test $-\frac{3}{2}$, $f\left(-\frac{3}{2}\right) = \left(-\frac{3}{2}\right)^2 + 3\left(-\frac{3}{2}\right) + 2 = -\frac{1}{4}$

C: Test 0, $f(0) = 0^2 + 3\cdot0 + 2 = 2$

The solution set for this portion of the conjunction is $\{x|x < -2 \text{ or } x > -1\}$, or $(-\infty, -2) \cup (-1, \infty)$.

Next we solve $x^2 + 3x - 4 < 0$.

Solve the related equation.

$x^2 + 3x - 4 = 0$

$(x + 4)(x - 1) = 0$

$x = -4$ or $x = 1$

Use the numbers -4 and 1 to divide the number line as shown.

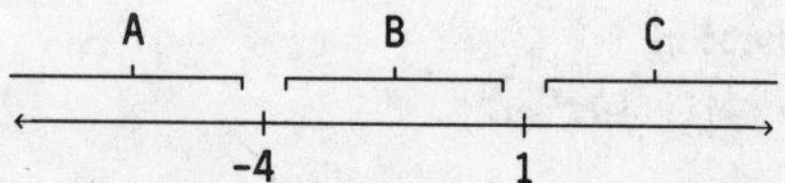

Try test numbers in each interval.

A: Test -5, $f(-5) = (-5)^2 + 3(-5) - 4 = 6$

B: Test 0, $f(0) = 0^2 + 3\cdot0 - 4 = -4$

C: Test 2, $f(2) = 2^2 + 3\cdot2 - 4 = 6$

The solution set for this portion of the conjunction is $\{x|-4 < x < 1\}$, or $(-4, 1)$.

The solution set for the original inequality is the intersection of the solution sets for the two portions of the disjunction.

$[(-\infty, -2) \cup (-1, \infty)] \cap (-4, 1) =$
$(-4, -2) \cup (-1, 1)$

62. $(-\infty, -1] \cup [1, 4] \cup [6, \infty)$

63. We first make a drawing.

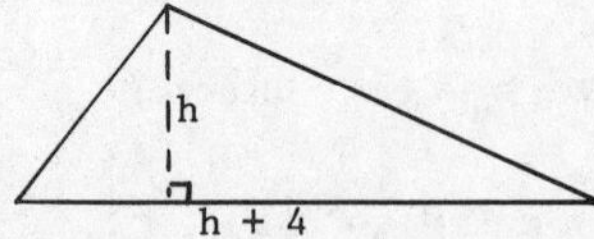

We let h represent the height and h + 4 the base. The area is $\frac{1}{2}(h + 4)h$. We now have an inequality:

$\frac{1}{2}h(h + 4) > 10$

$h(h + 4) > 20$

$h^2 + 4h > 20$

$h^2 + 4h - 20 > 0$

Consider the function $f(h) = h^2 + 4h - 20$. The inputs that produce outputs that are greater than 0 are the solutions of the inequality.

63. (continued)

Set $f(h) = 0$ and use the quadratic formula to find the x-intercepts.

$h^2 + 4h - 20 = 0$

$$h = \frac{-4 \pm \sqrt{4^2 - 4(1)(-20)}}{2\cdot 1} = \frac{-4 \pm \sqrt{16 + 80}}{2}$$

$$= \frac{-4 \pm \sqrt{96}}{2} = \frac{-4 \pm 4\sqrt{6}}{2} = -2 \pm 2\sqrt{6}$$

The x-intercepts are $(-2 - 2\sqrt{6}, 0)$ and $(-2 + 2\sqrt{6}, 0)$.

The graph of the function opens upwards. Function values will be greater than 0 when h is less than $-2 - 2\sqrt{6}$ and when h is greater than $-2 + 2\sqrt{6}$. Since the height can only be positive, we only consider $h > -2 + 2\sqrt{6}$.

Thus, $\{h \mid h > -2 + 2\sqrt{6} \text{ cm}\}$.

64. $\left\{w \middle| w > \frac{-3 + \sqrt{69}}{2} \text{ m}\right\}$

65. a) $-3x^2 + 630x - 6000 > 0$

$x^2 - 210x + 2000 < 0$ $\left[\text{Multiplying by } -\frac{1}{3}\right]$

$(x - 200)(x - 10) < 0$

The solutions of $(x - 200)(x - 10) = 0$ are 200 and 10. We use the numbers to divide the number line.

Sign of x - 200: - - - - | - - - - | + + + +

Sign of x - 10: - - - - | + + + + | + + + +

Sign of product: + + + + | - - - - | + + + +

10 200

The solution set is $\{x \mid 10 < x < 200\}$, or $(10, 200)$.

b) Use the diagram in part a). Note that x must be nonnegative in this problem. The solution set is $\{x \mid 0 \leqslant x < 10 \text{ or } x > 200\}$, or $[0, 10) \cup (200, \infty)$.

66. a) $\{t \mid 0 \text{ sec} < t < 2 \text{ sec}\}$, or $(0 \text{ sec}, 2 \text{ sec})$

b) $\{t \mid t > 10 \text{ sec}\}$, or $(10 \text{ sec}, \infty)$

67. $78 \leqslant \frac{n(n - 1)}{2} \leqslant 1225$

$156 \leqslant n^2 - n \leqslant 2450$

$156 \leqslant n^2 - n$ and $n^2 - n \leqslant 2450$

First we solve $156 \leqslant n^2 - n$.

$0 \leqslant n^2 - n - 156$

$0 \leqslant (n - 13)(n + 12)$

The solutions of $(n - 13)(n + 12) = 0$ are 13 and -12. They divide the number line as shown.

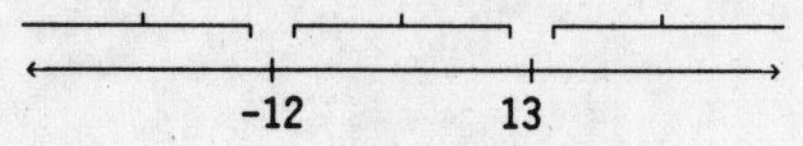

67. (continued)

However, only nonnegative values of n make sense in this exercise, so we need only consider the intervals shown below.

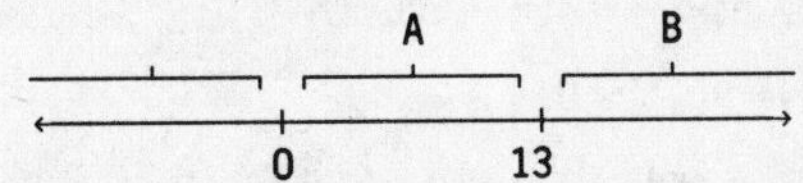

A: Test 1, $f(1) = 1^2 - 1 - 156 = -156$

B: Test 14, $f(14) = 14^2 - 14 - 156 = 26$

The solution set for this portion of the conjunction is $\{n \mid n \geqslant 13\}$, or $[13, \infty)$.

Next we consider $n^2 - n \leqslant 2450$.

$n^2 - n - 2450 \leqslant 0$

$(n - 50)(n + 49) \leqslant 0$

The solutions of $(n - 50)(n + 49) = 0$ are 50 and -49. They divide the number line as shown.

Again however, we need only consider nonnegative values of n.

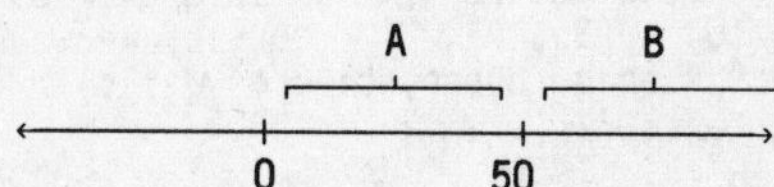

A: Test 1, $f(1) = 1^2 - 1 - 2450 = -2450$

B: Test 51, $f(51) = 51^2 - 51 - 2450 = 100$

The solution set for this portion of the conjunction is $\{n \mid 0 \leqslant n \leqslant 50\}$, or $[0, 50]$.

The solution set of the original inequality is $[13, \infty) \cap [0, 50] = [13, 50]$.

68. $\{n \mid 10 \leqslant n \leqslant 20\}$

69. a) Solve $R(x) = C(x)$.

$50x - x^2 = 5x + 350$

$0 = x^2 - 45x + 350$

$0 = (x - 35)(x - 10)$

$x = 35$ or $x = 10$

The break even values are 10 units and 35 units.

69. (continued)

b) $P(x) = R(x) - C(x)$

$P(x) = (50x - x^2) - (5x + 350)$

$P(x) = -x^2 + 45x - 350$

Solve $P(x) > 0$.

$-x^2 + 45x - 350 > 0$

$x^2 - 45x + 350 < 0$ (Multiplying by -1)

$(x - 35)(x - 10) < 0$

The solutions of $(x - 35)(x - 10) = 0$ are 35 and 10.

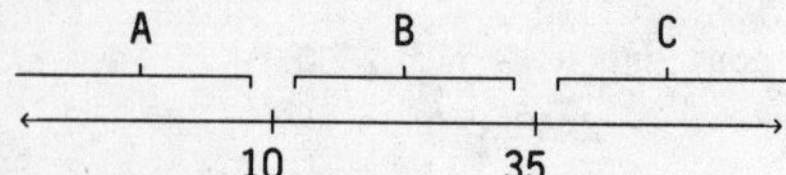

A: Test 1, $f(1) = 1^2 - 45\cdot 1 + 350 = 306$

B: Test 11, $f(11) = 11^2 - 45\cdot 11 + 350 = -24$

C: Test 36, $f(36) = 36^2 - 45\cdot 36 + 350 = 26$

There is a profit for $\{x|10 < x < 35\}$, or $(10, 35)$.

c) Solve $P(x) < 0$. From part b) we see that there is a loss for $\{x|x < 10 \text{ or } x > 35\}$, or $(-\infty, 10) \cup (35, \infty)$. (Of course, we would expect x to be nonnegative as well.)

70. a) 10, 60

b) $\{x|10 < x < 60\}$

c) $\{x|x < 10 \text{ or } x > 60\}$

71. The discriminant of $x^2 + kx + 1 = 0$ is $k^2 - 4$.

a) There are two real-number solutions when the discriminant is positive, so we solve $k^2 - 4 > 0$, or $(k + 2)(k - 2) > 0$.

The solutions of $(k + 2)(k - 2) = 0$ are -2 and 2. They divide the number line as shown.

Sign of $k + 2$: - - - - | + + + + | + + + +

Sign of $k - 2$: - - - - | - - - - | + + + +

Sign of product: + + + + | - - - - | + + + +

-2 2

The solution set is $\{k|k < -2 \text{ or } k > 2\}$, or $(-\infty, -2) \cup (2, \infty)$.

b) There is no real-number solution when the discriminant is negative, so we solve $k^2 - 4 < 0$. Using the diagram in part a), we see that the solution set is $\{k|-2 < k < 2\}$, or $(-2, 2)$.

72. a) $\left\{k \middle| k < -2\sqrt{2} \text{ or } k > 2\sqrt{2}\right\}$

b) $\left\{k \middle| -2\sqrt{2} < k < 2\sqrt{2}\right\}$

73. $f(x) = \sqrt{1 - x^2}$

The radicand must be nonnegative, so we solve $1 - x^2 \geqslant 0$, or $x^2 - 1 \leqslant 0$ (multiplying by -1). Using the graph in Exercise 7, we see that the domain is $\{x|-1 \leqslant x \leqslant 1\}$, or $[-1, 1]$.

74. $\{x|-1 < x < 1\}$, or $(-1, 1)$

75. $g(x) = \sqrt{x^2 + 2x - 3}$

The radicand must be nonnegative, so we solve $x^2 + 2x - 3 \geqslant 0$, or $(x + 3)(x - 1) \geqslant 0$. The solutions of $(x + 3)(x - 1) = 0$ are -3 and 1. They divide the number line as shown.

A B C

-3 1

A: Test -4, $f(-4) = (-4)^2 + 2(-4) - 3 = 5$

B: Test 0, $f(0) = 0^2 + 2\cdot 0 - 3 = -3$

C: Test 2, $f(2) = 2^2 + 2\cdot 2 - 3 = 5$

The domain is $\{x|x \leqslant -3 \text{ or } x \geqslant 1\}$, or $(-\infty, -3] \cup [1, \infty)$.

76. $\{x|-2 \leqslant x \leqslant 2\}$, or $[-2, 2]$

77. Use a compute or a graphic calculator to graph the function.

$f(x) = x^3 - 2x^2 - 5x + 6$

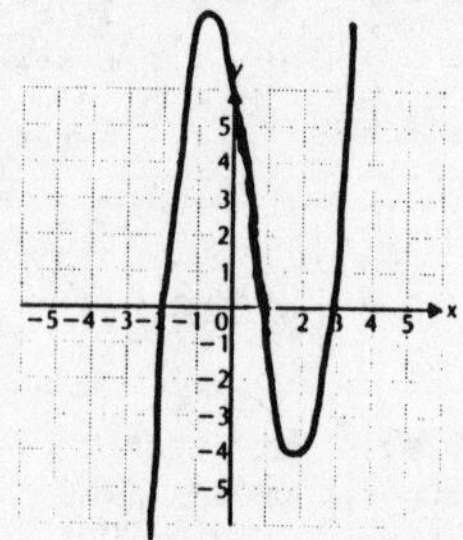

From the graph we determine the following:

The solutions of $f(x) = 0$ are -2, 1, and 3.

The solution of $f(x) < 0$ is $\{x|x < -2 \text{ or } 1 < x < 3\}$, or $(-\infty,-2) \cup (1,3)$.

The solution of $f(x) > 0$ is $\{x|-2 < x < 1 \text{ or } x > 3\}$, or $(-2,1) \cup (3,\infty)$.

78. $f(x) = 0$ for $x = -2$ and $x = 1$;

$f(x) < 0$ for $\{x|x < -2\}$, or $(-\infty,-2)$;

$f(x) > 0$ for $\{x|-2 < x < 1 \text{ or } x > 1\}$, or $(-2,1) \cup (1,\infty)$

79. Use a computer or a graphic calculator to graph the function.

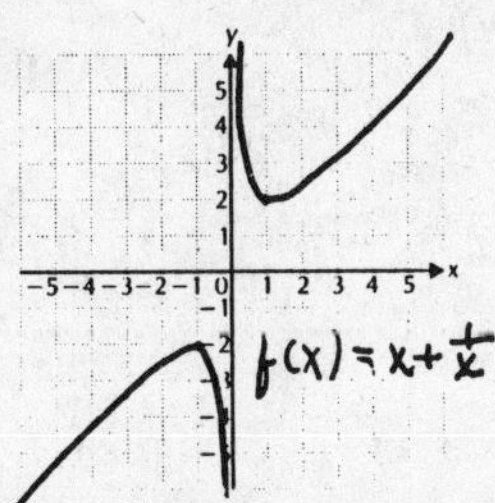

From the graph we determine the following:

f(x) has no zeros.

The solutions f(x) < 0 are {x|x < 0}, or (-∞,0).

The solutions of f(x) > 0 are {x|x > 0}, or (0,∞).

80. f(x) = 0 for x = 0 and x = 1;

f(x) < 0 for {x|0 < x < 1}, or (0,1);

f(x) > 0 for {x|x > 1}, or (1,∞)

81. Use a computer or a graphic calculator to graph the function.

$f(x) = x^4 - 4x^3 - x^2 + 16x - 12$

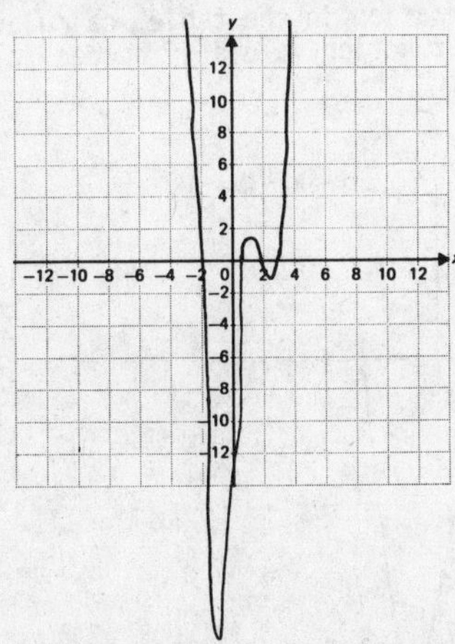

From the graph we determine the following:

The solutions of f(x) = 0 are -2, 1, 2, and 3.

The solutions of f(x) < 0 are {x|-2 < x < 1 or 2 < x < 3}, or (-2,1) ∪ (2,3).

The solutions of f(x) > 0 are {x|x < -2 or 1 < x < 2 or x > 3}, or (-∞,-2) ∪ (1,2) ∪ (3,∞).

82. f(x) = 0 for x = -2, x = 0, and x = 1;

f(x) < 0 for {x|x < -3 or -2 < x < 0 or 1 < x < 2}, or (-∞,-3) ∪ (-2,0) ∪ (1,2);

f(x) > 0 for {x|-3 < x < -2 or 0 < x < 1 or x > 2}, or (-3,-2) ∪ (0,1) ∪ (2,∞)

Exercise Set 4.8

1. $f(x) = \frac{1}{3}x^6$

The graph of $f(x) = \frac{1}{3}x^6$ has the same general shape as $f(x) = x^2$. This is an even function, so it is symmetric with respect to the y-axis. We can compute some function values and use symmetry to find others.

x	0	1	2	3	-1	-2	-3
f(x)	0	$\frac{1}{3}$	$\frac{64}{3}$	243	$\frac{1}{3}$	$\frac{64}{3}$	243

Found using symmetry

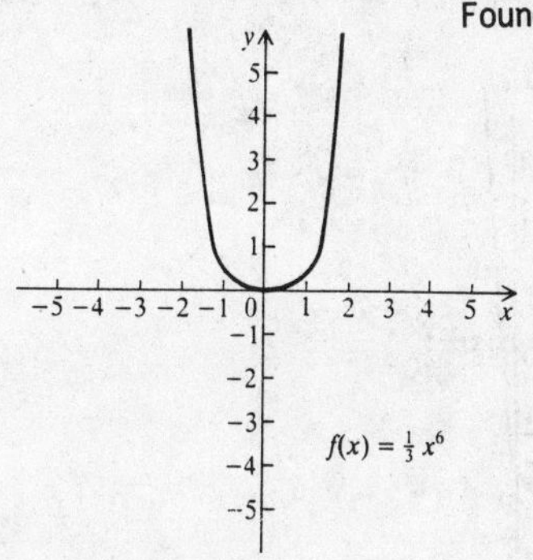

2.

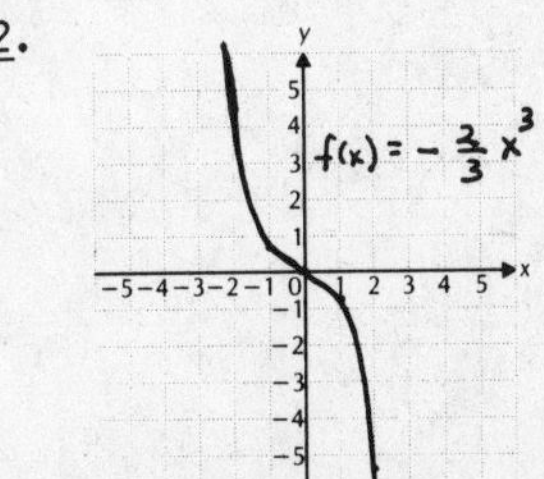

3. $f(x) = -0.6x^5$

The graph of $f(x) = -0.6x^5$ has the same general shape as $f(x) = x^3$. It is an odd function, so it is symmetric with respect to the origin. We can compute some function values and use symmetry to find others.

x	0	1	2	-1	-2
f(x)	0	-0.6	-19.2	0.6	19.2

Found using symmetry

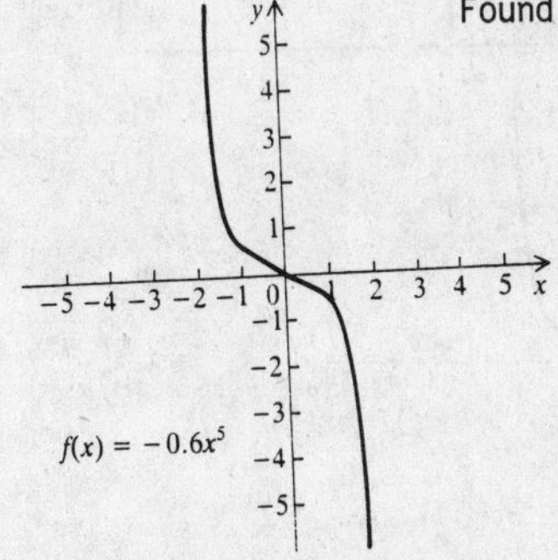

4.

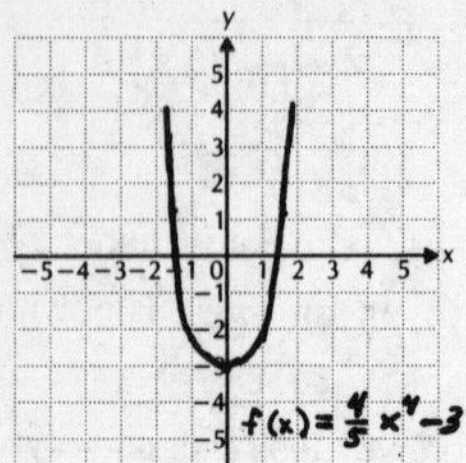

5. $f(x) = (x + 1)^5 - 4$

Using the results of Section 3.8, we see that the graph of $f(x) = (x + 1)^5 - 4$ is obtained from the graph of $f(x) = x^5$ (see page 265) by translating it 1 unit the left and 4 units down.

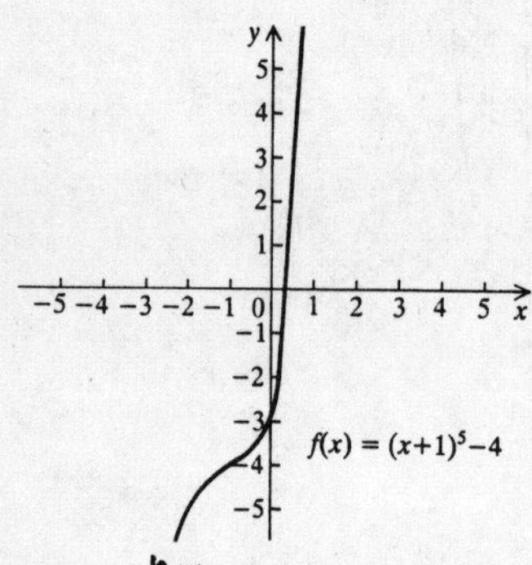

6.

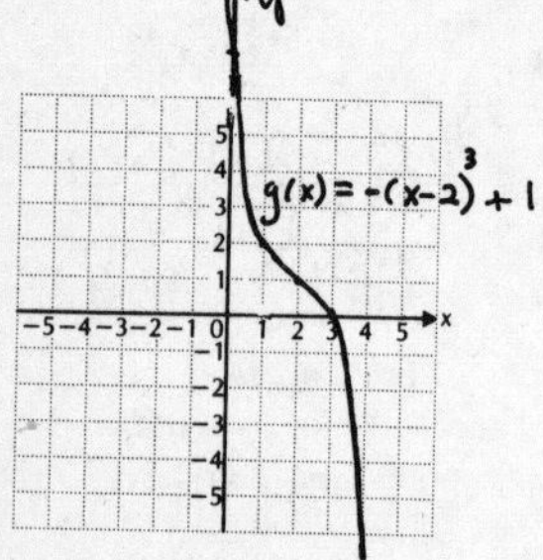

7. $f(x) = \frac{1}{4}(x + 1)^4$

This function is a transformation of $f(x) = x^4$, so its graph will have the same general shape as $f(x) = x^2$. We compute some function values.

x	-4	-3	-2	-1	0	1	2
f(x)	$\frac{81}{4}$	4	$\frac{1}{4}$	0	$\frac{1}{4}$	4	$\frac{81}{4}$

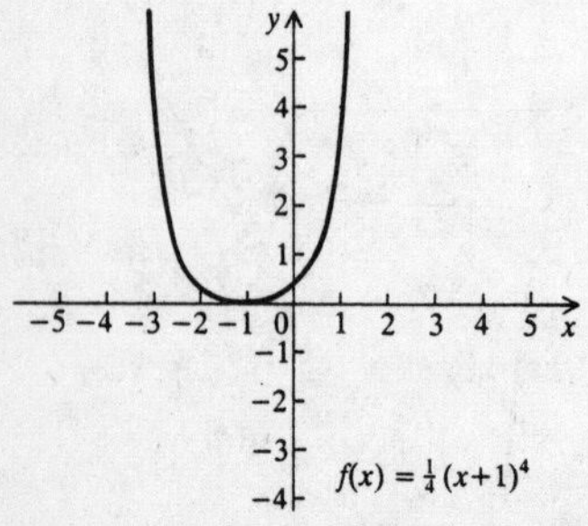

8.

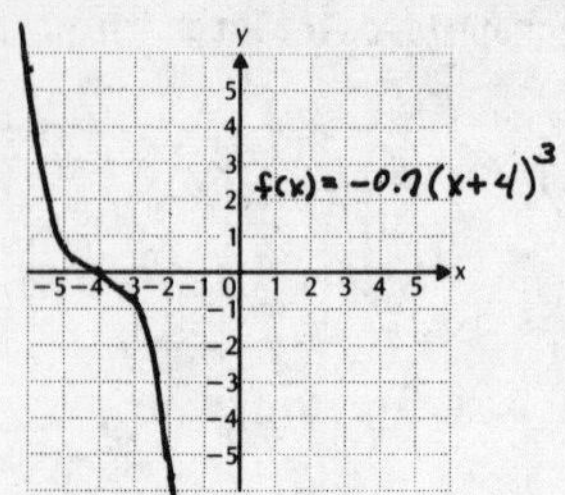

9. $f(x) = (x + 3)(x - 2)(x + 1)$

The zeros of this function are -3, 2, and -1. They divide the real-number line into 4 open intervals as indicated in the table below. We try a test value in each interval and determine the sign of the function for values of x in the interval.

Interval	$(-\infty, -3)$	$(-3, -1)$	$(-1, 2)$	$(2, \infty)$
Test value	$f(-4) = -18$	$f(-2) = 4$	$f(0) = -6$	$f(3) = 12$
Sign of f(x)	Negative	Positive	Negative	Positive
Location of points on graph	Below x-axis	Above x-axis	Below x-axis	Above x-axis

We use the information in the table, calculate some additional function values if needed, plot points, and sketch the graph.

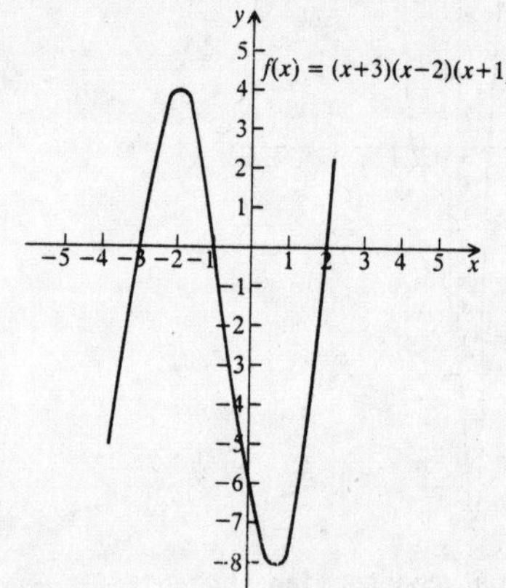

10.

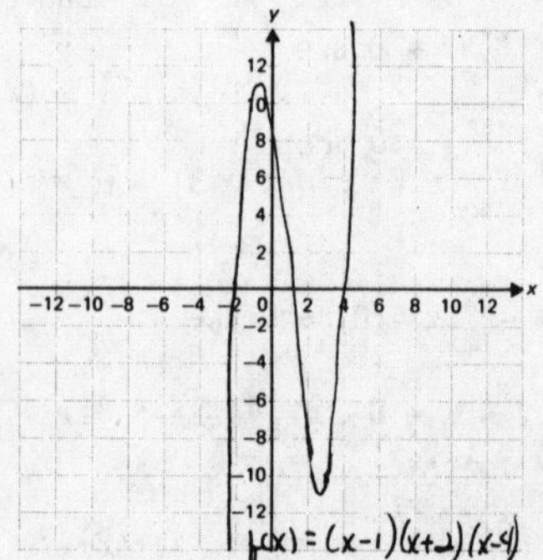

11. $f(x) = 9x^2 - x^4$
$= x^2(9 - x^2)$
$= x^2(3 + x)(3 - x)$

The zeros of the function are 0, -3, and 3.

Interval	$(-\infty, -3)$	$(-3, 0)$	$(0, 3)$	$(3, \infty)$
Test value	$f(-4) = -112$	$f(-1) = 8$	$f(1) = 8$	$f(4) = -112$
Sign of f(x)	Negative	Positive	Positive	Negative
Location of points on graph	Below x-axis	Above x-axis	Above x-axis	Below x-axis

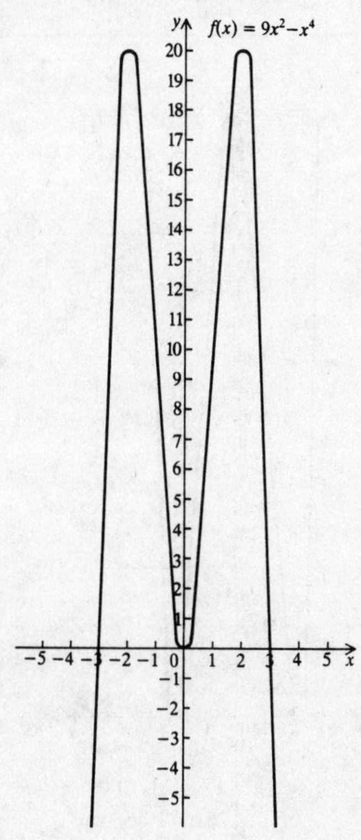

12.

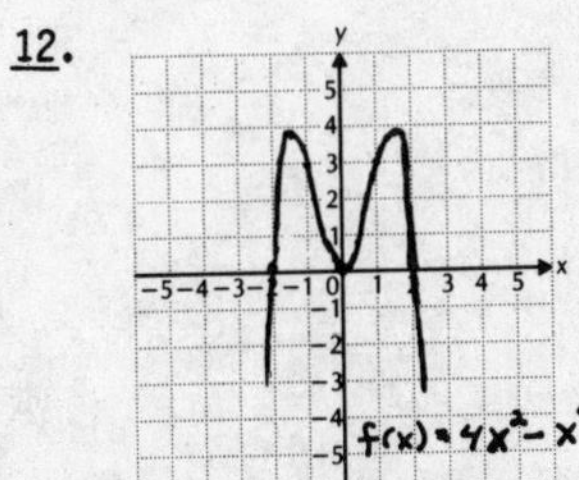

13. $f(x) = x^4 - x^3$
$= x^3(x - 1)$

The zeros of the function are 0 and 1.

Interval	$(-\infty, 0)$	$(0, 1)$	$(1, \infty)$
Test value	$f(-1) = 2$	$f\left(\frac{1}{2}\right) = -\frac{1}{16}$	$f(2) = 8$
Sign of f(x)	Positive	Negative	Positive
Location of points on graph	Above x-axis	Below x-axis	Above x-axis

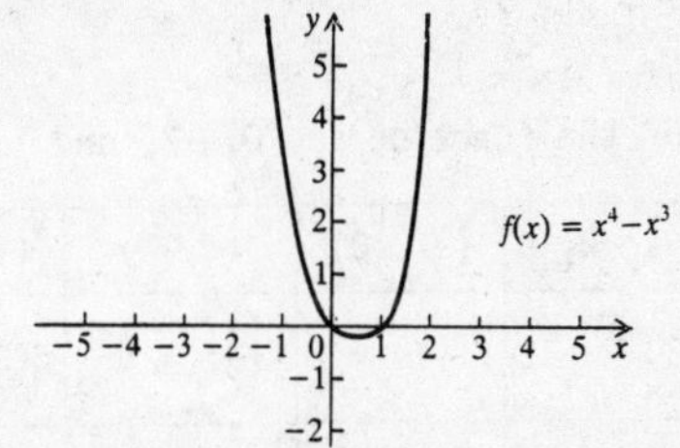

14.

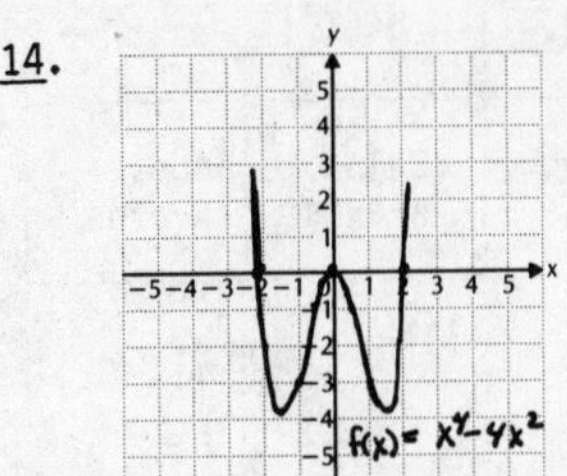

15. $f(x) = x^3 - 4x$
$= x(x^2 - 4)$
$= x(x + 2)(x - 2)$

The zeros of the function are 0, -2, and 2.

Interval	$(-\infty, -2)$	$(-2, 0)$	$(0, 2)$	$(2, \infty)$
Test value	$f(-3) = -15$	$f(-1) = 3$	$f(1) = -3$	$f(3) = 15$
Sign of f(x)	Negative	Positive	Negative	Positive
Location of points on graph	Below x-axis	Above x-axis	Below x-axis	Above x-axis

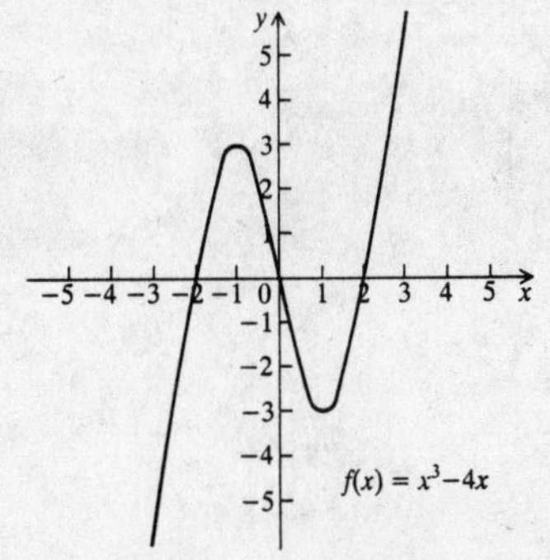

16.

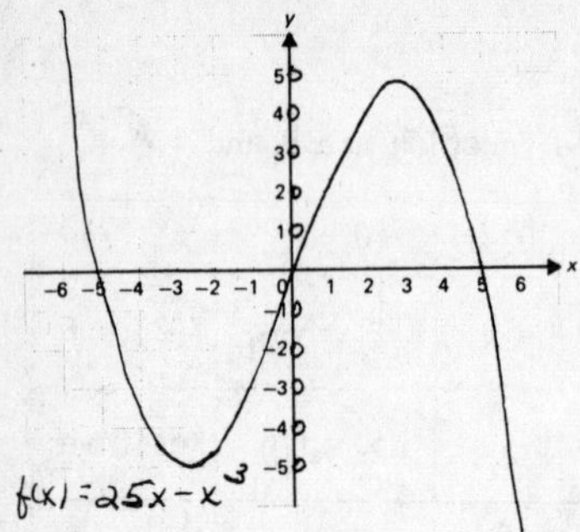

17. $f(x) = x^3 + x^2 - 2x$
$= x(x^2 + x - 2)$
$= x(x + 2)(x - 1)$

The zeros of the function are 0, -2, and 1.

Interval	$(-\infty, -2)$	$(-2, 0)$	$(0, 1)$	$(1, \infty)$
Test value	$f(-3) = -12$	$f(-1) = 2$	$f\left(\frac{1}{2}\right) = -\frac{5}{8}$	$f(2) = 8$
Sign of f(x)	Negative	Positive	Negative	Positive
Location of points on graph	Below x-axis	Above x-axis	Below x-axis	Above x-axis

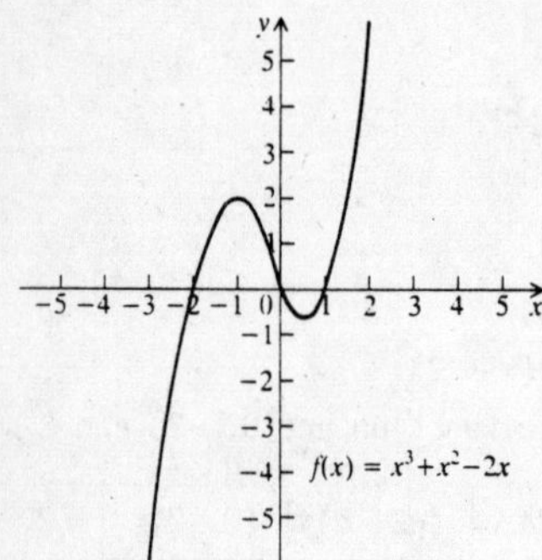

18.

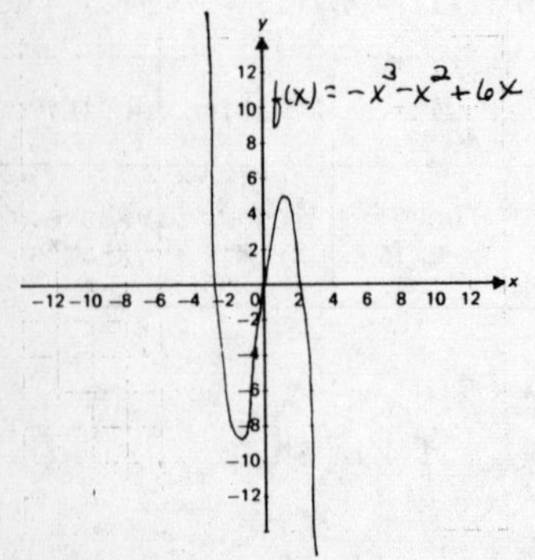

19. $f(x) = x^4 - 9x^2 + 20$
$= (x^2 - 4)(x^2 - 5)$
$= (x + 2)(x - 2)(x + \sqrt{5})(x - \sqrt{5})$

The zeros of the function are -2, 2, $-\sqrt{5}$, and $\sqrt{5}$.

Interval	$(-\infty,-\sqrt{5})$	$(-\sqrt{5},-2)$	$(-2, 2)$	$(2,\sqrt{5})$	$(\sqrt{5},\infty)$
Test value	$f(-3) = 20$	$f(-2.1) = -0.2419$	$f(0) = 20$	$f(2.1) = -0.2419$	$f(3) = 20$
Sign of f(x)	+	-	+	-	+
Location of points on graph	Above x-axis	Below x-axis	Above x-axis	Below x-axis	Above x-axis

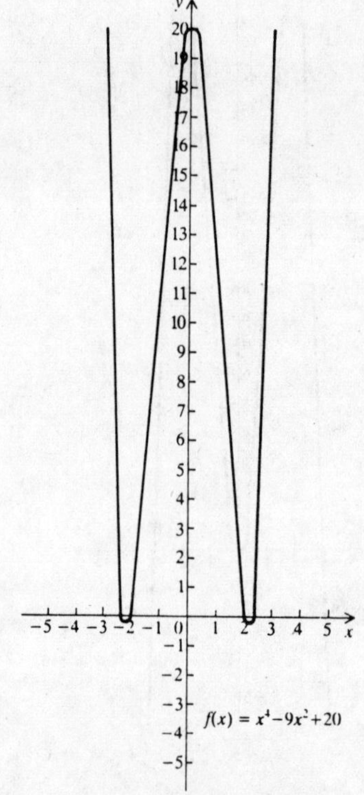

20.

f(x) = x⁴ − 3x² + 2

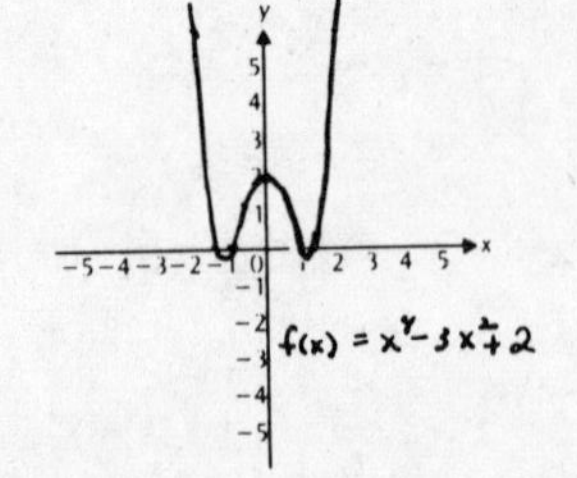

21. $f(x) = x^3 - 3x^2 - 4x + 12$

$= x^2(x - 3) - 4(x - 3)$

$= (x^2 - 4)(x - 3)$

$= (x + 2)(x - 2)(x - 3)$

The zeros of the function are -2, 2, and 3.

Interval	$(-\infty, -2)$	$(-2, 2)$	$(2, 3)$	$(3, \infty)$
Test value	$f(-3) = -30$	$f(0) = 12$	$f\left(\frac{5}{2}\right) = -\frac{9}{8}$	$f(4) = 12$
Sign of $f(x)$	Negative	Positive	Negative	Positive
Location of points on graph	Below x-axis	Above x-axis	Below x-axis	Above x-axis

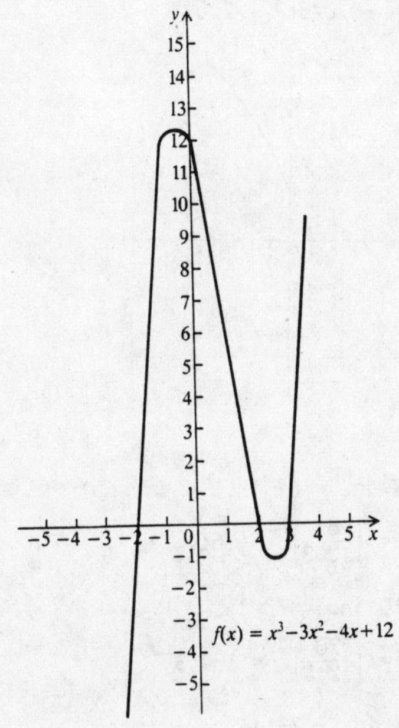

22.

$f(x) = x^3-2x^2-9x+18$

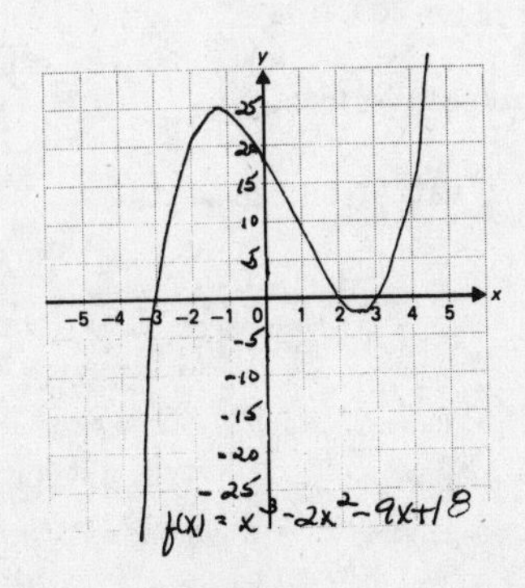

23. $f(x) = -x^4 - 3x^3 - 3x^2$

$= -x^2(x^2 + 3x + 3)$

The discriminant of $x^2 + 3x + 3 = 0$ is negative, so the only zero of the function is 0.

Interval	$(-\infty, 0)$	$(0, \infty)$
Test value	$f(-1) = -1$	$f(1) = -7$
Sign of $f(x)$	Negative	Negative
Location of points on graph	Below x-axis	Below x-axis

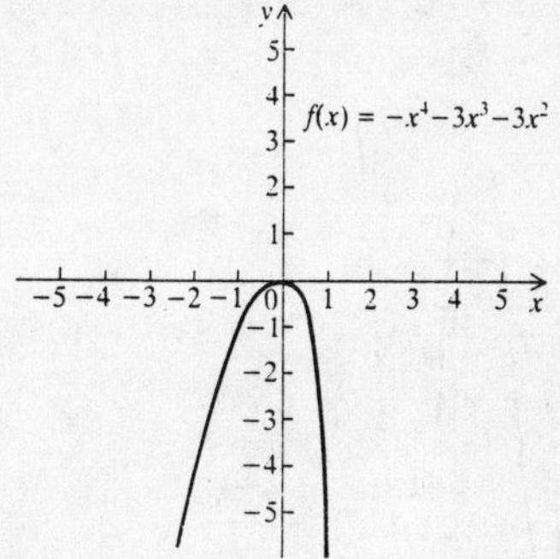

24.

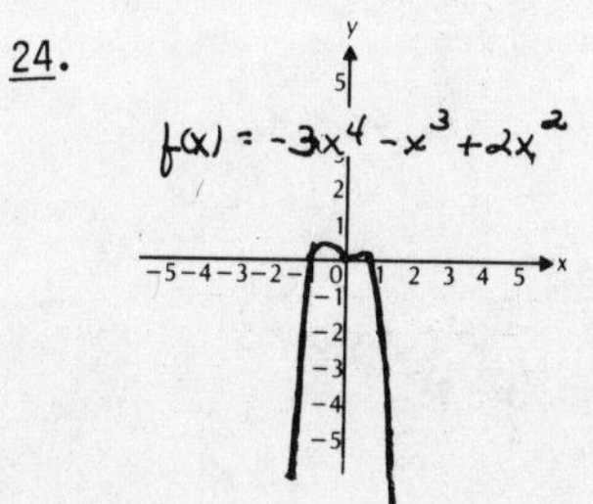

25. $f(x) = x(x - 2)(x + 1)(x + 3)$

The zeros of the function are 0, 2, -1, and -3.

Interval	$(-\infty,-3)$	$(-3,-1)$	$(-1,0)$	$(0,2)$	$(2,\infty)$
Test value	$f(-4) = 72$	$f(-2) = -8$	$f\left(-\frac{1}{2}\right) = \frac{25}{16}$	$f(1) = -8$	$f(3) = 72$
Sign of f(x)	+	-	+	-	+
Location of points on graph	Above x-axis	Below x-axis	Above x-axis	Below x-axis	Above x-axis

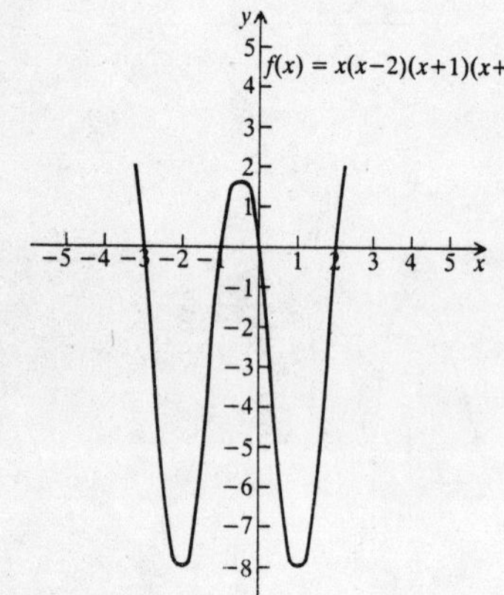

26.

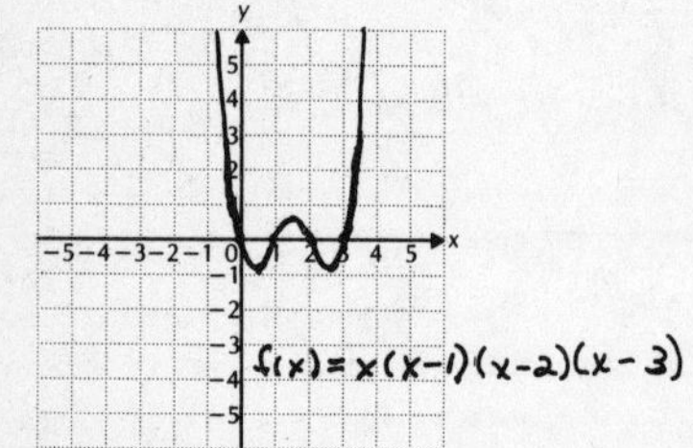

27. a) 1 unit: $y = \frac{1}{13}(1)^3 - \frac{1}{14}(1) \approx 0.0055$

2 units: $y = \frac{1}{13}(2)^3 - \frac{1}{14}(2) \approx 0.4725$

3 units: $y = \frac{1}{13}(3)^3 - \frac{1}{14}(3) \approx 1.8626$

6 units: $y = \frac{1}{13}(6)^3 - \frac{1}{14}(6) \approx 16.1868$

8 units: $y = \frac{1}{13}(8)^3 - \frac{1}{14}(8) \approx 38.8132$

9 units: $y = \frac{1}{13}(9)^3 - \frac{1}{14}(9) \approx 55.4341$

10 units: $y = \frac{1}{13}(10)^3 - \frac{1}{14}(10) \approx 76.2088$

27. (continued)

b)

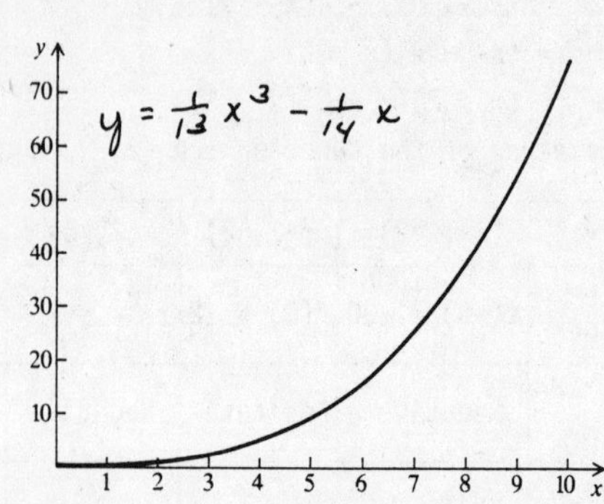

28. a) 4.03, 5.79, 7.00, 7.38, 6.65, 4.54, 0.78

b) $N(t) = -0.046t^3 + 2.08t + 2$

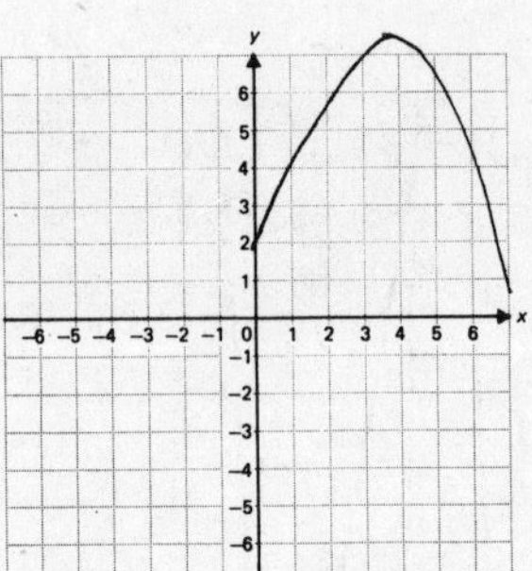

29. a) 5 ft 7 in. = 67 in.

$W(67) = \left(\frac{67}{12.3}\right)^3 \approx 161.6$ lb

5 ft 10 in. = 70 in.

$W(70) = \left(\frac{70}{12.3}\right)^3 \approx 184.3$ lb

b) 6 ft 1 in. = 73 in.

$W(73) = \left(\frac{73}{12.3}\right)^3 \approx 209.1$ lb

He should not gain weight.

30. a) 7.68 watts, 15 watts, 50.625 watts

b) 20 mph

31. a) We make a drawing.

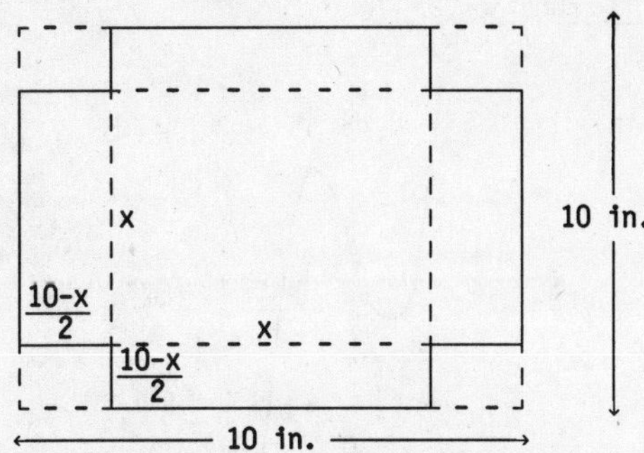

The length of the box is x, the width is x, the height is $\frac{10 - x}{2}$. Thus, the volume is given by

$V(x) = x \cdot x \cdot \left[\frac{10 - x}{2}\right] = x^2\left[\frac{10 - x}{2}\right]$, or

$V(x) = 5x^2 - \frac{x^3}{2}$.

b) The zeros of V(x) are 0 and 10.

Interval	$(-\infty, 0)$	$(0, 10)$	$(10, \infty)$
Test value	$V(-1) = \frac{11}{2}$	$V(1) = \frac{9}{2}$	$V(11) = -\frac{121}{5}$
Sign of f(x)	Positive	Positive	Negative
Location of points on graph	Above x-axis	Above x-axis	Below x-axis

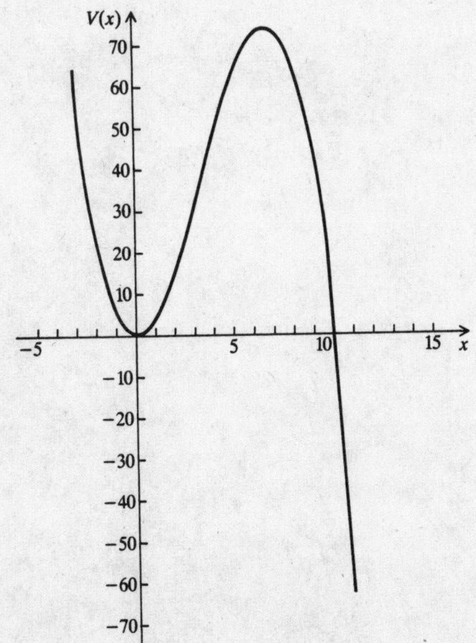

c) Using the table in part b), we see that the function is positive over $(-\infty, 0)$, and $(0, 10)$.

32. a) $V(x) = x^2\left[\frac{8 - x}{2}\right]$, or $V(x) = 4x^2 - \frac{x^3}{2}$

b)

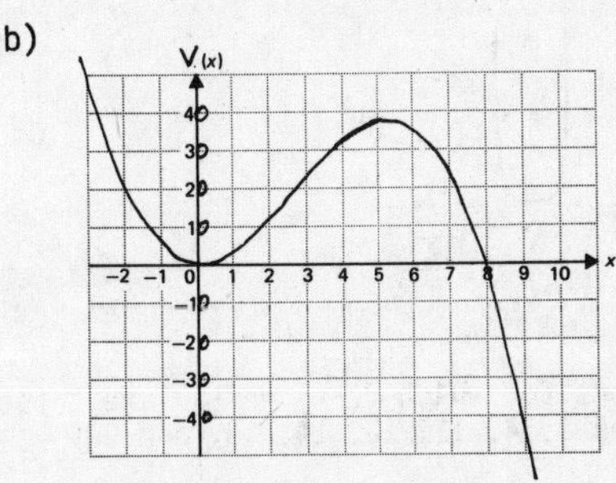

c) $(-\infty, 0) \cup (0, 8)$

33. a)

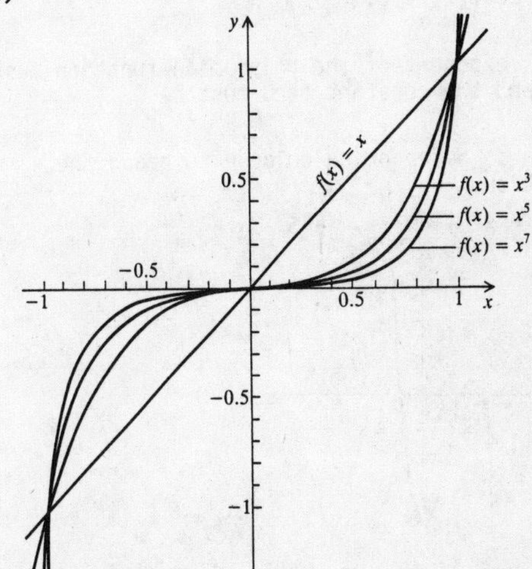

b) The graphs all contain the points (-1, -1), (0, 0), and (1, 1). In all cases, $f(x) > 0$ for $x > 0$ and $f(x) < 0$ for $x < 0$. As the exponent increases, the graph gets closer to the x-axis in [-1, 1] and gets steeper elsewhere. The graph of $f(x) = x^{13}$ contains (-1, -1), (0, 0), and (1, 1). Points on the graph lie below the x-axis for $x < 0$ and above the x-axis for $x > 0$. The graph will be close to the x-axis on [-1, 1] and will rise or fall steeply elsewhere.

34. a)

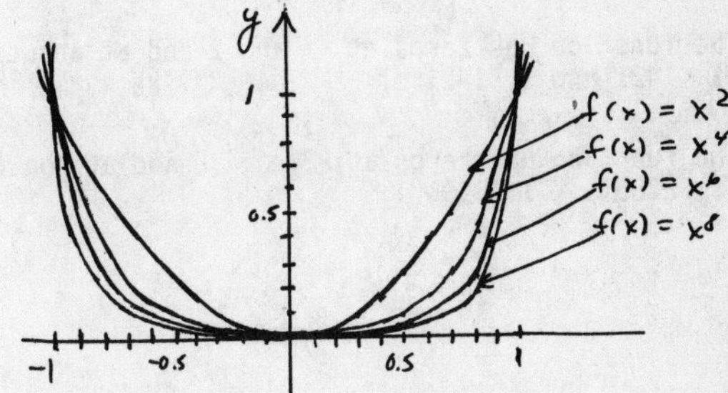

34. (continued)

b)

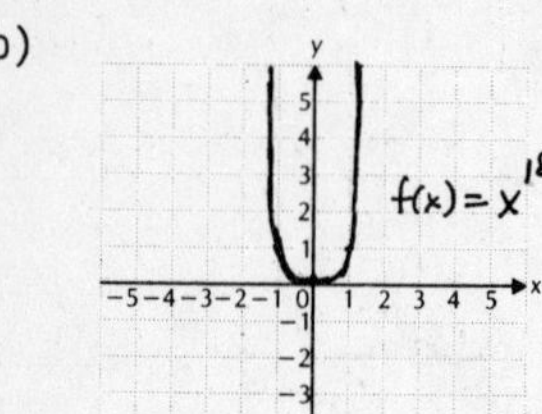

35. The functions for which $f(-x) = f(x)$ are those in Exercises 1, 4, 11, 12, 14, 19, and 20.

36. Exercise 2, 3, 15 and 16

37. Every exponent of the polynomial function must be even or $f(x) = 0$.

38. Every exponent of the polynomial function must be odd and the constant term must be 0.

39. Use a computer or calculator to graph the function.

$f(x) = x^3 + 4x^2 + x - 6$

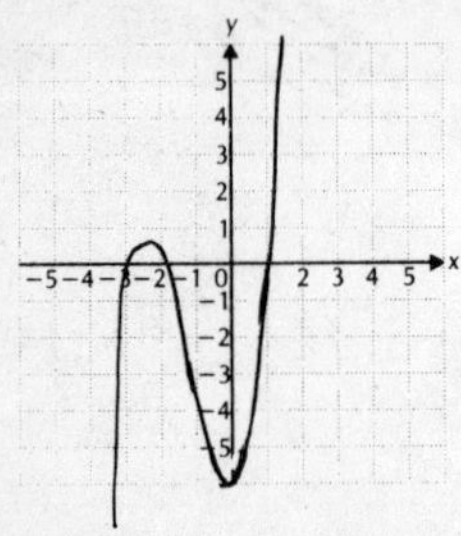

The zeros of the function are -3, -2, and 1.

40. The function has a zero at about -3.33691.

41. Use a computer or calculator to graph the function.

$f(x) = -x^4 + x^3 + 4x^2 - 2x - 4$

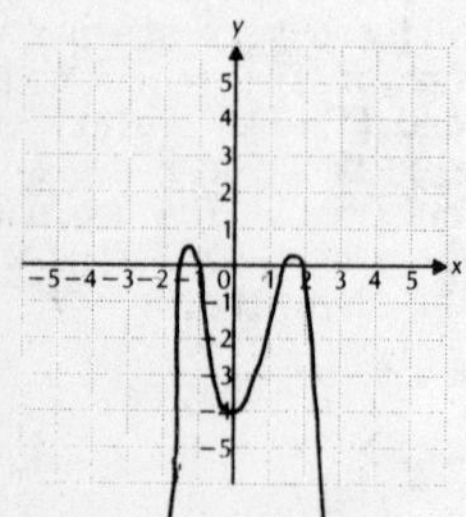

The function has zeros at -1 and 2 and at about -1.41421 and 1.41421.

42. The function has zeros at -2, 0, 3, and at about -1.73205 and 1.73205.

43. a) Use a computer or calculator to graph the function.

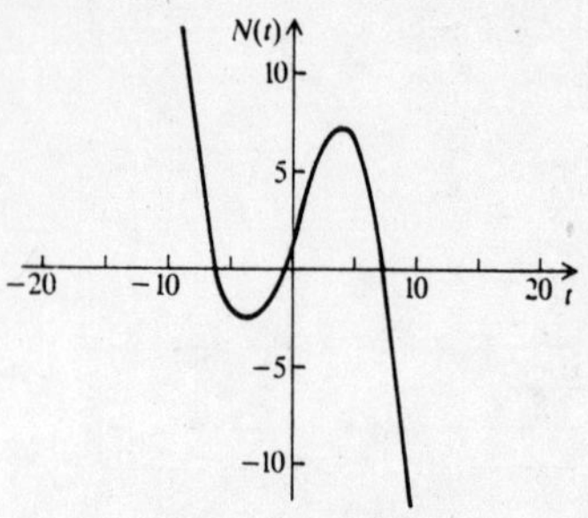

b) The zeros of the function are approximately -6.17908, -0.98251, and 7.16159.

c) The concentration will be 0 after about 7.2 hr.

Exercise Set 5.1

1. We interchange the first and second coordinates of each ordered pair to find the inverse of the relation. It is

   {(1, 0), (6, 5), (-4, -2)}.

2. {(3, -1), (5, 2), (5, -3), (0, 2)}

3. We interchange the first and second coordinates of each ordered pair to find the inverse of the relation. It is

   {(8, 7), (8, -2), (-4, 3), (-8, 8)}.

4. {(-1, -1), (4, -3)}

5. Interchange x and y.

   $y = 4x - 5$

   $\downarrow \quad \downarrow$

   $x = 4y - 5$ (Equation of the inverse relation)

6. $x = 3y + 5$

7. Interchange x and y.

   $x^2 - 3y^2 = 3$

   $\downarrow \quad \downarrow$

   $y^2 - 3x^2 = 3$ (Equation of the inverse relation)

8. $2y^2 + 5x^2 = 4$

9. Interchange x and y.

   $y = 3x^2 + 2$

   $\downarrow \quad \downarrow$

   $x = 3y^2 + 2$ (Equation of the inverse relation)

10. $x = 5y^2 - 4$

11. Interchange x and y.

    $xy = 7$

    $\downarrow\downarrow$

    $yx = 7$ (Equation of the inverse relation)

12. $yx = -5$

13. Graph $y = x^2 + 1$. A few solutions of the equation are (0, 1), (1, 2), (-1, 2), (2, 5), and (-2, 5). Plot these points and draw the curve. Then reflect the graph of $y = x^2 + 1$ across the line $y = x$ (----). A few solutions of the inverse relation, $x = y^2 + 1$, are (1, 0), (2, 1), (2, -1), (5, 2), and (5, -2). Both graphs are shown below.

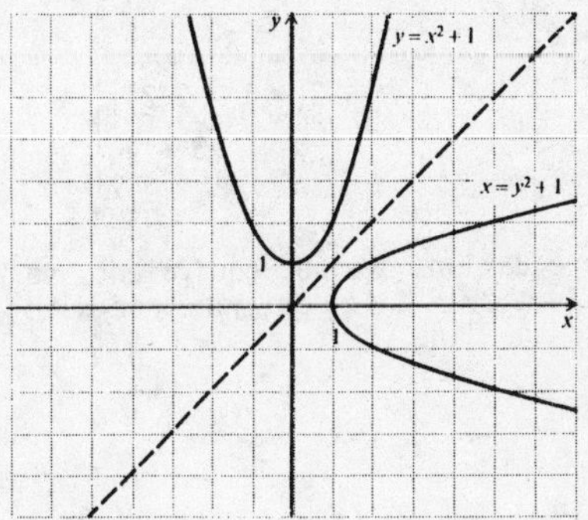

14.

$y = x^2 - 3$

$x = y^2 - 3$

15. Graph $x = |y|$. A few solutions of the equation are (0, 0), (2, 2), (2, -2), (5, 5), and (5, -5). Plot these points and draw the graph. Then reflect the graph across the line $y = x$ (----). A few solutions of the inverse relation, $y = |x|$, are (0, 0), (2, 2), (-2, 2), (5, 5), and (-5, 5). Both graphs are shown below.

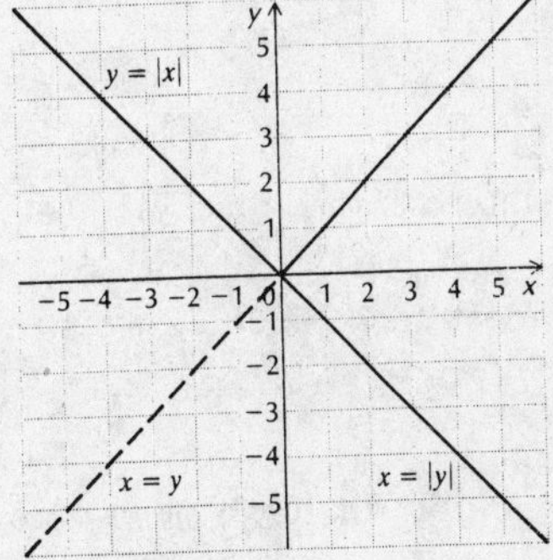

16.

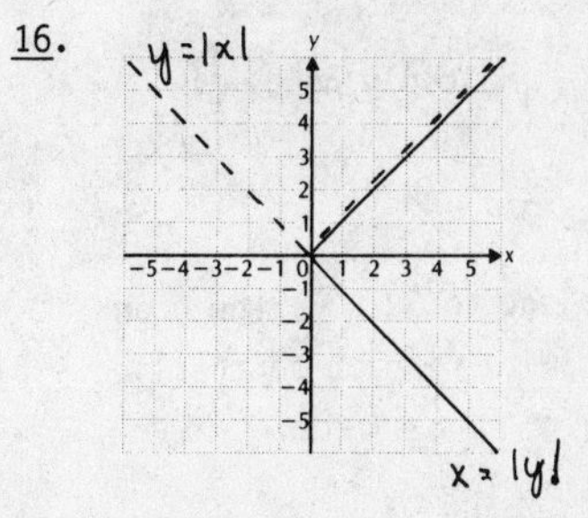

17. $3x + 2y = 4$

$\downarrow \quad \downarrow$

$3y + 2x = 4$ (Interchanging x and y)

Is the resulting equation equivalent to the original? Note that (2, -1) is a solution of $3x + 2y = 4$, but it is not a solution of $3y + 2x = 4$.

$3x + 2y = 4$	
$3(2) + 2(-1)$	4
$6 + -2$	
4	

$3y + 2x = 4$	
$3(-1) + 2(2)$	4
$-3 + 4$	
1	

Thus, the equations are not equivalent, so the graph of $3x + 2y = 4$ is not symmetric with respect to the line $y = x$.

18. No

19. $4x + 4y = 3$

$\downarrow \quad \downarrow$

$4y + 4x = 3$ (Interchanging x and y)

The commutative law of addition guarantees that the resulting equation is equivalent to the original. Thus the graph is symmetric with respect to the line $y = x$.

20. Yes

21. $xy = 10$

$\downarrow\downarrow$

$yx = 10$ (Interchanging x and y)

The commutative law of multiplication guarantees that the resulting equation is equivalent to the original. Thus the graph is symmetric with respect to the line $y = x$.

22. Yes

23. $3x = \frac{4}{y}$

$\downarrow \quad \downarrow$

$3y = \frac{4}{x}$ (Interchanging x and y)

Is $3y = \frac{4}{x}$ equivalent to $3x = \frac{4}{y}$?

$$3y = \frac{4}{x}$$

$$3y \cdot x = \frac{4}{x} \cdot x \qquad \text{(Multiplying by x)}$$

$$3yx = 4$$

$$3yx \cdot \frac{1}{y} = 4 \cdot \frac{1}{y} \qquad \left[\text{Multiplying by } \frac{1}{y}\right]$$

$$3x = \frac{4}{y}$$

Since $3y = \frac{4}{x}$ is equivalent to $3x = \frac{4}{y}$, the graph of $3x = \frac{4}{y}$ is symmetric with respect to the line $y = x$.

24. Yes

25. $y = |2x|$

$\downarrow \quad \downarrow$

$x = |2y|$ (Interchanging x and y)

Is the resulting equation equivalent to the original? Note that (-3, 6) is a solution of $y = |2x|$, but it is not a solution of $x = |2y|$.

$y = \vert 2x \vert$	
6	$\vert 2(-3) \vert$
	$\vert -6 \vert$
	6

$x = \vert 2y \vert$	
-3	$\vert 2 \cdot 6 \vert$
	$\vert 12 \vert$
	12

Thus, the equations are not equivalent, so the graph of $y = |2x|$ is not symmetric with respect to the line $y = x$.

26. No

27. $4x^2 + 4y^2 = 3$

$\downarrow \quad \downarrow$

$4y^2 + 4x^2 = 3$ (Interchanging x and y)

The commutative law of addition guarantees that the resulting equation is equivalent to the original. Thus the graph is symmetric with respect to the line $y = x$.

28. Yes

29. The graph of $f(x) = 5x - 8$ is shown below.

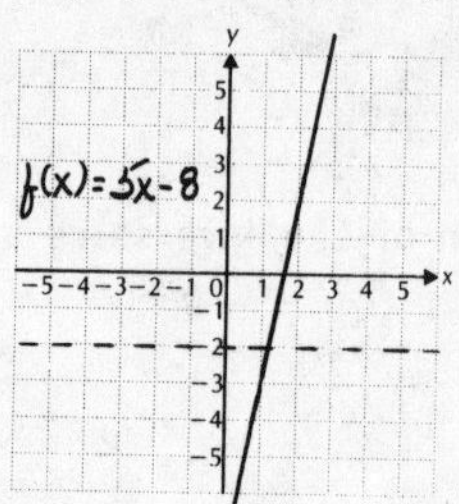

Since there is no horizontal line that crosses the graph more than once, the function is one-to-one.

30. Yes

31. The graph of $f(x) = x^2 - 7$ is shown below.

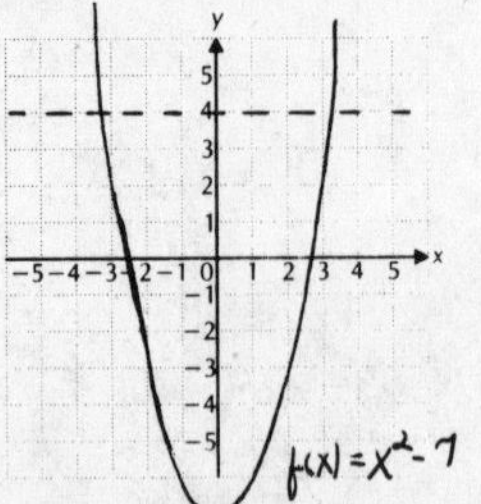

There are many horizontal lines that cross the graph more than once. In particular, the line $y = 4$ crosses the graph more than once. The function is not one-to-one.

32. No

33. The graph of $g(x) = 3$ is shown below.

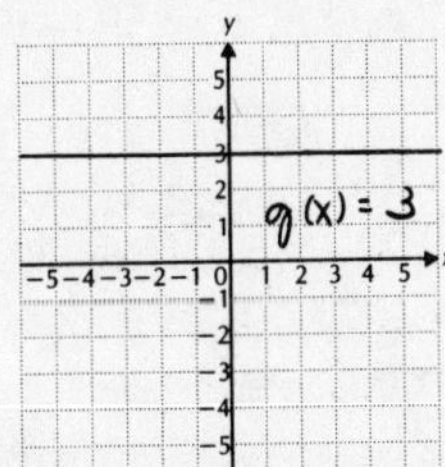

Since the horizontal line $y = 3$ crosses the graph more than once, the function is not one-to-one.

34. No

35. The graph of $g(x) = |x|$ is shown below.

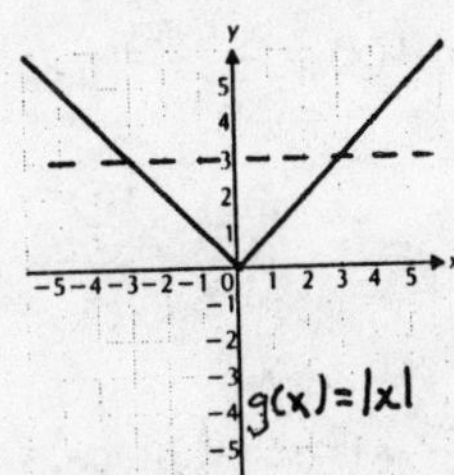

There are many horizontal lines that cross the graph more than once. In particular, the line $y = 3$ crosses the graph more than once. The function is not one-to-one.

36. No

37. The graph of $f(x) = |x + 1|$ is shown below.

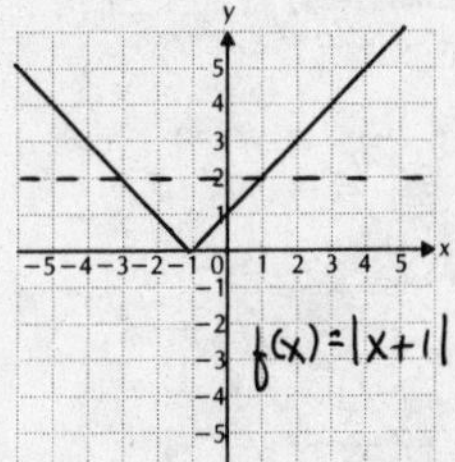

The line $y = 2$ is one of many lines that cross the graph more than once. The function is not one-to-one.

38. No

39. The graph of $g(x) = \frac{-4}{x}$ is shown below.

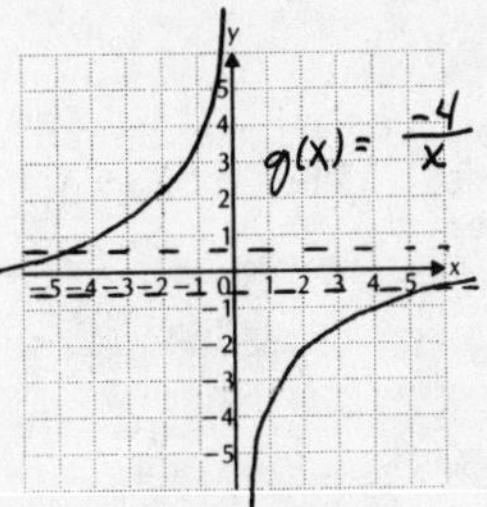

Since there is no horizontal line that crosses the graph more than once, the function is one-to-one.

40. Yes

41. a) The graph of $f(x) = x + 4$ is shown below. It passes the horizontal line test, so it is one-to-one.

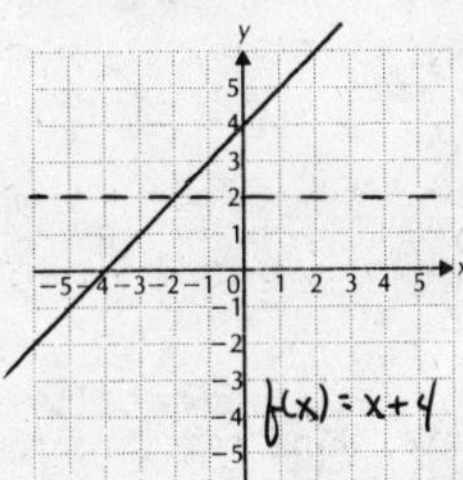

b) Replace $f(x)$ by y: $y = x + 4$

Solve for x: $y - 4 = x$

Interchange x and y: $x - 4 = y$

Replace y by $f^{-1}(x)$: $f^{-1}(x) = x - 4$

42. a) Yes

b) $f^{-1}(x) = x - 5$

43. a) The graph of $f(x) = 5 - x$ is shown below. It passes the horizontal line test, so the function is one-to-one.

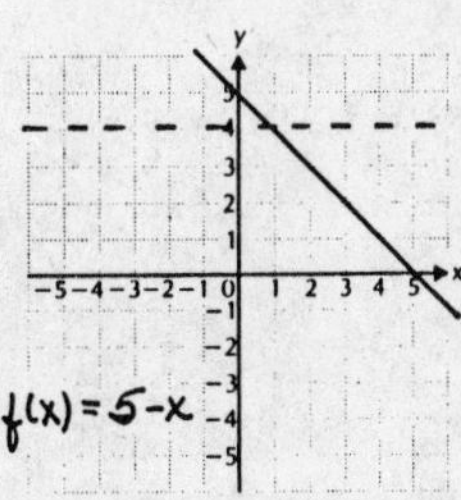

b) Replace $f(x)$ by y: $y = 5 - x$

Solve for x: $x = 5 - y$

Interchange x and y: $y = 5 - x$

Replace y by $f^{-1}(x)$: $f^{-1}(x) = 5 - x$

44. a) Yes

b) $f^{-1}(x) = 7 - x$

45. a) The graph of $g(x) = x - 3$ is shown below. It passes the horizontal line test, so the function is one-to-one.

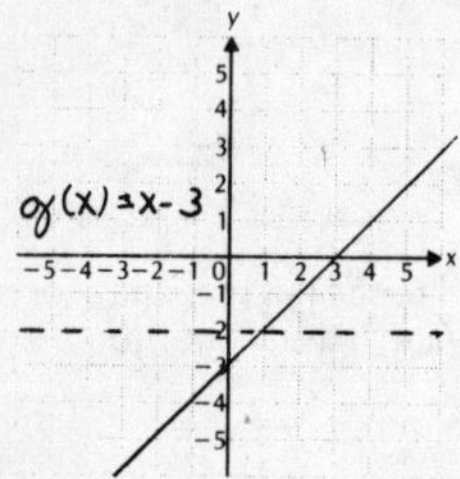

b) Replace g(x) by y: $y = x - 3$

Solve for x: $y + 3 = x$

Interchange x and y: $x + 3 = y$

Replace y by $g^{-1}(x)$: $g^{-1}(x) = x + 3$

46. a) Yes

b) $g^{-1}(x) = x + 10$

47. a) The graph of $f(x) = 2x$ is shown below. It passes the horizontal line test, so the function is one-to-one.

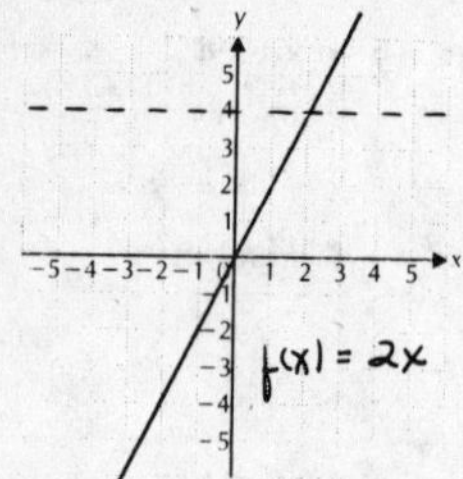

b) Replace f(x) by y: $y = 2x$

Solve for x: $\frac{y}{2} = x$

Interchange x and y: $\frac{x}{2} = y$

Replace y by $f^{-1}(x)$: $f^{-1}(x) = \frac{x}{2}$

48. a) Yes

b) $f^{-1}(x) = \frac{x}{5}$

49. a) The graph of $g(x) = 2x + 5$ is shown below. It passes the horizontal line test, so the function is one-to-one.

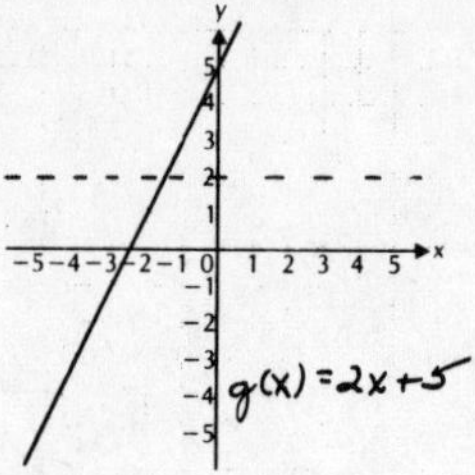

b) Replace g(x) by y: $y = 2x + 5$

Solve for x: $y - 5 = 2x$

$\frac{y - 5}{2} = x$

Interchange variables: $\frac{x - 5}{2} = y$

Replace y by $g^{-1}(x)$: $g^{-1}(x) = \frac{x - 5}{2}$

50. a) Yes

b) $g^{-1}(x) = \frac{x - 8}{5}$

51. a) The graph of $h(x) = \frac{4}{x + 7}$ is shown below. It passes the horizontal line test, so the function is one-to-one.

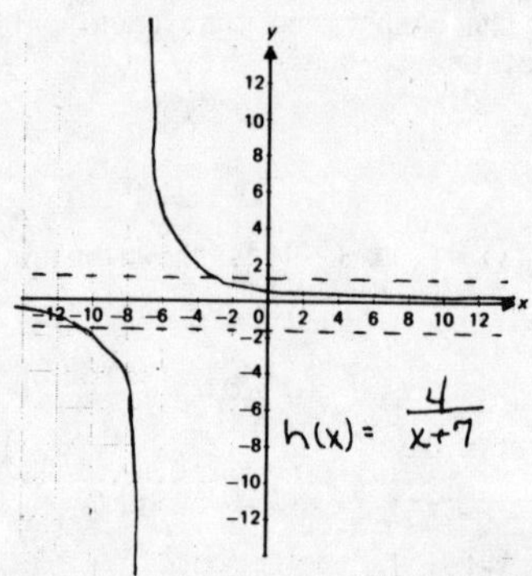

b) Replace h(x) by y: $y = \frac{4}{x + 7}$

Solve for x: $y(x + 7) = 4$

$x + 7 = \frac{4}{y}$

$x = \frac{4}{y} - 7$

Interchange x and y: $y = \frac{4}{x} - 7$

Replace y by $h^{-1}(x)$: $h^{-1}(x) = \frac{4}{x} - 7$

52. a) Yes

b) $h^{-1}(x) = \frac{1}{x} + 6$

53. a) The graph of $f(x) = \frac{1}{x}$ is shown below. It passes the horizontal line test, so the function is one-to-one.

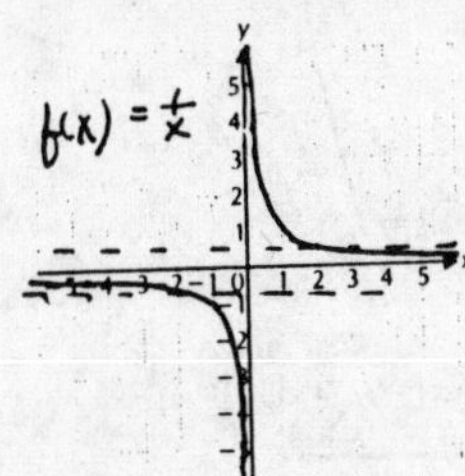

b) Replace f(x) by y: $y = \frac{1}{x}$

Solve for x: $xy = 1$

$x = \frac{1}{y}$

Interchange x and y: $y = \frac{1}{x}$

Replace y by $f^{-1}(x)$: $f^{-1}(x) = \frac{1}{x}$

54. a) Yes

b) $f^{-1}(x) = -\frac{4}{x}$

55. a) The graph of $f(x) = \frac{2x + 3}{4}$ is shown below. It passes the horizontal line test, so the function is one-to-one.

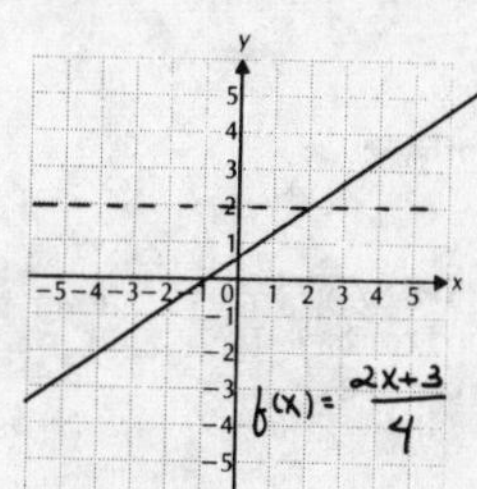

b) Replace f(x) by y: $y = \frac{2x + 3}{4}$

Solve for x: $4y = 2x + 3$

$4y - 3 = 2x$

$\frac{4y - 3}{2} = x$

Interchange x and y: $\frac{4x - 3}{2} = y$

Replace y by $f^{-1}(x)$: $f^{-1}(x) = \frac{4x - 3}{2}$

56. a) Yes

b) $f^{-1}(x) = \frac{4x + 5}{3}$

57. a) The graph of $g(x) = \frac{x + 4}{x - 3}$ is shown below. It passes the horizontal line test, so the function is one-to-one.

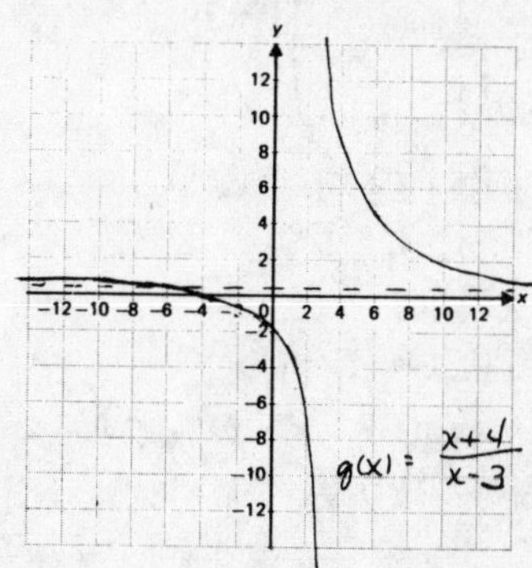

b) Replace g(x) by y: $y = \frac{x + 4}{x - 3}$

Solve for x: $(x - 3)y = x + 4$

$xy - 3y = x + 4$

$xy - x = 3y + 4$

$x(y - 1) = 3y + 4$

$x = \frac{3y + 4}{y - 1}$

Interchange x and y: $y = \frac{3x + 4}{x - 1}$

Replace y by $g^{-1}(x)$: $g^{-1}(x) = \frac{3x + 4}{x - 1}$

58. a) Yes

b) $g^{-1}(x) = \frac{x + 3}{5 - 2x}$

59. a) The graph of $f(x) = x^3 - 1$ is shown below. It passes the horizontal line test, so the function is one-to-one.

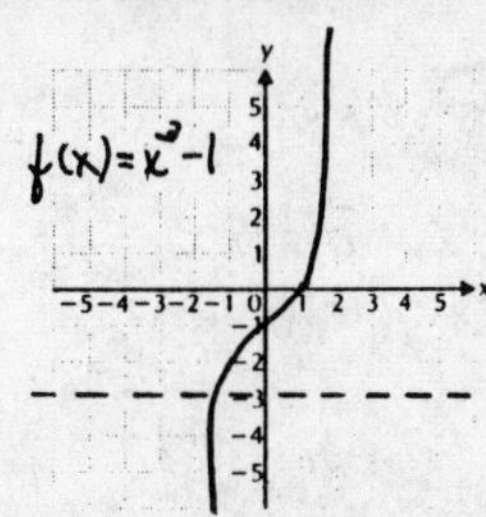

b) Replace f(x) by y: $y = x^3 - 1$

Solve for x: $y = x^3 - 1$

$y + 1 = x^3$

$\sqrt[3]{y + 1} = x$

Interchange x and y: $\sqrt[3]{x + 1} = y$

Replace y by $f^{-1}(x)$: $f^{-1}(x) = \sqrt[3]{x + 1}$

60. a) Yes

b) $f^{-1}(x) = \sqrt[3]{x - 7}$

61. a) The graph of $G(x) = (x - 4)^3$ is shown below. It passes the horizontal line test, so the function is one-to-one.

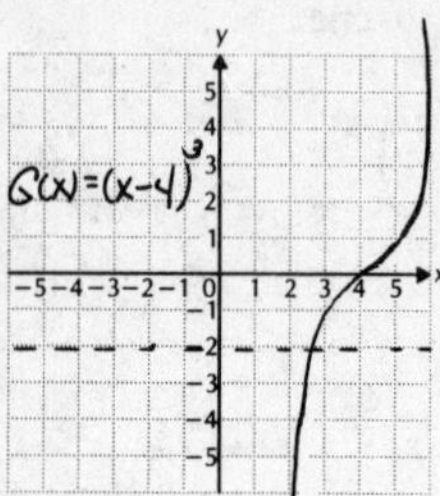

b) Replace $G(x)$ by y: $y = (x - 4)^3$

Solve for x: $\sqrt[3]{y} = x - 4$

$\sqrt[3]{y} + 4 = x$

Interchange x and y: $\sqrt[3]{x} + 4 = y$

Replace y by $G^{-1}(x)$: $G^{-1}(x) = \sqrt[3]{x} + 2$

62. a) Yes

b) $G^{-1}(x) = \sqrt[3]{x} - 5$

63. a) The graph of $f(x) = \sqrt[3]{x}$ is shown below. It passes the horizontal line test, so the function is one-to-one.

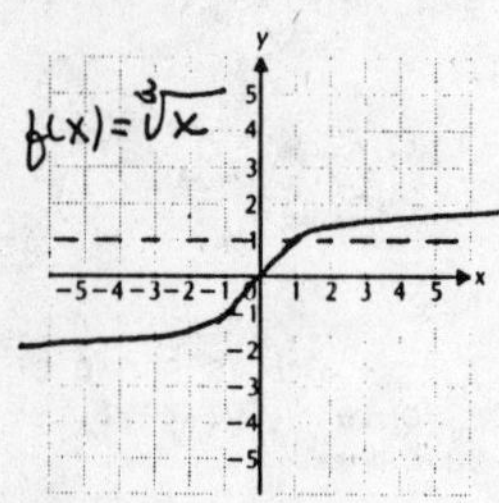

b) Replace $f(x)$ by y: $y = \sqrt[3]{x}$

Solve for x: $y^3 = x$

Interchange x and y: $x^3 = y$

Replace y by $f^{-1}(x)$: $f^{-1}(x) = x^3$

64. a) Yes

b) $f^{-1}(x) = x^3 + 8$

65. a) The graph of $f(x) = 4x^2 + 3$, $x > 3$, is shown below. It passes the horizontal line test, so the function is one-to-one.

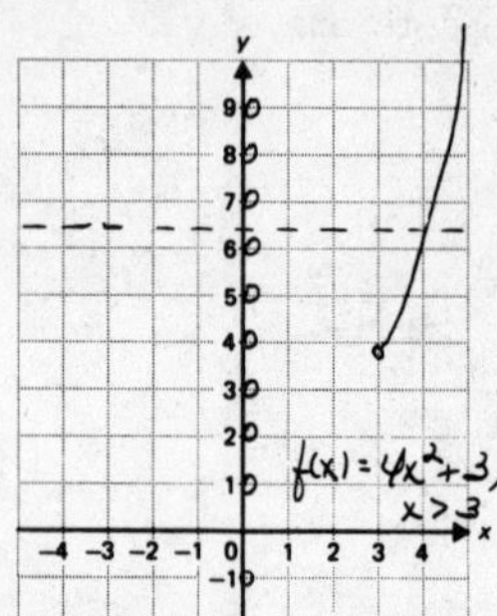

b) Replace $f(x)$ by y: $y = 4x^2 + 3$

Solve for x: $y - 3 = 4x^2$

$\frac{y - 3}{4} = x^2$

$\frac{\sqrt{y - 3}}{2} = x$

(We take the principal square root since $x > 3$.)

Interchange x and y: $\frac{\sqrt{x - 3}}{2} = y$

Replace y by $f^{-1}(x)$: $f^{-1}(x) = \frac{\sqrt{x - 3}}{2}$ for all x in the range of $f(x)$, or

$f^{-1}(x) = \frac{\sqrt{x - 3}}{2}$, $x > 39$

66. a) Yes

b) $f^{-1}(x) = \sqrt{\frac{x + 2}{5}}$, $x > 18$

67. a) The graph of $f(x) = \sqrt{x + 1}$ is shown below. It passes the horizontal line test, so the function is one-to-one.

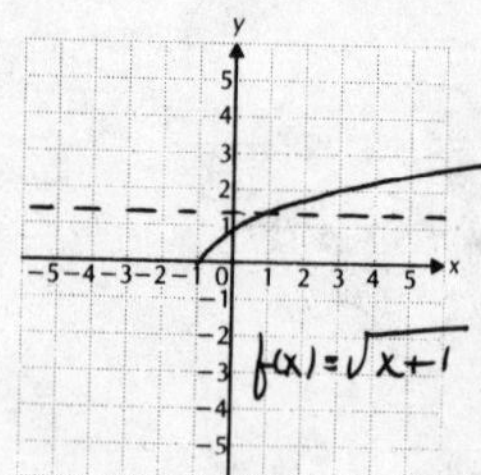

b) Replace $f(x)$ by y: $y = \sqrt{x + 1}$

Solve for x: $y^2 = x + 1$

$y^2 - 1 = x$

Interchange x and y: $x^2 - 1 = y$

Replace y by $f^{-1}(x)$: $f^{-1}(x) = x^2 - 1$ for all x in the range of $f(x)$, or $f^{-1}(x) = x^2 - 1$, $x \geqslant 0$.

68. a) Yes

b) $g^{-1}(x) = \frac{x^2 + 3}{2}$, $x \geqslant 0$

69. Only a) and c) pass the horizontal line test, and thus, have inverses that are functions.

70. (a), (d)

71. First graph $f(x) = \frac{1}{2}x - 4$. Then graph the inverse by flipping the graph of $f(x) = \frac{1}{2}x - 4$ over the line $y = x$. The graph of the inverse function can also be found by first finding a formula for the inverse and then substituting to find function values.

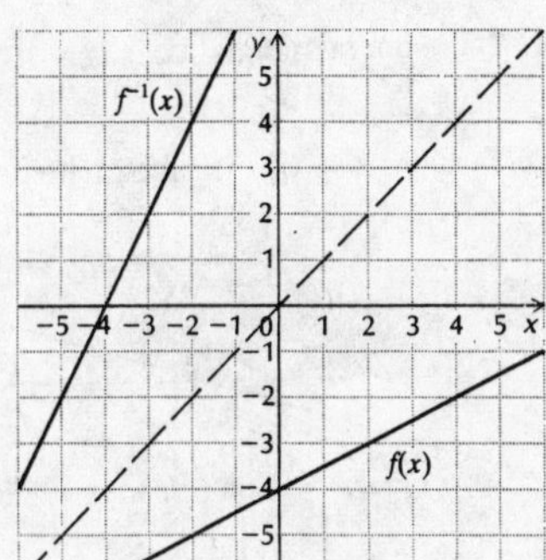

72.

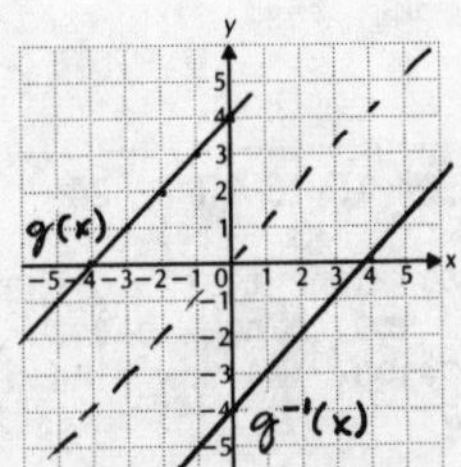

73. Use the procedure described in Exercise 71 to graph the function and its inverse.

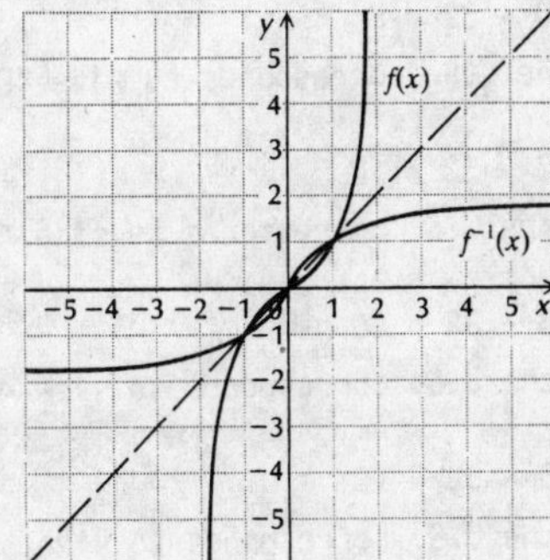

74.

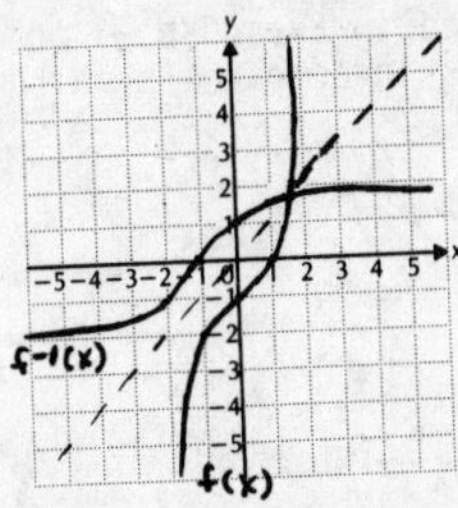

75. Use the procedure described in Exercise 71 to graph the function and its inverse.

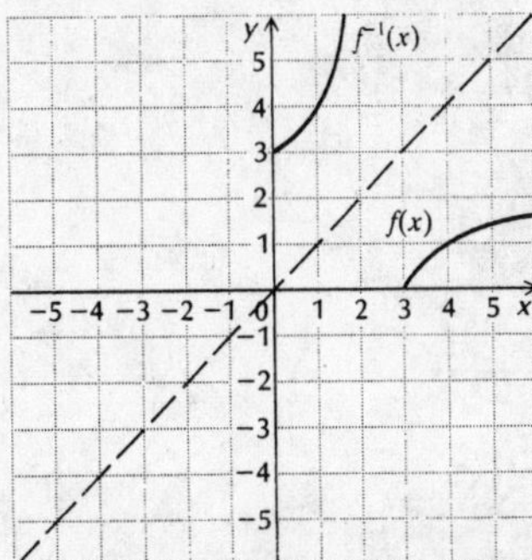

76.

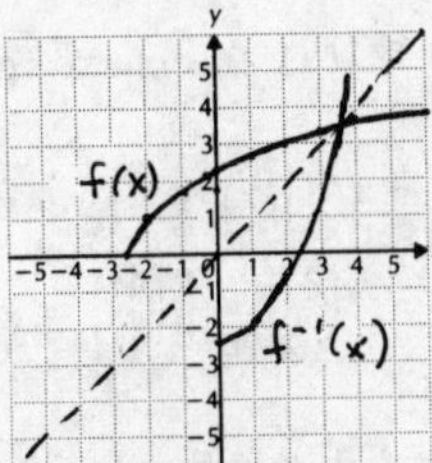

77. Use the procedure described in Exercise 71 to graph the function and its inverse.

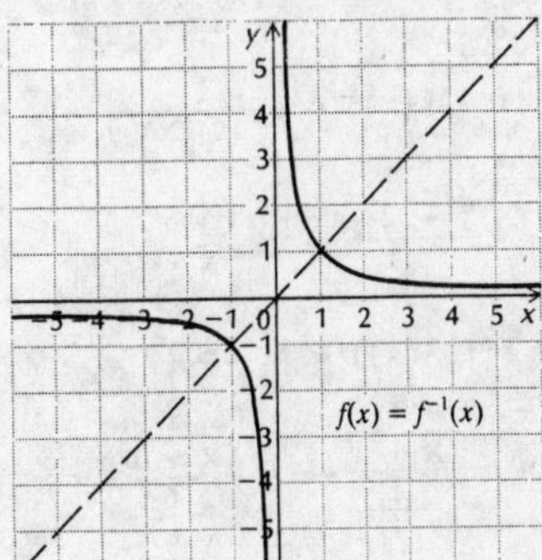

78.

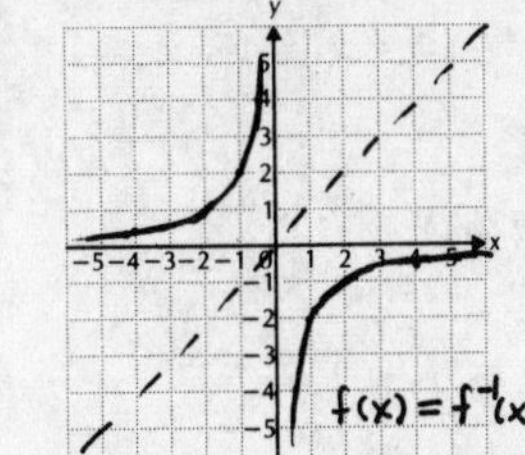

79. Use the procedure described in Exercise 71 to graph the function and its inverse.

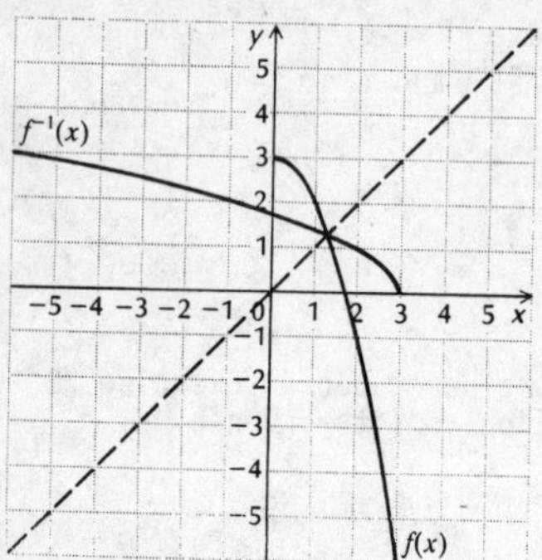

80.

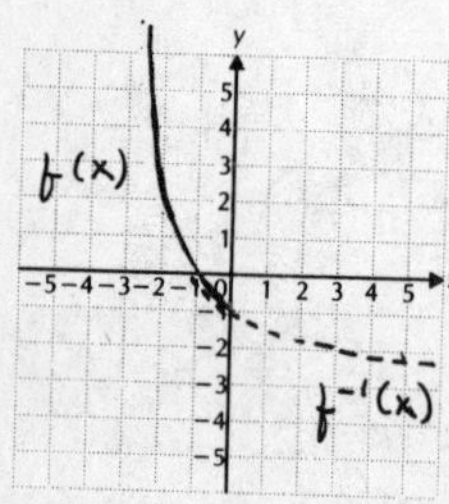

81. We find $f^{-1} \circ f(x)$ and $f \circ f^{-1}(x)$ and check to see that each is x.

a) $f^{-1} \circ f(x) = f^{-1}(f(x)) = f^{-1}\left(\frac{7}{8}x\right) = \frac{8}{7}\left(\frac{7}{8}x\right) = x$

b) $f \circ f^{-1}(x) = f(f^{-1}(x)) = f\left(\frac{8}{7}x\right) = \frac{7}{8}\left(\frac{8}{7}x\right) = x$

82. a) $f^{-1} \circ f(x) = 4\left(\frac{x + 5}{4}\right) - 5 = x + 5 - 5 = x$

b) $f \circ f^{-1}(x) = \frac{4x - 5 + 5}{4} = \frac{4x}{4} = x$

83. We find $f^{-1} \circ f(x)$ and $f \circ f^{-1}(x)$ and check to see that each is x.

a) $f^{-1} \circ f(x) = f^{-1}(f(x)) = f^{-1}\left(\frac{1 - x}{x}\right) =$

$$\frac{1}{\frac{1 - x}{x} + 1} = \frac{1}{\frac{1 - x + x}{x}} = \frac{1}{\frac{1}{x}} = x$$

b) $f \circ f^{-1}(x) = f(f^{-1}(x)) = f\left(\frac{1}{x + 1}\right) =$

$$\frac{1 - \frac{1}{x + 1}}{\frac{1}{x + 1}} = \frac{\frac{x + 1 - 1}{x + 1}}{\frac{1}{x + 1}} = \frac{\frac{x}{x + 1}}{\frac{1}{x + 1}} = x$$

84. a) $f^{-1} \circ f(x) = \sqrt[3]{x^3 - 4 + 4} = \sqrt[3]{x^3} = x$

b) $f \circ f^{-1}(x) = (\sqrt[3]{x + 4})^3 - 4 = x + 4 - 4 = x$

85. Using Theorem 1 we have $f^{-1}(f(3)) = 3$ and $f(f^{-1}(-125)) = -125$.

86. 5; -12

87. Using Theorem 1, we have $f^{-1}(f(12,053)) = 12,053$ and $f(f^{-1}(-17,243)) = -17,243$.

88. 489; -17,422

89. a) $f(8) = 8 + 32 = 40$

Size 40 in France corresponds to size 8 in the U.S.

$f(10) = 10 + 32 = 42$

Size 42 in France corresponds to size 10 in the U.S.

$f(14) = 14 + 32 = 46$

Size 46 in France corresponds to size 14 in the U.S.

$f(18) = 18 + 32 = 50$

Size 50 in France corresponds to size 18 in the U.S.

b) The graph of $f(x) = x + 32$ is shown below.

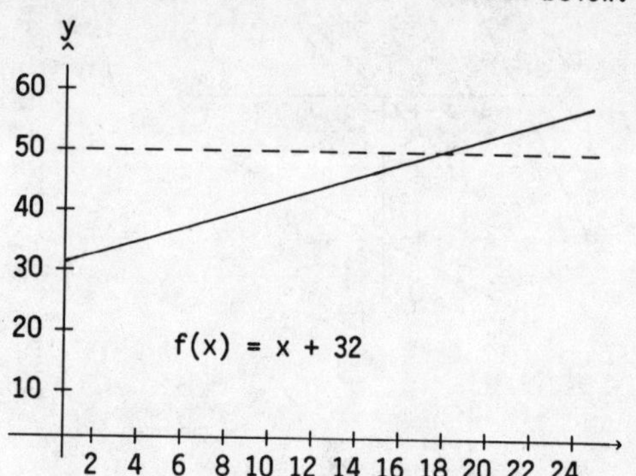

It passes the horizontal line test, so the function is one-to-one and, hence, has an inverse that is a function. We now find a formula for the inverse.

Replace f(x) by y: $y = x + 32$

Solve for x: $y - 32 = x$

Interchange x and y: $x - 32 = y$

Replace y by $f^{-1}(x)$: $f^{-1}(x) = x - 32$

c) $f^{-1}(40) = 40 - 32 = 8$

Size 8 in the U.S. corresponds to size 40 in France.

$f^{-1}(42) = 42 - 32 = 10$

Size 10 in the U.S. corresponds to size 42 in France.

$f^{-1}(46) = 46 - 32 = 14$

Size 14 in the U.S. corresponds to size 46 in France.

$f^{-1}(50) = 50 - 32 = 18$

Size 18 in the U.S. corresponds to size 50 in France.

90. a) 40, 44, 52, 60

b) Yes. $f^{-1}(x) = \frac{x - 24}{2}$, or $\frac{x}{2} - 12$

c) 8, 10, 14, 18

91. The graph of $f(x) = 5$ is shown below. Since the horizontal line $y = 5$ crosses the graph in more than one place, the function does not have an inverse that is a function.

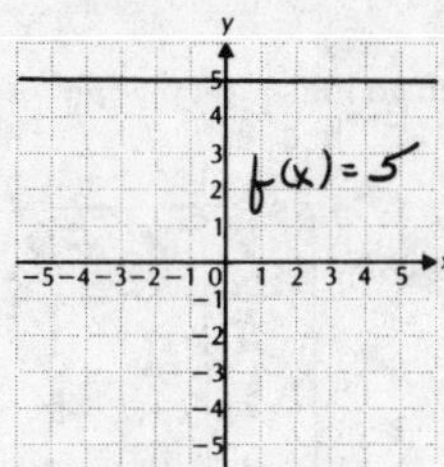

92. $C^{-1}(x) = \frac{100}{x - 5}$

$C^{-1}(x)$ gives the number of people in the group, where x is the cost per person, in dollars.

93. Check to see if $g \circ f(x) = x$ and $f \circ g(x) = x$.

a) $g \circ f(x) = g(f(x)) = g\left(\frac{2}{3}\right) = \frac{3}{2}$

Since $g \circ f(x) \neq x$, the functions are not inverses of each other.

94. Yes

95. Check to see if $g \circ f(x) = x$ and $f \circ g(x) = x$.

a) $g \circ f(x) = g(f(x)) = g(\sqrt[4]{x}) = (\sqrt[4]{x})^4 = x$

b) $f \circ g(x) = f(g(x)) = f(x^4) = \sqrt[4]{x^4} = |x|$

For $x < 0$, $|x| = -x$ and $f \circ g(x) \neq x$.

The functions are not inverses of each other.

96. No

97. Answers may vary. $f(x) = \frac{3}{x}$, $f(x) = 1 - x$, $f(x) = x$

98. Consider any two inputs a and b, $a < b$, of an increasing function $f(x)$. Then $f(a) < f(b)$. Thus, if $a \neq b$, $f(a) \neq f(b)$ and the function is one-to-one.

99. Graph $y = \frac{1}{x^2}$.

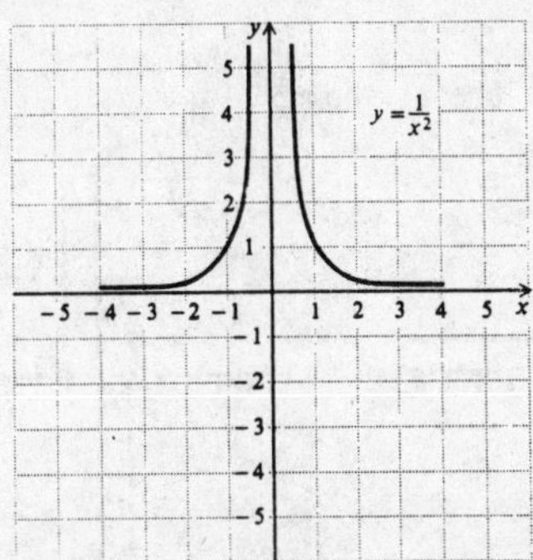

Test $y = \frac{1}{x^2}$ for symmetry with respect to the x-axis.

Replace y by -y: $-y = \frac{1}{x^2}$

Since $y = \frac{1}{x^2}$ is not equivalent to $-y = \frac{1}{x^2}$, $y = \frac{1}{x^2}$ is not symmetric with respect to the x-axis.

Test $y = \frac{1}{x^2}$ for symmetry with respect to the y-axis.

Replace x by -x: $y = \frac{1}{(-x)^2} = \frac{1}{x^2}$

Since the equations are equivalent, $y = \frac{1}{x^2}$ is symmetric with respect to the y-axis.

Test $y = \frac{1}{x^2}$ for symmetry with respect to the origin.

Replace x by -x and y by -y: $-y = \frac{1}{(-x)^2} = \frac{1}{x^2}$

Since $y = \frac{1}{x^2}$ is not equivalent to $-y = \frac{1}{x^2}$, $y = \frac{1}{x^2}$ is not symmetric with respect to the origin.

Test $y = \frac{1}{x^2}$ for symmetry with respect to the line $y = x$.

Interchange x and y: $x = \frac{1}{y^2}$

Since $y = \frac{1}{x^2}$ is not equivalent to $x = \frac{1}{y^2}$, $y = \frac{1}{x^2}$ is not symmetric with respect to the line $y = x$.

100. C and F are inverses.

101. See the answer section in the text.

102. a) The inverse of the given function $y = \frac{|x|}{x}$ is $x = \frac{|y|}{y}$.

The graphs are as shown:

$y = \frac{|x|}{x}$

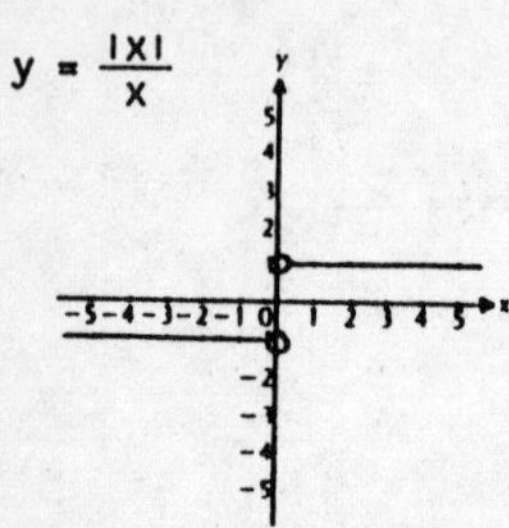

$x = \frac{|y|}{y}$

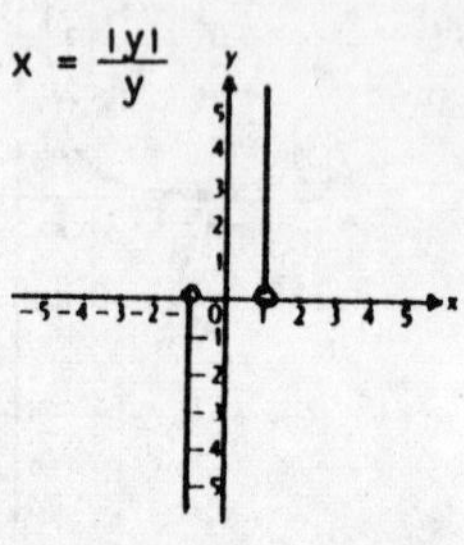

b) Each of the two graphs is symmetric with respect to the origin. Neither is symmetric with respect to the x-axis, y-axis, nor the line y = x.

Exercise Set 5.2

1. Graph: $y = 2^x$

We compute some function values, thinking of y as f(x), and keep the results in a table.

$f(0) = 2^0 = 1$

$f(1) = 2^1 = 2$

$f(2) = 2^2 = 4$

$f(-1) = 2^{-1} = \frac{1}{2^1} = \frac{1}{2}$

$f(-2) = 2^{-2} = \frac{1}{2^2} = \frac{1}{4}$

x	y, or f(x)
0	1
1	2
2	4
-1	$\frac{1}{2}$
-2	$\frac{1}{4}$

Next we plot these points and connect them with a smooth curve.

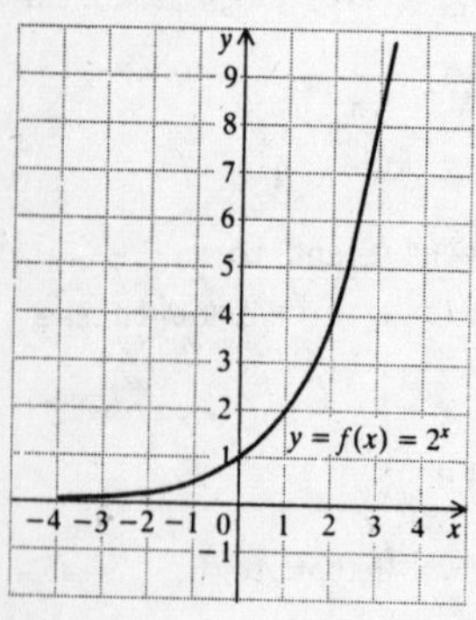

2.

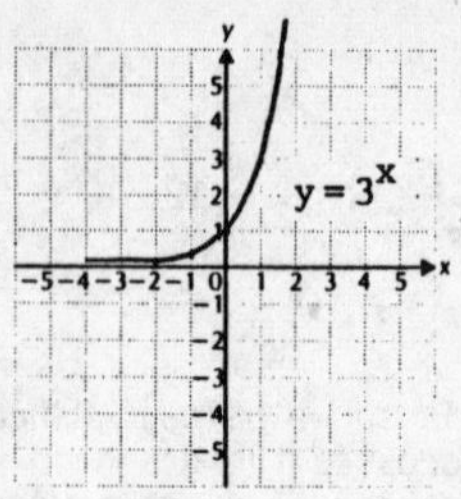

3. Graph: $y = 5^x$

We compute some function values, thinking of y as f(x), and keep the results in a table.

$f(0) = 5^0 = 1$

$f(1) = 5^1 = 5$

$f(2) = 5^2 = 25$

$f(-1) = 5^{-1} = \frac{1}{5^1} = \frac{1}{5}$

$f(-2) = 5^{-2} = \frac{1}{5^2} = \frac{1}{25}$

x	y, or f(x)
0	1
1	5
2	25
-1	$\frac{1}{5}$
-2	$\frac{1}{25}$

Next we plot these points and connect them with a smooth curve.

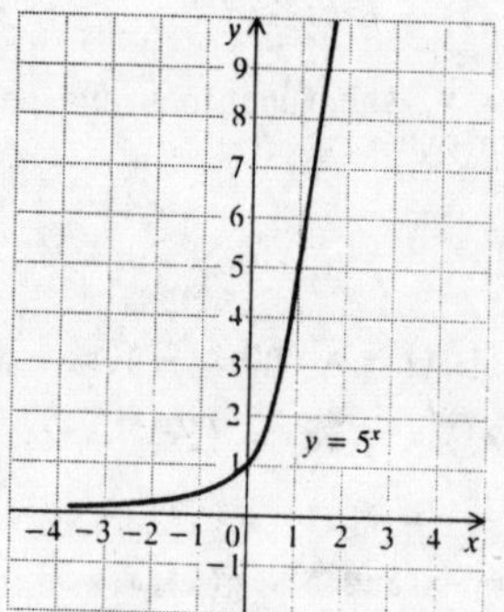

4.

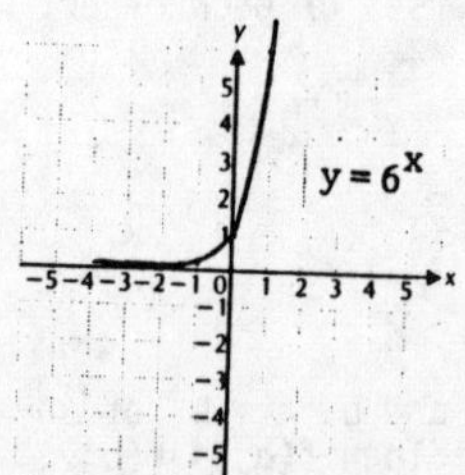

5. Graph: $y = 2^{x+1}$

We compute some function values, thinking of y as f(x), and keep the results in a table.

$f(0) = 2^{0+1} = 2^1 = 2$

$f(-1) = 2^{-1+1} = 2^0 = 1$

$f(-2) = 2^{-2+1} = 2^{-1} = \frac{1}{2^1} = \frac{1}{2}$

$f(-3) = 2^{-3+1} = 2^{-2} = \frac{1}{2^2} = \frac{1}{4}$

$f(1) = 2^{1+1} = 2^2 = 4$

$f(2) = 2^{2+1} = 2^3 = 8$

x	y, or f(x)
0	2
-1	1
-2	$\frac{1}{2}$
-3	$\frac{1}{4}$
1	4
2	8

Next we plot these points and connect them with a smooth curve.

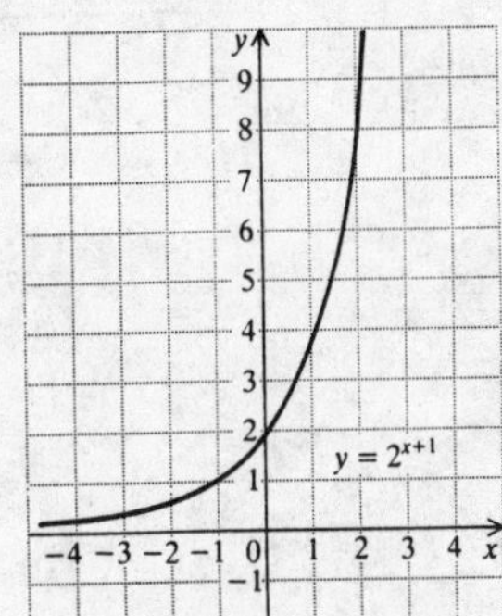

6.

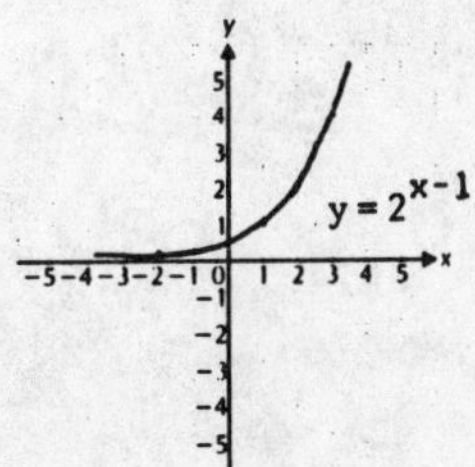

7. Graph: $y = 3^{x-2}$

We compute some function values, thinking of y as f(x), and keep the results in a table.

$f(0) = 3^{0-2} = 3^{-2} = \frac{1}{3^2} = \frac{1}{9}$

$f(1) = 3^{1-2} = 3^{-1} = \frac{1}{3^1} = \frac{1}{3}$

$f(2) = 3^{2-2} = 3^0 = 1$

$f(3) = 3^{3-2} = 3^1 = 3$

$f(4) = 3^{4-2} = 3^2 = 9$

$f(-1) = 3^{-1-2} = 3^{-3} = \frac{1}{3^3} = \frac{1}{27}$

$f(-2) = 3^{-2-2} = 3^{-4} = \frac{1}{3^4} = \frac{1}{81}$

x	y, or f(x)
0	$\frac{1}{9}$
1	$\frac{1}{3}$
2	1
3	3
4	9
-1	$\frac{1}{27}$
-2	$\frac{1}{81}$

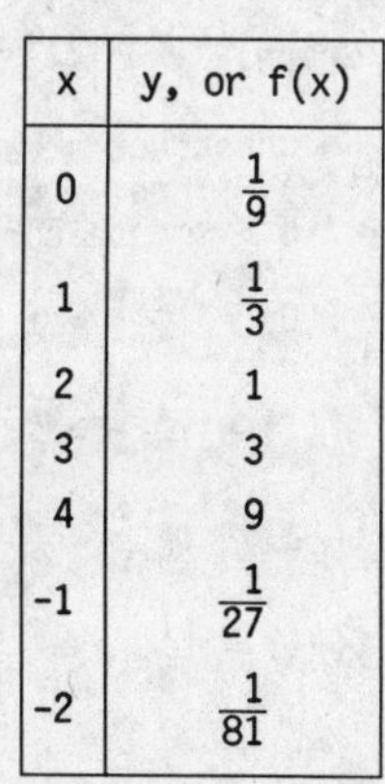

Next we plot these points and connect them with a smooth curve.

7. (continued)

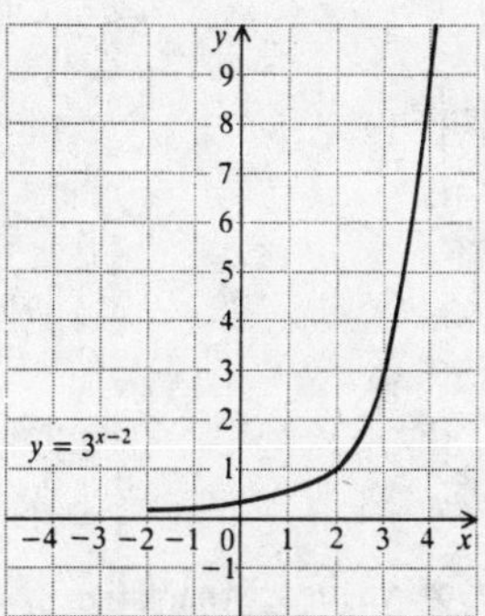

8.

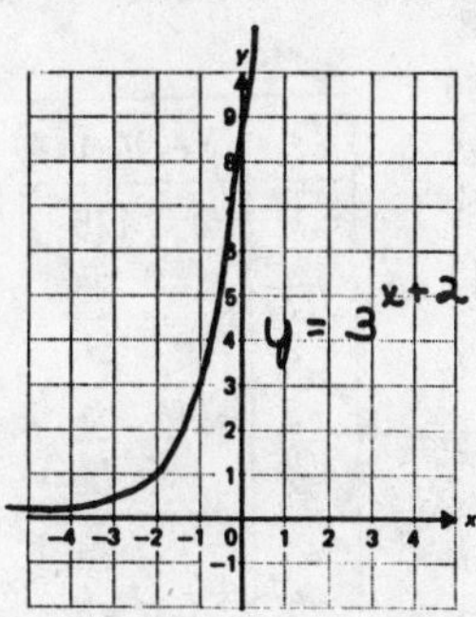

9. Graph: $y = 2^x - 3$

We construct a table of values, thinking of y as f(x). Then we plot the points and connect them with a smooth curve.

$f(0) = 2^0 - 3 = 1 - 3 = -2$

$f(1) = 2^1 - 3 = 2 - 3 = -1$

$f(2) = 2^2 - 3 = 4 - 3 = 1$

$f(3) = 2^3 - 3 = 8 - 3 = 5$

$f(-1) = 2^{-1} - 3 = \frac{1}{2} - 3 = -\frac{5}{2}$

$f(-2) = 2^{-2} - 3 = \frac{1}{4} - 3 = -\frac{11}{4}$

x	y, or f(x)
0	-2
1	-1
2	1
3	5
-1	$-\frac{5}{2}$
-2	$-\frac{11}{4}$

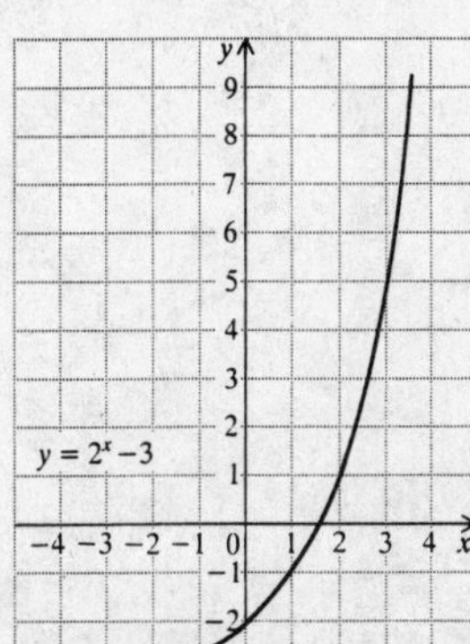

10.

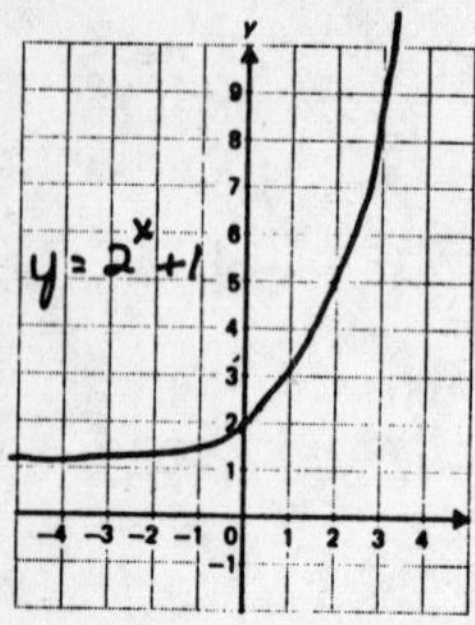

11. Graph: $y = 5^{x+3}$

We construct a table of values, thinking of y as f(x). Then we plot the points and connect them with a smooth curve.

$f(0) = 5^{0+3} = 5^3 = 125$

$f(-1) = 5^{-1+3} = 5^2 = 25$

$f(-2) = 5^{-2+3} = 5^1 = 5$

$f(-3) = 5^{-3+3} = 5^0 = 1$

$f(-4) = 5^{-4+3} = 5^{-1} = \frac{1}{5}$

$f(-5) = 5^{-5+3} = 5^{-2} = \frac{1}{25}$

x	y, or f(x)
0	125
-1	25
-2	5
-3	1
-4	$\frac{1}{5}$
-5	$\frac{1}{25}$

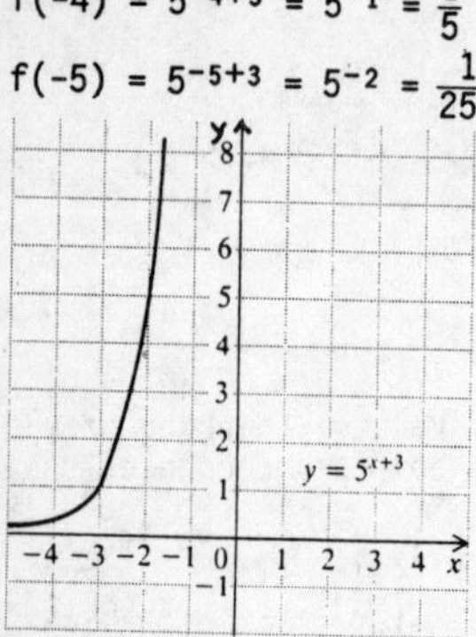

12.

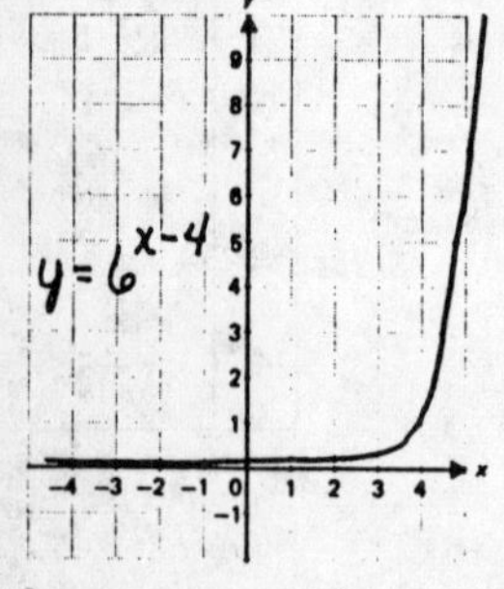

13. Graph: $y = \left(\frac{1}{2}\right)^x$

We construct a table of values, thinking of y as f(x). Then we plot the points and connect them with a smooth curve.

$f(0) = \left(\frac{1}{2}\right)^0 = 1$

$f(1) = \left(\frac{1}{2}\right)^1 = \frac{1}{2}$

$f(2) = \left(\frac{1}{2}\right)^2 = \frac{1}{4}$

$f(3) = \left(\frac{1}{2}\right)^3 = \frac{1}{8}$

$f(-1) = \left(\frac{1}{2}\right)^{-1} = \frac{1}{\left(\frac{1}{2}\right)^1} = \frac{1}{\frac{1}{2}} = 2$

$f(-2) = \left(\frac{1}{2}\right)^{-2} = \frac{1}{\left(\frac{1}{2}\right)^2} = \frac{1}{\frac{1}{4}} = 4$

$f(-3) = \left(\frac{1}{2}\right)^{-3} = \frac{1}{\left(\frac{1}{2}\right)^3} = \frac{1}{\frac{1}{8}} = 8$

x	y, or f(x)
0	1
1	$\frac{1}{2}$
2	$\frac{1}{4}$
3	$\frac{1}{8}$
-1	2
-2	4
-3	8

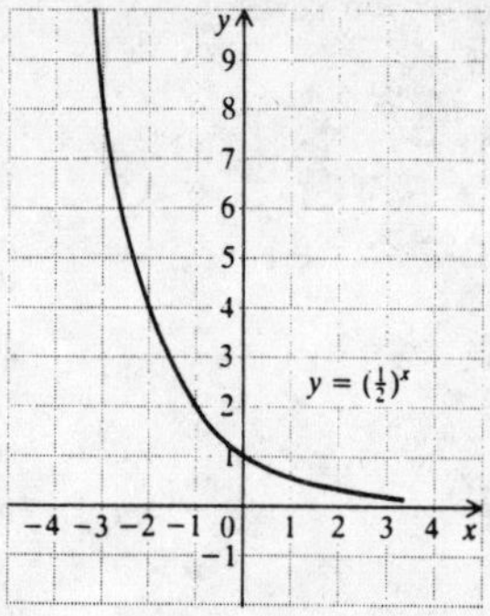

14.

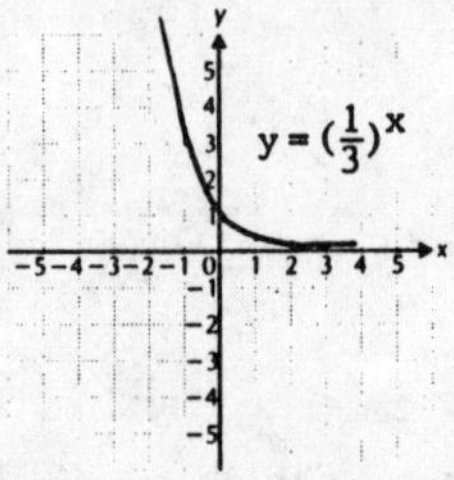

15. Graph: $y = \left(\frac{1}{5}\right)^x$

We construct a table of values, thinking of y as f(x). Then we plot the points and connect them with a smooth curve.

$f(0) = \left(\frac{1}{5}\right)^0 = 1$

$f(1) = \left(\frac{1}{5}\right)^1 = \frac{1}{5}$

$f(2) = \left(\frac{1}{5}\right)^2 = \frac{1}{25}$

$f(-1) = \left(\frac{1}{5}\right)^{-1} = \frac{1}{\frac{1}{5}} = 5$

$f(-2) = \left(\frac{1}{5}\right)^{-2} = \frac{1}{\frac{1}{25}} = 25$

x	y, or f(x)
0	1
1	$\frac{1}{5}$
2	$\frac{1}{25}$
-1	5
-2	25

15\. (continued)

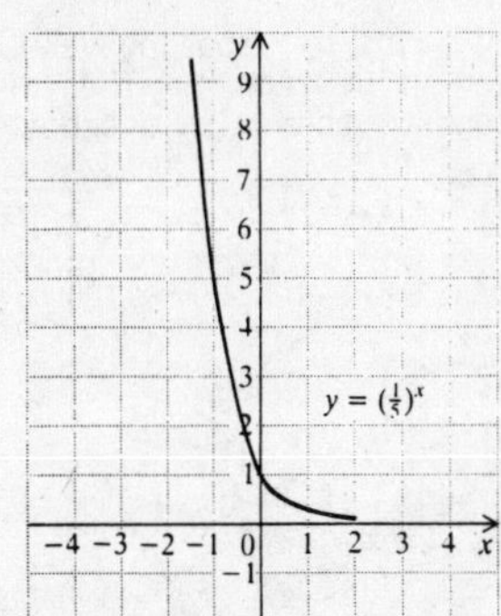

16\.

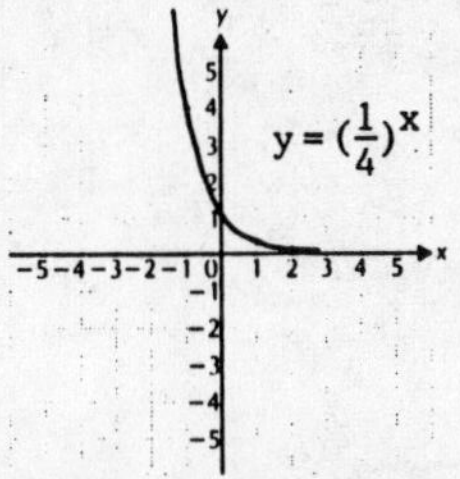

17\. Graph: $y = 2^{2x-1}$

We construct a table of values, thinking of y as f(x). Then we plot the points and connect them with a smooth curve.

$f(0) = 2^{2\cdot 0-1} = 2^{-1} = \frac{1}{2}$

$f(1) = 2^{2\cdot 1-1} = 2^{1} = 2$

$f(2) = 2^{2\cdot 2-1} = 2^{3} = 8$

$f(-1) = 2^{2(-1)-1} = 2^{-3} = \frac{1}{8}$

$f(-2) = 2^{2(-2)-1} = 2^{-5} = \frac{1}{32}$

x	y, or f(x)
0	$\frac{1}{2}$
1	2
2	8
-1	$\frac{1}{8}$
-2	$\frac{1}{32}$

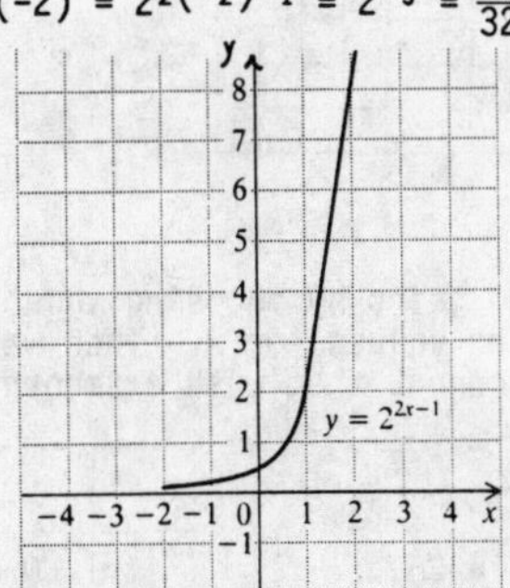

18\.

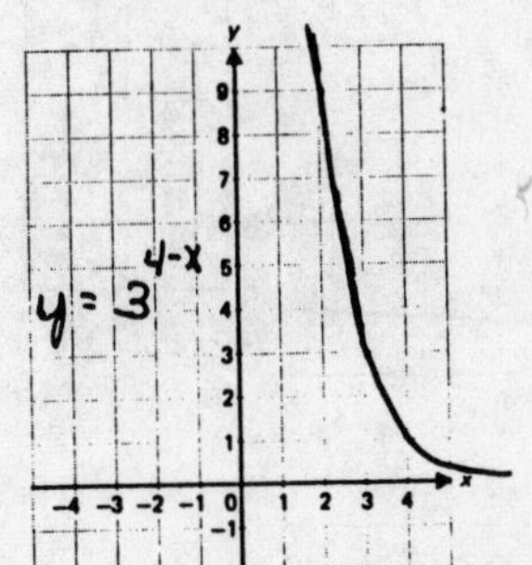

19\. Graph: $y = 2^{x-1} - 3$

We construct a table of values, thinking of y as f(x). Then we plot the points and connect them with a smooth curve.

$f(0) = 2^{0-1} - 3 = 2^{-1} - 3 = \frac{1}{2} - 3 = -\frac{5}{2}$

$f(1) = 2^{1-1} - 3 = 2^{0} - 3 = 1 - 3 = -2$

$f(2) = 2^{2-1} - 3 = 2^{1} - 3 = 2 - 3 = -1$

$f(3) = 2^{3-1} - 3 = 2^{2} - 3 = 4 - 3 = 1$

$f(4) = 2^{4-1} - 3 = 2^{3} - 3 = 8 - 3 = 5$

$f(-1) = 2^{-1-1} - 3 = 2^{-2} - 3 = \frac{1}{4} - 3 = -\frac{11}{4}$

$f(-2) = 2^{-2-1} - 3 = 2^{-3} - 3 = \frac{1}{8} - 3 = -\frac{23}{8}$

x	y, or f(x)
0	$-\frac{5}{2}$
1	-2
2	-1
3	1
4	5
-1	$-\frac{11}{4}$
-2	$-\frac{23}{8}$

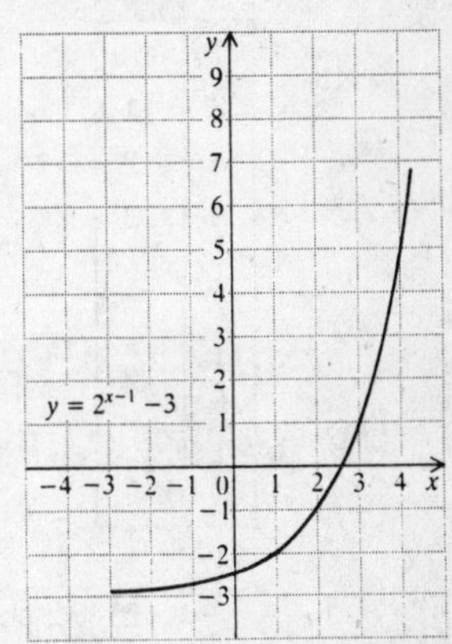

20\.

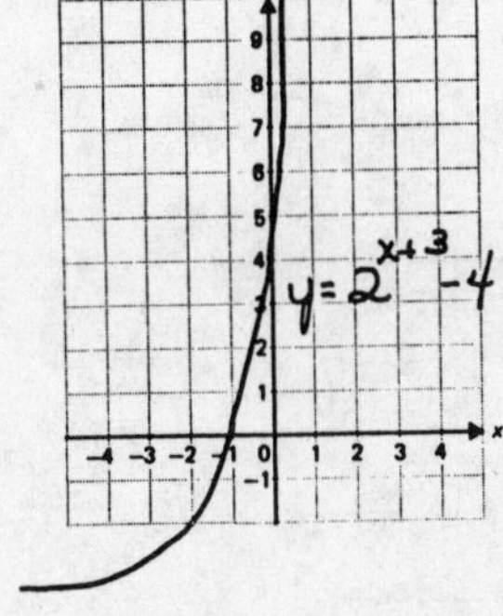

21. Graph: $x = 2^y$

We can find ordered pairs by choosing values for y and then computing values for x.

For $y = 0$, $x = 2^0 = 1$.

For $y = 1$, $x = 2^1 = 2$.

For $y = 2$, $x = 2^2 = 4$.

For $y = 3$, $x = 2^3 = 8$.

For $y = -1$, $x = 2^{-1} = \frac{1}{2^1} = \frac{1}{2}$.

For $y = -2$, $x = 2^{-2} = \frac{1}{2^2} = \frac{1}{4}$.

For $y = -3$, $x = 2^{-3} = \frac{1}{2^3} = \frac{1}{8}$.

x	y
1	0
2	1
4	2
8	3
$\frac{1}{2}$	-1
$\frac{1}{4}$	-2
$\frac{1}{8}$	-3

(1) Choose values for y.

(2) Compute values for x.

We plot these points and connect them with a smooth curve.

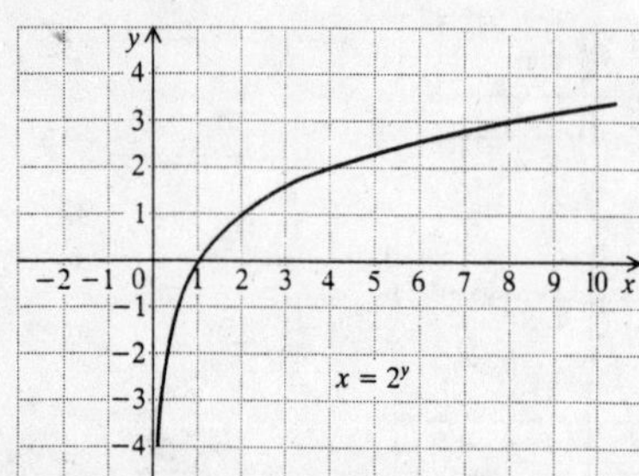

22.

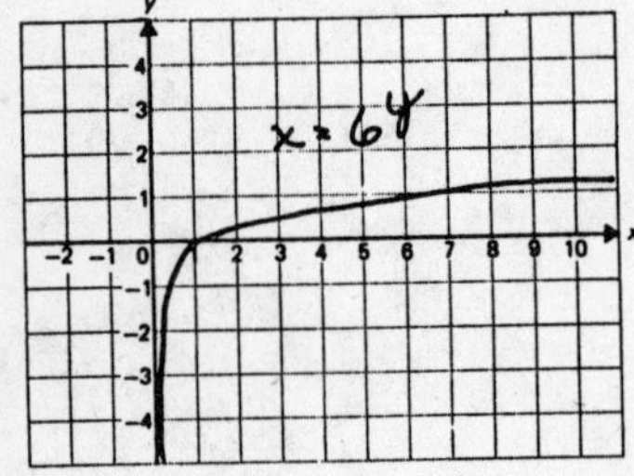

23. Graph: $x = \left(\frac{1}{2}\right)^y$

We can find ordered pairs by choosing values for y and then computing values for x. Then we plot these points and connect them with a smooth curve.

For $y = 0$, $x = \left(\frac{1}{2}\right)^0 = 1$.

For $y = 1$, $x = \left(\frac{1}{2}\right)^1 = \frac{1}{2}$.

For $y = 2$, $x = \left(\frac{1}{2}\right)^2 = \frac{1}{4}$.

For $y = 3$, $x = \left(\frac{1}{2}\right)^3 = \frac{1}{8}$.

For $y = -1$, $x = \left(\frac{1}{2}\right)^{-1} = \frac{1}{\frac{1}{2}} = 2$.

For $y = -2$, $x = \left(\frac{1}{2}\right)^{-2} = \frac{1}{\frac{1}{4}} = 4$.

For $y = -3$, $x = \left(\frac{1}{2}\right)^{-3} = \frac{1}{\frac{1}{8}} = 8$.

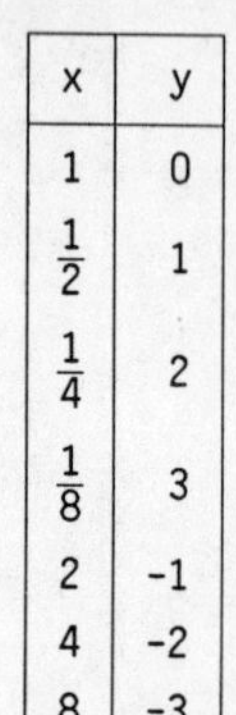

x	y
1	0
$\frac{1}{2}$	1
$\frac{1}{4}$	2
$\frac{1}{8}$	3
2	-1
4	-2
8	-3

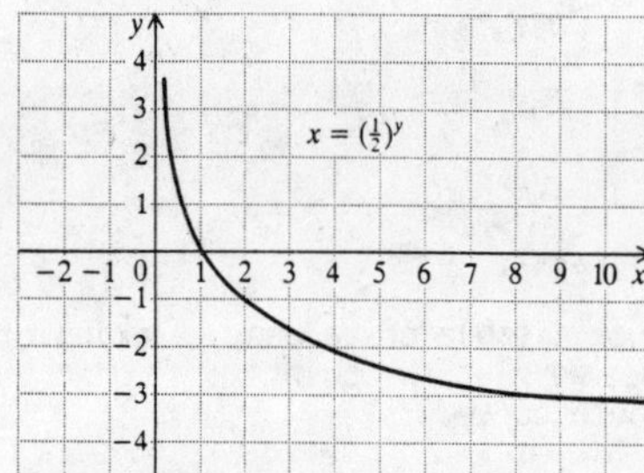

24.

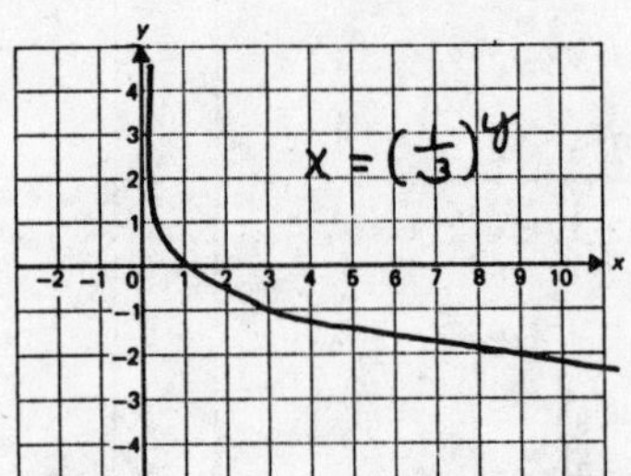

25. Graph: $x = 5^y$

We can find ordered pairs by choosing values for y and then computing values for x. Then we plot these points and connect them with a smooth curve.

For $y = 0$, $x = 5^0 = 1$.

For $y = 1$, $x = 5^1 = 5$.

For $y = 2$, $x = 5^2 = 25$.

For $y = -1$, $x = 5^{-1} = \frac{1}{5}$.

For $y = -2$, $x = 5^{-2} = \frac{1}{25}$.

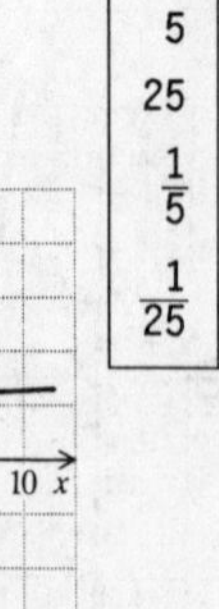

x	y
1	0
5	1
25	2
$\frac{1}{5}$	-1
$\frac{1}{25}$	-2

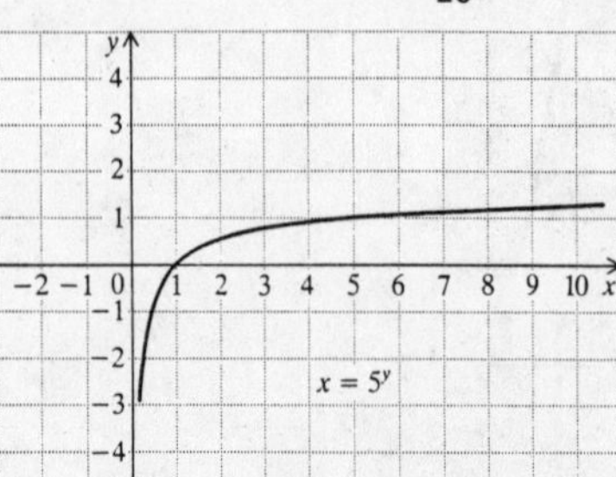

26.

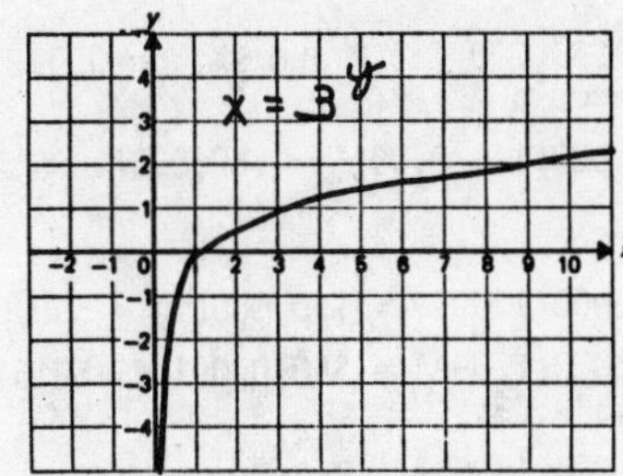

27. Graph: $x = \left(\frac{2}{3}\right)^y$

We can find ordered pairs by choosing values for y and then computing values for x. Then we plot these points and connect them with a smooth curve.

For y = 0, $x = \left(\frac{2}{3}\right)^0 = 1$.

For y = 1, $x = \left(\frac{2}{3}\right)^1 = \frac{2}{3}$.

For y = 2, $x = \left(\frac{2}{3}\right)^2 = \frac{4}{9}$.

For y = 3, $x = \left(\frac{2}{3}\right)^3 = \frac{8}{27}$.

For y = -1, $x = \left(\frac{2}{3}\right)^{-1} = \frac{1}{\frac{2}{3}} = \frac{3}{2}$.

For y = -2, $x = \left(\frac{2}{3}\right)^{-2} = \frac{1}{\frac{4}{9}} = \frac{9}{4}$.

For y = -3, $x = \left(\frac{2}{3}\right)^{-3} = \frac{1}{\frac{8}{27}} = \frac{27}{8}$.

x	y
1	0
$\frac{2}{3}$	1
$\frac{4}{9}$	2
$\frac{8}{27}$	3
$\frac{3}{2}$	-1
$\frac{9}{4}$	-2
$\frac{27}{8}$	-3

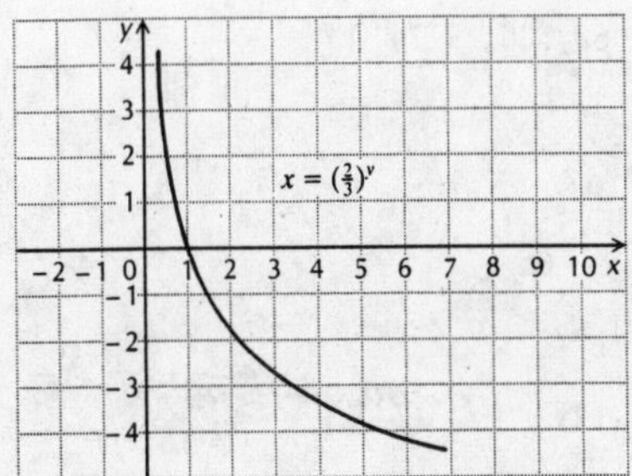

28.

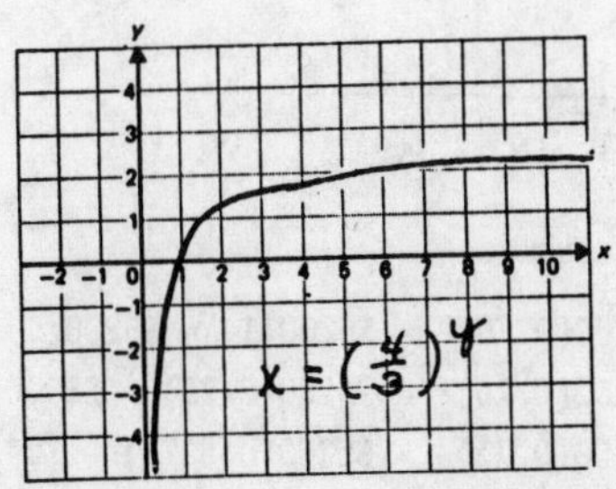

29. Graph $y = 2^x$ (see Exercise 1) and $x = 2^y$ (see Exercise 21) using the same set of axes.

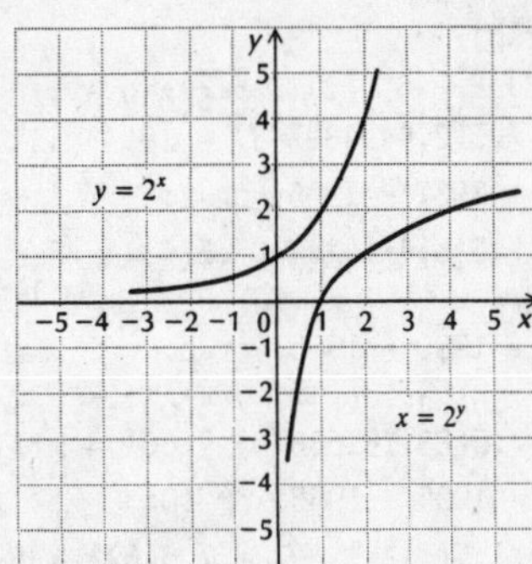

30.

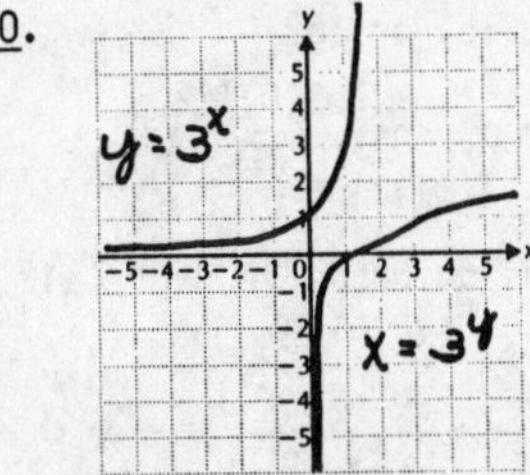

31. Graph $y = \left(\frac{1}{2}\right)^x$ (see Exercise 13) and $x = \left(\frac{1}{2}\right)^y$ (see Exercise 23) using the same set of axes.

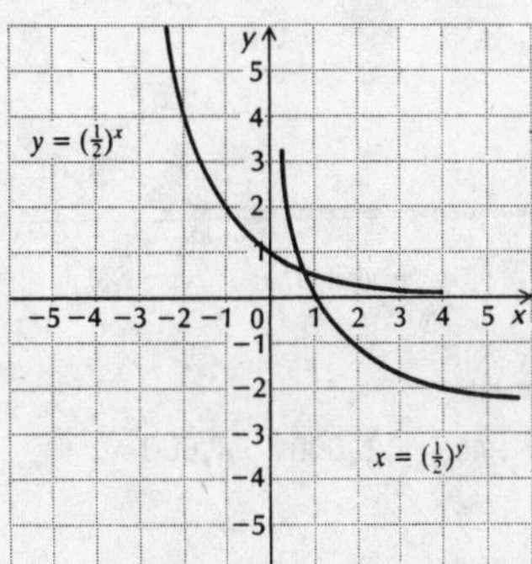

32.

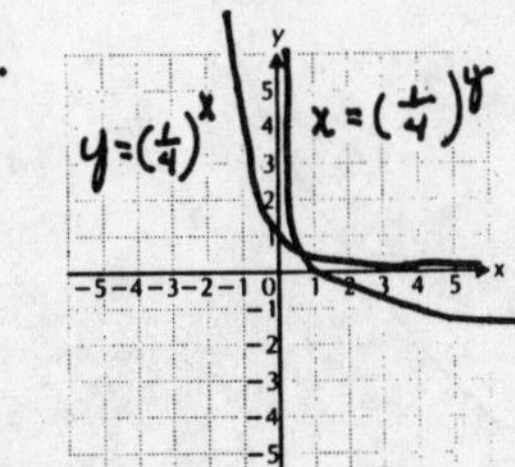

33. a) Keep in mind that t represents the number of years after 1985 and that N is given in millions.

For 1986, t = 1:

$N(1) = 7.5(6)^{0.5(1)} = 7.5(6)^{0.5} \approx 7.5(2.449489743) \approx 18.4$ million;

For 1987, t = 1987 - 1985, or 2:

$N(2) = 7.5(6)^{0.5(2)} = 7.5(6) = 45$ million;

For 1990, t = 1990 - 1985, or 5:

$N(5) = 7.5(6)^{0.5(5)} = 7.5(6)^{2.5} \approx 7.5(88.18163074) \approx 661.4$ million;

For 1995, t = 1995 - 1985, or 10:

$N(10) = 7.5(6)^{0.5(10)} = 7.5(6)^5 = 7.5(7776) = 58,320$ million;

For 2000, t = 2000 - 1985, or 15:

$N(15) = 7.5(6)^{0.5(15)} = 7.5(6)^{7.5} \approx 7.5(685,700.3606) \approx 5,142,752.7$ million

b) Use the function values computed in part (a) to draw the graph of the function.

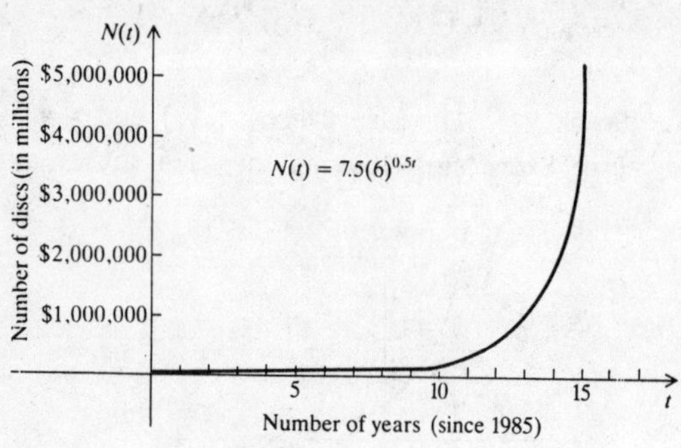

34. a) 4243, 6000, 8485, 12,000, 24,000

b)

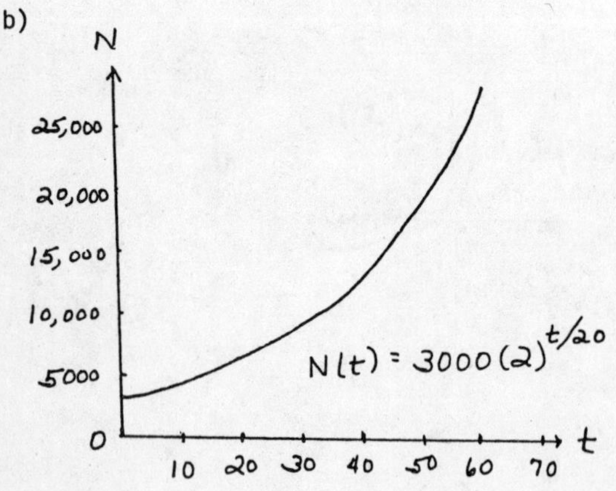

35. a) Substitute \$50,000 for P and 9%, or 0.09, for i in the formula $A = P(1 + i)^t$:

$A(t) = \$50,000(1 + 0.09)^t = \$50,000(1.09)^t$

b) Substitute for t.

$A(0) = \$50,000(1.09)^0 = \$50,000(1) = \$50,000;$

$A(4) = \$50,000(1.09)^4 = \$50,000(1.41158161) \approx \$70,579.08;$

$A(8) = \$50,000(1.09)^8 \approx \$50,000(1.992562642) \approx \$99,628.13;$

$A(10) = \$50,000(1.09)^{10} \approx \$50,000(2.367363675) \approx \$118,368.18$

c) We use the function values computed in part (b) to draw the graph. Note that the axes are scaled differently because of the large numbers.

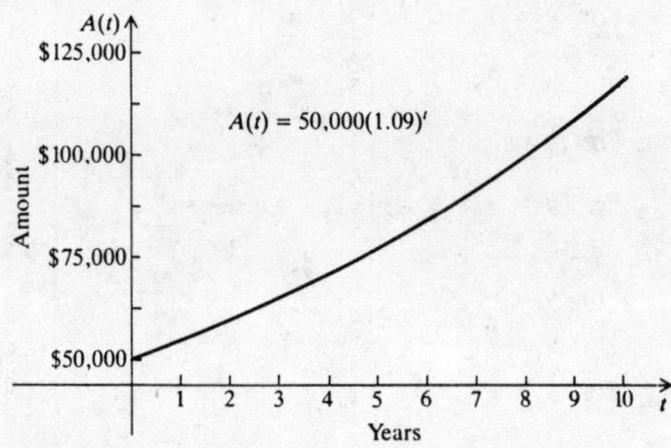

36. a) 250,000, 62,500, 977, 0

b)

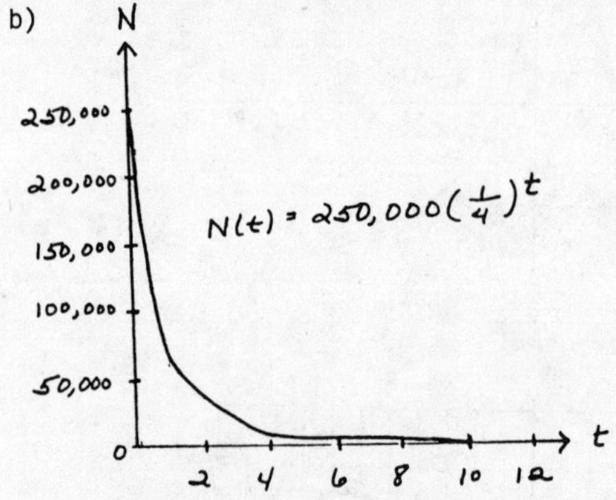

37. a) Substitute for t.

$V(0) = \$5200(0.75)^0 = \$5200(1) = \$5200;$

$V(1) = \$5200(0.75)^1 = \$5200(0.75) = \$3900;$

$V(2) = \$5200(0.75)^2 = \$5200(0.5625) = \$2925;$

$V(5) = \$5200(0.75)^5 \approx \$5200(0.237304687) \approx \$1233.98;$

$V(10) = \$5200(0.75)^{10} \approx \$5200(0.056313514) \approx \$292.83$

37. (continued)

b) Use the function values computed in part (a) to draw the graph of the function.

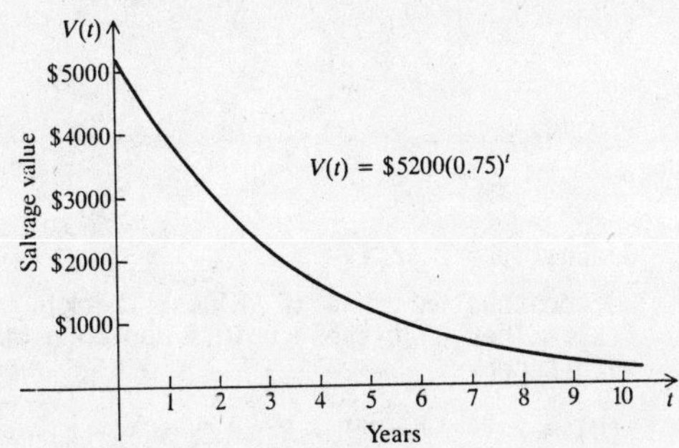

38. a) 2.6 lb, 3.7 lb, 10.9 lb, 13.1 lb, 29.1 lb

b) 29.1 lb

c)

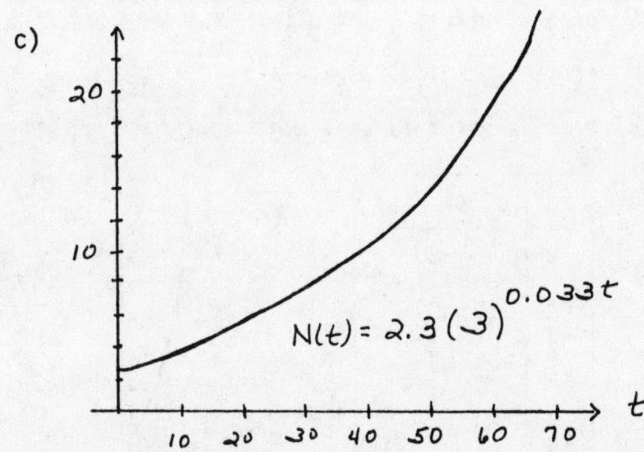

39. Use a calculator.

a) $7^3 = 343$

b) $7^{3.1} \approx 416.681217$

c) $7^{3.14} \approx 450.409815$

d) $7^{3.141} \approx 451.287125$

e) $7^{3.1415} \approx 451.726421$

f) $7^{3.14159} \approx 451.805540$

40. π^7

41. Since the bases are the same, the one with the larger exponent is the larger number. Thus $\pi^{3.2}$ is larger.

42. $4^{\sqrt{3}}$

43. Graph: $f(x) = (2.7)^x$

Use a calculator with a power key to construct a table of values. (We will round values to the nearest hundredth.) Then plot these points and connect them with a smooth curve.

x	f(x)
0	1
1	2.7
2	7.29
3	19.68
-1	0.37
-2	0.14

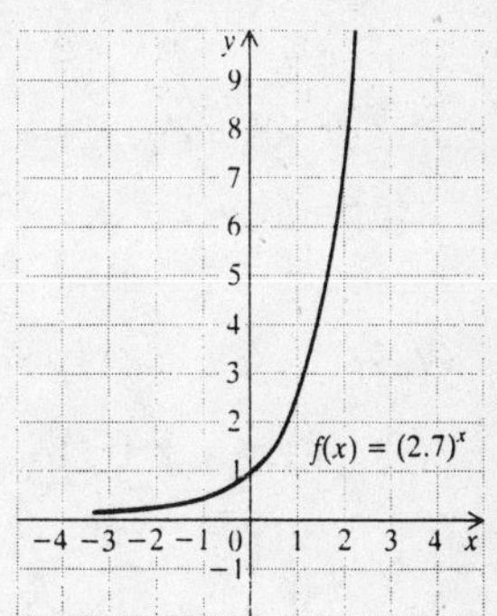

44.

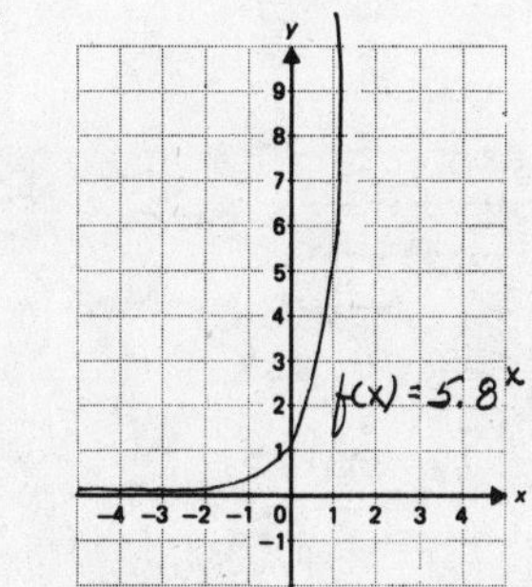

45. Graph: $g(x) = (0.745)^x$

Use the procedure described in Exercise 43, rounding values of g(x) to the nearest thousandth.

x	g(x)
0	1
1	0.745
2	0.555
3	0.413
-1	1.342
-2	1.802

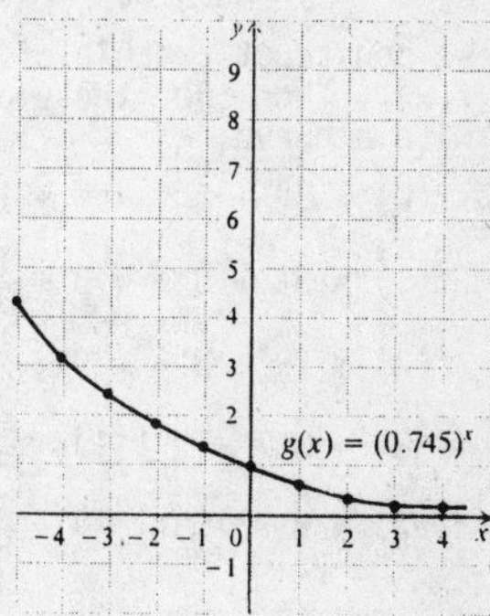

46.

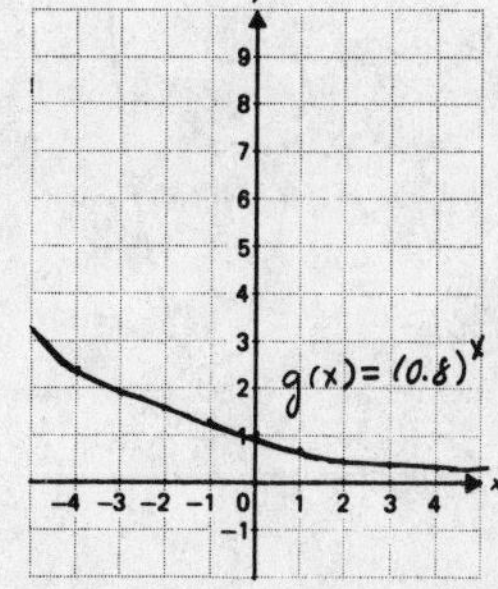

47. Graph: $y = 2^x + 2^{-x}$

Construct a table of values, thinking of y as f(x). Then plot these points and connect them with a curve.

$f(0) = 2^0 + 2^{-0} = 1 + 1 = 2$

$f(1) = 2^1 + 2^{-1} = 2 + \frac{1}{2} = 2\frac{1}{2}$

$f(2) = 2^2 + 2^{-2} = 4 + \frac{1}{4} = 4\frac{1}{4}$

$f(3) = 2^3 + 2^{-3} = 8 + \frac{1}{8} = 8\frac{1}{8}$

$f(-1) = 2^{-1} + 2^{-(-1)} = \frac{1}{2} + 2 = 2\frac{1}{2}$

$f(-2) = 2^{-2} + 2^{-(-2)} = \frac{1}{4} + 4 = 4\frac{1}{4}$

$f(-3) = 2^{-3} + 2^{-(-3)} = \frac{1}{8} + 8 = 8\frac{1}{8}$

x	y, or f(x)
0	2
1	$2\frac{1}{2}$
2	$4\frac{1}{4}$
3	$8\frac{1}{8}$
-1	$2\frac{1}{2}$
-2	$4\frac{1}{4}$
-3	$8\frac{1}{8}$

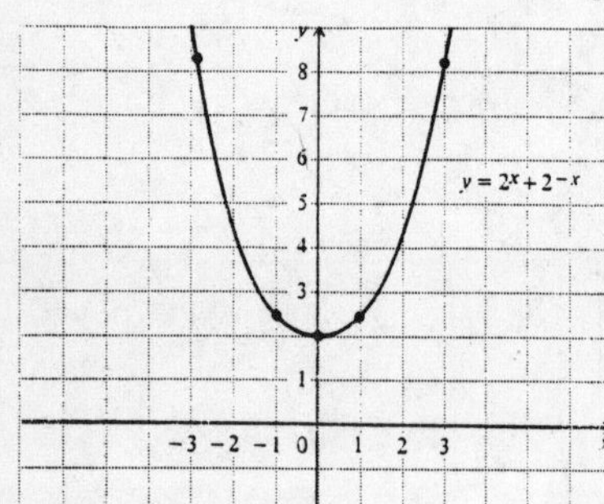

48.

$y = (\frac{1}{2})^x - 1$

49. Graph: $y = 3^x + 3^{-x}$

We construct a table of values, thinking of y as f(x). Then plot these points and connect them with a curve.

$f(0) = 3^0 + 3^{-0} = 1 + 1 = 2$

$f(1) = 3^1 + 3^{-1} = 3 + \frac{1}{3} = 3\frac{1}{3}$

$f(2) = 3^2 + 3^{-2} = 9 + \frac{1}{9} = 9\frac{1}{9}$

$f(-1) = 3^{-1} + 3^{-(-1)} = \frac{1}{3} + 3 = 3\frac{1}{3}$

$f(-2) = 3^{-2} + 3^{-(-2)} = \frac{1}{9} + 9 = 9\frac{1}{9}$

x	y, or f(x)
0	2
1	$3\frac{1}{3}$
2	$9\frac{1}{9}$
-1	$3\frac{1}{3}$
-2	$9\frac{1}{9}$

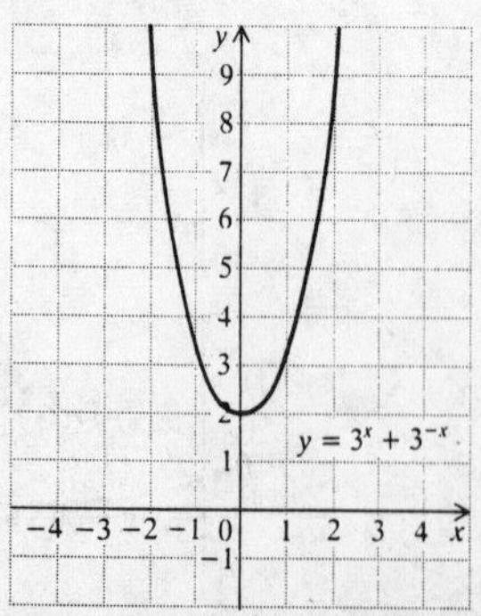

50.

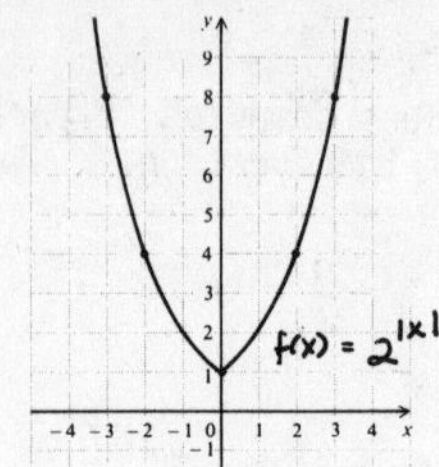

51. Graph: $y = 2^{-(x-1)}$

We construct a table of values, thinking of y as f(x). Then plot these points and connect them with a curve.

$f(0) = 2^{-(0-1)} = 2^1 = 2$

$f(1) = 2^{-(1-1)} = 2^0 = 1$

$f(2) = 2^{-(2-1)} = 2^{-1} = \frac{1}{2}$

$f(4) = 2^{-(4-1)} = 2^{-3} = \frac{1}{8}$

$f(-1) = 2^{-(-1-1)} = 2^2 = 4$

$f(-2) = 2^{-(-2-1)} = 2^3 = 8$

x	y, or f(x)
0	2
1	1
2	$\frac{1}{2}$
4	$\frac{1}{8}$
-1	4
-2	8

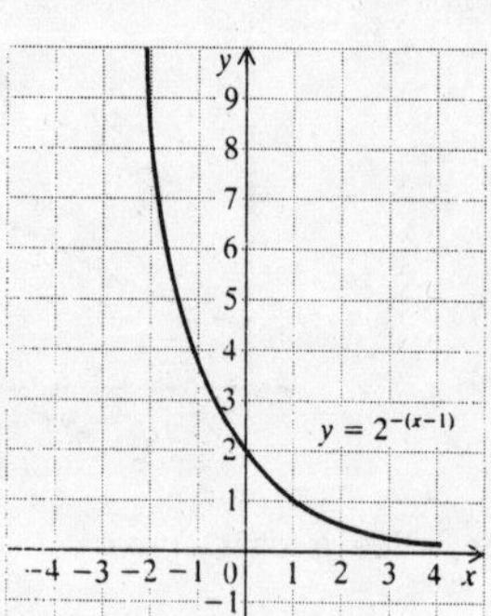

52.

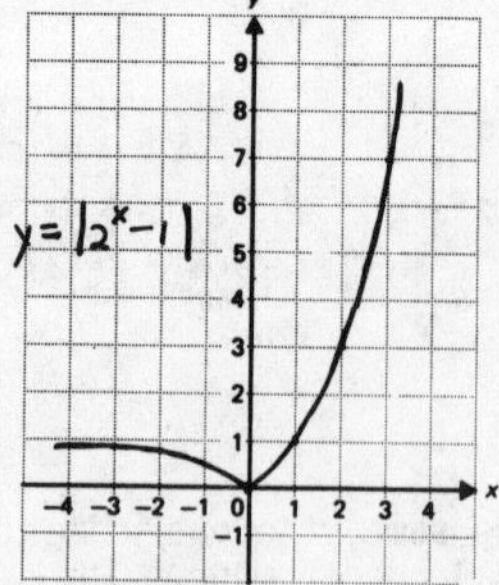

53. Graph $y = |2^x - 2|$

We construct a table of values, thinking of y as f(x). Then plot these points and connect them with a curve.

$f(0) = |2^0 - 2| = |1 - 2| = 1$

$f(1) = |2^1 - 2| = |2 - 2| = 0$

$f(2) = |2^2 - 2| = |4 - 2| = 2$

$f(3) = |2^3 - 2| = |8 - 2| = 6$

$f(-1) = |2^{-1} - 2| = \left|\frac{1}{2} - 2\right| = \frac{3}{2}$

$f(-2) = |2^{-2} - 2| = \left|\frac{1}{4} - 2\right| = \frac{7}{4}$

$f(-4) = |2^{-4} - 2| = \left|\frac{1}{16} - 2\right| = \frac{31}{16}$

x	y, or f(x)
0	1
1	0
2	2
3	6
-1	$\frac{3}{2}$
-2	$\frac{7}{4}$
-4	$\frac{31}{16}$

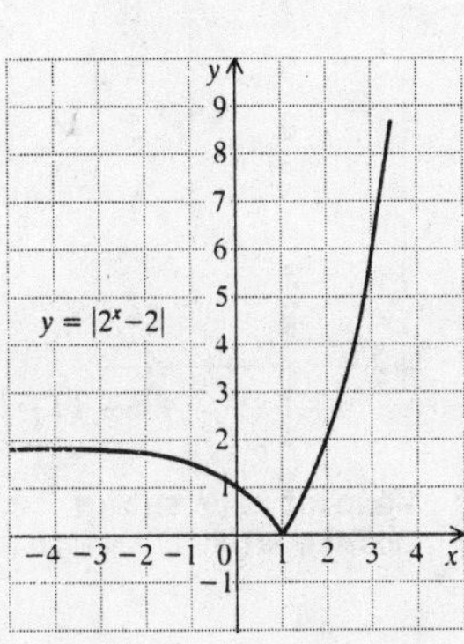

54.

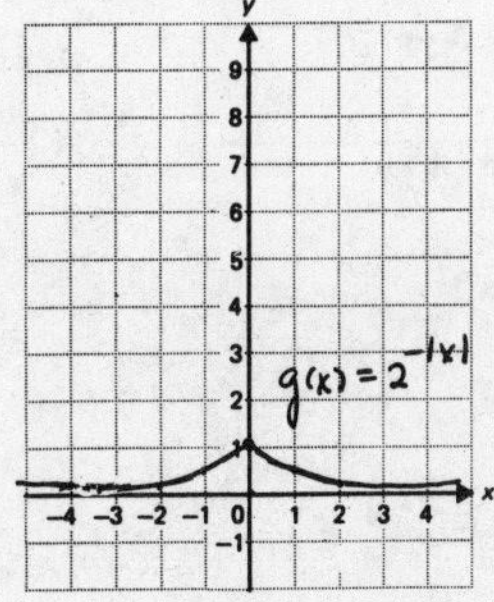

55. $y = 3^{-(x-1)}$ $x = 3^{-(y-1)}$

x	y
0	3
1	1
2	$\frac{1}{3}$
3	$\frac{1}{9}$
-1	9

x	y
3	0
1	1
$\frac{1}{3}$	2
$\frac{1}{9}$	3
9	-1

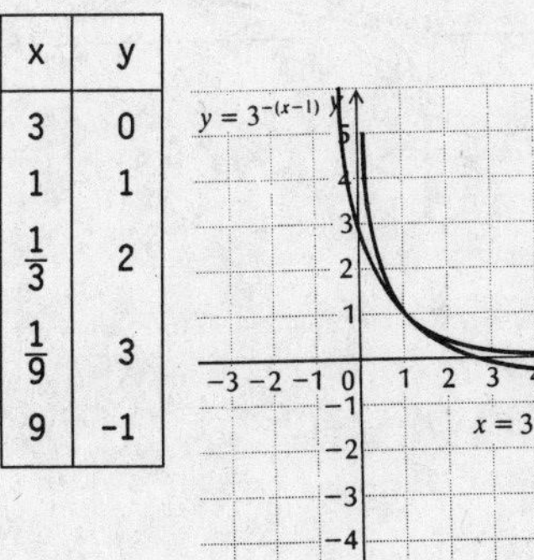

56.

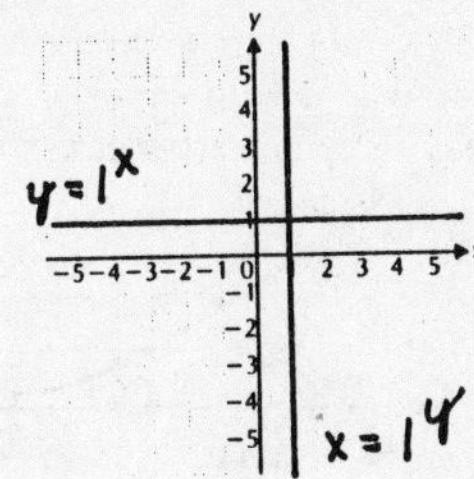

57. First we graph $y = 2^x$.

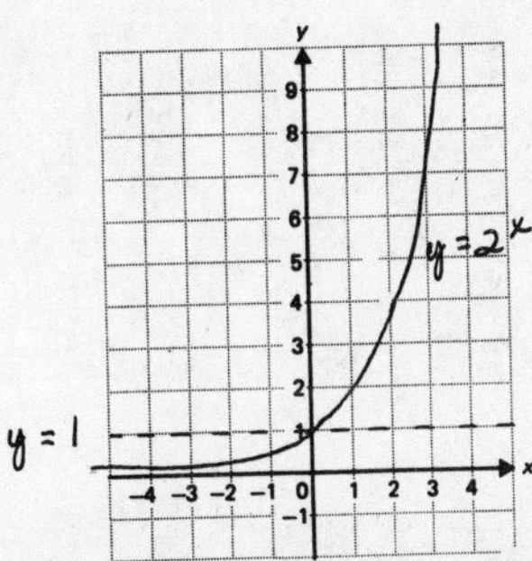

Then we study the graph to see which inputs give outputs greater than 1. (These are the inputs whose outputs are above the dashed line y = 1.) We see that all inputs greater than 0 give outputs greater than 1. The solution set is $\{x|x > 0\}$.

58. $\{x|x \leqslant 0\}$

59. a) We substitute for t.

$S(10) = 200[1 - (0.86)^{10}] \approx 155.7$ words per minute;

$S(20) = 200[1 - (0.86)^{20}] \approx 190.2$ words per minute;

$S(40) = 200[1 - (0.86)^{40}] \approx 199.5$ words per minute;

$S(100) = 200[1 - (0.86)^{100}] \approx 199.9999$ words per minute

b) Use the function values computed in part (a) to draw the graph of the function.

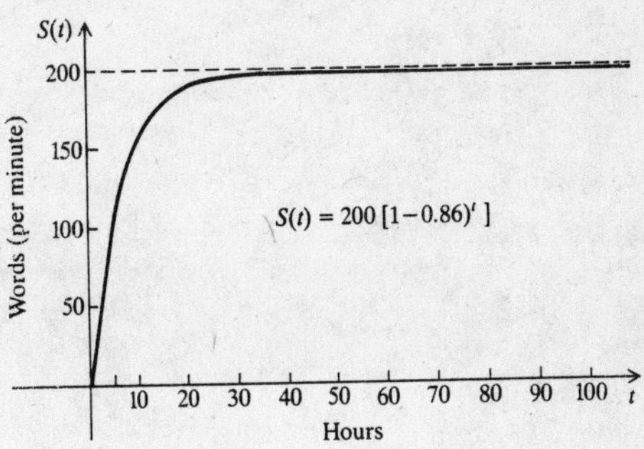

60.

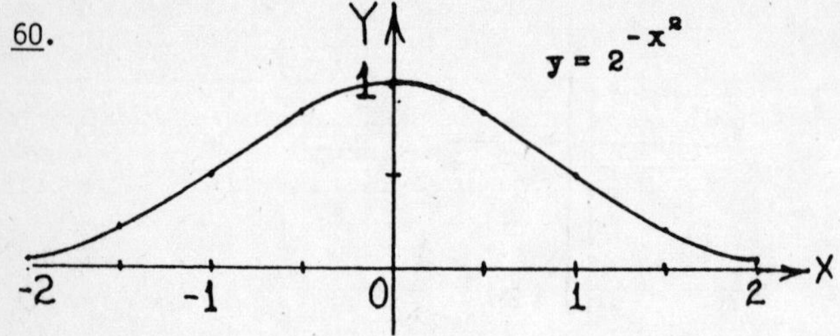

61. Graph: $y = f(x) = 3^{-(x+1)^2}$

$f(0) = 3^{-(0+1)^2} = 3^{-1} = \frac{1}{3}$

$f(1) = 3^{-(1+1)^2} = 3^{-4} = \frac{1}{81}$

$f(-1) = 3^{-(-1+1)^2} = 3^0 = 1$

$f(-2) = 3^{-(-2+1)^2} = 3^{-1} = \frac{1}{3}$

$f(-3) = 3^{-(-3+1)^2} = 3^{-4} = \frac{1}{81}$

x	y
0	$\frac{1}{3}$
1	$\frac{1}{81}$
-1	1
-2	$\frac{1}{3}$
-3	$\frac{1}{81}$

$y = 3^{-(x+1)^2}$

62. $y = |2x^2 - 8|$

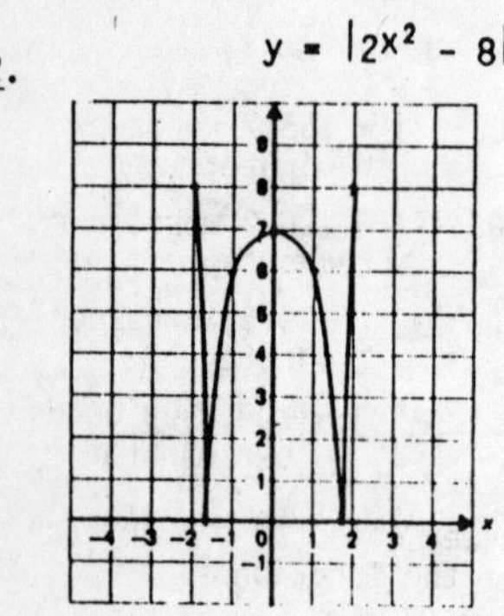

Exercise Set 5.3

1. Graph: $y = \log_3 x$

The equation $y = \log_3 x$ is equivalent to $3^y = x$. We can find ordered pairs by choosing values for y and computing the corresponding x-values.

For $y = 0$, $x = 3^0 = 1$.

For $y = 1$, $x = 3^1 = 3$.

For $y = 2$, $x = 3^2 = 9$.

For $y = -1$, $x = 3^{-1} = \frac{1}{3}$.

For $y = -2$, $x = 3^{-2} = \frac{1}{9}$.

x, or 3^y	y
1	0
3	1
9	2
$\frac{1}{3}$	-1
$\frac{1}{9}$	-2

(1) Select y.
(2) Compute x.

We plot the set of ordered pairs and connect the points with a smooth curve.

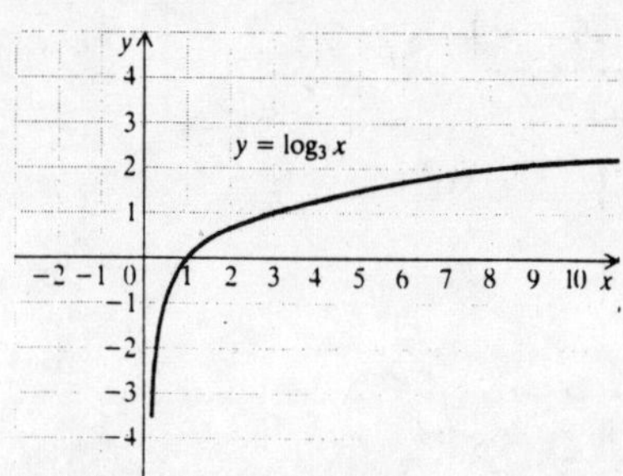

2.

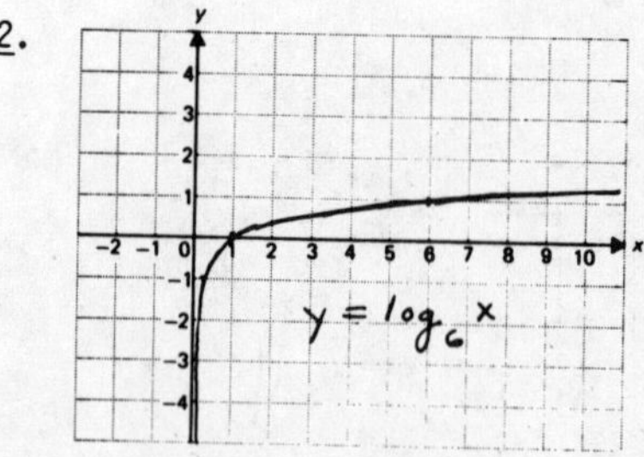

3. Graph: $y = \log_{10} x$

The equation $y = \log_{10} x$ is equivalent to $10^y = x$. We can find ordered pairs by choosing values for y and computing the corresponding x-values.

For $y = 0$, $x = 10^0 = 1$.

For $y = 1$, $x = 10^1 = 10$.

For $y = 2$, $x = 10^2 = 100$.

For $y = -1$, $x = 10^{-1} = \frac{1}{10}$.

For $y = -2$, $x = 10^{-2} = \frac{1}{100}$.

x, or 10^y	y
1	0
10	1
100	2
$\frac{1}{10}$	-1
$\frac{1}{100}$	-2

We plot the set of ordered pairs and connect the points with a smooth curve.

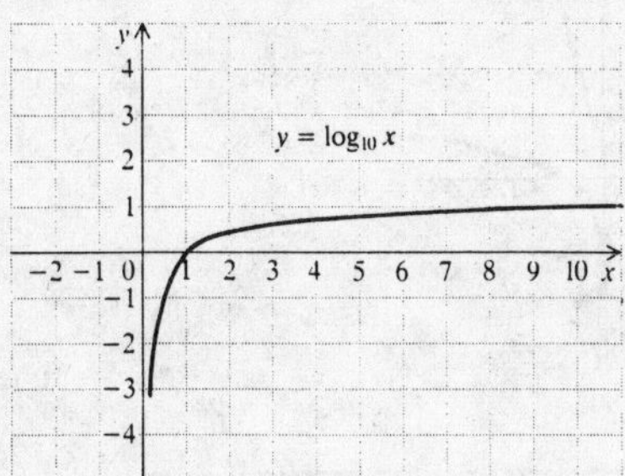

4.

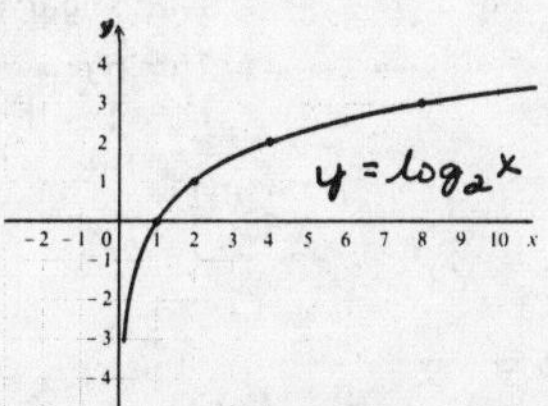

5. Graph: $f(x) = \log_4 x$

Think of f(x) as y. Then $y = \log_4 x$ is equivalent to $4^y = x$. We find ordered pairs by choosing values for x and computing the corresponding x-values. Then we plot the points and connect them with a smooth curve.

For $y = 0$, $x = 4^0 = 1$.

For $y = 1$, $x = 4^1 = 4$.

For $y = 2$, $x = 4^2 = 16$.

For $y = -1$, $x = 4^{-1} = \frac{1}{4}$.

For $y = -2$, $x = 4^{-2} = \frac{1}{16}$.

x, or 4^y	y
1	0
4	1
16	2
$\frac{1}{4}$	-1
$\frac{1}{16}$	-2

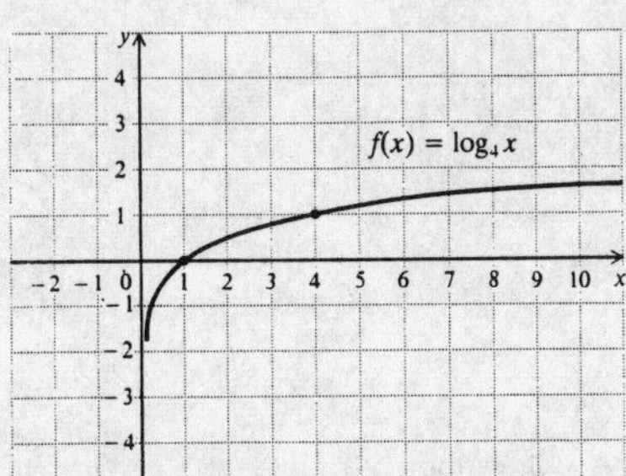

6.

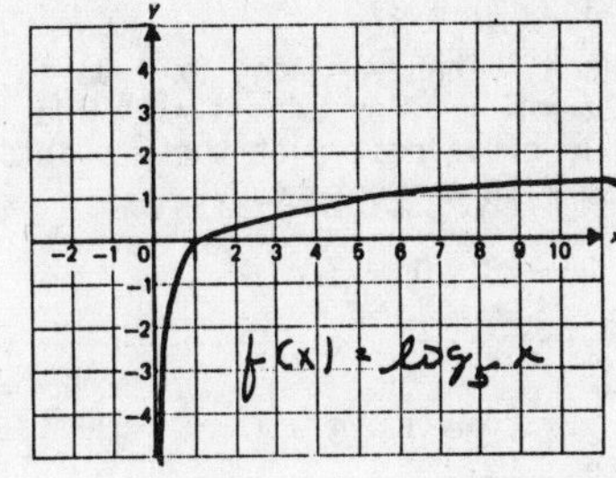

7. Graph: $f(x) = \log_{1/2} x$

Think of f(x) as y. Then $y = \log_{1/2} x$ is equivalent to $\left(\frac{1}{2}\right)^y = x$. We construct a table of values, plot these points, and connect them with a smooth curve.

For $y = 0$, $x = \left(\frac{1}{2}\right)^0 = 1$.

For $y = 1$, $x = \left(\frac{1}{2}\right)^1 = \frac{1}{2}$.

For $y = 2$, $x = \left(\frac{1}{2}\right)^2 = \frac{1}{4}$.

For $y = -1$, $x = \left(\frac{1}{2}\right)^{-1} = 2$.

For $y = -2$, $x = \left(\frac{1}{2}\right)^{-2} = 4$.

For $y = -3$, $x = \left(\frac{1}{2}\right)^{-3} = 8$.

x, or $\left(\frac{1}{2}\right)^y$	y
1	0
$\frac{1}{2}$	1
$\frac{1}{4}$	2
2	-1
4	-2
8	-3

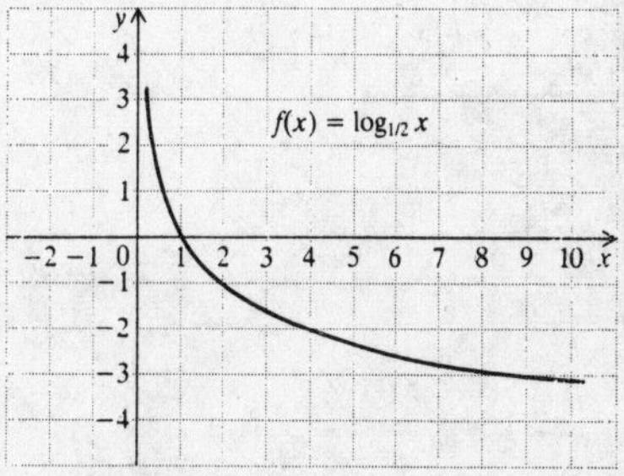

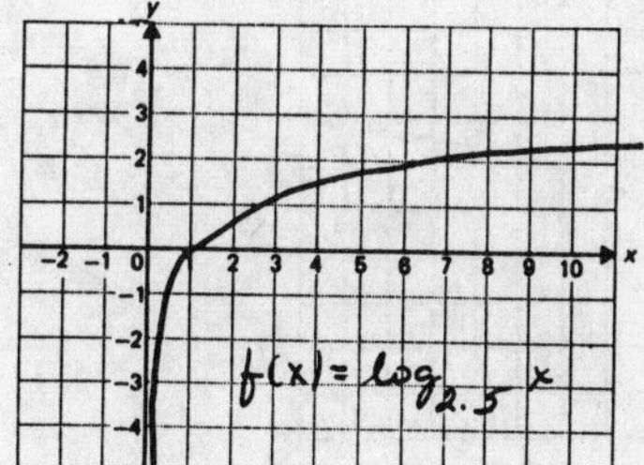

8.

f(x) = log₂.₅ x

9. Graph: $f(x) = \log_2 (x + 3)$

Think of $f(x)$ as y. Then $y = \log_2 (x + 3)$ is equivalent to $2^y = x + 3$, or $2^y - 3 = x$. We construct a table of values, plot these points, and connect them with a smooth curve.

For $y = 0$, $x = 2^0 - 3 = 1 - 3 = -2$

For $y = 1$, $x = 2^1 - 3 = 2 - 3 = -1$

For $y = 2$, $x = 2^2 - 3 = 4 - 3 = 1$

For $y = 3$, $x = 2^3 - 3 = 8 - 3 = 5$

For $y = -1$, $x = 2^{-1} - 3 = \frac{1}{2} - 3 = -\frac{5}{2}$

For $y = -2$, $x = 2^{-2} - 3 = \frac{1}{4} - 3 = -\frac{11}{4}$

x, or $2^y - 3$	y
-2	0
-1	1
1	2
5	3
$-\frac{5}{2}$	-1
$-\frac{11}{4}$	-2

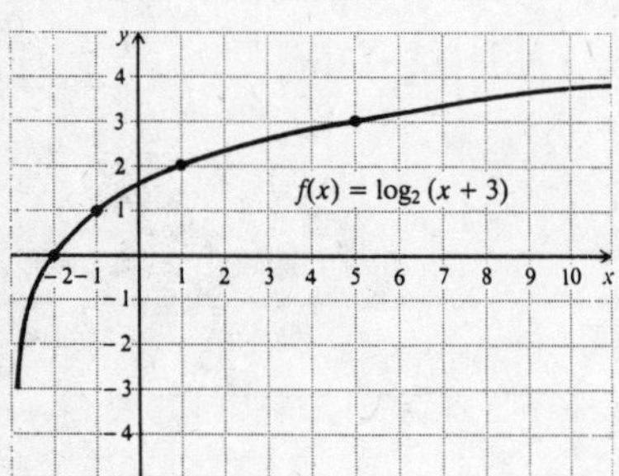

10. $y = \log_3 (x - 2)$

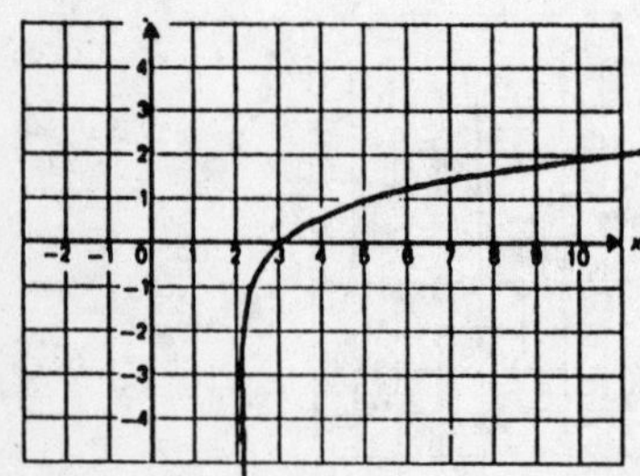

11. Graph $f(x) = 3^x$ (see Exercise Set 5.2, Exercise 2) and $f^{-1}(x) = \log_3 x$ (see Exercise 1 above) on the same set of axes.

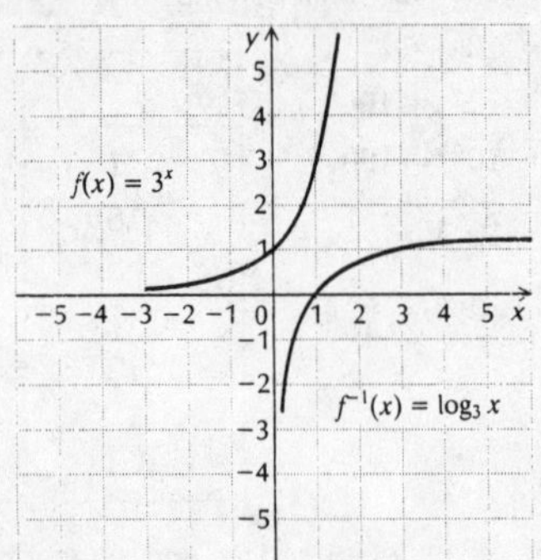

12.

13. $10^3 = 1000 \longrightarrow 3 = \log_{10} 1000$ The exponent is the logarithm. The base remains the same.

14. $2 = \log_{10} 100$

15. $5^{-3} = \frac{1}{125} \longrightarrow -3 = \log_5 \frac{1}{125}$ The exponent is the logarithm. The base remains the same.

16. $-5 = \log_4 \frac{1}{1024}$

17. $8^{1/3} = 2 \longrightarrow \frac{1}{3} = \log_8 2$

18. $\frac{1}{4} = \log_{16} 2$

19. $10^{0.3010} = 2 \longrightarrow 0.3010 = \log_{10} 2$

20. $0.4771 = \log_{10} 3$

21. $e^3 = t \longrightarrow 3 = \log_e t$

22. $k = \log_p 3$

23. $Q^t = x \longrightarrow t = \log_Q x$

24. $m = \log_p V$

25. $e^3 = 20.0855 \longrightarrow 3 = \log_e 20.0855$

26. $2 = \log_e 7.3891$

27. $e^{-1} = 0.3679 \longrightarrow -1 = \log_e 0.3679$

28. $-6 = \log_e 0.00247$

29. $t = \log_4 7 \longrightarrow 4^t = 7$ The logarithm is the exponent. The base remains the same.

30. $6^h = 29$

31. $\log_2 32 = 5 \longrightarrow 2^5 = 32$ The logarithm is the exponent. The base remains the same.

32. $5^1 = 5$

33. $\log_{10} 0.1 = -1 \longrightarrow 10^{-1} = 0.1$

34. $10^{-2} = 0.01$

35. $\log_{10} 7 = 0.845 \longrightarrow 10^{0.845} = 7$

36. $10^{0.4771} = 3$

37. $\log_e 30 = 3.4012 \longrightarrow e^{3.4012} = 30$

38. $e^{2.3026} = 10$

39. $\log_t Q = k \longrightarrow t^k = Q$

40. $m^a = P$

41. $\log_e 0.38 = -0.9676 \longrightarrow e^{-0.9676} = 0.38$

42. $e^{-0.0987} = 0.906$

43. $\log_r M = -x \longrightarrow r^{-x} = M$

44. $c^{-w} = W$

45. $\log_{10} x = 3$

$10^3 = x$ Converting to an exponential equation

$1000 = x$ Computing 10^3

46. 4

47. $\log_3 3 = x$

$3^x = 3$ Converting to an exponential equation

$3^x = 3^1$

$x = 1$ The exponents are the same.

48. -2

49. $\log_3 x = 2$

$3^2 = x$ Converting to an exponential equation

$9 = x$ Computing 3^2

50. 64

51. $\log_x 16 = 2$

$x^2 = 16$ Converting to an exponential equation

$x = 4$ or $x = -4$ Taking square roots

$\log_4 16 = 2$ because $4^2 = 16$. Thus, 4 is a solution. Since all logarithm bases must be positive, $\log_{-4} 16$ is not defined and -4 is not a solution.

52. 4

53. $\log_2 x = -1$

$2^{-1} = x$ Converting to an exponential equation

$\frac{1}{2} = x$ Simplifying

54. $\frac{1}{9}$

55. $\log_8 x = \frac{1}{3}$

$8^{1/3} = x$

$2 = x$

56. 2

57. Let $\log_{10} 1000 = x$.

Then $10^x = 1000$

$10^x = 10^3$

$x = 3.$

Thus, $\log_{10} 1000 = 3$.

58. 7

59. Let $\log_{10} 0.1 = x$.

Then $10^x = 0.1 = \frac{1}{10}$

$10^x = 10^{-1}$

$x = -1.$

Thus, $\log_{10} 0.1 = -1$.

60. -3

61. Let $\log_{10} 1 = x$.

Then $10^x = 1$

$10^x = 10^0$ $(10^0 = 1)$

$x = 0.$

Thus, $\log_{10} 1 = 0$.

62. 1

63. Let $\log_5 625 = x$.
Then $5^x = 625$
$5^x = 5^4$
$x = 4$.
Thus, $\log_5 625 = 4$.

64. 6

65. Let $\log_5 \frac{1}{25} = x$.
Then $5^x = \frac{1}{25}$
$5^x = 5^{-2}$
$x = -2$.
Thus, $\log_5 \frac{1}{25} = -2$.

66. -4

67. Let $\log_3 1 = x$.
Then $3^x = 1$
$3^x = 3^0$ ($3^0 = 1$)
$x = 0$.
Thus, $\log_3 1 = 0$.

68. 1

69. Find $\log_e 1$.
We know that $\log_e 1$ is the exponent to which e is raised to get 1. That exponent is 0. Therefore, $\log_e 1 = 0$.

70. 1

71. Find $\log_{81} 9$.
We know that $\log_{81} 9$ is the exponent to which 81 is raised to get 9. Since $9 = \sqrt{81} = 81^{1/2}$, it follows that the exponent is $\frac{1}{2}$. Therefore, $\log_{81} 9 = \frac{1}{2}$.

72. $\frac{1}{3}$

73. Find $\log_e e^5$.
We know that $\log_e e^5$ is the exponent to which e is raised to get e^5. That exponent is 5. Therefore, $\log_e e^5 = 5$.

74. -2

75. Find $\log_{10} 10^m$.
We know that $\log_{10} 10^m$ is the exponent to which 10 is raised to get 10^m. That exponent is m. Therefore, $\log_{10} 10^m = m$.

76. t

77. Graph: $y = \left(\frac{2}{3}\right)^x$

x	y, or $\left(\frac{2}{3}\right)^x$
0	1
1	$\frac{2}{3}$
2	$\frac{4}{9}$
3	$\frac{8}{27}$
-1	$\frac{3}{2}$
-2	$\frac{9}{4}$
-3	$\frac{27}{8}$

Graph: $y = \log_{2/3} x$, or $x = \left(\frac{2}{3}\right)^y$

x, or $\left(\frac{2}{3}\right)^y$	y
1	0
$\frac{2}{3}$	1
$\frac{4}{9}$	2
$\frac{8}{27}$	3
$\frac{3}{2}$	-1
$\frac{9}{4}$	-2
$\frac{27}{8}$	-3

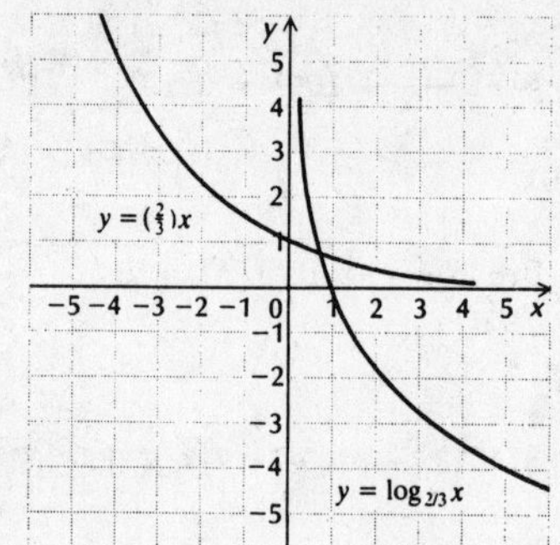

78.

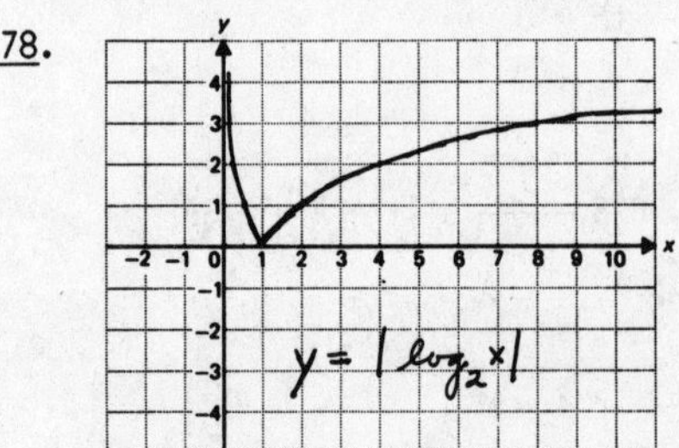

79. Graph: $y = \log_2 |x|$

x	$\pm\frac{1}{4}$	$\pm\frac{1}{2}$	± 1	± 2	± 4
y	-2	-1	0	1	2

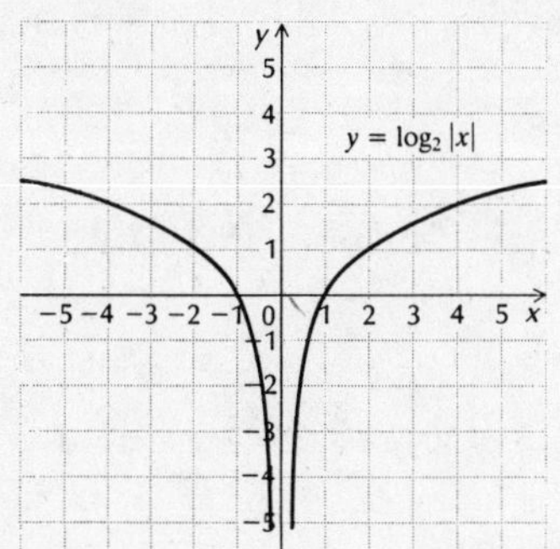

80. $27, \frac{1}{27}$

81. $\log_{125} x = \frac{2}{3}$
$$125^{2/3} = x$$
$$(5^3)^{2/3} = x$$
$$5^2 = x$$
$$25 = x$$

82. 4

83. $\log_{\sqrt{5}} x = -3$
$(\sqrt{5})^{-3} = x$, or $5^{-3/2} = x$

84. $0, \frac{1}{2}$

85. $\log_4 (3x - 2) = 2$
$$4^2 = 3x - 2$$
$$16 = 3x - 2$$
$$18 = 3x$$
$$6 = x$$

86. $\frac{25}{16}$

87. $\log_x \sqrt[5]{36} = \frac{1}{10}$
$$x^{1/10} = \sqrt[5]{36}$$
$$x^{1/10} = 36^{1/5}$$
$$(x^{1/10})^{10} = (36^{1/5})^{10}$$
$$x = 36^2$$
$$x = 1296$$

88. -25, 4

89. Let $\log_{1/4} \frac{1}{64} = x$.

Then $\left(\frac{1}{4}\right)^x = \frac{1}{64}$
$$\left(\frac{1}{4}\right)^x = \left(\frac{1}{4}\right)^3$$
$$x = 3.$$
Thus, $\log_{1/4} \frac{1}{64} = 3$.

90. 1

91. $\log_{10} (\log_4 (\log_3 81))$
$= \log_{10} (\log_4 4)$ $\quad (\log_3 81 = 4)$
$= \log_{10} 1$ $\quad (\log_4 4 = 1)$
$= 0$

92. 1

93. Let $\log_{\sqrt{3}} \frac{1}{81} = x$.

Then $(\sqrt{3})^x = \frac{1}{81}$
$$(3^{1/2})^x = \frac{1}{3^4}$$
$$3^{x/2} = 3^{-4}$$
$$\frac{x}{2} = -4$$
$$x = -8$$
Thus, $\log_{\sqrt{3}} \frac{1}{81} = -8$.

94. -2

95. $f(x) = 3^x$

3^x is defined for all real numbers. Thus the domain is the set of all real numbers.

96. $\{x|x > 0\}$

97. $f(x) = \log_a x^2$

The domain of a logarithmic function is the set of all positive real numbers. Since $x^2 > 0$ for all real numbers, except 0, the domain is $\{x|x \neq 0\}$.

98. $\{x|x > 0\}$

99. $f(x) = \log_{10} (3x - 4)$

The domain of a logarithmic function is the set of all positive real numbers. Thus, we solve
$$3x - 4 > 0$$
$$3x > 4$$
$$x > \frac{4}{3}$$
The domain is $\left\{x \middle| x > \frac{4}{3}\right\}$.

100. $\{x|x \neq 0\}$

101. $f(x) = \log_6 (x^2 - 9)$

The domain of a logarithmic function is the set of all positive real numbers. For $x^2 - 9$ to be greater than 0, x must be less than -3 or greater than 3. The domain is $\{x \mid x < -3 \text{ or } x > 3\}$.

102. $\{x \mid 0 < x < 1\}$

103. Solve $\log_2 x \geqslant 4$ by graphing. To do this we first graph $y = \log_2 x$ (or $2^y = x$).

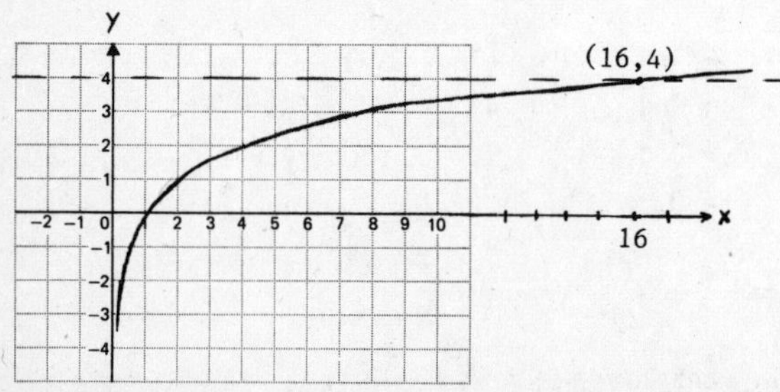

Next we study the graph to see for which inputs the outputs are greater than or equal to 4. We see that $\log_2 x \geqslant 4$ for those values of x for which $x \geqslant 16$. Thus the solution set of $\log_2 x \geqslant 4$ is $\{x \mid x \geqslant 16\}$.

104. $\{x \mid 3 < x \leqslant 35\}$

Exercise Set 5.4

1. $\log_2 (64 \cdot 8) = \log_2 64 + \log_2 8$ Property 1

2. $\log_3 81 + \log_3 27$

3. $\log_4 (32 \cdot 64) = \log_4 32 + \log_4 64$ Property 1

4. $\log_5 125 + \log_5 25$

5. $\log_c QP = \log_c Q + \log_c P$ Property 1

6. $\log_t 9Y = \log_t 9 + \log_t Y$

7. $\log_b 8 + \log_b 90 = \log_b (8 \cdot 90)$ Property 1
   $= \log_b 720$

8. $\log_a 150$

9. $\log_c P + \log_c Q = \log_c PQ$ Property 1

10. $\log_e MT$

11. $\log_a x^4 = 4 \log_a x$ Property 2

12. $3 \log_b t$

13. $\log_c y^5 = 5 \log_c y$ Property 2

14. $8 \log_{10} y$

15. $\log_b Q^{-6} = -6 \log_b Q$ Property 2

16. $-6 \log_c K$

17. $\log_a \frac{76}{13} = \log_a 76 - \log_a 13$ Property 3

18. $\log_t M - \log_t 8$

19. $\log_b \frac{5}{4} = \log_b 5 - \log_b 4$ Property 3

20. $\log_a x - \log_a y$

21. $\log_a 18 - \log_a 5 = \log_a \frac{18}{5}$ Property 3

22. $\log_b \frac{54}{6}$, or $\log_b 9$

23. $\log_a x^3y^2z = \log_a x^3 + \log_a y^2 + \log_a z$ Property 1
   $= 3 \log_a x + 2 \log_a y + \log_a z$ Property 2

24. $\log_a 6 + \log_a x + 5 \log_a y + 4 \log_a z$

25. $\log_b \frac{x^2y}{b^3} = \log_b x^2y - \log_b b^3$ Property 3
   $= \log_b x^2 + \log_b y - \log_b b^3$ Property 1
   $= \log_b x^2 + \log_b y - 3$ Property 4
   $= 2 \log_b x + \log_b y - 3$ Property 2

26. $2 \log_b p + 5 \log_b q - 4 \log_b m - 9$

27. $\log_c \sqrt[3]{\frac{x^4}{y^3z^2}}$
   $= \log_c \left(\frac{x^4}{y^3z^2}\right)^{1/3}$
   $= \frac{1}{3} \log_c \frac{x^4}{y^3z^2}$ Property 2
   $= \frac{1}{3} (\log_c x^4 - \log_c y^3z^2)$ Property 3
   $= \frac{1}{3}[\log_c x^4 - (\log_c y^3 + \log_c z^2)]$ Property 1
   $= \frac{1}{3}(\log_c x^4 - \log_c y^3 - \log_c z^2)$ Removing parentheses
   $= \frac{1}{3}(4 \log_c x - 3 \log_c y - 2 \log_c z)$ Property 2

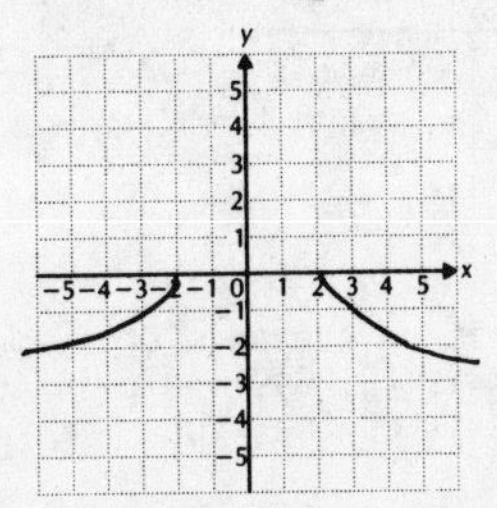

28. $\frac{1}{2}(6 \log_a x - 5 \log_a p - 8 \log_a q)$

29. $\log_a \sqrt[4]{\frac{m^8n^{12}}{a^3b^5}}$

$= \frac{1}{4}\log_a \frac{m^8n^{12}}{a^3b^5}$ Property 2

$= \frac{1}{4}(\log_a m^8n^{12} - \log_a a^3b^5)$ Property 3

$= \frac{1}{4}[\log_a m^8 + \log_a n^{12} - (\log_a a^3 + \log_a b^5)]$ Property 1

$= \frac{1}{4}(\log_a m^8 + \log_a n^{12} - \log_a a^3 - \log_a b^5)$ Removing parentheses

$= \frac{1}{4}(\log_a m^8 + \log_a n^{12} - 3 - \log_a b^5)$ Property 4

$= \frac{1}{4}(8\log_a m + 12\log_a n - 3 - 5\log_a b)$ Property 2

30. $2 + \frac{3}{2}\log_a b$

31. $\frac{2}{5}\log_a x - \frac{1}{3}\log_a y = \log_a x^{\frac{2}{5}} - \log_a y^{\frac{1}{3}}$ Property 2

$= \log_a \frac{x^{\frac{2}{5}}}{y^{\frac{1}{3}}}$ Property 3

32. $\log_a \frac{x^{\frac{1}{2}}y^4}{x^3}$, or $\log_a x^{-\frac{5}{2}}y^4$

33. $\log_a 2x + 3(\log_a x - \log_a y)$

$= \log_a 2x + 3\log_a x - 3\log_a y$

$= \log_a 2x + \log_a x^3 - \log_a y^3$ Property 2

$= \log_a 2x^4 - \log_a y^3$ Property 1

$= \log_a \frac{2x^4}{y^3}$ Property 3

34. $\log_a x$

35. $\log_a \frac{a}{\sqrt{x}} - \log_a \sqrt{ax}$

$= \log_a ax^{-1/2} - \log_a a^{1/2}x^{1/2}$

$= \log_a \frac{ax^{-1/2}}{a^{1/2}x^{1/2}}$ Property 3

$= \log_a \frac{a^{1/2}}{x}$

$= \log_a \frac{\sqrt{a}}{x}$

We can simplify this further as follows:

$\log_a \frac{\sqrt{a}}{x} = \log_a \sqrt{a} - \log_a x$ Property 3

$= \log_a a^{1/2} - \log_a x$ Writing exponential notation

$= \frac{1}{2} - \log_a x$ Property 4

36. $\log_a (x + 2)$

37. $\log_b 15 = \log_b (3\cdot 5)$

$= \log_b 3 + \log_b 5$ Property 1

$= 0.5283 + 0.7740$

$= 1.3023$

38. 0.2457

39. $\log_b \frac{3}{5} = \log_b 3 - \log_b 5$ Property 3

$= 0.5283 - 0.7740$

$= -0.2457$

40. -0.7740

41. $\log_b \frac{1}{3} = \log_b 1 - \log_b 3$ Property 3

$= 0 - 0.5283$ $(\log_b 1 = 0)$

$= -0.5283$

42. $\frac{1}{2}$

43. $\log_b \sqrt{b^3} = \log_b b^{3/2}$ Writing exponential notation

$= \frac{3}{2}$ Property 4

44. 1.7740

45. $\log_b 3b = \log_b 3 + \log_b b$ Property 1

$= 0.5283 + 1$ $(\log_b b = 1)$

$= 1.5283$

46. 1.0566

47. $\log_b 25 = \log_b 5^2$

$= 2\log_b 5$ Property 2

$= 2(0.7740) = 1.548$

48. 2.0763

49. $\log_t t^{11} = 11$ Property 4

50. 3

51. $\log_e e^{|x-4|} = |x - 4|$ Property 4

52. $\sqrt{5}$

53. $3^{\log_3 4x} = 4x$ Property 5

54. 4x - 3

55. $a^{\log_a Q} = Q$ Property 5

56. e^5

57. $a^{\log_a x} = 15$

$x = 15$ Property 5

58. 4

59. $\log_e e^x = -7$

$x = -7$ Property 4

60. 2.7

61. $\dfrac{\log_a M}{\log_a N} = \log_a M - \log_a N$, False

By Property 3,

$\log_a M - \log_a N = \log_a \frac{M}{N}$, not $\dfrac{\log_a M}{\log_a N}$.

62. False

63. $\dfrac{\log_a M}{c} = \dfrac{1}{c} \log_a M = \log_a M^{1/c}$

Thus, true by Property 2.

64. True

65. $\log_a 2x = 2 \log_a x$, False

By Property 1,

$\log_a 2x = \log_a 2 + \log_a x$, not $2 \log_a x$.

66. True

67. $\log_a (M + N) = \log_a M + \log_a N$, False

By Property 1,

$\log_a M + \log_a N = \log_a MN$, not $\log_a (M + N)$.

68. True

69. $\log_c a - \log_c b = \log_c \left[\frac{a}{b}\right]$ is true by Property 3.

70. False

71. $\log_a (x^8 - y^8) - \log_a (x^2 + y^2)$

$= \log_a \dfrac{x^8 - y^8}{x^2 + y^2}$ Property 3

$= \log_a \dfrac{(x^4 + y^4)(x^2 + y^2)(x + y)(x - y)}{x^2 + y^2}$ Factoring

$= \log_a [(x^4 + y^4)(x^2 - y^2)]$ Simplifying

$= \log_a (x^6 - x^4y^2 + x^2y^4 - y^6)$ Multiplying

72. $\log_a (x^3 + y^3)$

73. $\log_a \sqrt{4 - x^2}$

$= \log_a (4 - x^2)^{1/2}$

$= \frac{1}{2} \log_a (4 - x^2)$

$= \frac{1}{2} \log_a [(2 + x)(2 - x)]$

$= \frac{1}{2} [\log_a (2 + x) + \log_a (2 - x)]$

74. $\frac{1}{2} \log_a (x - y) - \frac{1}{2} \log_a (x + y)$

75. $\log_a \dfrac{\sqrt[3]{x^2z}}{\sqrt[3]{y^2z^{-2}}}$

$= \log_a \left[\dfrac{x^2z^3}{y^2}\right]^{1/3}$

$= \frac{1}{3} (\log_a x^2z^3 - \log_a y^2)$

$= \frac{1}{3} (2 \log_a x + 3 \log_a z - 2 \log_a y)$

$= \frac{1}{3} [2\cdot 2 + 3\cdot 4 - 2\cdot 3]$ Substituting

$= \frac{1}{3}(10)$

$= \frac{10}{3}$

76. 0, 5

77. $\log_a 3x = \log_a 3 + \log_a x$

$\log_a 3x = \log_a 3x$ Property 1

We get an equation that is true for all values of x for which the logarithm function is defined. Thus, the solution is $\{x|x > 0\}$, or $(0,\infty)$.

78. $\frac{1}{2}$

79. $3^{\log_3 (8x - 4)} = 5$

$8x - 4 = 5$ Property 5

$8x = 9$

$x = \frac{9}{8}$

80. $\sqrt{7}$

81. $8^{2 \log_8 x + \log_8 x} = 27$

$8^{3 \log_8 x} = 27$

$8^{\log_8 x^3} = 27$

$x^3 = 27$ Property 5

$x = 3$

82. $\{x|x > 0\}$

83. $\log_b \dfrac{5}{x + 2} = \log_b 5 - \log_b (x + 2)$

$\log_b \dfrac{5}{x + 2} = \log_b \dfrac{5}{x + 2}$ Property 3

The equation is true for all values of x for which the logarithm function is defined. That is, for all x for which $\dfrac{5}{x + 2} > 0$, or

$x + 2 > 0$

$x > -2$.

The solution set is $\{x|x > -2\}$, or $(-2, \infty)$.

84. -2

85. $\log_a x = 2$ Given

$a^2 = x$ Definition

Let $\log_{1/a} x = n$ and solve for n.

$\log_{1/a} a^2 = n$ Substituting a^2 for x

$\left(\frac{1}{a}\right)^n = a^2$

$(a^{-1})^n = a^2$

$a^{-n} = a^2$

$-n = 2$

$n = -2$

Thus, $\log_{1/a} x = -2$ when $\log_a x = 2$.

86. $\log_a \frac{1}{x} = \log 1 - \log_a x = -\log_a x$

87. $\log_a \frac{x + \sqrt{x^2 - 5}}{5} \cdot \frac{x - \sqrt{x^2 - 5}}{x - \sqrt{x^2 - 5}}$

$= \log_a \frac{5}{5(x - \sqrt{x^2 - 5})} = \log_a \frac{1}{x - \sqrt{x^2 - 5}}$

$= \log_a 1 - \log_a (x - \sqrt{x^2 - 5})$

$= -\log_a (x - \sqrt{x^2 - 5})$.

88. Let $M = \log_a \left(\frac{1}{x}\right)$ or $\frac{1}{x} = a^M$ or $x = \left(\frac{1}{a}\right)^M$

or $\log_{\frac{1}{a}} x = M$.

89. See the answer section in the text.

Exercise Set 5.5

1. 0.4771
2. 0.8451
3. 0.9031
4. 1.1139
5. 0.3692
6. 0.4871
7. 1.8129
8. 1.9243
9. 1.7952
10. 1.0334
11. 2.7259
12. 2.2923
13. 4.1271
14. 4.9689
15. -0.2441
16. -0.1612
17. -1.2840
18. -0.4123
19. -2.0084
20. -3.2905
21. 1000
22. 100,000
23. 501.1872
24. 6.3096
25. 3.0001
26. 1.1623
27. 0.2841
28. 0.4567
29. 0.0011
30. 79,104.2833
31. 5.7498
32. -11.3349
33. Does not exist
34. Does not exist
35. 1.0088
36. 1.0394
37. 1.0986
38. 0.6931
39. 2.0794
40. 2.5649
41. 4.4067
42. 3.9120
43. 9.0318

44. 6.6962

45. -5.1328

46. -7.9020

47. 36.7890

48. 138.5457

49. 0.0023

50. 0.1002

51. 1.0057

52. 1.0112

53. 5.8346×10^{14}

54. 2.0917×10^{24}

55. 66.5569

56. 28.3395

57. -30.6182

58. -116.3164

59. 1637.9488

60. 547.7396

61. 2.8044

62. 0.0017

63. 0.0000027

64. 54,515,738.62

65. Find $\log_4 100$. We will use common logarithms in the change-of-base formula:

$$\log_b M = \frac{\log_a M}{\log_a b}$$

Substitute 10 for a, 4 for b, and 100 for M.

$$\log_4 100 = \frac{\log_{10} 100}{\log_{10} 4}$$

$$\approx \frac{2.0000}{0.6021} \quad \text{(Using a calculator)}$$

$$\approx 3.3219$$

66. 2.7268

67. Find $\log_2 12$. We will use common logarithms in the change-of-base formula:

$$\log_b M = \frac{\log_a M}{\log_a b}$$

Substitute 10 for a, 2 for b, and 12 for M.

$$\log_2 12 = \frac{\log_{10} 12}{\log_{10} 2}$$

$$\approx \frac{1.0792}{0.3010} \quad \text{(Using a calculator)}$$

$$\approx 3.5850$$

68. 2.2920

69. Find $\log_{100} 0.3$. We will use common logarithms in the change-of-base formula:

$$\log_b M = \frac{\log_a M}{\log_a b}$$

Substitute 10 for a, 100 for b, and 0.3 for M.

$$\log_{100} 0.3 = \frac{\log_{10} 0.3}{\log_{10} 100}$$

$$\approx \frac{-0.52288}{2} \quad \text{(Using a calculator)}$$

$$= -0.26144$$

70. 0.7384

71. Find $\log_{0.5} 7$. We will use common logarithms in the change-of-base formula:

$$\log_b M = \frac{\log_a M}{\log_a b}$$

Substitute 10 for a, 0.5 for b and 7 for M.

$$\log_{0.5} 7 = \frac{\log_{10} 7}{\log_{10} 0.5}$$

$$\approx \frac{0.8451}{-0.3010} \quad \text{(Using a calculator)}$$

$$\approx -2.8074$$

72. -0.3010

73. Find $\log_3 0.3$. We will use common logarithms in the change-of-base formula:

$$\log_b M = \frac{\log_a M}{\log_b b}$$

Substitute 10 for a, 3 for b, and 0.3 for M.

$$\log_3 0.3 = \frac{\log_{10} 0.3}{\log_{10} 3}$$

$$\approx \frac{-0.5229}{0.4771} \quad \text{(Using a calculator)}$$

$$\approx -1.0959$$

74. -4.0589

75. Find $\log_\pi 100$. We will use common logarithms in the change-of-base formula:

$$\log_b M = \frac{\log_a M}{\log_a b}$$

Substitute 10 for a, π for b, and 100 for M.

$$\log_\pi 100 = \frac{\log_{10} 100}{\log_{10} \pi}$$

$$\approx \frac{2.0000}{0.4971} \quad \text{(Using a calculator)}$$

$$\approx 4.0229$$

76. 2.8119

77. See the answer section in the text.

78. $\log x = \frac{\ln x}{\ln 10} = \frac{1}{\ln 10} \cdot \ln x = 0.4343 \ln x$

79. Graph $y = f(x) = 10^x$

We use a calculator to find function values.

x	y, or f(x)
-1	0.1
0	1
0.5	3.1623
1	10
1.5	31.6228
2	100
2.5	316.2278
3	1000
3.1	1258.9254

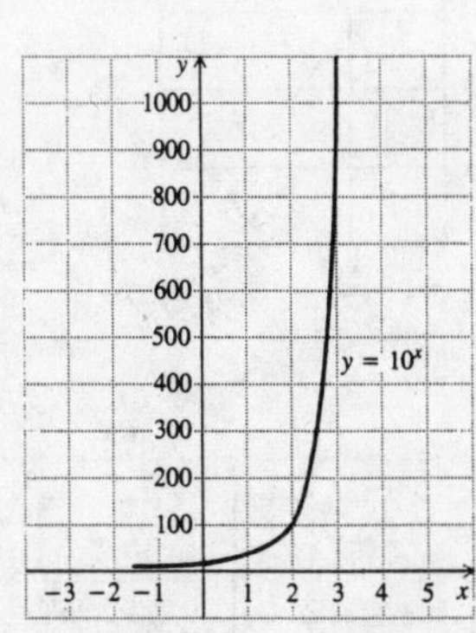

80.

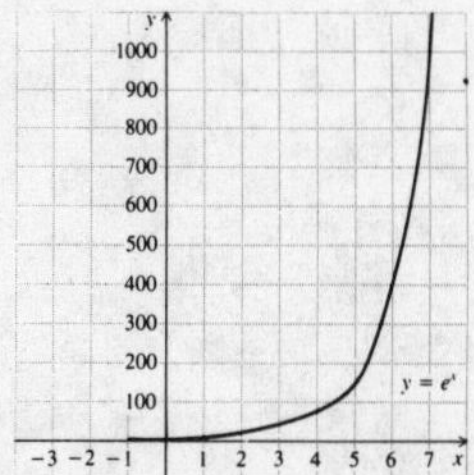

81. Graph $y = f^{-1}(x) = \log x$.

We can interchange the first and second coordinates in each pair found in Exercise 79 to obtain ordered pairs on the graph of $y = \log x$.

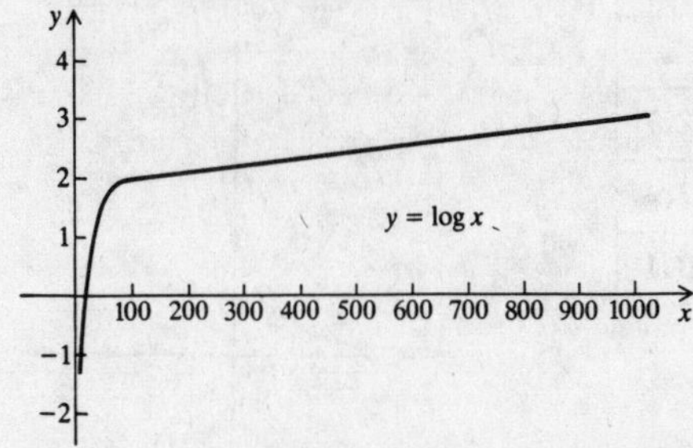

82. $y = \ln x$

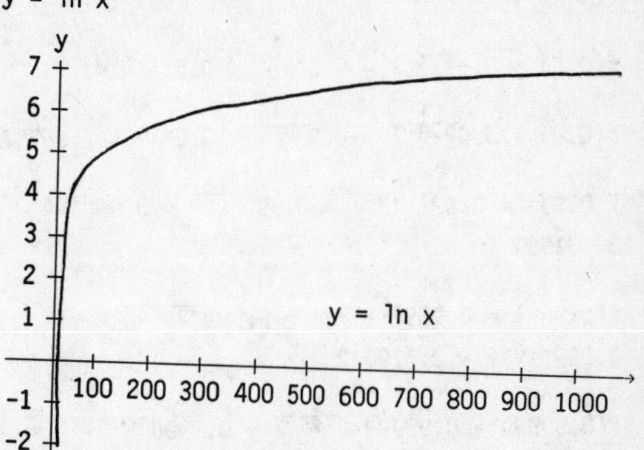

83. $\frac{\log_5 8}{\log_5 2} = \frac{\log_5 2^3}{\log_5 2} = \frac{3 \log_5 2}{\log_5 2} = 3$

84. $\frac{3}{2}$

85.

$$\log 872x = 5.3442$$

$$\log 872 + \log x = 5.3442$$

$$2.940516485 + \log x \approx 5.3442$$

$$\log x \approx 2.403683515$$

$$x = 253.3282 \quad \text{Finding the antilogarithm}$$

86. 1554.4107

87.

$$\log 784 + \log x = \log 2322$$

$$2.894316063 + \log x \approx 3.365862215$$

$$\log x \approx 0.471546152$$

$$x \approx 2.9617 \quad \text{Finding the antilogarithm}$$

88. 1.1799

89. Let a = 3, b = 10, M = e. Substitute in the change-of-base formula.

$$\log e = \frac{\ln e}{\ln 10} = \frac{1}{\ln 10}$$

90. Let a = e, b = 10, M = M. Substitute in the change-of-base formula.

$$\log M = \frac{\ln M}{\ln 10}$$

91. Let a = M, b = b, M = M. Substitute in the change-of-base formula.

$$\log_b M = \frac{\log_M M}{\log_M b} = \frac{1}{\log_M b}$$

92. Let a = 10, b = 3, M = M. Substitute in the change-of-base formula.

$$\ln M = \frac{\log M}{\log e}$$

93. See the answer section in the text.

94. 2, 2.25, 2.48832, 2.593742, 2.704814, 2.716924

95. $f(t) = t^{\frac{1}{t-1}}$

$f(0.5) = 0.5^{\frac{1}{0.5-1}} = 0.5^{\frac{1}{-0.5}} = 0.5^{-2} = 4$

$f(0.9) = 0.9^{\frac{1}{0.9-1}} = 0.9^{\frac{1}{-0.1}} = 0.9^{-10} \approx 2.867972$

$f(0.99) = 0.99^{\frac{1}{0.99-1}} = 0.99^{\frac{1}{-0.01}} = 0.99^{-100} \approx$ 2.731999

$f(0.999) = 0.999^{\frac{1}{0.999-1}} = 0.999^{\frac{1}{-0.001}} =$ $0.999^{-1000} \approx 2.719642$

$f(0.9999) = 0.9999^{\frac{1}{0.9999-1}} = 0.9999^{\frac{1}{-0.0001}} =$ $0.9999^{-10,000} \approx 2.718418$

96. e^{π}

97. $e^{\sqrt{\pi}} \approx 5.8853$, $\sqrt{e^{\pi}} \approx 4.8105$

$e^{\sqrt{\pi}}$ is larger.

98. Use a calculator to find antilogarithms in both base 10 and base e and compare with the value of x. Study graphs if possible.

Exercise Set 5.6

1. Graph $f(x) = 2e^x$.

Use a calculator to find approximate values of $2e^x$, and use these values to draw the graph.

x	$2e^x$
-2	0.3
-1	0.7
0	2
1	5.4
2	14.8

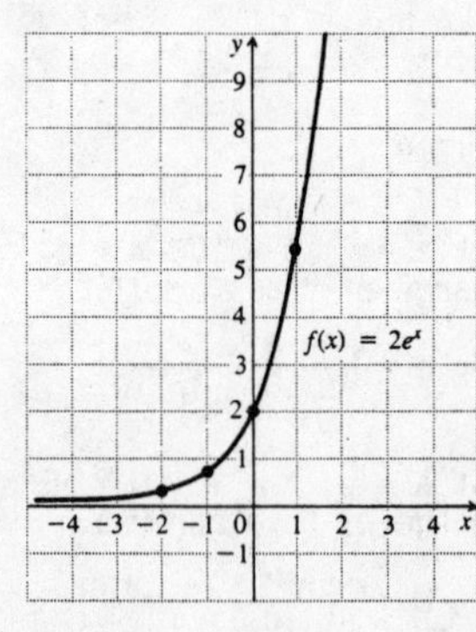

2.

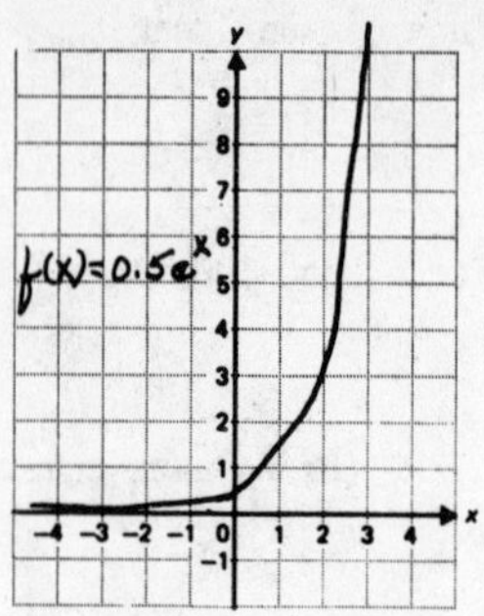

3. Graph $f(x) = e^{(1/2)x}$.

Use a calculator to find approximate values of $e^{(1/2)x}$, and use these values to draw the graph.

x	$e^{(1/2)x}$
-2	0.4
-1	0.6
0	1
1	1.6
2	2.7

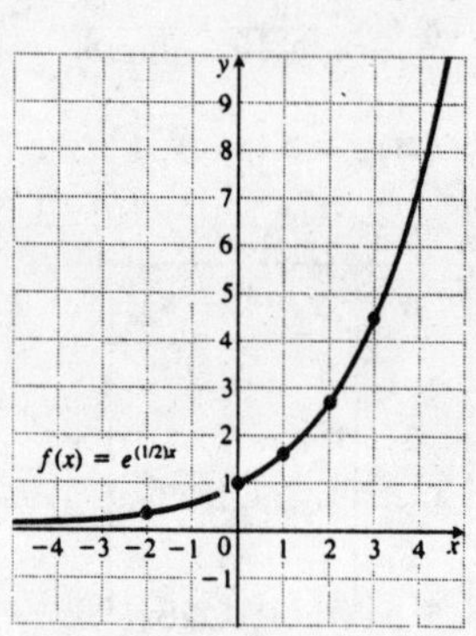

4.

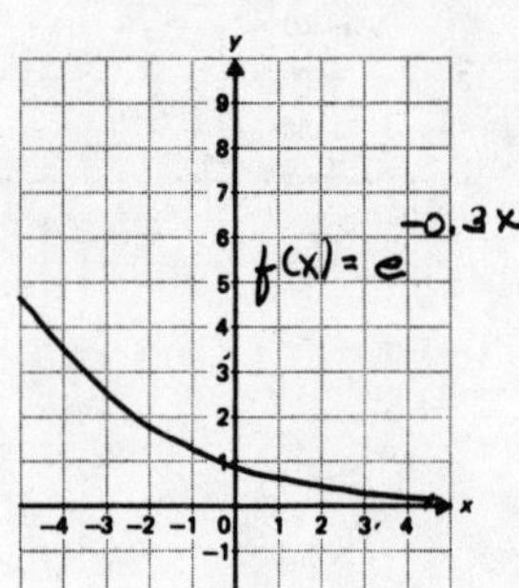

5. Graph $f(x) = e^{x+1}$.

Use a calculator to find approximate values of e^{x+1}, and use these values to draw the graph.

x	e^{x+1}
-2	0.4
-1	1
0	2.7
1	7.4
2	20.1

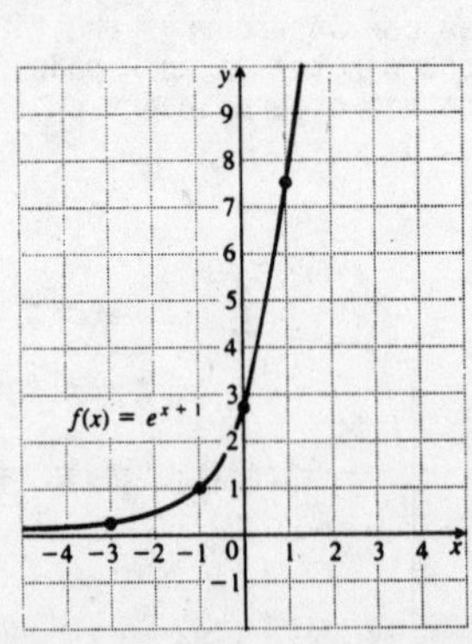

6.

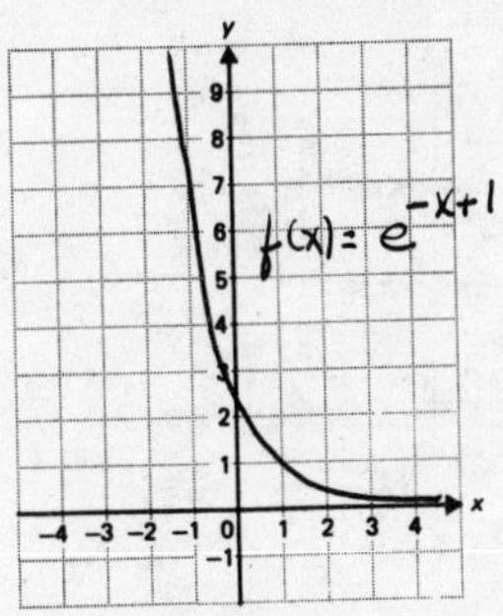

7. Graph $f(x) = e^{2x} + 1$.

Use a calculator to find approximate values of $e^{2x} + 1$, and use these values to draw the graph.

x	$e^{2x} + 1$
-2	1.0 +
-1	1.1
0	2
1	8.4
2	55.6

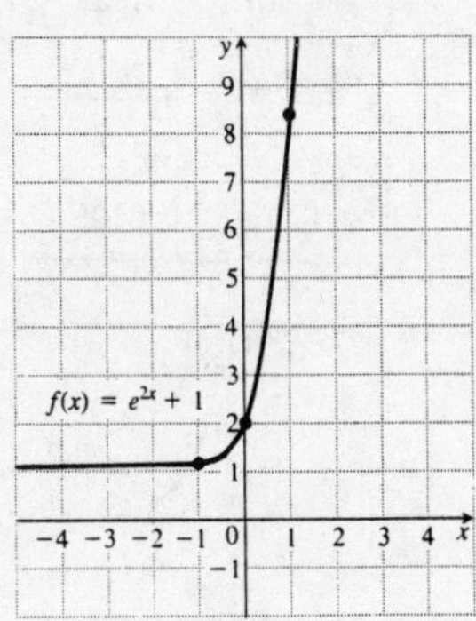

8.

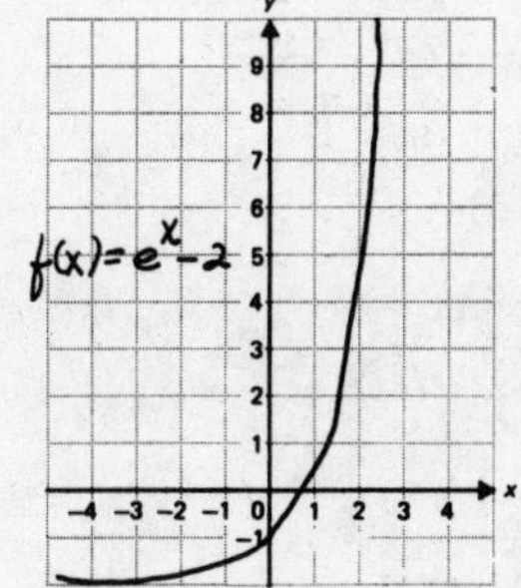

9. Graph $f(x) = 1 - e^{-0.01x}$, for nonnegative values of x. Use a calculator to find approximate values of $1 - e^{-0.01x}$ for nonnegative values of x, and use these values to draw the graph.

x	$1 - e^{-0.01x}$
0	0
100	0.63
200	0.86
300	0.95
400	0.98

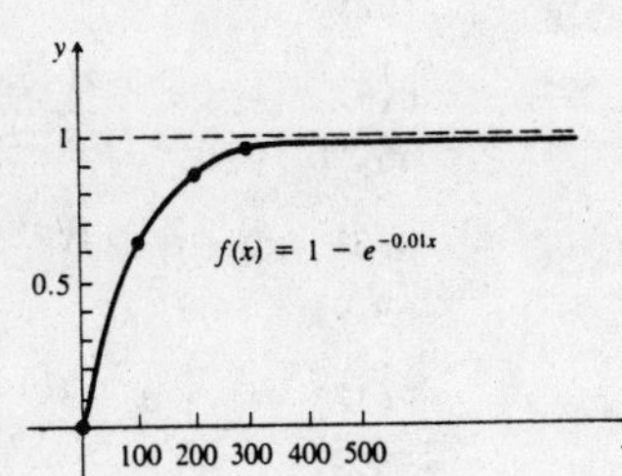

10.

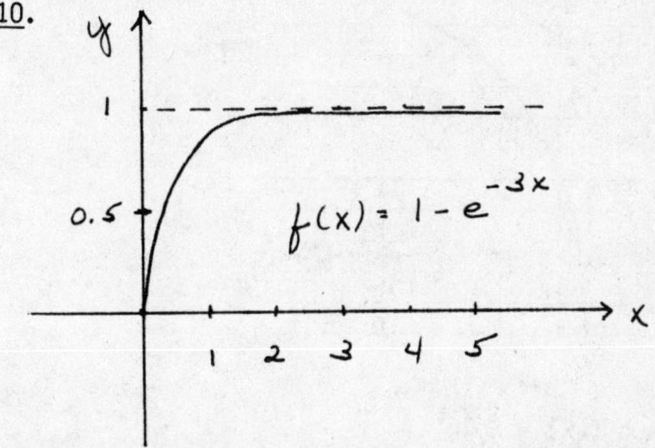

11. Graph $f(x) = 2(1 - e^{-x})$, for nonnegative values of x. Use a calculator to find approximate values of $2(1 - e^{-x})$ for nonnegative values of x, and use these values to draw the grpah.

x	$2(1 - e^{-x})$
0	0
1	1.26
2	1.73
3	1.90
4	1.96

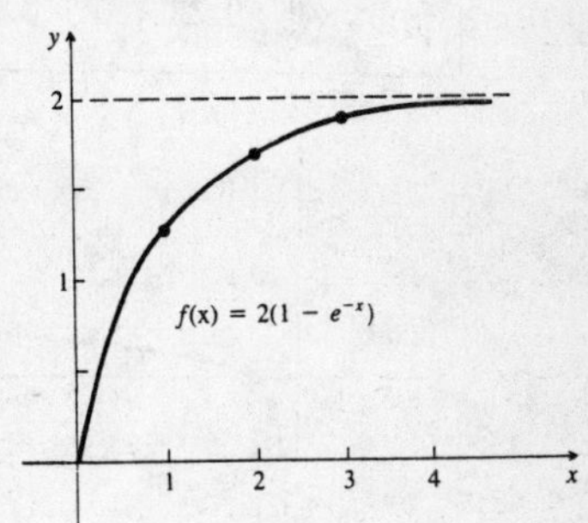

12.

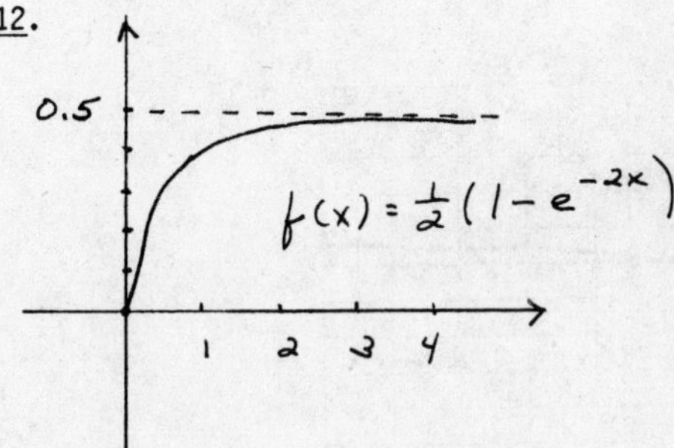

13. Graph $f(x) = 4 \ln x$.

Find some solutions with a calculator, plot them, and draw the graph.

x	0.25	0.5	1	1.5	2	3
4 ln x	-5.55	-2.77	0	1.62	2.77	4.39

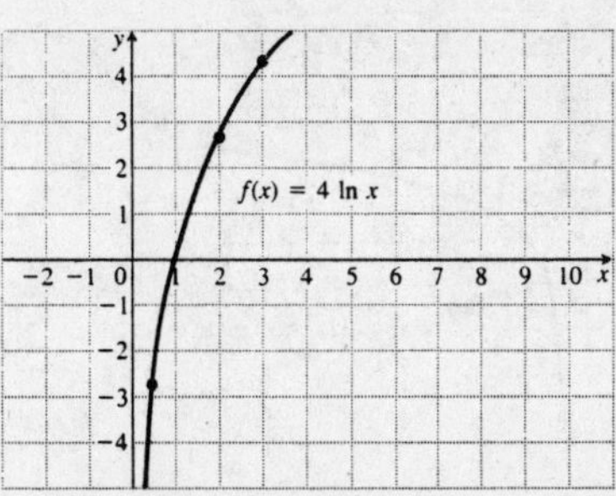

14.

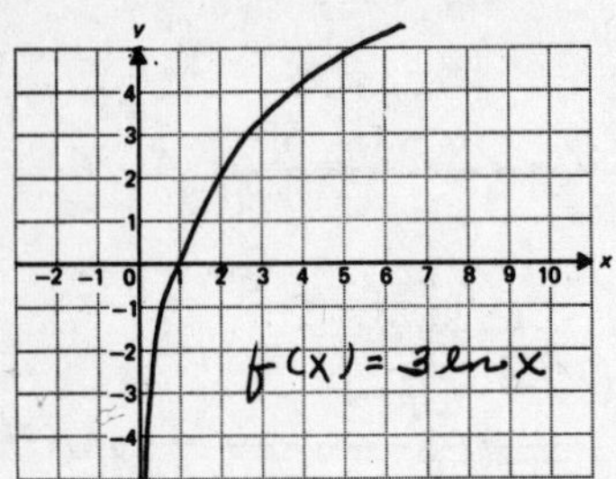

15. Graph $f(x) = \frac{1}{2} \ln x$.

Find some solutions with a calculator, plot them, and draw the graph.

x	0.5	1	2	4	7	8
$\frac{1}{2} \ln x$	-0.35	0	0.35	0.69	0.97	1.04

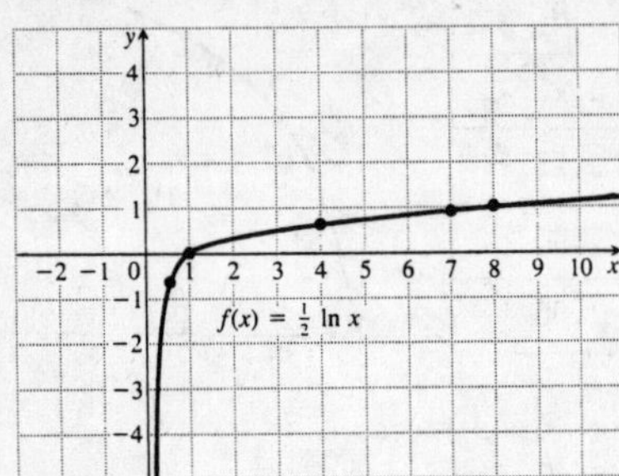

16.

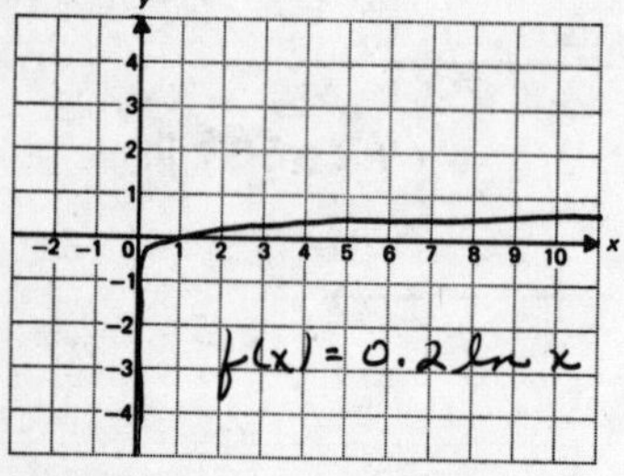

17. Graph $f(x) = \ln (x - 2)$.

Find some solutions using a calculator, plot them, and draw the graph. When x = 3, $f(x) = \ln (3 - 2) = \ln 1 = 0$, and so on.

x	2.5	3	4	6	8	10
ln (x-2)	-0.69	0	0.69	1.39	1.79	2.08

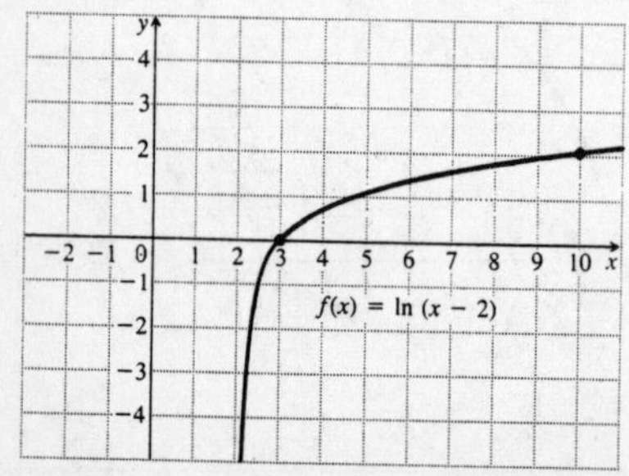

18.

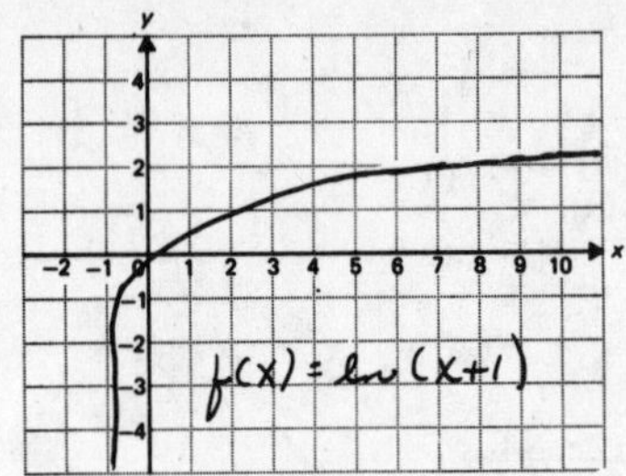

19. Graph $f(x) = 2 - \ln x$.

Find some solutions using a calculator, plot them, and draw the graph.

x	0.25	0.5	1	2	4	7	8	10
2 - ln x	3.39	2.69	2	1.31	0.61	0.05	-0.08	-0.30

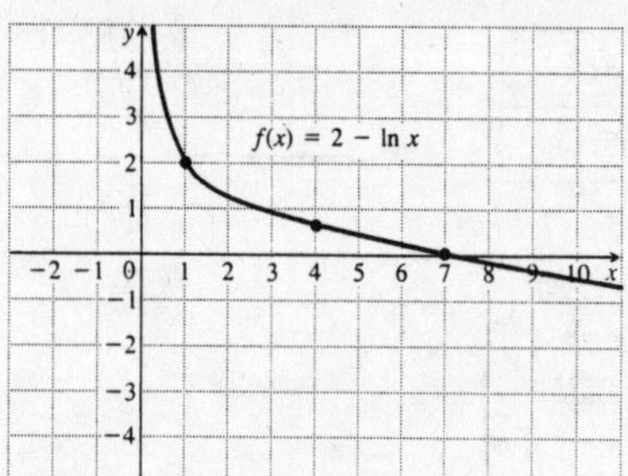

20.

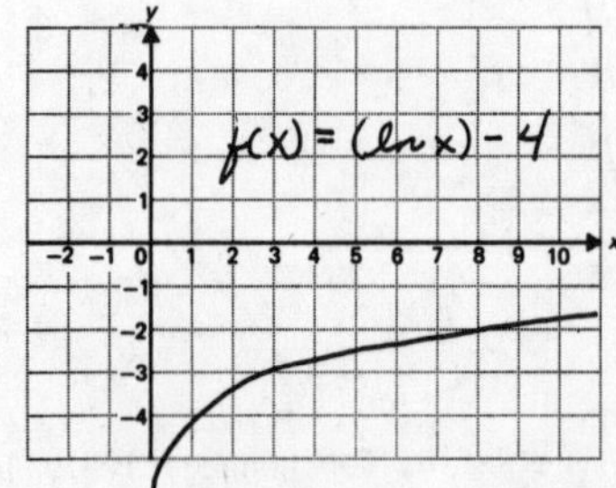

21. $P(t) = 1 - e^{-0.2t}$

a) $P(1) = 1 - e^{-0.2(1)} = 1 - e^{-0.2} \approx 0.181$, or 18.1%

$P(4) = 1 - e^{-0.2(4)} = 1 - e^{-0.8} \approx 0.551$, or 55.1%

$P(6) = 1 - e^{-0.2(6)} = 1 - e^{-1.2} \approx 0.699$, or 69.9%

$P(12) = 1 - e^{-0.2(12)} = 1 - e^{-2.4} \approx 0.909$, or 90.9%

21. (continued)

b) Plot the points (0, 0%), (1, 18.1%), (4, 55.1%), (6, 69.9%), and (12, 90.9%) and draw the graph.

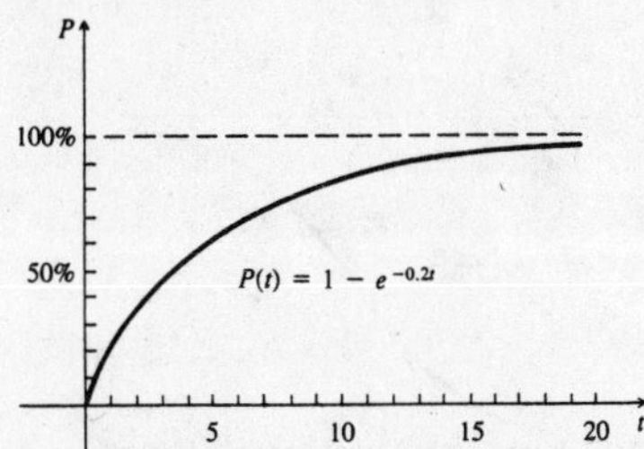

22. a) 26, 78, 91, 95

b)

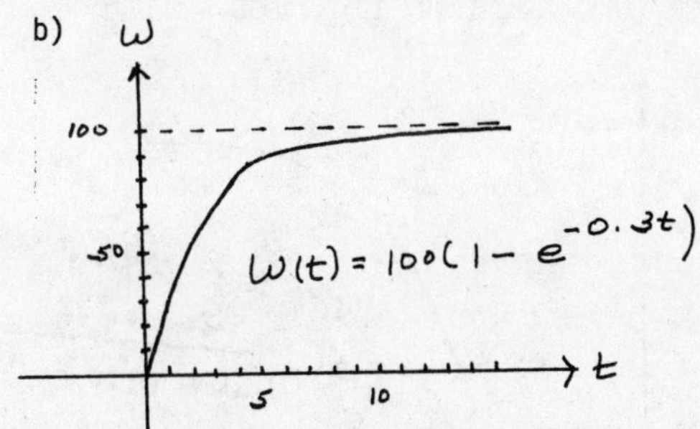

23. $V(t) = \$58(1 - e^{-1.1t}) + \$20$

a) $V(1) = \$58(1 - e^{-1.1(1)}) + \$20$

$= \$58(1 - e^{-1.1}) + \$20$

$\approx \$58.69$

$V(2) = \$58(1 - e^{-1.1(2)}) + \$20$

$= \$58(1 - e^{-2.2}) + \$20$

$\approx \$71.57$

$V(4) = \$58(1 - e^{-1.1(4)}) + \$20$

$= \$58(1 - e^{-4.4}) + \$20$

$\approx \$77.29$

$V(6) = \$58(1 - e^{-1.1(6)}) + \$20$

$= \$58(1 - e^{-6.6}) + \$20$

$\approx \$77.92$

$V(12) = \$58(1 - e^{-1.1(12)}) + \$20$

$= \$58(1 - e^{-13.2}) + \$20$

$\approx \$77.99 +$

b) Plot the points (0, \$20), (1, \$58.69), (2, \$71.57), (4, \$77.29), (6, \$77.92), and (12, \$77.99 +) and draw the graph.

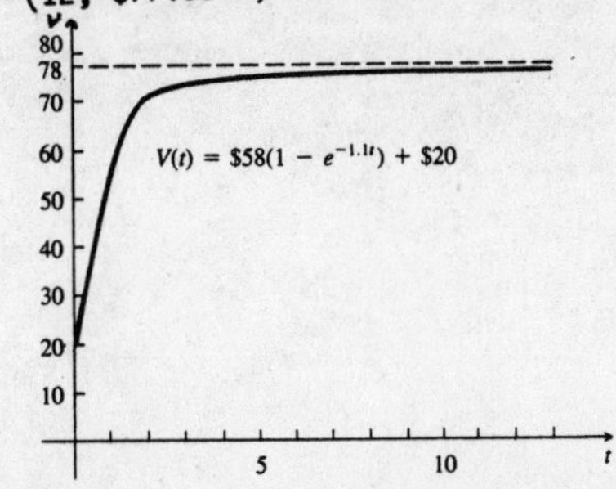

24. a) 25.9%, 59.3%, 83.5%, 97.3%

b)

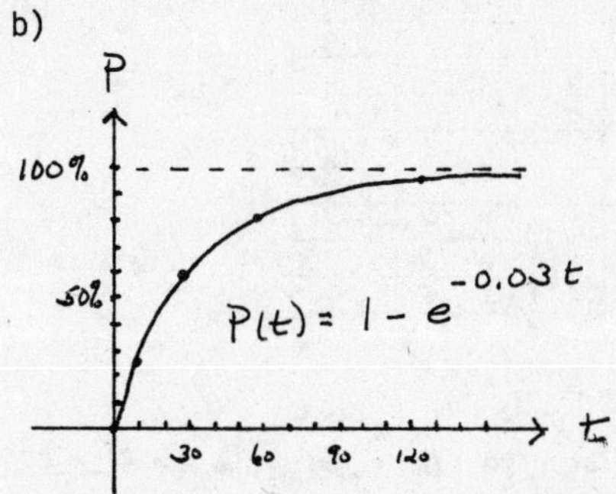

25. Graph $g(x) = e^{|x|}$.

Use a calculator to find some values of $e^{|x|}$, and use these values to draw the graph.

x	$e^{\vert x\vert}$
-3	20.09
-2	7.39
-1	2.72
0	1
1	2.72
2	7.39

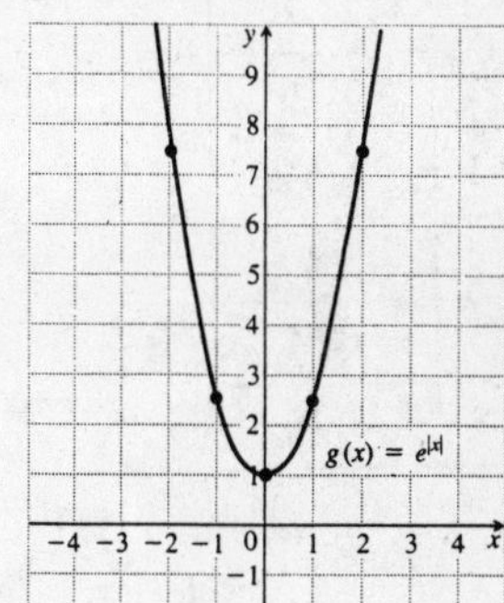

26.

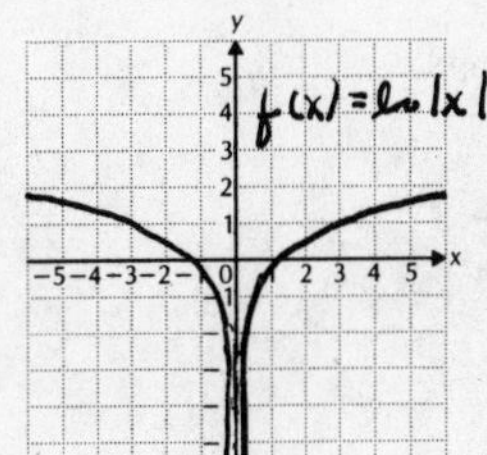

27. Graph $f(x) = |\ln x|$.

Find some solutions using a calculator, plot them, and draw the graph.

x	0.25	0.5	1	3	6	10
$\vert\ln x\vert$	1.39	0.69	0	1.10	1.79	2.30

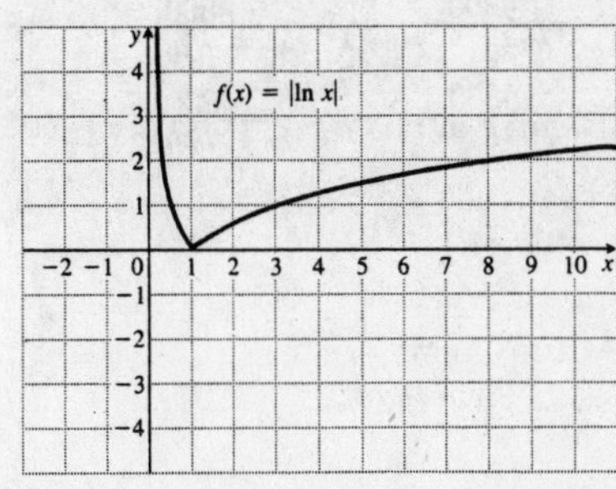

28.

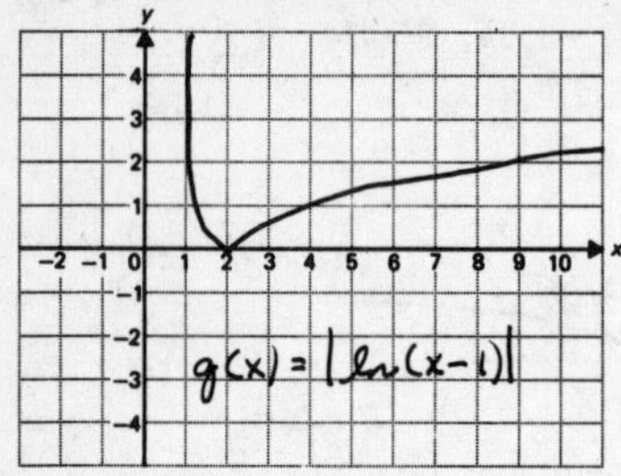

29. Graph $f(x) = \dfrac{e^x + e^{-x}}{2}$

Use a calculator to find some values of $\dfrac{e^x + e^{-x}}{2}$, and use these values to draw the graph.

x	$\dfrac{e^x + e^{-x}}{2}$
-3	13.76
-2	3.76
-1	1.54
0	1
1	1.54
2	3.76

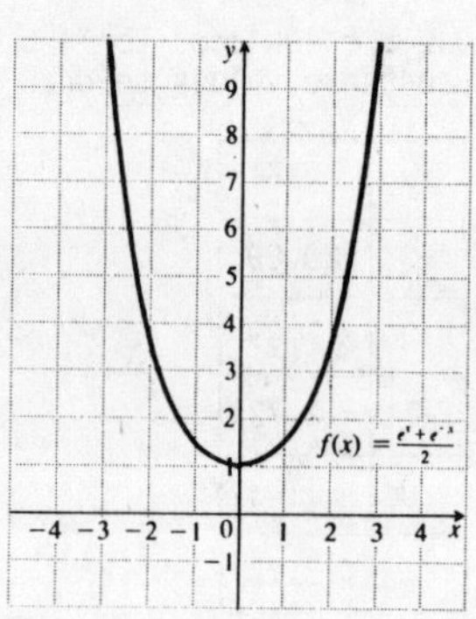

30.

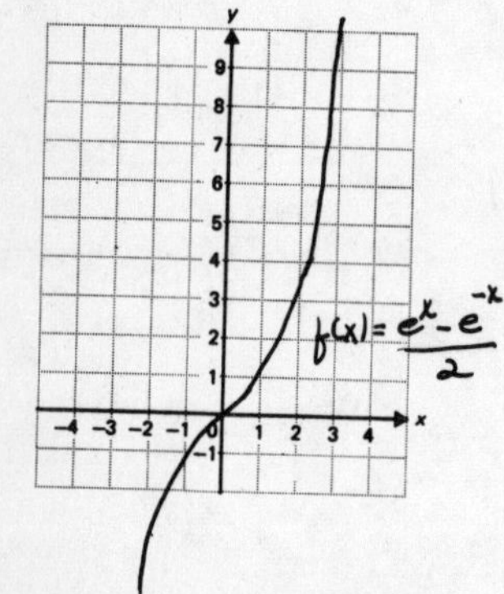

31. $N(t) = \dfrac{4800}{6 + 794e^{-0.4t}}$

a) $N(3) = \dfrac{4800}{6 + 794e^{-0.4(3)}} = \dfrac{4800}{6 + 794e^{-1.2}} \approx 20$

$N(5) = \dfrac{4800}{6 + 794e^{-0.4(5)}} = \dfrac{4800}{6 + 794e^{-2}} \approx 42$

$N(10) = \dfrac{4800}{6 + 794e^{-0.4(10)}} = \dfrac{4800}{6 + 794e^{-4}} \approx 234$

$N(15) = \dfrac{4800}{6 + 794e^{-0.4(15)}} = \dfrac{4800}{6 + 794e^{-6}} \approx 602$

31. (continued)

b) Plot the points (0, 6), (3, 20), (5, 42), (10, 234), and (15, 602) and draw the graph.

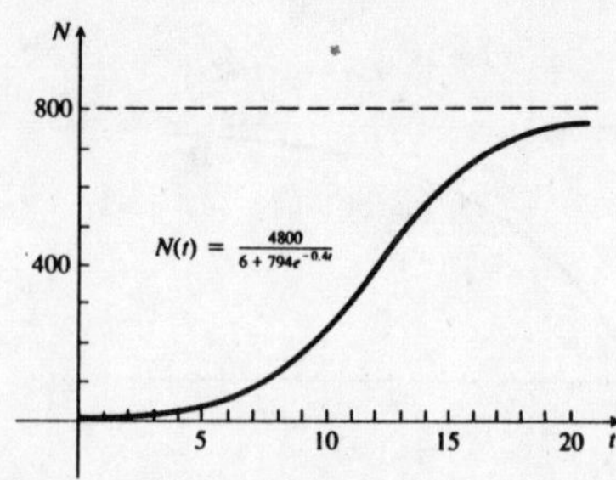

32. a) 33, 183, 758, 1339, 1952, 1998

b)

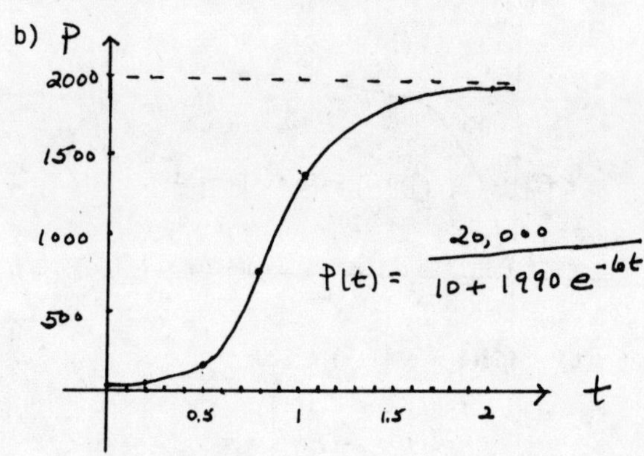

33. a)

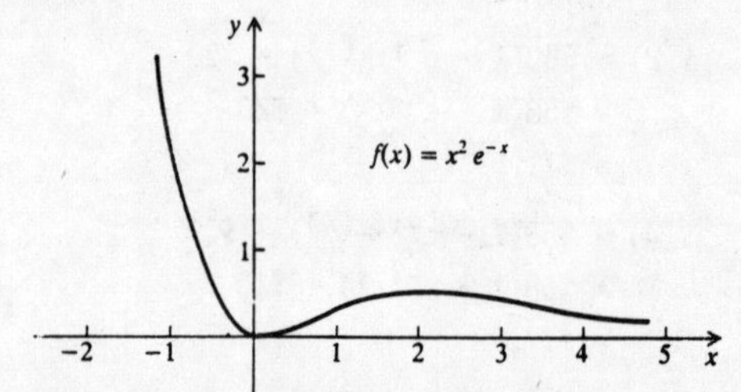

b) From the graph we see that 0 is the only zero of the function.

c) The function has no maximum value. The minimum value is 0.

34. a)

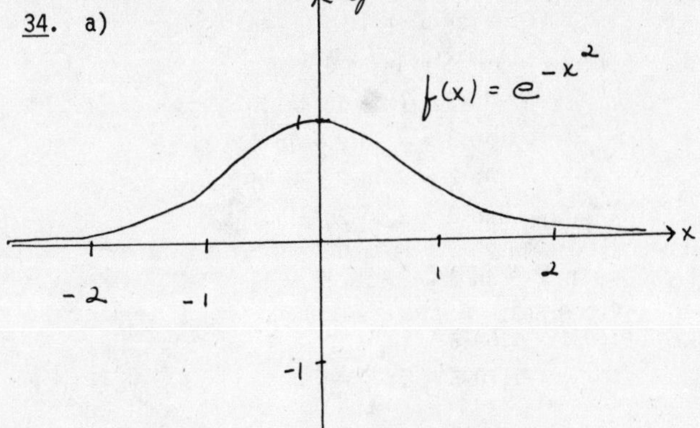

b) None

c) Maximum: 1; no minimum

35. a)

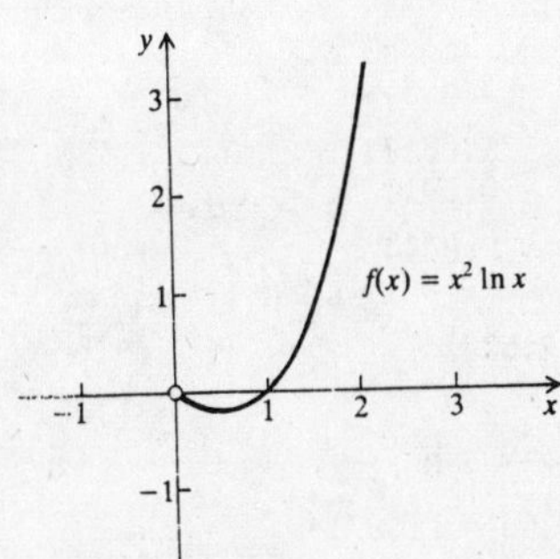

b) From the graph we see that 1 is the only zero of the function.

c) The function has no maximum value. The minimum value is about -0.2.

36. a)

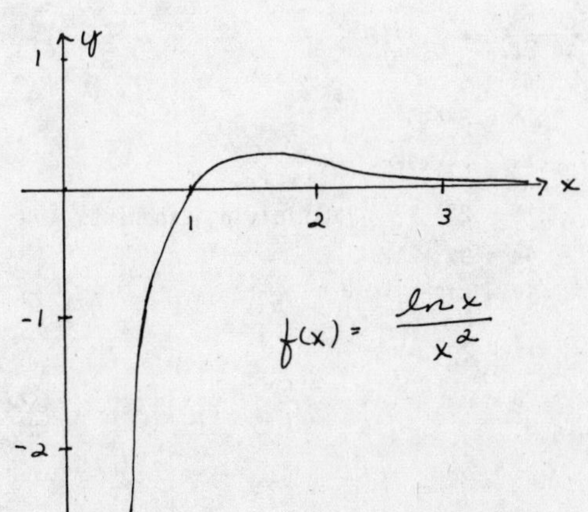

b) 1

c) Maximum: about 0.18; no minimum

Exercise Set 5.7

1. $2^x = 32$
$2^x = 2^5$
$x = 5$ The exponents are the same.

2. 4

3. $3^x = 81$
$3^x = 3^4$
$x = 4$ The exponents are the same.

4. 4

5. $2^{2x} = 8$
$2^{2x} = 2^3$
$2x = 3$ The exponents are the same.
$x = \frac{3}{2}$

6. $\frac{1}{2}$

7. $3^{7x} = 27$
$3^{7x} = 3^3$
$7x = 3$
$x = \frac{3}{7}$

8. 1

9. $2^x = 33$
$\log 2^x = \log 33$ Taking the common logarithm on both sides
$x \log 2 = \log 33$ Property 2
$x = \frac{\log 33}{\log 2}$
$x \approx \frac{1.5185}{0.3010}$
$x \approx 5.0444$

10. $\frac{\log 20}{\log 2} \approx 4.3219$

11. $2^x = 40$
$\log 2^x = \log 40$
$x \log 2 = \log 40$
$x = \frac{\log 40}{\log 2}$
$x \approx \frac{1.6021}{0.3010}$
$x \approx 5.3219$

12. $\frac{\log 19}{\log 2} \approx 4.2479$

13. $5^{4x-7} = 125$

$5^{4x-7} = 5^3$

$4x - 7 = 3$

$4x = 10$

$x = \frac{10}{4} = \frac{5}{2}$

14. -1

15. $3^{x^2+4x} = \frac{1}{27}$

$3^{x^2+4x} = 3^{-3}$

$x^2 + 4x = -3$

$x^2 + 4x + 3 = 0$

$(x + 3)(x + 1) = 0$

$x = -3$ or $x = -1$

16. $\frac{1}{2}$, -3

17. $84^x = 70$

$\log 84^x = \log 70$

$x \log 84 = \log 70$

$x = \frac{\log 70}{\log 84}$

$x \approx \frac{1.8451}{1.9243}$

$x \approx 0.9589$

18. $\frac{1}{\log 28} \approx 0.6910$

19. $e^t = 1000$

$\ln e^t = \ln 1000$

$t = \ln 1000$ Property 4

$t \approx 6.9078$

20. $\ln 100 \approx 4.6052$

21. $e^{-t} = 0.3$

$\ln e^{-t} = \ln 0.3$

$-t = \ln 0.3$

$t = -\ln 0.3$

$t \approx 1.2040$

22. $-\ln 0.04 \approx 3.2189$

23. $e^{-0.03t} = 0.08$

$\ln e^{-0.03t} = \ln 0.08$

$-0.03t = \ln 0.08$

$t = \frac{\ln 0.08}{0.03}$

$t \approx \frac{-2.5257}{-0.03}$

$t \approx 84.1910$

24. $\frac{\ln 4}{0.09} \approx 15.4033$

25. $3^x = 2^{x-1}$

$\ln 3^x = \ln 2^{x-1}$

$x \ln 3 = (x - 1) \ln 2$

$x \ln 3 = x \ln 2 - \ln 2$

$\ln 2 = x \ln 2 - x \ln 3$

$\ln 2 = x(\ln 2 - \ln 3)$

$\frac{\ln 2}{\ln 2 - \ln 3} = x$

$\frac{0.6931}{0.6931 - 1.0986} \approx x$

$-1.7095 \approx x$

26. $\frac{2 \log 5 + \log 4}{\log 4 - \log 5}$, or $\frac{2}{\log 4 - \log 5} \approx -20.6377$

27. $(3.9)^x = 48$

$\log (3.9)^x = \log 48$

$x \log 3.9 = \log 48$

$x = \frac{\log 48}{\log 3.9}$

$x \approx \frac{1.6812}{0.5911}$

$x \approx 2.8444$

28. $\frac{2}{\log 5.6} \approx 2.6731$

29. $250 - (1.87)^x = 0$

$250 = (1.87)^x$

$\log 250 = \log (1.87)^x$

$\log 250 = x \log 1.87$

$\frac{\log 250}{\log 1.87} = x$

$\frac{2.3979}{0.2718} \approx x$

$8.8211 \approx x$

30. $\frac{\log 4805}{\log 21.3} \approx 2.7716$

31. $4^{2x} = 8^{3x-4}$

$(2^2)^{2x} = (2^3)^{3x-4}$

$2^{4x} = 2^{9x-12}$ Multiplying exponents

$4x = 9x - 12$

$-5x = -12$

$x = \frac{12}{5}$

32. -16

33. $\frac{e^x - e^{-x}}{t} = 5$

$e^x - e^{-x} = 5t$ Multiplying by t

$e^x - \frac{1}{e^x} = 5t$ Rewriting e^{-x} with a positive exponent

$e^{2x} - 1 = 5te^x$ Multiplying by e^x

$(e^x)^2 - 5t \cdot e^x - 1 = 0$

This equation is reducible to quadratic with $u = e^x$. We use the quadratic formula with a = 1, b = -5t and c = -1.

$e^x = \frac{-(-5t) \pm \sqrt{(-5t)^2 - 4 \cdot 1 \cdot (-1)}}{2 \cdot 1}$

$= \frac{5t \pm \sqrt{25t^2 + 4}}{2}$

$\ln e^x = \ln \left(\frac{5t \pm \sqrt{25t^2 + 4}}{2}\right)$ Taking the natural logarithm on both sides

$x = \ln \left(\frac{5t \pm \sqrt{25t^2 + 4}}{2}\right)$ Property 4

34. $\ln \left(\frac{5 \pm \sqrt{21}}{2}\right) \approx \pm 1.5668$

35. $\frac{e^x + e^{-x}}{e^x - e^{-x}} = t$

$e^x + e^{-x} = te^x - te^{-x}$ Multiplying by $e^x - e^{-x}$

$e^x + \frac{1}{e^x} = te^x - \frac{t}{e^x}$

$e^{2x} + 1 = te^{2x} - t$ Multiplying by e^x

$t + 1 = te^{2x} - e^{2x}$

$t + 1 = e^{2x}(t - 1)$

$\frac{t + 1}{t - 1} = e^{2x}$

$\ln \left(\frac{t + 1}{t - 1}\right) = \ln e^{2x}$

$\ln \left(\frac{t + 1}{t - 1}\right) = 2x$

$\frac{1}{2} \ln \left(\frac{t + 1}{t - 1}\right) = x$

36. $\frac{1}{2} \log_5 \left(\frac{t + 1}{1 - t}\right)$

37. $\log_5 x = 4$

$x = 5^4$ Writing an equivalent exponential equation

$x = 625$

38. 4

39. $\log_5 x = -3$

$x = 5^{-3}$ Writing an equivalent exponential equation

$x = \frac{1}{125}$

40. 5

41. $\log x = 2$ The base is 10.

$x = 10^2$

$x = 100$

42. 10

43. $\log x = -3$ The base is 10.

$x = 10^{-3}$

$x = \frac{1}{1000}$, or 0.001

44. 10^{-4}, or 0.0001

45. $\ln x = 1$

$x = e^1 = e$

46. e^3

47. $\ln x = -2$

$x = e^{-2}$

48. e^{-1}

49. $\log_5 (8 - 7x) = 3$

$5^3 = 8 - 7x$ Writing an equivalent exponential equation

$125 = 8 - 7x$

$117 = -7x$

$-\frac{117}{7} = x$

The answer checks. The solution is $-\frac{117}{7}$.

50. $\frac{22}{3}$

51. $\log x + \log (x - 9) = 1$ The base is 10.

$\log_{10} [x(x - 9)] = 1$ Property 1

$x(x - 9) = 10^1$

$x^2 - 9x = 10$

$x^2 - 9x - 10 = 0$

$(x - 10)(x + 1) = 0$

$x = 10$ or $x = -1$

Check: For 10:

$\log x + \log (x - 9) = 1$	
$\log 10 + \log (10 - 9)$	1
$\log 10 + \log 1$	
$1 + 0$	
1	

For -1:

$\log x + \log (x - 9) = 1$	
$\log (-1) + \log (-1 - 9)$	1

The number -1 does not check, because negative numbers do not have logarithms. The solution is 10.

52. 1

53. $\log x - \log (x + 3) = -1$ The base is 10.

$\log_{10} \frac{x}{x + 3} = -1$ Property 3

$\frac{x}{x + 3} = 10^{-1}$

$\frac{x}{x + 3} = \frac{1}{10}$

$10x = x + 3$

$9x = 3$

$x = \frac{1}{3}$

The answer checks. The solution is $\frac{1}{3}$.

54. 1

55. $\log_2 (x + 1) + \log_2 (x - 1) = 3$

$\log_2 [(x + 1)(x - 1)] = 3$ Property 1

$(x + 1)(x - 1) = 2^3$

$x^2 - 1 = 8$

$x^2 = 9$

$x = \pm 3$

The number 3 checks, but -3 does not. The solution is 3.

56. $\frac{1}{63}$

57. $\log_8 (x + 1) - \log_8 x = \log_8 4$

$\log_8 \left(\frac{x + 1}{x}\right) = \log_8 4$ Property 3

$\frac{x + 1}{x} = 4$ Taking antilogarithms

$x + 1 = 4x$

$1 = 3x$

$\frac{1}{3} = x$

The answer checks. The solution is $\frac{1}{3}$.

58. $\frac{21}{8}$

59. $\log_4 (x + 3) + \log_4 (x - 3) = 2$

$\log_4 [(x + 3)(x - 3)] = 2$ Property 1

$(x + 3)(x - 3) = 4^2$

$x^2 - 9 = 16$

$x^2 = 25$

$x = \pm 5$

The number 5 checks, but -5 does not. The solution is 5.

60. $\sqrt{41}$

61. $\log \sqrt[4]{x} = \sqrt{\log x}$

$\log x^{1/4} = \sqrt{\log x}$ Writing exponential notation

$\frac{1}{4} \log x = \sqrt{\log x}$ Property 2

$\frac{1}{16} (\log x)^2 = \log x$ Squaring both sides

$\frac{1}{16} (\log x)^2 - \log x = 0$

Let $u = \log x$, substitute and solve for u.

$\frac{1}{16}u^2 - u = 0$

$u\left(\frac{1}{16}u - 1\right) = 0$

$u = 0$ or $u = 16$

$\log x = 0$ or $\log x = 16$

$x = 1$ or $x = 10^{16}$

Both numbers check. The solutions are 1 and 10^{16}.

62. 1, 109

63. $\log_5 \sqrt{x^2 + 1} = 1$

$\sqrt{x^2 + 1} = 5^1$

$x^2 + 1 = 25$ Squaring both sides

$x^2 = 24$

$x = \pm\sqrt{24}$, or $\pm 2\sqrt{6}$

Both numbers check. The solutions are $\pm 2\sqrt{6}$.

64. 1, 10,000

65. $\log x^2 = (\log x)^2$

$2 \log x = (\log x)^2$

$0 = (\log x)^2 - 2 \log x$

Let $u = \log x$.

$0 = u^2 - 2u$

$0 = u(u - 2)$

$u = 0$ or $u = 2$

$\log x = 0$ or $\log x = 2$

$x = 10^0$ or $x = 10^2$

$x = 1$ or $x = 100$

Both numbers check. The solutions are 1 and 100.

66. 27, $\frac{1}{3}$

67. $\log_3 (\log_4 x) = 0$

$\log_4 x = 3^0$ Writing an equivalent exponential equation

$\log_4 x = 1$

$x = 4^1$

$x = 4$

The answer checks. The solution is 4.

68. 10^{100}

69. $(\log_a x)^{-1} = \log_a x^{-1}$

$\frac{1}{\log_a x} = \log_a x^{-1}$ Writing $(\log_a x)^{-1}$ with a positive exponent

$\frac{1}{\log_a x} = -\log_a x$ Property 2

$1 = -(\log_a x)^2$ Multiplying by $\log_a x$

$-1 = (\log_a x)^2$

Since $(\log_a x)^2 \geqslant 0$ for all values of x, the equation has no real-number solution. The solution set is $\emptyset$.

70. $25, \frac{1}{25}$

71. $\log_7 \sqrt{x^2 - 9} = 1$

$\sqrt{x^2 - 9} = 7^1$

$x^2 - 9 = 49$ Squaring both sides

$x^2 = 58$

$x = \pm\sqrt{58}$

Both numbers check. The solutions are $\pm\sqrt{58}$.

72. -1

73. $\log (\log x) = 3$ The base is 10.

$\log x = 10^3 = 1000$

$x = 10^{1000}$

74. -3, -1

75. $\log_3 |x| = 2$

$|x| = 3^2 = 9$

$x = 9$ or $x = -9$

76. 125, -125

77. $\log x^{\log x} = 4$

$\log x (\log x) = 4$ Property 2

$(\log x)^2 = 4$

$\log x = 2$ or $\log x = -2$

$x = 10^2$ or $x = 10^{-2}$

$x = 100$ or $x = \frac{1}{100}$

Both answers check. The solutions are 100 and $\frac{1}{100}$.

78. $\frac{1}{2}$, 5000

79. $\log_a a^{x^2 + 5x} = 24$

$x^2 + 5x = 24$ Property 4

$x^2 + 5x - 24 = 0$

$(x + 8)(x - 3) = 0$

$x = -8$ or $x = 3$

80. 100, 10

81. $x^{\log_{10} x} = \frac{x^{-4}}{1000}$

$x^{\log_{10} x} \cdot x^4 = \frac{1}{1000}$ Multiplying by x^4

$x^{\log_{10} x + 4} = \frac{1}{1000}$ Adding exponents

$\log_{10} x^{\log_{10} x + 4} = \log_{10} \frac{1}{1000}$

$(\log_{10} x + 4)\log_{10} x = -3$

$(\log_{10} x)^2 + 4 \log_{10} x + 3 = 0$

Let $u = \log_{10} x$.

$u^2 + 4u + 3 = 0$

$(u + 3)(u + 1) = 0$

$u = -3$ or $u = -1$

$\log_{10} x = -3$ or $\log_{10} x = -1$

$x = 10^{-3}$ or $x = 10^{-1}$

$x = \frac{1}{1000}$ or $x = \frac{1}{10}$

Both numbers check. The solutions are $\frac{1}{1000}$ and $\frac{1}{10}$.

82. $\frac{\log 7}{\log 5} \approx 1.2091$, $\frac{\log 2}{\log 5} \approx 0.4307$

83. $(32^{x-2})(64^{x+1}) = 16^{2x-3}$

$(2^5)^{x-2}(2^6)^{x+1} = (2^4)^{2x-3}$

$(2^{5x-10})(2^{6x+6}) = 2^{8x-12}$ Multiplying exponents

$2^{11x-4} = 2^{8x-12}$ Simplifying

$11x - 4 = 8x - 12$

$3x = -8$

$x = -\frac{8}{3}$

84. $\frac{\log \left(\frac{5140}{2400}\right)}{\log 49} \approx 0.1957$

85. $4^{3x} - 4^{3x-1} = 48$

$4^{3x-1}(4 - 1) = 48$ Factoring

$(3)4^{3x-1} = 48$

$4^{3x-1} = 16$

$4^{3x-1} = 4^2$

$3x - 1 = 2$

$3x = 3$

$x = 1$

86. $100, \frac{1}{10}$

87. $\frac{(e^{3x+1})^2}{e^4} = e^{10x}$

$\frac{e^{6x+2}}{e^4} = e^{10x}$

$e^{6x-2} = e^{10x}$

$6x - 2 = 10x$

$-2 = 4x$

$-\frac{1}{2} = x$

88. $\frac{7}{4}$

89. $P = P_0 e^{kt}$

$\ln P = \ln (P_0 e^{kt})$

$\ln P = \ln P_0 + \ln e^{kt}$

$\ln P = \ln P_0 + kt$ Property 4

$\ln P - \ln P_0 = kt$

$\frac{\ln P - \ln P_0}{k} = t$

90. $t = \frac{\ln P_0 - \ln P}{k}$

91. $T = T_0 + (T_1 - T_0)\, e^{-kt}$

$T - T_0 = (T_1 - T_0)\, e^{-kt}$

$\frac{T - T_0}{T_1 - T_0} = e^{-kt}$

$\ln \left[\frac{T - T_0}{T_1 - T_0}\right] = \ln e^{-kt}$

$\ln \left[\frac{T - T_0}{T_1 - T_0}\right] = -kt$ Property 4

$\frac{1}{k} \ln \left[\frac{T - T_0}{T_1 - T_0}\right] = t$

92. $n = \log_v c - \log_v P$, or $\log_v \frac{c}{P}$

93. $\log_a Q = \frac{1}{3} \log_a y + b$

$\log_a Q = \log_a y^{\frac{1}{3}} + b$ Property 2

$\log_a Q - \log_a y^{\frac{1}{3}} = b$

$\log_a \frac{Q}{y^{\frac{1}{3}}} = b$ Property 3

$\frac{Q}{y^{\frac{1}{3}}} = a^b$

$Q = a^b y^{\frac{1}{3}}$, or $a^b \sqrt[3]{y}$

94. 38

95. $\log_5 125 = 3$ and $\log_{125} 5 = \frac{1}{3}$, so

$x = (\log_{125} 5)^{\log_5 125}$ is equivalent to

$x = \left(\frac{1}{3}\right)^3 = \frac{1}{27}$. Then $\log_3 x = \log_3 \frac{1}{27} = -3$.

96. 5

97. $|\log_a x| = \log_a |x|$

Case I: Assume $0 < a < 1$.

A. Assume $0 < x \leqslant 1$. Then $|\log_a x| = \log_a x$ and $|x| = x$. We have:

$\log_a x = \log_a x$

This is true for all values of x in the interval under consideration, $0 < x \leqslant 1$.

B. Assume $x > 1$. Then $|\log_a x| = -\log_a x$ and $|x| = x$. We have:

$-\log_a x = \log_a x$

$0 = 2 \log_a x$

$0 = \log_a x^2$

$a^0 = x^2$

$1 = x^2$

$\pm 1 = x$

This yields no solution, since we have assumed $x > 1$.

Case II. Assume $a > 1$.

A. Assume $0 < x < 1$. Then $|\log_a x| = -\log_a x$ and $|x| = x$. We have:

$-\log_a x = \log_a x$

$0 = 2 \log_a x$

$\pm 1 = x$ (See Case IB above)

This yields no solution, since we have assumed $0 < x < 1$.

B. Assume $x \geqslant 1$. Then $|\log_a x| = \log_a x$ and $|x| = x$. We have:

$\log_a x = \log_a x$

This is true for all values of x in the interval under consideration, $x \geqslant 1$.

The solution is as follows:

If $0 < a < 1$, $0 < x \leqslant 1$, and if $a > 1$, $x \geqslant 1$.

98. 2^{10}

99. $(0.5)^x < \frac{4}{5}$

$\log (0.5)^x < \log \frac{4}{5}$

$x \log 0.5 < \log 0.8$ Property 2; writing $\frac{4}{5}$ as 0.8

$x > \frac{\log 0.8}{\log 0.5}$ Reversing the inequality symbol ($\log 0.5 < 0$)

or $x > 0.3219$

100. 88

101. $2 \log_3 (x - 2y) = \log_3 x + \log_3 y$

$\log_3 (x - 2y)^2 = \log_3 (xy)$ Properties 2 and 1

$(x - 2y)^2 = xy$

$x^2 - 4xy + 4y^2 = xy$

$x^2 - 5xy + 4y^2 = 0$

$(x - 4y)(x - y) = 0$

$x - 4y = 0$ or $x - y = 0$

$x = 4y$ or $x = y$

$\frac{x}{y} = 4$ $\frac{x}{y} = 1$

102. The general solution is $\left(x, \frac{\frac{3}{2} \log 3 - x \log 2}{\log 3}\right)$.

When $x = 0$, we have $\left(0, \frac{3}{2}\right)$.

103. $a = \log_8 225$, so $8^a = 225 = 15^2$.

$b = \log_2 15$, so $2^b = 15$.

Then $8^a = (2^b)^2$

$(2^3)^a = 2^{2b}$

$2^{3a} = 2^{2b}$

$3a = 2b$

$a = \frac{2}{3}b$.

Exercise Set 5.8

1. a) One billion is 1000 million, so we set $N(t) = 1000$ and solve for t:

$$1000 = 7.5(6)^{0.5t}$$

$$\frac{1000}{7.5} = (6)^{0.5t}$$

$$\log \frac{1000}{7.5} = \log(6)^{0.5t}$$

$$\log 1000 - \log 7.5 = 0.5t \log 6$$

$$t = \frac{\log 1000 - \log 7.5}{0.5 \log 6}$$

$$t \approx \frac{3 - 0.87506}{0.5(0.77815)} \approx 5.5$$

After about 5.5 years, one billion compact discs will be sold in a year.

1. (continued)

b) When $t = 0$, $N(t) = 7.5(6)^{0.5(0)} = 7.5(6)^0 = 7.5(1) = 7.5$. Twice this initial number is 15, so we set $N(t) = 15$ and solve for t:

$$15 = 7.5(6)^{0.5t}$$

$$2 = (6)^{0.5t}$$

$$\log 2 = \log(6)^{0.5t}$$

$$\log 2 = 0.5t \log 6$$

$$t = \frac{\log 2}{0.5 \log 6} \approx \frac{0.30103}{0.5(0.77815)} \approx 0.8$$

The doubling time is about 0.8 year.

2. a) 86.4 minutes

b) 300.5 minutes

c) 20 minutes

3. a) We have

$$A(t) = \$50,000(1.09)^t.$$

b) We set $A(t) = \$450,000$ and solve for t:

$$450,000 = 50,000(1.09)^t$$

$$\frac{450,000}{50,000} = (1.09)^t$$

$$9 = (1.09)^t$$

$\log 9 = \log (1.09)^t$ Taking the common logarithm on both sides

$\log 9 = t \log 1.09$ Property 2

$$t = \frac{\log 9}{\log 1.09} \approx \frac{0.95424}{0.03743} \approx 25.5$$

It will take about 25.5 years for the \$50,000 to grow to \$450,000.

c) We set $A(t) = \$100,000$ and solve for t:

$$100,000 = 50,000(1.09)^t$$

$$2 = (1.09)^t$$

$\log 2 = \log (1.09)^t$

Taking the common logarithm on both sides

$\log 2 = t \log 1.09$ Property 2

$$t = \frac{\log 2}{\log 1.09} \approx \frac{0.30103}{0.03743} \approx 8.04$$

The doubling time is about 8.04 years.

4. a) 1.0 year

b) 7.3 years

5. a) We set V(t) = \$1200 and solve for t.

$$1200 = 5200(0.8)^t$$
$$\frac{3}{13} = (0.8)^t$$
$$\log \frac{3}{13} = \log (0.8)^t$$
$$\log \frac{3}{13} = t \log 0.8$$
$$t = \frac{\log \left(\frac{3}{13}\right)}{\log 0.8} \approx \frac{-0.63682}{-0.09691} \approx 6.6$$

After about 6.6 years the salvage value will be \$1200.

b) We set $V(t) = \frac{1}{2}(\$5200)$, or \$2600, and solve for t.

$$2600 = 5200(0.8)^t$$
$$0.5 = (0.8)^t$$
$$\log 0.5 = \log (0.8)^t$$
$$\log 0.5 = t \log 0.8$$
$$t = \frac{\log 0.5}{\log 0.8} \approx \frac{-0.30103}{-0.09691} \approx 3.1$$

After about 3.1 years the salvage value will be half of its original value.

6. a) 59.7 years

b) 19.1 years

7. a) $S(0) = 68 - 20 \log (0 + 1) = 68 - 20 \log 1 = 68 - 20(0) = 68\%$

b) $S(4) = 68 - 20 \log (4 + 1) = 68 - 20 \log 5 \approx 68 - 20(0.69897) \approx 54\%$

$S(24) = 68 - 20 \log (24 + 1) = 68 - 20 \log 25 \approx 68 - 20 (1.39794) \approx 40\%$

c) Using the values we computed in parts (a) and (b) and any others we wish to calculate, we sketch the graph:

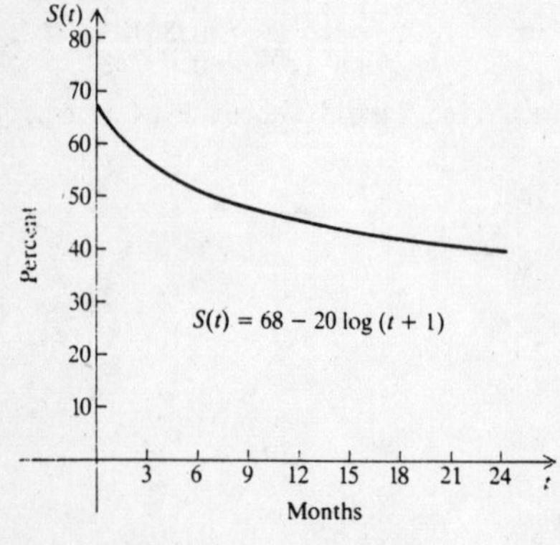

7. (continued)

d) We set S(t) = 50 and solve for t:

$$50 = 68 - 20 \log (t + 1)$$
$$-18 = -20 \log (t + 1)$$
$$0.9 = \log (t + 1)$$
$$10^{0.9} = t + 1$$ Using the definintion of logarithms or taking the antilogarithm
$$7.9 \approx t + 1$$
$$6.9 \approx t$$

After about 6.9 months, the average score was 50.

8. a) 78%

b) 68%, 57%

c)

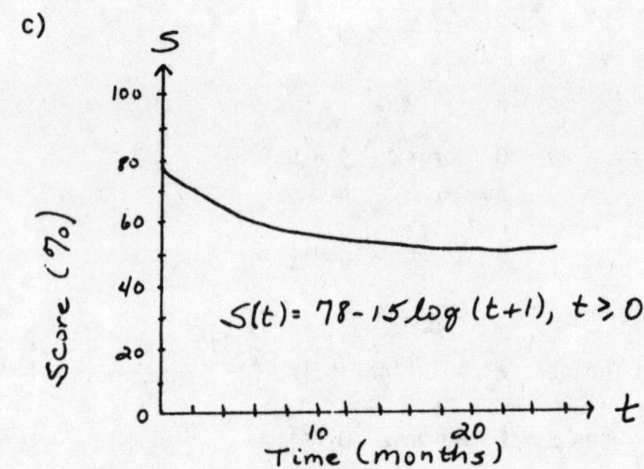

d) 1584 months

9. a) $N(1) = 1000 + 200 \log 1 = 1000 + 200(0) = 1000$ units

b) $N(5) = 1000 + 200 \log 5 \approx 1000 + 200(0.6990) \approx 1140$ units

c) Using the values computed in parts (a) and (b) and any others we wish to calculate, we can sketch the graph:

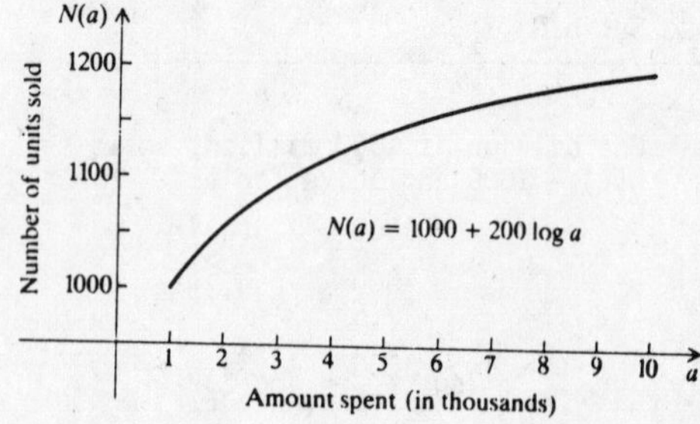

9. (continued)

d) Set N(a) = 1276 and solve for a.

$$1276 = 1000 + 200 \log a$$
$$276 = 200 \log a$$
$$1.38 = \log a$$
$$10^{1.38} = a$$ Using the definition of logarithms or taking the antilogarithm
$$23.98833 \approx a$$

About $23,988.33 would have to be spent.

10. a) 2000 units

b) 2452 units

c)

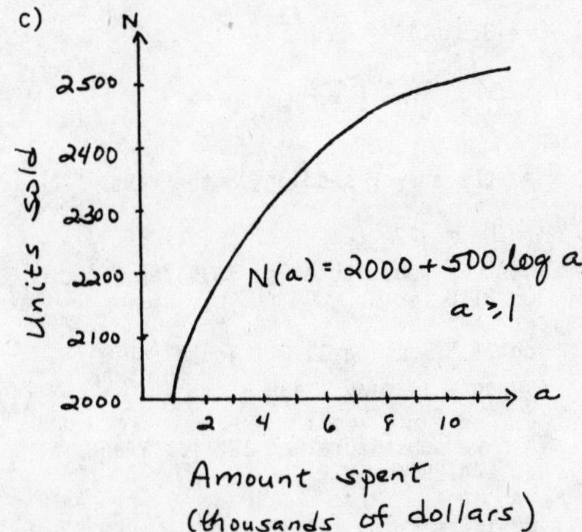

d) $1,000,000 thousand, or $1,000,000,000

11. We substitute 1.6×10^{-4} for H^+ in the formula for pH.

$pH = -\log [H^+] = -\log [1.6 \times 10^{-4}] =$

$-[\log 1.6 + \log 10^{-4}] \approx -[0.2041 - 4] =$

$-[-3.7959] = 3.7959 \approx 3.8$

The pH of pineapple juice is about 3.8.

12. 7.4

13. We substitute 6.3×10^{-5} for H^+ in the formula for p^H.

$pH = -\log [H^+] = -\log [6.3 \times 10^{-5}] =$

$-[\log 6.3 + \log 10^{-5}] \approx -[0.7993 - 5] =$

$-[-4.2007] = 4.2007 \approx 4.2$

The pH of a tomato is about 4.2.

14. 7.8

15. We substitute 5.4 for pH in the formula and solve for H^+.

$$5.4 = -\log [H^+]$$
$$-5.4 = \log [H^+]$$
$$10^{-5.4} = H^+$$ Using the definition of logarithm or taking the antilogarithm
$$0.0000040 \approx H^+$$
$$4.0 \times 10^{-6} \approx H^+$$

The hydrogen ion concentration of rainwater is about 4.0×10^{-6} moles/liter.

16. 10^{-7} moles/liter

17. We substitute 4.8 for pH in the formula and solve for H^+.

$$4.8 = -\log [H^+]$$
$$-4.8 = \log [H^+]$$
$$10^{-4.8} = H^+$$ Using the definition of logarithm or taking the antilogarithm
$$0.000016 \approx H^+$$
$$1.6 \times 10^{-5} \approx H^+$$

The hydrogen ion concentration of wine is about 1.6×10^{-5} moles/liter.

18. 6.3×10^{-4} moles/liter

19. We substitute into the formula.

$$R = \log \frac{10^{8.25} I_0}{I_0} = \log 10^{8.25} = 8.25$$

20. 5

21. We substitute into the formula.

$$L = 10 \log \frac{2510 I_0}{I_0} = 10 \log 2510 \approx 10(3.399674) \approx$$
34 decibels

22. 64 decibels

23. We substitute into the formula.

$$L = 10 \log \frac{10^6 I_0}{I_0} = 10 \log 10^6 = 10(6) =$$
60 decibels

24. 90 decibels

25. a) Substitute 100 for S(t) and solve for t.

$$100 = 200[1 - (0.86)^t]$$
$$0.5 = 1 - (0.86)^t$$
$$(0.86)^t = 0.5$$
$$\log(0.86)^t = \log 0.5$$
$$t \log 0.86 = \log 0.5$$
$$t = \frac{\log 0.5}{\log 0.86} \approx \frac{-0.3010}{-0.0655} \approx 4.6$$

The typist's speed will be 100 words per minute after she has studied typing for about 4.6 hours.

25. (continued)

b) Substitute 150 for S(t) and solve for t.

$$150 = 200[1 - (0.86)^t]$$
$$0.75 = 1 - (0.86)^t$$
$$(0.86)^t = 0.25$$
$$\log(0.86)^t = \log 0.25$$
$$t \log 0.86 = \log 0.25$$
$$t = \frac{\log 0.25}{\log 0.86} \approx \frac{-0.6021}{-0.0655} \approx 9.2$$

About 9.2 hours of studying occurred in the course.

26. a) $A = \$100{,}000(1.02)^{4t}$

b) 29.1 years

c) 8.8 years

27. $R = \log \frac{I}{I_0}$

$$\frac{I}{I_0} = 10^R$$
$$I = 10^R\, I_0$$

28. $I = 10^{0.1L}\, I_0$

29. $pH = -\log [H^+]$

$$-pH = \log [H^+]$$
$$[H^+] = 10^{-pH}$$

30. a) $L_2 - L_1 = 10 \log \frac{I_2}{I_0} - 10 \log \frac{I_1}{I_0}$

$$= 10 \log \frac{\frac{I_2}{I_0}}{\frac{I_1}{I_0}}$$
$$= 10 \log \frac{I_2}{I_1}$$

b) $L_1 + L_2 = 10 \log \left[\frac{I_1 I_2}{I_0^2}\right]$

31. a) We substitute -1.5 for M_1 and -0.3 for M_2.

$$-0.3 - (-1.5) = 2.5 \log \frac{I_1}{I_2}$$
$$1.2 = 2.5 \log \frac{I_1}{I_2}$$
$$0.48 = \log \frac{I_1}{I_2}$$
$$3 \approx \frac{I_1}{I_2} \quad \text{Taking the antilogarithm}$$

b) We substitute -1.5 for M_1 and 12.6 for M_2.

$$12.6 - (-1.5) = 2.5 \log \frac{I_1}{I_2}$$
$$14.1 = 2.5 \log \frac{I_1}{I_2}$$
$$5.64 = \log \frac{I_1}{I_2}$$
$$436{,}515.8 \approx \frac{I_1}{I_2} \quad \text{Taking the antilogarithm}$$

31. (continued)

c) We substitute 5 for $M_2 - M_1$.

$$5 = 2.5 \log \frac{I_1}{I_2}$$
$$2 = \log \frac{I_1}{I_2}$$
$$100 = \frac{I_1}{I_2}$$
$$100I_2 = I_1$$

One star is 100 times brighter than the other.

32. a) 3.4×10^{22} watts

b) 8.5×10^{27} watts

c) $L = 3.9 \times 10^{26} \times 10^{\frac{4.75-M}{2.5}}$, or

$$3.9 \times 10^{\frac{69.75-M}{2.5}} = L$$

33. a) See the answer section in the text.

b) $Y = kX + \log b$

First we substitute log 78.5 for Y and log 31.0 for X.

$\log 78.5 = k \log 31.0 + \log b$, or

$1.8949 = 1.4914k + \log b$ (1)

Then we substitute log 122 for Y and log 144.25 for X.

$\log 122 = k \log 144.25 + \log b$, or

$2.0864 = 2.1591k + \log b$ (2)

Solving the system of equations composed of equations (1) and (2), we find $k \approx 0.287$ and $\log b \approx 1.47$.

Thus, $Y = 0.287X + 1.47$.

c) Substitute log y for Y and log x for X and solve for y.

$$\log y = 0.287 \log x + 1.47$$
$$\log y = \log x^{0.287} + 1.47$$
$$\log y - \log x^{0.287} = 1.47$$
$$\log \frac{y}{x^{0.287}} = 1.47$$
$$\frac{y}{x^{0.287}} = 10^{1.47}$$
$$y = x^{0.287} 10^{1.47}$$

d) Substitute 131.0 mm for x and compute y.

$y = (131.0)^{0.287} 10^{1.47}$

$y \approx 119.6$ mm Using a calculator

Exercise Set 5.9

1. We substitute 7900 for P, since P is in thousands.

$R(P) = 0.37 \ln P + 0.05$

$R(7900) = 0.37 \ln 7900 + 0.05$

$\approx 0.37(8.9746) + 0.05$ Finding ln 7900 on a calculator

≈ 3.4 ft/sec

2. 2.5 ft/sec

3. We substitute 50.4 for P in the function, since P is in thousands.

$R(50.4) = 0.37 \ln 50.4 + 0.05$

$\approx 0.37(3.9200) + 0.05$

≈ 1.5 ft/sec

4. 1.8 ft/sec

5. a) We can use the function $C(t) = C_0 e^{kt}$ as a model. At t = 0(1962), the cost was 5¢, or \$0.05. We substitute 0.05 for C_0 and 9.7%, or 0.097, for k:

$C(t) = 0.05 e^{0.097t}$

b) In 1990, t = 1990 - 1962, or 28. We substitute 28 for t:

$C(28) = 0.05 e^{0.097(28)}$

$= 0.05 e^{2.716}$

$\approx 0.05(15.1197)$

≈ 0.76

In 1993 a Hershey bar will cost about \$0.76.

In 2000, t = 2000 - 1962, or 38. We substitute 38 for t:

$C(38) = 0.05 e^{0.097(38)}$

$= 0.05 e^{3.686}$

$\approx 0.05(39.8850)$

≈ 1.99

In 2000 a Hershey bar will cost about \$1.99.

c) We set C(t) = 5 and solve for t:

$5 = 0.05 e^{0.097t}$

$100 = e^{0.097t}$

$\ln 100 = \ln e^{0.097t}$

$\ln 100 = 0.097t$

$\frac{\ln 100}{0.097} = t$

$\frac{4.6052}{0.097} \approx t$

$47.5 \approx t$

A Hershey bar will cost \$5 about 47.5 years after 1962.

d) Using the ordered pairs found in parts (b) and (c) and any others we wish to compute, we draw the graph:

5. (continued)

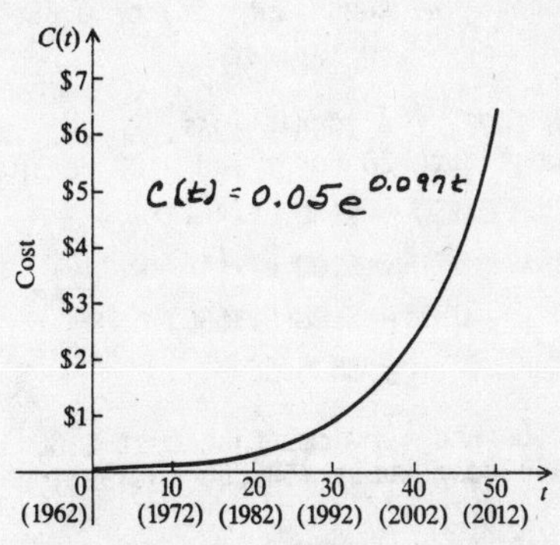

e) To find the doubling time, we set C(t) = 2(\$0.05), or \$0.10 and solve for t:

$0.1 = 0.05 e^{0.097t}$

$2 = e^{0.097t}$

$\ln 2 = \ln e^{0.097t}$

$\ln 2 = 0.097t$

$\frac{\ln 2}{0.097} = t$

(Note: We could also have used the expression relating the growth rate k and doubling time T: $T = \frac{\ln 2}{k}$.)

$\frac{0.6931}{0.097} \approx t$

$7.1 \approx t$

The doubling time is about 7.1 years.

6. a) $P(t) = 5e^{0.028t}$, where t is the number of years after 1987 and P is in billions.

b) 6.4 billion, 7.2 billion

c) 6.5 years after 1987

d)

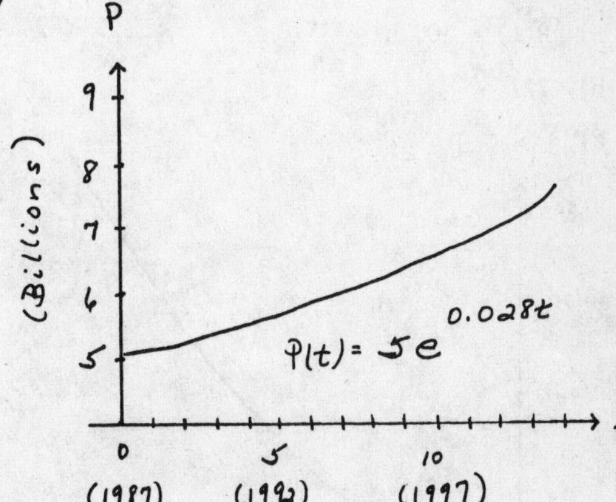

7. a) We can use the function $P(t) = P_0 e^{kt}$ as a model. We substitute 6%, or 0.06, for k.

$$P(t) = P_0 e^{0.06t}$$

b) In 1995, t = 1995 - 1967, or 28. We substitute 28 for t and \$100 for P_0.

$$C(28) = \$100 e^{0.06(28)}$$
$$= \$100 e^{1.68}$$
$$\approx \$100(5.3656)$$
$$\approx \$536.56$$

Goods and services that cost \$100 in 1967 will cost about \$536.56 in 1995.

c) In 2000, t = 2000 - 1967, or 33. We substitute 33 for t and \$100 for P_0.

$$C(33) = \$100 e^{0.06(33)}$$
$$= \$100 e^{1.98}$$
$$\approx \$100(7.2427)$$
$$\approx \$724.27$$

Goods and services that cost \$100 in 1967 will cost about \$724.27 in 2000.

d) Using the ordered pairs found in parts (b) and (c) and any others we wish to compute, we draw the graph:

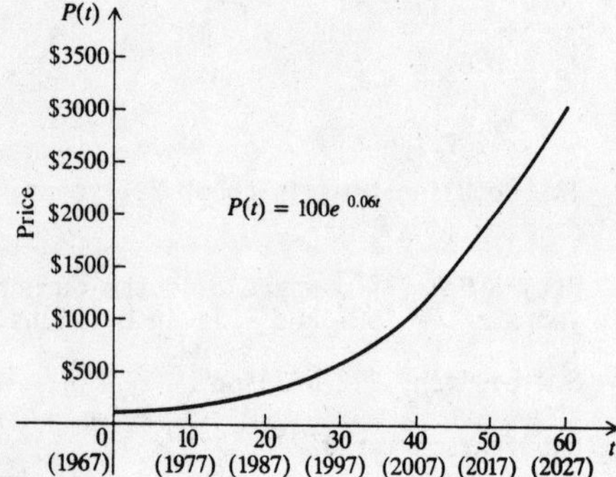

8. a) $P(t) = 100 e^{0.117t}$

b) 227

c)

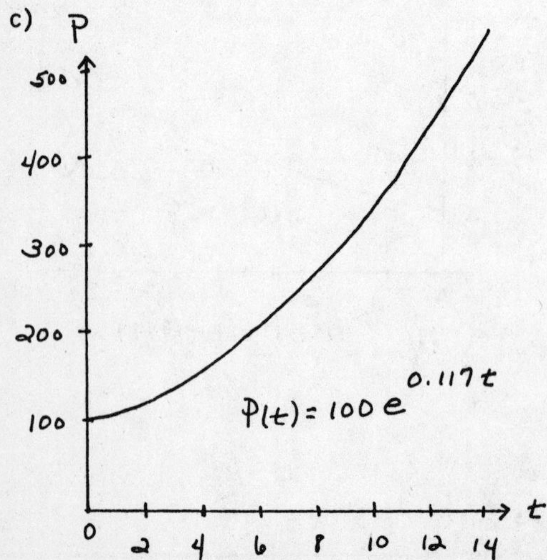

8. (continued)

d) 5.9 days

9. a) We plot the ordered pairs given in the table and sketch the following graph. We draw a curve as close to the data points as possible. It is very close to the graph of an exponential function.

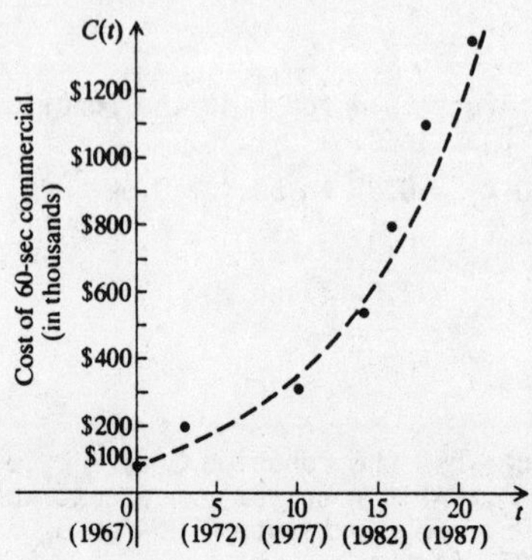

b) The exponential growth function is $C(t) = C_0 e^{kt}$. Substituting 80 for C_0, we have

$$C(t) = 80 e^{kt},$$

where t is the number of years after 1967.

To find k using the data point C(21) = \$1350 thousand, we substitute \$1350 for C(t) and 21 for t and solve for k:

$$1350 = 80 e^{k(21)}$$
$$16.875 = e^{21k}$$
$$\ln 16.875 = \ln e^{21k}$$
$$\ln 16.875 = 21k$$
$$\frac{\ln 16.875}{21} = k$$
$$\frac{2.8258}{21} \approx k$$
$$0.135 \approx k$$

The exponential growth function for the cost of a Super Bowl commercial is

$$C(t) = 80 e^{0.135t}.$$

c) The year 1995 is 28 years from 1967. We let t = 28 and find C(28):

$$C(28) = 80 e^{0.135(28)} \approx 3505$$

The cost of a 60-second commercial will be about \$3505 thousand, or \$3,505,000 in 1995.

9. (continued)

d) We set $C(t) = 3000$ (\$3000 thousand is \$3,000,000) and solve for t:

$$3000 = 80\, e^{0.135t}$$
$$37.5 = e^{0.135t}$$
$$\ln 37.5 = \ln e^{0.135t}$$
$$\ln 37.5 = 0.135t$$
$$\frac{\ln 37.5}{0.135} = t$$
$$\frac{3.6243}{0.135} \approx t$$
$$26.8 \approx t$$

The cost of a commercial will be \$3,000,000 about 26.8 years after 1967.

e) We can set $C(t) = 160$ and solve for t, or we can use the expression relating growth rate k and doubling time T, $T = \frac{\ln 2}{k}$. We will use the latter.

$$T = \frac{\ln 2}{0.135} \approx \frac{0.6931}{0.135} \approx 5.1 \text{ years}$$

10. a) $0.03 \approx k$, $P(t) = 52\, e^{0.03t}$

b) 107¢, or \$1.07

c) 23 years

d) About 58 years from 1970

11. a) Substitute 0.09 for k:

$$P(t) = P_0\, e^{0.09t}$$

b) To find the balance after one year, set $P_0 = 5000$ and $t = 1$. We find $P(1)$:

$P(1) = 5000\, e^{0.09(1)} = 5000\, e^{0.09} \approx 5000(1.094174284) \approx \$5470.87$

To find the balance after 2 years, set $P_0 = 5000$ and $t = 2$. We find $P(2)$:

$P(2) = 5000\, e^{0.09(2)} = 5000\, e^{0.18} \approx 5000(1.197217363) \approx \$5986.09$

c) We will use the expression relating growth rate k and doubling time T. (We could also set $P(t) = 10{,}000$ and solve for t.)

$$T = \frac{\ln 2}{0.09} \approx \frac{0.6931}{0.09} \approx 7.7 \text{ years}$$

12. a) $P(t) = P_0\, e^{0.1t}$

b) \$38,680.98, \$42,749.10

c) 6.9 years

13. We will use the expression relating growth rate k and doubling time T:

$$T = \frac{\ln 2}{k}.$$

Substitute 3.5%, or 0.035, for k:

$$T = \frac{\ln 2}{0.035} \approx \frac{0.6931}{0.035} \approx 19.8 \text{ years}$$

14. 69.3 years

15. We will use the expression relating growth rate k and doubling time T:

$$k = \frac{\ln 2}{T}$$
$$k = \frac{\ln 2}{7} \approx \frac{0.6931}{7} \approx 0.099$$

The annual interest rate is about 9.9%.

16. 12.8%

17. a) The exponential growth function is $V(t) = V_0\, e^{kt}$. We will express $V(t)$ in thousands of dollars and t as the number of years after 1947. Since $V_0 = 84$ thousand, we have

$$V(t) = 84\, e^{kt}.$$

In 1987 ($t = 40$), we know that $V(t) = 53{,}900$ thousand. We substitute and solve for k.

$$53{,}900 = 84\, e^{k(40)}$$
$$\frac{1925}{3} = e^{40k}$$
$$\ln \frac{1925}{3} = \ln e^{40k}$$
$$\ln \frac{1925}{3} = 40k$$
$$\frac{\ln \frac{1925}{3}}{40} = k$$
$$\frac{6.4641}{40} \approx k$$
$$0.16 \approx k$$

The exponential growth rate is about 0.16, or 16%. The exponential growth function is $V(t) = 84\, e^{0.16t}$, where V is in thousands of dollars and t is the number of years after 1947.

b) In 2007, $t = 60$. We find $V(60)$.

$V(60) = 84\, e^{0.16(60)} = 84\, e^{9.6} \approx 84(14{,}764.7816) \approx \$1{,}240{,}241.652$ thousand or \$1,240,241,652

c) We will use the expression relating growth rate k and doubling time T. (We could also set $V(t) = 168$ and solve for t.)

$$T = \frac{\ln 2}{k} \approx \frac{0.6931}{0.16} \approx 4.3 \text{ years}$$

17. (continued)

d) \$1 billion = \$1,000,000 thousand. We set $V(t) = 1,000,000$ and solve for t.

$$1,000,000 = 84\, e^{0.16t}$$

$$\frac{250,000}{21} = e^{0.16t}$$

$$\ell n\, \frac{250,000}{21} = \ell n\, e^{0.16t}$$

$$\ell n\, \frac{250,000}{21} = 0.16t$$

$$\frac{\ell n\, \frac{250,000}{21}}{0.16} = t$$

$$\frac{9.3847}{0.16} \approx t$$

$$58.7 \approx t$$

The value of the painting will be \$1 billion about 58.7 years after 1947.

18. a) $k \approx 0.31$, $V(t) = 7.75\, e^{0.31t}$, where t is the number of years after 1983.

b) \$319.80, \$1506.72

c) 2.2 years

d) 17.9 years

19. Using the expression relating growth rate k and doubling time T, we substitute 4%, or 0.04, for k:

$$T = \frac{\ell n\, 2}{0.04} \approx \frac{0.6931}{0.04} \approx 17.3 \text{ years from 1990}$$

20. 6.9 years from 1990

21. a) The exponential growth function with $P_0 = 2,812,000$ is $P(t) = 2,812,000\, e^{kt}$, where t is the number of years after 1970. To find k we substitue 14 for t and 3,097,000 for P and solve for k.

$$3,097,000 = 2,812,000\, e^{k(14)}$$

$$\frac{3,097,000}{2,812,000} = e^{14k}$$

$$\ell n\, \frac{3,097,000}{2,812,000} = \ell n\, e^{14k}$$

$$\ell n\, \frac{3,097,000}{2,812,000} = 14k$$

$$\frac{\ell n\, \frac{3,097,000}{2,812,000}}{14} = k$$

$$\frac{0.0965}{14} \approx k$$

$$0.007 \approx k$$

The value of k is about 0.007, or 0.7%. The exponential growth function is

$$P(t) = 2,812,000\, e^{0.007t},$$

where t is the number of years after 1970.

21. (continued)

b) In 1996, t = 1996 - 1970, or 26. We find P(26).

$P(26) = 2,812,000\, e^{0.007(26)} =$
$2,812,000\, e^{0.182} \approx 2,812,000(1.1996) \approx$
$3,373,315$

In 1996 the population of Los Angeles will be about 3,373,315.

22. a) $k \approx 0.018$, $P(t) = 786,000\, e^{0.018t}$, where t is the number of years after 1980.

b) 1,126,597

23. We will use the function derived in Example 7:

$$P(t) = P_0\, e^{-0.00012t}$$

If the mummy has lost 46% of its carbon-14, then 54% (P_0) is the amount present. To find the age of the mummy, we substitute 54% (P_0), or 0.54 P_0, for P(t) and solve for t.

$$0.54\, P_0 = P_0\, e^{-0.00012t}$$

$$0.54 = e^{-0.00012t}$$

$$\ell n\, 0.54 = \ell n\, e^{-0.00012t}$$

$$-0.6162 \approx -0.00012t$$

$$t \approx \frac{-0.6162}{-0.00012} \approx 5135 \text{ years}$$

24. 3590 years

25. We use the function $P(t) = P_0\, e^{-kt}$, $k > 0$. When $t = 3$, $P(t) = \frac{1}{2} P_0$. We substitute and solve for k.

$$\frac{1}{2} P_0 = P_0\, e^{-k(3)}, \text{ or } \frac{1}{2} = e^{-3k}$$

$$\ell n\, \frac{1}{2} = \ell n\, e^{-3k} = -3k$$

$$k = \frac{\ell n\, 0.5}{-3} \approx \frac{-0.6931}{-3} \approx 0.231$$

The decay rate is 0.231, or 23.1% per minute.

26. 3.2% per year

27. We use the function $P(t) = P_0\, e^{-kt}$, $k > 0$. For iodine-131, k = 9.6%, or 0.096. To find the half-life we substitute 0.096 for k and $\frac{1}{2} P_0$ for P(t), and solve for t.

$$\frac{1}{2} P_0 = P_0\, e^{-0.096t}, \text{ or } \frac{1}{2} = e^{-0.096t}$$

$$\ell n\, \frac{1}{2} = \ell n\, e^{-0.096t} = -0.096t$$

$$t = \frac{\ell n\, 0.5}{-0.096} \approx \frac{-0.6931}{-0.096} \approx 7.2 \text{ days}$$

28. 11 years

29. a) The value of k in the function is 0.007, so the animal loses 0.7% of its weight each day.

b) Find W(30).

$W(30) = W_0\ e^{-0.007(30)} = W_0\ e^{-0.21} \approx 0.811\ W_0$

After 30 days, about 0.811, or 81.1%, of its initial weight W_0 remains.

30. a) 6.7 watts

b) 115.5 days

c) 298.6 days

d) 60 watts

31. a) Substitute 14.7 for P_0 and 2000 for a.

$P = 14.7\ e^{-0.00005(2000)} = 14.7\ e^{-0.1} \approx$

$14.7(0.9048) \approx 13.3\ \text{lb/in}^2$

b) Substitute 14.7 for P_0 and 14,162 for a.

$P = 14.7\ e^{-0.00005(14,162)} = 14.7\ e^{-0.7081} \approx$

$14.7(0.4926) \approx 7.24\ \text{lb/in}^2$

c) Substitute 1.47 for P and 14.7 for P_0, and solve for a.

$1.47 = 14.7\ e^{-0.00005a}$

$0.1 = e^{-0.00005a}$

$\ln 0.1 = \ln e^{-0.00005a}$

$\ln 0.1 = -0.00005a$

$a = \dfrac{\ln 0.1}{-0.00005} \approx \dfrac{-2.3026}{-0.00005} \approx 46,052\ \text{ft}$

d) Substitute 14.7 for P_0 and 0.39 for P and solve for a.

$0.39 = 14.7\ e^{-0.00005a}$

$\dfrac{0.39}{14.7} = e^{-0.00005a}$

$\ln \dfrac{0.39}{14.7} = \ln e^{-0.00005a}$

$\ln \dfrac{0.39}{14.7} = -0.00005a$

$\dfrac{\ln \frac{0.39}{14.7}}{-0.00005} = a$

$\dfrac{-3.6295}{-0.00005} \approx a$

$a \approx 72,589\ \text{ft}$

32. a) $46,000

b) $6225.42

33. a) At 1 m: $I = I_0\ e^{-1.4(1)} \approx 0.247\ I_0$

24.7% of I_0 remains.

At 3 m: $I = I_0\ e^{-1.4(3)} \approx 0.015\ I_0$

1.5% of I_0 remains.

At 5 m: $I = I_0\ e^{-1.4(5)} \approx 0.0009\ I_0$

0.09% of I_0 remains.

At 50 m: $I = I_0\ e^{-1.4(50)} \approx (3.98 \times 10^{-31})\ I_0$

$3.98 \times 10^{-31} = (3.98 \times 10^{-2}) \times 10^{-29}$, so

(3.98×10^{-29})% remains.

b) $I = I_0\ e^{-1.4(10)} \approx 0.0000008\ I_0$

0.00008% remains

34. $11,846.39

35. $v = c \ln R$

$\dfrac{v}{c} = \ln R$

$R = e^{\frac{v}{c}}$

36. $t = -\dfrac{L}{R}\left[\ln\left[1 - \dfrac{iR}{V}\right]\right]$

37. Set S(x) = D(x) and solve for x.

$e^x = 163,000e^{-x}$

$e^{2x} = 163,000$ Multiplying by e^{-x}

$\ln e^{2x} = \ln 163,000$

$2x = \ln 163,000$

$x = \dfrac{\ln 163,000}{2} \approx \dfrac{12.0015}{2}$

$x \approx 6$

To find the second coordinate of the equilibrium point, find S(6) or D(6). We will find S(6).

$S(6) = e^6 \approx 403$

The equilibrium point is (6, $403). (Answers may vary slightly due to rounding differences.)

38. Measure the atmospheric pressure P at the top of building. Substitute that value in the equation $P = 14.7\ e^{-0.00005a}$, and solve for the height, or altitude, a. (Note: We assume that the base of the Empire State Building is essentially at sea level.)

39. To find k we substitute 105 for T_1, 32 for T_0, 5 for t, and 70 for T(t) and solve for k.

$70 = 32 + |105 - 32|\ e^{-k(5)}$

$38 = 73\ e^{-5k}$

$\dfrac{38}{73} = e^{-5k}$

$\ln \dfrac{38}{73} = \ln e^{-5k}$

$\ln \dfrac{38}{73} = -5k$

$k = \dfrac{\ln \frac{38}{73}}{-5} \approx 0.13$

39. (continued)

The function is

$$T(t) = 32 + 73\, e^{-0.13t}.$$

We find T(10).

$$T(10) = 32 + 73\, e^{-0.13(10)} \approx 51.9° \text{ F}$$

40. 9 AM

Exercise Set 6.1

1. See the answer section in the text.

2.

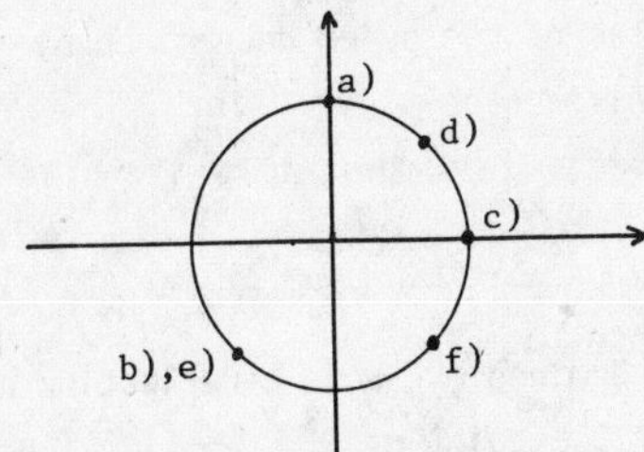

3. See the answer section in the text.

4.

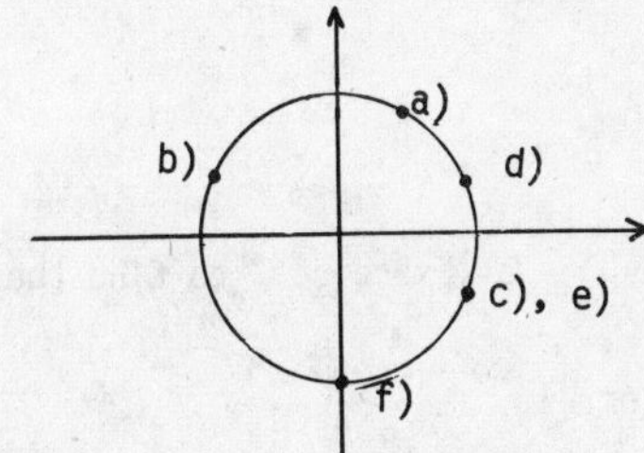

5. See the answer section in the text.

6.

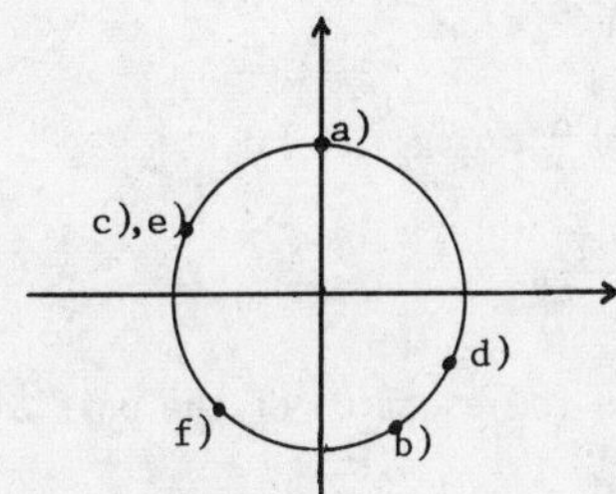

7.

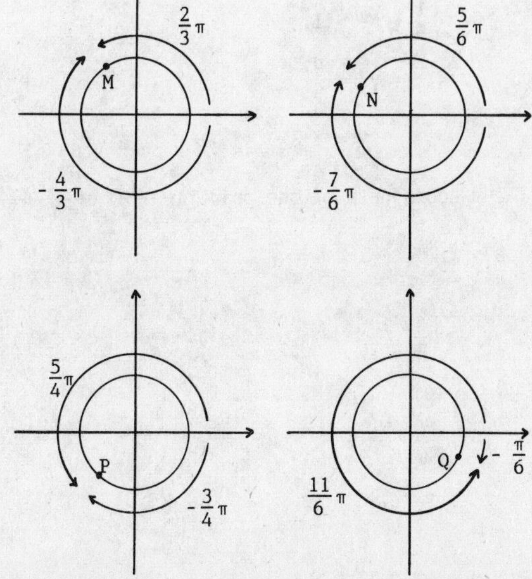

8. M: $\frac{\pi}{3}, -\frac{5}{3}\pi$; N: $\frac{7}{4}\pi, -\frac{\pi}{4}$

P: $\frac{4}{3}\pi, -\frac{2}{3}\pi$; Q: $\frac{7}{6}\pi, -\frac{5}{6}\pi$

9.

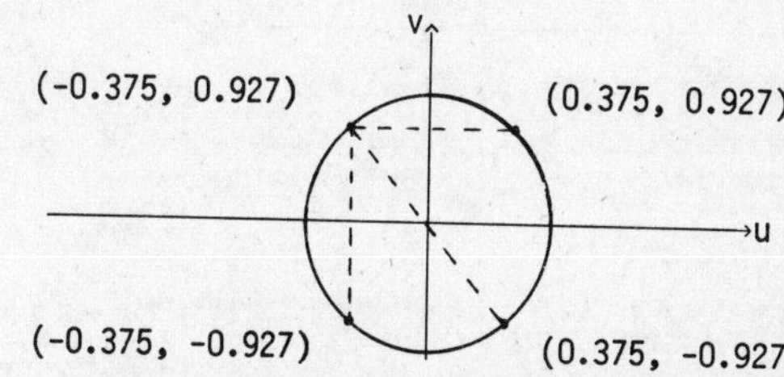

a) The reflection of (-0.375, 0.927) across the u-axis is (-0.375, -0.927). The first coordinates are the same, and the second coordinates are additive inverses of each other.

b) The reflection of (-0.375, 0.927) across the v-axis is (0.375, 0.927). The first coordinates are additive inverses of each other, and the second coordinates are the same.

c) The reflection of (-0.375, 0.927) across the origin is (0.375, -0.927). Both the first coordinates and the second coordinates are additive inverses of each other.

d) By symmetry, all these points are on the circle.

10. a) (0.625, 0.781) b) (-0.625, -0.781)
c) (-0.625, 0.781) d) Yes, by symmetry.

11.

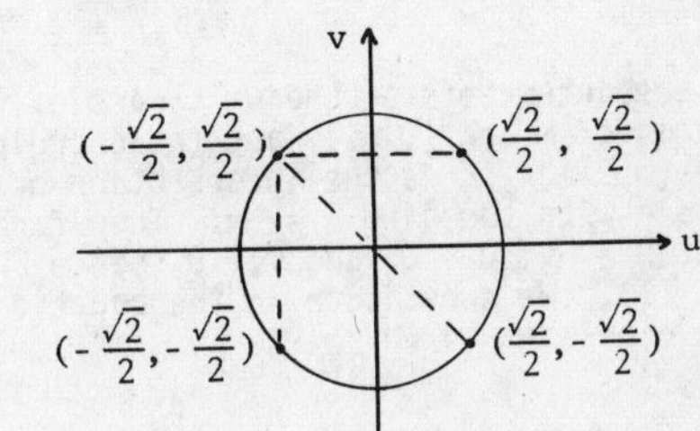

a) $\left(-\frac{\sqrt{2}}{2}, -\frac{\sqrt{2}}{2}\right)$, b) $\left(\frac{\sqrt{2}}{2}, \frac{\sqrt{2}}{2}\right)$,

c) $\left(\frac{\sqrt{2}}{2}, -\frac{\sqrt{2}}{2}\right)$

12. a) $\left(-\frac{3}{4}, -\frac{\sqrt{7}}{4}\right)$, b) $\left(\frac{3}{4}, \frac{\sqrt{7}}{4}\right)$, c) $\left(\frac{3}{4}, -\frac{\sqrt{7}}{4}\right)$

13.

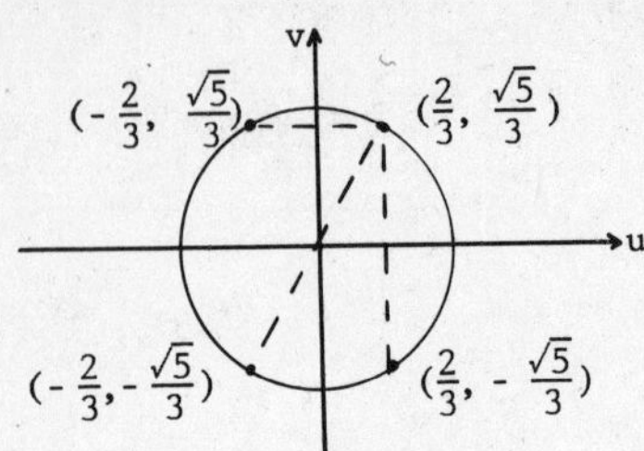

a) $\left(\frac{2}{3}, -\frac{\sqrt{5}}{3}\right)$, b) $\left(-\frac{2}{3}, \frac{\sqrt{5}}{3}\right)$, c) $\left(-\frac{2}{3}, -\frac{\sqrt{5}}{3}\right)$

14. a) $\left(-\frac{\sqrt{3}}{2}, \frac{1}{2}\right)$, b) $\left(\frac{\sqrt{3}}{2}, -\frac{1}{2}\right)$, c) $\left(\frac{\sqrt{3}}{2}, \frac{1}{2}\right)$

15.

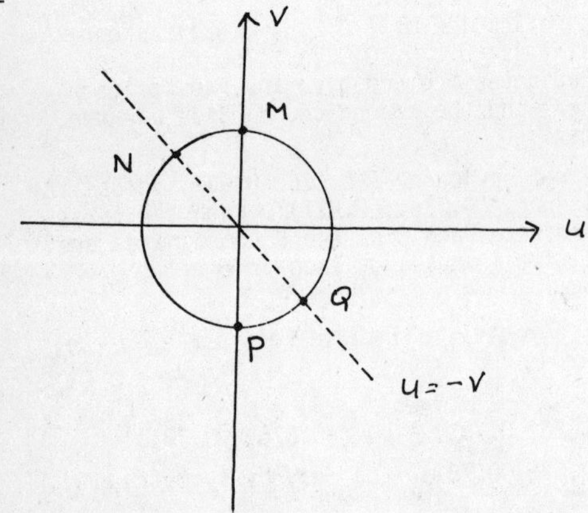

M and P are intercepts of the unit circle. The coordinates of M are (0, 1), and the coordinates of P are (0, -1). E is the intersection of the unit circle with the line u = -v. Then for E we know that u = -v, u < 0, and v > 0 (since E is in quadrant II). We substitute in the equation of the unit circle:

$$u^2 + v^2 = 1$$
$$(-v)^2 + v^2 = 1, \text{ or } 2v^2 = 1$$
$$v^2 = \frac{1}{2}$$
$$V = \sqrt{\frac{1}{2}}, \text{ or } \frac{\sqrt{2}}{2}$$

Then $u = -v = -\frac{\sqrt{2}}{2}$.

The coordinates of E are $\left(-\frac{\sqrt{2}}{2}, \frac{\sqrt{2}}{2}\right)$.

Q is the reflection of E across the origin. Hence its coordinates are $\left(\frac{\sqrt{2}}{2}, -\frac{\sqrt{2}}{2}\right)$.

16. M: (1, 0), N: $\left(\frac{\sqrt{2}}{2}, \frac{\sqrt{2}}{2}\right)$, P: (-1, 0),

Q: $\left(-\frac{\sqrt{2}}{2}, -\frac{\sqrt{2}}{2}\right)$

17. The point determined by $-\frac{\pi}{6}$ is a reflection across the u-axis of the point determined by $\frac{\pi}{6}$, $\left(\frac{\sqrt{3}}{2}, \frac{1}{2}\right)$. The first coordinates are the same, and the second coordinates are additive inverses of each other. The coordinates of the point determined by $-\frac{\pi}{6}$ are $\left(\frac{\sqrt{3}}{2}, -\frac{1}{2}\right)$.

18. $\left(\frac{1}{2}, -\frac{\sqrt{3}}{2}\right)$

19. The point determined by $-\alpha$ is a reflection across the u-axis of the point α, $\left(\frac{3}{4}, -\frac{\sqrt{7}}{4}\right)$. The first coordinates are the same, and the second coordinates are additive inverses of each other. The coordinates of the point determined by $-\alpha$ are $\left(\frac{3}{4}, \frac{\sqrt{7}}{4}\right)$.

20. $\left(-\frac{2}{3}, -\frac{\sqrt{5}}{3}\right)$

21. We use the results of Exercise 7 to find the numbers.

M: $\frac{2}{3}\pi + 2\pi$, or $\frac{8}{3}\pi$

N: $\frac{5}{6}\pi + 2\pi$, or $\frac{17}{6}\pi$

P: $\frac{5}{4}\pi + 2\pi$, or $\frac{13}{4}\pi$

Q: $\frac{11}{6}\pi + 2\pi$, or $\frac{23}{6}\pi$

22. M: $-\frac{11}{3}\pi$; N: $-\frac{9}{4}\pi$;

P: $-\frac{8}{3}\pi$; Q: $-\frac{17}{6}\pi$

23. We substitute in the equation of the unit circle.

$$u^2 + v^2 = 1$$
$$\left(-\frac{1}{3}\right)^2 + v^2 = 1$$
$$\frac{1}{9} + v^2 = 1$$
$$v^2 = \frac{8}{9}$$
$$v = \pm\frac{\sqrt{8}}{3} = \pm\frac{2\sqrt{2}}{3}$$

The v-coordinate of the point is $\frac{2\sqrt{2}}{3}$ or $-\frac{2\sqrt{2}}{3}$.

24. $\pm\frac{2}{5}$

25. We substitute in the equation of the unit circle.

$$u^2 + v^2 = 1$$
$$(0.25671)^2 + v^2 = 1$$
$$0.06590 + v^2 \approx 1$$
$$v^2 \approx 0.93410$$
$$v \approx \pm\ 0.96649$$

The v-coordinate of the point is 0.96649 or -0.96649.

26. $\pm\ 0.47421$

Exercise Set 6.2

1. The point $\frac{\pi}{4}$ has coordinates $\left(\frac{\sqrt{2}}{2}, \frac{\sqrt{2}}{2}\right)$ on the unit circle. The value of the sine function is the second coordinate.

$$\sin \frac{\pi}{4} = \frac{\sqrt{2}}{2}$$

2. 0

3. The point $\frac{\pi}{6}$ has coordinates $\left(\frac{\sqrt{3}}{2}, \frac{1}{2}\right)$ on the unit circle. The value of the sine function is the second coordinate.

$$\sin \frac{\pi}{6} = \frac{1}{2}$$

4. $-\frac{\sqrt{2}}{2}$

5. The point $\frac{3}{2}\pi$ has coordinates (0, -1) on the unit circle.

$$\sin \frac{3}{2}\pi = -1$$

6. 1

7. The point $\frac{\pi}{3}$ has coordinates $\left(\frac{1}{2}, \frac{\sqrt{3}}{2}\right)$ on the unit circle.

$$\sin \frac{\pi}{3} = \frac{\sqrt{3}}{2}$$

8. $\frac{\sqrt{2}}{2}$

9. The point $\frac{7}{4}\pi$ has coordinates $\left(\frac{\sqrt{2}}{2}, -\frac{\sqrt{2}}{2}\right)$ on the unit circle.

$$\sin \frac{7}{4}\pi = -\frac{\sqrt{2}}{2}$$

10. 0

11. The point $\frac{5}{6}\pi$ has coordinates $\left(-\frac{\sqrt{3}}{2}, \frac{1}{2}\right)$ on the unit circle.

$$\sin \frac{5}{6}\pi = \frac{1}{2}$$

12. $-\frac{\sqrt{2}}{2}$

13. The point $-\pi$ has coordinates (-1, 0) on the unit circle.

$$\sin (-\pi) = 0$$

14. -1

15. The point $-\frac{3}{4}\pi$ has coordinates $\left(-\frac{\sqrt{2}}{2}, -\frac{\sqrt{2}}{2}\right)$ on the unit circle.

$$\sin \left(-\frac{3}{4}\pi\right) = -\frac{\sqrt{2}}{2}$$

16. $\frac{\sqrt{2}}{2}$

17. The point $-\frac{5}{4}\pi$ has coordinates $\left(-\frac{\sqrt{2}}{2}, \frac{\sqrt{2}}{2}\right)$ on the unit circle.

$$\sin \left(-\frac{5}{4}\pi\right) = \frac{\sqrt{2}}{2}$$

18. 1

19. The point $-\frac{\pi}{3}$ has coordinates $\left(\frac{1}{2}, -\frac{\sqrt{3}}{2}\right)$ on the unit circle.

$$\sin \left(-\frac{\pi}{3}\right) = -\frac{\sqrt{3}}{2}$$

20. 0

21. The point $-\frac{2}{3}\pi$ has coordinates $\left(-\frac{1}{2}, -\frac{\sqrt{3}}{2}\right)$ on the unit circle.

$$\sin \left(-\frac{2}{3}\pi\right) = -\frac{\sqrt{3}}{2}$$

22. $-\frac{1}{2}$

23. The value of the cosine function of a number is the first coordinate of the point on the unit circle determined by the number. See Exercise 1 for the coordinates of the point $\frac{\pi}{4}$ on the unit circle.

$$\cos \frac{\pi}{4} = \frac{\sqrt{2}}{2}$$

24. -1

25. The point $\frac{5}{4}\pi$ has coordinates $\left(-\frac{\sqrt{2}}{2}, -\frac{\sqrt{2}}{2}\right)$ on the unit circle.

$\cos \frac{5}{4}\pi = -\frac{\sqrt{2}}{2}$

26. $\frac{\sqrt{3}}{2}$

27. The point $\frac{\pi}{2}$ has coordinates (0, 1) on the unit circle.

$\cos \frac{\pi}{2} = 0$

28. $\frac{1}{2}$

29. The point $\frac{3}{4}\pi$ has coordinates $\left(-\frac{\sqrt{2}}{2}, \frac{\sqrt{2}}{2}\right)$ on the unit circle.

$\cos \frac{3}{4}\pi = -\frac{\sqrt{2}}{2}$

30. $\frac{\sqrt{2}}{2}$

31. $\cos \frac{3}{2}\pi = 0$ (See Exercise 5.)

32. $-\frac{\sqrt{3}}{2}$

33. The point $-\frac{\pi}{4}$ has coordinates $\left(\frac{\sqrt{2}}{2}, -\frac{\sqrt{2}}{2}\right)$ on the unit circle.

$\cos\left(-\frac{\pi}{4}\right) = \frac{\sqrt{2}}{2}$

34. -1

35. The point 2π has coordinates (1, 0) on the unit circle.

$\cos 2\pi = 1$

36. 0

37. $\cos\left(-\frac{3}{4}\pi\right) = -\frac{\sqrt{2}}{2}$ (See Exercise 15.)

38. $-\frac{\sqrt{2}}{2}$

39. The point $-\frac{3}{2}\pi$ has coordinates (0, 1) on the unit circle.

$\cos\left(-\frac{3}{2}\pi\right) = 0$

40. $\frac{1}{2}$

41. The point $-\frac{\pi}{6}$ has coordinates $\left(\frac{\sqrt{3}}{2}, -\frac{1}{2}\right)$ on the unit circle.

$\cos\left(-\frac{\pi}{6}\right) = \frac{\sqrt{3}}{2}$

42. $\frac{\sqrt{2}}{2}$

43. $\cos\left(-\frac{2}{3}\pi\right) = -\frac{1}{2}$ (See Exercise 21.)

44. $-\frac{\sqrt{3}}{2}$

45. $\cos(-x) = \cos x$

46. $-\sin x$

47. $\sin(x + \pi) = -\sin x$

48. $-\sin x$

49. $\cos(\pi - x) = -\cos x$

50. $\sin x$

51. $\cos(x + 2k\pi) = \cos x$

52. $\sin x$

53. $\cos(x - \pi) = -\cos x$

54. $-\cos x$

55. See the answer section in the text.

56. a) See Margin Exercise 10.

b) Same as a).

c)

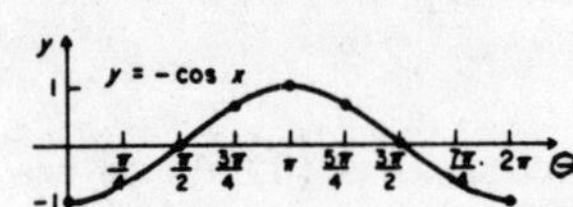

d) Same

57. See the answer section in the text.

58. a) See Margin Exercise 3.

b)

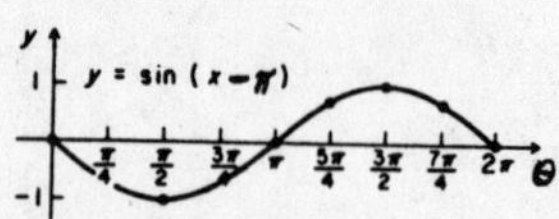

c) Same as b).

d) Same

59. See the answer section in the text.

60. a) See Margin Exercise 10.

b)

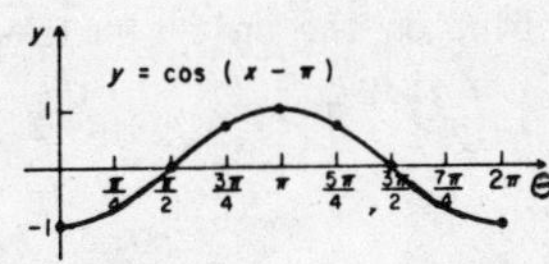

c) Same as b).

d) Same

61. a) $\sin \frac{\pi}{2} = 1$

$\sin\left(\frac{\pi}{2} + 2\pi\right) = 1$ $\qquad \sin\left(\frac{\pi}{2} - 2\pi\right) = 1$

$\sin\left(\frac{\pi}{2} + 2\cdot 2\pi\right) = 1$ $\qquad \sin\left(\frac{\pi}{2} - 2\cdot 2\pi\right) = 1$

$\sin\left(\frac{\pi}{2} + 3\cdot 2\pi\right) = 1$ $\qquad \sin\left(\frac{\pi}{2} - 3\cdot 2\pi\right) = 1$

$\sin\left(\frac{\pi}{2} + k\cdot 2\pi\right) = 1$, k any integer

Thus

$x = \frac{\pi}{2} + 2k\pi$, k any integer

b) $\sin \frac{3}{2}\pi = -1$

$\sin\left(\frac{3}{2}\pi + 2\pi\right) = -1$ $\qquad \sin\left(\frac{3}{2}\pi - 2\pi\right) = -1$

$\sin\left(\frac{3}{2}\pi + 2\cdot 2\pi\right) = -1$ $\qquad \sin\left(\frac{3}{2}\pi - 2\cdot 2\pi\right) = -1$

$\sin\left(\frac{3}{2}\pi + 3\cdot 2\pi\right) = -1$ $\qquad \sin\left(\frac{3}{2}\pi - 3\cdot 2\pi\right) = -1$

$\sin\left(\frac{3}{2}\pi + k\cdot 2\pi\right) = -1$, k any integer

Thus

$x = \frac{3}{2}\pi + 2k\pi$, k any integer

62. a) $2k\pi$, k any integer

b) $(2k + 1)\pi$, k any integer

63. For which numbers, x, is $\sin x = 0$?

$\sin 0 = 0$

$\sin \pi = 0$ $\qquad \sin(-\pi) = 0$

$\sin 2\pi = 0$ $\qquad \sin(-2\pi) = 0$

$\sin 3\pi = 0$ $\qquad \sin(-3\pi) = 0$

$\sin 4\pi = 0$ $\qquad \sin(-4\pi) = 0$

$\sin k\pi = 0$, k any integer

Thus

$x = k\pi$, k any integer

64. $x = \frac{\pi}{2} + k\pi$, k any integer

65. $f(x) = x^2 + 2x$ $\qquad g(x) = \cos x$

$f \circ g(x) = f(g(x)) = f(\cos x)$

$= (\cos x)^2 + 2\cos x$

$= \cos^2 x + 2\cos x$

$g \circ f(x) = g(f(x)) = g(x^2 + 2x)$

$= \cos(x^2 + 2x)$

66. a) 0.7071; b) 0.8660

67. a) $\cos \frac{\pi}{6} = \frac{\sqrt{3}}{2} \approx \frac{1.7320}{2} = 0.8660$

b) $\cos \frac{\pi}{4} = \frac{\sqrt{2}}{2} \approx \frac{1.4142}{2} = 0.7071$

68. For example,

$\sin\left(\frac{\pi}{4} + \frac{\pi}{4}\right) = \sin \frac{\pi}{2} = 1$, but

$\sin \frac{\pi}{4} + \sin \frac{\pi}{4} = \frac{\sqrt{2}}{2} + \frac{\sqrt{2}}{2} = \sqrt{2}$

69. See the answer section in the text.

70.

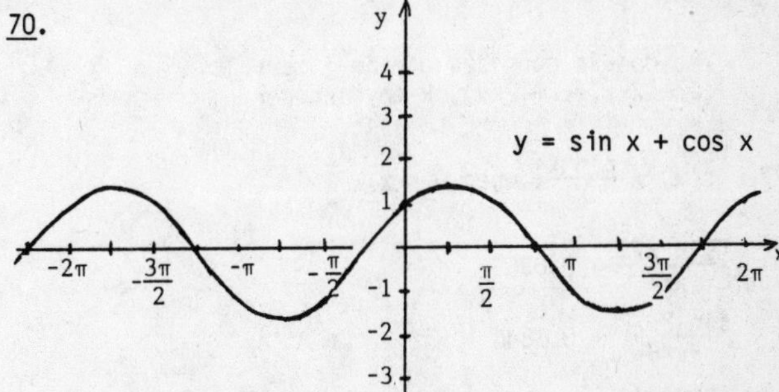

71. $y = \sin^2 x = \sin x \cdot \sin x$

It is helpful to graph the function.

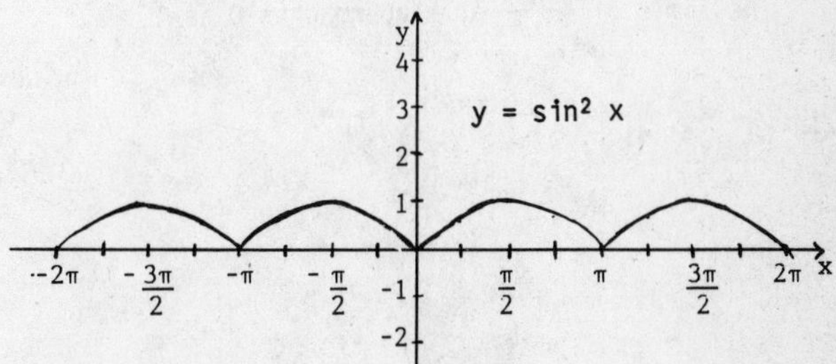

Domain: Set of all real numbers

Range: Set of all real numbers from 0 to 1 inclusive.

Period: π $\quad [\sin^2 x = \sin^2(x + k\pi)$, k any integer]

Amplitude: $\frac{1}{2}$ $\left(\frac{1}{2} \text{ of } 1 - 0 \text{ is } \frac{1}{2}\right)$

72. Domain: Set of all real numbers

Range: Set of all real numbers from 1 to 2 inclusive.

Period: π

Amplitude: $\frac{1}{2}$

73. $f(x) = \sqrt{\cos x}$

Solve $\cos x \geqslant 0$.

Since $\cos x \geqslant 0$ when x is in the intervals $\left[-\frac{\pi}{2} + 2k\pi, \frac{\pi}{2} + 2k\pi\right]$, the domain consists of the intervals $\left[-\frac{\pi}{2} + 2k\pi, \frac{\pi}{2} + 2k\pi\right]$.

74. The domain is the set of all real numbers except $k\pi$ for any integer k.

75. $f(x) = \frac{\sin x}{\cos x}$

Note that $\cos x \neq 0$ and $\cos x = 0$ when $x = \frac{\pi}{2} + k\pi$ for any integer k. Thus the domain is the set of all real numbers except $\frac{\pi}{2} + k\pi$ for any integer k.

76. The domain consists of the intervals $(0 + 2k\pi, \pi + 2k\pi)$, k any integer.

77. $f(x) = \frac{\sin x}{x}$, when $0 < x < \frac{\pi}{2}$

$\frac{\sin \pi/2}{\pi/2} \approx 0.6369$

$\frac{\sin 3\pi/8}{3\pi/8} \approx 0.7846$

$\frac{\sin \pi/4}{\pi/4} \approx 0.9008$

$\frac{\sin \pi/8}{\pi/8} \approx 0.9750$

The limit of $\frac{\sin x}{x}$ as x approaches 0 is 1.

78. $\left[-\frac{3\pi}{4}, \frac{\pi}{4}\right]$

Exercise Set 6.3

1. $\frac{\pi}{4}$ determines a point on the unit circle with coordinates $\left(\frac{\sqrt{2}}{2}, \frac{\sqrt{2}}{2}\right)$. Then

$$\cot \frac{\pi}{4} = \frac{\cos \frac{\pi}{4}}{\sin \frac{\pi}{4}} = \frac{\frac{\sqrt{2}}{2}}{\frac{\sqrt{2}}{2}} = 1.$$

2. -1

3. $\frac{\pi}{6}$ determines a point on the unit circle with coordinates $\left(\frac{\sqrt{3}}{2}, \frac{1}{2}\right)$. Then

$$\tan \frac{\pi}{6} = \frac{\sin \frac{\pi}{6}}{\cos \frac{\pi}{6}} = \frac{\frac{1}{2}}{\frac{\sqrt{3}}{2}} = \frac{1}{\sqrt{3}} = \frac{\sqrt{3}}{3}.$$

4. $-\sqrt{3}$

5. See Exercise 1.

$$\sec \frac{\pi}{4} = \frac{1}{\cos \frac{\pi}{4}} = \frac{1}{\frac{\sqrt{2}}{2}} = \frac{2}{\sqrt{2}} = \sqrt{2}$$

6. $\sqrt{2}$

7. $\frac{3\pi}{2}$ determines a point on the unit circle with coordinates (0, -1). Then

$$\tan \frac{3\pi}{2} = \frac{\sin \frac{3\pi}{2}}{\cos \frac{3\pi}{2}} = \frac{-1}{0}, \text{ which is undefined.}$$

8. Undefined

9. $\frac{2\pi}{3}$ determines a point on the unit circle with coordinates $\left(-\frac{1}{2}, \frac{\sqrt{3}}{2}\right)$.

$$\tan \frac{2\pi}{3} = \frac{\sin \frac{2\pi}{3}}{\cos \frac{2\pi}{3}} = \frac{\frac{\sqrt{3}}{2}}{-\frac{1}{2}} = -\sqrt{3}$$

10. $\frac{\sqrt{3}}{3}$

11. $-\frac{7\pi}{4}$ determines a point on the unit circle with coordinates $\left(\frac{\sqrt{2}}{2}, \frac{\sqrt{2}}{2}\right)$.

$$\sec\left(-\frac{7\pi}{4}\right) = \frac{1}{\cos\left(-\frac{7\pi}{4}\right)} = \frac{1}{\frac{\sqrt{2}}{2}} = \frac{2}{\sqrt{2}} = \sqrt{2}$$

12. $\frac{2\sqrt{3}}{3}$

13. $\frac{5\pi}{6}$ determines a point on the unit circle with coordinates $\left(-\frac{\sqrt{3}}{2}, \frac{1}{2}\right)$.

$$\tan \frac{5\pi}{6} = \frac{\sin \frac{5\pi}{6}}{\cos \frac{5\pi}{6}} = \frac{\frac{1}{2}}{-\frac{\sqrt{3}}{2}} = -\frac{1}{\sqrt{3}} = -\frac{\sqrt{3}}{3}$$

14. $\sqrt{3}$

15. $\sin(-x) = -\sin x$

The sine function is an odd function.

$\cos(-x) = \cos x$

The cosine function is an even function.

$$\tan(-x) = \frac{\sin(-x)}{\cos(-x)} = \frac{-\sin x}{\cos x} = -\frac{\sin x}{\cos x} = -\tan x$$

The tangent function is an odd function.

$$\cot(-x) = \frac{\cos(-x)}{\sin(-x)} = \frac{\cos x}{-\sin x} = -\frac{\cos x}{\sin x} = -\cot x$$

The cotangent function is an odd function.

$$\sec(-x) = \frac{1}{\cos(-x)} = \frac{1}{\cos x} = \sec x$$

The secant function is an even function.

$$\csc(-x) = \frac{1}{\sin(-x)} = \frac{1}{-\sin x} = -\frac{1}{\sin x} = -\csc x$$

The cosecant function is an odd function.

16. See Exercise 15.

17. See the graph and the list of properties for each function in Sections 6.2 and 6.3 in the text. The sin, cos, sec, and csc functions are periodic, with period 2π.

18. tangent, cotangent

19.

Function	I	II	III	IV
sin x	+	+	-	-
cos x	+	-	-	+
$\tan x = \frac{\sin x}{\cos x}$	+	-	+	-

The tangent function is positive in the first and third quadrants and negative in the second and fourth quadrants.

20. Positive: I and III

Negative: II and IV

21.

	I	II	III	IV
cos x	+	-	-	+
$\sec x = \frac{1}{\cos x}$	+	-	-	+

The secant function is positive in the first and fourth quadrants and negative in the second and third quadrants.

22. Positive: I and II

Negative: III and IV

23. See table in text.

$\frac{\pi}{16}$: $\tan \frac{\pi}{16} = \frac{\sin(\pi/16)}{\cos(\pi/16)} \approx \frac{0.19509}{0.98079} \approx 0.19891$

$\cot \frac{\pi}{16} = \frac{\cos(\pi/16)}{\sin(\pi/16)} \approx \frac{0.98079}{0.19509} \approx 5.02737$

$\sec \frac{\pi}{16} = \frac{1}{\cos(\pi/16)} \approx \frac{1}{0.98079} \approx 1.01959$

$\csc \frac{\pi}{16} = \frac{1}{\sin(\pi/16)} = \frac{1}{0.19509} = 5.12584$

$\frac{\pi}{6}$: $\sin \frac{\pi}{6} = \frac{1}{2} = 0.50000$

$\cos \frac{\pi}{6} = \frac{\sqrt{3}}{2} \approx \frac{1.73205}{2} \approx 0.86603$

$\tan \frac{\pi}{6} = \frac{\sin(\pi/6)}{\cos(\pi/6)} \approx \frac{0.50000}{0.86603} \approx 0.57735$

$\cot \frac{\pi}{6} = \frac{\cos(\pi/6)}{\sin(\pi/6)} \approx \frac{0.86603}{0.50000} \approx 1.73206$

$\sec \frac{\pi}{6} = \frac{1}{\cos(\pi/6)} \approx \frac{1}{0.86603} \approx 1.15469$

$\csc \frac{\pi}{6} = \frac{1}{\sin(\pi/6)} = \frac{1}{0.50000} = 2.00000$

$\frac{\pi}{4}$: $\sin \frac{\pi}{4} = \frac{\sqrt{2}}{2} \approx \frac{1.41421}{2} \approx 0.70711$

$\cos \frac{\pi}{4} = \frac{\sqrt{2}}{2} \approx 0.70711$

$\tan \frac{\pi}{4} = \frac{\sin(\pi/4)}{\cos(\pi/4)} \approx \frac{0.70711}{0.70711} = 1.00000$

$\cot \frac{\pi}{4} = \frac{\cos(\pi/4)}{\sin(\pi/4)} \approx \frac{0.70711}{0.70711} = 1.00000$

$\sec \frac{\pi}{4} = \frac{1}{\cos(\pi/4)} \approx \frac{1}{0.70711} = 1.41421$

$\csc \frac{\pi}{4} = \frac{1}{\sin(\pi/4)} \approx \frac{1}{0.70711} = 1.41421$

$\frac{3\pi}{8}$: $\tan \frac{3\pi}{8} = \frac{\sin(3\pi/8)}{\cos(3\pi/8)} = \frac{0.92388}{0.38268} \approx 2.41424$

$\cot \frac{3\pi}{8} = \frac{1}{\tan(3\pi/8)} \approx \frac{1}{2.41424} \approx 0.41421$

$\sec \frac{3\pi}{8} = \frac{1}{\cos(3\pi/8)} = \frac{1}{0.38268} \approx 2.61315$

$\csc \frac{3\pi}{8} = \frac{1}{\sin(3\pi/8)} = \frac{1}{0.92388} \approx 1.08239$

23. (continued)

$\frac{7\pi}{16}$: $\tan\frac{7\pi}{16} = \frac{\sin(7\pi/16)}{\cos(7\pi/16)} = \frac{0.98079}{0.19509} \approx 5.02737$

$$\cot\frac{7\pi}{16} = \frac{1}{\tan(7\pi/16)} \approx \frac{1}{5.02737} \approx 0.19891$$

$$\sec\frac{7\pi}{16} = \frac{1}{\cos(7\pi/16)} = \frac{1}{0.19509} \approx 5.12584$$

$$\csc\frac{7\pi}{16} = \frac{1}{\sin(7\pi/16)} = \frac{1}{0.98079} \approx 1.01959$$

24.

	$-\frac{\pi}{16}$	$-\frac{\pi}{8}$	$-\frac{\pi}{6}$	$-\frac{\pi}{4}$	$-\frac{\pi}{3}$
sin	-0.19509	-0.38268	-0.50000	-0.70711	-0.86603
cos	0.98079	0.92388	0.86603	0.70711	0.50000
tan	-0.19891	-0.41421	-0.57735	-1.00000	-1.73206
cot	-5.02737	-2.41424	-1.73206	-1.00000	-0.57735
sec	1.01959	1.08239	1.15469	1.41421	2.00000
csc	-5.12584	-2.61315	-2.00000	-1.41421	-1.15469

25. From Margin Exercise 34, we know that

$$\sin s = \pm\sqrt{1 - \cos^2 s},$$

where the sign of the radical depends on the quadrant in which s lies. Since s is in the first quadrant, the sine function is positive.

$$\sin s = \sqrt{1 - \cos^2 s} = \sqrt{1 - \left(\frac{1}{3}\right)^2} = \sqrt{1 - \frac{1}{9}} =$$

$$\sqrt{\frac{8}{9}} = \frac{\sqrt{8}}{3} = \frac{2\sqrt{2}}{3}$$

Now that we know sin s and cos s, we can find the other function values.

$$\tan s = \frac{\sin s}{\cos s} = \frac{\frac{2\sqrt{2}}{3}}{\frac{1}{3}} = 2\sqrt{2}$$

$$\cot s = \frac{1}{\tan s} = \frac{1}{2\sqrt{2}} = \frac{\sqrt{2}}{4}$$

$$\sec s = \frac{1}{\cos s} = \frac{1}{\frac{1}{3}} = 3$$

$$\csc s = \frac{1}{\sin s} = \frac{1}{\frac{2\sqrt{2}}{3}} = \frac{3}{2\sqrt{2}} = \frac{3\sqrt{2}}{4}$$

26. $\cos = \frac{\sqrt{5}}{3}$, $\tan s = \frac{2\sqrt{5}}{5}$, $\cot s = \frac{\sqrt{5}}{2}$, $\sec s = \frac{3\sqrt{5}}{5}$, $\csc s = \frac{3}{2}$

27. Solving the identity $1 + \tan^2 s = \sec^2 s$ for sec s we get

$$\sec s = \pm\sqrt{1 + \tan^2 s},$$

where the sign of the radical depends on the quadrant in which s lies. We know s is in the third quadrant, $\sec s = \frac{1}{\cos s}$, and the cosine function is negative in the third quadrant, so the secant function is also negative in the third quadrant. Thus,

$$\sec s = -\sqrt{1 + \tan^2 s} = -\sqrt{1 + 3^2} = -\sqrt{10}.$$

Next we find cos s:

$$\cos s = \frac{1}{\sec s} = \frac{1}{-\sqrt{10}} = -\frac{\sqrt{10}}{10}$$

Since the sine function is negative in the third quadrant, we have

$$\sin s = -\sqrt{1 - \cos^2 s} = -\sqrt{1 - \left(-\frac{\sqrt{10}}{10}\right)^2} =$$

$$-\sqrt{1 - \frac{10}{100}} = -\sqrt{\frac{9}{10}} = -\frac{3}{\sqrt{10}} = -\frac{3\sqrt{10}}{10}.$$

Next we find csc s:

$$\csc s = \frac{1}{\sin s} = \frac{1}{-\frac{3\sqrt{10}}{10}} = -\frac{\sqrt{10}}{3}$$

Finally we find cot s:

$$\cot s = \frac{1}{\tan s} = \frac{1}{3}$$

28. $\csc s = -\sqrt{17}$, $\sin s = -\frac{\sqrt{17}}{17}$, $\cos s = -\frac{4\sqrt{17}}{17}$, $\sec s = -\frac{\sqrt{17}}{4}$, and $\tan s = \frac{1}{4}$

29. First we find cos s:

$$\cos s = \frac{1}{\sec s} = \frac{1}{-\frac{5}{3}} = -\frac{3}{5}$$

Since the sine function is positive in the second quadrant, we have

$$\sin s = \sqrt{1 - \cos^2 s} = \sqrt{1 - \left(-\frac{3}{5}\right)^2} = \sqrt{1 - \frac{9}{25}} =$$

$$\sqrt{\frac{16}{25}} = \frac{4}{5}.$$

Now that we have sin s and cos s, we can find the remaining function values.

$$\tan s = \frac{\sin s}{\cos s} = \frac{\frac{4}{5}}{-\frac{3}{5}} = -\frac{4}{3}$$

$$\cot s = \frac{1}{\tan s} = \frac{1}{-\frac{4}{3}} = -\frac{3}{4}$$

$$\csc s = \frac{1}{\sin s} = \frac{1}{\frac{4}{5}} = \frac{5}{4}$$

30. $\sin s = -\frac{4}{5}$, $\cos s = \frac{3}{5}$, $\tan s = -\frac{4}{3}$,

$\cot s = -\frac{3}{4}$, and $\sec s = \frac{5}{3}$

31. Since the cosine function is negative in the third quadrant, we have

$$\cos s = -\sqrt{1 - \sin^2 s} = -\sqrt{1 - \left(-\frac{2}{5}\right)^2} =$$

$$-\sqrt{\frac{21}{25}} = -\frac{\sqrt{21}}{5}.$$

Now that we have sin s and cos s, we can find the other function values:

$$\tan s = \frac{\sin s}{\cos s} = \frac{-\frac{2}{5}}{-\frac{\sqrt{21}}{5}} = \frac{2}{\sqrt{21}} = \frac{2\sqrt{21}}{21}$$

$$\cot s = \frac{1}{\tan s} = \frac{1}{\frac{2\sqrt{21}}{21}} = \frac{\sqrt{21}}{2}$$

$$\sec s = \frac{1}{\cos s} = \frac{1}{-\frac{\sqrt{21}}{5}} = -\frac{5}{\sqrt{21}} = -\frac{5\sqrt{21}}{21}$$

$$\csc s = \frac{1}{\sin s} = \frac{1}{-\frac{2}{5}} = -\frac{5}{2}$$

32. $\sin s = -\frac{\sqrt{35}}{6}$, $\tan s = \sqrt{35}$, $\cot s = \frac{\sqrt{35}}{35}$,

$\sec s = -6$, and $\csc s = -\frac{6\sqrt{35}}{35}$

33. See the answer section in the text.

34. $\csc(-x) = \frac{1}{\sin(-x)} = \frac{1}{-\sin x} = -\csc x$

35. See the answer section in the text.

36. $\cot(x - \pi) = \frac{\cos(x - \pi)}{\sin(x - \pi)} = \frac{-\cos x}{-\sin x} = \cot x$

37. See the answer section in the text.

38. $\tan(\pi - x) = \frac{\sin(\pi - x)}{\cos(\pi - x)} = \frac{\sin x}{-\cos x} = -\tan x$

39. See the answer section in the text.

40. The graph of tan $(x + \pi)$ is like that of tan x, but moved π units to the left. This graph is identical to that of tan x.

41. See the answer section in the text.

42. If the graph of sec x were translated to the right $\frac{\pi}{2}$ units, the graph of csc x would be obtained. There are other descriptions.

43. Studying the graphs of the six circular functions we see that

$\sin x = 0$ when $x = 0, \pm\pi, \pm 2\pi, \pm 3\pi, \ldots$

$\cos x = 0$ when $x = \pm\frac{\pi}{2}, \pm\frac{3}{2}\pi, \pm\frac{5}{2}\pi, \ldots$

$\tan x = 0$ when $x = 0, \pm\pi, \pm 2\pi, \pm 3\pi, \ldots$

$\cot x = 0$ when $x = \pm\frac{\pi}{2}, \pm\frac{3}{2}\pi, \pm\frac{5}{2}\pi, \ldots$

sec x is never 0.

$\left[\text{Think: } \sec x = \frac{1}{\cos x}; \frac{1}{\cos x} \text{ is never } 0\right]$

csc x is never 0.

$\left[\text{Think: } \csc x = \frac{1}{\sin x}; \frac{1}{\sin x} \text{ is never } 0\right]$

In the sine and tangent functions, the same inputs give outputs of 0. Also in the cosine and cotangent functions, the same inputs give outputs of 0.

44. The tangent and secant functions have the same asymptotes. The cotangent and cosecant functions have the same asymptotes. The inputs producing zero outputs of the first two occur where the asymptotes of the latter occur, and conversely.

45. See the answer section in the text.

46.

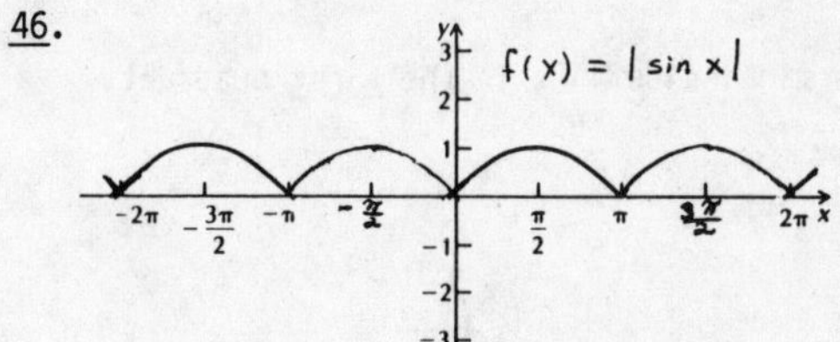

47. See the answer section in the text.

48.

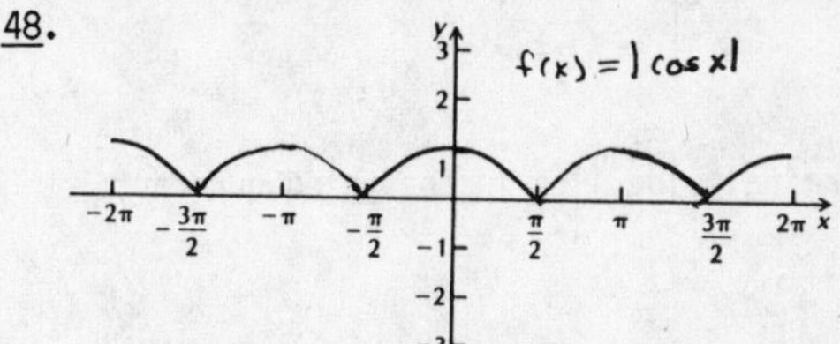

49. See the answer section in the text.

50.

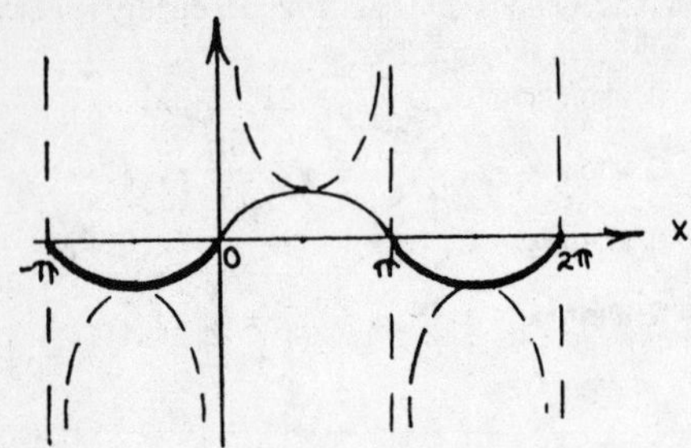

It is true that sin x > csc x for all x in the intervals $\left[\pi + k\cdot 2\pi, \frac{3}{2}\pi + k\cdot 2\pi\right)$ and $\left(\frac{3}{2}\pi + k\cdot 2\pi, 2\pi + k\cdot 2\pi\right]$.

Exercise Set 6.4

1.

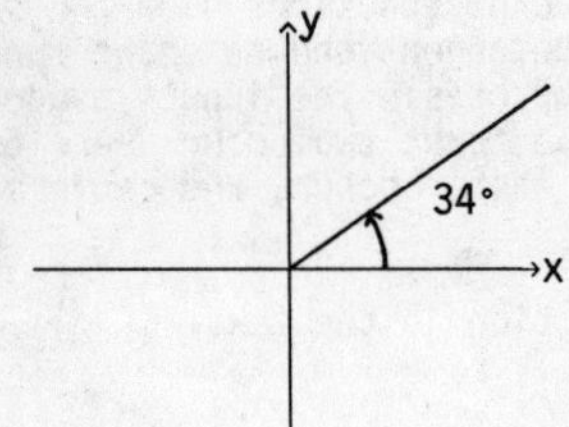

The terminal side lies in the first quadrant.

2. IV

3.

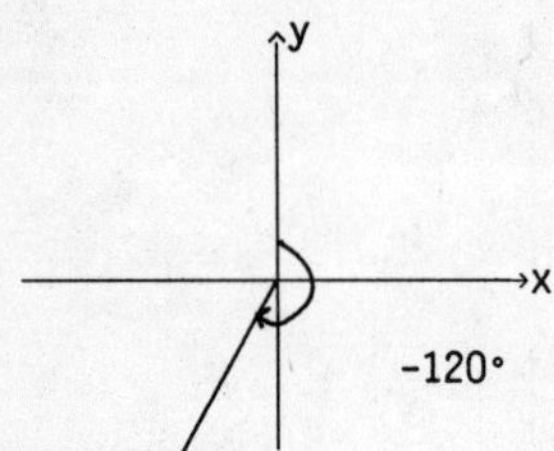

The terminal side lies in the third quadrant.

4. II

5.

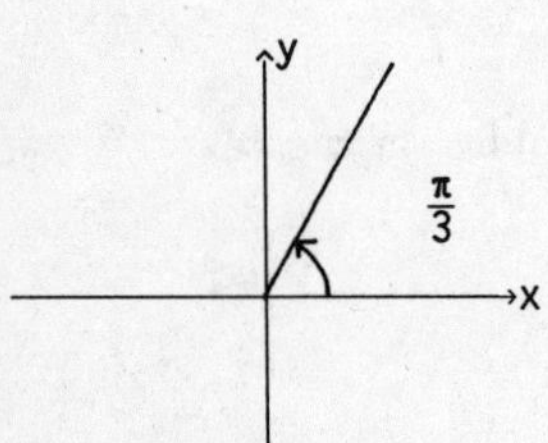

The terminal side lies in the first quadrant.

6. III

7.

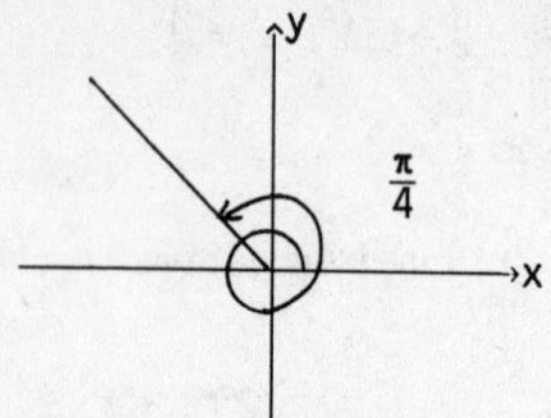

The terminal side lies in the second quadrant.

8. II

9. To get two positive angles coterminal with 58° we add 360° and 720°:

58° + 360° = 418° and 58° + 720° = 778°

To get two negative angles coterminal with 58° we add -360° and -720°:

58° + (-360°) = -302° and 58° + (-720°) = -662°

Answers may vary.

10. Positive: 680°, 1040°; negative: -40°, -400°. Answers may vary.

11. To get two positive angles coterminal with -120° we add 360° and 720°:

-120° + 360° = 240° and -120° + 720° = 600°

To get two negative angles coterminal with -120° we add -360° and -720°:

-120° + (-360°) = -480° and -120° + (-720°) = -840°

Answers may vary.

12. Positive: 145°, 505°; negative: -575°, -935°. Answers may vary.

13. 90° = 89°60'
 -57°23'
 32°37'

180° = 179°60'
 -57°23'
 122°37'

The complement of 57°23' is 32°37' and the supplement is 122°37'.

14. 42°22', 132°22'

15. 90° = 89°59'60"
 -73°45'11"
 16°14'49"

180° = 179°59'60"
 -73°45'11"
 106°14'49"

The complement of 73°45'11" is 16°14'49" and the supplement is 106°14'49".

16. 77°56'46", 167°56'46"

17. $90° - 67.31° = 22.69°$,
$180° - 67.31° = 112.69°$

The complement of 67.31° is 22.69° and the supplement is 112.69°.

18. 76.32°, 166.32°

19. $90° - 11.2344° = 78.7656°$,
$180° - 11.2344° = 168.7656°$

The complement of 11.2344° is 78.7656° and the supplement is 168.7656°.

20. 4.93688°, 94.93688°

21. $8.6° = 8° + 0.6 \times 1°$

Now substituting 60' for 1°,
$0.6 \times 1° = 0.6 \times 60' = 36'$.

So 8.6° = 8° 36'.

22. 47° 48'

23. $72.25° = 72° + 0.25 \times 1°$

Substituting 60' for 1°,
$0.25 \times 1° = 0.25 \times 60' = 15'$.

So 72.25° = 72° 15'.

24. 11° 45'

25. $$\begin{aligned} 46.38° &= 46° + 0.38 \times 1° \\ &= 46° + 0.38 \times 60' \qquad (1° = 60') \\ &= 46° + 22.8' \\ &\approx 46° 23' \end{aligned}$$

26. 85° 13'

27. $$\begin{aligned} 67.84° &= 67° + 0.84 \times 1° \\ &= 67° + 0.84 \times 60' \qquad (1° = 60') \\ &= 67° + 50.4' \\ &\approx 67° 50' \end{aligned}$$

28. 38° 29'

29. $$\begin{aligned} 87.3456° &= 87° + 0.3456 \times 1° \\ &= 87° + 0.3456 \times 60' \qquad (1° = 60') \\ &= 87° + 20.736' \\ &= 87° 20' + 0.736 \times 1' \\ &= 87° 20' + 0.736 \times 60'' \qquad (1' = 60'') \\ &= 87° 20' + 44.16'' \\ &\approx 87° 20' 44'' \end{aligned}$$

30. 11° 1' 32"

31. $$\begin{aligned} 48.02498° &= 48° + 0.02498 \times 1° \\ &= 48° + 0.02498 \times 60' \qquad (1° = 60') \\ &= 48° + 1.4988' \\ &= 48° 1' + 0.4988 \times 1' \\ &= 48° 1' + 0.4988 \times 60'' \qquad (1' = 60'') \\ &= 48° 1' + 29.928'' \\ &\approx 48° 1' 30'' \end{aligned}$$

32. 27° 53' 58"

33. $$\begin{aligned} 9° 45' &= 9° + \left(\frac{45}{60}\right)° \qquad \left(1' = \frac{1°}{60}\right) \\ &= 9° + 0.75° \\ &= 9.75° \end{aligned}$$

34. 52.25°

35. $$\begin{aligned} 35° 50' &= 35° + \left(\frac{50}{60}\right)° \qquad \left(1' = \frac{1°}{60}\right) \\ &\approx 35° + 0.83° \\ &= 35.83° \end{aligned}$$

36. 64.67°

37. $$\begin{aligned} 80° 33' &= 80° + \left(\frac{33}{60}\right)° \qquad \left(1' = \frac{1°}{60}\right) \\ &= 80° + 0.55° \\ &= 80.55° \end{aligned}$$

38. 27.32°

39. $$\begin{aligned} 3° 2' &= 3° + \left(\frac{2}{60}\right)° \qquad \left(1' = \frac{1°}{60}\right) \\ &\approx 3° + 0.03° \\ &= 3.03° \end{aligned}$$

40. 10.13°

41. $$\begin{aligned} 19° 47' 23'' &= 19° + \left(\frac{47}{60}\right)^0 + \left(\frac{23}{3600}\right)^0 \\ &\qquad \left(1' = \frac{1^0}{60} \text{ and } 1'' = \frac{1}{3600}^0\right) \\ &\approx 19° + 0.7833° + 0.0064° \\ &= 19.7897° \end{aligned}$$

42. 49.6461°

43. $$\begin{aligned} 31° 57' 55'' &= 31° + \left(\frac{57}{60}\right)^0 + \left(\frac{55}{3600}\right)^0 \\ &\qquad \left(1' = \frac{1^0}{60} \text{ and } 1'' = \frac{1}{3600}^0\right) \\ &\approx 31° + 0.95° + 0.0153° \\ &= 31.9653° \end{aligned}$$

44. 76.1928°

45. $30° = 30° \cdot \frac{\pi \text{ radians}}{180°} = \frac{\pi}{6}$ radians, or $\frac{\pi}{6}$.

46. $\frac{\pi}{12}$

47. $60° = 60° \cdot \frac{\pi \text{ radians}}{180°} = \frac{\pi}{3}$ radians, or $\frac{\pi}{3}$.

48. $\frac{10}{9}\pi$

49. $75° = 75° \cdot \frac{\pi \text{ radians}}{180°} = \frac{5}{12}\pi$ radians, or $\frac{5}{12}\pi$.

50. $\frac{5}{3}\pi$

51. $37.71° = 37.71° \cdot \frac{\pi \text{ radians}}{180°}$

$= 0.2095\pi$ radians, or 0.2095π

52. 0.0707π

53. $214.6° = 214.6° \cdot \frac{\pi \text{ radians}}{180°}$

$\approx 1.1922\pi$ radians, or 1.1922π

54. 0.4104π

55. $120° = 120° \cdot \frac{\pi \text{ radians}}{180°} = \frac{2}{3}\pi \approx 2.0933$

56. 4.187

57. $320° = 320° \cdot \frac{\pi \text{ radians}}{180°} = \frac{16}{9}\pi \approx 5.5822$

58. 1.308

59. $200° = 200° \cdot \frac{\pi \text{ radians}}{180°} = \frac{10}{9}\pi \approx 3.4889$

60. 5.233

61. $117.8° = 117.8° \cdot \frac{\pi \text{ radians}}{180°} = \frac{117.8}{180}\pi \approx 2.0550$

62. 4.0332

63. $1.354° = 1.354° \cdot \frac{\pi \text{ radians}}{180°} = \frac{1.354}{180}\pi \approx 0.0236$

64. 5.7200

65. $1 \text{ radian} = 1 \text{ radian} \cdot \frac{180°}{\pi \text{ radians}}$

$\approx \left(\frac{180}{3.14}\right)° \approx 57.32°$

66. 114.6°

67. $8\pi = 8\pi \text{ radians} \cdot \frac{180°}{\pi \text{ radians}}$

$= 8 \cdot 180° = 1440°$

68. -2160°

69. $\frac{3}{4}\pi = \frac{3}{4}\pi \text{ radians} \cdot \frac{180°}{\pi \text{ radians}}$

$= \frac{3}{4} \cdot 180° = 135°$

70. 225°

71. $1.303 = 1.303 \text{ radians} \cdot \frac{180°}{\pi \text{ radians}}$

$\approx \frac{1.303 \cdot 180°}{3.14} \approx 74.69°$

72. 134.5°

73. $0.7532\pi = 0.7532\pi \text{ radians} \cdot \frac{180°}{\pi \text{ radians}}$

$= 0.7532 \cdot 180° = 135.576°$

74. -216.9°

75. $0° = 0$ radians, $30° = \frac{\pi}{6}$, $60° = \frac{\pi}{3}$, $135° = \frac{3}{4}\pi$, $180° = \pi$, $225° = \frac{5}{4}\pi$, $270° = \frac{3}{2}\pi$, $315° = \frac{7}{4}\pi$

76. $-30° = -\frac{\pi}{6}$, $-60° = -\frac{\pi}{3}$, $-90° = -\frac{\pi}{2}$, $-135° = -\frac{3}{4}\pi$

$-225° = -\frac{5}{4}\pi$, $-270° = -\frac{3}{2}\pi$, $-315° = -\frac{7}{4}\pi$

77. $\theta = \frac{s}{r}$

θ is radian measure of central angle, s is arc length, and r is radius length.

$\theta = \frac{132 \text{ cm}}{120 \text{ cm}}$ (Substituting 132 cm for s and 120 cm for r)

$\theta = \frac{11}{10}$, or 1.1 (The unit is understood to be radians)

$1.1 \text{ radians} = 1.1 \text{ radians} \cdot \frac{180°}{\pi \text{ radians}} \approx 63°$.

78. 0.325 radians, 19°

79. In 60 minutes the minute hand rotates 2π radians. In 50 minutes it rotates

$\frac{50}{60} \cdot 2\pi$, or 5.233 radians.

80. 4526 radians

81. $\theta = \frac{s}{r}$

$1.6 = \frac{s}{10 \text{ m}}$ (Substituting 1.6 for θ and 10 m for r)

$1.6(10 \text{ m}) = s$

$16 \text{ m} = s$

82. 10.5 m

83. Since the linear speed must be in cm/min, the given angular speed, 7 radians/sec, must be changed to radians/min.

$$\omega = \frac{7 \text{ radians}}{1 \text{ sec}} = \frac{7}{1 \text{ sec}} \cdot \frac{60 \text{ sec}}{1 \text{ min}} = \frac{420}{1 \text{ min}}$$

$$r = \frac{d}{2} = \frac{15 \text{ cm}}{2} = 7.5 \text{ cm}$$

Using $v = r\omega$, we have

$$v = 7.5 \text{ cm} \cdot \frac{420}{1 \text{ min}} \qquad \text{(Substituting)}$$

$$= 3150 \frac{\text{cm}}{\text{min}}$$

84. 54 m/min

85. First change ω to radians per second.

$$\omega = 33\frac{1}{3} \frac{\text{rev}}{\text{min}} = 33\frac{1}{3} \cdot \frac{2\pi}{\text{min}} \qquad \text{(Substituting } 2\pi \text{ for 1 rev)}$$

$$= 33\frac{1}{3} \cdot \frac{2\pi}{\text{min}} \cdot \frac{1 \text{ min}}{60 \text{ sec}}$$

$$\approx 3.4888 \frac{\text{radians}}{\text{sec}}$$

Using $v = r\omega$, we have

$$v = 15 \text{ cm} \cdot \frac{3.4888}{1 \text{ sec}} \approx 52.33 \frac{\text{cm}}{\text{sec}}$$

86. 41 cm/sec

87. First find ω in radians per hour.

$$\omega = \frac{1 \text{ rev}}{24 \text{ hr}} = \frac{2\pi}{24 \text{ hr}} = \frac{\pi}{12} \frac{\text{radians}}{\text{hr}}$$

Using $v = r\omega$, we have

$$v = 4000 \text{ mi} \cdot \frac{\pi}{12 \text{ hr}} = 1047 \text{ mph}$$

88. 66,626 mph

89. The units of distance for v and r must be the same. Thus,

$$v = 11 \frac{\text{m}}{\text{s}} \cdot \frac{100 \text{ cm}}{1 \text{ m}} = 1100 \frac{\text{cm}}{\text{s}}$$

$$r = \frac{d}{2} = \frac{32 \text{ cm}}{2} = 16 \text{ cm}$$

Using $v = r\omega$, we have

$$1100 \frac{\text{cm}}{\text{s}} = 16 \cdot \text{cm} \cdot \omega$$

$$\frac{1100}{16} \cdot \frac{\text{radians}}{\text{s}} = \omega$$

$$68.75 \frac{\text{radians}}{\text{sec}} \approx \omega$$

90. 1429 rad/hr

91. First change ω to radians per hour.

$$\omega = 14 \frac{\text{rev}}{\text{min}} = 14 \cdot \frac{2\pi}{\text{min}} \cdot \frac{60 \text{ min}}{1 \text{ hr}} \approx 5275.2 \frac{\text{radians}}{\text{hr}}$$

Next change 10 ft to miles.

$$r = 10 \text{ ft} \cdot \frac{1 \text{ mi}}{5280 \text{ ft}} = \frac{1}{528} \text{ mi}$$

Using $v = r\omega$, we have

$$v = \frac{1}{528} \text{ mi} \cdot \frac{5275.2}{1 \text{ hr}} \approx 10 \text{ mph}$$

92. 11.4 mph

93. $$\omega = 12 \frac{\text{rev}}{\text{min}} = 12 \cdot \frac{2\pi}{\text{min}}$$

$$= 24\pi \frac{\text{radians}}{\text{min}} \approx 75.36 \frac{\text{radians}}{\text{min}}$$

$$r = \frac{d}{2} = \frac{24 \text{ in}}{2} = 12 \text{ in} = 1 \text{ ft}$$

Using $v = r\omega$, we have

$$v = 1 \text{ ft} \cdot \frac{75.36}{\text{min}} = 75.36 \frac{\text{ft}}{\text{min}}$$

The bike will travel 75.36 ft in 1 min.

94. 4710 ft

95. First find v in ft/sec.

$$v = 30 \frac{\text{mi}}{\text{hr}} = 30 \cdot \frac{5280 \text{ ft}}{3600 \text{ sec}} = 44 \frac{\text{ft}}{\text{sec}}$$

Next find r in ft.

$$r = 14 \text{ in} = 14 \text{ in} \cdot \frac{1 \text{ ft}}{12 \text{ in}} = \frac{7}{6} \text{ ft.}$$

Using $v = r\omega$, we have

$$44 \frac{\text{ft}}{\text{sec}} = \frac{7}{6} \text{ ft} \cdot \omega$$

$$\frac{6}{7} \cdot 44 \frac{\text{radians}}{\text{sec}} = \omega$$

$$37.7 \frac{\text{radians}}{\text{sec}} \approx \omega$$

If $\omega = 37.7 \frac{\text{radians}}{\text{sec}}$, then the angle through which a wheel rotates is $37.7 \frac{\text{radians}}{\text{sec}} \cdot 10$ sec, or 377 radians.

96. 563 radians

97. $$90° = \frac{\pi}{2} \text{ radians} = 100 \text{ grads}$$

Note the following:

$$1 = \frac{90°}{100 \text{ grads}}; \text{ also } \frac{100 \text{ grads}}{90°} = 1$$

$$1 = \frac{\frac{\pi}{2} \text{ radians}}{100 \text{ grads}} = \frac{\pi \text{ radians}}{200 \text{ grads}}; \text{ also } \frac{200 \text{ grads}}{\pi \text{ radians}} = 1$$

a) $$48° = 48° \cdot \frac{100 \text{ grads}}{90°}$$

$$= \frac{48}{90} \cdot 100 \text{ grads} = 53.3 \text{ grads}$$

97. (continued)

b) $153° = 153° \cdot \frac{100 \text{ grads}}{90°}$

$= \frac{153}{90} \cdot 100 \text{ grads} = 170 \text{ grads}$

c) $\frac{\pi}{8}$ radians $= \frac{\pi}{8}$ radians $\cdot \frac{200 \text{ grads}}{\pi \text{ radians}}$

$= \frac{200}{8} \text{ grads} = 25 \text{ grads}$

d) $\frac{5\pi}{7}$ radians $= \frac{5\pi}{7}$ radians $\cdot \frac{200 \text{ grads}}{\pi \text{ radians}}$

$= \frac{5}{7} \cdot 200 \text{ grads} \approx 142.9 \text{ grads}$

98. a) 5° 37' 30"

b) 19° 41' 15"

99. One degree of latitude is $\frac{1}{360}$ of the circumference of the earth.

$C = \pi d$, or $2\pi r$

When $r = 6400$ km, $C = 2\pi \cdot 6400 \approx 40{,}192$ km.
Thus 1° of latitude is $\frac{1}{360} \cdot 40{,}192$, or ≈ 112 km.

When $r = 4000$ mi, $C = 2\pi \cdot 4000 \approx 25{,}120$ mi.
Thus 1° of latitude is $\frac{1}{360} \cdot 25{,}120$, or ≈ 70 mi.

100. a) 21,600 NM

b) 3439 NM

101.

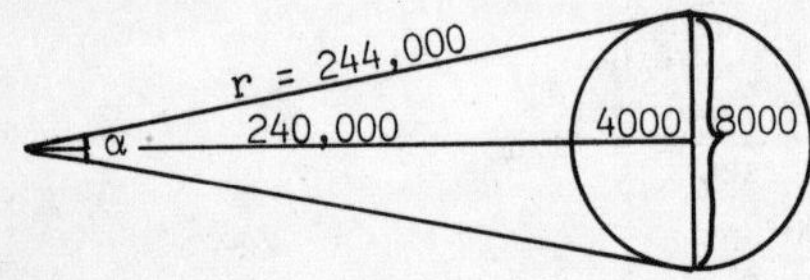

Use $\alpha = \frac{s}{r}$.

The arc, s, is approximately 8000 miles, the approximate length of the earth's diameter. The radius, r, is the sum of the earth's distance away from the astronaut, 240,000 miles, and the earth's radius, 4000 miles. Thus r is approximately 244,000 miles.

$\alpha = \frac{s}{r}$

$\alpha = \frac{8{,}000}{244{,}000}$ (Substituting 8000 for s and 244,000 for r)

$\alpha \approx 0.03$ radian

102. 25,000 mi

103.

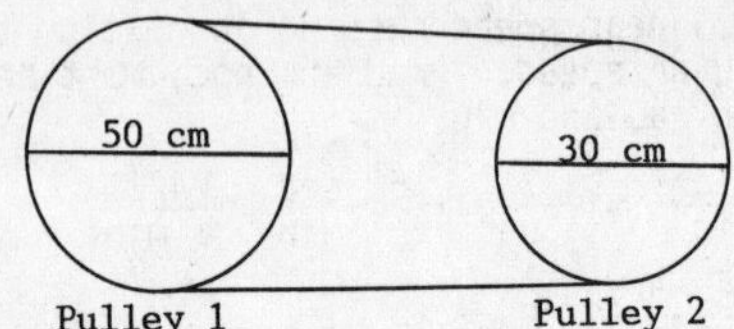

Since the pulleys are connected by the same belt, their linear speed v, will be the same.

$v = r_1\omega_1 = r_2\omega_2$

Then $\frac{r_1}{r_2} = \frac{\omega_2}{\omega_1}$

Find ω_1, the angular speed of the larger pulley, in radians/sec. The larger pulley makes 12 revolutions per minute (each revolution is 2π radians).

$\omega_1 = \frac{12 \cdot 2\pi \text{ radians}}{1 \text{ min}}$

$= \frac{24\pi \text{ radians}}{1 \text{ min}}$

$= \frac{24\pi \text{ radians}}{1 \text{ min}} \cdot \frac{1 \text{ min}}{60 \text{ sec}}$ (Multiplying by 1)

$= \frac{1.256 \text{ radians}}{1 \text{ sec}}$

$\frac{r_1}{r_2} = \frac{\omega_2}{\omega_1}$

$\frac{25}{15} = \frac{\omega_2}{1.256}$ (Substituting)

$1.256 \cdot \frac{25}{15} = \omega_2$

$2.093 = \omega_2$

The smaller pulley has an angular speed of 2.093 radians/sec.

104. 1.675 rad/sec

105. a) Angular acceleration $\approx \frac{(2500 - 800) \text{ rpm}}{4.3 \text{ sec}}$

≈ 395.35 rpm/sec

b) $395.35 \frac{\text{rpm}}{\text{sec}} = 395.35 \frac{\text{rev}}{\text{min}} \cdot \frac{1}{\text{sec}}$

$= 395.35 \frac{2\pi \text{ radians}}{\text{min}} \cdot \frac{1}{\text{sec}} \cdot \frac{1 \text{ min}}{60 \text{ sec}}$

$= \frac{395.35 \times 2}{60} \pi \cdot \frac{\text{radians}}{\text{sec}^2}$

≈ 41.38 radians/sec^2

106. a) 12.5 knots/sec

b) $1.01(10)^{-6}$ rad/sec^2

107.

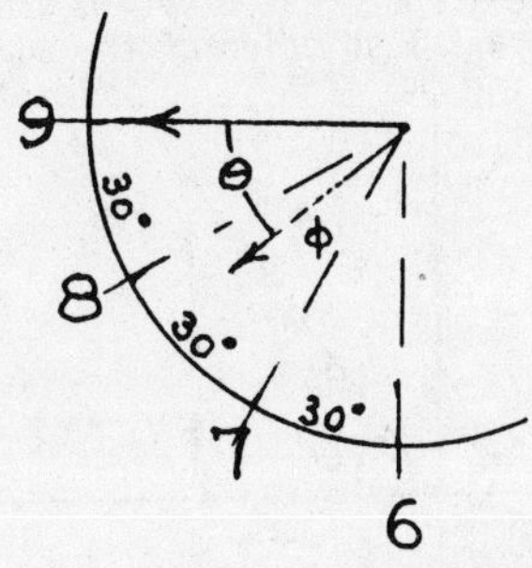

In going from 7:00 to 7:45, the hour hand moves $\frac{3}{4}(30°) = 22.5° = \phi$. Then $\theta = 60° - \phi = 37.5°$.

108. 12:16:22 (to the nearest second)

109. From Exercise 100, the earth's diameter $= 2(3439) = 6878$ NM.

Then $\frac{7926.4 \text{ statute mi}}{6878 \text{ NM}} \approx 1.15$ statute mi/NM.

110. 1.46 NM

Exercise Set 6.5

1. $\cos 180° = \cos \pi = -1$

2. Undefined

3. $\sin 45° = \sin \frac{\pi}{4} = \frac{\sqrt{2}}{2}$

4. 1

5. $\cos(-135°) = \cos\left(-\frac{3\pi}{4}\right) = -\frac{\sqrt{2}}{2}$

6. $\frac{1}{2}$

7. $\cot(-60°) = \cot\left(-\frac{\pi}{3}\right) = -\frac{1}{\sqrt{3}}$, or $-\frac{\sqrt{3}}{3}$

8. $\sqrt{3}$

9.

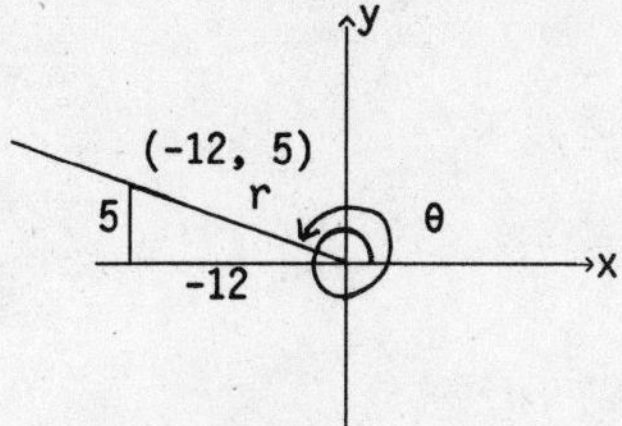

We first determine r.

$r = \sqrt{x^2 + y^2}$

$r = \sqrt{(-12)^2 + 5^2}$

$= \sqrt{144 + 25}$

$= \sqrt{169}$

$= 13$

Substituting -12 for x, 5 for y, and 13 for r, the trigonometric function values of θ are

$\sin\theta = \frac{y}{r} = \frac{5}{13}$

$\cos\theta = \frac{x}{r} = \frac{-12}{13} = -\frac{12}{13}$

$\tan\theta = \frac{y}{x} = \frac{5}{-12} = -\frac{5}{12}$

$\cot\theta = \frac{x}{y} = \frac{-12}{5} = -\frac{12}{5}$

$\sec\theta = \frac{r}{x} = \frac{13}{-12} = -\frac{13}{12}$

$\csc\theta = \frac{r}{y} = \frac{13}{5}$

10. $\sin\theta = -\frac{5}{13}$

$\cos\theta = -\frac{12}{13}$

$\tan\theta = \frac{5}{12}$

$\cot\theta = \frac{12}{5}$

$\sec\theta = -\frac{13}{12}$

$\csc\theta = -\frac{13}{5}$

11.

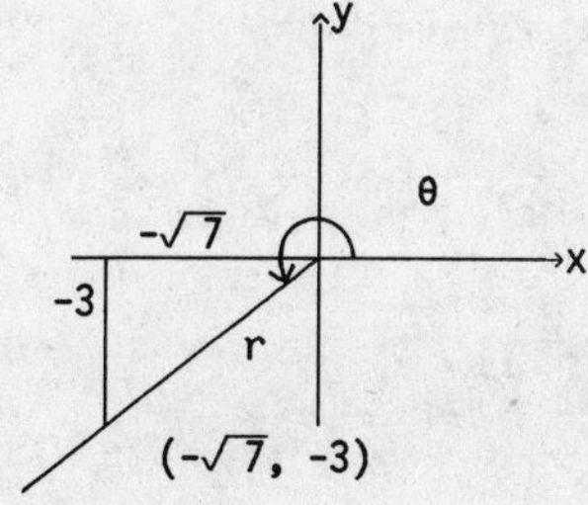

$r = \sqrt{x^2 + y^2}$

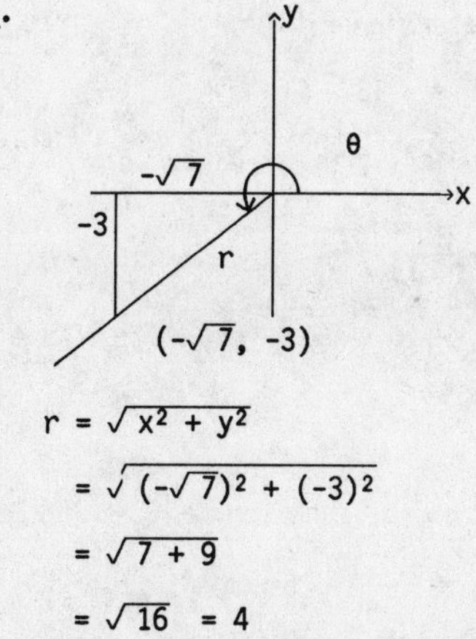

$= \sqrt{(-\sqrt{7})^2 + (-3)^2}$

$= \sqrt{7 + 9}$

$= \sqrt{16} = 4$

11. (continued)

$\sin\theta = \frac{y}{r} = \frac{-3}{4} = -\frac{3}{4}$

$\cos\theta = \frac{x}{r} = \frac{-\sqrt{7}}{4} = -\frac{\sqrt{7}}{4}$

$\tan\theta = \frac{y}{x} = \frac{-3}{-\sqrt{7}} = \frac{3}{\sqrt{7}}$, or $\frac{3\sqrt{7}}{7}$

$\cot\theta = \frac{x}{y} = \frac{-\sqrt{7}}{-3} = \frac{\sqrt{7}}{3}$

$\sec\theta = \frac{r}{x} = \frac{4}{-\sqrt{7}} = -\frac{4}{\sqrt{7}}$, or $-\frac{4\sqrt{7}}{7}$

$\csc\theta = \frac{r}{y} = \frac{4}{-3} = -\frac{4}{3}$

12. $\sin\theta = -\frac{3}{4}$

$\cos\theta = \frac{\sqrt{7}}{4}$

$\tan\theta = -\frac{3}{\sqrt{7}}$, or $-\frac{3\sqrt{7}}{7}$

$\cot\theta = -\frac{\sqrt{7}}{3}$

$\sec\theta = \frac{4}{\sqrt{7}}$, or $\frac{4\sqrt{7}}{7}$

$\csc\theta = -\frac{4}{3}$

13. First we draw the graph of $2x + 3y = 0$ and determine a quadrant IV solution of the equation.

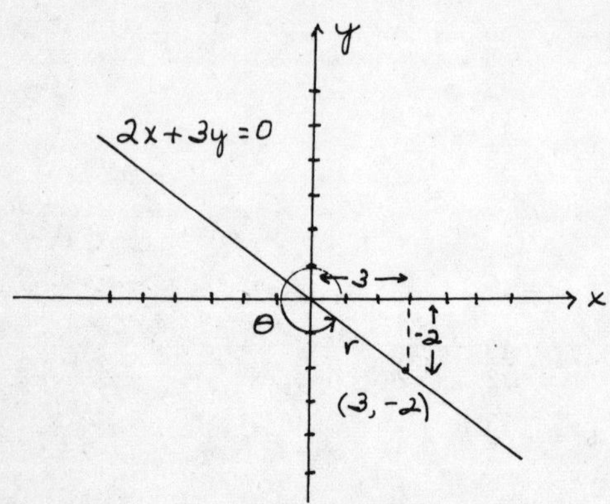

Using (3, -2), we determine r:

$r = \sqrt{3^2 + (-2)^2} = \sqrt{13}$

Then using $x = 3$, $y = -2$, and $r = \sqrt{13}$, we find

$\sin\theta = \frac{-2}{\sqrt{13}} = -\frac{2\sqrt{13}}{13}$,

$\cos\theta = \frac{3}{\sqrt{13}} = \frac{3\sqrt{13}}{13}$,

$\tan\theta = \frac{-2}{3} = -\frac{2}{3}$.

14. $\sin\theta = \frac{4\sqrt{17}}{17}$, $\cos\theta = -\frac{\sqrt{17}}{17}$, $\tan\theta = -4$

15. First we draw the graph of $5x - 4y = 0$ and determine a quadrant I solution of the equation.

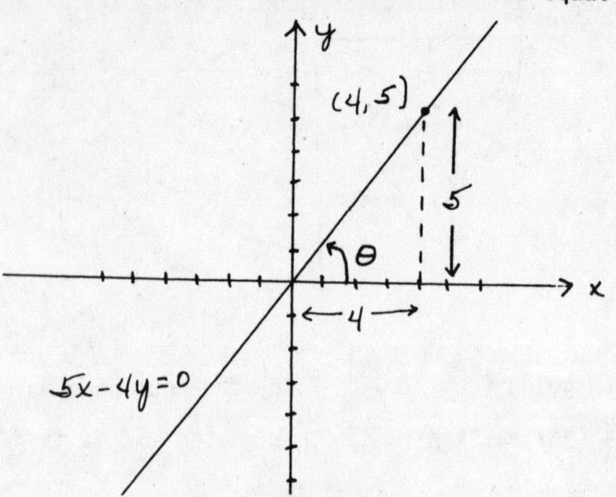

Using (4, 5), we determine r:

$r = \sqrt{4^2 + 5^2} = \sqrt{41}$

Then using $x = 4$, $y = 5$, and $r = \sqrt{41}$, we find

$\sin\theta = \frac{5}{\sqrt{41}} = \frac{5\sqrt{41}}{41}$,

$\cos\theta = \frac{4}{\sqrt{41}} = \frac{4\sqrt{41}}{41}$,

$\tan\theta = \frac{5}{4}$.

16. $\sin\theta = -\frac{4\sqrt{41}}{41}$, $\cos\theta = \frac{-5\sqrt{41}}{41}$, $\tan\theta = \frac{4}{5}$

17. $270° < 319° < 360°$, so P(x, y) is in the fourth quadrant. The cosine and secant are positive, and the other four function values are negative.

18. The cosine and secant are positive, and the other four function values are negative.

19.

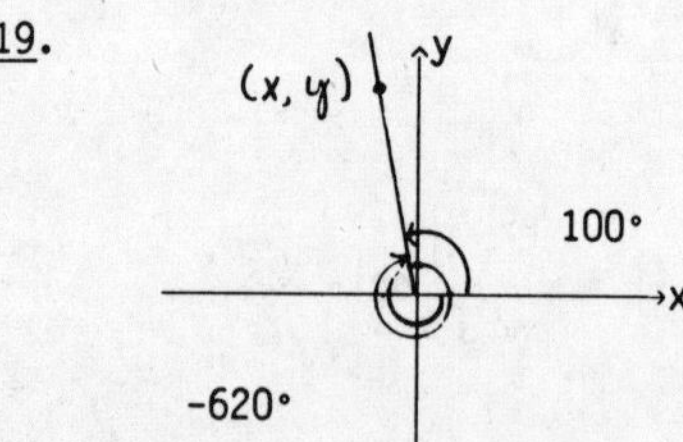

-620° has the same terminal side as 100°.

$90° < 100° < 180°$, so P(x, y) is in the second quadrant. The sine and cosecant are positive and the other four function values are negative.

20. The tangent and cotangent are positive, and the other four function values are negative.

21.

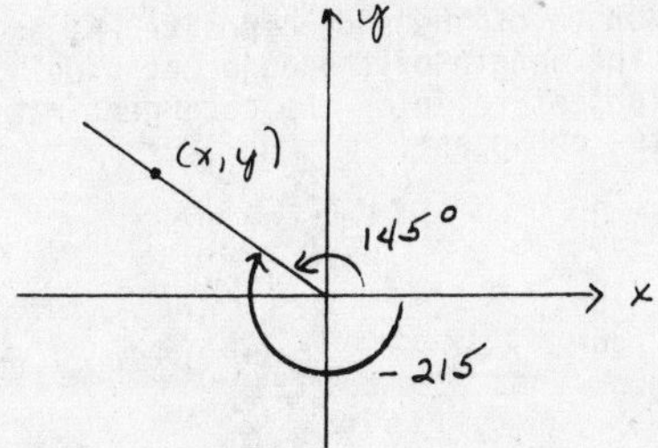

-215° has the same terminal side as 145°.

90° < 145° < 180°, so P(x, y) is in the second quadrant. The sine and cosecant are positive and the other four function values are negative.

22. The cosine and secant are positive and the other four function values are negative.

23. 90° < 91° < 180°, so P(x, y) is in the second quadrant. The sine and cosecant are positive and the other four function values are negative.

24. All six function values are positive.

25.

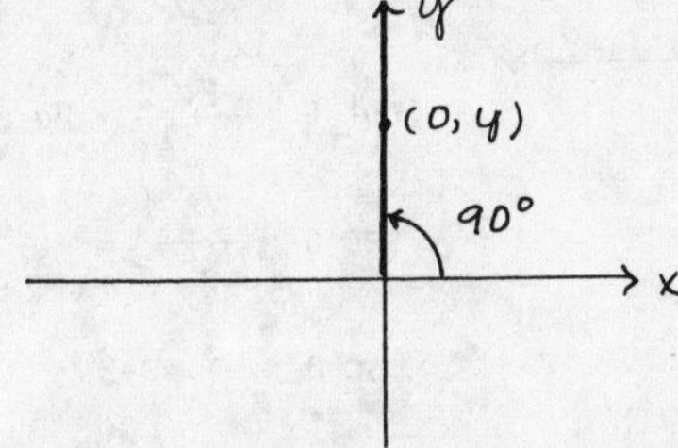

Let (0, y) represent any point, other than the vertex, on the terminal side of a 90° angle in standard position. We note that the first coordinate is 0, the second coordinate is positive, and r = y.

$\sin 90° = \frac{y}{r} = 1$ (Since r = y)

$\cos 90° = \frac{0}{r} = 0$

$\tan 90° = \frac{y}{0}$, tan 90° is undefined

$\cot 90° = \frac{0}{y} = 0$

$\sec 90° = \frac{r}{0}$, sec 90° is undefined

$\csc 90° = \frac{y}{r} = 1$ (Since r = y)

26. sin 360° = 0, cos 360° = 1, tan 360° = 0, cot 360° is undefined, sec 360° = 1, csc 360° is undefined

27.

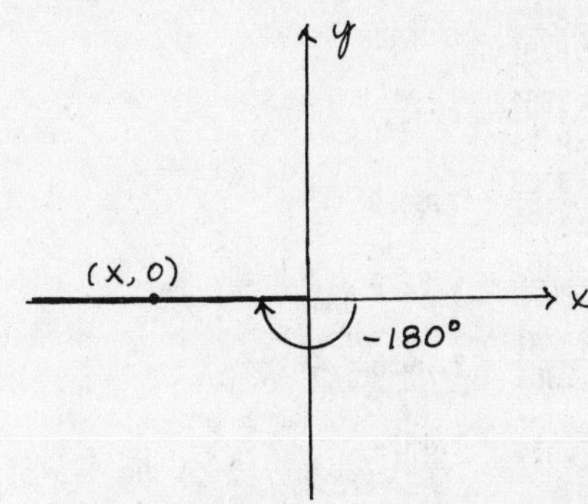

Let (x, 0) represent any point, other than the vertex, on the terminal side of a -180° angle in standard position. We note that the first coordinate is negative, the second coordinate is 0, and that x and r have the same absolute value.

$\sin (-180°) = \frac{0}{r} = 0$

$\cos (-180°) = \frac{x}{r} = -1$ (Since $|x| = |r|$, but x and r have opposite signs)

$\tan (-180°) = \frac{0}{x} = 0$

$\cot (-180°) = \frac{x}{0}$, cot (-180°) is undefined

$\sec (-180°) = \frac{r}{x} = -1$ (The reciprocal of cos (-180°))

$\csc (-180°) = \frac{r}{0}$, csc (-180°) is undefined

28. sin (-270°) = -1, cos (-270°) = 0, tan (-270°) is undefined, cot (-270°) = 0, sec (-270°) is undefined, csc (-270°) = -1.

29. $\sin \theta = \frac{\text{opposite}}{\text{hypotenuse}} = \frac{8}{17}$ $\csc \theta = \frac{\text{hypotenuse}}{\text{opposite}} = \frac{17}{8}$

$\cos \theta = \frac{\text{adjacent}}{\text{hypotenuse}} = \frac{15}{17}$ $\sec \theta = \frac{\text{hypotenuse}}{\text{adjacent}} = \frac{17}{15}$

$\tan \theta = \frac{\text{opposite}}{\text{adjacent}} = \frac{8}{15}$ $\cot \theta = \frac{\text{adjacent}}{\text{opposite}} = \frac{15}{8}$

30. $\sin \phi = \frac{15}{17}$, $\cos \phi = \frac{8}{17}$, $\tan \phi = \frac{15}{8}$

$\csc \phi = \frac{17}{15}$, $\sec \phi = \frac{17}{8}$, $\cot \phi = \frac{8}{15}$

31. $\sin \theta = \frac{\text{opposite}}{\text{hypotenuse}} = \frac{3}{h}$ $\csc \theta = \frac{\text{hypotenuse}}{\text{opposite}} = \frac{h}{3}$

$\cos \theta = \frac{\text{adjacent}}{\text{hypotenuse}} = \frac{7}{h}$ $\sec \theta = \frac{\text{hypotenuse}}{\text{adjacent}} = \frac{h}{7}$

$\tan \theta = \frac{\text{opposite}}{\text{adjacent}} = \frac{3}{7}$ $\cot \theta = \frac{\text{adjacent}}{\text{opposite}} = \frac{7}{3}$

32. $\sin \phi = \frac{7}{h}$ $\cos \phi = \frac{4}{h}$, $\tan \phi = \frac{7}{4}$

$\csc \phi = \frac{h}{7}$, $\sec \phi = \frac{h}{4}$, $\cot \phi = \frac{4}{7}$

33. $\sin\theta = \frac{7.8023}{8.8781} = 0.8788$

$\cos\theta = \frac{4.2361}{8.8781} = 0.4771$

$\tan\theta = \frac{7.8023}{4.2361} = 1.8419$

$\cot\theta = \frac{4.2361}{7.8023} = 0.5429$

$\sec\theta = \frac{8.8781}{4.2361} = 2.0958$

$\csc\theta = \frac{8.8781}{7.8023} = 1.1379$

34. $\sin\phi = 0.4771$, $\cos\phi = 0.8788$, $\tan\phi = 0.5429$
$\csc\phi = 2.0958$, $\sec\phi = 1.1379$, $\cot\phi = 1.8419$

35. $\sin 37.5° = \frac{\text{opp.}}{\text{hyp.}} = \frac{28}{\ell}$ $\quad \csc 37.5° = \frac{\text{hyp.}}{\text{opp.}} = \frac{\ell}{28}$

$\cos 37.5° = \frac{\text{adj.}}{\text{hyp.}} = \frac{d}{\ell}$ $\quad \sec 37.5° = \frac{\text{hyp.}}{\text{adj.}} = \frac{\ell}{d}$

$\tan 37.5° = \frac{\text{opp.}}{\text{adj.}} = \frac{28}{d}$ $\quad \cot 37.5° = \frac{\text{adj.}}{\text{opp.}} = \frac{d}{28}$

36. $\sin 36° = \frac{36}{c}$, $\cos 36° = \frac{a}{c}$, $\tan 36° = \frac{36}{a}$
$\csc 36° = \frac{c}{36}$, $\sec 36° = \frac{c}{a}$, $\cot 36° = \frac{a}{36}$

37.

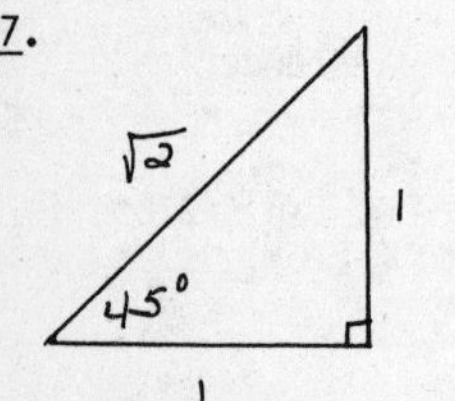

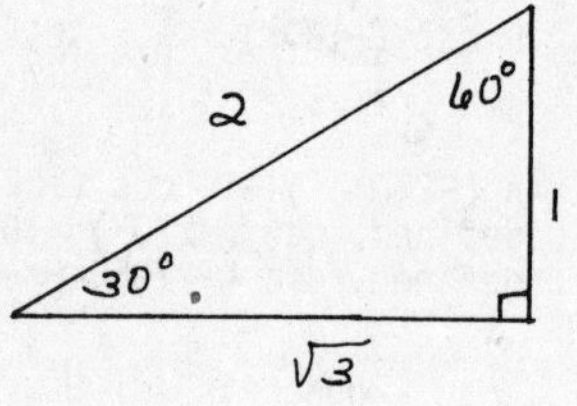

We use the triangles above to complete the table.

θ	sin	cos	tan	cot	sec	csc
45°	$\frac{\sqrt{2}}{2}$	$\frac{\sqrt{2}}{2}$	1	1	$\sqrt{2}$	$\sqrt{2}$
30°	$\frac{1}{2}$	$\frac{\sqrt{3}}{2}$	$\frac{\sqrt{3}}{3}$	$\sqrt{3}$	$\frac{2\sqrt{3}}{3}$	2
60°	$\frac{\sqrt{3}}{2}$	$\frac{1}{2}$	$\sqrt{3}$	$\frac{\sqrt{3}}{3}$	2	$\frac{2\sqrt{3}}{3}$

38.

θ	sin	cos	tan	cot	sec	csc
-45°	$-\frac{\sqrt{2}}{2}$	$\frac{\sqrt{2}}{2}$	-1	-1	$\sqrt{2}$	$-\sqrt{2}$
-30°	$-\frac{1}{2}$	$\frac{\sqrt{3}}{2}$	$-\frac{\sqrt{3}}{3}$	$-\sqrt{3}$	$\frac{2\sqrt{3}}{3}$	-2
-60°	$-\frac{\sqrt{3}}{2}$	$\frac{1}{2}$	$-\sqrt{3}$	$-\frac{\sqrt{3}}{3}$	2	$-\frac{2\sqrt{3}}{3}$

39. We know the length of the side opposite ∠B, and we want to find the length of the adjacent side. We can use the tangent ratio or the cotangent ratio. Here we use the cotangent:

$$\cot 30° = \frac{a}{36}$$
$$\sqrt{3} = \frac{a}{36}$$
$$36\sqrt{3} = a$$
$$62.4 \text{ m} \approx a$$

40. 32.3 ft

41. First we sketch a third quadrant triangle. Since $\sin\theta = -\frac{1}{3}$, we have a leg of length 1 and the hypotenuse of length 3. The other leg must then have length $\sqrt{8}$, or $2\sqrt{2}$. Now we can read off the appropriate ratios:

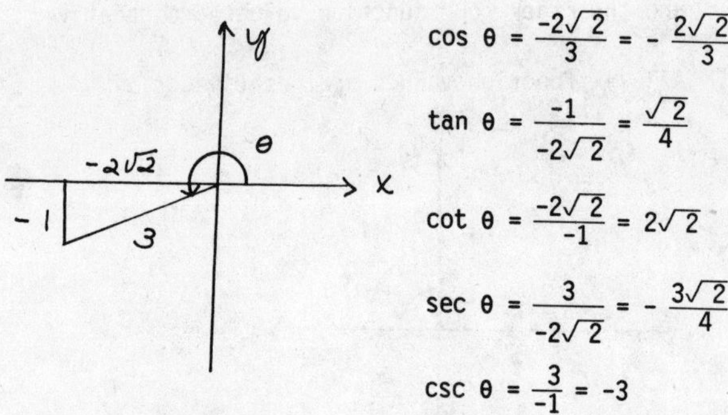

$\cos\theta = \frac{-2\sqrt{2}}{3} = -\frac{2\sqrt{2}}{3}$

$\tan\theta = \frac{-1}{-2\sqrt{2}} = \frac{\sqrt{2}}{4}$

$\cot\theta = \frac{-2\sqrt{2}}{-1} = 2\sqrt{2}$

$\sec\theta = \frac{3}{-2\sqrt{2}} = -\frac{3\sqrt{2}}{4}$

$\csc\theta = \frac{3}{-1} = -3$

42. $\cos\theta = \frac{2\sqrt{6}}{5}$, $\tan\theta = -\frac{\sqrt{6}}{12}$, $\cot\theta = -2\sqrt{6}$,
$\sec\theta = \frac{5\sqrt{6}}{12}$, $\csc\theta = -5$

43.

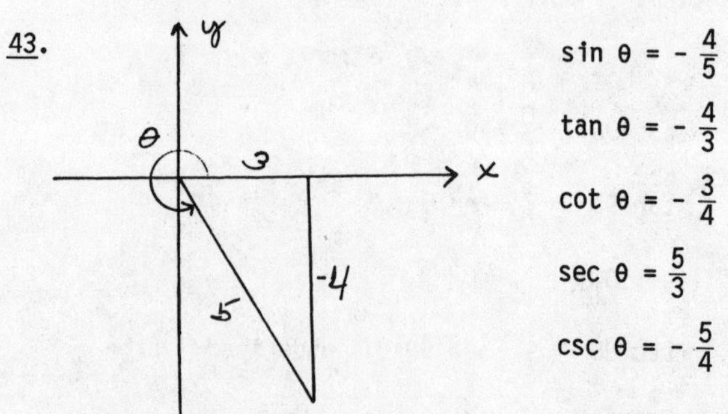

$\sin\theta = -\frac{4}{5}$

$\tan\theta = -\frac{4}{3}$

$\cot\theta = -\frac{3}{4}$

$\sec\theta = \frac{5}{3}$

$\csc\theta = -\frac{5}{4}$

44. $\sin\theta = \frac{3}{5}$, $\tan\theta = -\frac{3}{4}$, $\cot\theta = -\frac{4}{3}$, $\sec\theta = -\frac{5}{4}$,
$\csc\theta = \frac{5}{3}$

45.

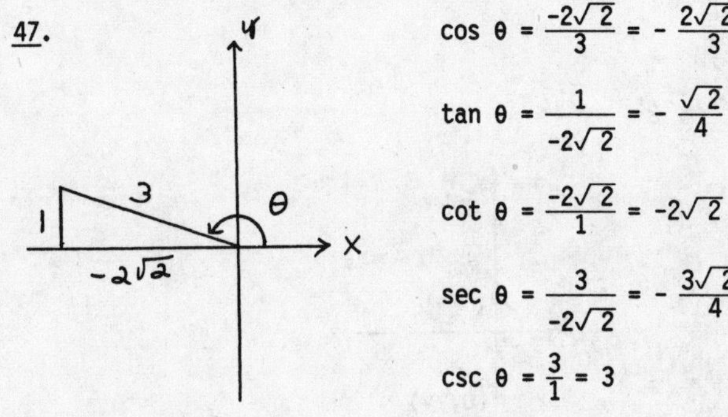

$\sin\theta = \frac{-1}{\sqrt{5}} = -\frac{\sqrt{5}}{5}$

$\cos\theta = \frac{2}{\sqrt{5}} = \frac{2\sqrt{5}}{5}$

$\tan\theta = \frac{-1}{2} = -\frac{1}{2}$

$\sec\theta = \frac{\sqrt{5}}{2}$

$\csc\theta = -\sqrt{5}$

46. $\sin\theta = -\frac{5\sqrt{26}}{26}$, $\cos\theta = -\frac{\sqrt{26}}{26}$, $\cot\theta = \frac{1}{5}$, $\sec\theta = -\sqrt{26}$, $\csc\theta = -\frac{\sqrt{26}}{5}$

47.

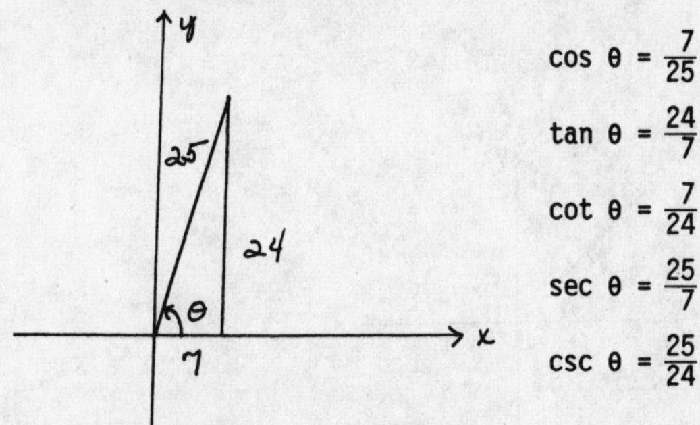

$\cos\theta = \frac{-2\sqrt{2}}{3} = -\frac{2\sqrt{2}}{3}$

$\tan\theta = \frac{1}{-2\sqrt{2}} = -\frac{\sqrt{2}}{4}$

$\cot\theta = \frac{-2\sqrt{2}}{1} = -2\sqrt{2}$

$\sec\theta = \frac{3}{-2\sqrt{2}} = -\frac{3\sqrt{2}}{4}$

$\csc\theta = \frac{3}{1} = 3$

48. $\sin\theta = -\frac{3}{5}$, $\tan\theta = -\frac{3}{4}$, $\cot\theta = -\frac{4}{3}$, $\sec\theta = \frac{5}{4}$, $\csc\theta = -\frac{5}{3}$

49. Since θ is an acute angle, we sketch a first quadrant triangle.

$\cos\theta = \frac{7}{25}$

$\tan\theta = \frac{24}{7}$

$\cot\theta = \frac{7}{24}$

$\sec\theta = \frac{25}{7}$

$\csc\theta = \frac{25}{24}$

50. $\sin\theta = \frac{\sqrt{51}}{10}$, $\tan\theta = \frac{\sqrt{51}}{7}$, $\cot\theta = \frac{7\sqrt{51}}{51}$, $\sec\theta = \frac{10}{7}$, $\csc\theta = \frac{10\sqrt{51}}{51}$

51. Since θ is an acute angle, we sketch a first quadrant triangle.

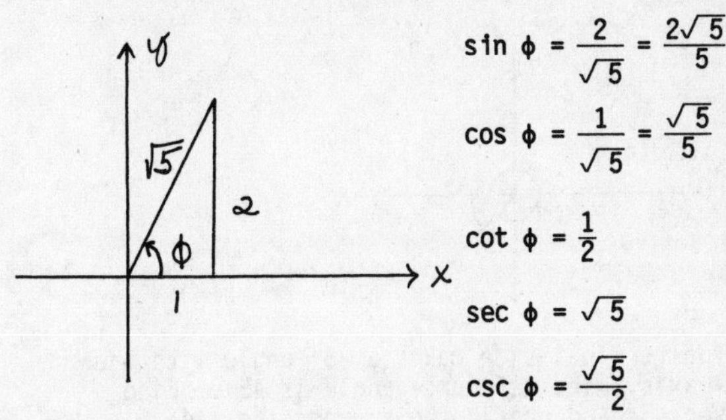

$\sin\phi = \frac{2}{\sqrt{5}} = \frac{2\sqrt{5}}{5}$

$\cos\phi = \frac{1}{\sqrt{5}} = \frac{\sqrt{5}}{5}$

$\cot\phi = \frac{1}{2}$

$\sec\phi = \sqrt{5}$

$\csc\phi = \frac{\sqrt{5}}{2}$

52. $\sin\phi = \frac{4\sqrt{17}}{17}$, $\cos\phi = \frac{\sqrt{17}}{17}$ $\tan\phi = 4$, $\cot\phi = \frac{1}{4}$, $\csc\phi = \frac{\sqrt{17}}{4}$

53. See the answer section in the text.

54. a) Δ OPA ~ Δ ODB

Thus, $\frac{AP}{OA} = \frac{BD}{OB}$

$\frac{\sin\theta}{\cos\theta} = \frac{BD}{1}$

$\tan\theta = BD$

b) Δ OPA ~ Δ ODB

$\frac{OD}{OP} = \frac{OB}{OA}$

$\frac{OD}{1} = \frac{1}{\cos\theta}$

$OD = \sec\theta$

c) Δ OAP ~ Δ ECO

$\frac{OE}{PO} = \frac{CO}{AP}$

$\frac{OE}{1} = \frac{1}{\sin\theta}$

$OE = \csc\theta$

d) Δ OAP ~ ECO

$\frac{CE}{AO} = \frac{CO}{AP}$

$\frac{CE}{\cos\theta} = \frac{1}{\sin\theta}$

$CE = \frac{\cos\theta}{\sin\theta}$

$CE = \cot\theta$

Exercise Set 6.6

1.

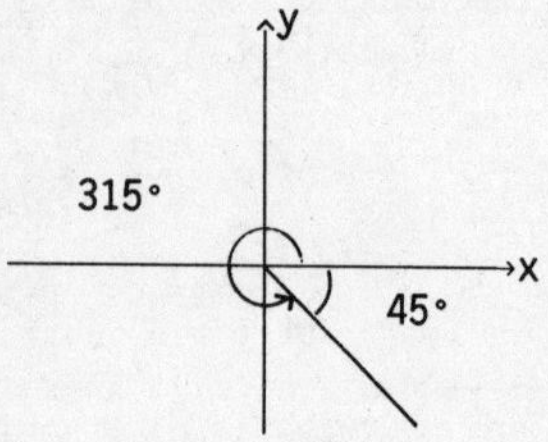

The terminal side makes a 45° angle with the x-axis. The reference angle is 45°. Find sec 45° and prefix the appropriate sign.

$\sec 45° = \frac{2}{\sqrt{2}}$, or $\sqrt{2}$

Since the secant function is also positive in the fourth quadrant,

$\sec 315° = \frac{2}{\sqrt{2}}$, or $\sqrt{2}$.

2. $-\frac{2}{\sqrt{2}}$, or $-\sqrt{2}$

3.

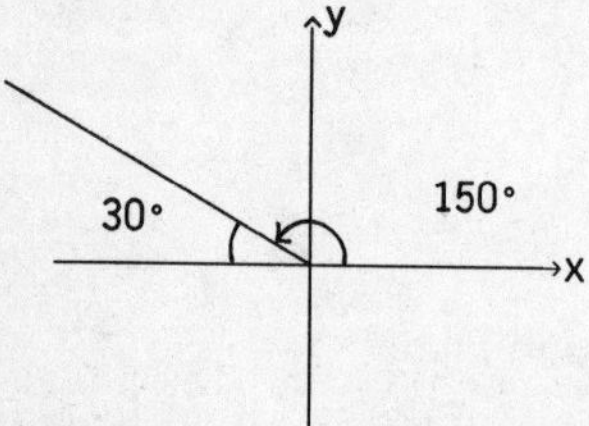

The terminal side makes a 30° angle with the x-axis. The reference angle is 30°. Find sin 30° and prefix the appropriate sign.

$\sin 30° = \frac{1}{2}$

Since the sine function is also positive in the second quadrant,

$\sin 150° = \frac{1}{2}$.

4. $-\frac{\sqrt{3}}{2}$

5.

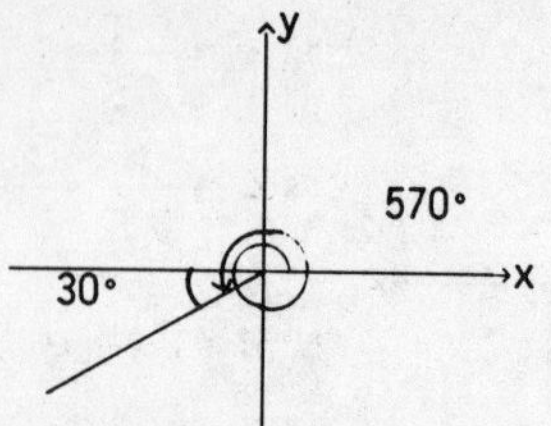

The terminal side makes a 30° angle with the x-axis. The reference angle is 30°. Find cot 30° and prefix the appropriate sign.

$\cot 30° = \frac{3}{\sqrt{3}}$, or $\sqrt{3}$

Since the cotangent function is also positive in the third quadrant,

$\cot 570° = \frac{3}{\sqrt{3}}$, or $\sqrt{3}$.

6. $\frac{1}{\sqrt{3}}$, or $\frac{\sqrt{3}}{3}$

7.

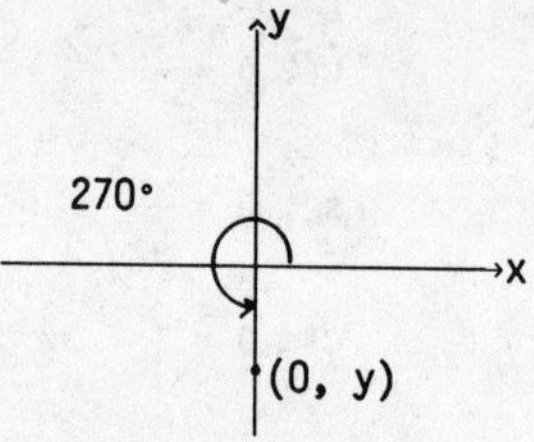

Note that the first coordinate is 0, the second coordinate is negative, and that $|y| = r$ (r is always positive).

$\csc \theta = \frac{r}{y}$

$\csc 270° = \frac{r}{y} = -1$ ($|y| = r$, y is negative)

8. -1

9.

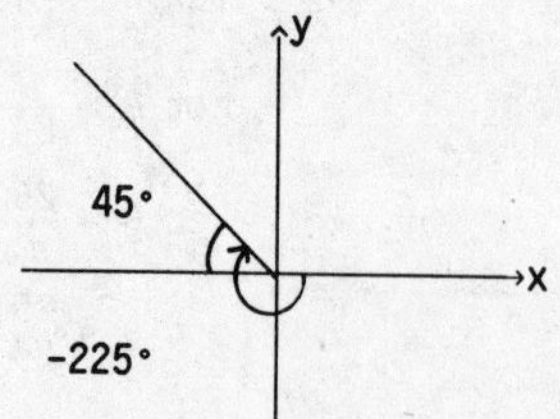

The terminal side makes a 45° angle with the x-axis. The reference angle is 45°. Find cot 45° and prefix the appropriate sign.

$\cot 45° = 1$

Since the cotangent function is negative in the second quadrant,

$\cot (-225°) = -1$.

10. $-\sqrt{2}$

11.

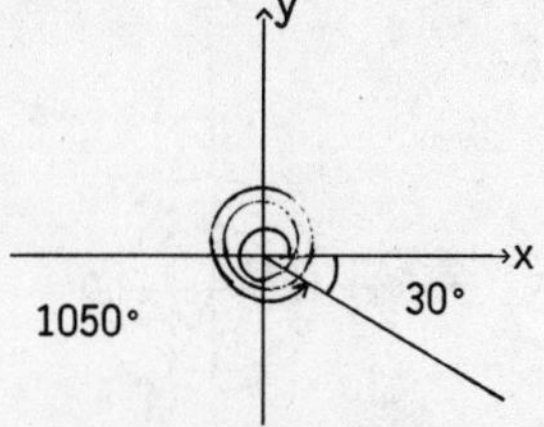

The terminal side makes a 30° angle with the x-axis. The reference angle is 30°. Find sin 30° and prefix the appropriate sign.

$\sin 30° = \frac{1}{2}$

Since the sine function is negative in the fourth quadrant,

$\sin 1050° = -\frac{1}{2}$.

12. $\frac{\sqrt{2}}{2}$

13.

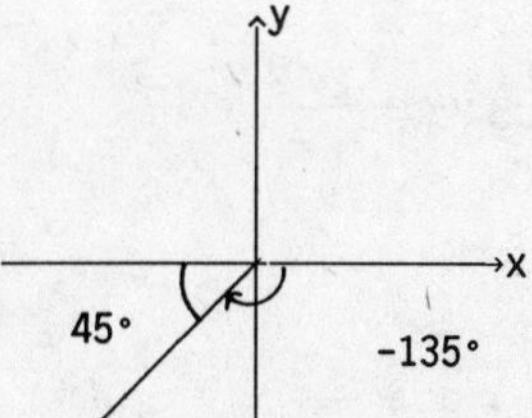

The terminal side makes a 45° angle with the x-axis. The reference angle is 45°. Find tan 45° and prefix the appropriate sign.

$\tan 45° = 1$

Since the tangent function is also positive in the third quadrant,

$\tan(-135°) = 1$.

14. $-\frac{\sqrt{2}}{2}$

15.

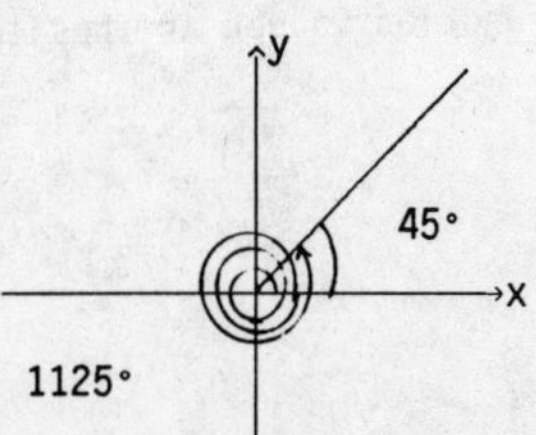

The terminal side makes a 45° angle with the x-axis. The reference angle is 45°. Find sec 45° and prefix the appropriate sign.

$\sec 45° = \frac{2}{\sqrt{2}}$, or $\sqrt{2}$

Note that the terminal side of 1125° also lies in the first quadrant. Thus,

$\sec 1125° = \frac{2}{\sqrt{2}}$, or $\sqrt{2}$.

16. $\sqrt{2}$

17.

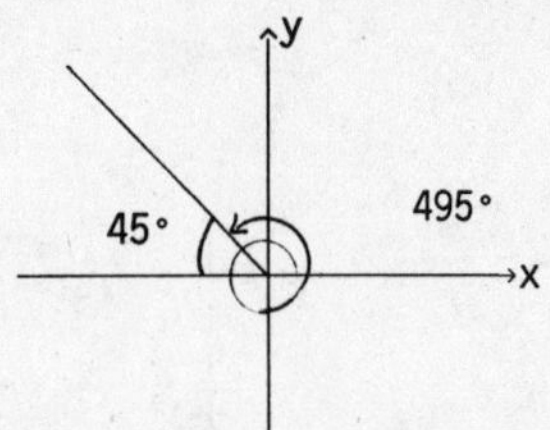

The terminal side makes a 45° angle with the x-axis. The reference angle is 45°. Find sin 45° and prefix the appropriate sign.

$\sin 45° = \frac{\sqrt{2}}{2}$

Since the sine function is also positive in the second quadrant,

$\sin 495° = \frac{\sqrt{2}}{2}$.

18. $-\sqrt{3}$

19. $\cos 5220° = \cos(5040° + 180°)$

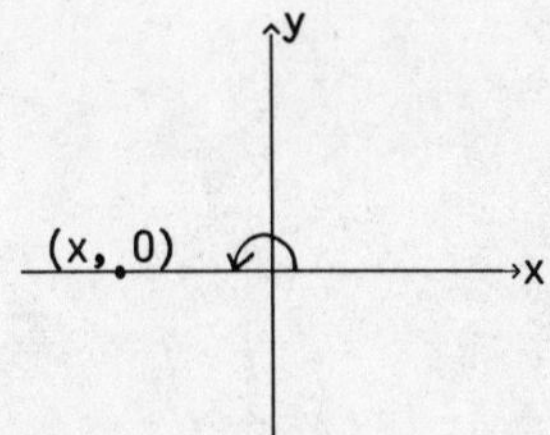

The first coordinate is negative, the second coordinate is 0, and $|x| = r$ (r is always positive).

$\cos \theta = \frac{x}{r}$

$\cos 5220° = \frac{x}{r} = -1$ ($|x| = r$, r is negative)

20. 0

21. $\frac{13\pi}{4} = \frac{12\pi}{4} + \frac{\pi}{4}$

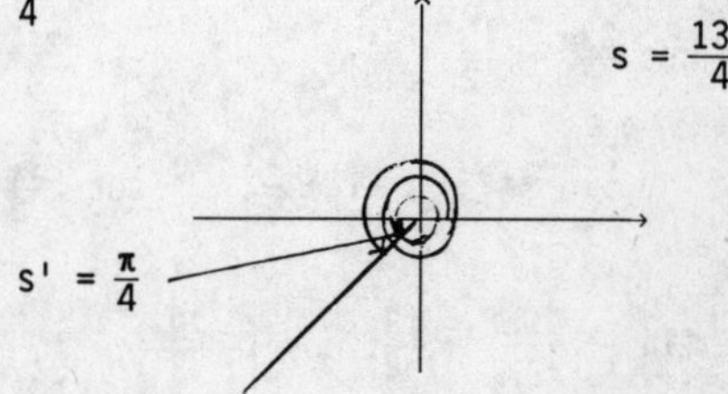

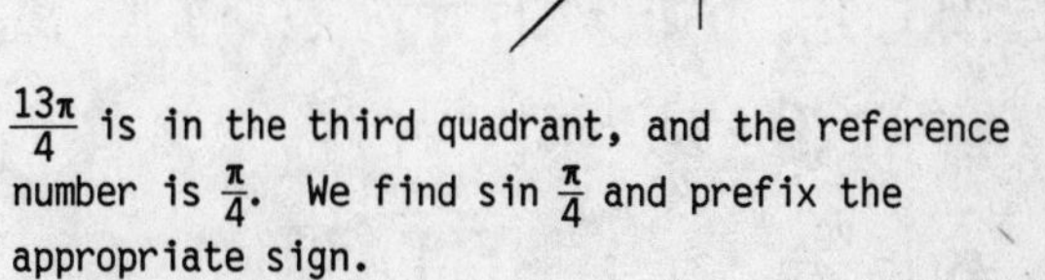

$\frac{13\pi}{4}$ is in the third quadrant, and the reference number is $\frac{\pi}{4}$. We find $\sin \frac{\pi}{4}$ and prefix the appropriate sign.

$\sin \frac{\pi}{4} = \frac{\sqrt{2}}{2}$

Since the sine function is negative in the third quadrant,

$\sin \frac{13\pi}{4} = -\frac{\sqrt{2}}{2}$.

22. $-\frac{\sqrt{2}}{2}$

23. $\frac{11\pi}{3} = \frac{9\pi}{3} + \frac{2\pi}{3}$

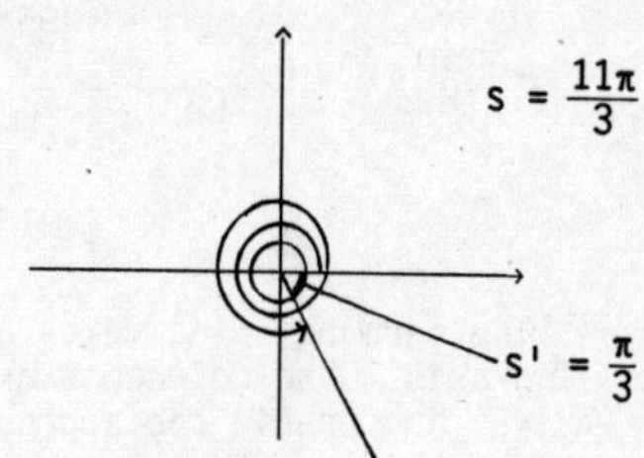

$\frac{11\pi}{3}$ is in the fourth quadrant, and the reference number is $\frac{\pi}{3}$. We find $\tan\frac{\pi}{3}$ and prefix the appropriate sign.

$\tan\frac{\pi}{3} = \sqrt{3}$

Since the tangent function is negative in the fourth quadrant,

$\tan\frac{\pi}{3} = -\sqrt{3}$.

24. 2

25. $\frac{23\pi}{6} = \frac{18\pi}{6} + \frac{5\pi}{6}$

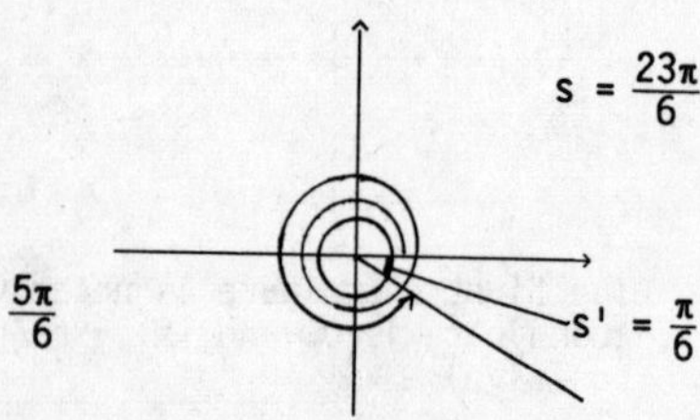

$\frac{23\pi}{6}$ is in the fourth quadrant, and the reference number is $\frac{\pi}{6}$. We know that in the fourth quadrant the cosine is positive and the sine and tangent are negative.

$\sin\frac{23\pi}{6} = -\frac{1}{2}$ $\qquad$ $\csc\frac{23\pi}{6} = -2$

$\cos\frac{23\pi}{6} = \frac{\sqrt{3}}{2}$ $\qquad$ $\sec\frac{23\pi}{6} = \frac{2}{\sqrt{3}}$, or $\frac{2\sqrt{3}}{3}$

$\tan\frac{23\pi}{6} = -\frac{1}{\sqrt{3}}$, or $-\frac{\sqrt{3}}{3}$ $\qquad$ $\cot\frac{23\pi}{6} = -\sqrt{3}$

26. $\sin\left(-\frac{29\pi}{3}\right) = \frac{\sqrt{3}}{2}$, $\cos\left(-\frac{29\pi}{3}\right) = \frac{1}{2}$,

$\tan\left(-\frac{29\pi}{3}\right) = \sqrt{3}$, $\cot\left(-\frac{29\pi}{3}\right) = \frac{\sqrt{3}}{3}$

$\sec\left(-\frac{29\pi}{3}\right) = 2$, $\csc\left(-\frac{29\pi}{3}\right) = \frac{2\sqrt{3}}{3}$

27. $-\frac{19\pi}{3} = -\left(\frac{18\pi}{3} + \frac{\pi}{3}\right)$

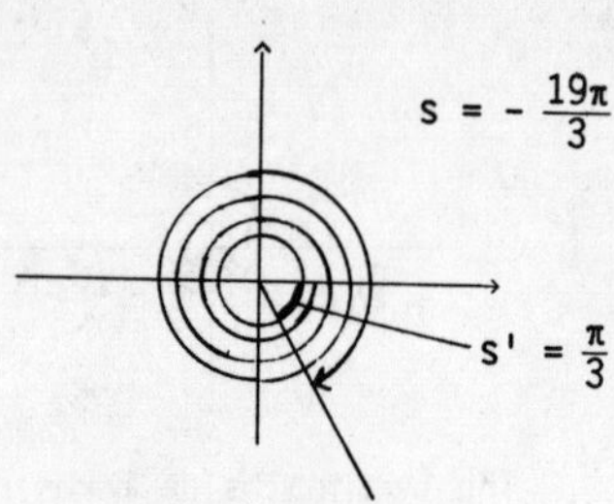

$-\frac{19\pi}{3}$ is in the fourth quadrant, and the reference number is $\frac{\pi}{3}$. We know that in the fourth quadrant, the cosine is positive and the sine and tangent are negative.

$\sin\left(-\frac{19\pi}{3}\right) = -\frac{\sqrt{3}}{2}$

$\cos\left(-\frac{19\pi}{3}\right) = \frac{1}{2}$

$\tan\left(-\frac{19\pi}{3}\right) = -\sqrt{3}$

$\csc\left(-\frac{19\pi}{3}\right) = -\frac{2}{\sqrt{3}}$, or $-\frac{2\sqrt{3}}{3}$

$\sec\left(-\frac{19\pi}{3}\right) = 2$

$\cot\left(-\frac{19\pi}{3}\right) = -\frac{1}{\sqrt{3}}$, or $-\frac{\sqrt{3}}{3}$

28. $\sin\frac{37\pi}{6} = \frac{1}{2}$, $\cos\frac{37\pi}{6} = \frac{\sqrt{3}}{2}$, $\tan\frac{37\pi}{6} = \frac{\sqrt{3}}{3}$,

$\cot\frac{37\pi}{6} = \sqrt{3}$, $\sec\frac{37\pi}{6} = \frac{2\sqrt{3}}{3}$, $\csc\frac{37\pi}{6} = 2$

29. $\frac{31\pi}{3} = \frac{30\pi}{3} + \frac{\pi}{3}$

We see that $\frac{31\pi}{3}$ is in the first quadrant, and the reference number is $\frac{\pi}{3}$. In the first quadrant, all the function values are positive.

$\sin\frac{31\pi}{3} = \frac{\sqrt{3}}{2}$

$\cos\frac{31\pi}{3} = \frac{1}{2}$

$\tan\frac{31\pi}{3} = \sqrt{3}$

$\csc\frac{31\pi}{3} = \frac{2}{\sqrt{3}}$, or $\frac{2\sqrt{3}}{3}$

$\sec\frac{31\pi}{3} = 2$

$\cot\frac{31\pi}{3} = \frac{1}{\sqrt{3}}$, or $\frac{\sqrt{3}}{3}$

30. $\sin\left(-\frac{25\pi}{6}\right) = -\frac{1}{2}$, $\cos\left(-\frac{25\pi}{6}\right) = \frac{\sqrt{3}}{2}$,

$\tan\left(-\frac{25\pi}{6}\right) = -\frac{\sqrt{3}}{3}$, $\cot\left(-\frac{25\pi}{6}\right) = -\sqrt{3}$,

$\sec\left(-\frac{25\pi}{6}\right) = \frac{2\sqrt{3}}{3}$, $\csc\left(-\frac{25\pi}{6}\right) = -2$

31.

y

-750°

x

30°

We see that -750° is in the fourth quadrant and the reference angle is 30°. In the fourth quadrant the cosine is positive and the sine and tangent are negative.

$\sin(-750°) = -\frac{1}{2}$

$\cos(-750°) = \frac{\sqrt{3}}{2}$

$\tan(-750°) = -\frac{1}{\sqrt{3}}$, or $-\frac{\sqrt{3}}{3}$

$\csc(-750°) = -2$

$\sec(-750°) = \frac{2}{\sqrt{3}}$, or $\frac{2\sqrt{3}}{3}$

$\cot(-750°) = -\sqrt{3}$

32. $\sin 1590° = \frac{1}{2}$, $\cos 1590° = -\frac{\sqrt{3}}{2}$,

$\tan 1590° = -\frac{\sqrt{3}}{3}$, $\cot 1590° = -\sqrt{3}$,

$\sec 1590° = -\frac{2\sqrt{3}}{3}$, $\csc 1590° = 2$

33. Use a calculator to find cos 18°.

Enter 18 and then press the [cos] key.

cos 18° ≈ 0.951057

34. 0.601815

35. Use a calculator to find tan 2.6°.

Enter 2.6 and then press the [tan] key.

tan 2.6° ≈ 0.045410

36. 0.821149

37. Use a calculator to find sin 62° 20'.

We first convert to degrees and decimal parts of degrees on a calculator.

$62°\ 20' = 62° + \left(\frac{20}{60}\right)° \approx 62.33333333°$

Then press the [sin] key.

sin 62° 20' ≈ 0.885664

38. 1.455009

39. Use a calculator to find cos 15° 35'.

We first convert to degrees and decimal parts of degrees on a calculator.

$15°\ 35' = 15° + \left(\frac{35}{60}\right)° \approx 15.58333333°$

Then press the [cos] key.

cos 15° 35' ≈ 0.963241

40. 0.654961

41. Use a calculator to find csc 29° 11' 36". We first convert to degrees and decimal parts of degrees on a calculator.

$29°\ 11'\ 36" = 29° + \left(\frac{11}{60}\right)° + \left(\frac{36}{3600}\right)°$

$= 29° + 0.18333333° + 0.01°$

$\approx 29.19333333°$

Now press the [sin] key, and then press the reciprocal key.

$\csc 29°\ 11'\ 36" = \frac{1}{\sin 29°\ 11'\ 36"}$

≈ 2.050197

42. 0.706781

43. Use a calculator to find sec 10° 31' 42".

We first convert to degrees and decimal parts of degrees.

$10°\ 31'\ 42" = 10° + \left(\frac{31}{60}\right)° + \left(\frac{42}{3600}\right)°$

$\approx 10° + 0.51666667° + 0.01166667°$

$\approx 10.52833333°$

Now press the [cos] key, and then press the reciprocal key.

$\sec 10°\ 31'\ 42" = \frac{1}{\cos 10°\ 31'\ 42"}$

≈ 1.017124

44. 1.407927

45. Use a calculator to find sin 561.2344°. Enter 561.2344 and then press the [sin] key.

sin 561.2344° ≈ -0.362184

46. -1.405422

47. Use a calculator to find cot (-900.23°). Enter -900.23, press the [tan] key, and then press the reciprocal key.

$$\cot(-900.23°) = \frac{1}{\tan(-900.23°)} \approx -249.1107$$

48. 1.000006

49. Use a calculator to find tan 295° 14'.

We first convert to degrees and decimal parts of degrees on a calculator.

$$295°\ 14' = 295° + \left(\frac{14}{60}\right)° \approx 29.23333333°$$

Then press the [tan] key.

tan 295° 14' ≈ -2.121903

50. -0.630902

51. Use a calculator to find sec 146.9°.

Enter 146.9, press the [cos] key, and then press the reciprocal key.

$$\sec 146.9° = \frac{1}{\cos 146.9°} \approx -1.193718$$

52. 0.989272

53. Use a calculator to find sin 756° 25'. We first convert to degrees and decimal parts of degrees on a calculator.

$$756°\ 25' = 756° + \left(\frac{25}{60}\right)° \approx 756.4166667°$$

Then press the [sin] key.

sin 756° 25' ≈ 0.593653

54. -0.188349

55. Use a calculator to find cos (-1000.85°).

Enter -1000.85 and press the [cos] key.

cos (-1000.85°) ≈ 0.188238

56. -0.107752

57. Find sin 37 using a calculator.

With the calculator in radian mode, enter 37 and then press the [sin] key.

sin 37 ≈ -0.643538

58. 0.862349

59. Find cos (-10) using a calculator.

With the calculator in radian mode, enter -10 and then press the [cos] key.

cos (-10) ≈ -0.839072

60. 0.756802

61. Find tan 5π using a calculator.

Using the [π] key, we find 5π ≈ 15.70796327. Then, with the calculator in radian mode, press the [tan] key.

tan 5π = 0

62. Undefined

63. Find cot 1000 using a calculator.

With the calculator in radian mode, enter 1000, press the [tan] key, and then press the reciprocal key.

$$\cot 1000 = \frac{1}{\tan 1000} \approx 0.680122$$

64. 0.463021

65. Find $\sec\left(-\frac{\pi}{5}\right)$ using a calculator.

Using the [π] key we find $-\frac{\pi}{5} \approx -0.62831853$.

Using radian mode, press the [cos] key and then press the reciprocal key.

$$\sec\left(-\frac{\pi}{5}\right) = \frac{1}{\cos\left(-\frac{\pi}{5}\right)} \approx 1.236068$$

66. -0.324920

67. Find $\sec\frac{10\pi}{7}$ using a calculator.

Using the [π] key, we find $\frac{10\pi}{7} \approx 4.487989505$. With the calculator in radian mode, press the [cos] key and then press the reciprocal key.

$$\sec\frac{10\pi}{7} = \frac{1}{\cos\frac{10\pi}{7}} \approx -4.493959$$

68. 0.797473

69. Find cos (-13π) using a calculator.

Using the [π] key, we find -13π ≈ -40.8407045. Then, with the calculator in radian mode, press the [cos] key.

cos (-13π) = -1

70. 0

71. Find tan 2.5 using a calculator.

With the calculator in radian mode, enter 2.5 and then press the [tan] key.

tan 2.5 ≈ -0.747022

72. -0.530323

73. With the calculator in degree mode, enter 0.5125 and press the $\boxed{\sin^{-1}}$ key.

$\theta = \sin^{-1} 0.5125 \approx 30.83°$.

With the calculator in radian mode, enter 0.5125 and press the $\boxed{\sin^{-1}}$ key.

$\theta = \sin^{-1} 0.5125 \approx 0.5381$

74. 34.50°, 0.6022

75. With the calculator in degree mode, enter 0.6512 and press the $\boxed{\cos^{-1}}$ key.

$\theta = \cos^{-1} 0.6512 \approx 49.37°$

With the calculator in radian mode, repeat this process.

$\theta = \cos^{-1} 0.6512 \approx 0.8616$

76. 55.29°, 0.9649

77. With the calculator in degree mode, enter 7.425 and press the $\boxed{\tan^{-1}}$ key.

$\theta = \tan^{-1} 7.425 \approx 82.33°$

With the calculator in radian mode, repeat this process.

$\theta = \tan^{-1} 7.425 \approx 1.4369$

78. 72.46°, 1.2646

79. $\csc \theta = \dfrac{1}{\sin \theta} = 6.277$

$\sin \theta = \dfrac{1}{6.277} \approx 0.1593$

With the calculator in degree mode, enter 0.1593 and press the $\boxed{\sin^{-1}}$ key.

$\theta = \sin^{-1} 0.1593 \approx 9.17°$

With the calculator in radian mode, repeat this process.

$\theta = \sin^{-1} 0.1593 \approx 0.1600$

80. 31.67°, 0.5528

81. $\sin \theta = -0.9956$, $270° < \theta < 360°$

We find the reference angle, ignoring the fact that $\sin \theta$ is negative. Enter 0.9956 and press the $\boxed{\sin^{-1}}$ key. The reference angle is 84.62°, or 84° 37'.

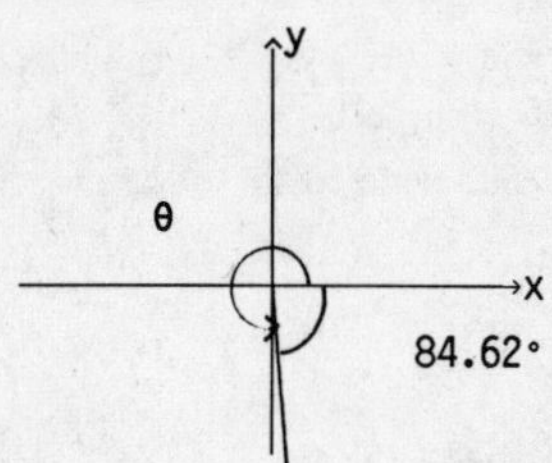

$\theta = 360° - 84.62° = 275.38°$, or 275° 23'.

82. 154.45°, or 154° 27'

83. $\cos \theta = -0.9388$, $180° < \theta < 270°$

We find the reference angle, ignoring the fact that $\cos \theta$ is negative. Enter 0.9388 and press the $\boxed{\cos^{-1}}$ key. The reference angle is 20.15°, or 20° 9'.

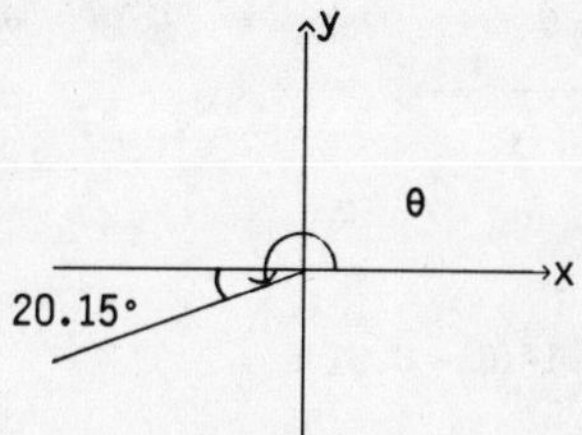

$\theta = 180° + 20.15° = 200.15°$, or 200° 9'.

84. 95.68°, or 95° 41'

85. $\tan \theta = 0.2460$, $180° < \theta < 270°$

Enter 0.2460 and press the $\boxed{\tan^{-1}}$ key. The reference angle is 13.82°, or 13° 49'.

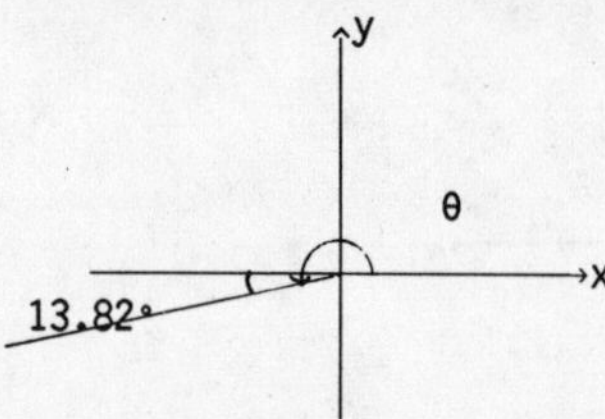

$\theta = 180° + 13.82° = 193.82°$, or 193° 49'.

86. 288.13°, or 288° 8'

87. $\sec \theta = -1.0485$, $90° < \theta < 180°$

Find the reference angle, ignoring the fact that $\sec \theta$ is negative. Enter 1.0485, press the reciprocal key, and then press the $\boxed{\cos^{-1}}$ key. The reference angle is 17.49°, or 17° 30'.

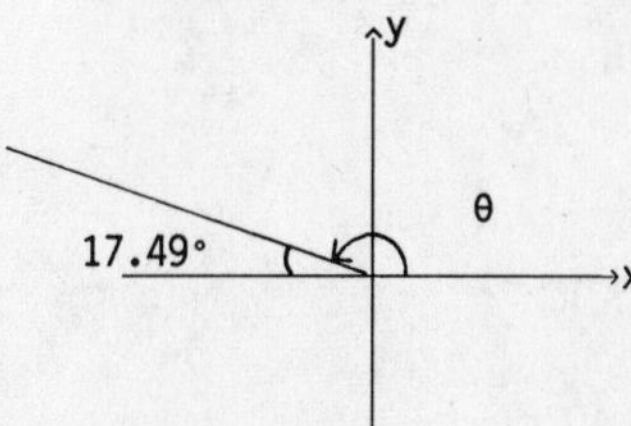

$\theta = 180° - 17.49° = 162.51°$, or 162° 30'.

88. 72.59°, or 72° 36'

89. $\sin\theta = 0.5766,\ \frac{\pi}{2} < \theta < \pi$

With the calculator in radian mode, enter 0.5766, and press the $\boxed{\sin^{-1}}$ key. The reference angle is 0.6146

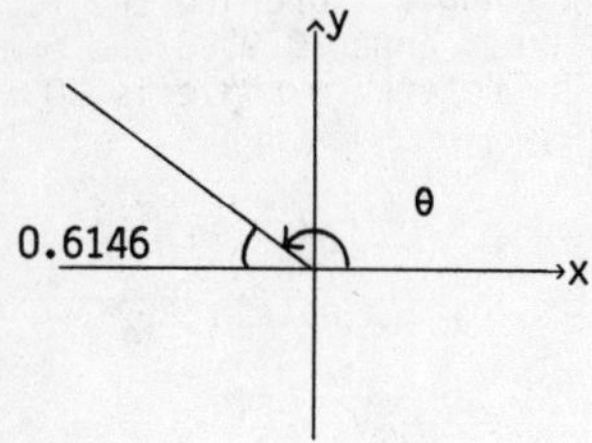

$\theta = \pi - 0.6146 \approx 3.1416 - 0.6146 = 2.5270$

90. 5.1917

91. $\cos\theta = -0.211499,\ \frac{\pi}{2} < \theta < \pi$

We find the reference angle, ignoring the fact that $\cos\theta$ is negative. With the calculator in radian mode, enter 0.211499 and press the $\boxed{\cos^{-1}}$ key. The reference angle is 1.3577.

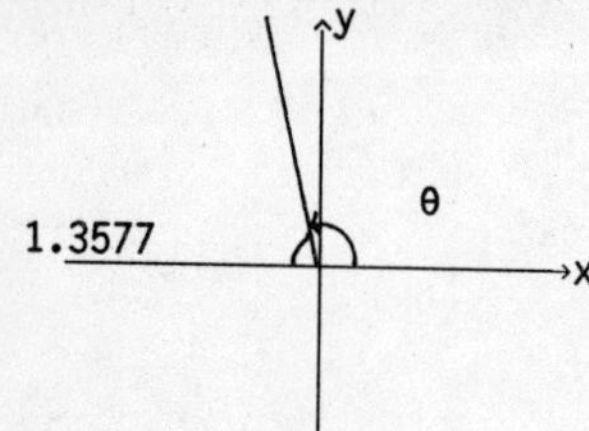

$\theta = \pi - 1.3577 \approx 3.1416 - 1.3577 = 1.7839$

92. 3.4933

93. $\tan\theta = -19.0541,\ \frac{3\pi}{2} < \theta < 2\pi$

We find the reference angle, ignoring the fact that $\tan\theta$ is negative. With the calculator in radian mode, enter 19.0541 and press the $\boxed{\tan^{-1}}$ key. The reference angle is 1.5184.

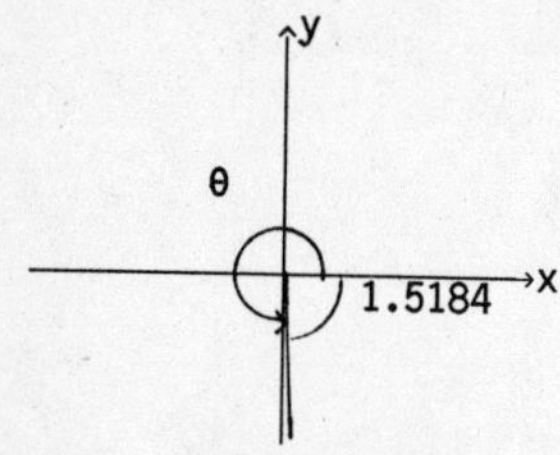

$\theta = 2\pi - 1.5184 \approx 6.2832 - 1.5184 = 4.7648$

94. 4.6814

95. $\csc\theta = \frac{1}{\sin\theta} = 11.21048,\ 0 < \theta < \frac{\pi}{2}.$

$\sin\theta = \frac{1}{11.21048} \approx 0.08920$

With the calculator in radian mode, enter 0.08920 and press the $\boxed{\sin^{-1}}$ key.

$\theta = 0.0893$

96. 1.8090

97. We draw the terminal side of an angle of 390°.

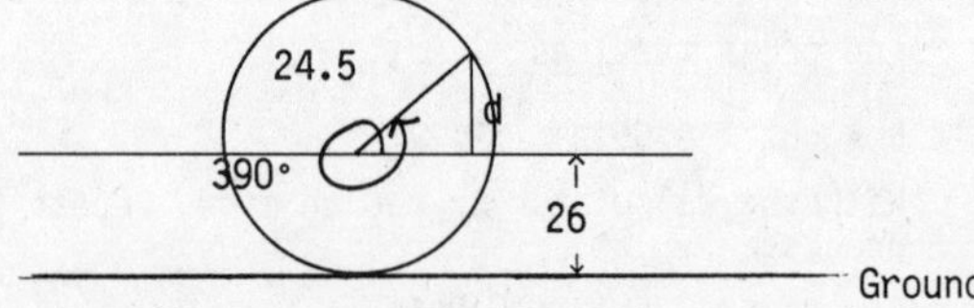

The reference angle is 30°.

Let d be the vertical distance of the valve cap above the center of the wheel.

$\frac{d}{24.5} = \sin 30°$

$d = 24.5 \sin 30°$

$= 24.5 \times \frac{1}{2}$

$= 12.25$

The distance above the ground is then

12.25 in. + 26 in. = 38.25 in.

98. 15.3 ft

99. Using Table 3 in the back of the book, we see that an angle must be 0.0436 radiants.

100. 0.0669

101. $\frac{\pi}{7} \approx 25.7143°$

$$\begin{aligned} 25.7143° &= 25° + 0.7143 \times 1° \\ &= 25° + 0.7143 \times 60' \\ &= 25° + 42.858' \\ &= 25°\ 42' + 0.858 \times 1' \\ &= 25°\ 42' + 0.858 \times 60'' \\ &= 25°\ 42' + 51'' \\ &= 25°\ 42'\ 51'' \end{aligned}$$

102. 61.63944°

103. $\sin 0.5 = 0.5 - \frac{(0.5)^3}{6} + \frac{(0.5)^5}{120}$

$\approx 0.5 - 0.02083 + 0.00026$

$= 0.47943$

104.

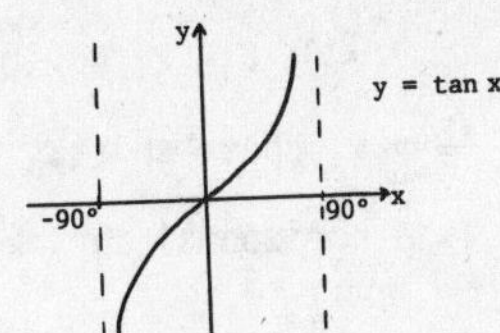

Domain: Set of all real numbers except odd multiples of 90°.

105. $\sin\theta \approx \theta$ when θ is small and when θ is in radian measure.

We convert 31' 59" to radian measure:

$$31'\ 59" = \left(\frac{31}{60}\right)^\circ + \left(\frac{59}{3600}\right)^\circ \approx 0.5331^\circ$$

$$0.5331^\circ \cdot \frac{\pi}{180^\circ} \approx 0.0093$$

We make a drawing. Let d represent the diameter of the sun.

Earth ——— 0.0093 ——— d

93,000,000 mi

$$\sin 0.0093 = \frac{d}{93{,}000{,}000}$$

$$0.0093 \approx \frac{d}{93{,}000{,}000} \quad (\sin 0.0093 \approx 0.0093)$$

$$865{,}000 \approx d$$

The diameter of the sun is about 865,000 miles.

106. 17.38°, or 17° 23'

107. a) When $y = 0$, the quadratic equation

$0 = -\frac{1}{2}gt^2 + (V\sin\theta)t + h$ has the positive root $t_1 = \dfrac{V\sin\theta + \sqrt{V^2\sin^2\theta + 2gh}}{g}$.

b) In $x = (V\cos\theta)t$, substitute the value for t_1 to obtain

$$x = V\cos\theta\,\frac{V\sin\theta + \sqrt{V^2\sin^2\theta + 2gh}}{g}.$$

c) Write the answer in part b) as follows:

$gx - V^2\sin\theta\cos\theta$

$= V\cos\theta\sqrt{V^2\sin^2\theta + 2gh}.$

Squaring both sides we obtain, after simplifying,

$V^2 = \dfrac{x^2g}{2\cos\theta\,(x\sin\theta + h\cos\theta)}$. Using the given data: $x = 21.6$, $h = 1.6$, $\theta = 44^\circ$, $g = 9.8$, we find $V \approx 14.026$ m/sec.

107. (continued)

d) The answer to b) shows that x is larger when h is larger; thus a taller shotputter can put the shot a greater distance.

e) In the given equations for x and y, solve the first equation for $V = \dfrac{x\sec\theta}{t}$ and substitute in the second equation to obtain

$y = -\frac{1}{2}gt^2 + x\tan\theta + h$. Using the given data, $16t^2 = 300\tan 20^\circ - 3.5$ so that $t \approx 2.57$ sec; then

$$V = \frac{300\sec 20^\circ}{2.57} \approx 124 \text{ ft/sec.}$$

Exercise Set 6.7

1. $y = 2 + \sin x$ $(A = 1, B = 2, C = 1, D = 0)$

Amplitude: 1, period: 2π, phase shift: 0

The graph of $y = 2 + \sin x$ is a translation of $y = \sin x$ upward 2 units.

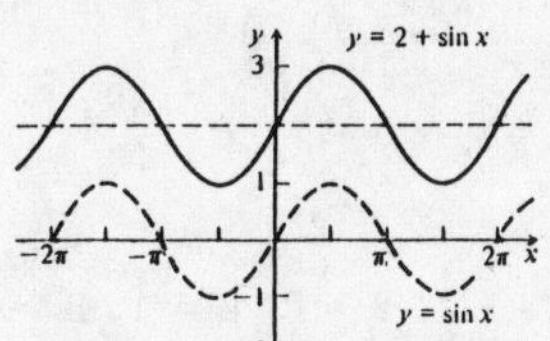

2. Amplitude: 1, period: 2π, phase shift: 0

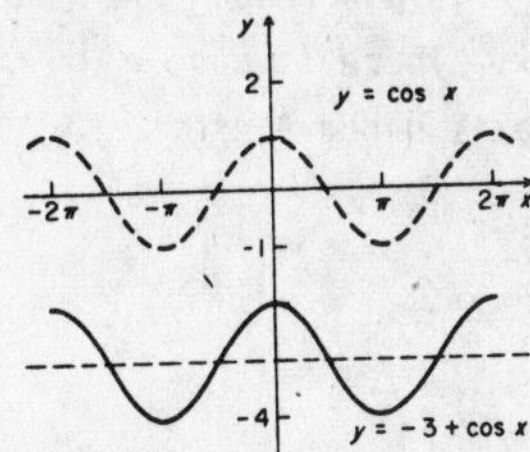

3. $y = \frac{1}{2}\sin x$ $\left[A = \frac{1}{2}, B = 0, C = 1, D = 0\right]$

Amplitude: $\frac{1}{2}$, period: 2π, phase shift: 0

The graph of $y = \frac{1}{2}\sin x$ is a vertical shrinking of the graph of $y = \sin x$.

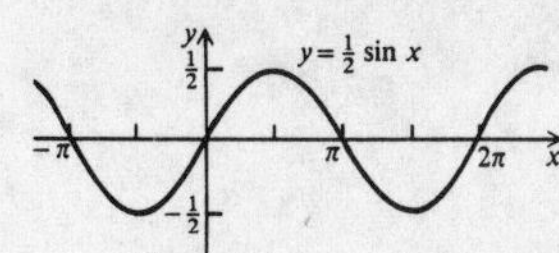

4. Amplitude: $\frac{1}{3}$, period: 2π, phase shift: 0

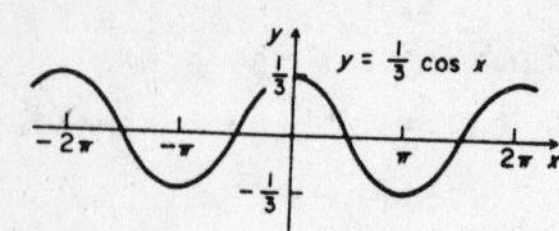

5. $y = 2 \cos x$ $(A = 2, B = 0, C = 1, D = 0)$

Amplitude: 2, period: 2π, phase shift: 0

The graph of $y = 2 \cos x$ is a vertical stretching of the graph of $y = \cos x$.

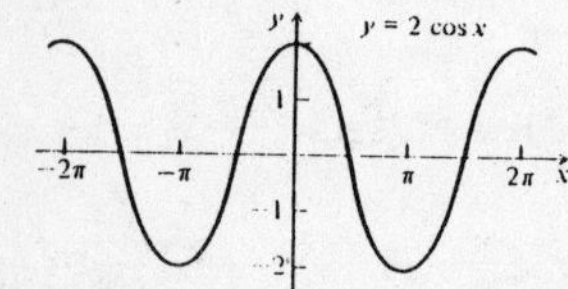

6. Amplitude: 3, period: 2π, phase shift: 0

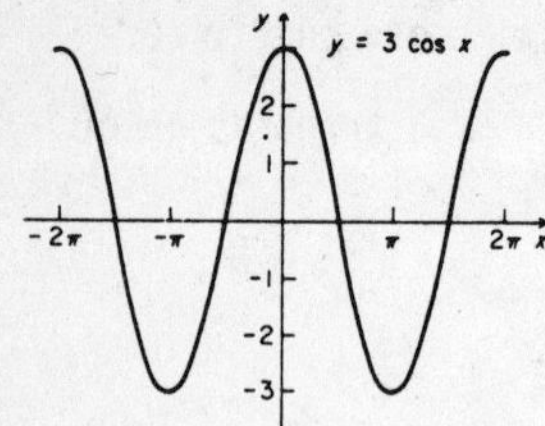

7. $y = -\frac{1}{2} \cos x$ $\left[A = -\frac{1}{2}, B = 0, C = 1, D = 0\right]$

Amplitude: $\left|-\frac{1}{2}\right| = \frac{1}{2}$, period: 2π, phase shift: 0

The graph of $y = -\frac{1}{2} \cos x$ is a vertical shrinking of the graph of $y = \cos x$. There is also a reflection across the x-axis since $A < 0$.

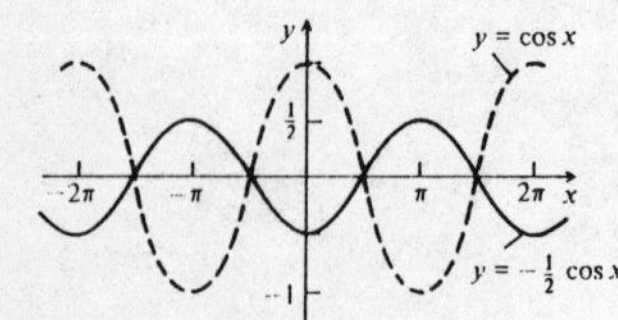

8. Amplitude: 2, period: 2π, phase shift: 0

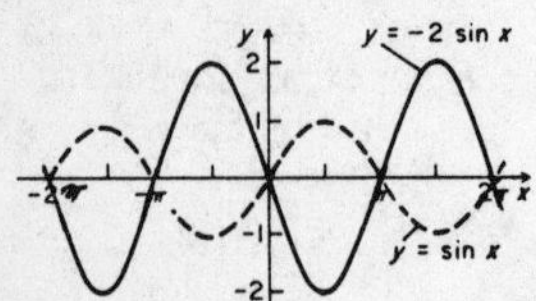

9. $y = \cos 2x$ $(A = 1, B = 0, C = 2, D = 0)$

Amplitude: 1, period: $\frac{2\pi}{2} = \pi$, phase shift: 0

The graph of $y = \cos 2x$ is a horizontal shrinking of the graph of $y = \cos x$.

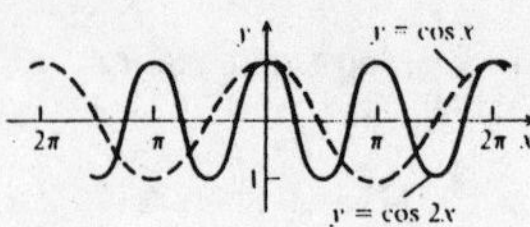

10. Amplitude: 1, period: $\frac{2\pi}{3}$, phase shift: 0

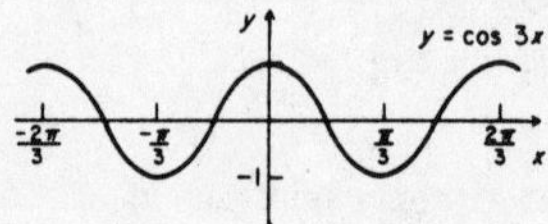

11. $y = \cos(-2x)$ $(A = 1, B = 0, C = -2, D = 0)$

Amplitude: 1, period: $\left|\frac{2\pi}{-2}\right| = \pi$,

phase shift: 0

The graph of $y = \cos(-2x)$ is a horizontal shrinking of the graph of $y = \cos x$. There is also a reflection across the y-axis since $C < 0$. Note that the graphs of $y = \cos 2x$ and $y = \cos(-2x)$ are identical. The cosine function is an even function, $\cos(-2x) = \cos 2x$.

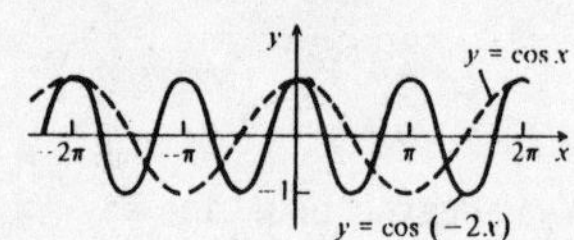

12. The graph of $y = \cos(-3x)$ is the same as the graph of $y = \cos 3x$. See the graph in Exercise 10 above.

13. $y = \sin \frac{1}{2} x$ $\left[A = 1, B = 0, C = \frac{1}{2}, D = 0\right]$

Amplitude: 1, period: $\frac{2\pi}{\frac{1}{2}} = 4\pi$, phase shift: 0

The graph of $y = \sin \frac{1}{2} x$ is a horizontal stretching of the graph of $y = \sin x$.

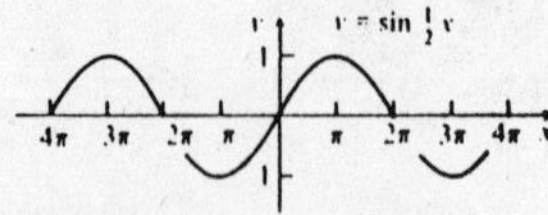

14. Amplitude: 1, period: 6π, phase shift: 0

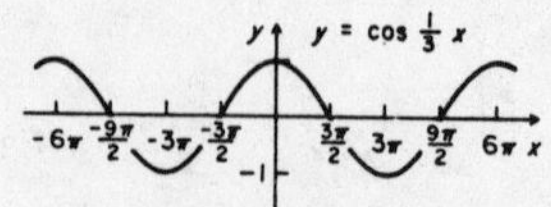

15. $y = \sin\left(-\frac{1}{2}x\right)$ $\left[A = 1, B = 0, C = -\frac{1}{2}, D = 0\right]$

Amplitude: 1, period: $\left|\frac{2\pi}{-\frac{1}{2}}\right| = 4\pi$,

phase shift: 0

The graph of $y = \sin\left(-\frac{1}{2}x\right)$ is a horizontal stretching of the graph of y = sin x. There is also a reflection across the y-axis since C < 0.

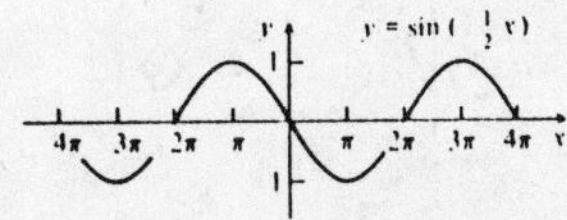

16. Amplitude: 1, period: 6π, phase shift: 0

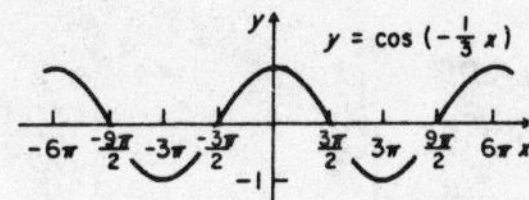

17. $y = \cos(2x - \pi)$, or $y = \cos 2\left(x - \frac{\pi}{2}\right)$

$(A = 1, B = 0, C = 2, D = \pi)$

Amplitude: 1, period: $\frac{2\pi}{2} = \pi$, phase shift: $\frac{\pi}{2}$

The $\frac{\pi}{2}$ translates the graph of y = cos 2x (see Exercise 9) a distance of $\frac{\pi}{2}$ to the right.

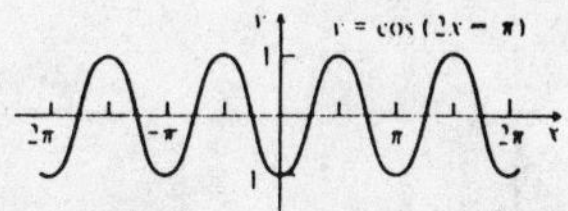

18. Amplitude: 1, period: π, phase shift: $-\frac{\pi}{2}$

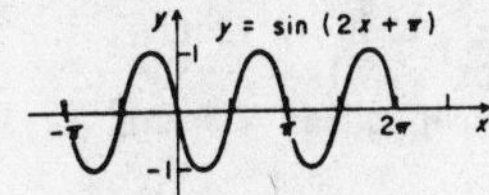

19. $y = 2\cos\left(\frac{1}{2}x - \frac{\pi}{2}\right)$, or $y = 2\cos\frac{1}{2}(x - \pi)$

$\left[A = 2, B = 0, C = \frac{1}{2}, D = \frac{\pi}{2}\right]$

The graph of $y = 2\cos\frac{1}{2}(x - \pi)$ has an amplitude of 2, a period of 4π $\left[2\pi \div \frac{1}{2} = 4\pi\right]$, and a phase shift of π.

First graph y = cos x.

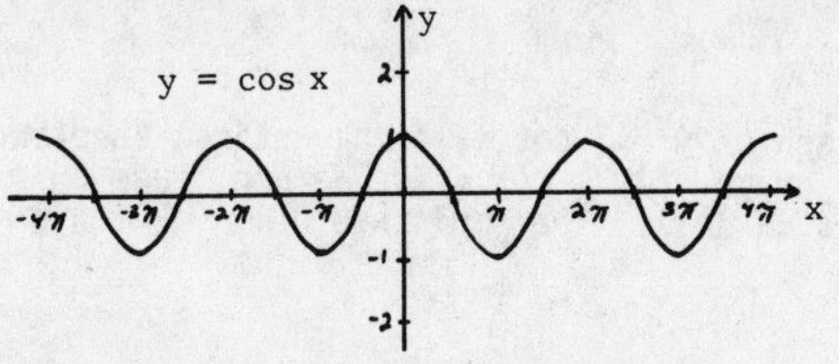

19. (continued)

The graph of $y = \cos\frac{1}{2}x$ is a horizontal stretching of the graph of y = cos x. The period of y = cos x is 2π. The period of $y = \cos\frac{1}{2}x$ is 4π.

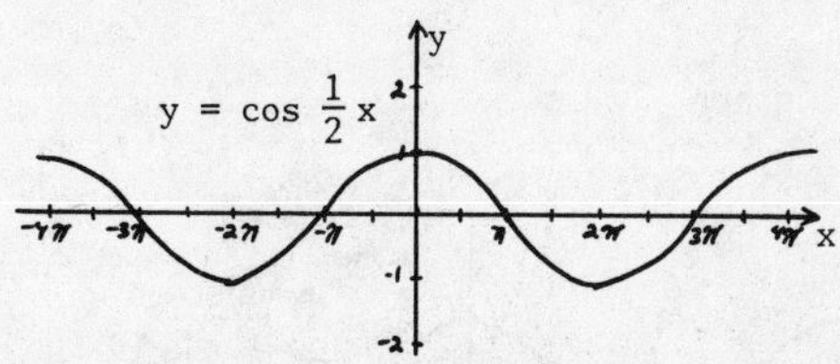

The graph of $y = 2\cos\frac{1}{2}x$ is a vertical stretching of the graph of $y = \cos\frac{1}{2}x$. The amplitude of $y = \cos\frac{1}{2}x$ is 1. The amplitude of $y = 2\cos\frac{1}{2}x$ is 2.

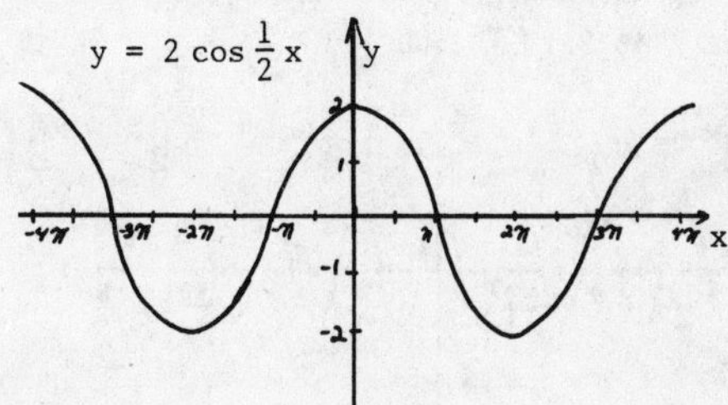

The graph of $y = 2\cos\frac{1}{2}(x - \pi)$ is a translation of the graph of $y = 2\cos\frac{1}{2}x$ a distance of π to the right.

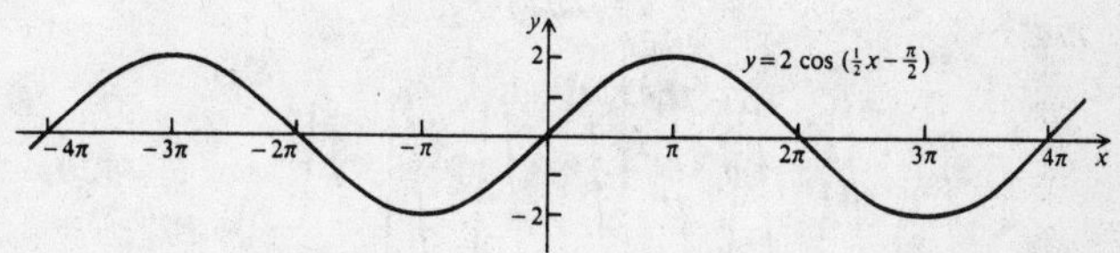

20. Amplitude: 4, period: 8π, phase shift: $-\frac{\pi}{2}$

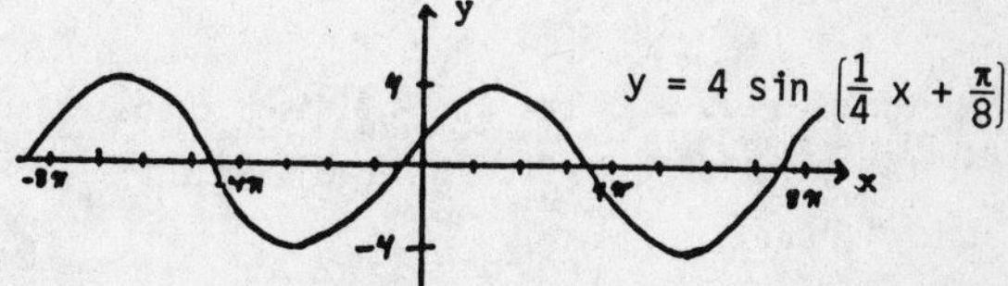

21. $y = -3\cos(4x - \pi)$, or $y = -3\cos 4\left[x - \frac{\pi}{4}\right]$

$(A = -3,\ B = 0,\ C = 4,\ D = \pi)$

The graph of $y = -3\cos 4\left[x - \frac{\pi}{4}\right]$ has an amplitude of 3, a period of $\frac{\pi}{2}$ $\left[2\pi \div 4 = \frac{\pi}{2}\right]$, and a phase shift of $\frac{\pi}{4}$.

First graph $y = \cos x$.

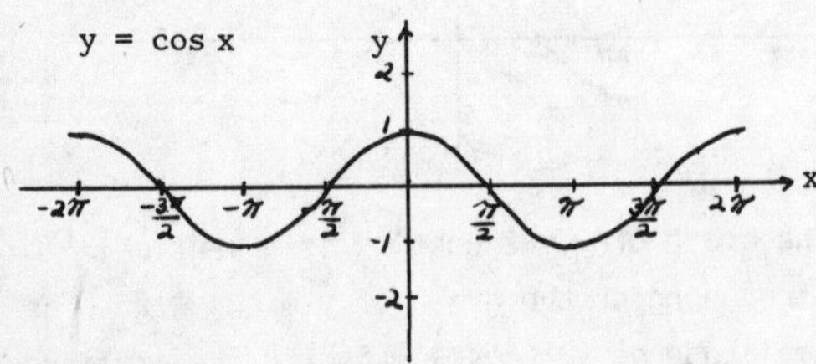

The graph of $y = \cos 4x$ is a horizontal shrinking of the graph of $y = \cos x$. The period of $y = \cos x$ is 2π. The period of $y = \cos 4x$ is $\frac{\pi}{2}$.

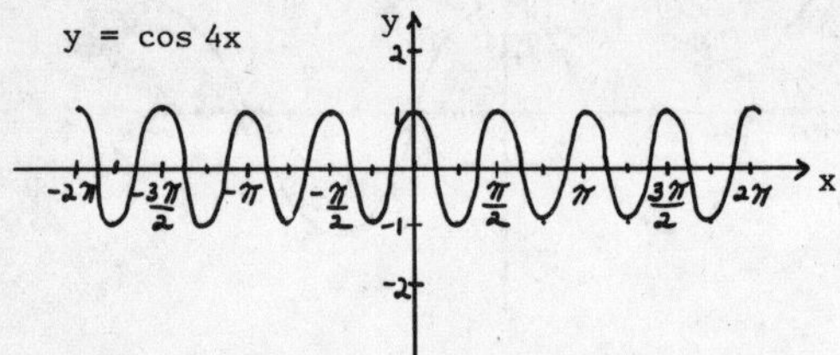

The graph of $y = -3\cos 4x$ is a vertical stretching of the graph of $y = \cos 4x$. The amplitude of $y = \cos 4x$ is 1. The amplitude of $y = -3\cos 4x$ is 3. The graph is also a reflection across the x-axis since B is negative.

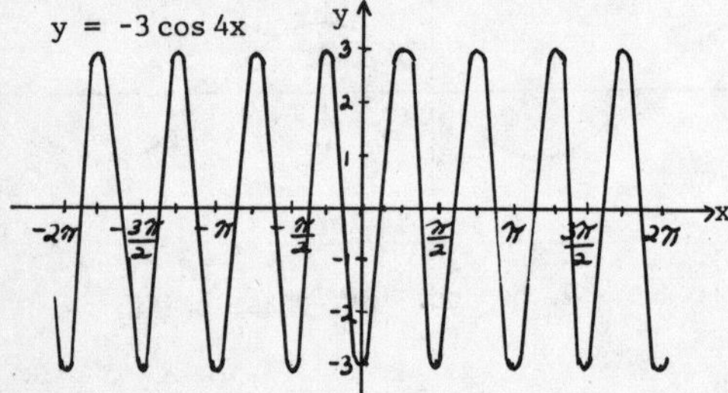

The graph of $y = -3\cos 4\left[x - \frac{\pi}{4}\right]$ is a translation of the graph of $y = -3\cos 4x$ a distance of $\frac{\pi}{4}$ to the right.

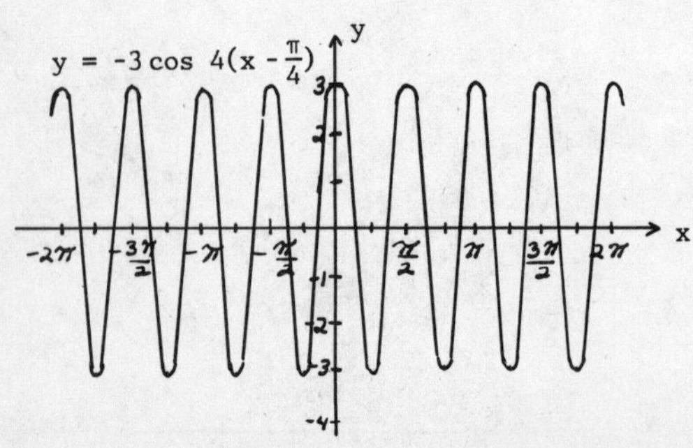

22. Amplitude: 3, period: π, phase shift: $-\frac{\pi}{4}$

$y = -3\sin\left[2x + \frac{\pi}{2}\right]$

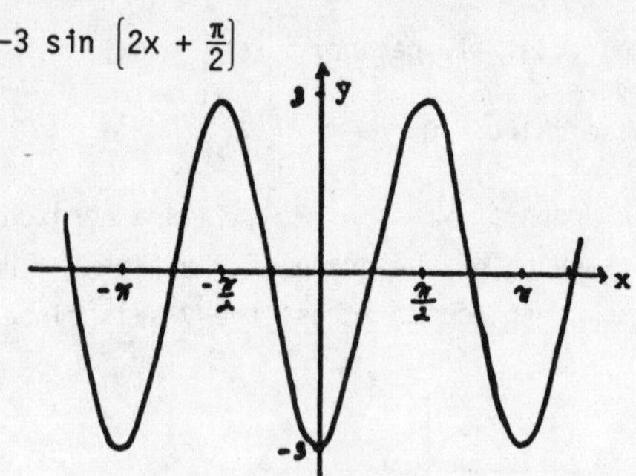

23. $y = 2 + 3\cos(\pi x - 3)$, or $y = 2 + 3\cos \pi\left[x - \frac{3}{\pi}\right]$

$(A = 3,\ B = 2,\ C = \pi,\ D = 3)$

The graph of $y = 2 + 3\cos \pi\left[x - \frac{3}{\pi}\right]$ has an amplitude of 3, a period of $2(2\pi \div \pi = 2)$, and a phase shift of $\frac{3}{\pi}$.

First graph $y = \cos x$.

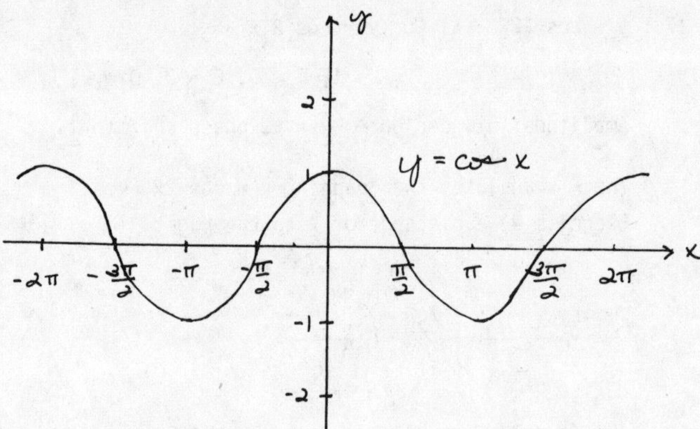

The graph of $y = \cos \pi x$ is a horizontal shrinking of the graph of $y = \cos x$. The period of $y = \cos x$ is 2π. The period of $y = \cos \pi x$ is 2.

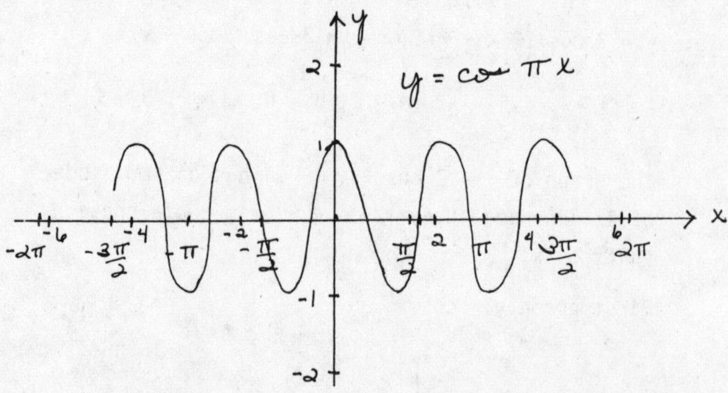

The graph of $y = 3\cos \pi x$ is a vertical stretching of the graph of $y = \cos \pi x$. The amplitude of $y = \cos \pi x$ is 1. The amplitude of $y = 3\cos \pi x$ is 3.

23. (continued)

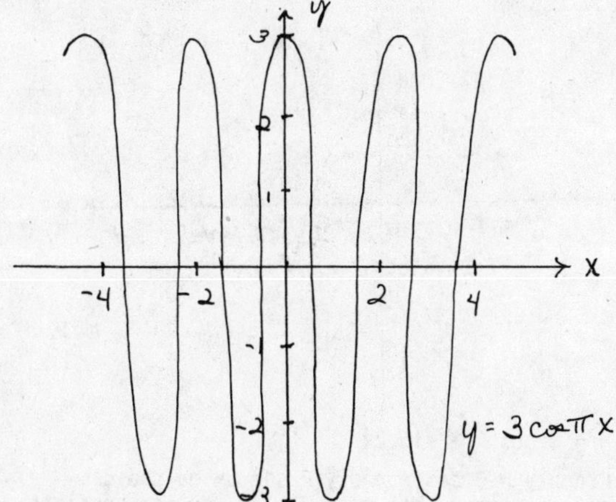

The graph of $y = 3 \cos \pi \left(x - \frac{3}{\pi}\right)$ is a translation of the graph of $y = 3 \cos \pi x$ a distance of $\frac{3}{\pi}$ to the right.

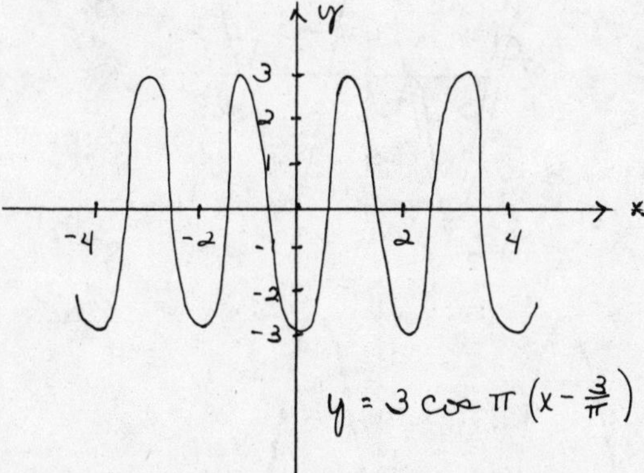

The graph of $y = 2 + 3 \cos \pi\left[x - \frac{3}{\pi}\right]$ is a translation of the graph of $y = 3 \cos \pi\left[x - \frac{3}{\pi}\right]$ up 2 units.

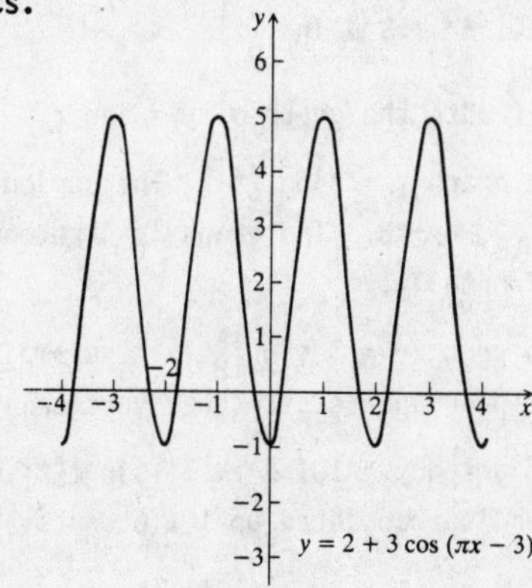

24. Amplitude: 2, period: 4, phase shift: -1

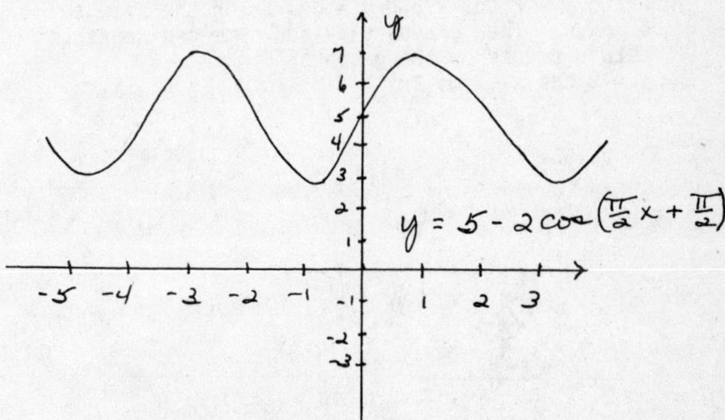

25. $y = 3 \cos \left[3x - \frac{\pi}{2}\right]$ $\left[A = 3, B = 0, C = 3, D = \frac{\pi}{2}\right]$

$y = 3 \cos 3\left[x - \frac{\pi}{6}\right]$

The amplitude is 3.

The period is $2\pi \div 3$, or $\frac{2}{3}\pi$.

The phase shift is $\frac{\pi}{6}$.

26. Amplitude: 4, Period: $\frac{\pi}{2}$, Phase shift: $\frac{\pi}{12}$

27. $y = -5 \cos \left[4x + \frac{\pi}{3}\right]$

$y = -5 \cos \left[4x - \left[-\frac{\pi}{3}\right]\right]$ $\left[A = -5, B = 0, C = 4, D = -\frac{\pi}{3}\right]$

$y = -5 \cos 4\left[x - \left[-\frac{\pi}{12}\right]\right]$

The amplitude is $|-5|$, or 5.

The period is $2\pi \div 4$, or $\frac{\pi}{2}$.

The phase shift is $-\frac{\pi}{12}$.

28. Amplitude: 4, Period: $\frac{2}{5}\pi$, Phase shift: $-\frac{\pi}{10}$

29. $y = \frac{1}{2} \sin (2\pi x + \pi)$

$y = \frac{1}{2} \sin [2\pi x - (-\pi)]$

$\left[A = \frac{1}{2}, B = 0, C = 2\pi, D = -\pi\right]$

$y = \frac{1}{2} \sin 2\pi\left[x - \left[-\frac{1}{2}\right]\right]$

The amplitude is $\frac{1}{2}$.

The period is $2\pi \div 2\pi$, or 1.

The phase shift is $-\frac{1}{2}$.

30. Amplitude: $\frac{1}{4}$, Period: 2, Phase shift: $\frac{4}{\pi}$

31. $y = 2 \cos x + \cos 2x$

Graph $y = 2 \cos x$ and $y = \cos 2x$ on the same set of axes. Then graphically add some ordinates to obtain points on the graph of $y = 2 \cos x + \cos 2x$.

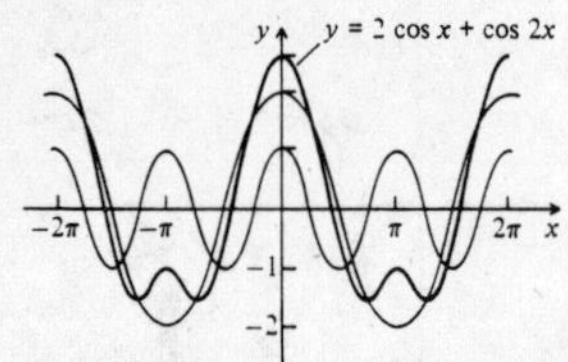

32.

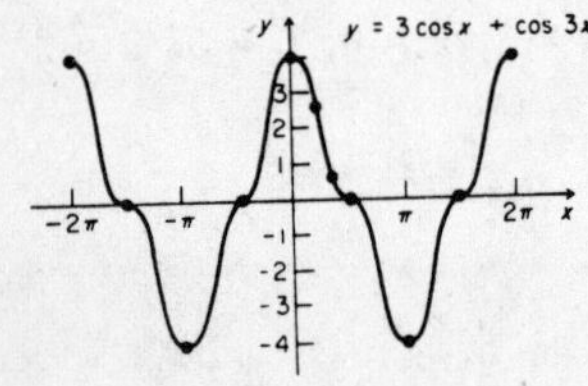

33. $y = \sin x + \cos 2x$

Graph $y = \sin x$ and $y = \cos 2x$ on the same set of axes. Then graphically add some ordinates to obtain points on the graph of $y = \sin x + \cos 2x$.

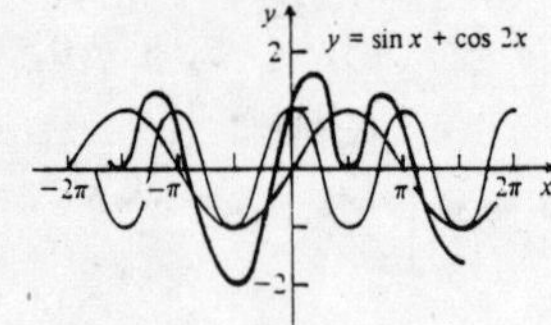

34. , graph $y = 2 \sin x$

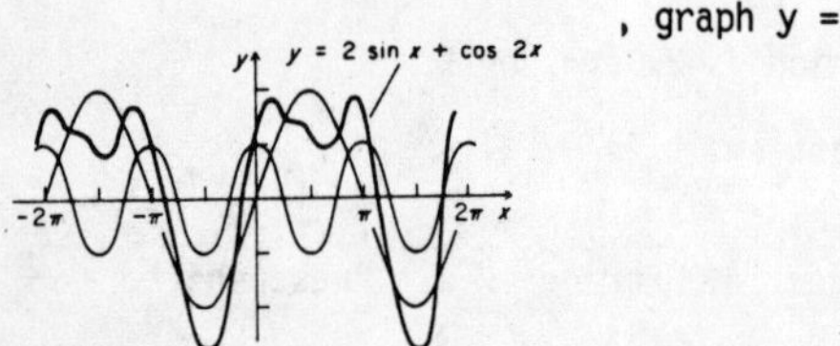

35. $y = \sin x - \cos x$

Graph $y = \sin x$ and $y = -\cos x$ on the same set of axes. Then graphically add some ordinates to obtain points on the graph of $y = \sin x - \cos x$.

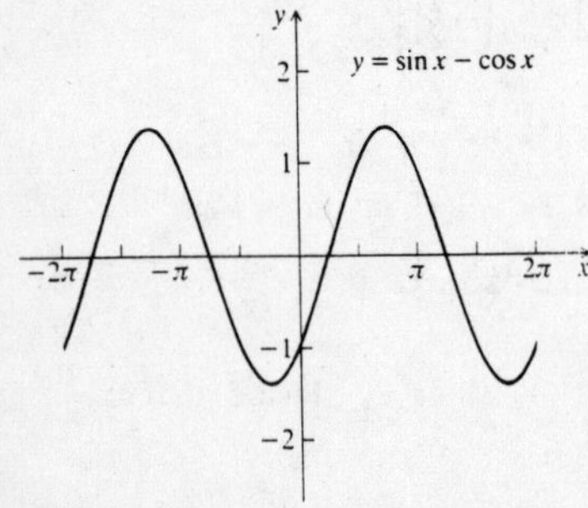

36.

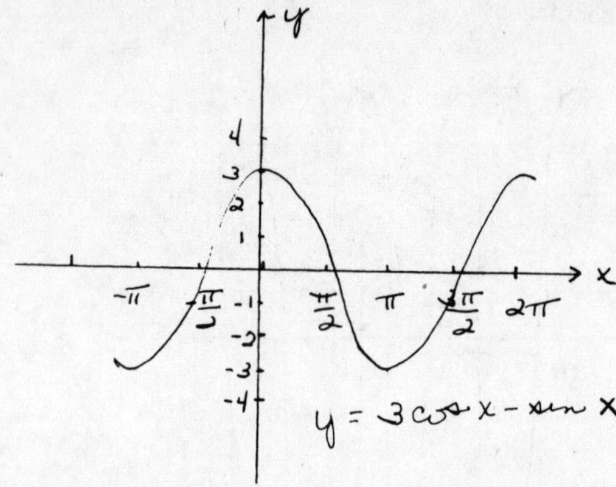

37. $y = 3 \cos x + \sin 2x$

Graph $y = 3 \cos x$ and $y = \sin 2x$ on the same set of axes. Then graphically add some ordinates to obtain points on the graph of $y = 3 \cos x + \sin 2x$.

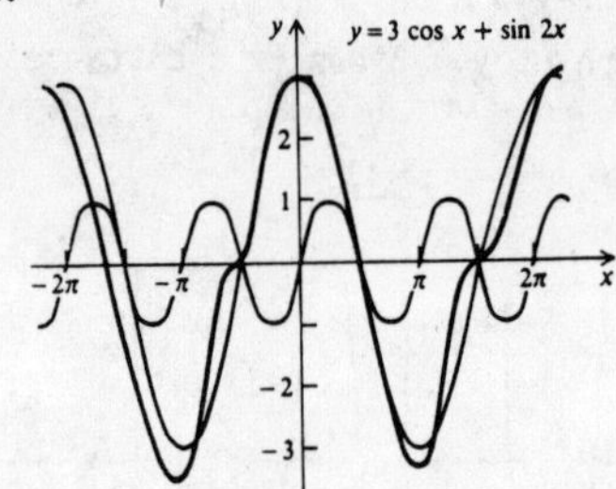

38.

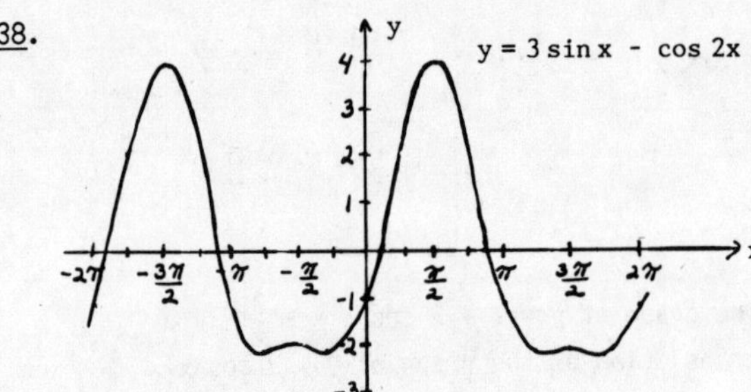

39. $T(t) = 101.6° + 3° \sin \left(\frac{\pi}{8}t\right)$

a) Start with the graph of $y = \sin t$.

Then graph $y = \sin \left(\frac{\pi}{8}t\right)$. The period is $2\pi \div \frac{\pi}{8}$, or 16. The graph is stretched horizontally.

Then graph $y = 3 \sin \left(\frac{\pi}{8}t\right)$. The amplitude is 3. The graph is stretched vertically.

Then graph $y = 101.6° + 3° \sin \left(\frac{\pi}{8}t\right)$. The graph is translated up 101.6 units.

39. (continued)

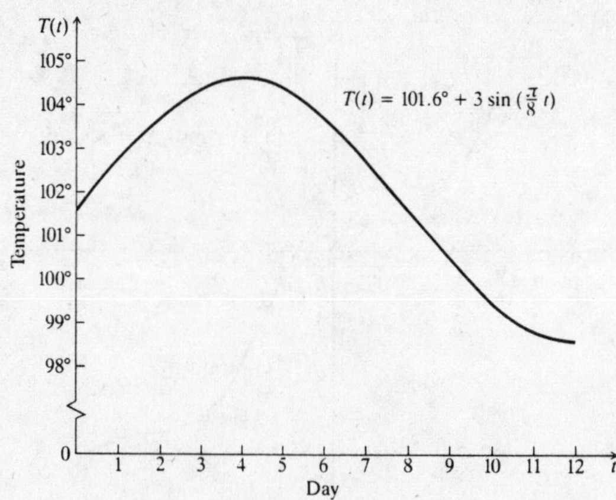

b) The maximum value occurs on day 4 when the sine function takes its maximum value, 1. It is $101.6° + 3 \cdot 1$, or $104.6°$.

The minimum value occurs on day 12 when the sine function takes its minimum value, -1. It is $101.6° + 3(-1)$, or $98.6°$.

40. a)

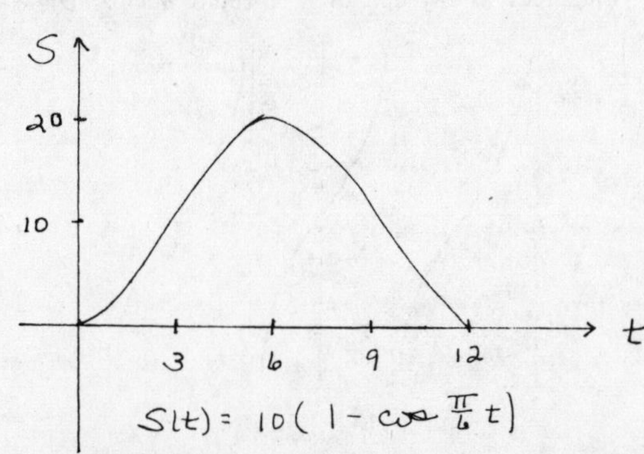

b) 12

c) \$0 in July (at t = 0 and t = 12)

d) \$20,000 in January (at t = 6)

41. $y = 3000\left[\cos \frac{\pi}{45}(t - 10)\right]$

a) Start with the graph of $y = \cos t$.

Then graph $y = \cos\left[\frac{\pi}{45}t\right]$. The period is $2\pi \div \frac{\pi}{45}$, or 90. The graph is stretched horizontally.

Then graph $y = 3000 \cos\left[\frac{\pi}{4}t\right]$. The amplitude is 3000. The graph is stretched vertically.

Then graph $y = 3000 \cos \frac{\pi}{45}(t - 10)$. The phase shift is 10. The graph is translated 10 units to the right.

41. (continued)

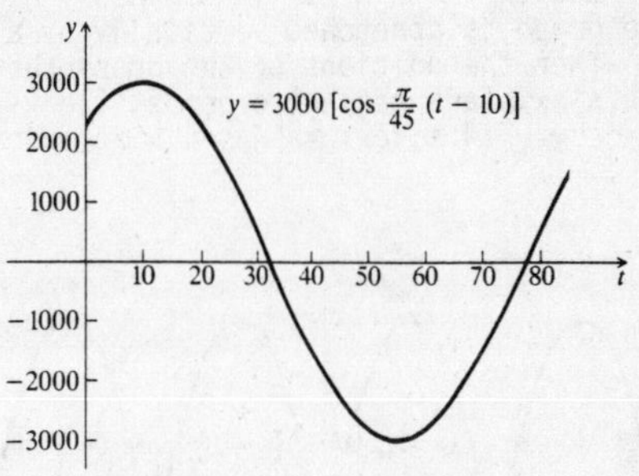

b) Amplitude: 3000, period: 90, phase shift: 10

42. a)

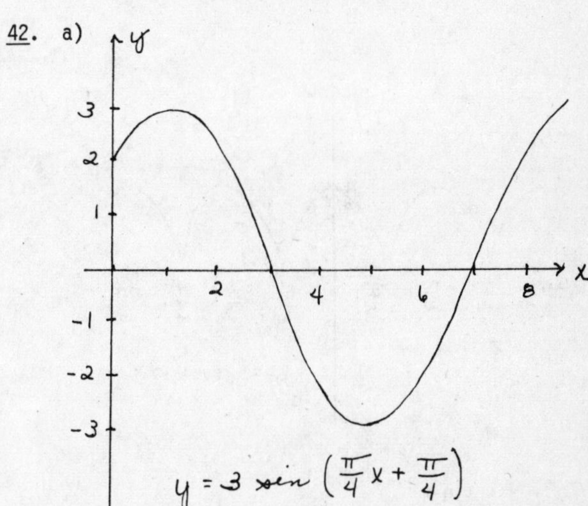

b) Amplitude: 3, period: 8, phase shift: -1

43. We use $y = B + A \sin\left[C\left(x - \frac{D}{C}\right)\right]$.

$A = 8.6$; $\frac{2\pi}{C} = 5$, or $C = \frac{2\pi}{5}$; $\frac{D}{C} = -11$; B = any real number. We choose B = 0. We have $y = 8.6 \sin\left[\frac{2\pi}{5}(x + 11)\right]$. Answers may vary.

44. $y = 6.7 \cos\left[2\left(x - \frac{4}{7}\right)\right]$. Answers may vary.

45. We use $y = B + A \cos\left[C\left(x - \frac{D}{C}\right)\right]$.

$A = 16$; $\frac{2\pi}{C} = \frac{\pi}{3}$, or $C = 6$; $\frac{D}{C} = -\frac{2}{\pi}$, B = any real number. We choose B = 0. We have $y = 16 \cos\left[6\left(x + \frac{2}{\pi}\right)\right]$. Answers may vary.

46. $y = 34 \sin\left[10\left(x + \frac{8}{\pi}\right)\right]$. Answers may vary.

47. $y = |4 \sin x|$

Start with $y = \sin x$. Then graph $y = 4 \sin x$. The graph is stretched vertically by a factor of 4. Then the portions of the graph that lie below the x-axis are reflected across the x-axis since $y = |4 \sin x| = -4 \sin x$ for $\sin x < 0$.

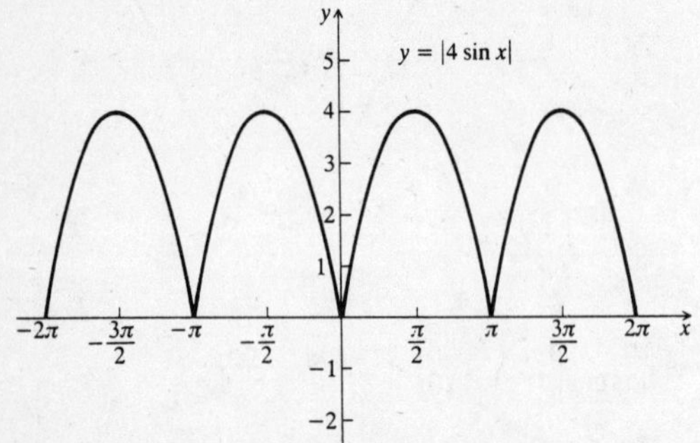

48.

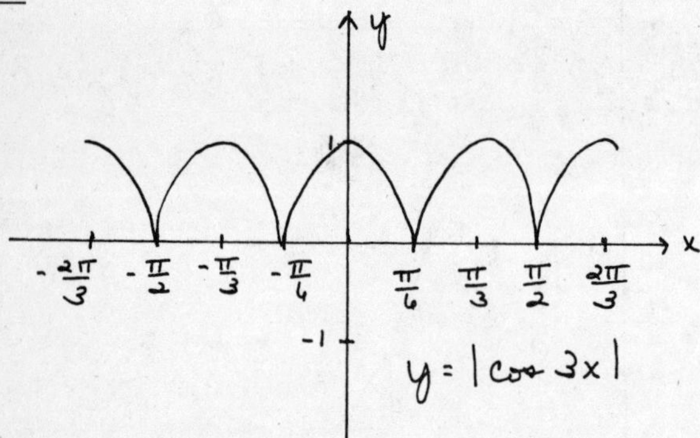

49. $y = x + \sin x$

Graph $y = x$ and $y = \sin x$ on the same set of axes. Then graphically add some ordinates to obtain points on the graph of $y = x + \sin x$.

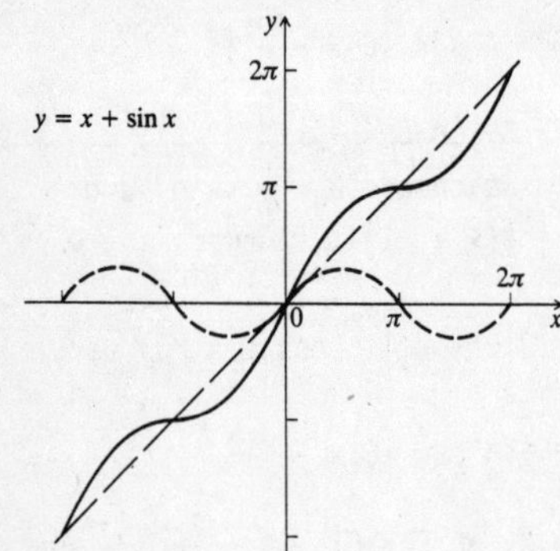

50.

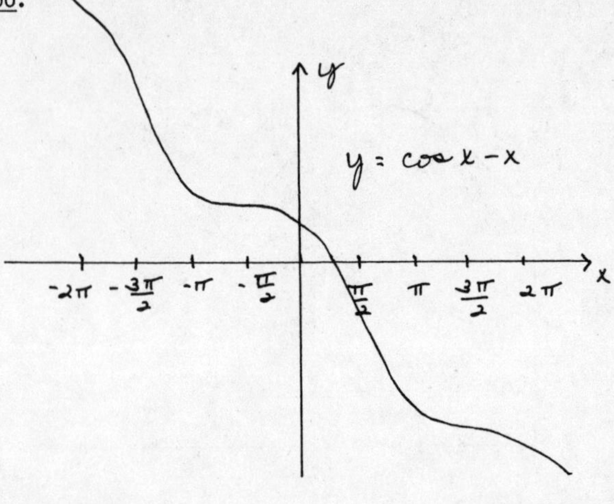

Exercise Set 6.8

1. $y = -\tan x$

Reflect the graph of $y = \tan x$ across the x-axis.

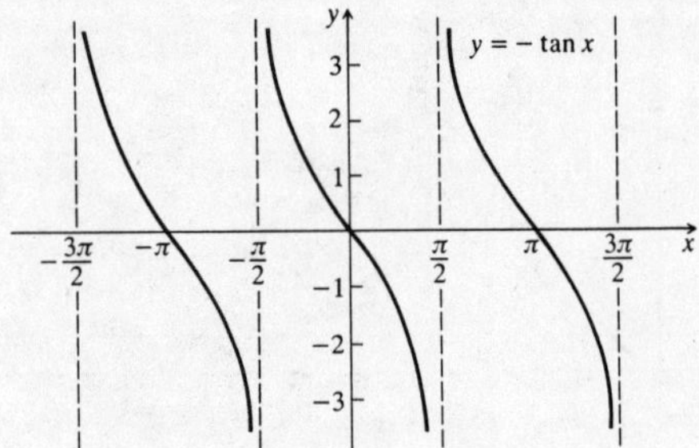

2.

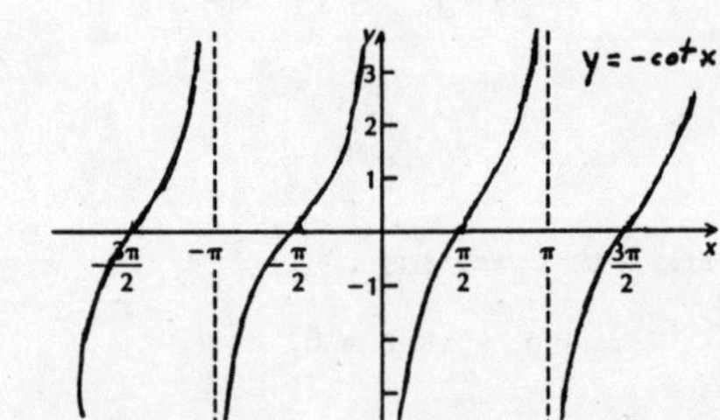

3. $y = -\csc x$

Reflect the graph of $y = \csc x$ across the x-axis.

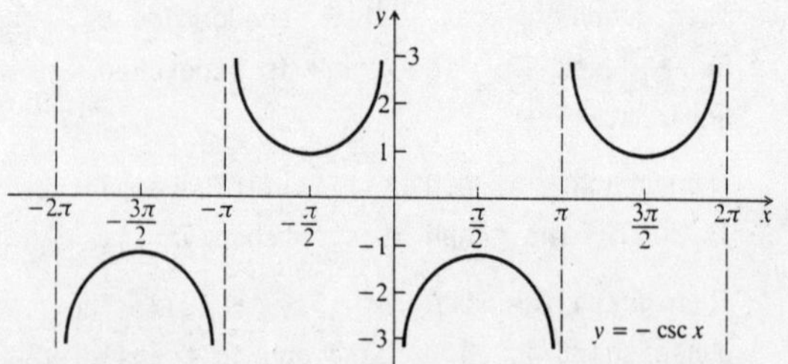

4.

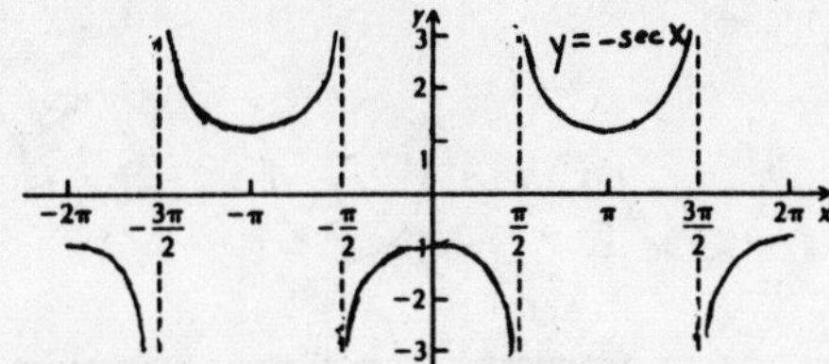

5. y = sec (-x)

sec (-x) = sec x (see Exercise 33, Exercise Set 6.3), so we graph y = sec x.

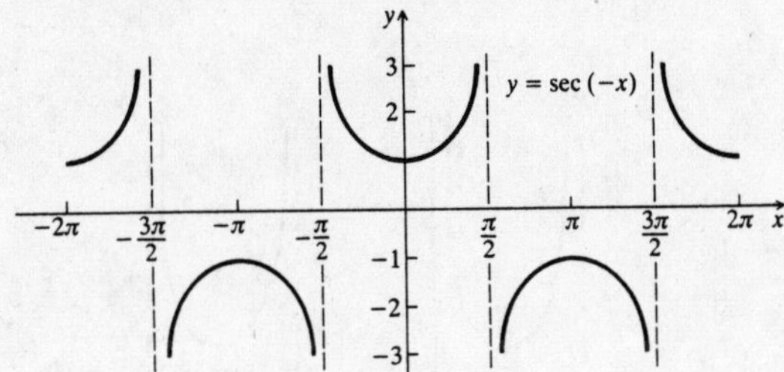

6.

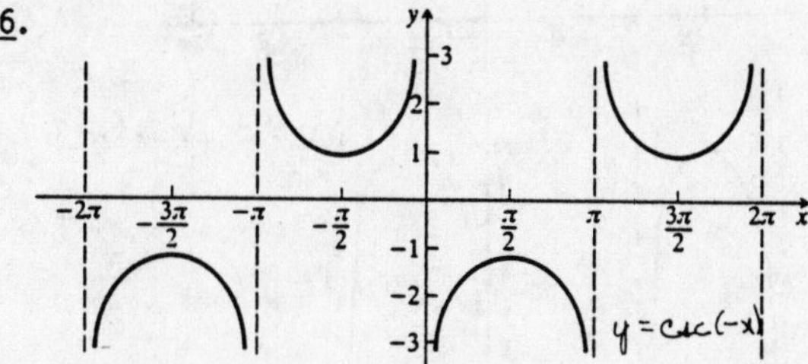

7. y = cot (-x)

cot (-x) = -cot x (see Margin Exercise 37, Exercise Set 6.3), so we graph y = -cot x. This is the reflection of the graph of y = cot x across the x-axis.

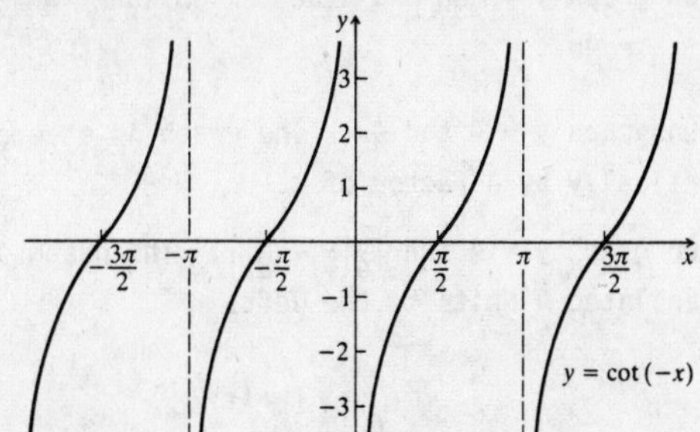

8.

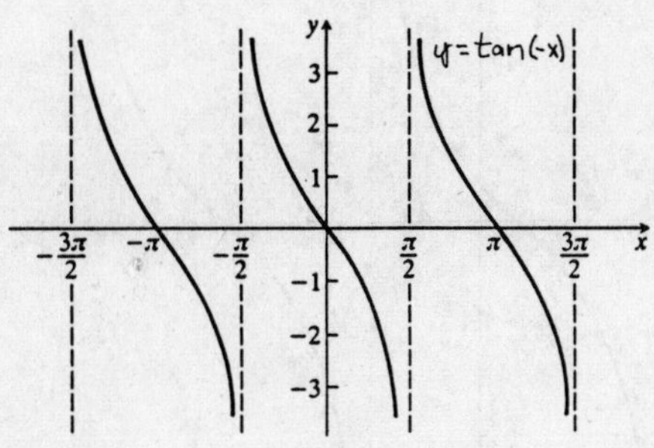

9. y = -2 + cot x

This is the graph of y = cot x translated down 2 units.

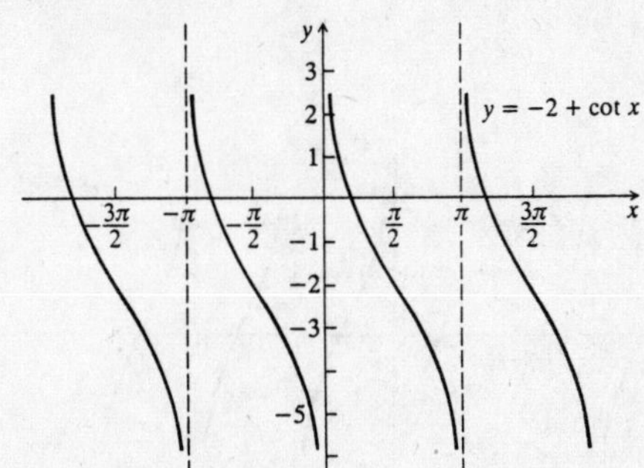

10.

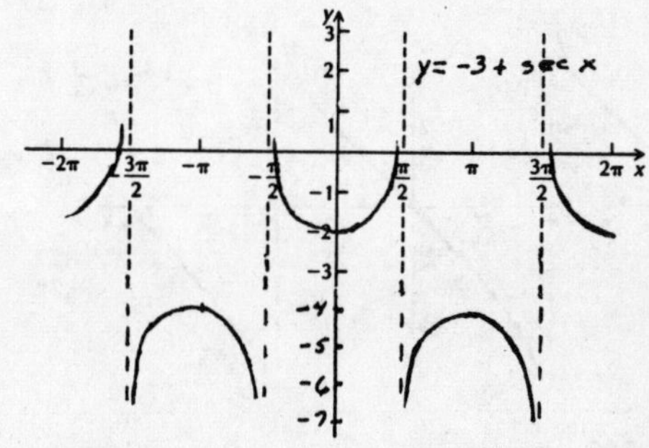

11. $y = -\frac{3}{2}\csc x$

The graph of y = csc x is stretched vertically by a factor of $\frac{3}{2}$ and is also reflected across the x-axis.

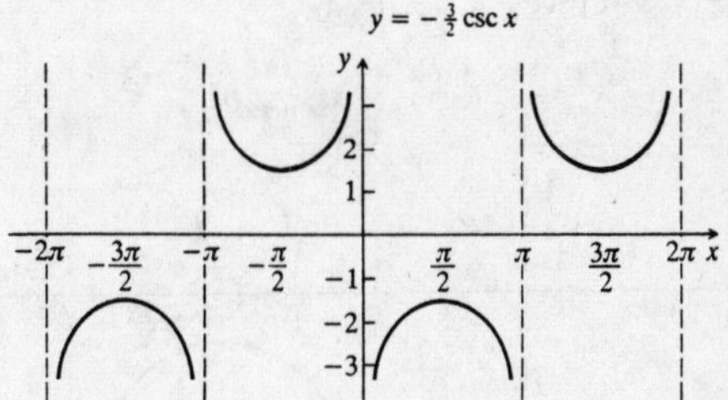

12.

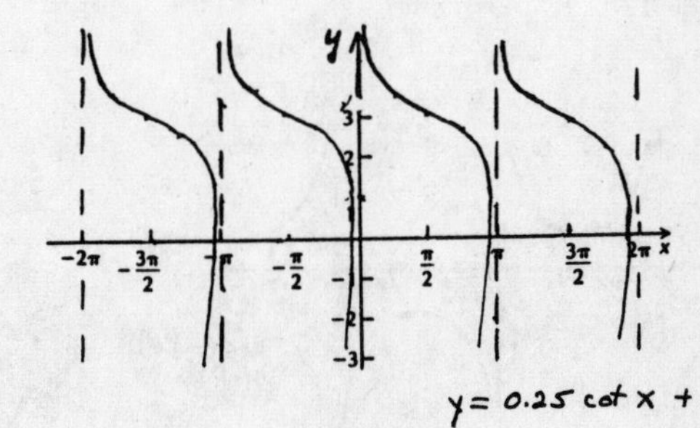

13. $y = \cot 2x$

Start with the graph of $y = \cot x$. The period of the cotangent function is π, so the period of $y = \cot 2x$ is $\frac{\pi}{2}$.

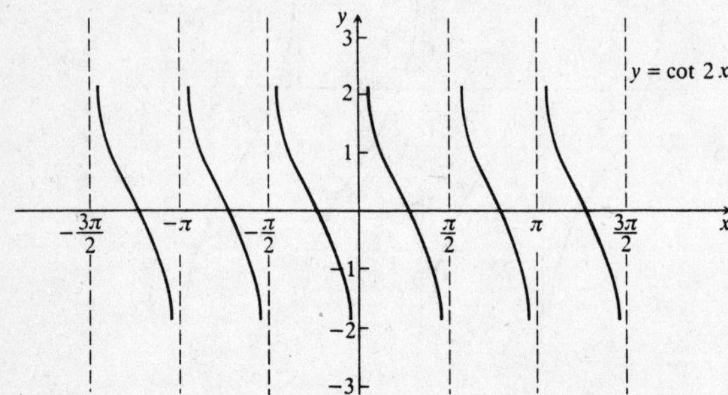

14.

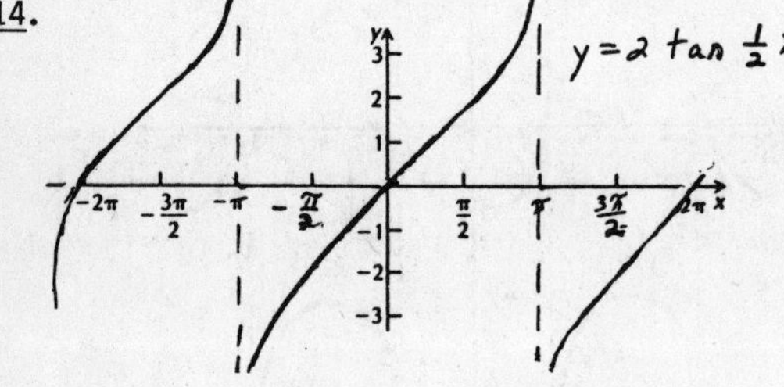

15. $y = \sec\left(-\frac{1}{2}x\right)$

$\sec\left(-\frac{1}{2}x\right) = \sec\left(\frac{1}{2}x\right)$ (see Exercise 33, Exercise Set 6.3), so we graph $y = \sec\left(\frac{1}{2}x\right)$. Start with the graph of $y = \sec x$. The period of the secant function is 2π, so the period of $y = \sec\left(\frac{1}{2}x\right)$ is $2\pi \div \frac{1}{2}$, or 4π.

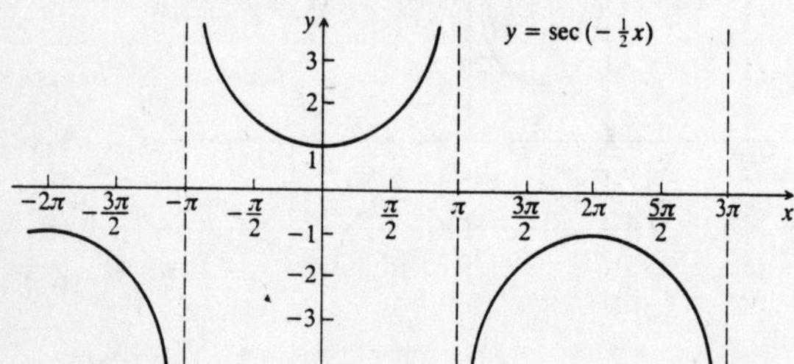

16.

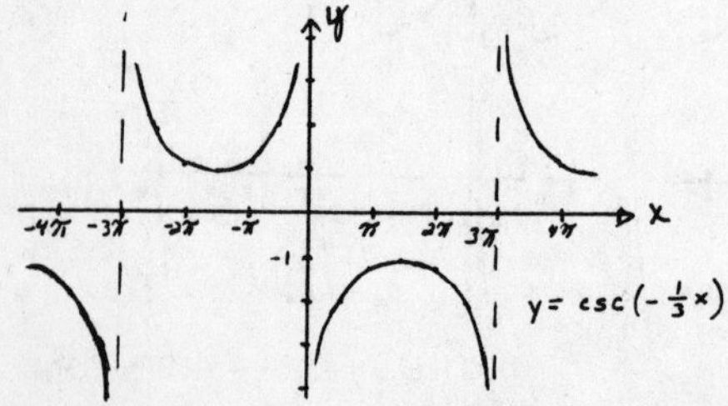

17. $y = 2 \sec (x - \pi)$

Start with the graph of $y = \sec x$.

Then graph $y = 2 \sec x$. The graph is stretched vertically by a factor of 2.

Then graph $y = 2 \sec (x - \pi)$. The graph is translated π units to the right.

17. (continued)

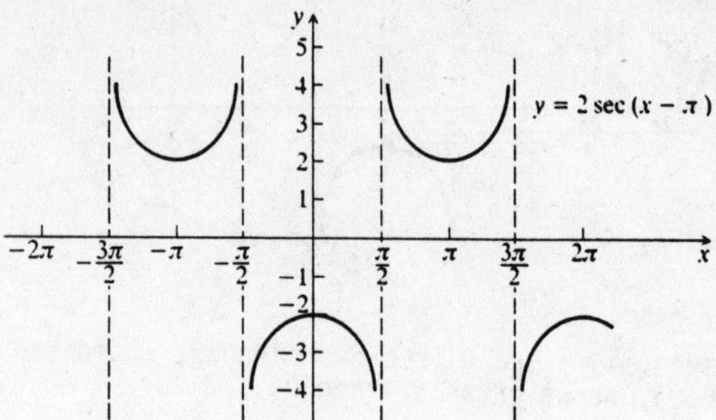

18.

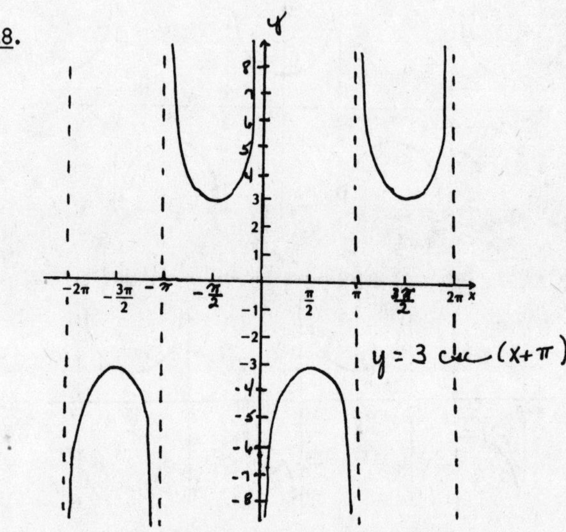

19. $y = 4 \tan\left(\frac{1}{4}x + \frac{\pi}{8}\right) = 4 \tan\left[\frac{1}{4}\left(x + \frac{\pi}{2}\right)\right]$

Start with the graph of $y = \tan x$.

Then graph $y = \tan \frac{1}{4}x$. The period is $\pi \div \frac{1}{4} = 4\pi$.

Then graph $y = 4 \tan \frac{1}{4}x$. The graph is stretched vertically by a factor of 4.

Then graph $y = 4 \tan\left[\frac{1}{4}\left(x + \frac{\pi}{2}\right)\right]$. The graph is translated $\frac{\pi}{2}$ units to the left.

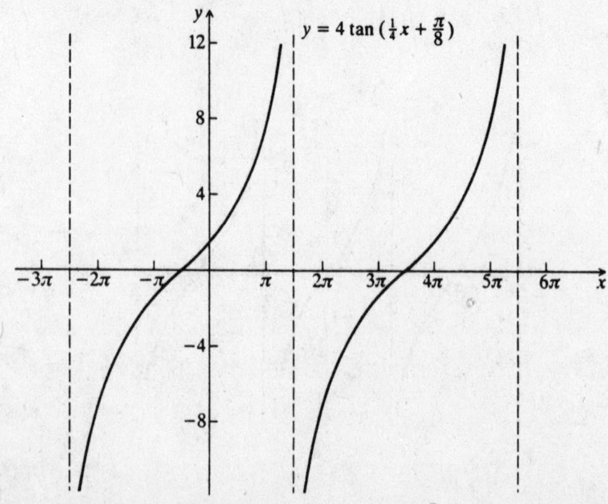

20.

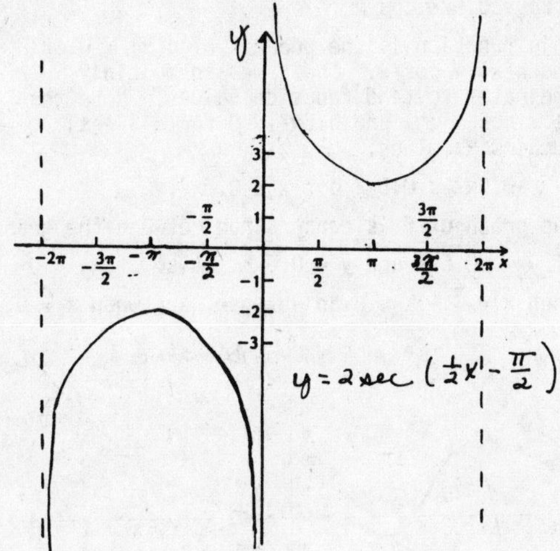

21. $y = -3 \cot \left[2x + \frac{\pi}{2}\right] = -3 \cot \left[2\left(x + \frac{\pi}{4}\right)\right]$

Start with the graph of y = cot x.

Then graph y = cot 2x. The period is $\frac{\pi}{2}$.

Then graph y = -3 cot 2x. The graph is stretched vertically by a factor of 3 and is reflected across the x-axis.

Then graph $y = -3 \cot \left[2\left(x + \frac{\pi}{4}\right)\right]$. The graph is translated $\frac{\pi}{4}$ units to the left.

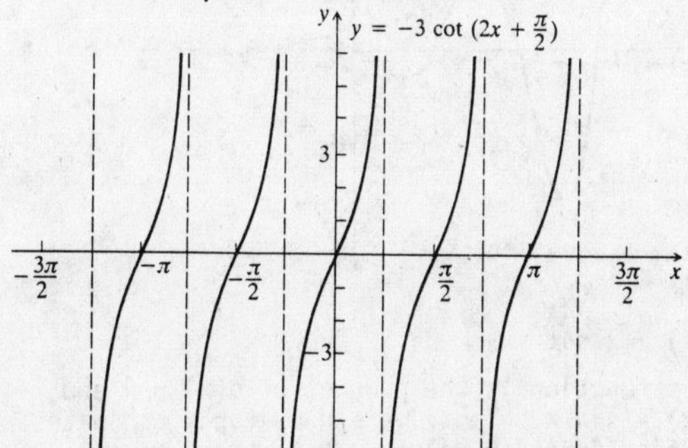

22.

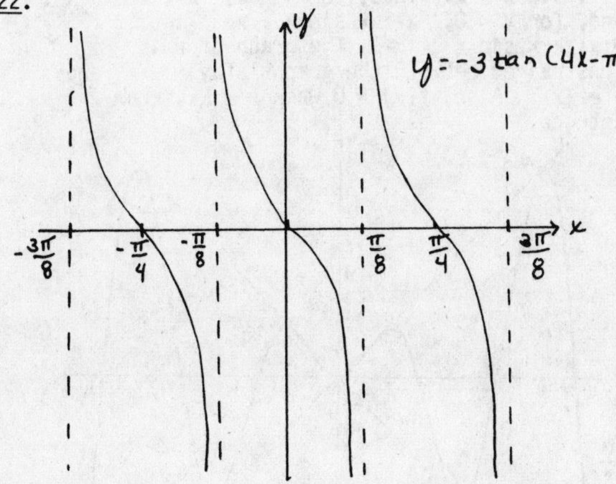

23. $y = 4 \sec (2x - \pi) = 4 \sec \left[2\left(x - \frac{\pi}{2}\right)\right]$

Start with the graph of y = sec x.

Then graph y = sec 2x. The period is $2\pi \div 2$, or π.

Then graph y = 4 sec 2x. The graph is stretched vertically by a factor of 4.

Then graph $y = 4 \sec \left[2\left(x - \frac{\pi}{2}\right)\right]$. The graph is translated $\frac{\pi}{2}$ units to the right.

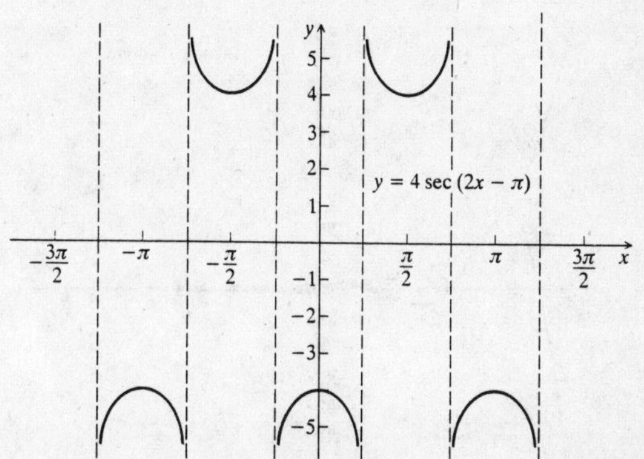

24.

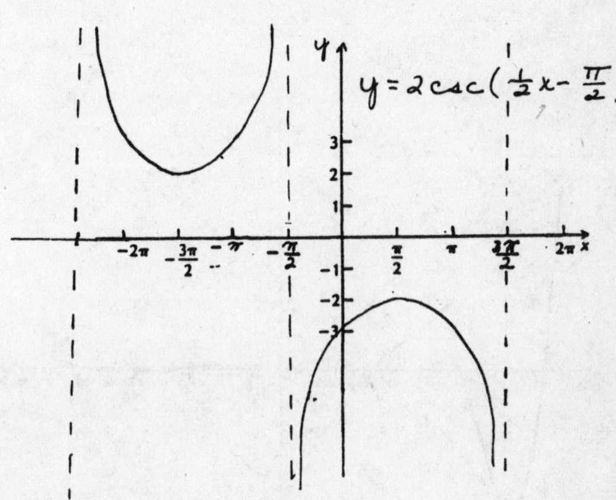

25. $f(x) = e^{-x/2} \cos x$

This function is the product of $g(x) = e^{-x/2}$ and $h(x) = \cos x$. Thus, we can multiply ordinates to find function values. Note that $-1 \leqslant \cos x \leqslant 1$ and $e^{-x/2} > 0$ for all real numbers x. Thus,

$$-e^{-x/2} \leqslant e^{-x/2} \cos x \leqslant e^{-x/2}.$$

The graph of f is constrained between the graphs of $y = -e^{-x/2}$ and $y = e^{-x/2}$. Also, $f(x) = 0$ when $x = \frac{\pi}{2} + k\pi$, k an integer.

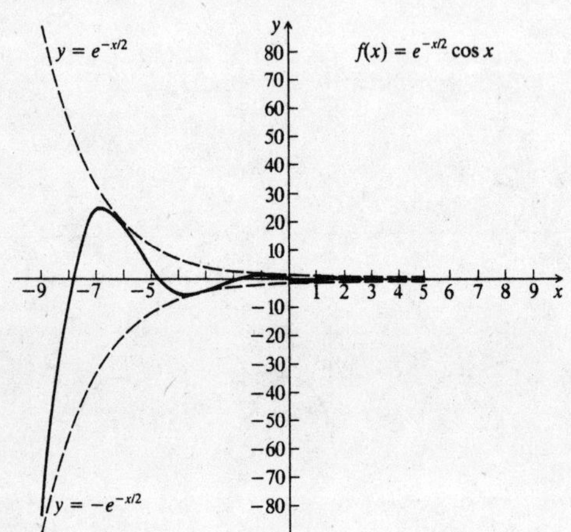

26.

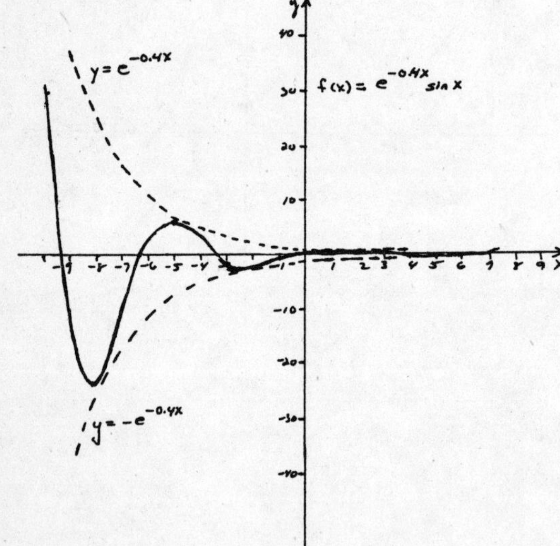

27. $f(x) = 0.6x^2 \cos x$

This function is the product of $g(x) = 0.6x^2$ and $h(x) = \cos x$. Thus, we can multiply ordinates to find function values. Note that $-1 \leqslant \cos x \leqslant 1$ and $0.6x^2 \geqslant 0$ for all real numbers x. Thus,

$$-0.6x^2 \leqslant 0.6x^2 \cos x \leqslant 0.6x^2.$$

The graph of f is constrained between the graphs of $y = -0.6x^2$ and $y = 0.6x^2$. Also, $f(x) = 0$ when $x = \frac{\pi}{2} + k\pi$, k an integer, and when $x = 0$.

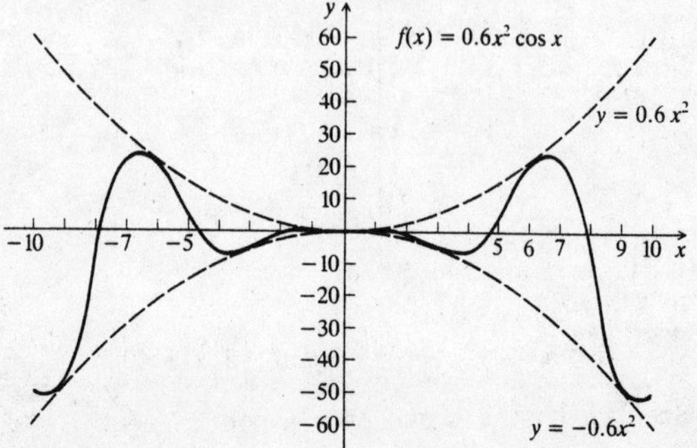

28.

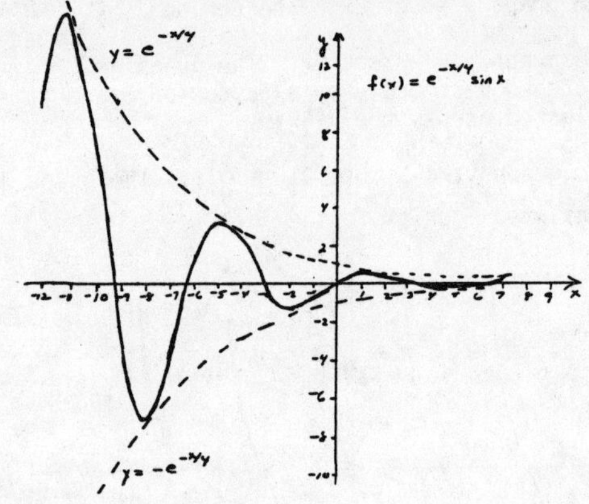

29. $f(x) = x \sin x$

This function is the product of $g(x) = x$ and $h(x) = \sin x$. Thus, we can multiply ordinates to find function values. Note that $-1 \leqslant \sin x \leqslant 1$. Thus, for $x \geqslant 0$, $-x \leqslant x \sin x \leqslant x$ and, for $x < 0$, $-x \geqslant x \sin x \geqslant x$. That is $-|x| \leqslant x \sin x \leqslant |x|$. The graph of f is constrained between the graphs of $y = -|x|$ and $y = |x|$. Also, $f(x) = 0$ when $x = k\pi$, k an integer.

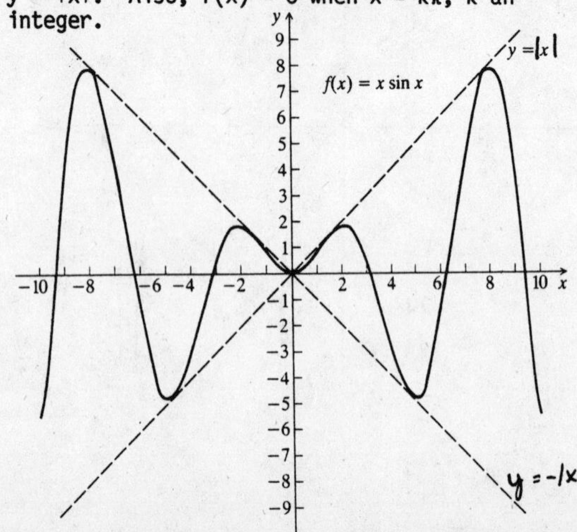

30.

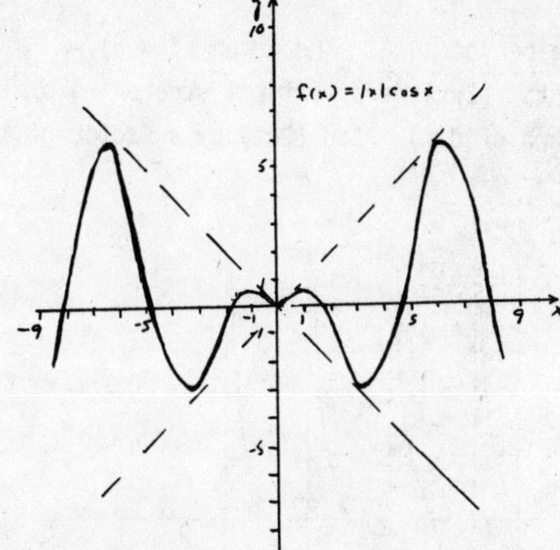

31. $f(x) = 2^{-x}\sin x$

This function is the product of $g(x) = 2^{-x}$ and $h(x) = \sin x$. Thus, we can multiply ordinates to find function values. Note that $-1 \leqslant \sin x \leqslant 1$ and $2^{-x} > 0$ for all real numbers x. Thus,

$$-2^{-x} \leqslant 2^{-x}\sin x \leqslant 2^{-x}.$$

The graph of f is constrained between the graphs of $y = -2^{-x}$ and $y = 2^{-x}$. Also, $f(x) = 0$ when $x = \frac{\pi}{2} + k\pi$, k an integer.

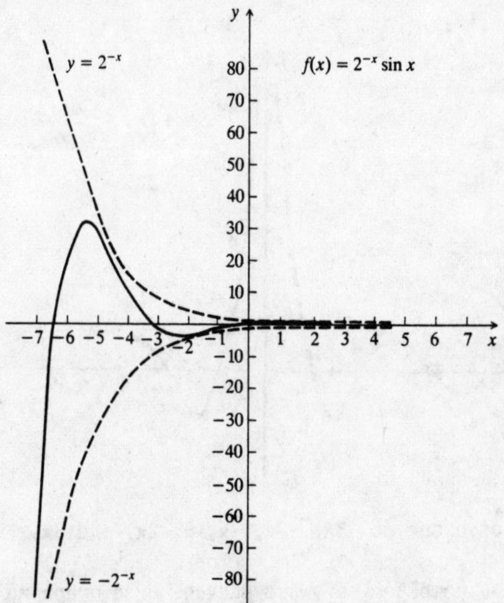

32.

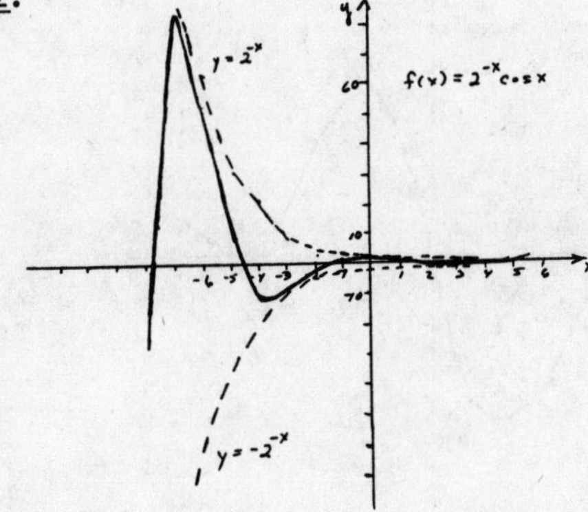

33. $f(x) = |\tan x|$

Start with the graph of $f(x) = \tan x$. Then reflect across the x-axis all points that lie below the x-axis, since $|\tan x| = -\tan x$ when $\tan x < 0$.

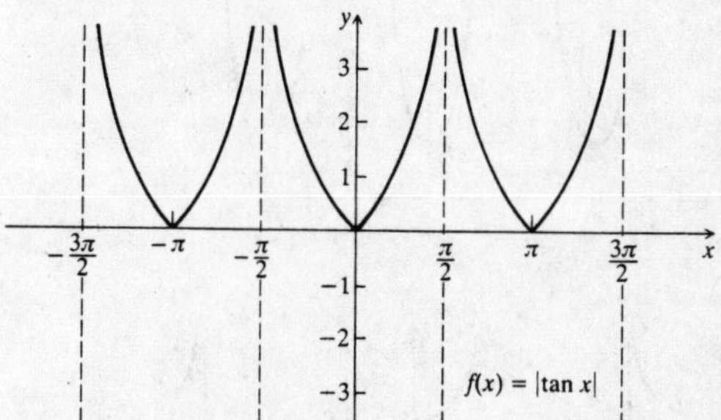

34.

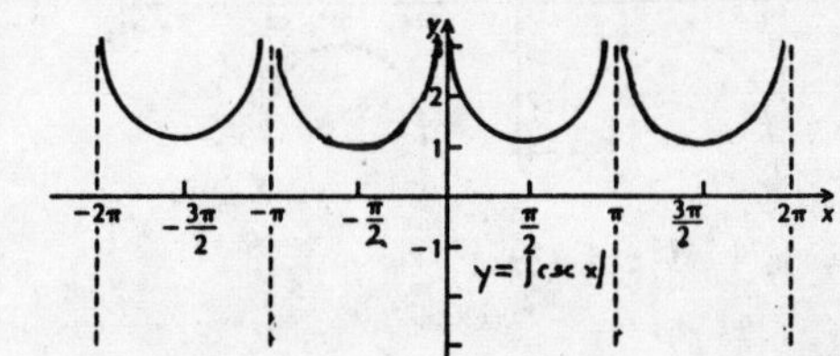

35. $f(x) = (\ell n\ x)(\sin x)$

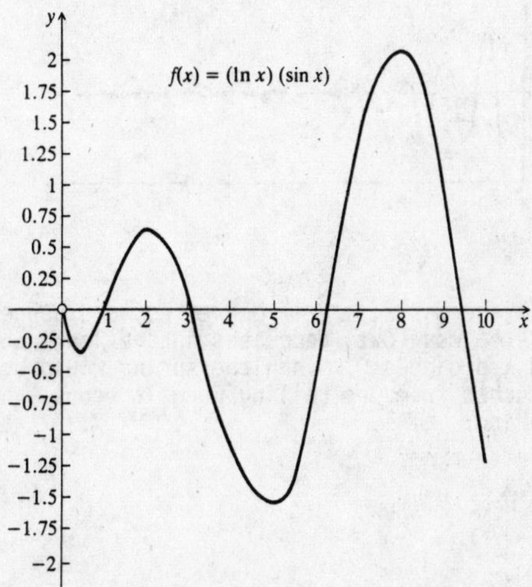

36.

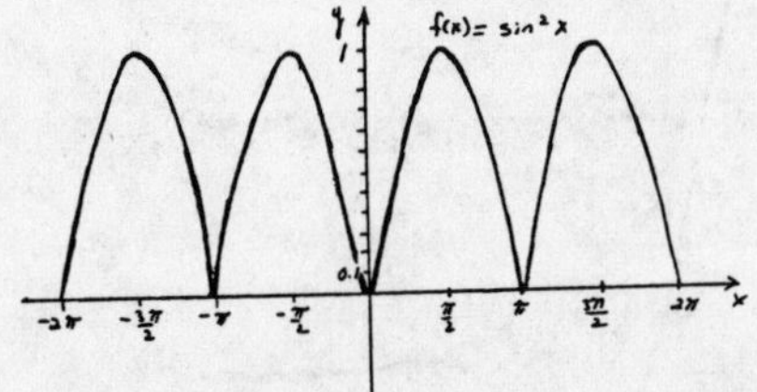

37. $f(x) = \sec^2 x$

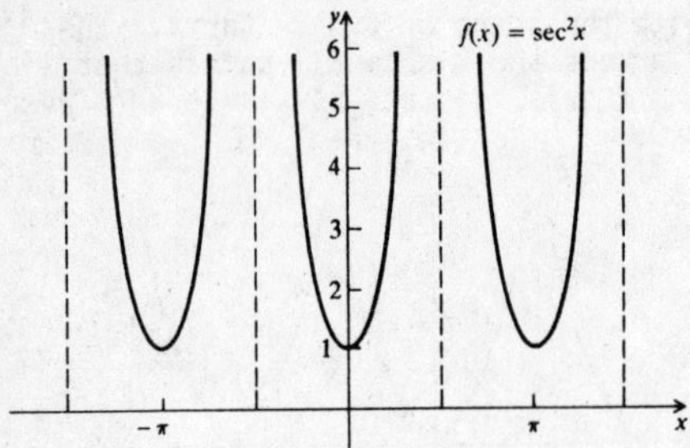

38.

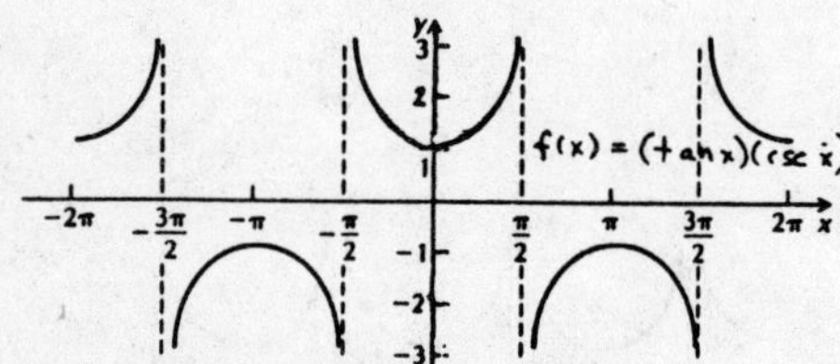

39. $d(t) = 6e^{-0.8t} \cos(6\pi t) + 4$

a)

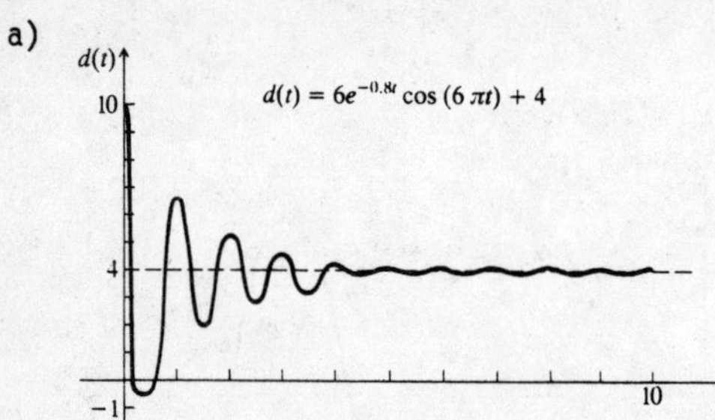

b) As t increases, $e^{-0.8t}$ decreases, and hence $6e^{-0.8t} \cos(6\pi t)$ decreases in the long run and approaches 0. Then the spring would be 4 inches from the ceiling when it stops bobbing.

40. a)

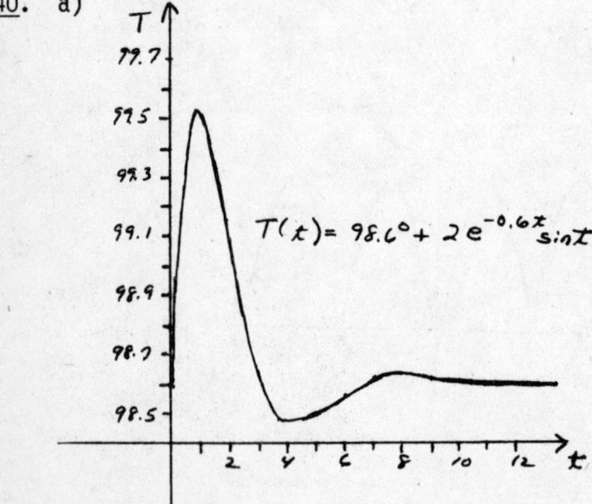

b) 98.6°

41. a) The period is $\frac{1}{2}$. The graph of $d(t) = 10 \tan(2\pi t)$ is a vertical stretching of the graph of $d(t) = \tan(2\pi t)$ by a factor of 10.

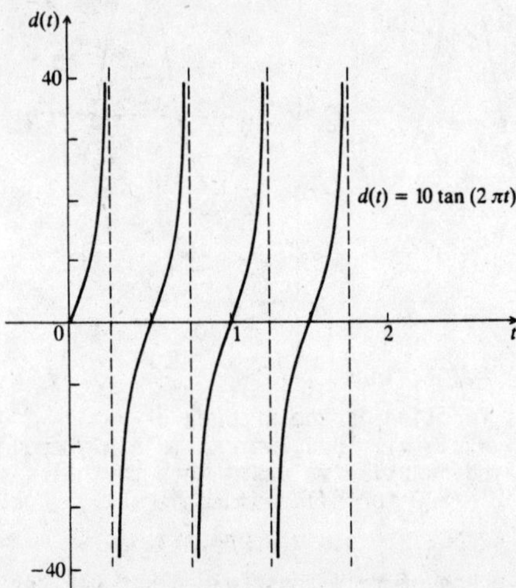

b) The values of t for which the function is undefined are those values for which the beam is parallel to the wall.

42.

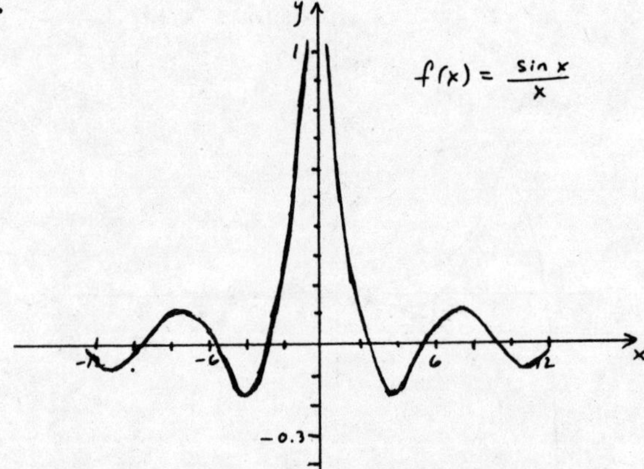

Zeros occur at -3π, -2π, $-\pi$, π, 2π, and 3π.

43. Use a computer software package or a graphing calculator to graph the function.

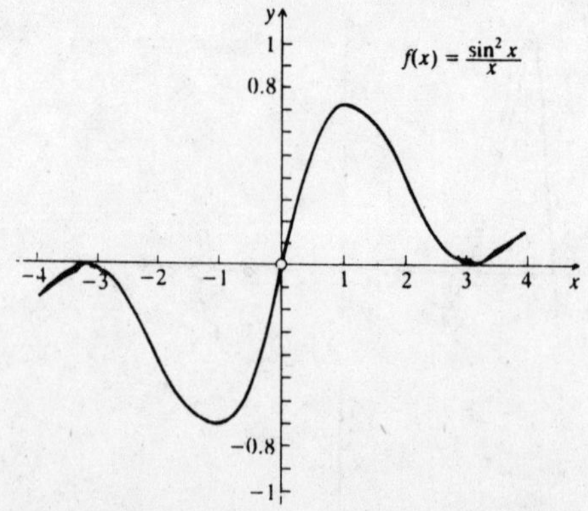

In the interval [-4, 4], the function has zeros at $x = -\pi$ and at $x = \pi$.

44.

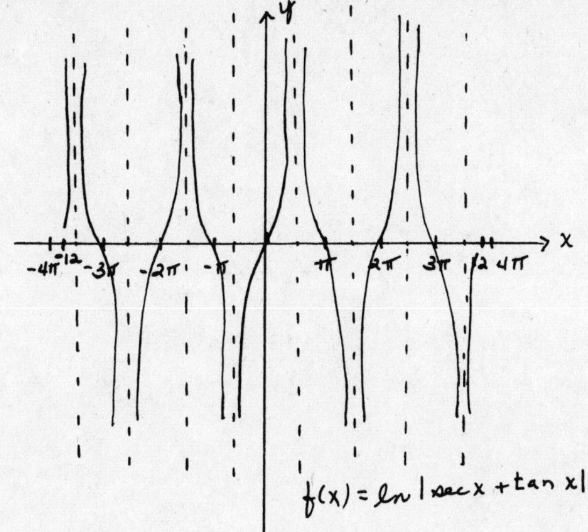

Zeros occur at -3π, -2π, $-\pi$, 0, π, 2π, and 3π.

45. Use a computer software package or a graphing calculator to graph the function.

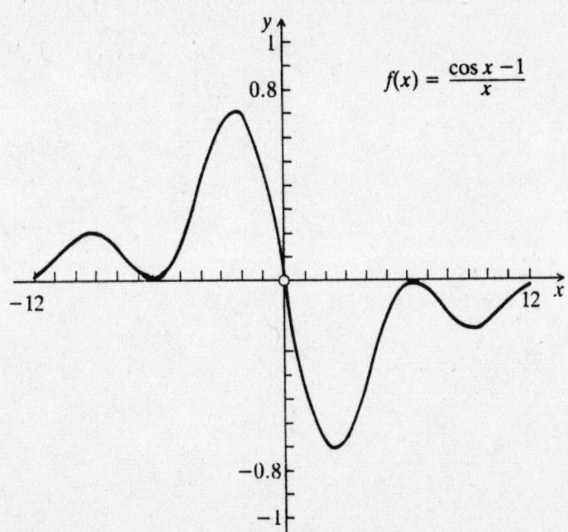

In the interval [-12, 12], the function has zeros at $x = -2\pi$ and at $x = 2\pi$.

Exercise Set 7.1

1. $(\sin x - \cos x)(\sin x + \cos x)$
$= \sin^2 x - \cos^2 x$

2. $\tan^2 y - \cot^2 y$

3. $\tan x\,(\cos x - \csc x)$
$= \dfrac{\sin x}{\cos x}\left[\cos x - \dfrac{1}{\sin x}\right]$
$= \sin x - \dfrac{1}{\cos x}$
$= \sin x - \sec x$

4. $\cos x + \csc x$

5. $\cos y \sin y\,(\sec y + \csc y)$
$= \cos y \sin y\left[\dfrac{1}{\cos y} + \dfrac{1}{\sin y}\right]$
$= \sin y + \cos y$

6. $\sin y - \tan y$

7. $(\sin x + \cos x)(\csc x - \sec x)$
$= \sin x \csc x - \sin x \sec x + \cos x \csc x - \cos x \sec x$
$= \sin x \cdot \dfrac{1}{\sin x} - \sin x \cdot \dfrac{1}{\cos x} + \cos x \cdot \dfrac{1}{\sin x} - \cos x \cdot \dfrac{1}{\cos x}$
$= 1 - \dfrac{\sin x}{\cos x} + \dfrac{\cos x}{\sin x} - 1$
$= \cot x - \tan x$

8. $2 + \tan x + \cot x$

9. $(\sin y - \cos y)^2$
$= \sin^2 y - 2 \sin y \cos y + \cos^2 y$
$= 1 - 2 \sin y \cos y \qquad (\sin^2 y + \cos^2 y = 1)$

10. $1 + 2 \sin y \cos y$

11. $(1 + \tan x)^2$
$= 1 + 2 \tan x + \tan^2 x$
$= \sec^2 x + 2 \tan x \qquad (1 + \tan^2 x = \sec^2 x)$

12. $2 \cot x + \csc^2 x$

13. $(\sin y - \csc y)^2$
$= \sin^2 y - 2 \sin y \csc y + \csc^2 y$
$= \sin^2 y - 2 \sin y \cdot \dfrac{1}{\sin y} + \csc^2 y$
$= \sin^2 y + \csc^2 y - 2$

14. $\cos^2 y + \sec^2 y + 2$

15. $(\cos x - \sec x)(\cos^2 x + \sec^2 x + 1)$
$= \cos^3 x + \cos x \sec^2 x + \cos x - \sec x \cos^2 x - \sec^3 x - \sec x$
$= \cos^3 x + \cos x \cdot \dfrac{1}{\cos^2 x} + \cos x - \dfrac{1}{\cos x} \cdot \cos^2 x - \sec^3 x - \dfrac{1}{\cos x}$
$= \cos^3 x + \dfrac{1}{\cos x} + \cos x - \cos x - \sec^3 x - \dfrac{1}{\cos x}$
$= \cos^3 x - \sec^3 x$

16. $\sin^3 x + \csc^3 x$

17. $(\cot x - \tan x)(\cot^2 x + 1 + \tan^2 x)$
$= \cot^3 x + \cot x + \cot x \tan^2 x - \tan x \cot^2 x - \tan x - \tan^3 x$
$= \cot^3 x + \cot x + \cot x \cdot \dfrac{1}{\cot^2 x} - \dfrac{1}{\cot x} \cdot \cot^2 x - \dfrac{1}{\cot x} - \tan^3 x$
$= \cot^3 x + \cot x + \dfrac{1}{\cot x} - \cot x - \dfrac{1}{\cot x} - \tan^3 x$
$= \cot^3 x - \tan^3 x$

18. $\cot^3 y + \tan^3 y$

19. $(1 - \sin x)(1 + \sin x)$
$= 1 - \sin^2 x$
$= \cos^2 x$

20. $\sin^2 x$

21. $\sin x \cos x + \cos^2 x$
$= \cos x\,(\sin x + \cos x)$

22. $\csc x\,(\sec x - \csc x)$

23. $\sin^2\theta - \cos^2\theta$
$= (\sin\theta + \cos\theta)(\sin\theta - \cos\theta)$

24. $(\tan\theta - \cot\theta)(\tan\theta + \cot\theta)$

25. $\tan x + \sin(\pi - x)$
$= \dfrac{\sin x}{\cos x} + \sin x \qquad [\sin(\pi - x) = \sin x]$
$= \sin x\left[\dfrac{1}{\cos x} + 1\right]$
$= \sin x\,(\sec x + 1)$

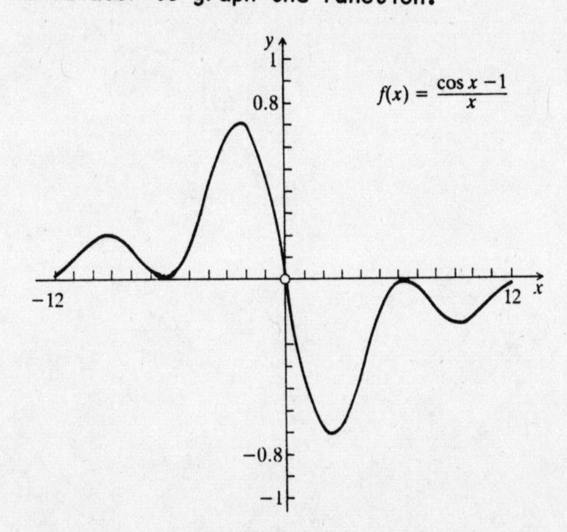

26. $\cos x\,(\csc x + 1)$

27. $\sin^4 x - \cos^4 x$
$= (\sin^2 x + \cos^2 x)(\sin^2 x - \cos^2 x)$
$= \sin^2 x - \cos^2 x$
$= (\sin x + \cos x)(\sin x - \cos x)$

28. $-1(\tan^2 x + \sec^2 x)$

29. $3 \cot^2 y + 6 \cot y + 3$
$= 3(\cot^2 y + 2 \cot y + 1)$
$= 3(\cot y + 1)^2$

30. $4(\sin y + 1)^2$

31. $\csc^4 x + 4 \csc^2 x - 5$
$= (\csc^2 x + 5)(\csc^2 x - 1)$
$= (\csc^2 x + 5)(\csc x + 1)(\csc x - 1)$

32. $(\tan^2 x + 2)(\tan x - 2)(\tan x + 2)$

33. $\sin^3 y + 27$
$= (\sin y)^3 + 3^3$
$= (\sin y + 3)(\sin^2 y - 3 \sin y + 9)$

34. $(1 - 5 \tan y)(1 + 5 \tan y + 25 \tan^2 y)$

35. $\sin^3 v - \csc^3 v$
$= (\sin v - \csc v)(\sin^2 v + \sin v \csc v + \csc^2 v)$
$= (\sin v - \csc v)(\sin^2 v + 1 + \csc^2 v)$
$\left[\sin v \csc v = \sin v \cdot \frac{1}{\sin v} = 1\right]$

36. $(\cos u - \sec u)(\cos^2 u + 1 + \sec^2 u)$

37. $\frac{\sin^2 x \cos x}{\cos^2 x \sin x}$
$= \frac{\sin x}{\cos x} \cdot \frac{\sin x \cos x}{\sin x \cos x}$
$= \frac{\sin x}{\cos x}$
$= \tan x$

38. $\cot x$

39. $\frac{4 \sin x \cos^3 x}{18 \sin^2 x \cos x}$
$= \frac{2}{9} \cdot \frac{\cos^2 x}{\sin x} \cdot \frac{\sin x \cos x}{\sin x \cos x}$
$= \frac{2 \cos^2 x}{9 \sin x}$

40. $\frac{5 \sin^2 x}{\cos x}$

41. $\frac{\cos^2 x - 2 \cos x + 1}{\cos x - 1}$
$= \frac{(\cos x - 1)^2}{\cos x - 1}$
$= \cos x - 1$

42. $\sin x + 1$

43. $\frac{\cos^2 \alpha - 1}{\cos \alpha - 1}$
$= \frac{(\cos \alpha + 1)(\cos \alpha - 1)}{\cos \alpha - 1}$
$= \cos \alpha + 1$

44. $\sin \alpha - 1$

45. $\frac{4 \tan x \sec x + 2 \sec x}{6 \sin x \sec x + 2 \sec x}$
$= \frac{2 \sec x\ (2 \tan x + 1)}{2 \sec x\ (3 \sin x + 1)}$
$= \frac{2 \tan x + 1}{3 \sin x + 1}$

46. $\frac{2 \tan x - 1}{3 \sin x + 1}$

47. $\frac{\csc(-\theta)}{\cot(-\theta)} = \frac{-\csc \theta}{-\cot \theta}$
$= \frac{\frac{1}{\sin \theta}}{\frac{\cos \theta}{\sin \theta}}$
$= \frac{1}{\sin \theta} \cdot \frac{\sin \theta}{\cos \theta}$
$= \frac{1}{\cos \theta}$
$= \sec \theta$

48. $-\frac{1}{\sin \beta \cos \beta}$, or $-\csc \beta \sec \beta$

49. $\frac{\sin^4 x - \cos^4 x}{\sin^2 x - \cos^2 x}$
$= \frac{(\sin^2 x + \cos^2 x)(\sin^2 x - \cos^2 x)}{\sin^2 x - \cos^2 x}$
$= 1$ $(\sin^2 x + \cos^2 x = 1)$

50. 1

51. $\frac{2 \sin^2 x}{\cos^3 x} \cdot \left[\frac{\cos x}{2 \sin x}\right]^2$
$= \frac{2 \sin^2 x}{\cos^3 x} \cdot \frac{\cos^2 x}{4 \sin^2 x}$
$= \frac{1}{2} \cdot \frac{1}{\cos x} \cdot \frac{\cos^2 x \sin^2 x}{\cos^2 x \sin^2 x}$
$= \frac{1}{2 \cos x}$, or $\frac{1}{2} \sec x$

52. $\frac{\cos x}{4}$

53. $\frac{3\sin x}{\cos^2 x}\cdot\frac{\cos^2 x+\cos x\sin x}{\cos^2 x-\sin^2 x}$

$=\frac{3\sin x}{\cos^2 x}\cdot\frac{\cos x\,(\cos x+\sin x)}{(\cos x-\sin x)(\cos x+\sin x)}$

$=3\cdot\frac{\sin x}{\cos x}\cdot\frac{1}{\cos x-\sin x}\cdot\frac{\cos x(\cos x+\sin x)}{\cos x(\cos x+\sin x)}$

$=\frac{3\tan x}{\cos x-\sin x}$

54. $\frac{5\cot x}{\sin x+\cos x}$

55. $\frac{\tan^2\gamma}{\sec\gamma}\div\frac{3\tan^3\gamma}{\sec\gamma}$

$=\frac{\tan^2\gamma}{\sec\gamma}\cdot\frac{\sec\gamma}{3\tan^3\gamma}$

$=\frac{1}{3\tan\gamma}\cdot\frac{\sec\gamma\tan^2\gamma}{\sec\gamma\tan^2\gamma}$

$=\frac{1}{3}\cot\gamma$

56. $\frac{1}{4}\cot\phi$

57. $\frac{1}{\sin^2 y-\cos^2 y}-\frac{2}{\cos y+\sin y}$

$=\frac{1}{\sin^2 y-\cos^2 y}-\frac{2}{\sin y+\cos y}\cdot\frac{\sin y-\cos y}{\sin y-\cos y}$

$=\frac{1-2\sin y+2\cos y}{\sin^2 y-\cos^2 y}$

58. $\frac{3\cos y+3\sin y+2}{\cos^2 y-\sin^2 y}$ or $\frac{-3\cos y-3\sin y-2}{\sin^2 y-\cos^2 y}$

59. $\left(\frac{\sin x}{\cos x}\right)^2-\frac{1}{\cos^2 x}$

$=\tan^2 x-\sec^2 x$

$=-1$ $\qquad(1+\tan^2 x\equiv\sec^2 x)$

60. 1

61. $\frac{\sin^2 x-9}{2\cos x+1}\cdot\frac{10\cos x+5}{3\sin x+9}$

$=\frac{(\sin x+3)(\sin x-3)}{2\cos x+1}\cdot\frac{5(2\cos x+1)}{3(\sin x+3)}$

$=\frac{5(\sin x-3)}{3}$

62. $\frac{(3\cos x+5)(\cos x+1)}{4}$

63. $\sqrt{\sin^2 x\cos x}\cdot\sqrt{\cos x}$

$=\sqrt{\sin^2 x\cos^2 x}$

$=\sin x\cos x$

64. $\cos x\sin x$

65. $\sqrt{\sin^3 y}+\sqrt{\sin y\cos^2 y}$

$=\sin y\sqrt{\sin y}+\cos y\sqrt{\sin y}$

$=\sqrt{\sin y}\,(\sin y+\cos y)$

66. $\sqrt{\cos y}\,(\sin y-\cos y)$

67. $\sqrt{\sin^2 x+2\cos x\sin x+\cos^2 x}$

$=\sqrt{(\sin x+\cos x)^2}$

$=\sin x+\cos x$

68. $\tan x-\sin x$

69. $(1-\sqrt{\sin y})(\sqrt{\sin y}+1)$

$=(1-\sqrt{\sin y})(1+\sqrt{\sin y})$

$=1-\sin y$ $\qquad[(\sqrt{\sin y})^2=\sin y]$

70. $4-\tan y$

71. $\sqrt{\sin x}\,(\sqrt{2\sin x}+\sqrt{\sin x\cos x})$

$=\sqrt{2\sin^2 x}+\sqrt{\sin^2 x\cos x}$

$=\sin x\cdot\sqrt{2}+\sin x\cdot\sqrt{\cos x}$

$=\sin x\,(\sqrt{2}+\sqrt{\cos x})$

72. $\cos x\,(\sqrt{3}-\sqrt{\sin x})$

73. $\sqrt{\frac{\sin x}{\cos x}}$

$=\sqrt{\frac{\sin x}{\cos x}\cdot\frac{\cos x}{\cos x}}$

$=\sqrt{\frac{\sin x\cos x}{\cos^2 x}}$

$=\frac{\sqrt{\sin x\cos x}}{\cos x}$

74. $\frac{\sqrt{\sin x\cos x}}{\sin x}$ or $\sqrt{\cot x}$

75. $\sqrt{\frac{\sin x}{\cot x}}$

$=\sqrt{\frac{\sin x}{\cot x}\cdot\frac{\cot x}{\cot x}}$

$=\sqrt{\frac{\sin x\cot x}{\cot^2 x}}$

$=\frac{\sqrt{\cos x}}{\cot x}$ $\qquad\left[\sqrt{\sin x\cot x}=\sqrt{\sin x\cdot\frac{\cos x}{\sin x}}=\sqrt{\cos x}\right]$

76. $\frac{\sqrt{\sin x}}{\tan x}$

77. $\sqrt{\dfrac{\cos^2 x}{2\sin^2 x}}$

$= \sqrt{\dfrac{\cot^2 x}{2}\cdot\dfrac{2}{2}}$

$= \dfrac{\sqrt{2}\cot x}{2}$

78. $\dfrac{\sqrt{3}\tan x}{3}$

79. $\sqrt{\dfrac{1+\sin x}{1-\sin x}}$

$= \sqrt{\dfrac{1+\sin x}{1-\sin x}\cdot\dfrac{1-\sin x}{1-\sin x}}$

$= \sqrt{\dfrac{1-\sin^2 x}{(1-\sin x)^2}}$

$= \dfrac{\sqrt{\cos^2 x}}{1-\sin x}$ $(1-\sin^2 x = \cos^2 x)$

$= \dfrac{\cos x}{1-\sin x}$

80. $\dfrac{\sin x}{1+\cos x}$

81. $\sqrt{\dfrac{\sin x}{\cos x}} = \sqrt{\dfrac{\sin x}{\cos x}\cdot\dfrac{\sin x}{\sin x}}$

$= \sqrt{\dfrac{\sin^2 x}{\cos x\sin x}}$

$= \dfrac{\sin x}{\sqrt{\cos x\sin x}}$

82. $\dfrac{\cos x}{\sqrt{\sin x\cos x}}$

83. $\sqrt{\dfrac{\sin x}{\cot x}} = \sqrt{\dfrac{\sin x}{\cot x}\cdot\dfrac{\sin x}{\sin x}}$

$= \sqrt{\dfrac{\sin^2 x}{\cos x}}$ $\left[\cot x\cdot\sin x = \dfrac{\cos x}{\sin x}\cdot\sin x = \cos x\right]$

$= \dfrac{\sin x}{\sqrt{\cos x}}$

84. $\dfrac{\cos x}{\sqrt{\sin x}}$

85. $\sqrt{\dfrac{\cos^2 x}{2\sin^2 x}} = \sqrt{\dfrac{\cot^2 x}{2}} = \dfrac{\cot x}{\sqrt{2}}$

86. $\dfrac{\tan x}{\sqrt{3}}$

87. $\sqrt{\dfrac{1+\sin x}{1-\sin x}} = \sqrt{\dfrac{1+\sin x}{1-\sin x}\cdot\dfrac{1+\sin x}{1+\sin x}}$

$= \sqrt{\dfrac{(1+\sin x)^2}{1-\sin^2 x}}$

$= \dfrac{1+\sin x}{\sqrt{\cos^2 x}}$

$= \dfrac{1+\sin x}{\cos x}$

88. $\dfrac{1-\cos x}{\sin x}$

89. $\sin^2\theta + \cos^2\theta = 1$

$\cos^2\theta = 1 - \sin^2\theta$

$\cos\theta = \pm\sqrt{1-\sin^2\theta}$

$\tan\theta = \dfrac{\sin\theta}{\cos\theta} = \pm\dfrac{\sin\theta}{\sqrt{1-\sin^2\theta}}$

$\cot\theta = \dfrac{1}{\tan\theta} = \pm\dfrac{\sqrt{1-\sin^2\theta}}{\sin\theta}$

$\sec\theta = \dfrac{1}{\cos\theta} = \pm\dfrac{1}{\sqrt{1-\sin^2\theta}}$

$\csc\theta = \dfrac{1}{\sin\theta}$

90. $\sin\theta = \pm\sqrt{1-\cos^2\theta}$, $\tan\theta = \pm\dfrac{\sqrt{1-\cos^2\theta}}{\cos\theta}$,

$\cot\theta = \pm\dfrac{\cos\theta}{\sqrt{1-\cos^2\theta}}$, $\sec\theta = \dfrac{1}{\cos\theta}$,

$\csc\theta = \pm\dfrac{1}{\sqrt{1-\cos^2\theta}}$

91. $\sec^2\theta = 1 + \tan^2\theta$

$\sec\theta = \pm\sqrt{1+\tan^2\theta}$

$\cos\theta = \dfrac{1}{\sec\theta} = \pm\dfrac{1}{\sqrt{1+\tan^2\theta}}$

$\tan\theta = \dfrac{\sin\theta}{\cos\theta}$, so $\sin\theta = \tan\theta\cos\theta =$

$\pm\dfrac{\tan\theta}{\sqrt{1+\tan^2\theta}}$

$\csc\theta = \dfrac{1}{\sin\theta} = \pm\dfrac{\sqrt{1+\tan^2\theta}}{\tan\theta}$

$\cot\theta = \dfrac{1}{\tan\theta}$

92. $\sin\theta = \pm\dfrac{1}{\sqrt{1+\cot^2\theta}}$, $\cos\theta = \pm\dfrac{\cot\theta}{\sqrt{1+\cot^2\theta}}$,

$\tan\theta = \dfrac{1}{\cot\theta}$, $\sec\theta = \pm\dfrac{\sqrt{1+\cot^2\theta}}{\cot\theta}$,

$\csc\theta = \pm\sqrt{1+\cot^2\theta}$

93. $\cos\theta = \dfrac{1}{\sec\theta}$

$1 + \tan^2\theta = \sec^2\theta$

$\tan^2\theta = \sec^2\theta - 1$

$\tan\theta = \pm\sqrt{\sec^2\theta - 1}$

$\tan\theta = \dfrac{\sin\theta}{\cos\theta}$, so $\sin\theta = \cos\theta\tan\theta = \pm\dfrac{\sqrt{\sec^2\theta - 1}}{\sec\theta}$

$\cot\theta = \dfrac{1}{\tan\theta} = \pm\dfrac{1}{\sqrt{\sec^2\theta - 1}}$

$\csc\theta = \dfrac{1}{\sin\theta} = \pm\dfrac{\sec\theta}{\sqrt{\sec^2\theta - 1}}$

94. $\sin\theta = \dfrac{1}{\csc\theta}$, $\cos\theta = \pm\dfrac{\sqrt{\csc^2\theta - 1}}{\csc\theta}$,

$\tan\theta = \pm\dfrac{1}{\sqrt{\csc^2\theta - 1}}$, $\cot\theta = \pm\sqrt{\csc^2\theta - 1}$,

$\sec\theta = \pm\dfrac{\csc\theta}{\sqrt{\csc^2\theta - 1}}$

95. $\sqrt{a^2 - x^2} = \sqrt{a^2 - (a\sin\theta)^2}$ (Substituting)

$= \sqrt{a^2 - a^2\sin^2\theta}$

$= \sqrt{a^2(1 - \sin^2\theta)}$

$= \sqrt{a^2\cos^2\theta}$

$= a\cos\theta$ $\left[a > 0 \text{ and } \cos\theta > 0 \text{ for } 0 < \theta < \frac{\pi}{2}\right]$

Then $\cos\theta = \dfrac{\sqrt{a^2 - x^2}}{a}$.

Also $x = a\sin\theta$, so $\sin\theta = \dfrac{x}{a}$. Then

$\tan\theta = \dfrac{\sin\theta}{\cos\theta} = \dfrac{\frac{x}{a}}{\frac{\sqrt{a^2 - x^2}}{a}} = \dfrac{x}{a}\cdot\dfrac{a}{\sqrt{a^2 - x^2}}$

$= \dfrac{x}{\sqrt{a^2 - x^2}}$.

96. $\cos x = \dfrac{2}{\sqrt{4 + x^2}}$, or $\dfrac{2\sqrt{4 + x^2}}{4 + x^2}$

$\sin x = \dfrac{x}{\sqrt{4 + x^2}}$, or $\dfrac{x\sqrt{4 + x^2}}{4 + x^2}$

97. $x = 3\sec\theta$

$x = \dfrac{3}{\cos\theta}$ $\left[\sec\theta = \dfrac{1}{\cos\theta}\right]$

$\cos\theta = \dfrac{3}{x}$

$\sqrt{x^2 - 9} = \sqrt{(3\sec\theta)^2 - 9}$ (Substituting)

$= \sqrt{9\sec^2\theta - 9}$

$= \sqrt{9(\sec^2\theta - 1)}$

$= \sqrt{9\tan^2\theta}$

$= 3\tan\theta$ $\left[\tan\theta > 0 \text{ for } 0 < \theta < \frac{\pi}{2}\right]$

Then $\tan\theta = \dfrac{\sqrt{x^2 - 9}}{3}$

$\dfrac{\sin\theta}{\cos\theta} = \dfrac{\sqrt{x^2 - 9}}{3}$

$\dfrac{\sin\theta}{\frac{3}{x}} = \dfrac{\sqrt{x^2 - 9}}{3}$ $\left[\cos\theta = \dfrac{3}{x}\right]$

$\sin\theta = \dfrac{3}{x}\cdot\dfrac{\sqrt{x^2 - 9}}{3}$

$\sin\theta = \dfrac{\sqrt{x^2 - 9}}{x}$.

98. $\sin\theta = \dfrac{\sqrt{x^2 - a^2}}{x}$, $\cos\theta = \dfrac{a}{x}$

99. $\dfrac{x^2}{\sqrt{1 - x^2}} = \dfrac{\sin^2\theta}{\sqrt{1 - \sin^2\theta}}$ (Substituting)

$= \dfrac{\sin^2\theta}{\sqrt{\cos^2\theta}}$

$= \dfrac{\sin^2\theta}{\cos\theta} = \sin\theta\cdot\dfrac{\sin\theta}{\cos\theta}$

$= \sin\theta\tan\theta$

100. $\dfrac{\sin\theta\cos\theta}{4}$

101. See the answer section in the text.

102. Let $\theta = \dfrac{3\pi}{2}$.

103. See the answer section in the text.

104. Let $x = \dfrac{3\pi}{4}$.

105. See the answer section in the text.

106. Let $\theta = \dfrac{\pi}{2}$.

107. See the answer section in the text.

108. Let $\theta = \dfrac{\pi}{4}$.

109. See the answer section in the text.

Exercise Set 7.2

1. $\sin 75° = \sin (45° + 30°)$
$= \sin 45° \cos 30° + \cos 45° \sin 30°$
$= \frac{\sqrt{2}}{2} \cdot \frac{\sqrt{3}}{2} + \frac{\sqrt{2}}{2} \cdot \frac{1}{2}$
$= \frac{\sqrt{6}}{4} + \frac{\sqrt{2}}{4} = \frac{\sqrt{6} + \sqrt{2}}{4}$

2. $\frac{\sqrt{6} - \sqrt{2}}{4}$

3. $\sin 15° = \sin (45° - 30°)$
$= \sin 45° \cos 30° - \cos 45° \sin 30°$
$= \frac{\sqrt{2}}{2} \cdot \frac{\sqrt{3}}{2} - \frac{\sqrt{2}}{2} \cdot \frac{1}{2}$
$= \frac{\sqrt{6}}{4} - \frac{\sqrt{2}}{4} = \frac{\sqrt{6} - \sqrt{2}}{4}$

4. $\frac{\sqrt{6} + \sqrt{2}}{4}$

5. $\sin 105° = \sin (75° + 30°)$
$= \sin 75° \cos 30° + \cos 75° \sin 30°$
$= \frac{\sqrt{6} + \sqrt{2}}{4} \cdot \frac{\sqrt{3}}{2} + \frac{\sqrt{6} - \sqrt{2}}{4} \cdot \frac{1}{2}$
$= \frac{3\sqrt{2} + \sqrt{6}}{8} + \frac{\sqrt{6} - \sqrt{2}}{8}$
$= \frac{2\sqrt{6} + 2\sqrt{2}}{8} = \frac{\sqrt{6} + \sqrt{2}}{4}$

6. $\frac{\sqrt{2} - \sqrt{6}}{4}$

7. $\tan 75° = \tan (45° + 30°)$
$= \frac{\tan 45° + \tan 30°}{1 - \tan 45° \tan 30°}$
$= \frac{1 + \frac{\sqrt{3}}{3}}{1 - 1 \cdot \frac{\sqrt{3}}{3}} = \frac{3 + \sqrt{3}}{3 - \sqrt{3}}$

8. $\frac{3 + \sqrt{3}}{3 - \sqrt{3}}$

9. $\sin 37° \cos 22° + \cos 37° \sin 22° = \sin(37° + 22°)$
$= \sin 59°$, or 0.8572

10. $\cos 59°$, or 0.5150

11. $\frac{\tan 20° + \tan 32°}{1 - \tan 20° \tan 32°} = \tan (20° + 32°)$
$= \tan 52°$, or 1.2799

12. $\tan 23°$, or 0.4245

Use the figures and function values below for Exercises 13 - 18.

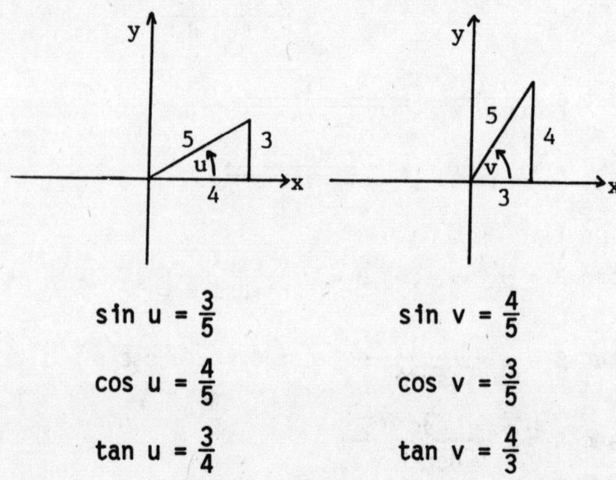

$\sin u = \frac{3}{5}$ $\sin v = \frac{4}{5}$

$\cos u = \frac{4}{5}$ $\cos v = \frac{3}{5}$

$\tan u = \frac{3}{4}$ $\tan v = \frac{4}{3}$

13. $\sin (u + v) = \sin u \cos v + \cos u \sin v$
$= \frac{3}{5} \cdot \frac{3}{5} + \frac{4}{5} \cdot \frac{4}{5}$
$= \frac{9}{25} + \frac{16}{25} = \frac{25}{25} = 1$

14. $-\frac{7}{25}$

15. Use drawings above Exercise 13.
$\cos (u + v) = \cos u \cos v - \sin u \sin v$
$= \frac{4}{5} \cdot \frac{3}{5} - \frac{3}{5} \cdot \frac{4}{5}$
$= \frac{12}{25} - \frac{12}{25} = 0$

16. $\frac{24}{25}$

17. Use drawings above Exercise 13.
$\tan (u + v) = \frac{\tan u + \tan v}{1 - \tan u \tan v}$
$= \frac{\frac{3}{4} + \frac{4}{3}}{1 - \frac{3}{4} \cdot \frac{4}{3}} = \frac{\frac{3}{4} + \frac{4}{3}}{0}$

Division by 0 is undefined. Thus $\tan (u + v)$ is undefined.

18. $-\frac{7}{24}$

19. Given: $\sin\theta = 0.6249$ and $\cos\phi = 0.1102$.

Find $\cos\theta$ and $\sin\phi$ using the following identity:

$$\sin^2 x + \cos^2 x = 1$$

If $\sin\theta = 0.6249$, then

$$(0.6249)^2 + \cos^2\theta = 1$$
$$\cos^2\theta = 1 - 0.3905 = 0.6095$$
$$\cos\theta = 0.7807$$

If $\cos\phi = 0.1102$, then

$$\sin^2\phi + (0.1102)^2 = 1$$
$$\sin^2\phi = 1 - 0.0121 = 0.9879$$
$$\sin\phi = 0.9939$$

$$\begin{aligned}\sin(\theta + \phi) &= \sin\theta\cos\phi + \cos\theta\sin\phi \\ &= (0.6249)(0.1102) + (0.7807)(0.9939) \\ &= 0.0689 + 0.7759 \\ &= 0.8448\end{aligned}$$

20. -0.5350

21. $\sin(\alpha + \beta) + \sin(\alpha - \beta)$

$$\begin{aligned}&= (\sin\alpha\cos\beta + \cos\alpha\sin\beta) + \\ &\qquad (\sin\alpha\cos\beta - \cos\alpha\sin\beta) \\ &= 2\sin\alpha\cos\beta\end{aligned}$$

22. $2\cos\alpha\sin\beta$

23. $\cos(\alpha + \beta) + \cos(\alpha - \beta)$

$$\begin{aligned}&= (\cos\alpha\cos\beta - \sin\alpha\sin\beta) + \\ &\qquad (\cos\alpha\cos\beta + \sin\alpha\sin\beta) \\ &= 2\cos\alpha\cos\beta\end{aligned}$$

24. $-2\sin\alpha\sin\beta$

25. $\cos(u + v)\cos v + \sin(u + v)\sin v$

$$\begin{aligned}&= (\cos u\cos v - \sin u\sin v)\cos v + \\ &\qquad (\sin u\cos v + \cos u\sin v)\sin v \\ &= \cos u\cos^2 v - \sin u\sin v\cos v + \\ &\qquad \sin u\cos v\sin v + \cos u\sin^2 v \\ &= \cos u(\cos^2 v + \sin^2 v) \\ &= \cos u\end{aligned}$$

26. $\sin u$

27. Solve each equation for y; then determine the slope of each line.

ℓ_1: $3y = \sqrt{3}x + 2$ $\qquad$ ℓ_2: $3y + \sqrt{3}x = -3$

$y = \frac{\sqrt{3}}{3}x + \frac{2}{3}$ $\qquad$ $y = -\frac{\sqrt{3}}{3}x - 1$

Thus $m_1 = \frac{\sqrt{3}}{3}$. $\qquad$ Thus $m_2 = -\frac{\sqrt{3}}{3}$.

27. (continued)

Let ϕ be the smallest angle from ℓ_1 to ℓ_2.

$$\tan\phi = \frac{-\frac{\sqrt{3}}{3} - \frac{\sqrt{3}}{3}}{1 + \left(-\frac{\sqrt{3}}{3}\right)\left(\frac{\sqrt{3}}{3}\right)} \qquad \left[\tan\phi = \frac{m_2 - m_1}{1 + m_2 m_1}\right]$$

$$= \frac{-\frac{2\sqrt{3}}{3}}{\frac{2}{3}} = -\sqrt{3}$$

Since $\tan\phi$ is negative, ϕ is obtuse. Hence ϕ is $\frac{2\pi}{3}$.

28. $\frac{\pi}{6}$

29. Solve each equation for y; then determine the slope of each line.

ℓ_1: $2x = 3 - 2y$ $\qquad$ ℓ_2: $x + y = 5$

$y = -x + \frac{3}{2}$ $\qquad$ $y = -x + 5$

Thus $m_1 = -1$ and the y-intercept is $\frac{3}{2}$. $\qquad$ Thus $m_2 = -1$ and the y-intercept is 5.

The lines do not form an angle. The lines are parallel. When the formula is used, the result is 0°.

30. The lines are parallel and do not form an angle.

31. Solve each equation for y; then determine the slope of each line.

ℓ_1: $2x - 5y + 1 = 0$

$$y = \frac{2}{5}x + \frac{1}{5}$$

Thus $m_1 = \frac{2}{5}$.

ℓ_2: $3x + y - 7 = 0$

$$y = -3x + 7$$

Thus $m_2 = -3$.

Let ϕ be the smallest angle from ℓ_1 to ℓ_2.

$$\tan\phi = \frac{-3 - \frac{2}{5}}{1 + (-3)\left(\frac{2}{5}\right)} \qquad \left[\tan\phi = \frac{m_2 - m_1}{1 + m_2 m_1}\right]$$

$$= \frac{-\frac{17}{5}}{-\frac{1}{5}} = 17$$

Since $\tan\phi$ is positive we know that ϕ is acute. Using a calculator we find $\phi \approx 86.63°$, or 86° 38'.

32. 126.87° or 126° 52'.

33. Find the slope of each line.

ℓ_1: $y = 3$ $(y = 0x + 3)$

Thus $m = 0$.

ℓ_2: $x + y = 5$

$y = -x + 5$

Thus $m = -1$.

Let ϕ be the smallest angle from ℓ_1 to ℓ_2.

$$\tan \phi = \frac{-1 - 0}{1 + (-1)(0)} \qquad \left[\tan \phi = \frac{m_2 - m_1}{1 + m_2 m_1}\right]$$

$$= \frac{-1}{1} = -1$$

Since $\tan \phi$ is negative, we know that ϕ is obtuse. Thus $\phi = \frac{3\pi}{4}$, or 135°.

34. 45°

35. $$\tan \phi = \frac{-0.79 - 1.25}{1 + (-0.79)(1.25)} \qquad \left[\tan \phi = \frac{m_2 - m_1}{1 + m_2\, m_1}\right]$$

$$= \frac{-2.04}{0.0125} = -163.2$$

Since $\tan \phi$ is negative we know that ϕ is obtuse. From a calculator, we find that an angle whose tangent is 1.632 is approximately 89.65°, or 89° 39'. This is a reference angle. Then ϕ = 180° - 89° 39' = 90° 21'.

36. 86.55°, or 86° 33'

37. $$\sin 2\theta = \sin (\theta + \theta)$$
$$= \sin \theta \cos \theta + \cos \theta \sin \theta$$
$$= 2 \sin \theta \cos \theta$$

38. $\cos^2 \theta - \sin^2 \theta$, or $1 - 2 \sin^2 \theta$, or $2 \cos^2 \theta - 1$

39. See the answer section in the text.

40. $\dfrac{\cot \alpha \cot \beta + 1}{\cot \beta - \cot \alpha}$

41. See the answer section in the text.

42. $$\cos \left(x - \frac{3\pi}{2}\right) = \cos x \cos \frac{3\pi}{2} + \sin x \sin \frac{3\pi}{2}$$
$$= (\cos x)(0) + (\sin x)(-1)$$
$$= -\sin x$$

43. See the answer section in the text.

44. $$\sin \left(x - \frac{3\pi}{2}\right) = \sin x \cos \frac{3\pi}{2} - \cos x \sin \frac{3\pi}{2}$$
$$= (\sin x)(0) - (\cos x)(-1)$$
$$= \cos x$$

45. See the answer section in the text.

46. $$\tan \left(x - \frac{\pi}{4}\right) = \frac{\tan x - \tan \frac{\pi}{4}}{1 + \tan x \tan \frac{\pi}{4}}$$
$$= \frac{\tan x - 1}{\tan x + 1}$$

47. See the answer section in the text.

48. $$\frac{\sin (\alpha + \beta)}{\sin (\alpha - \beta)} = \frac{\sin \alpha \cos \beta + \cos \alpha \sin \beta}{\sin \alpha \cos \beta - \cos \alpha \sin \beta}$$
$$= \frac{\dfrac{\sin \alpha \cos \beta + \cos \alpha \sin \beta}{\cos \alpha \cos \beta}}{\dfrac{\sin \alpha \cos \beta - \cos \alpha \sin \beta}{\cos \alpha \cos \beta}}$$
$$= \frac{\dfrac{\sin \alpha}{\cos \alpha} + \dfrac{\sin \beta}{\cos \beta}}{\dfrac{\sin \alpha}{\cos \alpha} - \dfrac{\sin \beta}{\cos \beta}} = \frac{\tan \alpha + \tan \beta}{\tan \alpha - \tan \beta}$$

49. See the answer section in the text.

50. $$\cos (\alpha + \beta) \cdot \cos (\alpha - \beta)$$
$$= (\cos \alpha \cos \beta - \sin \alpha \sin \beta)(\cos \alpha \cos \beta + \sin \alpha \sin \beta)$$
$$= \cos^2\alpha \cos^2\beta - \sin^2\alpha \sin^2\beta$$
$$= (1 - \sin^2\alpha)(1 - \sin^2\beta) - \sin^2\alpha \sin^2\beta$$
$$= 1 - \sin^2\beta - \sin^2\alpha + \sin^2\alpha \sin^2\beta - \sin^2\alpha \sin^2\beta$$
$$= 1 - \sin^2\beta - \sin^2\alpha$$
$$= \cos^2\alpha - \sin^2\beta$$

51. $$\tan \phi = \frac{m_2 - m_1}{1 + m_2 m_1}$$
$$\tan 45° = \frac{\frac{4}{3} - m_1}{1 + \frac{4}{3} \cdot m_1}$$
$$1 = \frac{4 - 3m_1}{3 + 4m_1}$$
$$3 + 4m_1 = 4 - 3m_1$$
$$7m_1 = 1$$
$$m_1 = \frac{1}{7}$$

52. $-\dfrac{11}{16}$

53. Find the slope of ℓ_1.

$$m_1 = \frac{-1 - 4}{5 - (-2)} = \frac{-5}{7} = -\frac{5}{7}$$

Use the following formula to determine the slope of ℓ_2.

$$\tan \phi = \frac{m_2 - m_1}{1 + m_2 m_1}$$

$$\tan 45° = \frac{m_2 - \left(-\frac{5}{7}\right)}{1 + m_2\left(-\frac{5}{7}\right)}$$

$$1 = \frac{7m_2 + 5}{7 - 5m_2}$$

$$7 - 5m_2 = 7m_2 + 5$$

$$2 = 12m_2$$

$$\frac{1}{6} = m_2$$

54. $-\frac{5}{2}$

55. Find the slope of each line.

$$m_1 = \frac{3.899 - (-0.9012)}{-2.123 - (-4.892)} = \frac{4.8002}{2.769} \approx 1.7336$$

$$m_2 = \frac{4.013 - (-3.814)}{5.925 - 0} = \frac{7.827}{5.925} \approx 1.3210$$

Let ϕ be the smallest angle from ℓ_1 to ℓ_2.

$$\tan \phi = \frac{1.3210 - 1.7336}{1 + (1.3210)(1.7336)}$$

$$\approx \frac{-0.4126}{3.2901} \approx -0.1254$$

Since $\tan \phi$ is negative, we know that ϕ is obtuse. Thus $\phi = 172.85°$, or $172° \; 51'$.

56. 168.7°

57.

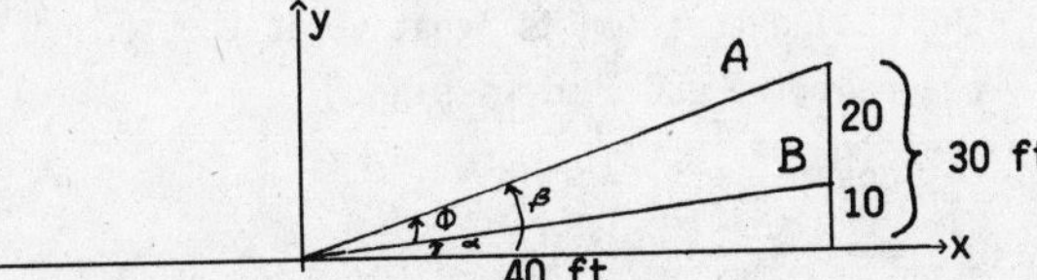

Position a set of coordinate axes as shown above. Let ℓ_1 be the line containing the points (0, 0) and (40, 10). Find its slope.

$$m_1 = \tan \alpha = \frac{10}{40} = \frac{1}{4}$$

Let ℓ_2 be the line containing the points (0, 0) and (40, 30). Find its slope.

$$m_2 = \tan \beta = \frac{30}{40} = \frac{3}{4}$$

Now ϕ is the smallest positive angle from ℓ_1 to ℓ_2.

$$\tan \phi = \frac{\frac{3}{4} - \frac{1}{4}}{1 + \frac{3}{4} \cdot \frac{1}{4}} = \frac{\frac{1}{2}}{\frac{19}{16}} = \frac{8}{19} \approx 0.4211$$

$$\phi \approx 22.83°, \text{ or } 22° \; 50'$$

58. $\dfrac{\sin (x + h) - \sin x}{h}$

$$= \frac{\sin x \cos h + \cos x \sin h - \sin x}{h}$$

$$= \frac{\sin x \cos h - \sin x}{h} + \frac{\cos x \sin h}{h}$$

$$= \sin x \left(\frac{\cos h - 1}{h}\right) + \cos x \left(\frac{\sin h}{h}\right)$$

59. See the answer section in the text.

60. $\dfrac{\tan (x + h) - \tan x}{h}$

$$= \frac{\frac{\tan x + \tan h}{1 - \tan x \tan h} - \tan x}{h}$$

$$= \frac{\tan x + \tan h - \tan x + \tan^2 x \tan h}{(1 - \tan x \tan h)h}$$

$$= \frac{\tan h\,(1 + \tan^2 x)}{(1 - \tan x \tan h)h} = \frac{\sec^2 x}{1 - \tan x \tan h}\left(\frac{\tan h}{h}\right)$$

61. See the answer section in the text.

62. $\sin \alpha \cos \beta \cos \gamma + \cos \alpha \sin \beta \cos \gamma +$
$\cos \alpha \cos \beta \sin \gamma - \sin \alpha \sin \beta \sin \gamma$

Exercise Set 7.3

1. Since 25° and 65° are complements, we have

$\sin 25° = \cos 65° = 0.4226$

$\cos 25° = \sin 65° = 0.9063$

$\tan 25° = \cot 65° = 0.4663$

$\cot 25° = \tan 65° = 2.145$

$\sec 25° = \csc 65° = 1.103$

$\csc 25° = \sec 65° = 2.366$

2. $\sin 58° = 0.8480$ $\qquad \csc 58° = 1.179$

$\cos 58° = 0.5299$ $\qquad \sec 58° = 1.887$

$\tan 58° = 1.600$ $\qquad \cot 58° = 0.6249$

3. a) $\tan 22.5° = \dfrac{\sin 22.5°}{\cos 22.5°} = \dfrac{\frac{\sqrt{2 - \sqrt{2}}}{2}}{\frac{\sqrt{2 + \sqrt{2}}}{2}}$

$$= \frac{\sqrt{2 - \sqrt{2}}}{\sqrt{2 + \sqrt{2}}}$$

$$\cot 22.5° = \frac{1}{\tan 22.5°} = \frac{\sqrt{2 + \sqrt{2}}}{\sqrt{2 - \sqrt{2}}}$$

$$\sec 22.5° = \frac{1}{\cos 22.5°} = \frac{2}{\sqrt{2 + \sqrt{2}}}$$

$$\csc 22.5° = \frac{1}{\sin 22.5°} = \frac{2}{\sqrt{2 - \sqrt{2}}}$$

3. (continued)

b) Since 67.5° and 22.5° are complements, we have

$\sin 67.5° = \cos 22.5° = \dfrac{\sqrt{2 + \sqrt{2}}}{2}$

$\cos 67.5° = \sin 22.5° = \dfrac{\sqrt{2 - \sqrt{2}}}{2}$

$\tan 67.5° = \cot 22.5° = \dfrac{\sqrt{2 + \sqrt{2}}}{\sqrt{2 - \sqrt{2}}}$

$\cot 67.5° = \tan 22.5° = \dfrac{\sqrt{2 - \sqrt{2}}}{\sqrt{2 + \sqrt{2}}}$

$\sec 67.5° = \csc 22.5° = \dfrac{2}{\sqrt{2 - \sqrt{2}}}$

$\csc 67.5° = \sec 22.5° = \dfrac{2}{\sqrt{2 + \sqrt{2}}}$

4. a) $\tan \frac{\pi}{12} = \dfrac{\sqrt{2 - \sqrt{3}}}{\sqrt{2 + \sqrt{3}}}$, $\cot \frac{\pi}{12} = \dfrac{\sqrt{2 + \sqrt{3}}}{\sqrt{2 - \sqrt{3}}}$,

$\csc \frac{\pi}{12} = \dfrac{2}{\sqrt{2 - \sqrt{3}}}$, $\sec \frac{\pi}{12} = \dfrac{2}{\sqrt{2 + \sqrt{3}}}$

b) $\sin \frac{5\pi}{12} = \dfrac{\sqrt{2 + \sqrt{3}}}{2}$, $\cos \frac{5\pi}{12} = \dfrac{\sqrt{2 - \sqrt{3}}}{2}$,

$\tan \frac{5\pi}{12} = \dfrac{\sqrt{2 + \sqrt{3}}}{\sqrt{2 - \sqrt{3}}}$, $\cot \frac{5\pi}{12} = \dfrac{\sqrt{2 - \sqrt{3}}}{\sqrt{2 + \sqrt{3}}}$,

$\csc \frac{5\pi}{12} = \dfrac{2}{\sqrt{2 + \sqrt{3}}}$, $\sec \frac{5\pi}{12} = \dfrac{2}{\sqrt{2 - \sqrt{3}}}$

5. See the answer section in the text.

6. $\cot \left(x - \frac{\pi}{2}\right) = \dfrac{\cos \left(x - \frac{\pi}{2}\right)}{\sin \left(x - \frac{\pi}{2}\right)} = \dfrac{\sin x}{-\cos x} = -\tan x$

7. See the answer section in the text.

8. $\sec \left(x + \frac{\pi}{2}\right) = \dfrac{1}{\cos \left(x + \frac{\pi}{2}\right)} = \dfrac{1}{-\sin x} = -\csc x$

9. See the answer section in the text.

10. $\cos (\theta - 270°) = \cos [90°(-3) + \theta] = -\sin \theta$

11. $\tan (\theta - 270°) = \tan [90°(-3) + \theta]$

Since -3 is odd, we know that $\tan (\theta - 270°) = \pm \cot \theta$. Suppose θ is in the first quadrant. Then $\theta - 270°$ is in the second quadrant. Since the tangent function is negative in the second quadrant, we use a minus sign.

$\tan (\theta - 270°) = -\cot \theta$

12. $\cos \theta$

13. $\sin (\theta + 450°) = \sin (90° \cdot 5 + \theta)$

Since 5 is odd, we know that $\sin (\theta + 450°) = \pm \cos \theta$. Suppose θ is in the first quadrant. Then $\theta + 450°$ is in the second quadrant. Since the sine function is positive in the second quadrant, we use a plus sign.

$\sin (\theta + 450°) = \cos \theta$

14. $\sin \theta$

15. $\sec (\pi - \theta) = \sec \left[\frac{\pi}{2} \cdot 2 - \theta\right]$

Since 2 is even, we know that $\sec (\pi - \theta) = \pm \sec \theta$. Suppose θ is in the first quadrant. Then $\pi - \theta$ is in the second quadrant. Since the secant function is negative in the second quadrant, we use a minus sign.

$\sec (\pi - \theta) = -\sec \theta$

16. $-\csc \theta$

17. $\tan (x - 4\pi) = \tan \left[\frac{\pi}{2} (-8) + x\right]$

Since -8 is even, we know that $\tan (x - 4\pi) = \pm \tan x$. Suppose x is in the first quadrant. Then $x - 4\pi$ is also in the first quadrant. Since the tangent function is positive in the first quadrant, we use a plus sign.

$\tan (x - 4\pi) = \tan x$

18. $\sec x$

19. $\cos \left(\frac{9\pi}{2} + x\right) = \cos \left(\frac{\pi}{2} \cdot 9 + x\right)$

Since 9 is odd, we know that $\cos \left(\frac{9\pi}{2} + x\right) = \pm \sin x$. Suppose x is in the first quadrant. Then $\frac{9\pi}{2} + x$ is in the second quadrant. Since the cosine function is negative in the second quadrant, we use a minus sign.

$\cos \left(\frac{9\pi}{2} + x\right) = -\sin x$

20. $\tan x$

21. $\csc (540° - \theta) = \csc (90° \cdot 6 - \theta)$

Since 6 is even, we know that $\csc (540° - \theta) = \pm \csc \theta$. Suppose θ is in the first quadrant. Then $540° - \theta$ is in the second quadrant. Since the cosecant function is positive in the second quadrant, we use a plus sign.

$\csc (540° - \theta) = \csc \theta$

22. $\tan \theta$

23. $\cot\left(x - \frac{3\pi}{2}\right) = \cot\left[\frac{\pi}{2}(-3) + x\right]$

Since -3 is odd, we know that $\cot\left(x - \frac{3\pi}{2}\right) = \pm\tan x$. Suppose x is in the first quadrant. Then $x - \frac{3\pi}{2}$ is in the second quadrant. Since the cotangent function is negative in the second quadrant, we use a minus sign.

$$\cot\left(x - \frac{3\pi}{2}\right) = -\tan x$$

24. $-\cos x$

25. Use $c \sin ax + d \cos ax = A \sin(ax + b)$, where $A = \sqrt{c^2 + d^2}$ and b is a number whose cosine is $\frac{c}{A}$ and whose sine is $\frac{d}{A}$.

$\sin 2x + \sqrt{3}\cos 2x$
$a = 2,\quad c = 1,\quad \text{and } d = \sqrt{3}$

Find A:

$A = \sqrt{c^2 + d^2} = \sqrt{1^2 + (\sqrt{3})^2} = \sqrt{4} = 2$

Find b:
We know that b is a number such that
$\cos b = \frac{1}{2}$ and $\sin b = \frac{\sqrt{3}}{2}$.

Therefore, we can use 60° or $\frac{\pi}{3}$ for b.

Thus, $\sin 2x + \sqrt{3}\cos 2x = 2\sin\left(2x + \frac{\pi}{3}\right)$.

26. $2\sin\left(3x - \frac{\pi}{6}\right)$

27. Use $c \sin ax + d \cos ax = A \sin(ax + b)$, where $A = \sqrt{c^2 + d^2}$ and b is a number whose cosine is $\frac{c}{A}$ and whose sine is $\frac{d}{A}$.

$4\sin x + 3\cos x$
$a = 1,\quad c = 4,\quad \text{and } d = 3$

Find A:

$A = \sqrt{c^2 + d^2} = \sqrt{4^2 + 3^2} = \sqrt{25} = 5$

Find b:
We know that b is a number such that
$\cos b = \frac{4}{5}$ and $\sin b = \frac{3}{5}$.

Thus, $4\sin x + 3\cos x = 5\sin(x + b)$, where b is a number whose cosine is $\frac{4}{5}$ and whose sine is $\frac{3}{5}$.

28. $5\sin(2x + b)$, where $b \approx -36.9°$

29. Use $c \sin ax + d \cos ax = A \sin(ax + b)$, where $A = \sqrt{c^2 + d^2}$ and b is a number whose cosine is $\frac{c}{A}$ and whose sine is $\frac{d}{A}$.

$6.75 \sin 0.374x + 4.08 \cos 0.374x$
$a = 0.374,\quad c = 6.75,\quad \text{and } d = 4.08$

Find A:

$A = \sqrt{(6.75)^2 + (4.08)^2} = \sqrt{45.5625 + 16.6464}$
$= \sqrt{62.2089} \approx 7.89$

Find b:
We know that b is a number such that
$\cos b = \frac{6.75}{7.89} \approx 0.85$ and $\sin b = \frac{4.08}{7.89} \approx 0.517$.

Therefore, we can use 31.2° for b.
$6.75 \sin 0.374x + 4.08 \cos 0.374x$
$= 7.89 \sin(0.374x + 31.2°)$

30. $97.93 \sin(0.8081x - 2.86°)$

31. Use $c \sin ax + d \cos ax = A \sin(ax + b)$, where $A = \sqrt{c^2 + d^2}$ and b is a number whose cosine is $\frac{c}{A}$ and whose sine is $\frac{d}{A}$.

$y = \sin 2x - \cos 2x$
$a = 2,\ c = 1, \text{ and } d = -1$
$A = \sqrt{1^2 + (-1)^2} = \sqrt{2}$

We know that b is a number such that $\cos b = \frac{1}{\sqrt{2}}$ and $\sin b = -\frac{1}{\sqrt{2}}$. Therefore, b is $-\frac{\pi}{4}$.

The equation is $y = \sqrt{2}\sin\left(2x - \frac{\pi}{4}\right)$.

We graph the equation using the procedures in Section 6.7.

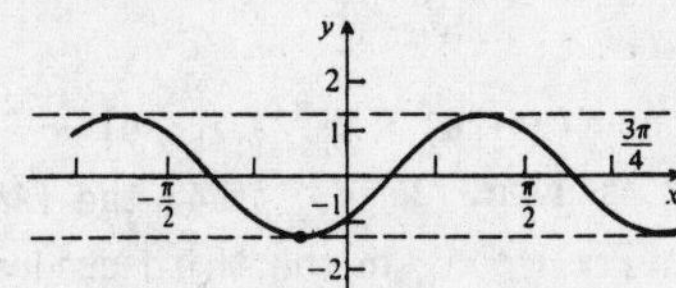

32.

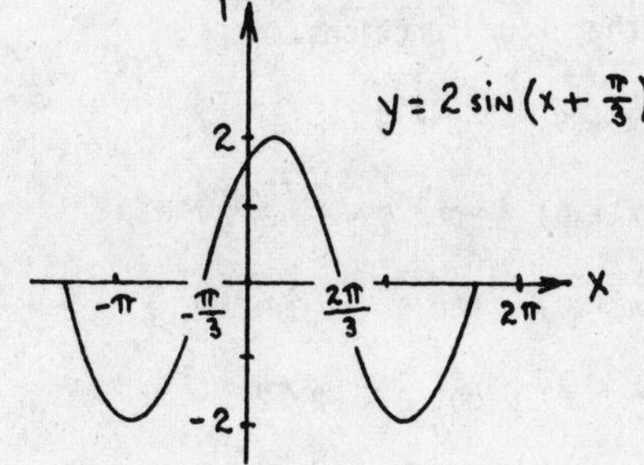

33. We sketch a second quadrant triangle.

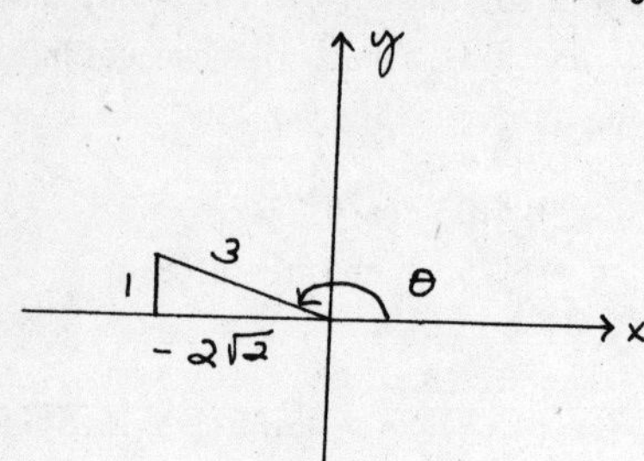

a) $\cos\theta = \frac{-2\sqrt{2}}{3} = -\frac{2\sqrt{2}}{3}$

$\tan\theta = \frac{1}{-2\sqrt{2}} = -\frac{\sqrt{2}}{4}$

$\cot\theta = \frac{-2\sqrt{2}}{1} = -2\sqrt{2}$

$\sec\theta = \frac{3}{-2\sqrt{2}} = -\frac{3\sqrt{2}}{4}$

$\csc\theta = \frac{3}{1} = 3$

b) Note that $f\left(\frac{\pi}{2} - \theta\right) = f\left(\frac{\pi}{2}\cdot 1 - \theta\right) = g(\theta)$, where g is the cofunction of f, since 1 is odd and $\frac{\pi}{2} - \theta$ is in the first quadrant when θ is in the first quadrant.

$\sin\left(\frac{\pi}{2} - \theta\right) = \cos\theta = -\frac{2\sqrt{2}}{3}$

$\cos\left(\frac{\pi}{2} - \theta\right) = \sin\theta = \frac{1}{3}$

$\tan\left(\frac{\pi}{2} - \theta\right) = \cot\theta = -2\sqrt{2}$

$\cot\left(\frac{\pi}{2} - \theta\right) = \tan\theta = -\frac{\sqrt{2}}{4}$

$\sec\left(\frac{\pi}{2} - \theta\right) = \csc\theta = 3$

$\csc\left(\frac{\pi}{2} - \theta\right) = \sec\theta = -\frac{3\sqrt{2}}{4}$

c) Note that $f(\pi + \theta) = f\left(\frac{\pi}{2}\cdot 2 + \theta\right) = \pm f(\theta)$ since 2 is even. When θ is in the first quadrant, $\pi + \theta$ is in the third quadrant. Thus, we will use a plus sign for the tangent and cotangent functions and a minus sign for the other four functions.

$\sin(\pi + \theta) = -\sin\theta = -\frac{1}{3}$

$\cos(\pi + \theta) = -\cos\theta = \frac{2\sqrt{2}}{3}$

$\tan(\pi + \theta) = \tan\theta = -\frac{\sqrt{2}}{4}$

$\cot(\pi + \theta) = \cot\theta = -2\sqrt{2}$

$\sec(\pi + \theta) = -\sec\theta = \frac{3\sqrt{2}}{4}$

$\csc(\pi + \theta) = -\csc\theta = -3$

33. (continued)

d) Note that $f(\pi - \theta) = f\left(\frac{\pi}{2}\cdot 2 - \theta\right) = \pm f(\theta)$ since 2 is even. When θ is in the first quadrant, $\pi - \theta$ is in the second quadrant. Thus, we will use a plus sign for the sine and cosecant functions and a minus sign for the other four functions.

$\sin(\pi - \theta) = \sin\theta = \frac{1}{3}$

$\cos(\pi - \theta) = -\cos\theta = \frac{2\sqrt{2}}{3}$

$\tan(\pi - \theta) = -\tan\theta = \frac{\sqrt{2}}{4}$

$\cot(\pi - \theta) = -\cot\theta = 2\sqrt{2}$

$\sec(\pi - \theta) = -\sec\theta = \frac{3\sqrt{2}}{4}$

$\csc(\pi - \theta) = \csc\theta = 3$

e) Note that $f(2\pi - \theta) = f\left(\frac{\pi}{2}\cdot 4 - \theta\right) = \pm f(\theta)$ since 4 is even. When θ is in the first quadrant, $2\pi - \theta$ is in the fourth quadrant. Thus, we will use a plus sign for the cosine and secant functions and a minus sign for the other four functions.

$\sin(2\pi - \theta) = -\sin\theta = -\frac{1}{3}$

$\cos(2\pi - \theta) = \cos\theta = -\frac{2\sqrt{2}}{3}$

$\tan(2\pi - \theta) = -\tan\theta = \frac{\sqrt{2}}{4}$

$\cot(2\pi - \theta) = -\cot\theta = 2\sqrt{2}$

$\sec(2\pi - \theta) = \sec\theta = -\frac{3\sqrt{2}}{4}$

$\csc(2\pi - \theta) = -\csc\theta = -3$

34. a) $\sin\theta = -\frac{3}{5}$, $\tan\theta = -\frac{3}{4}$, $\cot\theta = -\frac{4}{3}$, $\sec\theta = \frac{5}{4}$, $\csc\theta = -\frac{5}{3}$

b) $\sin\left(\frac{\pi}{2} + \theta\right) = \frac{4}{5}$, $\cos\left(\frac{\pi}{2} + \theta\right) = \frac{3}{5}$, $\tan\left(\frac{\pi}{2} + \theta\right) = \frac{4}{3}$, $\cot\left(\frac{\pi}{2} + \theta\right) = \frac{3}{4}$, $\sec\left(\frac{\pi}{2} + \theta\right) = \frac{5}{3}$, $\csc\left(\frac{\pi}{2} + \theta\right) = \frac{5}{4}$

c) $\sin(\pi + \theta) = \frac{3}{5}$, $\cos(\pi + \theta) = -\frac{4}{5}$, $\tan(\pi + \theta) = -\frac{3}{4}$, $\cot(\pi + \theta) = -\frac{4}{3}$, $\sec(\pi + \theta) = -\frac{5}{4}$, $\csc(\pi + \theta) = \frac{5}{3}$

34. (continued)

d) $\sin(\pi - \theta) = -\frac{3}{5}$, $\cos(\pi - \theta) = -\frac{4}{5}$,

$\tan(\pi - \theta) = \frac{3}{4}$, $\cot(\pi - \theta) = \frac{4}{3}$,

$\sec(\pi - \theta) = -\frac{5}{4}$, $\csc(\pi - \theta) = -\frac{5}{3}$

e) $\sin(2\pi - \theta) = \frac{3}{5}$, $\cos(2\pi - \theta) = \frac{4}{5}$,

$\tan(2\pi - \theta) = \frac{3}{4}$, $\cot(2\pi - \theta) = \frac{4}{3}$,

$\sec(2\pi - \theta) = \frac{5}{4}$, $\csc(2\pi - \theta) = \frac{5}{3}$

35. a) We sketch a first quadrant triangle. Note that $0.45399 = \frac{0.45399}{1}$. We can use the Pythagorean theorem to find the length of the other leg.

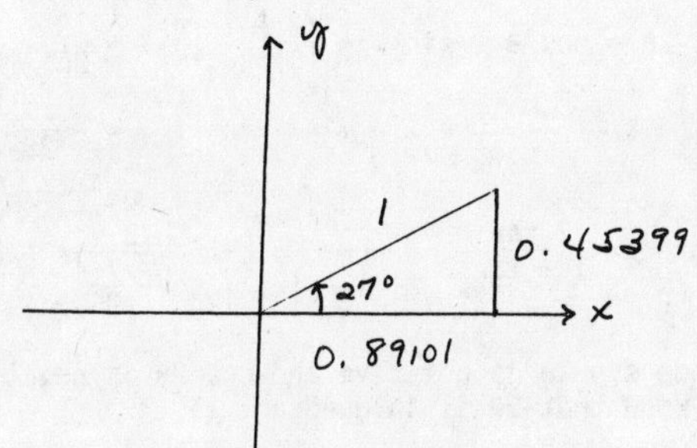

$\cos 27° = \frac{0.89101}{1} = 0.89101$

$\tan 27° = \frac{0.45399}{0.89101} \approx 0.50952$

$\cot 27° = \frac{0.89101}{0.45399} \approx 1.9626$

$\sec 27° = \frac{1}{0.89101} \approx 1.1223$

$\csc 27° = \frac{1}{0.45399} \approx 2.2027$

b) Since 63° and 27° are complements, we have

$\sin 63° = \cos 27° = 0.89101$

$\cos 63° = \sin 27° = 0.45399$

$\tan 63° = \cot 27° = 1.9626$

$\cot 63° = \tan 27° = 0.50952$

$\sec 63° = \csc 27° = 2.2027$

$\csc 63° = \sec 27° = 1.1223$

36. a) $\sin 54° = 0.80902$, $\cos 54° = 0.58779$, $\tan 54° = 1.3764$, $\sec 54° = 1.7013$, $\csc 54° = 1.2361$

b) $\sin 36° = 0.58779$, $\cos 36° = 0.80902$, $\tan 36° = 0.72654$, $\cot 36° = 1.3764$, $\sec 36° = 1.2361$, $\csc 36° = 1.7013$

37.

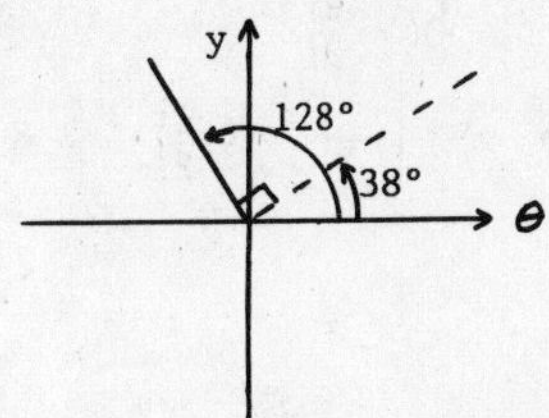

$\sin 128° = \sin(90° + 38°) = \sin(90° \cdot 1 + 38°)$
$= \cos 38°$
$= 0.78801$

Using $\cos^2\theta + \sin^2\theta = 1$, find $\cos 128°$.

$(\cos 128°)^2 + (\sin 128°)^2 = 1$

$(\cos 128°)^2 + (0.78801)^2 = 1$

$(\cos 128°)^2 \approx 1 - 0.62096$

$\cos 128° \approx -0.61566$

(cosine is negative in quadrant II)

$\tan 128° = \frac{0.78801}{-0.61566} \approx -1.27994$

$\cot 128° = \frac{1}{-1.27994} \approx -0.78129$

$\sec 128° = \frac{1}{-0.61566} \approx -1.62427$

$\csc 128° = \frac{1}{0.78801} \approx 1.26902$

38. $\sin 343° = -0.29237$ $\csc 343° = -3.42030$
$\cos 343° = 0.95631$ $\sec 343° = 1.04569$
$\tan 343° = -0.30573$ $\cot 343° = -3.27085$

39. $\sin\left(\frac{\pi}{2} - x\right)[\sec x - \cos x]$

$= -\sin\left(x - \frac{\pi}{2}\right)\left[\frac{1}{\cos x} - \cos x\right]$

$\left[\sin\left(\frac{\pi}{2} - x\right) = \sin\left[-\left(x - \frac{\pi}{2}\right)\right] = -\sin\left(x - \frac{\pi}{2}\right)\right]$

$= -(-\cos x)\left(\frac{1 - \cos^2 x}{\cos x}\right)$ $\left[\sin\left(x - \frac{\pi}{2}\right) = -\cos x\right]$

$= 1 - \cos^2 x$

$= \sin^2 x$

40. $\cos^2 x$

41. $\sin x \cos y - \cos\left(x + \frac{\pi}{2}\right)\tan y$

$= \sin x \cos y - (-\sin x)\tan y$

$\left[\cos\left(x + \frac{\pi}{2}\right) = -\sin x\right]$

$= \sin x \cos y + \sin x \tan y$

$= \sin x(\cos y + \tan y)$

42. $\sin x(\tan y + \cot y)$

43. $\cos(\pi - x) + \cot x \sin\left(x - \frac{\pi}{2}\right)$

$= -\cos x + \cot x\,(-\cos x)$ $\left[\sin\left(x - \frac{\pi}{2}\right) = -\cos x\right]$

$= -\cos x\,(1 + \cot x)$

44. $\sin x\,(1 - \tan x)$

45. $\dfrac{\cos^2 x - 1}{\sin\left(\frac{\pi}{2} - x\right) - 1}$

$= \dfrac{(\cos x + 1)(\cos x - 1)}{\cos x - 1}$

$\left[\sin\left(\frac{\pi}{2} - x\right) = -\sin\left(x - \frac{\pi}{2}\right) = \cos x\right]$

$= \cos x + 1$

46. $\sin x - 1$

47. $\dfrac{\sin x - \cos\left(x - \frac{\pi}{2}\right)\cos x}{-\sin x - \cos\left(x - \frac{\pi}{2}\right)\tan x}$

$= \dfrac{\sin x - \sin x\cos x}{-\sin x - \sin x\tan x}$ $\left[\cos\left(x - \frac{\pi}{2}\right) = \sin x\right]$

$= \dfrac{\sin x\,(1 - \cos x)}{-\sin x\,(1 + \tan x)}$

$= \dfrac{\cos x - 1}{1 + \tan x}$ $[-1(1 - \cos x) = \cos x - 1]$

48. $\dfrac{1 - \sin x}{1 + \tan x}$

49. $\dfrac{\cos^2 x + 2\sin\left(x - \frac{\pi}{2}\right) + 1}{\sin\left(\frac{\pi}{2} - x\right) - 1}$

$= \dfrac{\cos^2 x + 2(-\cos x) + 1}{-\sin\left(x - \frac{\pi}{2}\right) - 1}$ $\left[\sin\left(x - \frac{\pi}{2}\right) = -\cos x\right]$

$= \dfrac{\cos^2 x - 2\cos x + 1}{\cos x - 1}$

$= \dfrac{(\cos x - 1)^2}{\cos x - 1}$

$= \cos x - 1$

50. $\sin x - 1$

51. $\dfrac{\sin^2 y\cos\left(y + \frac{\pi}{2}\right)}{\cos^2 y\cos\left(\frac{\pi}{2} - y\right)}$

$= \dfrac{\sin^2 y\,(-\sin y)}{\cos^2 y\,(\sin y)}$ $\left[\cos\left(\frac{\pi}{2} - y\right) = \cos\left(y - \frac{\pi}{2}\right) = \sin y\right]$

$= -\tan^2 y$

52. $\cot^2 y$

Exercise Set 7.4

1. Sketch a first quadrant triangle. Since $\sin\theta = \frac{4}{5}$, label the leg parallel to the y-axis 4 and the hypotenuse 5. Using the Pythagorean theorem we know the other leg has length 3.

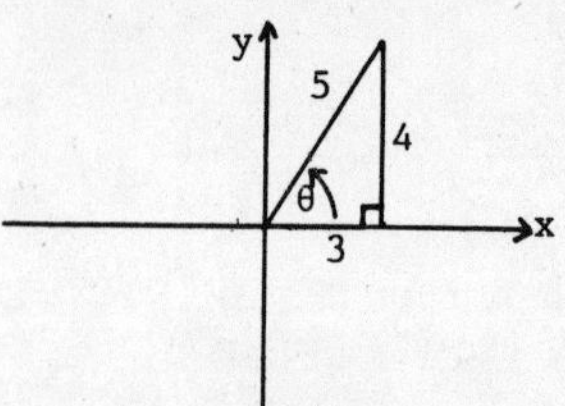

From this diagram we find that $\cos\theta = \frac{3}{5}$ and $\tan\theta = \frac{4}{3}$.

$\sin 2\theta = 2\sin\theta\cos\theta = 2\cdot\frac{4}{5}\cdot\frac{3}{5} = \frac{24}{25}$

$\cos 2\theta = \cos^2\theta - \sin^2\theta = \left(\frac{3}{5}\right)^2 - \left(\frac{4}{5}\right)^2 = \frac{9}{25} - \frac{16}{25}$

$= -\frac{7}{25}$

$\tan 2\theta = \dfrac{2\tan\theta}{1 - \tan^2\theta} = \dfrac{2\cdot\frac{4}{3}}{1 - \left(\frac{4}{3}\right)^2} = \dfrac{\frac{8}{3}}{-\frac{7}{9}} = -\frac{24}{7}$

Since $\sin 2\theta$ is positive and $\cos 2\theta$ is negative, we know that 2θ is in quadrant II.

2. $\sin 2\theta = \frac{120}{169}$, $\cos 2\theta = \frac{119}{169}$, $\tan 2\theta = \frac{120}{119}$;

2θ is in quadrant I

3. Sketch a third quadrant triangle. Since $\cos\theta = -\frac{4}{5}$, label the leg on the x-axis -4 and the hypotenuse 5. Using the Pythagorean theorem we know the other leg has length 3.

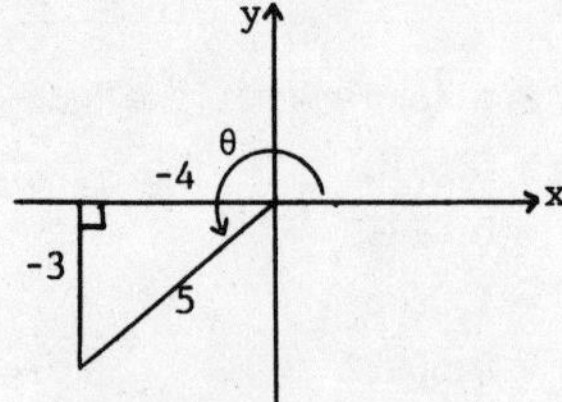

From this diagram we find that $\sin\theta = -\frac{3}{5}$ and $\tan\theta = \frac{3}{4}$.

$\sin 2\theta = 2\sin\theta\cos\theta = 2\left(-\frac{3}{5}\right)\left(-\frac{4}{5}\right) = \frac{24}{25}$

$\cos 2\theta = \cos^2\theta - \sin^2\theta$

$= \left(-\frac{4}{5}\right)^2 - \left(-\frac{3}{5}\right)^2 = \frac{16}{25} - \frac{9}{25} = \frac{7}{25}$

3. (continued)

$$\tan 2\theta = \frac{2 \tan \theta}{1 - \tan^2\theta}$$

$$= \frac{2 \cdot \frac{3}{4}}{1 - \left(\frac{3}{4}\right)^2} = \frac{\frac{3}{2}}{\frac{7}{16}} = \frac{24}{7}$$

Since both sin 2θ and cos 2θ are positive, we know that 2θ is in quadrant I.

4. $\sin 2\theta = \frac{24}{25}$, $\cos 2\theta = -\frac{7}{25}$, $\tan 2\theta = -\frac{24}{7}$;

2θ is in quadrant II

5. Sketch a third quadrant triangle. Since $\tan \theta = \frac{4}{3}$, label the leg on the x-axis -3 and the leg parallel to the y-axis -4. Using the Pythagorean theorem we know the hypotenuse has length 5.

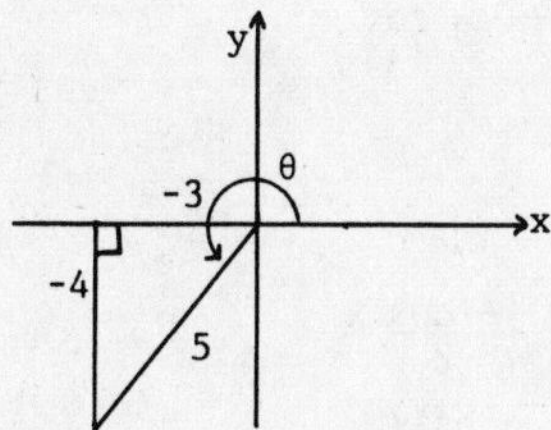

From this diagram we find that $\sin \theta = -\frac{4}{5}$ and $\cos \theta = -\frac{3}{5}$.

$$\sin 2\theta = 2 \sin \theta \cos \theta$$

$$= 2\left(-\frac{4}{5}\right)\left(-\frac{3}{5}\right) = \frac{24}{25}$$

$$\cos 2\theta = \cos^2\theta - \sin^2\theta$$

$$= \left(-\frac{3}{5}\right)^2 - \left(-\frac{4}{5}\right)^2 = \frac{9}{25} - \frac{16}{25} = -\frac{7}{25}$$

$$\tan 2\theta = \frac{2 \tan \theta}{1 - \tan^2\theta}$$

$$= \frac{2 \cdot \frac{4}{3}}{1 - \left(\frac{4}{3}\right)^2} = \frac{\frac{8}{3}}{-\frac{7}{9}} = -\frac{24}{7}$$

Since sin 2θ is positive and cos 2θ is negative, 2θ is in quadrant II.

6. Same as Exercise 3.

7.
$$\begin{aligned} \sin 4\theta &= \sin 2(2\theta) \\ &= 2 \sin 2\theta \cos 2\theta \\ &= 2(2 \sin \theta \cos \theta)(\cos^2\theta - \sin^2\theta) \\ &= 4 \sin \theta \cos^3\theta - 4 \sin^3\theta \cos \theta \end{aligned}$$

or

$$\begin{aligned} \sin 4\theta &= \sin 2(2\theta) \\ &= 2 \sin 2\theta \cos 2\theta \\ &= 2(2 \sin \theta \cos \theta)(2 \cos^2\theta - 1) \\ &= 8 \sin \theta \cos^3\theta - 4 \sin \theta \cos \theta \end{aligned}$$

or

$$\begin{aligned} \sin 4\theta &= \sin 2(2\theta) \\ &= 2 \sin 2\theta \cos 2\theta \\ &= 2(2 \sin \theta \cos \theta)(1 - 2 \sin^2\theta) \\ &= 4 \sin \theta \cos \theta - 8 \sin^3\theta \cos \theta \end{aligned}$$

8. $1 - 8 \sin^2 \theta \cos^2 \theta$, or
$\cos^4 \theta - 6 \sin^2 \theta \cos^2 \theta + \sin^4 \theta$, or
$8 \cos^4 \theta - 8 \cos^2 \theta + 1$

9.
$$\begin{aligned} \sin^4\theta &= \sin^2\theta \cdot \sin^2\theta \\ &= \frac{1 - \cos 2\theta}{2} \cdot \frac{1 - \cos 2\theta}{2} \\ &= \frac{1 - 2 \cos 2\theta + \cos^2 2\theta}{4} \\ &= \frac{1 - 2 \cos 2\theta + \frac{1 + \cos 4\theta}{2}}{4} \\ &= \frac{2 - 4 \cos 2\theta + 1 + \cos 4\theta}{8} \\ &= \frac{3 - 4 \cos 2\theta + \cos 4\theta}{8} \end{aligned}$$

10. $\frac{3 + 4 \cos 2\theta + \cos 4\theta}{8}$

11.
$$\begin{aligned} \sin 75^\circ &= \sin \frac{150^\circ}{2} \\ &= \sqrt{\frac{1 - \cos 150^\circ}{2}} = \sqrt{\frac{1 - (-\sqrt{3}/2)}{2}} \\ &= \sqrt{\frac{2 + \sqrt{3}}{4}} = \frac{\sqrt{2 + \sqrt{3}}}{2} \end{aligned}$$

The expression is positive because 75° is in the first quadrant.

12. $\frac{\sqrt{2 - \sqrt{3}}}{2}$

13.
$$\begin{aligned} \tan 75^\circ &= \tan \frac{150^\circ}{2} \\ &= \frac{\sin 150^\circ}{1 + \cos 150^\circ} = \frac{\frac{1}{2}}{1 + \left(-\frac{\sqrt{3}}{2}\right)} \\ &= \frac{1}{2 - \sqrt{3}}, \text{ or } 2 + \sqrt{3} \end{aligned}$$

14. $1 + \sqrt{2}$

15. $\sin \frac{5\pi}{8} = \sin \frac{\frac{5\pi}{4}}{2}$

$= \sqrt{\frac{1 - \cos (5\pi/4)}{2}} = \sqrt{\frac{1 - (-\sqrt{2}/2)}{2}}$

$= \sqrt{\frac{2 + \sqrt{2}}{4}} = \frac{\sqrt{2 + \sqrt{2}}}{2}$

The expression is positive because $\frac{5\pi}{8}$ is in the second quadrant.

16. $-\frac{\sqrt{2 - \sqrt{2}}}{2}$

17. $112.5° = \frac{5\pi}{8}$. This is the same as Exercise 15.

18. $\frac{\sqrt{2 + \sqrt{2}}}{2}$

19. $\cos \frac{\pi}{8} = \cos \frac{\frac{\pi}{4}}{2}$

$= \sqrt{\frac{1 + \cos (\pi/4)}{2}} = \sqrt{\frac{1 + (\sqrt{2}/2)}{2}}$

$= \sqrt{\frac{2 + \sqrt{2}}{4}} = \frac{\sqrt{2 + \sqrt{2}}}{2}$

The expression is positive because $\frac{\pi}{8}$ is in the first quadrant.

20. $\sqrt{\frac{2 - \sqrt{3}}{2 + \sqrt{3}}}$

21. $\sin 22.5° = \sin \frac{45°}{2}$

$= \sqrt{\frac{1 - \cos 45°}{2}} = \sqrt{\frac{1 - (\sqrt{2}/2)}{2}}$

$= \sqrt{\frac{2 - \sqrt{2}}{4}} = \frac{\sqrt{2 - \sqrt{2}}}{2}$

The expression is positive because 22.5° is in the first quadrant.

22. $-\sqrt{\frac{2 + \sqrt{2}}{2 - \sqrt{2}}}$

23. First find $\cos \theta$ using the following identity:

$\sin^2\theta + \cos^2\theta = 1$

$(0.3416)^2 + \cos^2\theta = 1$

$\cos^2\theta = 1 - 0.11669056,$

$\cos \theta = 0.9398$ (θ is in quadrant I)

$\sin 2\theta = 2 \sin \theta \cos \theta$

$= 2(0.3416)(0.9398)$

$= 0.6421$

24. 0.7666

25. $\sin 4\theta = 4 \sin \theta \cos^3\theta - 4 \sin^3\theta \cos \theta$

(See Exercise 7)

$= 4(0.3416)(0.9398)^3 - 4(0.3416)^3(0.9398)$

$= 1.1342 - 0.1498$

$= 0.9844$

26. 0.1754

27. $\sin \frac{\theta}{2} = \sqrt{\frac{1 - \cos \theta}{2}}$ $\left[\frac{\theta}{2} \text{ is in quadrant I}\right]$

$= \sqrt{\frac{1 - 0.9398}{2}}$

$= 0.1735$

28. 0.9848

29. $\frac{\sin 2x}{2 \sin x} = \frac{2 \sin x \cos x}{2 \sin x} = \cos x$

30. $\sin x$

31. $1 - 2 \sin^2 \frac{x}{2} = 1 - 2\left[\frac{1 - \cos x}{2}\right]$

$= 1 - 1 + \cos x$

$= \cos x$

32. $\cos x$

33. $2 \sin \frac{x}{2} \cos \frac{x}{2} = \sin 2\left[\frac{x}{2}\right] = \sin x$

34. $\sin 4x$

35. $\cos^2 \frac{x}{2} - \sin^2 \frac{x}{2} = \cos 2\left[\frac{x}{2}\right] = \cos x$

36. $\cos 2x$

37. $(\sin x + \cos x)^2 - \sin 2x$

$= \sin^2x + 2 \sin x \cos x + \cos^2x - \sin 2x$

$= \sin^2x + \sin 2x + \cos^2x - \sin 2x$

$= 1$

38. 1

39. $2 \sin^2 \frac{x}{2} + \cos x$

$= 2\left[\frac{1 - \cos x}{2}\right] + \cos x$

$= 1 - \cos x + \cos x$

$= 1$

40. 1

41. $(-4 \cos x \sin x + 2 \cos 2x)^2 + (2 \cos 2x + 4 \sin x \cos x)^2$
$= (4 \cos^2 2x - 16 \sin x \cos x \cos 2x + 16\cos^2 x \sin^2 x) + (4 \cos^2 2x + 16 \sin x \cos x \cos 2x + 16 \sin^2 x \cos^2 x)$
$= 8 \cos^2 2x + 32 \cos^2 x \sin^2 x$
$= 8(\cos^2 x - \sin^2 x)^2 + 32 \cos^2 x \sin^2 x$
$= 8(\cos^4 x - 2 \cos^2 x \sin^2 x + \sin^4 x) + 32 \cos^2 x \sin^2 x$
$= 8 \cos^4 x + 16 \cos^2 x \sin^2 x + 8 \sin^4 x$
$= 8(\cos^4 x + 2 \cos^2 x \sin^2 x + \sin^4 x)$
$= 8(\cos^2 x + \sin^2 x)$
$= 8$

42. 32

43. $2 \sin x \cos^3 x + 2 \sin^3 x \cos x$
$= 2 \sin x \cos x(\cos^2 x + \sin^2 x)$
$= 2 \sin x \cos x$
$= \sin 2x$

44. $\sin 2x \cos 2x$, or $\frac{1}{2} \sin 4x$

45. $(\sin x + \cos x)^2$
$= \sin^2 x + 2 \sin x \cos x + \cos^2 x$
$= 1 + \sin 2x$

Thus $(\sin x + \cos x)^2 = 1 + \sin 2x$.

46. $\cos^4 x - \sin^4 x = \cos 2x$

47. $\frac{2 \cot x}{\cot^2 x - 1} = \frac{2\frac{\cos x}{\sin x}}{\frac{\cos^2 x}{\sin^2 x} - 1}$
$= \frac{2 \cos x}{\sin x} \cdot \frac{\sin^2 x}{\cos^2 x - \sin^2 x}$
$= \frac{2 \cos x}{\sin x} \cdot \frac{\sin^2 x}{\cos 2x}$
$= \frac{2 \cos x \sin x}{\cos 2x}$
$= \frac{\sin 2x}{\cos 2x}$
$= \tan 2x$

Thus $\frac{2 \cot x}{\cot^2 x - 1} = \tan 2x$.

48. $\frac{2 - \sec^2 x}{\sec^2 x} = \cos 2x$

49. $2 \sin^2 2x + \cos 4x$
$= 2 \sin^2 2x + \cos 2(2x)$
$= 2 \sin^2 2x + (\cos^2 2x - \sin^2 2x)$
$= \sin^2 2x + \cos^2 2x$
$= 1$

Thus $2 \sin^2 2x + \cos 4x = 1$.

50. $\frac{1 + \sin 2x + \cos 2x}{1 + \sin 2x - \cos 2x} = \cot x$

51. $\frac{\pi}{2} \leqslant 2\theta \leqslant \pi$, so $\frac{\pi}{4} \leqslant \theta \leqslant \frac{\pi}{2}$ and all the function values for θ are positive.

We sketch a second quadrant triangle.

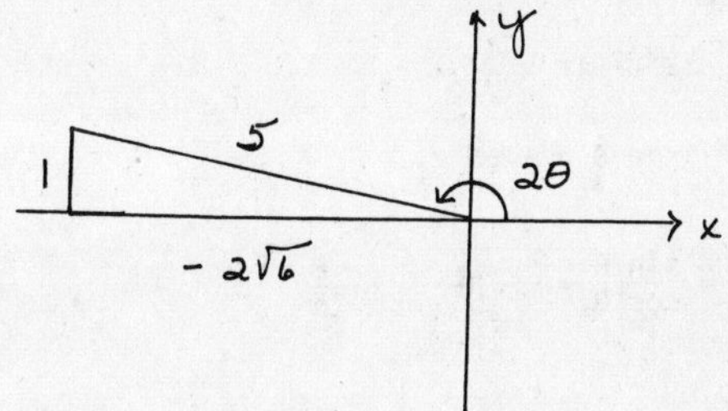

We see that $\cos 2\theta = -\frac{2\sqrt{6}}{5}$.

$\cos 2\theta = 1 - 2 \sin^2\theta$
$-\frac{2\sqrt{6}}{5} = 1 - 2 \sin^2\theta$
$2 \sin^2\theta = 1 + \frac{2\sqrt{6}}{5} = \frac{5 + 2\sqrt{6}}{5}$
$\sin^2\theta = \frac{5 + 2\sqrt{6}}{10}$
$\sin \theta = \sqrt{\frac{5 + 2\sqrt{6}}{10}}$, or $\sqrt{\frac{1}{2} + \frac{\sqrt{6}}{5}}$

$\cos 2\theta = 2 \cos^2\theta - 1$
$-\frac{2\sqrt{6}}{5} = 2 \cos^2\theta - 1$
$1 - \frac{2\sqrt{6}}{5} = 2 \cos^2\theta$
$\frac{5 - 2\sqrt{6}}{10} = \cos^2\theta$
$\sqrt{\frac{5 - 2\sqrt{6}}{10}} = \cos \theta$, or
$\sqrt{\frac{1}{2} - \frac{\sqrt{6}}{5}} = \cos \theta$

$\tan \theta = \frac{\sin \theta}{\cos \theta} = \frac{\sqrt{\frac{5 + 2\sqrt{6}}{10}}}{\sqrt{\frac{5 - 2\sqrt{6}}{10}}} = \sqrt{\frac{5 + 2\sqrt{6}}{5 - 2\sqrt{6}}}$

52. $\sin \theta = \frac{\sqrt{30}}{12}$, $\cos \theta = -\frac{\sqrt{114}}{12}$, $\tan \theta = -\frac{\sqrt{95}}{19}$

53. Since $\frac{3\pi}{2} \leqslant \theta \leqslant 2\pi$, cos θ is positive and sin θ and tan θ are negative.

$$\tan \frac{\theta}{2} = -\sqrt{\frac{1 - \cos \theta}{1 + \cos \theta}}$$

$$-\frac{1}{4} = -\sqrt{\frac{1 - \cos \theta}{1 + \cos \theta}}$$

$$\frac{1}{16} = \frac{1 - \cos \theta}{1 + \cos \theta}$$

$$1 + \cos \theta = 16 - 16 \cos \theta$$

$$17 \cos \theta = 15$$

$$\cos \theta = \frac{15}{17}$$

$$\sin^2\theta + \cos^2\theta = 1$$

$$\sin^2\theta + \left(\frac{15}{17}\right)^2 = 1$$

$$\sin^2\theta = \frac{64}{289}$$

$$\sin \theta = -\frac{8}{17}$$

$$\tan \theta = \frac{\sin \theta}{\cos \theta} = \frac{-\frac{8}{17}}{\frac{15}{17}} = -\frac{8}{15}$$

54. $\sin \theta = -\frac{15}{17}$, $\cos \theta = -\frac{8}{17}$, $\tan \theta = \frac{15}{8}$

55. a) We substitute 42° for φ.

$$N(42°) = 6066 - 31 \cos (2 \cdot 42°)$$
$$= 6066 - 31 \cos 84°$$
$$\approx 6066 - 31(0.104528463)$$
$$\approx 6062.76 \text{ ft}$$

b) We substitute 90° for φ.

$$N(90°) = 6066 - 31 \cos (2 \cdot 90°)$$
$$= 6066 - 31 \cos 180°$$
$$= 6066 - 31(-1) = 6066 + 31$$
$$= 6097 \text{ ft}$$

55. (continued)

c) We substitute $2 \cos^2\theta - 1$ for $\cos 2\phi$.

$$N(\phi) = 6066 - 31(2 \cos^2\phi - 1), \text{ or}$$
$$N(\phi) = 6097 - 62 \cos^2\phi$$

56. a) 9.80359 m/sec²

b) 9.80180 m/sec²

c) $g = 9.78049(1 + 0.005264 \sin^2\phi + 0.000024 \sin^4\phi)$

d) g is greatest at 90° N and 90° S (that is, at the poles).

g is least at 0° (that is, at the equator).

57. Since $\cos^2 x - \sin^2 x = \cos 2x$, we graph the equivalent function, $f(x) = \cos 2x$.

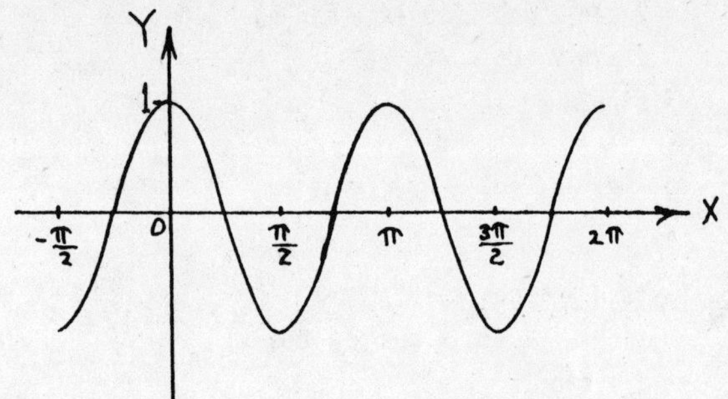

58.

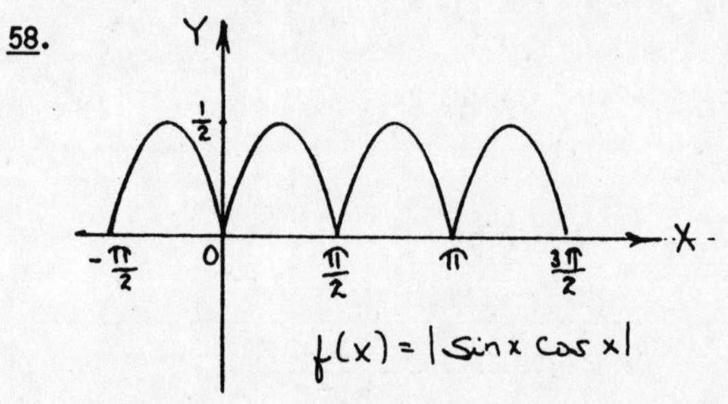

Exercise Set 7.5

Note: Answers for the odd-numbered exercises 1 - 81 are in the answer section in the text.

2.

sec x - sin x tan x	cos x
$\frac{1}{\cos x} - \sin x \cdot \frac{\sin x}{\cos x}$	
$\frac{1 - \sin^2 x}{\cos x}$	
$\frac{\cos^2 x}{\cos x}$	
cos x	

4.

$\frac{1}{\sin \theta \cos \theta} - \frac{\cos \theta}{\sin \theta}$	$\frac{\sin \theta \cos \theta}{1 - \sin^2\theta}$
$\frac{1 - \cos^2\theta}{\sin \theta \cos \theta}$	$\frac{\sin \theta \cos \theta}{\cos^2 \theta}$
$\frac{\sin^2\theta}{\sin \theta \cos \theta}$	$\frac{\sin \theta}{\cos \theta}$
$\frac{\sin \theta}{\cos \theta}$	

6.

$\dfrac{1 - \cos x}{\sin x}$	$\dfrac{\sin x}{1 + \cos x}$
	$\dfrac{\sin x}{1 + \cos x} \cdot \dfrac{1 - \cos x}{1 - \cos x}$
	$\dfrac{\sin x\,(1 - \cos x)}{1 - \cos^2 x}$
	$\dfrac{\sin x\,(1 - \cos x)}{\sin^2 x}$
	$\dfrac{1 - \cos x}{\sin x}$

8.

$\dfrac{\cot \theta - 1}{1 - \tan \theta}$	$\dfrac{\csc \theta}{\sec \theta}$
$\dfrac{\dfrac{\cos \theta}{\sin \theta} - 1}{1 - \dfrac{\sin \theta}{\cos \theta}}$	$\dfrac{\dfrac{1}{\sin \theta}}{\dfrac{1}{\cos \theta}}$
$\dfrac{\dfrac{\cos \theta - \sin \theta}{\sin \theta}}{\dfrac{\cos \theta - \sin \theta}{\cos \theta}}$	$\dfrac{\cos \theta}{\sin \theta}$
$\dfrac{\cos \theta}{\sin \theta}$	

10.

$\dfrac{\sin x - \cos x}{\sec x - \csc x}$	$\dfrac{\cos x}{\csc x}$
$\dfrac{\sin x - \cos x}{\dfrac{1}{\cos x} - \dfrac{1}{\sin x}}$	$\sin x \cos x$
$\dfrac{\sin x - \cos x}{\dfrac{\sin x - \cos x}{\sin x \cos x}}$	
$\sin x \cos x$	

12.

$\dfrac{\cos^2\theta + \cot \theta}{\cos^2\theta - \cot \theta}$	$\dfrac{\cos^2\theta \tan \theta + 1}{\cos^2\theta \tan \theta - 1}$
$\dfrac{\cos^2\theta + \dfrac{\cos \theta}{\sin \theta}}{\cos^2\theta - \dfrac{\cos \theta}{\sin \theta}}$	$\dfrac{\cos^2\theta\,\dfrac{\sin \theta}{\cos \theta} + 1}{\cos^2\theta\,\dfrac{\sin \theta}{\cos \theta} - 1}$
$\dfrac{\cos \theta\left(\cos \theta + \dfrac{1}{\sin \theta}\right)}{\cos \theta\left(\cos \theta - \dfrac{1}{\sin \theta}\right)}$	$\dfrac{\sin \theta \cos \theta + 1}{\sin \theta \cos \theta - 1}$
$\dfrac{\cos \theta + \dfrac{1}{\sin \theta}}{\cos \theta - \dfrac{1}{\sin \theta}}$	
$\dfrac{\dfrac{\sin \theta \cos \theta + 1}{\sin \theta}}{\dfrac{\sin \theta \cos \theta - 1}{\sin \theta}}$	
$\dfrac{\sin \theta \cos \theta + 1}{\sin \theta \cos \theta - 1}$	

14.

$\dfrac{2 \tan \theta}{1 + \tan^2\theta}$	$\sin 2\theta$
$\dfrac{2 \tan \theta}{\sec^2\theta}$	$2 \sin \theta \cos \theta$
$\dfrac{2 \sin \theta}{\cos \theta} \cdot \dfrac{\cos^2\theta}{1}$	
$2 \sin \theta \cos \theta$	

16.

$\cot 2\theta$	$\dfrac{\cot^2\theta - 1}{2 \cot \theta}$
$\dfrac{\cos 2\theta}{\sin 2\theta}$	$\dfrac{\dfrac{\cos^2\theta}{\sin^2\theta} - 1}{2\,\dfrac{\cos \theta}{\sin \theta}}$
$\dfrac{\cos^2\theta - \sin^2\theta}{2 \sin \theta \cos \theta}$	$\dfrac{1}{2}\left[\dfrac{\cos^2\theta - \sin^2\theta}{\sin^2\theta} \cdot \dfrac{\sin \theta}{\cos \theta}\right]$
	$\dfrac{\cos^2\theta - \sin^2\theta}{2 \sin \theta \cos \theta}$

18.

$\dfrac{\cos(\alpha - \beta)}{\cos \alpha \sin \beta}$	$\tan \alpha + \cot \beta$
$\dfrac{\cos \alpha \cos \beta + \sin \alpha \sin \beta}{\cos \alpha \sin \beta}$	
$\dfrac{\cos \alpha \cos \beta}{\cos \alpha \sin \beta} + \dfrac{\sin \alpha \sin \beta}{\cos \alpha \sin \beta}$	
$\dfrac{\cos \beta}{\sin \beta} + \dfrac{\sin \alpha}{\cos \alpha}$	
$\cot \beta + \tan \alpha$	

20.

$2 \sin \theta \cos^3\theta + 2 \sin^3\theta \cos \theta$	$\sin 2\theta$
$2 \sin \theta \cos \theta(\cos^2\theta + \sin^2\theta)$	$2 \sin \theta \cos \theta$
$2 \sin \theta \cos \theta$	

22.

$\dfrac{\tan \theta - \sin \theta}{2 \tan \theta}$	$\sin^2 \dfrac{\theta}{2}$
$\dfrac{1}{2}\left[\dfrac{\dfrac{\sin \theta}{\cos \theta} - \sin \theta}{\dfrac{\sin \theta}{\cos \theta}}\right]$	$\dfrac{1 - \cos \theta}{2}$
$\dfrac{1}{2}\,\dfrac{\sin \theta - \sin \theta \cos \theta}{\cos \theta} \cdot \dfrac{\cos \theta}{\sin \theta}$	
$\dfrac{1 - \cos \theta}{2}$	

24.

$\dfrac{\cos^4 x - \sin^4 x}{1 - \tan^4 x}$	$\cos^4 x$
$\dfrac{\cos^4 x - \sin^4 x}{1 - \dfrac{\sin^4 x}{\cos^4 x}}$	
$\dfrac{\cos^4 x - \sin^4 x}{\dfrac{\cos^4 x - \sin^4 x}{\cos^4 x}}$	
$\cos^4 x$	

26.

$\left(\dfrac{1 + \tan\theta}{1 - \tan\theta}\right)^2$	$\dfrac{1 + \sin 2\theta}{1 - \sin 2\theta}$
$\left[\dfrac{\frac{\cos\theta + \sin\theta}{\cos\theta}}{\frac{\cos\theta - \sin\theta}{\cos\theta}}\right]^2$	$\dfrac{1 + 2\sin\theta\cos\theta}{1 - 2\sin\theta\cos\theta}$
$\dfrac{\cos^2\theta + 2\sin\theta\cos\theta + \sin^2\theta}{\cos^2\theta - 2\sin\theta\cos\theta + \sin^2\theta}$	
$\dfrac{1 + 2\sin\theta\cos\theta}{1 - 2\sin\theta\cos\theta}$	

28.

$\dfrac{\sin^3 t + \cos^3 t}{\sin t + \cos t}$	$\dfrac{2 - \sin 2t}{2}$
$\dfrac{(\sin t + \cos t)(\sin^2 t - \sin t\cos t + \cos^2 t)}{\sin t + \cos t}$	$\dfrac{2 - 2\sin t\cos t}{2}$
$1 - \sin t\cos t$	$1 - \sin t\cos t$

30.

$\cos(\alpha + \beta)\cos(\alpha - \beta)$	$\cos^2\alpha - \sin^2\beta$
$(\cos\alpha\cos\beta - \sin\alpha\sin\beta)(\cos\alpha\cos\beta + \sin\alpha\sin\beta)$	
$\cos^2\alpha\cos^2\beta - \sin^2\alpha\sin^2\beta$	
$(1 - \sin^2\alpha)(1 - \sin^2\beta) - \sin^2\alpha\sin^2\beta$	
$1 - \sin^2\alpha - \sin^2\beta + \sin^2\alpha\sin^2\beta - \sin^2\alpha\sin^2\beta$	
$1 - \sin^2\alpha - \sin^2\beta$	
$\cos^2\alpha - \sin^2\beta$	

32.

$\sin(\alpha + \beta) + \sin(\alpha - \beta)$	$2\sin\alpha\cos\beta$
$\sin\alpha\cos\beta + \cos\alpha\sin\beta + \sin\alpha\cos\beta - \cos\alpha\sin\beta$	
$2\sin\alpha\cos\beta$	

34.

$\cos^2 x\,(1 - \sec^2 x)$	$-\sin^2 x$
$\cos^2 x\,(-\tan^2 x)$	
$\cos^2 x\left(-\dfrac{\sin^2 x}{\cos^2 x}\right)$	
$-\sin^2 x$	

36.

$\tan\theta\,(\tan\theta + \cot\theta)$	$\sec^2\theta$
$\tan^2\theta + \tan\theta\cot\theta$	
$\tan^2\theta + 1$	
$\sec^2\theta$	

38.

$\dfrac{\tan x + \sin x}{1 + \sec x}$	$\sin x$
$\dfrac{\dfrac{\sin x}{\cos x} + \dfrac{\sin x}{1}}{1 + \dfrac{1}{\cos x}}$	
$\dfrac{\dfrac{\sin x + \sin x \cos x}{\cos x}}{\dfrac{\cos x + 1}{\cos x}}$	
$\dfrac{\sin x\,(1 + \cos x)}{\cos x} \cdot \dfrac{\cos x}{\cos x + 1}$	
$\sin x$	

40.

$\dfrac{1 + \tan^2\theta}{\csc^2\theta}$	$\tan^2\theta$
$\dfrac{\sec^2\theta}{\csc^2\theta}$	$\dfrac{\sin^2\theta}{\cos^2\theta}$
$\dfrac{1}{\cos^2\theta} \cdot \dfrac{\sin^2\theta}{1}$	
$\dfrac{\sin^2\theta}{\cos^2\theta}$	

42.

$\dfrac{1 + \cos^2 x}{\sin^2 x}$	$2\csc^2 x - 1$
$\dfrac{1}{\sin^2 x} + \dfrac{\cos^2 x}{\sin^2 x}$	
$\csc^2 x + \cot^2 x$	
$\csc^2 x + \csc^2 x - 1$	
$2\csc^2 x - 1$	

44.

$\dfrac{\csc x - \sin x}{\cot x}$	$\cos x$
$\dfrac{\dfrac{1}{\sin x} - \sin x}{\dfrac{\cos x}{\sin x}}$	
$\dfrac{1 - \sin^2 x}{\sin x} \cdot \dfrac{\sin x}{\cos x}$	
$\dfrac{1 - \sin^2 x}{\cos x}$	
$\dfrac{\cos^2 x}{\cos x}$	
$\cos x$	

46.

$\dfrac{\csc\theta - \sin\theta}{\cos^2\theta}$	$\csc\theta$
$\dfrac{\dfrac{1}{\sin\theta} - \sin\theta}{\cos^2\theta}$	$\dfrac{1}{\sin\theta}$
$\dfrac{1 - \sin^2\theta}{\sin\theta} \cdot \dfrac{1}{\cos^2\theta}$	
$\dfrac{\cos^2\theta}{\sin\theta} \cdot \dfrac{1}{\cos^2\theta}$	
$\dfrac{1}{\sin\theta}$	

48.

$\dfrac{1 + \sin x}{1 - \sin x} + \dfrac{\sin x - 1}{1 + \sin x}$	$4 \sec x \tan x$
$\dfrac{(1 + \sin x)^2 - (1 - \sin x)^2}{1 - \sin^2 x}$	$4 \cdot \dfrac{1}{\cos x} \cdot \dfrac{\sin x}{\cos x}$
$\dfrac{(1+2\sin x+\sin^2 x)-(1-2\sin x+\sin^2 x)}{\cos^2 x}$	$\dfrac{4 \sin x}{\cos^2 x}$
$\dfrac{4 \sin x}{\cos^2 x}$	

50.

$\cos^2\theta \cot^2\theta$	$\cot^2\theta - \cos^2\theta$
$(1 - \sin^2\theta)\cot^2\theta$	
$\cot^2\theta - \sin^2\theta \cdot \dfrac{\cos^2\theta}{\sin^2\theta}$	
$\cot^2\theta - \cos^2\theta$	

52.

$\dfrac{\cos^2 x - 1}{1 - \sec^2 x}$	$\dfrac{1}{\tan^2 x + 1}$
$\dfrac{-\sin^2 x}{-\tan^2 x}$	$\dfrac{1}{\sec^2 x}$
$\sin^2 x \cdot \dfrac{\cos^2 x}{\sin^2 x}$	$\cos^2 x$
$\cos^2 x$	

54.

$\tan\theta - \cot\theta$	$(\sec\theta - \csc\theta)(\sin\theta + \cos\theta)$
$\dfrac{\sin\theta}{\cos\theta} - \dfrac{\cos\theta}{\sin\theta}$	$\left(\dfrac{1}{\cos\theta} - \dfrac{1}{\sin\theta}\right)(\sin\theta + \cos\theta)$
$\dfrac{\sin^2\theta - \cos^2\theta}{\cos\theta \sin\theta}$	$\left(\dfrac{\sin\theta - \cos\theta}{\sin\theta \cos\theta}\right)\left(\dfrac{\sin\theta + \cos\theta}{1}\right)$
	$\dfrac{\sin^2\theta - \cos^2\theta}{\sin\theta \cos\theta}$

56.

$\csc x - \cot x$	$\dfrac{1}{\csc x + \cot x}$
$\dfrac{1}{\sin x} - \dfrac{\cos x}{\sin x}$	$\dfrac{1}{\dfrac{1}{\sin x} + \dfrac{\cos x}{\sin x}}$
$\dfrac{1 - \cos x}{\sin x} \cdot \dfrac{1 + \cos x}{1 + \cos x}$	$\dfrac{1}{\dfrac{1 + \cos x}{\sin x}}$
$\dfrac{1 - \cos^2 x}{\sin x\,(1 + \cos x)}$	$\dfrac{\sin x}{1 + \cos x}$
$\dfrac{\sin^2 x}{\sin x\,(1 + \cos x)}$	
$\dfrac{\sin x}{1 + \cos x}$	

58.

$2\sin^2\theta\cos^2\theta + \cos^4\theta$	$1 - \sin^4\theta$
$\cos^2\theta(2\sin^2\theta + \cos^2\theta)$	$(1+\sin^2\theta)(1-\sin^2\theta)$
$\cos^2\theta(\sin^2\theta + \sin^2\theta + \cos^2\theta)$	$(1 + \sin^2\theta)(\cos^2\theta)$
$\cos^2\theta(\sin^2\theta + 1)$	

60.

$\dfrac{\cot\theta}{\csc\theta - 1}$	$\dfrac{\csc\theta + 1}{\cot\theta}$
$\dfrac{\cot\theta}{\csc\theta - 1}\cdot\dfrac{\csc\theta + 1}{\csc\theta + 1}$	
$\dfrac{\cot\theta(\csc\theta + 1)}{\csc^2\theta - 1}$	
$\dfrac{\cot\theta(\csc\theta + 1)}{\cot^2\theta}$	
$\dfrac{\csc\theta + 1}{\cot\theta}$	

62.

$\dfrac{1 + \sin x}{1 - \sin x}$	$(\sec x + \tan x)^2$
$\dfrac{1 + \sin x}{1 - \sin x}\cdot\dfrac{1 + \sin x}{1 + \sin x}$	$\left[\dfrac{1}{\cos x} + \dfrac{\sin x}{\cos x}\right]^2$
$\dfrac{(1 + \sin x)^2}{1 - \sin^2 x}$	$\dfrac{(1 + \sin x)^2}{\cos^2 x}$
$\dfrac{(1 + \sin x)^2}{\cos^2 x}$	

64.

$\sec^4\theta - \tan^2\theta$	$\tan^4\theta + \sec^2\theta$
$\sec^4\theta - (\sec^2\theta - 1)$	$(\tan^2\theta)^2 + \sec^2\theta$
$\sec^4\theta - \sec^2\theta + 1$	$(\sec^2\theta - 1)^2 + \sec^2\theta$
	$\sec^4\theta - 2\sec^2\theta + 1 + \sec^2\theta$
	$\sec^4\theta - \sec^2\theta + 1$

66.

$1 + \sec x$	$\csc x(\sin x + \tan x)$
$1 + \dfrac{1}{\cos x}$	$\dfrac{1}{\sin x}\left[\dfrac{\sin x}{1} + \dfrac{\sin x}{\cos x}\right]$
	$1 + \dfrac{1}{\cos x}$

68.

$\dfrac{\sin^3\theta - \cos^3\theta}{\sin\theta - \cos\theta}$	$\sin\theta\cos\theta + 1$
$\dfrac{(\sin\theta-\cos\theta)(\sin^2\theta+\sin\theta\cos\theta+\cos^2\theta)}{\sin\theta - \cos\theta}$	
$\sin^2\theta + \sin\theta\cos\theta + \cos^2\theta$	
$\sin\theta\cos\theta + 1$	

70.

$2\sec^2 x + \dfrac{\csc x}{1 - \csc x}$	$\dfrac{\csc x}{\csc x + 1}$
$2\cdot\dfrac{1}{\cos^2 x} + \dfrac{\frac{1}{\sin x}}{1 - \frac{1}{\sin x}}$	$\dfrac{\frac{1}{\sin x}}{\frac{1}{\sin x} + 1}$
$\dfrac{2}{\cos^2 x} + \dfrac{\frac{1}{\sin x}}{\frac{\sin x - 1}{\sin x}}$	$\dfrac{\frac{1}{\sin x}}{\frac{1 + \sin x}{\sin x}}$
$\dfrac{2}{\cos^2 x} + \dfrac{1}{\sin x - 1}$	$\dfrac{1}{1+\sin x}\cdot\dfrac{1-\sin x}{1-\sin x}$
$\dfrac{2}{\cos^2 x} + \dfrac{1}{\sin x - 1}\cdot\dfrac{\sin x + 1}{\sin x + 1}$	$\dfrac{1 - \sin x}{1 - \sin^2 x}$
$\dfrac{2}{\cos^2 x} + \dfrac{\sin x + 1}{\sin^2 x - 1}$	$\dfrac{1 - \sin x}{\cos^2 x}$
$\dfrac{2}{\cos^2 x} + \dfrac{\sin x + 1}{-\cos^2 x}$	
$\dfrac{1 - \sin x}{\cos^2 x}$	

72.

$\cos\theta + \dfrac{\sin\theta}{\cot\theta - 1}$	$\dfrac{\cos\theta}{1 - \tan\theta} - \sin\theta$
$\cos\theta + \dfrac{\sin\theta}{\frac{\cos\theta}{\sin\theta} - 1}$	$\dfrac{\cos\theta}{1 - \frac{\sin\theta}{\cos\theta}} - \sin\theta$
$\cos\theta + \dfrac{\sin\theta}{1}\cdot\dfrac{\sin\theta}{\cos\theta-\sin\theta}$	$\dfrac{\cos\theta}{1}\cdot\dfrac{\cos\theta}{\cos\theta-\sin\theta} - \sin\theta$
$\cos\theta + \dfrac{\sin^2\theta}{\cos\theta - \sin\theta}$	$\dfrac{\cos^2\theta}{\cos\theta - \sin\theta} - \sin\theta$
$\dfrac{\cos^2\theta-\cos\theta\sin\theta+\sin^2\theta}{\cos\theta - \sin\theta}$	$\dfrac{\cos^2\theta-\sin\theta\cos\theta+\sin^2\theta}{\cos\theta - \sin\theta}$

74.

$\dfrac{\cos x + \cot x}{1 + \csc x}$	$\cos x$
$\dfrac{\frac{\cos x}{1} + \frac{\cos x}{\sin x}}{1 + \frac{1}{\sin x}}$	
$\dfrac{\sin x\cos x + \cos x}{\sin x}\cdot\dfrac{\sin x}{\sin x + 1}$	
$\dfrac{\cos x(\sin x + 1)}{\sin x + 1}$	
$\cos x$	

76.

$\sec^2\theta - \csc^2\theta$	$\dfrac{\tan\theta - \cot\theta}{\sin\theta\cos\theta}$
$\dfrac{1}{\cos^2\theta} - \dfrac{1}{\sin^2\theta}$	$\dfrac{\frac{\sin\theta}{\cos\theta} - \frac{\cos\theta}{\sin\theta}}{\sin\theta\cos\theta}$
$\dfrac{\sin^2\theta - \cos^2\theta}{\cos^2\theta\sin^2\theta}$	$\dfrac{\sin^2\theta - \cos^2\theta}{\sin\theta\cos\theta}\cdot\dfrac{1}{\sin\theta\cos\theta}$
	$\dfrac{\sin^2\theta - \cos^2\theta}{\sin^2\theta\cos^2\theta}$

78.

$$\begin{array}{r|l} \sec^4 x - 4\tan^2 x & (1 - \tan^2 x)^2 \\ (\sec^2 x)^2 - 4\tan^2 x & 1 - 2\tan^2 x + \tan^4 x \\ (1 + \tan^2 x)^2 - 4\tan^2 x & \\ 1 + 2\tan^2 x + \tan^4 x - 4\tan^2 x & \\ 1 - 2\tan^2 x + \tan^4 x & \end{array}$$

80.

$$\begin{array}{c|l} \dfrac{\tan^2 x + \sec^2 x}{\sec^4 x} & 1 - \sin^4 x \\ \dfrac{\dfrac{\sin^2 x}{\cos^2 x} + \dfrac{1}{\cos^2 x}}{\dfrac{1}{\cos^4 x}} & (1 - \sin^2 x)(1 + \sin^2 x) \\ \dfrac{\dfrac{\sin^2 x + 1}{\cos^2 x}}{\dfrac{1}{\cos^4 x}} & \cos^2 x\,(\sin^2 x + 1) \\ \cos^2 x(\sin^2 x + 1) & \end{array}$$

82. By Exercise 81,

$$2 \sin \frac{x + y}{2} \cos \frac{x - y}{2}$$

$$= 2 \cdot \frac{1}{2}\left[\sin \left(\frac{x + y}{2} + \frac{x - y}{2}\right) + \sin \left(\frac{x + y}{2} - \frac{x - y}{2}\right)\right]$$

$$= \sin x + \sin y$$

Other formulas follow similarly.

83. We will use the identity

$$\cos u \cdot \sin v = \frac{1}{2} [\sin (u + v) - \sin (u - v)].$$

Here $u + v = 3\theta$ and $u - v = 5\theta$.

Note that $u = \dfrac{(u + v) + (u - v)}{2}$ and $v = \dfrac{(u + v) - (u - v)}{2}$.

$$\begin{aligned} \sin 3\theta - \sin 5\theta &= 2 \cos \left(\frac{3\theta + 5\theta}{2}\right) \cdot \sin \left(\frac{3\theta - 5\theta}{2}\right) \\ &= 2 \cos 4\theta \sin (-\theta) \\ &= -2 \cos 4\theta \sin \theta \qquad (\sin (-\theta) = -\sin \theta) \end{aligned}$$

84. $\sin 7x - \sin 4x = 2 \cos \dfrac{11x}{2} \sin \dfrac{3x}{2}$

85. We will use the identity

$$\sin u \cdot \cos v = \frac{1}{2} [\sin (u + v) + \sin (u - v)]$$

Here $u + v = 8\theta$ and $u - v = 5\theta$.

We find u and v as in Exercise 83.

$$\begin{aligned} \sin 8\theta + \sin 5\theta &= 2 \sin \left(\frac{8\theta + 5\theta}{2}\right) \cdot \cos \left(\frac{8\theta - 5\theta}{2}\right) \\ &= 2 \sin \frac{13\theta}{2} \cos \frac{3\theta}{2} \end{aligned}$$

86. $\cos \theta - \cos 7\theta = 2 \sin 4\theta \sin 3\theta$

87. We will use the identity

$$\sin u \cdot \sin v = \frac{1}{2}\,[\cos(u - v) - \cos(u + v)].$$

Here u is 7u and v is 5u.

$$\sin 7u \sin 5u = \frac{1}{2}\,[\cos(7u - 5u) - \cos(7u + 5u)]$$

$$= \frac{1}{2}\,(\cos 2u - \cos 12u)$$

88. $2 \sin 7\theta \cos 3\theta = \sin 10\theta + \sin 4\theta$

89. We will use the identity

$$\cos u \cdot \sin v = \frac{1}{2}\,[\sin(u + v) - \sin(u - v)]$$

Here u is θ and v is 7θ.

$$7 \cos\theta \sin 7\theta = 7 \cdot \frac{1}{2}\,[\sin(\theta + 7\theta) - \sin(\theta - 7\theta)]$$

$$= \frac{7}{2}\,[\sin 8\theta - \sin(-6\theta)]$$

$$= \frac{7}{2}\,(\sin 8\theta + \sin 6\theta) \qquad (\sin(-6\theta) = -\sin 6\theta)$$

90. $\cos 2t \sin t = \frac{1}{2}\,(\sin 3t - \sin t)$

91. We will use the identity

$$\cos u \cdot \sin v = \frac{1}{2}\,[\sin(u + v) - \sin(u - v)].$$

Here u is 55° and v is 25°.

$$\cos 55^\circ \sin 25^\circ = \frac{1}{2}\,[\sin(55^\circ + 25^\circ) - \sin(55^\circ - 25^\circ)]$$

$$= \frac{1}{2}\,(\sin 80^\circ - \sin 30^\circ), \text{ or}$$

$$\frac{1}{2}\left[\sin 80^\circ - \frac{1}{2}\right], \text{ or} \qquad \left[\sin 30^\circ = \frac{1}{2}\right]$$

$$\frac{1}{2}\sin 80^\circ - \frac{1}{4}$$

92. $7 \cos 5\theta \cos 7\theta = \frac{7}{2}\,(\cos 2\theta + \cos 12\theta)$

Answers for the odd-numbered exercises 93 - 111 are in the answer section in the text.

94.

$\tan 2x(\cos x + \cos 3x)$	$\sin x + \sin 3x$
$\frac{\sin 2x}{\cos 2x}\left[2 \cos \frac{4x}{2} \cos \frac{-2x}{2}\right]$	$2 \sin \frac{4x}{2} \cos \frac{-2x}{2}$
$\frac{\sin 2x}{\cos 2x}\,(2 \cos 2x \cos x)$	$2 \cos 2x \cos x$
$2 \sin 2x \cos x$	

96.

$\tan\frac{x+y}{2}$	$\frac{\sin x + \sin y}{\cos x + \cos y}$
$\frac{\sin\frac{x+y}{2}}{\cos\frac{x+y}{2}}$	$\frac{2\sin\frac{x+y}{2}\cos\frac{x-y}{2}}{2\cos\frac{x+y}{2}\cos\frac{x-y}{2}}$
	$\frac{\sin\frac{x+y}{2}}{\cos\frac{x+y}{2}}$

98.

$\tan\frac{\theta+\phi}{2}\tan\frac{\phi-\theta}{2}$	$\frac{\cos\theta - \cos\phi}{\cos\theta + \cos\phi}$
	$\frac{2\sin\frac{\phi+\theta}{2}\sin\frac{\phi-\theta}{2}}{2\cos\frac{\phi+\theta}{2}\cos\frac{\phi-\theta}{2}}$
	$\tan\frac{\theta+\phi}{2}\tan\frac{\phi-\theta}{2}$

100.

$\sin 2\theta + \sin 4\theta + \sin 6\theta$	$4\cos\theta\cos 2\theta\sin 3\theta$
$2\sin\frac{6\theta}{2}\cos\frac{-2\theta}{2} + \sin 6\theta$	
$2\sin 3\theta\cos\theta + 2\sin 3\theta\cos 3\theta$	
$2\sin 3\theta(\cos\theta + \cos 3\theta)$	
$2\sin 3\theta\left[2\cos\frac{4\theta}{2}\cos\frac{2\theta}{2}\right]$	
$2\sin 3\theta(2\cos 2\theta\cos\theta)$	
$4\cos\theta\cos 2\theta\sin 3\theta$	

102.

$\ln\lvert\tan x\rvert$	$-\ln\lvert\cot x\rvert$
$\ln\left\lvert\frac{1}{\cot x}\right\rvert$	
$\ln\lvert 1\rvert - \ln\lvert\cot x\rvert$	
$0 - \ln\lvert\cot x\rvert$	
$-\ln\lvert\cot x\rvert$	

104.

$\ln\lvert\csc x\rvert$	$-\ln\lvert\sin x\rvert$
$\ln\left\lvert\frac{1}{\sin x}\right\rvert$	
$\ln\lvert 1\rvert - \ln\lvert\sin x\rvert$	
$-\ln\lvert\sin x\rvert$	

106.

$e^{\ln\lvert\cos t\rvert}$	$\lvert\cos t\rvert$
$\lvert\cos t\rvert$	

(Property 5)

108.

$\ln \lvert \sec\theta + \tan\theta \rvert$	$-\ln \lvert \sec\theta - \tan\theta \rvert$
$\ln \left\lvert \sec\theta + \tan\theta \cdot \dfrac{\sec\theta - \tan\theta}{\sec\theta - \tan\theta} \right\rvert$	
$\ln \left\lvert \dfrac{\sec^2\theta - \tan^2\theta}{\sec\theta - \tan\theta} \right\rvert$	
$\ln \left\lvert \dfrac{1}{\sec\theta - \tan\theta} \right\rvert$	
$\ln \lvert 1 \rvert - \ln \lvert \sec\theta - \tan\theta \rvert$	
$0 - \ln \lvert \sec\theta - \tan\theta \rvert$	
$-\ln \lvert \sec\theta - \tan\theta \rvert$	

110. $\sin\theta = \cos\phi$

112. Using the given equations for E_1 and E_2 we have

(a) $$\frac{E_1 + E_2}{2} = \frac{\sqrt{2}\,E_t \cos\left(\theta + \frac{\pi}{P}\right) + \sqrt{2}\,E_t \cos\left(\theta - \frac{\pi}{P}\right)}{2}$$

$$= \frac{\sqrt{2}\,E_t \left[\left(\cos\theta\cos\frac{\pi}{P} - \sin\theta\sin\frac{\pi}{P}\right) + \left(\cos\theta\cos\frac{\pi}{P} + \sin\theta\sin\frac{\pi}{P}\right)\right]}{2}$$

$$= \frac{\sqrt{2}\,E_t \left(2\cos\theta\cos\frac{\pi}{P}\right)}{2}$$

$$= \sqrt{2}\,E_t \cos\theta\cos\frac{\pi}{P}$$

(b) $$\frac{E_1 - E_2}{2} = \frac{\sqrt{2}\,E_t \cos\left(\theta + \frac{\pi}{P}\right) - \sqrt{2}\,E_t \cos\left(\theta - \frac{\pi}{P}\right)}{2}$$

$$= \frac{\sqrt{2}\,E_t \left[\left(\cos\theta\cos\frac{\pi}{P} - \sin\theta\sin\frac{\pi}{P}\right) - \left(\cos\theta\cos\frac{\pi}{P} + \sin\theta\sin\frac{\pi}{P}\right)\right]}{2}$$

$$= \frac{\sqrt{2}\,E_t \left(-2\sin\theta\sin\frac{\pi}{P}\right)}{2}$$

$$= -\sqrt{2}\,E_t \sin\theta\sin\frac{\pi}{P}$$

Exercise Set 7.6

1. Find $\arcsin \frac{\sqrt{2}}{2}$.

   The only number in the restricted range $\left[-\frac{\pi}{2}, \frac{\pi}{2}\right]$ whose sine is $\frac{\sqrt{2}}{2}$ is $\frac{\pi}{4}$, or 45°.

2. $\frac{\pi}{3}$, or 60°

3. Find $\cos^{-1} \frac{\sqrt{2}}{2}$.

   The only number in the restricted range $[0, \pi]$ whose cosine is $\frac{\sqrt{2}}{2}$ is $\frac{\pi}{4}$, or 45°.

4. $\frac{\pi}{6}$, or 30°

5. Find $\sin^{-1}\left(-\frac{\sqrt{2}}{2}\right)$.

   The only number in the restricted range $\left[-\frac{\pi}{2}, \frac{\pi}{2}\right]$ whose sine is $-\frac{\sqrt{2}}{2}$ is $-\frac{\pi}{4}$, or -45°.

6. $-\frac{\pi}{3}$, or -60°

7. Find arccos $\left(-\frac{\sqrt{2}}{2}\right)$.

The only number in the restricted range $[0, \pi]$ whose cosine is $-\frac{\sqrt{2}}{2}$ is $\frac{3\pi}{4}$, or 135°.

8. $\frac{5\pi}{6}$, or 150°

9. Find arctan $\sqrt{3}$.

The only number in the restricted range $\left(-\frac{\pi}{2}, \frac{\pi}{2}\right)$ whose tangent is $\sqrt{3}$ is $\frac{\pi}{3}$, or 60°.

10. $\frac{\pi}{6}$, or 30°

11. Find $\cot^{-1} 1$.

The only number in the restricted range $(0, \pi)$ whose cotangent is 1 is $\frac{\pi}{4}$, or 45°.

12. $\frac{\pi}{6}$, or 30°

13. Find arctan $\left(-\frac{\sqrt{3}}{3}\right)$.

The only number in the restricted range $\left(-\frac{\pi}{2}, \frac{\pi}{2}\right)$ whose tangent is $-\frac{\sqrt{3}}{3}$ is $-\frac{\pi}{6}$, or -30°.

14. $-\frac{\pi}{3}$, or -60°

15. Find arccot (-1).

The only number in the restricted range $(0, \pi)$ whose cotangent is -1 is $\frac{3\pi}{4}$, or 135°.

16. $\frac{5\pi}{6}$, or 150°

17. Find arcsec 1.

The only number in the restricted range $\left[0, \frac{\pi}{2}\right) \cup \left[\pi, \frac{3\pi}{2}\right)$ whose secant is 1 is 0, or 0°.

18. $\frac{\pi}{3}$, or 60°

19. Find $\csc^{-1} 1$.

The only number in the restricted range $\left(0, \frac{\pi}{2}\right] \cup \left(-\pi, -\frac{\pi}{2}\right]$ whose cosecant is 1 is $\frac{\pi}{2}$, or 90°.

20. $\frac{\pi}{6}$, or 30°

21. Find arcsin $\left(-\frac{\sqrt{2}}{2}\right)$.

The only number in the restricted range $\left[-\frac{\pi}{2}, \frac{\pi}{2}\right]$ whose sine is $-\frac{\sqrt{2}}{2}$ is $-\frac{\pi}{4}$, or -45°.

22. $\frac{\pi}{6}$, or 30°

23. Find $\cos^{-1} \frac{1}{2}$.

The only number in the restricted range $[0, \pi]$ whose cosine is $\frac{1}{2}$ is $\frac{\pi}{3}$, or 60°.

24. $\frac{\pi}{4}$, or 45°

25. Find arcsin $\left(-\frac{\sqrt{3}}{2}\right)$.

The only number in the restricted range $\left[-\frac{\pi}{2}, \frac{\pi}{2}\right]$ whose sine is $-\frac{\sqrt{3}}{2}$ is $-\frac{\pi}{3}$, or -60°.

26. $-\frac{\pi}{6}$, or -30°

27. Same as Exercise 7.

28. Same as Exercise 8.

29. Same as Exercise 13.

30. Same as Exercise 14.

31. Find arccot $\left(-\frac{\sqrt{3}}{3}\right)$.

The only number in the restricted range $(0, \pi)$ whose cotangent is $-\frac{\sqrt{3}}{3}$ is $\frac{2\pi}{3}$, or 120°.

32. Same as Exercise 16.

33. 23.00°

34. 74.01°, or 74° 29"

35. -38.31°, or -38° 18'

36. -60.15°, or -60° 9'

37. 36.97°, or 36° 58'

38. 22.10°, or 22° 6'

39. 168.82°, or 168° 49'

40. 105.76°, or 105° 45'

41. 20.17°, or 20° 10'

42. 47.49°, or 47° 30'

43. 38.33°, or 38° 20'

44. 64.50°, or 64° 30'

45. 31.09°, or 31° 5'

46. 226.02°, or 226° 1'

47. -170.83°, or -170° 50'
Some calculators may give arccsc (-6.2774) = -9.17°. To find the inverse function value in the restricted range $\left[0, \frac{\pi}{2}\right] \cup \left[-\pi, -\frac{\pi}{2}\right]$, we compute -180° + 9.17° = -170.83°.

48. 64.15°, or 64° 9'

49. 13.50°, or 13° 30'

50. 26.83°, or 26° 50'

51. -39.50°, or -39° 30'

52. -55.00°, or -55° 17"

53. -62.70°, or -62° 42'

54. 107.07°, or 107° 4'

55. -22.23°, or -22° 14'

56. -13.55°, or -13° 33'

57. 170.44°, o0 170° 26'
(180° - 9.56°, or 170.44°, is in the restricted range.)

58. 142.83°, or 142° 50'

59. 177.51°, or 177° 31'
(180° - 2.49°, or 177.51°, is in the restricted range.)

60. 89.64°, or 89° 38'

61.

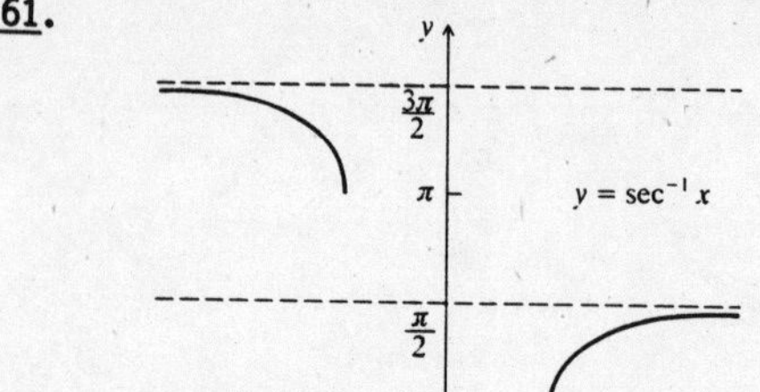

62.

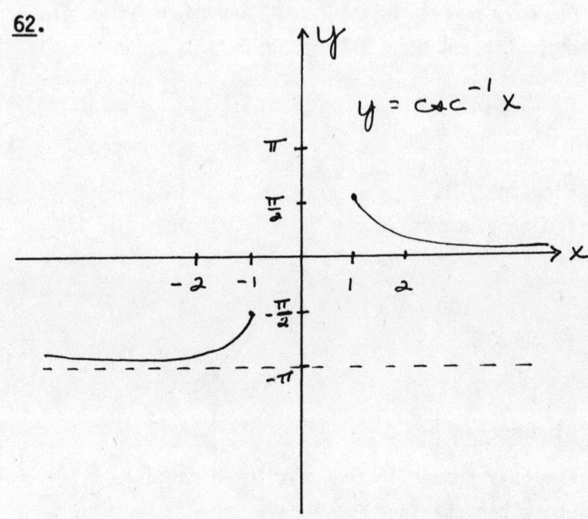

63. See the answer section in the text.

64. Let x = 1. Then $\cos^{-1} 1 = 0$, but $(\cos 1)^{-1} \approx (0.5403)^{-1} \approx 1.8508$.

65. See the answer section in the text.

66. Let x = 1. Then $\cot^{-1} 1 = \frac{\pi}{4} \approx 0.7854$, but $(\cot 1)^{-1} \approx (0.6421)^{-1} \approx 1.5574$.

67. See the answer section in the text.

68. Let x = 0. Then $(\cos^{-1} 0)^2 + (\sin^{-1} 0)^2 = \left(\frac{\pi}{2}\right)^2 + 0^2 \approx 2.4674 \neq 1$.

69. We see from the triangle in the text that $\tan \theta = \frac{\text{opposite side}}{\text{adjacent side}} = \frac{50}{b}$. That is, θ is the number whose tangent is $\frac{50}{b}$, and $-\frac{\pi}{2} < \theta < \frac{\pi}{2}$, so $\theta = \tan^{-1} \frac{50}{b}$.

70. $\sin \theta = \frac{2000}{h}$, $-\frac{\pi}{2} \leqslant \theta \leqslant \frac{\pi}{2}$, so $\theta = \sin^{-1} \left(\frac{2000}{h}\right)$.

71. a) Using a calculator, we find

$16 \tan^{-1} \frac{1}{5} - 4 \tan^{-1} \frac{1}{239} \approx 3.141592654$.

b) This expression seems to approximate π.

Exercise Set 7.7

1. Since 0.3 is in the interval [-1, 1], sin (arcsin 0.3) = 0.3.

2. 0.2

3. The domain of the inverse tangent function is $(-\infty, \infty)$, so $\tan [\tan^{-1} (-4.2)] = -4.2$

4. -1.5

5. $\arcsin \sin \frac{2\pi}{3}$

$= \arcsin \frac{\sqrt{3}}{2}$ $\left[\sin \frac{2\pi}{3} = \frac{\sqrt{3}}{2}\right]$

$= \frac{\pi}{3}$ [The restricted range of the arcsin function is $\left[-\frac{\pi}{2}, \frac{\pi}{2}\right]$]

Note: $\arcsin \sin \frac{2\pi}{3} \neq \frac{2\pi}{3}$ because $\frac{2\pi}{3}$ is not in the range of the arcsine function.

6. $\frac{\pi}{2}$

7. $\sin^{-1} \sin \left(-\frac{3\pi}{4}\right)$

$= \sin^{-1} \left(-\frac{\sqrt{2}}{2}\right)$ $\left[\sin \left(-\frac{3\pi}{4}\right) = -\frac{\sqrt{2}}{2}\right]$

$= -\frac{\pi}{4}$ [The restricted range of the arcsin function is $\left[-\frac{\pi}{2}, \frac{\pi}{2}\right]$]

Note: $\sin^{-1} \sin \left(-\frac{3\pi}{4}\right) \neq -\frac{3\pi}{4}$ because $-\frac{3\pi}{4}$ is not in the range of the arcsine function.

8. $\frac{\pi}{4}$

9. $\sin^{-1} \sin \frac{\pi}{5} = \frac{\pi}{5}$ because $\frac{\pi}{5}$ is in the range of the arcsine function.

10. $\frac{\pi}{7}$

11. $\tan^{-1} \tan \frac{2\pi}{3}$

$= \tan^{-1} (-\sqrt{3})$ $\left[\tan \frac{2\pi}{3} = -\sqrt{3}\right]$

$= -\frac{\pi}{3}$ [The restricted range of the arctangent function is $\left(-\frac{\pi}{2}, \frac{\pi}{2}\right)$.]

Note: $\tan^{-1} \tan \frac{2\pi}{3} \neq \frac{2\pi}{3}$ because $\frac{2\pi}{3}$ is not in the range of the arctangent function.

12. $\frac{2\pi}{3}$

13. $\sin \arctan \sqrt{3}$

$= \sin \frac{\pi}{3}$

$= \frac{\sqrt{3}}{2}$

14. $\frac{1}{2}$

15. $\cos \arcsin \frac{\sqrt{3}}{2}$

$= \cos \frac{\pi}{3}$

$= \frac{1}{2}$

16. $\frac{\sqrt{2}}{2}$

17. $\tan \cos^{-1} \frac{\sqrt{2}}{2}$

$= \tan \frac{\pi}{4}$

$= 1$

18. $\frac{\sqrt{3}}{3}$

19. $\cos^{-1} \sin \frac{\pi}{3}$

$= \cos^{-1} \frac{\sqrt{3}}{2}$

$= \frac{\pi}{6}$

20. $\frac{\pi}{2}$

21. $\arcsin \cos \frac{\pi}{6}$

$= \arcsin \frac{\sqrt{3}}{2}$

$= \frac{\pi}{3}$

22. $\frac{\pi}{4}$

23. $\sin^{-1} \tan \frac{\pi}{4}$

$= \sin^{-1} 1$

$= \frac{\pi}{2}$

24. $-\frac{\pi}{2}$

25. Find $\sin \left[\arctan \frac{x}{2}\right]$.

We draw two right triangles with sides x and 2 so that $\tan \theta = \frac{x}{2}$.

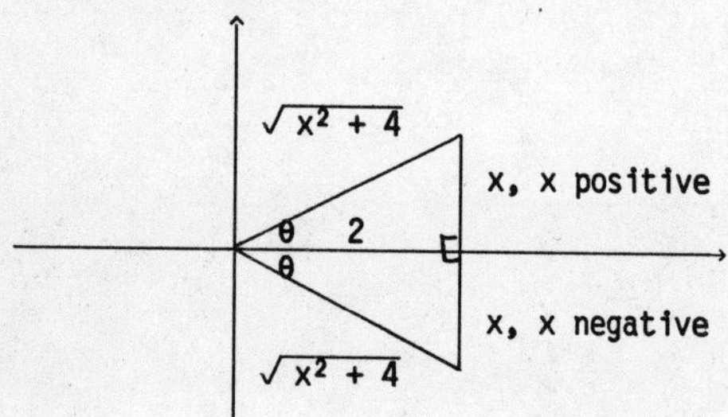

We find the length of the hypotenuse and then read off the sine ratio. In either case, we get

$$\sin \left[\arctan \frac{x}{2}\right] = \frac{x}{\sqrt{x^2 + 4}}.$$

26. $\frac{a}{\sqrt{a^2 + 9}}$

27. Find $\tan \left[\cos^{-1} \frac{3}{x}\right]$.

Since the restricted range of the inverse cosine function is $[0, \pi]$, we draw a first quadrant right triangle with one side 3 and hypotenuse x $(x > 0)$ so that $\cos \theta = \frac{3}{x}$.

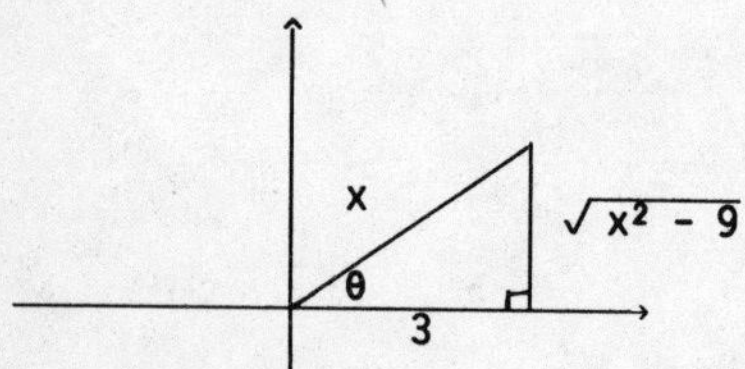

We find the length of the other leg and read off the tangent ratio.

$$\tan \left[\cos^{-1} \frac{3}{x}\right] = \frac{\sqrt{x^2 - 9}}{3}$$

28. $\frac{\sqrt{y^2 - 25}}{5}$

29. Find $\cot \left[\sin^{-1} \frac{a}{b}\right]$.

We draw two right triangles with one side a and hypotenuse of length b $(b > 0)$ so that $\sin \theta = \frac{a}{b}$.

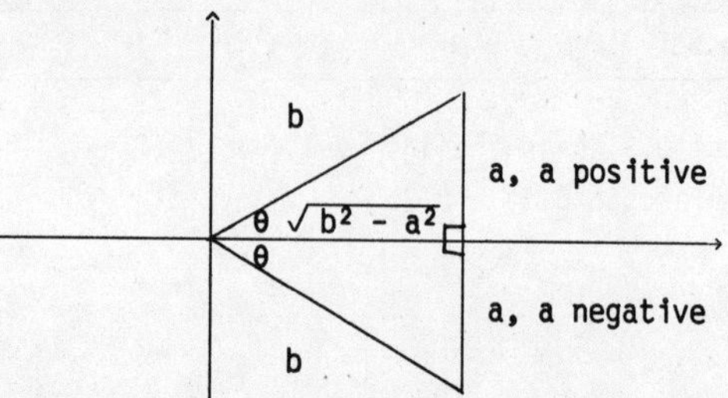

We find the length of the other side and then read off the cotangent ratio. In either case, we get

$$\cot \left[\sin^{-1} \frac{a}{b}\right] = \frac{\sqrt{b^2 - a^2}}{a}.$$

30. $\frac{\sqrt{q^2 - p^2}}{p}$

31. Find $\cos \tan^{-1} \frac{\sqrt{2}}{3}$.

We wish to find the cosine of an angle whose tangent is $\frac{\sqrt{2}}{3}$.

The length of the hypotenuse is $\sqrt{11}$.

Thus, $\cos \tan^{-1} \frac{\sqrt{2}}{3} = \frac{3}{\sqrt{11}}$.

32. $\cos \theta = \frac{4}{\sqrt{19}}$

33. Find tan arcsin 0.1

We wish to find the tangent of an angle whose sine is 0.1, or $\frac{1}{10}$.

The length of the other leg is $3\sqrt{11}$.

Thus, $\tan \arcsin 0.1 = \frac{1}{3\sqrt{11}}$.

34. $\frac{1}{2\sqrt{6}}$

35. Find $\cot \cos^{-1} (-0.2)$

We wish to find the cotangent of an angle whose cosine is -0.2.

The length of the other leg is $4\sqrt{6}$.

Thus, $\cot \cos^{-1} (-0.2) = \frac{-2}{4\sqrt{6}} = -\frac{1}{2\sqrt{6}}$.

36. $\frac{-3}{\sqrt{91}}$

37. Find sin (arccot y).

We draw two right triangles with sides y and 1 so that $\cot\theta = \frac{y}{1}$.

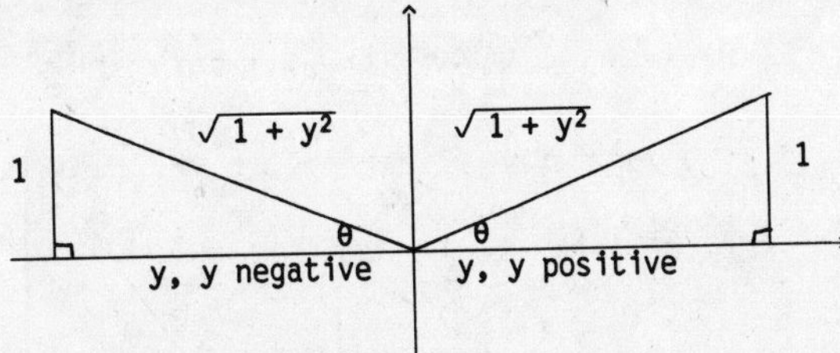

We find the length of the hypotenuse and then read off the sine ratio. In either case we get

$\sin(\text{arccot } y) = \frac{1}{\sqrt{1 + y^2}}$.

38. $\frac{1}{\sqrt{1 + x^2}}$

39. Find cos (arctan t).

We draw two right triangles with sides t and 1 so that $\tan\theta = \frac{t}{1}$.

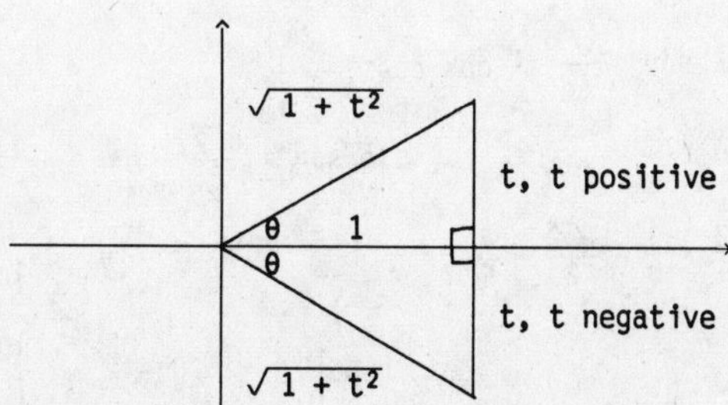

We find the length of the hypotenuse and then read off the cosine ratio. In either case we get

$\cos(\arctan t) = \frac{1}{\sqrt{1 + t^2}}$

40. $\frac{t}{\sqrt{1 + t^2}}$

41. Find cot ($\sin^{-1}$ y).

We draw two right triangles with one side y and hypotenuse 1 so that $\sin\theta = \frac{y}{1}$.

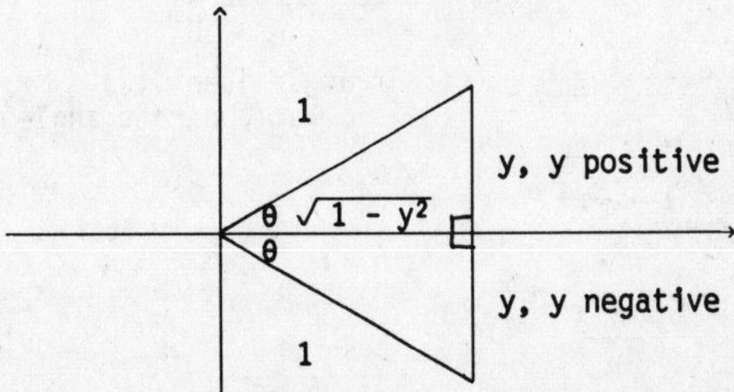

We find the length of the other side and then read off the cotangent ratio. In either case we get

$\cot(\sin^{-1} y) = \frac{\sqrt{1 - y^2}}{y}$.

42. $\frac{\sqrt{1 - y^2}}{y}$

43. Find sin ($\cos^{-1}$ x).

We draw two right triangles with one side x and hypotenuse 1 so that $\cos\theta = \frac{x}{1}$.

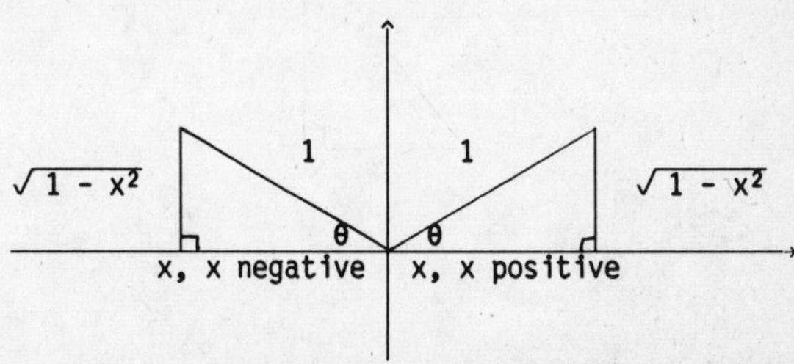

We find the length of the other side and then read off the sine ratio. In either case we get

$\sin(\cos^{-1} x) = \frac{\sqrt{1 - x^2}}{1}$, or $\sqrt{1 - x^2}$.

44. $\sqrt{1 - x^2}$

45. $\tan\left(\frac{1}{2}\arcsin\frac{4}{5}\right)$

$= \tan\frac{\theta}{2}$ $\left[\text{Letting } \theta = \arcsin\frac{4}{5}\right]$

$= \frac{\sin\theta}{1 + \cos\theta}$ (Half-angle identity)

$= \frac{\frac{4}{5}}{1 + \frac{3}{5}}$

$= \frac{\frac{4}{5}}{\frac{8}{5}}$

$= \frac{1}{2}$

5, 4, 3, θ

46. $2 - \sqrt{3}$

47. $\cos\left[\frac{1}{2}\arcsin\frac{1}{2}\right]$

$= \cos\frac{\theta}{2}$ $\quad\left[\text{Letting } \theta = \arcsin\frac{1}{2}\right]$

$= \sqrt{\frac{1 + \cos\theta}{2}}$ $\quad$ (Half-angle identity) (θ is a 1st quadrant angle)

$= \sqrt{\frac{1 + \frac{\sqrt{3}}{2}}{2}}$

$= \sqrt{\frac{2 + \sqrt{3}}{4}}$

$= \frac{\sqrt{2 + \sqrt{3}}}{2}$

48. $\frac{\sqrt{3}}{2}$

49. $\sin\left[2\cos^{-1}\frac{3}{5}\right]$

$= \sin 2\theta$ $\quad\left[\text{Letting } \theta = \cos^{-1}\frac{3}{5}\right]$

$= 2\sin\theta\cos\theta$ $\quad$ (Double-angle identity)

$= 2 \cdot \frac{4}{5} \cdot \frac{3}{5}$

$= \frac{24}{25}$

50. $\frac{\sqrt{3}}{2}$

51. $\cos\left[2\sin^{-1}\frac{5}{13}\right]$

$= \cos 2\theta$ $\quad\left[\text{Letting } \theta = \sin^{-1}\frac{5}{13}\right]$

$= 1 - 2\sin^2\theta$ $\quad$ (Double-angle identity)

$= 1 - 2 \cdot \left[\frac{5}{13}\right]^2$

$= 1 - \frac{50}{169}$

$= \frac{119}{169}$

52. $\frac{7}{25}$

53. $\sin\left[\sin^{-1}\frac{1}{2} + \cos^{-1}\frac{3}{5}\right]$

$= \sin(u + v)$ $\quad\left[\text{Letting } u = \sin^{-1}\frac{1}{2} \text{ and } v = \cos^{-1}\frac{3}{5}\right]$

$= \sin u\cos v + \cos u\sin v$ $\quad$ (Sum identity)

$= \sin\sin^{-1}\frac{1}{2} \cdot \cos\cos^{-1}\frac{3}{5} + \cos\sin^{-1}\frac{1}{2} \cdot \sin\cos^{-1}\frac{3}{5}$

$= \frac{1}{2} \cdot \frac{3}{5} + \frac{\sqrt{3}}{2} \cdot \frac{4}{5}$

$= \frac{3 + 4\sqrt{3}}{10}$

54. $\frac{4 - 3\sqrt{3}}{10}$

55. $\cos\left[\sin^{-1}\frac{\sqrt{2}}{2} + \cos^{-1}\frac{3}{5}\right]$

$= \cos(u + v)$ $\quad\left[\text{Letting } u = \sin^{-1}\frac{\sqrt{2}}{2} \text{ and } v = \cos^{-1}\frac{3}{5}\right]$

$= \cos u\cos v - \sin u\sin v$ $\quad$ (Sum identity)

$= \cos\sin^{-1}\frac{\sqrt{2}}{2} \cdot \cos\cos^{-1}\frac{3}{5} - \sin\sin^{-1}\frac{\sqrt{2}}{2} \cdot \sin\cos^{-1}\frac{3}{5}$

$= \frac{\sqrt{2}}{2} \cdot \frac{3}{5} - \frac{\sqrt{2}}{2} \cdot \frac{4}{5}$

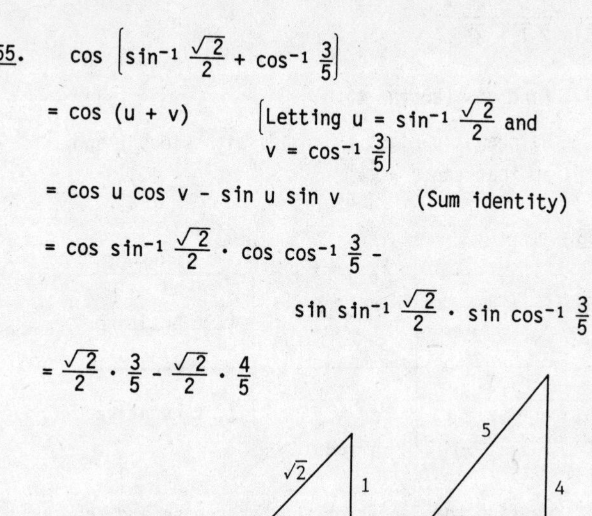

$= -\frac{\sqrt{2}}{10}$

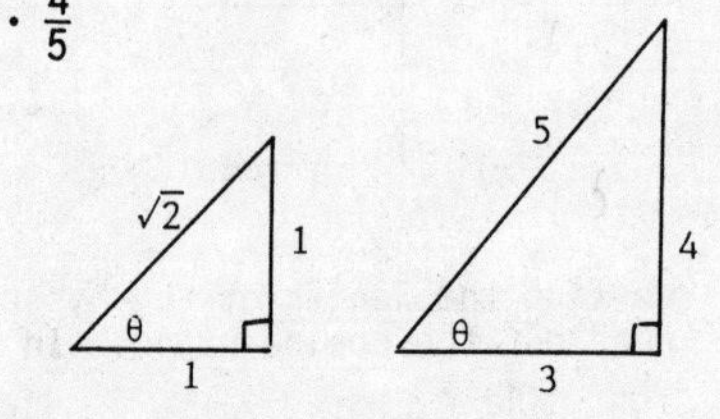

56. $\frac{3 + 4\sqrt{3}}{10}$

57.

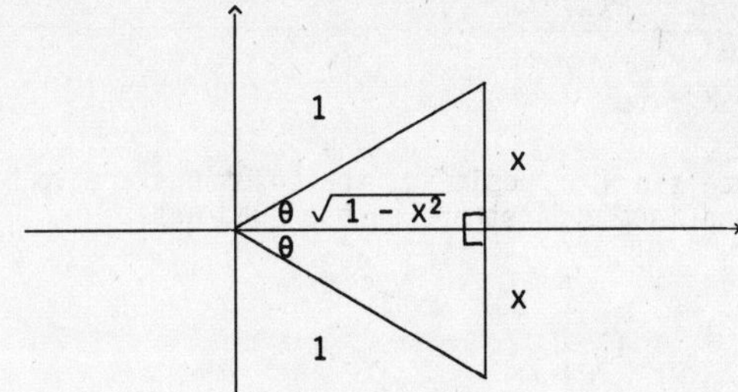

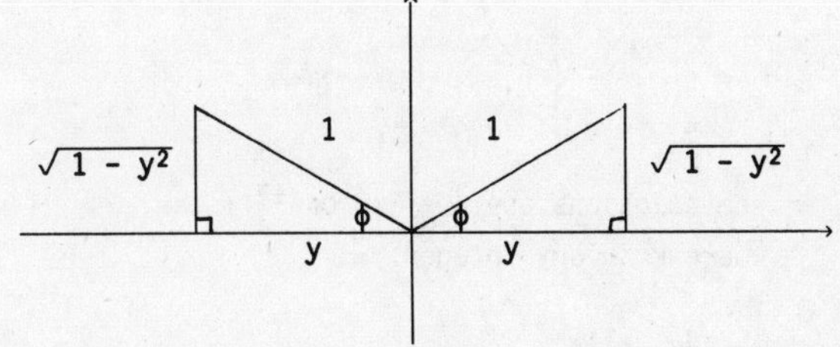

$\sin(\sin^{-1} x + \cos^{-1} y)$

$= \sin(u + v)$ (Letting $u = \sin^{-1} x$ and $v = \cos^{-1} y$)

$= \sin u \cos v + \cos u \sin v$ (Sum identity)

$= \sin \sin^{-1} x \cdot \cos \cos^{-1} y + \cos \sin^{-1} x \cdot \sin \cos^{-1} y$

$= x \cdot y + \sqrt{1 - x^2} \cdot \sqrt{1 - y^2}$

$= xy + \sqrt{(1 - x^2)(1 - y^2)}$

58. $xy - \sqrt{1 - x^2}\sqrt{1 - y^2}$

59. See the drawings in Exercise 57.

$\cos(\sin^{-1} x + \cos^{-1} y)$

$= \cos(u + v)$ (Letting $u = \sin^{-1} x$ and $v = \cos^{-1} y$)

$= \cos u \cos v - \sin u \sin v$ (Sum identity)

$= \cos \sin^{-1} x \cdot \cos \cos^{-1} y - \sin \sin^{-1} x \cdot \sin \cos^{-1} y$

$= \sqrt{1 - x^2} \cdot y - x \cdot \sqrt{1 - y^2}$

$= y\sqrt{1 - x^2} - x\sqrt{1 - y^2}$

60. $y\sqrt{1 - x^2} + x\sqrt{1 - y^2}$

61. $\sin(\sin^{-1} 0.6032 + \cos^{-1} 0.4621)$

$= \sin \sin^{-1} 0.6032 \cdot \cos \cos^{-1} 0.4621 + \cos \sin^{-1} 0.6032 \cdot \sin \cos^{-1} 0.4621$

$\approx 0.6032 \cdot 0.4621 + 0.7976 \cdot 0.8868$

$\approx 0.27874 + 0.70731$

≈ 0.9861

62. 0.9704

63.

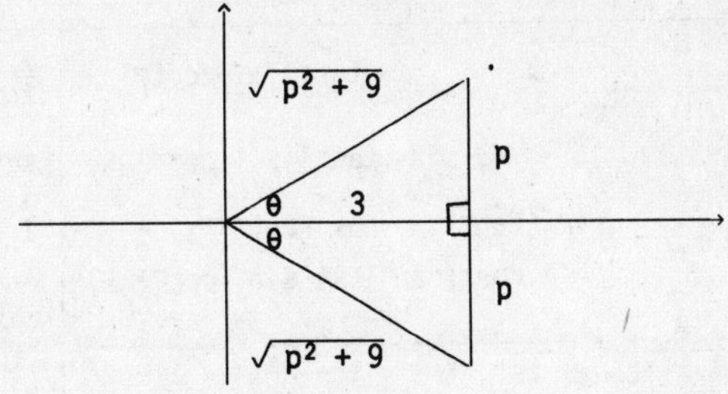

$\tan\left[\arcsin \frac{p}{\sqrt{p^2 + 9}}\right] = \frac{p}{3}$

64. $\frac{5}{\sqrt{25 - p^2}}$

65. See the first drawing in Exercise 57.

$\sin \theta = x$, $\cos \theta = \sqrt{1 - x^2}$, $\tan \theta = \frac{x}{\sqrt{1 - x^2}}$,

$\cos \theta = \frac{\sqrt{1 - x^2}}{x}$, $\sec \theta = \frac{1}{\sqrt{1 - x^2}}$, $\csc \theta = \frac{1}{x}$

66. $\sin \theta = \sqrt{1 - x^2}$, $\cos \theta = x$, $\tan \theta = \frac{\sqrt{1 - x^2}}{x}$,

$\cos \theta = \frac{x}{\sqrt{1 - x^2}}$, $\sec \theta = \frac{1}{x}$, $\csc \theta = \frac{1}{\sqrt{1 - x^2}}$

67.

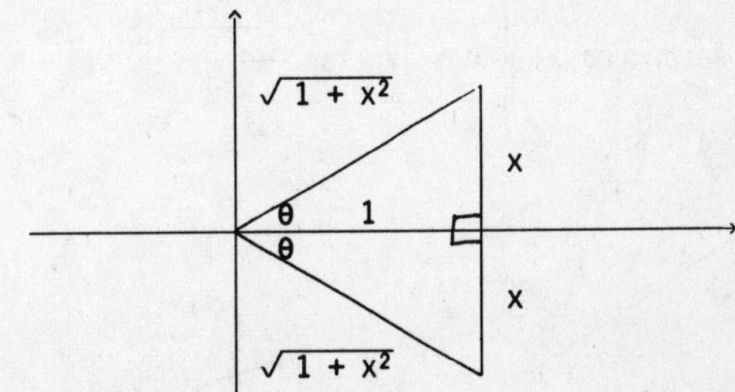

$\sin \theta = \frac{x}{\sqrt{1 + x^2}}$, $\cos \theta = \frac{1}{\sqrt{1 + x^2}}$, $\tan \theta = x$,

$\cot \theta = \frac{1}{x}$, $\sec \theta = \sqrt{1 + x^2}$, $\csc \theta = \frac{\sqrt{1 + x^2}}{x}$

68. $\sin \theta = \frac{1}{\sqrt{1 + x^2}}$, $\cos \theta = \frac{x}{\sqrt{1 + x^2}}$, $\tan \theta = \frac{1}{x}$,

$\cot \theta = x$, $\sec \theta = \frac{\sqrt{1 + x^2}}{x}$, $\csc \theta = \sqrt{1 + x^2}$

69. See the answer section in the text.

70.

$\tan^{-1} x + \cot^{-1} x$	$\frac{\pi}{2}$
$\sin(\tan^{-1} x + \cot^{-1} x)$	$\sin \frac{\pi}{2}$
$\sin(\tan^{-1} x)\cos(\cot^{-1} x) + \cos(\tan^{-1} x)\sin(\cot^{-1} x)$	1
$\frac{x}{\sqrt{1+x^2}} \cdot \frac{x}{\sqrt{1+x^2}} + \frac{1}{\sqrt{1+x^2}} \cdot \frac{1}{\sqrt{1+x^2}}$	
$\frac{x^2+1}{x^2+1}$	
1	

71. See the answer section in the text.

72.

$\tan^{-1} x$	$\sin^{-1} \frac{x}{\sqrt{x^2+1}}$
$\sin(\tan^{-1} x)$	$\sin\left[\sin^{-1} \frac{x}{\sqrt{x^2+1}}\right]$
$\frac{x}{\sqrt{x^2+1}}$	$\frac{x}{\sqrt{x^2+1}}$

73. See the answer section in the text.

74.

arccos x	$\arctan \frac{\sqrt{1-x^2}}{x}$
cos (arccos x)	$\cos\left[\arctan \frac{\sqrt{1-x^2}}{x}\right]$
x	x

75.

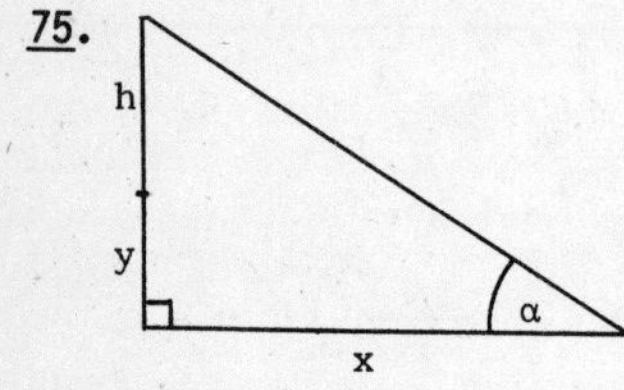

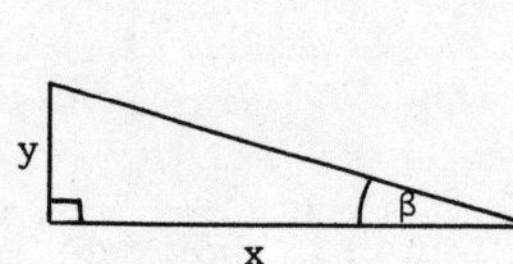

Let $\theta = \alpha - \beta$

$\tan \alpha = \frac{h+y}{x}$, $\alpha = \arctan \frac{h+y}{x}$

$\tan \beta = \frac{y}{x}$, $\beta = \arctan \frac{y}{x}$

Thus, $\theta = \arctan \frac{h+y}{x} - \arctan \frac{y}{x}$

76. 38.7°

Exercise Set 7.8

1. $\sin x = \frac{\sqrt{3}}{2}$

Since sin x is positive, the solutions are to be found in the first and second quadrants.

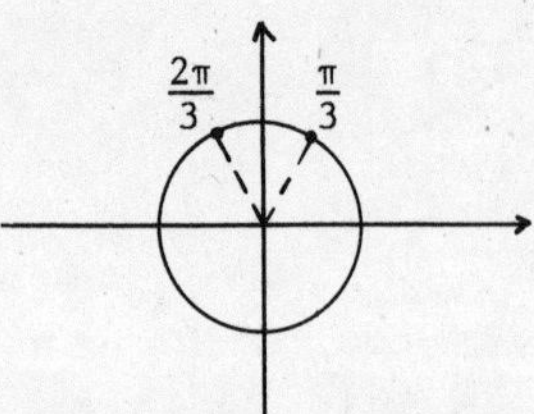

The solutions are $\frac{\pi}{3} + 2k\pi$ or $\frac{2\pi}{3} + 2k\pi$, where k is any integer.

2. $\frac{\pi}{6} + 2k\pi$, $\frac{11\pi}{6} + 2k\pi$

3. $\cos x = \frac{1}{\sqrt{2}}$, or $\frac{\sqrt{2}}{2}$

Since cos x is positive, the solutions are to be found in the first and fourth quadrants.

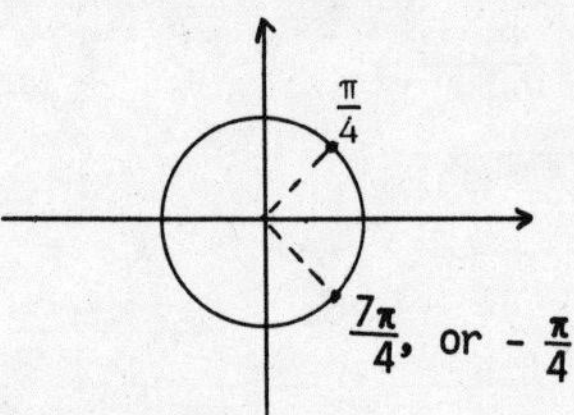

The solutions are $\frac{\pi}{4} + 2k\pi$ or $\frac{7\pi}{4} + 2k\pi$, where k is any integer; this can also be expressed as $\frac{\pi}{4} + 2k\pi$ or $-\frac{\pi}{4} + 2k\pi$ where k is any integer.

4. $\frac{\pi}{3} + k\pi$

5. sin x = 0.3448

Using a calculator, we find the reference angle is 20.17°, or 20° 10'. Since sin x is positive, the solutions are to be found in the first and second quadrants.

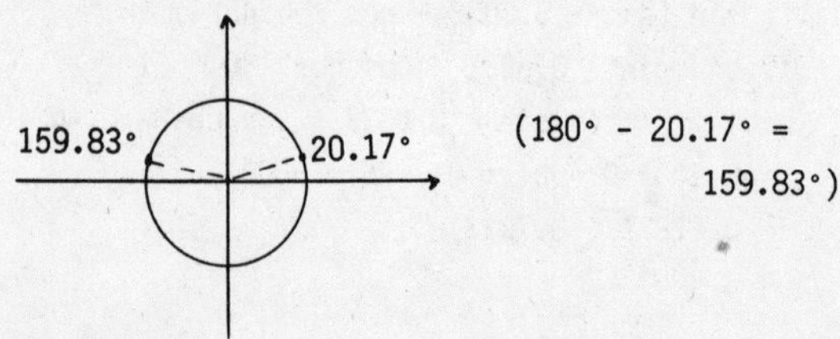

The solutions are 20.17° (or 20° 10') + k·360° or 159.83° (or 159° 50') + k·360°, where k is any integer.

6. 50.17° (or $50^\circ\ 10'$) + $k\cdot 360^\circ$,
309.83° (or $309^\circ\ 50'$) + $k\cdot 360^\circ$

7. $\cos x = -0.5495$

Using a calculator, we find the reference angle is 56.67°, or $56^\circ\ 40'$. Since cos x is negative, the solutions are in the second and third quadrants.

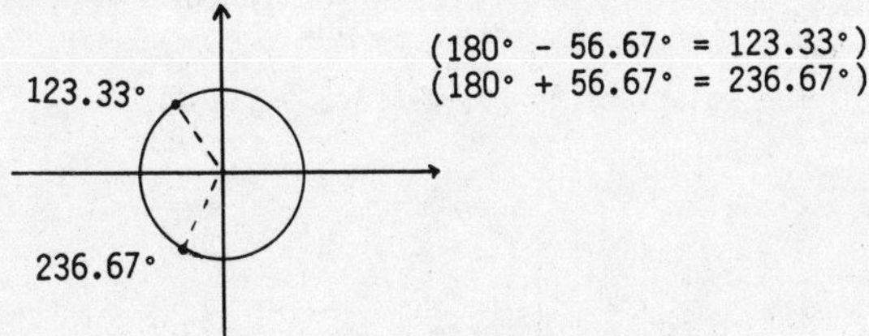

The solutions in $[0^\circ, 360^\circ)$ are 123.33° (or $123^\circ\ 20'$) and 236.67° (or $236^\circ\ 40'$).

8. 205.33° (or $205^\circ\ 20'$), 334.67° (or $334^\circ\ 40'$)

9. $2 \sin x + \sqrt{3} = 0$

$2 \sin x = -\sqrt{3}$

$\sin x = -\frac{\sqrt{3}}{2}$

There are two points on the unit circle for which the sine is $-\frac{\sqrt{3}}{2}$.

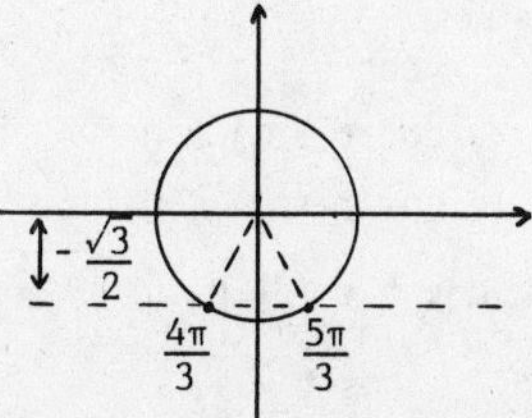

The solutions in $[0, 2\pi)$ are $\frac{4\pi}{3}$ and $\frac{5\pi}{3}$.

10. $\tan x = -\frac{1}{\sqrt{3}}$ gives $x = \frac{5\pi}{6}, \frac{11\pi}{6}$.

11. $2 \tan x + 3 = 0$

$2 \tan x = -3$

$\tan x = -1.5$

Using a calculator, we find the reference angle is 56.31°, or $56^\circ\ 19'$. Since tan x is negative, the solutions are in the second and fourth quadrants.

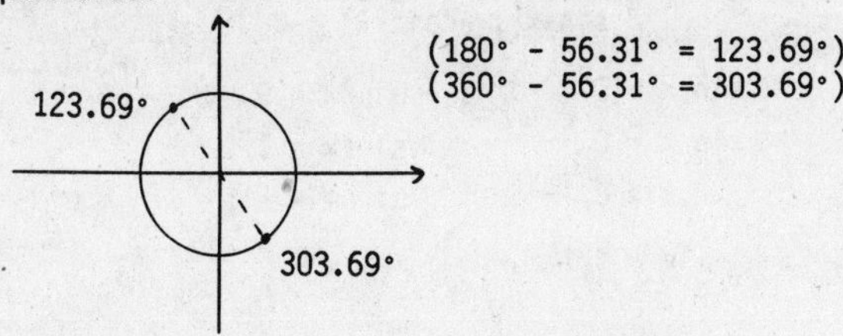

The solutions in $[0^\circ, 360^\circ)$ are 123.69° (or $123^\circ\ 41'$) and 303.69° (or $303^\circ\ 41'$).

12. 14.48° (or $14^\circ\ 29'$), 165.52° (or $165^\circ\ 31'$)

13. $4 \sin^2 x - 1 = 0$

$4 \sin^2 x = 1$

$\sin^2 x = \frac{1}{4}$

$\sin x = \pm \frac{1}{2}$

Use the unit circle to find those numbers having a sine of $\frac{1}{2}$ or $-\frac{1}{2}$.

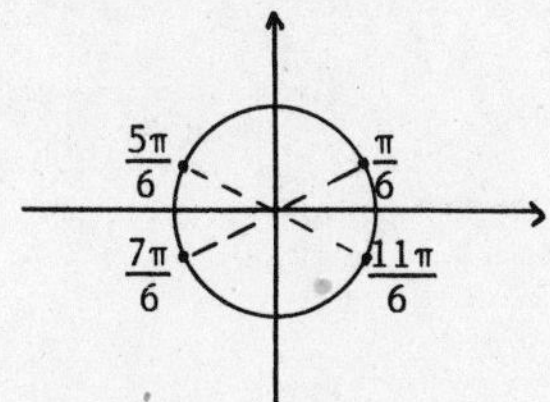

The solutions in $[0, 2\pi)$ are $\frac{\pi}{6}, \frac{5\pi}{6}, \frac{7\pi}{6}$, and $\frac{11\pi}{6}$.

14. $\frac{\pi}{4}, \frac{3\pi}{4}, \frac{5\pi}{4}, \frac{7\pi}{4}$

15. $\cot^2 x - 3 = 0$

$\cot^2 x = 3$

$\cot x = \pm \sqrt{3}$

Use the unit circle to find those numbers having a cotangent of $\sqrt{3}$ or $-\sqrt{3}$.

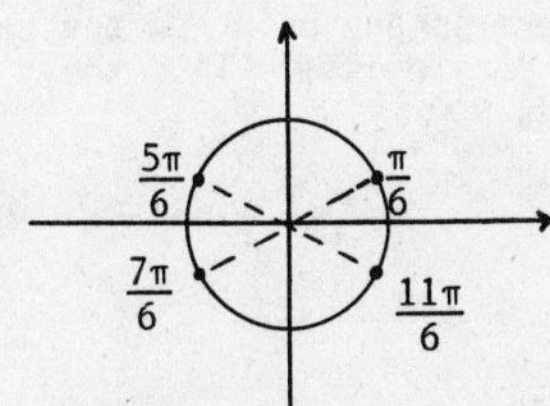

The solutions in $[0, 2\pi)$ are $\frac{\pi}{6}, \frac{5\pi}{6}, \frac{7\pi}{6}$, and $\frac{11\pi}{6}$.

16. $\frac{\pi}{6}, \frac{5\pi}{6}, \frac{7\pi}{6}, \frac{11\pi}{6}$

17.
$$2\sin^2 x + \sin x = 1$$
$$2\sin^2 x + \sin x - 1 = 0$$
$$(2\sin x - 1)(\sin x + 1) = 0$$

$2\sin x - 1 = 0$ or $\sin x + 1 = 0$

$2\sin x = 1$ or $\sin x = -1$

$\sin x = \frac{1}{2}$ or $\sin x = -1$

Use a unit circle to find those numbers having a sine of $\frac{1}{2}$ or -1.

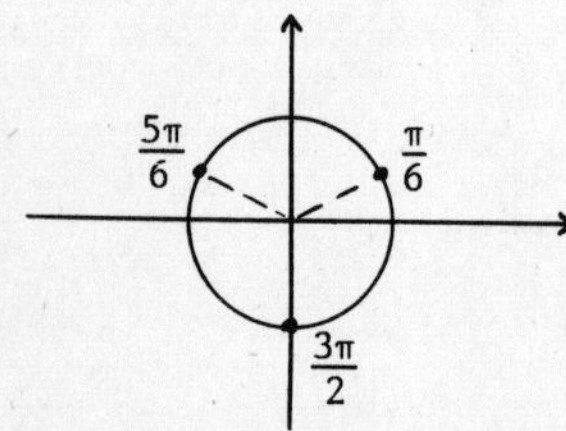

The solutions in $[0, 2\pi)$ are $\frac{\pi}{6}$, $\frac{5\pi}{6}$, and $\frac{3\pi}{2}$.

18. $\frac{2\pi}{3}, \frac{4\pi}{3}, \pi$

19.
$$\cos^2 x + 2\cos x = 3$$
$$\cos^2 x + 2\cos x - 3 = 0$$
$$(\cos x + 3)(\cos x - 1) = 0$$

$\cos x + 3 = 0$ or $\cos x - 1 = 0$

$\cos x = -3$ or $\cos x = 1$

Since cosines are never less than -1, $\cos x = -3$ has no solution. There is one point on the unit circle for which the cosine is 1. Thus the solution in $[0, 2\pi)$ is 0.

20. $\frac{3\pi}{2}$

21.
$$4\sin^3 x - \sin x = 0$$
$$\sin x(4\sin^2 x - 1) = 0$$
$$\sin x(2\sin x + 1)(2\sin x - 1) = 0$$

$\sin x = 0$ or $2\sin x + 1 = 0$

$\sin x = 0$ or $\sin x = -\frac{1}{2}$

$x = 0, \pi$ or $x = \frac{7\pi}{6}, \frac{11\pi}{6}$

or $2\sin x - 1 = 0$

or $\sin x = \frac{1}{2}$

or $x = \frac{\pi}{6}, \frac{5\pi}{6}$

The solutions in $[0, 2\pi)$ are 0, $\frac{\pi}{6}$, $\frac{5\pi}{6}$, π, $\frac{7\pi}{6}$, and $\frac{11\pi}{6}$.

22. $\frac{\pi}{2}, \frac{3\pi}{2}, \frac{\pi}{6}, \frac{11\pi}{6}$

23.
$$2\sin^2\theta + 7\sin\theta = 4$$
$$2\sin^2\theta + 7\sin\theta - 4 = 0$$
$$(2\sin\theta - 1)(\sin\theta + 4) = 0$$

$2\sin\theta - 1 = 0$ or $\sin\theta + 4 = 0$

$\sin\theta = \frac{1}{2}$ or $\sin\theta = -4$

$\theta = \frac{\pi}{6}, \frac{5\pi}{6}$ Since sines are never less than -1, $\sin\theta = -4$ has no solution.

The solutions in $[0, 2\pi)$ are $\frac{\pi}{6}$ and $\frac{5\pi}{6}$.

24. $\frac{\pi}{6}, \frac{5\pi}{6}$

25.
$$6\cos^2\phi + 5\cos\phi + 1 = 0$$
$$(3\cos\phi + 1)(2\cos\phi + 1) = 0$$

$3\cos\phi + 1 = 0$ or $2\cos\phi + 1 = 0$

$\cos\phi = -\frac{1}{3}$ or $\cos\phi = -\frac{1}{2}$

$\phi = 109.47°$ (or $109°28'$), $250.53°$ (or $250°32'$) or $\phi = 120°, 240°$

The solutions in $[0, 360°)$ are $109.47°$ (or $109°\ 28'$), $120°$, $240°$, $250.53°$ (or $250°\ 32'$).

26. $\frac{\pi}{6}, \frac{5\pi}{6}, \frac{3\pi}{2}$

27.
$$2\sin t\cos t + 2\sin t - \cos t - 1 = 0$$
$$2\sin t(\cos t + 1) - (\cot t + 1) = 0$$
$$(2\sin t - 1)(\cos t + 1) = 0$$

$2\sin t - 1 = 0$ or $\cos t + 1 = 0$

$\sin t = \frac{1}{2}$ or $\cos t = -1$

$t = \frac{\pi}{6}, \frac{5\pi}{6}$ or $t = \pi$

The solutions in $[0, 2\pi)$ are $\frac{\pi}{6}$, $\frac{5\pi}{6}$, and π.

28. $\frac{7\pi}{6}, \frac{11\pi}{6}, \frac{\pi}{4}, \frac{5\pi}{4}$

29.
$$\cos 2x\sin x + \sin x = 0$$
$$(1 - 2\sin^2 x)\sin x + \sin x = 0$$
$$\sin x - 2\sin^3 x + \sin x = 0$$
$$-2\sin^3 x + 2\sin x = 0$$
$$2\sin x(1 - \sin^2 x) = 0$$

$2\sin x = 0$ or $1 - \sin^2 x = 0$

$\sin x = 0$ or $\sin^2 x = 1$

$x = 0, \pi$ or $\sin x = \pm 1$

$x = 0, \pi$ or $x = \frac{\pi}{2}, \frac{3\pi}{2}$

The solutions in $[0, 2\pi)$ are 0, $\frac{\pi}{2}$, π, and $\frac{3\pi}{2}$.

30. $\frac{\pi}{2}, \frac{3\pi}{2}, \frac{\pi}{4}, \frac{5\pi}{4}$

31. $5 \sin^2 x - 8 \sin x = 3$

$5 \sin^2 x - 8 \sin x - 3 = 0$

$\sin x = \frac{8 \pm \sqrt{64 + 60}}{10}$ (Using the quadratic formula)

$= \frac{8 \pm \sqrt{124}}{10} = \frac{8 \pm 11.1355}{10}$

$\sin x = -0.3136$ or $\sin x = 1.9136$

Since sines are never greater than 1, the second equation has no solution. Using the first equation, we find the reference angle is 18.28°, or 18° 17'. Since sin x is negative, the solutions are to be found in the third and fourth quadrants. The solutions are 198.28° (or 198° 17') and 341.72° (or 341° 43').

32. 139.81° (or 139° 49'), 220.19° (or 220° 11')

33. $2 \tan^2 x = 3 \tan x + 7$

$2 \tan^2 x - 3 \tan x - 7 = 0$

$\tan x = \frac{3 \pm \sqrt{9 + 56}}{4}$ (Using the quadratic formula)

$= \frac{3 \pm \sqrt{65}}{4} = \frac{3 \pm 8.0623}{4}$

$\tan x = -1.2656$ or $\tan x = 2.7656$

Using the first equation, we find the reference angle is 51.69° (or 51° 41'). Since tan x is negative, the solutions will be found in the second and fourth quadrants. The solutions of tan x = -1.2656 are 128.31° (or 128° 19') and 308.31° (or 308° 19').

Using the second equation, we find the reference angle is 70.12° (or 70° 7'). Since tan x is positive, the solutions will be found in the first and third quadrants. The solutions of tan x = 2.7656 are 70.12° (70° 7') and 250.12° (or 250° 7').

34. 207.22° (or 207° 13'), 332.78° (or 332° 47')

35. $7 = \cot^2 x + 4 \cot x$

$0 = \cot^2 x + 4 \cot x - 7$

$\cot x = \frac{-4 \pm \sqrt{16 + 28}}{2}$ (Using the quadratic formula)

$= \frac{-4 \pm \sqrt{44}}{2} = \frac{-4 \pm 6.6332}{2}$

$\cot x = 1.3166$ or $\cot x = -5.3166$

Using the first equation, we find the reference angle is 37.22° (or 37° 13'). Since cot x is positive, the solutions will be found in the first and third quadrants. The solutions of cot x = 1.3166 are 37.22° (or 37° 13') and 217.22° (or 217° 13').

35. (continued)

Using the second equation, we find the reference angle is 10.65° (or 10° 39'). Since cot x is negative, the solutions will be found in the second and fourth quadrants. The solutions of cot x = -5.3166 are 169.35° (or 169° 21') and 349.35° (or 349° 21').

36. 50.89° (or 50° 53'), 230.89° (or 230° 53'), 117.80° (or 117° 48'), 297.80° (or 297° 48')

37. $|\sin x| = \frac{\sqrt{3}}{2}$

$\sin x = \pm \frac{\sqrt{3}}{2}$

$x = \frac{\pi}{3}, \frac{2\pi}{3}, \frac{4\pi}{3}, \frac{5\pi}{3}$

or 60°, 120°, 240°, 300°

38. $\frac{\pi}{3}, \frac{5\pi}{3}, \frac{2\pi}{3}, \frac{4\pi}{3}$

39. $\sqrt{\tan x} = \sqrt[4]{3}$

$(\tan x)^{1/2} = 3^{1/4}$

$\tan^2 x = 3$

$|\tan x| = \sqrt{3}$

$\tan x = \pm \sqrt{3}$

$x = \frac{\pi}{3}, \frac{2\pi}{3}, \frac{4\pi}{3}, \frac{5\pi}{3}$

When tan x is negative $\sqrt{\tan x}$ is undefined. Thus the only solutions are $\frac{\pi}{3}$ and $\frac{4\pi}{3}$ (or 60° and 240°).

40. 6.38°, 173.62°, 3.58°, 176.42°

41. $16 \cos^4 x - 16 \cos^2 x + 3 = 0$

$(4 \cos^2 x - 1)(4 \cos^2 x - 3) = 0$

$4 \cos^2 x - 1 = 0$ or $4 \cos^2 x - 3 = 0$

$\cos^2 x = \frac{1}{4}$ or $\cos^2 x = \frac{3}{4}$

$\cos x = \pm \frac{1}{2}$ or $\cos x = \pm \frac{\sqrt{3}}{2}$

$x = \frac{\pi}{3}, \frac{2\pi}{3}, \frac{4\pi}{3}, \frac{5\pi}{3}$ or $x = \frac{\pi}{6}, \frac{5\pi}{6}, \frac{7\pi}{6}, \frac{11\pi}{6}$

The solutions are

$\frac{\pi}{6}, \frac{\pi}{3}, \frac{2\pi}{3}, \frac{5\pi}{6}, \frac{7\pi}{6}, \frac{4\pi}{3}, \frac{5\pi}{3}$, and $\frac{11\pi}{6}$.

or 30°, 60°, 120°, 150°, 210°, 240°, 300°, and 330°.

42. 0

43. $e^{\sin x} = 1$

$\ln e^{\sin x} = \ln 1$

$\sin x = 0$

$x = 0, \pi$

44. $e^{3\pi/2}$

45. $e^{\ln \sin x} = 1$

$\sin x = 1$

$x = \frac{\pi}{2}$

46. a) See Exercise 69, Exercise Set 7.6.

b) 51.34° (or 51° 20')

47. a) We make a drawing.

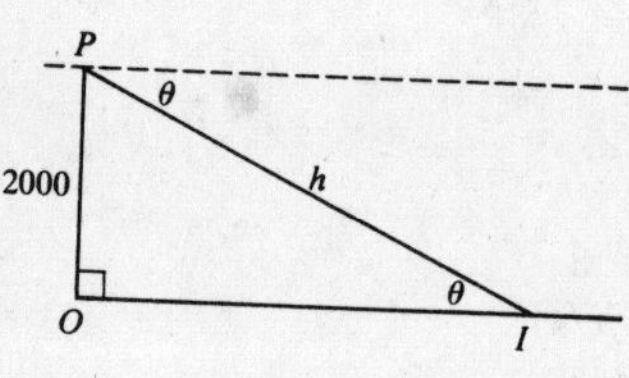

$\sin \theta = \frac{\text{opposite side}}{\text{hypotenuse}} = \frac{2000}{h}$

b) $\sin \theta = \frac{2000}{3000}$ (Substituting)

$\theta = \sin^{-1} \frac{2000}{3000} \approx \sin^{-1} 0.6667 \approx 41.81°$ (or 41° 49')

48. $0 + k\cdot 2\pi, \frac{\pi}{2} + k\cdot 2\pi, \frac{3\pi}{2} + k\cdot 2\pi$

Exercise Set 7.9

1. $\tan x \sin x - \tan x = 0$

$\tan x(\sin x - 1) = 0$

$\tan x = 0$ or $\sin x - 1 = 0$

$x = 0, \pi$ or $\sin x = 1$

$x = \frac{\pi}{2}$

The value $\frac{\pi}{2}$ does not check, but the other values do. Thus the solutions in $[0, 2\pi)$ are 0 and π.

2. $0, \pi, \frac{2\pi}{3}, \frac{4\pi}{3}$

3. $2 \sec x \tan x + 2 \sec x + \tan x + 1 = 0$

$2 \sec x(\tan x + 1) + (\tan x + 1) = 0$

$(2 \sec x + 1)(\tan x + 1) = 0$

$2 \sec x + 1 = 0$ or $\tan x + 1 = 0$

$\sec x = -\frac{1}{2}$ or $\tan x = -1$

$\cos x = -2$ $\quad x = \frac{3\pi}{4}, \frac{7\pi}{4}$

No solution

Both values check. The solutions in $[0, 2\pi)$ are $\frac{3\pi}{4}$ and $\frac{7\pi}{4}$.

4. $\frac{\pi}{3}, \frac{5\pi}{3}, \frac{\pi}{6}, \frac{5\pi}{6}$

5. $\sin 2x - \cos x = 0$

$2 \sin x \cos x - \cos x = 0$

$\cos x(2 \sin x - 1) = 0$

$\cos x = 0$ or $2 \sin x - 1 = 0$

$x = \frac{\pi}{2}, \frac{3\pi}{2}$ or $\sin x = \frac{1}{2}$

$x = \frac{\pi}{6}, \frac{5\pi}{6}$

All values check. The solutions in $[0, 2\pi)$ are $\frac{\pi}{6}, \frac{\pi}{2}, \frac{5\pi}{6}$, and $\frac{3\pi}{2}$.

6. $0, \pi, \frac{7\pi}{6}, \frac{11\pi}{6}$

7. $\sin 2x \sin x - \cos x = 0$

$2 \sin x \cos x \cdot \sin x - \cos x = 0$

$\cos x(2 \sin^2 x - 1) = 0$

$\cos x = 0$ or $2 \sin^2 x - 1 = 0$

$x = \frac{\pi}{2}, \frac{3\pi}{2}$ $\quad \sin^2 x = \frac{1}{2}$

$\sin x = \pm \frac{\sqrt{2}}{2}$

$x = \frac{\pi}{4}, \frac{3\pi}{4}, \frac{5\pi}{4}, \frac{7\pi}{4}$

All values check. The solutions in $[0, 2\pi)$ are $\frac{\pi}{4}, \frac{\pi}{2}, \frac{3\pi}{4}, \frac{5\pi}{4}, \frac{3\pi}{2}, \frac{7\pi}{4}$.

8. $0, \pi, \frac{\pi}{4}, \frac{7\pi}{4}, \frac{3\pi}{4}, \frac{5\pi}{4}$

9. $\sin 2x + 2 \sin x \cos x = 0$

$$\sin 2x + \sin 2x = 0$$
$$2 \sin 2x = 0$$
$$\sin 2x = 0$$
$$2x = 0, \pi, 2\pi, 3\pi$$
$$x = 0, \frac{\pi}{2}, \pi, \frac{3\pi}{2}$$

All values check. The solutions in $[0, 2\pi)$ are $0, \frac{\pi}{2}, \pi$, and $\frac{3\pi}{2}$.

10. $0, \pi, \frac{\pi}{2}, \frac{3\pi}{2}$

11.
$$\cos 2x \cos x + \sin 2x \sin x = 1$$
$$(1 - 2 \sin^2 x)\cos x + 2 \sin x \cos x \cdot \sin x = 1$$
$$(1 - 2 \sin^2 x)\cos x + 2 \sin^2 x \cos x = 1$$
$$\cos x = 1$$
$$x = 0$$

The value 0 checks. The only solution in $[0, 2\pi)$ is 0.

12. $0, \pi, \frac{\pi}{2}, \frac{3\pi}{2}$

13.
$$\sin 4x - 2 \sin 2x = 0$$
$$\sin 2(2x) - 2 \sin 2x = 0$$
$$2 \sin 2x \cos 2x - 2 \sin 2x = 0$$
$$\sin 2x(\cos 2x - 1) = 0$$

$\sin 2x = 0$ or $\cos 2x - 1 = 0$

$\sin 2x = 0$ or $\cos 2x = 1$

$2x = 0, \pi, 2\pi, 3\pi$ or $2x = 0, 2\pi$

$x = 0, \frac{\pi}{2}, \pi, \frac{3\pi}{2}$ $\quad x = 0, \pi$

All values check. The solutions in $[0, 2\pi)$ are $0, \frac{\pi}{2}, \pi$, and $\frac{3\pi}{2}$.

14. $0, \frac{\pi}{2}, \pi, \frac{3\pi}{2}$

15.
$$\sin 2x + 2 \sin x - \cos x - 1 = 0$$
$$2 \sin x \cos x + 2 \sin x - \cos x - 1 = 0$$
$$2 \sin x(\cos x + 1) - (\cos x + 1) = 0$$
$$(2 \sin x - 1)(\cos x + 1) = 0$$

$2 \sin x - 1 = 0$ or $\cos x + 1 = 0$

$\sin x = \frac{1}{2}$ or $\cos x = -1$

$x = \frac{\pi}{6}, \frac{5\pi}{6}$ or $x = \pi$

All values check. The solutions in $[0, 2\pi)$ are $\frac{\pi}{6}, \frac{5\pi}{6}$, and π.

16. $\frac{3\pi}{2}, \frac{2\pi}{3}, \frac{4\pi}{3}$

17. $\sec^2 x = 4 \tan^2 x$

$$\sec^2 x = 4(\sec^2 x - 1)$$
$$\sec^2 x = 4 \sec^2 x - 4$$
$$4 = 3 \sec^2 x$$
$$\frac{4}{3} = \sec^2 x$$
$$\pm \frac{2}{\sqrt{3}} = \sec x$$
$$\pm \frac{\sqrt{3}}{2} = \cos x$$
$$x = \frac{\pi}{6}, \frac{5\pi}{6}, \frac{7\pi}{6}, \frac{11\pi}{6}$$

All values check. The solutions in $[0, 2\pi)$ are $\frac{\pi}{6}, \frac{5\pi}{6}, \frac{7\pi}{6}$, and $\frac{11\pi}{6}$.

18. $\frac{\pi}{4}, \frac{5\pi}{4}, \frac{3\pi}{4}, \frac{7\pi}{4}$

19.
$$\sec^2 x + 3 \tan x - 11 = 0$$
$$(1 + \tan^2 x) + 3 \tan x - 11 = 0$$
$$\tan^2 x + 3 \tan x - 10 = 0$$
$$(\tan x + 5)(\tan x - 2) = 0$$

$\tan x + 5 = 0$ or $\tan x - 2 = 0$

$\tan x = -5$ or $\tan x = 2$

$x = 101.31°$ (or $101° 19'$), $281.31°$ (or $281° 19'$) or $x = 63.43°$ (or $63° 26'$), $243.43°$ (or $243° 26'$)

All values check. The solutions in $[0, 360°)$ are $63.43°$ (or $63° 26'$), $101.31°$ (or $101° 19'$), $243.43°$ (or $243° 26'$), and $281.31°$ (or $281° 19'$).

20. $45°, 225°, 116.57°$ (or $116° 34'$), $296.57°$ (or $296° 34'$).

21. $\cot x = \tan(2x - 3\pi)$

$\cot x = \dfrac{\tan 2x - \tan 3\pi}{1 + \tan 2x \tan 3\pi}$

$\cot x = \dfrac{\tan 2x - 0}{1 + \tan 2x \cdot 0}$

$\cot x = \tan 2x$

$\dfrac{1}{\tan x} = \dfrac{2 \tan x}{1 - \tan^2 x}$

(In this form of the equation, x cannot be $\pi/2$ and $3\pi/2$. Therefore $\pi/2$ and $3\pi/2$ must also be checked in the original equation.)

$1 - \tan^2 x = 2 \tan^2 x$

$1 = 3 \tan^2 x$

$\frac{1}{3} = \tan^2 x$

$\dfrac{1}{\sqrt{3}} = |\tan x|$

$\pm \dfrac{\sqrt{3}}{3} = \tan x$

$\frac{\pi}{6}, \frac{5\pi}{6}, \frac{7\pi}{6}, \frac{11\pi}{6} = x$

These values and also $\frac{\pi}{2}$ and $\frac{3\pi}{2}$ check. The solutions in $[0, 2\pi)$ are $\frac{\pi}{6}, \frac{\pi}{2}, \frac{5\pi}{6}, \frac{7\pi}{6}, \frac{3\pi}{2}$, and $\frac{11\pi}{6}$.

22. $\frac{\pi}{6}, \frac{7\pi}{7}, \frac{5\pi}{7}, \frac{11\pi}{7}$

23. $\cos(\pi - x) + \sin\left(x - \frac{\pi}{2}\right) = 1$

$(-\cos x) + (-\cos x) = 1$

$-2 \cos x = 1$

$\cos x = -\frac{1}{2}$

$x = \frac{2\pi}{3}, \frac{4\pi}{3}$

Both values check. The solutions in $[0, 2\pi)$ are $\frac{2\pi}{3}$ and $\frac{4\pi}{3}$.

24. $\frac{\pi}{6}, \frac{5\pi}{6}$

25. $\dfrac{\cos^2 x - 1}{\sin\left(\frac{\pi}{2} - x\right) - 1} = \dfrac{\sqrt{2}}{2} + 1$

$\dfrac{(\cos x + 1)(\cos x - 1)}{\cos x - 1} = \dfrac{\sqrt{2}}{2} + 1$

$\cos x + 1 = \dfrac{\sqrt{2}}{2} + 1$

$\cos x = \dfrac{\sqrt{2}}{2}$

$x = \frac{\pi}{4}, \frac{7\pi}{4}$

Both values check. The solutions in $[0, 2\pi)$ are $\frac{\pi}{4}$ and $\frac{7\pi}{4}$.

26. $\frac{\pi}{4}, \frac{3\pi}{4}$

27. $2 \cos x + 2 \sin x = \sqrt{6}$

$\cos x + \sin x = \dfrac{\sqrt{6}}{2}$

$\cos^2 x + 2 \sin x \cos x + \sin^2 x = \frac{6}{4}$

$\sin 2x + 1 = \frac{3}{2}$

$\sin 2x = \frac{1}{2}$

$2x = \frac{\pi}{6}, \frac{5\pi}{6}, \frac{13\pi}{6}, \frac{17\pi}{6}$

$x = \frac{\pi}{12}, \frac{5\pi}{12}, \frac{13\pi}{12}, \frac{17\pi}{12}$

The values $\frac{13\pi}{12}$ and $\frac{17\pi}{12}$ do not check, but the other values do. The solutions in $[0, 2\pi)$ are $\frac{\pi}{12}$ and $\frac{5\pi}{12}$.

28. $\frac{7\pi}{12}, \frac{23\pi}{12}$

29. $\sqrt{3} \cos x - \sin x = 1$

$\sqrt{3} \cos x = 1 + \sin x$

$3 \cos^2 x = 1 + 2 \sin x + \sin^2 x$

$3(1 - \sin^2 x) = 1 + 2 \sin x + \sin^2 x$

$3 - 3 \sin^2 x = 1 + 2 \sin x + \sin^2 x$

$0 = 4 \sin^2 x + 2 \sin x - 2$

$0 = 2 \sin^2 x + \sin x - 1$

$0 = (2 \sin x - 1)(\sin x + 1)$

$2 \sin x - 1 = 0$ or $\sin x + 1 = 0$

$\sin x = \frac{1}{2}$ or $\sin x = -1$

$x = \frac{\pi}{6}, \frac{5\pi}{6}$ or $x = \frac{3\pi}{2}$

The value $\frac{5\pi}{6}$ does not check, but the other values do check. The solutions in $[0, 2\pi)$ are $\frac{\pi}{6}$ and $\frac{3\pi}{2}$.

30. $\frac{7\pi}{4}$

31. $\sec^2 x + 2 \tan x = 6$

$1 + \tan^2 x + 2 \tan x = 6$

$\tan^2 x + 2 \tan x - 5 = 0$

$\tan x = \frac{-2 \pm \sqrt{4 + 20}}{2} = \frac{-2 \pm \sqrt{24}}{2} = \frac{-2 \pm 4.8990}{2}$

$\tan x = 1.4495$ or $\tan x = -3.4495$

Using the first equation, we find the reference angle is 0.9669 (radians). Thus the solutions of $\tan x = 1.4495$ are 0.9669 and $\pi + 0.9669$, or 4.1085.

Using the second equation, we find the reference angle is 1.2886. Thus the solutions of $\tan x = -3.4495$ are $\pi - 1.2886$, or 1.8530, and $2\pi - 1.2886$, or 4.9946.

All four values check.

32. 0.8639, 4.0055, 2.9724, 6.1140

33. $3 \cos 2x + \sin x = 1$

$3(1 - 2 \sin^2 x) + \sin x = 1$

$3 - 6 \sin^2 x + \sin x = 1$

$0 = 6 \sin^2 x - \sin x - 2$

$0 = (3 \sin x - 2)(2 \sin x + 1)$

$3 \sin x - 2 = 0$ or $2 \sin x + 1 = 0$

$\sin x = \frac{2}{3}$ or $\sin x = -\frac{1}{2}$

$x = 0.7297, 2.4119$ or $x = \frac{7\pi}{6}, \frac{11\pi}{6}$

(or 3.6652, 5.7596)

All four values check.

34. 0.3792, 2.7624, 3.4152, 6.0096

35. $T(t) = 101.6° + 3° \sin \left[\frac{\pi}{8} t\right], 0 \leqslant t \leqslant 12$

$103 = 101.6 + 3 \sin \left[\frac{\pi}{8} t\right]$ (Substituting)

$1.4 = 3 \sin \left[\frac{\pi}{8} t\right]$

$0.4667 \approx \sin \frac{\pi}{8} t$

$\frac{\pi}{8} t \approx 0.4855, 2.6561$ $\left[0 \leqslant t \leqslant 12, \text{ so } 0 \leqslant \frac{\pi}{8} t \leqslant \frac{3\pi}{2}\right]$

$t \approx 1.24, 6.76$

Both values check.

The patient's temperature was 103° for $t \approx 1.24$ days and $t \approx 6.76$ days.

36. 23.28 min, 113.28 min, 203.28 min

37. $N(\phi) = 6066 - 31 \cos 2\phi$

We consider ϕ in the interval [0°, 90°] since we want latitude north.

$6040 = 6066 - 31 \cos 2\phi$ (Substituting)

$-26 = -31 \cos 2\phi$

$0.8387 \approx \cos 2\phi$ $(0° \leqslant 2\phi \leqslant 180°)$

$2\phi \approx 33.0°$

$\phi \approx 16.5°$

The value checks.

At about 16.5° N the length of a British nautical mile is found to be 6040 ft.

38. 37.95615° N

39. Sketch a triangle having an angle θ whose cosine is $\frac{3}{5}$. Using the Pythagorean theorem we know the other leg has length 4. Then $\sin \theta = \frac{4}{5}$. Thus, $\arccos \frac{3}{5} = \arcsin \frac{4}{5}$.

$\arccos x = \arccos \frac{3}{5} - \arcsin \frac{4}{5}$

$\arccos x = 0$

$\cos 0 = x$

$1 = x$

40. $\frac{\sqrt{2}}{2}$

41. First graph $y = \sin x - \cos x$. Then graph $y = \cot x$ on the same set of axes and find the points of intersection of the graphs. The solutions are 1.15, 5.65, -0.63, -0.513, etc.

42. ± 1.12, ± 2.77, ± 6.44, ± 9.32, etc.

43. $\sin x = 5 \cos x$

$\frac{\sin x}{\cos x} = 5$

$\tan x = 5$

$x \approx 1.3734$ (Finding x in quadrant I)

Then $\sin x \cos x = \sin (1.3734) \cos (1.3734)$

$\approx 0.1923.$

Exercise Set 8.1

1.

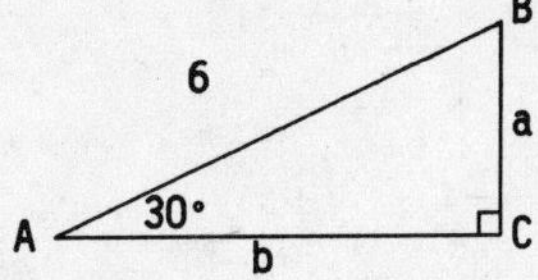

To solve this triangle find B, a, and b.

$B = 90° - 30° = 60°$

$\frac{a}{6} = \sin 30°$

$a = 6 \sin 30°$

$a = 6 \times \frac{1}{2}$

$a = 3$

$\frac{b}{6} = \cos 30°$

$b = 6 \cos 30°$

$b = 6 \times \frac{\sqrt{3}}{2}$

$b = 3\sqrt{3} \approx 5.20$

2. $B = 30°$, $b = \frac{27}{\sqrt{3}} \approx 15.59$, $C = \frac{54}{\sqrt{3}} \approx 31.18$

3.

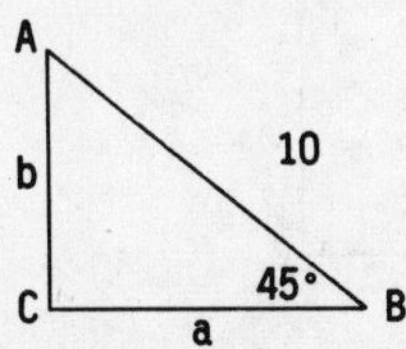

To solve this triangle find A, a, and b.

$A = 90° - 45° = 45°$

$\frac{a}{10} = \cos 45°$

$a = 10 \cos 45°$

$a = 10 \times \frac{\sqrt{2}}{2}$

$a = 5\sqrt{2} \approx 7.07$

$\frac{b}{10} = \sin 45°$

$b = 10 \sin 45°$

$b = 10 \times \frac{\sqrt{2}}{2}$

$b = 5\sqrt{2} \approx 7.07$

4. $B = 45°$, $a = 3$, $c = 3\sqrt{2} \approx 4.24$

5.

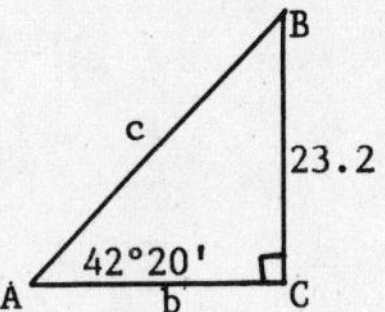

To solve this triangle find B, b, and c.

$B = 90° - 42°\ 20' = 47°\ 40'$

$\frac{b}{23.2} = \cot 42°\ 20'$

$b = 23.2 \cot 42°\ 20'$

$b = 23.2 \times 1.0977$

$b = 25.5$

$\frac{c}{23.2} = \csc 42°\ 20'$

$c = 23.2 \csc 42°\ 20'$

$c = 23.2 \times 1.4849$

$c = 34.4$

6. $B = 61°\ 30'$, $b = 31.9$, $c = 36.3$

7.

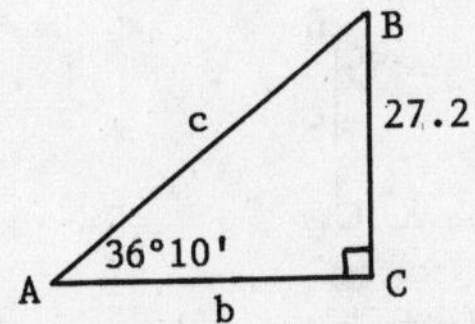

To solve this triangle find B, b, and c.

$B = 90° - 36°\ 10' = 53°\ 50'$

$\frac{b}{27.2} = \cot 36°\ 10'$

$b = 27.2 \cot 36°\ 10'$

$b = 27.2 \times 1.3680$

$b = 37.2$

$\frac{c}{27.2} = \csc 36°\ 10'$

$c = 27.2 \csc 36°\ 10'$

$c = 27.2 \times 1.6945$

$c = 46.1$

8. $B = 2°\ 20'$, $b = 0.40$, $c = 9.74$

9.

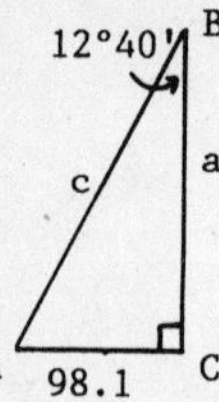

To solve this triangle find A, a, and c.
$A = 90° - 12° 40' = 77° 20'$

$\frac{a}{98.1} = \cot 12° 40'$

$a = 98.1 \cot 12° 40'$
$a = 98.1 \times 4.4494$
$a = 436.5$

$\frac{c}{98.1} = \csc 12° 40'$

$c = 98.1 \csc 12° 40'$
$c = 98.1 \times 4.5604$
$c = 447.4$

10. $A = 20° 10'$, $a = 46.6$, $c = 135$

11.

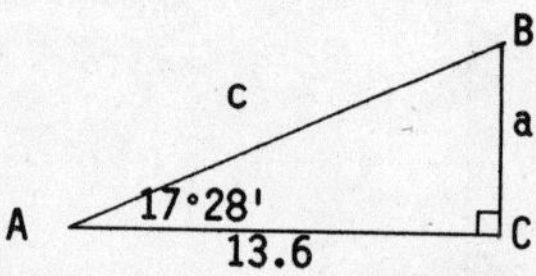

To solve this triangle find B, a, and c.
$B = 90° - 17° 28' = 72° 32'$

$\frac{a}{13.6} = \tan 17° 28'$

$a = 13.6 \tan 17° 28'$
$a = 13.6 \times 0.3147$
$a = 4.3$

$\frac{c}{13.6} = \sec 17° 28'$

$c = 13.6 \sec 17° 28'$
$c = 13.6 \times 1.048$
$c = 14.3$

12. $B = 11° 18'$, $a = 6706$, $c = 6839$

13.

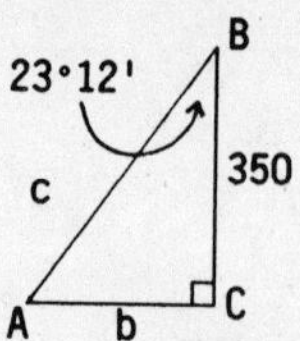

To solve this triangle find A, b, and c.
$A = 90° - 23° 12' = 66° 48'$

$\frac{b}{350} = \tan 23° 12'$

$b = 350 \tan 23° 12'$
$b = 350 \times 0.4286$
$b = 150$

$\frac{c}{350} = \sec 23° 12'$

$c = 350 \sec 23° 12'$
$c = 350 \times 1.088$
$c = 381$

14. $A = 20° 38'$, $b = 637$, $c = 681$

15.

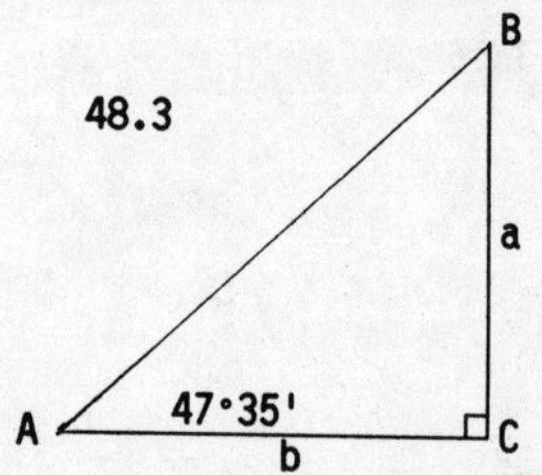

To solve this triangle find B, a, and b.
$B = 90° - 47° 35' = 42° 25'$

$\frac{a}{48.3} = \sin 47° 35'$

$a = 48.3 \sin 47° 35'$
$a = 48.3 \times 0.7383$
$a = 35.7$

$\frac{b}{48.3} = \cos 47° 35'$

$b = 48.3 \cos 47° 35'$
$b = 48.3 \times 0.6745$
$b = 32.6$

16. $B = 1° 5'$, $a = 3949$, $b = 75$

17.

To solve this triangle find A, a, and b.

$A = 90° - 82° 20' = 7° 40'$

$\frac{a}{0.982} = \cos 82° 20'$

$a = 0.982 \cos 82° 20'$

$a = 0.982 \times 0.1334$

$a = 0.131$

$\frac{b}{0.982} = \sin 82° 20'$

$b = 0.982 \sin 82° 20'$

$b = 0.982 \times 0.9911$

$b = 0.973$

18. $A = 33° 30'$, $a = 0.0247$, $b = 0.0372$

19.

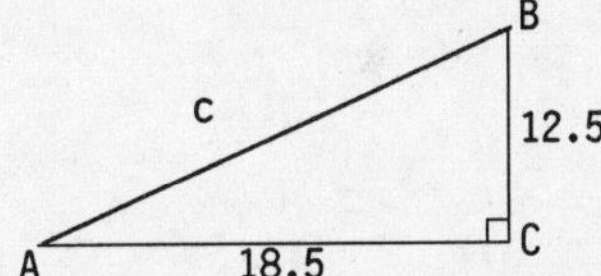

To solve this triangle find c, A, and B.

$c^2 = (12.5)^2 + (18.5)^2$

$c^2 = 156.25 + 342.25$

$c^2 = 498.5$

$c = 22.3$

$\tan A = \frac{12.5}{18.5} \approx 0.6757$

$A = 34.05°$, or $34° 3'$

$B = 90° - 34.05° = 55.95°$, or $55° 57'$

20. $A = 26.56°$, or $26° 34'$; $B = 63.44°$, or $63° 26'$; $c = 22.8$

21.

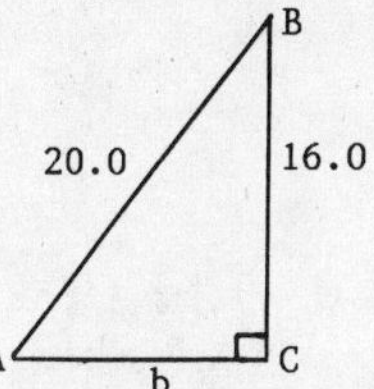

To solve this triangle find b, A, and B.

$b^2 = (20.0)^2 - (16.0)^2$

$b^2 = 400 - 256$

$b^2 = 144$

$b = 12$

$\sin A = \frac{16.0}{20.0} = 0.8000$

Thus $A = 53.13°$, or $53° 8'$

$B = 90° - 53.13° = 36.87°$, or $36° 52'$

22. $b = 42.4$, $A = 19.47°$, or $19° 28'$, $B = 70.53°$, or $70° 32'$

23.

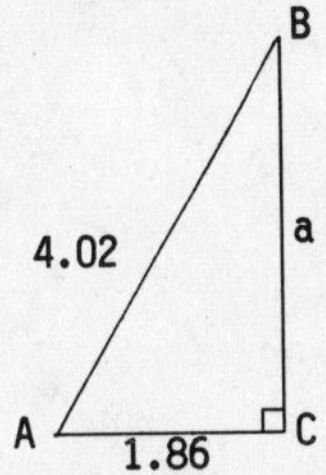

To solve this triangle find a, A, and B.

$a^2 + (1.86)^2 = (4.02)^2$

$a^2 + 3.4596 = 16.1604$

$a^2 = 12.7008$

$a = 3.56$

$\cos A = \frac{1.86}{4.02} \approx 0.4627$

$A = 62.44°$, or $62° 26'$

$B = 90° - 62.44° = 27.56°$ or $27° 34'$

24. $a = 439$, $A = 77.16°$, or $77° 10'$, $B = 12.84°$, or $12° 50'$

25.

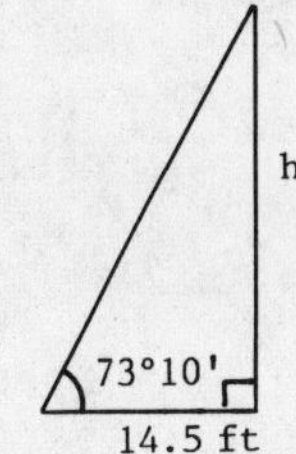

$\frac{h}{14.5} = \tan 73° 10'$

$h = 14.5 \tan 73° 10'$

$h = 14.5 \times 3.3052$

$h = 47.9$ ft

26. 9.59 ft

27.

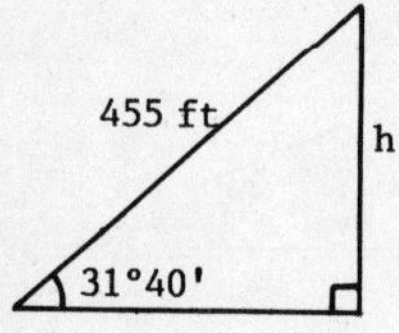

$\frac{h}{455} = \sin 31° 40'$

$h = 455 \sin 31° 40'$

$h = 455 \times 0.5250$

$h = 239$ ft

28. 171 ft

29.

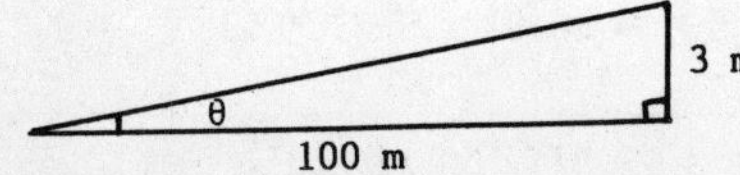

$\tan \theta = \frac{3}{100} = 0.03$

Thus $\theta = 1.72°$, or $1° 43'$.

30. 10.32°, or 10° 19'

31.

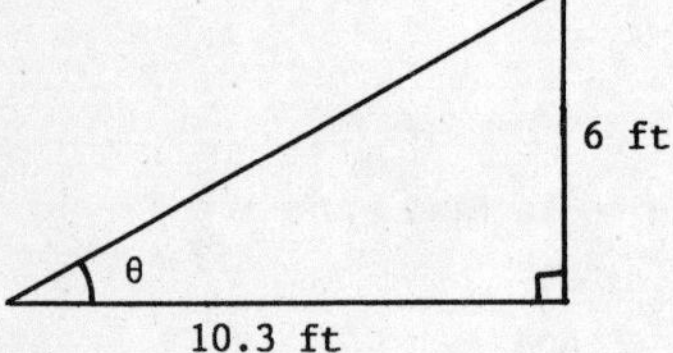

$\tan \theta = \frac{6}{10.3} = 0.5825$

Thus $\theta = 30.22°$, or $30° 13'$.

32. 60.26°, or 60° 15'

33.

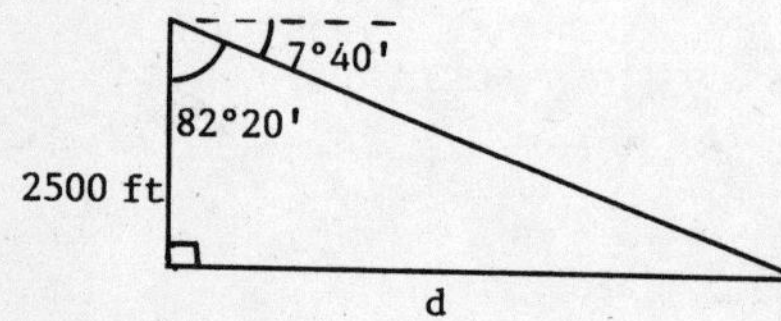

$\frac{d}{2500} = \tan 82° 20'$

$d = 2500 \tan 82° 20'$

$d = 2500 \times 7.4287$

$d = 18{,}571$ ft, or 3.52 mi

34. 274 ft

35.

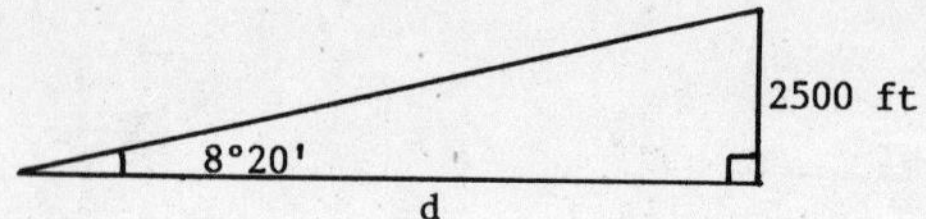

$\frac{d}{2500} = \cot 8° 20'$

$d = 2500 \cot 8° 20'$

$d = 2500 \times 6.8269$

$d = 17{,}067$ ft, or 3.23 mi

36. 328 ft

37.

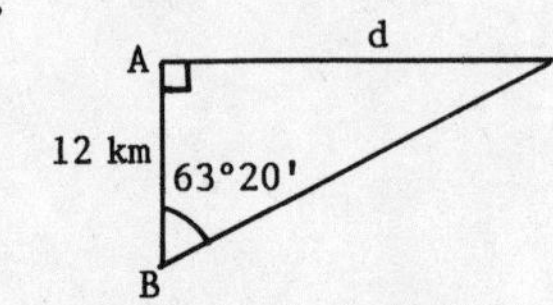

$\frac{d}{12} = \tan 63° 20'$

$d = 12 \tan 63° 20'$

$d = 12 \times 1.9912$

$d = 23.9$ km

38. 19.3 km

39.

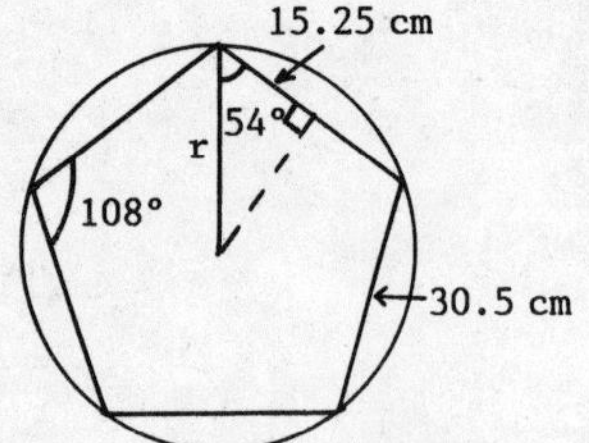

$\frac{r}{15.25} = \sec 54°$

$r = 15.25 \sec 54°$

$r = 15.25 \times 1.7013$

$r = 25.9$ cm

40. 36.4 cm

41.

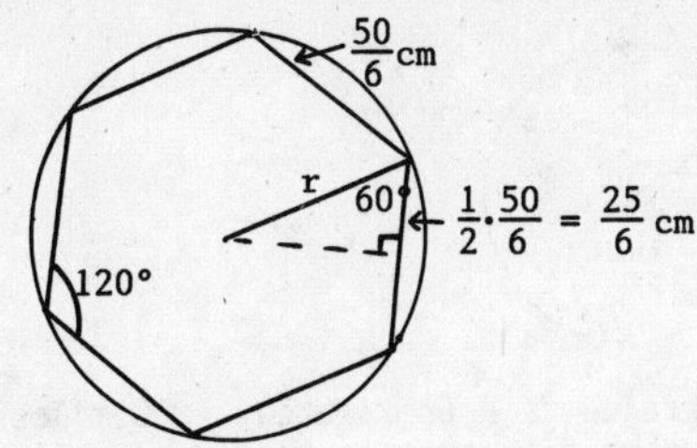

$$\frac{r}{\frac{25}{6}} = \sec 60°$$

$$r = \frac{25}{6} \sec 60°$$

$$r = \frac{25}{6} \times 2$$

$$r = \frac{25}{3}, \text{ or } 8.33 \text{ cm}$$

42. 96.7 cm

43.

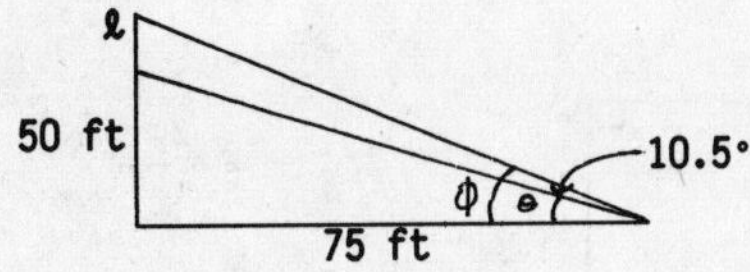

$$\tan \theta = \frac{50}{75} \approx 0.6667$$

$$\theta = 33.7°$$

$$\phi = \theta + 10.5° = 33.7° + 10.5° = 44.2°$$

$$\tan 44.2° = \frac{50 + \ell}{75}$$

$$75 \tan 44.2° = 50 + \ell$$

$$75 \times 0.9721 = 50 + \ell$$

$$72.9 = 50 + \ell$$

$$22.9 = \ell$$

The length of the antenna is 22.9 ft.

44. 44.9 ft

45.

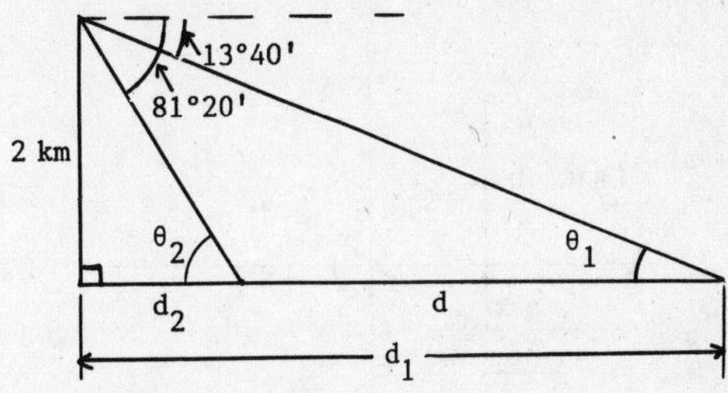

The distance to find is d, which is $d_1 - d_2$.

$$\frac{d_1}{2} = \cot \theta_1 \qquad \frac{d_2}{2} = \cot \theta_2$$

$$d_1 = 2 \cot 13° \, 40' \qquad d_2 = 2 \cot 81° \, 20'$$

$$\begin{aligned} d = d_1 - d_2 &= 2 \cot 13° \, 40' - 2 \cot 81° \, 20' \\ &= 2(\cot 13° \, 40' - \cot 81° \, 20') \\ &= 2(4.1126 - 0.1524) \\ &= 7.92 \text{ km} \end{aligned}$$

46. 4671 m

47.

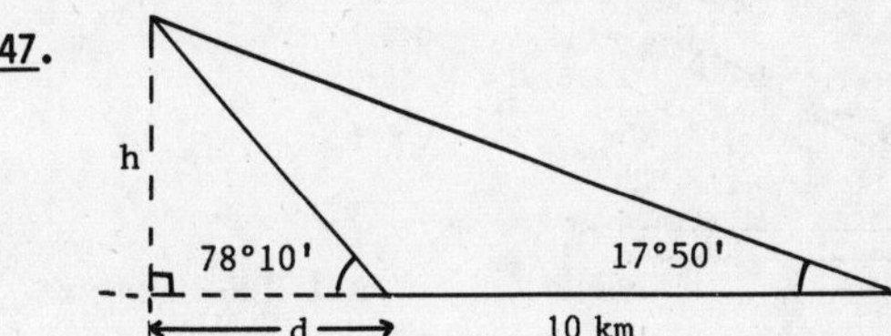

First find d. Then solve for h.

$$\frac{h}{d} = \tan 78° \, 10' \qquad \frac{h}{d + 10} = \tan 17° \, 50'$$

$$h = d \tan 78° \, 10' \qquad h = (d + 10) \tan 17° \, 50'$$

$$d \tan 78° \, 10' = d \tan 17° \, 50' + 10 \tan 17° \, 50'$$

$$d(\tan 78° \, 10' - \tan 17° \, 50') = 10 \tan 17° \, 50'$$

$$d = \frac{10 \tan 17° \, 50'}{\tan 78° \, 10' - \tan 17° \, 50'}$$

$$d = \frac{10 \times 0.3217}{4.7729 - 0.3217}$$

$$d = \frac{3.217}{4.4512}$$

$$d = 0.7227$$

$$\frac{h}{0.7227} = \tan 78° \, 10'$$

$$h = 0.7227 \tan 78° \, 10'$$

$$h = 0.7227(4.7729)$$

$$h = 3.45 \text{ km}$$

48. 225 ft

49.

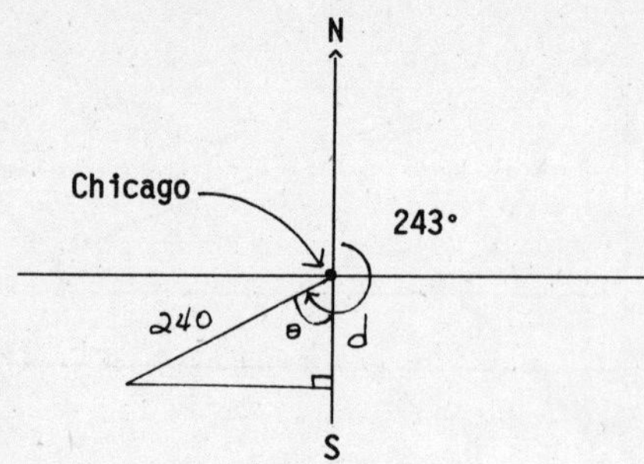

At 120 km/h, in 2 hr the plane travels 120·2, or 240 km.

$\theta = 243° - 180° = 63°$

$\frac{d}{240} = \cos 63°$

$d = 240 \cos 63°$

$d = 240 \times 0.4540$

$d \approx 109$ km

50. 201 km

51.

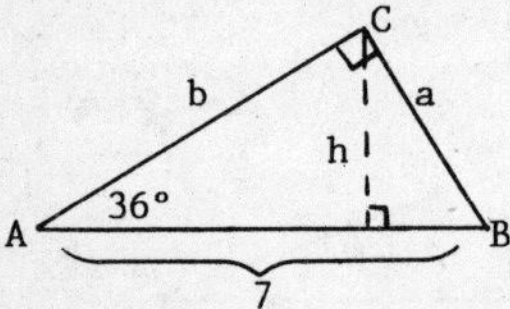

$\frac{b}{7} = \cos 36°$

$b = 7 \cos 36°$

$b = 7 \times 0.8090$

$b = 5.66$

$\frac{h}{5.66} = \sin 36°$

$h = 5.66 \sin 36°$

$h = 5.66 \times 0.5878$

$h = 3.33$

52. 5.88

53. See the answer section in the text.

54.

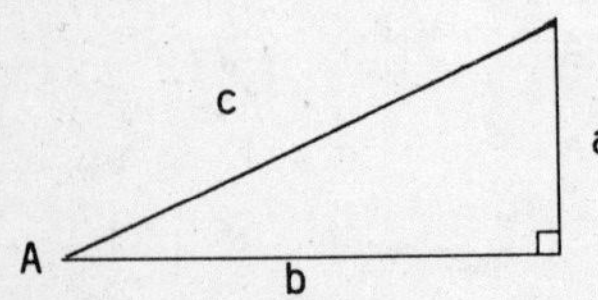

Area $= \frac{1}{2}$ b·a where a = c sin A, so

Area $= \frac{1}{2}$ bc sin A.

55.

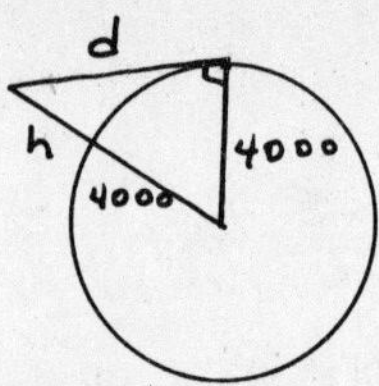

The earth's radius is approximately 4000 miles. Use the Pythagorean theorem to find d (in miles):

$d^2 + 4000^2 = (4000 + h)^2$

$d^2 + 4000^2 = 4000^2 + 8000h + h^2$

$d^2 = 8000h + h^2$

$d = \sqrt{8000h + h^2}$

When h = 1000 ft:

$1000 \text{ ft} \cdot \frac{1 \text{ mi}}{5280 \text{ ft}} = \frac{1000}{5280} \text{ mi} = \frac{25}{132} \text{ mi}$

$d = \sqrt{8000\left(\frac{25}{132}\right) + \left(\frac{25}{132}\right)^2}$

≈ 38.9 mi

56.

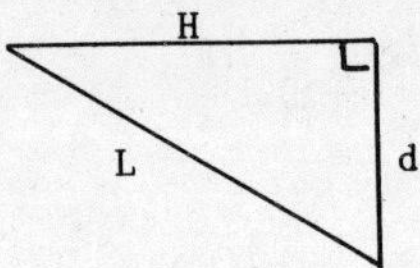

$C = L - H$

$d^2 = L^2 - H^2 = (L + H)(L - H) = C(L + H)$

$C = \frac{d^2}{L + H} \approx \frac{d^2}{2L}$

57. $\tan \theta = \frac{8}{1.5} \approx 5.3333$

$\theta = 79.38°$ or 79° 23'

58. a) 1.96 cm

b) 4.00 cm

c) $d = 1.02\ \overline{VP}$

d) $\overline{VP} = 0.98\ d$

59. $A = \frac{1}{2}bh$, where $h = a \sin \theta$

$A = \frac{1}{2}ab \sin \theta$

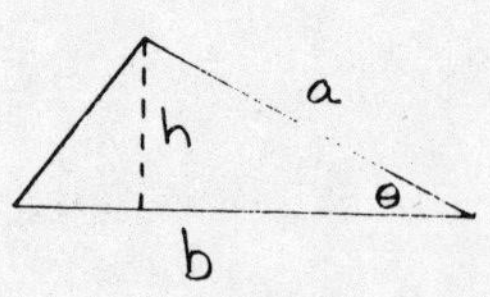

60. 27°

61. 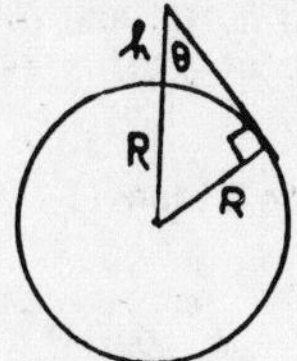

$h = 14{,}162 \text{ ft} \approx 2.682197 \text{ mi}$

$\theta = 87° \; 53'$

$\sin \theta = \frac{R}{R + h}$ gives $R = \frac{h \sin \theta}{1 - \sin \theta}$

$R = \frac{2.682197 \sin 87° \; 53'}{1 - \sin 87° \; 53'} \approx 3928 \text{ mi}$

62. 2.72°, or 2° 43'; 1.68°, or 1° 41'

Exercise Set 8.2

1. 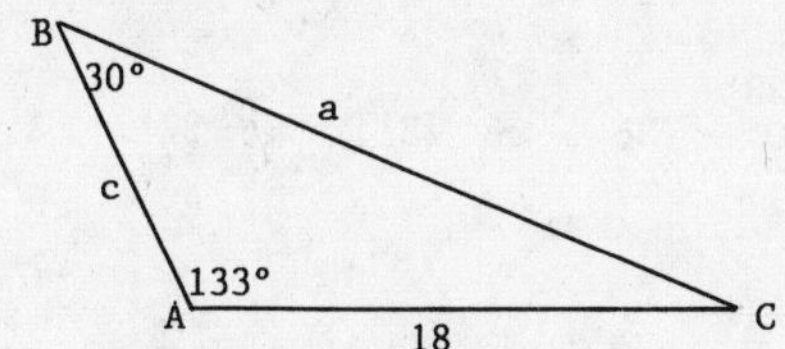

To solve this triangle find C, a, and c.

$C = 180° - (133° + 30°) = 17°$

Use the law of sines to find a and c.

Find a:

$\frac{a}{\sin A} = \frac{b}{\sin B}$

$a = \frac{b \sin A}{\sin B}$

$a = \frac{18 \sin 133°}{\sin 30°} = \frac{18 \times 0.7314}{0.5} = 26.3$

Find c:

$\frac{c}{\sin C} = \frac{b}{\sin B}$

$c = \frac{b \sin C}{\sin B}$

$c = \frac{18 \sin 17°}{\sin 30°} = \frac{18 \times 0.2924}{0.5} = 10.5$

2. $A = 30°$, $b = 16\sqrt{3} \approx 27.7$, $c = 16$

3. 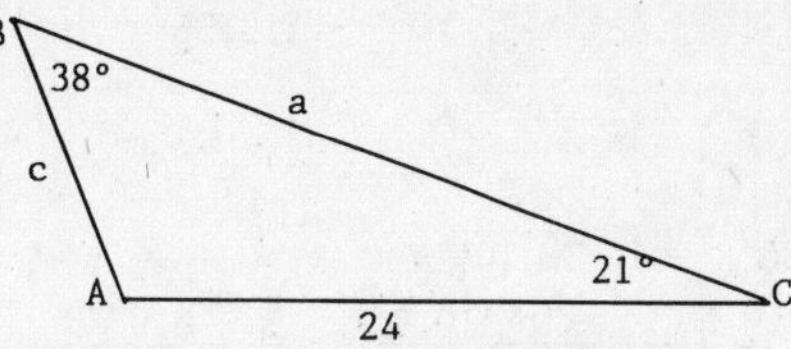

To solve this triangle find A, a, and c.

$A = 180° - (38° + 21°) = 121°$

Use the law of sines to find a and c.

Find a:

$\frac{a}{\sin A} = \frac{b}{\sin B}$

$a = \frac{b \sin A}{\sin B}$

$a = \frac{24 \sin 121°}{\sin 38°} = \frac{24 \times 0.8572}{0.6157} = 33.4$

Find c:

$\frac{c}{\sin C} = \frac{b}{\sin B}$

$c = \frac{b \sin C}{\sin B}$

$c = \frac{24 \sin 21°}{\sin 38°} = \frac{24 \times 0.3584}{0.6157} = 14.0$

4. $B = 26°$, $a = 17.2$, $c = 8.91$

5. 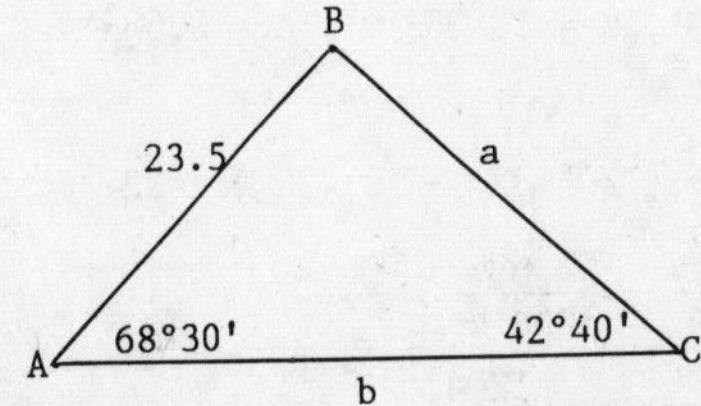

To solve this triangle find B, a, and b.

$B = 180° - (68° \; 30' + 42° \; 40')$

$= 180° - 111° \; 10'$

$= 68° \; 50'$

Use the law of sines to find a and b.

$\frac{a}{\sin A} = \frac{c}{\sin C}$

$a = \frac{c \sin A}{\sin C}$

$a = \frac{23.5 \sin 68° \; 30'}{\sin 42° \; 40'} = \frac{23.5 \times 0.9304}{0.6777} = 32.3 \text{ cm}$

$\frac{b}{\sin B} = \frac{c}{\sin C}$

$b = \frac{c \sin B}{\sin C}$

$b = \frac{23.5 \sin 68° \; 50'}{\sin 42° \; 40'} = \frac{23.5 \times 0.9325}{0.6777} = 32.3 \text{ cm}$

$b = 3.43 \text{ mi}$

6. $A = 16°$, $a = 13.2$ cm, $c = 34.2$ cm

7. $A = 180° - (37.48° + 32.16°) = 110.36°$

$$\frac{a}{\sin A} = \frac{c}{\sin C}$$

$$\frac{a}{\sin 110.36°} = \frac{3}{\sin 32.16°}$$

$$a = \frac{3 \times \sin 110.36°}{\sin 32.16°} = \frac{3 \times 0.9375}{0.5322}$$

$$a = 5.28 \text{ mi}$$

$$\frac{b}{\sin B} = \frac{c}{\sin C}$$

$$\frac{b}{\sin 37.48°} = \frac{3}{\sin 32.16°}$$

$$b = \frac{3 \times \sin 37.48°}{\sin 32.16°} = \frac{3 \times 0.6085}{0.5322}$$

$$b = 3.43 \text{ mi}$$

8. $B = 125.27°$, $b = 301.76$ m, $c = 138.28$ m

9. $C = 180° - (16° 56' + 59° 23') = 103° 41'$

$$\frac{a}{\sin A} = \frac{c}{\sin C}$$

$$\frac{a}{\sin 16° 56'} = \frac{6019}{\sin 103° 41'}$$

$$a = \frac{6019 \times \sin 16° 56'}{\sin 103° 41'} = \frac{6019 \times 0.2913}{0.9716}$$

$$a = 1804 \text{ km}$$

10. $B = 31° 56'$, $b = 10{,}068$ mi, $c = 8868$ mi

11. $B = 180° - (129° 32' + 18° 28') = 32°$

$$\frac{a}{\sin A} = \frac{b}{\sin B}$$

$$\frac{a}{\sin 129° 32'} = \frac{1204}{\sin 32°}$$

$$a = \frac{1204 \times \sin 129° 32'}{\sin 32°} = \frac{1204 \times 0.7713}{0.5299}$$

$$a = 1752 \text{ in.}$$

$$\frac{c}{\sin C} = \frac{b}{\sin B}$$

$$\frac{c}{\sin 18° 28'} = \frac{1204}{\sin 32°}$$

$$c = \frac{1204 \times \sin 18° 28'}{\sin 32°} = \frac{1204 \times 0.3168}{0.5299}$$

$$c = 720 \text{ in.}$$

12. $A = 122° 22'$, $a = 331$ mm, $c = 132$ mm

13. To solve this triangle we find c, B, and C. There will be two solutions. An arc of radius 24 meets the base at two points.

Solution I

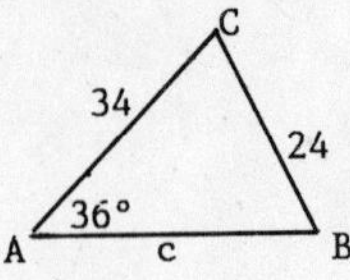

Solution II

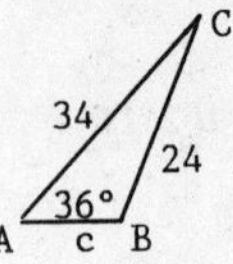

Find B:

$$\frac{b}{\sin B} = \frac{a}{\sin A}$$

$$\sin B = \frac{b \sin A}{a}$$

$$\sin B = \frac{34 \sin 36°}{24} = \frac{34 \times 0.5878}{24} = 0.8327$$

There are two angles less than 180° having a sine of 0.8327. They are 56° 23' and 123° 37'. This gives us two possible solutions.

Solution I

If $B = 56° 23'$
then $C = 180° - (36° + 56° 23') = 87° 37'$.

Find c:

$$\frac{c}{\sin C} = \frac{a}{\sin A}$$

$$c = \frac{a \sin C}{\sin A}$$

$$c = \frac{24 \sin 87° 37'}{\sin 36°} = \frac{24 \times 0.9991}{0.5878} = 40.8$$

Solution II

If $B = 123° 37'$,
then $C = 180° - (36° + 123° 37') = 20° 23'$.

Find c:

$$\frac{c}{\sin C} = \frac{a}{\sin A}$$

$$c = \frac{a \sin C}{\sin A}$$

$$c = \frac{24 \sin 20° 23'}{\sin 36°} = \frac{24 \times 0.3483}{0.5878} = 14.2$$

14. $B = 41° 7'$, $A = 95° 53'$, $a = 40.8$

15. 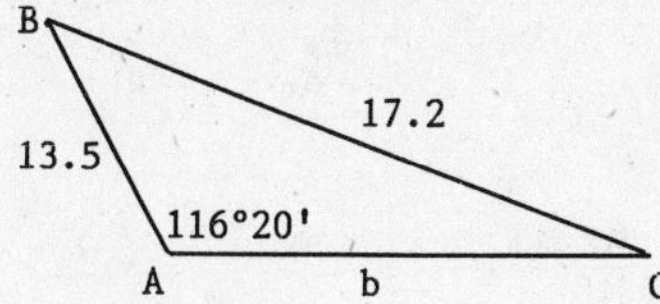

To solve this triangle find B, C, and b.

Find C:

$$\frac{c}{\sin C} = \frac{a}{\sin A}$$

$$\sin C = \frac{c \sin A}{a}$$

$$\sin C = \frac{13.5 \sin 116° 20'}{17.2} = \frac{13.5 \times 0.8962}{17.2} = 0.7034$$

Then C = 44° 42' or C = 135° 18'. An angle of 135° 18' cannot be an angle of this triangle because it already has an angle of 116° 20' and these two would total more than 180°. Thus C = 44° 42'.

Find B:

B = 180° - (116° 20' + 44° 42') = 18° 58'

Find b:

$$\frac{b}{\sin B} = \frac{a}{\sin A}$$

$$b = \frac{a \sin B}{\sin A}$$

$$b = \frac{17.2 \sin 18° 58'}{\sin 116° 20'} = \frac{17.2 \times 0.3250}{0.8962} = 6.24$$

16. B = 28° 28', C = 103° 42', c = 37.1

17. 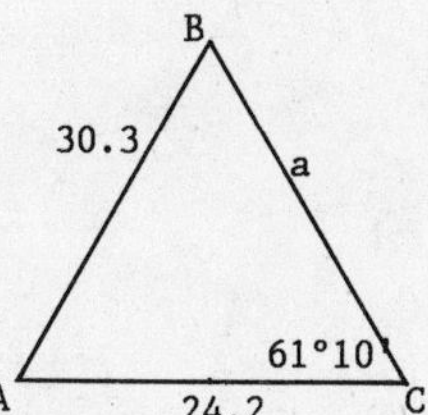

To solve this triangle find A, B, and a.

Find B:

$$\frac{b}{\sin B} = \frac{c}{\sin C}$$

$$\sin B = \frac{b \sin C}{c}$$

$$\sin B = \frac{24.2 \sin 61° 10'}{30.3} = \frac{24.2 \times 0.8760}{30.3} = 0.6996$$

Then B = 44° 24' or B = 135° 36'. An angle of 135° 36' cannot be an angle of this triangle because it already has an angle of 61° 10' and the two would total more than 180°. Thus B = 44° 24'.

Find A:

A = 180° - (61° 10' + 44° 24') = 74° 26'

17. (continued)

Find a:

$$\frac{a}{\sin A} = \frac{c}{\sin C}$$

$$a = \frac{c \sin A}{\sin C}$$

$$a = \frac{30.3 \sin 74° 26'}{\sin 61° 10'} = \frac{30.3 \times 0.9633}{0.8760} = 33.3$$

18. A = 41° 7', C = 80° 13', c = 37.6

19. $$\frac{b}{\sin B} = \frac{a}{\sin A}$$

$$\frac{2345}{\sin B} = \frac{2345}{\sin 124.67°}$$

$$\sin B = \frac{2345 \times \sin 124.67°}{2345} = \frac{2345 \times 0.8224}{2345}$$

sin B = 0.8224

Then B = 55.33° or B = 124.67°. An angle of 55.33° cannot be an angle of this triangle because it already has an angle of 124.67° and the two would total 180°. A second angle of 124.67° cannot be an angle of this triangle either because the two 124.67° angles would total more than 180°.

There is no solution.

20. B = 83° 47', A = 12° 26', a = 12 yd

21. $$\frac{b}{\sin B} = \frac{a}{\sin A}$$

$$\frac{6}{\sin B} = \frac{2}{\sin 30°}$$

$$\sin B = \frac{6 \times \sin 30°}{2} = \frac{6 \times 0.5}{2}$$

sin B = 1.5

Since there is no angle having a sine greater than 1, there is no solution.

22. No solution

23. $$\frac{b}{\sin B} = \frac{a}{\sin A}$$

$$\frac{8000}{\sin B} = \frac{4000}{\sin 32° 52'}$$

$$\sin B = \frac{8000 \times \sin 32° 52'}{4000} = \frac{8000 \times 0.5427}{4000}$$

sin B = 1.0854

Since there is no angle having a sine greater than 1, there is no solution.

24. B = 14.48°, C = 135.52°, c = 28 cm

25. $\frac{b}{\sin B} = \frac{c}{\sin C}$

$\frac{4.157}{\sin B} = \frac{3.446}{\sin 51° 48'}$

$\sin B = \frac{4.157 \times \sin 51° 48'}{3.446} = \frac{4.157 \times 0.7859}{3.446}$

$\sin B = 0.9480$

Then $B = 71° 26'$ or $B = 108° 34'$

Solution I

If $B = 71° 26'$, then
$A = 180° - (71° 26' + 51° 48') = 56° 46'$.

$\frac{a}{\sin A} = \frac{c}{\sin C}$

$\frac{a}{\sin 56° 46'} = \frac{3.446}{\sin 51° 48'}$

$a = \frac{3.446 \times \sin 56° 46'}{\sin 51° 48'}$

$= \frac{3.446 \times 0.8364}{0.7859}$

$a = 3.668$ km

Solution II

If $B = 108° 34'$,
then $A = 180° - (108° 34' + 51° 48') = 19° 38'$.

$\frac{a}{\sin A} = \frac{c}{\sin C}$

$\frac{a}{\sin 19° 38'} = \frac{3.446}{\sin 51° 48'}$

$a = \frac{3.446 \times \sin 19° 38'}{\sin 51° 48'}$

$= \frac{3.446 \times 0.3360}{0.7859}$

$a = 1.473$ km

26. $B = 57.68°$, $A = 73.56°$, $a = 49.57$ mm; or
$B = 122.32°$, $A = 8.92°$, $a = 8.01$ mm

27. $\frac{b}{\sin B} = \frac{a}{\sin A}$

$\frac{18.4}{\sin B} = \frac{15.6}{\sin 89°}$

$\sin B = \frac{18.4 \times \sin 89°}{15.6} = \frac{18.4 \times 0.9998}{15.6}$

$\sin B = 1.1793$

Since there is no angle having a sine greater than 1, there is no solution.

28. $C = 47.28°$, $A = 96.72°$, $a = 13.5$ ft; or
$C = 132.72°$, $A = 11.28°$, $a = 2.7$ ft

29. $\frac{c}{\sin C} = \frac{a}{\sin A}$

$\frac{110.0}{\sin C} = \frac{90.0}{\sin 41° 50'}$

$\sin C = \frac{110.0 \times \sin 41° 50'}{90.0} = \frac{110.0 \times 0.6670}{90.0}$

$\sin C = 0.8152$

Then $C = 54° 36'$ or $C = 125° 24'$.

Solution I

If $C = 54° 36'$, then
$B = 180° - (41° 50' + 54° 36') = 83° 34'$.

$\frac{b}{\sin B} = \frac{a}{\sin A}$

$\frac{b}{\sin 83° 34'} = \frac{90.0}{\sin 41° 50'}$

$b = \frac{90.0 \times \sin 83° 34'}{\sin 41° 50'} = \frac{90.0 \times 0.9937}{0.6670}$

$b = 134.1$ mi

Solution II

If $C = 125° 24'$, then
$B = 180° - (41° 50' + 125° 24') = 12° 46'$.

$\frac{b}{\sin B} = \frac{a}{\sin A}$

$\frac{b}{\sin 12° 46'} = \frac{90.0}{\sin 41° 50'}$

$b = \frac{90.0 \times \sin 12° 46'}{\sin 41° 50'} = \frac{90.0 \times 0.2210}{0.6670}$

$b = 29.8$ mi

30. No solution

31.

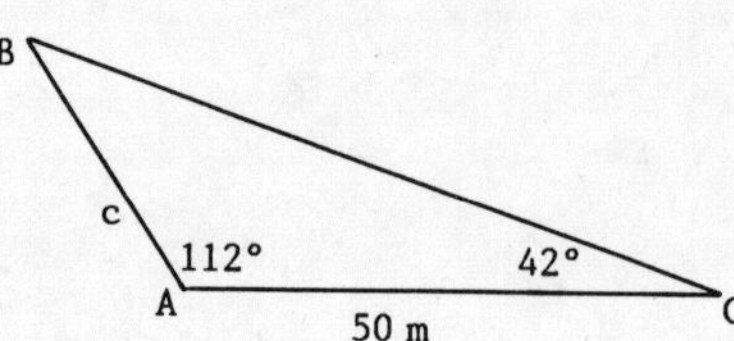

Find B:

$B = 180° - (112° + 42°) = 26°$

Find c:

$\frac{c}{\sin C} = \frac{b}{\sin B}$

$c = \frac{b \sin C}{\sin B}$

$c = \frac{50 \sin 42°}{\sin 26°} = \frac{50 \times 0.6991}{0.4384} = 76.3$

The width of the crater is 76.3 m.

32. 26.9 ft

33.

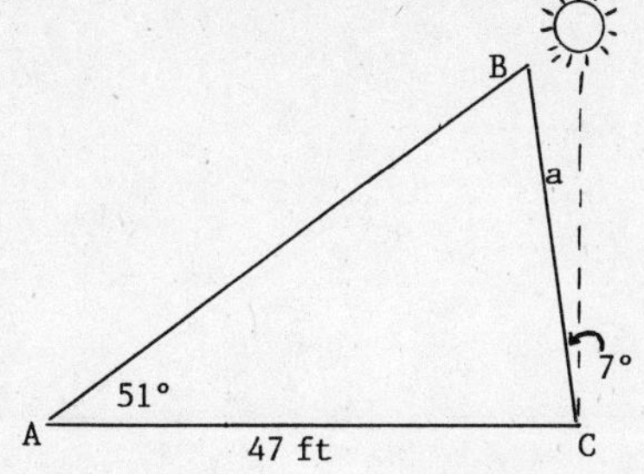

Find C: C = 90° - 7° = 83°

Find B: B = 180° - (51° + 83°) = 46°

Find a: (a is the length of the pole.)

$$\frac{a}{\sin A} = \frac{b}{\sin B}$$

$$a = \frac{b \sin A}{\sin B}$$

$$a = \frac{47 \sin 51°}{\sin 46°} = \frac{47 \times 0.7771}{0.7193} = 50.8$$

The pole is 50.8 ft long.

34. 9.29 ft

35.

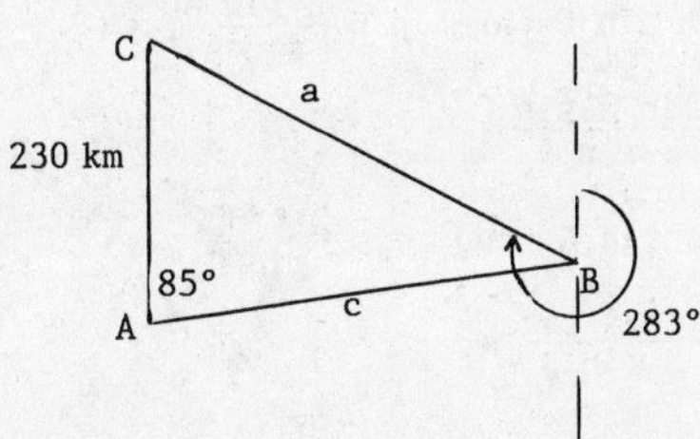

Find B: B = 283° - 180° - 85° = 18°

Find C: C = 180° - (85° + 18°) = 77°

Find a:

$$\frac{a}{\sin A} = \frac{b}{\sin B}$$

$$a = \frac{b \sin A}{\sin B}$$

$$a = \frac{230 \sin 85°}{\sin 18°} = \frac{230 \times 0.9962}{0.3090} = 742$$

Find c:

$$\frac{c}{\sin C} = \frac{b}{\sin B}$$

$$c = \frac{b \sin C}{\sin B}$$

$$c = \frac{230 \sin 77°}{\sin 18°} = \frac{230 \times 0.9744}{0.3090} = 725$$

The plane flew a total of 742 + 725, or 1467 km.

36. 24.7 km, 28.4 km

37.

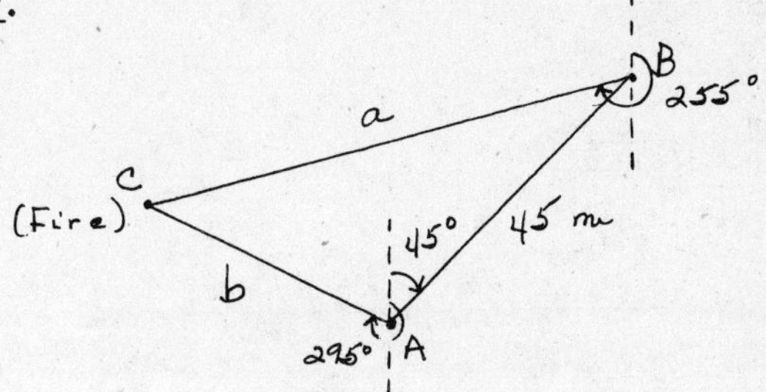

A = 45° + (360° - 295°) = 110°

B = 255° - 180° - 45° = 30°

C = 180° - (110° + 30°) = 40°

The distance from Tower A to the fire is b:

$$\frac{b}{\sin B} = \frac{c}{\sin C}$$

$$\frac{b}{\sin 30°} = \frac{45}{\sin 40°}$$

$$b = \frac{45 \times \sin 30°}{\sin 40°} = \frac{45 \times 0.5}{0.6428}$$

$$b \approx 35.0 \text{ mi}$$

The distance from Tower B to the fire is a:

$$\frac{a}{\sin A} = \frac{c}{\sin C}$$

$$\frac{a}{\sin 110°} = \frac{45}{\sin 40°}$$

$$a = \frac{45 \times \sin 110°}{\sin 40°} = \frac{45 \times 0.9397}{0.6428}$$

$$a \approx 65.8 \text{ mi}$$

38. 373 km

39.

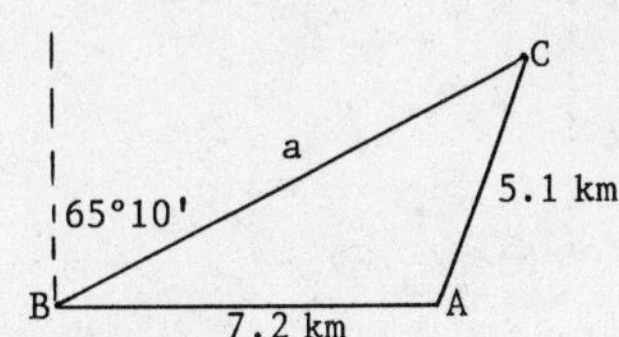

Find B: B = 90° - 65° 10' = 24° 50'

Find C:

$$\frac{c}{\sin C} = \frac{b}{\sin B}$$

$$\sin C = \frac{c \sin B}{b}$$

$$\sin C = \frac{7.2 \sin 24° 50'}{5.1} = \frac{7.2 \times 0.4200}{5.1} = 0.5929$$

Then C = 36° 20' or C = 143° 40'.

If C = 143° 40', then A = 180° - (24° 50' + 143° 40') = 11° 30'.

39. (continued)

Find a:

$$\frac{a}{\sin A} = \frac{b}{\sin B}$$

$$a = \frac{b \sin A}{\sin B}$$

$$a = \frac{5.1 \sin 11° 30'}{\sin 24° 50'} = \frac{5.1 \times 0.1994}{0.4200} = 2.42$$

If C = 36° 20',
then A = 180° - (24° 50' + 36° 20') = 118° 50'.

Find a:

$$\frac{a}{\sin A} = \frac{b}{\sin B}$$

$$a = \frac{b \sin A}{\sin B}$$

$$a = \frac{5.1 \sin 118° 50'}{\sin 24° 50'} = \frac{5.1 \times 0.8760}{0.4200} = 10.6$$

The boat is either 2.42 km or 10.6 km from lighthouse B.

40. 59.8 mi

41. Place the figure on a coordinate system as shown.

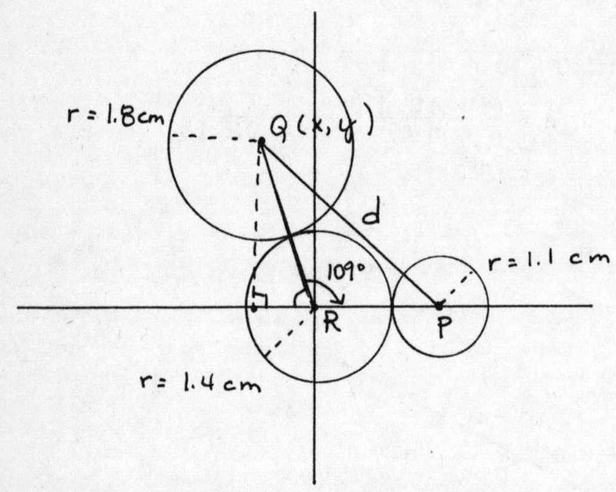

Let (x, y) be the coordinates of point Q. The coordinates of point P are (1.4 + 1.1, 0), or (2.5, 0). The length of QR is 1.8 + 1.4, or 3.2.

Then $\cos 109° = \frac{x}{3.2}$, or $x = 3.2 \cos 109° \approx -1.0418$

and $\sin 109° = \frac{y}{3.2}$, or $y = 3.2 \sin 109° \approx 3.0257$.

Now use the distance formula to find the distance d from P to Q.

$$d = \sqrt{(-1.0418 - 2.5)^2 + (3.0257 - 0)^2}$$

$$d \approx 4.7 \text{ cm}$$

42. 89.4°

43.

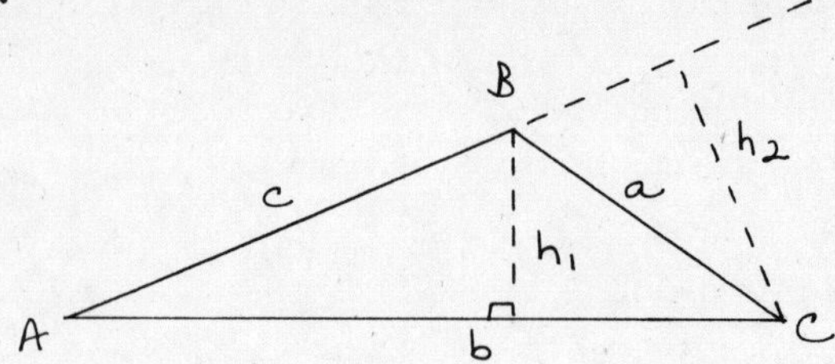

$K = \frac{1}{2} bh_1 = \frac{1}{2} bc \sin A$ where $h_1 = c \sin A$.

$K = \frac{1}{2} bh_1 = \frac{1}{2} ba \sin C$ where $h_1 = a \sin C$.

$K = \frac{1}{2} ch_2 = \frac{1}{2} ca \sin B$ where $h_2 = a \sin B$.

44. Use the results of Exercise 43 and the law of sines.

$K = \frac{1}{2} bc \sin A$ and $b = \frac{c \sin B}{\sin C}$, so

$K = \frac{c^2 \sin A \sin B}{2 \sin C}$.

$K = \frac{1}{2} ab \sin C$ and $b = \frac{a \sin B}{\sin A}$, so

$K = \frac{a^2 \sin B \sin C}{2 \sin A}$

$K = \frac{1}{2} bc \sin A$ and $c = \frac{b \sin C}{\sin B}$, so

$K = \frac{b^2 \sin A \sin C}{2 \sin B}$.

45. $\frac{a}{\sin A} = \frac{b}{\sin B}$, so $a = \frac{b \sin A}{\sin B}$.

$\frac{c}{\sin C} = \frac{b}{\sin B}$, so $c = \frac{b \sin C}{\sin B}$.

$$\frac{a + b}{c} = \frac{\frac{b \sin A}{\sin B} + b}{\frac{b \sin C}{\sin B}} \qquad \text{(Substituting)}$$

$$= \frac{\frac{b \sin A + b \sin B}{\sin B}}{\frac{b \sin C}{\sin B}}$$

$$= \frac{b(\sin A + \sin B)}{b \sin C}$$

$$= \frac{\sin A + \sin B}{\sin C}$$

$$= \frac{2 \sin \frac{1}{2}(A + B) \cos \frac{1}{2}(A - B)}{\sin C}$$

(Using a reduction formula)

$$= \frac{2 \sin \frac{1}{2}(A + B) \cos \frac{1}{2}(A - B)}{2 \sin \frac{1}{2} C \cos \frac{1}{2} C}$$

$$= \frac{\sin \frac{1}{2}(A + B) \cos \frac{1}{2}(A - B)}{\sin \frac{1}{2} C \cos \frac{1}{2} [\pi - (A + B)]}$$

$[C = \pi - (A + B)]$

45. (continued)

$$= \frac{\sin\frac{1}{2}(A+B)\cos\frac{1}{2}(A-B)}{\sin\frac{1}{2}C\cos\left[\frac{\pi}{2}-\frac{1}{2}(A+B)\right]}$$

$$= \frac{\sin\frac{1}{2}(A+B)\cos\frac{1}{2}(A-B)}{\sin\frac{1}{2}C\sin\frac{1}{2}(A+B)}$$

$$= \frac{\cos\frac{1}{2}(A-B)}{\sin\frac{1}{2}C}$$

46. We substitute as in Exercise 45.

$$\frac{a-b}{c} = \frac{\frac{b\sin A}{\sin B} - b}{\frac{b\sin C}{\sin B}}$$

$$= \frac{\frac{b\sin A - b\sin B}{\sin B}}{\frac{b\sin C}{\sin B}}$$

$$= \frac{\sin A - \sin B}{\sin C}$$

$$= \frac{2\cos\frac{1}{2}(A+B)\sin\frac{1}{2}(A-B)}{2\sin\frac{1}{2}C\cos\frac{1}{2}C}$$

$$= \frac{\cos\frac{1}{2}(A+B)\sin\frac{1}{2}(A-B)}{\sin\frac{1}{2}[\pi-(A+B)]\cos\frac{1}{2}C}$$

$$= \frac{\cos\frac{1}{2}(A+B)\sin\frac{1}{2}(A-B)}{\cos\frac{1}{2}(A+B)\cos\frac{1}{2}C}$$

$$= \frac{\sin\frac{1}{2}(A-B)}{\cos\frac{1}{2}C}$$

47. See the answer section in the text.

48.

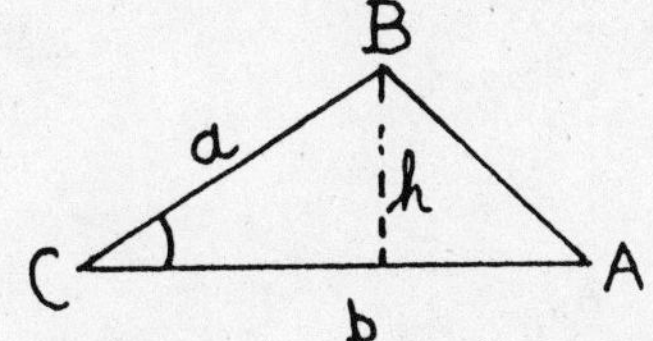

The area K of Δ ABC is $\frac{1}{2}bh$ where $h = a\sin C$; thus $K = \frac{1}{2}ab\sin C$; i.e., "one-half the product of two sides times the sine of the included angle."

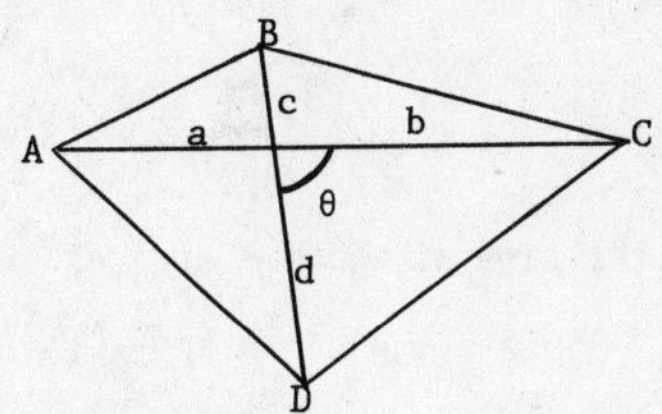

Then for the quadrilateral ABCD, we have

$$\text{Area} = \frac{1}{2}bd\sin\theta + \frac{1}{2}ac\sin\theta + \frac{1}{2}ad\sin(180° - \theta) + \frac{1}{2}bc\sin(180° - \theta)$$

$$= \frac{1}{2}(bd + ac + ad + bc)\sin\theta$$

$$= \frac{1}{2}(a+b)(c+d)\sin\theta$$

49. See the answer section in the text.

50. Suppose that objects at A and B are moving as shown, velocities v_a and v_b, paths intersecting at P. Let θ be the bearing from A to B, assumed constant.

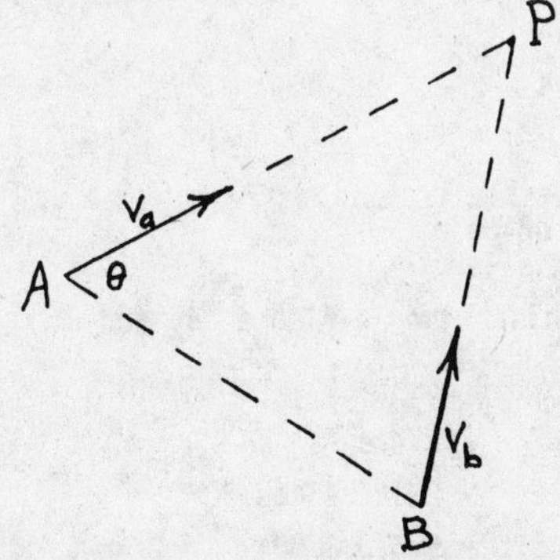

At a later time, t, A will have moved a distance of $v_a t$ and B a distance of $v_b t$. A'B' $\parallel$ AB because θ is constant. Let x = MN.

50. (continued)

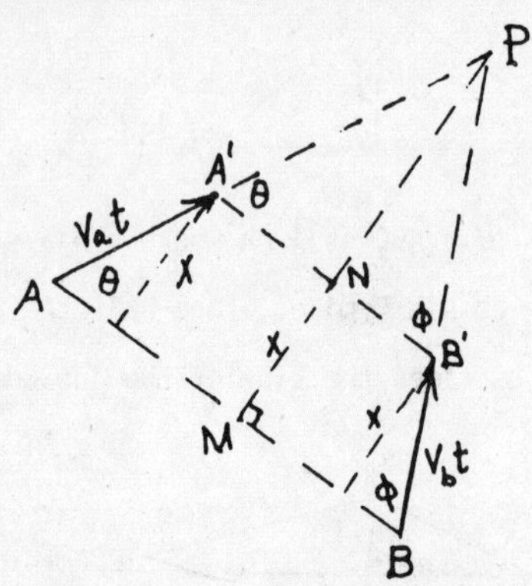

Let T be the time at which A arrives at P. Then $AP = v_aT$. Since $\frac{x}{AA'} = \sin\theta$ and $\frac{x}{BB'} = \sin\phi$, we have $AA'\sin\theta = BB'\sin\phi$, then $v_at\sin\theta = v_bt\sin\phi$. Therefore, $\frac{\sin\theta}{\sin\phi} = \frac{v_bt}{v_at}$, and at time T, $\frac{\sin\theta}{\sin\phi} = \frac{v_bT}{v_aT}$. Note that $\sin\theta = \frac{MP}{AP}$ and $\sin\phi = \frac{MP}{BP}$, so that $AP\sin\theta = BP\sin\phi$ and $\frac{\sin\theta}{\sin\phi} = \frac{BP}{AP}$. Therefore, $\frac{BP}{AP} = \frac{v_bT}{v_aT}$ so $BP = AP\cdot\frac{v_bT}{v_aT} = AP\cdot\frac{v_bT}{AP} = v_bT$. Thus at time T, the object that was at B will be at point P.

Exercise Set 8.3

1.

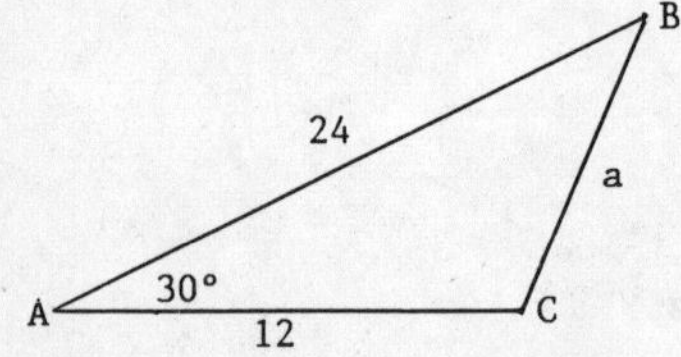

To solve this triangle find a, B, and C.

From the law of cosines,

$a^2 = b^2 + c^2 - 2bc\cos A$

$a^2 = 12^2 + 24^2 - 2\cdot12\cdot24\cos 30°$

$a^2 = 144 + 576 - 576(0.8660)$

$a^2 = 221$

Then $a = \sqrt{221} = 14.9$.

1. (continued)

Next we use the law of cosines again to find a second angle.

$b^2 = a^2 + c^2 - 2ac\cos B$

$12^2 = (14.9)^2 + 24^2 - 2(14.9)(24)\cos B$

$144 = 222.01 + 576 - 715.2\cos B$

$-654.01 = -715.2\cos B$

$\cos B = 0.9144$

Then B = 23° 53'. (Answers may vary due to rounding or law used.)

The third angle is easy to find.

C = 180° - (30° + 23° 53') = 126° 7'

2. c = 13.7, A = 71° 4', B = 48° 56'

3.

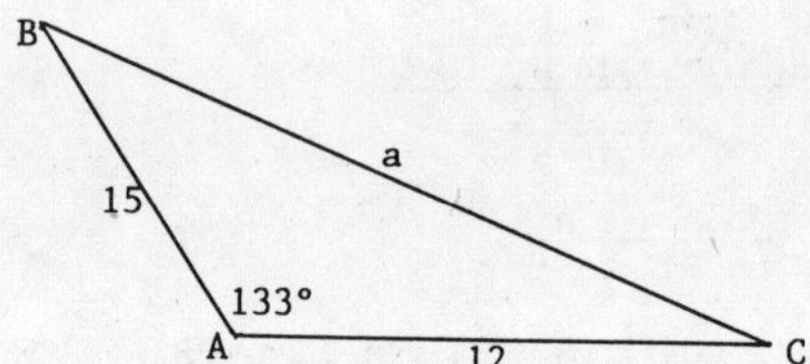

To solve this triangle find a, B, and C.

From the law of cosines,

$a^2 = b^2 + c^2 - 2bc\cos A$

$a^2 = 12^2 + 15^2 - 2\cdot12\cdot15\cos 133°$

$a^2 = 144 + 225 - 360(-0.6820)$

$a^2 = 615$

Then $a = \sqrt{615} = 24.8$.

Next we use the law of cosines again to find a second angle.

$b^2 = a^2 + c^2 - 2ac\cos B$

$12^2 = (24.8)^2 + 15^2 - 2(24.8)(15)\cos B$

$144 = 615.04 + 225 - 744\cos B$

$-696.04 = -744\cos B$

$\cos B = 0.9355$

Then B = 20° 41'. (Answers may vary due to rounding or law used.)

The third angle is easy to find.

C = 180° - (133° + 20° 41') = 26° 19'.

4. b = 47.6, A = 35° 49', C = 28° 11'

5. 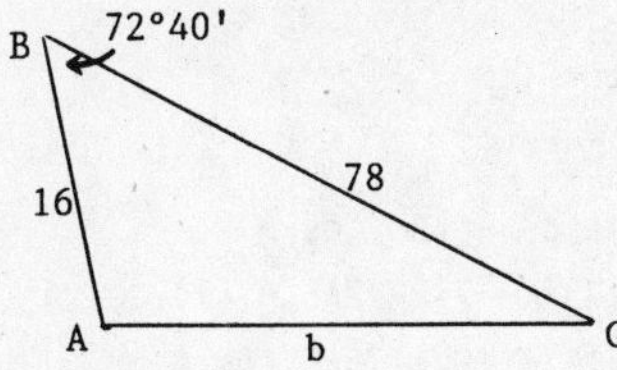

To solve this triangle find b, A, and C.

From the law of cosines,

$b^2 = a^2 + c^2 - 2ac \cos B$

$b^2 = 78^2 + 16^2 - 2 \cdot 78 \cdot 16 \cos 72° \, 40'$

$b^2 = 6084 + 256 - 2496(0.2979)$

$b^2 = 5596$

Then $b = \sqrt{5596} = 74.8$ m

Next we use the law of cosines again to find a second angle.

$c^2 = a^2 + b^2 - 2ab \cos C$

$16^2 = 78^2 + (74.8)^2 - 2(78)(74.8) \cos C$

$256 = 6084 + 5595.04 - 11{,}668.8 \cos C$

$-11{,}423.04 = -11{,}668.8 \cos C$

$\cos C = 0.9789$

Then $C = 11° \, 47'$.

The third angle is easy to find.

$A = 180° - (72° \, 40' + 11° \, 47') = 95° \, 33'$

6. $a = 55.6$ ft, $B = 149° \, 9'$, $C = 6° \, 21'$

7. $c^2 = a^2 + b^2 - 2ab \cos C$

$c^2 = (25.4)^2 + (73.8)^2 - 2(25.4)(73.8) \cos 22.28°$

$c^2 = 645.16 + 5446.44 - 3749.04(0.9253)$

$c^2 = 2622$

Then $c = \sqrt{2622} = 51.2$ cm.

Next we use the law of cosines again to find a second angle.

$a^2 = b^2 + c^2 - 2bc \cos A$

$(25.4)^2 = (73.8)^2 + (51.2)^2 - 2(73.8)(51.2) \cos A$

$645.16 = 5446.44 + 2621.44 - 7557.12 \cos A$

$-7422.72 = -7557.12 \cos A$

$\cos A = 0.9822$

Then $A = 10.82°$.
And $B = 180° - (10.82° + 22.28°) = 146.9°$.
(Answers may vary due to rounding and law used.)

8. $b = 29.38$ km, $A = 50.59°$, $C = 56.75°$

9. $a^2 = b^2 + c^2 - 2bc \cos A$

$a^2 = (15.8)^2 + (18.4)^2 - 2(15.8)(18.4) \cos (96° \, 13')$

$a^2 = 249.64 + 338.56 - 581.44(-0.1083)$

$a^2 = 651$

Then $a = \sqrt{651} = 25.5$ yd.

Next we use the law of cosines again to find a second angle.

$b^2 = a^2 + c^2 - 2ac \cos B$

$(15.8)^2 = (25.5)^2 + (18.4)^2 - 2(25.5)(18.4) \cos B$

$249.64 = 650.25 + 338.56 - 938.4 \cos B$

$-739.17 = -938.4 \cos B$

$\cos B = 0.7877$

Then $B = 38° \, 2'$.
And $C = 180° - (96° \, 13' + 38° \, 2') = 45° \, 45'$.
(Answers may vary due to rounding and law used.)

10. $c = 4.7$ mm, $A = 37° \, 32'$, $B = 113° \, 45'$

11. $c^2 = a^2 + b^2 - 2ab \cos C$

$c^2 = (60.12)^2 + (40.23)^2 - 2(60.12)(40.23) \cos 48.7°$

$c^2 = 3614.4144 + 1618.4529 - 4837.2552(0.6600)$

$c^2 = 2040$

Then $c = \sqrt{2040} = 45.17$ mi.

Next we use the law of cosines again to find a second angle.

$a^2 = b^2 + c^2 - 2bc \cos A$

$(60.12)^2 = (40.23)^2 + (45.17)^2 - 2(40.23)(45.17) \cos A$

$3614.4144 = 1618.4529 + 2040.3289 - 3634.3782 \cos A$

$-44.3674 = -3634.3782 \cos A$

$\cos A = 0.0122$

Then $A = 89.3°$.
And $B = 180° - (89.3° + 48.7°) = 42°$.

12. $a = 13.9$ in., $B = 36.127°$, $C = 90.417°$

13. 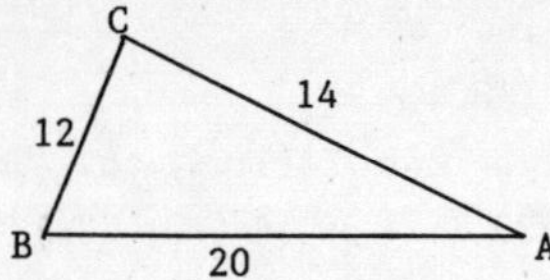

To solve this triangle find A, B, and C.

Find B:

$$b^2 = a^2 + c^2 - 2ac \cos B$$
$$14^2 = 12^2 + 20^2 - 2\cdot 12\cdot 20 \cos B$$
$$196 = 144 + 400 - 480 \cos B$$
$$\cos B = \frac{348}{480} = 0.7250$$

Thus B = 43° 32'.

Find A:

$$a^2 = b^2 + c^2 - 2bc \cos A$$
$$12^2 = 14^2 + 20^2 - 2\cdot 14\cdot 20 \cos A$$
$$144 = 196 + 400 - 560 \cos A$$
$$\cos A = \frac{452}{560} = 0.8071$$

Thus A = 36° 11'.

Then C = 180° - (43° 30' + 36° 11') = 100° 17'.

14. A = 37° 18', B = 37° 18', C = 105° 24'

15. Find A:

$$a^2 = b^2 + c^2 - 2bc \cos A$$
$$16^2 = 20^2 + 32^2 - 2\cdot 20\cdot 32 (\cos A)$$
$$256 = 400 + 1024 - 1280 \cos A$$
$$\cos A = \frac{1168}{1280} = 0.9125$$

Thus A = 24° 9'.

Find B:

$$b^2 = a^2 + c^2 - 2ac \cos B$$
$$20^2 = 16^2 + 32^2 - 2\cdot 16\cdot 32 (\cos B)$$
$$400 = 256 + 1024 - 1024 \cos B$$
$$\cos B = \frac{880}{1024} = 0.8594$$

Thus B = 30° 45'.
Then C = 180° - (24° 9' + 30° 45') = 125° 6'.

16. A = 25° 43', C = 28° 16', B = 126° 1'

17. Find A:

$$a^2 = b^2 + c^2 - 2bc \cos A$$
$$2^2 = 3^2 + 8^2 - 2\cdot 3\cdot 8 (\cos A)$$
$$4 = 9 + 64 - 48 \cos A$$
$$\cos A = \frac{69}{48} = 1.4375$$

Since there is no angle whose cosine is greater than 1, there is no solution.

18. No solution

19. Find A:

$$a^2 = b^2 + c^2 - 2bc \cos A$$
$$(11.2)^2 = (5.4)^2 + 7^2 - 2(5.4)(7) \cos A$$
$$125.44 = 29.16 + 49 - 75.6 \cos A$$
$$\cos A = -\frac{47.28}{75.6} = -0.6254$$

Thus A = 128.7°.

Find B:

$$b^2 = a^2 + c^2 - 2ac \cos B$$
$$(5.4)^2 = (11.2)^2 + 7^2 - 2(11.2)(7) \cos B$$
$$29.16 = 125.44 + 49 - 156.8 \cos B$$
$$\cos B = \frac{145.28}{156.8} = 0.9265$$

Thus B = 22.1°.
Then C = 180° - (128.7° + 22.1°) = 29.2°.

20. 79.9°, 53.6°, 46.5°

21. We are given two sides and the angle opposite one of them. The law of sines applies.

C = 180° - (70° + 12°) = 98°

$$\frac{a}{\sin A} = \frac{b}{\sin B}$$
$$\frac{a}{\sin 70°} = \frac{21.4}{\sin 12°}$$
$$a = \frac{21.4 \times \sin 70°}{\sin 12°} = \frac{21.4 \times 0.9397}{0.2079}$$
$$a = 96.7$$
$$\frac{c}{\sin C} = \frac{b}{\sin B}$$
$$\frac{c}{\sin 98°} = \frac{21.4}{\sin 12°}$$
$$c = \frac{21.4 \times \sin 98°}{\sin 12°} = \frac{21.4 \times 0.9903}{0.2079}$$
$$c = 101.9$$

22. Law of cosines; b = 13, A = 92.2°, C = 25.8°

23. We are given all three sides of the triangle. The law of cosines applies.

Find A:

$$a^2 = b^2 + c^2 - 2bc \cos A$$
$$(3.3)^2 = (2.7)^2 + (2.8)^2 - 2(2.7)(2.8) \cos A$$
$$10.89 = 7.29 + 7.84 - 15.12 \cos A$$
$$\cos A = \frac{4.24}{15.12} = 0.2804$$

Thus A = 73.7°.

23. (continued)

Find B:

$b^2 = a^2 + c^2 - 2ac \cos B$

$(2.7)^2 = (3.3)^2 + (2.8)^2 - 2(3.3)(2.8) \cos B$

$7.29 = 10.89 + 7.84 - 18.48 \cos B$

$\cos B = \frac{11.44}{18.48} = 0.6190$

Thus B = 51.8°.

Then C = 180° - (73.7° + 51.8°) = 54.5°.

24. Law of sines; no solution

25. We are given two sides of the triangle and the included angle. The law of cosines applies.

$c^2 = a^2 + b^2 - 2ab \cos C$

$c^2 = 60^2 + 40^2 - 2(60)(40) \cos 48°$

$c^2 = 3600 + 1600 - 4800(0.6691)$

$c^2 = 1988.32$

Then $c = \sqrt{1988.32} = 44.6$.

Next we use the law of cosines again to find a second angle.

$a^2 = b^2 + c^2 - 2bc \cos A$

$60^2 = 40^2 + (44.6)^2 - 2(40)(44.6) \cos A$

$3600 = 1600 + 1989.16 - 3568 \cos A$

$\cos A = -\frac{10.84}{3568} = -0.0030$

Then A ≈ 90° and B = 180° - (48° + 90°) = 42°.

26. Law of cosines; A = 33.7°, B = 107.1°, C = 39.2°

27. We are given two angles and a side of the triangle. The law of sines applies.

A = 180° - (110° 30' + 8° 10') = 61° 20'

$\frac{b}{\sin B} = \frac{c}{\sin C}$

$\frac{b}{\sin 110° 30'} = \frac{0.912}{\sin 8° 10'}$

$b = \frac{0.912 \times \sin 110° 30'}{\sin 8° 10'}$

$= \frac{0.912 \times 0.9367}{0.1421}$

$b = 6.014$

$\frac{a}{\sin A} = \frac{c}{\sin C}$

$\frac{a}{\sin 61° 20'} = \frac{0.912}{\sin 8° 10'}$

$a = \frac{0.912 \times \sin 61° 20'}{\sin 8° 10'} = \frac{0.912 \times 0.8774}{0.1421}$

$a = 5.631$

28. Law of sines; A = 113°, c = 19.8, a = 70.6

29.

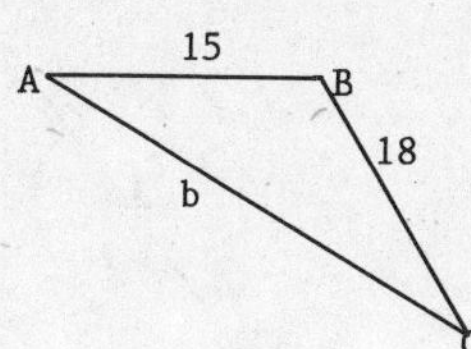

Find B: B = 90° + 27° = 117°

Find b:

$b^2 = a^2 + c^2 - 2ac \cos B$

$b^2 = 18^2 + 15^2 - 2 \cdot 18 \cdot 15 \cos 117°$

$= 324 + 225 - 540(-0.4540)$

$= 794.16$

Then $b = \sqrt{794.16} = 28.2$.

Find A:

$a^2 = b^2 + c^2 - 2bc \cos A$

$18^2 = (28.2)^2 + 15^2 - 2(28.2)(15) \cos A$

$324 = 795.24 + 225 - 846 \cos A$

$\cos A = \frac{696.24}{846} = 0.8229$

Thus A = 34° 40' and 90° - 34° 40' = 55° 20'.

The ship is 28.2 nautical mi from the harbor in a direction of S55° 20'E.

30. 315 km, 241°

31.

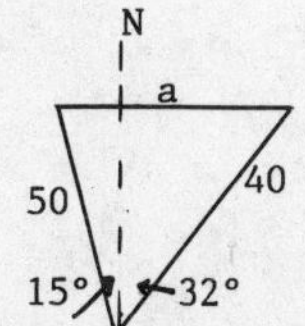

2 hr × 25 knots = 50 nautical mi

2 hr × 20 knots = 40 nautical mi

Find A: A = 15° + 32° = 47°

Find a:

$a^2 = b^2 + c^2 - 2bc \cos A$

$a^2 = 50^2 + 40^2 - 2 \cdot 50 \cdot 40 \cos 47°$

$= 2500 + 1600 - 4000(0.6820)$

$= 1372$

Then $a = \sqrt{1372} = 37.0$.

The ships are 37 nautical mi apart.

32. 912 km

33.

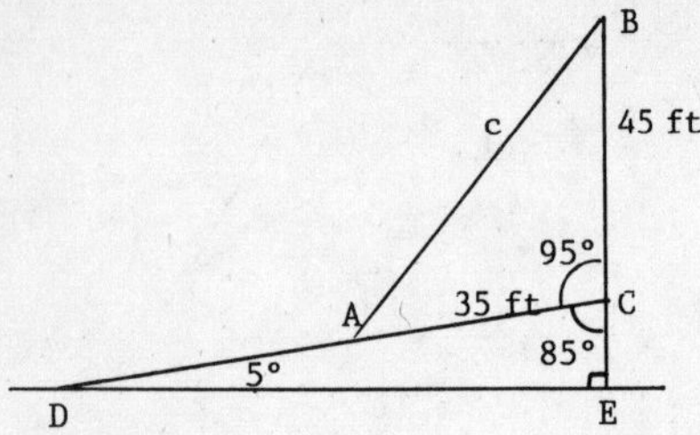

Since D = 5° and E = 90°, ∠DCE = 85° and ∠ACB = 95°.

Using the law of cosines,

$c^2 = 45^2 + 35^2 - 2 \cdot 45 \cdot 35 \cos 95°$

$= 2025 + 1225 - 3150(-0.0872)$

$= 3525$

Then $c = \sqrt{3525} = 59.4$.

The length of the rope must be 59.4 ft.

34. 87.4 ft

35. The perimeter is 5.5 m. The length of the third side is 5.5 - (1.5 + 2), or 2 m.

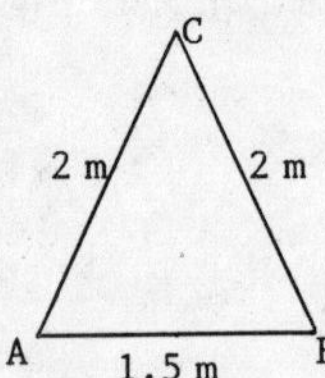

Find A:

$a^2 = b^2 + c^2 - 2bc \cos A$

$2^2 = (1.5)^2 + 2^2 - 2(1.5)(2) \cos A$

$4 = 2.25 + 4 - 6 \cos A$

$\cos A = \frac{2.25}{6} = 0.375$

Thus A = 68°. Since △ABC is isosceles B is also 68°.

Find C:

C = 180° - (68° + 68°) = 44°

The angles are 68°, 68°, and 44°.

36. 52° 50', 85° 30', 41° 40'

37.

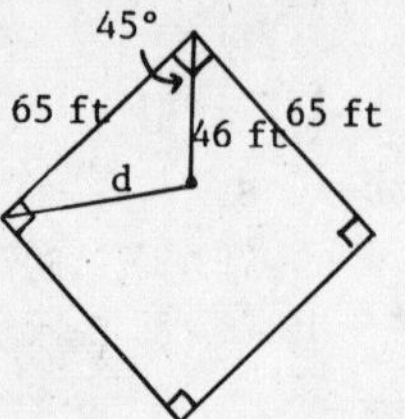

Using the law of cosines,

$d^2 = 65^2 + 46^2 - 2 \cdot 65 \cdot 46 \cdot \cos 45°$

$= 4225 + 2116 - 5980(0.7071)$

$= 2112.5$

Then $d = \sqrt{2112.5} = 45.96$ ft

The distance from the pitcher's mound to first base is 45.96 ft.

38. 63.7 ft

39.

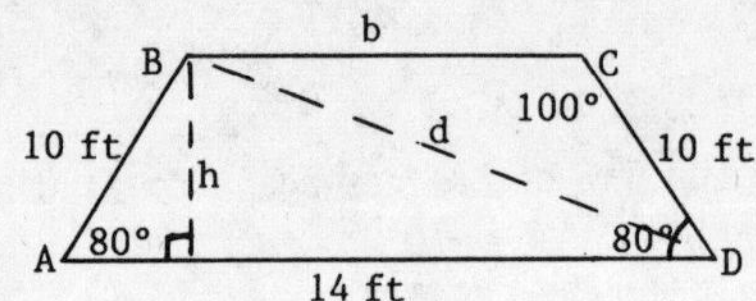

Let b represent the other base, d represent the diagonal, and h the height.

a) Find d:

Using the law of cosines,

$d^2 = 10^2 + 14^2 - 2 \cdot 10 \cdot 14 \cos 80°$

$= 100 + 196 - 280(0.1736)$

$= 247.4$

Then $d = \sqrt{247.4} = 15.7$.

The length of the diagonal is 15.7 ft.

b) Find h:

$\frac{h}{10} = \sin 80°$

$h = 10 \sin 80°$

$h = 10(0.9848)$

$h = 9.85$

Find ∠CBD:

Using the law of sines,

$\frac{10}{\sin \angle CBD} = \frac{15.7}{\sin 100°}$

$\sin \angle CBD = \frac{10 \sin 100°}{15.7} = \frac{10 \times 0.9848}{15.7} = 0.6273$

Thus ∠CBD = 38° 50'.

Then ∠CDB = 180° - (100° + 38° 50') = 41° 10'.

39. (continued)

Find b:

Using the law of sines,

$$\frac{b}{\sin 41^\circ\ 10'} = \frac{15.7}{\sin 100^\circ}$$

$$b = \frac{15.7 \sin 41^\circ\ 10'}{\sin 100^\circ} = \frac{15.7 \times 0.6583}{0.9848}$$

$$= 10.5$$

$$\text{Area} = \frac{1}{2} h(b_1 + b_2)$$

$$\text{Area} = \frac{1}{2} (9.85)(10.5 + 14)$$

$$= 121 \text{ ft}^2$$

40. 123 ft²

41.

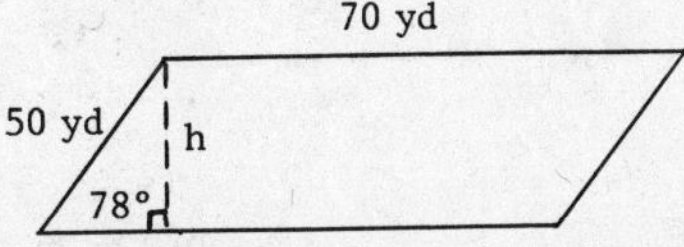

Let b = 70 yd; then find h.

$$\frac{h}{50} = \sin 78^\circ$$

$$h = 50 \sin 78^\circ$$

$$h = 50(0.9781)$$

$$h = 48.9$$

$$\text{Area} = b \cdot h$$

$$\text{Area} = 70 \cdot 48.9$$

$$= 3423 \text{ yd}^2$$

42. 13°

43.

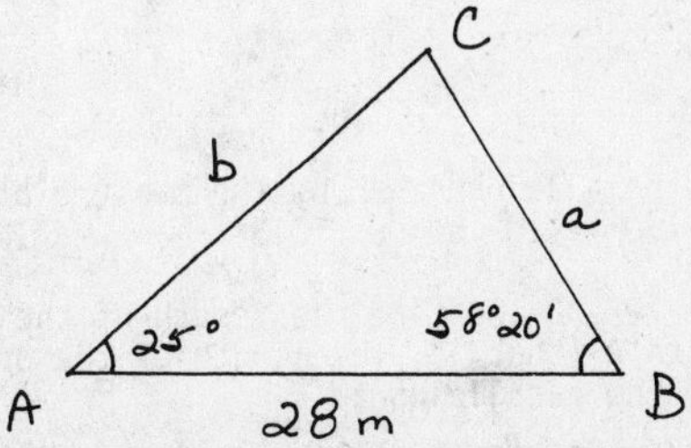

We are given two angles and a side of a triangle. The law of sines applies.

$$C = 180^\circ - (25^\circ + 58^\circ\ 20') = 96^\circ\ 40'$$

$$\frac{a}{\sin A} = \frac{c}{\sin C}$$

$$\frac{a}{\sin 25^\circ} = \frac{28}{\sin 96^\circ\ 40'}$$

$$a = \frac{28 \times \sin 25^\circ}{\sin 96^\circ\ 40'} = \frac{28 \times 0.4226}{0.9932}$$

$$a = 11.9$$

43. (continued)

$$\frac{b}{\sin B} = \frac{c}{\sin C}$$

$$\frac{b}{\sin 58^\circ\ 20'} = \frac{28}{\sin 96^\circ\ 40'}$$

$$b = \frac{28 \times \sin 58^\circ\ 20'}{\sin 96^\circ\ 40'} = \frac{28 \times 0.8511}{0.9932}$$

$$b = 24.0$$

The lengths of the other two sides are 11.9 m and 24.0 m.

44. 15.3 m

45.

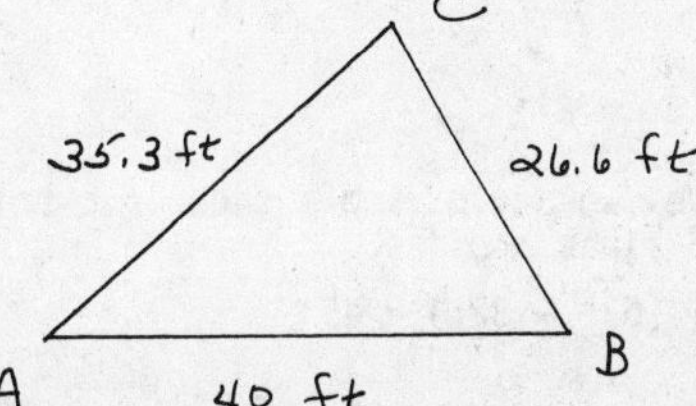

We are given all three sides of a triangle. The law of cosines applies.

Find A:

$$a^2 = b^2 + c^2 - 2bc \cos A$$

$$(26.6)^2 = (35.3)^2 + 40^2 - 2(35.3)(40) \cos A$$

$$707.56 = 1246.09 + 1600 - 2824 \cos A$$

$$-2138.53 = -2824 \cos A$$

$$\cos A = \frac{2138.53}{2824} = 0.7573$$

Then A = 40.8°.

Find B:

$$b^2 = a^2 + c^2 - 2ac \cos B$$

$$(35.3)^2 = (26.6)^2 + 40^2 - 2(26.6)(40) \cos B$$

$$1246.09 = 707.56 + 1600 - 2128 \cos B$$

$$-1061.47 = -2128 \cos B$$

$$\cos B = \frac{1061.47}{2128} = 0.4988$$

Then B = 60.1° and
C = 180° - (40.8° + 60.1°) = 79.1°,

46. 28.8 ft

47.

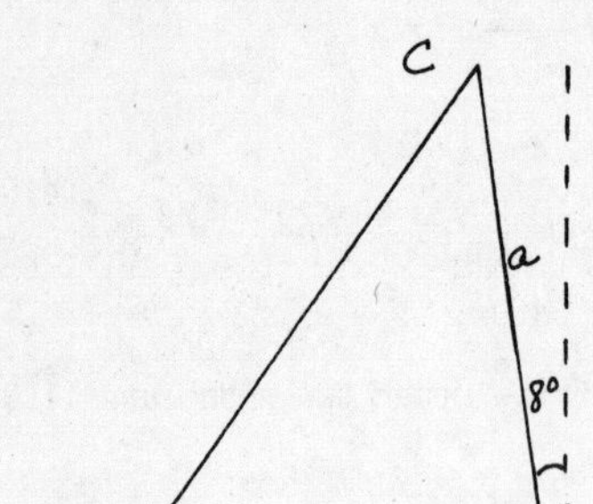

$B = 90° - 8° = 82°$

Now we have two angles and a side of a triangle. The law of sines applies.

$C = 180° - (53° + 82°) = 45°$

$$\frac{a}{\sin A} = \frac{c}{\sin C}$$

$$\frac{a}{\sin 53°} = \frac{44}{\sin 45°}$$

$$a = \frac{44 \times \sin 53°}{\sin 45°} = \frac{44 \times 0.7986}{0.7071}$$

$a = 49.7$ ft

48. 57.4 ft

49. $K = \frac{1}{2} bc \sin A$

$K^2 = \frac{1}{4} b^2c^2 \sin^2 A$ Squaring both sides

$K^2 = \frac{1}{4} b^2c^2 (1 - \cos^2 A)$

$K^2 = \frac{1}{4} b^2c^2 - \frac{1}{4} b^2c^2 \cos^2 A$

Now, the law of cosines gives us

$a^2 = b^2 + c^2 - 2bc \cos A$ so

$bc \cos A = \frac{-a^2 + b^2 + c^2}{2}$ and

$b^2c^2 \cos^2 A = \left[\frac{-a^2 + b^2 + c^2}{2}\right]^2.$

We substitute in the expression for K^2:

$$K^2 = \frac{1}{4} b^2c^2 - \frac{1}{4} \left[\frac{-a^2 + b^2 + c^2}{2}\right]^2$$

$$= \frac{1}{4} b^2c^2 - \frac{1}{4} \cdot \frac{1}{4} (-a^2 + b^2 + c^2)^2$$

$$= \frac{1}{16} [4b^2c^2 - (-a^2 + b^2 + c^2)^2]$$

$$= \frac{1}{16} [2bc+(-a^2 + b^2 + c^2)] [2bc-(-a^2 + b^2 + c^2)]$$

$$= \frac{1}{16} [(b^2 + 2bc + c^2)-a^2] [a^2-(b^2 - 2bc + c^2)]$$

$$= \frac{1}{16} [(b + c)^2 - a^2] [a^2 - (b - c)^2]$$

49. (continued)

$$= \frac{1}{16}(b + c + a)(b + c - a)(a - b + c)(a + b - c)$$

$$= \frac{1}{2} (a + b + c)\cdot\left[\frac{1}{2} (a + b + c) - a\right]\cdot \left[\frac{1}{2} (a + b + c) - b\right]\cdot\left[\frac{1}{2} (a + b + c) - c\right]$$

Thus, $K^2 = s(s - a)(s - b)(s - c)$, where $s = \frac{1}{2} (a + b + c)$.

Then $K = \sqrt{s(s - a)(s - b)(s - c)}$, where $s = \frac{1}{2} (a + b + c)$.

50. 461.1 ft^2

51.

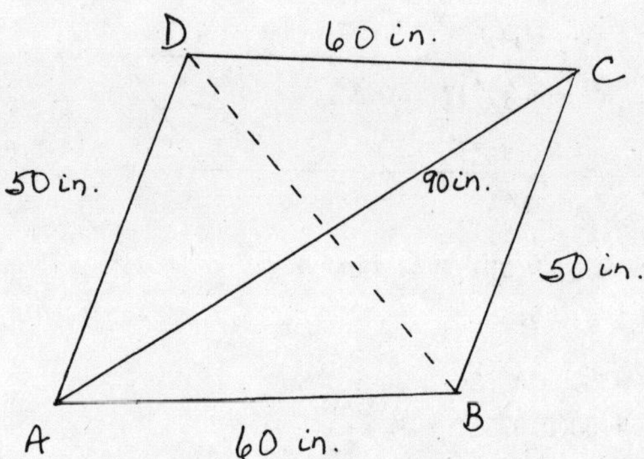

First consider Δ ABC and find B:

$b^2 = a^2 + c^2 - 2ac \cos B$

$90^2 = 50^2 + 60^2 - 2\cdot50\cdot60 (\cos B)$

$8100 = 2500 + 3600 - 6000 \cos B$

$2000 = -6000 \cos B$

$\cos B = - \frac{2000}{6000} = -0.3333$

$B = 109.5°$

Then in the parallelogram ABCD, the measure of angle A is 180° - 109.5°, or 70.5°.

Now consider Δ ABD and find the length of the side opposite A (that is, the length of the other diagonal of the parallelogram.)

$a^2 = b^2 + d^2 - 2bd \cos A$

$a^2 = 50^2 + 60^2 - 2\cdot50\cdot60 (\cos 70.5°)$

$a^2 = 2500 + 3600 - 6000(0.3338)$

$a^2 = 4097$

$a = \sqrt{4097} = 64$ in.

52. 62.3°

53.

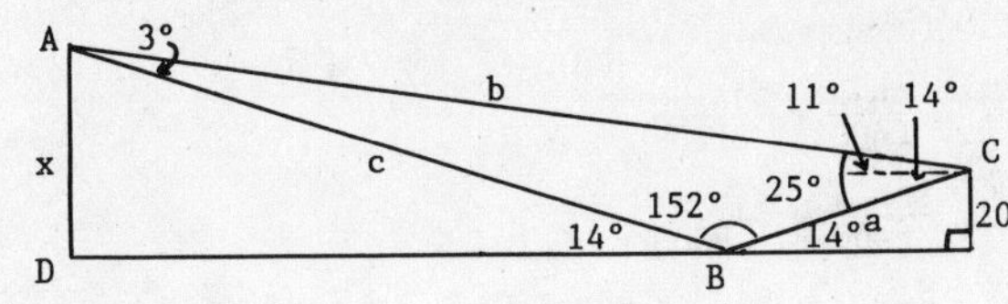

Find ∠ABC, ∠CAB, and ∠ACB:

The angle of incidence equals the angle of reflection. Thus ∠ABC = 180° - (14° + 14°), or 152°. Then ∠CAB = 180° - (25° + 152°), or 3°, ∠ACB = 11° + 14°, or 25°.

Find a:

$\frac{a}{20} = \csc 14°$

$a = 20 \csc 14° = 20 \times 4.1336 = 82.7$

Find c:

Using the law of sines,

$\frac{c}{\sin 25°} = \frac{82.7}{\sin 3°}$

$c = \frac{82.7 \sin 25°}{\sin 3°} = \frac{82.7 \times 0.4226}{0.0523} = 668$

Find x:

$\frac{x}{668} = \sin 14°$

$x = 668 \sin 14° = 668 \times 0.2419 = 162$ ft

The height of the tree is 162 ft.

54. 9386 ft

55.

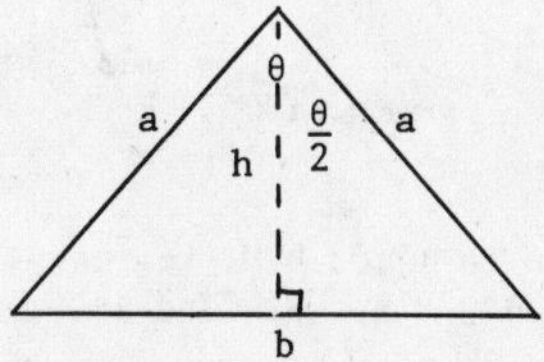

Let b represent the base, h the height, and θ the included angle.

Find h:

$\frac{h}{a} = \cos \frac{\theta}{2}$

$h = a \cos \frac{\theta}{2}$

Find b:

$b^2 = a^2 + a^2 - 2 \cdot a \cdot a \cos \theta$

$b^2 = 2a^2 - 2a^2 \cos \theta$

$b^2 = 2a^2(1 - \cos \theta)$

$b = \sqrt{2}\, a\sqrt{1 - \cos \theta}$

55. (continued)

Find A (area):

$A = \frac{1}{2}\, bh$

$A = \frac{1}{2}\left[\sqrt{2}\, a\sqrt{1 - \cos \theta}\right]\left[a \cos \frac{\theta}{2}\right]$

$= \left[a\sqrt{\frac{1 - \cos \theta}{2}}\right]\left[a\sqrt{\frac{1 + \cos \theta}{2}}\right]$

$= \frac{1}{2}\, a^2 \sqrt{1 - \cos^2\theta} = \frac{1}{2}\, a^2 \sqrt{\sin^2\theta}$

$= \frac{1}{2}\, a^2 \sin \theta$

The maximum area occurs when θ is 90°.

56. a) S 87° 36' E

b) 504.9 ft, N 70° 17' E

57. Let $S^2 = a^2 + b^2 + c^2$. Then

$a^2 = b^2 + c^2 - 2bc \cos A$

$b^2 = c^2 + a^2 - 2ca \cos B$

(+) $c^2 = a^2 + b^2 - 2ab \cos C$

$S^2 = S^2 + S^2 - 2(bc \cos A + ac \cos B + ab \cos C).$

$S^2 = a^2 + b^2 + c^2$

$= 2(bc \cos A + ac \cos B + ab \cos C)$

58. From Exercise 57 we have

$2bc \cos A + 2ac \cos B + 2ab \cos C = a^2 + b^2 + c^2.$

Multiplying each term on both sides of the equation by $\frac{1}{2abc}$ we obtain

$\frac{\cos A}{a} + \frac{\cos B}{b} + \frac{\cos C}{c} = \frac{a^2 + b^2 + c^2}{2abc}.$

59. We add labels to the drawing in the text.

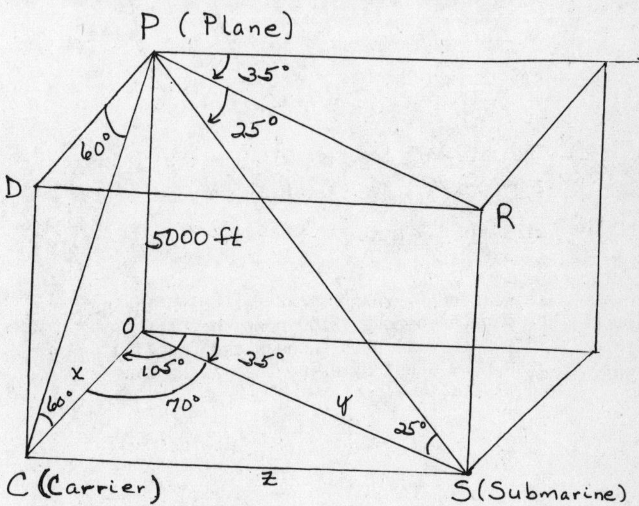

59. (continued)

We find x using Δ PCO:

$\cot 60° = \frac{x}{5000}$

$x = 5000 \cot 60°$

$x \approx 2886.75$

We find y using Δ PSO:

$\cot 25° = \frac{y}{5000}$

$y = 5000 \cot 25°$

$y \approx 10{,}722.53$

Finally, we use Δ OCS and the law of cosines to find z, the distance from the submarine to the carrier.

$z^2 = x^2 + y^2 - 2xy \cos 70°$

$z^2 = (2886.75)^2 + (10{,}722.53)^2 - 2(2886.75)(10{,}722.53) \cos 70°$

$z^2 \approx 102{,}132{,}695.9$

$z \approx 10{,}106$ ft

60. $\frac{rR}{R - r}$, where R is the radius of the sphere with center A and r is the radius of the sphere with center B

61.

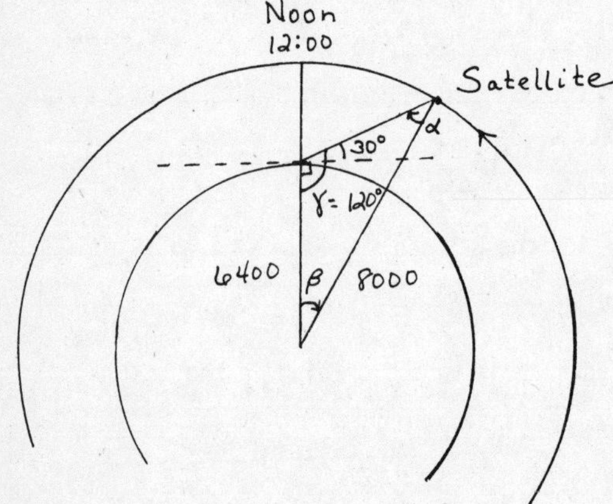

a) The satellite travels 360° (one orbit) in 2 hr, so it travels 1° in $\frac{1}{3}$ min

$\left(\frac{120 \text{ min}}{360°} = \frac{x \text{ min}}{1°},\ x = \frac{1}{3}\right)$.

If we find β, then we can determine the number of minutes before 12:00 noon that the satellite will pass through the tracking beam. We start by using the law of sines to find α.

61. (continued)

$\frac{\sin \alpha}{6400} = \frac{\sin \gamma}{8000}$

$\frac{\sin \alpha}{6400} = \frac{\sin 120°}{8000}$

$\sin \alpha = \frac{6400 \times \sin 120°}{8000} = \frac{6400 \times 0.8660}{8000}$

$\sin \alpha \approx 0.6928$

$\alpha \approx 43.8538°$

$\beta = 180° - (\alpha + \gamma) = 180° - (43.8538° + 120°)$
$= 16.1462°$

$\frac{\frac{1}{3} \text{ min}}{1°} = \frac{m \text{ min}}{16.1462°}$

$m = \frac{1}{3}(16.1462) \approx 5.3821$

Thus, the satellite passes through the tracking beam about 5 minutes before 12:00 noon, or at about 11:55 A.M.

b)

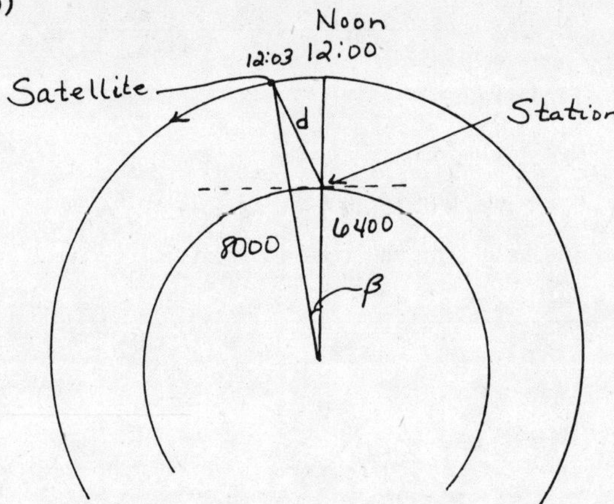

From 12:00 noon to 12:03 P.M. the satellite passes through angle β. We find β:

$\frac{\beta}{3 \text{ min}} = \frac{1°}{\frac{1}{3} \text{ min}}$ (See part (a))

$\beta = 9°$

Now use the law of cosines to find d:

$d^2 = 6400^2 + 8000^2 - 2(6400)(8000) \cos \beta$

$d^2 = 6400^2 + 8000^2 - 2(6400)(8000) \cos 9°$

$d^2 \approx 3{,}820{,}713.923$

$d \approx 1955$ km

62. 51.26°

63.

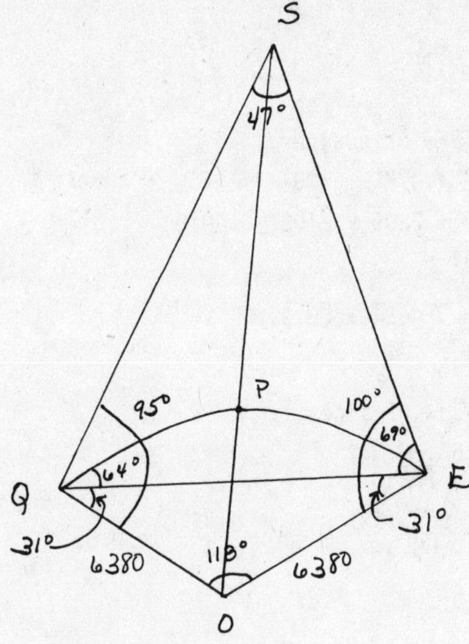

Consider Δ QEO:

$$\frac{\overline{QE}}{\sin 118°} = \frac{6380}{\sin 31°}$$

$$\overline{QE} = \frac{6380 \times \sin 118°}{\sin 31°} \approx 10,937.455$$

Consider Δ QES:

$$\frac{\overline{QS}}{\sin 69°} = \frac{\overline{QE}}{\sin 47°}$$

$$\overline{QS} = \frac{10,937.455 \times \sin 69°}{\sin 47°} \approx 13,961.772$$

Consider Δ OQS:

$$\overline{OS}^2 = 6380^2 + \overline{QS}^2 - 2 \cdot \overline{QS} \cdot 6380 \ (\cos 95°)$$

$$\overline{OS}^2 = 6380^2 + (13,961.772)^2 - 2(6380)(13,961.772) \cos 95°$$

$$\overline{OS}^2 \approx 251,162,467.4$$

$$\overline{OS} \approx 15,848$$

Then the distance of the satellite from the earth, $\overline{PS}$, is 15,848 - 6380 = 9468 km.

Exercise Set 8.4

1.

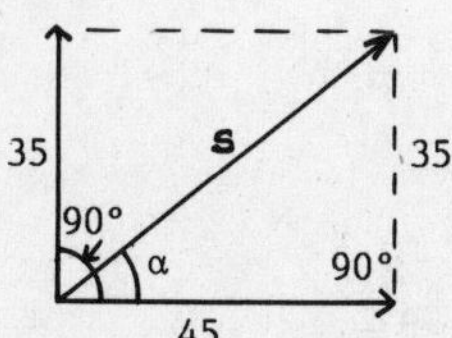

Find the resultant, s, and the angle, α, it makes with vector **a**. Consider the following triangle:

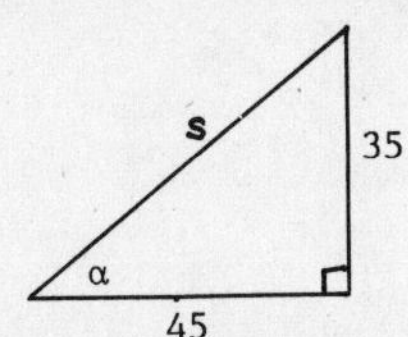

1. (continued)

Find $|s|$:

Using the Pythagorean theorem,

$|s|^2 = 35^2 + 45^2$ ($|s|$ denotes the length of s)

$= 1225 + 2025$

$= 3250$

Then $|s| = \sqrt{3250} = 57.0$

Find α:

$\tan \alpha = \frac{35}{45} = 0.7778$

Thus $\alpha = 38°$.

2. 69, 39°

3.

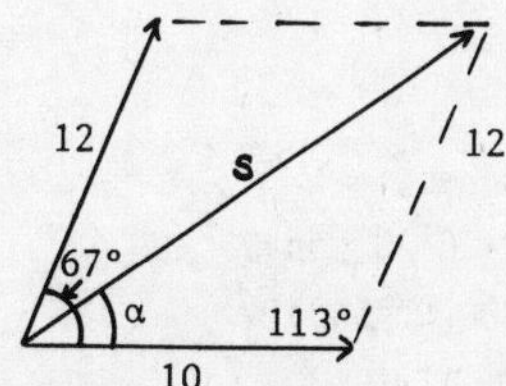

Find the resultant, s, and the angle, α, it makes with vector **a**. Consider the following triangle:

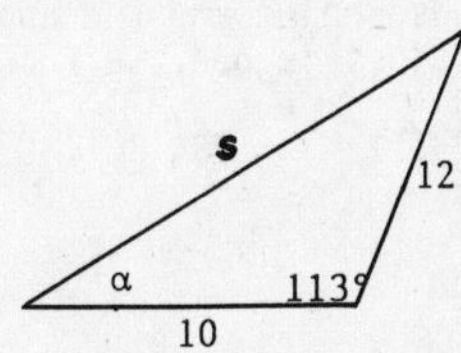

Find $|s|$:

Using the law of cosines,

$|s|^2 = 10^2 + 12^2 - 2 \cdot 10 \cdot 12 \cos 113°$

$= 100 + 144 - 240(-0.3907)$

$= 338$

Then $|s| = \sqrt{338} = 18.4$.

Find α:

Using the law of sines,

$$\frac{12}{\sin \alpha} = \frac{18.4}{\sin 113°}$$

$$\sin \alpha = \frac{12 \sin 113°}{18.4} = \frac{12 \times 0.9205}{18.4} = 0.6003$$

Thus $\alpha = 37°$.

4. 43.7, 42°

5. 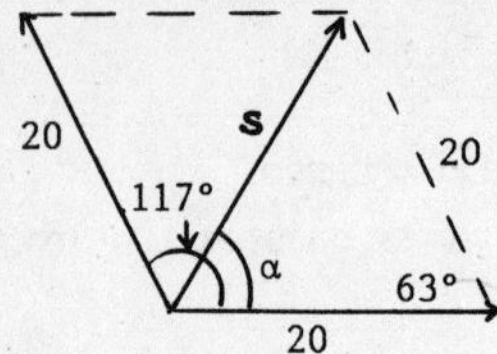

Find the resultant, s, and the angle, α, it makes with vector **a**. Consider the following triangle:

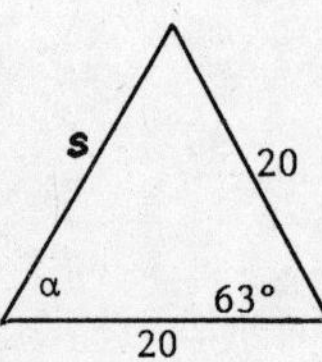

Find $|s|$:

Using the law of cosines,

$$|s|^2 = 20^2 + 20^2 - 2\cdot20\cdot20\cos 63^\circ$$
$$= 400 + 400 - 800(0.4540)$$
$$= 437$$

Then $|s| = \sqrt{437} = 20.9$.

Find α:

Since the triangle is isosceles and the angle between the congruent sides is 63°, we know that

$$\alpha = \frac{1}{2}(180^\circ - 63^\circ) = \frac{1}{2}(117^\circ) = 58^\circ\ 30'.$$

6. 28.6, 62°

7. 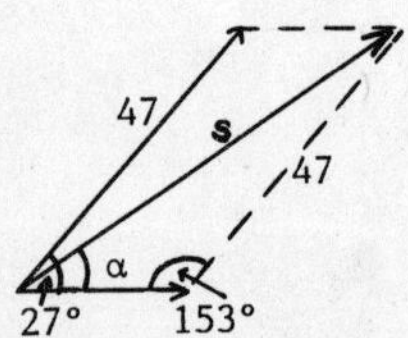

Find the resultant, s, and the angle, α, it makes with vector **a**. Consider the following triangle:

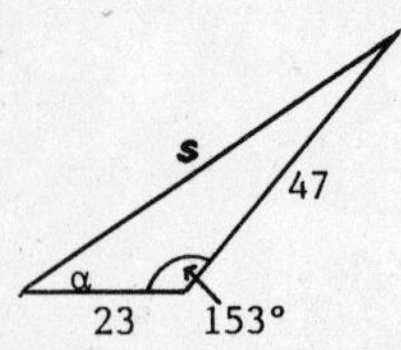

7. (continued)

Find $|s|$:

Using the law of cosines,

$$|s|^2 = 23^2 + 47^2 - 2\cdot23\cdot47\cos 153^\circ$$
$$= 529 + 2209 - 2162(-0.8910)$$
$$= 4664$$

Then $|s| = \sqrt{4664} = 68.3$.

Find α:

Using the law of sines,

$$\frac{47}{\sin\alpha} = \frac{68.3}{\sin 153^\circ}$$

$$\sin\alpha = \frac{47\sin 153^\circ}{68.3} = \frac{47 \times 0.4540}{68.3} = 0.3124$$

Thus $\alpha = 18^\circ$.

8. 89.2, 52°

9. 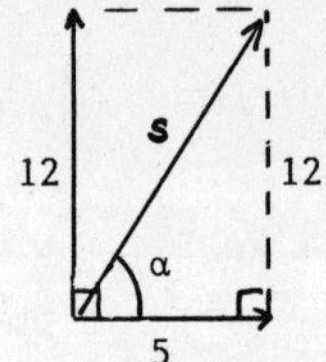

Find $|s|$:

Using the Pythagorean theorem,

$$|s|^2 = 12^2 + 5^2$$
$$= 144 + 25$$
$$= 169$$

Then $|s| = \sqrt{169} = 13$ kg.

Find α:

$$\tan\alpha = \frac{12}{5} = 2.4$$

Then $\alpha = 67^\circ$.

10. 50 kg, 53°

11. 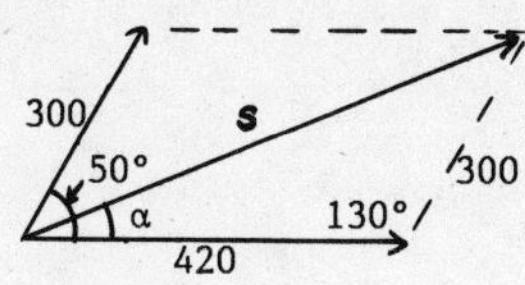

Find $|s|$:

Using the law of cosines,

$$|s|^2 = 420^2 + 300^2 - 2\cdot 420\cdot 300 \cos 130^\circ$$
$$= 176{,}400 + 90{,}000 - 252{,}000(-0.6428)$$
$$= 428{,}386$$

Then $|s| = \sqrt{428{,}386} = 655$ kg

Find α:

Using the law of sines,

$$\frac{300}{\sin\alpha} = \frac{655}{\sin 130^\circ}$$

$$\sin\alpha = \frac{300 \sin 130^\circ}{655} = \frac{300 \times 0.7660}{655} = 0.3508$$

Thus $\alpha = 21^\circ$.

12. 929 kg, 28°

13. 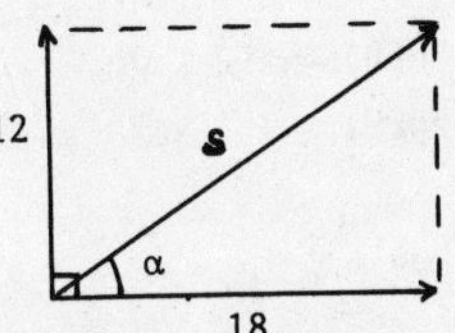

Find $|s|$, the speed of the balloon, and α, the angle it makes with the horizontal. Consider the following triangle:

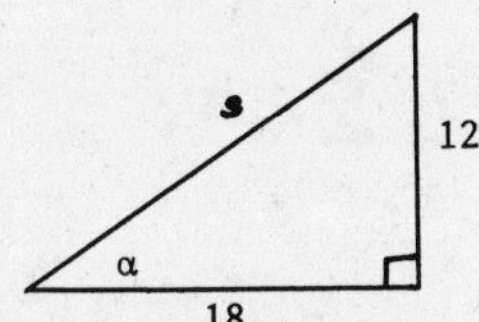

Find $|s|$:

$$|s|^2 = 18^2 + 12^2 = 324 + 144 = 468$$

Then $|s| = \sqrt{468} = 21.6$.

Find α:

$$\tan\alpha = \frac{12}{18} = 0.6667$$

Thus $\alpha = 34^\circ$.

The speed of the balloon is 21.6 ft/sec, and it makes a 34° angle with the horizontal.

14. 11.2 ft/sec, 63°

15. 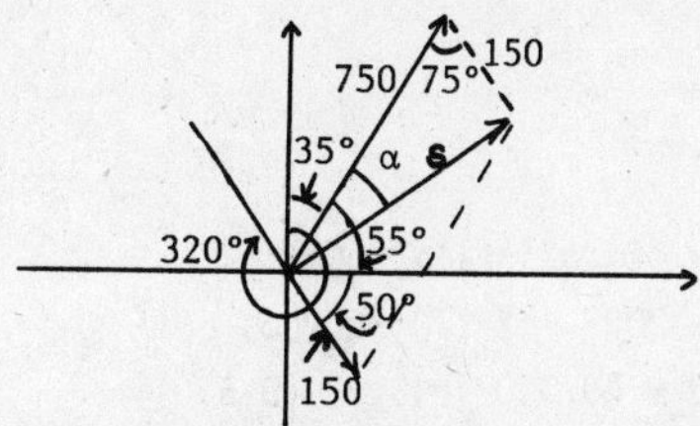

Find $|s|$:

Using the law of cosines,

$$|s|^2 = 750^2 + 150^2 - 2\cdot 750\cdot 150 \cos 75^\circ$$
$$= 562{,}500 + 22{,}500 - 225{,}000(0.2588)$$
$$= 526{,}770$$

Then $|s| = \sqrt{526{,}770} = 726$.

Find α:

Using the law of sines,

$$\frac{150}{\sin\alpha} = \frac{726}{\sin 75^\circ}$$

$$\sin\alpha = \frac{150 \sin 75^\circ}{726} = \frac{150 \times 0.9659}{726} = 0.1996$$

Thus $\alpha = 12^\circ$.

The resultant force is 726 lb, and the boat is moving in the direction 35° + 12°, or 47°.

16. 729 lb, 225°

17. 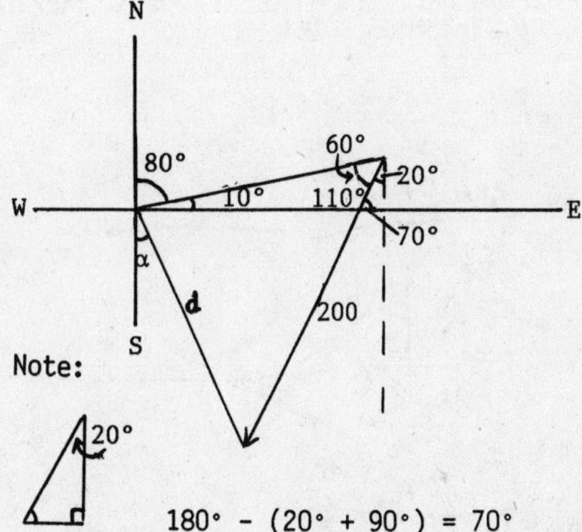

Note:

20° $\quad 180^\circ - (20^\circ + 90^\circ) = 70^\circ$

70° $\quad 180^\circ - 70^\circ = 110^\circ$

80° $\quad 90^\circ - 80^\circ = 10^\circ$

10° 110° $\quad 180^\circ - (10^\circ + 110^\circ) = 60^\circ$

17. (continued)

Find $|d|$:

Using the law of cosines,

$|d|^2 = 120^2 + 200^2 - 2\cdot120\cdot200 \cos 60°$

$= 14{,}400 + 40{,}000 - 48{,}000(0.5)$

$= 30{,}400$

Then $|d| = \sqrt{30{,}400} = 174$.

Use the law of sines to find θ in this triangle.

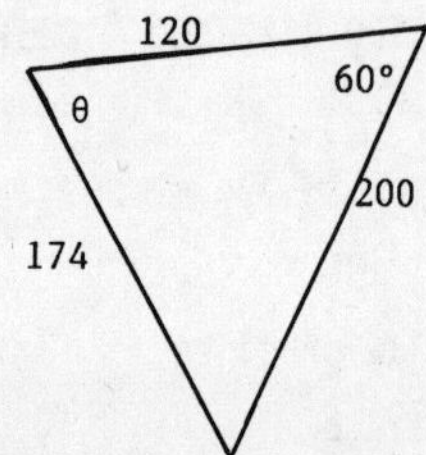

$$\frac{200}{\sin\theta} = \frac{174}{\sin 60°}$$

$$\sin\theta = \frac{200 \sin 60°}{174} = \frac{200(0.8660)}{174} = 0.9954$$

Thus θ = 85°.

Now we can find α.

$\alpha = 180° - (80° + 85°) = 15°$

The ship is 174 nautical miles from the starting point in the direction S 15° 30' E.

18. 215 km, 345°

19.

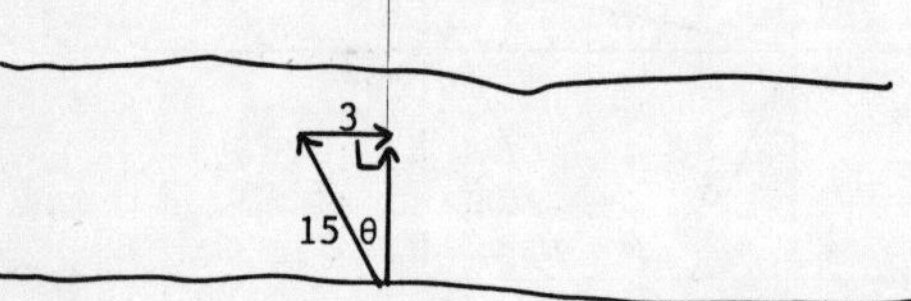

Find θ:

$$\sin\theta = \frac{3}{15} = 0.2000$$

Thus θ = 12°.

The boat should be pointed at an angle of 12° upstream.

20. 70.4°

21.

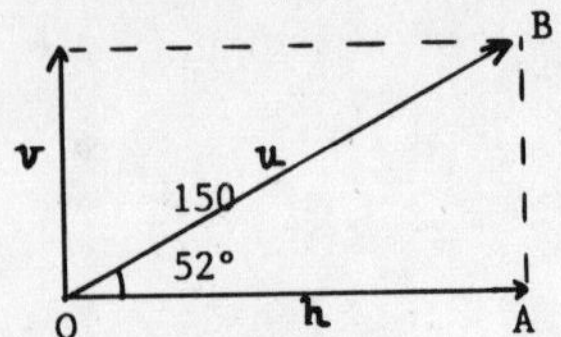

From Δ OAB, we find

$|h| = 150 \cos 52° = 150(0.6157) = 92.3$

$|v| = 150 \sin 52° = 150(0.7880) = 118.2$

22. Vert: 151.5 downward,
Hort: 77.2 to the right

23.

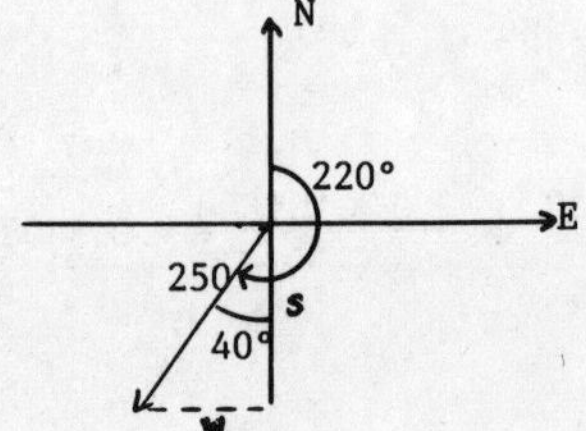

From the drawing we find

$|s| = 250 \cos 40° = 250(0.7660) = 192$ km/h

$|w| = 250 \sin 40° = 250(0.6428) = 161$ km/h

24. S: 16.1 mph, E: 19.2 mph

25.

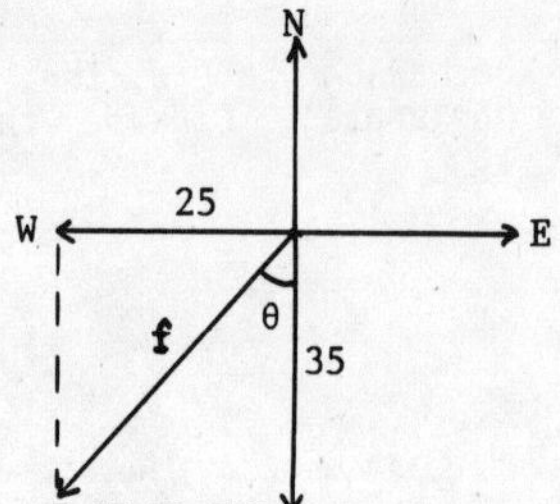

Find $|f|$:

$|f| = \sqrt{25^2 + 35^2} = \sqrt{1850} = 43.0$

Find θ:

$$\tan\theta = \frac{25}{35} = 0.7143$$

$\theta = 35°\ 32'$

The vector f has a magnitude of 43 kg and a direction of S 35° 32' W.

26. 111 kg, 35° 50' with horizontal

27.

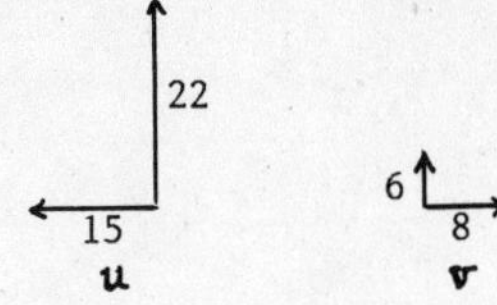

a) Adding the east-west components, we obtain 7 west. Adding the north-south components, we obtain 28 north. The components of **u** + **v** are 7 west and 28 north.

b)

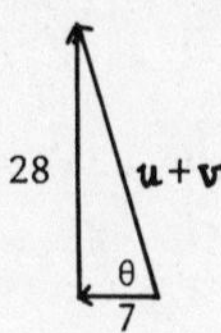

Find θ:

$\tan \theta = \frac{28}{7} = 4$

$\theta = 75° 58'$

Thus $90° - 75° 58' = 14° 2'$.

Find $|\mathbf{u} + \mathbf{v}|$:

$|\mathbf{u} + \mathbf{v}| = 28 \csc 75° 58' = 28(1.0308) = 28.9$

The vector **u** + **v** has a magnitude of 28.9 and a direction of N 14° 2' W.

28. a) <23, 13>

b) 26.4, 29° 29'

29. <3, 7> + <2, 9>
= <3 + 2, 7 + 9>
= <5, 16>

30. <2, -5>

31. <17, 7.6> + <-12.2, 6.1>
= <17 + (-12.2), 7.6 + 6.1>
= <4.8, 13.7>

32. <-7.3, 19.3>

33. <-650, -750> + <-12, 324>
= <-650 + (-12), -750 + 324>
= <-662, -426>

34. <-429, -717>

35. Let **v** = <3, 4>.

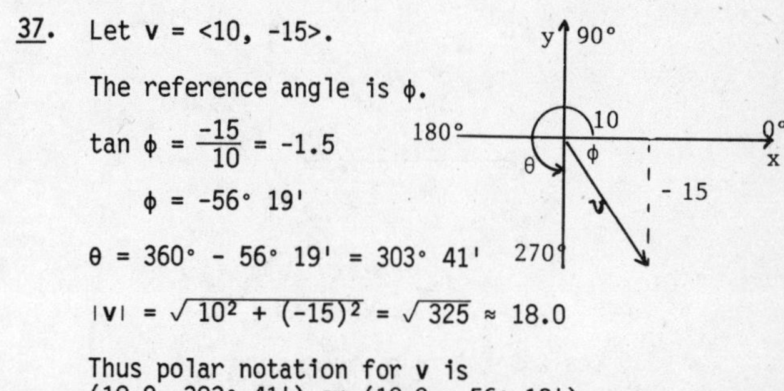

$\tan \theta = \frac{4}{3} = 1.333$

$\theta = 53° 8'$

$|\mathbf{v}| = \sqrt{3^2 + 4^2} = \sqrt{25} = 5$

Thus polar notation for **v** is (5, 53° 8').

36. (5, 36° 52')

37. Let **v** = <10, -15>.

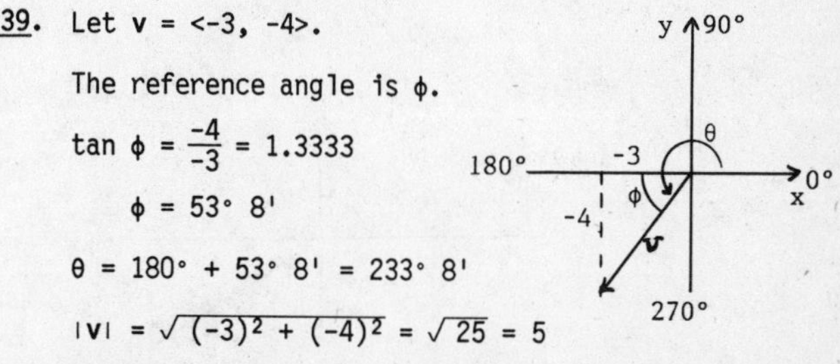

The reference angle is ϕ.

$\tan \phi = \frac{-15}{10} = -1.5$

$\phi = -56° 19'$

$\theta = 360° - 56° 19' = 303° 41'$

$|\mathbf{v}| = \sqrt{10^2 + (-15)^2} = \sqrt{325} \approx 18.0$

Thus polar notation for **v** is (18.0, 303° 41') or (18.0, -56° 19').

38. (19.7, 329° 32')

39. Let **v** = <-3, -4>.

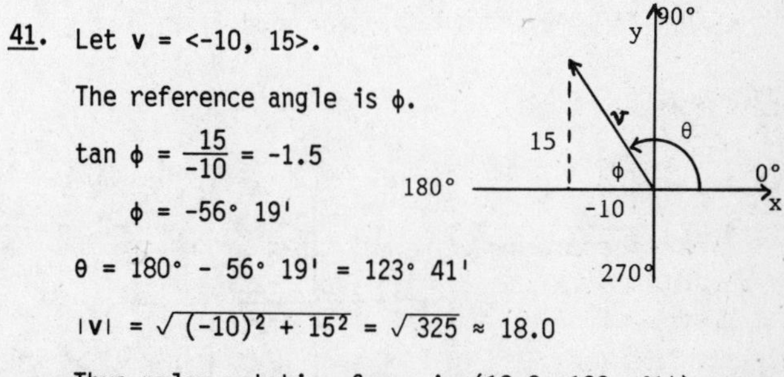

The reference angle is ϕ.

$\tan \phi = \frac{-4}{-3} = 1.3333$

$\phi = 53° 8'$

$\theta = 180° + 53° 8' = 233° 8'$

$|\mathbf{v}| = \sqrt{(-3)^2 + (-4)^2} = \sqrt{25} = 5$

Thus polar notation for **v** is (5, 233° 8').

40. (5, 216° 52')

41. Let **v** = <-10, 15>.

y 90° 15 v θ ϕ 180° -10 0° x 270°

The reference angle is ϕ.

$\tan \phi = \frac{15}{-10} = -1.5$

$\phi = -56° 19'$

$\theta = 180° - 56° 19' = 123° 41'$

$|\mathbf{v}| = \sqrt{(-10)^2 + 15^2} = \sqrt{325} \approx 18.0$

Thus polar notation for **v** is (18.0, 123° 41').

42. (19.7, 149° 32')

43. $v = (4, 30°)$

From the drawing we have

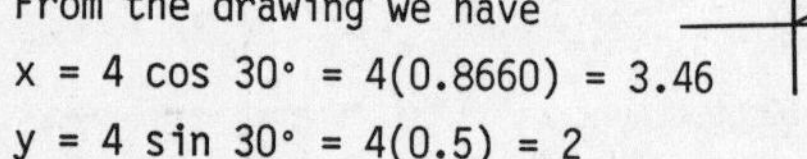

$x = 4 \cos 30° = 4(0.8660) = 3.46$

$y = 4 \sin 30° = 4(0.5) = 2$

Thus rectangular notation for **v** is $\langle 3.46, 2\rangle$.

44. $\langle 4, 6.93\rangle$

45. $v = (10, 235°)$

From the drawing we have

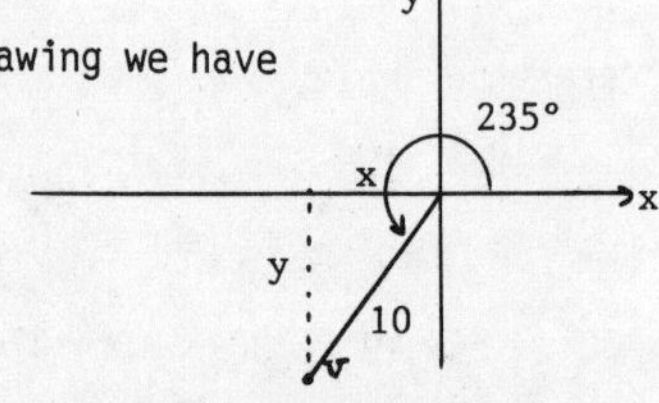

$x = 10 \cos 235° = 10(-0.5736) = -5.74$

$y = 10 \sin 235° = 10(-0.8192) = -8.19$

Thus rectangular notation for **v** is $\langle -5.74, -8.19\rangle$.

46. $\langle -13.0, -7.50\rangle$

47. $v = (20, 330°)$

From the drawing we have

$x = 20 \cos 330° = 20(0.8660) = 17.3$

$y = 20 \sin 330° = 20(-0.5) = -10$

Thus rectangular notation for **v** is $\langle 17.3, -10\rangle$.

48. $\langle -18.8, -6.84\rangle$

49. $v = (100, -45°)$

From the drawing we have

$x = 100 \cos (-45°) = 100(0.7071) = 70.7$

$y = 100 \sin (-45°) = 100(-0.7071) = -70.7$

Thus rectangular notation for **v** is $\langle 70.7, -70.7\rangle$.

50. $\langle 75, -130\rangle$

51. - 60.

Given: $\mathbf{u} = \langle 3, 4\rangle$, $\mathbf{v} = \langle 5, 12\rangle$, and $\mathbf{w} = \langle -6, 8\rangle$

51. $3\mathbf{u} + 2\mathbf{v} = 3\langle 3, 4\rangle + 2\langle 5, 12\rangle$
$= \langle 9, 12\rangle + \langle 10, 24\rangle$
$= \langle 9 + 10, 12 + 24\rangle$
$= \langle 19, 36\rangle$

52. $\langle -1, -12\rangle$

53. $(\mathbf{u} + \mathbf{v}) - \mathbf{w} = (\langle 3, 4\rangle + \langle 5, 12\rangle) - \langle -6, 8\rangle$
$= \langle 8, 16\rangle + \langle 6, -8\rangle$
$= \langle 14, 8\rangle$

54. $\langle 4, -16\rangle$

55. $|\mathbf{u}| = |\langle 3, 4\rangle| = \sqrt{3^2 + 4^2} = \sqrt{25} = 5$

$|\mathbf{v}| = |\langle 5, 12\rangle| = \sqrt{5^2 + 12^2} = \sqrt{169} = 13$

$|\mathbf{u}| + |\mathbf{v}| = 5 + 13 = 18$

56. -8

57. $|\mathbf{u} + \mathbf{v}| = |\langle 3, 4\rangle + \langle 5, 12\rangle|$
$= |\langle 8, 16\rangle|$
$= \sqrt{8^2 + 16^2} = \sqrt{64 + 256}$
$= \sqrt{320}$
$= 17.9$

58. 8.246

59. $2|\mathbf{u} + \mathbf{v}| = 2(17.9)$ (See Exercise 57.)
$= 35.8$

60. 36

61. Let $\mathbf{PQ} = \langle x, y\rangle$. Then $\mathbf{QP} = \langle -x, -y\rangle$ and $\mathbf{PQ} + \mathbf{QP} = \langle x, y\rangle + \langle -x, -y\rangle = \langle 0, 0\rangle = \mathbf{0}$

62. DEFINE: $\mathbf{u \cdot v} = uv \cos \theta$ where $u = |\mathbf{u}|$, $v = |\mathbf{v}|$, and θ is the angle between the vectors.

PROVE: **u** **v** if $\mathbf{u \cdot v} = 0$; and conversely.

i) By hypothesis, $uv(\cos \theta) = 0$. Hence $\theta = 90°$ or $270°$; in either case the vectors are perpendicular.

ii) Conversely, if the vectors are perpendicular, $\cos \theta = 0$ and then $\mathbf{u \cdot v} = 0$.

63. SHOW: If k is a scalar, then
k(u + v) = ku + kv.

In the following proof we will be using

(1) <m, n> + <p, q> = <m + p, n + q>

(2) r<x, y> = <rx, ry>

(3) k(r + s) = kr + ks.

Proof: Let u = <a, b> and v = <c, d>. Then

k(u + v) = k(<a, b> + <c, d>)
= k<a + c, b + d> by (1)
= <k(a + c), k(b + d)> by (2)
= <ka + kc, kb + kd> by (3)
= <ka, kb> + <kc, kd> by (1)
= k<a, b> + k<c, d> by (2)
= ku + kv

64. $\langle \frac{3}{5}, \frac{4}{5} \rangle$

65. PROVE: $|u + v| \leqslant |u| + |v|$

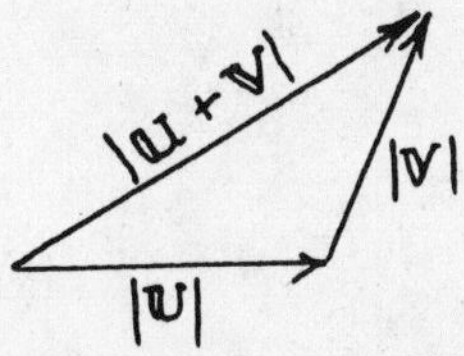

If the vectors are not collinear (or parallel) they form a triangle as shown. From plane geometry (Euclid) we know that (any) one side of a triangle is less than the sum of the other two sides.

If the vectors are collinear, then either

$|u + v| = |u| + |v|$

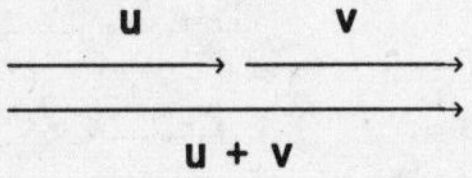

or

$|u + v| < |u| + |v|$

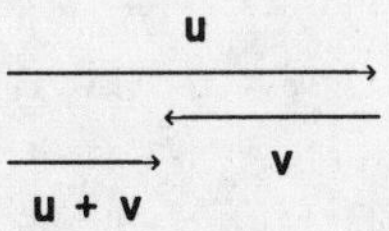

Exercise Set 8.5

1. At the end of the boom there are three forces acting at a point:

1) The 150-lb weight of the sign acting down.
2) The cable pulling up and to the left.
3) The boom pushing to the right.

It helps to draw a force diagram.

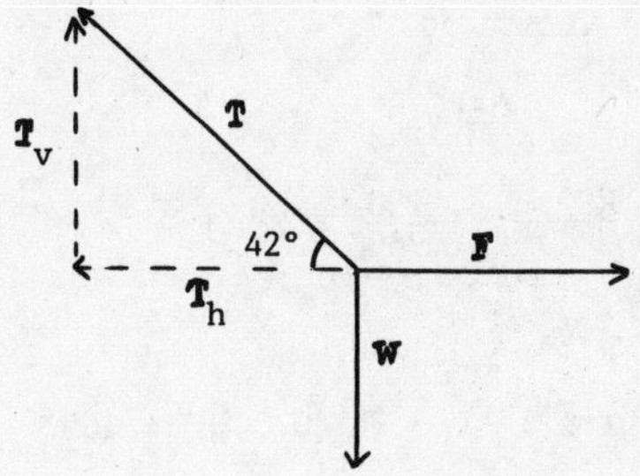

We let T represent the tension in the cable, F the compression in the boom, and W the weight. We resolve the tension in the cable into horizontal (T_h) and vertical (T_v) components. The sum of the horizontal components must be 0, and the sum of the vertical components must also be 0.

Thus, $|T_v| = 150$ lb and $|F| = |T_h|$.

$$\sin 42° = \frac{|T_v|}{|T|} = \frac{150}{|T|}$$

$$|T| = \frac{150}{\sin 42°} = \frac{150}{0.6691} = 224 \text{ lb}$$

$$\cos 42° = \frac{|T_h|}{|T|} = \frac{|T_h|}{224}$$

$$|T_h| = 224 \cos 42° = 224(0.7431) = 166$$

$$|F| = 166 \text{ lb}$$

Tension in the cable: 224 lb
Compression in the boom: 166 lb

2. 386 lb, 243 lb

3. There are three forces acting at point C:

1) The 200 kg weight acting down.
2) Rod AC pushing up and to the right.
3) Rod BC pulling to the left.

Draw a force diagram.

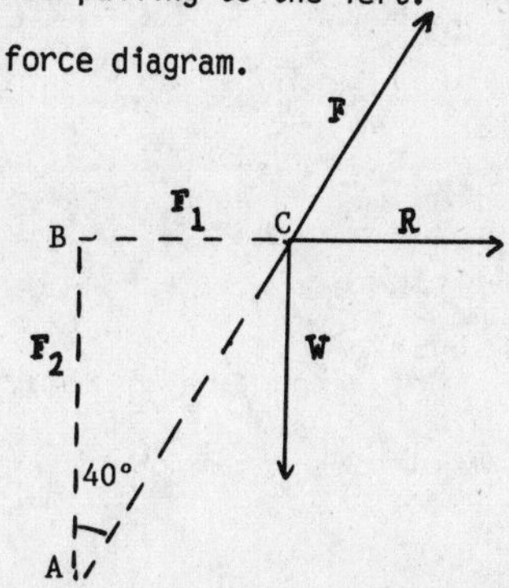

3. (continued)

F is the force exerted by rod AC.
R is the force exerted by rod BC.

Resolve the force in rod AC into horizontal and vertical components, F_1 and F_2. There is a balance so F + R + W = 0. The sum of the horizontal components (F_1 + R) is 0, and the sum of the vertical components (F_2 + W) must also be 0.

Thus, $|F_2| = 200$kg and $|R| = |F_1|$.

$$\cos 40° = \frac{|F_2|}{|F|} = \frac{200}{|F|}$$

$$|F| = \frac{200}{\cos 40°} = \frac{200}{0.7660} = 261 \text{ kg}$$

$$\sin 40° = \frac{|F_1|}{|F|} = \frac{|F_1|}{261}$$

$$|F_1| = 261 \sin 40° = 261(0.6428) = 168$$

$$|R| = 168 \text{ kg}$$

The force exerted by rod AC is 261 kg.
The force exerted by rod BC is 168 kg.

4. 467 kg, 358 kg

5.

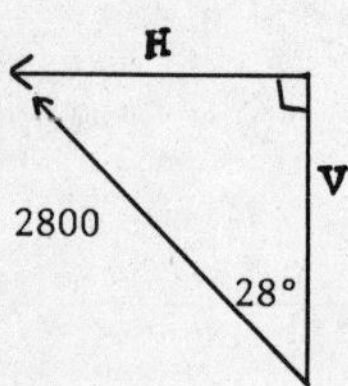

Let V represent the vertical component and H represent the horizontal component.

$$\sin 28° = \frac{|H|}{2800}$$

$$|H| = 2800 \sin 28° = 2800(0.4695) = 1315 \text{ lb}$$

$$\cos 28° = \frac{|V|}{2800}$$

$$|V| = 2800 \cos 28° = 2800(0.8829) = 2472 \text{ lb}$$

The lift is 2472 lb. The drag is 1315 lb.

6. 2968 lb, 1855 lb

7.

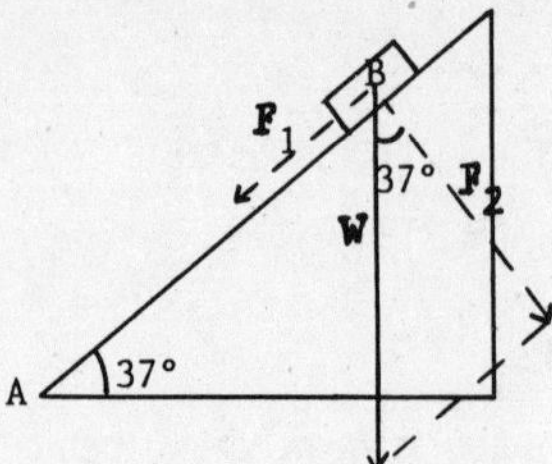

The weight W is acting downward. F_1 and F_2 are the components. The angle at B has the same measure as the angle at A because their sides are respectively perpendicular.

Find $|F_1|$.

$$\frac{|F_1|}{W} = \sin 37°$$

$$|F_1| = W \sin 37°$$

$$|F_1| = 100(0.6018) = 60$$

A force of 60 kg is needed to keep the block from sliding down.

8. 792 kg

9.

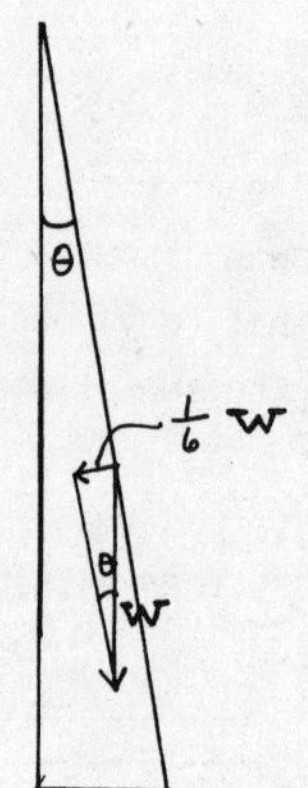

$$\sin \theta = \frac{\frac{1}{6} W}{W} = \frac{1}{6} = 0.16667$$

$$\theta \approx 9° \; 36'$$

10. 22°

11. First draw a force diagram.

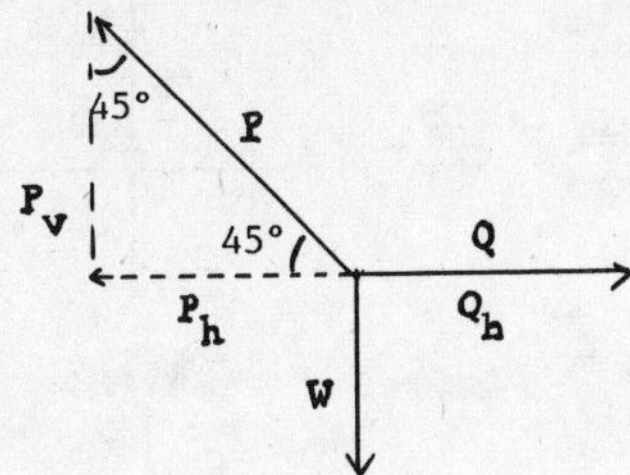

The forces exerted by the ropes are P and Q. The horizontal components of P and Q have magnitudes $|P| \cos 45°$ and $|Q| \cos 0°$. Since there is a balance, these must be the same.

$|P| \cos 45° = |Q| \cos 0°$

The vertical components of P and Q have magnitudes $|P| \sin 45°$ and $|Q| \sin 0°$. Since there is a balance, these must total 400 kg, the weight of the block.

$|P| \sin 45° + |Q| \sin 0° = 400$

We now have two equations with unknowns $|P|$ and $|Q|$.

$|P| \cos 45° - |Q| \cos 0° = 0 \quad (1)$

$|P| \sin 45° + |Q| \sin 0° = 400 \quad (2)$

Solve Equation (1) for $|P|$.

$|P| = |Q| \dfrac{\cos 0°}{\cos 45°} \quad (3)$

Substitute in Equation (2) and solve for $|Q|$.

$$|Q| \frac{\cos 0° \sin 45°}{\cos 45°} + |Q| \sin 0° = 400$$

$$|Q| \frac{\cos 0° \sin 45°}{\cos 45°} + |Q| \frac{\sin 0° \cos 45°}{\cos 45°} = 400$$

$$|Q|(\cos 0° \sin 45° + \sin 0° \cos 45°) = 400 \cos 45°$$

$$|Q|[\sin (45° + 0°)] = 400 \cos 45°$$

(Using an identity)

$$|Q|(\sin 45°) = 400 \cos 45°$$

$$|Q| = \frac{400 \cos 45°}{\sin 45°}$$

$$|Q| = 400 \tan 45°$$

$$|Q| = 400$$

Substituting in Equation (3), we obtain

$$|P| = 400 \frac{\cos 0°}{\cos 45°} = 400 \cdot \frac{1}{0.7071} = 566.$$

The tension in rope P is 566 kg, and the tension in rope Q is 400 kg.

12. 1732 lb

13. First draw force diagram.

The forces exerted by the ropes are P and Q. The horizontal components of P and Q have magnitudes $|P| \cos 30°$ and $|Q| \cos 30°$. Since there is a balance, these must be the same.

$|P| \cos 30° = |Q| \cos 30°$, or $|P| = |Q|$

The vertical components of P and Q have magnitudes $|P| \sin 30°$ and $|Q| \sin 30°$. Since there is a balance, these must total 2000 kg, the weight of the block.

$|P| \sin 30° + |Q| \sin 30° = 2000$

Substitute $|Q|$ for $|P|$ and solve for $|Q|$.

$$|Q| \sin 30° + |Q| \sin 30° = 2000$$

$$2|Q| \sin 30° = 2000$$

$$2|Q| \cdot \frac{1}{2} = 2000$$

$$|Q| = 2000$$

Since $|Q| = |P|$, the tension in each rope is 2000 kg.

14. 1061 lb

15. First draw a force diagram.

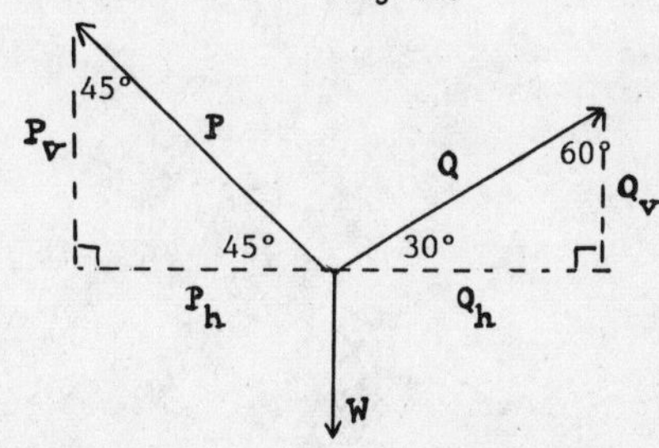

The forces exerted by the ropes are P and Q. The horizontal components of P and Q have magnitudes $|P| \cos 45°$ and $|Q| \cos 30°$. Since there is a balance, these must be the same.

$|P| \cos 45° = |Q| \cos 30°$

The vertical components of P and Q have magnitudes $|P| \sin 45°$ and $|Q| \sin 30°$. Since there is a balance, these must total 2500 kg, the weight of the block.

$|P| \sin 45° + |Q| \sin 30° = 2500$

We now have two equations with unknowns $|P|$ and $|Q|$.

$|P| \cos 45° - |Q| \cos 30° = 0 \quad (1)$

$|P| \sin 45° + |Q| \sin 30° = 2500 \quad (2)$

15. (continued)

Solve Equation (1) for $|P|$.

$$|P| = \frac{|Q| \cos 30°}{\cos 45°}$$

Substitute in Equation (2) and solve for $|Q|$.

$$\frac{|Q| \cos 30° \sin 45°}{\cos 45°} + |Q| \sin 30° = 2500$$

$$\frac{|Q| \cos 30° \sin 45°}{\cos 45°} + \frac{|Q| \sin 30° \cos 45°}{\cos 45°} = 2500$$

$$|Q|(\cos 30° \sin 45° + \sin 30° \cos 45°) = 2500 \cos 45°$$

$$|Q|[\sin(30° + 45°)] = 2500 \cos 45°$$

$$|Q| = \frac{2500 \cos 45°}{\sin 75°}$$

$$|Q| = \frac{2500(0.7071)}{0.9659}$$

$$|Q| = 1830 \text{ kg.}$$

Substituting in Equation (3), we obtain

$$|P| = \frac{1830 \cos 30°}{\cos 45°} = \frac{1830(0.8660)}{0.7071} = 2241 \text{ kg}$$

The tension in rope P is 2241 kg.
The tension in rope Q is 1830 kg.

16. 1464 kg, 1035 kg

Exercise Set 8.6

See text answer section for odd exercise answers 1.-23.

Even exercises 2. - 24.

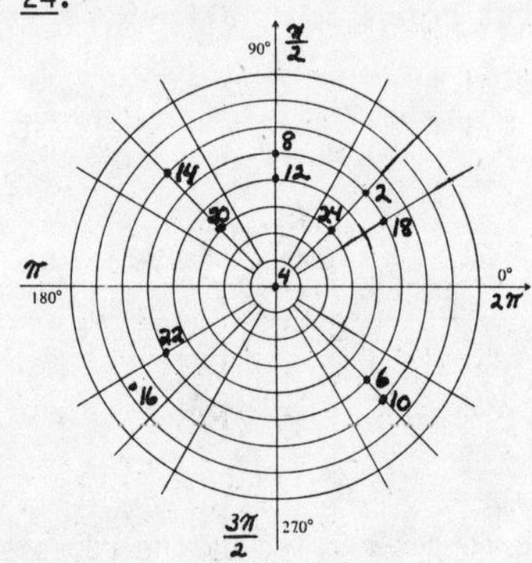

25. Find $|v|$:

$$|v| = \sqrt{4^2 + 4^2} = \sqrt{32} = 4\sqrt{2}$$

Find θ:

$$\tan \theta = \frac{4}{4} = 1$$

$$\theta = 45°$$

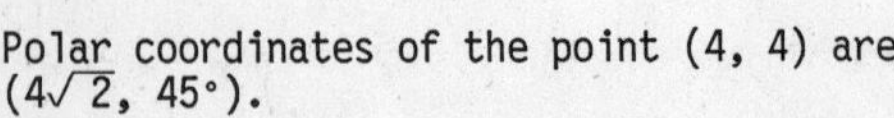

Polar coordinates of the point (4, 4) are $(4\sqrt{2}, 45°)$.

26. $(5\sqrt{2}, 45°)$

27. $|v| = 5$

$\theta = 90°$

Polar coordinates of the point (0, 5) are (5, 90°)

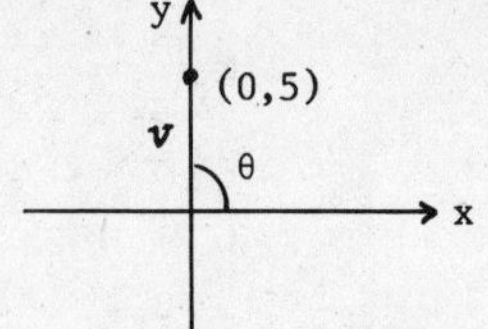

28. (3, 270°)

29. $|v| = 4$

$\theta = 0°$

Polar coordinates of the point (4, 0) are (4, 0°)

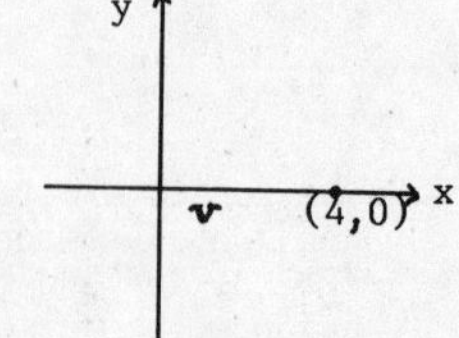

30. (5, 180°)

31. Find $|v|$:

$$|v| = \sqrt{3^2 + (3\sqrt{3})^2} = \sqrt{36} = 6$$

Find θ:

$$\tan \theta = \frac{3\sqrt{3}}{3} = \sqrt{3}$$

$$\theta = 60°$$

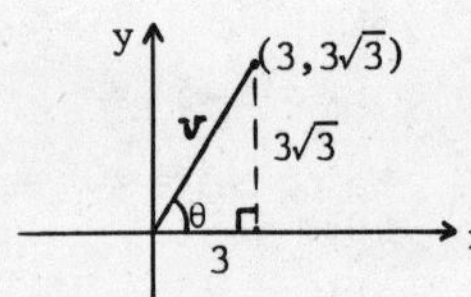

Polar coordinates of the point $(3, 3\sqrt{3})$ are (6, 60°)

32. (6, 240°)

33. Find $|v|$:

$$|v| = \sqrt{(\sqrt{3})^2 + 1^2} = \sqrt{4} = 2$$

Find θ:

$$\tan \theta = \frac{1}{\sqrt{3}}$$

$$\theta = 30°$$

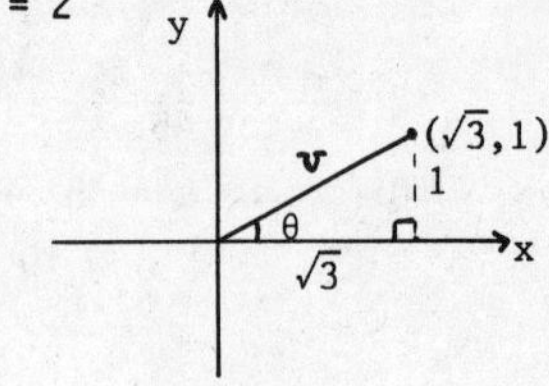

Polar coordinates of the point $(\sqrt{3}, 1)$ are (2, 30°).

34. (2, 150°)

35. Find $|v|$:

$$|v| = \sqrt{(3\sqrt{3})^2 + 3^2} = \sqrt{36} = 6$$

Find θ:

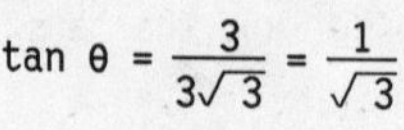

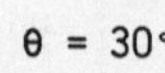

$$\tan \theta = \frac{3}{3\sqrt{3}} = \frac{1}{\sqrt{3}}$$

$$\theta = 30°$$

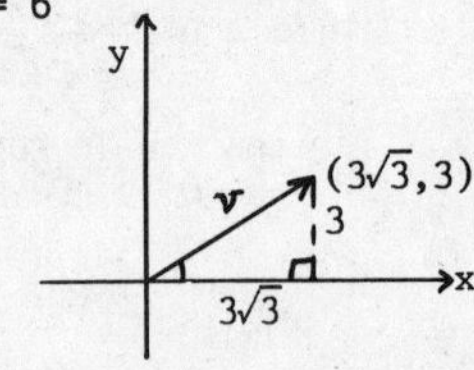

Polar coordinates of the point $(3\sqrt{3}, 3)$ are (6, 30°).

36. (8, 330°)

37. $x = 4 \cos 45° = 4 \cdot \frac{\sqrt{2}}{2} = 2\sqrt{2}$

$y = 4 \sin 45° = 4 \cdot \frac{\sqrt{2}}{2} = 2\sqrt{2}$

Cartesian coordinates of the point (4, 45°) are $(2\sqrt{2}, 2\sqrt{2})$.

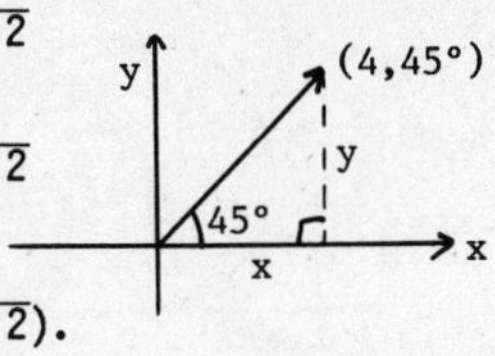

38. $\left(\frac{5}{\sqrt{2}}, \frac{5\sqrt{3}}{2}\right)$

39. $x = 0 \cos 23° = 0$
$y = 0 \sin 23° = 0$

Cartesian coordinates of the point (0, 23°) are (0, 0).

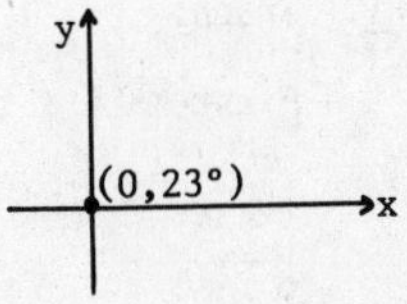

40. (0, 0)

41. $x = -3 \cos 45° = -3 \cdot \frac{\sqrt{2}}{2} = -\frac{3\sqrt{2}}{2}$

$y = -3 \sin 45° = -3 \cdot \frac{\sqrt{2}}{2} = -\frac{3\sqrt{2}}{2}$

Cartesian coordinates of the point (-3, 45°) are $\left(-\frac{3\sqrt{2}}{2}, -\frac{3\sqrt{2}}{2}\right)$.

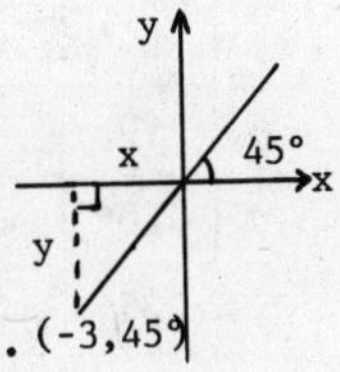

42. $\left(-\frac{5\sqrt{3}}{2}, -\frac{5}{2}\right)$

43. $x = 6 \cos(-60°) = 6 \cdot \frac{1}{2} = 3$

$y = 6 \sin(-60°) = 6\left(-\frac{\sqrt{3}}{2}\right) = -3\sqrt{3}$

Cartesian coordinates of the point (6, -60°) are $(3, -3\sqrt{3})$.

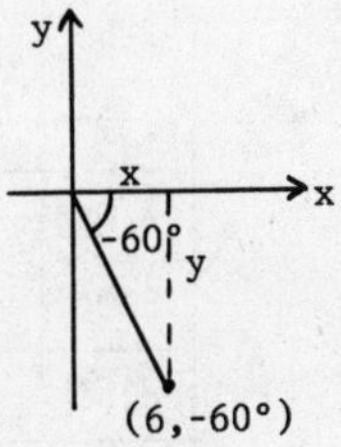

44. $\left(-\frac{3}{2}, -\frac{3\sqrt{3}}{2}\right)$

45. $x = 10 \cos \frac{\pi}{6} = 10 \cdot \frac{\sqrt{3}}{2} = 5\sqrt{3}$

$y = 10 \sin \frac{\pi}{6} = 10 \cdot \frac{1}{2} = 5$

Cartesian coordinates of the point $\left(10, \frac{\pi}{6}\right)$ are $(5\sqrt{3}, 5)$.

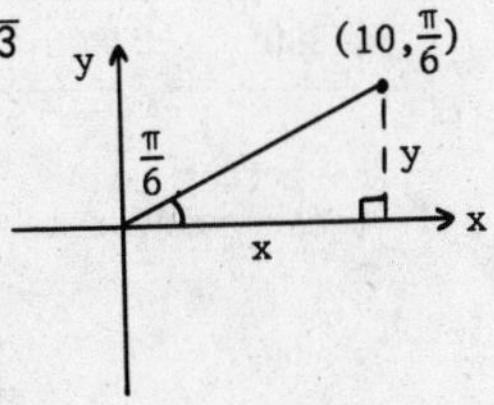

46. $(-6\sqrt{2}, 6\sqrt{2})$

47. $x = -5 \cos \frac{5\pi}{6} = -5\left(-\frac{\sqrt{3}}{2}\right) = \frac{5\sqrt{3}}{2}$

$y = -5 \sin \frac{5\pi}{6} = -5 \cdot \frac{1}{2} = -\frac{5}{2}$

Cartesian coordinates of the point $\left(-5, \frac{5\pi}{6}\right)$ are $\left(\frac{5\sqrt{3}}{2}, -\frac{5}{2}\right)$.

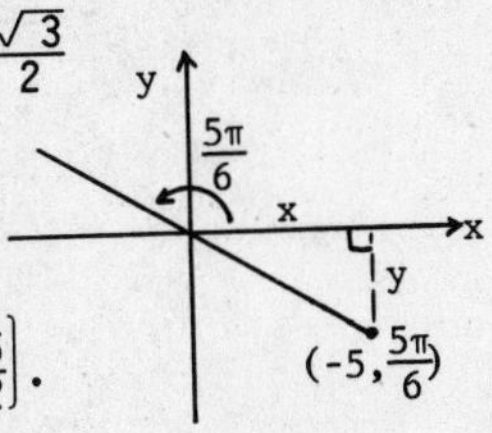

48. $(3\sqrt{2}, -3\sqrt{2})$

49. $3x + 4y = 5$

$3r \cos \theta + 4r \sin \theta = 5$ $(x = r \cos \theta, y = r \sin \theta)$

50. $r(5 \cos \theta + 3 \sin \theta) = 4$

51. $x = 5$

$r \cos \theta = 5$ $(x = r \cos \theta)$

52. $r \sin \theta = 4$

53. $x^2 + y^2 = 36$

$(r \cos \theta)^2 + (r \sin \theta)^2 = 36$ $(x = r \cos \theta, y = r \sin \theta)$

$r^2 \cos^2\theta + r^2 \sin^2\theta = 36$

$r^2(\cos^2\theta + \sin^2\theta) = 36$

$r^2 = 36$ $(\cos^2\theta + \sin^2\theta = 1)$

54. $r^2 = 16$, or $r = 4$, or $r = -4$

55. $x^2 - 4y^2 = 4$

$(r \cos \theta)^2 - 4(r \sin \theta)^2 = 4$

$r^2 \cos^2\theta - 4r^2 \sin^2\theta = 4$

$r^2(\cos^2\theta - 4 \sin^2\theta) = 4$

56. $r^2(\cos^2 \theta - 5 \sin^2 \theta) = 5$

57. $r = 5$

$+\sqrt{x^2 + y^2} = 5$ (Substituting for r)

$x^2 + y^2 = 25$ (Squaring)

58. $x^2 + y^2 = 64$

59. $\tan \theta = \frac{y}{x}$

$\tan \frac{\pi}{4} = \frac{y}{x}$ $\left(\theta = \frac{\pi}{4}\right)$

$1 = \frac{y}{x}$

$x = y$

60. $y = -x$

61. $r \sin \theta = 2$

$y = 2$ $(y = r \sin \theta)$

62. $x = 5$

63.
$$r = 4 \cos \theta$$
$$r^2 = 4r \cos \theta \quad \text{(Multiplying by r)}$$
$$x^2 + y^2 = 4x \quad (x^2 + y^2 = r^2,\ x = r \cos \theta)$$

64. $x^2 + y^2 = -3y$

65.
$$r - r \sin \theta = 2$$
$$r - y = 2 \quad (y = r \sin \theta)$$
$$r = 2 + y$$
$$r^2 = 4 + 4y + y^2 \quad \text{(Squaring)}$$
$$x^2 + y^2 = 4 + 4y + y^2 \quad (x^2 + y^2 = r^2)$$
$$x^2 - 4y = 4$$

66. $y^2 = -6x + 9$

67.
$$r - 2 \cos \theta = 3 \sin \theta$$
$$r^2 - 2r \cos \theta = 3r \sin \theta \quad \text{(Multiplying by r)}$$
$$x^2 + y^2 - 2x = 3y \quad (x^2 + y^2 = r^2,\ x = r \cos \theta,\ y = r \sin \theta)$$
$$x^2 + y^2 - 2x - 3y = 0$$

68. $x^2 - 7x + y^2 + 5y = 0$

69. Graph: $r = 4 \cos \theta$

First make a table of values.

θ	$\cos \theta$	$r = 4 \cos \theta$
0°	1	4
30°	0.866	3.46
45°	0.707	2.83
60°	0.5	2
90°	0	0
120°	-0.5	-2
135°	-0.707	-2.83
150°	-0.866	-3.46
180°	-1	-4
210°	-0.866	-3.46
225°	-0.707	-2.83

Note that points are beginning to repeat. Plot these points and draw the graph.

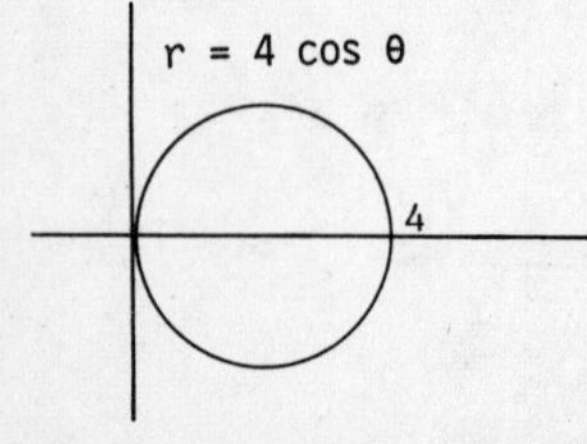

70.

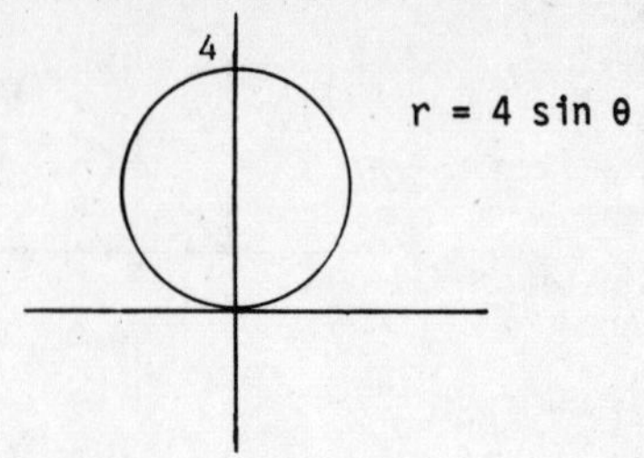

71. Graph: $r = 1 - \cos \theta$

First make a table of values.

θ	$\cos \theta$	$r = 1 - \cos \theta$
0°	1	0
30°	0.866	0.134
45°	0.707	0.293
60°	0.5	0.5
90°	0	1
120°	-0.5	1.5
135°	-0.707	1.707
150°	-0.866	1.866
180°	-1	2
210°	-0.866	1.866
225°	-0.707	1.707
240°	-0.5	1.5
270°	0	1
300°	0.5	0.5
315°	0.707	0.293
330°	0.866	0.134
360°	1	0
390°	0.866	0.134

Note that points are beginning to repeat. Plot these points and draw the graph.

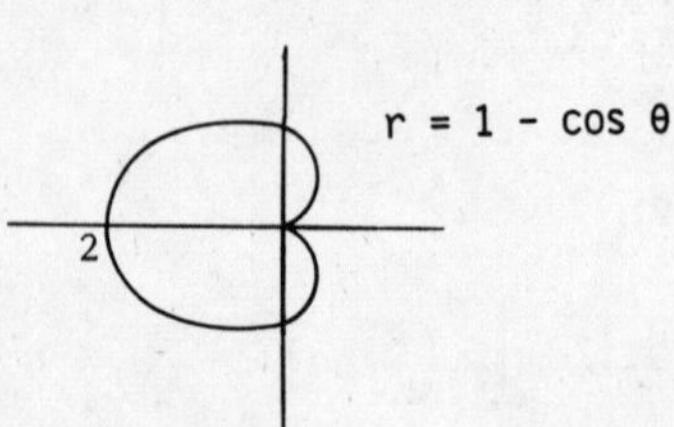

72.

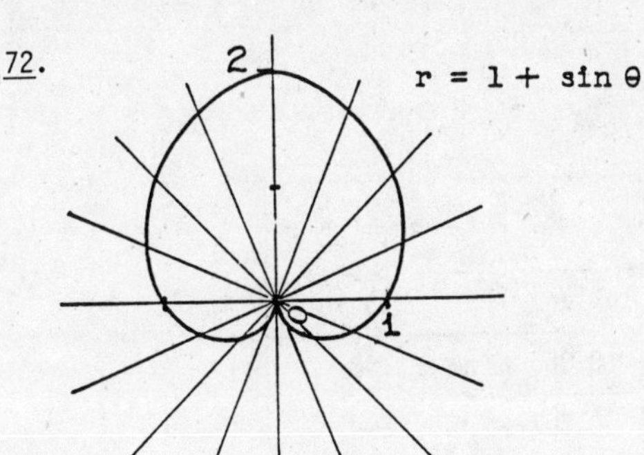

74. r = 3 cos 2θ

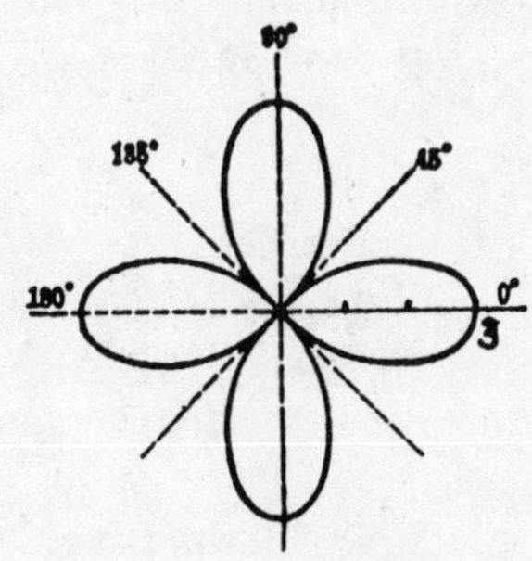

73. Graph: r = sin 2θ

First make a table of values.

θ	2θ	r = sin 2θ
0°	0°	0
30°	60°	0.866
45°	90°	1
60°	120°	0.866
90°	180°	0
120°	240°	-0.866
135°	270°	-1
150°	300°	-0.866
180°	360°	0
210°	420°	0.866
225°	450°	1
240°	480°	0.866
270°	540°	0
300°	600°	-0.866
315°	630°	-1
330°	660°	-0.866
360°	720°	0
390°	780°	0.866

Note that points are beginning to repeat. Plot these points and draw the graph.

r = sin 2θ

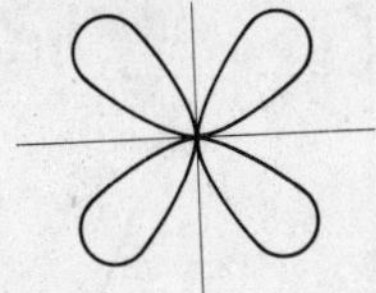

75. Graph: r = 2 cos 3θ

First make a table of values.

θ	3θ	r = 2 cos 3θ
0°	0°	2
30°	90°	0
45°	135°	-1.414
60°	180°	-2
90°	270°	0
120°	360°	2
135°	405°	1.414
150°	450°	0
180°	540°	-2
210°	630°	0
225°	675°	1.414
240°	720°	2
270°	810°	0
300°	900°	-2
315°	945°	-1.414
330°	990°	0
360°	1080°	2
390°	1170°	0

Note that points are beginning to repeat. Plot these points and draw the graph.

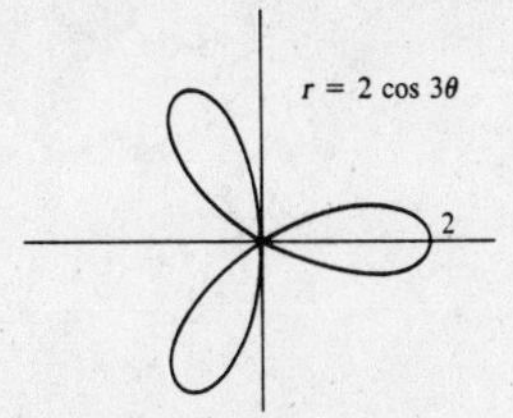

76.

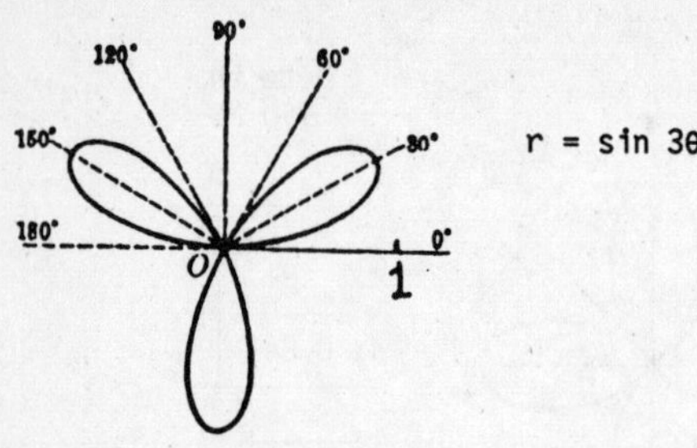

$r = \sin 3\theta$

77. Graph: $r \cos \theta = 4$

Express the function as $r = \dfrac{4}{\cos \theta}$. Then make a table of values.

θ	$\cos \theta$	$r = \dfrac{4}{\cos \theta}$
0°	1	4
30°	0.866	4.62
45°	0.707	5.66
60°	0.5	8
90°	0	Undefined
120°	-0.5	-8
135°	-0.707	-5.66
150°	-0.866	-4.62
180°	-1	-4
210°	-0.866	-4.62
225°	-0.707	-5.66

Note that points are begining to repeat. Plot these points and draw the graph.

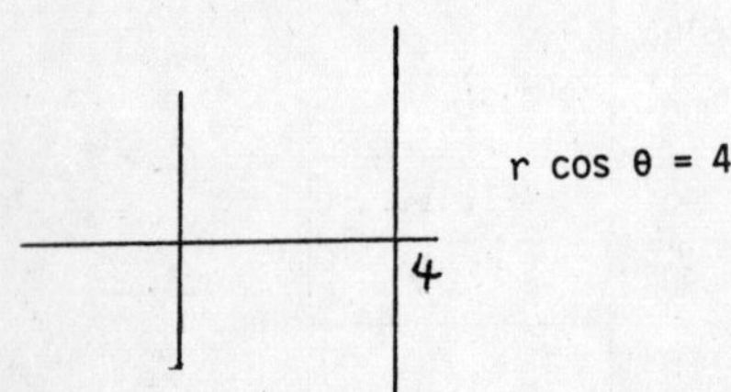

$r \cos \theta = 4$

78. $r \sin \theta = 6$

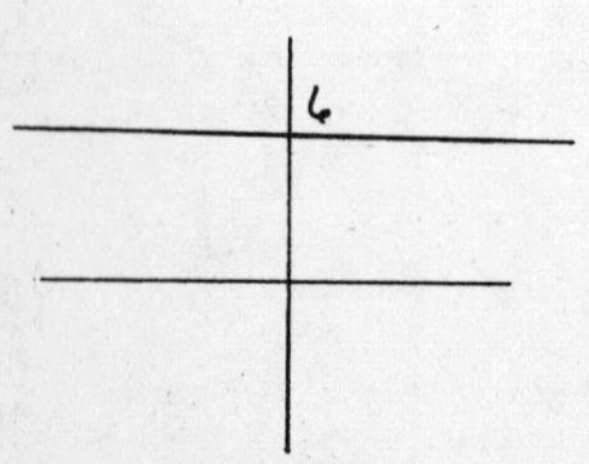

79. Graph: $r = \dfrac{5}{1 + \cos \theta}$

Make a table of values.

θ	$\cos \theta$	$r = \dfrac{5}{1 + \cos \theta}$
0°	1	2.5
30°	0.866	2.68
45°	0.707	2.93
60°	0.5	3.33
90°	0	5
120°	-0.5	10
135°	-0.707	17.07
150°	-0.866	37.32
180°	-1	Undefined
210°	-0.866	37.32
225°	-0.707	17.07
240°	-0.5	10
270°	0	5
300°	0.5	3.33
315°	0.707	2.93
330°	0.866	2.68
360°	1	2.5
390°	0.866	2.68

Note that points are beginning to repeat. Plot these points and draw the graph.

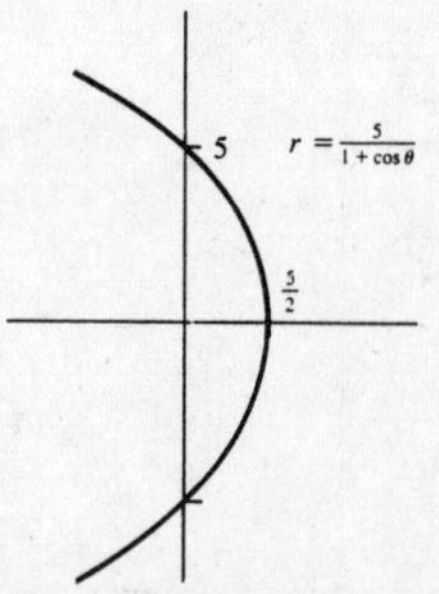

80. $r = \dfrac{3}{1 + \sin \theta}$

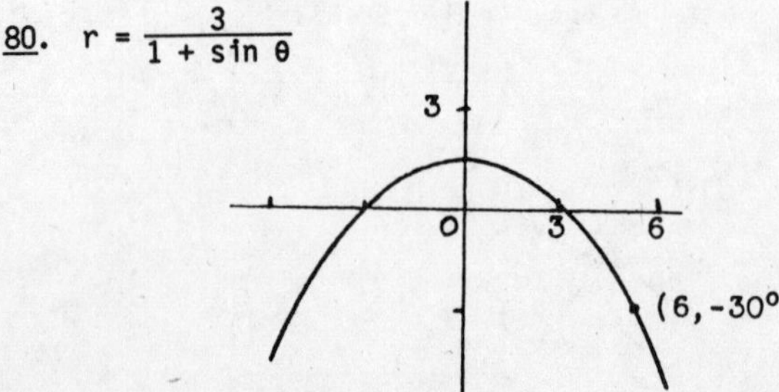

81. Graph: $r = \theta$

We make a table of values.

θ	r
0	0
$\frac{\pi}{4}$	$\frac{\pi}{4}$, or 0.785
$\frac{\pi}{2}$	$\frac{\pi}{2}$, or 1.571
$\frac{3\pi}{4}$	$\frac{3\pi}{4}$, or 2.356
π	π, or 3.142
$\frac{3\pi}{2}$	$\frac{3\pi}{2}$, or 4.712
2π	2π, or 6.283
$\frac{5\pi}{2}$	$\frac{5\pi}{2}$, or 7.854

Plot these points; take note of the fact that, as θ increases, r also increases; draw the graph.

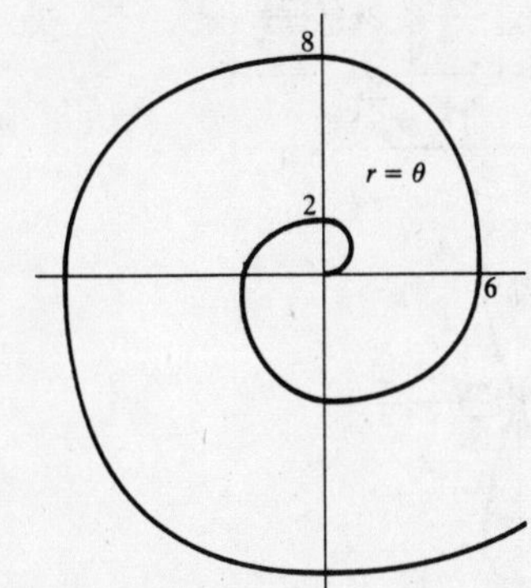

82.

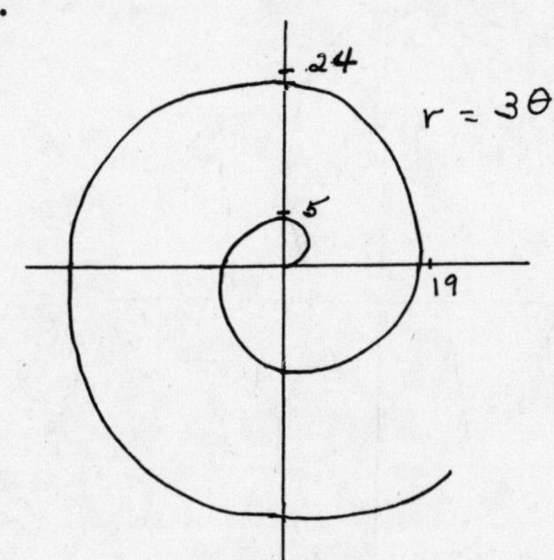

83. Graph: $r^2 = \sin 2\theta$

We make a table of values. Note that we cannot choose values of θ for which $\sin 2\theta$ is negative since r^2 is nonnegative.

θ	0	$\frac{\pi}{8}$	$\frac{\pi}{4}$	$\frac{3\pi}{8}$	$\frac{\pi}{2}$
r	0	$\pm$ 0.841	$\pm$ 1	$\pm$ 0.841	0

θ	π	$\frac{9\pi}{8}$	$\frac{5\pi}{4}$	$\frac{11\pi}{8}$	$\frac{3\pi}{2}$
r	0	$\pm$ 0.841	$\pm$ 1	$\pm$ 0.841	0

We only need to plot the points for $0 \leqslant \theta \leqslant \frac{\pi}{2}$.

The repetition has already begun for $\pi \leqslant \theta \leqslant \frac{3\pi}{2}$.

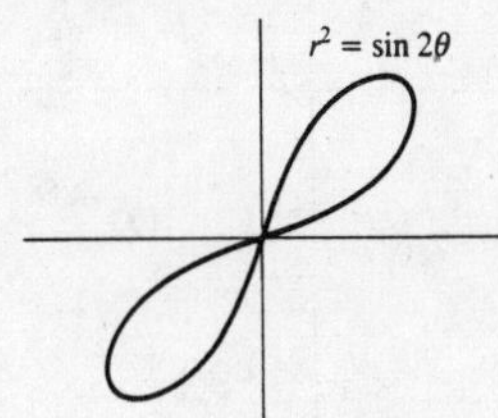

84.

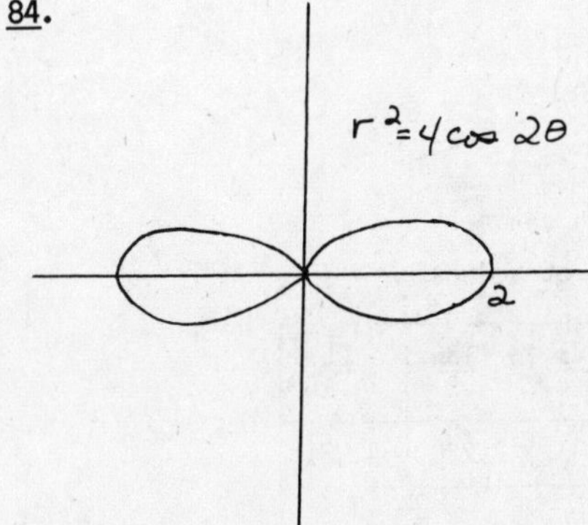

85. Graph: $r = e^{\theta/10}$ or $\ln r = \frac{\theta}{10}$

We make a table of values.

θ	$-\pi$	$-\frac{3\pi}{4}$	$-\frac{\pi}{2}$	$-\frac{\pi}{4}$
r	0.730	0.790	0.855	0.924

θ	0	$\frac{\pi}{4}$	$\frac{\pi}{2}$	$\frac{3\pi}{4}$	π
r	1	1.082	1.170	1.266	1.369

θ	$\frac{5\pi}{4}$	$\frac{3\pi}{2}$	$\frac{7\pi}{4}$	2π	$\frac{9\pi}{4}$
r	1.481	1.602	1.732	1.874	2.028

85. (continued)

Plot these points; take note of the fact that, as θ increases, r also increases; draw the graph.

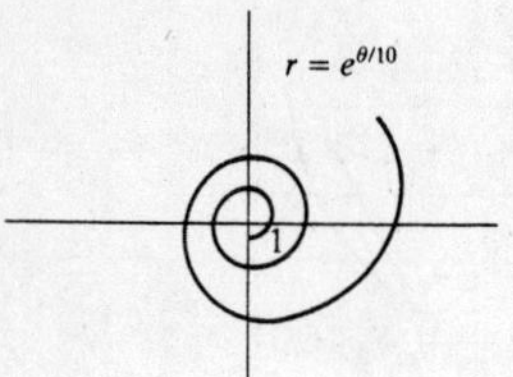

86. $r = 10^{2\theta}$, or
$\log r = 2\theta$

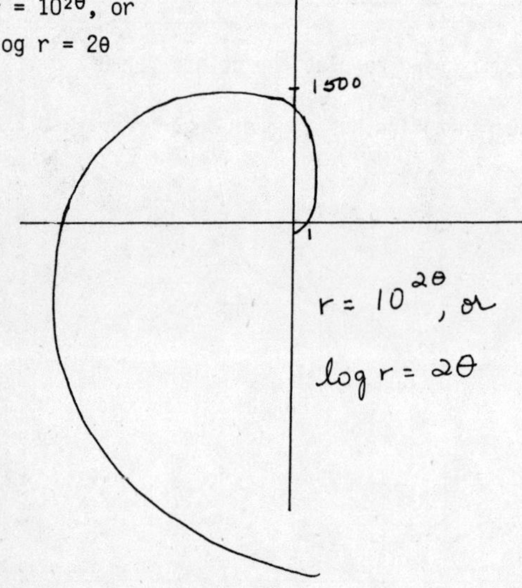

87. Graph: $r = \sin\theta\tan\theta$

We make a table of values.

θ	-2π	$-\frac{7\pi}{4}$	$-\frac{13\pi}{8}$	$-\frac{25\pi}{16}$
r	0	0.707	2.230	4.931

θ	$-\frac{7\pi}{16}$	$-\frac{3\pi}{8}$	$-\frac{\pi}{4}$	0
r	4.931	2.230	0.707	0

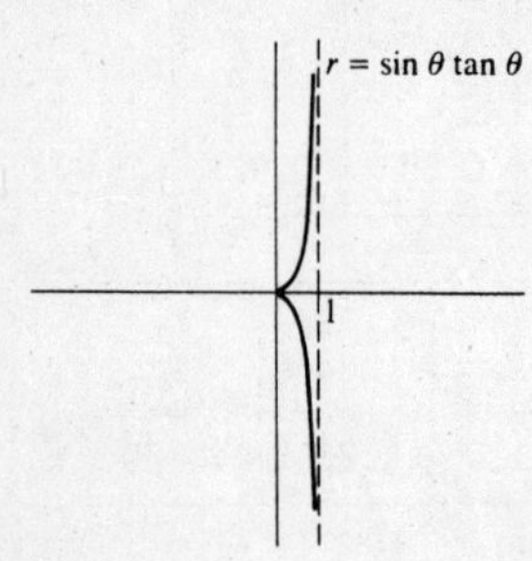

88.

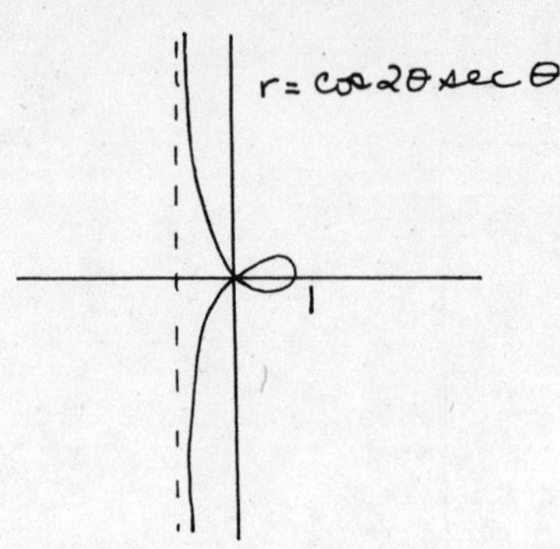

89. Graph: $r = 2\cos 2\theta - 1$

We make a table of values.

r	0	$\frac{\pi}{4}$	$\frac{\pi}{2}$	$\frac{3\pi}{4}$	π
θ	1	-1	-3	-1	1

r	$\frac{5\pi}{4}$	$\frac{3\pi}{2}$	$\frac{7\pi}{4}$	2π	$\frac{9\pi}{4}$
θ	-1	-3	-1	1	-1

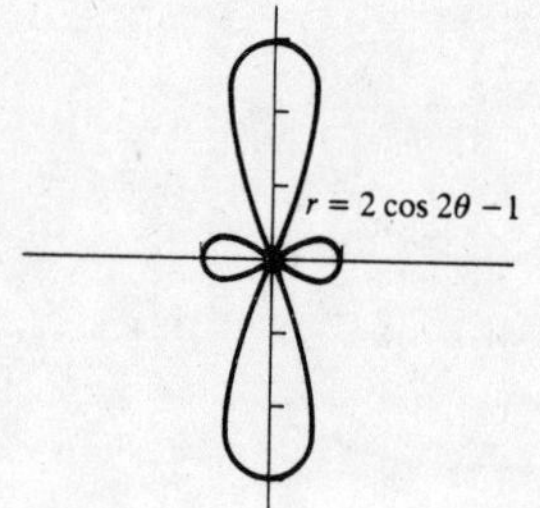

90.

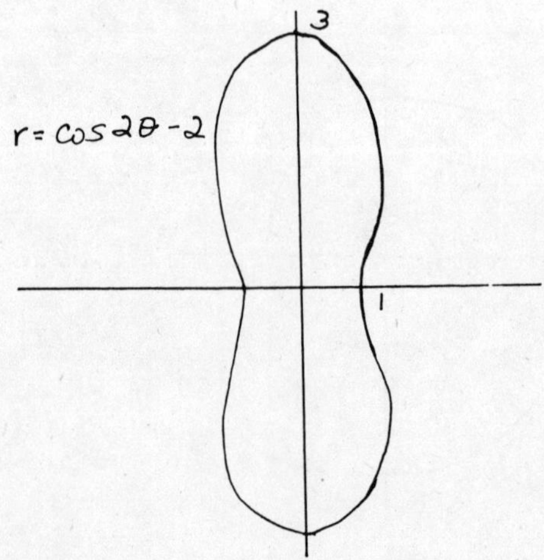

91. Graph: $r = \frac{1}{4}\tan^2\theta\sec\theta$

We make a table of values.

θ	-2π	$-\frac{7\pi}{4}$	$-\frac{13\pi}{8}$	$-\frac{25\pi}{16}$
r	0	0.354	3.808	32.388

θ	$-\frac{7\pi}{16}$	$-\frac{3\pi}{8}$	$-\frac{\pi}{4}$	0
r	32.388	3.808	0.354	0

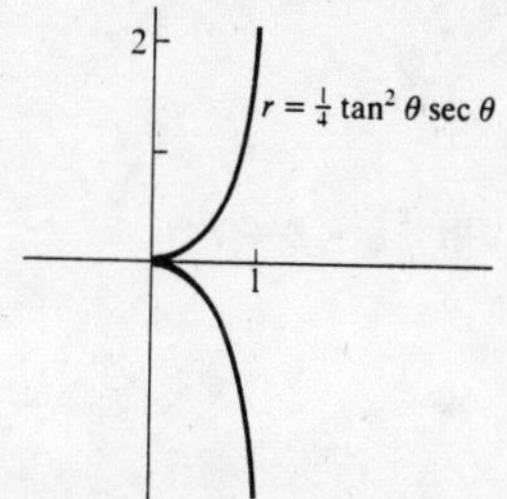

92.

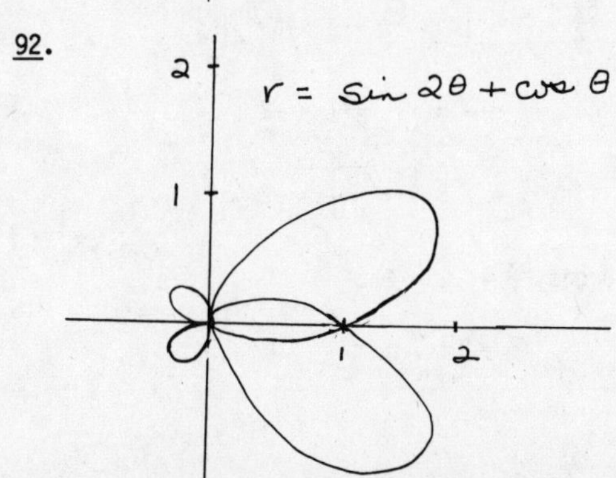

93. $r = \sec^2\frac{\theta}{2}$ or $r = \frac{1}{\cos^2\frac{\theta}{2}}$ or $r = \frac{1}{\frac{1 + \cos\theta}{2}}$

or

$$r = \frac{2}{1 + \cos\theta}$$

$$r + r\cos\theta = 2$$

$$\pm\sqrt{x^2 + y^2} + x = 2$$

$$x^2 + y^2 = (2 - x)^2$$

$$y^2 = -4x + 4$$

94.

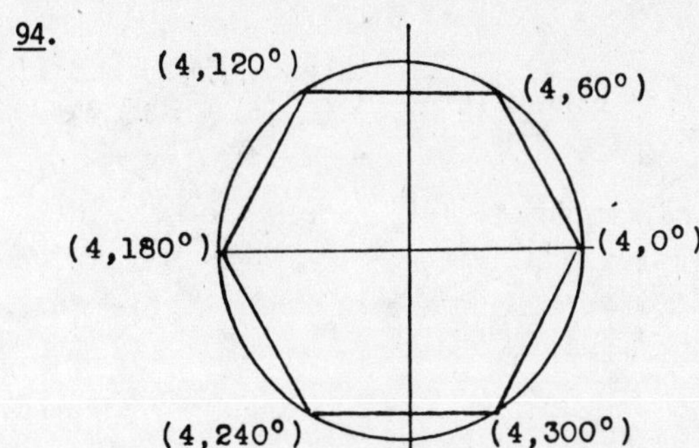

Exercise Set 8.7

1.

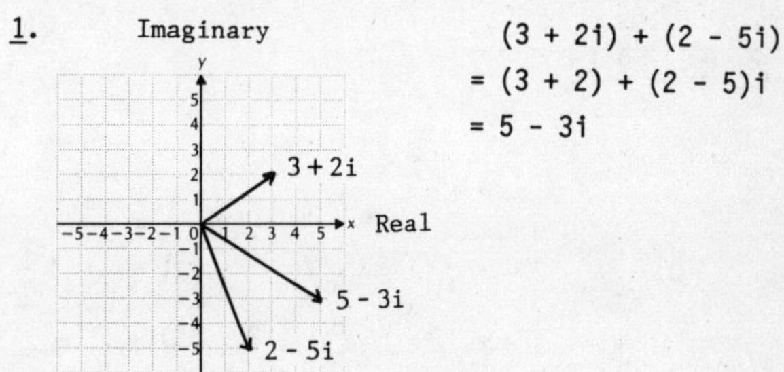

$(3 + 2i) + (2 - 5i)$
$= (3 + 2) + (2 - 5)i$
$= 5 - 3i$

2.

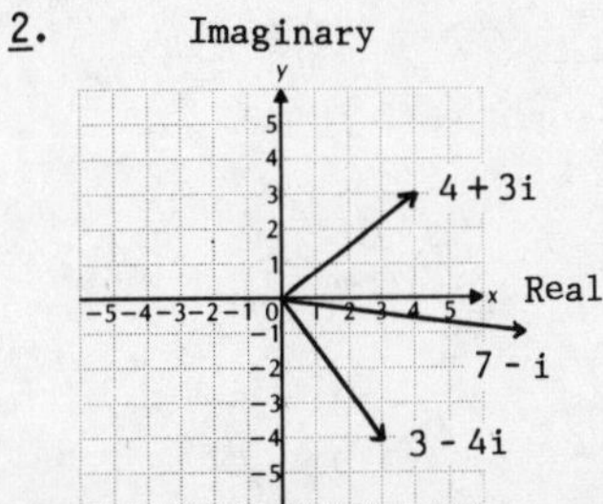

3.

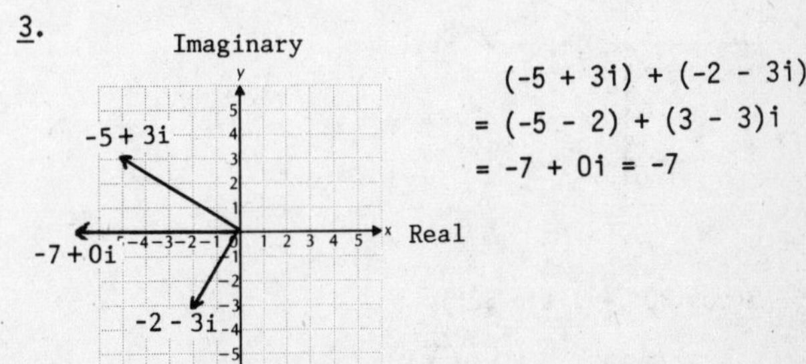

$(-5 + 3i) + (-2 - 3i)$
$= (-5 - 2) + (3 - 3)i$
$= -7 + 0i = -7$

4.

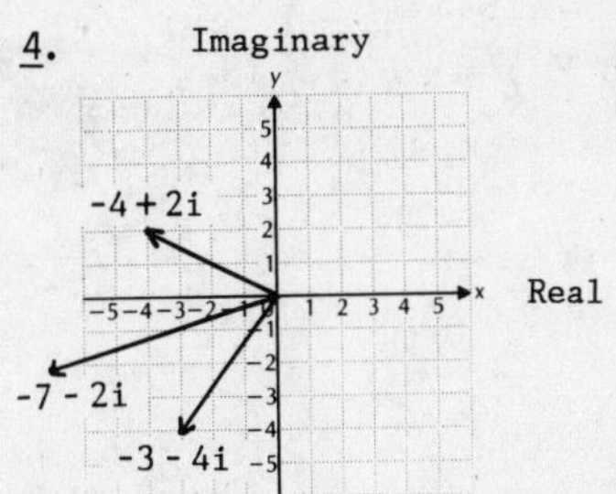

5. 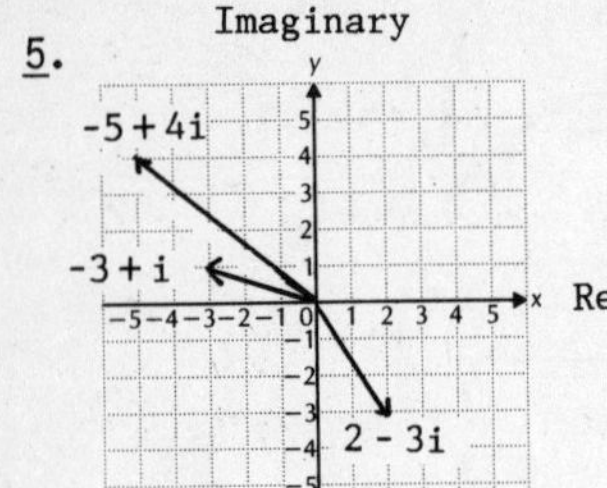

$(2 - 3i) + (-5 + 4i)$
$= (2 - 5) + (-3 + 4)i$
$= -3 + i$

6. 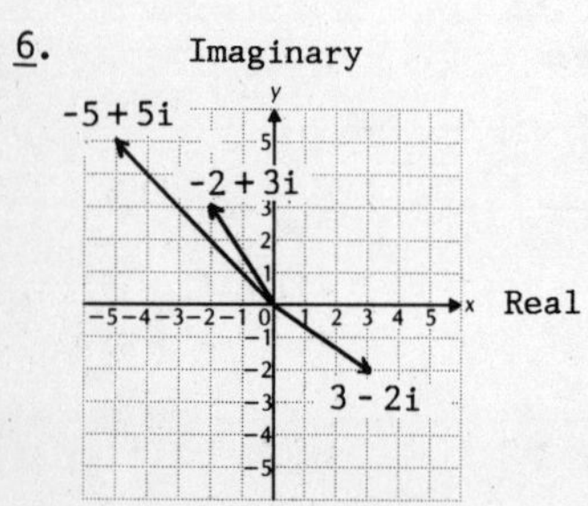

7. 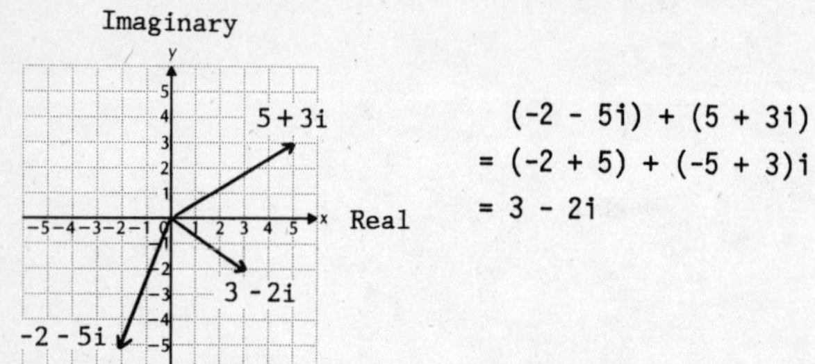

$(-2 - 5i) + (5 + 3i)$
$= (-2 + 5) + (-5 + 3)i$
$= 3 - 2i$

8. 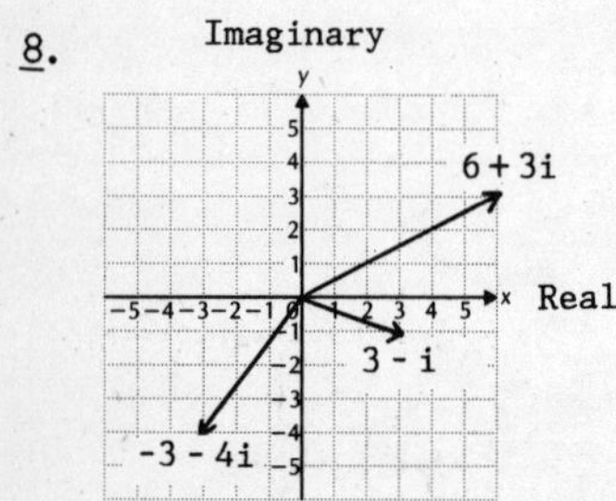

9. $3(\cos 30° + i \sin 30°)$

$a = 3 \cos 30° = 3 \cdot \frac{\sqrt{3}}{2} = \frac{3\sqrt{3}}{2}$

$b = 3 \sin 30° = 3 \cdot \frac{1}{2} = \frac{3}{2}$

Thus $3(\cos 30° + i \sin 30°) = \frac{3\sqrt{3}}{2} + \frac{3}{2} i$.

10. $-3\sqrt{3} + 3i$

11. $10 \text{ cis } 270° = 10(\cos 270° + i \sin 270°)$

$a = 10 \cos 270° = 10 \cdot 0 = 0$

$b = 10 \sin 270° = 10(-1) = -10$

Thus $10 \text{ cis } 270° = 0 + (-10)i = -10i$.

12. $\frac{5}{2} - \frac{5\sqrt{3}}{2} i$

13. $\sqrt{8} \left[\cos \frac{\pi}{4} + i \sin \frac{\pi}{4}\right]$

$a = \sqrt{8} \cos \frac{\pi}{4} = \sqrt{8} \cdot \frac{\sqrt{2}}{2} = 2$

$b = \sqrt{8} \sin \frac{\pi}{4} = \sqrt{8} \cdot \frac{\sqrt{2}}{2} = 2$

Thus $\sqrt{8} \left[\cos \frac{\pi}{4} + i \sin \frac{\pi}{4}\right] = 2 + 2i$

14. $\frac{5}{2} + \frac{5\sqrt{3}}{2} i$

15. $\sqrt{8} \text{ cis } \frac{5\pi}{4} = \sqrt{8} \left[\cos \frac{5\pi}{4} + i \sin \frac{5\pi}{4}\right]$

$a = \sqrt{8} \cos \frac{5\pi}{4} = \sqrt{8} \left[-\frac{\sqrt{2}}{2}\right] = -2$

$b = \sqrt{8} \sin \frac{5\pi}{4} = \sqrt{8} \left[-\frac{\sqrt{2}}{2}\right] = -2$

Thus $\sqrt{8} \text{ cis } \frac{5\pi}{4} = -2 - 2i$.

16. $2 - 2i$

17. $1 - i$

$a = 1$ and $b = -1$

$r = \sqrt{a^2 + b^2} = \sqrt{1^2 + (-1)^2} = \sqrt{2}$

$\sin \theta = \frac{b}{r} = \frac{-1}{\sqrt{2}} = -\frac{\sqrt{2}}{2}$

$\cos \theta = \frac{a}{r} = \frac{1}{\sqrt{2}} = \frac{\sqrt{2}}{2}$

Thus $\theta = \frac{7\pi}{4}$, or 315°, and we have

$1 - i = \sqrt{2} \text{ cis } \frac{7\pi}{4}$, or $\sqrt{2} \text{ cis } 315°$.

18. $2 \text{ cis } 30°$

19. $10\sqrt{3} - 10i$

$a = 10\sqrt{3}$ and $b = -10$

$r = \sqrt{a^2 + b^2} = \sqrt{(10\sqrt{3})^2 + (-10)^2} = \sqrt{400} = 20$

$\sin\theta = \frac{b}{r} = \frac{-10}{20} = -\frac{1}{2}$

$\cos\theta = \frac{a}{r} = \frac{10\sqrt{3}}{20} = \frac{\sqrt{3}}{2}$

Thus $\theta = \frac{11\pi}{6}$, or 330°, and we have

$10\sqrt{3} - 10i = 20 \text{ cis } \frac{11\pi}{6}$, or 20 cis 330°.

20. 20 cis 150°

21. -5 (Think $-5 + 0i$)

$a = -5$ and $b = 0$

$r = \sqrt{a^2 + b^2} = \sqrt{(-5)^2 + 0^2} = \sqrt{25} = 5$

$\sin\theta = \frac{b}{r} = \frac{0}{5} = 0$

$\cos\theta = \frac{-5}{5} = -1$

Thus $\theta = \pi$, or 180°, and we have

$-5 = 5 \text{ cis } \pi$, or 5 cis 180°.

22. 5 cis 270°

23. $(1 - i)(2 + 2i)$

Find polar notation for $1 - i$.

$a = 1$, $b = -1$, and $r = \sqrt{1^2 + (-1)^2} = \sqrt{2}$

$\sin\theta = \frac{-1}{\sqrt{2}} = -\frac{\sqrt{2}}{2}$, $\cos\theta = \frac{1}{\sqrt{2}} = \frac{\sqrt{2}}{2}$

Thus $\theta = \frac{7\pi}{4}$, or 315°, and $1 - i = \sqrt{2}$ cis 315°.

Find polar notation for $2 + 2i$.

$a = 2$, $b = 2$, and $r = \sqrt{2^2 + 2^2} = \sqrt{8} = 2\sqrt{2}$

$\sin\theta = \frac{2}{2\sqrt{2}} = \frac{\sqrt{2}}{2}$, $\cos\theta = \frac{2}{2\sqrt{2}} = \frac{\sqrt{2}}{2}$

Thus $\theta = \frac{\pi}{4}$, or 45°, and $2 + 2i = 2\sqrt{2}$ cis 45°

$(1 - i)(2 + 2i)$

$= \sqrt{2}$ cis 315° $\cdot$ $2\sqrt{2}$ cis 45°

$= \sqrt{2} \cdot 2\sqrt{2}$ cis (315° + 45°) (Theorem 3)

= 4 cis 0°, or 4

24. $2\sqrt{2}$ cis 105°

25. $(2\sqrt{3} + 2i)(2i)$

Find polar notation for $2\sqrt{3} + 2i$.

$a = 2\sqrt{3}$, $b = 2$, and $r = \sqrt{(2\sqrt{3})^2 + 2^2}$

$= \sqrt{16} = 4$

$\sin\theta = \frac{2}{4} = \frac{1}{2}$, $\cos\theta = \frac{2\sqrt{3}}{4} = \frac{\sqrt{3}}{2}$

Thus $\theta = \frac{\pi}{6}$, or 30°, and $2\sqrt{3} + 2i = 4$ cis 30°.

Find polar notation for $2i$.

$a = 0$, $b = 2$, and $r = \sqrt{0^2 + 2^2} = 2$

$\sin\theta = \frac{2}{2} = 1$, $\cos\theta = \frac{0}{2} = 0$

Thus $\theta = \frac{\pi}{2}$, or 90°, and $2i = 2$ cis 90°.

$(2\sqrt{3} + 2i)(2i)$

= 4 cis 30° $\cdot$ 2 cis 90°

= 4·2 cis (30° + 90°) (Theorem 3)

= 8 cis 120°

26. 12 cis 60°

27. $\frac{1 - i}{1 + i}$

Find polar notation for $1 - i$.

$a = 1$, $b = -1$, and $r = \sqrt{2}$

$\sin\theta = \frac{-1}{\sqrt{2}} = -\frac{\sqrt{2}}{2}$, $\cos\theta = \frac{1}{\sqrt{2}} = \frac{\sqrt{2}}{2}$

Thus $\theta = \frac{7\pi}{4}$, or 315°, and $1 - i = \sqrt{2}$ cis 315°.

Find polar notation for $1 + i$.

$a = 1$, $b = 1$, and $r = \sqrt{2}$

$\sin\theta = \frac{1}{\sqrt{2}} = \frac{\sqrt{2}}{2}$, $\cos\theta = \frac{1}{\sqrt{2}} = \frac{\sqrt{2}}{2}$

Thus $\theta = \frac{\pi}{4}$, or 45°, and $1 + i = \sqrt{2}$ cis 45°.

$\frac{1 - i}{1 + i} = \frac{\sqrt{2} \text{ cis } 315°}{\sqrt{2} \text{ cis } 45°}$

$= \frac{\sqrt{2}}{\sqrt{2}}$ cis (315° − 45°) (Theorem 4)

= cis 270°, or $-i$

28. $\frac{\sqrt{2}}{2}$ cis 345°

29. $\dfrac{2\sqrt{3} - 2i}{1 + \sqrt{3}\ i}$

Find polar notation for $2\sqrt{3} - 2i$.

$a = 2\sqrt{3}$, $b = -2$, and $r = \sqrt{(2\sqrt{3})^2 + (-2)^2}$

$= \sqrt{16} = 4$

$\sin \theta = \frac{-2}{4} = -\frac{1}{2}$, $\cos \theta = \frac{2\sqrt{3}}{4} = \frac{\sqrt{3}}{2}$

Thus $\theta = \frac{11\pi}{6}$, or 330°, and $2\sqrt{3} - 2i = 4 \text{ cis } 330°$.

Find polar notation for $1 + \sqrt{3}\ i$.

$a = 1$, $b = \sqrt{3}$, and $r = \sqrt{1^2 + (\sqrt{3})^2} = \sqrt{4} = 2$

$\sin \theta = \frac{\sqrt{3}}{2}$, $\cos \theta = \frac{1}{2}$

Thus $\theta = \frac{\pi}{3}$, or 60°, and $1 + \sqrt{3}\ i = 2 \text{ cis } 60°$.

$$\frac{2\sqrt{3} - 2i}{1 + \sqrt{3}\ i} = \frac{4 \text{ cis } 330°}{2 \text{ cis } 60°}$$
$$= \frac{4}{2} \text{ cis } (330° - 60°)$$
$$= 2 \text{ cis } 270°, \text{ or } -2i$$

30. 3 cis 330°

31. See the answer section in the text.

32. $z = a + bi$, $|z| = \sqrt{a^2 + b^2}$

$\bar{z} = a - bi$, $|\bar{z}| = \sqrt{a^2 + (-b)^2} = \sqrt{a^2 + b^2}$

$|z| = |\bar{z}|$

33. See the answer section in the text.

34. $|(a + bi)^2| = |a^2 - b^2 + 2abi|$

$= \sqrt{(a^2 - b^2)^2 + 4a^2b^2}$

$= \sqrt{a^4 + 2a^2b^2 + b^4}$

$= a^2 + b^2$

$|a + bi|^2 = (\sqrt{a^2 + b^2})^2 = a^2 + b^2$

35. See the answer section in the text.

36. $\frac{z}{w} = \frac{r_1 \text{cis } \theta_1}{r_2 \text{cis } \theta_2} = \frac{r_1}{r_2} \text{ cis } (\theta_1 - \theta_2)$

$$\left|\frac{z}{w}\right| = \sqrt{\left[\frac{r_1}{r_2} \cos (\theta_1 - \theta_2)\right]^2 + \left[\frac{r_1}{r_2} \sin (\theta_1 - \theta_2)\right]^2}$$

$$= \sqrt{\frac{r_1^2}{r_2^2}} = \frac{|r_1|}{|r_2|}$$

$|z| = \sqrt{(r_1 \cos \theta_1)^2 + (r_1 \sin \theta_1)^2}$

$= \sqrt{r_1^2} = |r_1|$

$|w| = \sqrt{(r_2 \cos \theta_2)^2 + (r_2 \sin \theta_2)^2}$

$= \sqrt{r_2^2} = |r_2|$

Then $\left|\frac{z}{w}\right| = \frac{|r_1|}{|r_2|} = \frac{|z|}{|w|}$.

37. Graph: $|z| = 1$

Let $z = a + bi$

Then $|a + bi| = 1$

$\sqrt{a^2 + b^2} = 1$ (Definition)

$a^2 + b^2 = 1$

The graph of $a^2 + b^2 = 1$ is a circle whose radius is 1 and whose center is (0, 0).

The graph is in the answer section in the text.

38.

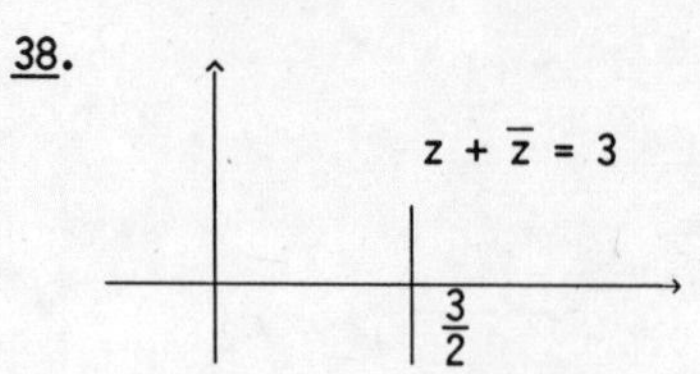

39. $(\cos \theta + i \sin \theta)^{-1}$

$= \frac{1}{\cos \theta + i \sin \theta} \cdot \frac{\cos \theta - i \sin \theta}{\cos \theta - i \sin \theta}$

$= \frac{\cos \theta - i \sin \theta}{\cos^2 \theta + \sin^2 \theta}$

$= \cos \theta - i \sin \theta$

40. 125 cis 60°

Exercise Set 8.8

1. $\left[2 \text{ cis } \frac{\pi}{3}\right]^3 = 2^3 \text{ cis } 3 \cdot \frac{\pi}{3}$ (Theorem 5)

$= 8 \text{ cis } \pi$

2. 81 cis 2π

3. $\left[2 \text{ cis } \frac{\pi}{6}\right]^6 = 2^6 \text{ cis } 6 \cdot \frac{\pi}{6}$ (Theorem 5)

$= 64 \text{ cis } \pi$

4. 32 cis π

5. $(1 + i)^6$

We first find polar notation for 1 + i. See Exercise 27 in Exercise Set 8.7.

$1 + i = \sqrt{2}\text{ cis } 45°$

$(1 + i)^6 = (\sqrt{2}\text{ cis } 45°)^6$

$= (\sqrt{2})^6\text{ cis } 6\cdot 45°$ (Theorem 5)

$= 8\text{ cis } 270°$

6. 8 cis 90°

7. $(2\text{ cis } 240°)^4 = 2^4\text{ cis } 4\cdot 240°$ (Theorem 5)

$= 16\text{ cis } 960°$

$= 16\text{ cis } 240°$

$= 16(\cos 240° + i\sin 240°)$

$= 16\left(-\frac{1}{2} - \frac{\sqrt{3}}{2}i\right)$

$= -8 - 8\sqrt{3}\,i$

8. $-8 + 8\sqrt{3}\,i$

9. $(1 + \sqrt{3}\,i)^4$

We first find polar notation for $1 + \sqrt{3}\,i$. See Exercise 29 in Exercise Set 8.7.

$1 + \sqrt{3}\,i = 2\text{ cis } 60°$

$(1 + \sqrt{3}\,i)^4 = (2\text{ cis } 60°)^4$

$= 2^4\text{ cis } 4\cdot 60°$ (Theorem 5)

$= 16\text{ cis } 240°$

$= 16(\cos 240° + i\sin 240°)$

$= 16\left(-\frac{1}{2} - \frac{\sqrt{3}}{2}i\right)$

$= -8 - 8\sqrt{3}\,i$

10. -64

11. $\left(\frac{1}{\sqrt{2}} + \frac{1}{\sqrt{2}}i\right)^{10}$

First find polar notation for $\frac{1}{\sqrt{2}} + \frac{1}{\sqrt{2}}i$.

$a = \frac{1}{\sqrt{2}}$, $b = \frac{1}{\sqrt{2}}$, and $r = \sqrt{\left(\frac{1}{\sqrt{2}}\right)^2 + \left(\frac{1}{\sqrt{2}}\right)^2}$

$= \sqrt{1} = 1$

$\sin\theta = \frac{\frac{1}{\sqrt{2}}}{1} = \frac{1}{\sqrt{2}}$, or $\frac{\sqrt{2}}{2}$

$\cos\theta = \frac{\frac{1}{\sqrt{2}}}{1} = \frac{1}{\sqrt{2}}$, or $\frac{\sqrt{2}}{2}$

Thus $\theta = \frac{\pi}{4}$, or 45° and $\frac{1}{\sqrt{2}} + \frac{1}{\sqrt{2}}i = 1\cdot\text{cis } 45°$.

11. (continued)

$\left(\frac{1}{\sqrt{2}} + \frac{1}{\sqrt{2}}i\right)^{10} = (\text{cis } 45°)^{10}$

$= \text{cis } 450°$ (Theorem 5)

$= \text{cis } 90°$

$= \cos 90° + i\sin 90°$

$= 0 + i$, or i

12. -1

13. $\left(\frac{\sqrt{3}}{2} + \frac{1}{2}i\right)^{12}$

First find polar notation for $\frac{\sqrt{3}}{2} + \frac{1}{2}i$.

$a = \frac{\sqrt{3}}{2}$, $b = \frac{1}{2}$, and $r = \sqrt{\left(\frac{\sqrt{3}}{2}\right)^2 + \left(\frac{1}{2}\right)^2} = \sqrt{1} = 1$

$\sin\theta = \frac{\frac{1}{2}}{1} = \frac{1}{2}$, $\cos\theta = \frac{\frac{\sqrt{3}}{2}}{1} = \frac{\sqrt{3}}{2}$

Thus $\theta = \frac{\pi}{6}$, or 30°, and $\frac{\sqrt{3}}{2} + \frac{1}{2}i = \text{cis } 30°$.

$\left(\frac{\sqrt{3}}{2} + \frac{1}{2}i\right)^{12} = (\text{cis } 30°)^{12}$

$= \text{cis } 12\cdot 30°$ (Theorem 5)

$= \text{cis } 360°$, or cis 0°

$= \cos 0° + i\sin 0°$

$= 1 + 0i$, or 1

14. $\frac{1}{2} - \frac{\sqrt{3}}{2}i$

15. $x^2 = i$

Find the square roots of i.

Polar notation for i is cis 90°.

Using Theorem 6, we know

$(\text{cis } 90°)^{1/2} = \text{cis}\left[\frac{90°}{2} + k\cdot\frac{360°}{2}\right]$, $k = 0, 1$

$= \text{cis }(45° + k\cdot 180°)$, $k = 0, 1$

Thus the roots are

cis 45° when k = 0 and cis 225° when k = 1.

Rectangular notation for these roots is as follows:

$\text{cis } 45° = \cos 45° + i\sin 45°$

$= \frac{\sqrt{2}}{2} + \frac{\sqrt{2}}{2}i$

$\text{cis } 225° = \cos 225° + i\sin 225°$

$= -\frac{\sqrt{2}}{2} - \frac{\sqrt{2}}{2}i$

The roots are $\frac{\sqrt{2}}{2} + \frac{\sqrt{2}}{2}i$ and $-\frac{\sqrt{2}}{2} - \frac{\sqrt{2}}{2}i$.

16. $-\frac{\sqrt{2}}{2} + \frac{\sqrt{2}}{2} i, \frac{\sqrt{2}}{2} - \frac{\sqrt{2}}{2} i$

17. $x^2 = 2\sqrt{2} - 2\sqrt{2}\, i$

Find the square roots of $2\sqrt{2} - 2\sqrt{2}\, i$.

Polar notation for $2\sqrt{2} - 2\sqrt{2}\, i$ is 4 cis 315°.

Using Theorem 6, we know

$$(4 \text{ cis } 315°)^{1/2} = 4^{1/2} \text{ cis } \left(\frac{315°}{2} + \frac{k \cdot 360°}{2}\right), \quad k = 0, 1$$

$$= 2 \text{ cis } (157.5° + k \cdot 180°), \quad k = 0, 1$$

Thus the roots are

2 cis 157.5° when k = 0 and 2 cis 337.5° when k = 1.

18. $\sqrt[4]{2}$ cis 22.5°, $\sqrt[4]{2}$ cis 202.5°

19. $x^2 = -1 + \sqrt{3}\, i$

Find the square roots of $-1 + \sqrt{3}\, i$.

Polar notation for $-1 + \sqrt{3}\, i$ is 2 cis 120°.

Using Theorem 6, we know

$$(2 \text{ cis } 120°)^{1/2} = 2^{1/2} \text{ cis } \left(\frac{120°}{2} + k \cdot \frac{360°}{2}\right), \quad k = 0, 1$$

$$= \sqrt{2} \text{ cis } (60° + k \cdot 180°), \quad k = 0, 1$$

Thus the roots are

$\sqrt{2}$ cis 60° when k = 0 and $\sqrt{2}$ cis 240° when k = 1.

Rectangular notation for these roots is as follows:

$$\sqrt{2} \text{ cis } 60° = \sqrt{2}\,(\cos 60° + i \sin 60°)$$

$$= \sqrt{2}\left(\frac{1}{2} + \frac{\sqrt{3}}{2} i\right) = \frac{\sqrt{2}}{2} + \frac{\sqrt{6}}{2} i$$

$$\sqrt{2} \text{ cis } 240° = \sqrt{2}\,(\cos 240° + i \sin 240°)$$

$$= \sqrt{2}\left(-\frac{1}{2} - \frac{\sqrt{3}}{2} i\right) = -\frac{\sqrt{2}}{2} - \frac{\sqrt{6}}{2} i$$

The roots are $\frac{\sqrt{2}}{2} + \frac{\sqrt{6}}{2} i$ and $-\frac{\sqrt{2}}{2} - \frac{\sqrt{6}}{2} i$.

20. $\sqrt{2}$ cis 105°, $\sqrt{2}$ cis 285°

21. $x^3 = i$

Find the cube roots of i.

Polar notation for i is cis 90°.

Using Theorem 6, we know

$$(\text{cis } 90°)^{1/3} = 1^{1/3} \text{ cis } \left(\frac{90°}{3} + k \cdot \frac{360°}{3}\right), \quad k = 0, 1, 2$$

$$= \text{cis } (30° + k \cdot 120°), \quad k = 0, 1, 2$$

Thus the roots are

cis 30° when k = 0, cis 150° when k = 1, and cis 270° when k = 2.

Rectangular notation for these roots is as follows:

$$\text{cis } 30° = \cos 30° + i \sin 30° = \frac{\sqrt{3}}{2} + \frac{1}{2} i$$

$$\text{cis } 150° = \cos 150° + i \sin 150° = -\frac{\sqrt{3}}{2} + \frac{1}{2} i$$

$$\text{cis } 270° = \cos 270° + i \sin 270° = -i$$

22. 4.09, -2.045 + 3.542i, -2.045 - 3.542i

23. Solve $x^4 = 16$.

Find the fourth roots of 16.

Polar notation for 16 is 16 cis 0°.

Using Theorem 6, we know

$$(16 \text{ cis } 0°)^{1/4} = 16^{1/4} \text{ cis } \left(\frac{0°}{4} + k \cdot \frac{360°}{4}\right), \quad k = 0, 1, 2, 3$$

$$= 2 \text{ cis } k \cdot 90°, \quad k = 0, 1, 2, 3$$

Thus the roots are

$2 \text{ cis } 0° = 2(\cos 0° + i \sin 0°) = 2(1 + 0i) = 2$ when k = 0

$2 \text{ cis } 90° = 2(\cos 90° + i \sin 90°) = 2(0 + i) = 2i$ when k = 1

$2 \text{ cis } 180° = 2(\cos 180° + i \sin 180°) = 2(-1 + 0i) = -2$ when k = 2

$2 \text{ cis } 270° = 2(\cos 270° + i \sin 270°) = 2(0 - i) = -2i$ when k = 3

The graph of ± 2 and ± 2i is as follows:

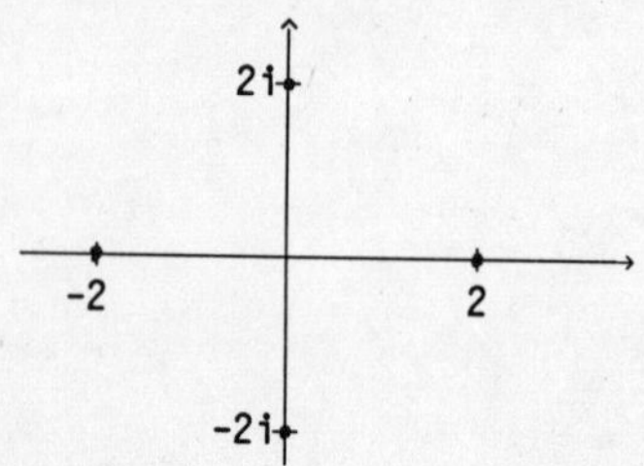

24. cis $22\frac{1}{2}°$, cis $112\frac{1}{2}°$, cis $202\frac{1}{2}°$, cis $292\frac{1}{2}°$

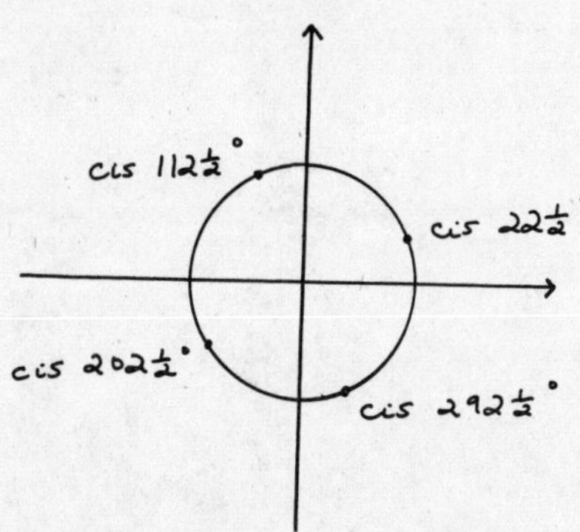

25. $-1 = 1$ cis 180°

$(1 \text{ cis } 180°)^{1/5} = 1^{1/5} \text{ cis } \left[\frac{180°}{5} + k \cdot \frac{360°}{5}\right],$

$k = 0, 1, 2, 3, 4$

The roots are 1 cis 36°, 1 cis 108°, 1 cis 180°, 1 cis 252° and 1 cis 324°, or cis 36°, cis 108°, cis 180°, cis 252°, cis 324°.

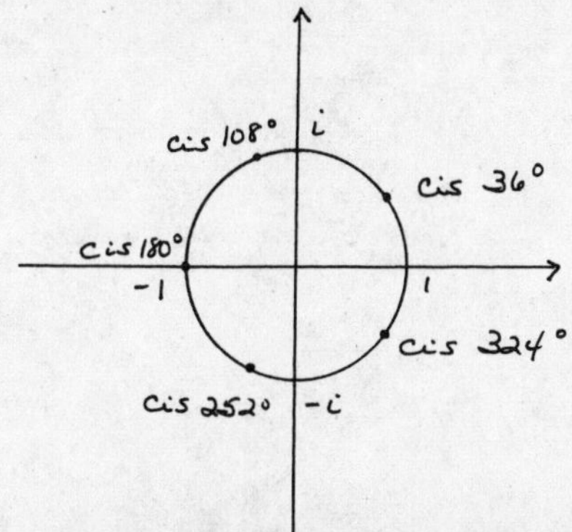

26. cis 0°, cis 60°, cis 120°, cis 180°, cis 240°, cis 300°, or 1, $\frac{1}{2} + \frac{\sqrt{3}}{2}i$, $-\frac{1}{2} + \frac{\sqrt{3}}{2}i$, -1, $-\frac{1}{2} - \frac{\sqrt{3}}{2}i$, $\frac{1}{2} - \frac{\sqrt{3}}{2}i$

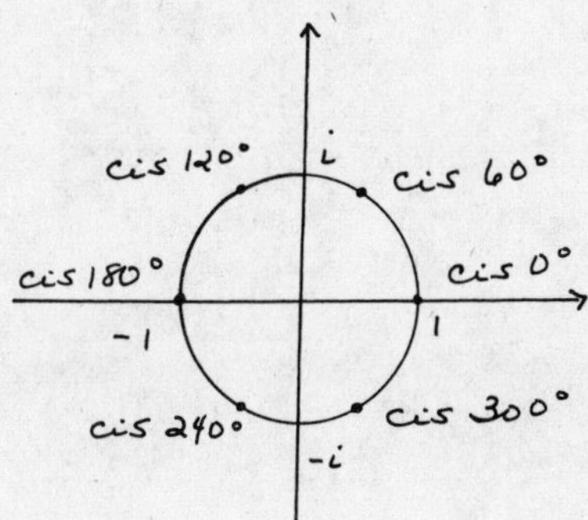

27. $-8 + i\,8\sqrt{3} = 16$ cis 120°

$(16 \text{ cis } 120°)^{1/4} = 16^{1/4} \text{ cis } \left[\frac{120°}{4} + k \cdot \frac{360°}{4}\right],$

$k = 0, 1, 2, 3$

The roots are 2 cis 30°, 2 cis 120°, 2 cis 210°, 2 cis 300°, or $\sqrt{3} + i$, $-1 + \sqrt{3}\,i$, $-\sqrt{3} - i$, $1 - \sqrt{3}\,i$.

28. 4 cis 90°, 4 cis 210°, 4 cis 330°, or $4i$, $-2\sqrt{3} - 2i$, $2\sqrt{3} - 2i$

29. $2\sqrt{3} - 2i = 4$ cis 330°

$(4 \text{ cis } 330°)^{1/3} = 4^{1/3} \text{ cis } \left[\frac{330°}{3} + k \cdot \frac{360°}{3}\right],$

$k = 0, 1, 2$

The roots are $\sqrt[3]{4}$ cis 110°, $\sqrt[3]{4}$ cis 230°, $\sqrt[3]{4}$ cis 350°.

30. $\sqrt[3]{2}$ cis 100°, $\sqrt[3]{2}$ cis 220°, $\sqrt[3]{2}$ cis 340°

31. $x^3 = 1$

The solutions of this equation are the cube roots of 1. These were found in Example 4 in the text.

32. cis 0°, cis 72°, cis 144°, cis 216°, cis 288°

33. $x^5 + 1 = 0$

$x^5 = -1$

The solutions of this equation are the fifth roots of -1. These were found in Exercise 25 above.

34. cis 67.5°, cis 157.5°, cis 247.5°, cis 337.5°

35. $x^5 + \sqrt{3} + i = 0$

$x^5 = -\sqrt{3} - i$

The solutions of this equation are the fifth roots of $-\sqrt{3} - i$.

$-\sqrt{3} - i = 2$ cis 210°

$(2 \text{ cis } 210°)^{1/5} = 2^{1/5} \text{ cis } \left[\frac{210°}{5} + k \cdot \frac{360°}{5}\right],$

$k = 0, 1, 2, 3, 4$

The solutions are $\sqrt[5]{2}$ cis 42°, $\sqrt[5]{2}$ cis 114°, $\sqrt[5]{2}$ cis 186°, $\sqrt[5]{2}$ cis 258°, $\sqrt[5]{2}$ cis 330°

36. 2 cis 40°, 2 cis 160°, 2 cis 280°

37. $x^4 + 81 = 0$

$x^4 = -81$

The solutions of this equation are the fourth roots of -81.

$-81 = 81 \text{ cis } 180°$

$(81 \text{ cis } 180°)^{1/4} = 81^{1/4} \text{ cis } \left(\frac{180°}{4} + k \cdot \frac{360°}{4}\right),$

$k = 0, 1, 2, 3$

The solutions are 3 cis 45°, 3 cis 135°, 3 cis 225°, 3 cis 315°, or $\frac{3\sqrt{2}}{2} + \frac{3\sqrt{2}}{2}i$, $-\frac{3\sqrt{2}}{2} + \frac{3\sqrt{2}}{2}i$, $-\frac{3\sqrt{2}}{2} - \frac{3\sqrt{2}}{2}i$, and $\frac{3\sqrt{2}}{2} - \frac{3\sqrt{2}}{2}i$.

38. cis 30°, cis 90°, cis 150°, cis 210°, cis 270°, cis 330°, or $\frac{\sqrt{3}}{2} + \frac{1}{2}i$, i, $-\frac{\sqrt{3}}{2} + \frac{1}{2}i$, $-\frac{\sqrt{3}}{2} - \frac{1}{2}i$, $-i$, $\frac{\sqrt{3}}{2} - \frac{1}{2}i$

39. $x^2 + (1 - i)x + i = 0$

$a = 1, \quad b = 1 - i, \quad c = 1$

$$x = \frac{-(1 - i) \pm \sqrt{(1 - i)^2 - 4 \cdot 1 \cdot i}}{2 \cdot 1}$$

$$= \frac{-1 + i \pm \sqrt{1 - 2i + i^2 - 4i}}{2}$$

$$= \frac{-1 + i \pm \sqrt{-6i}}{2}$$

Let $z = \sqrt{-6i} = [6 \text{ cis } 270°]^{1/2}$

$= 6^{1/2} \text{ cis } \left(\frac{270°}{2} + k \cdot \frac{360°}{2}\right),\ k = 0, 1.$

Hence $z_1 = \sqrt{6} \text{ cis } 135° = -\sqrt{3} + \sqrt{3}\, i$

$z_2 = \sqrt{6} \text{ cis } 315° = \sqrt{3} - \sqrt{3}\, i.$

Thus, substituting for $\sqrt{-6i}$, we get

$$x_1 = \frac{-1 + i - \sqrt{3} + \sqrt{3}\, i}{2} = -1.366 + 1.366i$$

$$x_2 = \frac{-1 + i + \sqrt{3} - \sqrt{3}\, i}{2} = 0.366 - 0.366i$$

40. $0.12 + 0.39i$, $-0.45 - 1.06i$

Exercise Set 9.1

1. We replace x by $\frac{1}{2}$ and y by 1.

$$\begin{array}{c|c} \multicolumn{2}{c}{3x + y = \frac{5}{2}} \\ \hline 3 \cdot \frac{1}{2} + 1 & \frac{5}{2} \\ \frac{3}{2} + \frac{2}{2} & \\ \frac{5}{2} & \end{array} \qquad \begin{array}{c|c} \multicolumn{2}{c}{2x - y = \frac{1}{4}} \\ \hline 2 \cdot \frac{1}{2} - 1 & \frac{1}{4} \\ 1 - 1 & \\ 0 & \end{array}$$

The ordered pair $\left(\frac{1}{2}, 1\right)$ is not a solution of $2x - y = \frac{1}{4}$. Therefore it is not a solution of the system of equations.

2. Yes

3. Graph both lines on the same set of axes. The x and y-intercepts of the line x + y = 2 are (2, 0) and (0, 2). Plot these points and draw the line they determine. Next graph the line 3x + y = 0. Three of the points on the line are (0, 0), (-1, 3), and (2, -6). Plot these points and draw the line they determine.

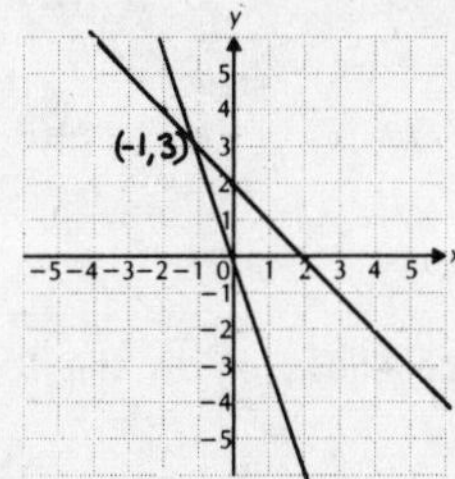

The solution (point of intersection) seems to be the point (-1, 3).

Check:

$$\begin{array}{c|c} \multicolumn{2}{c}{x + y = 2} \\ \hline -1 + 3 & 2 \\ 2 & \end{array} \qquad \begin{array}{c|c} \multicolumn{2}{c}{3x + y = 0} \\ \hline 3(-1) + 3 & 0 \\ -3 + 3 & \\ 0 & \end{array}$$

The solution is (-1, 3).

4. (3, -2)

5. Graph both lines on the same set of axes. The intercepts of y + 1 = 2x are $\left(\frac{1}{2}, 0\right)$ and (0, -1). The intercepts of y - 1 = 2x are $\left(-\frac{1}{2}, 0\right)$ and (0, 1).

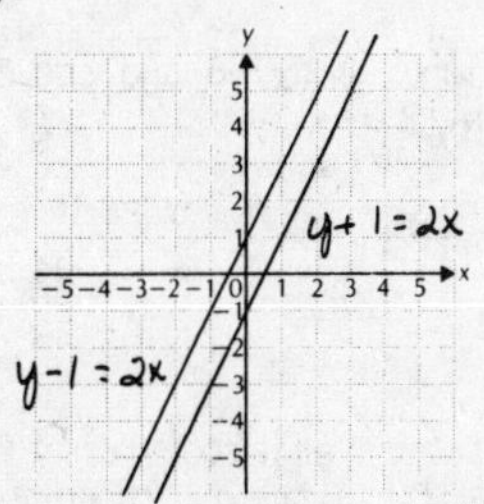

The lines have no point in common. The system has no solution.

6. Infinitely many solutions

7. $x - 5y = 4,$ (1)

$2x + y = 7$ (2)

Solve equation (1) for x.

$$x - 5y = 4$$
$$x = 5y + 4$$

Substitute 5y + 4 for x in equation (2) and solve for y.

$$2x + y = 7$$
$$2(5y + 4) + y = 7$$
$$10y + 8 + y = 7$$
$$11y + 8 = 7$$
$$11y = -1$$
$$y = -\frac{1}{11}$$

Now substitute $-\frac{1}{11}$ for y in either equation (1) or (2) and solve for x. It's easier to use equation (1).

$$x - 5y = 4$$
$$x - 5\left(-\frac{1}{11}\right) = 4$$
$$x + \frac{5}{11} = \frac{44}{11}$$
$$x = \frac{39}{11}$$

Check:

$$\begin{array}{c|c} \multicolumn{2}{c}{x - 5y = 4} \\ \hline \frac{39}{11} - 5\left(-\frac{1}{11}\right) & 4 \\ \frac{39}{11} + \frac{5}{11} & \\ \frac{44}{11} & \\ 4 & \end{array} \qquad \begin{array}{c|c} \multicolumn{2}{c}{2x + y = 7} \\ \hline 2\left(\frac{39}{11}\right) + \left(-\frac{1}{11}\right) & 7 \\ \frac{78}{11} - \frac{1}{11} & \\ \frac{77}{11} & \\ 7 & \end{array}$$

The solution is $\left(\frac{39}{11}, -\frac{1}{11}\right)$.

8. $\left(\frac{11}{8}, -\frac{7}{8}\right)$

9. $x - 3y = 2,$ (1)
$6x + 5y = -34$ (2)

We multiply equation (1) by -6 and add the result to equation (2).

$x - 3y = 2,$ $\quad -6x + 18y = -12$
$23y = -46 \leftarrow$ $\quad \underline{6x + 5y = -34}$
$\quad 23y = -46$

Now we solve the second equation for y. Then we substitute the result in the first equation to find x.

$-23y = -46$ $\quad x - 3(-2) = 2$
$y = -2$ $\quad x + 6 = 2$
$\quad x = -4$

The solution is (-4, -2).

10. (3, -1)

11. $0.3x + 0.2y = -0.9,$
$0.2x - 0.3y = -0.6$

Multiply each equation by 10 to eliminate decimals.

$3x + 2y = -9,$ (1)
$2x - 3y = -6$ (2)

Multiply equation (2) by 3 to make the x-coefficient a multiple of 3.

$3x + 2y = -9,$
$6x - 9y = -18$

Then multiply equation (1) by -2 and add the result to equation (2).

$3x + 2y = -9$ $\quad -6x - 4y = 18$
$-13y = 0 \leftarrow$ $\quad \underline{6x - 9y = -18}$
$\quad -13y = 0$

Now solve the second equation for y and substitute in the first equation to find x.

$-13y = 0$ $\quad 3x + 2\cdot 0 = -9$
$y = 0$ $\quad 3x = -9$
$\quad x = -3$

The solution is (-3, 0).

12. $\left(\frac{1}{2}, -\frac{1}{3}\right)$

13. $\frac{1}{5}x + \frac{1}{2}y = 6,$
$\frac{3}{5}x - \frac{1}{2}y = 2$

Multiply each equation by 10 to clear of fractions and then add.

$2x + 5y = 60,$ (1)
$6x - 5y = 20$ (2)

Multiply equation (1) by -3 and add the result to equation (2).

$2x + 5y = 60,$ $\quad -6x - 15y = -180$
$-20y = -160 \leftarrow$ $\quad \underline{6x - 5y = 20}$
$\quad -20y = -160$

Solve the second equation for y and substitute in the first equation to find x.

$-20y = -160$ $\quad 2x + 5\cdot 8 = 60$
$y = 8$ $\quad 2x + 40 = 60$
$\quad 2x = 20$
$\quad x = 10$

The solution is (10, 8).

14. (-12, -15)

15. $2a = 5 - 3b,$
$4a = 11 - 7b$

First write each equation in the form Ax + By = C.

$2a + 3b = 5,$ (1)
$4a + 7b = 11$ (2)

Multiply equation (1) by -2 and add the result to equation (2).

$2a + 3b = 5,$ $\quad -4a - 6b = -10$
$b = 1 \leftarrow$ $\quad \underline{4a + 7b = 11}$
$\quad b = 1$

Substitute in the first equation to find a.

$2a + 3\cdot 1 = 5$
$2a + 3 = 5$
$2a = 2$
$a = 1$

The solution is (1, 1).

16. (3, 1)

17. Familiarize and Translate:

Let x and y represent the numbers. A system of equations results directly from the statements in the problem.

$x + y = -10,$ (The sum is -10.)

$x - y = 1$ (The difference is 1.)

Carry out: We solve the system.

Multiply the first equation by -1 and add the result to the second equation.

$x + y = -10,$

$-2y = 11$

Solve the second equation for y and then substitute into the first equation to find x.

$-2y = 11$ $\quad x + \left(-\frac{11}{2}\right) = -10$

$y = -\frac{11}{2}$ $\quad x = -\frac{20}{2} + \frac{11}{2} = -\frac{9}{2}$

Check:

The sum of $-\frac{9}{2}$ and $-\frac{11}{2}$ is $-\frac{20}{2}$, or -10. The difference is $-\frac{9}{2} - \left(-\frac{11}{2}\right)$, or $\frac{2}{2}$ which is 1.

State:

Thus the numbers are $-\frac{9}{2}$ and $-\frac{11}{2}$.

18. $\frac{9}{2}$ and $-\frac{11}{2}$

19. Familiarize:

It helps to make a drawing. Then organize the information in a table. Let b represent the speed of the boat and s represent the speed of the stream. The speed upstream is b - s. The speed downstream is b + s.

$d_1 = 46$ $\quad t_1 = 2$ $\quad r_1 = b + s$

Downstream

$d_2 = 51$ $\quad t_2 = 3$ $\quad r_2 = b - s$

Upstream

	Distance	Speed	Time
Downstream	46	b + s	2
Upstream	51	b - s	3

Translate:

Using d = rt in each row of the table, we get a system of equations.

$46 = (b + s)2 = 2b + 2s,$ or $b + s = 23,$

$51 = (b - s)3 = 3b - 3s$ $\quad b - s = 17$

Carry out:

Multiply the first equation by -1 and add the result to the second equation.

$b + s = 23,$

$-2s = -6$

19. (continued)

Solve the second equation for s and then substitute in the first equation to find b.

$-2s = -6$ $\quad b + 3 = 23$

$s = 3$ $\quad b = 20$

Check:

The speed downstream is 20 + 3, or 23. The distance downstream is 23·2, or 46. The speed upstream is 20 - 3, or 17. The distance upstream is 17·3, or 51. The values check.

State:

The speed of the boat is 20 km/h. The speed of the stream is 3 km/h.

20. Plane: 875 km/h, wind: 125 km/h

21. Familiarize:

We organize the information in a table. We let x represent the number of liters of antifreeze A and y represent the number of liters of antifreeze B.

	Amount of antifreeze	Percent of alcohol	Amount of alcohol in antifreeze
A	x liters	18%	18%x
B	y liters	10%	10%y
Mixture	20 liters	15%	0.15 × 20, or 3

Translate:

If we add x and y in the first column, we get 20.

$x + y = 20$

If we add the amounts of alcohol in the third column, we get 3.

$18\%x + 10\%y = 3$

After changing percents to decimals and clearing, we have the following system:

$x + y = 20,$

$18x + 10y = 300$

Carry out:

Multiply -18 times the first equation and add the result to the second equation.

$x + y = 20,$

$-8y = -60$

Solve the second equation for y and substitute in the first equation to find x.

$-8y = -60$ $\quad x + 7.5 = 20$

$y = 7.5$ $\quad x = 12.5$

21. (continued)

Check:

We add the amounts of antifreeze: 12.5 L + 7.5 L = 20 L. Thus the amount of antifreeze checks. Next, we check the amount of alcohol: 18%(12.5) + 10%(7.5) = 2.25 + 0.75, or 3 L. The amount of alcohol also checks.

State:

The solution of the problem is 12.5 L of A and 7.5 L of B.

22. A: 15 L, B: 35 L

23. Familiarize:

We first make a drawing.

t hours 80 km/h 96 km/h t hours

d kilometers 528 - d kilometers

528 km

Then we organize the information in a table. Each car travels the same amount of time. Let t represent the time and d_1 and d_2 represent the distances of the slow car and the fast car respectively.

	Distance	Speed	Time
Slow car	d	80	t
Fast car	528 - d	96	t

Translate:

Using d = rt in each row of the table, we get

$d = 80t$, $528 - d = 96t$ or $d - 80t = 0$, $d + 96t = 528$.

Carry out.

Multiply the first equation by -1 and add the result to the second equation.

$d - 80t = 0$,
$176t = 528$

Solve the second equation for t. (We do not need to find d.)

$176t = 528$
$t = 3$

Check:

In 3 hours, one car travels 80·3, or 240 km, and the other car travels 96·3, or 288 km. The sum of the distances is 240 + 288, or 528 km. The value checks.

State:

In 3 hours the cars will be 528 km apart.

24. 375 km

25. Familiarize:

We first make a drawing.

t hours 190 km/h 200 km/h t hours

d kilometers 780 - d kilometers

780 km

Then we organize the information in a table. The time is the same for each plane. Let t represent the time and d_1 and d_2 represent the distances of the planes.

	Distance	Speed	Time
Slow plane	d	190	t
Fast plane	780 - d	200	t

Translate:

Using d = rt in each row of the table, we get

$d = 190t$, $780 - d = 200t$ or $d - 190t = 0$, $d + 200t = 780$.

Carry out:

Multiply the first equation by -1 and add the result to the second equation.

$d - 190t = 0$,
$390t = 780$

Solve the second equation for t. (We do not need to find d.)

$390t = 780$
$t = 2$

Check:

In 2 hours, one plane travels 2·190, or 380 km, and the other plane travels 2·200, or 400 km. The total distance is 380 + 400, or 780 km. The value checks.

State:

The planes will meet in 2 hours.

26. $1\frac{3}{4}$ hours

27. Familiarize and Translate:

Let x represent the number of white scarves and y represent the number of printed scarves. Thus, \$4.95x worth of white ones and \$7.95y worth of printed ones were sold. The total number of scarves was 40.

$x + y = 40$

The total sales were \$282.

$4.95x + 7.95y = 282$, or $495x + 795y = 28,200$

We solve the following system of equations:

$x + y = 40$,
$495x + 795y = 28,200$

27. (continued)

Carry out:

Multiply -495 times the first equation and add the result to the second equation.

$x + y = 40,$

$300y = 8400$

Solve the second equation for y and then substitute in the first equation to find x.

$300y = 8400 \qquad x + 28 = 40$

$y = 28 \qquad x = 12$

Check:

The total number of scarves was 12 + 28, or 40. Total sales were 4.95(12) + 7.95(28), or 59.40 + 222.60, or $282. The values check.

State:

12 white scarves and 28 printed ones were sold.

28. White: 8, yellow: 22

29. Familiarize and Translate:

Let x represent Paula's age now and y represent Bob's age now. It helps to organize the information in a table.

	Age now	Age four years from now
Paula	x	x + 4
Bob	y	y + 4

Paula is 12 years older than Bob.

$x = y + 12, \quad \text{or} \quad x - y = 12$

Four years from now, Bob will be $\frac{2}{3}$ as old as Paula.

$y + 4 = \frac{2}{3}(x + 4)$

or

$3(y + 4) = 2(x + 4)$

$3y + 12 = 2x + 8$

$4 = 2x - 3y$

We now have a system of equations.

$x - y = 12,$

$2x - 3y = 4$

Carry out:

Multiply the first equation by -2 and add the result to the second equation.

$x - y = 12,$

$-y = -20$

Complete the solution.

$-y = -20 \qquad x - 20 = 12$

$y = 20 \qquad x = 32$

29. (continued)

Check:

If Paula is 32 and Bob 20, Paula is 12 years older than Bob. In four years, Paula will be 36 and Bob 24. Thus Bob will be $\frac{2}{3}$ as old as Paula $\left[24 = \frac{2}{3} \cdot 36\right]$. The ages check.

State:

Paula is 32, and Bob is 20.

30. Marcia: 20, Carlos: 28

31. Familiarize:

If helps to make a drawing. We let ℓ represent the length and w represent the width.

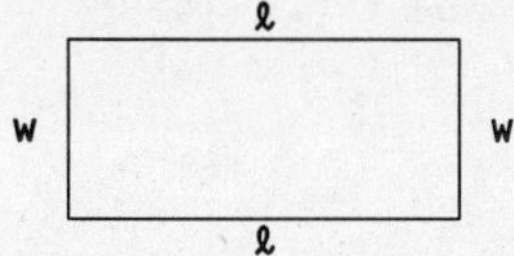

The formula for the perimeter is $P = 2\ell + 2w$.

Translate:

The perimeter is 190 m.

$2\ell + 2w = 190, \quad \text{or} \quad \ell + w = 95$

The width is one-fourth the length.

$w = \frac{1}{4}\ell$

The resulting system is

$\ell + w = 95,$

$w = \frac{1}{4}\ell.$

Carry out:

Substitute $\frac{1}{4}\ell$ for w in the first equation and solve for ℓ.

$\ell + w = 95$

$\ell + \frac{1}{4}\ell = 95$

$\frac{5}{4}\ell = 95$

$\frac{4}{5} \cdot \frac{5}{4}\ell = \frac{4}{5} \cdot 95$

$\ell = 76$

Then substitute 76 for ℓ in one of the equations of the system and solve for w.

$\ell + w = 95$

$76 + w = 95$

$w = 19$

Check:

If $\ell = 76$ m and $w = 19$ m, then the width is one-fourth the length. The perimeter is $2 \cdot 76 + 2 \cdot 19$, or 190 m. The values check.

State:

The length is 76 m; the width is 19 m.

32. Length: 160 m, width: 154 m

33. Familiarize:

We first make a drawing.

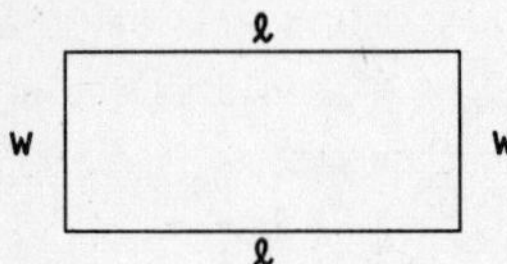

We let ℓ represent the length and w represent the width. The formula for the perimeter is $P = 2\ell + 2w$.

Translate:

The perimeter is 384 m.

$2\ell + 2w = 384$, or $\ell + w = 192$

The length is 82 m greater than the width.

$\ell = w + 82$

The resulting system is

$\ell + w = 192,$

$\ell = w + 82.$

Carry out:

Substitute $w + 82$ for ℓ in the first equation and solve for w.

$$\begin{aligned} \ell + w &= 192 \\ w + 82 + w &= 192 \\ 2w + 82 &= 192 \\ 2w &= 110 \\ w &= 55 \end{aligned}$$

Then substitute 55 for w in one of the original equations and solve for ℓ.

$$\begin{aligned} \ell &= w + 82 \\ \ell &= 55 + 82 \\ \ell &= 137 \end{aligned}$$

Check:

The length is 82 greater than the width ($137 = 55 + 82$). The perimeter is $2\cdot137 + 2\cdot55$, or 384 m. The values check.

State:

The length is 137 m; the width is 55 m.

34. 372 cm^2

35. Familiarize:

Let x represent the amount invested at 9%, and let y represent the amount invested at 10%. The interest earned by these investments is 9%x, or 0.09x, and 10%y, or 0.1y, respectively.

Translate:

The total investment is $15,000.

$x + y = 15{,}000$

The total interest is $1432.

$0.09x + 0.1y = 1432$, or $9x + 10y = 143{,}200$

The resulting system is

$x + y = 15{,}000,$

$9x + 10y = 143{,}200.$

Carry out:

Multiply the first equation by -9 and add.

$$\begin{aligned} x + y &= 15{,}000, \\ y &= 8200 \end{aligned}$$

Complete the solution.

$$\begin{aligned} x + 8200 &= 15{,}000 \\ x &= 6800 \end{aligned}$$

Check:

The total investment is $8200 + $6800, or $15,000.

The total interest is 0.09($6800) + 0.1($8200), or $612 + $820, or $1432.

State:

$6800 is invested at 9%, and $8200 is invested at 10%.

36. 5%: $30,000, 6%: $40,000

37. $\frac{x + y}{4} - \frac{x - y}{3} = 1$

$\frac{x - y}{2} + \frac{x + y}{4} = -9$

First multiply each equation by the LCM to clear fractions.

$3(x + y) - 4(x - y) = 12$ (Multiplying by 12)

$2(x - y) + (x + y) = -36$ (Multiplying by 4)

After simplifying, the resulting system is

$-x + 7y = 12,$

$3x - y = -36.$

Multiply the first equation by 3 and add the result to the second equation.

$$\begin{aligned} -x + 7y &= 12 \\ 20y &= 0 \end{aligned}$$

Complete the solution.

$$\begin{aligned} 20y &= 0 \\ y &= 0 \end{aligned} \qquad \begin{aligned} -x + 7\cdot0 &= 12 \\ -x &= 12 \\ x &= -12 \end{aligned}$$

The solution is (-12, 0).

38. (0, 0)

39. $2.35x - 3.18y = 4.82,$

$1.92x + 6.77y = -3.87$

It is helpful to multiply each equation by 100 to clear of decimal points.

$235x - 318y = 482,$

$192x + 677y = -387$

Multiply the second equation by 235 to make the x-coefficient a multiple of 235.

$235x - 318y = 482,$

$45{,}120x + 159{,}095y = -90{,}945$

Then multiply the first equation by -192 and add.

$235x - 318y = 482,$

$220{,}151y = -183{,}489$

Complete the solution.

$220{,}151y = -183{,}489 \qquad 235x - 318(-0.833) = 482$

$y \approx -0.833 \qquad 235x + 264.894 = 482$

$235x = 217.106$

$x \approx 0.924$

The solution is (0.924, -0.833).

40. (0.13122, -1.10518)

41. Familiarize:

We first make a drawing.

Jogging 8 km/h t_1 hours Walking t_2 hours

Home University

6 km

It helps to organize the information in a table. Using $d = rt$, we let $8t_1$ represent the distance jogging and $4t_2$ represent the distance walking.

	Distance	Speed	Time
Jogging	$8t_1$	8	t_1
Walking	$4t_2$	4	t_2
Total	6		1

Translate:

The total distance is 6 km. The total time is 1 hour. A system of equations results from the columns in the table.

$t_1 + t_2 = 1,$

$8t_1 + 4t_2 = 6$

Carry out:

Multiply the first equation by -8 and add the result to the second equation.

$t_1 + t_2 = 1$

$-4t_2 = -2$

41. (continued)

Complete the solution.

$-4t_2 = -2 \qquad t_1 + t_2 = 1$

$t_2 = \frac{1}{2} \qquad t_1 = \frac{1}{2}$

The jogging distance, $8t_1$, is $8 \cdot \frac{1}{2}$, or 4 km.

The walking distance, $4t_2$, is $4 \cdot \frac{1}{2}$, or 2 km.

Check:

The total distance is 4 + 2, or 6 km. The total time is $\frac{1}{2} + \frac{1}{2}$, or 1 hour. The values check.

State:

She jogs 4 km on each trip.

42. James: 32 yr, Joan: 14 yr

43. Familiarize:

Let x represent the number of one-book orders. Let y represent the number of two-book orders. Then 12x represents the amount taken in from the one-book orders and 20y represents the amount taken in from the two-book orders. It helps to organize the information in a table.

	Number of orders	Number of books	Amount taken in
One-book orders	x	x	12x
Two-book orders	y	2y	20y
Total		880	9840

Translate:

The total number of books sold was 880.

$x + 2y = 880$

The total amount of money taken in was $9840.

$12x + 20y = 9840, \quad \text{or} \quad 3x + 5y = 2460$

The resulting system of equations is

$x + 2y = 880,$

$3x + 5y = 2460.$

Multiply the first equation by -3 and add the result to the second equation.

$x + 2y = 880,$

$-y = -180$

Complete the solution.

$-y = -180 \qquad x + 2\cdot 180 = 880$

$y = 180 \qquad x + 360 = 880$

$x = 520$

Check:

The total number of books sold was $520 + 2\cdot 180$, or 880. The total amount taken in was $12\cdot 520 + 20\cdot 180$, or $9840. The values check.

43. (continued)

State:

180 members ordered two books.

44. 82

45. Familiarize and Translate:

Let x represent the numerator and y the denominator.

The numerator is 12 more than the denominator.

$x = y + 12$, or $x - y = 12$

The sum of the numerator and the denominator is 5 more than three times the denominator.

$x + y = 3y + 5$, or $x - 2y = 5$

The resulting system is

$x - y = 12$,

$x - 2y = 5$.

Carry out:

Multiply the first equation by -1 and add the result to the second equation.

$x - y = 12$,

$-y = -7$

Complete the solution.

$-y = -7$ $\quad$ $x - 7 = 12$

$y = 7$ $\quad$ $x = 19$

Check:

If $x = 19$ and $y = 7$, the fraction is $\frac{19}{7}$. The numerator is 12 more than the denominator. The sum of the numerator and the denominator, $19 + 7$ (or 26), is 5 more than three times the denominator ($26 = 5 + 3\cdot7$). The numbers check.

State:

The reciprocal of $\frac{19}{7}$ is $\frac{7}{19}$.

46. 137°

47. Familiarize and Translate:

First situation:

3 hours $\quad r_1$ km/h $\quad\quad r_2$ km/h $\quad$ 2 hours

Union $\quad\quad$ Central

216 km

Second situation:

1.5 hours $\quad r_1$ km/h $\quad\quad r_2$ km/h $\quad$ 3 hours

Union $\quad\quad$ Central

216 km

We let r_1 and r_2 represent the speeds of the trains and list the information in a table. Using $d = rt$, we let $3r_1$, $2r_2$, $1.5r_1$, and $3r_2$ represent the distances the trains travel.

47. (continued)

	Distance traveled in first situation	Distance traveled in second situation
$Train_1$ (from Union to Central)	$3r_1$	$1.5r_1$
$Train_2$ (from Central to Union)	$2r_2$	$3r_2$
Total	216	216

The total distance in each situation is 216 km. Thus, we have a system of equations.

$3\ r_1 + 2r_2 = 216$,

$1.5r_1 + 3r_2 = 216$.

Carry out:

Multiply the second equation by 2 to make the x-coefficient a multiple of 3.

$3r_1 + 2r_2 = 216$,

$3r_1 + 6r_2 = 432$

Now multiply the first equation by -1 and add.

$3r_1 + 2r_2 = 216$,

$4r_2 = 216$

Complete the solution.

$4r_2 = 216$ $\quad$ $3r_1 + 2\cdot54 = 216$

$r_2 = 54$ $\quad$ $3r_1 + 108 = 216$

$3r_1 = 108$

$r_1 = 36$

Check:

If $r_1 = 36$ and $r_2 = 54$, the total distance the trains travel in the first situation is $3\cdot36 + 2\cdot54$, or 216 km. The total distance they travel in the second situation is $1.5\cdot36 + 3\cdot54$, or 216 km. The values check.

State:

The speed of the first train is 36 km/h.
The speed of the second train is 54 km/h.

48. $4\frac{4}{7}$

49. Familiarize:

Let x represent the value of the horse.

Let y represent the value of the compensation the stablehand was to receive after one year.

Translate:

If the stablehand had worked for one year, he would have received \$240 and one horse.

$y = 240 + x$

\$100 and one horse is $\frac{7}{12}$ of one year's compensation.

$100 + x = \frac{7}{12}y$

49. (continued)

We have a system of equations:

$y = 240 + x,$

$100 + x = \frac{7}{12}y$

Carry out:

We substitute 240 + x for y in the second equation.

$100 + x = \frac{7}{12}(240 + x)$

$100 + x = 140 + \frac{7}{12}x$

$\frac{5}{12}x = 40$

$x = 96$

Check:

If the value of the horse is \$96, the compensation after one year would have been \$240 + \$96, or \$336. Since $\frac{7}{12} \cdot \$336 = \$196$, or \$100 + \$96, the value checks.

State:

The value of the horse is \$96.

50. 5

51. Familiarize and Translate:

Let x represent the number of girls in the family and y represent the number of boys. Then Phil has x sisters and y - 1 brothers, and Phyllis has x - 1 sisters and y brothers.

	Brothers	Sisters
Phil	y - 1	x
Phyllis	y	x - 1

Phil has the same number of brothers and sisters.

$y - 1 = x,$ or $x - y = -1$

Phyllis has twice as many brothers as sisters.

$y = 2(x - 1),$ or $2x - y = 2$

We now have a system of equations.

$x - y = -1,$

$2x - y = 2$

Carry out:

Multiply the first equation by -2 and add the result to the second equation.

$x - y = -1,$

$y = 4$

Complete the solution.

$x - 4 = -1$

$x = 3$

51. (continued)

Check:

If there are 3 girls and 4 boys in the family, Phil has 3 brothers and 3 sisters and Phyllis has 2 sisters and 4 brothers. Thus, Phil has the same number of brothers and sisters, and Phyllis has twice as many brothers as sisters.

State:

There are 3 girls and 4 boys in the family.

52. City: 261 mi, highway: 204 mi

53. Substituting the given solutions in the equation $y = mx + b$, we get a system of equations.

$3 = -2m + b,$ or $-2m + b = 3,$

$-5 = 4m + b$ $\quad$ $4m + b = -5$

Multiply the first equation by 2 and add.

$-2m + b = 3,$

$3b = 1$

Complete the solution.

$3b = 1$ $\quad$ $-2m + \frac{1}{3} = 3$

$b = \frac{1}{3}$ $\quad$ $-2m = \frac{8}{3}$

$m = -\frac{4}{3}$

54. $A = \frac{1}{10}$, $B = -\frac{7}{10}$

55. Solve the system of equations.

$x + 43p = 800,$

$x - 16p = 210.$

Multiply the first equation by -1 and add.

$x + 43p = 800,$

$-59p = -590$

Complete the solution.

$-59p = -590$ $\quad$ $x + 43 \cdot 10 = 800$

$p = 10$ $\quad$ $x + 430 = 800$

$x = 370$

The equilibrium point is (370, \$10).

56. (7600, \$40)

57. Solve the system of equations.

$x = 760 - 13p,$ or $x + 13p = 760$

$x = 430 + 2p$ $\quad x - 2p = 430.$

Multiply the first equation by -1 and add.

$x + 13p = 760,$

$-15p = -330$

Complete the solution.

$-15p = -330 \qquad x + 13\cdot 22 = 760$

$p = 22 \qquad x + 286 = 760$

$x = 474$

The equilibrium point is (474, \$22).

58. Solve $x + 60p = 2000,$

$x - 94p = 460.$

$x + 60p = 2000$

$-154p = -1540$

$p = 10,\ x = 1400$

The equilibrium point is (1400, \$10).

59. $\frac{1}{x} - \frac{3}{y} = 2$

$\frac{6}{x} + \frac{5}{y} = -34$

Substitute u for $\frac{1}{x}$ and v for $\frac{1}{y}$ to obtain a linear system.

$u - 3v = 2,$

$6u + 5v = -34$

Multiply the first equation by -6 and add the result to the second equation.

$u - 3v = 2$

$23v = -46$

Complete the solution for u and v.

$23v = -46 \qquad u - 3(-2) = 2$

$v = -2 \qquad u + 6 = 2$

$u = -4$

If $v = -2$, then $\frac{1}{y} = -2$, or $y = -\frac{1}{2}$.

If $u = -4$, then $\frac{1}{x} = -4$, or $x = -\frac{1}{4}$.

Both values check. The solution of the original system is $\left(-\frac{1}{4}, -\frac{1}{2}\right)$.

60. (-125, 100)

61. $3|x| + 5|y| = 30$

$5|x| + 3|y| = 34$

Substitute u for $|x|$ and v for $|y|$ to obtain a linear system.

$3u + 5v = 30,$

$5u + 3v = 34$

Multiply the second equation by 3 to make the x-coefficient a multiple of 3.

$3u + 5v = 30,$

$15u + 9v = 102$

Now multiply the first equation by -5 and add.

$3u + 5v = 30,$

$-16v = -48$

Complete the solution for u and v.

$-16v = -48 \qquad 3u + 5\cdot 3 = 30$

$v = 3 \qquad 3u + 15 = 30$

$3u = 15$

$u = 5$

If $v = 3$, then $|y| = 3$ and $y = 3$ or -3.

If $u = 5$, then $|x| = 5$ and $x = 5$ or -5.

Thus, the solution set for the original system is $\{(5, 3), (5, -3), (-5, 3), (-5, -3)\}$.

62. $(0, \sqrt[3]{3})$

63. Familiarize:

First we convert the given distances to miles:

$300 \text{ ft} = \frac{300}{5280} \text{ mi} = \frac{5}{88} \text{ mi},$

$500 \text{ ft} = \frac{500}{5280} \text{ mi} = \frac{25}{264} \text{ mi}$

Then at 10 mph, the student can run to point P in $\frac{\frac{5}{88}}{10}$, or $\frac{1}{176}$ hr, and she can run to point Q in $\frac{\frac{25}{264}}{10}$, or $\frac{5}{528}$ hr $\left[\text{using } d = rt, \text{ or } \frac{d}{r} = t\right]$.

Let d represent the distance, in miles, from the train to point P in the drawing in the text, and let r represent the speed of the train, in miles per hour.

We organize the information in a table.

	Distance	Rate	Time
Going to P	d	r	$\frac{1}{176}$
Going to Q	$d + \frac{5}{88} + \frac{25}{264}$	r	$\frac{5}{528}$

63. (continued)

Translate:

Using d = rt in each row of the table, we get a system of equations.

$$d = r\left(\frac{1}{176}\right),$$

$$d + \frac{5}{88} + \frac{25}{264} = r\left(\frac{5}{528}\right)$$

Carry out:

We substitute $\frac{r}{176}$ for d in the second equation.

$$\frac{r}{176} + \frac{5}{88} + \frac{25}{264} = \frac{5r}{528}$$

$$3r + 30 + 50 = 5r \quad \text{(Multiplying by 528)}$$

$$80 = 2r$$

$$40 = r$$

Check:

The check is left to the student.

State:

The train is traveling 40 mph.

Exercise Set 9.2

1. We substitute (-1, 1, 0) in each of the three equations.

$2x + 3y - 5z = 1$	
$2(-1) + 3\cdot1 - 5\cdot0$	1
$-2 + 3 - 0$	
1	

$6x - 6y + 10z = 3$	
$6(-1) - 6\cdot1 + 10\cdot0$	3
$-6 - 6 + 0$	
-12	

We do not need to proceed. Since (-1, 1, 0) is not a solution of one of the equations, it is not a solution of the system.

2. Yes

3. $x + y + z = 2$ (P1)
$6x - 4y + 5z = 31$ (P2)
$5x + 2y + 2z = 13$ (P3)

Multiply (P1) by -6 and add the result to (P2). We also multiply (P1) by -5 and add the result to (P3).

$x + y + z = 2$ (P1)
$-10y - z = 19$ (P2)
$-3y - 3z = 3$ (P3)

Multiply (P3) by 10 to make the y-coefficient a multiple of the y-coefficient in (P2).

$x + y + z = 2$ (P1)
$-10y - z = 19$ (P2)
$-30y - 30z = 30$ (P3)

3. (continued)

Multiply (P2) by -3 and add the result to (P3).

$x + y + z = 2$ (P1)
$-10y - z = 19$ (P2)
$-27z = -27$ (P3)

Solve (P3) for z.

$-27z = -27$

$z = 1$

Back-substitute 1 for z in (P2) and solve for y.

$-10y - z = 19$

$-10y - 1 = 19$

$-10y = 20$

$y = -2$

Back-substitute 1 for z and -2 for y in (P1) and solve for x.

$x + y + z = 2$

$x + (-2) + 1 = 2$

$x - 1 = 2$

$x = 3$

The solution is (3, -2, 1).

4. (-2, -1, 4)

5. $x - y + 2z = -3$ (P1)
$x + 2y + 3z = 4$ (P2)
$2x + y + z = -3$ (P3)

Multiply (P1) by -1 and add the result to (P2). Also multiply (P1) by -2 and add the result to (P3).

$x - y + 2z = -3$ (P1)
$3y + z = 7$ (P2)
$3y - 3z = 3$ (P3)

Multiply (P2) by -1 and add the result to (P3).

$x - y + 2z = -3$ (P1)
$3y + z = 7$ (P2)
$-4z = -4$ (P3)

Solve (P3) for z.

$-4z = -4$

$z = 1$

Back-substitute 1 for z in (P2) and solve for y.

$3y + z = 7$

$3y + 1 = 7$

$3y = 6$

$y = 2$

Back-substitute 1 for z and 2 for y in (P1) and solve for x.

$x - y + 2z = -3$

$x - 2 + 2\cdot1 = -3$

$x = -3$

The solution is (-3, 2, 1).

6. (1, 2, 3)

7. $4a + 9b = 8$ (P1)
$8a + 6c = -1$ (P2)
$6b + 6c = -1$ (P3)

Multiply (P1) by -2 and add the result to (P2).

$4a + 9b = 8$ (P1)
$-18b + 6c = -17$ (P2)
$6b + 6c = -1$ (P3)

Multiply (P3) by 3 to make the b-coefficient a multiple of the b-coefficient in (P2).

$4a + 9b = 8$ (P1)
$-18b + 6c = -17$ (P2)
$18b + 18c = -3$ (P3)

Add (P2) to (P3).

$4a + 9b = 8$ (P1)
$-18b + 6c = -17$ (P2)
$24c = -20$ (P3)

Solve (P3) for c.

$24c = -20$

$c = -\frac{20}{24} = -\frac{5}{6}$

Back-substitute $-\frac{5}{6}$ for c in (P2) and solve for b.

$-18b + 6c = -17$

$-18b + 6\left(-\frac{5}{6}\right) = -17$

$-18b - 5 = -17$

$-18b = -12$

$b = \frac{12}{18} = \frac{2}{3}$

Back-substitute $\frac{2}{3}$ for b in (P1) and solve for a.

$4a + 9b = 8$

$4a + 9 \cdot \frac{2}{3} = 8$

$4a + 6 = 8$

$4a = 2$

$a = \frac{1}{2}$

The solution is $\left(\frac{1}{2}, \frac{2}{3}, -\frac{5}{6}\right)$.

8. $\left(4, \frac{1}{2}, -\frac{1}{2}\right)$

9. $w + x + y + z = 2$ (P1)
$w + 2x + 2y + 4x = 1$ (P2)
$-w + x - y - z = -6$ (P3)
$-w + 3x + y - z = -2$ (P4)

Multiply (P1) by - 1 and add to (P2). Add (P1) to (P3) and to (P4).

$w + x + y + z = 2$ (P1)
$x + y + 3z = -1$ (P2)
$2x = -4$ (P3)
$4x + 2y = 0$ (P4)

Solve (P3) for x.

$2x = -4$

$x = -2$

Back-substitute -2 for x in (P4) and solve for y.

$4(-2) + 2y = 0$

$-8 + 2y = 0$

$2y = 8$

$y = 4$

Back-substitute -2 for x and 4 for y in (P2) and solve for z.

$-2 + 4 + 3z = -1$

$3z = -3$

$z = -1$

Back-substitute -2 for x, 4 for y, and -1 for z in (P1) and solve for w.

$w - 2 + 4 - 1 = 2$

$w = 1$

The solution is (1, -2, 4, -1).

10. (-3, -1, 0, 4)

11. Familiarize and Translate:

We let x, y, z represent the first, second, and third numbers respectively.

The sum of the three numbers is 26.

$x + y + z = 26$

Twice the first minus the second is 2 less than the third.

$2x - y = z - 2$, or $2x - y - z = -2$

The third is the second minus three times the first.

$z = y - 3x$, or $3x - y + z = 0$

We now have a system of three equations.

$x + y + z = 26$,
$2x - y - z = -2$,
$3x - y + z = 0$

Carry out:

We solve the system. The solution is (8, 21, -3).

11. (continued)

Check:

The sum of the numbers is 8 + 21 + (-3), or 26. Twice the first minus the second (2·8 - 21 = -5) is two less than the third (-3 - 2 = -5). The third (-3) is the second minus three times the first (21 - 3·8 = -3). The numbers check.

State:

The numbers are 8, 21, and -3.

12. 4, 2, and -1

13. Familiarize:

We make a drawing and use x, y, and z for the measures of the angles.

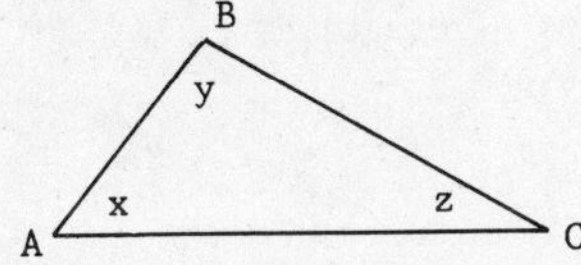

We must use the fact that the measures of the angles of a triangle add up to 180°.

Translate:

The sum of the angle measures is 180°.

$x + y + z = 180$

The measure of angle B is three times the measure of angle A.

$y = 3x$

The measure of angle C is 30° greater than the measure of angle A.

$z = x + 30$

We now have a system of equations.

$x + y + z = 180,$

$y = 3x,$

$z = x + 30$

Carry out:

We solve the system. The solution is (30, 90, 60).

Check:

The sum of the angle measures is 30 + 90 + 60, or 180°. The measure of angle B, 90°, is three times 30°, the measure of angle A. The measure of angle C, 60°, is 30° greater than the measure of angle A, 30°. The values check.

State:

The measures of angles A, B, and C are 30°, 90°, and 60°, respectively.

14. A = 34°, B = 104°, and C = 42°.

15. Familiarize and Translate:

Let x, y, and z represent the number of quarts Pat picked on Monday, Tuesday, and Wednesday, respectively.

He picked a total of 87 quarts.

$x + y + z = 87$

On Tuesday he picked 15 quarts more than on Monday.

$y = x + 15$, or $x - y = -15$

On Wednesday he picked 3 quarts fewer than on Tuesday.

$z = y - 3$, or $y - z = 3$

We now have a system of equations.

$x + y + z = 87,$

$x - y = -15,$

$y - z = 3$

Carry out:

We solve the system. The solution is (20, 35, 32).

Check:

The total quarts picked in three days was 20 + 35 + 32, or 87 quarts. On Tuesday Pat picked 35 quarts which is 15 more quarts than on Monday (20 + 15 = 35). On Wednesday he picked 32 quarts which is 3 quarts less than on Tuesday (35 - 3 = 32). The amounts check.

State:

Pat picked 20 quarts on Monday, 35 quarts on Tuesday, and 32 quarts on Wednesday.

16. Thursday: \$21, Friday: \$18, Saturday: \$27

17. Familiarize and Translate:

Let x, y, and z represent the number of board-feet of lumber produced per day by sawmills A, B, and C, respectively.

All three together can produce 7400 board-feet in a day.

$x + y + z = 7400$

A and B together can produce 4700 board-feet in a day.

$x + y = 4700$

B and C together can produce 5200 board-feet in a day.

$y + z = 5200$

We now have a system of equations.

$x + y + z = 7400$

$x + y = 4700$

$y + z = 5200$

Carry out:

We solve the system. The solution is (2200, 2500, 2700).

17. (continued)

Check:

All three can produce 2200 + 2500 + 2700, or 7400 board-feet per day. A and B together can produce 2200 + 2500, or 4700 board-feet. B and C together can produce 2500 + 2700, or 5200 board-feet. All values check.

State:

In a day, sawmill A can produce 2200 board-feet, sawmill B can produce 2500 board-feet, and sawmill C can produce 2700 board-feet.

18. A: 1500, B: 1900, C: 2300

19. Familiarize and Translate:

Let x, y, and z represent the number of linear feet per hour welders A, B, and C can weld, respectively.

Together A, B, and C can weld 37 linear feet.

$x + y + z = 37$

A and B together can weld 22 linear feet.

$x + y = 22$

A and C together can weld 25 linear feet.

$x + z = 25$

We now have a system of equations.

$$\begin{aligned} x + y + z &= 37, \\ x + y \quad &= 22, \\ x \quad + z &= 25 \end{aligned}$$

Carry out:

We solve the system. The solution is (10, 12, 15).

Check:

Together all three can weld 10 + 12 + 15, or 37 linear feet per hour. A and B together can weld 10 + 12, or 22 linear feet. A and C together can weld 10 + 15, or 25 linear feet. The values check.

State:

Welder A can weld 10 linear feet per hour.
Welder B can weld 12 linear feet per hour.
Welder C can weld 15 linear feet per hour.

20. A: 900 gal/hr, B: 1300 gal/hr, C: 1500 gal/hr

21. Familiarize:

Let x, y, and z represent the number of par-3, par-4, and par-5 holes, respectively. A golfer who shoots par on every hole has 3x from the par-3 holes, 4y from the par-4 holes, and 5z from the par-5 holes.

Translate:

The total number of holes is 18.

$x + y + z = 18$

A golfer who shoots par on every hole has a total of 72.

$3x + 4y + 5z = 72$

The sum of the number of par-3 holes and the number of par-5 holes is 8.

$x + z = 8$

We have a system of equations.

$$\begin{aligned} x + y + z &= 18, \\ 3x + 4y + 5z &= 72, \\ x \quad + z &= 8 \end{aligned}$$

Carry out:

We solve the system. The solution is (4, 10, 4).

Check:

The total number of holes is 4 + 10 + 4, or 18. A golfer who shoots par on every hole has $3 \cdot 4 + 4 \cdot 10 + 5 \cdot 4$, or 72. The sum of the number of par-3 holes and the number of par-5 holes is 4 + 4, or 8. The values check.

State:

There are 4 par-3 holes, 10 par-4 holes, and 4 par-5 holes.

22. Par-3: 6, par-4: 8, par-5: 4

23. Familiarize:

Let x, y, and z represent the amounts invested at 7%, 8%, and 9%, respectively. Then the interest from each investment is 7%x or 0.07x, 8%y or 0.08y, and 9%z or 0.09z.

Translate:

The total investment is $2500.

$x + y + z = 2500$

The total interest is $212.

$0.07x + 0.08y + 0.09z = 212$, or

$7x + 8y + 9z = 21{,}200$

The amount invested at 9% is $1100 more than the amount invested at 8%.

$z = y + 1100$

We have a system of equations.

$$\begin{aligned} x + y + z &= 2500, \\ 7x + 8y + 9z &= 21{,}200, \\ -y + z &= 1100 \end{aligned}$$

Carry out:

We solve the system. The solution is (400, 500, 1600).

23. (continued)

Check:

The total investment is \$400 + \$500 + \$1600, or \$2500. The total interest is 0.07(\$400) + 0.08(\$500) + 0.09(\$1600), or \$28 + \$40 + \$144, or \$212. The amount invested at 9%, \$1600, is \$1100 more than \$500, the amount invested at 8%. The values check.

State:

\$400 is invested at 7%, \$500 is invested at 8%, and \$1600 is invested at 9%.

24. 8%: \$300, 9%: \$300, 10%: \$2900

25. We wish to find a quadratic function

$$f(x) = ax^2 + bx + c$$

containing the three given points.

When we substitute, we get

for (1, 4) $\quad 4 = a\cdot 1^2 + b\cdot 1 + c$

for (-1, -2) $\quad -2 = a\cdot(-1)^2 + b\cdot(-1) + c$

for (2, 13) $\quad 13 = a\cdot 2^2 + b\cdot 2 + c$

We now have a system of equations in three unknowns, a, b, and c:

$$a + b + c = 4,$$
$$a - b + c = -2,$$
$$4a + 2b + c = 13$$

We solve this system, obtaining (2, 3, -1). Thus the function we are looking for is

$f(x) = 2x^2 + 3x - 1.$

26. $f(x) = 3x^2 - x + 2.$

27. a) We wish to find a quadratic function

$$E(t) = at^2 + bt + c$$

containing the three given points.

When we substitute, we get

for (1, 38) $\quad 38 = a\cdot 1^2 + b\cdot 1 + c$

for (2, 66) $\quad 66 = a\cdot 2^2 + b\cdot 2 + c$

for (3, 86) $\quad 86 = a\cdot 3^2 + b\cdot 3 + c$

We now have a system of equations in three unknowns, a, b, and c.

$$a + b + c = 38,$$
$$4a + 2b + c = 66,$$
$$9a + 3b + c = 86$$

We solve the system, obtaining (-4, 40, 2). Thus, the function we are looking for is $E(t) = -4t^2 + 40t + 2$.

b) $E(t) = -4t^2 + 40t + 2$

$$E(4) = -4\cdot 4^2 + 40\cdot 4 + 2$$
$$= -64 + 160 + 2$$
$$= \$98$$

28. a) $E(t) = 2500t^2 - 6500t + 5000$

b) \$19,000

29. a) We wish to find a quadratic function

$$f(x) = ax^2 + bx + c$$

containing the points (5, 1121), (7, 626), and (9, 967).

When we substitute, we get

$$1121 = 25a + 5b + c,$$
$$626 = 49a + 7b + c,$$
$$967 = 81a + 9b + c.$$

We solve this system of equations, obtaining (104.5, -1501.5, 6016). Thus,

$$f(x) = 104.5x^2 - 1501.5x + 6016.$$

b) $f(4) = 104.5(4)^2 - 1501.5(4) + 6016 = 1682$

$f(6) = 104.5(6)^2 - 1501.5(6) + 6016 = 769$

$f(10) = 104.5(10)^2 - 1501.5(10) + 6016 = 1451$

30. a) $f(x) = \frac{1}{300{,}000}x^2 + \frac{7}{3000}x$

b) 2.925 hr

c) 407

31. a) We wish to find a quadratic function

$$f(x) = ax^2 + bx + c$$

containing the points (8, 72), (11, 82), and (14, 69).

$72 = 64a + 8b + c,$

$82 = 121a + 11b + c,$ (Subsbituting)

$69 = 196a + 14b + c$

We solve the system of equations, obtaining $\left(-\frac{23}{18}, \frac{497}{18}, -\frac{604}{9}\right)$. Thus,

$$f(x) = -\frac{23}{18}x^2 + \frac{497}{18}x - \frac{604}{9}.$$

b) $f(10) = -\frac{23}{18}(10)^2 + \frac{497}{18}(10) - \frac{604}{9} \approx 81.2$ years

c) Solve $79 = -\frac{23}{18}x^2 + \frac{497}{18}x - \frac{604}{9}$, or

$\frac{23}{18}x^2 - \frac{497}{18}x + \frac{1315}{9} = 0.$

$23x^2 - 497x + 2630 = 0$ (Multiplying by 18)

$x = \frac{-(-497) \pm \sqrt{(-497)^2 - 4(23)(2630)}}{2\cdot 23}$ Using the quadratic formula

$x \approx 9.3$ or $x \approx 12.3$

The man's shoe size is about 9 or 12.

32. $\left(-\frac{1}{2}, -1, -\frac{1}{3}\right)$

33. $\frac{2}{x} - \frac{1}{y} - \frac{3}{z} = -1,$

$\frac{2}{x} - \frac{1}{y} + \frac{1}{z} = -9,$

$\frac{1}{x} + \frac{2}{y} - \frac{4}{z} = 17$

First substitute u for $\frac{1}{x}$, v for $\frac{1}{y}$, and w for $\frac{1}{z}$ and solve for u, v, and w.

$2u - v - 3w = -1,$

$2u - v + w = -9,$

$u + 2v - 4w = 17$

Solving this system we get (-1, 5, -2).

If u = -1 and $u = \frac{1}{x}$, then $\frac{1}{x} = -1$, or x = -1.

If v = 5 and $v = \frac{1}{y}$, then $\frac{1}{y} = 5$, or $y = \frac{1}{5}$.

If w = -2 and $w = \frac{1}{z}$, then $\frac{1}{z} = -2$, or $z = -\frac{1}{2}$.

The solution of the original system is $\left(-1, \frac{1}{5}, -\frac{1}{2}\right)$.

34. A: 24 hr, B: 12 hr, C: $4\frac{4}{5}$

35. Familiarize:

Let a, b, and c represent the time it would take A, B, and C, working alone, to do the job, respectively. Then A can do $\frac{1}{a}$ of the job in 1 hr, $\frac{2}{a}$ of the job in 2 hr and so on.

Translate:

Working together, A, B, and C can do a job in 2 hr.

$\frac{2}{a} + \frac{2}{b} + \frac{2}{c} = 1$

Working together, B and C can do the job in 4 hr.

$\frac{4}{b} + \frac{4}{c} = 1$

Working together, A and B can do the job in $\frac{12}{5}$ hr.

$\frac{\frac{12}{5}}{a} + \frac{\frac{12}{5}}{b} = 1$, or $\frac{12}{a} + \frac{12}{b} = 5$

We have a system of equations.

$\frac{2}{a} + \frac{2}{b} + \frac{2}{c} = 1,$

$\frac{4}{b} + \frac{4}{c} = 1,$

$\frac{12}{a} + \frac{12}{b} = 5$

35. (continued)

Carry out:

Substitute x for $\frac{1}{a}$, y for $\frac{1}{b}$, and z for $\frac{1}{c}$ and solve for x, y, and z.

$2x + 2y + 2z = 1,$

$4y + 4z = 1,$

$12x + 12y = 5$

Solving this system we get $\left(\frac{1}{4}, \frac{1}{6}, \frac{1}{12}\right)$.

If $x = \frac{1}{4}$ and $x = \frac{1}{a}$, then $\frac{1}{a} = \frac{1}{4}$ or a = 4.

If $y = \frac{1}{6}$ and $y = \frac{1}{b}$, then $\frac{1}{b} = \frac{1}{6}$ or b = 6.

If $z = \frac{1}{12}$ and $z = \frac{1}{c}$, then $\frac{1}{c} = \frac{1}{12}$ or c = 12.

The solution of the original system is (4, 6, 12).

Check:

The check is left to the student.

State:

Working alone, A can do the job in 4 hr, B can do the job in 6 hr, and C can do the job in 12 hr.

36. 1869

37. Label the angle measures at the tips of the stars a, b, c, d, and e. Also label the angles of the pentagon 1, 2, 3, 4, and 5.

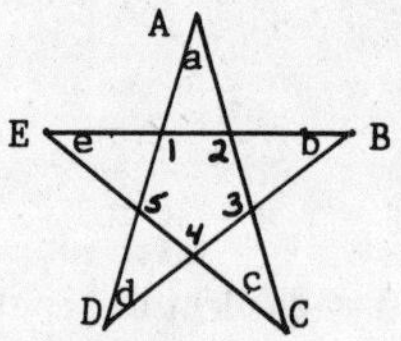

Using the geometric fact that the sum of the angle measures of a triangle is 180°, we get 5 equations.

$1 + b + d = 180$

$2 + c + e = 180$

$3 + a + d = 180$

$4 + b + e = 180$

$5 + a + c = 180$

Adding these equations, we get

$(1 + 2 + 3 + 4 + 5) + 2a + 2b + 2c + 2d + 2e = 5(180)$

The sum of the angle measures of any convex polygon with n sides is given by the formula $S = (n - 2)180$. Thus $1 + 2 + 3 + 4 + 5 = (5 - 2)180$, or 540. We substitute and solve for a + b + c + d + e.

$540 + 2(a + b + c + d + e) = 900$

$2(a + b + c + d + e) = 360$

$a + b + c + d + e = 180$

37. (continued)

The sum of the angle measures at the tips of the star is 180°.

38. $B = 2, M = \frac{1}{2}, N = \frac{1}{4}; y = 2 - \frac{1}{2}x - \frac{1}{4}z$

39. Substituting, we get

$A + \frac{3}{4}B + 3C = 12,$

$\frac{4}{3}A + B + 2C = 12,$

$2A + B + C = 12,$ or

$4A + 3B + 12C = 48,$

$4A + 3B + 6C = 36,$ (Clearing fractions)

$2A + B + C = 12.$

Solving the system, we get A = 3, B = 4, and C = 2. The equation is $3x + 4y + 2z = 12$.

40. Men: 5, women: 1, children: 94

41. First we clear decimals.

$3120x + 2140y - 988z = 3790,$ (P1)

$384x - 353y + 196z = 780,$ (P2)

$4630x - 1080y + 11z = 6340$ (P3)

Multiply (P2) by 195 and multiply (P3) by 312 to make the x-coefficients multiples of 3120.

$3120x + 2140y - 988z = 3790,$ (P1)

$74,880x - 68,835y + 38,220z = 152,100,$ (P2)

$1,444,560x - 336,960y + 3432z = 1,978,080$ (P3)

Multiply (P1) by -24 and add it to (P2). Also multiply (P1) by -463 and add it to (P3).

$3120x + 2140y - 988z = 3790,$

$-120,195y + 61,932z = 61,140,$

$-1,327,780y + 460,876z = -68,444,220$

Continuing in this manner we find the solution (1.49, 0.54, 2.03).

42. (-2.76, 2.19, 1.05)

43. Let k, m denote positive integers; and denote the scores by a, b, c, d, f.

GIVEN: (1) a is prime; f is 3rd
(2) Each score ⩽ 200
(3) a - c = 23; d = k·10
(4) a + b + c + d + f = 885; b = m·8
(5) a - b < 10; d - b = 14.

The diagram below follows easily.

c b=m·8 f a=prime d=k·10
0 200
< 10
14
23

43. (continued)

The greatest score d must be at least $\frac{1}{5}$ of the total 885.

Thus $d \geqslant \frac{885}{5} = 177$; d = 180 or 190 or 200, and k = 18 or 19 or 20, since d = 10·k.

Now d - b = 14 or

(10k) - (8m) = 14 so that

$m = \frac{5k - 7}{4}$ which requires an odd k = 19;

then m = 22.

Thus b = 8m = 176; and d = 10·k = 190.
From a - b < 10 we have a < 186. Thus a is a prime such that 176 = b < a < 186. There are only two primes between 176 and 186: 179 and 181.

If a = 179, then it follows incorrectly that f is in second place. Thus a = 181 and we have b = 176, c = 158, d = 190, f = 180.

Exercise Set 9.3

Note: The answers to Exercises 19 and 20 are included in the answers to Exercises 1 - 18.

1. $9x - 3y = 15,$ (1)
$6x - 2y = 10$ (2)

Multiply equation (2) by 3 to make the x-coefficient a multiple of 9.

$9x - 3y = 15,$

$18x - 6y = 30$

Multiply equation (1) by -2 and add the result to equation (2).

Now we have

$9x - 3y = 15,$

$0 = 0.$

which illustrates that the original system of equations is a dependent system of equations. For this particular system, there is an infinite number of solutions. The system is consistent.

These solutions can be described by expressing one variable in terms of the other. Since 0 = 0 contributes nothing, we solve 9x - 3y = 15 for x and obtain

$x = \frac{3y + 15}{9}$, or $\frac{y + 5}{3}$

The ordered pairs in the solution set can be described in terms of y only.

$\left(\frac{y + 5}{3}, y\right)$

Any value chosen for y gives a value for x, and thus an ordered pair in the solution set.

1. (continued)

We could have solved $9x - 3y = 15$ for y obtaining $y = \frac{9x - 15}{3}$ or $3x - 5$ and described the solution set in terms of x only.

$(x, 3x - 5)$

Then any value chosen for x would give a value for y, and thus an ordered pair in the solution set.

A few of the solutions are (0, -5), (1, -2), and (-1, -8).

2. Ø; inconsistent, independent

3. $5c + 2d = 24$, (1)

$30c + 12d = 10$ (2)

Multiply equation (1) by -6 and add the result to equation (2).

This gives us

$5c + 2d = 24,$

$0 = -134.$

The second equation says that $0 \cdot x + 0 \cdot y = -134$. There are no numbers x and y for which this is true. Thus, the system has no solutions. The solution set is Ø. The system is inconsistent. This system is not equivalent to a system of fewer than 2 linear equations. The system is independent.

4. Ø; inconsistent, independent

5. $3x + 2y = 5,$

$4y = 10 - 6x$

Rewrite each equation in the form $Ax + By = C$.

$3x + 2y = 5,$ (1)

$6x + 4y = 10$ (2)

Multiply equation (1) by -2 and add.

$3x + 2y = 5,$

$0 = 0$

The system is dependent. The equation $3x + 2y = 5$ has infinitely many solutions, so the system is consistent.

If we solve $3x + 2y = 5$ for x obtaining

$x = \frac{5 - 2y}{3},$

the ordered pairs in the solution set can be described in terms of y only.

$\left(\frac{5 - 2y}{3}, y\right)$

If we solve $3x + 2y = 5$ for y obtaining

$y = \frac{5 - 3x}{2},$

the ordered pairs in the solution set can be described in terms of x only.

$\left(x, \frac{5 - 3x}{2}\right)$

5. (continued)

A few of the solutions are $\left(0, \frac{5}{2}\right)$, (3, -2), and (-1, 4).

6. $\left(\frac{7y - 2}{5}, y\right)$ or $\left(x, \frac{5x + 2}{7}\right)$; (1, 1), $\left(0, \frac{2}{7}\right)$, (8,6), etc.; dependent, consistent

7. $12y - 8x = 6,$

$4x + 3 = 6y$

Rewrite each equation in the form $Ax + By = C$.

$-8x + 12y = 6,$ (1)

$4x - 6y = -3$ (2)

Multiply equation (2) by 2 to make the x-coefficient a multiple of -8.

$-8x + 12y = 6,$

$8x - 12y = -6$

Now add equation (1) to equation (2).

$-8x + 12y = 6,$

$0 = 0$

The system is dependent. The equation $-8x + 12y = 6$ has infinitely many solutions, so the system is consistent.

If we solve $4x - 6y = -3$ for x obtaining

$x = \frac{6y - 3}{4},$

the ordered pairs in the solution set can be described in terms of y only.

$\left(\frac{6y - 3}{4}, y\right)$

If we solve $4x - 6y = -3$ for y obtaining

$y = \frac{4x + 3}{6},$

the ordered pairs in the solution set can be described in terms of x only.

$\left(x, \frac{4x + 3}{6}\right)$

A few of the solutions are $\left(-\frac{3}{4}, 0\right)$, $\left(\frac{9}{4}, 2\right)$ and $\left(-\frac{9}{4}, -1\right)$.

8. $\left(\frac{6y + 5}{8}, y\right)$ or $\left(x, \frac{8x - 5}{6}\right)$; $\left(0, -\frac{5}{6}\right)$, $\left(1, \frac{1}{2}\right)$, $\left(-1, -\frac{13}{6}\right)$, etc.; dependent, consistent

9. $x + 2y - z = -8,$ (P1)

$2x - y + z = 4,$ (P2)

$8x + y + z = 2$ (P3)

Multiply (P1) by -2 and add the result to (P2).
Multiply (P1) by -8 and add the result to (P3).

$x + 2y - z = -8,$ (P1)

$-5y + 3z = 20,$ (P2)

$-15y + 9z = 66$ (P3)

Multiply (P2) by -3 and add the result to (P3).

$x + 2y - z = -8,$ (P1)

$-5y + 3z = 20,$ (P2)

$0 = 6$ (P3)

Since in (P3) we obtain a false equation 0 = 6, the system has no solution. The solution set is ∅. The system is <u>inconsistent</u>. This system is not equivalent to a system of fewer than 3 equations. The system is <u>independent</u>.

10. $\left(\frac{z + 28}{11}, \frac{5z + 8}{11}, z\right)$; $\left(\frac{28}{11}, \frac{8}{11}, 0\right)$, $\left(\frac{29}{11}, \frac{13}{11}, 1\right)$, $\left(\frac{27}{11}, \frac{3}{11}, -1\right)$, etc.; dependent, consistent

11. $2x + y - 3z = 1,$

$x - 4y + z = 6,$

$4x - 16y + 4z = 24$

First interchange the first two equations so all the x-coefficients are multiples of the first.

$x - 4y + z = 6,$ (P1)

$2x + y - 3z = 1,$ (P2)

$4x - 16y + 4z = 24$ (P3)

Multiply (P1) by -2 and add the result to (P2).
Multiply (P1) by -4 and add the result to (P3).

$x - 4y + z = 6,$ (P1)

$9y - 5z = -11,$ (P2)

$0 = 0$ (P3)

Now we know the system is <u>dependent</u>.

Solve (P2) for y.

$$9y - 5z = -11$$
$$9y = 5z - 11$$
$$y = \frac{5z - 11}{9}$$

Substitute $\frac{5z - 11}{9}$ for y in (P1) and solve for x.

$$x - 4y + z = 6$$
$$x - 4\left(\frac{5z - 11}{9}\right) + z = 6$$
$$9x - 4(5z - 11) + 9z = 54$$
$$9x - 20z + 44 + 9z = 54$$
$$9x = 11z + 10$$
$$x = \frac{11z + 10}{9}$$

11. (continued)

The ordered triples in the solution set can be described as follows:

$\left(\frac{11z + 10}{9}, \frac{5z - 11}{9}, z\right)$

There is an infinite number of solutions. The system is <u>consistent</u>.

A few of the solutions are $\left(\frac{10}{9}, -\frac{11}{9}, 0\right)$, $\left(\frac{7}{3}, -\frac{2}{3}, 1\right)$, and $\left(\frac{32}{9}, -\frac{1}{9}, 2\right)$.

12. $\left(\frac{z + 5}{5}, -\frac{7}{5}z, z\right)$; (1, 0, 0), (2, -7, 5), $\left(\frac{6}{5}, -\frac{7}{5}, 1\right)$, etc.; dependent, consistent

13. $2x + y - 3z = 0,$

$x - 4y + z = 0,$

$4x - 16y + 4z = 0$

Note that this is a system of homogeneous equations. The trivial solution is (0, 0, 0). There may or may not be other solutions.

First we interchange the first two equations so all the x-coefficients are multiples of the first:

$x - 4y + z = 0,$ (P1)

$2x + y - 3z = 0,$ (P2)

$4x - 16y + 4z = 0$ (P3)

Multiply (P1) by -2 and add to (P2).
Multiply (P1) by -4 and add to (P3).

$x - 4y + z = 0,$ (P1)

$9y - 5z = 0,$ (P2)

$0 = 0$ (P3)

Now we know that the system is <u>dependent</u>.

Solve (P2) for y.

$$9y - 5z = 0$$
$$9y = 5z$$
$$y = \frac{5}{9}z$$

Substitute $\frac{5}{9}z$ for y in (P1) and solve for x.

$$x - 4y + z = 0$$
$$x - 4\left(\frac{5}{9}z\right) + z = 0$$
$$x - \frac{20}{9}z + \frac{9}{9}z = 0$$
$$x - \frac{11}{9}z = 0$$
$$x = \frac{11}{9}z$$

The ordered triples in the solution set can be described as follows:

$\left(\frac{11}{9}z, \frac{5}{9}z, z\right)$

13. (continued)

There is an infinite number of solutions. The system is <u>consistent</u>.

A few of the solutions are (0, 0, 0), $\left(\frac{11}{8}, \frac{5}{18}, \frac{1}{2}\right)$ and $\left(-\frac{11}{9}, -\frac{5}{9}, -1\right)$.

14. $\left(\frac{1}{5}z, -\frac{7}{5}z, z\right)$; (0, 0, 0), (1, -7, 5), $\left(\frac{1}{5}, -\frac{7}{5}, 1\right)$, etc.; dependent, consistent

15. $x + y - z = -3,$ (P1)
$x + 2y + 2z = -1$ (P2)

Multiply (P1) by -1 and add to (P2).

$x + y - z = -3,$
$y + 3z = 2$

Solve (P2) for y.

$y + 3z = 2$
$y = -3z + 2$

Substitute -3z + 2 for y in (P1) and solve for x.

$x + y - z = -3$
$x + (-3z + 2) - z = -3$
$x - 4z + 2 = -3$
$x = 4z - 5$

The system has an infinite set of solutions. The system is <u>consistent</u>. The ordered triples in the solutions can be described as follows:

(4z - 5, -3z + 2, z)

A few of the solutions are (-5, 2, 0), (-9, 5, -1), and (3, -4, 2).

The system is <u>independent</u>.

16. $\left(-\frac{7}{2}z, -\frac{19}{2}z, z\right)$; (0, 0, 0), (-7, -19, 2), $\left(-\frac{7}{2}, -\frac{19}{2}, 1\right)$, etc.; independent, consistent

17. $2x + y + z = 0,$
$x + y - z = 0,$
$x + 2y + 2z = 0$

Any homogeneous system like this always has a solution, because (0, 0, 0) is a solution. There may or may not be other solutions.

Interchange the first two equations.

$x + y - z = 0,$ (P1)
$2x + y + z = 0,$ (P2)
$x + 2y + 2z = 0$ (P3)

Multiply (P1) by -2 and add to (P2). Multiply (P1) by -1 and add to (P3).

$x + y - z = 0,$ (P1)
$-y + 3z = 0,$ (P2)
$y + 3z = 0$ (P3)

17. (continued)

Add (P2) to (P3).

$x + y - z = 0,$ (P1)
$-y + 3z = 0,$ (P2)
$6z = 0$ (P3)

Solve (P3) for z.

$6z = 0$
$z = 0$

Substitute 0 for z in (P2) and solve for y.

$-y + 3z = 0$
$-y + 3\cdot 0 = 0$
$-y = 0$
$y = 0$

Substitute 0 for z and 0 for y in (P1) and solve for x.

$x + y - z = 0$
$x + 0 - 0 = 0$
$x = 0$

The only solution is (0, 0, 0). The system is <u>consistent</u>. The system is not equivalent to a system of fewer than 3 equations. Thus, the system is <u>independent</u>.

18. (0, 0, 0); independnet, consistent

19. - 20. Answers in Exercises 1 - 18.

21. $4.026x - 1.448y = 18.32,$
$0.724y = -9.16 + 2.013x$

Multiply each equation by 1000 and rewrite in the form Ax + By = C.

$4026x - 1448y = 18{,}320,$
$2013x - 724y = 9160$

Interchange the equations.

$2013x - 724y = 9160,$ (1)
$4026x - 1448y = 18{,}320$ (2)

Multiply equation (1) by -2 and add.

$2013x - 724y = 9160,$
$0 = 0$

The system is dependent.

If we solve 2013x - 724y = 9160 for x obtaining

$x = \frac{9160 + 724y}{2013}$,

the ordered pairs in the solution set can be described in terms of y only.

$\left(\frac{9160 + 724y}{2013}, y\right)$

21. (continued)

If we solve $2013x - 724y = 9160$ for y obtaining

$$y = \frac{2013x - 9160}{724},$$

the ordered pairs in the solution set can be described in terms of x only.

$$\left(x, \frac{2013x - 9160}{724}\right)$$

22. $\left(\frac{5260 - 142y}{40,570}, y\right)$ or $\left(x, \frac{5260 - 40,570x}{142}\right)$

23. a) $w + x + y + z = 4,$ (P1)
$w + x + y + z = 3,$ (P2)
$w + x + y + z = 3$ (P3)

Multiply (P1) by -1 and add the result to (P2) and to (P3).

$w + x + y + z = 4,$ (P1)
$0 = -1,$ (P2)
$0 = -1$ (P3)

Since we obtain the false equation $0 = -1$, the solution set is Ø.

b) Since there is no solution, the system is inconsistent.

c) Since (P2) and (P3) are identical, the system is equivalent to a system of fewer than 3 equations. It is dependent.

24. a) $(2x + 5y, x, y, 3x - 4y)$

b) Consistent

c) Dependent

25. $6x - 9y = -3,$ (1)
$-4x + 6y = k$ (2)

Multiply equation (2) by 3 to make the x-coefficient a multiple of 6.

$6x - 9y = -3,$ (1)
$-12x + 18y = 3k$ (2)

Multiply equation (1) by 2 and add.

$6x - 9y = -3,$
$0 = -6 + 3k$

If the system is to be dependent, it must be true that

$-6 + 3k = 0$ or
$k = 2.$

26. 25

27. Let a = the number of par 3 holes,
b = the number of par 4 holes, and
c = the number of par 5 holes.

Then

$$\left.\begin{aligned} a + b + c &= 18 \\ 3a + 4b + 5c &= 72 \\ a &= c \end{aligned}\right\} \text{(A)},$$

or, equivalently,

$$\left.\begin{aligned} a + b + c &= 18 \\ b + 2c &= 18 \\ 0 &= 0 \end{aligned}\right\} \text{(B)}.$$

Then $S = \{(a, 18 - 2a, a)\}$ where $a \leqslant 1 \leqslant 8$, a is an integer.

All permissible triples (a, b, c) are: (1, 16, 1), (2, 14, 2), (3, 12, 3), (4, 10, 4), (5, 8, 5), (6, 6, 6), (7, 4, 7), (8, 2, 8).

28. 12, 24, 36, or 48

Exercise Set 9.4

1. $4x + 2y = 11,$
$3x - y = 2$

Write a matrix using only the constants.

$$\begin{bmatrix} 4 & 2 & 11 \\ 3 & -1 & 2 \end{bmatrix}$$

Multiply row 2 by 4 to make the first number in row 2 a multiple of 4.

$$\begin{bmatrix} 4 & 2 & 11 \\ 12 & -4 & 8 \end{bmatrix}$$

Multiply row 1 by -3 and add it to row 2.

$$\begin{bmatrix} 4 & 2 & 11 \\ 0 & -10 & -25 \end{bmatrix}$$

Putting the variables back in, we have

$4x + 2y = 11,$ (1)
$-10y = -25$ (2)

Solve (2) for y.

$-10y = -25$

$y = \frac{5}{2}$

1. (continued)

Substitute $\frac{5}{2}$ for y in (1) and solve for x.

$$4x + 2y = 11$$
$$4x + 2 \cdot \frac{5}{2} = 11$$
$$4x + 5 = 11$$
$$4x = 6$$
$$x = \frac{3}{2}$$

The solution is $\left(\frac{3}{2}, \frac{5}{2}\right)$.

2. $\left(-\frac{1}{3}, -4\right)$

3. $x + 2y - 3z = 9,$
$2x - y + 2z = -8,$
$3x - y - 4z = 3$

Write a matrix using only the constants.

$$\begin{bmatrix} 1 & 2 & -3 & 9 \\ 2 & -1 & 2 & -8 \\ 3 & -1 & -4 & 3 \end{bmatrix}$$

Multiply row 1 by -2 and add it to row 2.
Multiply row 1 by -3 and add it to row 3.

$$\begin{bmatrix} 1 & 2 & -3 & 9 \\ 0 & -5 & 8 & -26 \\ 0 & -7 & 5 & -24 \end{bmatrix}$$

Multiply row 3 by 5.

$$\begin{bmatrix} 1 & 2 & -3 & 9 \\ 0 & -5 & 8 & -26 \\ 0 & -35 & 25 & -120 \end{bmatrix}$$

Multiply row 2 by -7 and add to row 3.

$$\begin{bmatrix} 1 & 2 & -3 & 9 \\ 0 & -5 & 8 & -26 \\ 0 & 0 & -31 & 62 \end{bmatrix}$$

Putting the variables back in, we have

$x + 2y - 3z = 9,$ (1)
$-5y + 8z = -26,$ (2)
$-31z = 62$ (3)

Solve (3) for z.

$-31z = 62$
$z = -2$

3. (continued)

Substitute -2 for z in (2) and solve for y.

$$-5y + 8z = -26$$
$$-5y + 8(-2) = -26$$
$$-5y - 16 = -26$$
$$-5y = -10$$
$$y = 2$$

Substitute -2 for z and 2 for y in (1) and solve for x.

$$x + 2y - 3z = 9$$
$$x + 2(2) - 3(-2) = 9$$
$$x + 10 = 9$$
$$x = -1$$

The solution is (-1, 2, -2).

4. (0, 2, 1)

5. $5x - 3y = -2,$
$4x + 2y = 5$

Write a matrix using only the constants.

$$\begin{bmatrix} 5 & -3 & -2 \\ 4 & 2 & 5 \end{bmatrix}$$

Multiply row 2 by 5.

$$\begin{bmatrix} 5 & -3 & -2 \\ 20 & 10 & 25 \end{bmatrix}$$

Multiply row 1 by -4 and add it to row 2.

$$\begin{bmatrix} 5 & -3 & -2 \\ 0 & 22 & 33 \end{bmatrix}$$

Putting the variables back in, we have

$5x - 3y = -2,$ (1)
$22y = 33$ (2)

Solve (2) for y.

$22y = 33$
$y = \frac{3}{2}$

Substitute $\frac{3}{2}$ for y in (1) and solve for x.

$$5x - 3y = -2$$
$$5x - 3 \cdot \frac{3}{2} = -2$$
$$5x - \frac{9}{2} = -\frac{4}{2}$$
$$5x = \frac{5}{2}$$
$$x = \frac{1}{2}$$

The solution is $\left(\frac{1}{2}, \frac{3}{2}\right)$.

6. $\left(-1, \frac{5}{2}\right)$

7. $4x - y - 3z = 1,$
$8x + y - z = 5,$
$2x + y + 2z = 5$

Write a matrix using only the constants.

$$\begin{bmatrix} 4 & -1 & -3 & 1 \\ 8 & 1 & -1 & 5 \\ 2 & 1 & 2 & 5 \end{bmatrix}$$

First interchange rows 1 and 3 so that each number below the first number in the first row is a multiple of that number.

$$\begin{bmatrix} 2 & 1 & 2 & 5 \\ 8 & 1 & -1 & 5 \\ 4 & -1 & -3 & 1 \end{bmatrix}$$

Multiply row 1 by -4 and add it to row 2.
Multiply row 1 by -2 and add it to row 3.

$$\begin{bmatrix} 2 & 1 & 2 & 5 \\ 0 & -3 & -9 & -15 \\ 0 & -3 & -7 & -9 \end{bmatrix}$$

Multiply row 2 by -1 and add it to row 3.

$$\begin{bmatrix} 2 & 1 & 2 & 5 \\ 0 & -3 & -9 & -15 \\ 0 & 0 & 2 & 6 \end{bmatrix}$$

Putting the variables back in, we have

$2x + y + 2z = 5,$ (1)
$-3y - 9z = -15,$ (2)
$2z = 6$ (3)

Solve (3) for z.

$2z = 6$
$z = 3$

Substitute 3 for z in (2) and solve for y.

$-3y - 9z = -15$
$-3y - 9(3) = -15$
$-3y - 27 = -15$
$-3y = 12$
$y = -4$

Substitute 3 for z and -4 for y in (1) and solve for x.

$2x + y + 2z = 5$
$2x + (-4) + 2(3) = 5$
$2x - 4 + 6 = 5$
$2x = 3$
$x = \frac{3}{2}$

The solution is $\left(\frac{3}{2}, -4, 3\right)$.

8. $\left(2, \frac{1}{2}, -2\right)$

9. $p + q + r = 1,$
$p + 2q + 3r = 4,$
$4p + 5q + 6r = 7$

Write a matrix using only the constants.

$$\begin{bmatrix} 1 & 1 & 1 & 1 \\ 1 & 2 & 3 & 4 \\ 4 & 5 & 6 & 7 \end{bmatrix}$$

Multiply row 1 by -1 and add it to row 2.
Multiply row 1 by -4 and add it to row 3.

$$\begin{bmatrix} 1 & 1 & 1 & 1 \\ 0 & 1 & 2 & 3 \\ 0 & 1 & 2 & 3 \end{bmatrix}$$

Multiply row 2 by -1 and add it to row 3.

$$\begin{bmatrix} 1 & 1 & 1 & 1 \\ 0 & 1 & 2 & 3 \\ 0 & 0 & 0 & 0 \end{bmatrix}$$

Putting the variables back in, we have

$p + q + r = 1,$
$q + 2r = 3,$
$0 = 0$

This system is dependent. In this case it has an infinite set of solutions.

Solve $q + 2r = 3$ for q.

$q = -2r + 3$

Substitute $-2r + 3$ for q in the first equation and solve for p.

$p + q + r = 1$
$p + (-2r + 3) + r = 1$
$p - r + 3 = 1$
$p - r = -2$
$p = r - 2$

The solutions can be described as follows:

$(r - 2, -2r + 3, r)$

10. Ø

11. $-2w + 2x + 2y - 2z = -10,$

$w + x + y + z = -5,$

$3w + x - y + 4z = -2,$

$w + 3x - 2y + 2z = -6$

Write a matrix using only the constants.

$$\begin{bmatrix} -2 & 2 & 2 & -2 & -10 \\ 1 & 1 & 1 & 1 & -5 \\ 3 & 1 & -1 & 4 & -2 \\ 1 & 3 & -2 & 2 & -6 \end{bmatrix}$$

Interchange rows 1 and 2.

$$\begin{bmatrix} 1 & 1 & 1 & 1 & -5 \\ -2 & 2 & 2 & -2 & -10 \\ 3 & 1 & -1 & 4 & -2 \\ 1 & 3 & -2 & 2 & -6 \end{bmatrix}$$

Multiply row 1 by 2 and add it to row 2.
Multiply row 1 by -3 and add it to row 3.
Multiply row 1 by -1 and add it to row 4.

$$\begin{bmatrix} 1 & 1 & 1 & 1 & -5 \\ 0 & 4 & 4 & 0 & -20 \\ 0 & -2 & -4 & 1 & 13 \\ 0 & 2 & -3 & 1 & -1 \end{bmatrix}$$

Interchange rows 2 and 3.

$$\begin{bmatrix} 1 & 1 & 1 & 1 & -5 \\ 0 & -2 & -4 & 1 & 13 \\ 0 & 4 & 4 & 0 & -20 \\ 0 & 2 & -3 & 1 & -1 \end{bmatrix}$$

Multiply row 2 by 2 and add it to row 3.
Add row 2 to row 4.

$$\begin{bmatrix} 1 & 1 & 1 & 1 & -5 \\ 0 & -2 & -4 & 1 & 13 \\ 0 & 0 & -4 & 2 & 6 \\ 0 & 0 & -7 & 2 & 12 \end{bmatrix}$$

Multiply row 4 by 4.

$$\begin{bmatrix} 1 & 1 & 1 & 1 & -5 \\ 0 & -2 & -4 & 1 & 13 \\ 0 & 0 & -4 & 2 & 6 \\ 0 & 0 & -28 & 8 & 48 \end{bmatrix}$$

Multiply row 3 by -7 and add it to row 4.

$$\begin{bmatrix} 1 & 1 & 1 & 1 & -5 \\ 0 & -2 & -4 & 1 & 13 \\ 0 & 0 & -4 & 2 & 6 \\ 0 & 0 & 0 & -6 & 6 \end{bmatrix}$$

11. (continued)

Putting the variables back in, we get

$w + x + y + z = -5,$ (1)

$-2x - 4y + z = 13,$ (2)

$-4y + 2z = 6,$ (3)

$-6z = 6$ (4)

Solve (4) for z.

$-6z = 6$

$z = -1$

Substitute -1 for z in (3) and solve for y.

$-4y + 2(-1) = 6$

$-4y = 8$

$y = -2$

Substitute -2 for y and -1 for z in (2) and solve for x.

$-2x - 4(-2) + (-1) = 13$

$-2x + 8 - 1 = 13$

$-2x = 6$

$x = -3$

Substitute -3 for x, -2 for y, and -1 for z in (1) and solve for w.

$w + (-3) + (-2) + (-1) = -5$

$w - 6 = -5$

$w = 1$

The solution is (1, -3, -2, -1).

12. (7, 4, 5, 6)

13. Let d represent the number of dimes and n represent the number of nickels. Translate to a system of equations.

$d + n = 34,$

$0.10d + 0.05n = 1.90$

Multiply the second equation by 100 to eliminate the decimal points.

$d + n = 34$

$10d + 5n = 190$

Use matrices to solve this system.

$$\begin{bmatrix} 1 & 1 & 34 \\ 10 & 5 & 190 \end{bmatrix}$$

Multiply row 1 by -10 and add it to row 2.

$$\begin{bmatrix} 1 & 1 & 34 \\ 0 & -5 & -150 \end{bmatrix}$$

Putting the variables back in, we have

$d + n = 34,$ (1)

$-5n = -150$ (2)

Solve (2) for n.

$-5n = -150$

$n = 30$

13. (continued)

Substitute 30 for n in (1) and solve for d.

$d + n = 34$

$d + 30 = 34$

$d = 4$

The solution is (4, 30). Thus there are 4 dimes and 30 nickels.

14. Dimes: 21, quarters: 22

15. Let x represent the number of nickels, y represent the number of dimes, and z represent the number of quarters. Translate to a system of equations.

$x + y + z = 22,$

$x + 2y + 5z = 58,$ ($2.90 = 290¢) ($5x + 10y + 25z = 290$)

$x - y = 6$ ($x = y + 6$)

Use matrices to solve this system.

$$\begin{bmatrix} 1 & 1 & 1 & 22 \\ 1 & 2 & 5 & 58 \\ 1 & -1 & 0 & 6 \end{bmatrix}$$

Multiply row 1 by -1 and add it to rows 2 and 3.

$$\begin{bmatrix} 1 & 1 & 1 & 22 \\ 0 & 1 & 4 & 36 \\ 0 & -2 & -1 & -16 \end{bmatrix}$$

Multiply row 2 by 2 and add it to row 3.

$$\begin{bmatrix} 1 & 1 & 1 & 22 \\ 0 & 1 & 4 & 36 \\ 0 & 0 & 7 & 56 \end{bmatrix}$$

Putting the variables back in, we have

$x + y + z = 22,$ (1)

$y + 4z = 36,$ (2)

$7z = 56$ (3)

Solve (3) for z.

$7z = 56$

$z = 8$

Substitute 8 for z in (2) and solve for y.

$y + 4z = 36$

$y + 4\cdot 8 = 36$

$y = 4$

Substitute 8 for z and 4 for y in (1) and solve for x.

$x + y + z = 22$

$x + 4 + 8 = 22$

$x = 10$

The solution is (10, 4, 8). Thus there are 10 nickels, 4 dimes, and 8 quarters.

16. Nickels: 6, dimes: 5, quarters: 7

17. Let x represent the number of pounds of tobacco that sells for $4.05 per pound. Let y represent the number of pounds of tobacco that sells for $2.70 per pound.

Translate to a system of equations.

$x + y = 15,$

$4.05x + 2.70y = 15(3.15)$

Multiply the second equation by 100 to eliminate the decimal points.

$x + y = 15$

$405x + 270y = 4725$

Use matrices to solve this system:

$$\begin{bmatrix} 1 & 1 & 15 \\ 405 & 270 & 4725 \end{bmatrix}$$

Multiply row 1 by -405 and add to row 2.

$$\begin{bmatrix} 1 & 1 & 15 \\ 0 & -135 & -1350 \end{bmatrix}$$

Putting the variables back in, we have

$x + y = 15,$ (1)

$-135y = -1350$ (2)

Solve (2) for y.

$-135y = -1350$

$y = 10$

Substitute 10 for y in (1) and solve for x.

$x + y = 15$

$x + 10 = 15$

$x = 5$

The solution is (5, 10). Thus, 5 lb of the $4.05 per lb tobacco and 10 lb of the $2.70 per lb tobacco should be used.

18. Candy: 14 lb, nuts: 6 lb

19. Let x represent the amount invested at 12½% and y represent the amount invested at 13%. Translate to a system of equations.

$-x + y = 10{,}000,$ ($y = x + 10{,}000$)

$125x + 130y = 8{,}950{,}000$ ($12\frac{1}{2}\%x + 13\%y = 8950$)

Use matrices to solve this system.

$$\begin{bmatrix} -1 & 1 & 10{,}000 \\ 125 & 130 & 8{,}950{,}000 \end{bmatrix}$$

Multiply row 1 by 125 and add to row 2.

$$\begin{bmatrix} -1 & 1 & 10{,}000 \\ 0 & 255 & 10{,}200{,}000 \end{bmatrix}$$

19. (continued)

Putting the variables back in, we have

$-x + y = 10,000,$ (1)

$225y = 10,200,000$ (2)

Solve (2) for y.

$225y = 10,200,000$

$y = 40,000$

Substitute 40,000 for y in (1) and solve for x.

$-x + y = 10,000$

$-x + 40,000 = 10,000$

$-x = -30,000$

$x = 30,000$

The solution is (30,000, 40,000). Thus, \$30,000 was invested at 12½%, and \$40,000 was invested at 13%.

20. $13\frac{1}{2}$%: \$10,000, $13\frac{3}{4}$%: \$14,000

21. $4.83x + 9.06y = -39.42,$

$-1.35x + 6.67y = -33.99$

First multiply each equation by 100 to clear of decimal points.

$483x + 906y = -3942$

$-135x + 667y = -3399$

Use matrices to solve this system.

$$\begin{bmatrix} 483 & 906 & -3942 \\ -135 & 667 & -3399 \end{bmatrix}$$

Multiply row 2 by 483.

$$\begin{bmatrix} 483 & 906 & -3942 \\ -65,205 & 322,161 & -1,641,717 \end{bmatrix}$$

Multiply row 1 by 135 and add it to row 2.

$$\begin{bmatrix} 483 & 906 & -3942 \\ 0 & 444,471 & -2,173,887 \end{bmatrix}$$

Putting the variables back in, we have

$483x + 906y = -3942,$ (1)

$444,471y = -2,173,887$ (2)

Solve (2) for y.

$444,471y = -2,173,887$

$y = -4.89095$

Substitute -4.89095 for y in (1) and solve for x.

$483x + 906y = -3942$

$483x + 906(-4.89095) = -3942$

$483x - 4431.2007 = -3942$

$483x = 489.2007$

$x = 1.01284$

The solution is (1.01284, -4.89095).

22. (-6.235, 2.451)

23. $3.55x - 1.35y + 1.03z = 9.16,$

$-2.14x + 4.12y + 3.61z = -4.50,$

$5.48x - 2.44y - 5.86z = 0.813$

Multiply each equation by 100.

$355x - 135y + 103z = 916$

$-214x + 412y + 361z = -450$

$548x - 244y - 586z = 81.3$

Use matrices to solve the system.

$$\begin{bmatrix} 355 & -135 & 103 & 916 \\ -214 & 412 & 361 & -450 \\ 548 & -244 & -586 & 81.3 \end{bmatrix}$$

Multiply rows 2 and 3 by 355.

$$\begin{bmatrix} 355 & -135 & 103 & 916 \\ -75,970 & 146,260 & 128,155 & -159,750 \\ 194,540 & -86,620 & -208,030 & 28,861.5 \end{bmatrix}$$

Multiply row 1 by 214 and add to row 2.
Multiply row 1 by -548 and add to row 3.

$$\begin{bmatrix} 355 & -135 & 103 & 916 \\ 0 & 117,370 & 150,197 & 36,274 \\ 0 & -12,640 & -264,474 & -473,106.5 \end{bmatrix}$$

Multiply row 3 by 11,737.

$$\begin{bmatrix} 355 & -135 & 103 & 916 \\ 0 & 117,370 & 150,197 & 36,274 \\ 0 & -148,355,680 & -3,104,131,338 & -5,552,850,991 \end{bmatrix}$$

Multiply row 2 by 1264 and add to row 3.

$$\begin{bmatrix} 355 & -135 & 103 & 916 \\ 0 & 117,370 & 150,197 & 36,274 \\ 0 & 0 & -2,914,282,330 & -5,507,000,655 \end{bmatrix}$$

Putting the variables back in, we have

$355x - 135y + 103z = 916,$ (1)

$117,370y + 150,197z = 36,274,$ (2)

$-2,914,282,330z = -5,507,000,655$ (3)

Solve (3) for z.

$-2,914,282,330z = -5,507,000,655$

$z \approx 1.8897$

23. (continued)

Substitute 1.8897 for z in (2) and solve for y.

$$117{,}370y + 150{,}197z = 36{,}274$$
$$117{,}370y + 150{,}197(1.8897) = 36{,}274$$
$$117{,}370y + 283{,}827.2709 = 36{,}274$$
$$117{,}370y = -247{,}553.2709$$
$$y \approx -2.1092$$

Substitute 1.8897 for z and -2.1092 for y in (1) and solve for x.

$$355x - 135y + 103z = 916$$
$$355x - 135(-2.1092) + 103(1.8897) = 916$$
$$355x + 284.742 + 194.6391 = 916$$
$$355x + 479.3811 = 916$$
$$355x = 436.6189$$
$$x \approx 1.2299$$

The solution is (1.2299, -2.1092, 1.8897).

24. (81.2, 39.7, -17.6)

25. $\sqrt{2}x + \pi y = 3,$

$\pi x - \sqrt{2}y = 1$

Write a matrix using only the constants.

$$\begin{bmatrix} \sqrt{2} & \pi & 3 \\ \pi & -\sqrt{2} & 1 \end{bmatrix}$$

Multiply row 2 by $\sqrt{2}$.

$$\begin{bmatrix} \sqrt{2} & \pi & 3 \\ \pi\sqrt{2} & -2 & \sqrt{2} \end{bmatrix}$$

25. (continued)

Multiply row 1 by $-\pi$ and add it to row 2.

$$\begin{bmatrix} \sqrt{2} & \pi & 3 \\ 0 & -\pi^2 - 2 & -3\pi + \sqrt{2} \end{bmatrix}$$

Putting the variables back in, we have

$\sqrt{2}x + \pi y = 3$ (1)

$(-\pi^2 - 2)y = -3\pi + \sqrt{2}$ (2)

Solve (2) for y.

$(-\pi^2 - 2)y = -3\pi + \sqrt{2}$

$$y = \frac{-3\pi + \sqrt{2}}{-\pi^2 - 2}, \text{ or } \frac{3\pi - \sqrt{2}}{\pi^2 + 2}$$

Substitute for y in (1) and solve for x.

$$\sqrt{2}x + \pi\left(\frac{3\pi - \sqrt{2}}{\pi^2 + 2}\right) = 3$$
$$\sqrt{2}x = 3 - \frac{3\pi^2 - \pi\sqrt{2}}{\pi^2 + 2}$$
$$\sqrt{2}x = \frac{3\pi^2 + 6 - 3\pi^2 + \pi\sqrt{2}}{\pi^2 + 2}$$
$$x = \frac{6 + \pi\sqrt{2}}{\sqrt{2}(\pi^2 + 2)}$$
$$x = \frac{3\sqrt{2} + \pi}{\pi^2 + 2}$$

The solution is $\left(\frac{3\sqrt{2} + \pi}{\pi^2 + 2}, \frac{3\pi - \sqrt{2}}{\pi^2 + 2}\right)$.

26. $\left(\frac{ce - bf}{bd - ac}, \frac{cd - af}{bd - ae}\right)$

Exercise Set 9.5

1. $$A + B = \begin{bmatrix} 1 & 2 \\ 4 & 3 \end{bmatrix} + \begin{bmatrix} -3 & 5 \\ 2 & -1 \end{bmatrix} = \begin{bmatrix} 1 + (-3) & 2 + 5 \\ 4 + 2 & 3 + (-1) \end{bmatrix} = \begin{bmatrix} -2 & 7 \\ 6 & 2 \end{bmatrix}$$

2. $$\begin{bmatrix} -2 & 7 \\ 6 & 2 \end{bmatrix}$$

3. $$E + 0 = \begin{bmatrix} 1 & 3 \\ 2 & 6 \end{bmatrix} + \begin{bmatrix} 0 & 0 \\ 0 & 0 \end{bmatrix} = \begin{bmatrix} 1 + 0 & 3 + 0 \\ 2 + 0 & 6 + 0 \end{bmatrix} = \begin{bmatrix} 1 & 3 \\ 2 & 6 \end{bmatrix}$$

4. $$\begin{bmatrix} 2 & 4 \\ 8 & 6 \end{bmatrix}$$

5. $3F = 3\begin{bmatrix} 3 & 3 \\ -1 & -1 \end{bmatrix} = \begin{bmatrix} 3\cdot 3 & 3\cdot 3 \\ 3\cdot(-1) & 3\cdot(-1) \end{bmatrix} = \begin{bmatrix} 9 & 9 \\ -3 & -3 \end{bmatrix}$

6. $\begin{bmatrix} -1 & -1 \\ -1 & -1 \end{bmatrix}$

7. $3F = 3\begin{bmatrix} 3 & 3 \\ -1 & -1 \end{bmatrix} = \begin{bmatrix} 9 & 9 \\ -3 & -3 \end{bmatrix}$, $2A = 2\begin{bmatrix} 1 & 2 \\ 4 & 3 \end{bmatrix} = \begin{bmatrix} 2 & 4 \\ 8 & 6 \end{bmatrix}$

Then
$3F + 2A = \begin{bmatrix} 9 & 9 \\ -3 & -3 \end{bmatrix} + \begin{bmatrix} 2 & 4 \\ 8 & 6 \end{bmatrix} = \begin{bmatrix} 9+2 & 9+4 \\ -3+8 & -3+6 \end{bmatrix} = \begin{bmatrix} 11 & 13 \\ 5 & 3 \end{bmatrix}$

8. $\begin{bmatrix} 4 & -3 \\ 2 & 4 \end{bmatrix}$

9. $B - A = \begin{bmatrix} -3 & 5 \\ 2 & -1 \end{bmatrix} - \begin{bmatrix} 1 & 2 \\ 4 & 3 \end{bmatrix} = \begin{bmatrix} -3 & 5 \\ 2 & -1 \end{bmatrix} + \begin{bmatrix} -1 & -2 \\ -4 & -3 \end{bmatrix}$ [B - A = B + (-A)]

$= \begin{bmatrix} -3+(-1) & 5+(-2) \\ 2+(-4) & -1+(-3) \end{bmatrix} = \begin{bmatrix} -4 & 3 \\ -2 & -4 \end{bmatrix}$

10. $\begin{bmatrix} 1 & 3 \\ -6 & 17 \end{bmatrix}$

11. $BA = \begin{bmatrix} -3 & 5 \\ 2 & -1 \end{bmatrix}\begin{bmatrix} 1 & 2 \\ 4 & 3 \end{bmatrix} = \begin{bmatrix} -3\cdot 1 + 5\cdot 4 & -3\cdot 2 + 5\cdot 3 \\ 2\cdot 1 + (-1)4 & 2\cdot 2 + (-1)3 \end{bmatrix} = \begin{bmatrix} 17 & 9 \\ -2 & 1 \end{bmatrix}$

12. $\begin{bmatrix} 0 & 0 \\ 0 & 0 \end{bmatrix}$

13. $CD = \begin{bmatrix} 1 & -1 \\ -1 & 1 \end{bmatrix}\begin{bmatrix} 1 & 1 \\ 1 & 1 \end{bmatrix} = \begin{bmatrix} 1\cdot 1 + (-1)\cdot 1 & 1\cdot 1 + (-1)\cdot 1 \\ -1\cdot 1 + 1\cdot 1 & -1\cdot 1 + 1\cdot 1 \end{bmatrix} = \begin{bmatrix} 0 & 0 \\ 0 & 0 \end{bmatrix}$

14. $\begin{bmatrix} 0 & 0 \\ 0 & 0 \end{bmatrix}$

15. $AI = \begin{bmatrix} 1 & 2 \\ 4 & 3 \end{bmatrix}\begin{bmatrix} 1 & 0 \\ 0 & 1 \end{bmatrix} = \begin{bmatrix} 1\cdot 1 + 2\cdot 0 & 1\cdot 0 + 2\cdot 1 \\ 4\cdot 1 + 3\cdot 0 & 4\cdot 0 + 3\cdot 1 \end{bmatrix} = \begin{bmatrix} 1 & 2 \\ 4 & 3 \end{bmatrix}$ (Note: AI = A)

16. $\begin{bmatrix} 1 & 2 \\ 4 & 3 \end{bmatrix}$ or A

17.

$$AB = \begin{bmatrix} 1 & 0 & -2 \\ 0 & -1 & 3 \\ 3 & 2 & 4 \end{bmatrix}\begin{bmatrix} -1 & -2 & 5 \\ 1 & 0 & -1 \\ 2 & -3 & 1 \end{bmatrix}$$

$$= \begin{bmatrix} 1(-1) + 0\cdot1 + (-2)2 & 1(-2) + 0\cdot0 + (-2)(-3) & 1\cdot5 + 0(-1) + (-2)1 \\ 0(-1) + (-1)1 + 3\cdot2 & 0(-2) + (-1)0 + 3(-3) & 0\cdot5 + (-1)(-1) + 3\cdot1 \\ 3(-1) + 2\cdot1 + 4\cdot2 & 3(-2) + 2\cdot0 + 4(-3) & 3\cdot5 + 2(-1) + 4\cdot1 \end{bmatrix}$$

$$= \begin{bmatrix} -5 & 4 & 3 \\ 5 & -9 & 4 \\ 7 & -18 & 17 \end{bmatrix}$$

18. $\begin{bmatrix} 14 & 12 & 16 \\ -2 & -2 & -6 \\ 5 & 5 & -9 \end{bmatrix}$

19.

$$CI = \begin{bmatrix} -2 & 9 & 6 \\ -3 & 3 & 4 \\ 2 & -2 & 1 \end{bmatrix}\begin{bmatrix} 1 & 0 & 0 \\ 0 & 1 & 0 \\ 0 & 0 & 1 \end{bmatrix}$$

$$= \begin{bmatrix} -2\cdot1 + 9\cdot0 + 6\cdot0 & -2\cdot0 + 9\cdot1 + 6\cdot0 & -2\cdot0 + 9\cdot0 + 6\cdot1 \\ -3\cdot1 + 3\cdot0 + 4\cdot0 & -3\cdot0 + 3\cdot1 + 4\cdot0 & -3\cdot0 + 3\cdot0 + 4\cdot1 \\ 2\cdot1 + (-2)\cdot0 + 1\cdot0 & 2\cdot0 + (-2)1 + 1\cdot0 & 2\cdot0 + (-2)\cdot0 + 1\cdot1 \end{bmatrix}$$

$$= \begin{bmatrix} -2 & 9 & 6 \\ -3 & 3 & 4 \\ 2 & -2 & 1 \end{bmatrix} \quad \text{(Note: } CI = C\text{)}$$

20. $\begin{bmatrix} -2 & 9 & 6 \\ -3 & 3 & 4 \\ 2 & -2 & 1 \end{bmatrix}$ or C

21.

$$\begin{bmatrix} -3 & 2 \end{bmatrix}\begin{bmatrix} 4 \\ -2 \end{bmatrix} = [-3\cdot4 + 2(-2)] = [-16]$$

22. $[-14]$

23.

$$[-5 \quad 1 \quad 2]\begin{bmatrix} 1 & 3 \\ -1 & 0 \\ 4 & -2 \end{bmatrix} = [-5\cdot1 + 1(-1) + 2\cdot4 \quad -5\cdot3 + 1\cdot0 + 2(-2)] = [2 \quad -19]$$

24. $\begin{bmatrix} -8 \\ 2 \\ -6 \end{bmatrix}$

25. $$\begin{bmatrix} 3 & -2 & 4 \\ 2 & 1 & -5 \end{bmatrix} \begin{bmatrix} x \\ y \\ z \end{bmatrix} = \begin{bmatrix} 17 \\ 13 \end{bmatrix}$$

26. $$\begin{bmatrix} 3 & 2 & 5 \\ 4 & -3 & 2 \end{bmatrix} \begin{bmatrix} x \\ y \\ z \end{bmatrix} = \begin{bmatrix} 9 \\ 10 \end{bmatrix}$$

27. $$\begin{bmatrix} 1 & -1 & 2 & -4 \\ 2 & -1 & -1 & 1 \\ 1 & 4 & -3 & -1 \\ 3 & 5 & -7 & 2 \end{bmatrix} \begin{bmatrix} x \\ y \\ z \\ w \end{bmatrix} = \begin{bmatrix} 12 \\ 0 \\ 1 \\ 9 \end{bmatrix}$$

28. $$\begin{bmatrix} 2 & 4 & -5 & 12 \\ 4 & -1 & 12 & -1 \\ -1 & 4 & 0 & 2 \\ 2 & 10 & 1 & 0 \end{bmatrix} \begin{bmatrix} x \\ y \\ z \\ w \end{bmatrix} = \begin{bmatrix} 2 \\ 5 \\ 13 \\ 5 \end{bmatrix}$$

29. Let $\begin{bmatrix} 3.61 & -2.14 & 16.7 \\ -4.33 & 7.03 & 12.9 \\ 5.82 & -6.95 & 2.34 \end{bmatrix} \begin{bmatrix} 3.05 & 0.402 & -1.34 \\ 1.84 & -1.13 & 0.024 \\ -2.83 & 2.04 & 8.81 \end{bmatrix} = \begin{bmatrix} a & b & c \\ d & e & f \\ g & h & i \end{bmatrix}$

Each matrix is a 3 × 3 matrix. The product is also a 3 × 3 matrix. We calculate each element in the product.

a) 3.61(3.05) + (-2.14)(1.84) + 16.7(-2.83) = -40.1881 ≈ -40.19

b) 3.61(0.402) + (-2.14)(-1.13) + 16.7(2.04) = 37.93742 ≈ 37.94

c) 3.61(-1.34) + (-2.14)(0.024) + 16.7(8.81) = 142.23824 ≈ 142.24

d) -4.33(3.05) + 7.03(1.84) + 12.9(-2.83) = -36.7783 ≈ -36.78

e) -4.33(0.402) + 7.03(-1.13) + 12.9(2.04) = 16.63144 ≈ 16.63

f) -4.33(-1.34) + 7.03(0.024) + 12.9(8.81) = 119.61992 ≈ 119.62

g) 5.82(3.05) + (-6.95)(1.84) + 2.34(-2.83) = -1.6592 ≈ -1.66

h) 5.82(0.402) + (-6.95)(-1.13) + 2.34(2.04) = 14.96674 ≈ 14.97

i) 5.82(-1.34) + (-6.95)(0.024) + 2.34(8.81) = 12.6498 ≈ 12.65

The product is $\begin{bmatrix} -40.19 & 37.94 & 142.24 \\ -36.78 & 16.63 & 119.62 \\ -1.66 & 14.97 & 12.65 \end{bmatrix}$

30. $\begin{bmatrix} 142.1 & -66.62 & -136.5 \\ 257.2 & 1038.8 & 2694.1 \\ 182.4 & 169.6 & 452.8 \end{bmatrix}$

31. $A = \begin{bmatrix} -1 & 0 \\ 2 & 1 \end{bmatrix}, \quad B = \begin{bmatrix} 1 & -1 \\ 0 & 2 \end{bmatrix}$

$(A + B)(A - B) = \begin{bmatrix} 0 & -1 \\ 2 & 3 \end{bmatrix}\begin{bmatrix} -2 & 1 \\ 2 & -1 \end{bmatrix} = \begin{bmatrix} -2 & 1 \\ 2 & -1 \end{bmatrix}$

$A^2 - B^2 = \begin{bmatrix} -1 & 0 \\ 2 & 1 \end{bmatrix}\begin{bmatrix} -1 & 0 \\ 2 & 1 \end{bmatrix} - \begin{bmatrix} 1 & -1 \\ 0 & 2 \end{bmatrix}\begin{bmatrix} 1 & -1 \\ 0 & 2 \end{bmatrix}$

$= \begin{bmatrix} 1 & 0 \\ 0 & 1 \end{bmatrix} - \begin{bmatrix} 1 & -3 \\ 0 & 4 \end{bmatrix} = \begin{bmatrix} 0 & 3 \\ 0 & -3 \end{bmatrix}$

32. $(A + B)(A + B) = \begin{bmatrix} -2 & -3 \\ 6 & 7 \end{bmatrix}, \ A^2 + 2AB + B^2 = \begin{bmatrix} 0 & -1 \\ 4 & 5 \end{bmatrix}$

33. $(A + B)(A - B) = \begin{bmatrix} -2 & 1 \\ 2 & -1 \end{bmatrix}$ (See Exercise 31.)

$A^2 = \begin{bmatrix} -1 & 0 \\ 2 & 1 \end{bmatrix}\begin{bmatrix} -1 & 0 \\ 2 & 1 \end{bmatrix} = \begin{bmatrix} 1 & 0 \\ 0 & 1 \end{bmatrix}$

$BA = \begin{bmatrix} 1 & -1 \\ 0 & 2 \end{bmatrix}\begin{bmatrix} -1 & 0 \\ 2 & 1 \end{bmatrix} = \begin{bmatrix} -3 & -1 \\ 4 & 2 \end{bmatrix}$

$AB = \begin{bmatrix} -1 & 0 \\ 2 & 1 \end{bmatrix}\begin{bmatrix} 1 & -1 \\ 0 & 2 \end{bmatrix} = \begin{bmatrix} -1 & 1 \\ 2 & 0 \end{bmatrix}$

$B^2 = \begin{bmatrix} 1 & -1 \\ 0 & 2 \end{bmatrix}\begin{bmatrix} 1 & -1 \\ 0 & 2 \end{bmatrix} = \begin{bmatrix} 1 & -3 \\ 0 & 4 \end{bmatrix}$

$A^2 + BA - AB - B^2 = \begin{bmatrix} 1 & 0 \\ 0 & 1 \end{bmatrix} + \begin{bmatrix} -3 & -1 \\ 4 & 2 \end{bmatrix} - \begin{bmatrix} -1 & 1 \\ 2 & 0 \end{bmatrix} - \begin{bmatrix} 1 & -3 \\ 0 & 4 \end{bmatrix}$

$= \begin{bmatrix} -2 & 1 \\ 2 & -1 \end{bmatrix}$

Therefore, $(A + B)(A - B) = A^2 + BA - AB - B^2$.

34. $(A + B)(A + B) = \begin{bmatrix} -2 & -3 \\ 6 & 7 \end{bmatrix} = A^2 + BA + AB + B^2$

35.
$$\begin{bmatrix} \cos x & \sin x \\ -\sin x & \cos x \end{bmatrix}\begin{bmatrix} \cos y & \sin y \\ -\sin y & \cos y \end{bmatrix}$$

$$= \begin{bmatrix} \cos x \cos y - \sin x \sin y & \cos x \sin y + \sin x \cos y \\ -\sin x \cos y - \cos x \sin y & -\sin x \sin y + \cos x \cos y \end{bmatrix}$$

$$= \begin{bmatrix} \cos (x + y) & \sin (x + y) \\ -\sin (x + y) & \cos (x + y) \end{bmatrix}$$ (Using sum and difference identities)

36.
$$A + B = \begin{bmatrix} a_{11} + b_{11} & a_{12} + b_{12} \\ a_{21} + b_{21} & a_{22} + b_{22} \end{bmatrix}$$

$$= \begin{bmatrix} b_{11} + a_{11} & b_{12} + a_{12} \\ b_{21} + a_{21} & b_{22} + a_{22} \end{bmatrix}$$ (By commutativity of addition of real numbers)

$= B + A$

37. $A + (B + C) = (A + B) + C$, is similar to Exercise 36, but uses associativity of addition of real numbers.

38.
$$k(A + B) = \begin{bmatrix} k(a_{11} + b_{11}) & k(a_{12} + b_{12}) \\ k(a_{21} + b_{21}) & k(a_{22} + b_{22}) \end{bmatrix}$$

$$= \begin{bmatrix} ka_{11} + kb_{11} & ka_{12} + kb_{12} \\ ka_{21} + kb_{21} & ka_{22} + kb_{22} \end{bmatrix}$$ (By the distributive law of real numbers)

$$= \begin{bmatrix} ka_{11} & ka_{12} \\ ka_{21} & ka_{22} \end{bmatrix} + \begin{bmatrix} kb_{11} & kb_{12} \\ kb_{21} & kb_{22} \end{bmatrix}$$

$= kA + kB$

39.
$$(k + m)A = \begin{bmatrix} (k + m)a_{11} & (k + m)a_{12} \\ (k + m)a_{21} & (k + m)a_{22} \end{bmatrix}$$

$$= \begin{bmatrix} ka_{11} + ma_{11} & ka_{12} + ma_{12} \\ ka_{21} + ma_{21} & ka_{22} + ma_{22} \end{bmatrix}$$

$$= \begin{bmatrix} ka_{11} & ka_{12} \\ ka_{21} & ka_{22} \end{bmatrix} + \begin{bmatrix} ma_{11} & ma_{12} \\ ma_{21} & ma_{22} \end{bmatrix}$$

$= kA + mA$

40. Find BC. Then A(BC). Find AB. Then (AB)C. Then compare A(BC) and (AB)C.

41. Find AI, IA, and compare with A.

Exercise Set 9.6

1. $\begin{vmatrix} -2 & -\sqrt{5} \\ -\sqrt{5} & 3 \end{vmatrix}$

$= -2\cdot3 - (-\sqrt{5})(-\sqrt{5}) = -6 - 5 = -11$

2. $2\sqrt{5} + 12$

3. $\begin{vmatrix} x & 4 \\ x & x^2 \end{vmatrix} = x\cdot x^2 - x\cdot 4 = x^3 - 4x$

4. $3y^2 + 2y$

5. $\begin{vmatrix} 3 & 1 & 2 \\ -2 & 3 & 1 \\ 3 & 4 & -6 \end{vmatrix}$

$= 3\begin{vmatrix} 3 & 1 \\ 4 & -6 \end{vmatrix} - (-2)\begin{vmatrix} 1 & 2 \\ 4 & -6 \end{vmatrix} + 3\begin{vmatrix} 1 & 2 \\ 3 & 1 \end{vmatrix}$

$= 3(-22) + 2(-14) + 3(-5)$

$= -66 - 28 - 15$

$= -109$

6. -9

7. $\begin{vmatrix} x & 0 & -1 \\ 2 & x & x^2 \\ -3 & x & 1 \end{vmatrix}$

$= x\begin{vmatrix} x & x^2 \\ x & 1 \end{vmatrix} - 2\begin{vmatrix} 0 & -1 \\ x & 1 \end{vmatrix} + (-3)\begin{vmatrix} 0 & -1 \\ x & x^2 \end{vmatrix}$

$= x(x - x^3) - 2(x) - 3(x)$

$= x^2 - x^4 - 2x - 3x$

$= -x^4 + x^2 - 5x$

8. $-2x^3$

9. $-2x + 4y = 3$
$3x - 7y = 1$

$$x = \frac{\begin{vmatrix} 3 & 4 \\ 1 & -7 \end{vmatrix}}{\begin{vmatrix} -2 & 4 \\ 3 & -7 \end{vmatrix}} = \frac{3(-7) - 1(4)}{-2(-7) - 3(4)} = \frac{-25}{2}$$

$$y = \frac{\begin{vmatrix} -2 & 3 \\ 3 & 1 \end{vmatrix}}{\begin{vmatrix} -2 & 4 \\ 3 & -7 \end{vmatrix}} = \frac{-2(1) - 3(3)}{-2(-7) - 3(4)} = \frac{-11}{2}$$

The solution is $\left(-\frac{25}{2}, -\frac{11}{2}\right)$.

10. $\left(\frac{9}{19}, \frac{51}{28}\right)$

11. $\sqrt{3}x + \pi y = -5$
$\pi x - \sqrt{3}y = 4$

$$x = \frac{\begin{vmatrix} -5 & \pi \\ 4 & -\sqrt{3} \end{vmatrix}}{\begin{vmatrix} \sqrt{3} & \pi \\ \pi & -\sqrt{3} \end{vmatrix}} = \frac{5\sqrt{3} - 4\pi}{-3 - \pi^2} = \frac{-5\sqrt{3} + 4\pi}{3 + \pi^2}$$

$$y = \frac{\begin{vmatrix} \sqrt{3} & -5 \\ \pi & 4 \end{vmatrix}}{\begin{vmatrix} \sqrt{3} & \pi \\ \pi & -\sqrt{3} \end{vmatrix}} = \frac{4\sqrt{3} + 5\pi}{-3 - \pi^2} = \frac{-4\sqrt{3} - 5\pi}{3 + \pi^2}$$

The solution is $\left(\frac{-5\sqrt{3} + 4\pi}{3 + \pi^2}, \frac{-4\sqrt{3} - 5\pi}{3 + \pi^2}\right)$, or $\left(\frac{-5\sqrt{3} + 4\pi}{3 + \pi^2}, \frac{4\sqrt{3} + 5\pi}{-3 - \pi^2}\right)$.

12. $\left(\frac{2\pi - 3\sqrt{5}}{\pi^2 + 5}, \frac{-3\pi - 2\sqrt{5}}{\pi^2 + 5}\right)$

13. $3x + 2y - z = 4$
$3x - 2y + z = 5$
$4x - 5y - z = -1$

$$x = \frac{\begin{vmatrix} 4 & 2 & -1 \\ 5 & -2 & 1 \\ -1 & -5 & -1 \end{vmatrix}}{\begin{vmatrix} 3 & 2 & -1 \\ 3 & -2 & 1 \\ 4 & -5 & -1 \end{vmatrix}}$$

$$= \frac{4\begin{vmatrix} -2 & 1 \\ -5 & -1 \end{vmatrix} - 5\begin{vmatrix} 2 & -1 \\ -5 & -1 \end{vmatrix} + (-1)\begin{vmatrix} 2 & -1 \\ -2 & 1 \end{vmatrix}}{3\begin{vmatrix} -2 & 1 \\ -5 & -1 \end{vmatrix} - 3\begin{vmatrix} 2 & -1 \\ -5 & -1 \end{vmatrix} + 4\begin{vmatrix} 2 & -1 \\ -2 & 1 \end{vmatrix}}$$

$$= \frac{4(7) - 5(-7) - 1(0)}{3(7) - 3(-7) + 4(0)}$$

$$= \frac{28 + 35}{21 + 21} = \frac{63}{42} = \frac{3}{2}$$

$$y = \frac{\begin{vmatrix} 3 & 4 & -1 \\ 3 & 5 & 1 \\ 4 & -1 & -1 \end{vmatrix}}{42}$$

$$= \frac{3\begin{vmatrix} 5 & 1 \\ -1 & -1 \end{vmatrix} - 3\begin{vmatrix} 4 & -1 \\ -1 & -1 \end{vmatrix} + 4\begin{vmatrix} 4 & -1 \\ 5 & 1 \end{vmatrix}}{42}$$

$$= \frac{3(-4) - 3(-5) + 4(9)}{42}$$

13. (continued)

$$= \frac{-12 + 15 + 36}{42} = \frac{39}{42} = \frac{13}{14}$$

$$z = \frac{\begin{vmatrix} 3 & 2 & 4 \\ 3 & -2 & 5 \\ 4 & -5 & -1 \end{vmatrix}}{42}$$

$$= \frac{3\begin{vmatrix} -2 & 5 \\ -5 & -1 \end{vmatrix} - 3\begin{vmatrix} 2 & 4 \\ -5 & -1 \end{vmatrix} + 4\begin{vmatrix} 2 & 4 \\ -2 & 5 \end{vmatrix}}{42}$$

$$= \frac{3(27) - 3(18) + 4(18)}{42}$$

$$= \frac{81 - 54 + 72}{42} = \frac{99}{42} = \frac{33}{14}$$

The solution is $\left(\frac{3}{2}, \frac{13}{14}, \frac{33}{14}\right)$.

14. $\left(-1, -\frac{6}{7}, \frac{11}{7}\right)$

15. $0x + 6y + 6z = -1$
$8x + 0y + 6z = -1$
$4x + 9y + 0z = 8$

$$x = \frac{\begin{vmatrix} -1 & 6 & 6 \\ -1 & 0 & 6 \\ 8 & 9 & 0 \end{vmatrix}}{\begin{vmatrix} 0 & 6 & 6 \\ 8 & 0 & 6 \\ 4 & 9 & 0 \end{vmatrix}}$$

$$= \frac{-1\begin{vmatrix} 0 & 6 \\ 9 & 0 \end{vmatrix} - (-1)\begin{vmatrix} 6 & 6 \\ 9 & 0 \end{vmatrix} + 8\begin{vmatrix} 6 & 6 \\ 0 & 6 \end{vmatrix}}{0\begin{vmatrix} 0 & 6 \\ 9 & 0 \end{vmatrix} - 8\begin{vmatrix} 6 & 6 \\ 9 & 0 \end{vmatrix} + 4\begin{vmatrix} 6 & 6 \\ 0 & 6 \end{vmatrix}}$$

$$= \frac{-1(-54) + 1(-54) + 8(36)}{0(-54) - 8(-54) + 4(36)} = \frac{54 - 54 + 288}{0 + 432 + 144}$$

$$= \frac{288}{576} = \frac{1}{2}$$

$$y = \frac{\begin{vmatrix} 0 & -1 & 6 \\ 8 & -1 & 6 \\ 4 & 8 & 0 \end{vmatrix}}{576} = \frac{384}{576} = \frac{2}{3}$$

$$z = \frac{\begin{vmatrix} 0 & 6 & -1 \\ 8 & 0 & -1 \\ 4 & 9 & 8 \end{vmatrix}}{576} = \frac{-480}{576} = -\frac{5}{6}$$

The solution is $\left(\frac{1}{2}, \frac{2}{3}, -\frac{5}{6}\right)$.

16. $\left(-\frac{31}{16}, \frac{25}{16}, -\frac{58}{16}\right)$

17. $\begin{vmatrix} x & 5 \\ -4 & x \end{vmatrix} = 24$

$$x^2 + 20 = 24$$
$$x^2 = 4$$
$$x = \pm 2$$

18. 3, -2

19. $\begin{vmatrix} x & -3 \\ -1 & x \end{vmatrix} \geqslant 0$

$$x^2 - 3 \geqslant 0$$

The solutions of $f(x) = x^2 - 3 = 0$ are $-\sqrt{3}$ and $\sqrt{3}$. They divide the real-number line as shown.

A B C

$-\sqrt{3}$ $\sqrt{3}$

We try test numbers in each interval.

$f(-2) = (-2)^2 - 3 = 4 - 3 = 1$
$f(0) = 0^2 - 3 = -3$
$f(2) = 2^2 - 3 = 4 - 3 = 1$

Since the inequality is $\geqslant$, we include $\pm\sqrt{3}$ in the solution set. The solution set is $\{x \mid x \leqslant -\sqrt{3} \text{ or } x \geqslant \sqrt{3}\}$.

20. $\{y \mid -\sqrt{10} < y < \sqrt{10}\}$

21. $\begin{vmatrix} x+3 & 4 \\ x-3 & 5 \end{vmatrix} = -7$

$$(x + 3)(5) - (x - 3)4 = -7$$
$$5x + 15 - 4x + 12 = -7$$
$$x + 27 = -7$$
$$x = -34$$

22. 3

23. $\begin{vmatrix} 2 & x & 1 \\ 1 & 2 & -1 \\ 3 & 4 & -2 \end{vmatrix} = -6$

$$2\begin{vmatrix} 2 & -1 \\ 4 & -2 \end{vmatrix} - 1\begin{vmatrix} x & 1 \\ 4 & -2 \end{vmatrix} + 3\begin{vmatrix} x & 1 \\ 2 & -1 \end{vmatrix} = -6$$

$$2(-4 + 4) - 1(-2x - 4) + 3(-x - 2) = -6$$
$$2x + 4 - 3x - 6 = -6$$
$$-x - 2 = -6$$
$$-x = -4$$
$$x = 4$$

24. 0

25. - 30. Answers may vary

25. $2L + 2W = \begin{vmatrix} L & -W \\ 2 & 2 \end{vmatrix}$

26. $\begin{vmatrix} \pi & -h \\ \pi & r \end{vmatrix}$

27. $a^2 + b^2 = \begin{vmatrix} a & b \\ -b & a \end{vmatrix}$

28. $\begin{vmatrix} \frac{1}{2}h & -b \\ \frac{1}{2}h & a \end{vmatrix}$

29. $2\pi r^2 + 2\pi rh = \begin{vmatrix} 2\pi r & 2\pi r \\ -h & r \end{vmatrix}$

30. $\begin{vmatrix} x^2 & 1 \\ Q^2 & y^2 \end{vmatrix}$

31. $\begin{vmatrix} \cos x & \sin x \\ -\sin x & \cos x \end{vmatrix} = \cos^2 x - (-\sin^2 x)$

$= \cos^2 x + \sin^2 x$

$= 1$

$\begin{vmatrix} \cos x & -\sin x \\ \sin x & \cos x \end{vmatrix} = \cos^2 x - (-\sin^2 x)$

$= \cos^2 x + \sin^2 x$

$= 1$

Therefore, $\begin{vmatrix} \cos x & \sin x \\ -\sin x & \cos x \end{vmatrix} = \begin{vmatrix} \cos x & -\sin x \\ \sin x & \cos x \end{vmatrix}$

$= 1.$

Exercise Set 9.7

1. a_{11} is the element in the 1st row and 1st column. $a_{11} = 7$
   a_{32} is the element in the 3rd row and 2nd column. $a_{32} = 2$
   a_{22} is the element in the 2nd row and 2nd column. $a_{22} = 0$

2. $a_{13} = -6$, $a_{31} = 1$, $a_{23} = -3$

3. To find M_{11} we delete the 1st row and the 1st column.

$$A = \begin{bmatrix} 7 & -4 & -6 \\ 2 & 0 & -3 \\ 1 & 2 & -5 \end{bmatrix}$$

We calculate the determinant of the matrix formed by the remaining elements.

$$M_{11} = \begin{vmatrix} 0 & -3 \\ 2 & -5 \end{vmatrix} = 0(-5) - 2(-3) = 0 + 6 = 6$$

To find M_{32} we delete the 3rd row and the 2nd column.

$$A = \begin{bmatrix} 7 & -4 & -6 \\ 2 & 0 & -3 \\ 1 & 2 & -5 \end{bmatrix}$$

We calculate the determinant of the matrix formed by the remaining elements.

$$M_{32} = \begin{vmatrix} 7 & -6 \\ 2 & -3 \end{vmatrix} = 7(-3) - 2(-6) = -21 + 12 = -9$$

3. (continued)

To find M_{22} we delete the 2nd row and the 2nd column.

$$A = \begin{bmatrix} 7 & -4 & -6 \\ 2 & 0 & -3 \\ 1 & 2 & -5 \end{bmatrix}$$

To calculate the determinant of the matrix formed by the remaining elements.

$$M_{22} = \begin{vmatrix} 7 & -6 \\ 1 & -5 \end{vmatrix} = 7(-5) - 1(-6) = -35 + 6 = -29.$$

4. $M_{13} = 4, \quad M_{31} = 12, \quad M_{23} = 18$

5.
$$A_{11} = (-1)^{1+1}\ M_{11} = (-1)^2 \cdot \begin{vmatrix} 0 & -3 \\ 2 & -5 \end{vmatrix} = 1 \cdot 6 = 6$$

$$A_{32} = (-1)^{3+2}\ M_{32} = (-1)^5 \cdot \begin{vmatrix} 7 & -6 \\ 2 & -3 \end{vmatrix} = -1 \cdot (-9) = 9$$

$$A_{22} = (-1)^{2+2}\ M_{22} = (-1)^4 \cdot \begin{vmatrix} 7 & -6 \\ 1 & -5 \end{vmatrix} = 1 \cdot (-29) = -29$$

6. $A_{13} = 4, \quad A_{31} = 12, \quad A_{23} = -18$

7.
$$A = \begin{bmatrix} 7 & -4 & -6 \\ \boxed{2 \quad 0 \quad -3} \\ 1 & 2 & -5 \end{bmatrix}$$

Find $|A|$ by expanding about the second row.

$$|A| = (2)(-1)^{2+1} \cdot \begin{vmatrix} -4 & -6 \\ 2 & -5 \end{vmatrix} + (0)(-1)^{2+2} \cdot \begin{vmatrix} 7 & -6 \\ 1 & -5 \end{vmatrix} + (-3)(-1)^{2+3} \cdot \begin{vmatrix} 7 & -4 \\ 1 & 2 \end{vmatrix}$$

$$= 2(-1)(32) + 0 + (-3)(-1)(18)$$
$$= -64 + 54 = -10$$

8. -10

9.
$$A = \begin{bmatrix} 7 & -4 & \boxed{-6} \\ 2 & 0 & -3 \\ 1 & 2 & -5 \end{bmatrix}$$

Find $|A|$ by expanding about the third column.

$$|A| = (-6)(-1)^{1+3} \cdot \begin{vmatrix} 2 & 0 \\ 1 & 2 \end{vmatrix} + (-3)(-1)^{2+3} \cdot \begin{vmatrix} 7 & -4 \\ 1 & 2 \end{vmatrix} + (-5)(-1)^{3+3} \cdot \begin{vmatrix} 7 & -4 \\ 2 & 0 \end{vmatrix}$$

$$= (-6)(1)(2 \cdot 2 - 1 \cdot 0) + (-3)(-1)[7 \cdot 2 - 1(-4)] + (-5)(1)[7 \cdot 0 - 2(-4)]$$
$$= (-6)(4) + (3)(18) + (-5)(8)$$
$$= -24 + 54 - 40 = -10$$

10. -10

11. To find M_{41} we delete the 4th row and the 1st column.

$$A = \begin{bmatrix} 1 & 0 & 0 & -2 \\ 4 & 1 & 0 & 0 \\ 5 & 6 & 7 & 8 \\ -2 & -3 & -1 & 0 \end{bmatrix}$$

We calculate the determinant of the matrix formed by the remaining elements.

$$M_{41} = \begin{vmatrix} 0 & 0 & -2 \\ 1 & 0 & 0 \\ 6 & 7 & 8 \end{vmatrix} = 0 + 0 + (-2)(-1)^{1+3} \cdot \begin{vmatrix} 1 & 0 \\ 6 & 7 \end{vmatrix} = -2 \cdot 1 \cdot 7 = -14$$

To find M_{33} we delete the 3rd row and the 3rd column.

$$A = \begin{bmatrix} 1 & 0 & 0 & -2 \\ 4 & 1 & 0 & 0 \\ 5 & 6 & 7 & 8 \\ -2 & -3 & -1 & 0 \end{bmatrix}$$

We calculate the determinant of the matrix formed by the remaining elements.

$$M_{33} = \begin{vmatrix} 1 & 0 & -2 \\ 4 & 1 & 0 \\ -2 & -3 & 0 \end{vmatrix} = -2(-1)^{1+3} \cdot \begin{vmatrix} 4 & 1 \\ -2 & -3 \end{vmatrix} + 0 + 0 = -2 \cdot 1 \cdot (-10) = 20$$

12. $M_{12} = 32, \quad M_{44} = 7$

13.

$$A = \begin{bmatrix} 1 & 0 & 0 & -2 \\ 4 & 1 & 0 & 0 \\ 5 & 6 & 7 & 8 \\ -2 & -3 & -1 & 0 \end{bmatrix} \qquad (a_{24} \text{ is } 0; \quad a_{43} \text{ is } -1)$$

A_{24} is the cofactor of the element a_{24}.

$$A_{24} = (-1)^{2+4} M_{24} \qquad (M_{24} \text{ is the minor of } a_{24})$$

$$= (-1)^6 \begin{vmatrix} 1 & 0 & 0 \\ 5 & 6 & 7 \\ -2 & -3 & -1 \end{vmatrix} \qquad \text{Think: } A = \begin{bmatrix} 1 & 0 & 0 & -2 \\ 4 & 1 & 0 & 0 \\ 5 & 6 & 7 & 8 \\ -2 & -3 & -1 & 0 \end{bmatrix}$$

$$= (1)\left[1(-1)^{1+1} \cdot \begin{vmatrix} 6 & 7 \\ -3 & -1 \end{vmatrix} + 0(-1)^{1+2} \cdot \begin{vmatrix} 5 & 7 \\ -2 & -1 \end{vmatrix} + 0(-1)^{1+3} \cdot \begin{vmatrix} 5 & 6 \\ -2 & -3 \end{vmatrix}\right]$$

(Here we expanded about the first row.)

$$= 1(1)[6(-1) - (-3)7] + 0 + 0 = -6 + 21 = 15$$

13. (continued)

A_{43} is the cofactor of the element a_{43}.

$A_{43} = (-1)^{4+3} M_{43}$ (M_{43} is the minor of a_{43})

$$= (-1)^7 \begin{vmatrix} 1 & 0 & -2 \\ 4 & 1 & 0 \\ 5 & 6 & 8 \end{vmatrix} \qquad \text{Think: } A = \begin{bmatrix} 1 & 0 & 0 & -2 \\ 4 & 1 & 0 & 0 \\ 5 & 6 & 7 & 8 \\ -2 & -3 & -1 & 0 \end{bmatrix}$$

$$= (-1)\left[4(-1)^{2+1} \cdot \begin{vmatrix} 0 & -2 \\ 6 & 8 \end{vmatrix} + 1(-1)^{2+2} \cdot \begin{vmatrix} 1 & -2 \\ 5 & 8 \end{vmatrix} + 0(-1)^{2+3} \cdot \begin{vmatrix} 1 & 0 \\ 5 & 6 \end{vmatrix}\right]$$

(Here we expanded about the second row.)

$= (-1)[4(-1)(0 \cdot 8 - 6(-2)) + 1(1 \cdot 8 - 5(-2)) + 0]$

$= (-1)[-4(12) + 18] = (-1)(-30) = 30$

14. $A_{22} = -10$, $A_{34} = 1$

15.

$$A = \begin{bmatrix} 1 & 0 & 0 & -2 \\ 4 & 1 & 0 & 0 \\ 5 & 6 & 7 & 8 \\ -2 & -3 & -1 & 0 \end{bmatrix}$$

Find $|A|$ by expanding about the first row.

$$|A| = (1)(-1)^{1+1} \cdot \begin{vmatrix} 1 & 0 & 0 \\ 6 & 7 & 8 \\ -3 & -1 & 0 \end{vmatrix} + 0 + 0 + (-2)(-1)^{1+4} \cdot \begin{vmatrix} 4 & 1 & 0 \\ 5 & 6 & 7 \\ -2 & -3 & -1 \end{vmatrix}$$

$$= 1 \cdot \left[1(-1)^{1+1} \cdot \begin{vmatrix} 7 & 8 \\ -1 & 0 \end{vmatrix} + 0 + 0\right] + 2 \cdot \left[4(-1)^{1+1} \cdot \begin{vmatrix} 6 & 7 \\ -3 & -1 \end{vmatrix} + 1(-1)^{1+2} \cdot \begin{vmatrix} 5 & 7 \\ -2 & -1 \end{vmatrix} + 0\right]$$

$= 1[1(0 - (-8))] + 2[4(-6 - (-21)) + (-1)(-5 - (-14))]$

$= 1 \cdot 8 + 2 \cdot (60 - 9) = 8 + 102 = 110$

16. 110

17. $\begin{vmatrix} \boxed{5} & -4 & 2 & -2 \\ 3 & -3 & -4 & 7 \\ -2 & 3 & 2 & 4 \\ -8 & 9 & 5 & -5 \end{vmatrix}$ (Here we will expand about the first column.)

$$= 5(-1)^{1+1} \cdot \begin{vmatrix} -3 & -4 & 7 \\ 3 & 2 & 4 \\ 9 & 5 & -5 \end{vmatrix} + 3(-1)^{2+1} \cdot \begin{vmatrix} -4 & 2 & -2 \\ 3 & 2 & 4 \\ 9 & 5 & -5 \end{vmatrix} +$$

$$(-2)(-1)^{3+1} \cdot \begin{vmatrix} -4 & 2 & -2 \\ -3 & -4 & 7 \\ 9 & 5 & -5 \end{vmatrix} + (-8)(-1)^{4+1} \cdot \begin{vmatrix} -4 & 2 & -2 \\ -3 & -4 & 7 \\ 3 & 2 & 4 \end{vmatrix}$$

$$= 5\left[-3(-1)^{1+1} \cdot \begin{vmatrix} 2 & 4 \\ 5 & -5 \end{vmatrix} + 3(-1)^{2+1} \cdot \begin{vmatrix} -4 & 7 \\ 5 & -5 \end{vmatrix} + 9(-1)^{3+1} \cdot \begin{vmatrix} -4 & 7 \\ 2 & 4 \end{vmatrix}\right] +$$

$$(-3)\left[-4(-1)^{1+1} \cdot \begin{vmatrix} 2 & 4 \\ 5 & -5 \end{vmatrix} + 3(-1)^{2+1} \cdot \begin{vmatrix} 2 & -2 \\ 5 & -5 \end{vmatrix} + 9(-1)^{3+1} \cdot \begin{vmatrix} 2 & -2 \\ 2 & 4 \end{vmatrix}\right] +$$

$$(-2)\left[-4(-1)^{1+1} \cdot \begin{vmatrix} -4 & 7 \\ 5 & -5 \end{vmatrix} + (-3)(-1)^{2+1} \cdot \begin{vmatrix} 2 & -2 \\ 5 & -5 \end{vmatrix} + 9(-1)^{3+1} \cdot \begin{vmatrix} 2 & -2 \\ -4 & 7 \end{vmatrix}\right] +$$

$$8\left[-4(-1)^{1+1} \cdot \begin{vmatrix} -4 & 7 \\ 2 & 4 \end{vmatrix} + (-3)(-1)^{2+1} \cdot \begin{vmatrix} 2 & -2 \\ 2 & 4 \end{vmatrix} + 3(-1)^{3+1} \cdot \begin{vmatrix} 2 & -2 \\ -4 & 7 \end{vmatrix}\right]$$

$= 5[-3(-30) + (-3)(-15) + 9(-30)] + (-3)[-4(-30) + (-3)(0) + 9(12)] +$
$(-2)[-4(-15) + 3(0) + 9(6)] + 8[-4(-30) + 3(12) + 3(6)]$

$= 5(90 + 45 - 270) + (-3)(120 + 0 + 108) + (-2)(60 + 0 + 54) +$
$8(120 + 36 + 18)$

$= 5(-135) + (-3)(228) + (-2)(114) + 8(174) = -675 - 684 - 228 + 1392 = -195$

18. xyzw

19. $\begin{vmatrix} -4 & 5 \\ 6 & 10 \end{vmatrix}$

$= 5\begin{vmatrix} -4 & 1 \\ 6 & 2 \end{vmatrix}$ (Theorem 9, factoring 5 out of the second column)

$= 5 \cdot 2\begin{vmatrix} -4 & 1 \\ 3 & 1 \end{vmatrix}$ (Theorem 9, factoring 2 out of the second row)

$= 5 \cdot 2\begin{vmatrix} -4 & 1 \\ 7 & 0 \end{vmatrix}$ (Theorem 10, multiplying each element of the first row by -1 and adding the products to the corresponding elements of the second row)

$= 5 \cdot 2(0 - 7)$

$= 10(-7) = -70$

NOTATION FOR EVEN-NUMBERED EXERCISES 20 - 30:

D is the given determinant.

$C_2 + 3C_1$ means that 3 times column 1 is added to column 2; etc.

20. -6

21. $\begin{vmatrix} 2 & 1 & 1 \\ 2 & -3 & -1 \\ -4 & 5 & 2 \end{vmatrix}$

$= \begin{vmatrix} 2 & 1 & 1 \\ 4 & -2 & 0 \\ -8 & 3 & 0 \end{vmatrix}$ (Theorem 10, adding the first row to the second row; also multiplying each element of the first row by -2 and adding the products to the corresponding elements of the third row)

$= 1(-1)^{1+3}\begin{vmatrix} 4 & -2 \\ -8 & 3 \end{vmatrix} + 0 + 0$ (Expanding the determinant about the third column)

$= 1(12 - 16) = -4$

22. -9

23. $\begin{vmatrix} 11 & -15 & 20 \\ 16 & 24 & -8 \\ 6 & 9 & 15 \end{vmatrix}$

$= 8\cdot\begin{vmatrix} 11 & -15 & 20 \\ 2 & 3 & -1 \\ 6 & 9 & 15 \end{vmatrix} = 8\cdot 3\begin{vmatrix} 11 & -15 & 20 \\ 2 & 3 & -1 \\ 2 & 3 & 5 \end{vmatrix}$ (Theorem 9, factoring 8 out of the second row and 3 out of the third row)

$= 8\cdot 3\cdot 3\begin{vmatrix} 11 & -5 & 20 \\ 2 & 1 & -1 \\ 2 & 1 & 5 \end{vmatrix}$ (Theorem 9, factoring 3 out of the second column)

$= 8\cdot 3\cdot 3\begin{vmatrix} 11 & -5 & 20 \\ 2 & 1 & -1 \\ 0 & 0 & 6 \end{vmatrix}$ (Theorem 10, multiplying each element of the second row by -1 and adding the products to the corresponding elements of the third row)

$= 8\cdot 3\cdot 3\cdot 6\begin{vmatrix} 11 & -5 & 20 \\ 2 & 1 & -1 \\ 0 & 0 & 1 \end{vmatrix}$ (Theorem 9 factoring 6 out of the third row)

$= 8\cdot 3\cdot 3\cdot 6\left[0 + 0 + 1(-1)^{3+3}\begin{vmatrix} 11 & -5 \\ 2 & 1 \end{vmatrix}\right]$ (Expanding the determinant about the third row)

$= 8\cdot 3\cdot 3\cdot 6\cdot 21 = 9072$

24. -546

25. $\begin{vmatrix} -3 & 0 & 2 & 6 \\ 2 & 4 & 0 & -1 \\ -1 & 0 & -5 & 2 \\ 0 & -1 & -2 & -3 \end{vmatrix}$

$= \begin{vmatrix} 0 & 0 & 17 & 0 \\ 2 & 4 & 0 & -1 \\ -1 & 0 & -5 & 2 \\ 0 & -1 & -2 & -3 \end{vmatrix}$ (Theorem 10, multiplying each element of the third row by -3 and adding the products to the corresponding elements of the first row)

$= 17\begin{vmatrix} 0 & 0 & 1 & 0 \\ 2 & 4 & 0 & -1 \\ -1 & 0 & -5 & 2 \\ 0 & -1 & -2 & -3 \end{vmatrix}$ (Theorem 9, factoring 17 out of the first row)

$= 17\left(0 + 0 + 1(-1)^{1+3}\begin{vmatrix} 2 & 4 & -1 \\ -1 & 0 & 2 \\ 0 & -1 & -3 \end{vmatrix} + 0\right)$ (Expanding the determinant about the first row)

$= -17\begin{vmatrix} 2 & 4 & -1 \\ 1 & 0 & -2 \\ 0 & -1 & -3 \end{vmatrix}$ (Theorem 9, factoring -1 out of the second row)

$= -17\begin{vmatrix} 0 & 4 & 3 \\ 1 & 0 & -2 \\ 0 & -1 & -3 \end{vmatrix}$ (Theorem 10, multiplying each element of the second row by -2 and adding the products to the corresponding elements in the first row)

$= -17\left[0 + 1(-1)^{2+1}\begin{vmatrix} 4 & 3 \\ -1 & -3 \end{vmatrix} + 0\right]$ (Expanding the determinant about the first column)

$= -17\cdot(-1)\cdot(-12 + 3) = 17(-9) = -153$

26. 746

27. $\begin{vmatrix} x & y & z \\ 0 & 0 & 0 \\ p & q & r \end{vmatrix} = 0$ (Theorem 6)

28. 0

29. $\begin{vmatrix} 2a & t & -7a \\ 2b & u & -7b \\ 2c & v & -7c \end{vmatrix}$

$= 2(-7)\begin{vmatrix} a & t & a \\ b & u & b \\ c & v & c \end{vmatrix}$ (Theorem 9, factoring 2 out of the first column and -7 out of the third column)

$= 2\cdot(-7)\cdot 0 = 0$ (Theorem 8; columns 1 and 3 are the same)

30. 0

31. $\begin{vmatrix} x^2 & x & 1 \\ y^2 & y & 1 \\ z^2 & z & 1 \end{vmatrix}$

$= \begin{vmatrix} x^2 - y^2 & x - y & 0 \\ y^2 & y & 1 \\ z^2 - y^2 & z - y & 0 \end{vmatrix}$ (Theorem 10, multiplying each element of the second row by -1 and adding the products to the corresponding elements of the first and third rows)

$= (x - y)(z - y)\begin{vmatrix} x + y & 1 & 0 \\ y^2 & y & 1 \\ z + y & 1 & 0 \end{vmatrix}$ (Theorem 9, factoring x - y out of the first row and z - y out of the third row)

$= (x - y)(z - y)\begin{vmatrix} x - z & 0 & 0 \\ y^2 & y & 1 \\ z + y & 1 & 0 \end{vmatrix}$ (Theorem 10, multiplying each element of the third row by -1 and adding the products to the corresponding elements of the first row)

$= (x - y)(z - y)(x - z)(-1)$ (Expanding the determinant about the first row)

$= (x - y)(z - y)(z - x)$, or $(x - y)(y - z)(x - z)$, or $(y - x)(z - y)(x - z)$

32. $(a - b)(b - c)(c - a)$

33. $\begin{vmatrix} x & x^2 & x^3 \\ y & y^2 & y^3 \\ z & z^2 & z^3 \end{vmatrix}$

$= xyz\cdot\begin{vmatrix} 1 & x & x^2 \\ 1 & y & y^2 \\ 1 & z & z^2 \end{vmatrix}$ (Theorem 9, factoring x out of the first row, y out of the second row, and z out of the third row)

$= xyz\cdot\begin{vmatrix} 0 & x - z & x^2 - z^2 \\ 0 & y - z & y^2 - z^2 \\ 1 & z & z^2 \end{vmatrix}$ (Theorem 10, multiplying each element of the third row by -1 and adding the products to the corresponding elements of of the first and second rows)

$= xyz(x - z)(y - z)\cdot\begin{vmatrix} 0 & 1 & x + z \\ 0 & 1 & y + z \\ 1 & z & z^2 \end{vmatrix}$ (Theorem 9, factoring x - z out of the first row and y - z out of the second row)

$= xyz(x - z)(y - z)\cdot\begin{vmatrix} 0 & 0 & x - y \\ 0 & 1 & y + z \\ 1 & z & z^2 \end{vmatrix}$ (Theorem 10, multiplying each element of the second row by -1 and adding the products to the corresponding elements of the first row)

$= xyz(x - z)(y - z)(x - y)\cdot\begin{vmatrix} 0 & 0 & 1 \\ 0 & 1 & y + z \\ 1 & z & z^2 \end{vmatrix}$ (Theorem 9, factoring x - y out of the first row)

$= xyz(x - z)(y - z)(x - y)\cdot\left[0 + 0 + 1(-1)^{1+3}\begin{vmatrix} 0 & 1 \\ 1 & z \end{vmatrix}\right]$ (Expanding the determinant about the first row)

$= xyz(x - z)(y - z)(x - y)(-1)$ $\left[\begin{vmatrix} 0 & 1 \\ 1 & z \end{vmatrix} = 0 - 1 = -1\right]$

33. (continued)

$= -xyz(x - z)(y - z)(x - y)$
or $xyz(z - x)(y - z)(x - y)$
or $xyz(x - z)(z - y)(x - y)$
or $xyz(x - z)(y - z)(y - x)$

34. $(a - b)(b - c)(c - a)(a + b + c)$

35. $\begin{vmatrix} x & y & 1 \\ x_1 & y_1 & 1 \\ x_2 & y_2 & 1 \end{vmatrix} = 0$

We first evaluate the determinant. Here we evaluate by expanding about the 1st row.

$$x(-1)^{1+1} \cdot \begin{vmatrix} y_1 & 1 \\ y_2 & 1 \end{vmatrix} + y(-1)^{1+2} \cdot \begin{vmatrix} x_1 & 1 \\ x_2 & 1 \end{vmatrix} + 1(-1)^{1+3} \cdot \begin{vmatrix} x_1 & y_1 \\ x_2 & y_2 \end{vmatrix} = 0$$

$$x(y_1 - y_2) - y(x_1 - x_2) + 1(x_1y_2 - x_2y_1) = 0$$

$$xy_1 - xy_2 - yx_1 + yx_2 + x_1y_2 - x_2y_1 = 0$$

The two-point equation of the line that contains the points (x_1, y_1) and (x_2, y_2) is

$y - y_1 = \frac{y_2 - y_1}{x_2 - x_1}(x - x_1)$

$(y - y_1)(x_2 - x_1) = (y_2 - y_1)(x - x_1)$ (Multiplying by $x_2 - x_1$)

$yx_2 - yx_1 - y_1x_2 + y_1x_1 = y_2x - y_2x_1 - y_1x + y_1x_1$ (Using FOIL)

$yx_2 - yx_1 - y_1x_2 + y_1x_1 - y_2x + y_2x_1 + y_1x - y_1x_1 = 0$

$xy_1 - xy_2 - yx_1 + yx_2 + x_1y_2 - x_2y_1 = 0$

Thus $\begin{vmatrix} x & y & 1 \\ x_1 & y_1 & 1 \\ x_2 & y_2 & 1 \end{vmatrix} = 0$ and $y - y_1 = \frac{y_2 - y_1}{x_2 - x_1}(x - x_1)$ are equivalent.

36. From Exercise 35, we know that the equation of line ℓ through (x_2, y_2), (x_3, y_3) is $\begin{vmatrix} x & y & 1 \\ x_2 & y_2 & 1 \\ x_3 & y_3 & 1 \end{vmatrix} = 0.$

Then (x_1, y_1) also lies on ℓ if and only if

$\begin{vmatrix} x_1 & y_1 & 1 \\ x_2 & y_2 & 1 \\ x_3 & y_3 & 1 \end{vmatrix} = 0.$

37. See the answer section in the text.

38. NOTICE: Parallel lines, by definition, are not coincident;
E_1, E_2 are the given equations;
m_1, m_2 are slopes;
D, D_x, D_y are the given determinants.

(I) If D = 0, then (3) $a_1b_2 - a_2b_1 = 0$. If $b_1, b_2 \neq 0$ then $\frac{-a_1}{b_1} = \frac{-a_2}{b_2}$ and $m_1 = m_2$. Thus, ℓ_1 and ℓ_2 are either parallel or coincident. If either of b_1, b_2 is 0, so is the other, by (3); and both lines are vertical and either parallel or coincident.

(II) If the lines coincide, $E_2 = k \cdot E_1$, $k \neq 0$; and
ℓ_1: $a_1x + b_1y = c_1$,
ℓ_2: $ka_1x + kb_1y = kc_1$.

Then $D_x = 0$ and $D_y = 0$. Thus a sufficient condition for distinct lines is $D_x \neq 0$ or $D_y \neq 0$.

In conclusion, sufficient conditions for two lines to be parallel are: $D = 0$ and $D_x \neq 0$ or $D_y \neq 0$.

Exercise Set 9.8

1.
$$A = \begin{bmatrix} 3 & 2 \\ 5 & 3 \end{bmatrix}$$

$$\begin{bmatrix} 3 & 2 & 1 & 0 \\ 5 & 3 & 0 & 1 \end{bmatrix} = \begin{bmatrix} 3 & 2 & 1 & 0 \\ 15 & 9 & 0 & 3 \end{bmatrix} = \begin{bmatrix} 3 & 2 & 1 & 0 \\ 0 & -1 & -5 & 3 \end{bmatrix} =$$

$$\begin{bmatrix} 3 & 2 & 1 & 0 \\ 0 & 1 & 5 & -3 \end{bmatrix} = \begin{bmatrix} 3 & 0 & -9 & 6 \\ 0 & 1 & 5 & -3 \end{bmatrix} = \begin{bmatrix} 1 & 0 & -3 & 2 \\ 0 & 1 & 5 & -3 \end{bmatrix}$$

$$A^{-1} = \begin{bmatrix} -3 & 2 \\ 5 & -3 \end{bmatrix}$$

2.
$$\begin{bmatrix} 2 & -5 \\ -1 & 3 \end{bmatrix}$$

3.
$$A = \begin{bmatrix} 11 & 3 \\ 7 & 2 \end{bmatrix}$$

$$\begin{bmatrix} 11 & 3 & 1 & 0 \\ 7 & 2 & 0 & 1 \end{bmatrix} = \begin{bmatrix} 7 & 2 & 0 & 1 \\ 11 & 3 & 1 & 0 \end{bmatrix} = \begin{bmatrix} 7 & 2 & 0 & 1 \\ 77 & 21 & 7 & 0 \end{bmatrix} =$$

$$\begin{bmatrix} 7 & 2 & 0 & 1 \\ 0 & -1 & 7 & -11 \end{bmatrix} = \begin{bmatrix} 7 & 2 & 0 & 1 \\ 0 & 1 & -7 & 11 \end{bmatrix} = \begin{bmatrix} 7 & 0 & 14 & -21 \\ 0 & 1 & -7 & 11 \end{bmatrix} =$$

$$\begin{bmatrix} 1 & 0 & 2 & -3 \\ 0 & 1 & -7 & 11 \end{bmatrix} \qquad A^{-1} = \begin{bmatrix} 2 & -3 \\ -7 & 11 \end{bmatrix}$$

4. $\begin{bmatrix} -3 & 5 \\ 5 & -8 \end{bmatrix}$

5. $A = \begin{bmatrix} 4 & -3 \\ 1 & 2 \end{bmatrix}$

$$\begin{bmatrix} 4 & -3 & 1 & 0 \\ 1 & 2 & 0 & 1 \end{bmatrix} = \begin{bmatrix} 1 & 2 & 0 & 1 \\ 4 & -3 & 1 & 0 \end{bmatrix} = \begin{bmatrix} 1 & 2 & 0 & 1 \\ 0 & -11 & 1 & -4 \end{bmatrix} =$$

$$\begin{bmatrix} 1 & 2 & 0 & 1 \\ 0 & 11 & -1 & 4 \end{bmatrix} = \begin{bmatrix} 11 & 22 & 0 & 11 \\ 0 & 11 & -1 & 4 \end{bmatrix} = \begin{bmatrix} 11 & 0 & 2 & 3 \\ 0 & 11 & -1 & 4 \end{bmatrix} =$$

$$\begin{bmatrix} 1 & 0 & \frac{2}{11} & \frac{3}{11} \\ 0 & 1 & -\frac{1}{11} & \frac{4}{11} \end{bmatrix}$$

$$A^{-1} = \begin{bmatrix} \frac{2}{11} & \frac{3}{11} \\ -\frac{1}{11} & \frac{4}{11} \end{bmatrix}$$

6. $\begin{bmatrix} 0 & 1 \\ -1 & 0 \end{bmatrix}$

7. $A = \begin{bmatrix} 3 & 1 & 0 \\ 1 & 1 & 1 \\ 1 & -1 & 2 \end{bmatrix}$

$$\begin{bmatrix} 3 & 1 & 0 & 1 & 0 & 0 \\ 1 & 1 & 1 & 0 & 1 & 0 \\ 1 & -1 & 2 & 0 & 0 & 1 \end{bmatrix} = \begin{bmatrix} 1 & 1 & 1 & 0 & 1 & 0 \\ 3 & 1 & 0 & 1 & 0 & 0 \\ 1 & -1 & 2 & 0 & 0 & 1 \end{bmatrix} =$$

$$\begin{bmatrix} 1 & 1 & 1 & 0 & 1 & 0 \\ 0 & -2 & -3 & 1 & -3 & 0 \\ 0 & -2 & 1 & 0 & -1 & 1 \end{bmatrix} = \begin{bmatrix} 1 & 1 & 1 & 0 & 1 & 0 \\ 0 & 2 & 3 & -1 & 3 & 0 \\ 0 & -2 & 1 & 0 & -1 & 1 \end{bmatrix} =$$

$$\begin{bmatrix} 1 & 1 & 1 & 0 & 1 & 0 \\ 0 & 2 & 3 & -1 & 3 & 0 \\ 0 & 0 & 4 & -1 & 2 & 1 \end{bmatrix} = \begin{bmatrix} 4 & 4 & 4 & 0 & 4 & 0 \\ 0 & 8 & 12 & -4 & 12 & 0 \\ 0 & 0 & 4 & -1 & 2 & 1 \end{bmatrix} =$$

$$\begin{bmatrix} 4 & 4 & 0 & 1 & 2 & -1 \\ 0 & 8 & 0 & -1 & 6 & -3 \\ 0 & 0 & 4 & -1 & 2 & 1 \end{bmatrix} = \begin{bmatrix} 8 & 8 & 0 & 2 & 4 & -2 \\ 0 & 8 & 0 & -1 & 6 & -3 \\ 0 & 0 & 4 & -1 & 2 & 1 \end{bmatrix} =$$

7. (continued)

$$\begin{bmatrix} 8 & 0 & 0 & 3 & -2 & 1 \\ 0 & 8 & 0 & -1 & 6 & -3 \\ 0 & 0 & 4 & -1 & 2 & 1 \end{bmatrix} = \begin{bmatrix} 1 & 0 & 0 & \frac{3}{8} & -\frac{1}{4} & \frac{1}{8} \\ 0 & 1 & 0 & -\frac{1}{8} & \frac{3}{4} & -\frac{3}{8} \\ 0 & 0 & 1 & -\frac{1}{4} & \frac{1}{2} & \frac{1}{4} \end{bmatrix}$$

$$A^{-1} = \begin{bmatrix} \frac{3}{8} & -\frac{1}{4} & \frac{1}{8} \\ -\frac{1}{8} & \frac{3}{4} & -\frac{3}{8} \\ -\frac{1}{4} & \frac{1}{2} & \frac{1}{4} \end{bmatrix}$$

8. $$\begin{bmatrix} -\frac{1}{2} & \frac{1}{2} & \frac{1}{2} \\ 1 & 0 & -1 \\ \frac{3}{2} & -\frac{1}{2} & -\frac{1}{2} \end{bmatrix}$$

9. $$A = \begin{bmatrix} 1 & -1 & 2 \\ 0 & 1 & 3 \\ 2 & 1 & -2 \end{bmatrix}$$

$$\begin{bmatrix} 1 & -1 & 2 & 1 & 0 & 0 \\ 0 & 1 & 3 & 0 & 1 & 0 \\ 2 & 1 & -2 & 0 & 0 & 1 \end{bmatrix} = \begin{bmatrix} 1 & -1 & 2 & 1 & 0 & 0 \\ 0 & 1 & 3 & 0 & 1 & 0 \\ 0 & 3 & -6 & -2 & 0 & 1 \end{bmatrix} =$$

$$\begin{bmatrix} 1 & -1 & 2 & 1 & 0 & 0 \\ 0 & 1 & 3 & 0 & 1 & 0 \\ 0 & 0 & -15 & -2 & -3 & 1 \end{bmatrix} = \begin{bmatrix} 15 & -15 & 30 & 15 & 0 & 0 \\ 0 & 5 & 15 & 0 & 5 & 0 \\ 0 & 0 & 15 & 2 & 3 & -1 \end{bmatrix} =$$

$$\begin{bmatrix} 15 & -15 & 0 & 11 & -6 & 2 \\ 0 & 5 & 0 & -2 & 2 & 1 \\ 0 & 0 & 15 & 2 & 3 & -1 \end{bmatrix} = \begin{bmatrix} 15 & 0 & 0 & 5 & 0 & 5 \\ 0 & 5 & 0 & -2 & 2 & 1 \\ 0 & 0 & 15 & 2 & 3 & -1 \end{bmatrix} =$$

$$\begin{bmatrix} 1 & 0 & 0 & \frac{1}{3} & 0 & \frac{1}{3} \\ 0 & 1 & 0 & -\frac{2}{5} & \frac{2}{5} & \frac{1}{5} \\ 0 & 0 & 1 & \frac{2}{15} & \frac{3}{15} & -\frac{1}{15} \end{bmatrix}$$

$$A^{-1} = \begin{bmatrix} \frac{1}{3} & 0 & \frac{1}{3} \\ -\frac{2}{5} & \frac{2}{5} & \frac{1}{5} \\ \frac{2}{15} & \frac{1}{5} & -\frac{1}{15} \end{bmatrix}$$

10. $\begin{bmatrix} -1 & 5 & 2 \\ -1 & 3 & 1 \\ \frac{1}{2} & -1 & -\frac{1}{2} \end{bmatrix}$

11. We cannot obtain the identity matrix on the left using the Gauss-Jordan reduction method. Thus, A^{-1} does not exist.

12. Does not exist

13.
$$A = \begin{bmatrix} 1 & 2 & 3 & 4 \\ 0 & 1 & 3 & -5 \\ 0 & 0 & 1 & -2 \\ 0 & 0 & 0 & -1 \end{bmatrix}$$

$$\begin{bmatrix} 1 & 2 & 3 & 4 & 1 & 0 & 0 & 0 \\ 0 & 1 & 3 & -5 & 0 & 1 & 0 & 0 \\ 0 & 0 & 1 & -2 & 0 & 0 & 1 & 0 \\ 0 & 0 & 0 & -1 & 0 & 0 & 0 & 1 \end{bmatrix} = \begin{bmatrix} 1 & 2 & 3 & 4 & 1 & 0 & 0 & 0 \\ 0 & 1 & 3 & -5 & 0 & 1 & 0 & 0 \\ 0 & 0 & 1 & -2 & 0 & 0 & 1 & 0 \\ 0 & 0 & 0 & 1 & 0 & 0 & 0 & -1 \end{bmatrix} =$$

$$\begin{bmatrix} 1 & 2 & 3 & 0 & 1 & 0 & 0 & 4 \\ 0 & 1 & 3 & 0 & 0 & 1 & 0 & -5 \\ 0 & 0 & 1 & 0 & 0 & 0 & 1 & -2 \\ 0 & 0 & 0 & 1 & 0 & 0 & 0 & -1 \end{bmatrix} = \begin{bmatrix} 1 & 2 & 0 & 0 & 1 & 0 & -3 & 10 \\ 0 & 1 & 0 & 0 & 0 & 1 & -3 & 1 \\ 0 & 0 & 1 & 0 & 0 & 0 & 1 & -2 \\ 0 & 0 & 0 & 1 & 0 & 0 & 0 & -1 \end{bmatrix} =$$

$$\begin{bmatrix} 1 & 0 & 0 & 0 & 1 & -2 & 3 & 8 \\ 0 & 1 & 0 & 0 & 0 & 1 & -3 & 1 \\ 0 & 0 & 1 & 0 & 0 & 0 & 1 & -2 \\ 0 & 0 & 0 & 1 & 0 & 0 & 0 & -1 \end{bmatrix}$$

$$A^{-1} = \begin{bmatrix} 1 & -2 & 3 & 8 \\ 0 & 1 & -3 & 1 \\ 0 & 0 & 1 & -2 \\ 0 & 0 & 0 & -1 \end{bmatrix}$$

14. Does not exist

15. We cannot obtain the identity matrix on the left using the Gauss-Jordan reduction method. Then, A^{-1} does not exist.

16. Does not exist

17. $11x + 3y = -4,$
$7x + 2y = 5$

We write a matrix equation equivalent to this system.

$$\begin{bmatrix} 11 & 3 \\ 7 & 2 \end{bmatrix}\begin{bmatrix} x \\ y \end{bmatrix} = \begin{bmatrix} -4 \\ 5 \end{bmatrix}$$

We solve the matrix equation by multiplying by A^{-1} on the left on each side.

$$\begin{bmatrix} 2 & -3 \\ -7 & 11 \end{bmatrix}\begin{bmatrix} 11 & 3 \\ 7 & 2 \end{bmatrix}\begin{bmatrix} x \\ y \end{bmatrix} = \begin{bmatrix} 2 & -3 \\ -7 & 11 \end{bmatrix}\begin{bmatrix} -4 \\ 5 \end{bmatrix}$$

$$\begin{bmatrix} 1 & 0 \\ 0 & 1 \end{bmatrix}\begin{bmatrix} x \\ y \end{bmatrix} = \begin{bmatrix} -23 \\ 83 \end{bmatrix}$$

$$\begin{bmatrix} x \\ y \end{bmatrix} = \begin{bmatrix} -23 \\ 83 \end{bmatrix}$$

The solution of the system of equations is (-23, 83).

18. (28, -46)

19. $3x + y = 2,$
$2x - y + 2z = -5,$
$x + y + z = 5$

We write a matrix equation equivalent to this system.

$$\begin{bmatrix} 3 & 1 & 0 \\ 2 & -1 & 2 \\ 1 & 1 & 1 \end{bmatrix}\begin{bmatrix} x \\ y \\ z \end{bmatrix} = \begin{bmatrix} 2 \\ -5 \\ 5 \end{bmatrix}$$

We solve the matrix equation by multiplying by A^{-1} on the left on each side.

$$\frac{1}{9}\begin{bmatrix} 3 & 1 & -2 \\ 0 & -3 & 6 \\ -3 & 2 & 5 \end{bmatrix}\begin{bmatrix} 3 & 1 & 0 \\ 2 & -1 & 2 \\ 1 & 1 & 1 \end{bmatrix}\begin{bmatrix} x \\ y \\ z \end{bmatrix} = \frac{1}{9}\begin{bmatrix} 3 & 1 & -2 \\ 0 & -3 & 6 \\ -3 & 2 & 5 \end{bmatrix}\begin{bmatrix} 2 \\ -5 \\ 5 \end{bmatrix}$$

$$\begin{bmatrix} 1 & 0 & 0 \\ 0 & 1 & 0 \\ 0 & 0 & 1 \end{bmatrix}\begin{bmatrix} x \\ y \\ z \end{bmatrix} = \frac{1}{9}\begin{bmatrix} -9 \\ 45 \\ 9 \end{bmatrix}$$

$$\begin{bmatrix} x \\ y \\ z \end{bmatrix} = \begin{bmatrix} -1 \\ 5 \\ 1 \end{bmatrix}$$

The solution of the system of equations is (-1, 5, 1).

20. (1, -7, -3)

21. $4x - 3y = 2,$ $\begin{bmatrix} 4 & -3 \\ 1 & 2 \end{bmatrix}\begin{bmatrix} x \\ y \end{bmatrix} = \begin{bmatrix} 2 \\ 1 \end{bmatrix}$
$x + 2y = -1$

The coefficient matrix is the matrix in Exercise 5. Its inverse is

$$\begin{bmatrix} \frac{2}{11} & \frac{3}{11} \\ -\frac{1}{11} & \frac{4}{11} \end{bmatrix}.$$

$$\begin{bmatrix} x \\ y \end{bmatrix} = \begin{bmatrix} \frac{2}{11} & \frac{3}{11} \\ -\frac{1}{11} & \frac{4}{11} \end{bmatrix}\begin{bmatrix} 2 \\ -1 \end{bmatrix} = \begin{bmatrix} \frac{1}{11} \\ -\frac{6}{11} \end{bmatrix}$$

The solution of the system is $\left(\frac{1}{11}, -\frac{6}{11}\right)$.

22.
$$\begin{bmatrix} 3 & 5 \\ 2 & 4 \end{bmatrix}\begin{bmatrix} x \\ y \end{bmatrix} = \begin{bmatrix} -4 \\ -2 \end{bmatrix}, A^{-1} = \begin{bmatrix} 2 & -\frac{5}{2} \\ -1 & \frac{3}{2} \end{bmatrix}, (-3, 1)$$

23. $7x - 2y = -3,$ $\begin{bmatrix} 7 & -2 \\ 9 & 3 \end{bmatrix}\begin{bmatrix} x \\ y \end{bmatrix} = \begin{bmatrix} -3 \\ 4 \end{bmatrix}$
$9x + 3y = 4$

The inverse of the coefficient matrix is $\begin{bmatrix} \frac{1}{13} & \frac{2}{39} \\ -\frac{3}{13} & \frac{7}{39} \end{bmatrix}$.

$$\begin{bmatrix} x \\ y \end{bmatrix} = \begin{bmatrix} \frac{1}{13} & \frac{2}{39} \\ -\frac{3}{13} & \frac{7}{39} \end{bmatrix}\begin{bmatrix} -3 \\ 4 \end{bmatrix} = \begin{bmatrix} -\frac{1}{39} \\ \frac{55}{39} \end{bmatrix}$$

The solution of the system of equations is $\left(-\frac{1}{39}, \frac{55}{39}\right)$.

24.
$$\begin{bmatrix} 5 & 3 \\ 4 & -1 \end{bmatrix}\begin{bmatrix} x \\ y \end{bmatrix} = \begin{bmatrix} -2 \\ 1 \end{bmatrix}, A^{-1} = \frac{1}{17}\begin{bmatrix} 1 & 3 \\ 4 & -5 \end{bmatrix}, \left(\frac{1}{17}, -\frac{13}{17}\right)$$

25. $x + z = 1,$
$2x + y = 3,$
$x - y + z = 4$

$$\begin{bmatrix} 1 & 0 & 1 \\ 2 & 1 & 0 \\ 1 & -1 & 1 \end{bmatrix}\begin{bmatrix} x \\ y \\ z \end{bmatrix} = \begin{bmatrix} 1 \\ 3 \\ 4 \end{bmatrix}$$

The inverse of the coefficient matrix is

$$\frac{1}{2}\begin{bmatrix} -1 & 1 & 1 \\ 2 & 0 & -2 \\ 3 & -1 & -1 \end{bmatrix}.$$

$$\begin{bmatrix} x \\ y \\ z \end{bmatrix} = \frac{1}{2}\begin{bmatrix} -1 & 1 & 1 \\ 2 & 0 & -2 \\ 3 & -1 & -1 \end{bmatrix}\begin{bmatrix} 1 \\ 3 \\ 4 \end{bmatrix} = \begin{bmatrix} 3 \\ -3 \\ -2 \end{bmatrix}$$

The solution of the system of equations is (3, -3, -2).

26. $\begin{bmatrix} 1 & 2 & 3 \\ 2 & -3 & 4 \\ -3 & 5 & -6 \end{bmatrix}\begin{bmatrix} x \\ y \\ z \end{bmatrix} = \begin{bmatrix} -1 \\ 2 \\ 4 \end{bmatrix}$, $A^{-1} = \begin{bmatrix} -2 & 27 & 17 \\ 0 & 3 & 2 \\ 1 & -11 & -7 \end{bmatrix}$,

(124, 14, -51)

27. $2w - 3x + 4y - 5z = 0,$
$3w - 2x + 7y - 3z = 2,$
$w + x - y + z = 1,$
$-w - 3x - 6y + 4z = 6$

$$\begin{bmatrix} 2 & -3 & 4 & -5 \\ 3 & -2 & 7 & -3 \\ 1 & 1 & -1 & 1 \\ -1 & -3 & -6 & 4 \end{bmatrix}\begin{bmatrix} w \\ x \\ y \\ z \end{bmatrix} = \begin{bmatrix} 0 \\ 2 \\ 1 \\ 6 \end{bmatrix}$$

The inverse of the coefficient matrix is

$$\frac{1}{11,165}\begin{bmatrix} 1430 & 605 & 6985 & 495 \\ 440 & -1045 & 2145 & -1870 \\ -2035 & 2145 & -2640 & -275 \\ -3025 & 2585 & -605 & 1100 \end{bmatrix}.$$

$$\begin{bmatrix} w \\ x \\ y \\ z \end{bmatrix} = \frac{1}{11,165}\begin{bmatrix} 1430 & 605 & 6985 & 495 \\ 440 & -1045 & 2145 & -1870 \\ -2035 & 2145 & -2640 & -275 \\ -3025 & 2585 & -605 & 1100 \end{bmatrix}\begin{bmatrix} 0 \\ 2 \\ 1 \\ 6 \end{bmatrix}$$

$$= \begin{bmatrix} 1 \\ -1 \\ 0 \\ 1 \end{bmatrix}$$

The solution of the system of equations is (1, -1, 0, 1).

28. $$\begin{bmatrix} 5 & -4 & 3 & -2 \\ 1 & 4 & -2 & 3 \\ 2 & -3 & 6 & -9 \\ 3 & -5 & 2 & -4 \end{bmatrix}\begin{bmatrix} w \\ x \\ y \\ z \end{bmatrix} = \begin{bmatrix} -6 \\ -5 \\ 14 \\ -3 \end{bmatrix}$$

$$A^{-1} = \frac{1}{4,218,336}\begin{bmatrix} 251,424 & 1,047,600 & 13,968 & 628,560 \\ -139,680 & 824,112 & 460,944 & -349,200 \\ 1,564,416 & -1,215,216 & 321,264 & -2,416,464 \\ 1,145,376 & -852,048 & -405,072 & 1,354,896 \end{bmatrix}$$

The solution of the system of equations is (-2, 1, 2, -1).

29. $$\begin{bmatrix} a & b & c \\ d & e & f \\ g & h & i \end{bmatrix}\begin{bmatrix} 1 & 0 & 0 \\ 0 & 1 & 0 \\ 0 & 0 & 1 \end{bmatrix} = \begin{bmatrix} a & b & c \\ d & e & f \\ g & h & i \end{bmatrix}, \qquad AI = A$$

$$\begin{bmatrix} 1 & 0 & 0 \\ 0 & 1 & 0 \\ 0 & 0 & 1 \end{bmatrix}\begin{bmatrix} a & b & c \\ d & e & f \\ g & h & i \end{bmatrix} = \begin{bmatrix} a & b & c \\ d & e & f \\ g & h & i \end{bmatrix} \qquad IA = A$$

30. A^{-1} exists if and only if $x \neq 0$.

$A^{-1} = \left[\frac{1}{x}\right]$

31.
$$A = \begin{bmatrix} x & 0 \\ 0 & y \end{bmatrix}$$

A^{-1} exists if and only if $xy \neq 0$.

$$\left[\begin{array}{cccc} x & 0 & 1 & 0 \\ 0 & y & 0 & 1 \end{array}\right]$$

Multiply the first row by $\frac{1}{x}$ and the second row by $\frac{1}{y}$.

$$\left[\begin{array}{cccc} 1 & 0 & \frac{1}{x} & 0 \\ 0 & 1 & 0 & \frac{1}{y} \end{array}\right]$$

$$A^{-1} = \begin{bmatrix} \frac{1}{x} & 0 \\ 0 & \frac{1}{y} \end{bmatrix}$$

32. A^{-1} exists if and only if $xyz \neq 0$.

$$A^{-1} = \begin{bmatrix} 0 & 0 & \frac{1}{z} \\ 0 & \frac{1}{y} & 0 \\ \frac{1}{x} & 0 & 0 \end{bmatrix}$$

33.
$$A = \begin{bmatrix} x & 1 & 1 & 1 \\ 0 & y & 0 & 0 \\ 0 & 0 & z & 0 \\ 0 & 0 & 0 & w \end{bmatrix}$$

A^{-1} exists if and only if $xyzw \neq 0$.

$$\left[\begin{array}{cccccccc} x & 1 & 1 & 1 & 1 & 0 & 0 & 0 \\ 0 & y & 0 & 0 & 0 & 1 & 0 & 0 \\ 0 & 0 & z & 0 & 0 & 0 & 1 & 0 \\ 0 & 0 & 0 & w & 0 & 0 & 0 & 1 \end{array}\right]$$

Multiply the first row by $-w$.

$$\left[\begin{array}{cccccccc} -xw & -w & -w & -w & -w & 0 & 0 & 0 \\ 0 & y & 0 & 0 & 0 & 1 & 0 & 0 \\ 0 & 0 & z & 0 & 0 & 0 & 1 & 0 \\ 0 & 0 & 0 & w & 0 & 0 & 0 & 1 \end{array}\right]$$

Add the fourth row to the first row.

$$\left[\begin{array}{cccccccc} -xw & -w & -w & 0 & -w & 0 & 0 & 1 \\ 0 & y & 0 & 0 & 0 & 1 & 0 & 0 \\ 0 & 0 & z & 0 & 0 & 0 & 1 & 0 \\ 0 & 0 & 0 & w & 0 & 0 & 0 & 1 \end{array}\right]$$

33. (continued)

Multiply the first row by -z.

$$\begin{bmatrix} xzw & zw & -zw & 0 & zw & 0 & 0 & -z \\ 0 & y & 0 & 0 & 0 & 1 & 0 & 0 \\ 0 & 0 & z & 0 & 0 & 0 & 1 & 0 \\ 0 & 0 & 0 & w & 0 & 0 & 0 & 1 \end{bmatrix}$$

Multiply the third row by w and add it to the first row.

$$\begin{bmatrix} xzw & zw & 0 & 0 & zw & 0 & w & -z \\ 0 & y & 0 & 0 & 0 & 1 & 0 & 0 \\ 0 & 0 & z & 0 & 0 & 0 & 1 & 0 \\ 0 & 0 & 0 & w & 0 & 0 & 0 & 1 \end{bmatrix}$$

Multiply the first row by -y.

$$\begin{bmatrix} -xzyw & -zyw & 0 & 0 & -zyw & 0 & -wy & yz \\ 0 & y & 0 & 0 & 0 & 1 & 0 & 0 \\ 0 & 0 & z & 0 & 0 & 0 & 1 & 0 \\ 0 & 0 & 0 & w & 0 & 0 & 0 & 1 \end{bmatrix}$$

Multiply the second row by zw and add it to the first row.

$$\begin{bmatrix} -xzyw & 0 & 0 & 0 & -zyw & zw & -wy & yz \\ 0 & y & 0 & 0 & 0 & 1 & 0 & 0 \\ 0 & 0 & z & 0 & 0 & 0 & 1 & 0 \\ 0 & 0 & 0 & w & 0 & 0 & 0 & 1 \end{bmatrix}$$

Multiply each row by the appropriate factor to get ones on the main diagonal.

$$\begin{bmatrix} 1 & 0 & 0 & 0 & \frac{1}{x} & -\frac{1}{xy} & \frac{1}{xz} & -\frac{1}{xw} \\ 0 & 1 & 0 & 0 & 0 & \frac{1}{y} & 0 & 0 \\ 0 & 0 & 1 & 0 & 0 & 0 & \frac{1}{z} & 0 \\ 0 & 0 & 0 & 1 & 0 & 0 & 0 & \frac{1}{w} \end{bmatrix}$$

$$A^{-1} = \begin{bmatrix} \frac{1}{x} & -\frac{1}{xy} & \frac{1}{xz} & -\frac{1}{xw} \\ 0 & \frac{1}{y} & 0 & 0 \\ 0 & 0 & \frac{1}{z} & 0 \\ 0 & 0 & 0 & \frac{1}{w} \end{bmatrix}$$

34. $\begin{bmatrix} a_{11} & a_{12} \\ a_{21} & a_{22} \end{bmatrix} \begin{bmatrix} x \\ y \end{bmatrix} = \begin{bmatrix} c_1 \\ c_2 \end{bmatrix}$ or $A \cdot X = C$. Then $X = A^{-1}C$ or

$$X = \begin{bmatrix} \frac{a_{22}}{|A|} & -\frac{a_{12}}{|A|} \\ -\frac{a_{21}}{|A|} & \frac{a_{11}}{|A|} \end{bmatrix} \begin{bmatrix} c_1 \\ c_2 \end{bmatrix} = \begin{bmatrix} \frac{a_{22}c_1 - a_{12}c_2}{|A|} \\ \frac{-a_{21}c_1 + a_{11}c_2}{|A|} \end{bmatrix} \quad \text{where } |A| = D.$$

$$\text{Thus } x = \frac{a_{22}c_1 - a_{12}c_1}{|A|} = \frac{\begin{vmatrix} c_1 & a_{12} \\ c_2 & a_{22} \end{vmatrix}}{D} = \frac{D_x}{D}; \text{ and}$$

$$y = \frac{a_{11}c_2 - a_{21}c_1}{|A|} = \frac{\begin{vmatrix} a_{11} & c_1 \\ a_{21} & c_2 \end{vmatrix}}{D} = \frac{D_y}{D}.$$

Exercise Set 9.9

1. We replace x by -4 and y by 2.

2x + y > -5	
2(-4) + 2	-5
-8 + 2	
-6	

Since -6 > -5 is false, (-4,2) is not a solution.

2. Yes

3. We replace x by 8 and y by 14.

2y - 3x ⩾ 5	
2·14 - 3·8	5
28 - 24	
4	

Since 4 ⩾ 5 is false, (8,14) is not a solution.

4. No

5. Graph: y > 2x

We first graph the line y = 2x. We draw the line dashed since the inequality symbol is >. To determine which half-plane to shade, test a point not on the line. We try (1,1) and substitute:

y > 2x	
1	2·1
	2

Since 1 > 2 is false, (1,1) is not a solution, nor are any points in the half-plane containing (1,1). The points in the opposite half-plane are solutions, so we shade that half-plane and obtain the graph.

5. (continued)

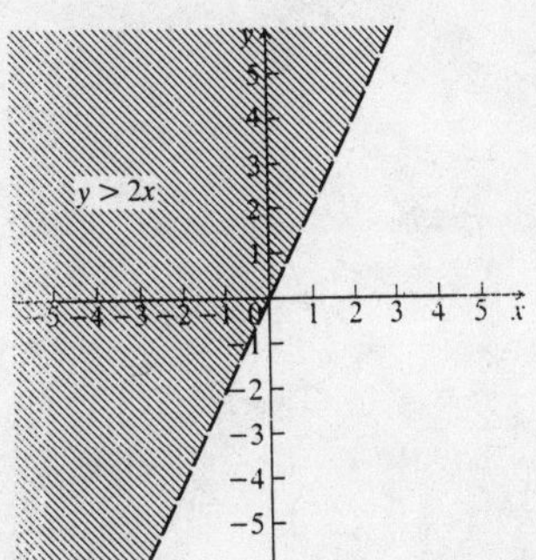

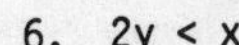
6. 2y < x

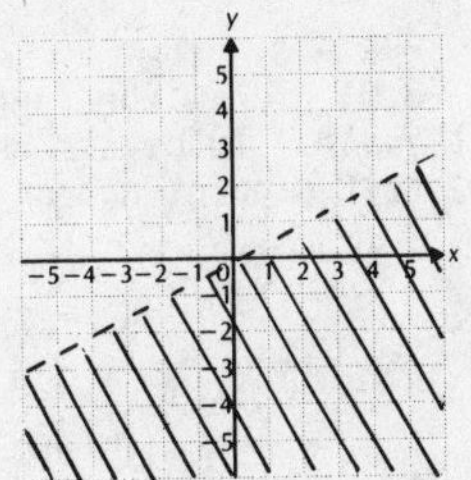

7. Graph: $y + x \geqslant 0$

First graph the equation $y + x = 0$. A few solutions of this equation are (-2, 2), (0, 0), and (3, -3). Plot these points and draw the line solid since the inequality is $\geqslant$.

Next determine which half-plane to shade by trying some point off the line. Here we use (2, 2) as a check.

$y + x \geqslant 0$	
$2 + 2$	0
4	

Since $4 \geqslant 0$ is true, (2, 2) is a solution. Thus shade the half-plane containing (2, 2).

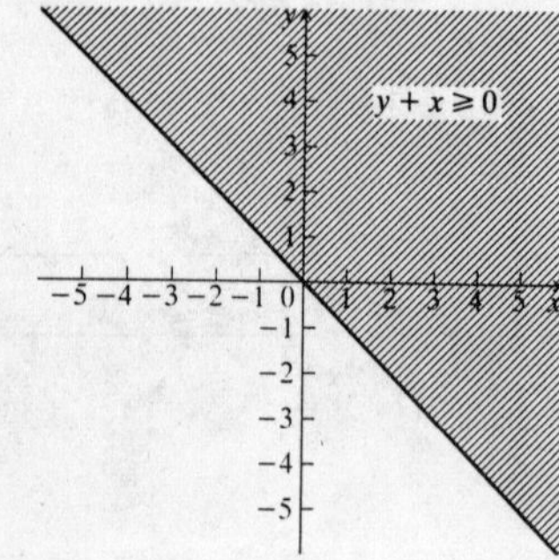

8. $y + x < 0$

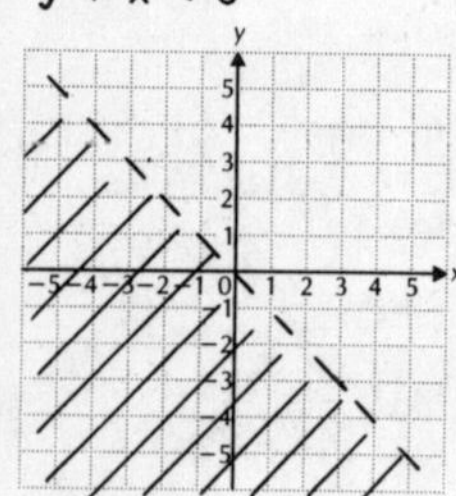

9. Graph: $y > x - 3$

First graph the equation $y = x - 3$. The intercepts are (0,-3) and (3,0). Draw the line dashed since the inequality is >. To determine which half-plane to shade, test a point not on the line. We try (0, 0).

$y > x - 3$	
0	$0 - 3$
	-3

Since $0 > -3$ is true, (0,0) is a solution. Thus we shade the half-plane containing (0,0).

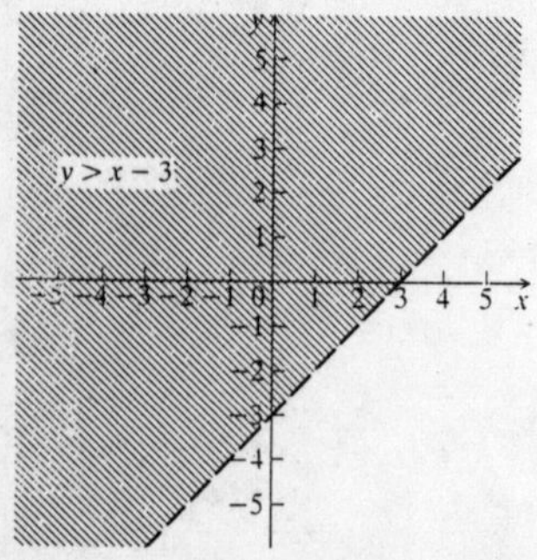

10. $y \leqslant x + 4$

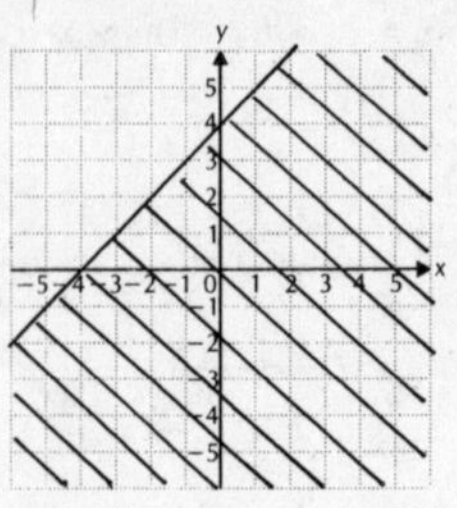

11. Graph: $x + y < 4$

First graph the equation $x + y = 4$. The intercepts are (0,4) and (4,0). Draw the line dashed since the inequality is <. To determine which half-plane to shade, test a point not on the line. We try (0,0).

$x + y < 4$	
$0 + 0$	4
0	

Since $0 < 4$ is true, (0,0) is a solution. Thus we shade the half-plane containing (0,0).

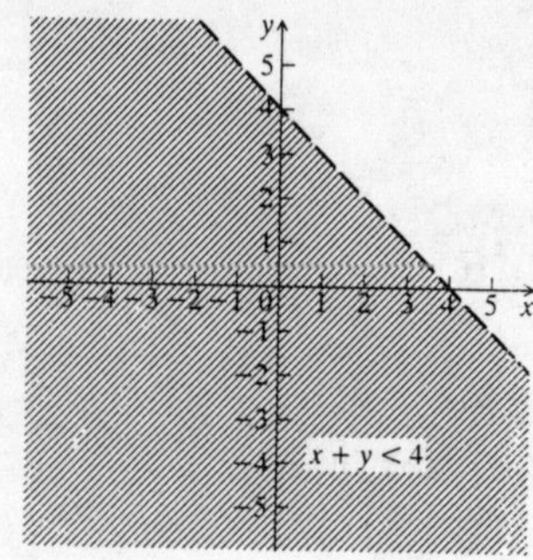

12. $x - y \geqslant 5$

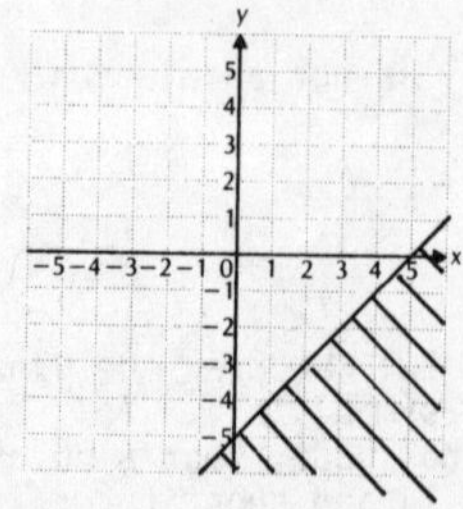

13. Graph: $3x - 2y \leqslant 6$

First graph the equation $3x - 2y = 6$. The y-intercept is (0, -3), and the x-intercept is (2, 0). Plot these points and draw the line solid since the inequality is $\leqslant$.

Next complete the graph by shading the correct half-plane. Since the line does not contain the origin, check the point (0, 0) in the inequality to see if we get a true sentence.

$3x - 2y \leqslant 6$	
$3(0) - 2(0)$	6
0	6

Since $0 \leqslant 6$ is a true sentence, (0, 0) is a solution. Thus shade the half-plane containing (0, 0).

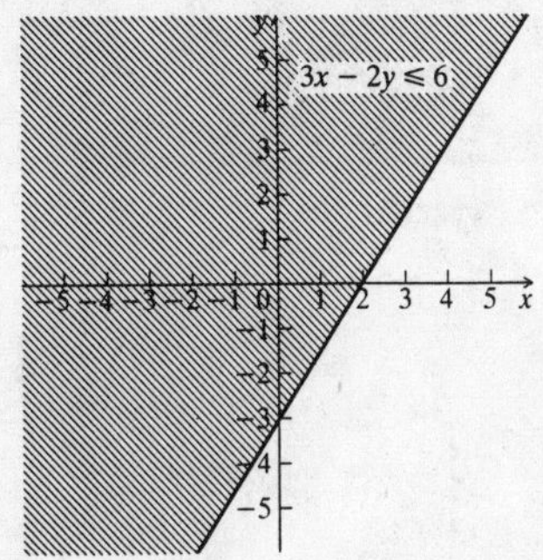

14. $2x - 5y < 10$

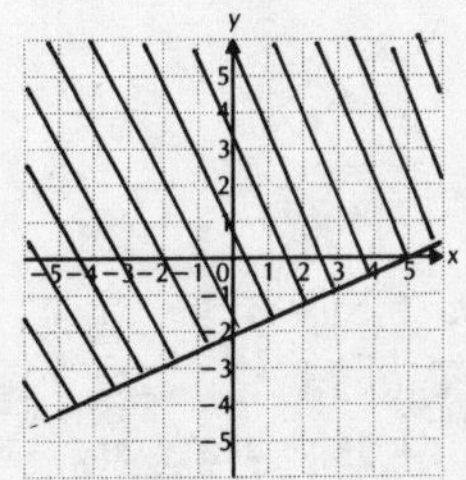

15. Graph: $3y + 2x \geqslant 6$

First graph the equation $3y + 2x = 6$. The y-intercept is (0, 2), and the x-intercept is (3, 0). Plot these points and draw the line solid since the inequality is $\geqslant$.

Next complete the graph by shading the correct half-plane. Since the line does not contain the origin, check the point (0, 0) in the inequality to see if we get a true sentence.

$3y + 2x \geqslant 6$	
$3\cdot0 + 2\cdot0$	6
0	

Since $0 \geqslant 6$ is false, (0, 0) is not a solution. We shade the half-plane which does not contain (0, 0).

15. (continued)

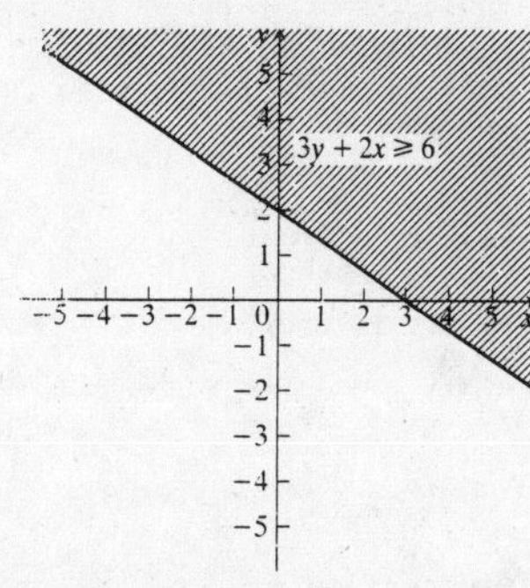

16. $2y + x \leqslant 4$

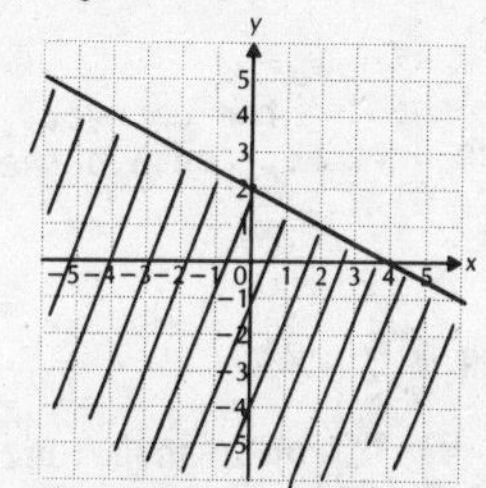

17. Graph: $3x - 2 \leqslant 5x + y$

$-2 \leqslant 2x + y$ (Adding -3x)

First graph $2x + y = -2$. The y-intercept is (0, -2), and the x-intercept is (-1, 0). Plot these points and draw the line solid since the inequality is $\leqslant$.

Next complete the graph by shading the correct half-plane. Since the line does not contain the origin, check the point (0, 0).

$2x + y \geqslant -2$	
$2(0) + 0$	-2
0	-2

Since $0 \geqslant -2$ is a true sentence, (0, 0) is a solution. Thus shade the half-plane containing the origin.

$3x - 2 \leqslant 5x + y$, or $2x + y \geqslant -2$

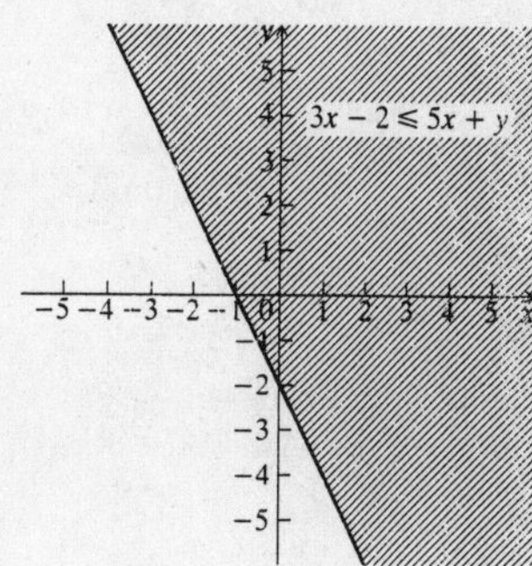

18. $2x - 6y \geqslant 8 + 4y$, or $2x - 10y \geqslant 8$

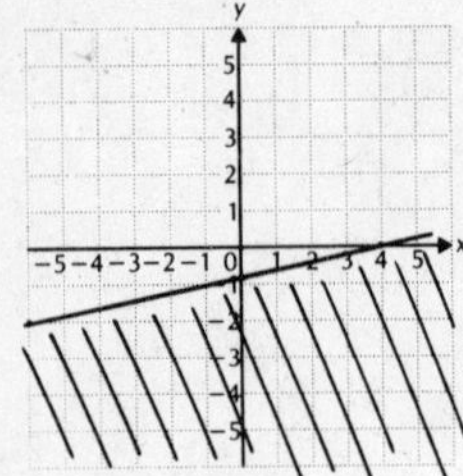

19. Graph: $x < -4$

We first graph the equation $x = -4$. The line is parallel to the y-axis with x-intercept $(-4, 0)$. We draw the line dashed since the inequality is $<$ and not $\leqslant$.

Next complete the graph by shading the correct half-plane. Since the line does not contain the origin, check the point (0, 0).

$$\begin{array}{c|c} x & < -4 \\ \hline 0 & -4 \end{array} \qquad \text{(Substituting 0 for x)}$$

Since $0 < -4$ is false, (0, 0) is not a solution. Thus, we shade the half-plane which does not contain the origin.

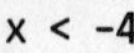
$x < -4$

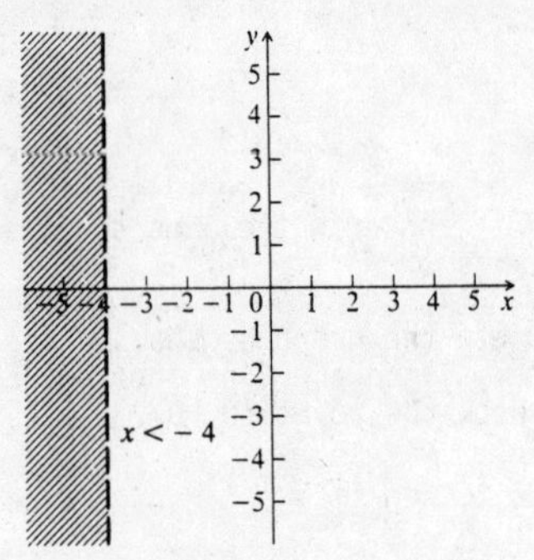

20. $y \geqslant 5$

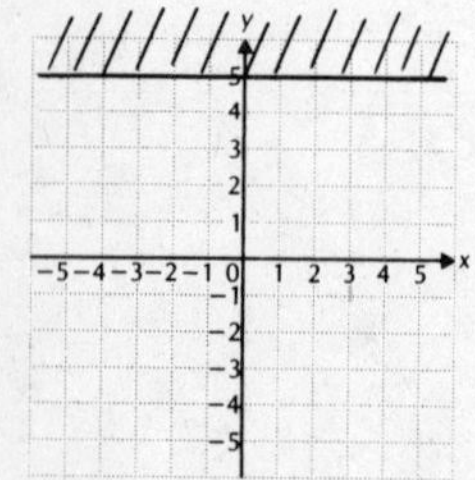

21. Graph: $y > -3$

First we graph the equation $y = -3$. We draw the line dashed since the inequality is $>$. To determine which half-plane to shade we test a point not on the line. We try (0,0).

$$\begin{array}{c|c} y & > -3 \\ \hline 0 & -3 \end{array}$$

Since $0 > -3$ is true, (0,0) is a solution. We shade the half-plane containing (0,0).

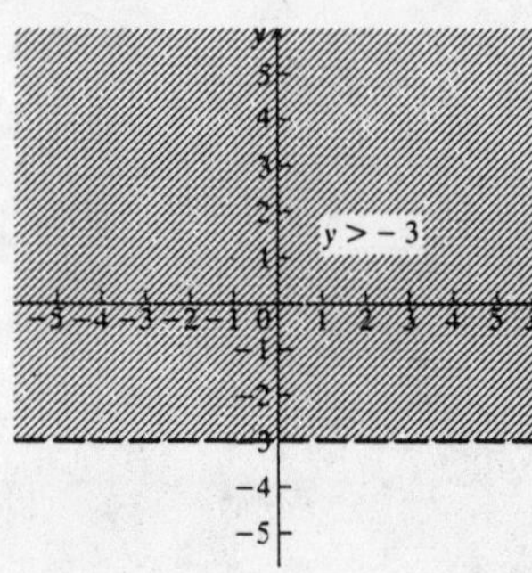

22. $x \leqslant 5$

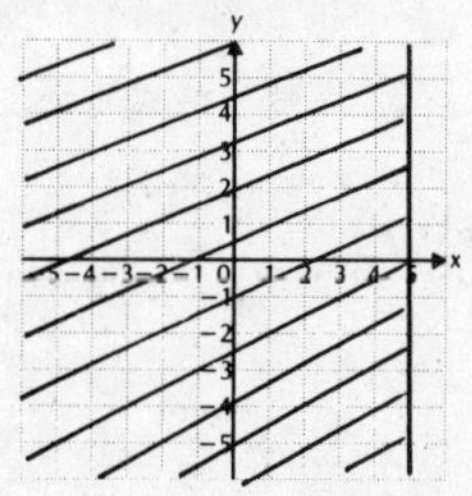

23. Graph: $-4 < y < -1$

This is a conjunction of two inequalities

$-4 < y$ and $y < -1$.

We can graph $-4 < y$ and $y < -1$ separately and then graph the intersection.

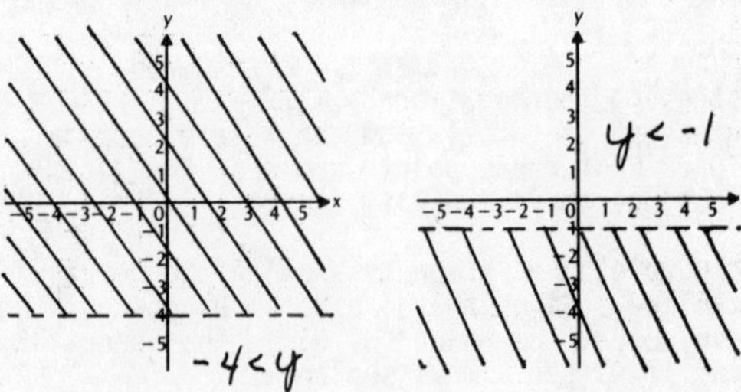

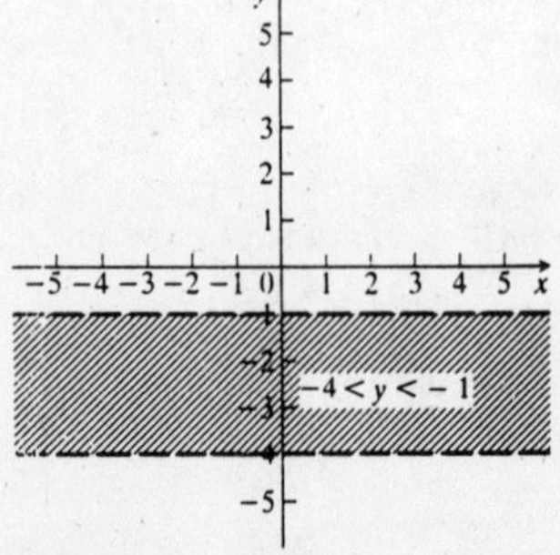

24. $-1 < y < 4$

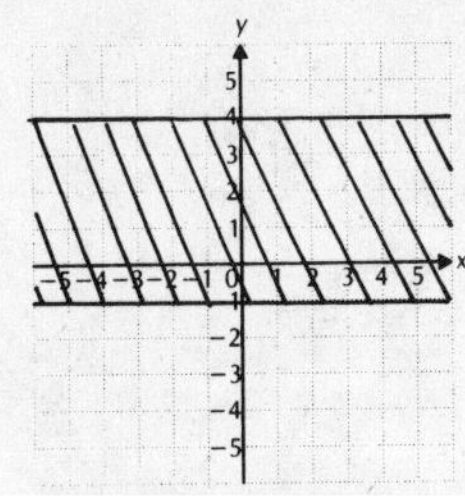

25. Graph: $-4 \leqslant x \leqslant 4$

This is a conjunction of inequalities

$-4 \leqslant x$ and $x \leqslant 4$.

We can graph $-4 \leqslant x$ and $x \leqslant 4$ separately and then graph the intersection.

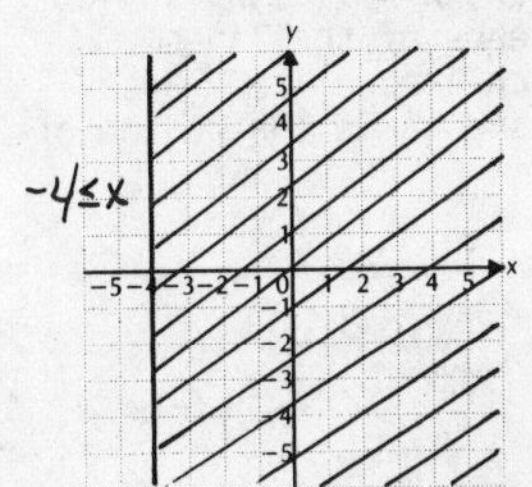

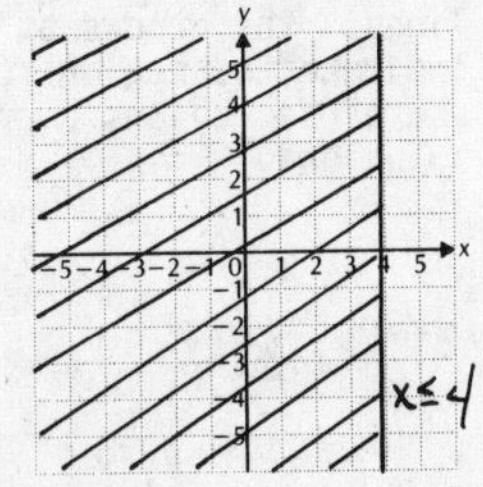

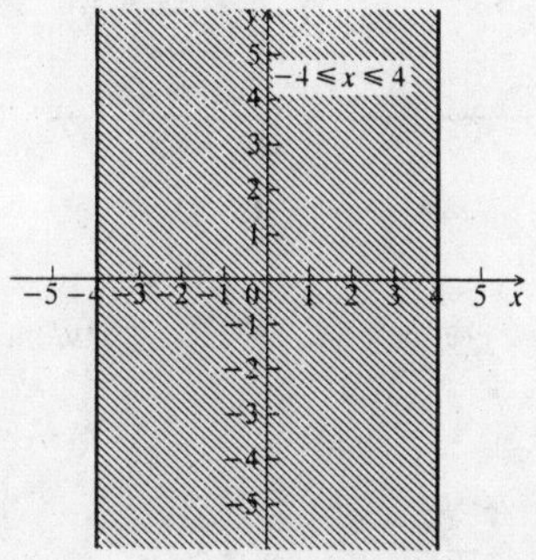

26. $-3 \leqslant x \leqslant 3$

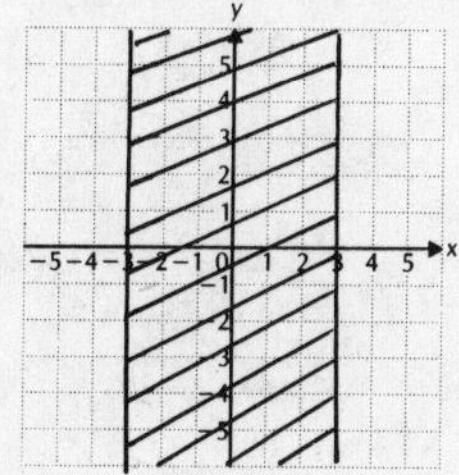

27. Graph: $y \geqslant |x|$

First graph $y = |x|$. A few solutions of this equation are (0, 0), (1, 1), (-1, 1), (4, 4), and (-4, 4). Plot these points and draw the graph solid since the inequality is $\geqslant$.

$y = |x|$

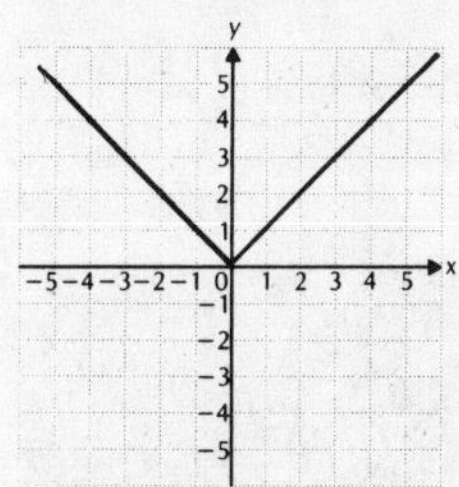

Note that in $y \geqslant |x|$, y is by itself. We interpret the graph of the inequality as the set of all ordered pairs (x, y) where the second coordinate y is greater than or equal to the absolute value of the first. Decide below which pairs satisfy the inequality and which do not.

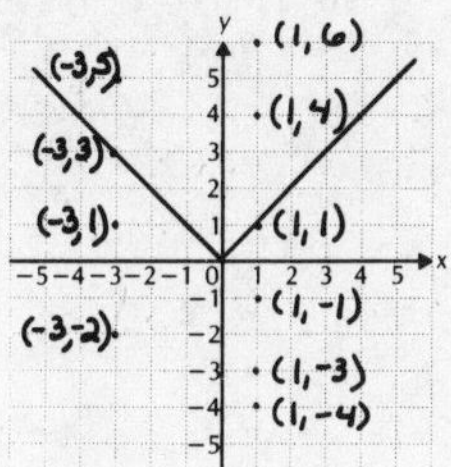

We see that any pair above the graph of $y = |x|$ is a solution as well as those in the graph of $y = |x|$. The graph is as follows.

$y \geqslant |x|$

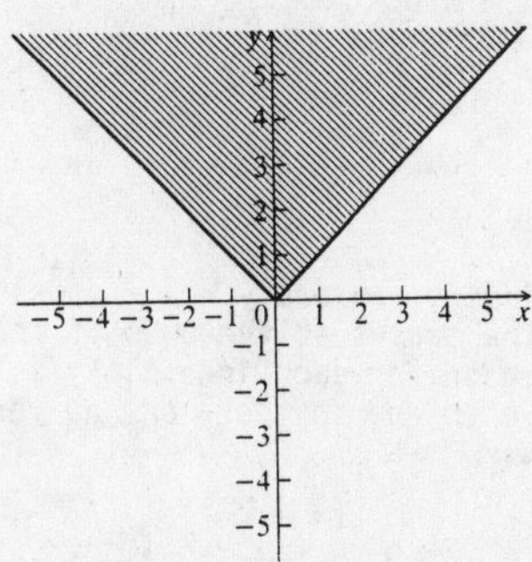

28. $y \leqslant |x|$

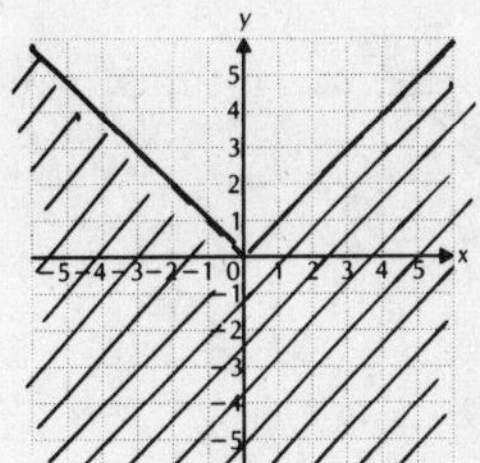

29. Graph: $y \leqslant x$,
$y \geqslant 3 - x$

We graph the lines $y = x$ and $y = 3 - x$ using solid lines. The arrows at the ends of the lines indicate the region for each inequality. Note where the regions overlap and shade the region of solutions.

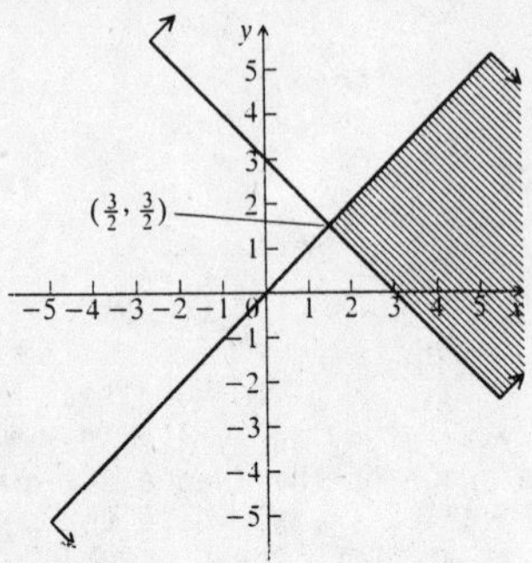

We find the vertex $\left(\frac{3}{2}, \frac{3}{2}\right)$ by solving the system

$y = x$,
$y = 3 - x$.

30. Graph: $y \geqslant x$,
$y \leqslant x - 5$

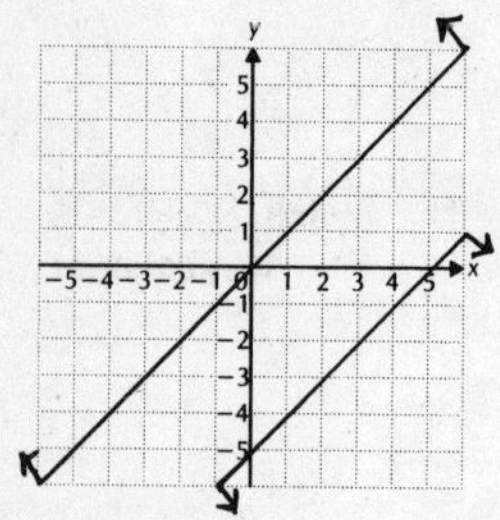

31. Graph $y \geqslant x$,
$y \leqslant x - 4$

We graph the lines $y = x$ and $y = x - 4$ using solid lines. The arrows at the ends of the lines indicate the region for each inequality. The regions do not overlap, so there are no vertices.

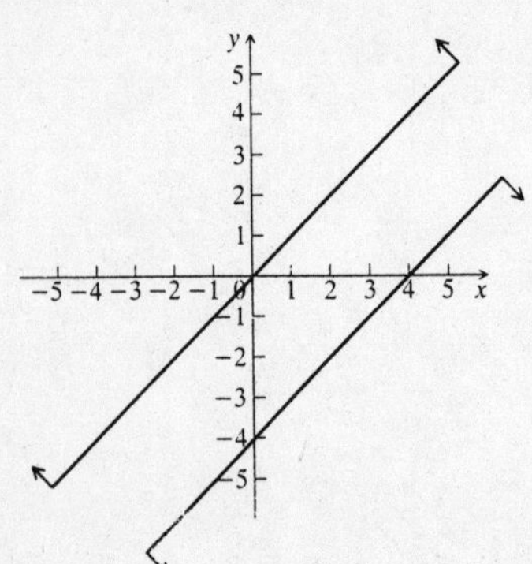

32. Graph: $y \geqslant x$,
$y \leqslant 2 - x$

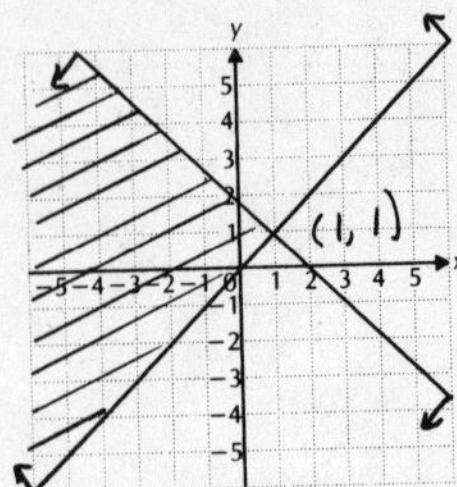

33. Graph: $y \geqslant -3$,
$x \geqslant 1$

We graph the lines $y = -3$ and $x = 1$ using solid lines. The arrows at the ends of the lines indicate the region for each inequality. Note where the regions overlap and shade the region of solutions.

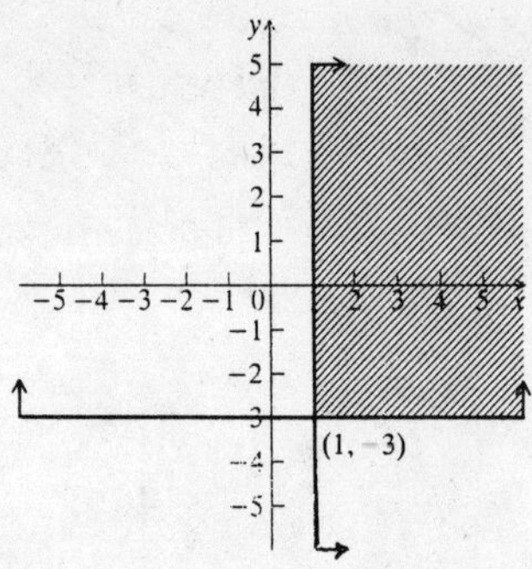

We find the vertex (1, -3) by solving the system

$y = -3$,
$x = 1$.

34. Graph: $y \leqslant -2$,
$x \geqslant 2$

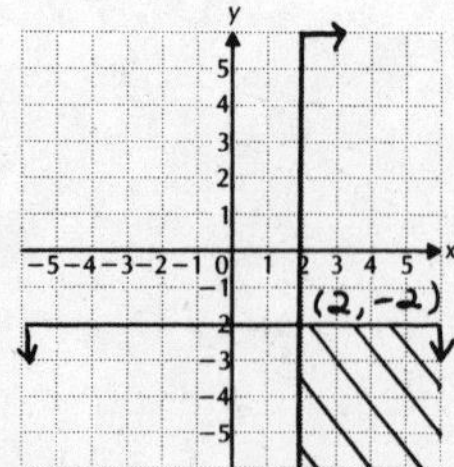

35. Graph: $x \leqslant 3$,

$y \geqslant 2 - 3x$

We graph the lines $x = 3$ and $y = 2 - 3x$ using solid lines. The arrows at the ends of the lines indicate the region for each inequality. Note where the regions overlap and shade the region of solutions.

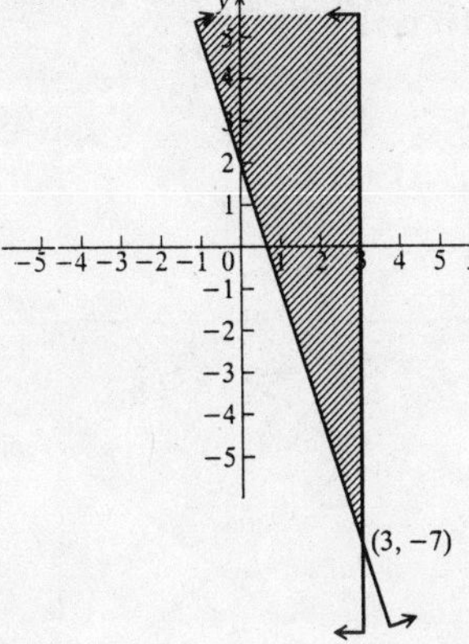

We find the vertex $(3,-7)$ by solving the system

$x = 3$,

$y = 2 - 3x$.

36. Graph: $x \geqslant -2$,

$y \leqslant 3 - 2x$

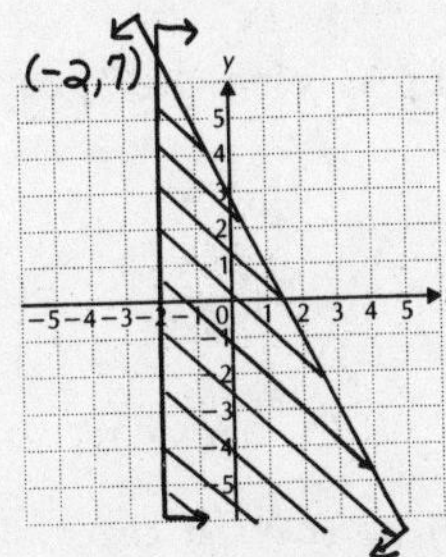

37. Graph: $x + y \leqslant 1$,

$x - y \leqslant 2$

We graph the lines $x + y = 1$ and $x - y = 2$ using solid lines. The arrows at the ends of the lines indicate the region for each inequality. Note where the regions overlap and shade the region of solutions.

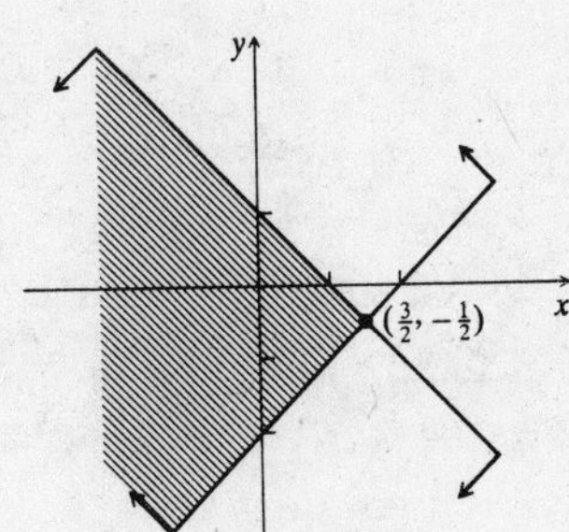

We find the vertex $\left(\frac{3}{2}, -\frac{1}{2}\right)$ by solving the system

$x + y = 1$,

$x - y = 2$.

38. Graph: $y + 3x \geqslant 0$,

$y + 3x \leqslant 2$

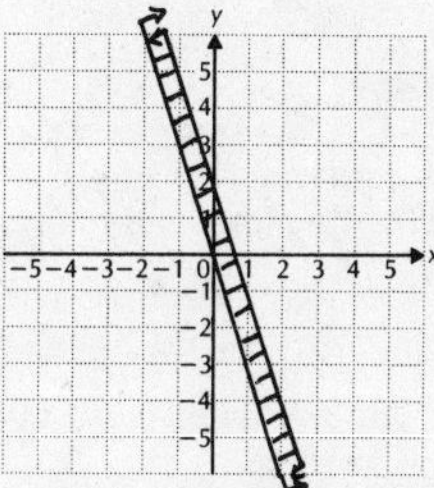

39. Graph: $2y - x \leqslant 2$,

$y + 3x \geqslant -1$

We graph the lines $2y - x = 2$ and $y + 3x = -1$ using solid lines. The arrows at the ends of the lines indicate the region for each inequality. Note where the regions overlap and shade the region of solutions.

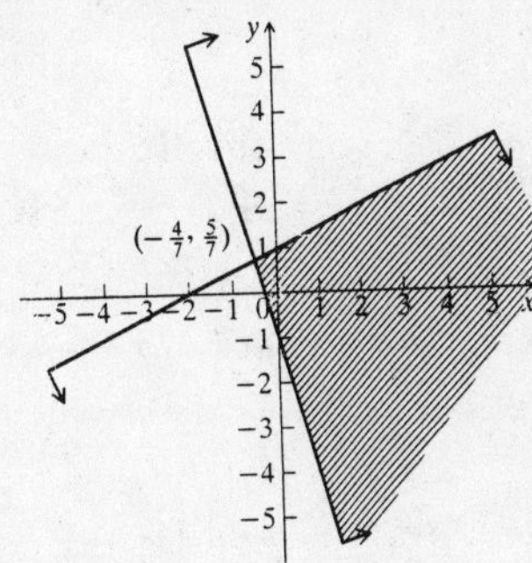

We find the vertex $\left(-\frac{4}{7}, \frac{5}{7}\right)$ by solving the system

$2y - x = 2$,

$y + 3x = -1$.

40. Graph: $y \leqslant 2x + 1$,

$y \geqslant -2x + 1$,

$x \leqslant 2$

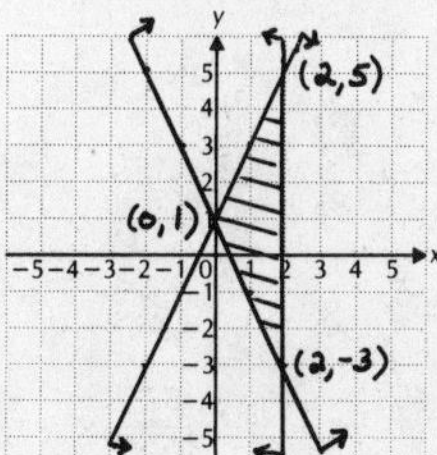

41. Graph: $x - y \leqslant 2$,
$x + 2y \geqslant 8$,
$y \leqslant 4$

We graph the lines $x - y = 2$, $x + 2y = 8$, and $y = 4$ using solid lines. The arrows at the ends of the lines indicate the region for each inequality. Note where the regions overlap and shade the region of solutions.

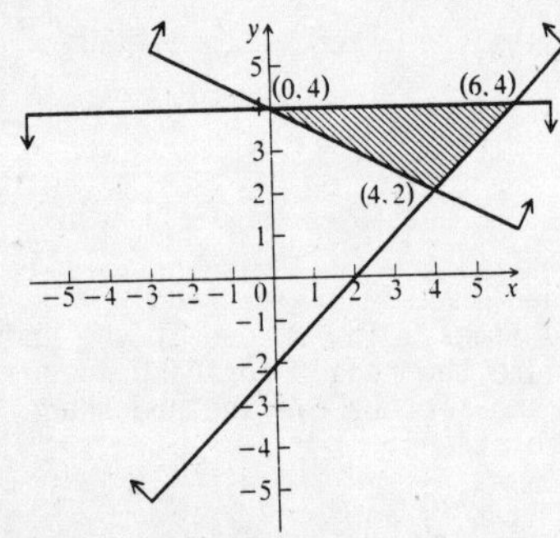

We find the vertex (0,4) by solving the system
$x + 2y = 8$,
$y = 4$.

We find the vertex (6,4) by solving the system
$x - y = 2$,
$y = 4$.

We find the vertex (4,2) by solving the system
$x - y = 2$,
$x + 2y = 8$.

42. Graph: $x + 2y \leqslant 12$,
$2x + y \leqslant 12$,
$x \geqslant 0$,
$y \geqslant 0$

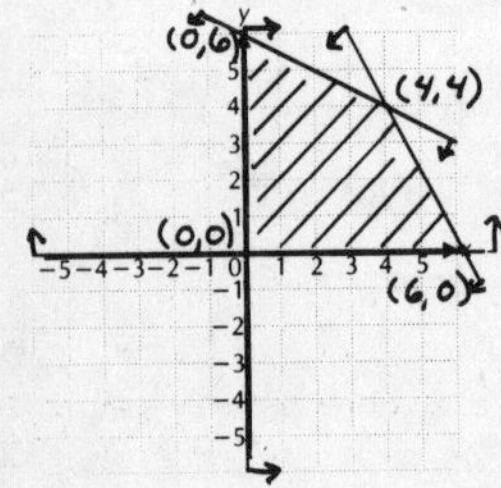

43. Graph: $4y - 3x \geqslant -12$,
$4y + 3x \geqslant -36$,
$y \leqslant 0$,
$x \leqslant 0$

Shade the intersection of the graphs of the four inequalities.

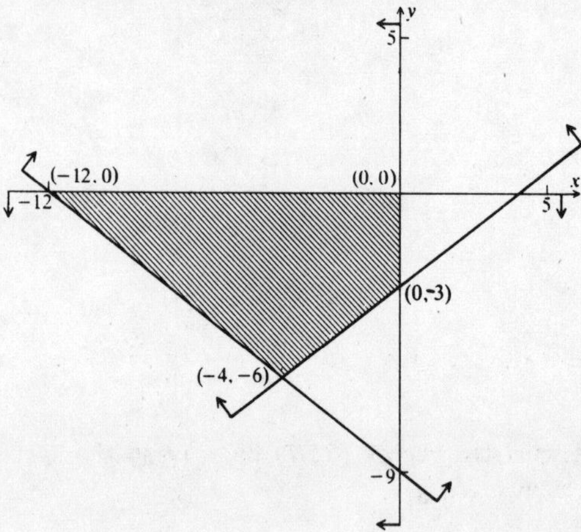

We find the vertex (-12, 0) by solving the system
$4y + 3x = -36$,
$y = 0$.

We find the vertex (0,0) by solving the system
$y = 0$,
$x = 0$.

We find the vertex (0,-3) by solving the system
$4y - 3x = -12$
$x = 0$.

We find the vertex (-4,-6) by solving the system
$4y - 3x = -12$,
$4y + 3x = -36$.

44. Graph: $8x + 5y \leqslant 40$,
$x + 2y \leqslant 8$,
$x \geqslant 0$,
$y \geqslant 0$

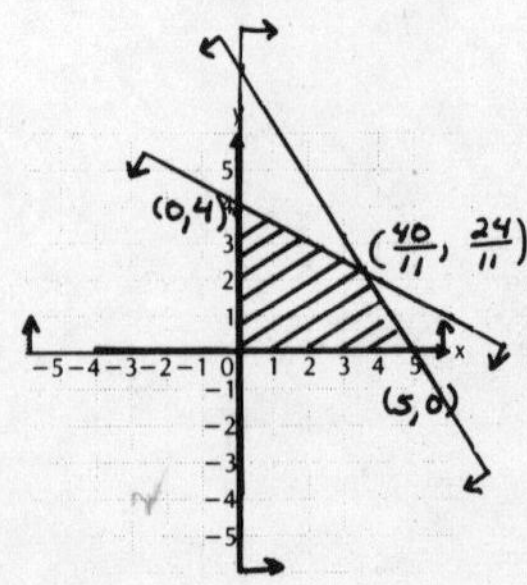

45. Graph: $3x + 4y \geqslant 12$,
$5x + 6y \leqslant 30$,
$1 \leqslant x \leqslant 3$

Shade the intersection of the graphs of the given inequalities.

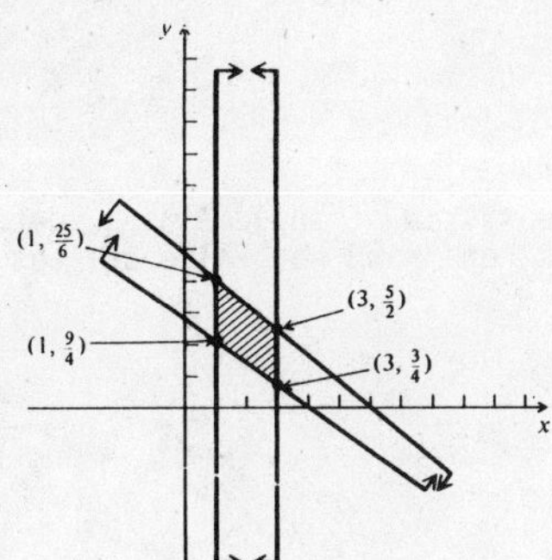

We find the vertex $\left(1, \frac{25}{6}\right)$ by solving the system

$$5x + 6y = 30,$$
$$x = 1.$$

We find the vertex $\left(3, \frac{5}{2}\right)$ by solving the system

$$5x + 6y = 30,$$
$$x = 3.$$

We find the vertex $\left(3, \frac{3}{4}\right)$ by solving the system

$$3x + 4y = 12,$$
$$x = 3.$$

We find the vertex $\left(1, \frac{9}{4}\right)$ by solving the system

$$3x + 4y = 12,$$
$$x = 1.$$

46. Graph: $y - x \geqslant 1$,
$y - x \leqslant 3$,
$2 \leqslant x \leqslant 5$

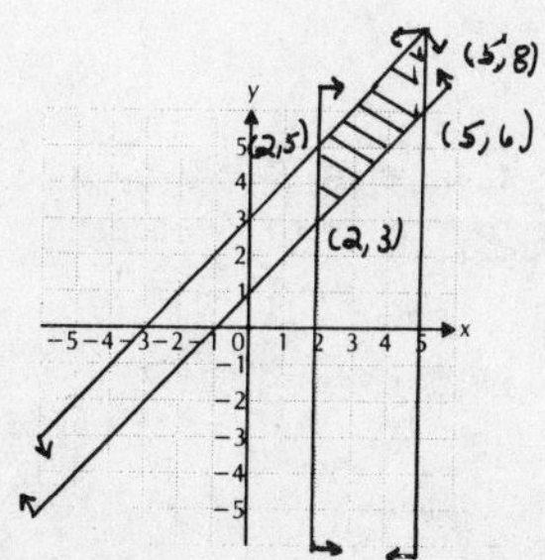

47. The graph of the domain is as follows:

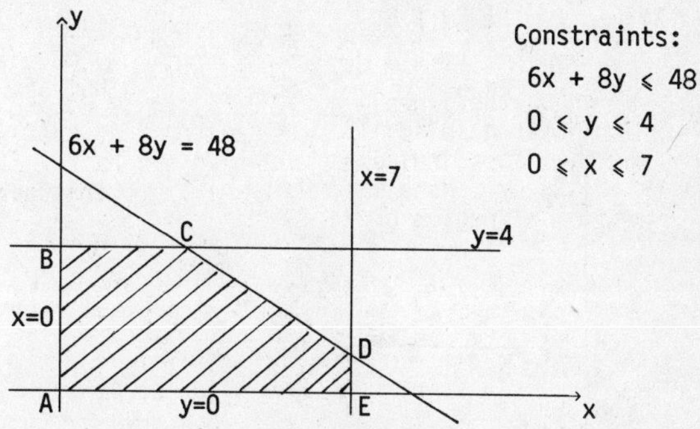

We need to find the coordinate of each vertex.

Vertex A: (0, 0)

Vertex B:

We solve the system $x = 0$ and $y = 4$.
The coordinates of point B are (0, 4).

Vertex C:

We solve the system $6x + 8y = 48$ and $y = 4$.
The coordinates of point C are $\left(\frac{8}{3}, 4\right)$.

Vertex D:

We solve the system $6x + 8y = 48$ and $x = 7$.
The coordinates of point D are $\left(7, \frac{3}{4}\right)$.

Vertex E:

We solve the system $x = 7$ and $y = 0$.
The coordinates of point E are (7, 0).

We compute the value of P for each vertex.

Vertex	$P = 17x - 3y + 60$
A(0, 0)	$17\cdot 0 - 3\cdot 0 + 60 = 60$
B(0, 4)	$17\cdot 0 - 3\cdot 4 + 60 = 48$
$C\left(\frac{8}{3}, 4\right)$	$17 \cdot \frac{8}{3} - 3\cdot 4 + 60 = 66\frac{2}{3}$
$D\left(7, \frac{3}{4}\right)$	$17\cdot 7 - 3 \cdot \frac{3}{4} + 60 = 176\frac{3}{4}$
E(7, 0)	$17\cdot 7 - 3\cdot 0 + 60 = 179$

The maximum value of P is 179 when $x = 7$ and $y = 0$.

The minimum value of P is 48 when $x = 0$ and $y = 4$.

48. Maximum: 151 when $x = 3$ and $y = \frac{5}{4}$;
minimum: $78\frac{2}{5}$ when $x = \frac{4}{5}$ and $y = 4$

49. The graph of the domain is as follows:

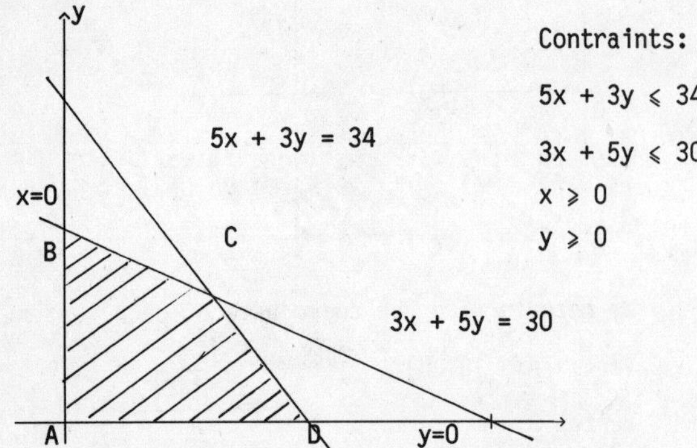

We need to find the coordinates of each vertex.

Vertex A: (0, 0)

Vertex B:

We solve the system $3x + 5y = 30$ and $x = 0$. The coordinates of point B are (0, 6).

Vertex C:

We solve the system $5x + 3y = 34$ and $3x + 5y = 30$. The coordinates of point C are (5, 3).

Vertex D:

We solve the system $5x + 3y = 34$ and $y = 0$. The coordinates of point D are $\left(\frac{34}{5}, 0\right)$.

We compute the value of F for each vertex.

Vertex	$F = 5x + 36y$
A(0, 0)	$5\cdot0 + 36\cdot0 = 0$
B(0, 6)	$5\cdot0 + 36\cdot6 = 216$
C(5, 3)	$5\cdot5 + 36\cdot3 = 133$
$D\left(\frac{34}{5}, 0\right)$	$5\cdot\frac{34}{5} + 36\cdot0 = 34$

The maximum value of F is 216 when $x = 0$ and $y = 6$.

The minimum value of F is 0 when $x = 0$ and $y = 0$.

50. Maximum: 81.2 when $x = 0$ and $y = 5.8$;
minimum: 0 when $x = 0$ and $y = 0$

51. Let x = the number of Biscuit Jumbos and y = the number of Mitimite Biscuits to be made per day. The income I is given by

$$I = \$0.10x + \$0.08y$$

subject to the constraints

$$x + y \leqslant 200,$$
$$2x + y \leqslant 300$$
$$x \geqslant 0,$$
$$y \geqslant 0.$$

We graph the system of inequalities, determine the vertices, and find the value of I at each vertex.

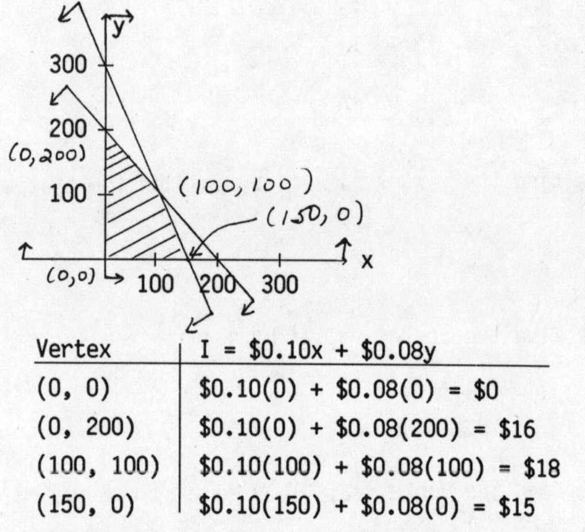

Vertex	$I = \$0.10x + \$0.08y$
(0, 0)	$\$0.10(0) + \$0.08(0) = \$0$
(0, 200)	$\$0.10(0) + \$0.08(200) = \$16$
(100, 100)	$\$0.10(100) + \$0.08(100) = \$18$
(150, 0)	$\$0.10(150) + \$0.08(0) = \$15$

The company will have a maximum income of \$18 when 100 of each type of biscuit is made.

52. The maximum number of miles is 480 when the car uses 9 gal and the moped uses 3 gal.

53. Let x = the number of type A questions and y = the number of type B questions you answer. The score S is given by

$$S = 10x + 25y$$

subject to the constraints

$$3 \leqslant x \leqslant 12,$$
$$4 \leqslant y \leqslant 15,$$
$$x + y \leqslant 20.$$

We graph the system of inequalities, determine the vertices, and find the value of S at each vertex.

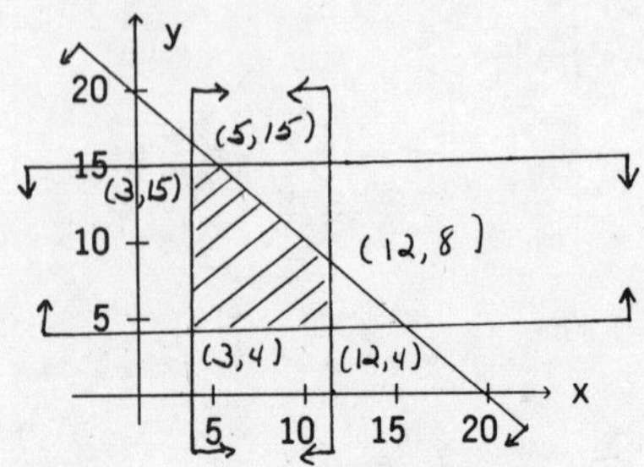

53. (continued)

Vertex	$S = 10x + 25y$
(3, 4)	$10\cdot3 + 25\cdot4 = 130$
(3, 15)	$10\cdot3 + 25\cdot15 = 405$
(5, 15)	$10\cdot5 + 25\cdot15 = 425$
(12, 8)	$10\cdot12 + 25\cdot8 = 320$
(12, 4)	$10\cdot12 + 25\cdot4 = 220$

The maximum score is 425 points when 5 type A questions and 15 type B questions are answered.

54. The maximum test score is 102 when 8 questions of type A and 10 questions of type B are answered.

55. Let x = the number of units of lumber and y = the number of units of plywood produced per week. The profit P is given by

$$P = \$20x + \$30y$$

subject to the constraints

$$x + y \leq 400,$$
$$x \geq 100,$$
$$y \geq 150.$$

We graph the system of inequalities, determine the vertices and find the value of P at each vertex.

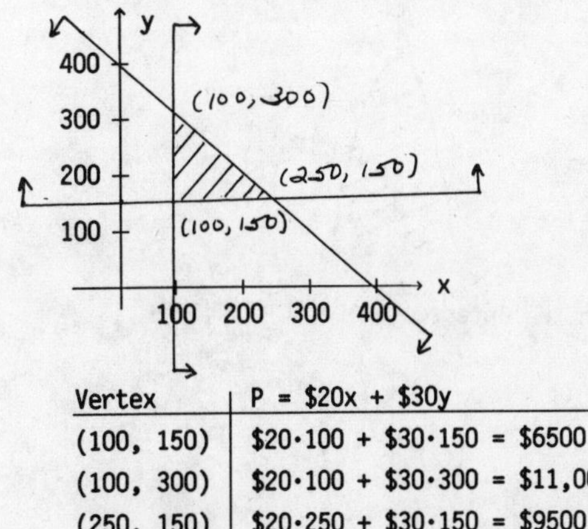

Vertex	$P = \$20x + \$30y$
(100, 150)	$\$20\cdot100 + \$30\cdot150 = \$6500$
(100, 300)	$\$20\cdot100 + \$30\cdot300 = \$11{,}000$
(250, 150)	$\$20\cdot250 + \$30\cdot150 = \$9500$

The maximum profit of $11,000 is achieved by producing 100 units of lumber and 300 units of plywood.

56. The maximum profit of $8000 occurs by planting 80 acres of corn and 160 acres of oats.

57. Let x = the amount invested in corporate bonds and y = the amount invested in municipal bonds. The income I is given by

$$I = 0.08x + 0.075y$$

subject to the constraints

$$x + y \leq \$40{,}000,$$
$$\$6000 \leq x \leq \$22{,}000,$$
$$y \leq \$30{,}000.$$

We graph the system of inequalities, determine the vertices, and find the value of I at each vertex.

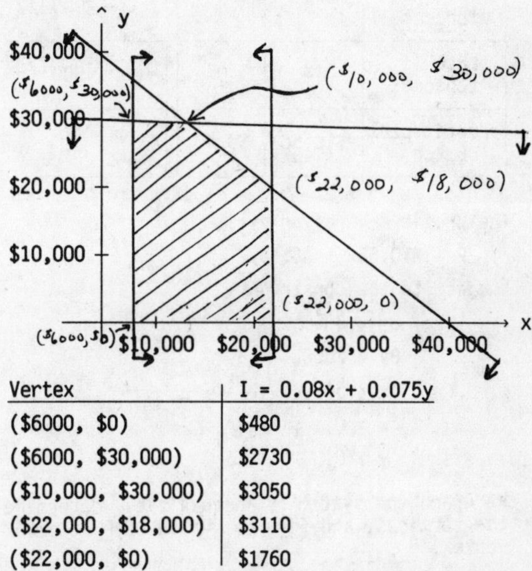

Vertex	$I = 0.08x + 0.075y$
($6000, $0)	$480
($6000, $30,000)	$2730
($10,000, $30,000)	$3050
($22,000, $18,000)	$3110
($22,000, $0)	$1760

The maximum income of $3110 occurs when $22,000 is invested in corporate bonds and $18,000 is invested in municipal bonds.

58. The maximum interest income is $1395 when $7000 is invested in bank X and $15,000 is invested in bank Y.

59. Let x = the number of batches of Smello and y = the number of batches of Roppo to be made. We organize the information in a table.

	Composition		Number of lbs Available
	Smello	Roppo	
Number of Batches	x	y	
English tobacco	12	8	3000
Virginia tobacco	0	8	2000
Latakia tobacco	4	0	500
Profit per Batch	\$10.56	\$6.40	

The profit P is given by

$$P = \$10.56x + \$6.40y$$

subject to the constraints

$$12x + 8y \leqslant 3000,$$
$$8y \leqslant 2000,$$
$$4x \leqslant 500,$$
$$x \geqslant 0,$$
$$y \geqslant 0.$$

We graph the system of inequalities, determine the vertices, and find the value of P at each vertex.

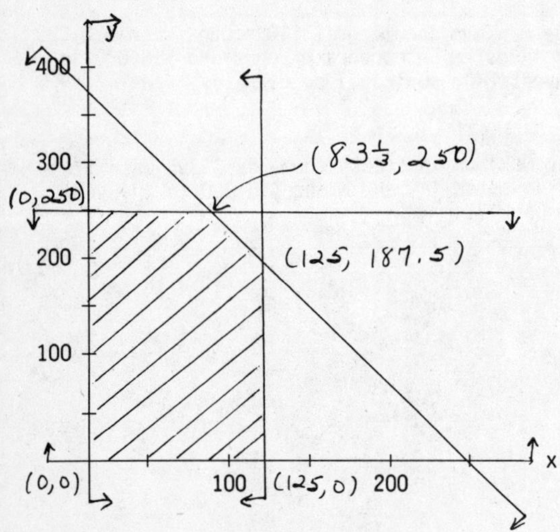

Vertex	P = \$10.56x + \$6.40y
(0, 0)	\$0
(125, 0)	\$1320
(125, 187.5)	\$2520
$\left(83\frac{1}{3}, 250\right)$	\$2480
(0, 250)	\$1600

The maximum profit of \$2520 occurs when 125 batches of Smello and 187.5 batches of Roppo are made.

60. The maximum profit per day is \$192 when 2 knit suits and 4 worsted suits are made.

61. Graph: $y \geqslant x^2 - 2$

$y \leqslant 2 - x^2$

Graph the equation $y = x^2 - 2$. A few solutions are (0, -2), (1, -1), (-1, -1), (2, 2), and (-2, 2). Plot these points and draw the graph solid since the inequality is $\geqslant$. Since for any point above the graph y is greater than $x^2 - 2$, we shade above $y = x^2 - 2$.

$y \geqslant x^2 - 2$

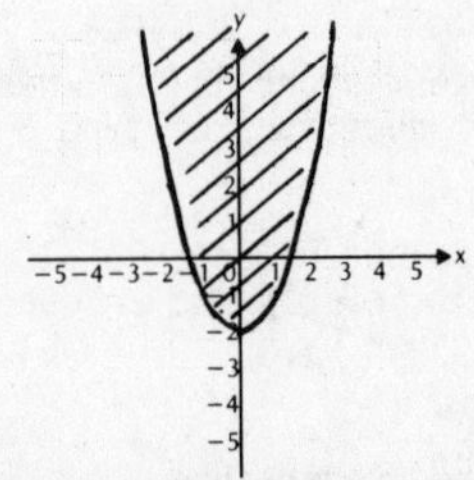

Graph the equation $y = 2 - x^2$. A few solutions are (0, 2), (1, 1), (-1, 1), (2, -2), and (-2, -2). Plot these points and draw the graph solid since the inequality is $\leqslant$. Since for any point below the graph y is less than $2 - x^2$, we shade below $y = 2 - x^2$.

$y \leqslant 2 - x^2$

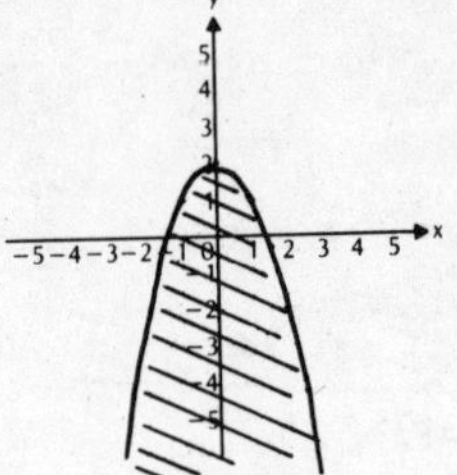

Now graph the intersection of the graphs.

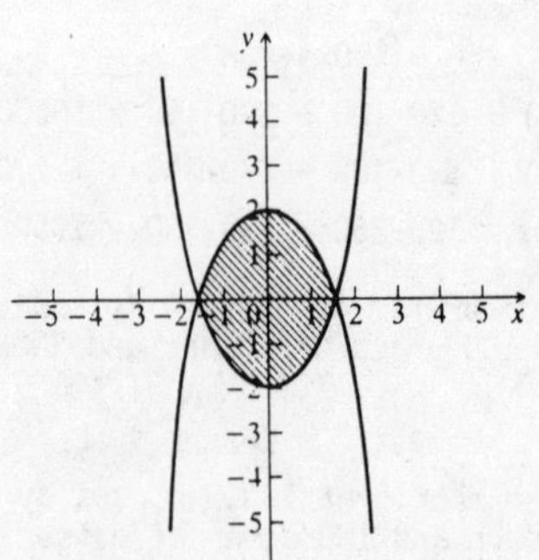

62. $y < x + 1$
$y \geqslant x^2$

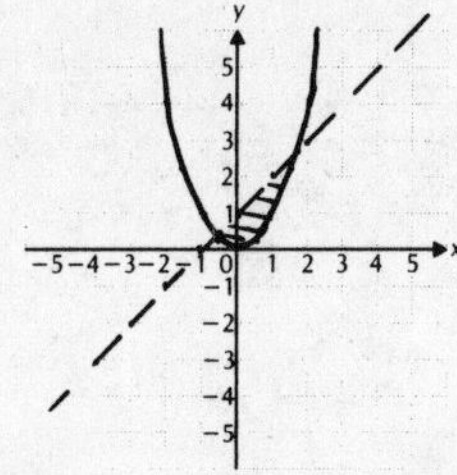

63. The team gets 2w points from w wins and t points from t ties. The number of wins and ties must each be nonnegative.

We have

$$2w + t \geqslant 60,$$
$$w \geqslant 0,$$
$$t \geqslant 0.$$

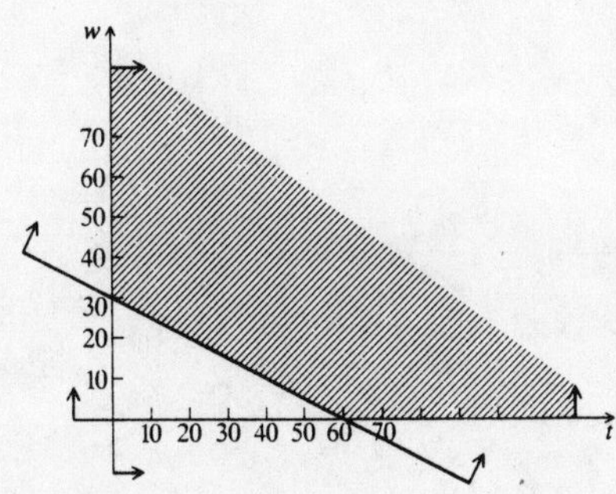

64. $75a + 35c > 1000,$
$a \geqslant 0,$
$c \geqslant 0$

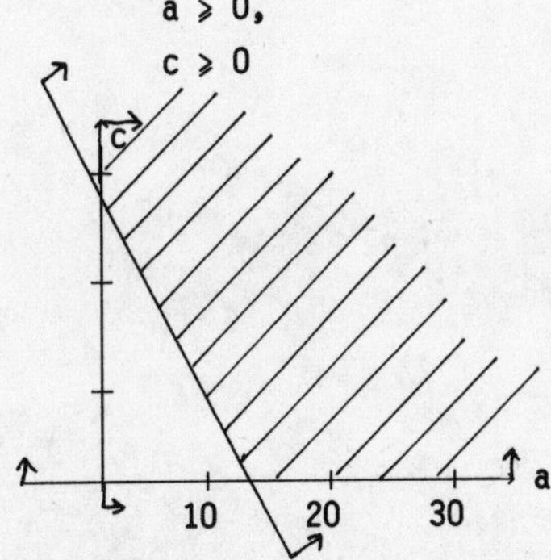

65. Graph: $0 \leqslant L \leqslant 74,$
$0 \leqslant W \leqslant 50$

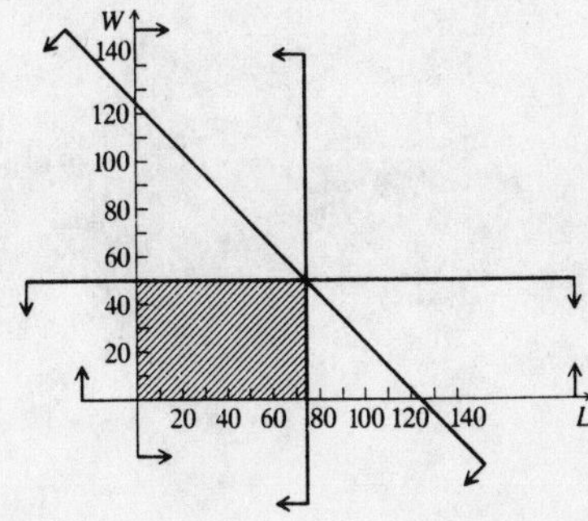

66. $|x| + |y| \leqslant 1$

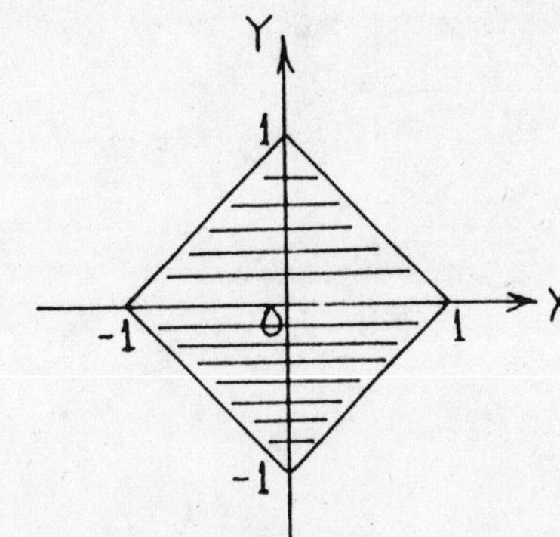

67.

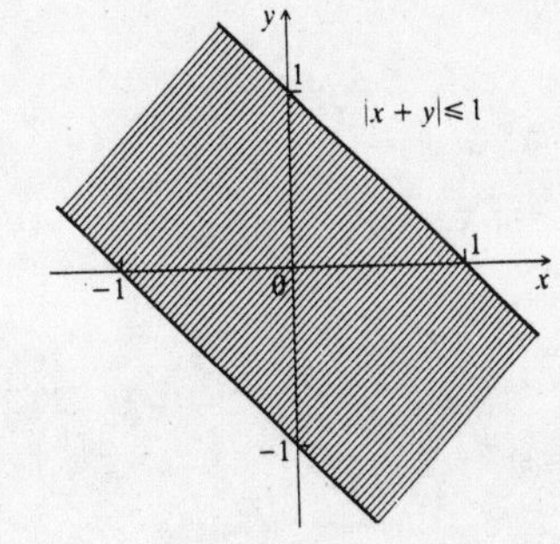

68. $|x - y| > 0$

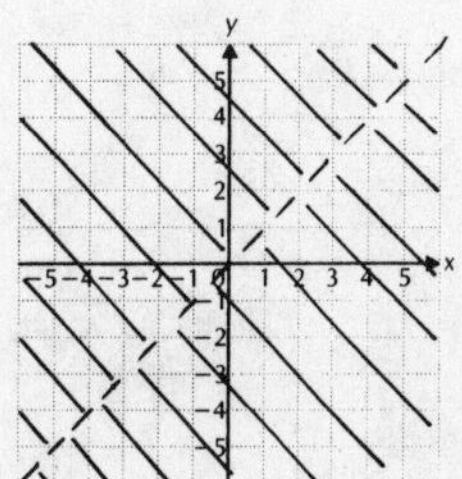

69.

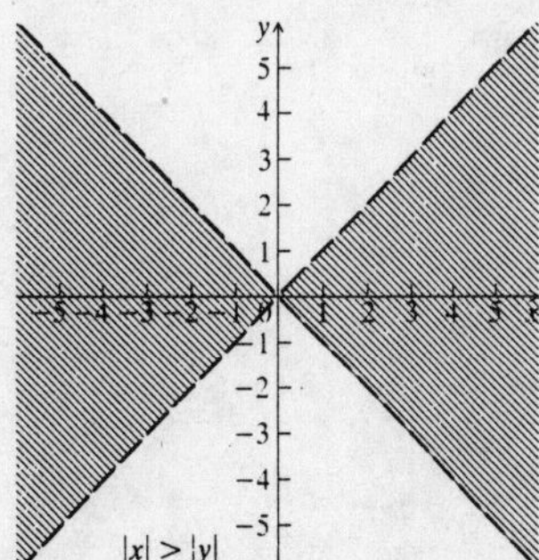

70. The maximum income of \$3350 occurs when 25 chairs and 9.5 sofas are made. (A more practical answer is that the maximum income of \$3200 is achieved when 25 chairs and 9 sofas are made.)

Exercise Set 10.1

1. $x^2 - y^2 = 0$

$(x + y)(x - y) = 0$

$x + y = 0$ or $x - y = 0$

$y = -x$ or $y = x$

The graph consists of two intersecting lines.

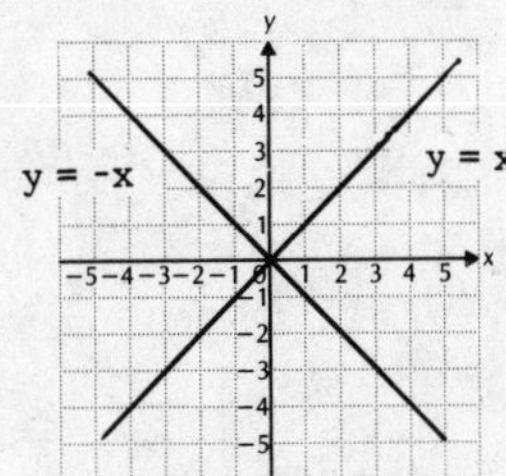

2.

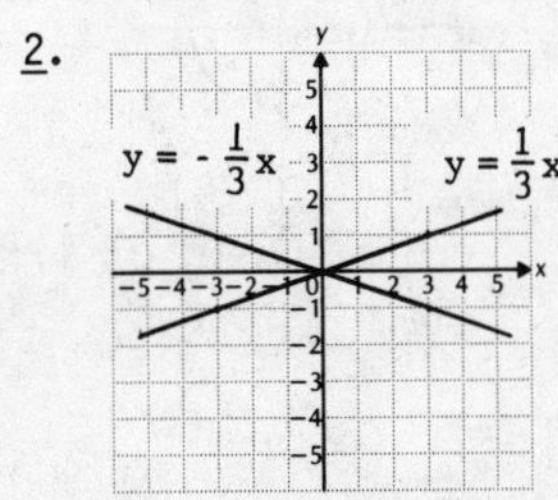

3. $3x^2 + xy - 2y^2 = 0$

$(3x - 2y)(x + y) = 0$

$3x - 2y = 0$ or $x + y = 0$

$y = \frac{3}{2}x$ or $y = -x$

The graph consists of two intersecting lines.

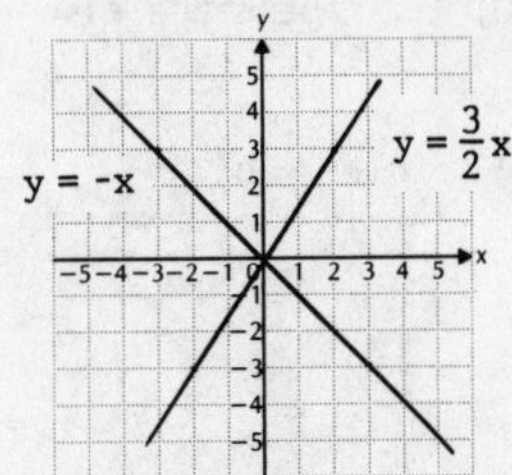

4.

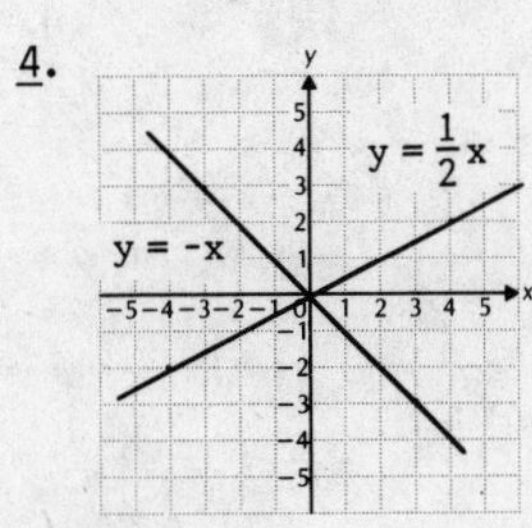

5. $2x^2 + y^2 = 0$

The expression $2x^2 + y^2$ is not factorable in the real-number system. The only real-number solution of the equation is (0, 0). The graph is the point (0, 0).

6. $5x^2 + y^2 = -3$

Since squares of numbers are never negative, the left side of the equation can never be negative. The equation has no real-number solution, hence has no graph.

7. $\frac{x^2}{4} + \frac{y^2}{1} = 1$

$\frac{x^2}{2^2} + \frac{y^2}{1^2} = 1$

The center of the ellipse is (0, 0); a = 2 and b = 1.

Two of the vertices are (-2, 0) and (2, 0). These are also the x-intercepts. The other two vertices are (0, -1) and (0, 1). These are also the y-intercepts.

Since a > b, we find c using $c^2 = a^2 - b^2$.

$c^2 = a^2 - b^2 = 4 - 1 = 3$

Thus $c = \sqrt{3}$.

The foci are on the x-axis. They are $(-\sqrt{3}, 0)$ and $(\sqrt{3}, 0)$.

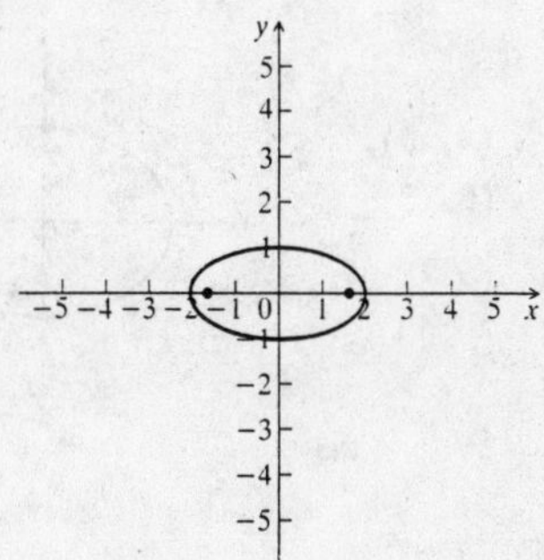

8. V: (-1, 0), (1, 0), (0, -2), (0, 2)

F: $(0, -\sqrt{3})$, $(0, \sqrt{3})$

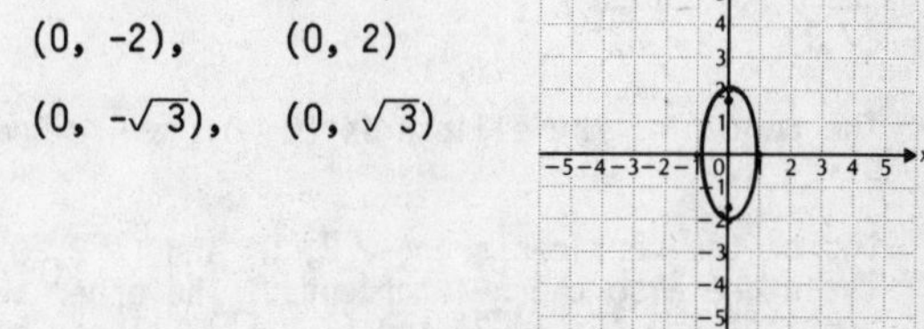

9. $16x^2 + 9y^2 = 144$

$\frac{x^2}{9} + \frac{y^2}{16} = 1$ [Multiplying by $\frac{1}{144}$]

$\frac{x^2}{3^2} + \frac{y^2}{4^2} = 1$

The center of the ellipse is (0, 0); a = 3 and b = 4.

Two of the vertices are (-3, 0) and (3, 0). These are also the x-intercepts. The other two vertices are (0, -4) and (0, 4). These are also the y-intercepts.

Since b > a, we find c using $c^2 = b^2 - a^2$.

$c^2 = b^2 - a^2 = 16 - 9 = 7$

Thus $c = \sqrt{7}$.

The foci are on the y-axis. They are $(0, -\sqrt{7})$ and $(0, \sqrt{7})$.

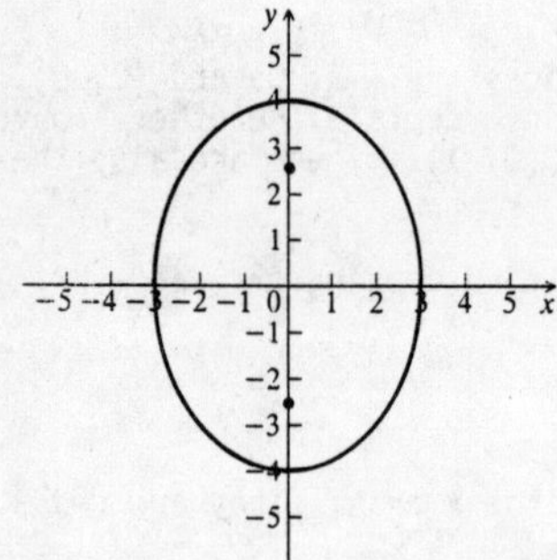

10. V: (-4, 0), (4, 0),
(0, -3), (0, 3)

F: $(-\sqrt{7}, 0)$, $(\sqrt{7}, 0)$

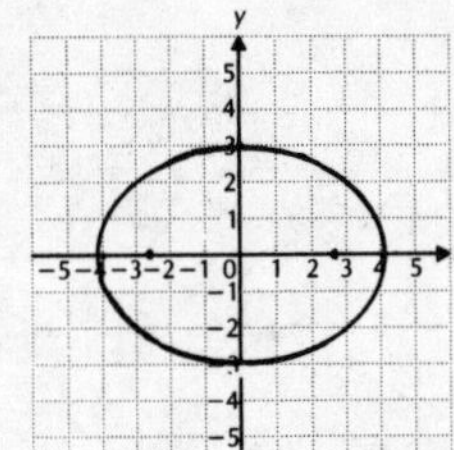

11. $2x^2 + 3y^2 = 6$

$\frac{x^2}{3} + \frac{y^2}{2} = 1$ [Multiplying by $\frac{1}{6}$]

$\frac{x^2}{(\sqrt{3})^2} + \frac{y^2}{(\sqrt{2})^2} = 1$

The center of the ellipse is (0, 0); $a = \sqrt{3}$ and $b = \sqrt{2}$.

Two of the vertices are $(-\sqrt{3}, 0)$ and $(\sqrt{3}, 0)$. These are also the x-intercepts. The other two vertices are $(0, -\sqrt{2})$ and $(0, \sqrt{2})$. These are also the y-intercepts.

Since a > b, we find c using $c^2 = a^2 - b^2$.

$c^2 = a^2 - b^2 = 3 - 2 = 1$

Thus c = 1.

The foci are on the x-axis. They are (-1, 0) and (1, 0).

11. (continued)

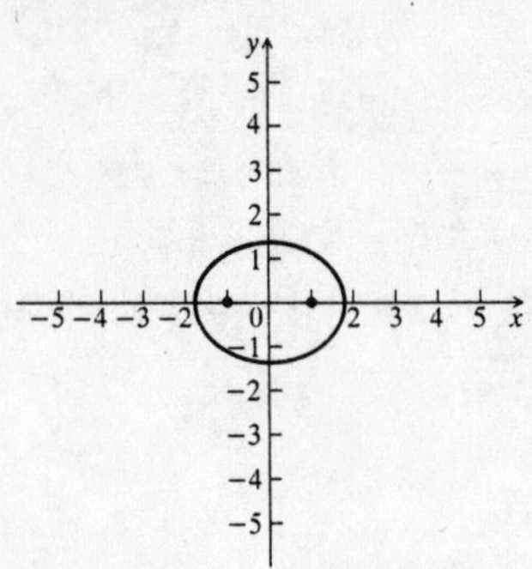

12. V: $(-\sqrt{7}, 0)$, $(\sqrt{7}, 0)$
$(0, -\sqrt{5})$, $(0, \sqrt{5})$

F: $(-\sqrt{2}, 0)$, $(\sqrt{2}, 0)$

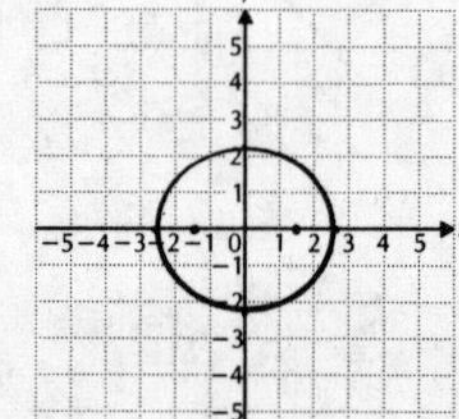

13. $4x^2 + 9y^2 = 1$

$\frac{x^2}{\frac{1}{4}} + \frac{y^2}{\frac{1}{9}} = 1$

$\frac{x^2}{\left(\frac{1}{2}\right)^2} + \frac{y^2}{\left(\frac{1}{3}\right)^2} = 1$

The center of the ellipse is (0, 0); $a = \frac{1}{2}$ and $b = \frac{1}{3}$.

Two of the vertices are $\left(-\frac{1}{2}, 0\right)$ and $\left(\frac{1}{2}, 0\right)$. These are also the x-intercepts. The other two vertices are $\left(0, -\frac{1}{3}\right)$ and $\left(0, \frac{1}{3}\right)$. These are also the y-intercepts.

Since a > b, we find c using $c^2 = a^2 - b^2$.

$c^2 = a^2 - b^2 = \frac{1}{4} - \frac{1}{9} = \frac{5}{36}$

Thus $c = \frac{\sqrt{5}}{6}$.

The foci are on the x-axis. They are $\left(-\frac{\sqrt{5}}{6}, 0\right)$ and $\left(\frac{\sqrt{5}}{6}, 0\right)$.

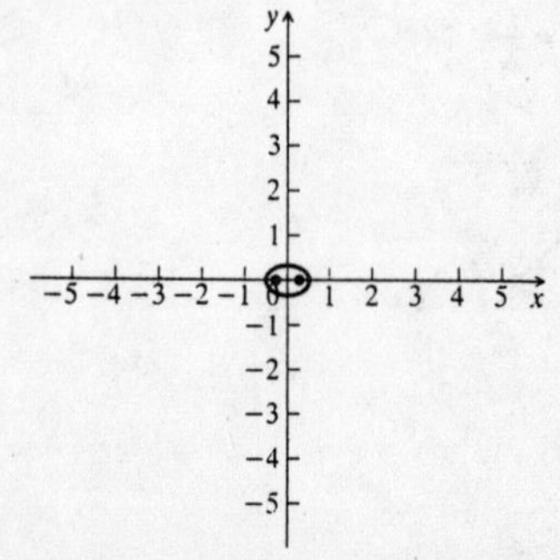

14. V: $\left[-\frac{1}{5}, 0\right]$, $\left[\frac{1}{5}, 0\right]$,

$\left[0, -\frac{1}{4}\right]$, $\left[0, \frac{1}{4}\right]$

F: $\left[0, -\frac{3}{20}\right]$, $\left[0, \frac{3}{20}\right]$

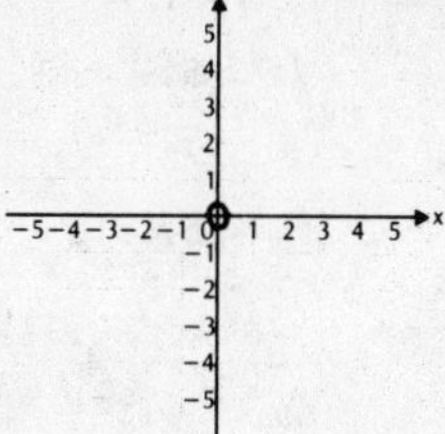

15. $\frac{(x-1)^2}{4} + \frac{(y-2)^2}{1} = 1$

$\frac{(x-1)^2}{2^2} + \frac{(y-2)^2}{1^2} = 1$

The center is (1, 2); a = 2 and b = 1. Since a > b, we find c using $c^2 = a^2 - b^2$.

$c^2 = a^2 - b^2 = 4 - 1 = 3$

Thus $c = \sqrt{3}$.

Consider the ellipse $\frac{x^2}{2^2} + \frac{y^2}{1^2} = 1$.

The center is (0, 0); a = 2, b = 1, and $c = \sqrt{3}$.

The vertices are (-2, 0), (2, 0), (0, -1), and (0, 1).

The foci are $(-\sqrt{3}, 0)$ and $(\sqrt{3}, 0)$.

The vertices and foci of the translated ellipse are found by translation in the same way in which the center has been translated.

The vertices are

(-2 + 1, 0 + 2), (2 + 1, 0 + 2), (0 + 1, -1 + 2), and (0 + 1, 1 + 2).

or

(-1, 2), (3, 2), (1, 1), and (1, 3).

The foci are

$(-\sqrt{3} + 1, 0 + 2)$ and $(\sqrt{3} + 1, 0 + 2)$

or

$(1 - \sqrt{3}, 2)$ and $(1 + \sqrt{3}, 2)$.

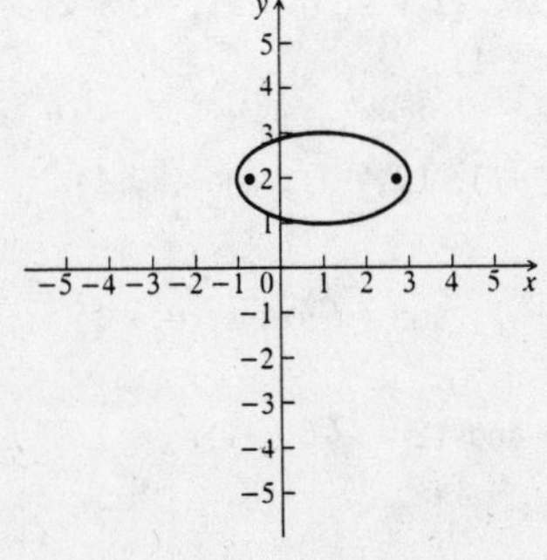

16. C: (1, 2)

V: (0, 2), (2, 2), (1, 0), (1, 4)

F: $(1, 2 - \sqrt{3})$, $(1, 2 + \sqrt{3})$

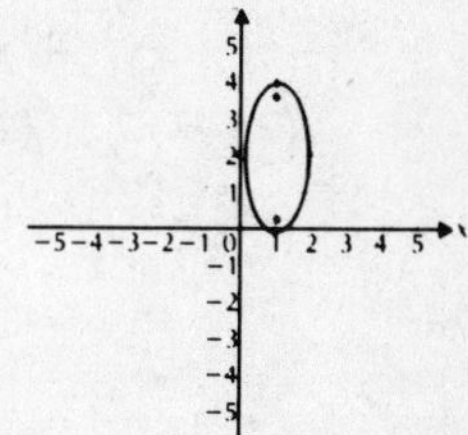

17. $\frac{(x+3)^2}{25} + \frac{(y-2)^2}{16} = 1$

$\frac{[x-(-3)]^2}{5^2} + \frac{(y-2)^2}{4^2} = 1$

The center is (-3, 2); a = 5 and b = 4. Since a > b, we find c using $c^2 = a^2 - b^2$.

$c^2 = a^2 - b^2 = 25 - 16 = 9$

Thus c = 3.

Consider the ellipse $\frac{x^2}{5^2} + \frac{y^2}{4^2} = 1$.

The center is (0, 0); a = 5, b = 4, and c = 3.

The vertices are (-5, 0), (5, 0), (0, -4) and (0, 4).

The foci are (-3, 0) and (3, 0).

The vertices and foci of the translated ellipse are found by translation in the same way in which the center has been translated.

The vertices are

(-5 - 3, 0 + 2), (5 - 3, 0 + 2), (0 - 3, -4 + 2), and (0 - 3, 4 + 2)

or

(-8, 2), (2, 2), (-3, -2), and (-3, 6).

The foci are

(-3 - 3, 0 + 2) and (3 - 3, 0 + 2)

or

(-6, 2) and (0, 2).

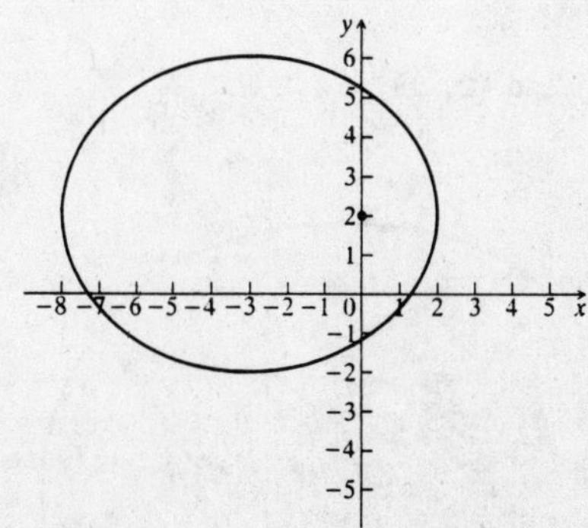

18. C: (2, -3)

V: (-3, -3), (7, -3), (2, -7), (2, 1)

F: (-1, -3), (5, -3)

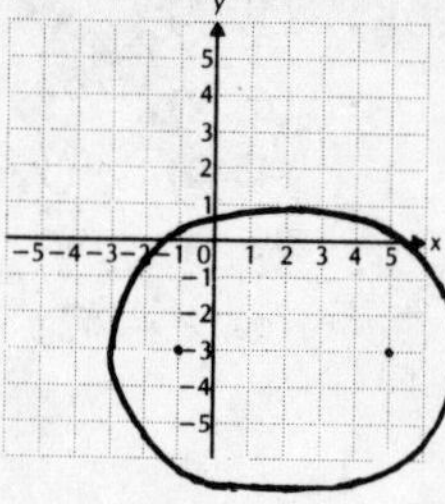

19. $3(x + 2)^2 + 4(y - 1)^2 = 192$

$$\frac{(x + 2)^2}{64} + \frac{(y - 1)^2}{48} = 1 \quad \left[\text{Multiplying by } \frac{1}{192}\right]$$

$$\frac{(x + 2)^2}{8^2} + \frac{(y - 1)^2}{(4\sqrt{3})^2} = 1$$

The center is (-2, 1); $a = 8$ and $b = 4\sqrt{3}$. Since $a > b$, we find c using $c^2 = a^2 - b^2$.

$c^2 = a^2 - b^2 = 64 - 48 = 16$

Thus $c = 4$.

Consider the ellipse $\frac{x^2}{8^2} + \frac{y^2}{(4\sqrt{3})^2} = 1$

The center is (0, 0); $a = 8$, $b = 4\sqrt{3}$, and $c = 4$.
The vertices are (-8, 0), (8, 0), $(0, -4\sqrt{3})$ and $(0, 4\sqrt{3})$.
The foci are (-4, 0) and (4, 0).

The vertices and foci of the translated ellipse are found by translation in the same way in which the center has been translated.

The vertices are

$(-8 - 2, 0 + 1)$, $(8 - 2, 0 + 1)$, $(0 - 2, -4\sqrt{3} + 1)$ and $(0 - 2, 4\sqrt{3} + 1)$

or

(-10, 1), (6, 1), $(-2, 1 - 4\sqrt{3})$, and $(-2, 1 + 4\sqrt{3})$.

The foci are

$(-4 - 2, 0 + 1)$ and $(4 - 2, 0 + 1)$

or

(-6, 1) and (2, 1).

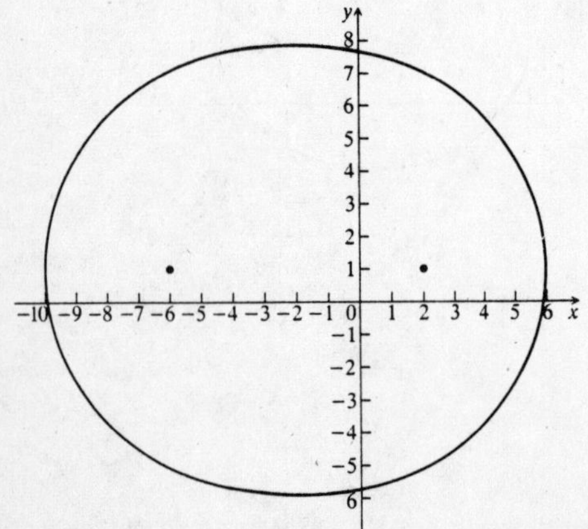

20. C: (5, 5)

V: $(5 - 4\sqrt{3}, 5)$, $(5 + 4\sqrt{3}, 5)$, (5, -3), (5, 13)

F: (5, 1), (5, 9)

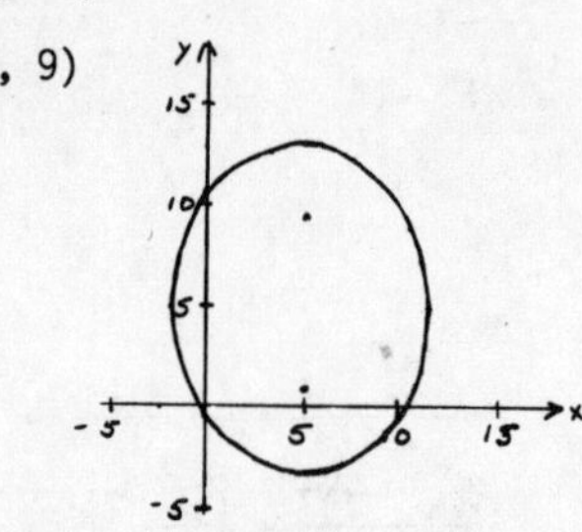

21. $4x^2 + 9y^2 - 16x + 18y - 11 = 0$

$$(4x^2 - 16x) + (9y^2 + 18y) = 11$$

$$4(x^2 - 4x) + 9(y^2 + 2y) = 11$$

$$4(x^2 - 4x + 4) + 9(y^2 + 2y + 1) = 11 + 16 + 9$$

$$4(x - 2)^2 + 9(y + 1)^2 = 36$$

$$\frac{(x - 2)^2}{9} + \frac{(y + 1)^2}{4} = 1$$

$$\frac{(x - 2)^2}{3^2} + \frac{[y - (-1)]^2}{2^2} = 1$$

The center is (2, -1); $a = 3$ and $b = 2$. Since $a > b$, we find c using $c^2 = a^2 - b^2$.

$c^2 = a^2 - b^2 = 9 - 4 = 5$

Thus $c = \sqrt{5}$.

Consider the ellipse $\frac{x^2}{3^2} + \frac{y^2}{2^2} = 1$.

The center is (0, 0); $a = 3$, $b = 2$, and $c = \sqrt{5}$.
The vertices are (-3, 0), (3, 0), (0, -2), and (0, 2).
The foci are $(-\sqrt{5}, 0)$ and $(\sqrt{5}, 0)$.

The vertices and foci of the translated ellipse are found by translation in the same way in which the center has been translated.

The vertices are

$(-3 + 2, 0 - 1)$, $(3 + 2, 0 - 1)$, $(0 + 2, -2 - 1)$, and $(0 + 2, 2 - 1)$

or

(-1, -1), (5, -1), (2, -3), and (2, 1).

The foci are

$(-\sqrt{5} + 2, 0 - 1)$ and $(\sqrt{5} + 2, 0 - 1)$

or

$(2 - \sqrt{5}, -1)$ and $(2 + \sqrt{5}, -1)$.

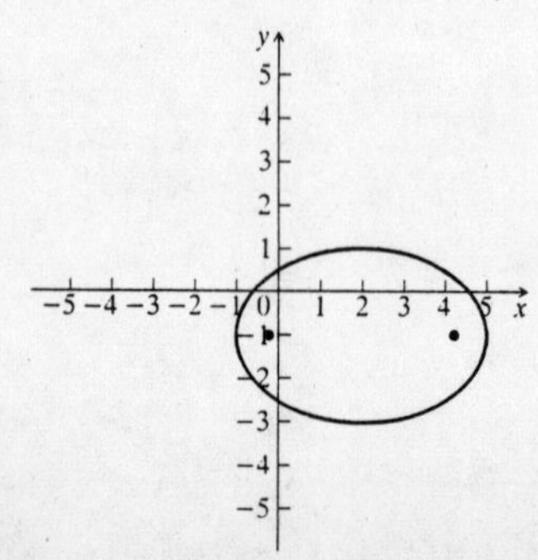

22. C: (5, -2)

V: (3, -2), (7, -2),

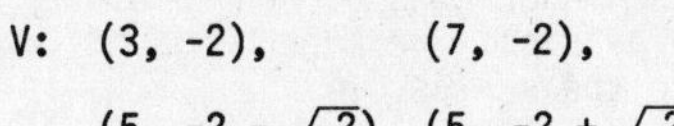

$(5, -2 - \sqrt{2})$, $(5, -2 + \sqrt{2})$

F: $(5 - \sqrt{2}, -2)$, $(5 + \sqrt{2}, -2)$

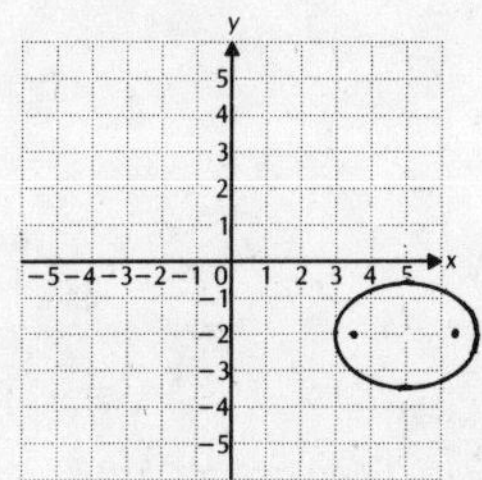

23.

$$4x^2 + y^2 - 8x - 2y + 1 = 0$$
$$(4x^2 - 8x) + (y^2 - 2y) = -1$$
$$4(x^2 - 2x) + (y^2 - 2y) = -1$$
$$4(x^2 - 2x + 1) + (y^2 - 2y + 1) = -1 + 4 + 1$$
$$4(x - 1)^2 + (y - 1)^2 = 4$$
$$\frac{(x - 1)^2}{1^2} + \frac{(y - 1)^2}{2^2} = 1$$

The center is (1, 1); a = 1 and b = 2. Since b > a, we find c using $c^2 = b^2 - a^2$.

$c^2 = b^2 - a^2 = 4 - 1 = 3$

Thus $c = \sqrt{3}$.

Consider the ellipse $\frac{x^2}{1^2} + \frac{y^2}{2^2} = 1$.

The center is (0, 0); a = 1, b = 2, and $c = \sqrt{3}$.

The vertices are (-1, 0), (1, 0), (0, -2), and (0, 2).

The foci are $(0, -\sqrt{3})$ and $(0, \sqrt{3})$.

The vertices and foci of the translated ellipse are found by translation in the same way in which the center has been translated.

The vertices are

(-1 + 1, 0 + 1), (1 + 1, 0 + 1), (0 + 1, -2 + 1), and (0 + 1, 2 + 1)

or

(0, 1), (2, 1), (1, -1), and (1, 3).

The foci are

$(0 + 1, -\sqrt{3} + 1)$ and $(0 + 1, \sqrt{3} + 1)$

or

$(1, 1 - \sqrt{3})$ and $(1, 1 + \sqrt{3})$.

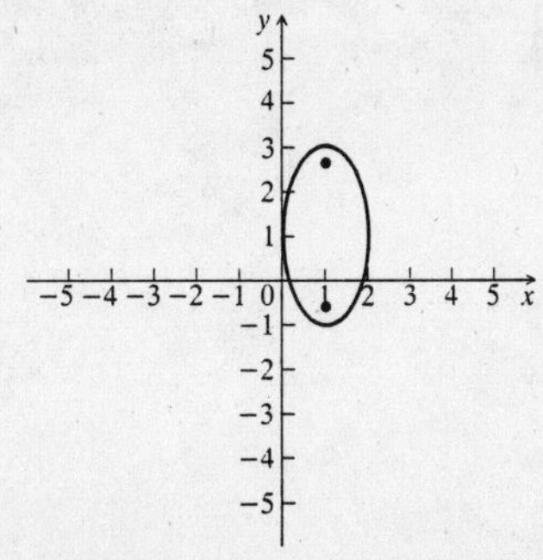

24. C: (-3, 1)

V: (-5, 1), (-1, 1)
(-3, -2), (-3, 4)

F: $(-3, 1 - \sqrt{5})$,
$(-3, 1 + \sqrt{5})$

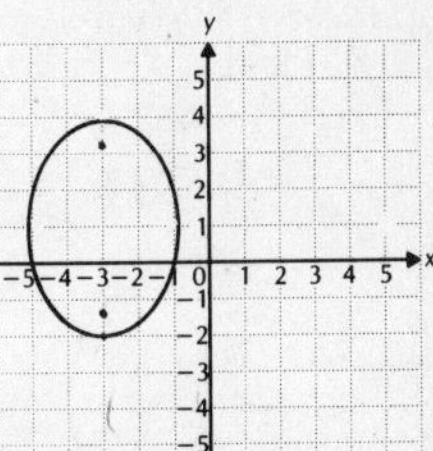

25.

$$4x^2 + 9y^2 - 16.025x + 18.0927y - 11.346 = 0$$
$$(4x^2 - 16.025x) + (9y^2 + 18.0927y) = 11.346$$
$$4(x^2 - 4.00625x) + 9(y^2 + 2.0103y) = 11.346$$
$$4(x^2 - 4.00625x + 4.0125098) + 9(y^2 + 2.0103y + 1.0103265) = 11.346 + 16.0500392 + 9.0929385$$
$$4(x - 2.003125)^2 + 9(y + 1.00515)^2 = 36.4889777$$
$$\frac{(x - 2.003125)^2}{9.122244425} + \frac{(y + 1.00515)^2}{4.054330856} = 1$$

Consider the ellipse $\frac{x^2}{(3.020305)^2} + \frac{y^2}{(2.013537)^2} = 1$.

The center is (0, 0). The vertices are (-3.020305, 0), (3.020305, 0), (0, -2.013537), and (0, 2.013537).

The center of the translated ellipse is (2.003125, -1.00515).

The vertices of the translated ellipse are:

(-3.020305 + 2.003125, 0 - 1.00515),
(3.020305 + 2.003125, 0 - 1.00515),
(0 + 2.003125, -2.013537 - 1.00515),
(0 + 2.003125, 2.013537 - 1.00515)

or

(-1.01718, -1.00515), (5.02343, -1.00515),
(2.003125, -3.018687), (2.003125, 1.008387).

26. C: (-3.0035, 1.002)

V: (-3.0035, -1.97008), (-3.0035, 3.97408)
(-1.02211, 1.002), (-4.98489, 1.002)

27. Graph the vertices and sketch the axes of the ellipse.

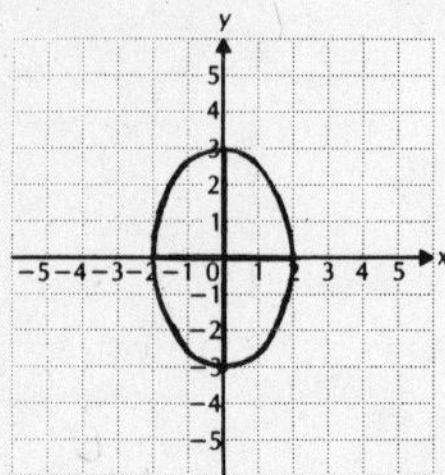

The intersection of the axes, which is (0, 0), is the center of the ellipse. Note that a = 2 and b = 3. Now find an equation of the ellipse (in standard form).

$$\frac{(x - h)^2}{a^2} + \frac{(y - k)^2}{b^2} = 1$$

$$\frac{(x - 0)^2}{2^2} + \frac{(y - 0)^2}{3^2} = 1 \quad \text{(Substituting)}$$

$$\frac{x^2}{4} + \frac{y^2}{9} = 1$$

28. $x^2 + \frac{y^2}{16} = 1$

29. Graph the vertices and sketch the axes of the ellipse.

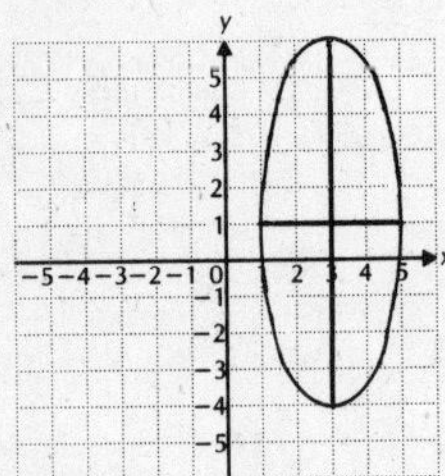

The intersection of the axes, which is (3, 1), is the center of the ellipse. Note that a = 2 and b = 5. Now find an equation of the ellipse (in standard form).

$$\frac{(x - h)^2}{a^2} + \frac{(y - k)^2}{b^2} = 1$$

$$\frac{(x - 3)^2}{2^2} + \frac{(y - 1)^2}{5^2} = 1 \quad \text{(Substituting)}$$

$$\frac{(x - 3)^2}{4} + \frac{(y - 1)^2}{25} = 1$$

30. $\frac{(x + 1)^2}{4} + \frac{(y - 2)^2}{9} = 1$

31. Graph the center, (-2, 3), and sketch the axes of the ellipse. The major axis of length 4 is parallel to the y-axis. The minor axis of length 1 is parallel to the x-axis.

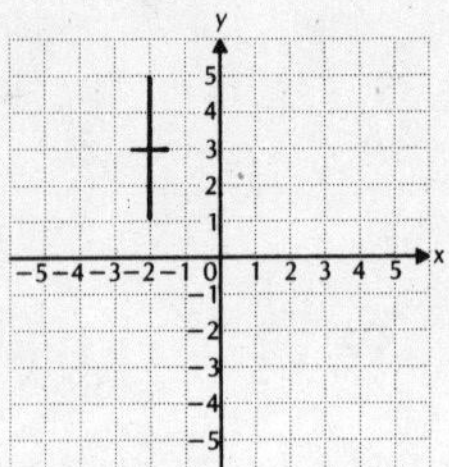

Note that $a = \frac{1}{2}$ and b = 2. Find an equation (in standard form) of the ellipse.

$$\frac{(x - h)^2}{a^2} + \frac{(y - k)^2}{b^2} = 1$$

$$\frac{[x - (-2)]^2}{\left(\frac{1}{2}\right)^2} + \frac{(y - 3)^2}{2^2} = 1 \quad \text{(Substituting)}$$

$$\frac{(x + 2)^2}{\frac{1}{4}} + \frac{(y - 3)^2}{4} = 1$$

32. $\frac{x^2}{9} + \frac{5y^2}{484} = 1$

33. a)

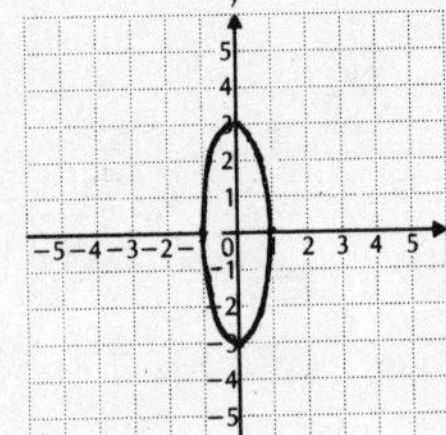

The relation is not a function. It does not pass the vertical line test.

b) $9x^2 + y^2 = 9$

$y^2 = 9 - 9x^2$

$y = \pm \sqrt{9 - 9x^2}$

$y = \pm 3\sqrt{1 - x^2}$

c)

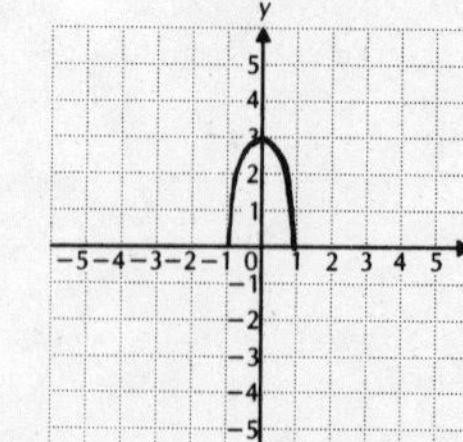

The relation is a function. It does pass the vertical line test.

Domain: $\{x | -1 \leqslant x \leqslant 1\}$

Range: $\{y | 0 \leqslant y \leqslant 3\}$

33. (continued)

d)

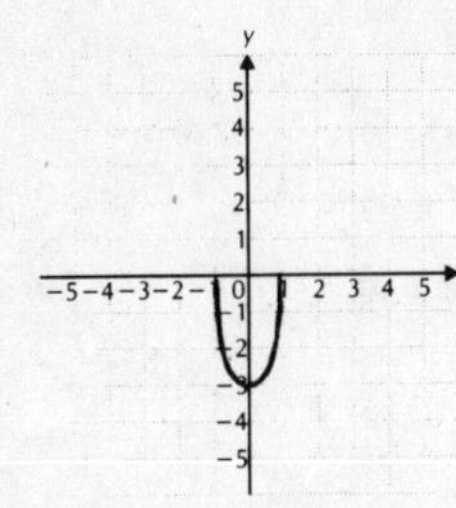

The relation is a function. It does pass the vertical line test.

Domain: $\{x \mid -1 \leqslant x \leqslant 1\}$

Range: $\{y \mid -3 \leqslant y \leqslant 0\}$

34. Circle with center (0, 0) and radius = a.

35.

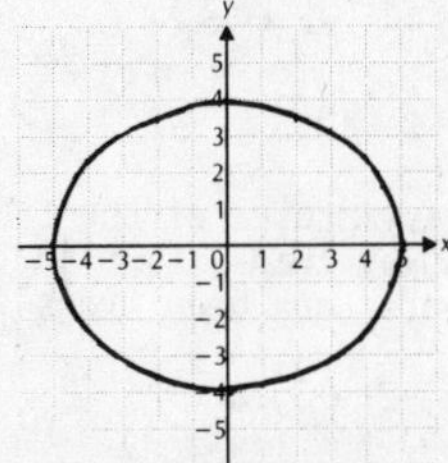

$$\frac{x^2}{25} + \frac{y^2}{36} = 1$$

36. 2×10^6 mi

37. The vertices are (-50, 0), (50, 0), (0, -12), and (0, 12). The equation of the ellipse is

$$\frac{x^2}{50^2} + \frac{y^2}{12^2} = 1, \text{ or } \frac{x^2}{2500} + \frac{y^2}{144} = 1.$$

38. $\frac{x^2}{524^2} + \frac{y^2}{449^2} = 1$, or $\frac{x^2}{274{,}576} + \frac{y^2}{201{,}601} = 1$

39.

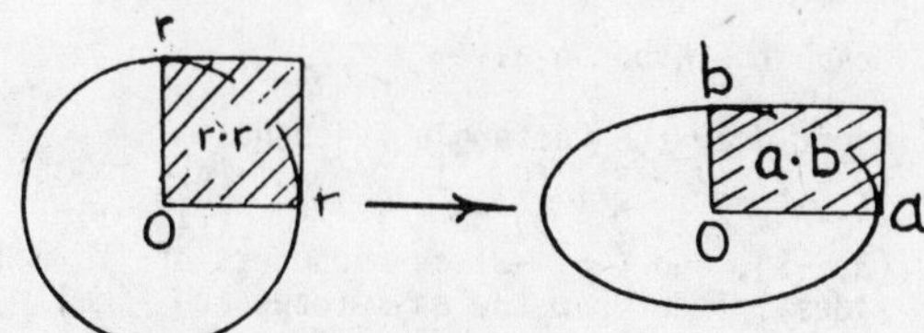

Stretching and shrinking, the area of the square, r·r, becomes the area of the rectangle, a·b. Hence, the area of the circle, π·r·r, becomes the area of the ellipse, π·a·b.

b) a = 4, b = 5

Area = $\pi \cdot 4 \cdot 5 = 20\pi$

c) a = 524, b = 449

Area = $\pi \cdot 524 \cdot 449 \approx 739{,}141.4$ ft^2

40. C = C(x, y)

θ is the angle which BC makes with the positive x-axis. Let AC = a, BC = b. Then, in all quadrants, $\frac{x}{a} = \cos\theta$ and $\frac{y}{b} = \sin\theta$.

Hence $\frac{x^2}{a^2} + \frac{y^2}{b^2} = 1$ which is the equation of an ellipse.

Exercise Set 10.2

1. $$\frac{x^2}{9} - \frac{y^2}{1} = 1$$

$$\frac{(x - 0)^2}{3^2} - \frac{(y - 0)^2}{1^2} = 1$$

The center is (0, 0); a = 3 and b = 1.

The vertices are (-3, 0) and (3, 0).

Since $c^2 = a^2 + b^2$, $c = \sqrt{a^2 + b^2} = \sqrt{9 + 1} = \sqrt{10}$.

The foci are $(-\sqrt{10}, 0)$ and $(\sqrt{10}, 0)$.

The asymptotes are $y = -\frac{1}{3}x$ and $y = \frac{1}{3}x$.

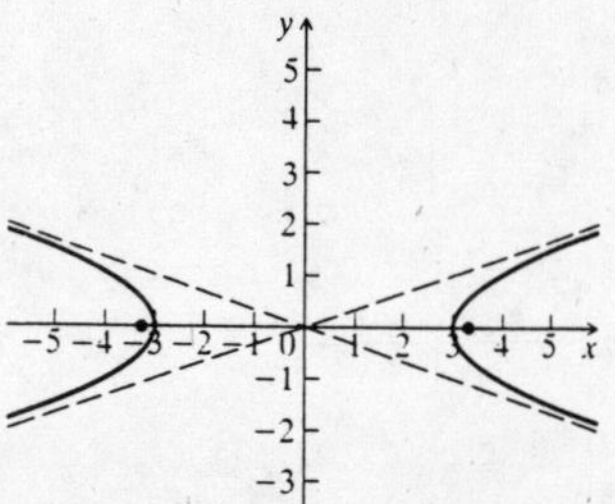

2. C: (0, 0)

V: (-1, 0), (1, 0)

F: $(-\sqrt{10}, 0)$, $(\sqrt{10}, 0)$

A: y = 3x, y = -3x

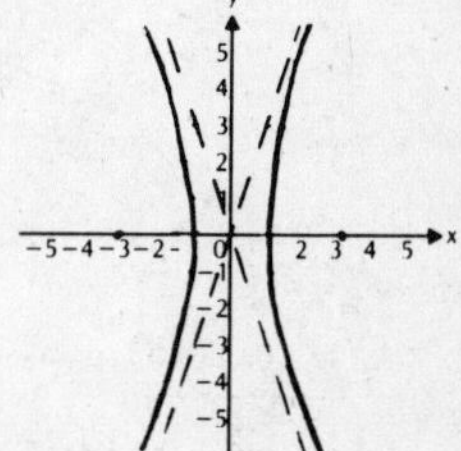

3. $$\frac{(x - 2)^2}{9} - \frac{(y + 5)^2}{1} = 1$$

$$\frac{(x - 2)^2}{3^2} - \frac{[y - (-5)]^2}{1^2} = 1$$

The center is (2, -5); a = 3 and b = 1.

The transverse axis is parallel to the x-axis.

Since $c^2 = a^2 + b^2$, $c = \sqrt{a^2 + b^2} = \sqrt{9 + 1} = \sqrt{10}$.

Consider the hyperbola $\frac{x^2}{3^2} - \frac{y^2}{1^2} = 1$.

The center is (0, 0); a = 3, b = 1, and c = $\sqrt{10}$.

The vertices are (-3, 0) and (3, 0).

The foci are $(-\sqrt{10}, 0)$ and $(\sqrt{10}, 0)$.

The asymptotes are $y = -\frac{1}{3}x$ and $y = \frac{1}{3}x$.

The vertices, foci, and asymptotes of the translated hyperbola are found by translation in the same way in which the center has been translated.

The vertices are (-3 + 2, 0 - 5) and (3 + 2, 0 - 5), or (-1, -5) and (5, -5).

The foci are $(-\sqrt{10} + 2, 0 - 5)$ and $(\sqrt{10} + 2, 0 - 5)$, or $(2 - \sqrt{10}, -5)$ and $(2 + \sqrt{10}, -5)$.

3. (continued)

The asymptotes are

$y - (-5) = -\frac{1}{3}(x - 2)$ and $y - (-5) = \frac{1}{3}(x - 2)$

or $y = -\frac{1}{3}x - \frac{13}{3}$ and $y = \frac{1}{3}x - \frac{17}{3}$.

Graph the hyperpola.

First draw the rectangle which has (-3 + 2, 0 - 5), (3 + 2, 0 - 5), (0 + 2, -1 - 5), and (0 + 2, 1 - 5) or (-1, -5), (5, -5), (2, -6), and (2, -4) as midpoints of its four sides. Then draw the asymptotes and finally the branches of the hyperbola outward from the vertices toward the asymptotes.

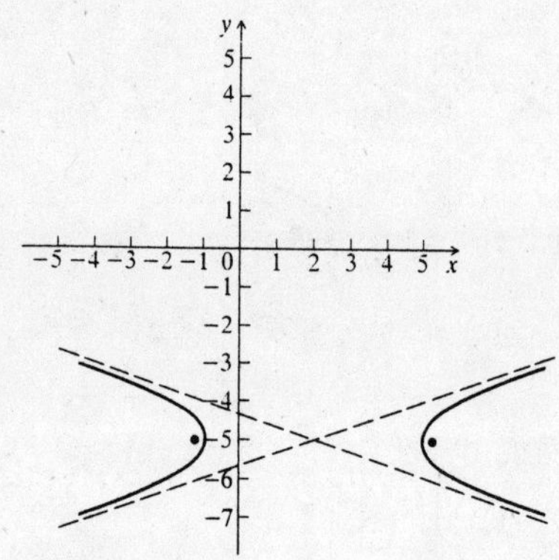

4. C: (2, -5)

V: (1, -5), (3, -5)

F: $(2 - \sqrt{10}, -5)$, $(2 + \sqrt{10}, -5)$

A: $y = -3x + 1$, $y = 3x - 11$

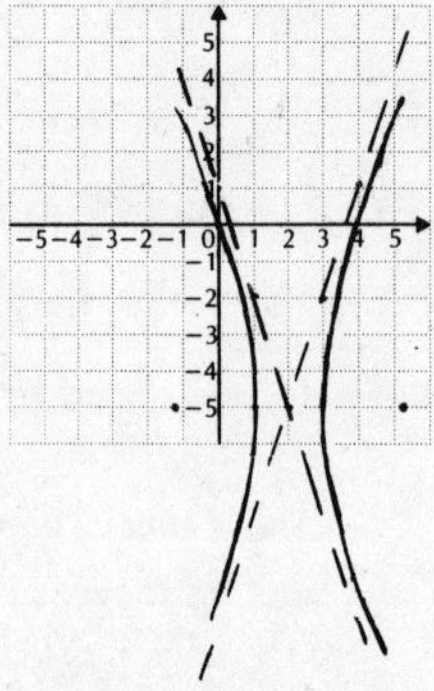

5. $\frac{(y + 3)^2}{4} - \frac{(x + 1)^2}{16} = 1$

$\frac{[y - (-3)]^2}{2^2} - \frac{[x - (-1)]^2}{4^2} = 1$

The center is (-1, -3); a = 4 and b = 2.

The transverse axis is parallel to the y-axis.

Since $c^2 = a^2 + b^2$, $c = \sqrt{a^2 + b^2} = \sqrt{16 + 4} = 2\sqrt{5}$.

Consider the hyperbola $\frac{y^2}{2^2} - \frac{x^2}{4^2} = 1$.

The center is (0, 0); a = 4, b = 2, and $c = 2\sqrt{5}$.

The vertices are (0, -2) and (0, 2).

The foci are $(0, -2\sqrt{5})$ and $(0, 2\sqrt{5})$.

The asymptotes are $y = -\frac{1}{2}x$ and $y = \frac{1}{2}x$.

The vertices, foci, and asymptotes of the translated hyperbola are found by translation in the same way in which the center has been translated.

The vertices are (0 - 1, -2 - 3) and (0 - 1, 2 - 3) or (-1, -5) and (-1, -1).

The foci are $(0 - 1, -2\sqrt{5} - 3)$ and $(0 - 1, 2\sqrt{5} - 3)$ or $(-1, -3 - 2\sqrt{5})$ and $(-1, -3 + 2\sqrt{5})$.

The asymptotes are

$y - (-3) = -\frac{1}{2}[x - (-1)]$ and

$y - (-3) = \frac{1}{2}[x - (-1)]$ or $y = -\frac{1}{2}x - \frac{7}{2}$

and $y = \frac{1}{2}x - \frac{5}{2}$.

Graph the hyperbola.

First draw the rectangle which has (0 - 1, -2 - 3), (0 - 1, 2 - 3), (4 - 1, 0 - 3), and (-4 - 1, 0 - 3) or (-1, -5), (-1, -1), (3, -3), and (-5, -3) as midpoints of its four sides. Then draw the asymptotes and finally the branches of the hyperbola outward from the vertices toward the asymptotes.

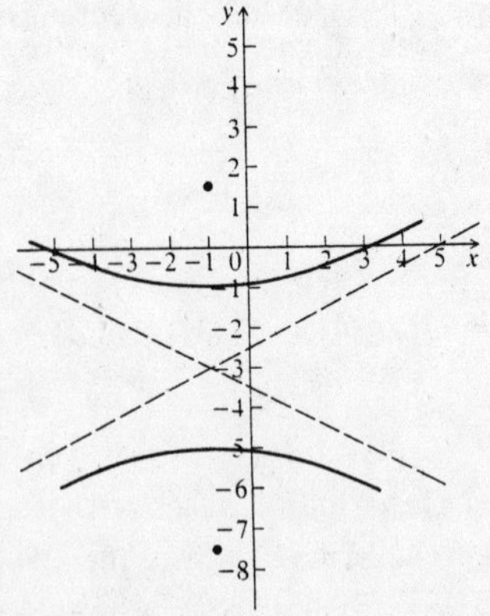

6. C: (-1, -3)

V: (-1, -8), (-1, 2)

F: $(-1, -3 - \sqrt{41})$, $(-1, -3 + \sqrt{41})$

A: $y = -\frac{5}{4}x - \frac{17}{4}$, $y = \frac{5}{4}x - \frac{7}{4}$

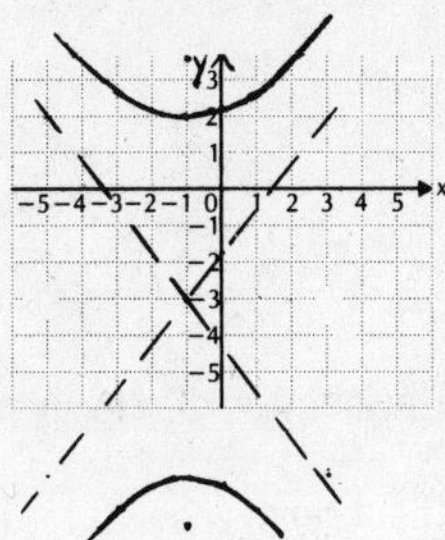

7. $$x^2 - 4y^2 = 4$$
$$\frac{x^2}{4} - \frac{y^2}{1} = 1$$
$$\frac{(x-0)^2}{2^2} - \frac{(y-0)^2}{1^2} = 1$$

The center is (0, 0); a = 2 and b = 1.
The vertices are (-2, 0) and (2, 0).

Since $c^2 = a^2 + b^2$, $c = \sqrt{a^2 + b^2} = \sqrt{4 + 1} = \sqrt{5}$.

The foci are $(-\sqrt{5}, 0)$ and $(\sqrt{5}, 0)$.

The asymptotes are $y = -\frac{1}{2}x$ and $y = \frac{1}{2}x$.

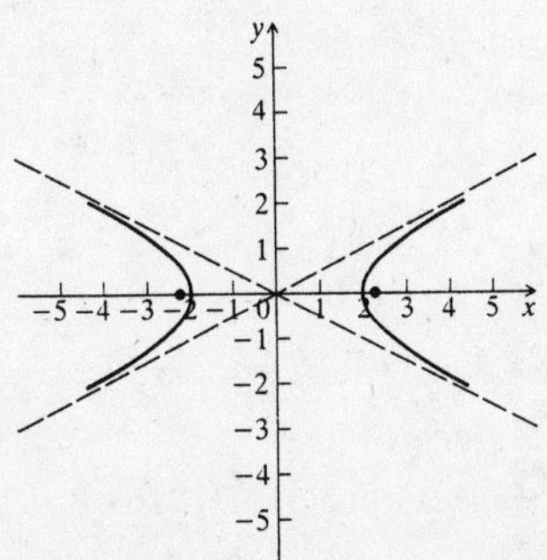

8. C: (0, 0)

V: (-1, 0), (1, 0)

F: $(-\sqrt{5}, 0)$, $(\sqrt{5}, 0)$

A: y = -2x, y = 2x

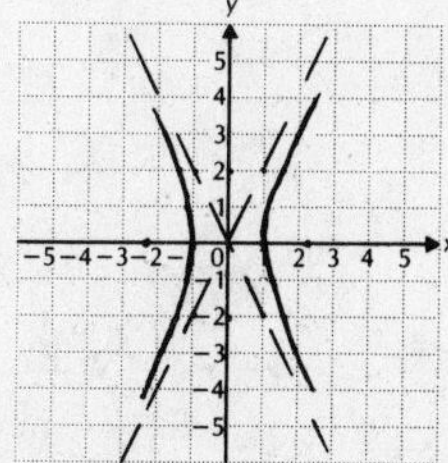

9. $$4y^2 - x^2 = 4$$
$$y^2 - \frac{x^2}{4} = 1$$
$$\frac{(y-0)^2}{1^2} - \frac{(x-0)^2}{2^2} = 1$$

The center is (0, 0); a = 2 and b = 1.
The vertices are (0, -1) and (0, 1).

Since $c^2 = a^2 + b^2$, $c = \sqrt{a^2 + b^2} = \sqrt{4 + 1} = \sqrt{5}$.

The foci are $(0, -\sqrt{5})$ and $(0, \sqrt{5})$.

The asymptotes are $y = -\frac{1}{2}x$ and $y = \frac{1}{2}x$.

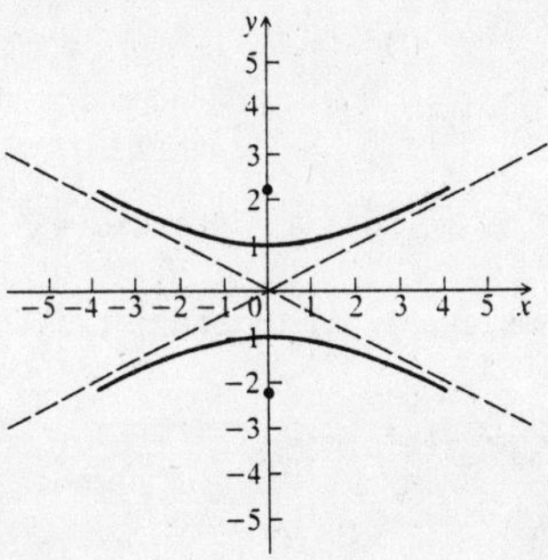

10. C: (0, 0)

V: (0, -2), (0, 2)

F: $(0, -\sqrt{5})$, $(0, \sqrt{5})$

A: y = -2x, y = 2x

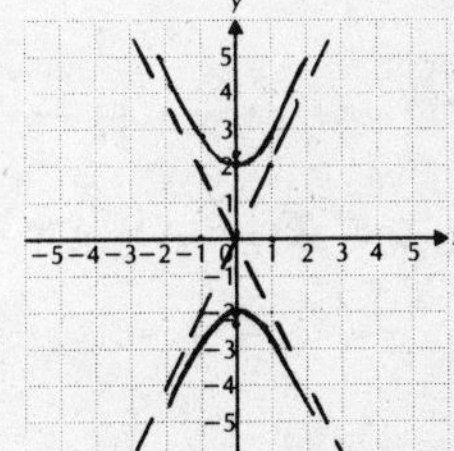

11. $$x^2 - y^2 = 2$$
$$\frac{x^2}{2} - \frac{y^2}{2} = 1$$
$$\frac{(x-0)^2}{(\sqrt{2})^2} - \frac{(y-0)^2}{(\sqrt{2})^2} = 1$$

The center is (0, 0); $a = \sqrt{2}$ and $b = \sqrt{2}$.

The vertices are $(-\sqrt{2}, 0)$ and $(\sqrt{2}, 0)$.

Since $c^2 = a^2 + b^2$, $c = \sqrt{a^2 + b^2} = \sqrt{2 + 2} = \sqrt{4} = 2$.

The foci are (-2, 0) and (2, 0).

The asymptotes are y = -x and y = x.

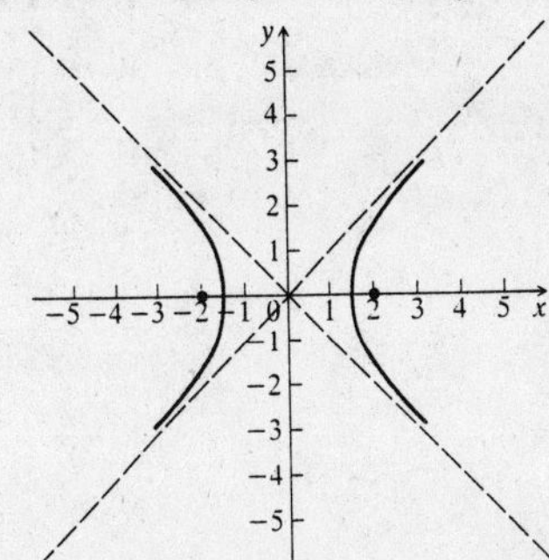

12. C: (0, 0)

V: $(-\sqrt{3}, 0)$, $(\sqrt{3}, 0)$

F: $(-\sqrt{6}, 0)$, $(\sqrt{6}, 0)$

A: $y = -x$, $y = x$

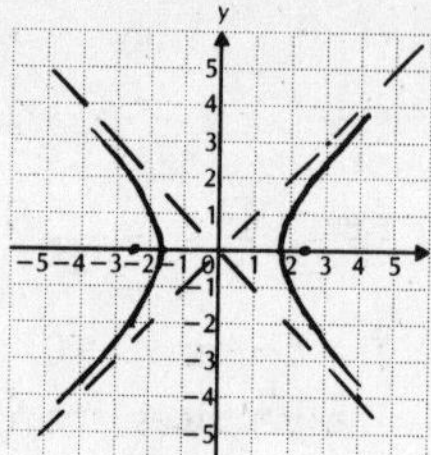

13.
$$x^2 - y^2 = \frac{1}{4}$$
$$\frac{x^2}{\frac{1}{4}} - \frac{y^2}{\frac{1}{4}} = 1$$
$$\frac{(x - 0)^2}{\left(\frac{1}{2}\right)^2} - \frac{(y - 0)^2}{\left(\frac{1}{2}\right)^2} = 1$$

The center is (0, 0); $a = \frac{1}{2}$ and $b = \frac{1}{2}$.

The vertices are $\left(-\frac{1}{2}, 0\right)$ and $\left(\frac{1}{2}, 0\right)$.

Since $c^2 = a^2 + b^2$, $c = \sqrt{a^2 + b^2} = \sqrt{\frac{1}{4} + \frac{1}{4}}$

$= \sqrt{\frac{1}{2}} = \frac{\sqrt{2}}{2}$.

The foci are $\left(-\frac{\sqrt{2}}{2}, 0\right)$ and $\left(\frac{\sqrt{2}}{2}, 0\right)$.

The asymptotes are $y = -x$ and $y = x$.

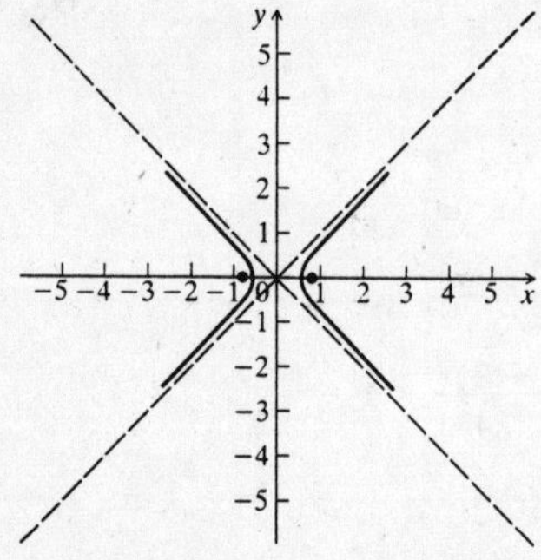

14. C: (0, 0)

V: $\left(-\frac{1}{3}, 0\right)$, $\left(\frac{1}{3}, 0\right)$

F: $\left(-\frac{\sqrt{2}}{3}, 0\right)$, $\left(\frac{\sqrt{2}}{3}, 0\right)$

A: $y = -x$, $y = x$

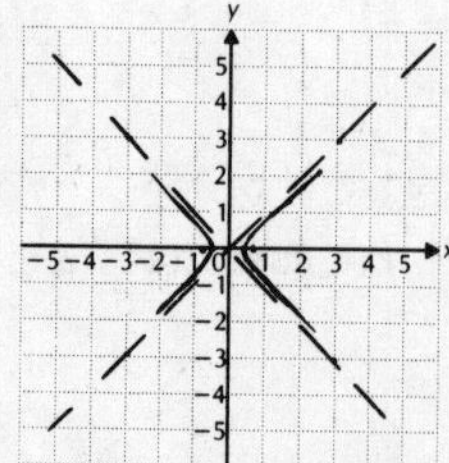

15.
$$x^2 - y^2 - 2x - 4y - 4 = 0$$
$$(x^2 - 2x) - (y^2 + 4y) = 4$$
$$(x^2 - 2x + 1) - (y^2 + 4y + 4) = 4 + 1 - 4$$
$$(x - 1)^2 - (y + 2)^2 = 1$$
$$\frac{(x - 1)^2}{1^2} - \frac{[y - (-2)]^2}{1^2} = 1$$

The center of the hyperbola is (1, -2); $a = 1$ and $b = 1$.

The transverse axis is parallel to the x-axis.

Since $c^2 = a^2 + b^2$, $c = \sqrt{a^2 + b^2} = \sqrt{1 + 1}$
$= \sqrt{2}$.

Consider the hyperbola $\frac{x^2}{1^2} - \frac{y^2}{1^2} = 1$

The center is (0, 0); $a = 1$, $b = 1$, and $c = \sqrt{2}$.

The vertices are (-1, 0) and (1, 0).

The foci are $(-\sqrt{2}, 0)$ and $(\sqrt{2}, 0)$.

The asymptotes are $y = -x$ and $y = x$.

The vertices, foci, and asymptotes of the translated hyperbola are found in the same way in which the center has been translated.

The vertices are

(-1 + 1, 0 - 2) and (1 + 1, 0 - 2)

or

(0, -2) and (2, -2).

The foci are

$(-\sqrt{2} + 1, 0 - 2)$ and $(\sqrt{2} + 1, 0 - 2)$

or

$(1 - \sqrt{2}, -2)$ and $(1 + \sqrt{2}, -2)$.

The asymptotes are

$y - (-2) = -(x - 1)$ and $y - (-2) = x - 1$

or

$y = -x - 1$ and $y = x - 3$.

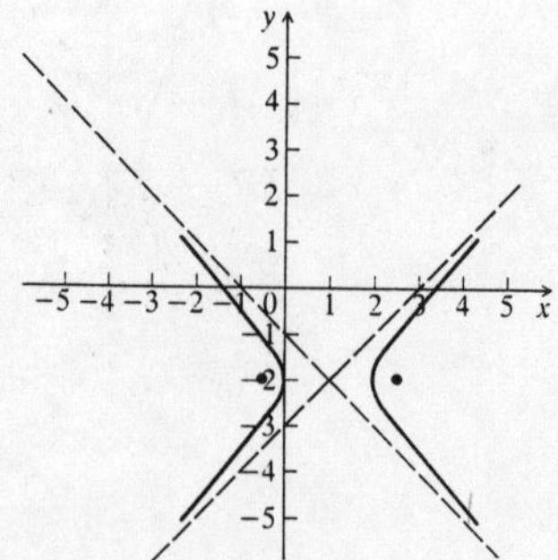

16. C: (-1, -2)

V: (-2, -2), (0, -2)

F: $(-1 - \sqrt{5}, -2)$,
$(-1 + \sqrt{5}, -2)$

A: $y = -2x - 4$, $y = 2x$

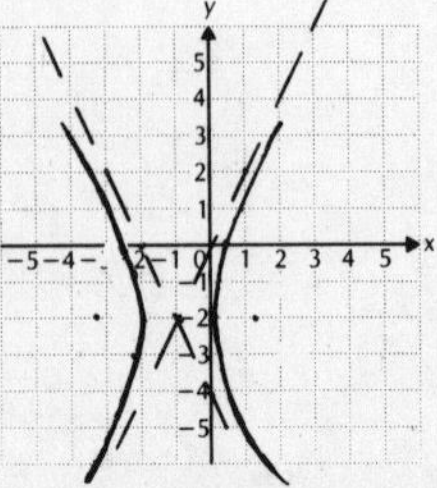

17. $$36x^2 - y^2 - 24x + 6y - 41 = 0$$
$$(36x^2 - 24x) - (y^2 - 6y) = 41$$
$$36\left(x^2 - \frac{2}{3}x\right) - (y^2 - 6y) = 41$$
$$36\left(x^2 - \frac{2}{3}x + \frac{1}{9}\right) - (y^2 - 6y + 9) = 41 + 4 - 9$$
$$36\left(x - \frac{1}{3}\right)^2 - (y - 3)^2 = 36$$
$$\frac{\left(x - \frac{1}{3}\right)^2}{1} - \frac{(y - 3)^2}{36} = 1$$
$$\frac{\left(x - \frac{1}{3}\right)^2}{1^2} - \frac{(y - 3)^2}{6^2} = 1$$

The center of the hyperbola is $\left(\frac{1}{3}, 3\right)$; a = 1 and b = 6.

The transverse axis is parallel to the x-axis.

Since $c^2 = a^2 + b^2$, $c = \sqrt{a^2 + b^2} = \sqrt{1 + 36}$ $= \sqrt{37}$.

Consider the hyperbola $\frac{x^2}{1^2} - \frac{y^2}{6^2} = 1$.

The center is (0, 0); a = 1, b = 6, and $c = \sqrt{37}$.
The vertices are (-1, 0) and (1, 0).
The foci are $(-\sqrt{37}, 0)$ and $(\sqrt{37}, 0)$.
The asymptotes are y = -6x and y = 6x.

The vertices, foci, and asymptotes of the translated hyperbola are found in the same way in which the center has been translated.

The vertices are

$\left(-1 + \frac{1}{3}, 0 + 3\right)$ and $\left(1 + \frac{1}{3}, 0 + 3\right)$

or

$\left(-\frac{2}{3}, 3\right)$ and $\left(\frac{4}{3}, 3\right)$.

The foci are

$\left(-\sqrt{37} + \frac{1}{3}, 0 + 3\right)$ and $\left(\sqrt{37} + \frac{1}{3}, 0 + 3\right)$

or

$\left(\frac{1}{3} - \sqrt{37}, 3\right)$ and $\left(\frac{1}{3} + \sqrt{37}, 3\right)$.

The asymptotes are

$y - 3 = -6\left(x - \frac{1}{3}\right)$ and $y - 3 = 6\left(x - \frac{1}{3}\right)$

or

y = -6x + 5 and y = 6x + 1.

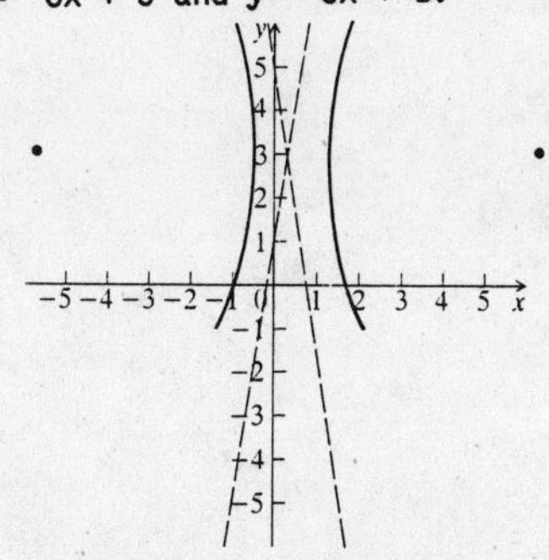

18. C: (-3, 1)

V: $\left(-3 + \frac{4\sqrt{2}}{3}, 1\right)$

$\left(-3 - \frac{4\sqrt{2}}{3}, 1\right)$

F: $\left(-3 + \frac{2\sqrt{26}}{3}, 1\right)$,

$\left(-3 - \frac{2\sqrt{26}}{3}, 1\right)$

A: $y = -\frac{3}{2}x - \frac{7}{2}$,

$y = \frac{3}{2}x + \frac{11}{2}$

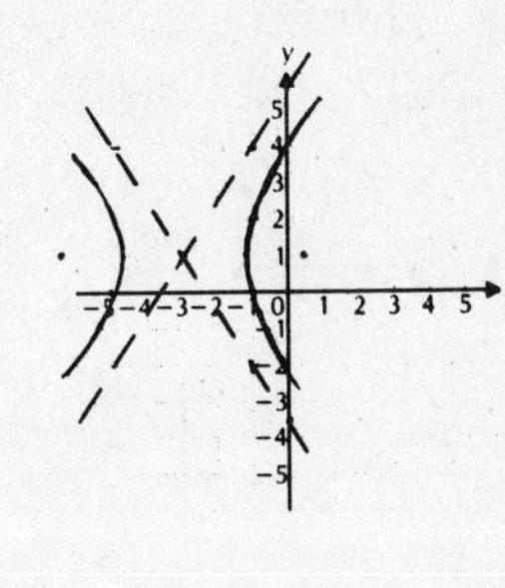

19. $$9y^2 - 4x^2 - 18y + 24x - 63 = 0$$
$$9(y^2 - 2y) - 4(x^2 - 6x) = 63$$
$$9(y^2 - 2y + 1) - 4(x^2 - 6x + 9) = 63 + 9 - 36$$
$$9(y - 1)^2 - 4(x - 3)^2 = 36$$
$$\frac{(y - 1)^2}{4} - \frac{(x - 3)^2}{9} = 1$$

The center of the hyperbola is (3, 1); a = 3 and b = 2.

The transverse axis is parallel to the y-axis.
Since $c^2 = a^2 + b^2$, $c = \sqrt{a^2 + b^2} = \sqrt{9 + 4}$ $= \sqrt{13}$.

Consider the hyperbola $\frac{y^2}{4} - \frac{x^2}{9} = 1$.

The center is (0, 0); a = 3, b = 2, and $c = \sqrt{13}$. The vertices are (0, -2) and (0, 2). The foci are $(0, -\sqrt{13})$ and $(0, \sqrt{13})$. The asymptotes are $y = -\frac{2}{3}x$ and $y = \frac{2}{3}x$.

The vertices, foci, and asymptotes of the translated hyperbola are found in the same way in which the center has been translated.

The vertices are (0 + 3, -2 + 1) and (0 + 3, 2 + 1) or (3, -1) and (3, 3).

The foci are $(0 + 3, -\sqrt{13} + 1)$ and $(0 + 3, \sqrt{13} + 1)$ or $(3, 1 - \sqrt{13})$ and $(3, 1 + \sqrt{13})$.

The asymptotes are

$y - 1 = -\frac{2}{3}(x - 3)$ and $y - 1 = \frac{2}{3}(x - 3)$ or

$y = -\frac{2}{3}x + 3$ and $y = \frac{2}{3}x - 1$.

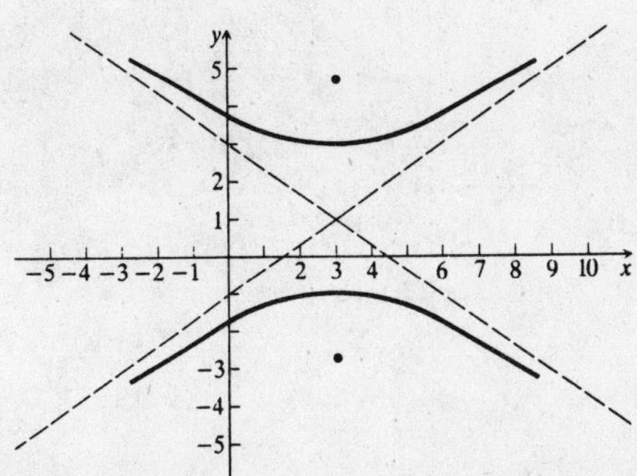

20. C: $(-3, -1)$

V: $(2, -1)$, $(-8, -1)$

F: $(-3 + \sqrt{26}, -1)$, $(-3 - \sqrt{26}, -1)$

A: $y = -\frac{1}{5}x - \frac{8}{5}$, $y = \frac{1}{5}x - \frac{2}{5}$

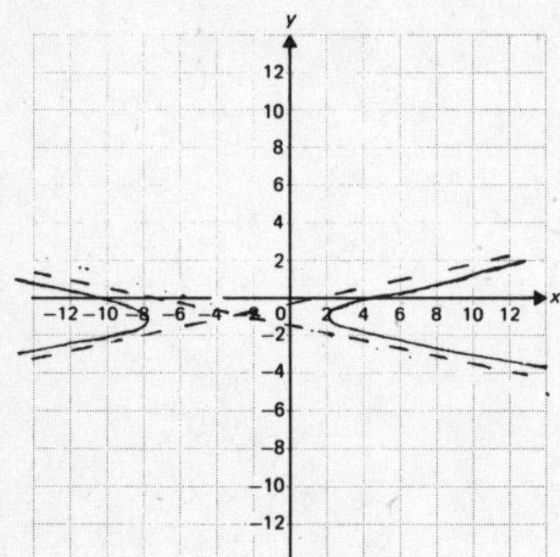

21.
$$x^2 - y^2 - 2x - 4y = 4$$
$$(x^2 - 2x + 1) - (y^2 + 4y + 4) = 4 + 1 - 4$$
$$(x - 1)^2 - (y + 2)^2 = 1, \text{ or}$$
$$\frac{(x - 1)^2}{1} - \frac{(y + 2)^2}{1} = 1$$

The center of the hyperbola is $(1, -2)$, $a = 1$ and $b = 1$.

The transverse axis is parallel to the x-axis. Since $c^2 = a^2 + b^2$, $c = \sqrt{a^2 + b^2} = \sqrt{1 + 1} = \sqrt{2}$.

Consider the hyperbola $\frac{x^2}{1} - \frac{y^2}{1} = 1$. The center is $(0, 0)$; $a = 1$, $b = 1$, and $c = \sqrt{2}$. The vertices are $(-1, 0)$ and $(1, 0)$. The foci are $(-\sqrt{2}, 0)$ and $(\sqrt{2}, 0)$. The asymptotes are $y = -x$ and $y = x$.

The vertices, foci, and asymptotes of the translated hyperbola are found in the same way in which the center has been translated.

The vertices are $(-1 + 1, 0 - 2)$ and $(1 + 1, 0 - 2)$ or $(0, -2)$ and $(2, -2)$.

The foci are $(-\sqrt{2} + 1, 0 - 2)$ and $(\sqrt{2} + 1, 0 - 2)$ or $(1 - \sqrt{2}, -2)$ and $(1 + \sqrt{2}, -2)$.

The asymptotes are

$y + 2 = -(x - 1)$ and $y + 2 = x - 1$, or

$y = -x - 1$ and $y = x - 3$.

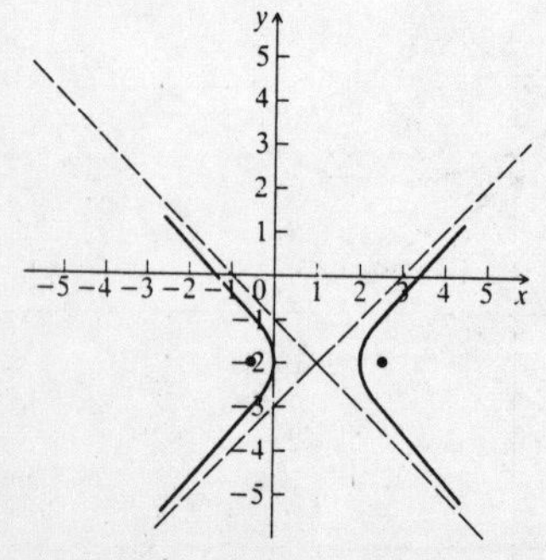

22. C: $(-1, 3)$

V: $(-1, 5)$, $(-1, 1)$

F: $(-1, 3 + \sqrt{13})$, $(-1, 3 - \sqrt{13})$

A: $y = -\frac{2}{3}x + \frac{7}{3}$, $y = \frac{2}{3}x + \frac{11}{3}$

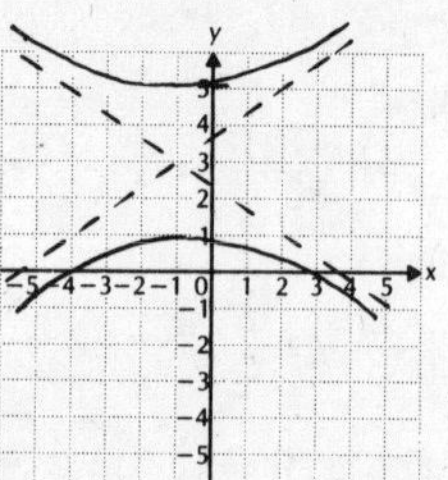

23.
$$y^2 - x^2 - 6x - 8y - 29 = 0$$
$$(y^2 - 8y) - (x^2 + 6x) = 29$$
$$(y^2 - 8y + 16) - (x^2 + 6x + 9) = 29 + 16 - 9$$
$$(y - 4)^2 - (x + 3)^2 = 36$$
$$\frac{(y - 4)^2}{36} - \frac{(x + 3)^2}{36} = 1$$

The center of the hyperbola is $(-3, 4)$; $a = 6$ and $b = 6$.

The transverse axis is parallel to the y-axis. Since $c^2 = a^2 + b^2$, $c = \sqrt{a^2 + b^2} = \sqrt{36 + 36} = 6\sqrt{2}$.

Consider the hyperbola $\frac{y^2}{36} - \frac{x^2}{36} = 1$. The center is is $(0, 0)$; $a = 6$, $b = 6$, and $c = 6\sqrt{2}$. The vertices are $(0, -6)$ and $(0, 6)$. The foci are $(0, -6\sqrt{2})$ and $(0, 6\sqrt{2})$. The asymptotes are $y = -x$ and $y = x$.

The vertices, foci, and asymptotes of the translated hyperbola are found in the same way in which the center was translated.

The vertices are $(0 - 3, 6 + 4)$ and $(0 - 3, -6 + 4)$ or $(-3, 10)$ and $(-3, -2)$.

The foci are $(0 - 3, 6\sqrt{2} + 4)$ and $(0 - 3, -6\sqrt{2} + 4)$ or $(-3, 4 + 6\sqrt{2})$ and $(-3, 4 - 6\sqrt{2})$.

The asymptotes are

$y - 4 = -(x + 3)$ and $y - 4 = x + 3$ or

$y = -x + 1$ and $y = x + 7$.

23. (continued)

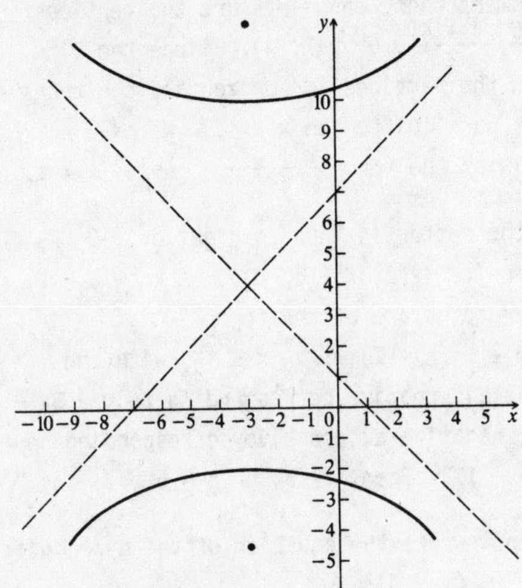

24. C: (4, 1)

V: $(4 + \sqrt{2}, 1)$, $(4 - \sqrt{2}, 1)$

F: (6, 1), (2, 1)

A: y = -x + 5, y = x - 3

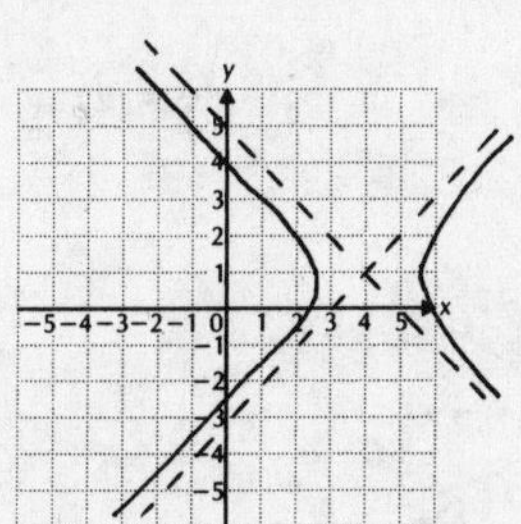

25. xy = 1

Since 1 > 0, the branches of the hyperbola lie in the first and third quadrants. The coordinate axes are its asymptotes.

x	y
$\frac{1}{4}$	4
$\frac{1}{2}$	2
1	1
$\frac{3}{2}$	$\frac{2}{3}$
3	$\frac{1}{3}$
4	$\frac{1}{4}$

x	y
$-\frac{1}{4}$	-4
$-\frac{1}{2}$	-2
-1	-1
$-\frac{3}{2}$	$-\frac{2}{3}$
-3	$-\frac{1}{3}$
-4	$-\frac{1}{4}$

xy = 1

26.

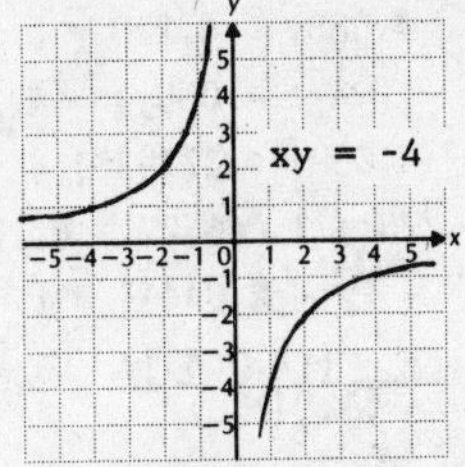

27. xy = -8

Since -8 < 0, the branches of the hyperbola lie in the second and fourth quadrants. The coordinates axes are its asymptotes.

x	y
1	-8
2	-4
4	-2
8	-1

x	y
-1	8
-2	4
-4	2
-8	1

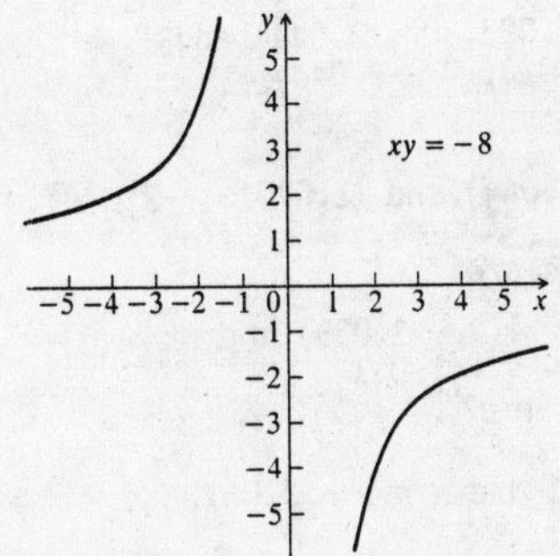

28.

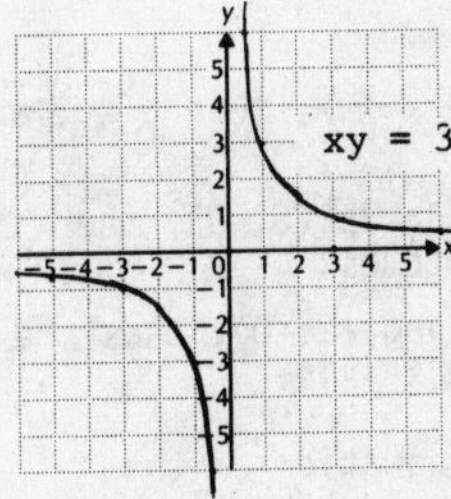

29. $$x^2 - y^2 - 2.046x - 4.088y - 4.228 = 0$$
$$(x^2 - 2.046x) - (y^2 + 4.088y) = 4.228$$
$$(x^2-2.046x+1.046529)-(y^2+4.088y+4.177936) =$$
$$4.228 + 1.046529 - 4.177936$$
$$(x - 1.023)^2 - (y + 2.044)^2 = 1.096593$$
$$\frac{(x - 1.023)^2}{(1.04718)^2} - \frac{[y - (-2.044)]^2}{(1.04718)^2} = 1$$
$$(\sqrt{1.096593} \approx 1.04718)$$

The center is (1.023, -2.044); a = b = 1.04718.

Consider the hyperbola $\frac{x^2}{(1.04718)^2} - \frac{y^2}{(1.04718)^2} = 1$.

The center is (0, 0); a = b = 1.04718.

The vertices are (-1.04718, 0) and (1.04718, 0).

The asymptotes are y = -x and y = x.

The vertices and asymptotes of the translated hyperbola are found by translation in the same way in which the center has been translated.

The vertices are

(-1.04718 + 1.023, 0 - 2.044) and
(1.04718 + 1.023, 0 - 2.044)

or

(-0.02418, -2.044) and (2.07018, -2.044).

The asymptotes are

y - (-2.044) = -(x - 1.023) and
y - (-2.044) = x - 1.023

or

y = -x - 1.021 and y = x - 3.067.

30. $x^2 - \frac{y^2}{3} = 1$

31. If the asymptotes are $y = \pm \frac{3}{2}x$, then a = 2 and b = 3.

The center of the hyperbola is (0, 0), the intersection of the asymptotes.

The vertices are (-2, 0) and (2, 0); thus the transverse axis is on the x-axis.

The standard form of the equation of the hyperbola must be in the form $\frac{x^2}{a^2} - \frac{y^2}{b^2} = 1$.

The equation is $\frac{x^2}{4} - \frac{y^2}{9} = 1$.

32. $$\frac{(y + 8)^2}{\left[\frac{11}{2}\right]^2} - \frac{(x - 3)^2}{3^2} = 1, \text{ or}$$
$$\frac{(y + 8)^2}{\frac{121}{4}} - \frac{(x - 3)^2}{9} = 1$$

33. The center of the hyperbola is the midpoint of the segment whose endpoints are the vertices: $\left[\frac{-9 - 5}{2}, \frac{4 + 4}{2}\right]$, or (-7, 4). Since the line through the vertices is horizontal, the transverse axis is parallel to the x-axis. We have a parabola of the form $\frac{(x + 7)^2}{a^2} - \frac{(y - 4)^2}{b^2} = 1$.

Since the center is 2 units right of (-9, 4) (and 2 units left of (-5, 4)), we know a = 2. The asymptotes are of the form $y - 4 = \pm \frac{b}{2}(x + 7) = \pm \frac{b}{2}x \pm \frac{7b}{2}$, with the positive alternative corresponding to y = 3x + 25 and the negative alternative corresponding to y = -3x - 17. Then $\frac{b}{2} = 3$, or b = 6.

We can now write the equation of the hyperbola:
$$\frac{(x + 7)^2}{4} - \frac{(y - 4)^2}{36} = 1$$

34. $$\frac{(y + 7)^2}{4} - \frac{(x - 4)^2}{36} = 1$$

V: (4, -5), (4, -9)

C: (4, -7)

35. a) Graph: $x^2 - 4y^2 = 4$, or $\frac{x^2}{4} - \frac{y^2}{1} = 1$

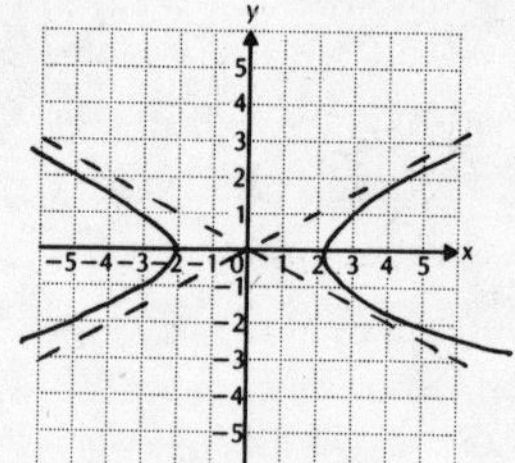

The relation is not a function, because it fails the vertical line test.

b) $$x^2 - 4y^2 = 4$$
$$x^2 - 4 = 4y^2$$
$$\frac{x^2 - 4}{4} = y^2$$
$$\pm \frac{1}{2}\sqrt{x^2 - 4} = y$$

c) Graph: $y = \frac{1}{2}\sqrt{x^2 - 4}$

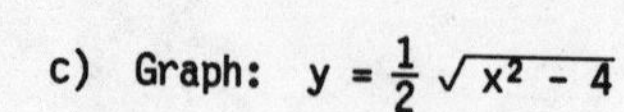

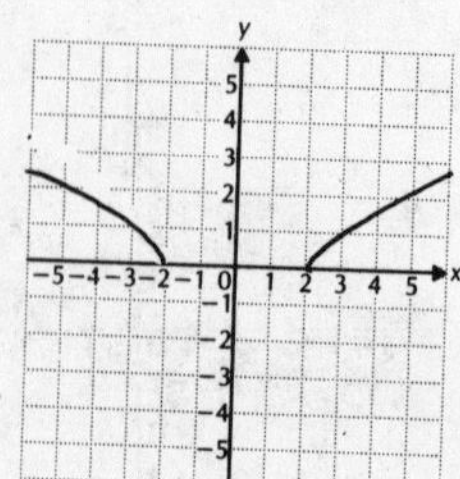

35. (continued)

The relation is a function because it passes the vertical line test. The domain is $\{x|x \leqslant -2 \text{ or } x \geqslant 2\}$. The range is $\{y|y \geqslant 0\}$.

d) Graph: $y = -\frac{1}{2}\sqrt{x^2 - 4}$

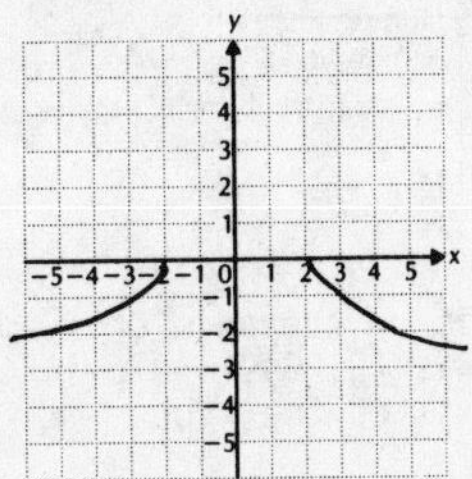

The relation is a function, because it passes the vertical line test. The domain is $\{x|x \leqslant -2 \text{ or } x \geqslant 2\}$. The range is $\{y|y \leqslant 0\}$.

36. $\left(\frac{x-h}{a}\right)\left(\frac{x-h}{a}\right) - \left(\frac{y-k}{b}\right)\left(\frac{y-k}{b}\right) = 1$, so

$$\frac{(x-h)^2}{a^2} - \frac{(y-k)^2}{b^2} = 1$$

37.

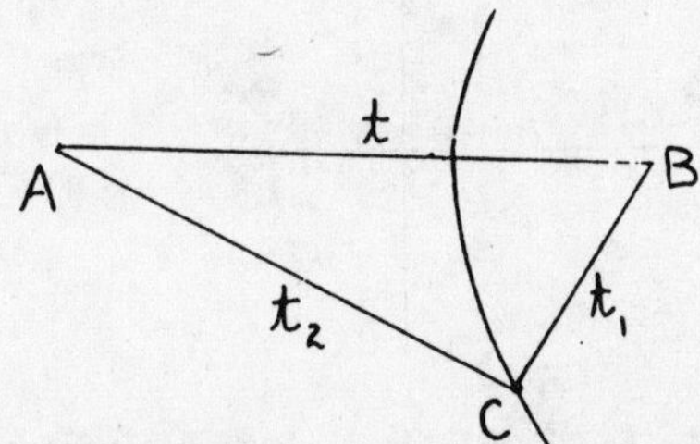

Let the constant b = the velocity of the bullet; let the constant s = the velocity of sound.

Then, by hypothesis,

$t_2 = t + t_1$. Now use: time $= \frac{\text{distance}}{\text{velocity}}$, then

$\frac{\overline{AC}}{s} = \frac{\overline{AB}}{b} + \frac{\overline{BC}}{s}$ or

$\overline{AC} - \overline{BC} = \frac{s}{b}\,\overline{AB}$, a constant.

Thus the (absolute value of the) difference of the distances from C to two fixed points A and B is a constant. By definition such a locus is a hyperbola.

38. Hyperbola

Exercise Set 10.3

1. $x^2 = 8y$

$x^2 = 4 \cdot 2 \cdot y$ (Writing $x^2 = 4py$)

Vertex: (0, 0)

Focus: (0, 2) [(0, p)]

Directrix: $y = -2$ $(y = -p)$

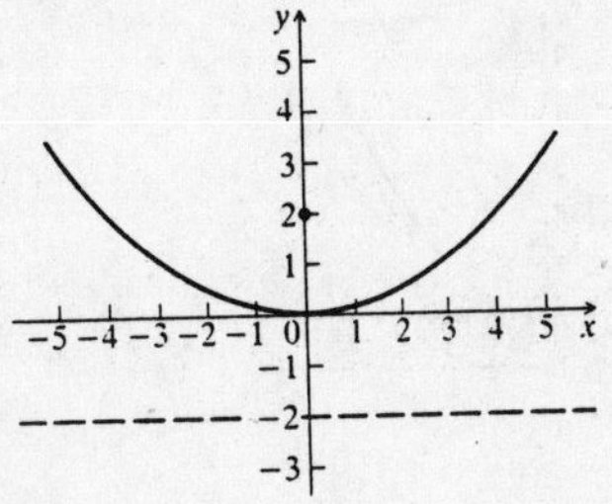

2. V: (0, 0)

F: (0, 4)

Directrix: $y = -4$

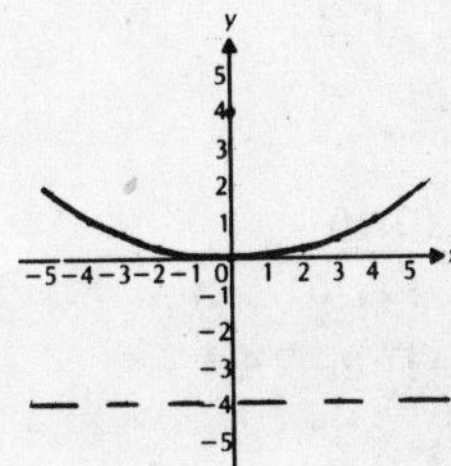

3. $y^2 = -6x$

$y^2 = 4\left(-\frac{3}{2}\right)x$ (Writing $y^2 = 4px$)

Vertex: (0, 0)

Focus: $\left(-\frac{3}{2}, 0\right)$ [(p, 0)]

Directrix: $x = -\left(-\frac{3}{2}\right) = \frac{3}{2}$ $(x = -p)$

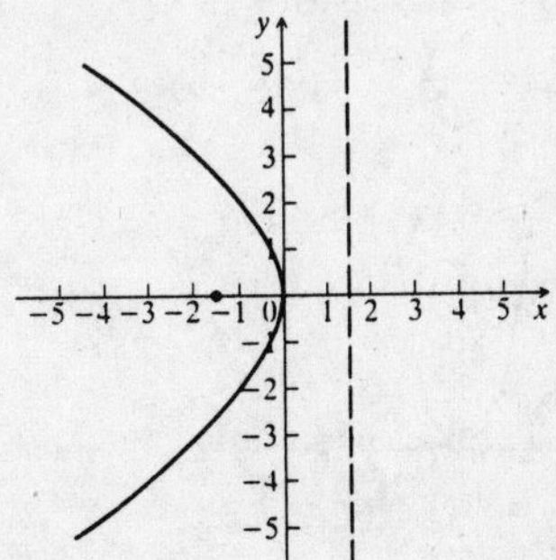

4. V: (0, 0)

F: $\left(-\frac{1}{2}, 0\right)$

Directrix: $x = \frac{1}{2}$

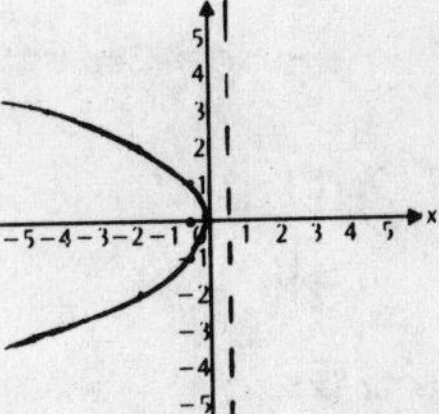

5. $x^2 - 4y = 0$

$x^2 = 4y$

$x^2 = 4 \cdot 1 \cdot y$ (Writing $x^2 = 4py$)

Vertex: (0, 0)

Focus: (0, 1) [(0, p)]

Directrix: $y = -1$ $(y = -p)$

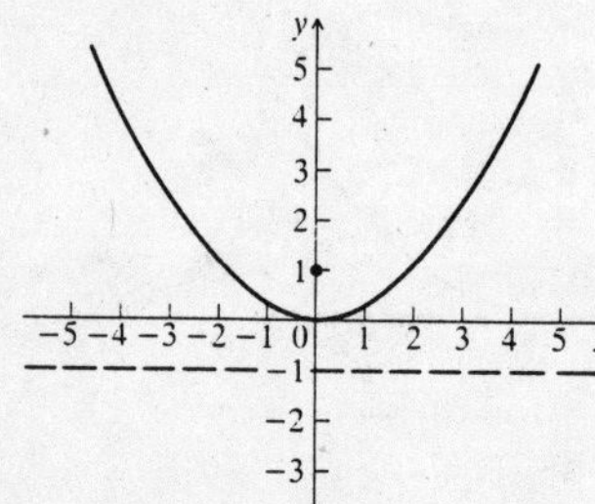

6. V: (0, 0)

F: (-1, 0)

Directrix: $x = 1$

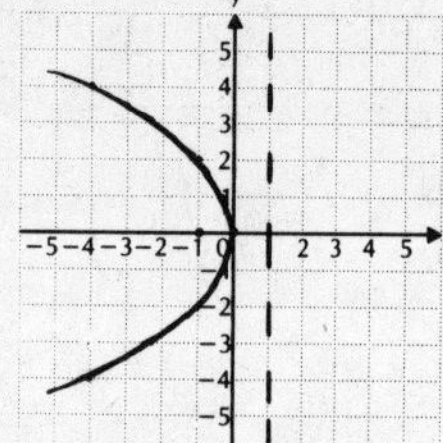

7. $y = 2x^2$

$x^2 = \frac{1}{2} y$

$x^2 = 4 \cdot \frac{1}{8} \cdot y$ (Writing $x^2 = 4py$)

Vertex: (0, 0)

Focus: $\left[0, \frac{1}{8}\right]$ [(0, p)]

Directrix: $y = -\frac{1}{8}$ $(y = -p)$

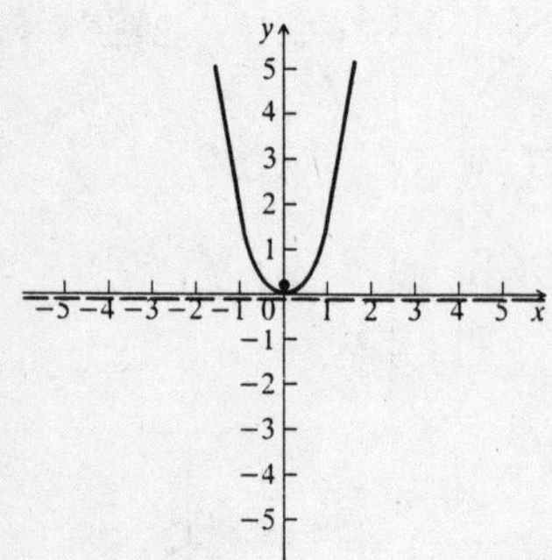

8. V: (0, 0)

F: $\left[0, \frac{1}{2}\right]$

Directrix: $y = -\frac{1}{2}$

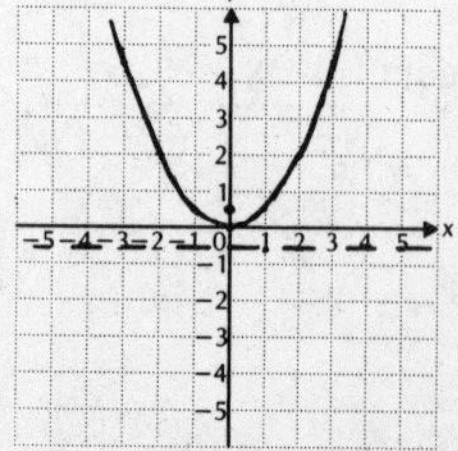

9. It helps to sketch a graph of the focus and the directrix.

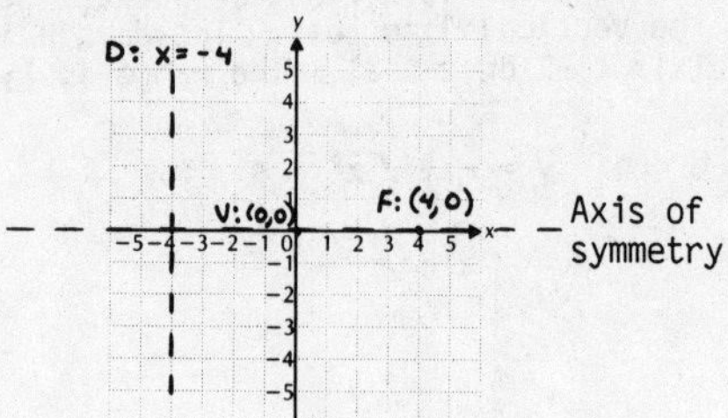

The focus, (4, 0), is on the axis of symmetry, the x-axis. The vertex is (0, 0), and p is 4 - 0, or 4. The equation is of the type

$y^2 = 4px$

$y^2 = 4 \cdot 4 \cdot x$ (Substituting 4 for p)

$y^2 = 16x$

10. $x^2 = y$

11. It is helpful to sketch a graph of the focus and the directrix.

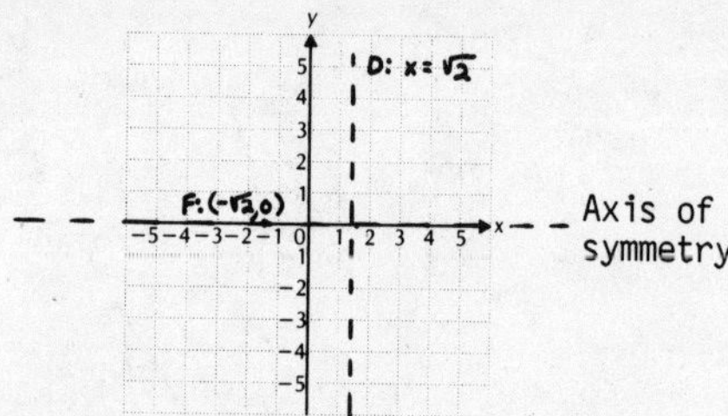

The focus, $(-\sqrt{2}, 0)$, is on the axis of symmetry, the x-axis. The vertex is (0, 0), and $p = -\sqrt{2} - 0$, or $-\sqrt{2}$. The equation is of the type

$y^2 = 4px$

$y^2 = 4(-\sqrt{2})x$ (Substituting $-\sqrt{2}$ for p)

$y^2 = -4\sqrt{2}\, x$

12. $x^2 = -4\pi y$

13. It is helpful to sketch a graph of the focus and the directrix.

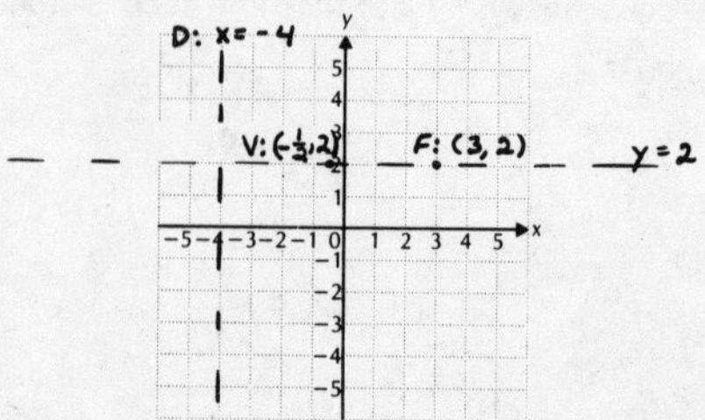

13. (continued)

The focus, (3, 2), is on the axis of symmetry, $y = 2$, which is parallel to the x-axis. The vertex is $\left(-\frac{1}{2}, 2\right)$, and $p = 3 - \left(-\frac{1}{2}\right)$, or $\frac{7}{2}$. The equation is of the type

$(y - k)^2 = 4p(x - h)$

$(y - 2)^2 = 4 \cdot \frac{7}{2}\left[x - \left(-\frac{1}{2}\right)\right]$ (Substituting)

$(y - 2)^2 = 14\left(x + \frac{1}{2}\right)$

14. $(x + 2)^2 = 12y$

15. $(x + 2)^2 = -6(y - 1)$

$[x - (-2)]^2 = 4\left(-\frac{3}{2}\right)(y - 1)$ $[(x - h)^2 = 4p(y-k)]$

Vertex: $(-2, 1)$ $[(h, k)]$

Focus: $\left(-2, 1 + \left(-\frac{3}{2}\right)\right)$, or $\left(-2, -\frac{1}{2}\right)$ $[(h, k + p)]$

Directrix: $y = 1 - \left(-\frac{3}{2}\right) = \frac{5}{2}$ $(y = k - p)$

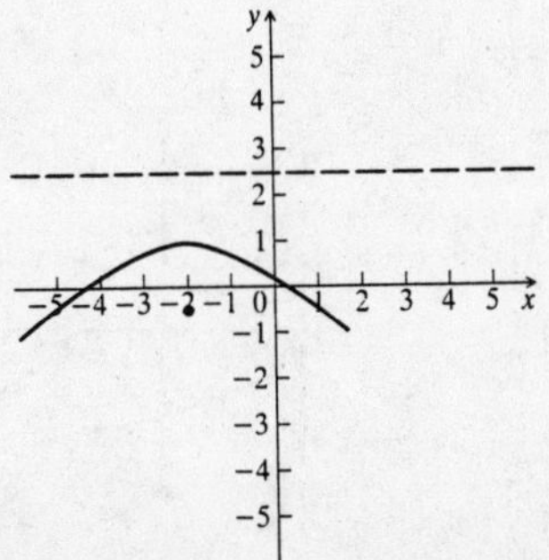

16. V: $(-2, 3)$
F: $(-7, 3)$
Directrix: $x = 3$

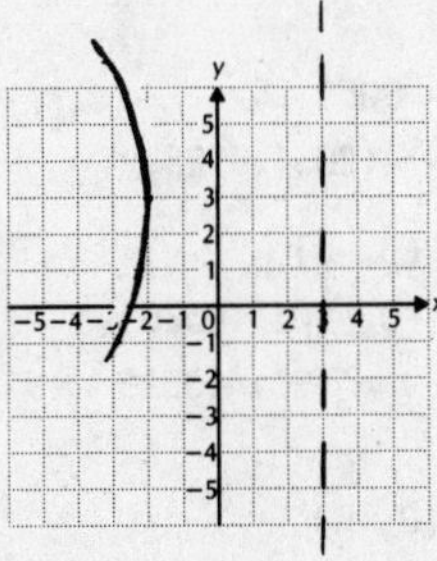

17. $x^2 + 2x + 2y + 7 = 0$

$x^2 + 2x = -2y - 7$

$(x^2 + 2x + 1) = -2y - 7 + 1 = -2y - 6$

$(x + 1)^2 = -2(y + 3)$

$[x - (-1)]^2 = 4\left(-\frac{1}{2}\right)[y - (-3)]$

$[(x - h)^2 = 4p(y - k)]$

Vertex: $(-1, -3)$ $[(h, k)]$

Focus: $\left(-1, -3 + \left(-\frac{1}{2}\right)\right)$, or $\left(-1, -\frac{7}{2}\right)$ $[(h, k+p)]$

Directrix: $y = -3 - \left(-\frac{1}{2}\right) = -\frac{5}{2}$ $(y = k - p)$

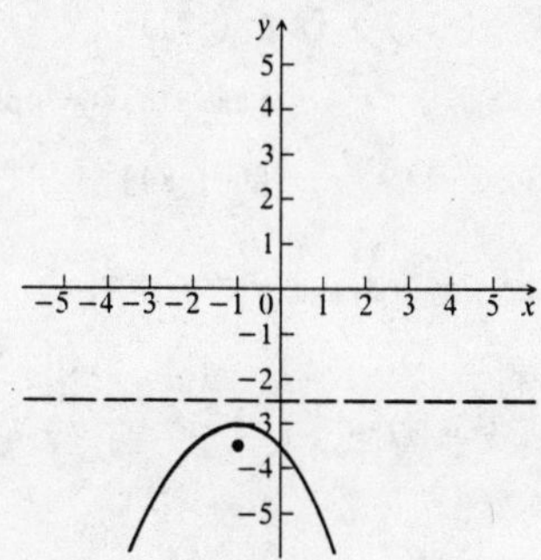

18. V: $(7, -3)$
F: $\left(\frac{29}{4}, -3\right)$
Directrix: $x = \frac{27}{4}$

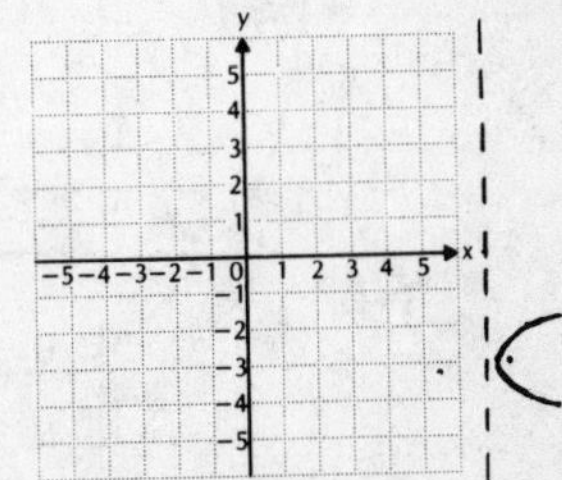

19. $x^2 - y - 2 = 0$

$x^2 = y + 2$

$(x - 0)^2 = 4 \cdot \frac{1}{4} \cdot [y - (-2)]$

$[(x - h)^2 = 4p(y - k)]$

Vertex: $(0, -2)$ $[(h, k)]$

Focus: $\left(0, -2 + \frac{1}{4}\right)$, or $\left(0, -\frac{7}{4}\right)$ $[(h, k + p)]$

Directrix: $y = -2 - \frac{1}{4} = -\frac{9}{4}$ $(y = k - p)$

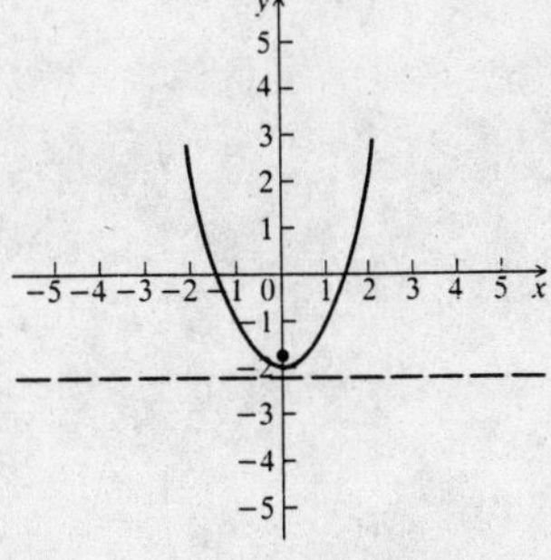

20. V: $(2, -2)$

F: $\left(2, -\frac{3}{2}\right)$

Directrix: $y = -\frac{5}{2}$

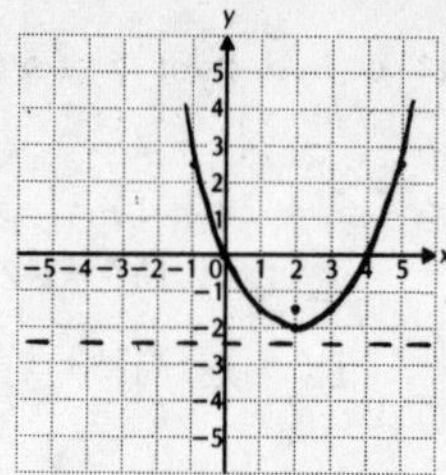

21.

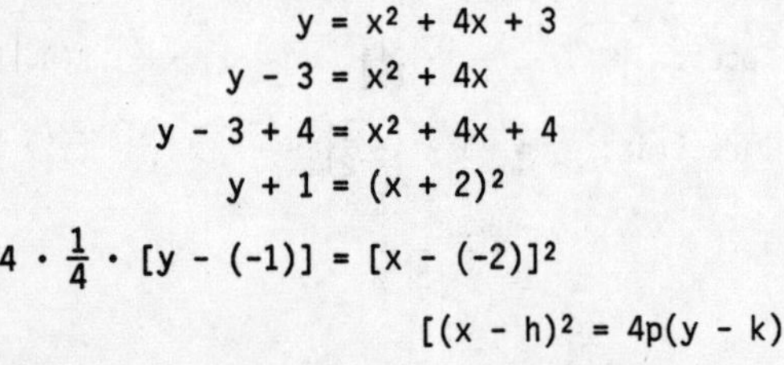

$$y = x^2 + 4x + 3$$
$$y - 3 = x^2 + 4x$$
$$y - 3 + 4 = x^2 + 4x + 4$$
$$y + 1 = (x + 2)^2$$
$$4 \cdot \frac{1}{4} \cdot [y - (-1)] = [x - (-2)]^2$$
$$[(x - h)^2 = 4p(y - k)]$$

Vertex: $(-2, -1)$ $[(h, k)]$

Focus: $\left(-2, -1 + \frac{1}{4}\right)$, or $\left(-2, -\frac{3}{4}\right)$ $[(h, k + p)]$

Directrix: $y = -1 - \frac{1}{4} = -\frac{5}{4}$ $(y = k - p)$

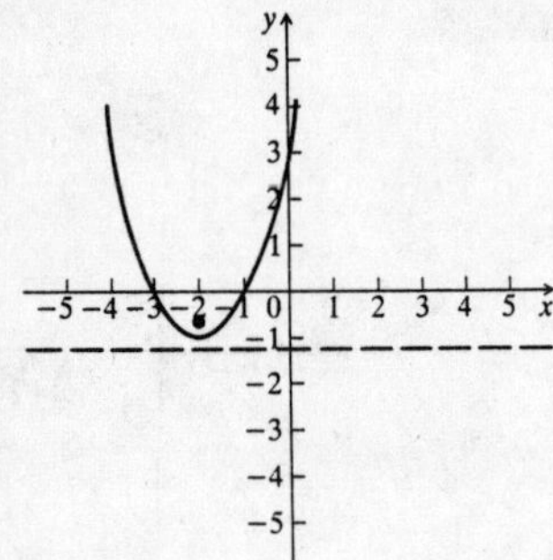

22. V: $(-3, 1)$

F: $\left(-3, \frac{5}{4}\right)$

Directrix: $y = \frac{3}{4}$

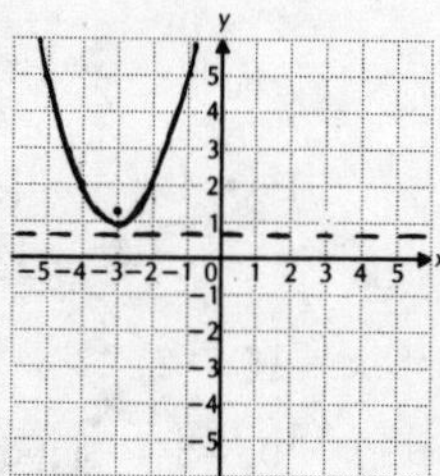

23. $4y^2 - 4y - 4x + 24 = 0$

$$y^2 - y - x + 6 = 0 \qquad \left(\text{Multiplying by } \frac{1}{4}\right)$$
$$y^2 - y = x - 6$$
$$y^2 - y + \frac{1}{4} = x - 6 + \frac{1}{4}$$
$$\left(y - \frac{1}{2}\right)^2 = x - \frac{23}{4}$$
$$\left(y - \frac{1}{2}\right)^2 = 4 \cdot \frac{1}{4}\left(x - \frac{23}{4}\right)$$
$$[(y - k)^2 = 4p(x - h)]$$

V: $\left(\frac{23}{4}, \frac{1}{2}\right)$ $[(h, k)]$

F: $\left(\frac{23}{4} + \frac{1}{4}, \frac{1}{2}\right)$, or $\left(6, \frac{1}{2}\right)$ $[(h + p, k)]$

D: $x = \frac{23}{4} - \frac{1}{4} = \frac{22}{4}$, or $\frac{11}{2}$ $(x = h - p)$

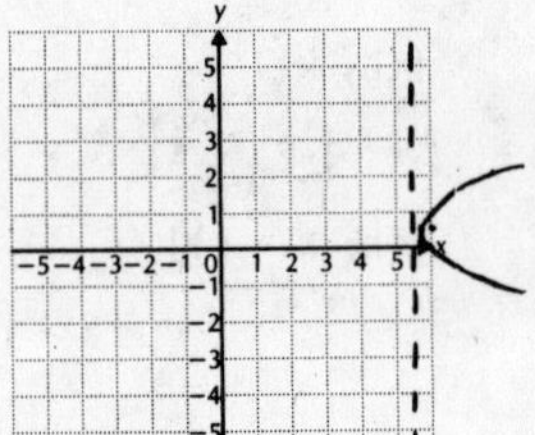

24. V: $\left(-\frac{17}{4}, -\frac{1}{2}\right)$

F: $\left(-4, -\frac{1}{2}\right)$

D: $x = -\frac{9}{2}$

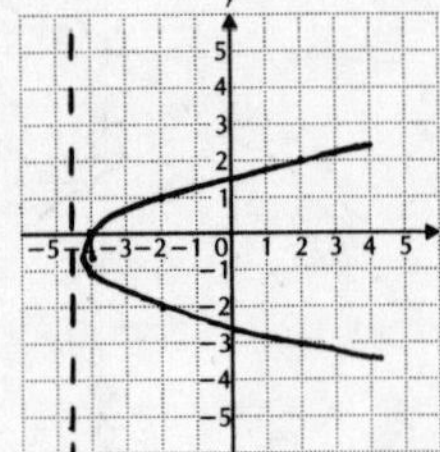

25. $x^2 = 8056.25y$

$x^2 = 4(2014.0625)y$ $(x^2 = 4py)$

V: $(0, 0)$

F: $(0, 2014.0625)$ $[(0, p)]$

D: $y = -2014.0625$ $(y = -p)$

26. V: $(0, 0)$

F: $(-1911.47, 0)$

Directrix: $x = 1911.47$

27. $x + 1 = 2y^2$

Only one of the variables is squared, so the graph is not a circle, an ellipse, or a hyperbola. We find an equivalent equation:

$$(y - 0)^2 = 4\left(\frac{1}{8}\right)(x + 1)$$

We have a parabola.

28. Circle

29. $$4y^2 + 25x^2 + 4 = 8y + 100x$$
$$25x^2 - 100x + 4y^2 - 8y = -4$$
$$25(x^2 - 4x + 4) + 4(y^2 - 2y + 1) = -4 + 100 + 4$$
$$25(x - 2)^2 + 4(y - 1)^2 = 100$$
$$\frac{(x - 2)^2}{4} + \frac{(y - 1)^2}{25} = 1$$

We have an ellipse.

30. Hyperbola

31. $$2x + 13 + y^2 = 8y - x^2$$
$$(x^2 + 2x + 1) + (y^2 - 8y + 16) = -13 + 1 + 16$$
$$(x + 1)^2 + (y - 4)^2 = 4$$

We have a circle.

32. Hyperbola

33. $$x = -16y + y^2 + 7$$
$$x - 7 + 64 = y^2 - 16y + 64$$
$$x + 57 = (y - 8)^2$$
$$4\left(\frac{1}{4}\right)(x + 57) = (y - 8)^2$$

We have a parabola.

34. Ellipse

35. $xy + 5x^2 = 9 + 7x^2 - 2x^2$
$xy = 9$

We have a hyperbola.

36. Ellipse

37.

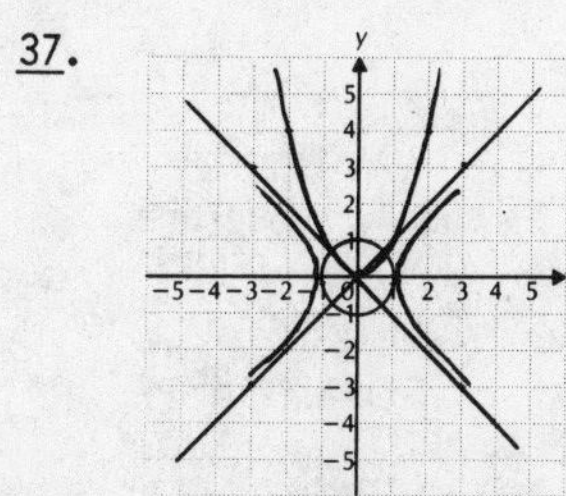

The graph of $x^2 - y^2 = 0$ is the union of the lines $x = y$ and $x = -y$. The others are respectively, a hyperbola, a circle, and a parabola.

38.

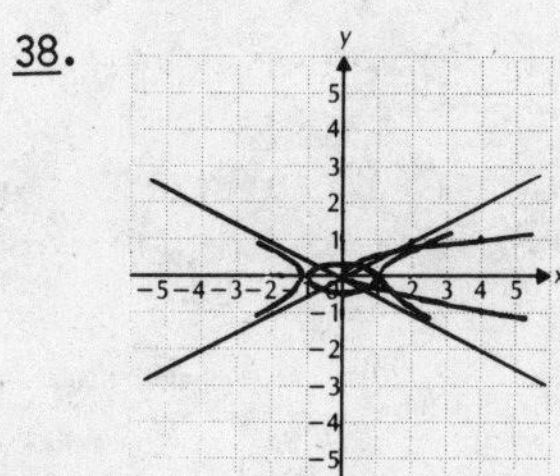

The graph of $x^2 - 4y^2 = 0$ is the union of the lines $x = 2y$ and $x = -2y$. The others are respectively, a hyperbola, an ellipse, and a parabola.

39. If the line of symmetry is parallel to the y-axis and the vertex (h, k) is (-1, 2), then the equation is of the type

$(x - h)^2 = 4p(y - k)$.

Solve for p substituting (-1, 2) for (h, k) and (-3, 1) for (x, y).

$$[-3 - (-1)]^2 = 4p(1 - 2)$$
$$4 = -4p$$
$$-1 = p$$

The equation of the parabola is

$[x - (-1)]^2 = 4(-1)(y - 2)$

or

$(x + 1)^2 = -4(y - 2)$.

40. a) Not a function

b) Not a function unless $p = 0$

41. If the line of symmetry is horizontal and the vertex (h, k) is (-2, 1), then the equation is of the type $(y - k)^2 = 4p(x - h)$.

Find p by substituting (-2, 1) for (h, k) and (-3, 5) for (x, y).

$$(5 - 1)^2 = 4p[-3 - (-2)]$$
$$16 = 4p(-1)$$
$$16 = -4p$$
$$-4 = p$$

The equation of the parabola is

$(y - 1)^2 = 4(-4)[x - (-2)]$

or $(y - 1)^2 = -16(x + 2)$.

42. 10 ft, 11.6 ft, 16.4 ft, 24.4 ft, 35.6 ft, 50 ft

43. See the answer section in the text.

44.

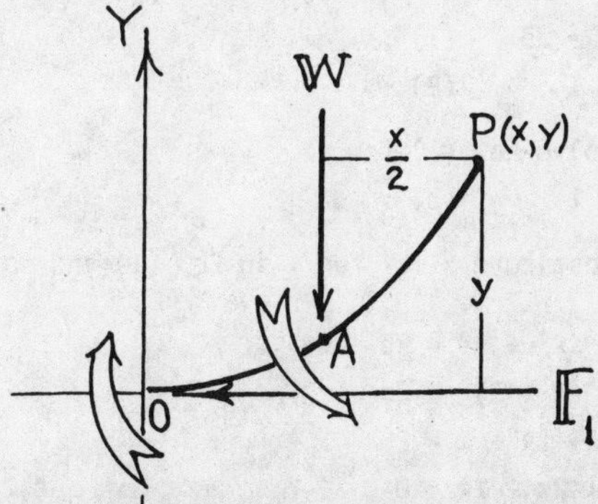

If a load W is distributed uniformly horizontally, this means that $W = kx$; that is, if the horizontal dimension of the cable is doubled, then the load is doubled, etc. We think of the entire load W as being concentrated at point A $\left(\text{whose abscissa is } \frac{x}{2}\right)$, roughly the midpoint of arc OP.

44. (continued)

A lever is in rotational equilibrium when the TORQUE(S) [or moment of force defined to be (FORCE) × (DISTANCE FROM FULCRUM)] tending to turn the lever clockwise about its fulcrum equals (in magnitude) the TORQUE(S) tending to turn the lever counter-clockwise.

Thinking of OP as a (rigid) lever with fulcrum at fixed point P, the torque due to F_1 (applied at O) tends to turn OP clockwise about P; the torque due to W (applied at A) tends to turn the lever counter-clockwise about P. These two torques are equal (in magnitude):

$F_1y = W \cdot \frac{x}{2}$ or

$F_1y = (kx) \cdot \frac{x}{2}$. Thus

$y = \frac{k}{2F_1} \cdot x^2$, a parabola.

[Since F_2 is applied at the fulcrum P, its torque is zero.]

Exercise Set 10.4

1.

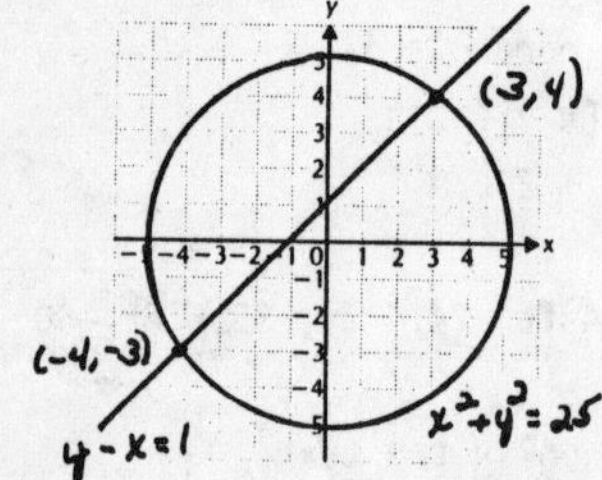

$x^2 + y^2 = 25$, (1)

$y - x = 1$ (2)

First solve Eq. (2) for y.

$y = x + 1$ (3)

Then substitute x + 1 for y in Eq. (1) and solve for x.

$$x^2 + y^2 = 25$$
$$x^2 + (x + 1)^2 = 25$$
$$x^2 + x^2 + 2x7+ 1 = 25$$
$$2x^2 + 2x - 24 = 0$$
$$x^2 + x - 12 = 0 \quad \text{Multiplying by } \frac{1}{2}$$
$$(x + 4)(x - 3) = 0 \quad \text{Factoring}$$

$x + 4 = 0$ or $x - 3 = 0$ Principle of zero products

$x = -4$ or $x = 3$

1. (continued)

Now substitute these numbers into Eq. (3) and solve for y.

$y = -4 + 1 = -3$

$y = 3 + 1 = 4$

The pairs (-4,-3) and (3,4) check, so they are the solutions.

2. (-8, -6), (6, 8)

3.

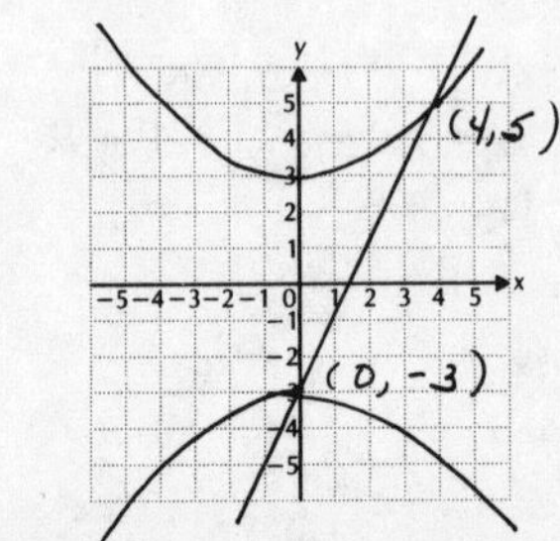

$y^2 - x^2 = 9$ (1)

$2x - 3 = y$ (2)

Substitute 2x - 3 for y in Eq. (1) and solve for x.

$$y^2 - x^2 = 9$$
$$(2x - 3)^2 - x^2 = 9$$
$$4x^2 - 12x + 9 - x^2 = 9$$
$$3x^2 - 12x = 0$$
$$x^2 - 4x = 0$$
$$x(x - 4) = 0$$

$x = 0$ or $x - 4 = 0$

$x = 0$ or $x = 4$

Now substitute these numbers into the linear Eq. (2) and solve for y.

If $x = 0$, $y = 2 \cdot 0 - 3 = -3$.

If $x = 4$, $y = 2 \cdot 4 - 3 = 5$.

The pairs (0, -3) and (4, 5) check, hence are solutions.

4. (-7, 1), (1, -7)

5. 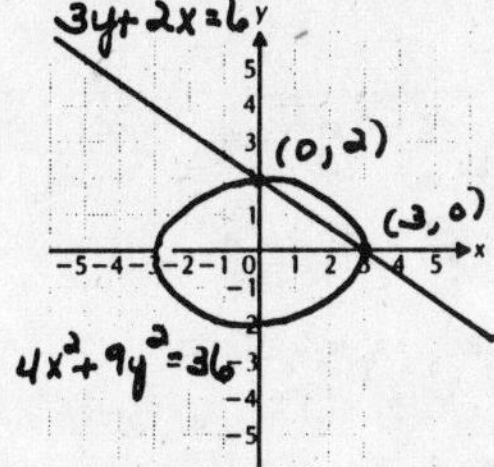

$4x^2 + 9y^2 = 36,$ (1)
$3y + 2x = 6$ (2)

First solve Eq. (2) for y.

$3y = -2x + 6$

$y = -\frac{2}{3}x + 2$ (3)

Then substitute $-\frac{2}{3}x + 2$ for y in Eq. (1) and solve for x.

$$4x^2 + 9y^2 = 36$$
$$4x^2 + 9\left(-\frac{2}{3}x + 2\right)^2 = 36$$
$$4x^2 + 9\left(\frac{4}{9}x^2 - \frac{8}{3}x + 4\right) = 36$$
$$4x^2 + 4x^2 - 24x + 36 = 36$$
$$8x^2 - 24x = 0$$
$$x^2 - 3x = 0$$
$$x(x - 3) = 0$$

$x = 0$ or $x = 3$

Now substitute these numbers in Eq. (3) and solve for y.

$y = -\frac{2}{3} \cdot 0 + 2 = 2$

$y = -\frac{2}{3} \cdot 3 + 2 = 0$

The pairs (0,2) and (3,0) check, so they are the solutions.

6. (2, 0), (0, 3)

7. 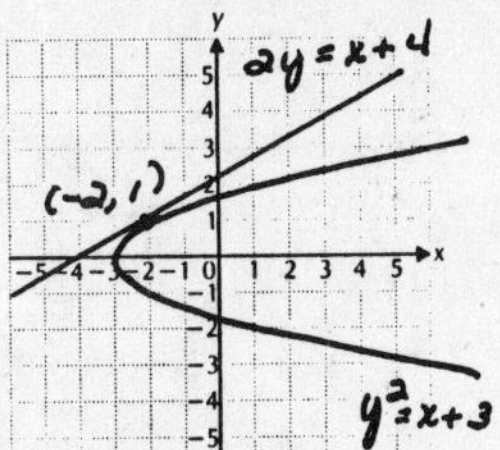

$y^2 = x + 3,$ (1)
$2y = x + 4$ (2)

First solve Eq. (2) for x.

$2y - 4 = x$ (3)

Then substitute 2y - 4 for x in Eq. (1) and solve for y.

$$y^2 = x + 3$$
$$y^2 = (2y - 4) + 3$$
$$y^2 = 2y - 1$$
$$y^2 - 2y + 1 = 0$$
$$(y - 1)(y - 1) = 0$$

$y = 1$ or $y = 1$

Now substitute 1 for y in Eq. (3) and solve for x.

$2 \cdot 1 - 4 = x$

$-2 = x$

The pair (-2,1) checks. It is the solution.

8. (2, 4), (1, 1)

9. $x^2 + 4y^2 = 25,$ (1)
$x + 2y = 7$ (2)

First solve Eq. (2) for x.

$x = -2y + 7$ (3)

Then substitute -2y + 7 for x in Eq. (1) and solve for y.

$$x^2 + 4y^2 = 25$$
$$(-2y + 7)^2 + 4y^2 = 25$$
$$4y^2 - 28y + 49 + 4y^2 = 25$$
$$8y^2 - 28y + 24 = 0$$
$$2y^2 - 7y + 6 = 0$$
$$(2y - 3)(y - 2) = 0$$

$y = \frac{3}{2}$ or $y = 2$

Now substitute these numbers in Eq. (3) and solve for x.

$x = -2 \cdot \frac{3}{2} + 7 = 4$

$x = -2 \cdot 2 + 7 = 3$

The pairs $\left(4, \frac{3}{2}\right)$ and (3,2) check, so they are the solutions.

10. $\left(-\frac{5}{3}, -\frac{13}{3}\right)$, (3, 5)

11. $x^2 - xy + 3y^2 = 27$, (1)

$x - y = 2$ (2)

First solve Eq. (2) for y.

$x - 2 = y$ (3)

Then substitute x - 2 for y in Eq. (1) and solve for x.

$$x^2 - xy + 3y^2 = 27$$
$$x^2 - x(x - 2) + 3(x - 2)^2 = 27$$
$$x^2 - x^2 + 2x + 3x^2 - 12x + 12 = 27$$
$$3x^2 - 10x - 15 = 0$$

$$x = \frac{-(-10) \pm \sqrt{(-10)^2 - 4(3)(-15)}}{2\cdot 3}$$
$$= \frac{10 \pm \sqrt{100 + 180}}{6}$$
$$= \frac{10 \pm \sqrt{280}}{6}$$
$$= \frac{10 \pm 2\sqrt{70}}{6}$$
$$= \frac{5 \pm \sqrt{70}}{3}$$

Now substitute these numbers in Eq. (3) and solve for y.

$$y = \frac{5 + \sqrt{70}}{3} - 2 = \frac{-1 + \sqrt{70}}{3}$$
$$y = \frac{5 - \sqrt{70}}{3} - 2 = \frac{-1 - \sqrt{70}}{3}$$

The pairs $\left(\frac{5 + \sqrt{70}}{3}, \frac{-1 + \sqrt{70}}{3}\right)$ and $\left(\frac{5 - \sqrt{70}}{3}, \frac{-1 - \sqrt{70}}{3}\right)$ check, so they are the solutions.

12. $\left(\frac{11}{4}, -\frac{9}{8}\right)$, (1, -2)

13. $3x + y = 7$ (1)

$4x^2 + 5y = 56$ (2)

First solve Eq. (1) for y.

$$3x + y = 7$$
$$y = 7 - 3x$$

Next substitute 7 - 3x for y in Eq. (2) for y and solve for x.

$$4x^2 + 5y = 56$$
$$4x^2 + 5(7 - 3x) = 56$$
$$4x^2 + 35 - 15x = 56$$
$$4x^2 - 15x - 21 = 0$$

Using the quadratic formula, we find that

$$x = \frac{15 - \sqrt{561}}{8} \text{ or } x = \frac{15 + \sqrt{561}}{8}.$$

Now substitute these numbers into Eq. (3) and solve for y.

If $x = \frac{15 - \sqrt{561}}{8}$, $y = 7 - 3\left(\frac{15 - \sqrt{561}}{8}\right)$, or $\frac{11 + 3\sqrt{561}}{8}$.

13. (continued)

If $x = \frac{15 + \sqrt{561}}{8}$, $y = 7 - 3\left(\frac{15 + \sqrt{561}}{8}\right)$, or $\frac{11 - 3\sqrt{561}}{8}$.

The pairs $\left(\frac{15 - \sqrt{561}}{8}, \frac{11 + 3\sqrt{561}}{8}\right)$ and $\left(\frac{15 + \sqrt{561}}{8}, \frac{11 - 3\sqrt{561}}{8}\right)$ check and are the solutions.

14. $\left(-3, \frac{5}{2}\right)$, (3, 1)

15. $a + b = 7$, (1)

$ab = 4$ (2)

First solve Eq. (1) for a.

$a = -b + 7$ (3)

Then substitute -b + 7 for a in Eq. (2) and solve for b.

$$(-b + 7)b = 4$$
$$-b^2 + 7b = 4$$
$$0 = b^2 - 7b + 4$$

$$b = \frac{-(-7) \pm \sqrt{(-7)^2 - 4\cdot 1\cdot 4}}{2\cdot 1}$$
$$b = \frac{7 \pm \sqrt{33}}{2}$$

Now substitute these numbers in Eq. (3) and solve for a.

$$a = -\left(\frac{7 + \sqrt{33}}{2}\right) + 7 = \frac{7 - \sqrt{33}}{2}$$
$$a = -\left(\frac{7 - \sqrt{33}}{2}\right) + 7 = \frac{7 + \sqrt{33}}{2}$$

The pairs $\left(\frac{7 - \sqrt{33}}{2}, \frac{7 + \sqrt{33}}{2}\right)$ and $\left(\frac{7 + \sqrt{33}}{2}, \frac{7 - \sqrt{33}}{2}\right)$ check, so they are the solutions.

16. (1, -7), (-7, 1)

17. $2a + b = 1$, (1)

$b = 4 - a^2$ (2)

Eq. (2) is already solved for b. Substitute $4 - a^2$ for b in Eq. (1) and solve for a.

$$2a + 4 - a^2 = 1$$
$$0 = a^2 - 2a - 3$$
$$0 = (a - 3)(a + 1)$$

a = 3 or a = -1

Substitute these numbers in Eq. (2) and solve for b.

$b = 4 - 3^2 = -5$

$b = 4 - (-1)^2 = 3$

The pairs (3,-5) and (-1,3) check.

18. $(3, 0), \left(-\frac{9}{5}, \frac{8}{5}\right)$

19. $a^2 + b^2 = 89$, (1)

$a - b = 3$ (2)

First solve Eq. (2) for a.

$a = b + 3$ (3)

Then substitute b + 3 for a in Eq. (1) and solve for b.

$(b + 3)^2 + b^2 = 89$

$b^2 + 6b + 9 + b^2 = 89$

$2b^2 + 6b - 80 = 0$

$b^2 + 3b - 40 = 0$

$(b + 8)(b - 5) = 0$

$b = -8$ or $b = 5$

Substitute these numbers in Eq. (3) and solve for a.

$a = -8 + 3 = -5$

$a = 5 + 3 = 8$

The pairs (-5,-8) and (8,5) check.

20. (1, 4), (4, 1)

21. $x^2 + y^2 = 5$, (1)

$x - y = 8$ (2)

First solve Eq. (2) for x.

$x = y + 8$ (3)

Then substitute y + 8 for x in Eq. (1) and solve for y.

$(y + 8)^2 + y^2 = 5$

$y^2 + 16y + 64 + y^2 = 5$

$2y^2 + 16y + 59 = 0$

$$y = \frac{-16 \pm \sqrt{(16)^2 - 4(2)(59)}}{2\cdot 2}$$

$$y = \frac{-16 \pm \sqrt{-216}}{4}$$

$$y = \frac{-16 \pm 6i\sqrt{6}}{4}$$

$$y = -4 \pm \frac{3}{2}i\sqrt{6}$$

Now substitute these numbers in Eq. (3) and solve for x.

$$x = -4 + \frac{3}{2}i\sqrt{6} + 8 = 4 + \frac{3}{2}i\sqrt{6}$$

$$x = -4 - \frac{3}{2}i\sqrt{6} + 8 = 4 - \frac{3}{2}i\sqrt{6}$$

The pairs $\left(4 + \frac{3}{2}i\sqrt{6}, -4 + \frac{3}{2}i\sqrt{6}\right)$ and $\left(4 - \frac{3}{2}i\sqrt{6}, -4 - \frac{3}{2}i\sqrt{6}\right)$ check.

22. $\left(-\frac{72}{13} + \frac{6}{13}i\sqrt{51}, \frac{32}{13} + \frac{6}{13}i\sqrt{51}\right)$,
$\left(-\frac{72}{13} - \frac{6}{13}i\sqrt{51}, \frac{32}{13} - \frac{6}{13}i\sqrt{51}\right)$

23. $x^2 + y^2 = 19{,}380{,}570.36$, (1)

$27{,}942.25x - 6.125y = 0$ (2)

Solve Eq. (2) for y.

$$\frac{27{,}942.25}{6.125}x = y$$

$4562x = y$ (3)

Substitute for y in Eq. (1) and solve for x.

$x^2 + (4562x)^2 = 19{,}380{,}510.36$

$20{,}811{,}845x^2 = 19{,}380{,}510.36$

$x^2 = 0.931225$

$x = \pm\ 0.965$

Now substitute for x in Eq. (3) and solve for y.

$y = 4562(\pm\ 0.965) = \pm\ 4402.33$

The pairs (0.965, 4402.33) and (-0.965, -4402.33) check.

24. (785, 45), (45, 785)

25. Familiarize: Let x = one number and y = the other number.

Translate: We translate to a system of equations.

The sum of two numbers is 12.

$x + y = 12$

The sum of their squares is 90.

$x^2 + y^2 = 90$

Carry out: We solve the system:

$x + y = 12$, (1)

$x^2 + y^2 = 90$ (2)

First solve Eq. (1) for y.

$y = 12 - x$ (3)

Then substitute 12 - x for y in Eq. (2) and solve for x.

$x^2 + y^2 = 90$

$x^2 + (12 - x)^2 = 90$

$x^2 + 144 - 24x + x^2 = 90$

$2x^2 - 24x + 54 = 0$

$x^2 - 12x + 27 = 0$

$(x - 9)(x - 3) = 0$

$x = 9$ or $x = 3$

Now substitute these numbers in Eq. (3) and solve for y.

$y = 12 - 9 = 3$

$y = 12 - 3 = 9$

Check: If the numbers are 9 and 3, the sum is 9 + 3, or 12. The sum of their squares is 81 + 9, or 90. The numbers check. The pair (3,9) does not give us another solution.

State: The numbers are 9 and 3.

26. 8 and 7

27. Familiarize: We first make a drawing. We let ℓ and w represent the length and width, respectively.

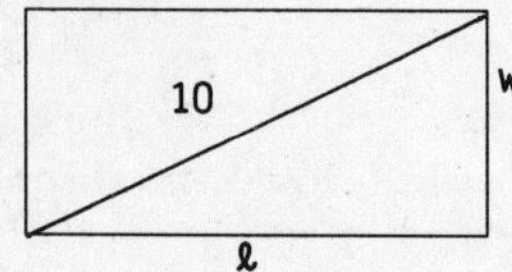

Translate: The perimeter is 28 cm.

$2\ell + 2w = 28$, or $\ell + w = 14$

Using the Pythagorean property we have another equation.

$\ell^2 + w^2 = 10^2$, or $\ell^2 + w^2 = 100$

Carry out: We solve the system:

$\ell + w = 14,$ (1)

$\ell^2 + w^2 = 100$ (2)

First solve Eq. (1) for w.

$w = 14 - \ell$ (3)

Then substitute $14 - \ell$ for w in Eq. (2) and solve for ℓ.

$$\begin{aligned} \ell^2 + w^2 &= 100 \\ \ell^2 + (14 - \ell)^2 &= 100 \\ \ell^2 + 196 - 28\ell + \ell^2 &= 100 \\ 2\ell^2 - 28\ell + 96 &= 0 \\ \ell^2 - 14\ell + 48 &= 0 \\ (\ell - 8)(\ell - 6) &= 0 \end{aligned}$$

$\ell = 8$ or $\ell = 6$

If $\ell = 8$, then $w = 14 - 8$, or 6. If $\ell = 6$, then $w = 14 - 6$, or 8. Since the length is usually considered to be longer than the width, we have the solution $\ell = 8$ and $w = 6$, or (8,6).

Check: If $\ell = 8$ and $w = 6$, then the perimeter is $2\cdot8 + 2\cdot6$, or 28. The length of a diagonal is $\sqrt{8^2 + 6^2}$, or $\sqrt{100}$, or 10. The numbers check.

State: The length is 8 cm, and the width is 6 cm.

28. Length: 2 m, width: 1 m

29. Familiarize: We first make a drawing. Let ℓ = the length and w = the width of the rectangle.

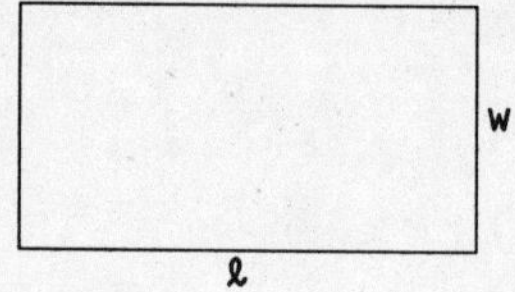

Translate:

Area: $\ell w = 20$

Perimeter: $2\ell + 2w = 18$, or $\ell + w = 9$

Carry out: We solve the system:

Solve the second equation for ℓ: $\ell = 9 - w$

Substitute $9 - w$ for ℓ in the first equation and solve for w.

$$\begin{aligned} (9 - w)w &= 20 \\ 9w - w^2 &= 20 \\ 0 &= w^2 - 9w + 20 \\ 0 &= (w - 5)(w - 4) \end{aligned}$$

$w = 5$ or $w = 4$

If $w = 5$, then $\ell = 9 - w$, or 4. If $w = 4$, then $\ell = 9 - 4$, or 5. Since length is usually considered to be longer than width, we have the solution $\ell = 5$ and $w = 4$, or (5,4).

Check: If $\ell = 5$ and $w = 4$, the area is $5\cdot4$, or 20. The perimeter is $2\cdot5 + 2\cdot4$, or 18. The numbers check.

State: The length is 5 in. and the width is 4 in.

30. Length: 2 yd, width: 1 yd

31. Familiarize: We make a drawing of the field. Let ℓ = the length and w = the width.

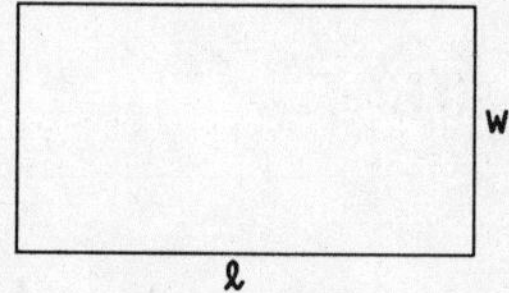

Since it takes 210 yd of fencing to enclose the field, we know that the perimeter is 210 yd.

Translate:

Perimeter: $2\ell + 2w = 210$, or $\ell + w = 105$

Area: $\ell w = 2250$

Carry out: We solve the system:

Solve the first equation for ℓ: $\ell = 105 - w$

Substitute $105 - w$ for ℓ in the second equation and solve for w.

$$\begin{aligned} (105 - w)w &= 2250 \\ 105w - w^2 &= 2250 \\ 0 &= w^2 - 105w + 2250 \\ 0 &= (w - 30)(w - 75) \end{aligned}$$

$w = 30$ or $w = 75$

31. (continued)

If $w = 30$, then $\ell = 105 - 30$, or 75. If $w = 75$, then $\ell = 105 - 75$, or 30. Since length is usually considered to be longer than width, we have the solution $\ell = 75$ and $w = 30$, or (75,30).

Check: If $\ell = 75$ and $w = 30$, the perimeter is $2 \cdot 75 + 2 \cdot 30$, or 210. The area is 75(30), or 2250. The numbers check.

State: The length is 75 yd and the width is 30 yd.

32. Length: 12 ft, width: 5 ft

33. Familiarize: Let x and y represent the numbers.

Translate: The product of the numbers is 2.

$$xy = 2$$

The sum of the reciprocals is $\frac{33}{8}$.

$$\frac{1}{x} + \frac{1}{y} = \frac{33}{8}$$

Carry out: We solve the system.

Solve the first equation for y: $y = \frac{2}{x}$

Substitute $\frac{2}{x}$ for y in the second equation and solve for x.

$$\frac{1}{x} + \frac{1}{\frac{2}{x}} = \frac{33}{8}$$

$$\frac{1}{x} + \frac{x}{2} = \frac{33}{8}$$

$$8 + 4x^2 = 33x \qquad \text{(Multiplying by 8x)}$$

$$4x^2 - 33x + 8 = 0$$

$$(4x - 1)(x - 8) = 0$$

$$x = \frac{1}{4} \quad \text{or} \quad x = 8$$

If $x = \frac{1}{4}$, then $y = \frac{2}{\frac{1}{4}} = 8$. If $x = 8$, then $y = \frac{2}{8} = \frac{1}{4}$. In either case we get the pair of numbers $\frac{1}{4}$ and 8.

Check: $\frac{1}{4} \cdot 8 = 2$ and $\frac{1}{\frac{1}{4}} + \frac{1}{8} = 4\,\frac{1}{8} = \frac{33}{8}$. The numbers check.

State: The numbers are $\frac{1}{4}$ and 8.

34. $(x + 2)^2 + (y - 1)^2 = 4$

35. Familiarize: Let x = the length of the longer piece and y = the length of the shorter piece. Then the lengths of the sides of the squares are $\frac{x}{4}$ and $\frac{y}{4}$.

Translate: The total length of the wire is 100 cm.

$$x + y = 100$$

The area of one square is 144 cm^2 greater than that of the other square.

$$\left(\frac{x}{4}\right)^2 = \left(\frac{y}{4}\right)^2 + 144$$

Carry out: We solve the system.

Solve the first equation for y: $y = 100 - x$

Substitute for y in the second equation and solve for x.

$$\left(\frac{x}{4}\right)^2 = \left(\frac{100 - x}{4}\right)^2 + 144$$

$$\frac{x^2}{16} = \frac{10{,}000 - 200x + x^2}{16} + 144$$

$$x^2 = 10{,}000 - 200x + x^2 + 2304$$

$$200x = 12{,}304$$

$$x = 61.52$$

If $x = 61.52$, $y = 100 - 61.52 = 38.48$.

Check: The total length of the wire is $61.52 + 38.48 = 100$ cm. The area of the larger square is $\left(\frac{61.52}{4}\right)^2$, or 236.5444 cm^2. This is 144 cm^2 greater than the area of the smaller square, $\left(\frac{38.48}{4}\right)^2$, or 92.5444 cm^2.

State: The wire should be cut into 61.52 cm and 38.48 cm lengths.

36. -2

37. The equation of the ellipse is of the form

$$\frac{x^2}{a^2} + \frac{y^2}{b^2} = 1.$$

Substitute $\left(1, \frac{\sqrt{3}}{2}\right)$ and $\left(\sqrt{3}, \frac{1}{2}\right)$ for (x, y) to get two equations.

$$\frac{1^2}{a^2} + \frac{\left(\frac{\sqrt{3}}{2}\right)^2}{b^2} = 1, \text{ or } \frac{1}{a^2} + \frac{3}{4b^2} = 1$$

$$\frac{(\sqrt{3})^2}{a^2} + \frac{\left(\frac{1}{2}\right)^2}{b^2} = 1, \text{ or } \frac{3}{a^2} + \frac{1}{4b^2} = 1$$

Substitute u for $\frac{1}{a^2}$ and v for $\frac{1}{b^2}$.

$$u + \frac{3}{4}v = 1, \qquad 4u + 3v = 4,$$
$$\text{or}$$
$$3u + \frac{1}{4}v = 1 \qquad 12u + v = 4$$

Solving for u and v, we get $u = \frac{1}{4}$, $v = 1$.

37. (continued)

Then $u = \frac{1}{a^2} = \frac{1}{4}$, so $a^2 = 4$; $v = \frac{1}{b^2} = 1$, so $b^2 = 1$.

Then the equation of the ellipse is

$$\frac{x^2}{4} + \frac{y^2}{1} = 1.$$

38. $\frac{x^2}{\frac{9}{4}} - \frac{y^2}{\frac{15}{4}} = 1$

39. $x - y = a + 2b,$ (1)

$x^2 - y^2 = a^2 + 2ab + b^2$ (2)

Solve Eq. (1) for x and substitute in Eq. (2). Then solve for y.

$x = y + a + 2b$ (3)

$$(y + a + 2b)^2 - y^2 = a^2 + 2ab + b^2$$
$$y^2 + 2ay + 4by + a^2 + 4ab + 4b^2 - y^2 = a^2 + 2ab + b^2$$
$$2ay + 4by = -2ab - 3b^2$$
$$y(2a + 4b) = -2ab - 3b^2$$
$$y = \frac{-2ab - 3b^2}{2a + 4b}$$

Substitute for y in Eq. (3) and solve for x.

$$x = \frac{-2ab - 3b^2}{2a + 4b} + a + 2b$$
$$x = \frac{-2ab - 3b^2 + 2a^2 + 8ab + 8b^2}{2a + 4b}$$
$$x = \frac{2a^2 + 6ab + 5b^2}{2a + 4b}$$

The solution is $\left(\frac{2a^2 + 6ab + 5b^2}{2a + 4b}, \frac{-2ab - 3b^2}{2a + 4b}\right)$.

40. $(a - b, 0)$, $\left(\frac{(a - b)(a^2 + b^2)}{2ab}, -\frac{(a^2 - b^2)(a - b)}{2ab}\right)$

41. Solve the system: $2L + 2W = P$

$LW = A$

See the answer section in the text.

42. There is no number x such that

$\frac{x^2}{a^2} - \frac{\left(\frac{b}{a}x\right)^2}{b^2} = 1$, because the left side simplifies to $\frac{x^2}{a^2} - \frac{x^2}{a^2}$ which is 0.

43. $(x - h)^2 + (y - k)^2 = r^2$

If (2, 4) is a point on the circle, then $(2 - h)^2 + (4 - k)^2 = r^2$.

If (3, 3) is a point on the circle, then $(3 - h)^2 + (3 - k)^2 = r^2$.

Thus

$$(2 - h)^2 + (4 - k)^2 = (3 - h)^2 + (3 - k)^2$$
$$4 - 4h + h^2 + 16 - 8k + k^2 = 9 - 6h + h^2 + 9 - 6k + k^2$$
$$-4h - 8k + 20 = -6h - 6k + 18$$
$$2h - 2k = -2$$
$$h - k = -1$$

If the center (h, k) is on the line $3x - y = 3$, then $3h - k = 3$.

Solving the system

$h - k = -1$

$3h - k = 3$

we find that $(h, k) = (2, 3)$.

Find r, substituting (2, 3) for (h, k) and (2, 4) for (x, y). We could also use (3, 3) for (x, y).

$$(x - h)^2 + (y - k)^2 = r^2$$
$$(2 - 2)^2 + (4 - 3)^2 = r^2$$
$$0 + 1 = r^2$$
$$1 = r^2$$
$$1 = r$$

The equation of the circle whose center is (2, 3) and whose radius is 1 is $(x - 2)^2 + (y - 3)^2 = 1$.

44. $(x + 1)^2 + (y + 3)^2 = 10^2$

45. $x^3 + y^3 = 72,$ (1)

$x + y = 6$ (2)

Solve Eq. (2) for y: $y = 6 - x$

Substitute for y in Eq. (1) and solve for x.

$$x^3 + (6 - x)^3 = 72$$
$$x^3 + 216 - 108x + 18x^2 - x^3 = 72$$
$$18x^2 - 108x + 144 = 0$$
$$x^2 - 6x + 8 = 0$$

$\left[\text{Multiplying by } \frac{1}{18}\right]$

$$(x - 4)(x - 2) = 0$$

$x = 4$ or $x = 2$

If $x = 4$, then $y = 6 - 4 = 2$.

If $x = 2$, then $y = 6 - 2 = 4$.

The pairs (4, 2) and (2, 4) check.

46. $h \approx 217.386816$, $k \approx 54.2592$, $\lambda \approx 0.000016956$

Exercise Set 10.5

1.

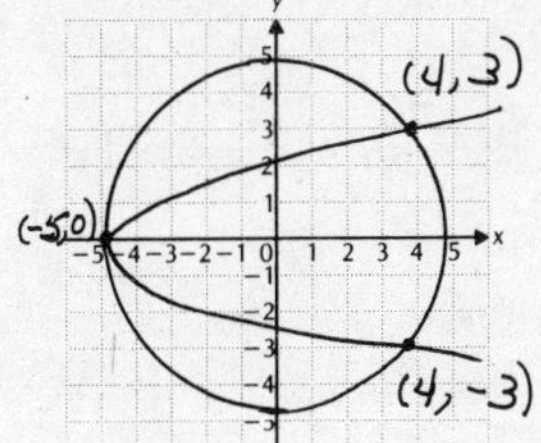

$x^2 + y^2 = 25,$ (1)

$y^2 = x + 5$ (2)

We substitute x + 5 for y^2 in Eq. (1) and solve for x.

$$x^2 + y^2 = 25$$
$$x^2 + (x + 5) = 25$$
$$x^2 + x - 20 = 0$$
$$(x + 5)(x - 4) = 0$$

$x + 5 = 0$ or $x - 4 = 0$

$x = -5$ or $x = 4$

Next we substitute these numbers for x in either Eq. (1) or Eq. (2) and solve for y. Here we use Eq. (2).

$y^2 = -5 + 5 = 0$ and $y = 0$.

$y^2 = 4 + 5 = 9$ and $y = \pm 3$.

The possible solutions are (-5,0), (4,3), and (4,-3).

Check:

For (-5,0):

$x^2 + y^2 = 25$	
$(-5)^2 + 0^2$	25
$25 + 0$	
25	

$y^2 = x + 5$	
0^2	$-5 + 5$
0	0

For (4,3):

$x^2 + y^2 = 25$	
$4^2 + 3^2$	25
$16 + 9$	
25	

$y^2 = x + 5$	
3^2	$4 + 5$
9	9

For (4,-3):

$x^2 + y^2 = 25$	
$4^2 + (-3)^2$	25
$16 + 9$	
25	

$y^2 = x + 5$	
$(-3)^2$	$4 + 5$
9	9

The solutions are (-5,0), (4,3), and (4,-3).

2. (0, 0), (1, 1)

3.

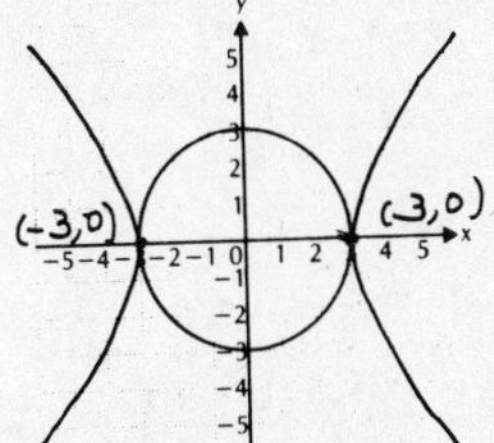

$x^2 + y^2 = 9,$ (1)

$x^2 - y^2 = 9$ (2)

Here we use the addition method.

$$x^2 + y^2 = 9$$
$$\underline{x^2 - y^2 = 9}$$
$$2x^2 = 18 \quad \text{Adding}$$
$$x^2 = 9$$
$$x = \pm 3$$

If x = 3, $x^2 = 9$, and if x = -3, $x^2 = 9$, so substituting 3 or -3 in Eq. (1) give us

$$x^2 + y^2 = 9$$
$$9 + y^2 = 9$$
$$y^2 = 0$$
$$y = 0.$$

The possible solutions are (3,0) and (-3,0).

Check:

$x^2 + y^2 = 9$	
$(\pm 3)^2 + (0)^2$	9
$9 + 0$	
9	

$x^2 - y^2 = 9$	
$(\pm 3)^2 - (0)^2$	9
$9 - 0$	
9	

The solutions are (3,0) and (-3,0).

4. (0, 2), (0, -2)

5.

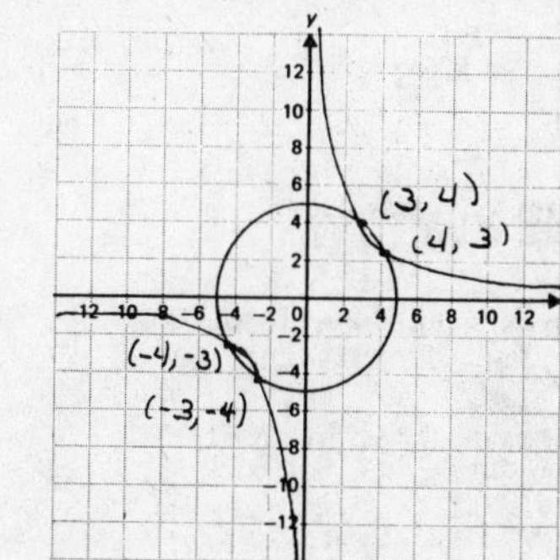

$x^2 + y^2 = 25,$ (1)

$xy = 12$ (2)

First we solve Eq. (2) for y.

$$xy = 12$$
$$y = \frac{12}{x}$$

Then we substitute $\frac{12}{x}$ for y in Eq. (1) and solve for x.

5. (continued)

$$x^2 + y^2 = 25$$

$$x^2 + \left(\frac{12}{x}\right)^2 = 25$$

$$x^2 + \frac{144}{x^2} = 25$$

$$x^4 + 144 = 25x^2 \quad \text{Multiplying by } x^2$$

$$x^4 - 25x^2 + 144 = 0$$

$$u^2 - 25u + 144 = 0 \quad \text{Letting } u = x^2$$

$$(u - 9)(u - 16) = 0$$

$$u = 9 \text{ or } u = 16$$

We now substitute x^2 for u and solve for x.

$$x^2 = 9 \text{ or } x^2 = 16$$

$$x = \pm 3 \text{ or } x = \pm 4$$

Since $y = 12/x$, if $x = 3$, $y = 4$; if $x = -3$, $y = -4$; if $x = 4$, $y = 3$; and if $x = -4$, $y = -3$. The pairs (3,4), (-3,-4), (4,3), (-4,-3) check. They are the solutions.

6. (-5, 3), (-5, -3), (4, 0)

7.

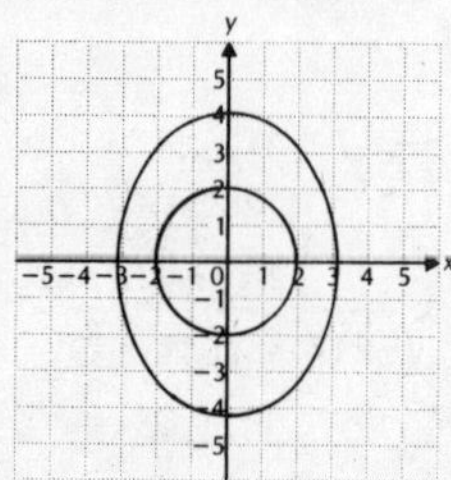

$$x^2 + y^2 = 4, \quad (1)$$

$$16x^2 + 9y^2 = 144 \quad (2)$$

$$-9x^2 - 9y^2 = -36 \quad \text{Multiplying (1) by -9}$$

$$16x^2 + 9y^2 = 144$$

$$7x^2 = 108 \quad \text{Adding}$$

$$x^2 = \frac{108}{7}$$

$$x = \pm\sqrt{\frac{108}{7}} = \pm 6\sqrt{\frac{3}{7}}$$

$$x = \pm\frac{6\sqrt{21}}{7} \quad \text{Rationalizing the denominator}$$

Substituting $\frac{6\sqrt{21}}{7}$ or $-\frac{6\sqrt{21}}{7}$ for x in Eq. (1) gives us

$$\frac{36\cdot 21}{49} + y^2 = 4$$

$$y^2 = 4 - \frac{108}{7}$$

$$y^2 = -\frac{80}{7}$$

$$y = \pm\sqrt{-\frac{80}{7}} = \pm 4i\sqrt{\frac{5}{7}}$$

$$y = \pm\frac{4i\sqrt{35}}{7}. \quad \text{Rationalizing the denominator}$$

7. (continued)

The pairs $\left(\frac{6\sqrt{21}}{7}, \frac{4i\sqrt{35}}{7}\right)$, $\left(\frac{6\sqrt{21}}{7}, -\frac{4i\sqrt{35}}{7}\right)$, $\left(-\frac{6\sqrt{21}}{7}, \frac{4i\sqrt{35}}{7}\right)$, and $\left(-\frac{6\sqrt{21}}{7}, -\frac{4i\sqrt{35}}{7}\right)$ check. They are the solutions.

8. (0, 5), (0, -5)

9. $x^2 + y^2 = 16,$ or $x^2 + y^2 = 16, \quad (1)$

$y^2 - 2x^2 = 10$ $\qquad -2x^2 + y^2 = 10 \quad (2)$

Here we use the addition method.

$$2x^2 + 2y^2 = 32 \quad \text{Multiplying (1) by 2}$$

$$-2x^2 + y^2 = 10$$

$$3y^2 = 42 \quad \text{Adding}$$

$$y^2 = 14$$

$$y = \pm\sqrt{14}$$

Substituting $\sqrt{14}$ or $-\sqrt{14}$ for y in Eq. (1) gives us

$$x^2 + 14 = 16$$

$$x^2 = 2$$

$$x = \pm\sqrt{2}$$

The pairs $(-\sqrt{2},-\sqrt{14})$, $(-\sqrt{2},\sqrt{14})$, $(\sqrt{2},-\sqrt{14})$, and $(\sqrt{2},\sqrt{14})$ check. They are the solutions.

10. $(-3, -\sqrt{5})$, $(-3, \sqrt{5})$, $(3, -\sqrt{5})$, $(3, \sqrt{5})$

11. $x^2 + y^2 = 5, \quad (1)$

$xy = 2 \quad (2)$

First we solve Eq. (2) for y.

$$xy = 2$$

$$y = \frac{2}{x}$$

Then we substitute $\frac{2}{x}$ for y in Eq. (1) and solve for x.

$$x^2 + y^2 = 5$$

$$x^2 + \left(\frac{2}{x}\right)^2 = 5$$

$$x^2 + \frac{4}{x^2} = 5$$

$$x^4 + 4 = 5x^2 \quad \text{Multiplying by } x^2$$

$$x^4 - 5x^2 + 4 = 0$$

$$u^2 - 5u + 4 = 0 \quad \text{Letting } u = x^2$$

$$(u - 4)(u - 1) = 0$$

$$u = 4 \text{ or } u = 1$$

We now substitute x^2 for u and solve for x.

$$x^2 = 4 \text{ or } x^2 = 1$$

$$x = \pm 2 \qquad x = \pm 1$$

Since $y = 2/x$, if $x = 2$, $y = 1$; if $x = -2$, $y = -1$; if $x = 1$, $y = 2$; and if $x = -1$, $y = -2$. The pairs (2,1), (-2,-1), (1,2), (-1,-2) check. They are the solutions.

12. (4, 2), (-4, -2), (2, 4), (-2, -4)

13. $x^2 + y^2 = 13$, (1)

$xy = 6$ (2)

First we solve Eq. (2) for y.

$xy = 6$

$y = \frac{6}{x}$

Then we substitute $\frac{6}{x}$ for y in Eq. (1) and solve for x.

$$x^2 + y^2 = 13$$

$$x^2 + \left(\frac{6}{x}\right)^2 = 13$$

$$x^2 + \frac{36}{x^2} = 13$$

$$x^4 + 36 = 13x^2 \quad \text{Multiplying by } x^2$$

$$x^4 - 13x^2 + 36 = 0$$

$$u^2 - 13u + 36 = 0 \quad \text{Letting } u = x^2$$

$$(u - 9)(u - 4) = 0$$

$u = 9$ or $u = 4$

We now substitute x^2 for u and solve for x.

$x^2 = 9$ or $x^2 = 4$

$x = \pm 3$ $\quad x = \pm 2$

Since $y = 6/x$, if $x = 3$, $y = 2$; if $x = -3$, $y = -2$; if $x = 2$, $y = 3$; and if $x = -2$, $y = -3$. The pairs (3,2), (-3,-2), (2,3), (-2,-3) check. They are the solutions.

14. (4, 1), (-4, -1), (2, 2), (-2, -2)

15. $x^2 + y^2 + 6y + 5 = 0$ (1)

$x^2 + y^2 - 2x - 8 = 0$ (2)

Using the addition method, multiply Eq. (2) by -1 and add the result to Eq. (1).

$$x^2 + y^2 + 6y + 5 = 0 \quad (1)$$

$$\underline{-x^2 - y^2 + 2x + 8 = 0} \quad (2)$$

$$2x + 6y + 13 = 0 \quad (3)$$

Solve Eq. (3) for x.

$$2x + 6y + 13 = 0$$

$$2x = -6y - 13$$

$$x = \frac{-6y - 13}{2}$$

Substitute $\frac{-6y - 13}{2}$ for x in Eq. (1) and solve for y.

$$x^2 + y^2 + 6y + 5 = 0$$

$$\left(\frac{-6y - 13}{2}\right)^2 + y^2 + 6y + 5 = 0$$

$$\frac{36y^2 + 156y + 169}{4} + y^2 + 6y + 5 = 0$$

$$36y^2 + 156y + 169 + 4y^2 + 24y + 20 = 0$$

$$40y^2 + 180y + 189 = 0$$

15. (continued)

Using the quadratic formula, we find that $y = \frac{-45 \pm 3\sqrt{15}}{20}$. Substitute $\frac{-45 \pm 3\sqrt{15}}{20}$ for y in $x = \frac{-6y - 13}{2}$ and solve for x.

If $y = \frac{-45 + 3\sqrt{15}}{20}$, then $x = \frac{-6\left(\frac{-45 + 3\sqrt{15}}{20}\right) - 13}{2}$

$$= \frac{5 - 9\sqrt{15}}{20}.$$

If $y = \frac{-45 - 3\sqrt{15}}{20}$, then $x = \frac{-6\left(\frac{-45 - 3\sqrt{15}}{20}\right) - 13}{2}$

$$= \frac{5 + 9\sqrt{15}}{20}.$$

The pairs $\left(\frac{5 + 9\sqrt{15}}{20}, \frac{-45 - 3\sqrt{15}}{20}\right)$ and $\left(\frac{5 - 9\sqrt{15}}{20}, \frac{-45 + 3\sqrt{15}}{20}\right)$ check and are the solutions.

16. (2, 1), (-2, -1)

17. $xy - y^2 = 2$, (1)

$2xy - 3y^2 = 0$ (2)

$$-2xy + 2y^2 = -4 \quad \text{Multiplying (1) by -2}$$

$$\underline{2xy - 3y^2 = 0}$$

$$-y^2 = -4$$

$$y^2 = 4$$

$$y = \pm 2$$

We substitute for y in Eq. (1) and solve for x.

When $y = 2$: $x \cdot 2 - 2^2 = 2$

$$2x - 4 = 2$$

$$2x = 6$$

$$x = 3$$

When $y = -2$: $x(-2) - (-2)^2 = 2$

$$-2x - 4 = 2$$

$$-2x = 6$$

$$x = -3$$

The pairs (3,2) and (-3,-2) check. They are the solutions.

18. $\left(2, -\frac{4}{5}\right)$, $\left(-2, -\frac{4}{5}\right)$, (5, 2), (-5, 2)

19. $m^2 - 3mn + n^2 + 1 = 0,$ (1)
$3m^2 - mn + 3n^2 = 13$ (2)

$m^2 - 3mn + n^2 = -1,$ (3) Rewriting (1)
$3m^2 - mn + 3n^2 = 13$ (2)

$-3m^2 + 9mn - 3n^2 = 3$ Multiplying (3) by -3
$3m^2 - mn + 3n^2 = 13$

$8mn = 16$
$mn = 2$
$n = \frac{2}{m}$ (4)

Substitute $\frac{2}{m}$ for n in Eq. (1) and solve for m.

$$m^2 - 3m\left(\frac{2}{m}\right) + \left(\frac{2}{m}\right)^2 + 1 = 0$$
$$m^2 - 6 + \frac{4}{m^2} + 1 = 0$$
$$m^2 - 5 + \frac{4}{m^2} = 0$$
$$m^4 - 5m^2 + 4 = 0 \quad \text{Multiplying by } m^2$$

Substitute u for m^2.

$u^2 - 5u + 4 = 0$
$(u - 4)(u - 1) = 0$

$u = 4$ or $u = 1$
$m^2 = 4$ or $m^2 = 1$
$m = \pm 2$ or $m = \pm 1$

Substitute for m in Eq. (4) and solve for n.

When $m = 2$, $n = \frac{2}{2} = 1$.

When $m = -2$, $n = \frac{2}{-2} = -1$.

When $m = 1$, $n = \frac{2}{1} = 2$.

When $m = -1$, $n = \frac{2}{-1} = -2$.

The pairs (2, 1), (-2, -1), (1, 2), and (-1, -2) check. They are the solutions.

20. $(-\sqrt{2}, \sqrt{2})$, $(\sqrt{2}, -\sqrt{2})$

21. $a^2 + b^2 = 14,$ (1)
$ab = 3\sqrt{5}$ (2)

Solve Eq. (1) for b.

$$b = \frac{3\sqrt{5}}{a}$$

Substitute $\frac{3\sqrt{5}}{a}$ for b in Eq. (1) and solve for a.

$$a^2 + \left(\frac{3\sqrt{5}}{a}\right)^2 = 14$$
$$a^2 + \frac{45}{a^2} = 14$$
$$a^4 + 45 = 14a^2$$
$$a^4 - 14a^2 + 45 = 0$$
$$u^2 - 14u + 45 = 0 \quad \text{Letting } u = a^2$$
$$(u - 9)(u - 5) = 0$$

$u = 9$ or $u = 5$
$a^2 = 9$ or $a^2 = 5$
$a = \pm 3$ or $a = \pm\sqrt{5}$

Since $b = 3\sqrt{5}/a$, if $a = 3$, $b = \sqrt{5}$; if $a = -3$, $b = -\sqrt{5}$; if $a = \sqrt{5}$, $b = 3$; and if $a = -\sqrt{5}$, $b = -3$. The pairs $(3,\sqrt{5})$, $(-3,-\sqrt{5})$, $(\sqrt{5},3)$, $(-\sqrt{5},-3)$ check. They are the solutions.

22. $\left(i\sqrt{3}, -\frac{8i\sqrt{3}}{3}\right)$, $\left(-i\sqrt{3}, \frac{8i\sqrt{3}}{3}\right)$

23. $x^2 + y^2 = 25,$ (1)
$9x^2 + 4y^2 = 36$ (2)

$-4x^2 - 4y^2 = -100$ Multiplying (1) by -4
$9x^2 + 4y^2 = 36$

$5x^2 = -64$

$$x^2 = -\frac{64}{5}$$
$$x = \pm\sqrt{\frac{-64}{5}} = \pm\frac{8i}{\sqrt{5}}$$
$$x = \pm\frac{8i\sqrt{5}}{5} \quad \text{Rationalizing the denominator}$$

Substituting $\frac{8i\sqrt{5}}{5}$ or $-\frac{8i\sqrt{5}}{5}$ for x in Eq. (1) and solving for y gives us

$$-\frac{64}{5} + y^2 = 25$$
$$y^2 = \frac{189}{5}$$
$$y = \pm\sqrt{\frac{189}{5}} = \pm 3\sqrt{\frac{21}{5}}$$
$$y = \pm\frac{3\sqrt{105}}{5}. \quad \text{Rationalizing the denominator}$$

The pairs $\left(\frac{8i\sqrt{5}}{5}, \frac{3\sqrt{105}}{5}\right)$, $\left(-\frac{8i\sqrt{5}}{5}, \frac{3\sqrt{105}}{5}\right)$, $\left(\frac{8i\sqrt{5}}{5}, -\frac{3\sqrt{105}}{5}\right)$, and $\left(-\frac{8i\sqrt{5}}{5}, -\frac{3\sqrt{105}}{5}\right)$ check. They are the solutions.

24. $\left(\frac{4\sqrt{10}}{5}, \frac{3\sqrt{15}}{5}\right)$, $\left(\frac{4\sqrt{10}}{5}, -\frac{3\sqrt{15}}{5}\right)$, $\left(-\frac{4\sqrt{10}}{5}, \frac{3\sqrt{15}}{5}\right)$, $\left(-\frac{4\sqrt{10}}{5}, -\frac{3\sqrt{15}}{5}\right)$

25.
$$\begin{array}{rll} 18.465x^2 + 788.723y^2 &= 6408 & (1) \\ \underline{106.535x^2 - 788.723y^2} &= \underline{2692} & (2) \\ 125x^2 &= 9100 & \\ x^2 &= 72.8 & \\ x &= \pm 8.53 & \end{array}$$

Substitute these values for x in either Eq. (1) or (2) and solve for y. Here we use equation (1).

$$\begin{aligned} 18.465x^2 + 788.723y^2 &= 6408 \\ 18.465(\pm 8.53)^2 + 788.723y^2 &= 6408 \\ 1344.252 + 788.723y^2 &= 6408 \\ 788.723y^2 &= 5063.748 \\ y^2 &= 6.42019 \\ y &= \pm 2.53 \end{aligned}$$

The pairs (-8.53, -2.53), (-8.53, 2.53), (8.53, -2.53) and (8.53, 2.53) check and are the solutions.

26. (400, 1.43), (400, -1.43), (-400, 1.43), (-400, -1.43)

27. Familiarize: Let x and y represent the numbers.

Translate:

The product of two numbers is 156.

$$xy = 156 \qquad (1)$$

The sum of their squares is 313.

$$x^2 + y^2 = 313 \qquad (2)$$

Carry out: We solve the system of equations.

First solve Eq. (1) for y.

$$xy = 156$$

$$y = \frac{156}{x}$$

Then we substitute $\frac{156}{x}$ for y in Eq. (2) and solve for x.

$$\begin{aligned} x^2 + y^2 &= 313 \qquad (2) \\ x^2 + \left(\frac{156}{x}\right)^2 &= 313 \\ x^2 + \frac{24,336}{x^2} &= 313 \\ x^4 + 24,336 &= 313x^2 \\ x^4 - 313x^2 + 24,336 &= 0 \\ u^2 - 313u + 24,336 &= 0 \qquad \text{Letting } u = x^2 \\ (u - 169)(u - 144) &= 0 \\ u = 169 \quad \text{or} \quad u &= 144 \end{aligned}$$

We now substitute x^2 for u and solve for x.

$$x^2 = 169 \quad \text{or} \quad x^2 = 144$$

$$x = \pm 13 \quad \text{or} \quad x = \pm 12$$

27. (continued)

Since y = 156/x, if x = 13, y = 12; if x = -13, y = -12; if x = 12, y = 13; and if x = -12, y = -13. The possible solutions are (13,12), (-13,-12), (12,13), and (-12,-13).

Check: If x = 13 and y = 12, their product is 156. If x = -13 and y = -12, their product is 156. The sum of the squares in either case is $(\pm 13)^2 + (\pm 12)^2 = 169 + 144 = 313$. The pairs (12,13) and (-12,-13) do not give us any other solutions.

State: The numbers are 13 and 12 or -13 and -12.

28. 6 and 10 or -6 and -10

29. Familiarize: We first make a drawing. Let ℓ = the length and w = the width.

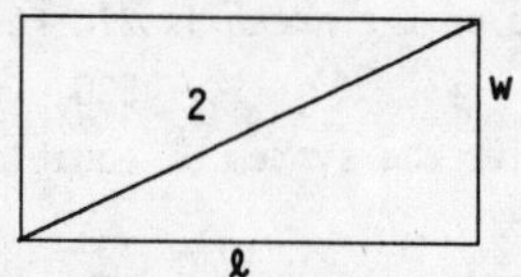

Translate:

Area: $\ell w = \sqrt{3}$ (1)

From the Pythagorean theorem: $\ell^2 + w^2 = 2^2$ (2)

Carry out: We solve the system of equations.

We first solve Eq. (1) for w.

$$\ell w = \sqrt{3}$$

$$w = \frac{\sqrt{3}}{\ell}$$

Then we substitute $\frac{\sqrt{3}}{\ell}$ for w in Eq. (2) and solve for ℓ.

$$\begin{aligned} \ell^2 + \left(\frac{\sqrt{3}}{\ell}\right)^2 &= 4 \\ \ell^2 + \frac{3}{\ell^2} &= 4 \\ \ell^4 + 3 &= 4\ell^2 \\ \ell^4 - 4\ell^2 + 3 &= 0 \\ u^2 - 4u + 3 &= 0 \qquad \text{Letting } u = \ell^2 \\ (u - 3)(u - 1) &= 0 \\ u = 3 \quad \text{or} \quad u &= 1 \end{aligned}$$

We now substitute ℓ^2 for u and solve for ℓ.

$$\ell^2 = 3 \quad \text{or} \quad \ell^2 = 1$$

$$\ell = \pm\sqrt{3} \quad \text{or} \quad \ell = \pm 1$$

Length cannot be negative, so we only need to consider $\ell = \sqrt{3}$ and $\ell = 1$. Since $w = \sqrt{3}/\ell$, if $\ell = \sqrt{3}$, w = 1 and if $\ell = 1$, $w = \sqrt{3}$. Length is usually considered to be longer than width, so we have the solution $\ell = \sqrt{3}$ and w = 1, or $(\sqrt{3},1)$.

Check: If $\ell = \sqrt{3}$ and w = 1, the area is $\sqrt{3}\cdot 1 = \sqrt{3}$. Also $(\sqrt{3})^2 + 1^2 = 3 + 1 = 4 = 2^2$. The numbers check.

State: The length is $\sqrt{3}$ m, and the width is 1 m.

30. Length: $\sqrt{2}$ m, width: 1 m

31. Familiarize: We let x = the length of a side of one peanut bed and y = the length of a side of the other peanut bed. Make a drawing.

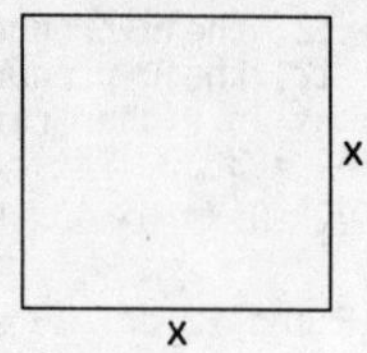

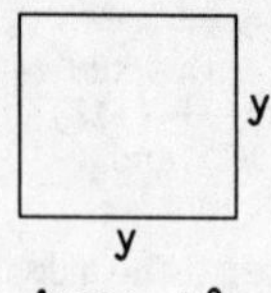

Area: x^2 Area: y^2

Translate:

The sum of the areas is 832 ft².

$$x^2 + y^2 = 832$$

The difference of the areas is 320 ft².

$$x^2 - y^2 = 320$$

Carry out: We solve the system of equations.

$$\begin{aligned} x^2 + y^2 &= 832 \\ x^2 - y^2 &= 320 \\ \hline 2x^2 &= 1152 \quad \text{Adding} \\ x^2 &= 576 \\ x &= \pm 24 \end{aligned}$$

Since length cannot be negative, we consider only x = 24. Substitute 24 for x in the first equation and solve for y.

$$\begin{aligned} 24^2 + y^2 &= 832 \\ 576 + y^2 &= 832 \\ y^2 &= 256 \\ y &= \pm 16 \end{aligned}$$

Again, we consider only the positive value, 16. The possible solution is (24,16).

Check: The areas of the peanut beds are 24^2, or 576, and 16^2, or 256. The sum of the areas is 576 + 256, or 832. The difference of the areas is 576 - 256, or 320. The values check.

State: The lengths of the beds are 24 ft and 16 ft.

32. Principal: \$125, interest rate: 6%

33. $(x - h)^2 + (y - k)^2 = r^2$ (Standard form)

Substitute (4, 6), (-6, 2), and (1, -3) for (x, y).

$$\begin{aligned} (4 - h)^2 + (6 - k)^2 &= r^2 \quad (1) \\ (-6 - h)^2 + (2 - k)^2 &= r^2 \quad (2) \\ (1 - h)^2 + (-3 - k)^2 &= r^2 \quad (3) \end{aligned}$$

Thus

$(4 - h)^2 + (6 - k)^2 = (-6 - h)^2 + (2 - k)^2$

or $5h + 2k = 3$

$(4 - h)^2 + (6 - k)^2 = (1 - h)^2 + (-3 - k)^2$

or $h + 3k = 7$

We solve the system $\begin{aligned} 5h + 2k &= 3 \\ h + 3k &= 7 \end{aligned}$ for (h, k).

Solving we get $h = -\frac{5}{13}$ and $k = \frac{32}{13}$. Substituting these values in equation (1), (2), or (3), we find that $r^2 = \frac{5365}{169}$.

The equation of the circle is

$$\left(x + \frac{5}{13}\right)^2 + \left(y - \frac{32}{13}\right)^2 = \frac{5365}{169}.$$

34. $(x - 10)^2 + (y + 3)^2 = 10^2$

35. Familiarize: Let x and y represent the numbers.

Translate:

The square of a certain number exceeds twice the square of another number by $\frac{1}{8}$.

$$x^2 = 2y^2 + \frac{1}{8}$$

The sum of the squares is $\frac{5}{16}$.

$$x^2 + y^2 = \frac{5}{16}$$

Carry out: We solve the system.

$$\begin{aligned} x^2 - 2y^2 &= \frac{1}{8}, \quad (1) \\ x^2 + y^2 &= \frac{5}{16} \quad (2) \end{aligned}$$

$$\begin{aligned} x^2 - 2y^2 &= \frac{1}{8}, \\ 2x^2 + 2y^2 &= \frac{5}{8} \quad \text{Multiplying (2) by 2} \\ \hline 3x^2 &= \frac{6}{8} \\ x^2 &= \frac{1}{4} \\ x &= \pm \frac{1}{2} \end{aligned}$$

35. (continued)

Substitute $\pm \frac{1}{2}$ for x in (2) and solve for y.

$$\left(\pm \frac{1}{2}\right)^2 + y^2 = \frac{5}{16}$$

$$\frac{1}{4} + y^2 = \frac{5}{16}$$

$$y^2 = \frac{1}{16}$$

$$y = \pm \frac{1}{4}$$

We get $\left(\frac{1}{2}, \frac{1}{4}\right)$, $\left(-\frac{1}{2}, \frac{1}{4}\right)$, $\left(\frac{1}{2}, -\frac{1}{4}\right)$ and $\left(-\frac{1}{2}, -\frac{1}{4}\right)$.

Check: It is true that $\left(\pm \frac{1}{2}\right)^2$ exceeds twice $\left(\pm \frac{1}{4}\right)^2$ by $\frac{1}{8}$: $\frac{1}{4} = 2\left(\frac{1}{16}\right) + \frac{1}{8}$

Also $\left(\pm \frac{1}{2}\right)^2 + \left(\pm \frac{1}{4}\right)^2 = \frac{5}{16}$. The pairs check.

State: The numbers are $\frac{1}{2}$ and $\frac{1}{4}$ or $-\frac{1}{2}$ and $\frac{1}{4}$ or $\frac{1}{2}$ and $-\frac{1}{4}$ or $-\frac{1}{2}$ and $-\frac{1}{4}$.

36. 10 in. by 7 in. by 5 in.

37. $x^2 + xy = a,$ (1)

$y^2 + xy = b$ (2)

Solve Eq. (1) for y.

$$xy = a - x^2$$

$$y = \frac{a - x^2}{x} \quad (3)$$

Substitute for y in Eq. (2) and solve for x.

$$\left(\frac{a - x^2}{x}\right)^2 + x\left(\frac{a - x^2}{x}\right) = b$$

$$\frac{a^2 - 2ax^2 + x^4}{x^2} + a - x^2 = b$$

$$a^2 - 2ax^2 + x^4 + ax^2 - x^4 = bx^2 \quad \text{Clearing the fraction}$$

$$a^2 = ax^2 + bx^2$$

$$a^2 = x^2(a + b)$$

$$\frac{a^2}{a + b} = x^2$$

$$\pm \frac{a}{\sqrt{a + b}} = x$$

$$\pm \frac{a\sqrt{a + b}}{a + b} = x \quad \text{Rationalizing the denominator}$$

Substitute for x in Eq. (3) and solve for y.

37. (continued)

When $x = \frac{a\sqrt{a + b}}{a + b}$: $y = \dfrac{a - \left(\frac{a\sqrt{a + b}}{a + b}\right)^2}{\frac{a\sqrt{a + b}}{a + b}} =$

$$\frac{a - \frac{a^2}{a + b}}{\frac{a\sqrt{a + b}}{a + b}} = \frac{a^2 + ab - a^2}{a\sqrt{a + b}} = \frac{ab}{a\sqrt{a + b}} = \frac{b}{\sqrt{a + b}} =$$

$$\frac{b\sqrt{a + b}}{a + b}$$

When $x = -\frac{a\sqrt{a + b}}{a + b}$: $y = \dfrac{a - \left(-\frac{a\sqrt{a + b}}{a + b}\right)^2}{-\frac{a\sqrt{a + b}}{a + b}} =$

$$\frac{a - \frac{a^2}{a + b}}{-\frac{a\sqrt{a + b}}{a + b}} = \frac{a^2 + ab - a^2}{-a\sqrt{a + b}} = \frac{ab}{-a\sqrt{a + b}} = -\frac{b}{\sqrt{a + b}} =$$

$$-\frac{b\sqrt{a + b}}{a + b}$$

The pairs $\left(\frac{a\sqrt{a + b}}{a + b}, \frac{b\sqrt{a + b}}{a + b}\right)$ and $\left(-\frac{a\sqrt{a + b}}{a + b}, -\frac{b\sqrt{a + b}}{a + b}\right)$ check. They are the solutions.

38. (a, b)

39. $p^2 + q^2 = 13$ (1)

$\frac{1}{pq} = -\frac{1}{6}$ (2)

Solve Eq. (2) for p.

$$\frac{1}{q} = -\frac{p}{6}$$

$$-\frac{6}{q} = p$$

Substitute -6/q for p in Eq. (1) and solve for q.

$$\left(-\frac{6}{q}\right)^2 + q^2 = 13$$

$$\frac{36}{q^2} + q^2 = 13$$

$$36 + q^4 = 13q^2$$

$$q^4 - 13q^2 + 36 = 0$$

$$u^2 - 13u + 36 = 0 \quad \text{Letting } u = q^2$$

$$(u - 9)(u - 4) = 0$$

$$u = 9 \quad \text{or} \quad u = 4$$

$$x^2 = 9 \quad \text{or} \quad x^2 = 4$$

$$x = \pm 3 \quad \text{or} \quad x = \pm 2$$

Since p = -6/q, if q = 3, p = -2; if q = -3, p = 2; if q = 2, p = -3, and if q = -2, p = 3. The pairs (-2,3), (2,-3), (-3,2), and (3,-2) check. They are the solutions.

40. $\left(\frac{1}{3}, \frac{1}{2}\right)$, $\left(\frac{1}{2}, \frac{1}{3}\right)$

41. $x^2 + y^2 = 4,$ (1)
$(x - 1)^2 + y^2 = 4$ (2)

Solve Eq. (1) for y^2:
$y^2 = 4 - x^2$ (3)

Substitute $4 - x^2$ for y^2 in Eq. (2) and solve for x.

$$(x - 1)^2 + (4 - x^2) = 4$$
$$x^2 - 2x + 1 + 4 - x^2 = 4$$
$$-2x = -1$$
$$x = \frac{1}{2}$$

Substitute $\frac{1}{2}$ for x in Eq. (3) and solve for y.

$$y^2 = 4 - \left(\frac{1}{2}\right)^2$$
$$y^2 = 4 - \frac{1}{4}$$
$$y^2 = \frac{15}{4}$$
$$y = \pm \frac{\sqrt{15}}{2}$$

The pairs $\left(\frac{1}{2}, \frac{\sqrt{15}}{2}\right)$ and $\left(\frac{1}{2}, -\frac{\sqrt{15}}{2}\right)$ check. They are the solutions.

42. $\left(\frac{a\sqrt{3}}{3}, \frac{b\sqrt{3}}{3}\right)$, $\left(\frac{a\sqrt{3}}{3}, -\frac{b\sqrt{3}}{3}\right)$, $\left(-\frac{a\sqrt{3}}{3}, \frac{b\sqrt{3}}{3}\right)$, $\left(-\frac{a\sqrt{3}}{3}, -\frac{b\sqrt{3}}{3}\right)$ where $a > 0$, $b > 0$

43. $(x - 2)^4 - (x - 2) = 0,$ (1)
$x^2 - kx + k = 0$ (2)

In Eq. (1) let $(x - 2) = z$. Then $z^4 - z = 0$, or $z(z - 1)(z^2 + z + 1) = 0$; so that $z = 0, 1, \frac{-1 \pm i\sqrt{3}}{2}$ and then $x = 2, 3, \frac{3 \pm i\sqrt{3}}{2}$.

The roots of Eq. (2) are $\frac{k \pm \sqrt{k^2 - 4k}}{2}$, and comparing this with $\frac{3 \pm i\sqrt{3}}{2}$ we have $k = 3$.

[If 2 were a root of Eq. (2) there would be no other common root; likewise for 3.]

44. $\left(\frac{2\sqrt{6}}{3}, \frac{\sqrt{6}}{3}\right)$, $\left(-\frac{2\sqrt{6}}{3}, -\frac{\sqrt{6}}{3}\right)$, $(1, -2)$, $(-1, 2)$

45. $5^{x+y} = 100,$
$3^{2x-y} = 1000$

$(x + y) \log 5 = 2,$
$(2x - y) \log 3 = 3$ Taking logarithms and simplifying

$x \log 5 + y \log 5 = 2,$ (1)
$2x \log 3 - y \log 3 = 3$ (2)

Multiply Eq. (1) by log 3 and Eq. (2) by log 5.

$$x \log 3 \cdot \log 5 + y \log 3 \cdot \log 5 = 2 \log 3$$
$$2x \log 3 \cdot \log 5 - y \log 3 \cdot \log 5 = 3 \log 5$$
$$3x \log 3 \cdot \log 5 = 2 \log 3 + 3 \log 5$$
$$x = \frac{2 \log 3 + 3 \log 5}{3 \log 3 \cdot \log 5}$$

Substitute in (1) to find y.

$$\frac{2 \log 3 + 3 \log 5}{3 \log 3 \cdot \log 5} \cdot \log 5 + y \log 5 = 2$$
$$y \log 5 = 2 - \frac{2 \log 3 + 3 \log 5}{3 \log 3}$$
$$y \log 5 = \frac{6 \log 3 - 2 \log 3 - 3 \log 5}{3 \log 3}$$
$$y \log 5 = \frac{4 \log 3 - 3 \log 5}{3 \log 3}$$
$$y = \frac{4 \log 3 - 3 \log 5}{3 \log 3 \cdot \log 5}$$

The pair $\left(\frac{2 \log 3 + 3 \log 5}{3 \log 3 \cdot \log 5}, \frac{4 \log 3 - 3 \log 5}{3 \log 3 \cdot \log 5}\right)$ checks. It is the solution.

46. (0, 0)

47. Familiarize: Let x and y represent the numbers. Let n represent the sum, product and sum of the squares of the numbers.

Translate: We have a system of equations.

$x + y = n,$ (1)
$xy = n,$ (2)
$x^2 + y^2 = n$ (3)

Carry out: We solve the system.

Solve Eq. (2) for y: $y = \frac{n}{x}$

Substitute in Eq. (1) and solve for n.

$$x + \frac{n}{x} = n$$
$$x^2 + n = nx$$
$$x^2 = nx - n$$
$$x^2 = n(x - 1)$$
$$\frac{x^2}{x - 1} = n$$

Then $y = \frac{n}{x} = \frac{\frac{x^2}{x - 1}}{x} = \frac{x}{x - 1}$.

47. (continued)

Substitute for y and n in Eq. (3) and solve for x.

$$x^2 + \left(\frac{x}{x - 1}\right)^2 = \frac{x^2}{x - 1}$$

$$x^2 + \frac{x^2}{(x - 1)^2} = \frac{x^2}{x - 1}$$

$$x^2(x - 1)^2 + x^2 = x^2(x - 1)$$ Multiplying by $(x - 1)^2$

$$x^4 - 2x^3 + x^2 + x^2 = x^3 - x^2$$

$$x^4 - 3x^3 + 3x^2 = 0$$

$$x^2(x^2 - 3x + 3) = 0$$

$x^2 = 0$ or $x^2 - 3x + 3 = 0$

Solving $x^2 = 0$, we get $x = 0$.
Solving $x^2 - 3x + 3 = 0$, we get

$$x = \frac{3 \pm \sqrt{9 - 12}}{2} = \frac{3 \pm i\sqrt{3}}{2}.$$

When $x = 0$, $y = \frac{0}{0 - 1} = 0$.

When $x = \frac{3 + i\sqrt{3}}{2}$, $y = \dfrac{\frac{3 + i\sqrt{3}}{2}}{\frac{3 + i\sqrt{3}}{2} - 1} = \frac{3 - i\sqrt{3}}{2}$.

When $x = \frac{3 - i\sqrt{3}}{2}$, $y = \dfrac{\frac{3 - i\sqrt{3}}{2}}{\frac{3 - i\sqrt{3}}{2} - 1} = \frac{3 + i\sqrt{3}}{2}$.

The pairs (0, 0), $\left(\frac{3 + i\sqrt{3}}{2}, \frac{3 - i\sqrt{3}}{2}\right)$, and $\left(\frac{3 - i\sqrt{3}}{2}, \frac{3 + i\sqrt{3}}{2}\right)$ check. They are the solutions.

Exercise Set 11.1

1. The degree of the polynomial, $\underline{x^4} - 3x^2 + 1$, is 4.

2. 5

3. The degree of the polynomial, $\underline{-2x} + 5$ (or $-2x^1 + 5$), is 1.

4. 1

5. The degree of the polynomial, $\underline{2x^2} - 3x + 4$, is 2.

6. 2

7. The degree of the polynomial, 3 (or $3x^0$), is 0.

8. No degree

9. $P(x) = x^3 + 6x^2 - x - 30$

$P(2) = 2^3 + 6(2)^2 - 2 - 30$ (Substituting)
$= 8 + 24 - 2 - 30$
$= 0$

Since $P(2) = 0$, 2 is a root, or zero, of the polynomial.

$P(3) = 3^3 + 6(3)^2 - 3 - 30$ (Substituting)
$= 27 + 54 - 3 - 30$
$= 48$

Since $P(3) \neq 0$, 3 is not a root, or zero, of the polynomial.

$P(-1) = (-1)^3 + 6(-1)^2 - (-1) - 30$
$= -1 + 6 + 1 - 30$
$= -24$

Since $P(-1) \neq 0$, -1 is not a root, or zero, of the polynomial.

10. 2 No, 3 No, -1 No

11. a)

$$\begin{array}{r}
x^2 + 8x + 15 \\
x - 2 \,\overline{)\, x^3 + 6x^2 - x - 30} \\
\underline{x^3 - 2x^2} \qquad\qquad \\
8x^2 - x \qquad \\
\underline{8x^2 - 16x} \qquad \\
15x - 30 \\
\underline{15x - 30} \\
0
\end{array}$$

Since the remainder is 0, we know that $x - 2$ is a factor of $x^3 + 6x^2 - x - 30$.

11. (continued)

b)

$$\begin{array}{r}
x^2 + 9x + 26 \\
x - 3 \,\overline{)\, x^3 + 6x^2 - x - 30} \\
\underline{x^3 - 3x^2} \qquad\qquad \\
9x^2 - x \qquad \\
\underline{9x^2 - 27x} \qquad \\
26x - 30 \\
\underline{26x - 78} \\
48
\end{array}$$

Since the remainder is not 0, we know that $x - 3$ is not a factor of $x^3 + 6x^2 - x - 30$.

c)

$$\begin{array}{r}
x^2 + 5x - 6 \\
x + 1 \,\overline{)\, x^3 + 6x^2 - x - 30} \\
\underline{x^3 + x^2} \qquad\qquad \\
5x^2 - x \qquad \\
\underline{5x^2 + 5x} \qquad \\
-6x - 30 \\
\underline{-6x - 6} \\
-24
\end{array}$$

Since the remainder is not 0, we know that $x + 1$ is not a factor of $x^3 + 6x^2 - x - 30$.

12. a) No, b) No, c) No

13. See work in Exercise 11 a).

$P(x) = d(x) \cdot Q(x) + R(x)$

$x^3 + 6x^2 - x - 30 = (x - 2)(x^2 + 8x + 15) + 0$

14. $Q(x) = 2x^2 + x + 3$, $R(x) = 5$

$P(x) = (x - 2)(2x^2 + x + 3) + 5$

15. See work in Exercise 11 b).

$P(x) = d(x) \cdot Q(x) + R(x)$

$x^3 + 6x^2 - x - 30 = (x - 3)(x^2 + 9x + 26) + 48$

16. $Q(x) = 2x^2 + 3x + 10$, $R(x) = 29$

$P(x) = (x - 3)(2x^2 + 3x + 10) + 29$

17.

$$\begin{array}{r}
x^2 - 2x + 4 \\
x + 2 \,\overline{)\, x^3 + 0x^2 + 0x - 8} \\
\underline{x^3 + 2x^2} \qquad\qquad \\
-2x^2 + 0x \qquad \\
\underline{-2x^2 - 4x} \qquad \\
4x - 8 \\
\underline{4x + 8} \\
-16
\end{array}$$

$P(x) = d(x) \cdot Q(x) + R(x)$

$x^3 - 8 = (x + 2)(x^2 - 2x + 4) - 16$

18. $Q(x) = x^2 - x + 1$, $R(x) = 26$

$P(x) = (x + 1)(x^2 - x + 1) + 26$

19.
$$\begin{array}{r} x^2 + 5 \\ x^2 + 4\,\overline{)\,x^4 + 9x^2 + 20} \\ \underline{x^4 + 4x^2} \\ 5x^2 + 20 \\ \underline{5x^2 + 20} \\ 0 \end{array}$$

$P(x) = d(x) \cdot Q(x) + R(x)$

$x^4 + 9x^2 + 20 = (x^2 + 4)(x^2 + 5) + 0$

20. $Q(x) = x^2 - x + 1$, $R(x) = 1$
$P(x) = (x^2 + x + 1)(x^2 - x + 1) + 1$

21.
$$\begin{array}{r} \frac{5}{2}x^5 + \frac{5}{4}x^4 - \frac{5}{8}x^3 - \frac{39}{16}x^2 - \frac{29}{32}x + \frac{113}{64} \\ 2x^2 - x + 1\,\overline{)\,5x^7 + 0x^6 + 0x^5 - 3x^4 + 0x^3 + 2x^2 + 0x - 3} \\ \underline{5x^7 - \frac{5}{2}x^6 + \frac{5}{2}x^5} \\ \frac{5}{2}x^6 - \frac{5}{2}x^5 - 3x^4 \\ \underline{\frac{5}{2}x^6 - \frac{5}{4}x^5 + \frac{5}{4}x^4} \\ -\frac{5}{4}x^5 + \frac{17}{4}x^4 + 0x^3 \\ \underline{-\frac{5}{4}x^5 - \frac{5}{8}x^4 - \frac{5}{8}x^3} \\ -\frac{39}{8}x^4 + \frac{5}{8}x^3 + 2x^2 \\ \underline{-\frac{39}{8}x^4 + \frac{39}{16}x^3 - \frac{39}{16}x^2} \\ -\frac{29}{16}x^3 + \frac{71}{16}x^2 + 0x \\ \underline{-\frac{29}{16}x^3 + \frac{29}{32}x^2 - \frac{29}{32}x} \\ \frac{113}{32}x^2 + \frac{29}{32}x - 3 \\ \underline{\frac{113}{32}x^2 - \frac{113}{64}x + \frac{113}{64}} \\ \frac{171}{64}x - \frac{305}{64} \end{array}$$

$P(x) = d(x) \cdot Q(x) + R(x)$

$5x^7 - 3x^4 + 2x^2 - 3$

$= (2x^2 - x + 1)\left[\frac{5}{2}x^5 + \frac{5}{4}x^4 - \frac{5}{8}x^3 - \frac{39}{16}x^2 - \frac{29}{32}x + \frac{113}{64}\right] + \left[\frac{171}{64}x - \frac{305}{64}\right]$

22. $P(x) = (3x^2 + 2x - 1)\left[2x^3 + \frac{2}{3}x - \frac{13}{9}\right] + \frac{41x - 31}{9}$

23. $P(x) = x^5 - 64$

a) $P(2) = 2^5 - 64 = 32 - 64 = -32$

b)
$$\begin{array}{r} x^4 + 2x^3 + 4x^2 + 8x + 16 \\ x - 2\,\overline{)\,x^5 + 0x^4 + 0x^3 + 0x^2 + 0x - 64} \\ \underline{x^5 - 2x^4} \\ 2x^4 + 0x^3 \\ \underline{2x^4 - 4x^3} \\ 4x^3 + 0x^2 \\ \underline{4x^3 - 8x^2} \\ 8x^2 + 0x \\ \underline{8x^2 - 16x} \\ 16x - 64 \\ \underline{16x - 32} \\ -32 \end{array}$$

The remainder is -32.

c) $P(-1) = (-1)^5 - 64 = -1 - 64 = -65$

d)
$$\begin{array}{r} x^4 - x^3 + x^2 - x + 1 \\ x + 1\,\overline{)\,x^5 + 0x^4 + 0x^3 + 0x^2 + 0x - 64} \\ \underline{x^5 + x^4} \\ -x^4 + 0x^3 \\ \underline{-x^4 - x^3} \\ x^3 + 0x^2 \\ \underline{x^3 + x^2} \\ -x^2 + 0x \\ \underline{-x^2 - x} \\ x - 64 \\ \underline{x + 1} \\ -65 \end{array}$$

The remainder is -65.

24. a) 0, b) 0, c) 12, d) 12

25. We solve $f(n) = 0$.

$\frac{1}{2}(n^2 - n) = 0$

$n^2 - n = 0$ (Multiplying by 2)

$n(n - 1) = 0$

$n = 0$ or $n = 1$

The zeros of the function are 0 and 1.

26. Approximately 7.16

27. We solve $W(x) = 0$, $0 \leqslant x < \infty$.

$\left(\frac{h}{12.3}\right)^3 = 0$

$\frac{h^3}{(12.3)^3} = 0$

$h^3 = 0$ [Multiplying by $(12.3)^3$]

$h = 0$

The root of the polynomial is 0.

28. 0 (This is the only zero in [0,2].)

29. $P(x) = 2x^2 - ix + 1$

a) $P(-i) = 2(-i)^2 - i(-i) + 1 = -2 - 1 + 1 = -2$

b)
$$\begin{array}{r} 2x \ - 3i \\ x + i \overline{\big)\, 2x^2 - ix + 1} \\ \underline{2x^2 + 2ix} \\ -3ix + 1 \\ \underline{-3ix + 3} \\ -2 \end{array}$$

The remainder is -2.

30. a) $-3 - i$

b) $-3 - i$

31. Degree of the product is the sum of the degrees of the factors.

32. The degree of the sum does not exceed the degree of either addend. Example: if $f(x) = x^3 + 3x^2 + 5x + 7$ and $g(x) = -x^3 - x^2 + 2x - 3$ then $f(x) + g(x) = 2x^2 + 7x + 4$ whose degree is actually less than the degree of either addend.

Exercise Set 11.2

1. $(2x^4 + 7x^3 + x - 12) \div (x + 3)$
$= (2x^4 + 7x^3 + 0x^2 + x - 12) \div [x - (-3)]$

$$\begin{array}{r|rrrrr} -3 & 2 & 7 & 0 & 1 & -12 \\ & & -6 & -3 & 9 & -30 \\ \hline & 2 & 1 & -3 & 10 & -42 \end{array}$$

The quotient is $2x^3 + x^2 - 3x + 10$.
The remainder is -42.

2. $Q(x) = x^2 - 5x + 3$, $R(x) = 9$

3. $(x^3 - 2x^2 - 8) \div (x + 2)$
$= (x^3 - 2x^2 + 0x - 8) \div [x - (-2)]$

$$\begin{array}{r|rrrr} -2 & 1 & -2 & 0 & -8 \\ & & -2 & 8 & -16 \\ \hline & 1 & -4 & 8 & -24 \end{array}$$

The quotient is $x^2 - 4x + 8$.
The remainder is -24.

4. $Q(x) = x^2 + 2x + 1$, $R(x) = 12$

5. $(x^4 - 1) \div (x - 1)$
$= (x^4 + 0x^3 + 0x^2 + 0x - 1) \div (x - 1)$

$$\begin{array}{r|rrrrr} 1 & 1 & 0 & 0 & 0 & -1 \\ & & 1 & 1 & 1 & 1 \\ \hline & 1 & 1 & 1 & 1 & 0 \end{array}$$

The quotient is $x^3 + x^2 + x + 1$.
The remainder is 0.

6. $Q(x) = x^4 - 2x^3 + 4x^2 - 8x + 16$, $R(x) = 0$

7. $(2x^4 + 3x^2 - 1) \div \left(x - \frac{1}{2}\right)$
$= (2x^4 + 0x^3 + 3x^2 + 0x - 1) \div \left(x - \frac{1}{2}\right)$

$$\begin{array}{r|rrrrr} \frac{1}{2} & 2 & 0 & 3 & 0 & -1 \\ & & 1 & \frac{1}{2} & \frac{7}{4} & \frac{7}{8} \\ \hline & 2 & 1 & \frac{7}{2} & \frac{7}{4} & -\frac{1}{8} \end{array}$$

The quotient is $2x^3 + x^2 + \frac{7}{2}x + \frac{7}{4}$.

The remainder is $-\frac{1}{8}$.

8. $Q(x) = 3x^3 + \frac{3}{4}x^2 - \frac{29}{16}x - \frac{29}{64}$, $R(x) = \frac{483}{256}$

9. $(x^4 - y^4) \div (x - y)$
$= (x^4 + 0x^3 + 0x^2 + 0x - y^4) \div (x - y)$

$$\begin{array}{r|rrrrr} y & 1 & 0 & 0 & 0 & -y^4 \\ & & y & y^2 & y^3 & y^4 \\ \hline & 1 & y & y^2 & y^3 & 0 \end{array}$$

The quotient is $x^3 + x^2y + xy^2 + y^3$.
The remainder is 0.

10. $Q(x) = x^2 + 2ix + (2 - 4i)$, $R(x) = -6 - 2i$

11. $P(x) = x^3 - 6x^2 + 11x - 6$

Find P(1).

$$\begin{array}{r|rrrr} 1 & 1 & -6 & 11 & -6 \\ & & 1 & -5 & 6 \\ \hline & 1 & -5 & 6 & 0 \end{array}$$

$P(1) = 0$

Find P(-2).

$$\begin{array}{r|rrrr} -2 & 1 & -6 & 11 & -6 \\ & & -2 & 16 & -54 \\ \hline & 1 & -8 & 27 & -60 \end{array}$$

$P(-2) = -60$

Find P(3).

$$\begin{array}{r|rrrr} 3 & 1 & -6 & 11 & -6 \\ & & 3 & -9 & 6 \\ \hline & 1 & -3 & 2 & 0 \end{array}$$

$P(3) = 0$

12. $P(-3) = 69$, $P(-2) = 41$, $P(1) = -7$

13. $P(x) = 2x^5 - 3x^4 + 2x^3 - x + 8$

Find $P(20)$.

$$\begin{array}{r|rrrrrr} 20 & 2 & -3 & 2 & 0 & -1 & 8 \\ & & 40 & 740 & 14{,}840 & 296{,}800 & 5{,}935{,}980 \\ \hline & 2 & 37 & 742 & 14{,}840 & 296{,}799 & 5{,}935{,}988 \end{array}$$

$P(20) = 5{,}935{,}988$

Find $P(-3)$.

$$\begin{array}{r|rrrrrr} -3 & 2 & -3 & 2 & 0 & -1 & 8 \\ & & -6 & 27 & -87 & 261 & -780 \\ \hline & 2 & -9 & 29 & -87 & 260 & -772 \end{array}$$

$P(-3) = -772$

14. $P(-10) = -220{,}050$, $P(5) = -750$

15. $P(x) = x^4 - 16$

Find $P(2)$.

$$\begin{array}{r|rrrrr} 2 & 1 & 0 & 0 & 0 & -16 \\ & & 2 & 4 & 8 & 16 \\ \hline & 1 & 2 & 4 & 8 & 0 \end{array}$$

$P(2) = 0$

Find $P(-2)$.

$$\begin{array}{r|rrrrr} -2 & 1 & 0 & 0 & 0 & -16 \\ & & -2 & 4 & -8 & 16 \\ \hline & 1 & -2 & 4 & -8 & 0 \end{array}$$

$P(-2) = 0$.

Find $P(3)$.

$$\begin{array}{r|rrrrr} 3 & 1 & 0 & 0 & 0 & -16 \\ & & 3 & 9 & 27 & 81 \\ \hline & 1 & 3 & 9 & 27 & 65 \end{array}$$

$P(3) = 65$.

16. $P(2) = 64$, $P(-2) = 0$, $P(3) = 275$

17. $P(x) = 3x^3 + 5x^2 - 6x + 18$

If -3 is a root of $P(x)$, then $P(-3) = 0$.
Find $P(-3)$ using synthetic division.

$$\begin{array}{r|rrrr} -3 & 3 & 5 & -6 & 18 \\ & & -9 & 12 & -18 \\ \hline & 3 & -4 & 6 & 0 \end{array}$$

Since $P(-3) = 0$, -3 is a root of $P(x)$.

If 2 is a root of $P(x)$, then $P(2) = 0$.
Find $P(2)$ using synthetic division.

$$\begin{array}{r|rrrr} 2 & 3 & 5 & -6 & 18 \\ & & 6 & 22 & 32 \\ \hline & 3 & 11 & 16 & 50 \end{array}$$

Since $P(2) \neq 0$, 2 is not a root of $P(x)$.

18. -4 Yes, 2 No

19. $P(x) = x^3 - \frac{7}{2}x^2 + x - \frac{3}{2}$

If -3 is a root of $P(x)$, then $P(-3) = 0$.
Find $P(-3)$ using synthetic division.

$$\begin{array}{r|rrrr} -3 & 1 & -\frac{7}{2} & 1 & -\frac{3}{2} \\ & & -3 & \frac{39}{2} & -\frac{123}{2} \\ \hline & 1 & -\frac{13}{2} & \frac{41}{2} & -63 \end{array}$$

Since $P(-3) \neq 0$, -3 is not a root of $P(x)$.

If $\frac{1}{2}$ is a root of $P(x)$, then $P\left(\frac{1}{2}\right) = 0$.

Find $P\left(\frac{1}{2}\right)$ using synthetic division.

$$\begin{array}{r|rrrr} \frac{1}{2} & 1 & -\frac{7}{2} & 1 & -\frac{3}{2} \\ & & \frac{1}{2} & -\frac{3}{2} & -\frac{1}{4} \\ \hline & 1 & -3 & -\frac{1}{2} & -\frac{7}{4} \end{array}$$

Since $P\left(\frac{1}{2}\right) \neq 0$, $\frac{1}{2}$ is not a root of $P(x)$.

20. i Yes, $-i$ Yes, -2 Yes

21. $P(x) = x^3 + 4x^2 + x - 6$

Try $x - 1$. Use synthetic division to see whether $P(1) = 0$.

$$\begin{array}{r|rrrr} 1 & 1 & 4 & 1 & -6 \\ & & 1 & 5 & 6 \\ \hline & 1 & 5 & 6 & 0 \end{array}$$

Since $P(1) = 0$, $x - 1$ is a factor of $P(x)$.
Thus $P(x) = (x - 1)(x^2 + 5x + 6)$.

Factoring the trinomial we get
$P(x) = (x - 1)(x + 2)(x + 3)$.

To solve the equation $P(x) = 0$, use the principle of zero products.

$(x - 1)(x + 2)(x + 3) = 0$

$x - 1 = 0$ or $x + 2 = 0$ or $x + 3 = 0$

$x = 1$ or $x = -2$ or $x = -3$

The solutions are 1, -2, and -3.

22. $P(x) = (x - 2)(x + 3)(x + 4)$; 2, -3, -4

23. $P(x) = x^3 - 6x^2 + 3x + 10$

Try x - 1. Use synthetic division to see whether P(1) = 0.

1	1	-6	3	10
		1	-5	-2
	1	-5	-2	8

Since P(1) ≠ 0, x - 1 <u>is not</u> a factor of P(x).

Try x + 1. Use synthetic division to see whether P(-1) = 0.

-1	1	-6	3	10
		-1	7	-10
	1	-7	10	0

Since P(-1) = 0, x + 1 <u>is</u> a factor of P(x).
Thus $P(x) = (x + 1)(x^2 - 7x + 10)$.

Factoring the trinomial we get

P(x) = (x + 1)(x - 2)(x - 5).

To solve the equation P(x) = 0, use the principle of zero products.

(x + 1)(x - 2)(x - 5) = 0

x + 1 = 0 or x - 2 = 0 or x - 5 = 0

x = -1 or x = 2 or x = 5

The solutions are -1, 2, and 5.

24. P(x) = (x - 1)(x - 2)(x + 5); 1, 2, -5

25. $P(x) = x^3 - x^2 - 14x + 24$

Try x + 1, x - 1, and x + 2. Using synthetic division we find that P(-1) ≠ 0, P(1) ≠ 0, and P(-2) ≠ 0. Thus x + 1, x - 2, and x + 2 are not factors of P(x).

Try x - 2. Use synthetic division to see whether P(2) = 0.

2	1	-1	-14	24
		2	2	-24
	1	1	-12	0

Since P(2) = 0, x - 2 <u>is</u> a factor of P(x).
Thus $P(x) = (x - 2)(x^2 + x - 12)$.

Factoring the trinomial we get

P(x) = (x - 2)(x + 4)(x - 3).

To solve the equation P(x) = 0, use the principle of zero products.

(x - 2)(x + 4)(x - 3) = 0

x - 2 = 0 or x + 4 = 0 or x - 3 = 0

x = 2 or x = -4 or x = 3

The solutions are 2, -4, and 3.

26. P(x) = (x - 2)(x - 4)(x + 3); 2, 4, -3

27. $P(x) = x^4 - x^3 - 19x^2 + 49x - 30$

Try x - 1. Use synthetic division to see whether P(1) = 0.

1	1	-1	-19	49	-30
		1	0	-19	30
	1	0	-19	30	0

Since P(1) = 0, x - 1 <u>is</u> a factor of P(x).
Thus $P(x) = (x - 1)(x^3 - 19x + 30)$.

We continue to use synthetic division to factor $x^3 - 19x + 30$. Trying x - 1, x + 1, and x + 2 we find that P(1) ≠ 0, P(-1) ≠ 0, and P(-2) ≠ 0. Thus x - 1, x + 1, and x + 2 are not factors of $x^3 - 19x + 30$. Try x - 2.

2	1	0	-19	30
		2	4	-30
	1	2	-15	0

Since P(2) = 0, x - 2 <u>is</u> a factor of $x^3 - 19x + 30$.

Thus $P(x) = (x - 1)(x - 2)(x^2 + 2x - 15)$.

Factoring the trinomial we get

P(x) = (x -1)(x - 2)(x - 3)(x + 5).

To solve the equation P(x) = 0, use the principle of zero products.

(x - 1)(x - 2)(x - 3)(x + 5) = 0

x - 1 = 0 or x - 2 = 0 or x - 3 = 0 or x + 5 = 0

x = 1 or x = 2 or x = 3 or x = -5

The solutions are 1, 2, 3, and -5.

28. P(x) = (x + 1)(x + 2)(x + 3)(x + 5);
-1, -2, -3, -5

29. $\frac{6x^2}{x^2 + 11} + \frac{60}{x^3 - 7x^2 + 11x - 77} = \frac{1}{x - 7}$

LCM = $(x^2 + 11)(x - 7)$

$$(x^2+11)(x-7)\left[\frac{6x^2}{x^2+11}+\frac{60}{(x^2+11)(x-7)}\right]=(x^2+11)(x-7)\cdot\frac{1}{x-7}$$

$$6x^2(x - 7) + 60 = x^2 + 11$$
$$6x^3 - 42x^2 + 60 = x^2 + 11$$
$$6x^3 - 43x^2 + 49 = 0$$

Use synthetic division to find factors of $P(x) = 6x^3 - 43x^2 + 49$. Try $x + 1$. Use synthetic division to see whether $P(-1) = 0$.

```
-1 | 6   -43    0    49
   |      -6   49   -49
   -------------------
     6   -49   49  |  0
```

Since $P(-1) = 0$, $x + 1$ is a factor of $P(x)$. Thus $P(x) = (x + 1)(6x^2 - 49x + 49)$. Factoring the trinomial we get $P(x) = (x + 1)(6x - 7)(x - 7)$.

To solve $P(x) = 0$, use the principle of zero products.

$(x + 1)(6x - 7)(x - 7) = 0$

$x + 1 = 0$ or $6x - 7 = 0$ or $x - 7 = 0$

$x = -1$ or $6x = 7$ or $x = 7$

$x = -1$ or $x = \frac{7}{6}$ or $x = 7$

The value $x = 7$ does not check, but $x = -1$ and $x = \frac{7}{6}$ do. The solutions are -1 and $\frac{7}{6}$.

30. $-1 \pm \sqrt{7}$

31. To solve $x^3 + 2x^2 - 13x + 10 > 0$ we first factor and solve $f(x) = x^3 + 2x^2 - 13x + 10 = 0$. Using synthetic division we find the solutions to be -5, 1, and 2. They are not solutions of the inequality, but they divide the number line as shown.

```
     A            B         C        D
 ---------|  |---------|  |----|  |--------
 ---------------+----------+----+----------
               -5          1    2
```

Try a test number in each interval.

A: $f(-6) = (-6)^3 + 2(-6)^2 - 13(-6) + 10 = -56$

B: $f(0) = 0^3 + 2\cdot0^2 - 13\cdot0 + 10 = 10$

C: $f\left(\frac{3}{2}\right) = \left(\frac{3}{2}\right)^3 + 2\left(\frac{3}{2}\right)^2 - 13\left(\frac{3}{2}\right) + 10 = -\frac{13}{8}$

D: $f(3) = 3^3 + 2\cdot3^2 - 13\cdot3 + 10 = 16$

The solution set is $\{x | -5 < x < 1 \text{ or } x > 2\}$.

32. $\{x | -5 < x < 1 \text{ or } 2 < x < 3\}$

33. $P(x) = x^3 - kx^2 + 3x + 7k$

Think of $x + 2$ as $x - (-2)$.

Find $P(-2)$.

```
-2 | 1    -k       3        7k
   |      -2     2k + 4   -4k - 14
   --------------------------------
     1  -k - 2   2k + 7  |  3k - 14
```

Thus $P(-2) = 3k - 14$.

We know that if $x + 2$ is a factor of $P(x)$, then $P(-2) = 0$.

We solve $0 = 3k - 14$ for k.

$0 = 3k - 14$

$14 = 3k$

$\frac{14}{3} = k$

34. a) -85.1587, b) -485.1587

35. Divide $x^2 + kx + 4$ by $x - 1$.

```
1 | 1    k       4
  |      1     k + 1
  ------------------
    1  k + 1 | k + 5
```

The remainder is $k + 5$.

Divide $x^2 + kx + 4$ by $x + 1$.

```
-1 | 1    k       4
   |     -1    -k + 1
   -------------------
     1  k - 1 | -k + 5
```

The remainder is $-k + 5$.

Set $k + 5 = -k + 5$ and solve for k.

$k + 5 = -k + 5$

$2k = 0$

$k = 0$

36. $k = -\frac{3}{2}$

37. See the answer section in the text.

38. Let $f(x) = x^n - a^n$. Then $f(a) = 0$ so that $x - a$ is a factor.

Exercise Set 11.3

1. $P(x) = (x + 3)^2(x - 1) = (x + 3)(x + 3)(x - 1)$

The factor $x + 3$ occurs twice. Thus the root -3 has a multiplicity of two.

The factor $x - 1$ occurs only one time. Thus the root 1 has a mulitplicity of one.

2. 3, Multiplicity 2; -4, Multiplicity 3; 0, Multiplicity 4

3. $x^3(x - 1)^2(x + 4) = 0$

   $x \cdot x \cdot x(x - 1)(x - 1)(x + 4) = 0$

   The factor x occurs three times. Thus the root 0 has a multiplicity of three.

   The factor x - 1 occurs twice. Thus the root 1 has a multiplicity of two.

   The factor x + 4 occurs only one time. Thus the root -4 has a multiplicity of one.

4. 3, Multiplicity 2; 2, Multiplicity 2

5. $P(x) = x^4 - 4x^2 + 3$

   We factor as follows:

   $P(x) = (x^2 - 3)(x^2 - 1)$

   $= (x - \sqrt{3})(x + \sqrt{3})(x - 1)(x + 1)$

   The roots of the polynomial are $\sqrt{3}$, $-\sqrt{3}$, 1, and -1. Each has multiplicity of one.

6. ± 3, ± 1; each has multiplicity of 1

7. $P(x) = x^3 + 3x^2 - x - 3$

   We factor by grouping:

   $P(x) = x^2(x + 3) - (x + 3)$

   $= (x^2 - 1)(x + 3)$

   $= (x - 1)(x + 1)(x + 3)$

   The roots of the polynomial are 1, -1, and -3. Each has multiplicity of one.

8. $\sqrt{2}$, $-\sqrt{2}$, 1; each has multiplicity of 1.

9. Find a polynomial of degree 3 with -2, 3, and 5 as roots.

   By Theorem 2 such a polynomial has factors x + 2, x - 3, and x - 5, so we have

   $P(x) = a_n(x + 2)(x - 3)(x - 5)$.

   The number a_n can be any nonzero number. The simplest polynomial will be obtained if we let it be 1. Multiplying the factors, we obtain

   $P(x) = (x + 2)(x - 3)(x - 5)$

   $= (x^2 - x - 6)(x - 5)$

   $= x^3 - 6x^2 - x + 30$

10. $x^3 - 2x^2 + x - 2$

11. Find a polynomial of degree 3 with -3, 2i, and -2i as roots.

    By Theorem 2 such a polynomial has factors x + 3, x - 2i, and x + 2i, so we have

    $P(x) = a_n(x + 3)(x - 2i)(x + 2i)$.

    The number a_n can be any nonzero number. The simplest polynomial will be obtained if we let it be 1. Multiplying the factors, we obtain

    $P(x) = (x + 3)(x - 2i)(x + 2i)$

    $= (x + 3)(x^2 + 4)$

    $= x^3 + 3x^2 + 4x + 12$

12. $x^3 - x^2 + 15x + 17$

13. Find a polynomial of degree 3 with $\sqrt{2}$, $-\sqrt{2}$, and $\sqrt{3}$ as roots.

    By Theorem 2 such a polynomial has factors $x - \sqrt{2}$, $x + \sqrt{2}$, and $x - \sqrt{3}$, so we have

    $P(x) = a_n(x - \sqrt{2})(x + \sqrt{2})(x - \sqrt{3})$.

    The number a_n can be any nonzero number. The simplest polynomial will be obtained if we let it be 1. Multiplying the factors, we obtain

    $P(x) = (x - \sqrt{2})(x + \sqrt{2})(x - \sqrt{3})$

    $= (x^2 - 2)(x - \sqrt{3})$

    $= x^3 - \sqrt{3}\,x^2 - 2x + 2\sqrt{3}$

14. $x^4 - 3x^3 - 7x^2 + 15x + 18$

15. A polynomial or polynomial equation of degree 5 has at most 5 roots. Three of the roots are 6, -3 + 4i, and $4 - \sqrt{5}$. Using Theorems 6 and 7 we know that -3 - 4i and $4 + \sqrt{5}$ are also roots.

16. 1 + i

17. Find a polynomial of lowest degree with rational coefficients that has 1 + i and 2 as some of its roots.

    1 - i is also a root. (Theorem 6)

    Thus the polynomial is

    $a_n(x - 2)[x - (1 + i)][x - (1 - i)]$.

    If we let $a_n = 1$, we obtain

    $(x - 2)[(x - 1) - i][(x - 1) + i]$

    $= (x - 2)[(x - 1)^2 - i^2]$

    $= (x - 2)(x^2 - 2x + 1 + 1)$

    $= (x - 2)(x^2 - 2x + 2)$

    $= x^3 - 4x^2 + 6x - 4$

18. $x^3 - 3x^2 + x + 5$

19. Find a polynomial of lowest degree with rational coefficients that has -4i and 5 as some of its roots.

    4i is also a root. (Theorem 6)

    Thus the polynomial is

    $a_n(x - 5)(x + 4i)(x - 4i)$.

    If we let $a_n = 1$, we obtain

    $(x - 5)[x^2 - (4i)^2]$

    $= (x - 5)(x^2 + 16)$

    $= x^3 - 5x^2 + 16x - 80$

20. $x^4 - 6x^3 + 11x^2 - 10x + 2$

21. Find a polynomial of lowest degree with rational coefficients that has $\sqrt{5}$ and $-3i$ as some of its roots.

$-\sqrt{5}$ is also a root. (Theorem 7)

$3i$ is also a root. (Theorem 6)

Thus the polynomial is

$a_n(x - \sqrt{5})(x + \sqrt{5})(x + 3i)(x - 3i)$.

If we let $a_n = 1$, we obtain

$(x^2 - 5)(x^2 + 9)$

$= x^4 + 4x^2 - 45$

22. $x^4 + 14x^2 - 32$

23. If $-i$ is a root of $x^4 - 5x^3 + 7x^2 - 5x + 6$, i is also a root (Theorem 6). Thus $x + i$ and $x - i$ or $(x + i)(x - i)$ which is $x^2 + 1$ are factors of the polynomial. Divide $x^4 - 5x^3 + 7x^2 - 5x + 6$ by $x^2 + 1$ to find the other factors.

$$\begin{array}{r|l} & x^2 - 5x + 6 \\ \hline x^2 + 1 & x^4 - 5x^3 + 7x^2 - 5x + 6 \\ & \underline{x^4 \qquad\quad + x^2} \\ & -5x^3 + 6x^2 - 5x \\ & \underline{-5x^3 \qquad\quad - 5x} \\ & 6x^2 \qquad + 6 \\ & \underline{6x^2 \qquad + 6} \\ & 0 \end{array}$$

Thus

$x^4 - 5x^3 + 7x^2 - 5x + 6 = (x + i)(x - i)(x^2 - 5x + 6)$

$= (x + i)(x - i)(x - 2)(x - 3)$

Using the principle of zero products we find the other roots to be i, 2, and 3.

24. $-2i$, 2, -2

25. $x^3 - 6x^2 + 13x - 20 = 0$

If 4 is a root, then $x - 4$ is a factor. Use synthetic division to find another factor.

$$\begin{array}{r|rrrr} 4 & 1 & -6 & 13 & -20 \\ & & 4 & -8 & 20 \\ \hline & 1 & -2 & 5 & 0 \end{array}$$

$(x - 4)(x^2 - 2x + 5) = 0$

$x - 4 = 0$ or $x^2 - 2x + 5 = 0$ (Principle of zero products)

$x = 4$ or $x = \dfrac{2 \pm \sqrt{4 - 20}}{2}$ (Quadratic formula)

$x = 4$ or $x = \dfrac{2 \pm 4i}{2} = 1 \pm 2i$

The other roots are $1 + 2i$ and $1 - 2i$.

26. $-1 + \sqrt{3}i$, $-1 - \sqrt{3}i$

27. $x^3 - 4x^2 + x - 4 = 0$

Using synthetic division we find that i is a root.

$$\begin{array}{r|rrrr} i & 1 & -4 & 1 & -4 \\ & & i & -4i - 1 & 4 \\ \hline & 1 & -4 + i & -4i & 0 \end{array}$$

Using Theorem 6 we know that if i is a root then $-i$ is also a root. Thus $(x - i)(x + i)$, or $x^2 + 1$, is a factor of $x^3 - 4x^2 + x - 4$. Using division we find that $x - 4$ is also a factor.

$$\begin{array}{r|l} & x - 4 \\ \hline x^2 + 1 & x^3 - 4x^2 + x - 4 \\ & \underline{x^3 \qquad\quad + x} \\ & -4x^2 \qquad - 4 \\ & \underline{-4x^2 \qquad - 4} \\ & 0 \end{array}$$

Therefore, $(x - i)(x + i)(x - 4) = 0$

Using the principle of zero products we find that the solutions are i, $-i$, and 4.

28. -3, $2 + i$, $2 - i$

29. $x^4 - 2x^3 - 2x - 1 = 0$

Using synthetic division we find that i is a root.

$$\begin{array}{r|rrrrr} i & 1 & -2 & 0 & -2 & -1 \\ & & i & -2i - 1 & 2 - i & 1 \\ \hline & 1 & -2 + i & -2i - 1 & -i & 0 \end{array}$$

Using Theorem 6 we know that if i is a root then $-i$ is also a root. Thus $(x - i)(x + i)$, or $x^2 + 1$, is a factor of $x^4 - 2x^3 - 2x - 1$. Using division we find that $x^2 - 2x - 1$ is also a factor.

$$\begin{array}{r|l} & x^2 - 2x - 1 \\ \hline x^2 + 1 & x^4 - 2x^3 + 0x^2 - 2x - 1 \\ & \underline{x^4 \qquad\quad + x^2} \\ & -2x^3 - x^2 - 2x \\ & \underline{-2x^3 \qquad\quad - 2x} \\ & -x^2 \qquad - 1 \\ & \underline{-x^2 \qquad - 1} \\ & 0 \end{array}$$

Therefore, $(x - i)(x + i)(x^2 - 2x - 1) = 0$. Using the principle of zero products and the quadratic formula, we find that the roots are i, $-i$, $1 + \sqrt{2}$, and $1 - \sqrt{2}$.

30. $\pm 4i$, $\pm \sqrt{7}$

31. If the equation has a double root, the discriminant must be 0.

$(2a)^2 - 4\cdot 1b = 0$ gives $b = a^2$.

Then $x^2 + 2ax + (a^2) = 0$ has the double root, $-a$.

32. For $P(x) = a_nx^n + a_{n-1}x^{n-1} + \ldots + a_0$ with a_i positive, consider any positive x. Every term will be positive, hence $P(x) > 0$.

33. Since $P(x)$ of odd degree n has n linear factors, of which an even number correspond to all the pairs of conjugate nonreal roots, there is at least one other factor $(x - a)$ which gives a real root a.

34. Each pair of nonreal roots, $a \pm bi$, gives a quadratic factor $x^2 - 2ax + a^2 + b^2$ with real coefficients. Each real root c gives a linear factor $(x - c)$. Hence $P(x)$ is composed of linear and quadratic factors with real coefficients.

35. See the answer section in the text.

36. Assume $P(x) = \cos x$. Then by Theorem 5, the polynomial has at least one root and at most n roots. But $\cos x = 0$ has infinitely many roots. Therefore, there is no polynomial P that defines the cosine function.

Exercise Set 11.4

1. $P(x) = x^5 - 3x^2 + 1$

Since the leading coefficient is 1, the only possibilities for rational roots (Theorem 8) are the factors of the last coefficient 1: 1 and -1.

2. $\pm (1, 2, 3, 4, 6, 12)$

3. $P(x) = 15x^6 + 47x^2 + 2$

By Theorem 8, if c/d, a rational number in lowest terms, is a root of $P(x)$, then c must be a factor of 2 and d must be a factor of 15.

The possibilities for c and d are

c: 1, -1, 2, -2   d: 1, -1, 3, -3, 5, -5, 15, -15

The resulting possibilities for c/d are

$\frac{c}{d}$: $1, -1, \frac{1}{3}, -\frac{1}{3}, \frac{1}{5}, -\frac{1}{5}, \frac{1}{15}, -\frac{1}{15}$

$2, -2, \frac{2}{3}, -\frac{2}{3}, \frac{2}{5}, -\frac{2}{5}, \frac{2}{15}, -\frac{2}{15}$

4. $\pm \left[1, 2, 3, 6, \frac{1}{10}, \frac{1}{5}, \frac{3}{10}, \frac{3}{5}, \frac{1}{2}, \frac{3}{2}, \frac{2}{5}, \frac{6}{5}\right]$

5. $P(x) = x^3 + 3x^2 - 2x - 6$

Using Theorem 8, the only possible rational roots are 1, -1, 2, -2, 3, -3, 6 and -6. Of these eight possibilities we know that at most three of them could be roots because $P(x)$ is of degree 3. Use synthetic division to determine which are roots.

1	1	3	-2	-6
		1	4	2
	1	4	2	-4

$P(1) \neq 0$, so 1 is not a root.

-1	1	3	-2	-6
		-1	-2	4
	1	2	-4	-2

$P(-1) \neq 0$, so -1 is not a root.

5. (continued)

2	1	3	-2	-6
		2	10	16
	1	5	8	10

$P(2) \neq 0$, so 2 is not a root.

-2	1	3	-2	-6
		-2	-2	8
	1	1	-4	2

$P(-2) \neq 0$, so -2 is not a root.

3	1	3	-2	-6
		3	18	48
	1	6	16	42

$P(3) \neq 0$, so 3 is not a root.

-3	1	3	-2	-6
		-3	0	6
	1	0	-2	0

$P(-3) = 0$, so -3 is a root.

We can now express $P(x)$ as follows:

$P(x) = (x + 3)(x^2 - 2)$.

To find the other roots we solve the quadratic equation $x^2 - 2 = 0$. The other roots are $\sqrt{2}$ and $-\sqrt{2}$. These are irrational numbers. Thus the only rational root is -3.

We write the polynomial in factored form:

$P(x) = (x + 3)(x - \sqrt{2})(x + \sqrt{2})$

6. $1, \sqrt{3}, -\sqrt{3}$; $(x - 1)(x - \sqrt{3})(x + \sqrt{3}) = 0$

7. $x^3 - 3x + 2 = 0$

Using Theorem 8, the only possible rational roots are 1, -1, 2, and -2. At most three of these can be roots because the polynomial is of degree 3. Use synthetic division to determine which are roots.

1	1	0	-3	2
		1	1	-2
	1	1	-2	0

1 is a root

We can now express the equation as follows:

$(x - 1)(x^2 + x - 2) = 0$

To find the other roots we solve the following quadratic equation:

$x^2 + x - 2 = 0$

$(x + 2)(x - 1) = 0$

$x = -2$ or $x = 1$

The rational roots are 1 and -2. (One is a root with multiplicity of 2.)

We write the polynomial in factored form:

$(x - 1)^2(x + 2) = 0$

8. There are no rational roots.

9. $P(x) = x^3 - 5x^2 + 11x - 19$

Using Theorem 8, the only possible rational roots are 1, -1, 19, and -19. At most three of these can be roots because P(x) is of degree 3. Use synthetic division to determine which are roots.

1	1	-5	11	-19
		1	-4	7
	1	-4	7	-12

$P(1) \neq 0$, so 1 is not a root.

-1	1	-5	11	-19
		-1	6	-17
	1	-6	17	-36

$P(-1) \neq 0$, so -1 is not a root.

19	1	-5	11	-19
		19	266	5263
	1	14	277	5244

$P(19) \neq 0$, so 19 is not a root.

-19	1	-5	11	-19
		-19	456	-8873
	1	-24	467	-8892

$P(-19) \neq 0$, so -19 is not a root.

There are no rational roots.

10. There are no rational roots.

11. $P(x) = 5x^4 - 4x^3 + 19x^2 - 16x - 4$

Using Theorem 8, the only possible rational roots are $\pm\left(1, 2, 4, \frac{1}{5}, \frac{2}{5}, \frac{4}{5}\right)$. Of these twelve possibilities we know that at most four of them could be roots because P(x) is of degree 4. Using synthetic division we determine that the only rational roots are 1 and $-\frac{1}{5}$.

1	5	-4	19	-16	-4
		5	1	20	4
	5	1	20	4	0

$P(1) = 0$, so 1 is a root.

We can now express P(x) as follows:
$P(x) = (x - 1)(5x^3 + x^2 + 20x + 4)$.

We now use $5x^3 + x^2 + 20x + 4$ and check to see if 1 is a double root and check for other possible rational roots. The only other rational root is $-\frac{1}{5}$.

$-\frac{1}{5}$	5	1	20	4
		-1	0	-4
	5	0	20	0

$P\left(-\frac{1}{5}\right) = 0$, so $-\frac{1}{5}$ is a root.

We can now express P(x) as follows:
$P(x) = (x - 1)\left(x + \frac{1}{5}\right)(5x^2 + 20)$

11. (continued)

To find the other roots we solve the quadratic equation $5x^2 + 20 = 0$.

$5x^2 + 20 = 0$
$5x^2 = -20$
$x^2 = -4$
$x = \pm 2i$

The other roots are 2i and -2i, neither of which is rational. The only rational roots are 1 and $-\frac{1}{5}$.

We write the polynomial in factored form:
$P(x) = (x - 1)\left(x + \frac{1}{5}\right)(x - 2i)(x + 2i)$

12. $\frac{1}{3}$, -1, 1 - i, 1 + i;
$P(x) = (x + 1)\left(x - \frac{1}{3}\right)(x - 1 + i)(x - 1 - i)$

13. $P(x) = x^4 - 3x^3 - 20x^2 - 24x - 8$

Using Theorem 8, the only possible rational roots are 1, -1, 2, -2, 4, -4, 8, and -8. Of these eight possibilities, we know that at most four of them could be roots because P(x) is of degree 4. Using synthetic division we determine that the only rational roots are -1 and -2.

-1	1	-3	-20	-24	-8
		-1	4	16	8
	1	-4	-16	-8	0

$P(-1) = 0$, so -1 is a root.

We can now express P(x) as follows:
$P(x) = (x + 1)(x^3 - 4x^2 - 16x - 8)$.

We now use $x^3 - 4x^2 - 16x - 8$ and check to see if -1 is a double root and check for other possible rational roots. The only other rational root is -2.

-2	1	-4	-16	-8
		-2	12	8
	1	-6	-4	0

$P(-2) = 0$, so -2 is a root.

We can now express P(x) as follows:
$P(x) = (x + 1)(x + 2)(x^2 - 6x - 4)$.

Since the factor $x^2 - 6x - 4$ is quadratic, use the quadratic formula to find the other roots, $3 + \sqrt{13}$ and $3 - \sqrt{13}$. These are irrational numbers. The only rational roots are -1 and -2.

We write the polynomial in factored form:
$P(x) = (x + 1)(x + 2)(x - 3 - \sqrt{13})(x - 3 + \sqrt{13})$

14. 1, 2, $-4 \pm \sqrt{21}$;
$P(x) = (x - 1)(x - 2)(x + 4 - \sqrt{21})(x + 4 + \sqrt{21})$

15. $P(x) = x^3 - 4x^2 + 2x + 4$

Using Theorem 8, the only possible rational roots are 1, -1, 2, -2, 4, and -4. Of these six possibilities, we know that at most three of them could be roots, because P(x) is of degree 3. Use synthetic division to determine which are roots.

1	1	-4	2	4
		1	-3	-1
	1	-3	-1	3

$P(1) \neq 0$, so 1 is not a root.

-1	1	-4	2	4
		-1	5	-7
	1	-5	7	-3

$P(-1) \neq 0$, so -1 is not a root.

2	1	-4	2	4
		2	-4	-4
	1	-2	-2	0

$P(2) = 0$, so 2 is a root.

We can now express P(x) as

$P(x) = (x - 2)(x^2 - 2x - 2)$.

To find the other roots we use the quadratic formula to solve the equation $x^2 - 2x - 2 = 0$. The other roots are $1 + \sqrt{3}$ and $1 - \sqrt{3}$. These are irrational roots.

The only rational root is 2.

We write the polynomial in factored form:

$P(x) = (x - 2)(x - 1 - \sqrt{3})(x - 1 + \sqrt{3})$

16. $4, 2 \pm \sqrt{3}$;

$P(x) = (x - 4)(x - 2 - \sqrt{3})(x - 2 + \sqrt{3})$

17. $P(x) = x^3 + 8$

Using Theorem 8, the only possible rational roots are 1, -1, 2, -2, 4, -4, 8, and -8. Of these eight possibilities we know that at most three of them could be roots because P(x) is of degree 3. Using synthetic division we determine that -2 is the only rational root.

-2	1	0	0	8
		-2	4	-8
	1	-2	4	0

$P(-2) = 0$, so -2 is a root.

We can now express P(x) as follows:

$P(x) = (x + 2)(x^2 - 2x + 4)$

Since the factor $x^2 - 2x + 4$ is quadratic, use the quadratic formula to find the other roots, $1 + \sqrt{3}i$ and $1 - \sqrt{3}i$. These are irrational numbers. Thus the only rational root is -2.

We write the polynomial in factored form:

$P(x) = (x + 2)(x - 1 - \sqrt{3}i)(x - 1 + \sqrt{3}i)$

18. $2, -1 + i\sqrt{3}, -1 - i\sqrt{3}$;

$P(x) = (x - 2)(x + 1 - i\sqrt{3})(x + 1 + i\sqrt{3})$

19. $P(x) = \frac{1}{3}x^3 - \frac{1}{2}x^2 - \frac{1}{6}x + \frac{1}{6}$

$6P(x) = 2x^3 - 3x^2 - x + 1$

We multiplied by 6, the LCM of the denominators. This equation is equivalent to the first, and all coefficients on the right are integers. Thus any root of 6P(x) is a root of P(x).

The possible rational roots of $2x^3 - 3x^2 - x + 1$ are $1, -1, \frac{1}{2}$, and $-\frac{1}{2}$. Of these four possibilities we know that at most three of them could be roots because the degree of 6P(x) is 3. Using synthetic division we determine that $\frac{1}{2}$ is a root.

$\frac{1}{2}$	2	-3	-1	1
		1	-1	-1
	2	-2	-2	0

We can now express 6P(x) as follows:

$$6P(x) = \left[x - \frac{1}{2}\right](2x^2 - 2x - 2)$$

$$= 2\left[x - \frac{1}{2}\right](x^2 - x - 1)$$

Since the factor $x^2 - x - 1$ is quadratic, use the quadratic formula to find the other roots, $\frac{1 + \sqrt{5}}{2}$ and $\frac{1 - \sqrt{5}}{2}$. These are irrational numbers. Thus the only rational root is $\frac{1}{2}$.

We write the polynomial in factored form:

$$6P(x) = 2\left[x - \frac{1}{2}\right]\left[x - \frac{1 + \sqrt{5}}{2}\right]\left[x - \frac{1 - \sqrt{5}}{2}\right], \text{ so}$$

$$P(x) = \frac{1}{3}\left[x - \frac{1}{2}\right]\left[x - \frac{1 + \sqrt{5}}{2}\right]\left[x - \frac{1 - \sqrt{5}}{2}\right]$$

20. $\frac{3}{4}, i, -i$; $P(x) = \frac{2}{3}\left[x - \frac{3}{4}\right](x - i)(x + i)$

21. $P(x) = x^4 + 32$

Using Theorem 8, the only possible rational roots are ± (1, 2, 4, 8, 16, 32). Of these twelve possibilities we know that at most four of them could be roots because P(x) is of degree 4. Use synthetic division to check each possibility.

1	1	0	0	0	32
		1	1	1	1
	1	1	1	1	33

-1	1	0	0	0	32
		-1	1	-1	1
	1	-1	1	-1	33

2	1	0	0	0	32
		2	4	8	16
	1	2	4	8	48

-2	1	0	0	0	32
		-2	4	-8	16
	1	-2	4	-8	48

$P(1) \neq 0$, $P(-1) \neq 0$, $P(2) \neq 0$, $P(-2) \neq 0$; therefore 1, -1, 2, and -2 are not roots. Similarly we can show that ± (4, 8, 16, 32) are not roots. Thus there are no rational roots.

22. None

23. $x^3 - x^2 - 4x + 3 = 0$

The possible rational roots are 1, -1, 3, and -3. Of these four possibilities, at most three of them could be roots because the polynomial equation is of degree 3. Use synthetic division to check each possibility.

1	1	-1	-4	3
		1	0	-4
	1	0	-4	-1

-1	1	-1	-4	3
		-1	2	2
	1	-2	-2	5

3	1	-1	-4	3
		3	6	6
	1	2	2	9

-3	1	-1	-4	3
		-3	12	-24
	1	-4	8	-21

$P(1) \neq 0$, $P(-1) \neq 0$, $P(3) \neq 0$, $P(-3) \neq 0$; therefore 1, -1, 3, and -3 are not roots. Thus there are no rational roots.

24. $-\frac{3}{2}$

25. $x^4 + 2x^3 + 2x^2 - 4x - 8 = 0$

The possible rational roots are $\pm$ (1, 2, 4, 8). Of these eight possibilities, at most four of them could be roots because the polynomial equation is of degree 4. Use synthetic division to check each possibility.

1	1	2	2	-4	-8
		1	3	5	1
	1	3	5	1	-7

-1	1	2	2	-4	-8
		-1	-1	-1	5
	1	1	1	-5	-3

2	1	2	2	-4	-8
		2	8	20	32
	1	4	10	16	24

-2	1	2	2	-4	-8
		-2	0	-4	16
	1	0	2	-8	8

$P(1) \neq 0$, $P(-1) \neq 0$, $P(2) \neq 0$, $P(-2) \neq 0$; therefore 1, -1, 2, and -2 are not roots. Similarly we can show that 4, -4, 8, and -8 are not roots. Thus there are no rational roots.

26. None

27. $P(x) = x^5 - 5x^4 + 5x^3 + 15x^2 - 36x + 20$

The possible rational roots are $\pm$ (1, 2, 4, 5, 10, 20). Of these twelve possibilities at most five of them could be roots because $P(x)$ is of degree 5. Using synthetic division we determine the only rational roots are -2, 1, and 2.

-2	1	-5	5	15	-36	20
		-2	14	-38	46	-20
	1	-7	19	-23	10	0

$P(-2) = 0$, so -2 is a root.

27. (continued)

1	1	-5	5	15	-36	20
		1	-4	1	16	-20
	1	-4	1	16	-20	0

$P(1) = 0$, so 1 is a root.

2	1	-5	5	15	-36	20
		2	-6	-2	26	-20
	1	-3	-1	13	-10	0

$P(2) = 0$, so 2 is a root.

Similarly we can show that $P(-1) \neq 0$, $P(-4) \neq 0$, $P(4) \neq 0$, $P(-5) \neq 0$, $P(5) \neq 0$, $P(-10) \neq 0$, $P(10) \neq 0$, $P(-20) \neq 0$, and $P(20) \neq 0$. Thus -1, -4, 4, -5, 5, -10, 10, -20, and 20 are not roots. The only rational roots are -2, 1, and 2.

28. 2, -2, 3

29. Solve $x^3 - 64 = 0$

One root is 4. We use synthetic division and express the polynomial in factored form.

4	1	0	0	-64
		4	16	64
	1	4	16	0

$x^3 - 64 = (x - 4)(x^2 + 4x + 16) = 0$

The discriminant of $x^2 + 4x + 16$ is -48, so the other roots of the polynomial are not real numbers. Thus, the length of a side is 4 cm.

30. 5 cm

31.

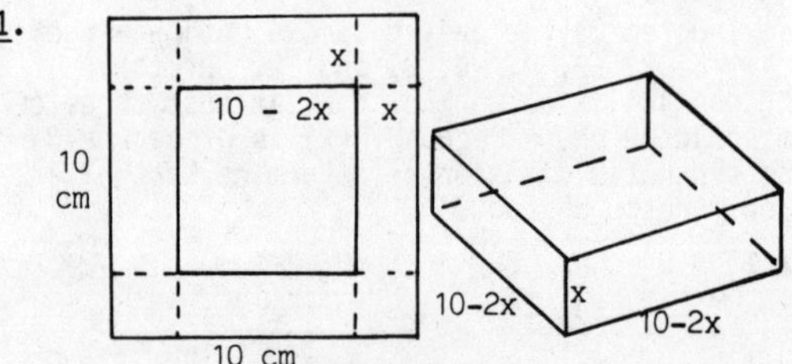

$V = \ell wh$ (Volume formula for rectangular prism)

$48 = (10 - 2x)(10 - 2x)(x)$

(Substituting $10 - 2x$ for ℓ and w, x for h, and 48 for V)

$48 = 100x - 40x^2 + 4x^3$

$12 = 25x - 10x^2 + x^3$

$0 = x^3 - 10x^2 + 25x - 12$

The possible rational roots of this equation are $\pm$ (1, 2, 3, 4, 6, 12). Using synthetic division we find that 3 is the only rational root.

3	1	-10	25	-12
		3	-21	12
	1	-7	4	0

31. (continued)

We can now express the polynomial equation as follows:

$0 = (x - 3)(x^2 - 7x + 4)$

Since the factor $x^2 - 7x + 4$ is quadratic, we use the quadratic formula to find the other roots, $\frac{7 + \sqrt{33}}{2}$ and $\frac{7 - \sqrt{33}}{2}$. Since the length, x, must be positive and less then 5, $\frac{7 + \sqrt{33}}{2}$ is not a possible solution. The length of a side of the square can be 3 cm or $\frac{7 - \sqrt{33}}{2}$ cm.

32. 5 cm or $\frac{15 - 5\sqrt{5}}{2}$ cm

33. $P(x) = x^3 - 31x^2 + 230x - 200$

a) Solve $x^3 - 31x^2 + 230x - 200 = 0$

The possible rational roots are ± 1, ± 2, ± 4, ± 5, ± 8, ± 10, ± 20, ± 25, ± 40, ± 50, ± 100, ± 200.

$$\begin{array}{r|rrrr} 1 & 1 & -31 & 230 & -200 \\ & & 1 & -30 & 200 \\ \hline & 1 & -30 & 200 & 0 \end{array}$$

We see that 1 is a root.

Then $(x - 1)(x^2 - 30x + 200) = 0$.

Solve $x^2 - 30x + 200 = 0$.

$(x - 10)(x - 20) = 0$

$x = 10$ or $x = 20$

The break-even values are 1, 10, and 20.

b) Solve $x^3 - 31x^2 + 230x - 200 > 0$, $x \geqslant 0$.

The numbers 1, 10, and 20 divide the nonnegative portion of the real number line as shown.

A B C D

0 1 10 20

Try a test number in each interval.

A: $P\left(\frac{1}{2}\right) = \left(\frac{1}{2}\right)^3 - 31\left(\frac{1}{2}\right)^2 + 230\left(\frac{1}{2}\right) - 200 = -\frac{741}{8}$

B: $P(2) = 2^3 - 31(2)^2 + 230 \cdot 2 - 200 = 144$

C: $P(11) = 11^3 - 31(11)^2 + 230 \cdot 11 - 200 = -90$

D: $P(21) = 21^3 - 31(21)^2 + 230 \cdot 21 - 200 = 220$

The nonnegative values of x for which the company makes a profit are $\{x \mid 1 < x < 10 \text{ or } x > 20\}$.

c) Using the results of part (b) we see that the nonnegative values of x for which the company has a loss are $\{x \mid 0 < x < 1 \text{ or } 10 < x < 20\}$.

34. $\frac{9}{2}\text{ m} \times \frac{3}{2}\text{ m} \times \frac{1}{2}\text{ m}$

35. $\sqrt{5}$ is a root of $x^2 - 5 = 0$ which has no rational roots (since ± 1, ± 5 are not roots). Hence $\sqrt{5}$ must be irrational.

36. $\sqrt{N}$ is rational if and only if $x^2 - N = 0$ has rational roots, where N is a positive integer.

37. $x^3 + 7 = 0$

The possible rational roots are 1, -1, 7, and -7. Synthetic division shows that none of these is a root. Since $x^3 + 7 = 0$, we know $x^3 = -7$ and one root is $x = -\sqrt[3]{7}$. We use synthetic division.

$$\begin{array}{r|rrrr} -\sqrt[3]{7} & 1 & 0 & 0 & 7 \\ & & -\sqrt[3]{7} & \sqrt[3]{7^2} & -7 \\ \hline & 1 & -\sqrt[3]{7} & \sqrt[3]{7^2} & 0 \end{array}$$

We can write the equation in factored form.

$(x + \sqrt[3]{7})(x^2 - \sqrt[3]{7}x + \sqrt[3]{7^2}) = 0$

Solve $x^2 - \sqrt[3]{7}x - \sqrt[3]{7^2}$ using the quadratic formula.

$$x = \frac{\sqrt[3]{7} \pm \sqrt{\sqrt[3]{7^2} - 4\sqrt[3]{7^2}}}{2}$$

$$x = \frac{\sqrt[3]{7} \pm \sqrt{-3\sqrt[3]{7^2}}}{2} = \frac{\sqrt[3]{7} \pm \sqrt{-3 \cdot 7^{2/3}}}{2}$$

$$x = \frac{\sqrt[3]{7} \pm \sqrt[3]{7}\sqrt{-3}}{2} \qquad (\sqrt{7^{2/3}} = (7^{2/3})^{1/2} = 7^{1/3})$$

$$x = \sqrt[3]{7}\left[\frac{1 \pm i\sqrt{3}}{2}\right]$$

The roots are $-\sqrt[3]{7}$, $\sqrt[3]{7}\left[\frac{1}{2} + \frac{\sqrt{3}}{2}i\right]$, and $\sqrt[3]{7}\left[\frac{1}{2} - \frac{\sqrt{3}}{2}i\right]$.

38. $\sqrt[3]{5}$, $\sqrt[3]{5}\left[-\frac{1}{2} + \frac{\sqrt{3}}{2}i\right]$, $\sqrt[3]{5}\left[-\frac{1}{2} - \frac{\sqrt{3}}{2}i\right]$

Exercise Set 11.5

1. $3x^5 - 2x^2 + x - 1$

a) b) c)

a) From positive to negative: a variation
b) From negative to positive: a variation
c) From positive to negative: a variation

The number of variations of sign is three. Therefore, the number of positive real roots is either 3 or 1 (Theorem 10).

2. 3 or 1

3. $6x^7 + 2x^2 + 5x + 4 = 0$

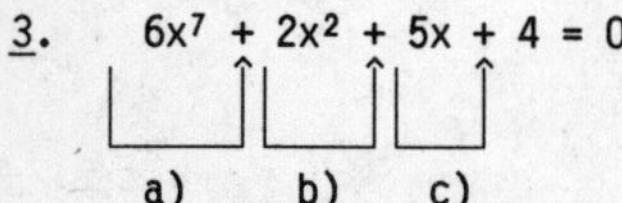

a) b) c)

a) From positive to positive: no variation
b) From positive to positive: no variation
c) From positive to positive: no variation

There are no variations of sign. Thus, there are no positive real roots (Theorem 10).

4. None

5. $3p^{18} + 2p^4 - 5p^2 + p + 3$

a) b) c) d)

a) From positive to positive: no variation
b) From positive to negative: a variation
c) From negative to positive: a variation
d) From positive to positive: no variation

The number of variations of sign is two. Therefore, the number of positive real roots is either 2 or 0 (Theorem 10).

6. 2 or 0

7. $P(x) = 3x^5 - 2x^2 + x - 1$

Replace x by -x.

$P(-x) = 3(-x)^5 - 2(-x)^2 + (-x) - 1$
$= -3x^5 - 2x^2 - x - 1$

a) b) c)

a) From negative to negative: no variation
b) From negative to negative: no variation
c) From negative to negative: no variation

There are no variations of sign. Thus, there are no negative real roots of P(x) (Theorem 11).

8. None

9. $6x^7 + 2x^2 + 5x + 4 = 0$

Replace x by -x.

$6(-x)^7 + 2(-x)^2 + 5(-x) + 4 = 0$
$-6x^7 + 2x^2 - 5x + 4 = 0$

a) b) c)

a) From negative to positive: a variation
b) From positive to negative: a variation
c) From negative to positive: a variation

The number of variations of signs is three. Therefore, the number of negative real roots is 3 or 1 (Theorem 11).

10. 1

11. $P(p) = 3p^{18} + 2p^3 - 5p^2 + p + 3$

Replace p with -p.

$P(-p) = 3(-p)^{18} + 2(-p)^3 - 5(-p)^2 + (-p) + 3$
$= 3p^{18} - 2p^3 - 5p - p + 3$

a) b) c) d)

a) From positive to negative: a variation
b) From negative to negative: no variation
c) From negative to negative: no variation
d) From negative to positive: a variation

The number of variations of sign is two. Therefore, the number of negative real roots of P(p) is 2 or 0 (Theorem 11).

12. 3 or 1

13. $P(x) = 3x^4 - 15x^2 + 2x - 3$

Try 1:

1	3	0	-15	2	-3
		3	3	-12	-10
	3	3	-12	-10	-13

Since some of the coefficients of the quotient and the remainder are negative, there is no guarantee that 1 is an upper bound (Theorem 12).

Try 2:

2	3	0	-15	2	-3
		6	12	-6	-8
	3	6	-3	-4	-11

Again there is no guarantee that 2 is an upper bound.

Try 3:

3	3	0	-15	2	-3
		9	27	36	114
	3	9	12	38	111

13. (continued)

Since all of the numbers in the bottom row are nonnegative, 3 is an upper bound (Theorem 12).

Also, 2 does not satisfy the requirements of Theorem 12. Thus, 3 is the smallest positive integer that is guaranteed by Theorem 12 to be an upper bound to the roots of P(x).

14. 2

15. $P(x) = 6x^3 - 17x^2 - 3x - 1$

Try 2:

2	6	-17	-3	-1
		12	-10	-26
	6	-5	-13	-27

Since some of the coefficients of the quotient and the remainder are negative, there is no guarantee that 2 is an upper bound.

Try 3:

3	6	-17	-3	-1
		18	3	0
	6	1	0	-1

Since the remainder is negative, there is no guarantee that 3 is an upper bound.

Try 4:

4	6	-17	-3	-1
		24	28	100
	6	7	25	99

Since all of the numbers in the bottom row are nonnegative, 4 is an upper bound.

Also, 3 does not satisfy the requirements of Theorem 12. Thus, 4 is the smallest positive integer that is guaranteed by Theorem 12 to be an upper bound to the roots of P(x).

16. 3

17. $P(x) = 3x^4 - 15x^3 + 2x - 3$

We will use Theorem 14, which is a corollary to Theorem 13.

Try -1:

-1	3	-15	0	2	-3
		-3	18	-18	16
	3	-18	18	-16	13

Since the odd-numbered coefficients (from left to right) are nonnegative and the even-numbered ones are nonpositive, -1 is a lower bound. And since -1 is the largest negative integer, -1 is the largest negative integer guaranteed by Theorem 13 to be a lower bound to the roots of P(x).

18. -1

19. $P(x) = 6x^3 + 15x^2 + 3x - 1$

We will use Theorem 14, which is a corollary of Theorem 13.

Try -2:

-2	6	15	3	-1
		-12	-6	6
	6	3	-3	5

We do not know whether or not -2 is a lower bound.

Try -3:

-3	6	15	3	-1
		-18	9	-36
	6	-3	12	-37

Since the odd-numbered coefficients (from left to right) are nonnegative and the even-numbered ones are nonpositive, -3 is a lower bound. Also, since -2 did not meet the requirements of Theorem 14, -3 is the largest negative integer guaranteed by Theorem 13 to be a lower bound to the roots of P(x).

20. -2 (Answers may vary.)

21. $P(x) = x^4 - 2x^2 + 12x - 8$

There are three variations of sign, so the number of positive real roots is 3 or 1 (Theorem 10).

$$P(-x) = (-x)^4 - 2(-x)^2 + 12(-x) - 8$$
$$= x^4 - 2x^2 - 12x - 8$$

There is one variation of sign, so the number of negative real roots of P(x) is 1 (Theorem 11).

Look for an upper bound (Use Theorem 12).

Try 1:

1	1	0	-2	12	-8
		1	1	-1	11
	1	1	-1	11	3

Since one of the coefficients is negative, we do not know whether or not 1 is an upper bound.

Try 2:

2	1	0	-2	12	-8
		2	4	4	32
	1	2	2	16	24

Since all of the numbers in the bottom row are nonnegative, 2 is an upper bound.

Look for a lower bound (Use Theorem 14).

Try -2:

-2	1	0	-2	12	-8
		-2	4	-4	-16
	1	-2	2	8	-24

We do not know whether or not -2 is a lower bound.

21. (continued)

Try -3:

-3	1	0	-2	12	-8
		-3	9	-21	27
	1	-3	7	-9	19

Since the odd-numbered coefficients (from left to right) are nonnegative and the even-numbered ones are nonpositive, -3 is a lower bound.

22. 3 or 1 positive, 1 negative,
Upper bound: 3, Lower bound: -4

23. $P(x) = x^4 - 2x^2 - 8$

There is one variation of sign, so the number of positive real roots is 1 (Theorem 10).

$P(-x) = (-x)^4 - 2(-x)^2 - 8$ (Replacing x by -x)

$= x^4 - 2x^2 - 8$

There is one variation of sign, so the number of negative real roots is 1 (Theorem 11). Since the degree of the equation is 4 and there is only one positive real root and one negative real root, there must be two nonreal roots.

Look for an upper bound (Use Theorem 12).

Try 1:

1	1	0	-2	0	-8
		1	1	-1	-1
	1	1	-1	-1	-9

Since some of the coefficients of the quotient and the remainder are negative, we do not know whether or not 1 is an upper bound.

Try 2:

2	1	0	-2	0	-8
		2	4	4	8
	1	2	2	4	0

Since all of the numbers in the bottom row are nonnegative, 2 is an upper bound. Also note that 2 is a root, $P(2) = 0$.

Look for a lower bound (Use Theorem 14).

Try -1:

-1	1	0	-2	0	-8
		-1	1	1	-1
	1	-1	-1	1	-9

We do not know whether or not -1 is a lower bound.

Try -2:

-2	1	0	-2	0	-8
		-2	4	-4	8
	1	-2	2	-4	0

Since the odd-numbered coefficients (from left to right) are nonnegative and the even-numbered ones are nonpositive, -2 is a lower bound.

24. 1 positive, 1 negative,
Upper bound: 2, Lower bound: -2

25. $P(x) = x^4 - 9x^2 - 6x + 4$

There are two variations of sign, so the number of positive real roots is 2 or 0 (Theorem 10).

$P(-x) = (-x)^4 - 9(-x)^2 - 6(-x) + 4$

$= x^4 - 9x^2 + 6x + 4$

There are two variations of sign, so the number of negative real roots of P(x) is 2 or 0 (Theorem 11).

Look for an upper bound (Use Theorem 12).

Try 3:

3	1	0	-9	-6	4
		3	9	0	-18
	1	3	0	-6	-14

Since one of the coefficients and the remainder are negative, we do not know whether or not 3 is an upper bound.

Try 4:

4	1	0	-9	-6	4
		4	16	28	88
	1	4	7	22	92

Since all of the numbers in the bottom row are nonnegative, 4 is an upper bound.

Look for a lower bound (Use Theorem 14).

Try -2:

-2	1	0	-9	-6	4
		-2	4	10	-8
	1	-2	-5	4	-4

We do not know whether or not -2 is a lower bound.

Try -3:

-3	1	0	-9	-6	4
		-3	9	0	18
	1	-3	0	-6	22

Since the odd-numbered coefficients (from left to right) are nonnegative and the even-numbered ones are nonpositive, -3 is a lower bound.

26. 2 or 0 positive, 2 or 0 negative,
Upper bound: 5, Lower bound: -5

27. $P(x) = x^4 + 3x^2 + 2$

There are no variations of sign. Thus there are no positive real roots (Theorem 10).

$P(-x) = (-x)^4 + 3(-x)^2 + 2$

$= x^4 + 3x^2 + 2$

There are no variations of sign. Thus there are no negative real roots (Theorem 11). Since the degree of the equation is 4, there are 4 nonreal roots.

Since there are no real roots, there is no upper bound and no lower bound.

28. No positive or negative roots

29. $P(x) = x^3 - 3x - 2$

The possible rational roots are 1, -1, 2, and -2. Synthetic division shows that -1 is a root with multiplicity of 2 and 2 is also a root. Thus all the roots are rational.

30. -0.9, 1.3, 2.5

31. $x^3 - 3x - 4 = 0$

The possible rational roots are 1, -1, 2, -2, 4, and -4. Synthetic division shows that none of these is a root. Sketch a graph of the polynomial $P(x) = x^3 - 3x - 4$.

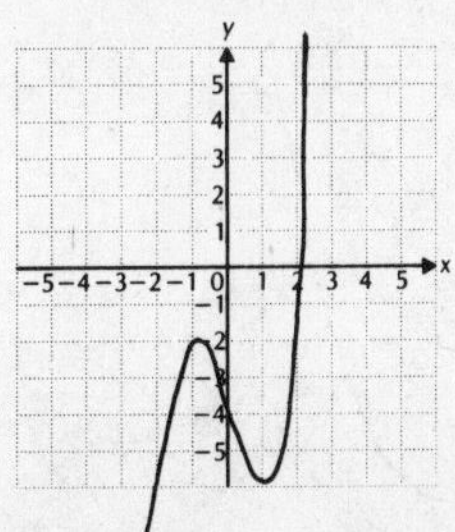

We see from the graph that there seems to be a root between 2 and 3. Checking function values confirms this: P(2) = -2 and P(3) = 14. To find a better approximation we compute additional function values.

x	P(x)	
2	-2	
2.1	-1.039	Since P(2.1) < 0, there is no root between 2 and 2.1.
2.2	0.048	Since P(2.1) < 0 and P(2.2) > 0, there is a root between 2.1 and 2.2.

We have the accuracy desired, so we stop. Since 0.048 is closer to 0 than -1.039, we accept 2.2 as a better approximation to the nearest tenth.

32. -1.1

33. $x^4 + x^2 + 1 = 0$

$P(x) = x^4 + x^2 + 1$ has no variations in sign and hence no positive real roots. $P(-x) = (-x)^4 + (-x)^2 + 1 = x^4 + x^2 + 1$ also has no variations in sign and hence P(x) has no negative real roots. Thus the equation has no real roots.

34. No real roots.

35. $x^4 - 6x^2 + 8 = 0$

The possible rational roots are ± 1, ± 2, ± 4, and ± 8. Using synthetic division we find that 2 and -2 are roots and we can factor the polynomial as follows: $(x - 2)(x + 2)(x^2 - 2)$. Solving $x^2 - 2 = 0$, we find that the irrational roots are $\sqrt{2}$ and $-\sqrt{2}$, or 1.4 and -1.4 to the nearest tenth.

36. -1.8, -0.8, 0.8, 1.8

37. $x^5 + x^4 - x^3 - x^2 - 2x - 2 = 0$

The possible rational roots are ± 1 and ± 2. Using synthetic division we find that -1 is a root and we can factor the polynomial as follows: $(x + 1)(x^4 - x^2 - 2)$. We can solve $x^4 - x^2 - 2 = 0$ by substituting u for x^2, factoring, and solving for u and then for x. We get $x = \pm\sqrt{2}$ or $x = \pm i$. The irrational roots are $\sqrt{2}$ and $-\sqrt{2}$, or 1.4 and -1.4 to the nearest tenth.

38. -1.7, 1.7

39. $P(x) = 2x^5 + 2x^3 - x^2 - 1$

x	P(x)	
0	-1	
1	2	
0.5	-0.9375	Since P(0.5) < 0, there is a root between 0.5 and 1.
0.7	-0.46786	Since P(0.7) < 0, there is a root between 0.7 and 1.
0.8	0.03936	Since P(0.8) > 0, there is a root between 0.7 and 0.8.
0.75	-0.24414	Since P(0.75) < 0, there is a root between 0.75 and 0.8.
0.79	-0.02261	Since P(0.79) < 0, there is a root between 0.79 and 0.8.

We have the accuracy we desire, so we stop. Since -0.02261 is closer to 0 than 0.03936, we accept 0.79 as a better approximation to the nearest hundredth.

40. 1.41

41. $P(x) = x^3 - 2x^2 - x + 4$

The possible rational roots are ± 1, ± 2, and ± 4. Synthetic division shows that none of these is a root. Sketch a graph of the polynomial.

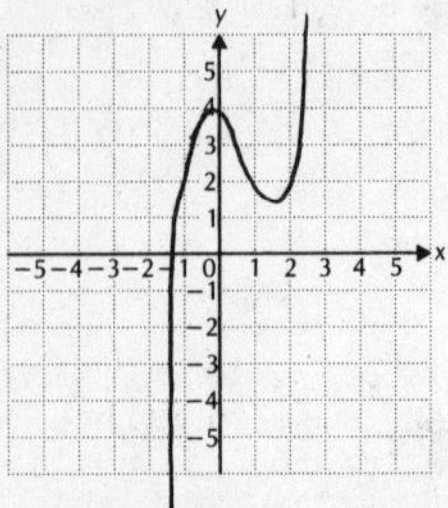

The graph shows that there seems to be a root between -2 and -1. We check function values to confirm this: $P(-2) = -10$, $P(-1) = 2$. We do further computations to find a better approximation.

x	P(x)	
-1	2	
-1.1	1.349	Since $P(-1.1) > 0$, there is no root between -1.1 and -1.
-1.2	0.592	Since $P(-1.2) > 0$, there is no root between -1.2 and -1.1.
-1.3	-0.277	Since $P(-1.3) < 0$, there is a root between -1.3 and -1.2.
-1.25	0.1719	Since $P(-1.25) > 0$, there is a root between -1.3 and -1.25.
-1.27	-0.0042	Since $P(-1.27) < 0$, there is a root between -1.27 and -1.25.
-1.26	0.0844	Since $P(-1.26) > 0$, there is a root between -1.27 and -1.26.

We have the desired accuracy, so we stop. Since -0.0042 is closer to 0 than 0.0844, we accept -1.27 as the better approximation to the nearest hundredth.

42. -0.70, 1.24, 3.46

43. See the answer section in the text.

44. $f(x)$ has one variation in sign; hence one positive root. $f(-x)$ has no variation in sign; hence no negative roots. Zero is not a root. Hence there is exactly one real root.

45. $f(x) = x^3 - 9x^2 + 27x + 50$

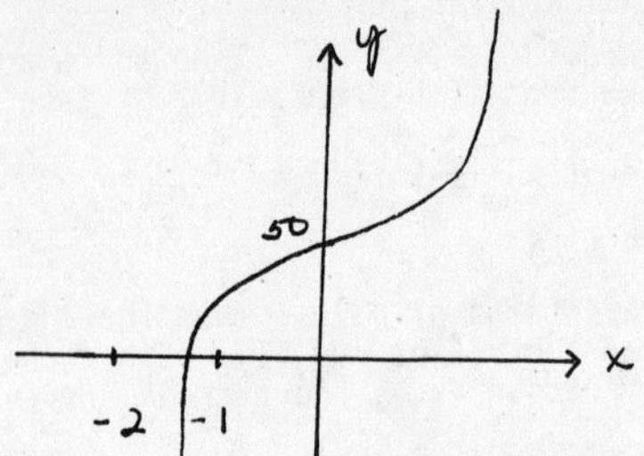

Using a computer software package or graphing calculator to graph the function, we see that there appears to be a root between -2 and -1. Examining the graph on increasingly smaller portions of the interval [-2, -1], we see that there is a root at approximately -1.3.

46. -1.88, 0.35, 1.53

47. $f(x) = x^4 + 4x^3 - 36x^2 - 160x + 300$

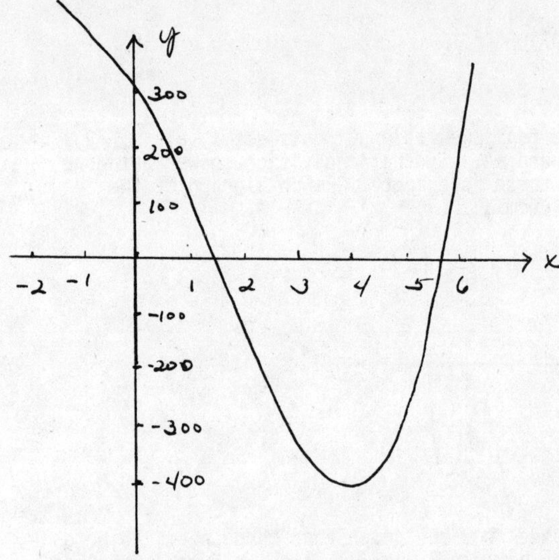

Using a computer software package or graphing calculator to graph the function, we see that there appear to be roots between 1 and 2 and between 5 and 6. Examining the graph on increasingly smaller portions of the intervals [1, 2] and [5, 6], we see that there are roots at approximately 1.5 and 5.7.

48. -7, -5, -2, 1, 3, 6

49. Approximate the root(s) of $N(t) = -0.046t^3 + 2.08t + 2$ in the interval [0, 8].

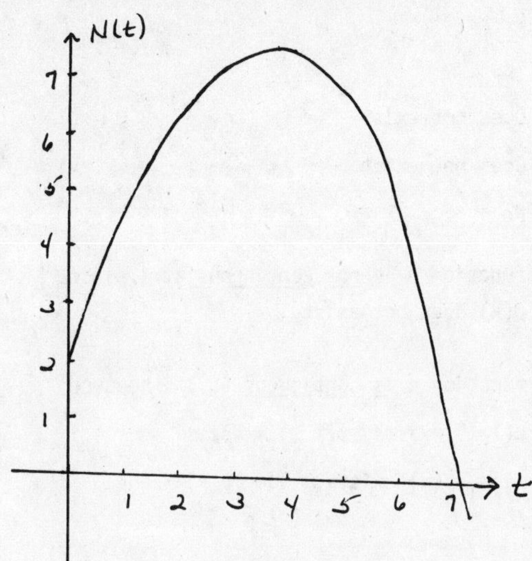

Use a computer software package or graphing calculator to graph the function. There appears to be a root between 7 and 8. Examining the graph on increasingly smaller portions of the interval [7, 8], we see that there is a root at approximately 7.16. Thus, the concentration will be 0 after about 7.16 hr.

50. 8%

51. See the answer section in the text.

52. Since $0 < k < 1$, we have f(r): + + + - - -. Thus the equation has one real root.

Exercise Set 11.6

1. The function is not continuous, because $\lim_{x \to 1} f(x)$ does not exist. ($\lim_{x \to 1^-} f(x) = 2$ and $\lim_{x \to 1^+} f(x) = -1$)

   Intuitively, we might say that the function is not continuous, because its graph cannot be traced without lifting a pencil from the paper.

2. No

3. The function is continuous, because it is continuous at each point in the domain. Intuitively, we might say that the function is continuous, because its graph can be traced without lifting a pencil from the paper.

4. No

5. a) As inputs x approach 1 from the right, outputs f(x) approach -1. Thus, the limit from the right is -1.

   $\lim_{x \to 1^+} f(x) = -1$

   As inputs x approach 1 from the left, outputs f(x) approach 2. Thus, the limit from the left is 2.

   $\lim_{x \to 1^-} f(x) = 2$

   Since the limit from the left, 2, is not the same as the limit from the right, -1, we say $\lim_{x \to 1} f(x)$ does not exist.

   b) When the input is 1, the output, f(1), is -1.

   $f(1) = -1$

   c) Since the limit at x = 1 does not exist, the function is not continuous at x = 1.

   d) As inputs x approach -2 from the left, outputs f(x) approach 3. Thus, the limit from the left is 3.

   $\lim_{x \to -2^-} f(x) = 3$

   As inputs x approach -2 from the right, outputs f(x) approach 3. Thus, the limit from the right is 3.

   $\lim_{x \to -2^+} f(x) = 3$

   Since the limit from the left, 3, is the same as the limit from the right, 3, we have

   $\lim_{x \to -2} f(x) = 3$.

   e) When the input is -2, the output, f(-2), is 3.

   $f(-2) = 3$

   f) The function f(x) is continuous at x = -2 because

   1) f(-2) exists, $f(-2) = 3$,

   2) $\lim_{x \to -2} f(x)$ exists, $\lim_{x \to -2} f(x) = 3$,

   and

   3) $\lim_{x \to -2} f(x) = f(-2) = 3$.

6. a) -2, -2, -2

   b) -2

   c) Yes

   d) Does not exist

   e) -3

   f) No

7. a) As inputs x approach 1 from the left, outputs h(x) approach 2. Thus, the limit from the left is 2.

$\lim_{x \to 1^-} h(x) = 2$

As inputs x approach 1 from the right, outputs h(x) approach 2. Thus, the limit from the right is 2.

$\lim_{x \to 1^+} h(x) = 2$

Since the limit from the left, 2, is the same as the limit from the right, 2, we have

$\lim_{x \to 1} h(x) = 2.$

b) When the input is 1, the output, h(1), is 2.

h(1) = 2

c) The function h(x) is continuous at x = 1 because

1) h(1) exists, h(1) = 2,

2) $\lim_{x \to 1} h(x)$ exists, $\lim_{x \to 1} h(x) = 2$,

and

3) $\lim_{x \to 1} h(x) = h(1) = 2.$

d) As inputs x approach -2 from the left, outputs h(x) approach 0. Thus, the limit from the left is 0.

$\lim_{x \to -2^-} h(x) = 0.$

As inputs x approach -2 from the right, outputs h(x) approach 0. Thus, the limit from the right is 0.

$\lim_{x \to -2^+} h(x) = 0.$

Since the limit from the left, 0, is the same as the limit from the right, 0, we have

$\lim_{x \to -2} h(x) = 0.$

e) When the input is -2, the output, h(-2), is 0.

h(-2) = 0

f) The function h(x) is continuous at x = -2 because

1) h(-2) exists, h(-2) = 0,

2) $\lim_{x \to -2} h(x)$ exists, $\lim_{x \to -2} h(x) = 0$,

and

3) $\lim_{x \to -2} h(x) = h(-2) = 0.$

8. a) $\frac{1}{2}$

b) $\frac{1}{2}$

c) Yes

d) Does not exist

e) Does not exist

f) No

9. The function p is not continuous at 1 since $\lim_{x \to 1} p(x)$ does not exist.

The function p is continuous at $1\frac{1}{2}$ because

1) $p(1\frac{1}{2})$ exists, $p(1\frac{1}{2}) = 45¢$,

2) $\lim_{x \to 1\frac{1}{2}} p(x)$ exists, $\lim_{x \to 1\frac{1}{2}} p(x) = 45¢$,

and

3) $\lim_{x \to 1\frac{1}{2}} p(x) = p(1\frac{1}{2}) = 45¢.$

The function p is not continuous at x = 2 since $\lim_{x \to 2} p(x)$ does not exist.

The function p is continuous at 2.53 because

1) p(2.53) exists, p(2.53) = 65¢,

2) $\lim_{x \to 2.53} p(x)$ exists, $\lim_{x \to 2.53} p(x) = 65¢$,

and

3) $\lim_{x \to 2.53} p(x) = p(2.53) = 65¢.$

10. No, yes, no, yes

11. As inputs x approach 1 from the left, outputs, p(x), approach 25¢. Thus the limit from the left is 25¢.

$\lim_{x \to 1^-} p(x) = 25¢$

As inputs x approach 1 from the right, outputs, p(x), approach 45¢. Thus the limit from the right is 45¢.

$\lim_{x \to 1^+} p(x) = 45¢$

Since the limit from the left, 25¢, is not the same as the limit from the right, 45¢, we have $\lim_{x \to 1} p(x)$ does not exist.

12. 45¢, 65¢, does not exist

13. As inputs x approach 2.3 from the left, outputs, p(x), approach 65¢. Thus the limit from the left is 65¢.

$\lim_{x \to 2.3^-} p(x) = 65¢$

As inputs x approach 2.3 from the right, outputs, p(x), approach 65¢. Thus the limit from the right is 65¢.

$\lim_{x \to 2.3^+} p(x) = 65¢$

Since the limit from the left, 65¢, is the same as the limit from the right, 65¢, we have

$\lim_{x \to 2.3} p(x) = 65¢.$

14. 25¢

15. Find $\lim_{x \to 1} (x^2 - 3)$.

From the continuity principles it follows that $x^2 - 3$ is continuous. Thus the limit can be found by substitution.

$$\begin{aligned} \lim_{x \to 1} (x^2 - 3) &= 1^2 - 3 \quad \text{(Substituting)} \\ &= 1 - 3 \\ &= -2 \end{aligned}$$

16. 5

17. Find $\lim_{x \to 0} \frac{3}{x}$.

The function $\frac{3}{x}$ is not continuous at x = 0, and there is no further algebraic simplification for $\frac{3}{x}$. Thus, direct substitution is not possible. We can use input-output tables or a graph. Here we use a graph.

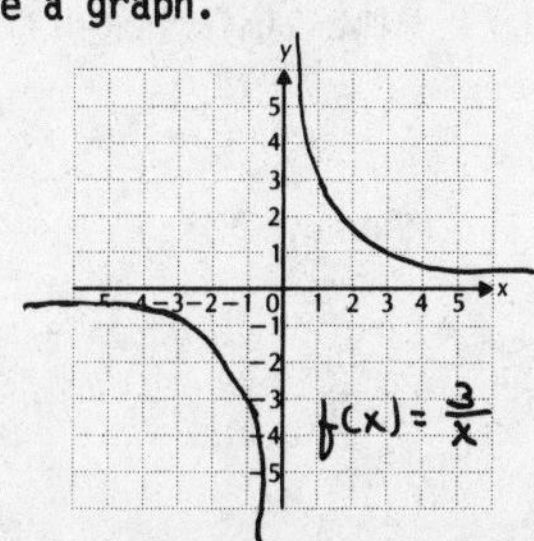

As x approaches 0 from the left, the outputs get smaller and smaller without bound. These numbers do not approach any real number, though it might be said that "the limit from the left is $-\infty$ (negative infinity)." As x approaches 0 from the right, the outputs get larger and larger without bound. These numbers do not approach any real number, though it might be said that "the limit from the right is ∞ (infinity)." For a limit to exist, the limits from the left and right must both exist and be the same.

Thus, $\lim_{x \to 0} \frac{3}{x}$ does not exist.

18. Does not exist

19. Find $\lim_{x \to 3} (2x + 5)$.

From the continuity principles it follows that 2x + 5 is continuous. Thus the limit can be found by substitution.

$$\begin{aligned} \lim_{x \to 3} (2x + 5) &= 2 \cdot 3 + 5 \quad \text{(Substituting)} \\ &= 6 + 5 \\ &= 11 \end{aligned}$$

20. -7

21. Find $\lim_{x \to -5} \frac{x^2 - 25}{x + 5}$.

The function $\frac{x^2 - 25}{x + 5}$ is not continuous at x = -5. We use some algebraic simplification and then some limit principles.

$$\begin{aligned} &\lim_{x \to -5} \frac{x^2 - 25}{x + 5} \\ &= \lim_{x \to -5} \frac{(x + 5)(x - 5)}{x + 5} \quad \text{(Factoring the numerator)} \\ &= \lim_{x \to -5} (x - 5) \quad \text{(Simplifying, assuming } x \neq -5\text{)} \\ &= \lim_{x \to -5} x - \lim_{x \to -5} 5 \quad \text{(By L3)} \\ &= -5 - 5 \quad \text{(By L2 and L1)} \\ &= -10 \end{aligned}$$

22. -8

23. Find $\lim_{x \to -2} \frac{5}{x}$.

The function $\frac{5}{x}$ is continuous at all real numbers except 0. Since $\frac{5}{x}$ is continuous at x = -2, we can substitute to find the limit.

$$\begin{aligned} \lim_{x \to -2} \frac{5}{x} &= \frac{5}{-2} \quad \text{(Substituting)} \\ &= -\frac{5}{2} \end{aligned}$$

24. $\frac{2}{5}$

25. Find $\lim_{x \to 2} \frac{x^2 + x - 6}{x - 2}$.

The function $\frac{x^2 + x - 6}{x - 2}$ is not continuous at $x = 2$. We use some algebraic simplification and then some limit principles.

$\lim_{x \to 2} \frac{x^2 + x - 6}{x - 2}$

$= \lim_{x \to 2} \frac{(x - 2)(x + 3)}{x - 2}$ (Factoring the numerator)

$= \lim_{x \to 2} (x + 3)$ (Simplifying, assuming $x \neq 2$)

$= \lim_{x \to 2} x + \lim_{x \to 2} 3$ (By L3)

$= 2 + 3$ (By L2 and L1)

$= 5$

26. -9

27. Find $\lim_{x \to 5} \sqrt[3]{x^2 - 17}$.

From the continuity principles it follows that $\sqrt[3]{x^2 - 17}$ is continuous. Thus the limit can be found by substitution.

$\lim_{x \to 5} \sqrt[3]{x^2 - 17} = \sqrt[3]{5^2 - 17}$ (Substituting)

$= \sqrt[3]{25 - 17}$

$= \sqrt[3]{8}$

$= 2$

28. 3

29. Find $\lim_{x \to 1} (x^4 - x^3 + x^2 + x + 1)$.

From the continuity principles it follows that $x^4 - x^3 + x^2 + x + 1$ is continuous. Thus the limit can be found by substitution.

$\lim_{x \to 1} (x^4 - x^3 + x^2 + x + 1)$

$= 1^4 - 1^3 + 1^2 + 1 + 1$ (Substituting)

$= 1 - 1 + 1 + 1 + 1$

$= 3$

30. 19

31. Find $\lim_{x \to 2} \frac{1}{x - 2}$.

The function $\frac{1}{x - 2}$ is not continuous at $x = 2$, and there is no further algebraic simplification for $\frac{1}{x - 2}$. Thus direct substitution is not possible. We can use input-output tables or a graph. Here we use a graph.

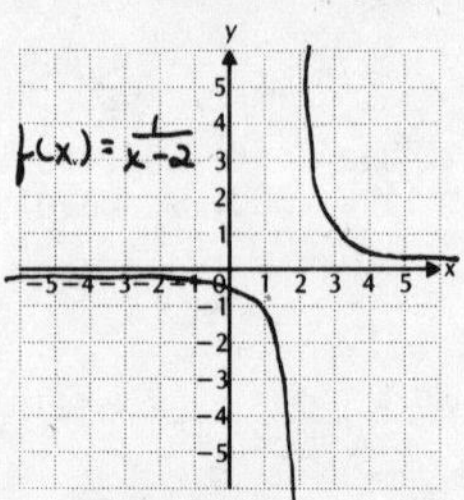

As x approaches 2 from the left, the outputs get smaller and smaller without bound. As x approaches 2 from the right, the outputs get larger and larger without bound. We say that

$\lim_{x \to 2} \frac{1}{x - 2}$ does not exist.

Remember: For a limit to exist, the limits from the left and right must both exist and be the same.

32. Does not exist

33. Find $\lim_{x \to 2} \frac{3x^2 - 4x + 2}{7x^2 - 5x + 3}$.

The rational function $\frac{3x^2 - 4x + 2}{7x^2 - 5x + 3}$ is continuous at all real numbers except those for which $7x^2 - 5x + 3 = 0$. Since $7x^2 - 5x + 3 \neq 0$ when $x = 2$, the function is continuous at $x = 2$. Thus we can substitute to find the limit.

$\lim_{x \to 2} \frac{3x^2 - 4x + 2}{7x^2 - 5x + 3}$

$= \frac{3 \cdot 2^2 - 4 \cdot 2 + 2}{7 \cdot 2^2 - 5 \cdot 2 + 3}$ (Substituting)

$= \frac{12 - 8 + 2}{28 - 10 + 3}$

$= \frac{6}{21}$

$= \frac{2}{7}$ (Simplifying)

34. $-\frac{8}{7}$

35. Find $\lim_{x\to 2} \frac{x^2 + x - 6}{x^2 - 4}$.

The function $\frac{x^2 + x - 6}{x^2 - 4}$ is not continuous at $x = 2$ ($x^2 - 4 = 0$ when $x = 2$). We first use some algebraic simplification.

$$\lim_{x\to 2} \frac{x^2 + x - 6}{x^2 - 4}$$

$$= \lim_{x\to 2} \frac{(x - 2)(x + 3)}{(x - 2)(x + 2)} \quad \text{(Factoring numerator and denominator)}$$

$$= \lim_{x\to 2} \frac{x + 3}{x + 2} \quad \text{(Simplifying, assuming } x \neq 2\text{)}$$

Using input-output tables, we see that as inputs x approach 2 from the left, outputs $\frac{x + 3}{x + 2}$ approach $\frac{5}{4}$. As inputs x approach 2 from the right, outputs $\frac{x + 3}{x + 2}$ approach $\frac{5}{4}$. Since the limit from the left, $\frac{5}{4}$, is the same as the limit from the right, $\frac{5}{4}$, we have

$$\lim_{x\to 2} \frac{x + 3}{x + 2} = \frac{5}{4}.$$

Thus, $\lim_{x\to 2} \frac{x^2 + x - 6}{x^2 - 4}$ is $\frac{5}{4}$.

Note that limit principles could have been used as an alternative to input-output tables in finding the limit.

36. $\frac{8}{7}$

37. Find $\lim_{x\to 1} \frac{1 - \sqrt{x}}{1 - x}$.

The function $\frac{1 - \sqrt{x}}{1 - x}$ is not continuous at $x = 1$ and, assuming we see no further algebraic simplification for $\frac{1 - \sqrt{x}}{1 - x}$, direct substitution is not possible. We can use an input-output table or a graph.

Here we use an input-output table.

x	$\frac{1 - \sqrt{x}}{1 - x}$
0	1
0.5	0.586
0.7	0.544
0.9	0.513
0.95	0.506
0.99	0.501
0.999	0.500

These inputs approach 1 from the left.

37. (continued)

x	$\frac{1 - \sqrt{x}}{1 - x}$
2	0.414
1.5	0.449
1.3	0.467
1.1	0.488
1.05	0.494
1.01	0.499
1.001	0.500

These inputs approach 1 from the right.

From the tables we see that as inputs x approach 1 from the left, outputs $\frac{1 - \sqrt{x}}{1 - x}$ approach 0.5. As inputs x approach 1 from the right, outputs $\frac{1 - \sqrt{x}}{1 - x}$ approach 0.5. Since the limit from the left is the same as the limit from the right, we have

$$\lim_{x\to 1} \frac{1 - \sqrt{x}}{1 - x} = 0.5, \text{ or } \frac{1}{2}.$$

Note that algebraic simplification of $\frac{1 - \sqrt{x}}{1 - x}$, although perhaps not obvious, <u>is</u> possible. A second method for finding this limit uses such simplification and some limit principles.

$$\lim_{x\to 1} \frac{1 - \sqrt{x}}{1 - x}$$

$$= \lim_{x\to 1} \frac{1 - \sqrt{x}}{1 - (\sqrt{x})^2} \quad (x = (\sqrt{x})^2)$$

$$= \lim_{x\to 1} \frac{1 - \sqrt{x}}{(1 - \sqrt{x})(1 + \sqrt{x})} \quad \text{(Factoring the denominator)}$$

$$= \lim_{x\to 1} \frac{1}{1 + \sqrt{x}} \quad \text{(Simplifying, assuming } x \neq 1\text{)}$$

$$= \frac{1}{\lim_{x\to 1} (1 + \sqrt{x})} \quad \text{(By L4)}$$

$$= \frac{1}{\lim_{x\to 1} 1 + \lim_{x\to 1} \sqrt{x}} \quad \text{(By L3)}$$

$$= \frac{1}{1 + 1} \quad \text{(By L1 and L2)}$$

$$= \frac{1}{2}$$

Thus, $\lim_{x\to 1} \frac{1 - \sqrt{x}}{1 - x} = \frac{1}{2}$, or 0.5.

38. $\frac{1}{4}$

39. Find $\lim_{x\to\infty} \frac{2x - 4}{5x}$.

We will use some algebra and the fact that as $x \to \infty$, $\frac{b}{ax^n} \to 0$, for any positive integer n.

$\lim_{x\to\infty} \frac{2x - 4}{5x}$

$= \lim_{x\to\infty} \frac{2x - 4}{5x} \cdot \frac{(1/x)}{(1/x)}$ (Multiplying by a form of 1)

$= \lim_{x\to\infty} \frac{2x \cdot \frac{1}{x} - 4 \cdot \frac{1}{x}}{5x \cdot \frac{1}{x}}$

$= \lim_{x\to\infty} \frac{2 - \frac{4}{x}}{5}$

$= \frac{2 - 0}{5}$ (As $x \to \infty$, $\frac{4}{x} \to 0$)

$= \frac{2}{5}$

40. $\frac{3}{4}$

41. Find $\lim_{x\to\infty} (5 - \frac{2}{x})$.

We will use the fact that as $x \to \infty$, $\frac{b}{ax^n} \to 0$, for any positive integer n.

$\lim_{x\to\infty} (5 - \frac{2}{x})$

$= 5 - 0$ (As $x \to \infty$, $\frac{2}{x} \to 0$)

$= 5$

42. 7

43. Find $\lim_{x\to\infty} \frac{2x - 5}{4x + 3}$.

We will use some algebra and the fact that as $x \to \infty$, $\frac{b}{ax^n} \to 0$, for any positive integer n.

$\lim_{x\to\infty} \frac{2x - 5}{4x + 3}$

$= \lim_{x\to\infty} \frac{2x - 5}{4x + 3} \cdot \frac{(1/x)}{(1/x)}$ (Multiplying by a form of 1)

$= \lim_{x\to\infty} \frac{2x \cdot \frac{1}{x} - 5 \cdot \frac{1}{x}}{4x \cdot \frac{1}{x} + 3 \cdot \frac{1}{x}}$

$= \lim_{x\to\infty} \frac{2 - \frac{5}{x}}{4 + \frac{3}{x}}$

$= \frac{2 - 0}{4 + 0}$ (As $x \to \infty$, $\frac{5}{x} \to 0$ and $\frac{3}{x} \to 0$)

$= \frac{2}{4}$

$= \frac{1}{2}$

44. $\frac{6}{5}$

45. Find $\lim_{x\to\infty} \frac{2x^2 - 5}{3x^2 - x + 7}$.

We will use some algebra and the fact that as $x \to \infty$, $\frac{b}{ax^n} \to 0$, for any positive integer n.

$\lim_{x\to\infty} \frac{2x^2 - 5}{3x^2 - x + 7}$

$= \lim_{x\to\infty} \frac{2x^2 - 5}{3x^2 - x + 7} \cdot \frac{(1/x^2)}{(1/x^2)}$ (Multiplying by a form of 1)

$= \lim_{x\to\infty} \frac{2x^2 \cdot \frac{1}{x^2} - 5 \cdot \frac{1}{x^2}}{3x^2 \cdot \frac{1}{x^2} - x \cdot \frac{1}{x^2} + 7 \cdot \frac{1}{x^2}}$

$= \lim_{x\to\infty} \frac{2 - \frac{5}{x^2}}{3 - \frac{1}{x} + \frac{7}{x^2}}$

$= \frac{2 - 0}{3 - 0 + 0}$ (As $x \to \infty$, $\frac{5}{x^2} \to 0$, $\frac{1}{x} \to 0$, and $\frac{7}{x^2} \to 0$)

$= \frac{2}{3}$

46. -4

47. It is helpful to graph the function.

$$f(x) = \begin{cases} 1, & \text{for } x \neq 2 \\ -1, & \text{for } x = 2 \end{cases}$$

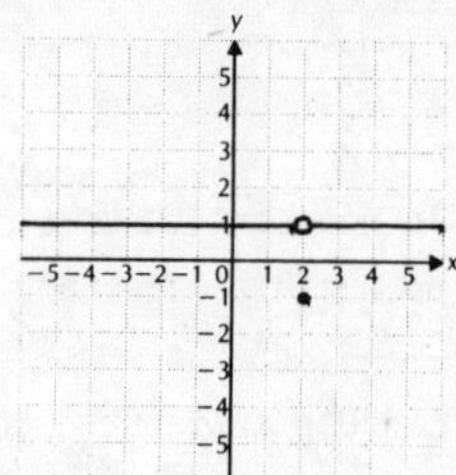

a) Find $\lim_{x\to 0} f(x)$.

As inputs x approach 0 from the left, outputs f(x) approach 1. Thus, the limit from the left is 1. As inputs x approach 0 from the right, outputs f(x) approach 1, so the limit from the right is 1. Since the limit from the left, 1, is the same as the limit from the right, 1, we have

$\lim_{x\to 0} f(x) = 1$.

b) Find $\lim_{x\to 2^-} f(x)$.

As inputs x approach 2 from the left, outputs, f(x), approach 1. Thus the limit from the left is 1.

$\lim_{x\to 2^-} f(x) = 1$

47. (continued)

c) Find $\lim_{x \to 2^+} f(x)$.

As inputs x approach 2 from the right, outputs, f(x), approach 1. Thus the limit from the right is 1.

$\lim_{x \to 2^+} f(x) = 1$

d) Since the limit from the left, 1, is the same as the limit from the right, 1, we have

$\lim_{x \to 2} f(x) = 1$.

e) The function f(x) is continuous at x = 0 because

1) f(0) exists, f(0) = 1,

2) $\lim_{x \to 0} f(x)$ exists, $\lim_{x \to 0} f(x) = 1$,

and

3) $\lim_{x \to 0} f(x) = f(0) = 1$.

The function f(x) is not continuous at x = 2 because the $\lim_{x \to 2} f(x)$ is not the same as the function value at x = 2.

$\lim_{x \to 2} f(x) = 1$

$f(2) = -1$

$\lim_{x \to 2} f(x) \neq f(2)$

48. a) 11

b) 11

c) 11

d) 9

e) No, yes

49. a) The value at the beginning of the first year is $10,000.

The depreciation during the first year is 8%·10,000, or $800.

The value at the beginning of the second year is 10,000 - 800, or $9200.

The depreciation during the second year is 8%·$9200, or $736.

The value at the beginning of the third year is 9200 - 736, or $8464.

The depreciation during the third year is 8%·8464, or $677.12.

The value at the beginning of the fourth year is 8464 - 677.12, or $7786.88.

The depreciation during the fourth year is 8%·7786.88, or $622.95.

The value at the beginning of the fifth year is 7786.88 - 622.95, or $7163.93.

49. (continued)

The depreciation during the fifth year is 8%·7163.93, or $573.11.

b) The value at the beginning of the sixth year is 7163.93 - 573.11, or $6590.82.

The depreciation during the sixth year is 8%·6590.82, or $527.27.

The value at the beginning of the seventh year is 6590.82 - 527.27, or $6063.55.

The depreciation during the seventh year is 8%·6063.55, or $485.08.

The value at the beginning of the eighth year is 6063.55 - 485.08, or $5578.47.

The depreciation during the eighth year is 8%·5578.47, or $446.28.

The value at the beginning of the ninth year is 5578.47 - 446.28, or $5132.19.

The depreciation during the ninth year is 8%·5132.19, or $410.58.

The value at the beginning of the tenth year is 5132.19 - 410.58, or $4721.61.

The depreciation during the tenth year is 8%·4721.61, or $377.73.

The total depreciation at the end of ten years is $800 + $736 + $677.12 + $622.95 + $573.11 + $527.27 + $485.08 + $446.28 + $410.58 + $377.73, or $5656.12.

c) The sum of the annual depreciation costs cannot exceed the new cost which was $10,000. The limit of the sum is $10,000.

50. a) $1800, $1260, $882, $617.40, $432.18

b) $5830.50

c) $6000

51. The limit of the distance of the offensive team from the goal is 0, but the limit is never reached. Thus, the offensive team cannot score a touchdown in this manner.

52. Does not exist

53. Find $\lim_{x\to 1} \frac{2 - \sqrt{x + 3}}{x - 1}$.

The function $\frac{2 - \sqrt{x + 3}}{x - 1}$ is not continuous at x - 1, and there is no further algebraic simplification for $\frac{2 - \sqrt{x + 3}}{x - 1}$. Thus direct substitution is not possible. Here we use an input-output table.

x	$\frac{2 - \sqrt{x + 3}}{x - 1}$
0	-0.268
0.5	-0.258
0.7	-0.255
0.8	-0.253
0.9	-0.252
0.99	-0.250

These inputs approach 1 from the left.

x	$\frac{2 - \sqrt{x + 3}}{x - 1}$
2	-0.236
1.5	-0.243
1.3	-0.245
1.2	-0.247
1.1	-0.248
1.01	-0.250

These inputs approach 1 from the right.

From the table we see that as inputs x approach 1 from the left, outputs approach -0.25. As inputs x approach 1 from the right, outputs approach -0.25. Since the limit from the left is the same as the limit from the right, we have

$$\lim_{x\to 1} \frac{2 - \sqrt{x + 3}}{x - 1} = -0.25.$$

54. $\frac{3}{2}$

55. Find $\lim_{x\to\infty} \frac{4 - 3x}{5 - 2x^2}$.

We will use some algebra and the fact that as $x \to \infty$, $\frac{b}{ax^n} \to 0$, for any positive integer n.

$$\lim_{x\to\infty} \frac{4 - 3x}{5 - 2x^2}$$

$$= \lim_{x\to\infty} \frac{4 - 3x}{5 - 2x^2} \cdot \frac{1/x^2}{1/x^2} \quad \text{(Multiplying by 1)}$$

$$= \lim_{x\to\infty} \frac{\frac{4}{x^2} - \frac{3x}{x^2}}{\frac{5}{x^2} - \frac{2x^2}{x^2}}$$

$$= \lim_{x\to\infty} \frac{\frac{4}{x^2} - \frac{3}{x}}{\frac{5}{x^2} - 2}$$

$$= \frac{0 - 0}{0 - 2} \quad \left(\text{As } x \to 0,\ \frac{4}{x^2} \to 0,\ \frac{3}{x} \to 0, \text{ and } \frac{5}{x^2} \to 0\right)$$

$$= \frac{0}{-2} = 0$$

56. Does not exist

Exercise Set 11.7

1. $f(x) = \frac{1}{x - 3}$

a) The function is neither even nor odd and no symmetries are apparent.

b) The degree of the numerator is less than the degree of the denominator, so the x-axis is an asymptote.

c) The zero of the denominator is 3, so x = 3 is a vertical asymptote. The function has no zeros.

d) The vertical asymptote divides the x-axis into intervals. To find where the function is positive or negative, we try a test point in each interval.

Interval	$(-\infty, 3)$	$(3, \infty)$
Test value	$f(0) = -\frac{1}{3}$	$f(4) = 1$
Sign of f(x)	Negative	Positive
Location of points on graph	Below x-axis	Above x-axis

e) We make a table of function values.

x	$3\frac{1}{8}$	$3\frac{1}{2}$	4	5	8	$2\frac{3}{4}$	2	1	0	-2
f(x)	8	2	1	$\frac{1}{2}$	$\frac{1}{5}$	-4	-1	$-\frac{1}{2}$	$-\frac{1}{3}$	$-\frac{1}{5}$

f) Using all available information, we draw the graph.

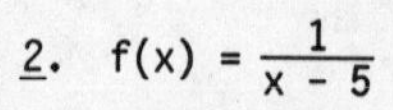

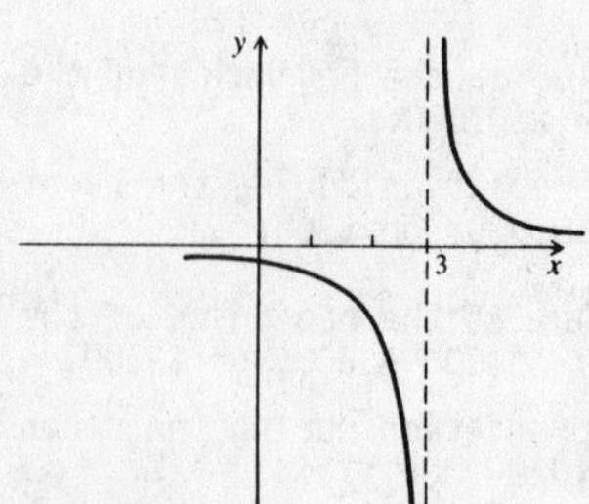

2. $f(x) = \frac{1}{x - 5}$

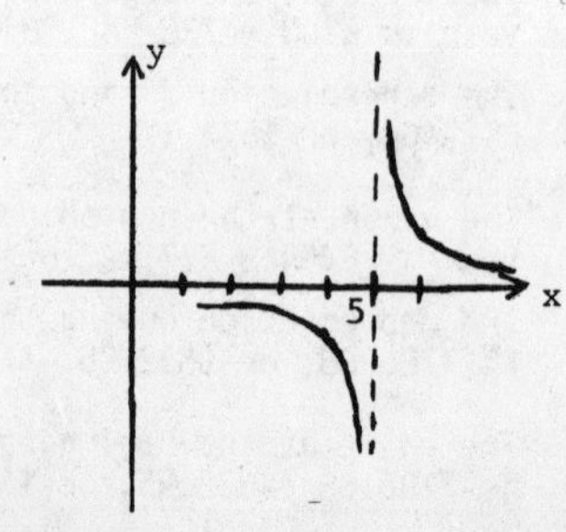

3. $f(x) = \frac{-2}{x - 5}$

a) The function is neither even nor odd and no symmetries are apparent.

b) The degree of the numerator is less than the degree of the denominator, so the x-axis is an asymptote.

c) The zero of the denominator is 5, so x = 5 is a vertical asymptote. The function has no zero.

d) The vertical asymptote divides the x-axis into intervals. To find where the function is positive or negative, we try a test point in each interval.

Interval	$(-\infty, 5)$	$(5, \infty)$
Test value	$f(0) = \frac{2}{5}$	$f(6) = -2$
Sign of f(x)	Positive	Negative
Location of points on graph	Above x-axis	Below x-axis

e) We make a table of function values.

x	-3	0	2	4	$4\frac{2}{3}$	$5\frac{1}{3}$	6	8	11
f(x)	$\frac{1}{4}$	$\frac{2}{5}$	$\frac{2}{3}$	2	6	-6	-2	$-\frac{2}{3}$	$-\frac{1}{3}$

f) Using all available information, we draw the graph.

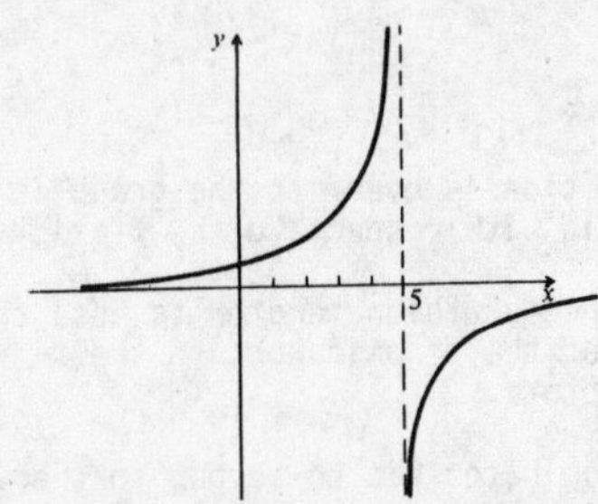

4. $f(x) = \frac{-3}{x - 3}$

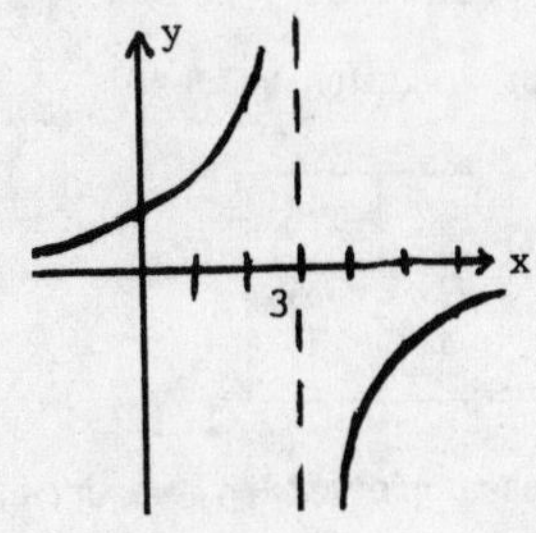

5. $f(x) = \frac{2x + 1}{x}$

a) The function is neither even nor odd and no symmetries are apparent.

b) Since the numerator and denominator have the same degree, the horizontal asymptote can be determined by dividing the leading coefficients of the two polynomials. Thus $2 \div 1 = 2$, and we know that y = 2 is a horizontal asymptote.

c) The zero of the denominator is 0, so x = 0 (the y-axis) is a vertical asymptote. The zero of the function is $-\frac{1}{2}$.

d) The vertical asymptote and the zero of the function divide the x-axis into intervals. To find where the function is positive or negative, we try a test number in each interval.

Interval	$\left(-\infty, -\frac{1}{2}\right)$	$\left(-\frac{1}{2}, 0\right)$	$(0, \infty)$
Test value	$f(-1) = 1$	$f\left(-\frac{1}{4}\right) = -8$	$f(1) = 3$
Sign of f(x)	Positive	Negative	Positive
Location of points on graph	Above x-axis	Below x-axis	Above x-axis

e) We make a table of function values.

x	-100	-10	-3	$-\frac{1}{2}$	$-\frac{1}{10}$	1	4	10	100
f(x)	$1\frac{99}{100}$	$1\frac{9}{10}$	$1\frac{2}{3}$	0	-8	3	$2\frac{1}{4}$	$2\frac{1}{10}$	$2\frac{1}{100}$

f) Using all available information, we draw the graph.

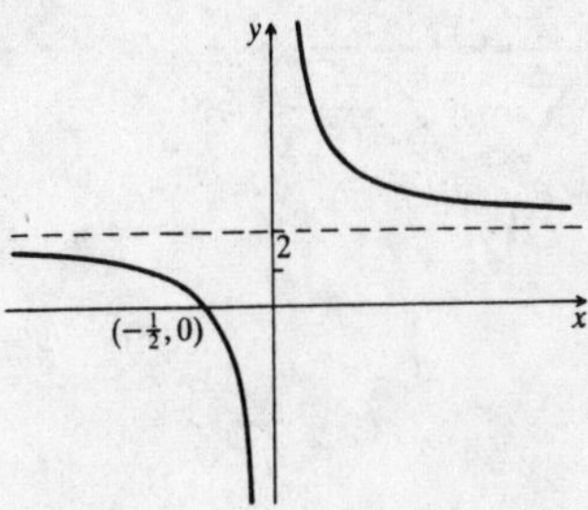

6. $f(x) = \frac{3x - 1}{x}$

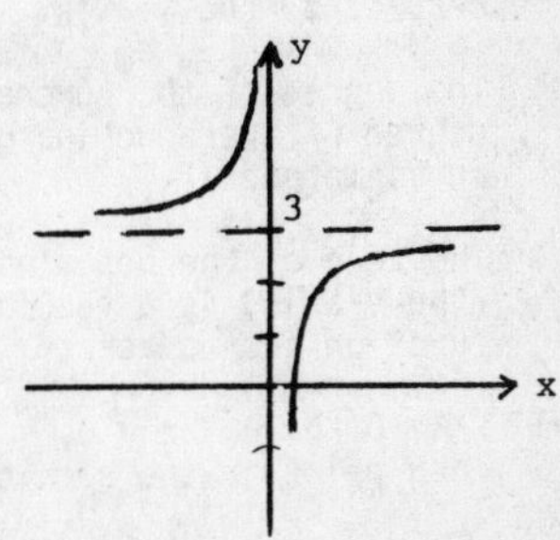

7. $f(x) = \frac{1}{(x - 2)^2}$

a) The function is neither even nor odd and no symmetries are apparent.

b) The degree of the numerator is less than the degree of the denominator, so the x-axis is an asymptote.

c) The zero of the denominator is 2, so x = 2 is a vertical asymptote. The function has no zeros.

d) $(x - 2)^2 > 0$ for all $x \neq 2$, so f(x) is positive at each point in the domain.

e) We make a table of function values.

x	$2\frac{1}{2}$	3	4	5	$1\frac{1}{2}$	1	0	-1
f(x)	4	1	$\frac{1}{4}$	$\frac{1}{9}$	4	1	$\frac{1}{4}$	$\frac{1}{9}$

f) Using all available information, we draw the graph.

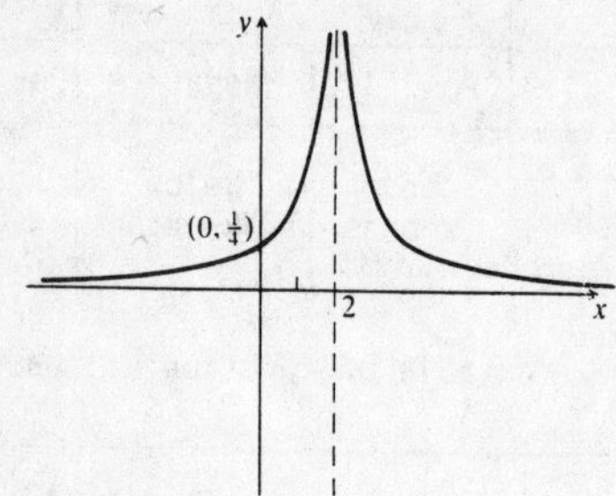

8. $f(x) = \frac{-2}{(x - 3)^2}$

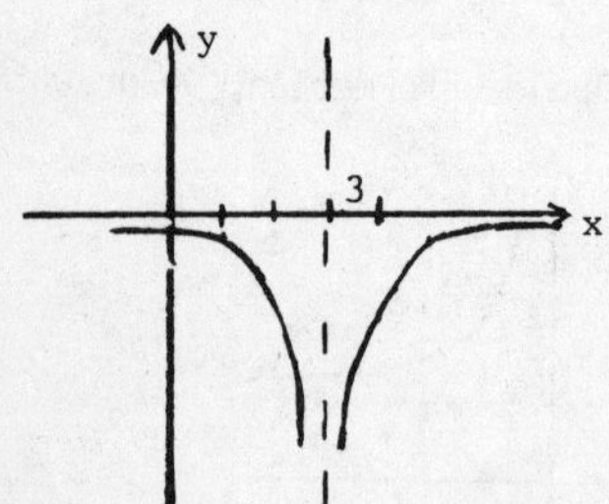

9. $f(x) = \frac{2}{x^2}$

a) The function is even, so it is symmetric with respect to the y-axis.

b) The degree of the numerator is less than the degree of the denominator, so the x-axis is an asymptote.

c) The zero of the denominator is 0, so x = 0 (the y-axis) is a vertical asymptote. The function has no zeros.

d) $x^2 > 0$ for all $x \neq 0$, so f(x) is positive at each point in the domain.

9. (continued)

e) We make a table of function values.

x	$\pm \frac{1}{2}$	± 1	± 2	± 3	± 4
f(x)	8	2	$\frac{1}{2}$	$\frac{2}{9}$	$\frac{1}{8}$

f) Using all available information, we draw the graph.

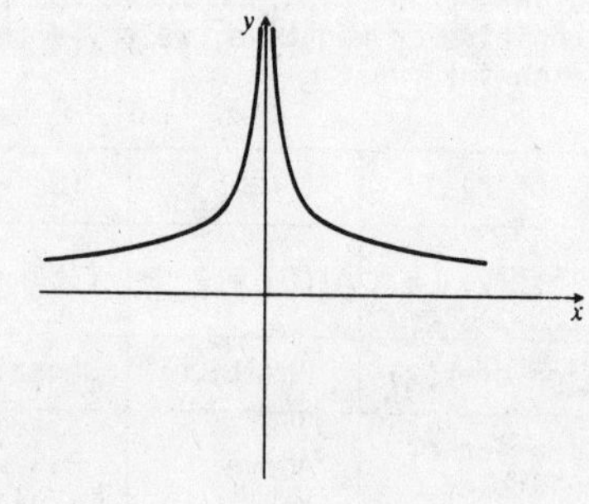

10. $f(x) = \frac{1}{3x^2}$

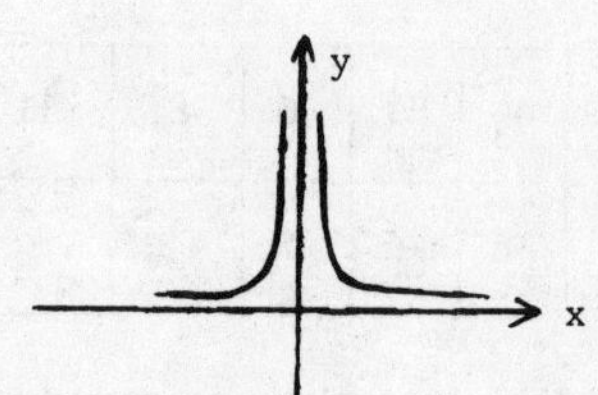

11. $f(x) = \frac{1}{x^2 + 3}$

a) The function is even, so the graph is symmetric with respect to the y-axis.

b) The degree of the numerator is less than the degree of the denominator, so the x-axis is an asymptote.

c) The denominator has no zeros, so there are no vertical asymptotes. The function has no zeros.

d) $x^2 + 3 > 0$ for all real numbers x, so f(x) is positive for all real numbers x.

e) We make a table of function values.

x	0	± 1	± 2	± 3
f(x)	$\frac{1}{3}$	$\frac{1}{4}$	$\frac{1}{7}$	$\frac{1}{12}$

f) Using all available information, we draw the graph.

11. (continued)

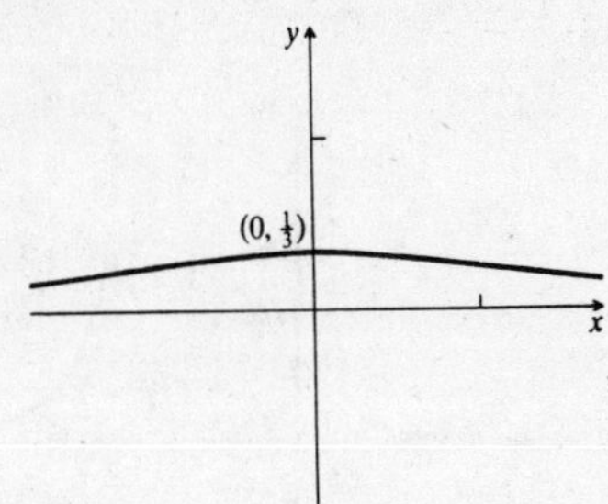

12. $f(x) = \frac{-1}{x^2 + 2}$

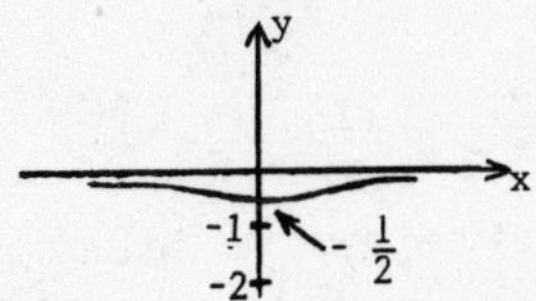

13. $f(x) = \frac{x - 1}{x + 2}$

a) The function is neither even nor odd and no symmetries are apparent.

b) Since the numerator and denominator have the same degree, the horizontal asymptote can be determined by dividing the leading coefficients of the two polynomials. Thus $1 \div 1 = 1$, and we know that $y = 1$ is a horizontal asymptote.

c) The zero of the denominator is -2, so $x = -2$ is a vertical asymptote. The zero of the function is 1.

d) The vertical asymptote and the zero of the function divide the x-axis into intervals. To find where the function is positive or negative, we try a test point in each interval.

Interval	$(-\infty, -2)$	$(-2, 1)$	$(1, \infty)$
Test value	$f(-3) = 4$	$f(0) = -\frac{1}{2}$	$f(2) = \frac{1}{4}$
Sign of f(x)	Positive	Negative	Positive
Location of points on graph	Above x-axis	Below x-axis	Above x-axis

e) We make a table of function values.

x	-6	-4	-3	$-2\frac{1}{2}$	$-1\frac{1}{2}$	-1	0	1	2	4	6	20
f(x)	$\frac{7}{4}$	$\frac{5}{2}$	4	7	-5	-2	$-\frac{1}{2}$	0	$\frac{1}{4}$	$\frac{1}{2}$	$\frac{5}{8}$	$\frac{19}{22}$

13. (continued)

f) Using all available information, we draw the graph.

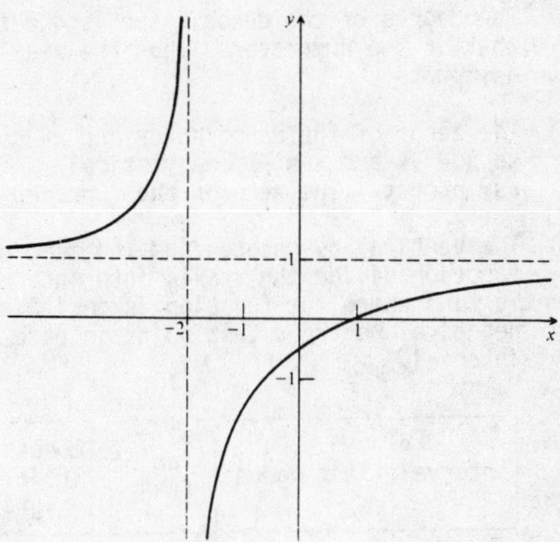

14. $f(x) = \frac{x - 2}{x + 1}$

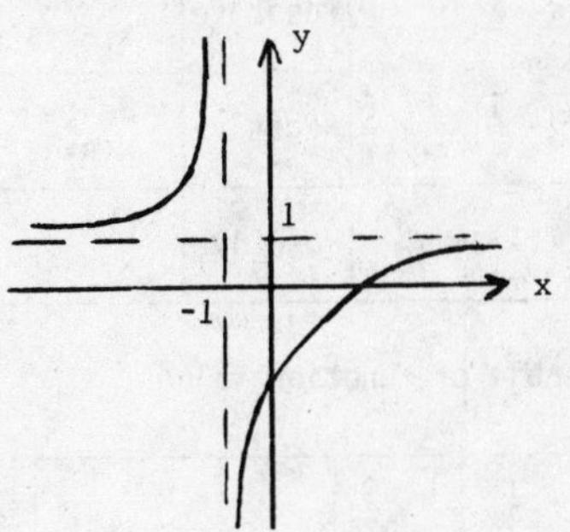

15. $f(x) = \frac{3x}{x^2 + 5x + 4} = \frac{3x}{(x + 4)(x + 1)}$

a) The function is neither even nor odd and no symmetries are apparent.

b) The degree of the denominator is greater than that of the numerator. Thus the x-axis is an asymptote.

c) The zeros of the denominator are -4 and -1, so x = -4 and x = -1 are vertical asymptotes. The zero of the function is 0.

d) The vertical asymptotes and the zero of the function divide the x-axis into intervals. To find where the function is positive or negative, we try a test point in each interval.

Interval	Test value	Sign of f(x)	Location of points on graph
$(-\infty, -4)$	$f(-5) = -\frac{15}{4}$	Negative	Below x-axis
$(-4, -1)$	$f(-2) = 3$	Positive	Above x-axis
$(-1, 0)$	$f\left(-\frac{1}{2}\right) = -\frac{6}{7}$	Negative	Below x-axis
$(0, \infty)$	$f(1) = \frac{3}{10}$	Positive	Above x-axis

e) We make a table of function values.

x	-6	-5	$-\frac{9}{2}$	$-\frac{7}{2}$	-3	-2	$-\frac{3}{2}$
f(x)	$-\frac{9}{5}$	$-\frac{15}{4}$	$-\frac{54}{7}$	$\frac{42}{5}$	$\frac{9}{2}$	3	$\frac{18}{5}$

x	$-\frac{3}{4}$	$-\frac{1}{2}$	0	1	2	3
f(x)	$-\frac{36}{13}$	$-\frac{6}{7}$	0	$\frac{3}{10}$	$\frac{1}{3}$	$\frac{7}{28}$

f) Using all the information, we draw the graph.

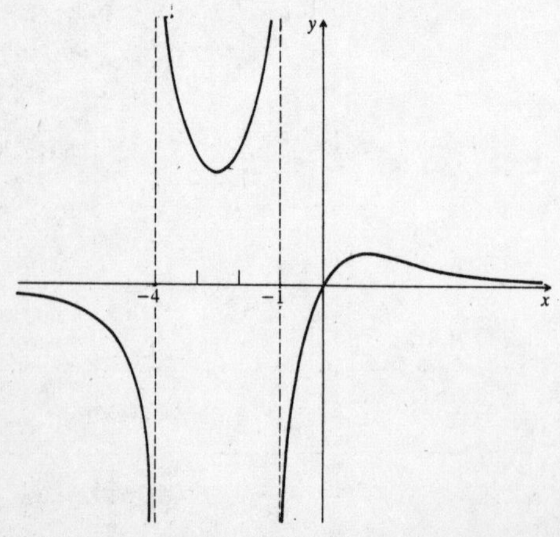

16. $f(x) = \frac{x + 3}{2x^2 - 5x - 3}$

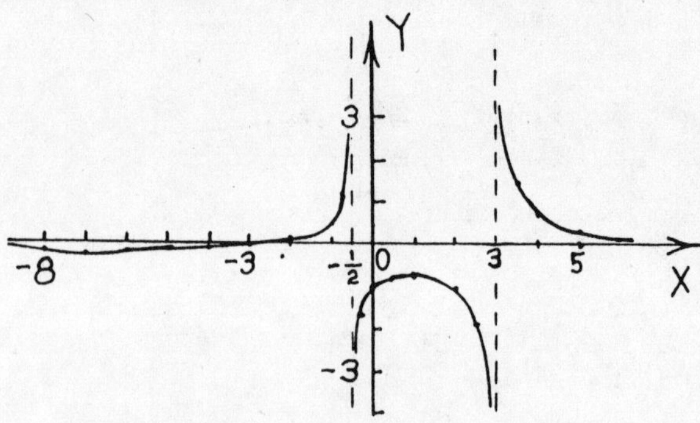

17. $f(x) = \frac{x^2 - 4}{x - 1} = \frac{(x + 2)(x - 2)}{x - 1}$

a) The function is neither even nor odd and no symmetries are apparent.

b) The degree of the numerator is one greater than that of the denominator. Dividing numerator by denominator will show that y = x + 1 is an oblique asymptote.

c) The zero of the denominator is 1, so x = 1 is a vertical asymptote. The zeros of the function are -2 and 2.

d) The vertical asymptote and the zeros of the function divide the x-axis into intervals.

To find where the function is positive or negative, we try a test point in each interval.

Interval	Test value	Sign of f(x)	Location of points on graph
$(-\infty, -2)$	$f(-3) = -\frac{5}{4}$	Negative	Below x-axis
$(-2, 1)$	$f(0) = 4$	Positive	Above x-axis
$(1, 2)$	$f\left(\frac{3}{2}\right) = -\frac{7}{2}$	Negative	Below x-axis
$(2, \infty)$	$f(3) = \frac{5}{2}$	Positive	Above x-axis

e) We make a table of function values.

x	-3	-2	-1	0	$\frac{1}{2}$	$\frac{3}{2}$	2	3
f(x)	$-1\frac{1}{4}$	0	$1\frac{1}{2}$	4	$7\frac{1}{2}$	$-3\frac{1}{2}$	9	$2\frac{1}{2}$

f) Using all available information, we draw the graph.

17. (continued)

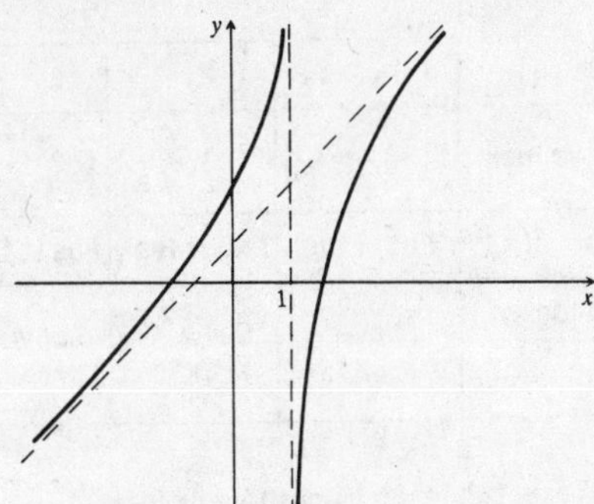

18. $f(x) = \frac{x^2 - 9}{x + 1}$

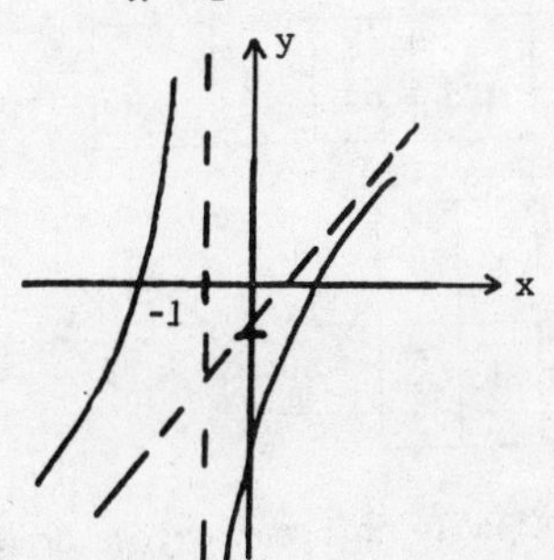

19. $f(x) = \frac{x^2 + x - 2}{2x^2 + 1} = \frac{(x + 2)(x - 1)}{2x^2 + 1}$

a) The function is neither even nor odd and no symmetries are apparent.

b) Since the numerator and denominator have the same degree, the horizontal asymptote can be determined by dividing the leading coefficients of the two polynomials. Thus $1 \div 2 = 1/2$, and we know that $y = 1/2$ is a horizontal asymptote.

c) The denominator has no zeros, so there are no vertical asymptotes. The zeros of the function are -2 and 1.

d) The zeros of the function divide the x-axis into intervals. To find where the function is positive or negative, we try a test point in each interval.

Interval	$(-\infty, -2)$	$(-2, 1)$	$(1, \infty)$
Test value	$f(-3) = \frac{4}{19}$	$f(0) = -2$	$f(2) = \frac{4}{9}$
Sign of f(x)	Positive	Negative	Positive
Location of points on graph	Above x-axis	Below x-axis	Above x-axis

e) We make a table of function values.

x	-5	-3	-2	-1	0	1	2	3
f(x)	$\frac{6}{17}$	$\frac{4}{19}$	0	$-\frac{2}{3}$	-2	0	$\frac{4}{9}$	$\frac{10}{19}$

19. (continued)

f) Using all the information, we draw the graph.

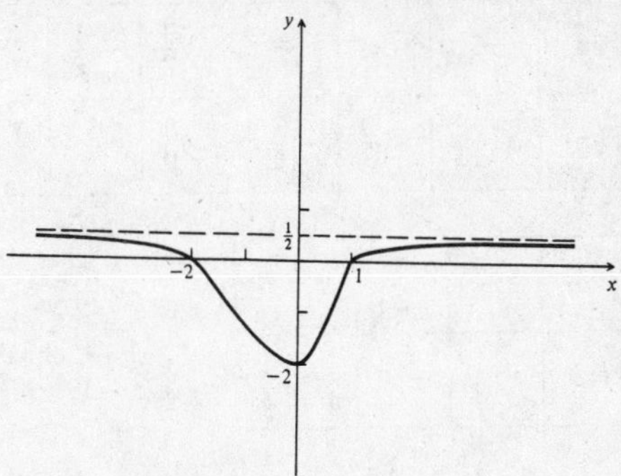

20. $f(x) = \frac{x^2 - 2x - 3}{3x^2 + 2}$

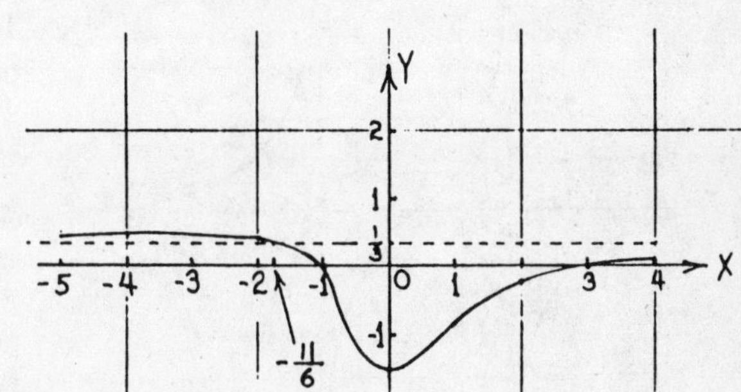

21. $f(x) = \frac{x - 1}{x^2 - 2x - 3} = \frac{x - 1}{(x - 3)(x + 1)}$

a) The function is neither even nor odd and no symmetries are apparent.

b) The degree of the denominator is greater than the numerator. Thus the x-axis is an asymptote.

c) The zeros of the denominator are 3 and -1, so $x = 3$ and $x = -1$ are vertical asymptotes. The zero of the function is 1.

d) The vertical asymptotes and the zeros of the function divide the x-axis into intervals. To find where the function is positive or negative, we try a test point in each interval.

Interval	Test value	Sign of f(x)	Location of points on graph
$(-\infty, -1)$	$f(-2) = -\frac{3}{5}$	Negative	Below x-axis
$(-1, 1)$	$f(0) = \frac{1}{3}$	Positive	Above x-axis
$(1, 3)$	$f(2) = -\frac{1}{3}$	Negative	Below x-axis
$(3, \infty)$	$f(4) = \frac{3}{5}$	Positive	Above x-axis

21. (continued)

e) We make a table of function values.

x	-5	-3	-2	$-\frac{3}{2}$	$-\frac{1}{2}$	0	1
f(x)	$-\frac{3}{16}$	$-\frac{1}{3}$	$-\frac{3}{5}$	$-\frac{10}{9}$	$\frac{6}{7}$	$\frac{1}{3}$	0

x	2	$\frac{5}{2}$	4	5
f(x)	$-\frac{1}{3}$	$-\frac{6}{7}$	$\frac{3}{5}$	$\frac{1}{3}$

f) Using all available information, we draw the graph.

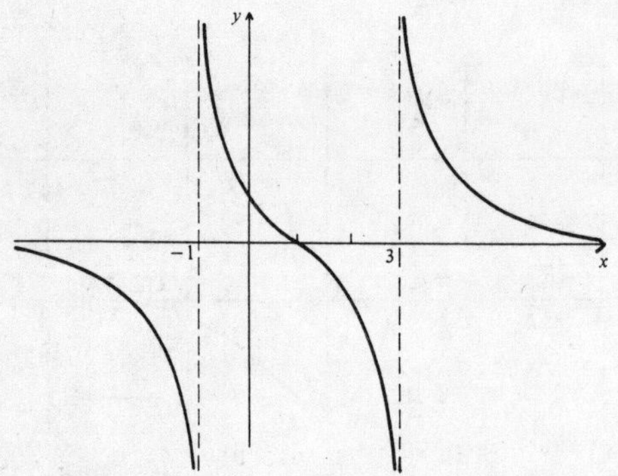

22. $f(x) = \frac{x + 2}{x^2 + 2x - 15}$

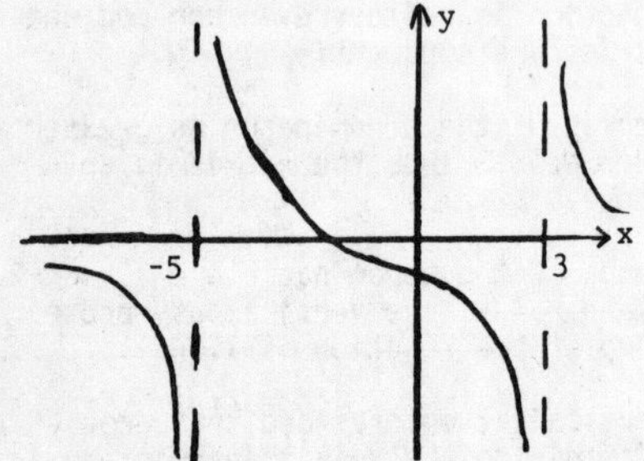

23. $f(x) = \frac{x + 2}{(x - 1)^3}$

a) The function is neither even nor odd and no symmetries are apparent.

b) The degree of the denominator is greater than the numerator. Thus the x-axis is an asymptote.

c) The zero of the denominator is 1, so x = 1 is a vertical asymptote. The zero of the function is -2.

d) The vertical asymptote and the zero of the function divide the x-axis into intervals. To find where the function is positive or negative, we try a test point in each interval.

23. (continued)

Interval	$(-\infty, -2)$	$(-2, 1)$	$(1, \infty)$
Test value	$f(-3) = \frac{1}{64}$	$f(0) = -2$	$f(2) = 4$
Sign of f(x)	Positive	Negative	Positive
Location of points on graph	Above x-axis	Below x-axis	Above x-axis

e) Make a table of function values.

x	-5	-4	-3	-2	-1	0	$\frac{1}{2}$
f(x)	$\frac{1}{72}$	$\frac{2}{125}$	$\frac{1}{64}$	0	$-\frac{1}{8}$	-2	-20

x	$\frac{3}{2}$	2	3
f(x)	28	4	$\frac{5}{8}$

f) Using all available information, we draw the graph.

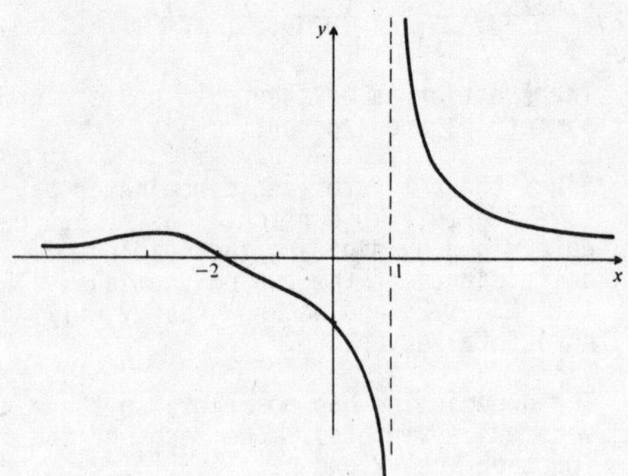

24. $f(x) = \frac{x - 3}{(x + 1)^3}$

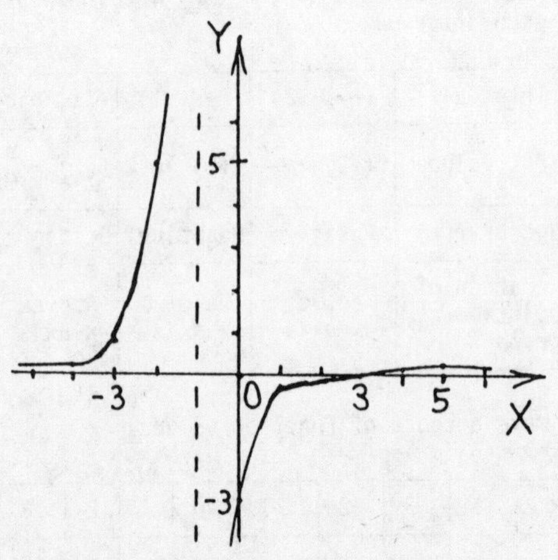

25. $f(x) = \frac{x^3 + 1}{x} = \frac{(x + 1)(x^2 - x + 1)}{x}$

a) The function is neither even nor odd and no symmetries are apparent.

b) The degree of the numerator is two greater than the denominator. There are no horizontal or oblique asymptotes.

c) The zero of the denominator is 0, so x = 0 (the y-axis) is a vertical asymptote. The only real-number zero of the function is -1.

d) The vertical asymptote and the real zero of the function divide the x-axis into intervals. To find where the function is positive or negative, we try a test point in each interval.

Interval	$(-\infty, -1)$	$(-1, 0)$	$(0, \infty)$
Test value	$f(-2) = \frac{7}{2}$	$f\left(-\frac{1}{2}\right) = -\frac{7}{4}$	$f(1) = 2$
Sign of f(x)	Positive	Negative	Positive
Location of points on graph	Above x-axis	Below x-axis	Above x-axis

e) Make a table of function values.

x	-3	-2	-1	$-\frac{1}{2}$	$\frac{1}{2}$	1	2	3
f(x)	$\frac{26}{3}$	$\frac{7}{2}$	0	$-\frac{7}{4}$	$\frac{9}{4}$	2	$\frac{9}{2}$	$\frac{28}{3}$

f) Draw the graph using all the known information.

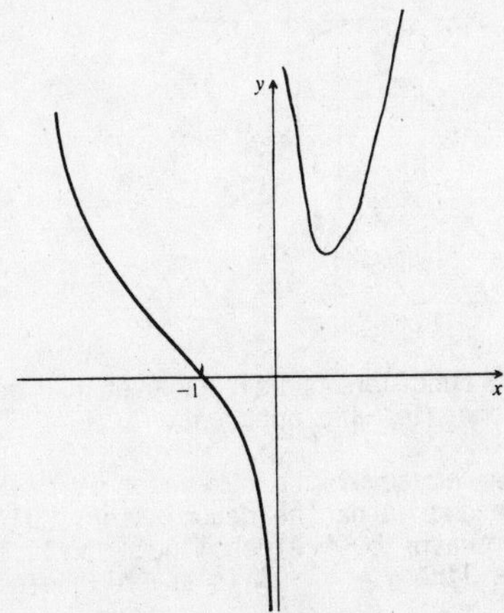

26. $f(x) = \frac{x^3 - 1}{x}$

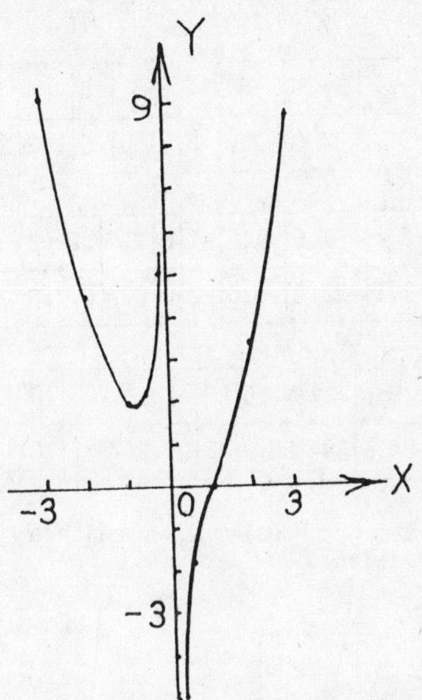

27. $f(x) = \frac{x^3 + 2x^2 - 15x}{x^2 - 5x - 14} = \frac{x(x + 5)(x - 3)}{(x + 2)(x - 7)}$

a) The function is neither even nor odd and no symmetries are apparent.

b) The degree of the numerator is one greater than that of the denominator. Dividing numerator by denominator will show that y = x + 7 is an oblique asymptote.

c) The zeros of the denominator are -2 and 7, so x = -2 and x = 7 are vertical asymptotes. The zeros of the function are 0, -5, and 3.

d) The vertical asymptotes and the zeros of the function divide the x-axis into intervals. To find where the function is positive or negative, we try a test value in each interval.

Interval	Test value	Sign of f(x)	Location of points on graph
$(-\infty, -5)$	$f(-6) = -\frac{27}{26}$	Negative	Below x-axis
$(-5, -2)$	$f(-3) = \frac{18}{5}$	Positive	Above x-axis
$(-2, 0)$	$f(-1) = -2$	Negative	Below x-axis
$(0, 3)$	$f(1) = \frac{2}{3}$	Positive	Above x-axis
$(3, 7)$	$f(4) = -2$	Negative	Below x-axis
$(7, \infty)$	$f(8) = 52$	Positive	Above x-axis

27. (continued)

e) Make a table of function values.

x	-100	-10	-8	-7	-6	-5	-4	-3	-2.5
f(x)	-93.2	-4.8	-2.9	-2	-1.0	0	1.3	3.6	7.2

x	-2.1	-1.9	-1.5	-1	0	1	2	3	4	5
f(x)	11.1	-32.4	-5.6	-2	0	0.7	0.7	0	-2	-7.1

x	6	7.1	7.5	8	10	20	100
f(x)	-24.8	394	88.8	42	29.2	29.7	107

f) Draw the graph using all available information.

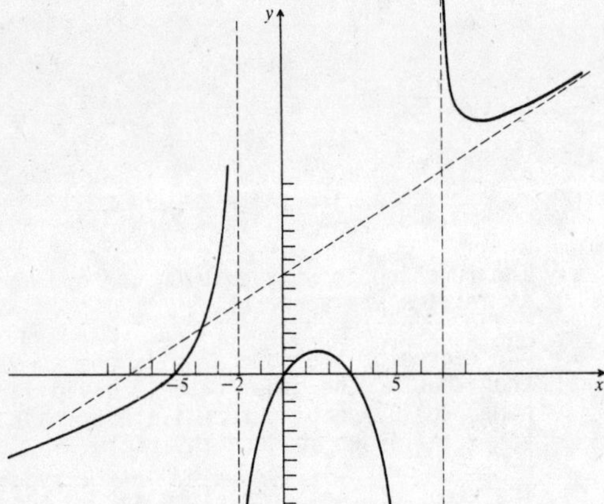

28. $f(x) = \frac{x^3 + 2x^2 - 3x}{x^2 - 25}$

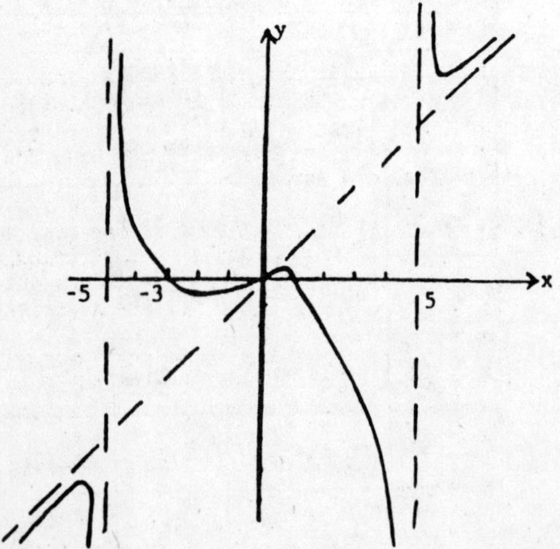

29. $f(x) = \frac{5x^4}{x^4 + 1}$

a) The function is even, so the graph is symmetric about the y-axis.

b) The degree of the numerator equals the degree of the denominator, so $y = \frac{5}{1}$, or $y = 5$, is a horizontal asymptote.

c) The denominator has no zeros, so there are no vertical asymptotes. The zero of the function is 0.

d) $5x^4 \geqslant 0$ for all real x and $x^4 + 1 > 0$ for all real x, so $f(x) \geqslant 0$ for all real values of x.

e) We make a table of function values.

x	± 5	± 3	± 1	0
f(x)	4.99	4.94	2.5	0

f) Using all available information, we draw the graph.

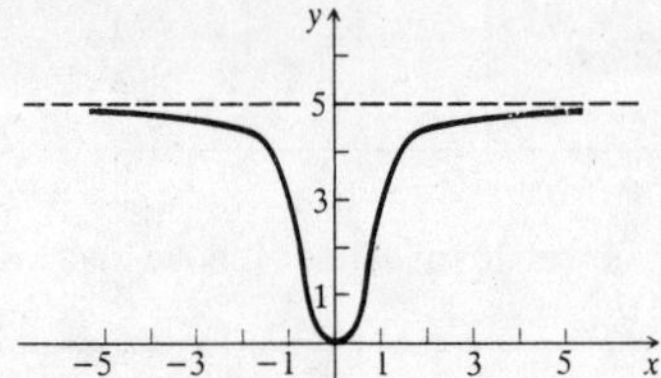

30. $f(x) = \frac{x + 1}{x^2 + x - 6}$

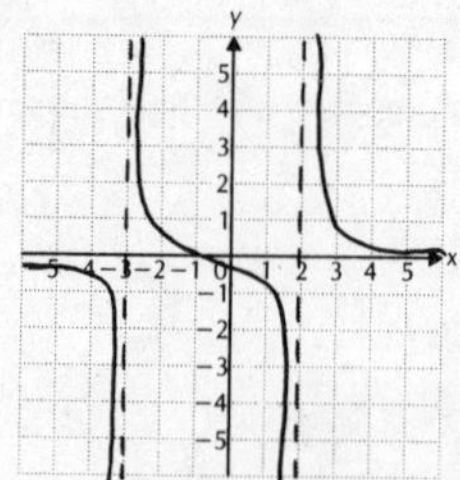

31. $f(x) = \frac{x^2 - x - 2}{x + 2} = \frac{(x - 2)(x + 1)}{x + 2}$

a) The function is neither even nor odd, and no symmetries are apparent.

b) The degree of the numerator is one more than the degree of the denominator. Dividing the numerator by the denominator will show that the line $y = x - 3$ is an oblique asymptote.

c) The zero of the denominator is -2, so $x = -2$ is a vertical asymptote. The zeros of the function are 2 and -1.

d) The vertical asymptote and the zeros of the function divide the x-axis into intervals. To find where the function is positive or negative, we try a test point in each interval.

31. (continued)

Interval	Test value	Sign of f(x)	Location of points on graph
$(-\infty, -2)$	$f(-3) = -10$	Negative	Below x-axis
$(-2, -1)$	$f\left(-\frac{3}{2}\right) = \frac{7}{2}$	Positive	Above x-axis
$(-1, 2)$	$f(0) = -1$	Negative	Below x-axis
$(2, \infty)$	$f(3) = \frac{4}{5}$	Positive	Above x-axis

e) Make a table of function values.

x	-5	-3	$-\frac{3}{2}$	-1	0	1	2
f(x)	$-\frac{28}{3}$	-10	$\frac{7}{2}$	0	-1	$-\frac{2}{3}$	0

x	3	5
f(x)	$\frac{4}{5}$	$\frac{18}{7}$

f) Using all available information, we draw the graph.

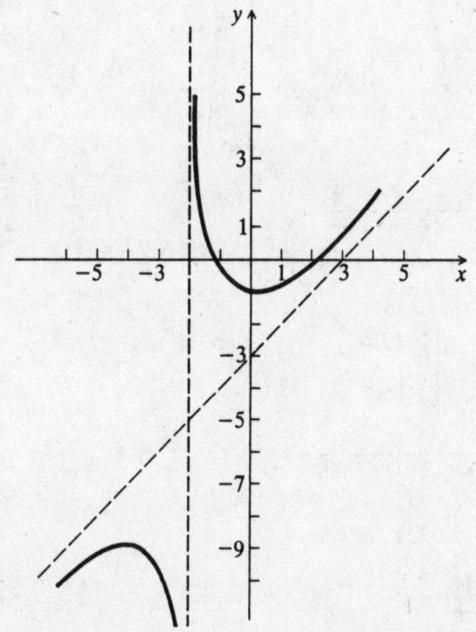

32. $f(x) = \dfrac{x^2}{x^2 - x - 2}$

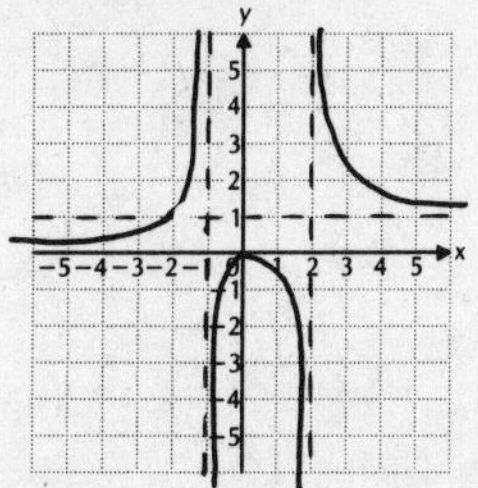

33. a) We use the formula $d = rt$:

$$d = rt$$

$$t = \frac{d}{r} \qquad \text{(Solving for t)}$$

$$t = \frac{500}{r} \qquad \text{(Substituting 500 for d)}$$

b) The graph over $(0, \infty)$ is one branch of a hyperbola.

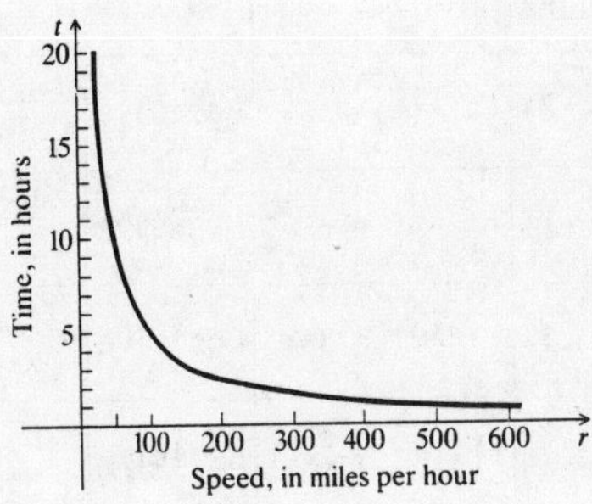

34. a) $R = \dfrac{10r}{10 + r}$

b)

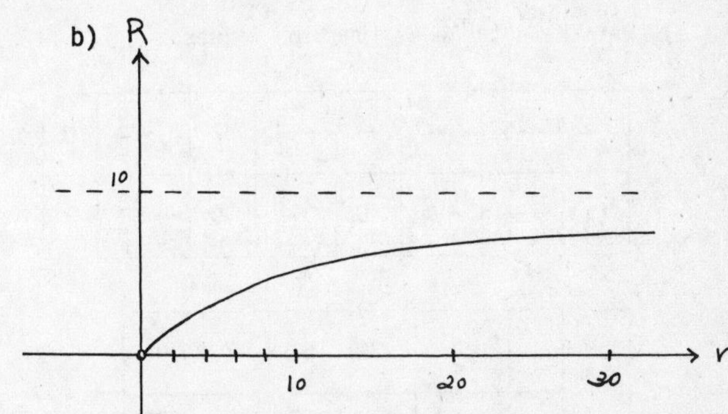

c) Domain: $(0, \infty)$, range: $(0, 10)$

d) 10

35. $f(x) = \dfrac{x^3 + 4x^2 + x - 6}{x^2 - x - 2} = \dfrac{(x + 3)(x + 2)(x - 1)}{(x - 2)(x + 1)}$

(We can use the rational roots theorem and synthetic division to factor the numerator.)

a) The function is neither even nor odd, and no symmetries are apparent.

b) The degree of the numerator is one more than the degree of the denominator. Dividing the numerator by the denominator will show that the line $y = x + 5$ is an oblique asymptote.

c) The zeros of the denominator are 2 and -1, so $x = 2$ and $x = -1$ are vertical asymptotes. The zeros of the function are -3, -2, and 1.

35. (continued)

d) We find where the function is positive or negative.

Interval	Test value	Sign of f(x)	Location of points on graph
$(-\infty, -3)$	$f(-4) = -\frac{5}{9}$	Negative	Below x-axis
$(-3, -2)$	$f\left(-\frac{5}{2}\right) = \frac{7}{54}$	Positive	Above x-axis
$(-2, -1)$	$f\left(-\frac{3}{2}\right) = -\frac{15}{14}$	Negative	Below x-axis
$(-1, 1)$	$f(0) = 3$	Positive	Above x-axis
$(1, 2)$	$f\left(\frac{3}{2}\right) = -\frac{63}{10}$	Negative	Below x-axis
$(2, \infty)$	$f(3) = 15$	Positive	Above x-axis

e) We make a table of function values.

x	-5	-4	-3	$-\frac{5}{2}$	-2	$-\frac{3}{2}$	0
f(x)	$-\frac{9}{7}$	$-\frac{5}{9}$	0	$\frac{7}{54}$	0	$-\frac{15}{14}$	3

x	1	$\frac{3}{2}$	3	5	7	9
f(x)	0	$-\frac{63}{10}$	15	$\frac{112}{9}$	$\frac{27}{2}$	$\frac{528}{35}$

f) Using all available information, we draw the graph.

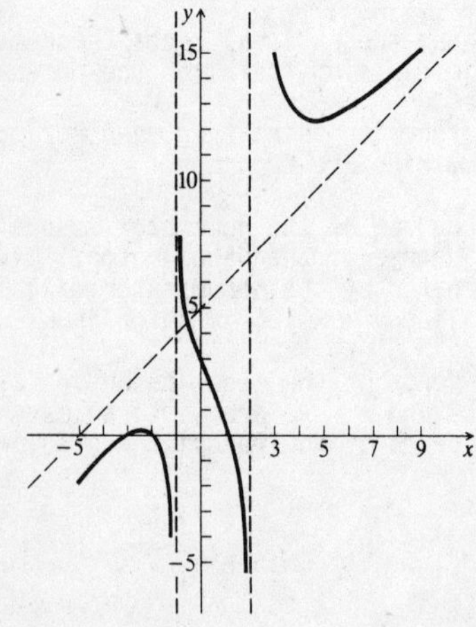

36. $f(x) = \dfrac{2x^3 + x^2 - 8x - 4}{x^3 + x^2 - 9x - 9}$

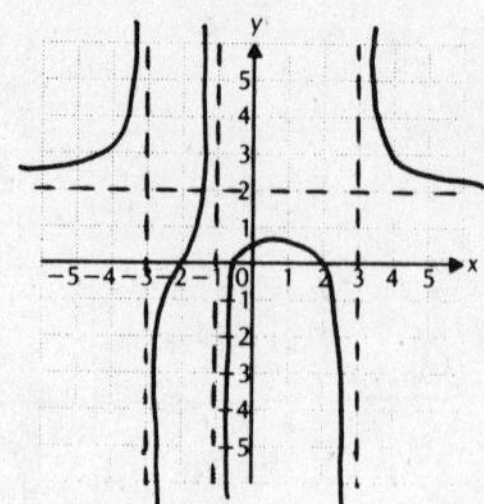

37. Use a computer software package or a graphing calculator to sketch the graph of

$f(x) = \dfrac{x^4 + 3x^3 + 21x^2 - 50x + 80}{x^4 + 8x^3 - x^2 + 20x - 10}$.

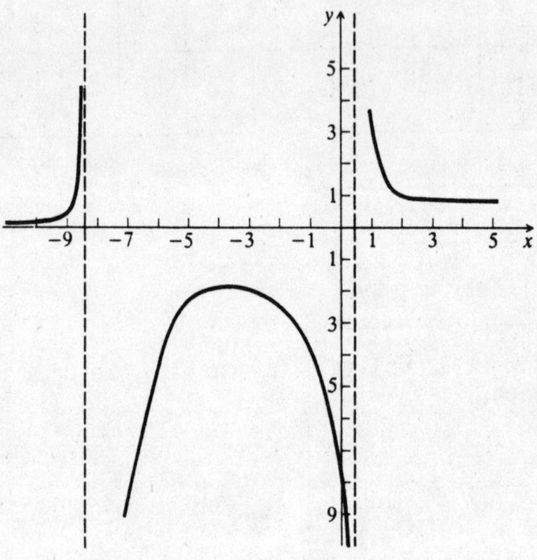

38. $f(x) = \dfrac{x^3 + x^2 + x}{x^2 - 1}$

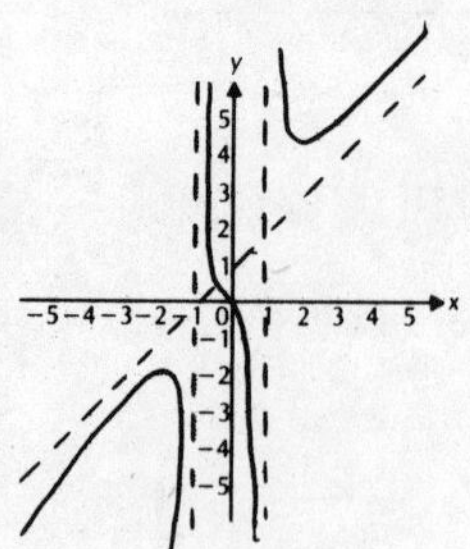

39. Use a computer software package or a graphing calculator to graph $f(x) = \frac{x^3}{x^2 - 1}$.

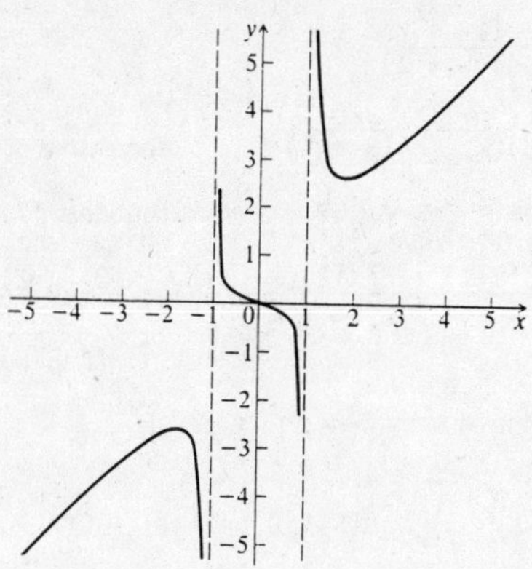

40. $f(x) = \frac{x^3 + 4}{x}$

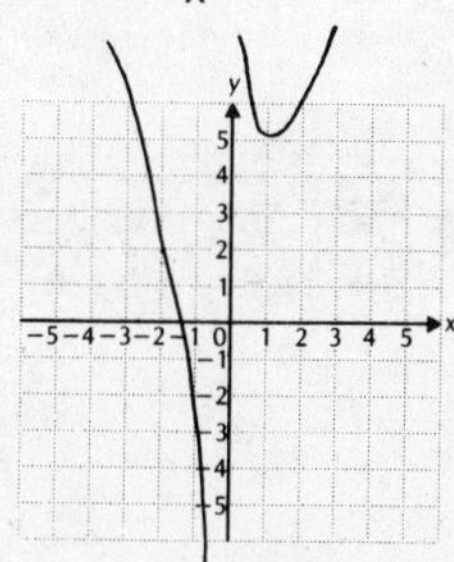

41. a) The distance of the round trip is 2·300, or 600 km, so $t_0 = \frac{600 \text{ km}}{200 \text{ km/h}} = 3$ hr.

We substitute:

$$T = \frac{(200)^2(3)}{(200)^2 - w^2} = \frac{120{,}000}{40{,}000 - w^2}$$

b) For w = 5 km/h:

$$T = \frac{120{,}000}{40{,}000 - 5^2} \approx 3.001876 \text{ hr}$$

For w = 10 km/h:

$$T = \frac{120{,}000}{40{,}000 - 10^2} \approx 3.007519 \text{ hr}$$

For w = 20 km/h:

$$T = \frac{120{,}000}{40{,}000 - 20^2} \approx 3.030303 \text{ hr}$$

c) The wind speed must be nonnegative, and it must also be less than the speed of the plane in still air. Thus the domain is $\{w \mid 0 \leqslant w < 200\}$.

For w = 0, T = 3; for other values of w in the domain, T > 3. Thus the range is $\{T \mid T \geqslant 3\}$.

41. (continued)

d)

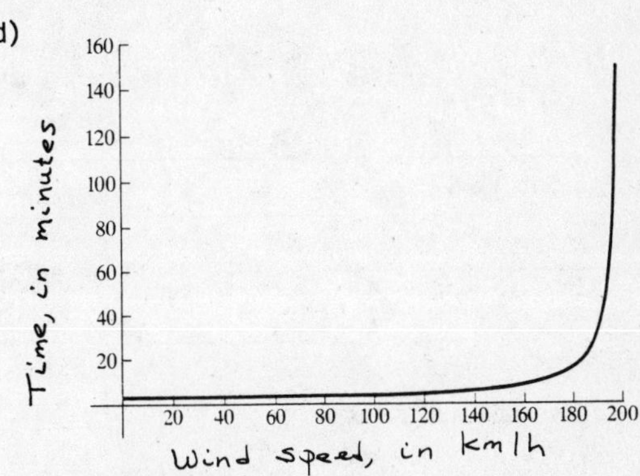

42. a) $W = \left(\frac{3965}{3965 + h}\right)^2(150)$, or

$$W = \frac{2{,}358{,}183{,}750}{(3975 + h)^2}$$

b)

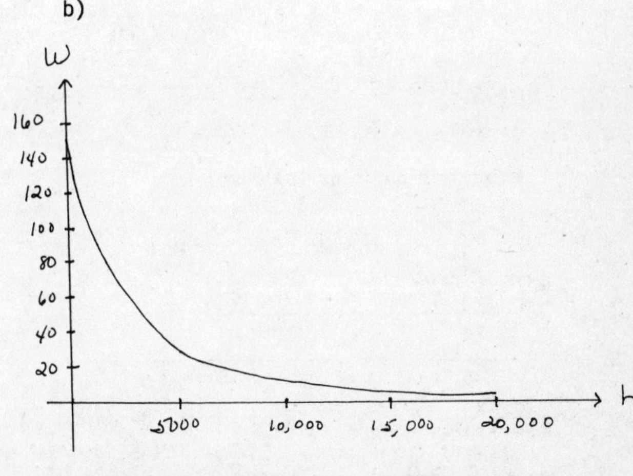

c) 2903 mi

43. a) S = area of base + 4·area of a side

$= x^2 + 4xy$

Now we express y in terms of x:

Volume = 108 = x^2y

$$\frac{108}{x^2} = y$$

Thus, $S = x^2 + 4x\left(\frac{108}{x^2}\right)$, or

$$S = x^2 + \frac{432}{x}.$$

b)

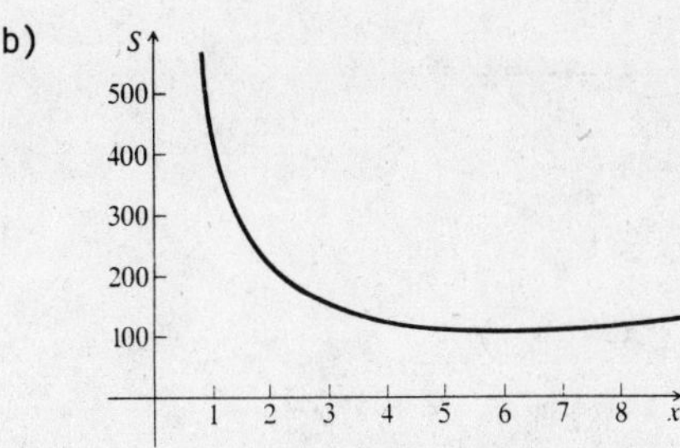

43. (continued)

c) Using the graph, we estimate that the minimum surface area is 108 cm². This occurs when x = 6 cm.

Exercise Set 11.8

1. $\frac{x + 7}{(x - 3)(x + 2)} = \frac{A}{x - 3} + \frac{B}{x + 2}$ (Theorem 17)

$\frac{x + 7}{(x - 3)(x + 2)} = \frac{A(x + 2) + B(x - 3)}{(x - 3)(x + 2)}$ (Adding)

Equate the numerators:

$x + 7 = A(x + 2) + B(x - 3)$

Let $x + 2 = 0$, or $x = -2$. Then we get

$-2 + 7 = 0 + B(-2 - 3)$

$5 = -5B$

$-1 = B$

Next let $x - 3 = 0$, or $x = 3$. Then we get

$3 + 7 = A(3 + 2) + 0$

$10 = 5A$

$2 = A$

The decomposition is as follows:

$\frac{2}{x - 3} - \frac{1}{x + 2}$

2. $\frac{1}{x + 1} + \frac{1}{x - 1}$

3. $\frac{7x - 1}{6x^2 - 5x + 1}$

$= \frac{7x - 1}{(3x - 1)(2x - 1)}$ (Factoring the denominator)

$= \frac{A}{3x - 1} + \frac{B}{2x - 1}$ (Theorem 17)

$= \frac{A(2x - 1) + B(3x - 1)}{(3x - 1)(2x - 1)}$ (Adding)

Equate the numerators:

$7x - 1 = A(2x - 1) + B(3x - 1)$

Let $2x - 1 = 0$, or $x = \frac{1}{2}$. Then we get

$7\left(\frac{1}{2}\right) - 1 = 0 + B\left(3 \cdot \frac{1}{2} - 1\right)$

$\frac{5}{2} = \frac{1}{2} B$

$5 = B$

Next let $3x - 1 = 0$, or $x = \frac{1}{3}$. We get

$7\left(\frac{1}{3}\right) - 1 = A\left(2 \cdot \frac{1}{3} - 1\right) + 0$

$\frac{7}{3} - 1 = A\left(\frac{2}{3} - 1\right)$

$\frac{4}{3} = -\frac{1}{3} A$

$-4 = A$

The decomposition is as follows:

$\frac{-4}{3x - 1} + \frac{5}{2x - 1}$

4. $\frac{-5}{4x + 3} + \frac{7}{3x - 5}$

5. $\frac{3x^2 - 11x - 26}{(x^2 - 4)(x + 1)}$

$= \frac{3x^2 - 11x - 26}{(x + 2)(x - 2)(x + 1)}$ (Factoring the denominator)

$= \frac{A}{x + 2} + \frac{B}{x - 2} + \frac{C}{x + 1}$ (Theorem 17)

$= \frac{A(x-2)(x+1) + B(x+2)(x+1) + C(x+2)(x-2)}{(x + 2)(x - 2)(x + 1)}$ (Adding)

Equate the numerators:

$3x^2 - 11x - 26 = A(x - 2)(x + 1) + B(x + 2)(x + 1) + C(x + 2)(x - 2)$

Let $x + 2 = 0$, or $x = -2$. Then, we get

$3(-2)^2 - 11(-2) - 26 = A(-2 - 2)(-2 + 1) + 0 + 0$

$12 + 22 - 26 = A(-4)(-1)$

$8 = 4A$

$2 = A$

Next let $x - 2 = 0$, or $x = 2$. Then, we get

$3 \cdot 2^2 - 11 \cdot 2 - 26 = 0 + B(2 + 2)(2 + 1) + 0$

$12 - 22 - 26 = B \cdot 4 \cdot 3$

$-36 = 12B$

$-3 = B$

Finally let $x + 1 = 0$, or $x = -1$. We get

$3(-1)^2 - 11(-1) - 26 = 0 + 0 + C(-1 + 2)(-1 - 2)$

$3 + 11 - 26 = C(1)(-3)$

$-12 = -3C$

$4 = C$

The decomposition is as follows:

$\frac{2}{x + 2} - \frac{3}{x - 2} + \frac{4}{x + 1}$

6. $\frac{6}{x - 4} + \frac{3}{x - 2} - \frac{4}{x + 1}$

7. $\frac{9}{(x + 2)^2(x - 1)}$

$= \frac{A}{x + 2} + \frac{B}{(x + 2)^2} + \frac{C}{x - 1}$ (Theorem 17)

$= \frac{A(x + 2)(x - 1) + B(x - 1) + C(x + 2)^2}{(x + 2)^2(x - 1)}$ (Adding)

Equate the numerators:

$9 = A(x + 2)(x - 1) + B(x - 1) + C(x + 2)^2$

Let $x - 1 = 0$, or $x = 1$. Then, we get

$9 = 0 + 0 + C(1 + 2)^2$

$9 = 9C$

$1 = C$

7. (continued)

Next let $x + 2 = 0$, or $x = -2$. Then, we get

$$9 = 0 + B(-1) + 0$$
$$9 = -3B$$
$$-3 = B$$

To find A we equate the coefficients of x^2.

Consider $9 = 0x^2 + 0x + 9$. Then

$$0 = A + C$$

Substituting 1 for C, we get $A = -1$.

The decomposition is as follows:

$$\frac{-1}{x + 2} - \frac{3}{(x + 2)^2} + \frac{1}{x - 1}$$

8. $\frac{1}{x - 2} + \frac{3}{(x - 2)^2} - \frac{2}{(x - 2)^3}$

9. $\frac{2x^2 + 3x + 1}{(x^2 - 1)(2x - 1)}$

$= \frac{2x^2 + 3x + 1}{(x + 1)(x - 1)(2x - 1)}$ (Factoring the denominator)

$= \frac{A}{x + 1} + \frac{B}{x - 1} + \frac{C}{2x - 1}$ (Theorem 17)

$= \frac{A(x-1)(2x-1) + B(x+1)(2x-1) + C(x+1)(x-1)}{(x + 1)(x -1)(2x - 1)}$ (Adding)

Equate the numerators:

$$2x^2 + 3x + 1 = A(x - 1)(2x - 1) + B(x + 1)(2x - 1) + C(x + 1)(x - 1)$$

Let $x + 1 = 0$, or $x = -1$. Then, we get

$$2(-1)^2 + 3(-1) + 1 = A(-1 - 1)[2(-1) - 1] + 0 + 0$$
$$2 - 3 + 1 = A(-2)(-3)$$
$$0 = 6A$$
$$0 = A$$

Next let $x - 1 = 0$, or $x = 1$. Then, we get

$$2\cdot 1^2 + 3\cdot 1 + 1 = 0 + B(1 + 1)(2\cdot 1 - 1) + 0$$
$$2 + 3 + 1 = B\cdot 2\cdot 1$$
$$6 = 2B$$
$$3 = B$$

Finally we let $2x -1 = 0$, or $x = \frac{1}{2}$. We get

$$2\left(\frac{1}{2}\right)^2 + 3\left(\frac{1}{2}\right) + 1 = 0 + 0 + C\left(\frac{1}{2} + 1\right)\left(\frac{1}{2} - 1\right)$$
$$\frac{1}{2} + \frac{3}{2} + 1 = C \cdot \frac{3}{2} \cdot \left(-\frac{1}{2}\right)$$
$$3 = -\frac{3}{4}C$$
$$-4 = C$$

The decomposition is as follows:

$$\frac{3}{x - 1} - \frac{4}{2x - 1}$$

10. $\frac{-4}{x - 3} + \frac{3}{x - 2} + \frac{2}{x - 1}$

11. $\frac{x^4 - 3x^3 - 3x^2 + 10}{(x + 1)^2(x - 3)}$

$= \frac{x^4 - 3x^3 - 3x^2 + 10}{x^3 - x^2 - 5x - 3}$ (Multiplying the denominator)

Since the degree of the numerator is greater than the degree of the denominator, we divide and then use Theorem 14.

$$\begin{array}{r} x - 2 \\ x^3 - x^2 - 5x - 3 \overline{\big) x^4 - 3x^3 - 3x^2 + 0x + 10} \\ \underline{x^4 - x^3 - 5x^2 - 3x } \\ -2x^3 + 2x^2 + 3x + 10 \\ \underline{-2x^3 + 2x^2 + 10x + 6} \\ -7x + 4 \end{array}$$

The original expression is thus equivalent to the following:

$$x - 2 + \frac{-7x + 4}{x^3 - x^2 - 5x - 3}$$

We proceed to decompose the fraction.

$\frac{-7x + 4}{(x + 1)^2(x - 3)}$

$= \frac{A}{x + 1} + \frac{B}{(x + 1)^2} + \frac{C}{x - 3}$ (Theorem 17)

$= \frac{A(x + 1)(x - 3) + B(x - 3) + C(x + 1)^2}{(x + 1)^2(x - 3)}$ (Adding)

Equate the numerators:

$$-7x + 4 = A(x + 1)(x - 3) + B(x - 3) + C(x + 1)^2$$

Let $x - 3 = 0$, or $x = 3$. Then, we get

$$-7\cdot 3 + 4 = 0 + 0 + C(3 + 1)^2$$
$$-17 = 16C$$
$$-\frac{17}{16} = C$$

Let $x + 1 = 0$, or $x = -1$. Then, we get

$$-7(-1) + 4 = 0 + B(-1 - 3) + 0$$
$$11 = -4B$$
$$-\frac{11}{4} = B$$

To find A we equate the coefficients of x^2.

Consider $-7x + 4 = 0x^2 - 7x + 4$.

$$0 = A + C$$

Substituting $-\frac{17}{16}$ for C, we get $A = \frac{17}{16}$.

The decomposition is as follows:

$$\frac{17/16}{x + 1} - \frac{11/4}{(x + 1)^2} - \frac{17/16}{x - 3}$$

The original expression is equivalent to the following:

$$x - 2 + \frac{17/16}{x + 1} - \frac{11/4}{(x + 1)^2} - \frac{17/16}{x - 3}$$

12. $10x - 5 + \frac{6}{x - 3} + \frac{14}{x + 2}$

13. $\dfrac{-x^2 + 2x - 13}{(x^2 + 2)(x - 1)}$

$= \dfrac{Ax + B}{x^2 + 2} + \dfrac{C}{x - 1}$ (Theorem 17)

$= \dfrac{(Ax + B)(x - 1) + C(x^2 + 2)}{(x^2 + 2)(x - 1)}$ (Adding)

Equate the numerators:

$-x^2 + 2x - 13 = (Ax + B)(x - 1) + C(x^2 + 2)$

Let $x - 1 = 0$, or $x = 1$. Then we get

$-1 + 2\cdot1 - 13 = 0 + C(1^2 + 2)$

$-1 + 2 - 13 = C(1 + 2)$

$-12 = 3C$

$-4 = C$

Equate the coefficients of x^2:

$-1 = A + C$

Substituting -4 for C, we get A = 3.

Equate the constant terms:

$-13 = -B + 2C$

Substituting -4 for C, we get B = 5.

The decomposition is as follows:

$\dfrac{3x + 5}{x^2 + 2} + \dfrac{-4}{x - 1}$

14. $\dfrac{41x + 3}{x^2 + 1} - \dfrac{15}{x + 5}$

15. $\dfrac{6 + 26x - x^2}{(2x - 1)(x + 2)^2}$

$= \dfrac{A}{2x - 1} + \dfrac{B}{x + 2} + \dfrac{C}{(x + 2)^2}$ (Theorem 17)

$= \dfrac{A(x + 2)^2 + B(2x - 1)(x + 2) + C(2x - 1)}{(2x - 1)(x + 2)^2}$

(Adding)

Equate the numerators:

$6 + 26x - x^2 = A(x + 2)^2 + B(2x - 1)(x + 2) + C(2x - 1)$

Let $2x - 1 = 0$, or $x = \frac{1}{2}$. Then, we get

$6 + 26 \cdot \frac{1}{2} - \left(\frac{1}{2}\right)^2 = A\left(\frac{1}{2} + 2\right)^2 + 0 + 0$

$6 + 13 - \frac{1}{4} = A\left(\frac{5}{2}\right)^2$

$\frac{75}{4} = \frac{25}{4} A$

$3 = A$

Let $x + 2 = 0$, or $x = -2$. We get

$6 + 26(-2) - (-2)^2 = 0 + 0 + C[2(-2) - 1]$

$6 - 52 - 4 = -5C$

$-50 = -5C$

$10 = C$

15. (continued)

Equate the coefficients of x^2:

$-1 = A + 2B$

Substituting 3 for A, we obtain B = -2.

The decomposition is as follows:

$\dfrac{3}{2x - 1} - \dfrac{2}{x + 2} + \dfrac{10}{(x + 2)^2}$

16. $\dfrac{1}{x - 1} - \dfrac{1}{x + 1} + \dfrac{3}{(x + 1)^2} - \dfrac{3}{(x + 1)^3} + \dfrac{2}{(x + 1)^4}$

17. $\dfrac{6x^3 + 5x^2 + 6x - 2}{2x^2 + x - 1}$

Since the degree of the numerator is greater than the degree of the denominator, we divide and then use Theorem 14.

$$\begin{array}{r} 3x + 1 \\ 2x^2 + x - 1 \overline{) 6x^3 + 5x^2 + 6x - 2} \\ \underline{6x^3 + 3x^2 - 3x} \\ 2x^2 + 9x - 2 \\ \underline{2x^2 + x - 1} \\ 8x - 1 \end{array}$$

The original expression is equivalent to

$3x + 1 + \dfrac{8x - 1}{2x^2 + x - 1}$

We proceed to decompose the fraction

$\dfrac{8x - 1}{2x^2 + x - 1} = \dfrac{8x - 1}{(2x - 1)(x + 1)}$ (Factoring the denominator)

$= \dfrac{A}{2x - 1} + \dfrac{B}{x + 1}$ (Theorem 17)

$= \dfrac{A(x + 1) + B(2x - 1)}{(2x - 1)(x + 1)}$ (Adding)

Equate the numerators:

$8x - 1 = A(x + 1) + B(2x - 1)$

Let $x + 1 = 0$, or $x = -1$. Then we get

$8(-1) - 1 = 0 + B[2(-1) - 1]$

$-8 - 1 = B(-2 - 1)$

$-9 = -3B$

$3 = B$

Next let $2x - 1 = 0$, or $x = \frac{1}{2}$. We get

$8\left(\frac{1}{2}\right) - 1 = A\left(\frac{1}{2} + 1\right) + 0$

$4 - 1 = A\left(\frac{3}{2}\right)$

$3 = \frac{3}{2} A$

$2 = A$

The decomposition is

$\dfrac{2}{2x - 1} + \dfrac{3}{x + 1}$.

The original expression is equivalent to

$3x + 1 + \dfrac{2}{2x - 1} + \dfrac{3}{x + 1}$.

18. $2x - 1 + \frac{1}{x + 3} + \frac{-4}{x - 1}$

19. $\frac{2x^2 - 11x + 5}{(x - 3)(x^2 + 2x - 5)}$

$= \frac{A}{x - 3} + \frac{Bx + C}{x^2 + 2x - 5}$ (Theorem 17)

$= \frac{A(x^2 + 2x - 5) + (Bx + C)(x - 3)}{(x - 3)(x^2 + 2x - 5)}$ (Adding)

Equate the numerators:

$2x^2 - 11x + 5 = A(x^2 + 2x - 5) + (Bx + C)(x - 3)$

Let $x - 3 = 0$, or $x = 3$. Then, we get

$2\cdot 3^2 - 11\cdot 3 + 5 = A(3^2 + 2\cdot 3 - 5) + 0$

$18 - 33 + 5 = A(9 + 6 - 5)$

$-10 = 10A$

$-1 = A$

Equate the coefficients of x^2:

$2 = A + B$

Substituting -1 for A, we get B = 3.

Equate the constant terms:

$5 = -5A - 3C$

Substituting -1 for A, we get C = 0.

The decomposition is as follows:

$\frac{-1}{x - 3} + \frac{3x}{x^2 + 2x - 5}$

20. $\frac{2}{x - 5} + \frac{x}{x^2 + x - 4}$

21. $\frac{-4x^2 - 2x + 10}{(3x + 5)(x + 1)^2}$

The decomposition looks like

$\frac{A}{3x + 5} + \frac{B}{x + 1} + \frac{C}{(x + 1)^2}$.

Add and equate the numerators.

$-4x^2 - 2x + 10 = A(x + 1)^2 + B(3x + 5)(x + 1) + C(3x + 5)$

$= A(x^2 + 2x + 1) + B(3x^2 + 8x + 5) + C(3x + 5)$

or

$-4x^2 - 2x + 10 = (A + 3B)x^2 + (2A + 8B + 3C)x + (A + 5B + 5C)$

Then equate corresponding coefficients.

$-4 = A + 3B$ (Coefficients of x^2-terms)

$-2 = 2A + 8B + 3C$ (Coefficients of x-terms)

$10 = A + 5B + 5C$ (Constant terms)

We solve this system of three equations and find A = 5, B = -3, C = 4.

The decomposition is

$\frac{5}{3x + 5} - \frac{3}{x + 1} + \frac{4}{(x + 1)^2}$.

22. $\frac{6}{x - 4} + \frac{1}{2x - 1} - \frac{3}{(2x - 1)^2}$

23. $\frac{36x + 1}{12x^2 - 7x - 10} = \frac{36x + 1}{(4x - 5)(3x + 2)}$

The decomposition looks like

$\frac{A}{4x - 5} + \frac{B}{3x + 2}$.

Add and equate the numerators.

$36x + 1 = A(3x + 2) + B(4x - 5)$

or $36x + 1 = (3A + 4B)x + (2A - 5B)$

Then equate corresponding coefficients.

$36 = 3A + 4B$ (Coefficients of x-terms)

$1 = 2A - 5B$ (Constant terms)

We solve this system of equations and find A = 8 and B = 3.

The decomposition is

$\frac{8}{4x - 5} + \frac{3}{3x + 2}$.

24. $\frac{7}{6x - 3} - \frac{4}{x + 7}$

25. $\frac{-4x^2 - 9x + 8}{(3x^2 + 1)(x - 2)}$

The decomposition looks like

$\frac{Ax + B}{3x^2 + 1} + \frac{C}{x - 2}$.

Add and equate the numerators.

$-4x^2 - 9x + 8 = (Ax + B)(x - 2) + C(3x^2 + 1)$

$= Ax^2 - 2Ax + Bx - 2B + 3Cx^2 + C$

or $-4x^2 - 9x + 8 = (A + 3C)x^2 + (-2A + B)x + (-2B + C)$

Then equate corresponding coefficients.

$-4 = A + 3C$ (Coefficients of x^2-terms)

$-9 = -2A + B$ (Coefficients of x-terms)

$8 = -2B + C$ (Constant terms)

We solve this system of equations and find A = 2, B = -5, C = -2.

The decomposition is

$\frac{2x - 5}{3x^2 + 1} - \frac{2}{x - 2}$.

26. $\frac{5x + 1}{x^2 + 4} + \frac{6}{x - 8}$

27. $\frac{x}{x^4 - a^4}$

$= \frac{x}{(x^2 + a^2)(x + a)(x - 1)}$ (Factoring the denominator)

$= \frac{Ax + B}{x^2 + a^2} + \frac{C}{x + a} + \frac{D}{x - a}$ (Theorem 17)

$= \frac{(Ax-B)(x+a)(x-a) + C(x^2+a^2)(x-a) + D(x^2+a^2)(x+a)}{(x^2 + a^2)(x + a)(x - a)}$

Equate the numerators:

$x = (Ax + B)(x + a)(x - a) + C(x^2 + a^2)(x - a) + D(x^2 + a^2)(x + a)$

Let $x - a = 0$, or $x = a$. Then, we get

$a = 0 + 0 + D(a^2 + a^2)(a + a)$

$a = D(2a^2)(2a)$

$a = 4a^3D$

$\frac{1}{4a^3} = D$

Let $x + a = 0$, or $x = -a$. We get

$-a = 0 + C[(-a)^2 + a^2](-a - a) + 0$

$-a = C(2a^2)(-2a)$

$-a = -4a^3C$

$\frac{1}{4a^2} = C$

Equate the coefficients of x^3:

$0 = A + C + D$

Substituting $\frac{1}{4a^2}$ for C and for D, we get $A = -\frac{1}{2a^2}$.

Equate the constant terms:

$0 = -Ba^2 - Ca^3 + Da^3$

Substitute $\frac{1}{4a^2}$ for C and for D. Then solve for B.

$0 = -Ba^2 - \frac{1}{4a^2} \cdot a^3 + \frac{1}{4a^2} \cdot a^3$

$0 = -Ba^2$

$0 = B$

The decomposition is as follows:

$$\frac{-\frac{1}{2a^2}x}{x^2 + a^2} + \frac{\frac{1}{4a^2}}{x + a} + \frac{\frac{1}{4a^2}}{x - a}$$

28. $\frac{-1}{x + 1} + \frac{1}{x - 2} + \frac{6}{(x - 2)^2} + \frac{12}{(x - 2)^3} + \frac{24}{(x - 2)^4}$

29. $y = \frac{3x}{x^2 + 5x + 4} = \frac{3x}{(x + 4)(x + 1)}$

Using Theorem 14 and adding, we get

$\frac{3x}{(x + 4)(x + 1)} = \frac{A}{x + 4} + \frac{B}{x + 1}$

$= \frac{A(x + 1) + B(x + 4)}{(x + 4)(x + 1)}$

Equate the numerators:

$3x = A(x + 1) + B(x + 4)$

Let $x + 1 = 0$, or $x = -1$. We get

$3(-1) = 0 + B(-1 + 4)$

$-3 = 3B$

$-1 = B$

Next let $x + 4 = 0$, or $x = -4$. Then we get

$3(-4) = A(-4 + 1) + 0$

$-12 = -3A$

$4 = A$

The decomposition is $y = \frac{4}{x + 4} - \frac{1}{x + 1}$.

We graph by addition of ordinates (adding respective fraction values.)

If $x = 4$, $y = \frac{4}{4 + 4} - \frac{1}{4 + 1} = \frac{1}{2} - \frac{1}{5} = \frac{3}{10}$.

If $x = 2$, $y = \frac{4}{2 + 4} - \frac{1}{2 + 1} = \frac{2}{3} - \frac{1}{3} = \frac{1}{3}$.

If $x = 0$, $y = \frac{4}{0 + 4} - \frac{1}{0 + 1} = 1 - 1 = 0$.

If $x = -2$, $y = \frac{4}{-2 + 4} - \frac{1}{-2 + 1} = 2 + 1 = 3$.

Continue adding fraction values as above. Then plot these points and draw the graph. Note that $x = -4$ and $x = -1$ are vertical asymptotes.

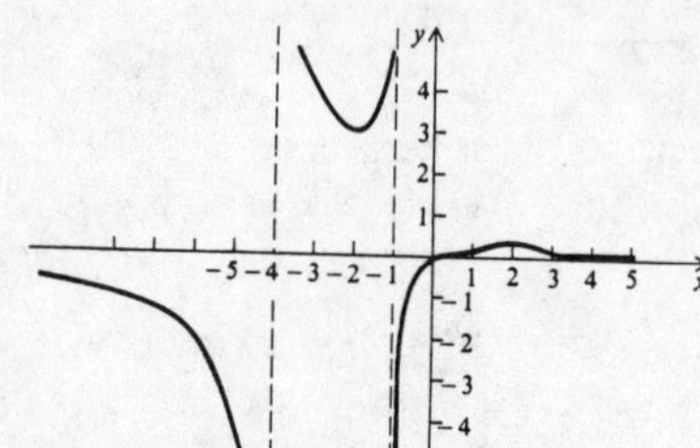

30. $\frac{\frac{1}{2}}{x - 3} + \frac{\frac{1}{2}}{x + 1}$

31. $\dfrac{1 + \ln x^2}{(\ln x + 2)(\ln x - 3)^2} = \dfrac{1 + 2 \ln x}{(\ln x + 2)(\ln x - 3)^2}$

Let $u = \ln x$. Then we have:

$$\frac{1 + 2u}{(u + 2)(u - 3)^2} = \frac{A}{u + 2} + \frac{B}{(u - 3)} + \frac{C}{(u - 3)^2}$$

$$= \frac{A(u-3)^2 + B(u+2)(u-3) + C(u+2)}{(u + 2)(u - 3)^2}$$

Equate the numerators:

$1 + 2u = A(u - 3)^2 + B(u + 2)(u - 3) + C(u + 2)$

Let $u - 3 = 0$, or $u = 3$.

$$1 + 2\cdot 3 = 0 + 0 + C(5)$$
$$7 = 5C$$
$$\frac{7}{5} = C$$

Let $u + 2 = 0$, or $u = -2$.

$$1 + 2(-2) = A(-2 - 3)^2 + 0 + 0$$
$$-3 = 25A$$
$$-\frac{3}{25} = A$$

To find B, we equate the coefficients of u^2:

$$0 = A + B$$

Substituting $-\frac{3}{25}$ for A and solving for B, we get $B = \frac{3}{25}$.

The decomposition of $\dfrac{1 + 2u}{(u + 2)(u - 3)^2}$ is as follows:

$$-\frac{3}{25(u + 2)} + \frac{3}{25(u - 3)} + \frac{7}{5(u - 3)^2}$$

Substituting $\ln x$ for u we get

$$-\frac{3}{25(\ln x + 2)} + \frac{3}{25(\ln x - 3)} + \frac{7}{5(\ln x - 3)^2}.$$

32. $\dfrac{1}{e^x + 1} - \dfrac{1}{2e^x + 1}$

Exercise Set 12.1

1. $a_n = 4n - 1$
$a_1 = 4\cdot1 - 1 = 3$, $a_4 = 4\cdot4 - 1 = 15$;
$a_2 = 4\cdot2 - 1 = 7$, $a_{10} = 4\cdot10 - 1 = 39$;
$a_3 = 4\cdot3 - 1 = 11$, $a_{15} = 4\cdot15 - 1 = 59$

2. $a_1 = 0$, $a_2 = 0$, $a_3 = 0$, $a_4 = 6$; $a_{10} = 504$;
$a_{15} = 2184$

3. $a_n = \frac{n}{n - 1}$, $n \geqslant 2$
The first 4 terms are a_2, a_3, a_4, and a_5:
$a_2 = \frac{2}{2 - 1} = 2$, $a_4 = \frac{4}{4 - 1} = \frac{4}{3}$,
$a_3 = \frac{3}{3 - 1} = \frac{3}{2}$, $a_5 = \frac{5}{5 - 1} = \frac{5}{4}$.
$a_{10} = \frac{10}{10 - 1} = \frac{10}{9}$, $a_{15} = \frac{15}{15 - 1} = \frac{15}{14}$

4. $a_1 = 0$, $a_2 = 3$, $a_3 = 8$, $a_4 = 15$; $a_{10} = 99$;
$a_{15} = 224$

5. $a_n = n^2 + 2n$
$a_1 = 1^2 + 2\cdot1 = 3$, $a_4 = 4^2 + 2\cdot4 = 24$;
$a_2 = 2^2 + 2\cdot2 = 8$, $a_{10} = 10^2 + 2\cdot10 = 120$;
$a_3 = 3^2 + 2\cdot3 = 15$, $a_{15} = 15^2 + 2\cdot15 = 255$

6. $a_1 = 0$, $a_2 = \frac{3}{5}$, $a_3 = \frac{4}{5}$, $a_4 = \frac{15}{17}$; $a_{10} = \frac{99}{101}$;
$a_{15} = \frac{112}{113}$

7. $a_n = n + \frac{1}{n}$
$a_1 = 1 + \frac{1}{1} = 2$, $a_4 = 4 + \frac{1}{4} = 4\frac{1}{4}$;
$a_2 = 2 + \frac{1}{2} = 2\frac{1}{2}$, $a_{10} = 10 + \frac{1}{10} = 10\frac{1}{10}$;
$a_3 = 3 + \frac{1}{3} = 3\frac{1}{3}$, $a_{15} = 15 + \frac{1}{15} = 15\frac{1}{15}$

8. $a_1 = 1$, $a_2 = -\frac{1}{2}$, $a_3 = \frac{1}{4}$, $a_4 = -\frac{1}{8}$; $a_{10} = -\frac{1}{512}$;
$a_{15} = \frac{1}{16{,}384}$

9. $a_n = (-1)^n n^2$
$a_1 = (-1)^1 1^2 = -1$, $a_4 = (-1)^4 4^2 = 16$;
$a_2 = (-1)^2 2^2 = 4$, $a_{10} = (-1)^{10} 10^2 = 100$;
$a_3 = (-1)^3 3^2 = -9$, $a_{15} = (-1)^{15} 15^2 = -225$

10. $a_1 = -4$, $a_2 = 5$, $a_3 = -6$, $a_4 = 7$; $a_{10} = 13$;
$a_{15} = -18$

11. $a_n = (-1)^{n+1}(3n - 5)$
$a_1 = (-1)^{1+1}(3\cdot1 - 5) = -2$,
$a_2 = (-1)^{2+1}(3\cdot2 - 5) = -1$,
$a_3 = (-1)^{3+1}(3\cdot3 - 5) = 4$,
$a_4 = (-1)^{4+1}(3\cdot4 - 5) = -7$;
$a_{10} = (-1)^{10+1}(3\cdot10 - 5) = -25$;
$a_{15} = (-1)^{15+1}(3\cdot15 - 5) = 40$

12. $a_1 = 0$, $a_2 = 7$, $a_3 = -26$, $a_4 = 63$; $a_{10} = 999$;
$a_{15} = -3374$

13. $a_n = \frac{n + 2}{n + 5}$
$a_1 = \frac{1 + 2}{1 + 5} = \frac{3}{6} = \frac{1}{2}$, $a_4 = \frac{4 + 2}{4 + 5} = \frac{6}{9} = \frac{2}{3}$;
$a_2 = \frac{2 + 2}{2 + 5} = \frac{4}{7}$, $a_{10} = \frac{10 + 2}{10 + 5} = \frac{12}{15} = \frac{4}{5}$;
$a_3 = \frac{3 + 2}{3 + 5} = \frac{5}{8}$, $a_{15} = \frac{15 + 2}{15 + 5} = \frac{17}{20}$

14. $a_1 = -1$, $a_2 = \frac{3}{2}$, $a_3 = 1$, $a_4 = \frac{7}{8}$; $a_{10} = \frac{19}{26}$;
$a_{15} = \frac{29}{41}$

15. $a_n = 5n - 6$
$a_8 = 5\cdot8 - 6 = 40 - 6 = 34$

16. 37

17. $a_n = (3n - 4)(2n + 5)$
$a_7 = (3\cdot7 - 4)(2\cdot7 + 5) = 17\cdot19 = 323$

18. 81

19. $a_n = (-1)^{n-1}(4.6n - 18.3)$
$a_{12} = (-1)^{12-1}[4.6(12) - 18.3] = -36.9$

20. -52,135,198.72

21. $a_n = 5n^2(4n - 100)$
$a_{11} = 5(11)^2(4\cdot11 - 100) = 5(121)(-56) = -33{,}880$

22. 528,528

23. $a_n = \left(1 + \frac{1}{n}\right)^2$
$a_{20} = \left(1 + \frac{1}{20}\right)^2 = \left(\frac{21}{20}\right)^2 = \frac{441}{400}$

24. $\frac{2744}{3375}$

25. $a_n = \log 10^n$
$a_{43} = \log 10^{43} = 43$

26. 67

27. $a_n = 1 + \frac{1}{n^2}$

$a_{38} = 1 + \frac{1}{38^2} = 1\frac{1}{1444}$, or $\frac{1445}{1444}$

28. -8

29. 1, 3, 5, 7, 9, . . .
These are odd integers, so the general term may be $2n - 1$.

30. 3^n

31. -2, 6, -18, 54, . . .
We can see a pattern if we write the sequence as
$-1\cdot2\cdot1$, $1\cdot2\cdot3$, $-1\cdot2\cdot9$, $1\cdot2\cdot27$, . . .
The general term may be $(-1)^n 2(3)^{n-1}$.

32. $5n - 7$

33. $\frac{2}{3}, \frac{3}{4}, \frac{4}{5}, \frac{5}{6}, \frac{6}{7}, \ldots$

These are fractions in which the denominator is 1 greater than the numerator. Also, each numerator is 1 greater than the preceding numerator. The general term may be $\frac{n+1}{n+2}$.

34. $\sqrt{2n}$

35. $\sqrt{3}, 3, 3\sqrt{3}, 9, 9\sqrt{3}, \ldots$
These are powers of $\sqrt{3}$. The general term may be $(\sqrt{3})^n$, or $3^{n/2}$.

36. $n(n + 1)$

37. -1, -4, -7, -10, -13, . . .
Each term is 3 less than the preceding term. The general term may be $-1 - 3(n - 1)$. After removing parentheses and simplifying, we can express the general term as $-3n + 2$, or $-(3n - 2)$.

38. $\log 10^{n-1}$, or $n - 1$

39. 1, 2, 3, 4, 5, 6, 7, . . .
$S_7 = 1 + 2 + 3 + 4 + 5 + 6 + 7 = 28$

40. -8

41. 2, 4, 6, 8, . . .
$S_5 = 2 + 4 + 6 + 8 + 10 = 30$

42. $\frac{5269}{3600}$

43. $\sum_{k=1}^{5} \frac{1}{2k} = \frac{1}{2\cdot1} + \frac{1}{2\cdot2} + \frac{1}{2\cdot3} + \frac{1}{2\cdot4} + \frac{1}{2\cdot5}$

$= \frac{1}{2} + \frac{1}{4} + \frac{1}{6} + \frac{1}{8} + \frac{1}{10}$

$= \frac{60}{120} + \frac{30}{120} + \frac{20}{120} + \frac{15}{120} + \frac{12}{120}$

$= \frac{137}{120}$

44. $\frac{1}{3} + \frac{1}{5} + \frac{1}{7} + \frac{1}{9} + \frac{1}{11} + \frac{1}{13} = \frac{43,024}{45,045}$

45. $\sum_{k=0}^{5} 2^k = 2^0 + 2^1 + 2^2 + 2^3 + 2^4 + 2^5$

$= 1 + 2 + 4 + 8 + 16 + 32$

$= 63$

46. $\sqrt{7} + \sqrt{9} + \sqrt{11} + \sqrt{13} \approx 12.5679$

47. $\sum_{k=7}^{10} \log k = \log 7 + \log 8 + \log 9 + \log 10 \approx 3.7024$

48. $0 + \pi + 2\pi + 3\pi + 4\pi \approx 31.4159$

49. $\sum_{k=1}^{8} \frac{k}{k+1} = \frac{1}{1+1} + \frac{2}{2+1} + \frac{3}{3+1} + \frac{4}{4+1} +$
$\frac{5}{5+1} + \frac{6}{6+1} + \frac{7}{7+1} + \frac{8}{8+1}$

$= \frac{1}{2} + \frac{2}{3} + \frac{3}{4} + \frac{4}{5} + \frac{5}{6} + \frac{6}{7} + \frac{7}{8} + \frac{8}{9}$

$= \frac{15,551}{2520}$

50. $0 + \frac{1}{5} + \frac{1}{3} + \frac{3}{7} = \frac{101}{105}$

51. $\sum_{k=1}^{5} (-1)^k = (-1)^1 + (-1)^2 + (-1)^3 + (-1)^4 + (-1)^5$

$= -1 + 1 - 1 + 1 - 1$

$= -1$

52. $1 - 1 + 1 - 1 + 1 = 1$

53. $\sum_{k=1}^{8} (-1)^{k+1}3k = (-1)^2 3\cdot1 + (-1)^3 3\cdot2 + (-1)^4 3\cdot3 +$
$(-1)^5 3\cdot4 + (-1)^6 3\cdot5 + (-1)^7 3\cdot6 +$
$(-1)^8 3\cdot7 + (-1)^9 3\cdot8$

$= 3 - 6 + 9 - 12 + 15 - 18 + 21 - 24$

$= -12$

54. $-4^2 + 4^3 - 4^4 + 4^5 - 4^6 + 4^7 - 4^8 = -52,432$

55. $\sum_{k=1}^{6} = \frac{2}{k^2+1} = \frac{2}{1^2+1} + \frac{2}{2^2+1} + \frac{2}{3^2+1} + \frac{2}{4^2+1} + \frac{2}{5^2+1} + \frac{2}{6^2+1}$

$= \frac{2}{2} + \frac{2}{5} + \frac{2}{10} + \frac{2}{17} + \frac{2}{26} + \frac{2}{37}$

$= 1 + \frac{2}{5} + \frac{1}{5} + \frac{2}{17} + \frac{1}{13} + \frac{2}{37}$

$= \frac{75{,}581}{40{,}885}$

56. $1\cdot2 + 2\cdot3 + 3\cdot4 + 4\cdot5 + 5\cdot6 + 6\cdot7 + 7\cdot8 + 8\cdot9 + 9\cdot10 + 10\cdot11 = 440$

57. $\sum_{k=0}^{5} (k^2 - 2k + 3) = (0^2 - 2\cdot0 + 3) + (1^2 - 2\cdot1 + 3) + (2^2 - 2\cdot2 + 3) + (3^2 - 2\cdot3 + 3) + (4^2 - 2\cdot4 + 3) + (5^2 - 2\cdot5 + 3)$

$= 3 + 2 + 3 + 6 + 11 + 18$

$= 43$

58. $4 + 2 + 2 + 4 + 8 + 14 = 34$

59. $\sum_{k=1}^{10} \frac{1}{k(k+1)} = \frac{1}{1(1+1)} + \frac{1}{2(2+1)} + \frac{1}{3(3+1)} + \frac{1}{4(4+1)} + \frac{1}{5(5+1)} + \frac{1}{6(6+1)} + \frac{1}{7(7+1)} + \frac{1}{8(8+1)} + \frac{1}{9(9+1)} + \frac{1}{10(10+1)}$

$= \frac{1}{2} + \frac{1}{6} + \frac{1}{12} + \frac{1}{20} + \frac{1}{30} + \frac{1}{42} + \frac{1}{56} + \frac{1}{72} + \frac{1}{90} + \frac{1}{110}$

$= \frac{10}{11}$

60. $\frac{2}{3} + \frac{4}{5} + \frac{8}{9} + \frac{16}{17} + \frac{32}{33} + \frac{64}{65} + \frac{128}{129} + \frac{256}{257} + \frac{512}{513} + \frac{1024}{1025} \approx 9.2365$

61. $\frac{1}{2} + \frac{2}{3} + \frac{3}{4} + \frac{4}{5} + \frac{5}{6} + \frac{6}{7}$

This is a sum of fractions in which the denominator is one greater than the numerator. Also, each numerator is 1 greater than the preceding numerator. Sigma notation is

$$\sum_{k=1}^{6} \frac{k}{k+1}.$$

62. $\sum_{k=1}^{5} 3k$

63. $-2 + 4 - 8 + 16 - 32 + 64$

This is a sum of powers of 2 with alternating signs. Sigma notation is

$$\sum_{k=1}^{6} (-1)^k 2^k, \text{ or } \sum_{k=1}^{6} (-2)^k.$$

64. $\sum_{k=1}^{5} \frac{1}{k^2}$

65. $4 - 9 + 16 - 25 + \ldots + (-1)^n n^2$

This is a sum of terms of the form $(-1)^k k^2$, beginning with $k = 2$ and continuing through $k = n$. Sigma notation is

$$\sum_{k=2}^{n} (-1)^k k^2.$$

66. $\sum_{k=3}^{n} (-1)^{k+1} k^2$

67. $5 + 10 + 15 + 20 + 25 + \ldots$

This is a sum of multiples of 5, and it is an infinite series. Sigma notation is

$$\sum_{k=1}^{\infty} 5k.$$

68. $\sum_{k=1}^{\infty} 7k$

69. $\frac{1}{1\cdot2} + \frac{1}{2\cdot3} + \frac{1}{3\cdot4} + \frac{1}{4\cdot5} + \ldots$

This is a sum of fractions in which the numerator is 1 and the denominator is a product of two consecutive integers. The larger integer in each product is the smaller integer in the succeeding product. It is an infinite series. Sigma notation is

$$\sum_{k=1}^{\infty} \frac{1}{k(k+1)}.$$

70. $\sum_{k=1}^{\infty} \frac{1}{k(k+1)^2}$

71. $a_1 = 4 \qquad a_{k+1} = 1 + \frac{1}{a_k}$

$a_2 = 1 + \frac{1}{4} = \frac{5}{4}$

$a_3 = 1 + \frac{1}{\frac{5}{4}} = 1 + \frac{4}{5} = \frac{9}{5}$

$a_4 = 1 + \frac{1}{\frac{9}{5}} = 1 + \frac{5}{9} = \frac{14}{9}$

72. $a_1 = 256,\ a_2 = 16,\ a_3 = 4,\ a_4 = 2$

73. $a_1 = 6561$ $a_{k+1} = (-1)^k \sqrt{a_k}$

$a_2 = (-1)^1 \sqrt{6561} = -81$

$a_3 = (-1)^2 \sqrt{-81} = 9i$

$a_4 = (-1)^3 \sqrt{9i} = -3\sqrt{i}$

74. $a_1 = e^Q$, $a_2 = \ln e^Q = Q$, $a_3 = \ln Q$, $a_4 = \ln (\ln Q)$

75. $a_1 = 2$ $a_{k+1} = a_k + a_{k-1}$

$a_2 = 3$

$a_3 = 3 + 2 = 5$

$a_4 = 5 + 3 = 8$

76. $a_1 = -10$, $a_2 = 8$, $a_3 = 18$, $a_4 = 10$

77. $a_n = \frac{1}{2^n} \log 1000^n$

$a_1 = \frac{1}{2} \log 1000 = \frac{1}{2} \cdot 3 = \frac{3}{2}$

$a_2 = \frac{1}{2^2} \log 1000^2 = \frac{1}{4} \cdot 6 = \frac{3}{2}$

$a_3 = \frac{1}{2^3} \log 1000^3 = \frac{1}{8} \cdot 9 = \frac{9}{8}$

$a_4 = \frac{1}{2^4} \log 1000^4 = \frac{1}{16} \cdot 12 = \frac{3}{4}$

$a_5 = \frac{1}{2^5} \log 1000^5 = \frac{1}{32} \cdot 15 = \frac{15}{32}$

$S_5 = \frac{3}{2} + \frac{3}{2} + \frac{9}{8} + \frac{3}{4} + \frac{15}{32}$

$= \frac{48}{32} + \frac{48}{32} + \frac{36}{32} + \frac{24}{32} + \frac{15}{32}$

$= \frac{171}{32}$

78. $i, -1, -i, 1, i; i$

79. $a_n = \ln (1 \cdot 2 \cdot 3 \cdots n)$

$a_1 = \ln 1 = 0$

$a_2 = \ln (1 \cdot 2) = \ln 2 \approx 0.693$

$a_3 = \ln (1 \cdot 2 \cdot 3) = \ln 6 \approx 1.792$

$a_4 = \ln (1 \cdot 2 \cdot 3 \cdot 4) = \ln 24 \approx 3.178$

$a_5 = \ln (1 \cdot 2 \cdot 3 \cdot 4 \cdot 5) = \ln 120 \approx 4.787$

$S_5 \approx 0 + 0.693 + 1.792 + 3.178 + 4.787 = 10.450$

80. 1, 0, -1, 0, 1; 1

81. $a_n = \cos^{-1} (-1)^n$

$a_1 = \cos^{-1} (-1)^1 = \cos^{-1} (-1) = \pi$

$a_2 = \cos^{-1} (-1)^2 = \cos^{-1} 1 = 0$

$a_3 = \cos^{-1} (-1)^3 = \cos^{-1} (-1) = \pi$

$a_4 = \cos^{-1} (-1)^4 = \cos^{-1} 1 = 0$

$a_5 = \cos^{-1} (-1)^5 = \cos^{-1} (-1) = \pi$

$S_5 = \pi + 0 + \pi + 0 + \pi = 3\pi$

82. $|\cos x|$, $|\cos x|$, $|\cos x|$, $|\cos x|$, $|\cos x|$; $5|\cos x|$

83. a) $a_n = n^2 - n + 41$

$a_1 = 1 - 1 + 41 = 41,$

$a_2 = 2^2 - 2 + 41 = 43,$

$a_3 = 3^2 - 3 + 41 = 47,$

$a_4 = 4^2 - 4 + 41 = 53,$

$a_5 = 5^2 - 5 + 41 = 61,$

$a_6 = 6^2 - 6 + 41 = 71$

b) All the terms are prime numbers.

c) $a_{41} = 41^2 - 41 + 41 = 41^2 = 1681$

The pattern does not hold since 1681 is not prime ($1681 = 41 \cdot 41$).

84. 2, 2.25, 2.370370, 2.441406, 2.488320, 2.521626

85. $a_n = \sqrt{n + 1} - \sqrt{n}$

$a_1 = \sqrt{1 + 1} - \sqrt{1} = \sqrt{2} - 1 \approx 0.414214$

$a_2 = \sqrt{2 + 1} - \sqrt{2} = \sqrt{3} - \sqrt{2} \approx 0.317837$

$a_3 = \sqrt{3 + 1} - \sqrt{3} = 2 - \sqrt{3} \approx 0.267949$

$a_4 = \sqrt{4 + 1} - \sqrt{4} = \sqrt{5} - 2 \approx 0.236068$

$a_5 = \sqrt{5 + 1} - \sqrt{5} = \sqrt{6} - \sqrt{5} \approx 0.213422$

$a_6 = \sqrt{6 + 1} - \sqrt{6} = \sqrt{7} - \sqrt{6} \approx 0.196262$

86. 2, 1.553774, 1.498834, 1.491398, 1.490378, 1.490238

87. $a_1 = 2,$ $a_{k+1} = \frac{1}{2} \left(a_k + \frac{2}{a_k}\right)$

$a_2 = \frac{1}{2} \left(2 + \frac{2}{2}\right) = 1.5$

$a_3 = \frac{1}{2} \left(1.5 + \frac{2}{1.5}\right) \approx 1.416667$

$a_4 = \frac{1}{2} \left(1.416667 + \frac{2}{1.416667}\right) \approx 1.414216$

$a_5 = \frac{1}{2} \left(1.414216 + \frac{2}{1.414216}\right) \approx 1.414214$

$a_6 = \frac{1}{2} \left(1.414214 + \frac{2}{1.414214}\right) \approx 1.414214$

88. 2, 4, 8, 16, 32, 64, 128, 256, 512, 1024, 2048, 4096, 8192, 16,384, 32,768, 65,536

89. Find each term by multiplying the preceding term by 0.75:

$3900, $2925, $2193.75, $1645.31, $1233.98, $925.49, $694.12, $520.59, $390.44, $292.83

90. $4.20, $4.35, $4.50, $4.65, $4.80, $4.95, $5.10, $5.25, $5.40, $5.55

91. $S_n = \ln 1 + \ln 2 + \ln 3 + \cdots + \ln n$

$= \ln (1\cdot2\cdot3 \cdots n) = \ln (n!)$.

92. $\dfrac{n}{n + 1}$

Exercise Set 12.2

1. 3, 8, 13, 18, . . .

$a_1 = 3$

$d = 5 \quad (8 - 3 = 5,\ 13 - 8 = 5,\ 18 - 13 = 5)$

2. $a_1 = 1.08$, $d = 0.08$

3. 9, 5, 1, -3, . . .

$a_1 = 9$

$d = -4 \quad (5 - 9 = -4,\ 1 - 5 = -4,\ -3 - 1 = -4)$

4. $a_1 = -8$, $d = 3$

5. $\frac{3}{2}, \frac{9}{4}, 3, \frac{15}{4}, \ldots$

$a_1 = \frac{3}{2}$

$d = \frac{3}{4} \quad \left[\frac{9}{4} - \frac{3}{2} = \frac{3}{4},\ 3 - \frac{9}{4} = \frac{3}{4}\right]$

6. $a_1 = \frac{3}{5}$, $d = -\frac{1}{2}$

7. $1.07, $1.14, $1.21, $1.28, . . .

$a_1 = \$1.07$

$d = \$0.07 \quad (\$1.14 - \$1.07 = \$0.07,\ \$1.21 - \$1.14 = \$0.07)$

8. $a_1 = \$316$, $d = -\$3$

9. 2, 6, 10, . . .

$a_1 = 2$, $d = 4$, and $n = 12$

$a_n = a_1 + (n - 1)d$

$a_{12} = 2 + (12 - 1)4 = 2 + 11\cdot4 = 2 + 44 = 46$

10. 0.57

11. 7, 4, 1, . . .

$a_1 = 7$, $d = -3$, and $n = 17$

$a_n = a_1 + (n - 1)d$

$a_{17} = 7 + (17 - 1)(-3) = 7 + 16(-3) = 7 - 48 = -41$

12. $-\frac{17}{3}$

13. $1200, $964.32, $728.64, . . .

$a_1 = \$1200$, $d = \$964.32 - \$1200 = -\$235.68$, and $n = 13$

$a_n = a_1 + (n - 1)d$

$a_{13} = \$1200 + (13 - 1)(-\$235.68) = \$1200 + 12(-\$235.68) = \$1200 - \$2828.16 = -\$1628.16$

14. $7941.62

15. $a_1 = 2$, $d = 4$

$a_n = a_1 + (n - 1)d$

Let $a_n = 106$, and solve for n.

$106 = 2 + (n - 1)(4)$

$106 = 2 + 4n - 4$

$108 = 4n$

$27 = n$

The 27th term is 106.

16. 33rd term

17. $a_1 = 7$, $d = -3$

$a_n = a_1 + (n - 1)d$

$-296 = 7 + (n - 1)(-3)$

$-296 = 7 - 3n + 3$

$-306 = -3n$

$102 = n$

The 102nd term is -296.

18. 46th term

19. $a_n = a_1 + (n - 1)d$

$a_{17} = 5 + (17 - 1)6$ Substituting 17 for n, 5 for a_1, and 6 for d

$= 5 + 16\cdot6$

$= 5 + 96$

$= 101$

20. -43

21. $a_n = a_1 + (n - 1)d$

$33 = a_1 + (8 - 1)4$ Substituting 33 for a_8, 8 for n, and 4 for d

$33 = a_1 + 28$

$5 = a_1$

(Note that this procedure is equivalent to subtracting d from a_8 seven times to get a_1: $33 - 7(4) = 33 - 28 = 5$)

22. 1.8

23. $a_n = a_1 + (n - 1)d$

$-76 = 5 + (n - 1)(-3)$ Substituting -76 for a_n, 5 for a_1 and -3 for d

$-76 = 5 - 3n + 3$

$-76 = 8 - 3n$

$-84 = -3n$

$28 = n$

24. 39

25. We know that $a_{17} = -40$ and $a_{28} = -73$. We would have to add d eleven times to get from a_{17} to a_{28}. That is,

$-40 + 11d = -73$

$11d = -33$

$d = -3.$

Since $a_{17} = -40$, we subtract d sixteen times to get to a_1.

$a_1 = -40 - 16(-3) = -40 + 48 = 8$

We write the first five terms of the sequence:

8, 5, 2, -1, -4

26. $\frac{1}{3}, \frac{5}{6}, \frac{4}{3}, \frac{11}{6}, \frac{7}{3}$

27. $5 + 8 + 11 + 14 + \ldots$

Note that $a_1 = 5$, $d = 3$, and $n = 20$. We use Formula 3.

$$S_n = \frac{n}{2}\left[2a_1 + (n - 1)d\right]$$

$$S_{20} = \frac{20}{2}[2\cdot 5 + (20 - 1)3] = 10[10 + 57] = 10\cdot 67 = 670$$

28. -210

29. The sum is $1 + 2 + 3 + \ldots + 299 + 300$. This is the sum of the arithmetic sequence for which $a_1 = 1$, $a_n = 300$, and $n = 300$. We use Formula 2.

$$S_n = \frac{n}{2}(a_1 + a_n)$$

$$S_{300} = \frac{300}{2}(1 + 300) = 150(301) = 45,150$$

30. 80,200

31. The sum is $2 + 4 + 6 + \ldots + 98 + 100$. This is the sum of the arithmetic sequence for which $a_1 = 2$, $a_n = 100$, and $n = 50$. We use Formula 2.

$$S_n = \frac{n}{2}(a_1 + a_n)$$

$$S_{50} = \frac{50}{2}(2 + 100) = 25(102) = 2550$$

32. 2500

33. The sum is $7 + 14 + 21 + \ldots + 91 + 98$. This is the sum of the arithmetic sequence for which $a_1 = 7$, $a_n = 98$, and $n = 14$. We use Formula 2.

$$S_n = \frac{n}{2}(a_1 + a_n)$$

$$S_{14} = \frac{14}{2}(7 + 98) = 7(105) = 735$$

34. 34,036

35. $S_n = \frac{n}{2}\left[2a_1 + (n - 1)d\right]$

$S_{20} = \frac{20}{2}[2\cdot 2 + (20 - 1)5]$ Substituting 20 for n, 2 for a_1, and 5 for d

$= 10[4 + 19\cdot 5]$

$= 10[4 + 95]$

$= 10\cdot 99$

$= 990$

36. -1264

37. We must find the smallest positive term of the sequence 35, 31, 27, ... It is an arithmetic sequence with $a_1 = 35$ and $d = -4$. We find the largest integer x for which $35 + x(-4) > 0$. Then we evaluate the expression $35 - 4x$ for that value of x.

$35 - 4x > 0$

$35 > 4x$

$\frac{35}{4} > x$

$8\frac{3}{4} > x$

The integer we are looking for is 8. Then $35 - 4x = 35 - 4(8) = 3$.

There will be 3 plants in the last row.

38. 62, 950

39. We go from 50 poles in a row, down to one pole in the top row, so there must be 50 rows. We want the sum 50 + 49 + 48 + . . . + 1. Thus we want the sum of an arithmetic sequence with $a_1 = 50$, $a_n = 1$, and $n = 50$. We will use the formula $S_n = \frac{n}{2}(a_1 + a_n)$.

$$S_{50} = \frac{50}{2}(50 + 1).$$

$$S_{50} = 25(51) = 1275$$

There will be 1275 poles in the pile.

40. 4960¢, or \$49.60

41. We want to find the sum of an arithmetic sequence with a_1 = \$600, d = \$100, and n = 20. We will use the formula

$$S_n = \frac{n}{2}[2a_1 + (n - 1)d].$$

$$S_{20} = \frac{20}{2}[2\cdot\$600 + (20 - 1)\$100].$$

$$S_{20} = 10(\$3100) = \$31{,}000$$

They save \$31,000 (disregarding interest).

42. \$10,230

43. We want to find the sum of an arithmetic sequence with $a_1 = 28$, $d = 4$, and $n = 50$. We will use the formula

$$S_n = \frac{n}{2}[2a_1 + (n - 1)d].$$

$$S_{50} = \frac{50}{2}[2\cdot 28 + (50 - 1)4].$$

$$S_{50} = 25(252) = 6300$$

There are 6300 seats.

44. \$462,500

45. 4, m_1, m_2, m_3, m_4, 13

We look for m_1, m_2, m_3, and m_4 such that 4, m_1, m_2, m_3, m_4, 13 is an arithmetic sequence. In this case $a_1 = 4$, $n = 6$, and $a_6 = 13$. We use Theorem 1.

$$a_n = a_1 + (n - 1)d$$

$$13 = 4 + (6 - 1)d$$

$$9 = 5d$$

$$1\frac{4}{5} = d$$

Thus,

$$m_1 = a_1 + d = 4 + 1\frac{4}{5} = 5\frac{4}{5}$$

$$m_2 = m_1 + d = 5\frac{4}{5} + 1\frac{4}{5} = 6\frac{8}{5} = 7\frac{3}{5}$$

$$m_3 = m_2 + d = 7\frac{3}{5} + 1\frac{4}{5} = 8\frac{7}{5} = 9\frac{2}{5}$$

$$m_4 = m_3 + d = 9\frac{2}{5} + 1\frac{4}{5} = 10\frac{6}{5} = 11\frac{1}{5}$$

46. -1, 1, 3

47. 1 + 3 + 5 + 7 + ... + n

$$S_n = \frac{n}{2}[2a_1 + (n - 1)d]$$

Note that $a_1 = 1$ and $d = 2$.

$$S_n = \frac{n}{2}[2\cdot 1 + (n - 1)2] = \frac{n}{2}(2 + 2n - 2)$$

$$= \frac{n}{2}\cdot 2n$$

$$= n^2$$

48. $\frac{n}{2}(1 + n)$, or $\frac{n(n + 1)}{2}$

49. Let x represent the first number in the sequence, and let d represent the common difference. Then the three numbers in the sequence are x, x + d, and x + 2d.

We write a system of equations:

The sum of the first and third numbers is 10.

$$x + x + 2d = 10$$

The product of the first and second numbers is 15.

$$x(x + d) = 15$$

Solving the system of equations we get x = 3 and d = 2. Thus the numbers are 3, 5, and 7.

50. -5p - 5q + 60, 5p + 2q - 20

51. $a_1 = \$8760$
$a_2 = \$8760 + (-\$798.23) = \$7961.67$
$a_3 = \$8760 + 2(-\$798.23) = \$7163.54$
$a_4 = \$8760 + 3(-\$798.23) = \$6365.31$
$a_5 = \$8760 + 4(-\$798.23) = \$5567.08$
$a_6 = \$8760 + 5(-\$798.23) = \$4768.85$
$a_7 = \$8760 + 6(-\$798.23) = \$3970.62$
$a_8 = \$8760 + 7(-\$798.23) = \$3172.39$
$a_9 = \$8760 + 8(-\$798.23) = \$2374.16$
$a_{10} = \$8760 + 9(-\$798.23) = \$1575.93$

52. \$51,679.65

53. $P(x) = x^4 + 4x^3 - 84x^2 - 176x + 640$ has at most 4 zeros because $P(x)$ is of degree 4. By the rational roots theorem, the possible zeros are

$\pm1, \pm2, \pm4, \pm5, \pm8, \pm10, \pm16, \pm20, \pm32, \pm40, \pm64, \pm80, \pm128, \pm160, \pm320, \pm640$

We find two of the zeros using synthetic division.

2	1	4	-84	-176	640
		2	12	-144	-640
	1	6	-72	-320	0

-4	1	6	-72	-320
		-4	-8	320
	1	2	-80	0

Also by synthetic division we determine that ±1 and -2 are not zeros. Therefore we determine that $d = 6$ in the arithmetic sequence.

Possible arithmetic sequences:

a)			-4	2	8	14
b)		-10	-4	2	8
c)	-16	-10	-4	2	

The solution cannot be a) because 14 is not a possible zero. Checking -16 by synthetic division we find that -16 is not a zero. Thus b) is the only arithmetic sequence which contains all four zeros. The zeros are -10, -4, 2, and 8.

54. Insert 16 arithmetic means, where $d = \frac{49}{17}$.

55. The sides are a, $a + d$, and $a + 2d$. By the Pythagorean theorem, we get $a^2 + (a + d)^2 = (a + 2d)^2$. Solving for d, we get $d = \frac{a}{3}$. Thus the sides are a, $a + \frac{a}{3}$, and $a + \frac{2a}{3}$, or a, $\frac{4a}{3}$, and $\frac{5a}{3}$ which are in the ratio 3:4:5.

56. a) $a_t = \$5200 - \$512.50t$

b) \$5200, \$4687.50, \$4175, \$3662.50, \$3150, \$1612.50, \$1100

57. $a_n = a_1 + (n - 1)d$

$a_n = f(n) = d \cdot n + (a_1 - d)$ where $f(n)$ is a linear function of n with slope d and y-intercept $(a_1 - d)$.

58. 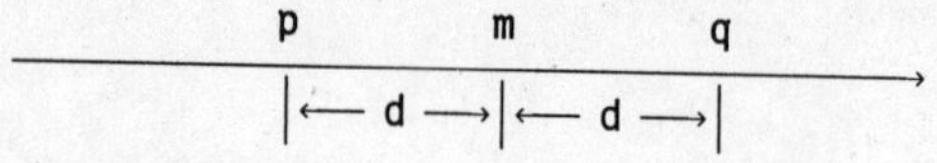

$$m = p + d$$
$$(+) \quad m = q - d$$
$$2m = p + q$$
$$m = \frac{p + q}{2}$$

Exercise Set 12.3

1. 2, 4, 8, 16, . . .
$\frac{4}{2} = 2, \ \frac{8}{4} = 2, \ \frac{16}{8} = 2$
$r = 2$

2. $-\frac{1}{3}$

3. 1, -1, 1, -1, . . .
$\frac{-1}{1} = -1, \ \frac{1}{-1} = -1, \ \frac{-1}{1} = -1$
$r = -1$

4. 0.1

5. $\frac{1}{2}, -\frac{1}{4}, \frac{1}{8}, -\frac{1}{16}, \ldots$

$$\frac{-\frac{1}{4}}{\frac{1}{2}} = -\frac{1}{4} \cdot \frac{2}{1} = -\frac{2}{4} = -\frac{1}{2}$$

$$\frac{\frac{1}{8}}{-\frac{1}{4}} = \frac{1}{8} \cdot \left(-\frac{4}{1}\right) = -\frac{4}{8} = -\frac{1}{2}$$

$r = -\frac{1}{2}$

6. -2

7. 75, 15, 3, $\frac{3}{5}$, . . .

$$\frac{15}{75} = \frac{1}{5}, \ \frac{3}{15} = \frac{1}{5}, \ \frac{\frac{3}{5}}{3} = \frac{3}{5} \cdot \frac{1}{3} = \frac{1}{5}$$

$r = \frac{1}{5}$

8. 0.1

9. $\frac{1}{x}, \frac{1}{x^2}, \frac{1}{x^3}, \ldots$

$$\frac{\frac{1}{x^2}}{\frac{1}{x}} = \frac{1}{x^2} \cdot \frac{x}{1} = \frac{x}{x^2} = \frac{1}{x}$$

$$\frac{\frac{1}{x^3}}{\frac{1}{x^2}} = \frac{1}{x^3} \cdot \frac{x^2}{1} = \frac{x^2}{x^3} = \frac{1}{x}$$

$r = \frac{1}{x}$

10. $\frac{m}{2}$

11. \$780, \$858, \$943.80, \$1038.18, . . .

$\frac{\$858}{\$780} = 1.1, \quad \frac{\$943.80}{\$858} = 1.1,$

$\frac{\$1038.18}{\$943.80} = 1.1$

$r = 1.1$

12. 0.95

13. 2, 4, 8, 16, . . .

$a_1 = 2$, $n = 6$, and $r = \frac{4}{2}$, or 2.

We use the formula $a_n = a_1r^{n-1}$.

$a_6 = 2(2)^{6-1} = 2 \cdot 2^5 = 2 \cdot 32 = 64$

14. 781,250

15. $2, 2\sqrt{3}, 6, \ldots$

$a_1 = 2$, $n = 9$, and $r = \frac{2\sqrt{3}}{2}$, or $\sqrt{3}$

$a_n = a_1r^{n-1}$

$a_9 = 2(\sqrt{3})^{9-1} = 2(\sqrt{3})^8 = 2 \cdot 81 = 162$

16. 1

17. $\frac{8}{243}, \frac{8}{81}, \frac{8}{27}, \ldots$

$a_1 = \frac{8}{243}$, $n = 10$, and $r = \frac{\frac{8}{81}}{\frac{8}{243}} = \frac{8}{81} \cdot \frac{243}{8} = 3$

$a_n = a_1r^{n-1}$

$a_{10} = \frac{8}{243}(3)^{10-1} = \frac{8}{243}(3)^8 = \frac{8}{243} \cdot 19{,}683 = 648$

18. $7(25)^{20}$

19. \$1000, \$1080, \$1166.40, . . .

$a_1 = \$1000$, $n = 5$, and $r = \frac{\$1080}{\$1000} = 1.08$

$a_n = a_1r^{n-1}$

$a_5 = \$1000(1.08)^{5-1} \approx \$1000(1.36048896) \approx$ \$1360.49

20. \$1402.55

21. 1, 3, 9, . . .

$a_1 = 1$ and $r = \frac{3}{1}$, or 3

$a_n = a_1r^{n-1}$

$a_n = 1(3)^{n-1} = 3^{n-1}$

22. 5^{3-n}

23. 1, -1, 1, -1, . . .

$a_1 = 1$ and $r = \frac{-1}{1} = -1$

$a_n = a_1r^{n-1}$

$a_n = 1(-1)^{n-1} = (-1)^{n-1}$

24. 2^n

25. $\frac{1}{x}, \frac{1}{x^2}, \frac{1}{x^3}, \ldots$

$a_1 = \frac{1}{x}$ and $r = \frac{1}{x}$ (see Exercise 9)

$a_n = a_1r^{n-1}$

$a_n = \frac{1}{x}\left(\frac{1}{x}\right)^{n-1} = \frac{1}{x} \cdot \frac{1}{x^{n-1}} = \frac{1}{x^{1+n-1}} = \frac{1}{x^n}$

26. $5\left(\frac{m}{2}\right)^{n-1}$

27. 6 + 12 + 24 + . . .

$a_1 = 6$, $n = 7$, and $r = \frac{12}{6}$, or 2

$S_n = \frac{a_1(r^n - 1)}{r - 1}$

$S_7 = \frac{6(2^7 - 1)}{2 - 1} = \frac{6(128 - 1)}{1} = 6 \cdot 127 = 762$

28. $\frac{21}{2}$, or 10.5

29. $\frac{1}{18} - \frac{1}{6} + \frac{1}{2} - \ldots$

$a_1 = \frac{1}{18}$, $n = 7$, and $r = \frac{-\frac{1}{6}}{\frac{1}{18}} = -\frac{1}{6} \cdot \frac{18}{1} = -3$

$S_n = \frac{a_1(r^n - 1)}{r - 1}$

$S_7 = \frac{\frac{1}{18}[(-3)^7 - 1]}{-3 - 1} = \frac{\frac{1}{18}[-2187 - 1]}{-4} = \frac{\frac{1}{18}(-2188)}{-4} =$

$\frac{1}{18}(-2188)\left(-\frac{1}{4}\right) = \frac{547}{18}$

30. $-\frac{171}{32}$

31. $1 + x + x^2 + x^3 + \ldots$

$a_1 = 1$, $n = 8$, and $r = \frac{x}{1}$, or x

$S_n = \frac{a_1(r^n - 1)}{r - 1}$

$S_8 = \frac{1(x^8 - 1)}{x - 1} = \frac{(x^4 + 1)(x^4 - 1)}{x - 1} =$

$\frac{(x^4 + 1)(x^2 + 1)(x^2 - 1)}{x - 1} =$

$\frac{(x^4 + 1)(x^2 + 1)(x + 1)(x - 1)}{x - 1} =$

$(x^4 + 1)(x^2 + 1)(x + 1)$

32. $\frac{x^{20} - 1}{x^2 - 1}$

33. \$200, \$200(1.06), \$200(1.06)², . . .

$a_1 = \$200$, $n = 16$, and $r = \frac{\$200(1.06)}{\$200} = 1.06$

$S_n = \frac{a_1(r^n - 1)}{r - 1}$

$S_{16} = \frac{\$200(1.06^{16} - 1)}{1.06 - 1} \approx \frac{\$200(2.540351685 - 1)}{0.06} \approx$

\$5134.51

34. \$60,893.30

35. $\sum_{k=1}^{\infty} \left(\frac{1}{2}\right)^{k-1}$

$a_1 = 1$, $r = \frac{1}{2}$

$\sum_{k=1}^{\infty} \left(\frac{1}{2}\right)^{k-1} = \lim_{n\to\infty} \frac{a_1(r^n - 1)}{r - 1} = \lim_{n\to\infty} \frac{1\left[\left(\frac{1}{2}\right)^n - 1\right]}{\frac{1}{2} - 1} =$

$\frac{1(0 - 1)}{-\frac{1}{2}} = 2$

36. Does not exist

37. $4 + 2 + 1 + \ldots$

$|r| = \left|\frac{2}{4}\right| = \left|\frac{1}{2}\right| = \frac{1}{2}$, and since $|r| < 1$, the series does have a sum.

$S_\infty = \frac{a_1}{1 - r} = \frac{4}{1 - \frac{1}{2}} = \frac{4}{\frac{1}{2}} = 4 \cdot \frac{2}{1} = 8$

38. $\frac{49}{4}$

39. $25 + 20 + 16 + \ldots$

$|r| = \left|\frac{20}{25}\right| = \left|\frac{4}{5}\right| = \frac{4}{5}$, and since $|r| < 1$, the series does have a sum.

$S_\infty = \frac{a_1}{1 - r} = \frac{25}{1 - \frac{4}{5}} = \frac{25}{\frac{1}{5}} = 25 \cdot \frac{5}{1} = 125$

40. 48

41. $100 - 10 + 1 - \frac{1}{10} + \ldots$

$|r| = \left|\frac{-10}{100}\right| = \left|-\frac{1}{10}\right| = \frac{1}{10}$, and since $|r| < 1$, the series does have a sum.

$S_\infty = \frac{a_1}{1 - r} = \frac{100}{1 - \left(-\frac{1}{10}\right)} = \frac{100}{\frac{11}{10}} = 100 \cdot \frac{10}{11} = \frac{1000}{11}$

42. Does not have a sum

43. $8 + 40 + 200 + \ldots$

$|r| = \left|\frac{40}{8}\right| = |5| = 5$, and since $|r| \not< 1$ the series does not have a sum.

44. −4

45. $0.6 + 0.06 + 0.006 + \ldots$

$|r| = \left|\frac{0.06}{0.6}\right| = |0.1| = 0.1$, and since $|r| < 1$, the series does have a sum.

$S_\infty = \frac{a_1}{1 - r} = \frac{0.6}{1 - 0.1} = \frac{0.6}{0.9} = \frac{6}{9} = \frac{2}{3}$

46. $\frac{37}{99}$

47. $\$500(1.11)^{-1} + \$500(1.11)^{-2} + \$500(1.11)^{-3} + \ldots$

$|r| = \left|\frac{\$500(1.11)^{-2}}{\$500(1.11)^{-1}}\right| = |(1.11)^{-1}| = (1.11)^{-1}$, or $\frac{1}{1.11}$, and since $|r| < 1$, the series does have a sum.

$S_\infty = \frac{a_1}{1 - r} = \frac{\$500(1.11)^{-1}}{1 - \left(\frac{1}{1.11}\right)} = \frac{\frac{\$500}{1.11}}{\frac{0.11}{1.11}} =$

$\frac{\$500}{1.11} \cdot \frac{1.11}{0.11} \approx \$4545.45$

48. \$12,500

49. $\sum_{k=1}^{\infty} 16(0.1)^{k-1}$

$a_1 = 16(0.1)^{1-1} = 16 \cdot 1 = 16$

$a_2 = 16(0.1)^{2-1} = 16(0.1) = 1.6$

$r = \frac{1.6}{16} = 0.1$

$S_\infty = \frac{a_1}{1 - r}$

$S_\infty = \frac{16}{1 - 0.1} = \frac{16}{0.9} = \frac{160}{9}$

50. 10

51. $\sum_{k=1}^{\infty} \frac{1}{2^{k-1}}$

$a_1 = \frac{1}{2^{1-1}} = \frac{1}{1} = 1$

$a_2 = \frac{1}{2^{2-1}} = \frac{1}{2^1} = \frac{1}{2}$

$r = \frac{\frac{1}{2}}{1} = \frac{1}{2}$

$S_\infty = \frac{a_1}{1 - r}$

$S_\infty = \frac{1}{1 - \frac{1}{2}} = \frac{1}{\frac{1}{2}} = 2$

52. $\frac{16}{3}$

53. $0.777\overline{7} = 0.7 + 0.07 + 0.007 + 0.0007 + \ldots$

Note that $a_1 = 0.7$ and $r = 0.1$

$S_\infty = \frac{a_1}{1 - r}$

$S_\infty = \frac{0.7}{1 - 0.1} = \frac{0.7}{0.9} = \frac{7}{9}$

54. 9

55. $0.533\overline{3} = 0.5 + 0.033\overline{3}$

First we consider the repeating part:

$0.033\overline{3} = 0.03 + 0.003 + 0.0003 + \cdots$

$a_1 = 0.03$ and $r = 0.1$

$S_\infty = \frac{a_1}{1 - r} = \frac{0.03}{1 - 0.1} = \frac{0.03}{0.9} = \frac{3}{90} = \frac{1}{30}$

Thus

$0.5333 = 0.5 + 0.033\overline{3} = \frac{1}{2} + \frac{1}{30} = \frac{16}{30} = \frac{8}{15}$

56. $\frac{29}{45}$

57. $5.1515\overline{15} = 5 + 0.151515$

First we consider the repeating part.

$0.1515\overline{15} = 0.15 + 0.0015 + 0.000015 + \ldots$

$a_1 = 0.15$ and $r = 0.01$

$S_\infty = \frac{a_1}{1 - r} = \frac{0.15}{1 - 0.01} = \frac{0.15}{0.99} = \frac{15}{99} = \frac{5}{33}$

Thus

$5.1515\overline{15} = 5 + \frac{5}{33} = 5\,\frac{5}{33}$, or $\frac{170}{33}$

58. $\frac{4121}{9990}$

59. The rebound distances form a geometric sequence:

$\frac{1}{4} \times 16,\quad \left(\frac{1}{4}\right)^2 \times 16,\quad \left(\frac{1}{4}\right)^3 \times 16, \ldots,$

or $4,\quad \frac{1}{4} \times 4,\quad \left(\frac{1}{4}\right)^2 \times 4, \ldots$

The height of the 6th rebound is the 6th term of the sequence.

We will use the formula $a_n = a_1r^{n-1}$, with $a_1 = 4$, $r = \frac{1}{4}$, and $n = 6$:

$a_6 = 4\left(\frac{1}{4}\right)^{6-1}$

$a_6 = \frac{4}{4^5} = \frac{1}{4^4} = \frac{1}{256}$

The ball rebounds $\frac{1}{256}$ ft the 6th time.

60. $5\frac{1}{3}$ ft

61. In one year, the population will be $100{,}000 + 0.03(100{,}000)$, or $(1.03)100{,}000$. In two years, the population will be $(1.03)100{,}000 + 0.03(1.03)100{,}000$, or $(1.03)^2100{,}000$. Thus the populations form a geometric sequence:

$100{,}000,\quad (1.03)100{,}000,\quad (1.03)^2100{,}000, \ldots$

The population in 15 years will be the 16th term of the sequence.

We will use the formula $a_n = a_1r^{n-1}$ with $a_1 = 100{,}000$, $r = 1.03$, and $n = 16$:

$a_{16} = 100{,}000(1.03)^{16-1}$

$a_{16} \approx 100{,}000(1.557967417) \approx 155{,}797$

In 15 years the population will be about 155,797.

62. About 24 years

63. The amounts owed at the beginning of successive years form a geometric sequence:

$\$1200,\quad (1.12)\$1200,\quad (1.12)^2\$1200,$

$(1.12)^3\$1200, \ldots$

The amount to be repaid at the end of 13 years is the amount owed at the beginning of the 14th year.

We use the formula $a_n = a_1r^{n-1}$ with $a_1 = \$1200$, $r = 1.12$, and $n = 14$:

$a_{14} = \$1200(1.12)^{14-1}$

$a_{14} \approx \$1200(4.363493112) \approx \$5236.19$

At the end of 13 years, \$5236.19 will be repaid.

64. 10,485.76 in.

65. The lengths of the falls form a geometric sequence:

$556, \left(\frac{3}{4}\right)556, \left(\frac{3}{4}\right)^2 556, \left(\frac{3}{4}\right)^3 556, \ldots$

The total length of the first 6 falls is the sum of the first six terms of this sequence. The heights of the rebounds also form a geometric sequence:

$\left(\frac{3}{4}\right)556, \left(\frac{3}{4}\right)^2 556, \left(\frac{3}{4}\right)^3 556, \ldots,$ or

$417, \left(\frac{3}{4}\right)417, \left(\frac{3}{4}\right)^2 417, \ldots$

When the ball hits the ground for the 6th time, it will have rebounded 5 times. Thus the total length of the rebounds is the sum of the first five terms of this sequence.

We use the formula $S_n = \frac{a_1(r^n - 1)}{r - 1}$ twice, once with $a_1 = 556$, $r = \frac{3}{4}$, and $n = 6$ and a second time with $a_1 = 417$, $r = \frac{3}{4}$, and $n = 5$.

$$D = \text{Length of falls} + \text{length of rebounds}$$
$$= \frac{556\left[\left(\frac{3}{4}\right)^6 - 1\right]}{\frac{3}{4} - 1} + \frac{417\left[\left(\frac{3}{4}\right)^5 - 1\right]}{\frac{3}{4} - 1}.$$

We use a calculator to obtain $D \approx 3100$.

The ball will have traveled about 3100 ft.

66. 3892 ft

67. The amounts form a geometric series:
$0.01, $0.01(2), $0.01(2)², $0.01(2³) + . . . + ($0.01)(2)²⁷

$\$0.01, \$0.01(2), \$0.01(2)^2, \$0.01(2^3) + \ldots + (\$0.01)(2)^{27}$

We use the formula $S_n = \frac{a_1(r^n - 1)}{r - 1}$ to find the sum of the geometric series with $a_1 = 0.01$, $r = 2$, and $n = 28$:

$$S_{28} = \frac{0.01(2^{28} - 1)}{2 - 1}$$

We use a calculator to obtain $S_{28} = \$2,684,355$.

You would earn $2,684,355.

68. $645,826.93

69. The total effect on the economy is $\$13,000,000,000 + \$13,000,000,000(0.85) + \$13,000,000,000(0.85)^2 + \$13,000,000,000(0.85)^3 + \ldots$ which is a geometric series with $a_1 = \$13,000,000,000$ and $r = 0.85$.

$$S_\infty = \frac{a_1}{1 - r}$$
$$S_\infty = \frac{\$13,000,000,000}{1 - 0.85} = \frac{\$13,000,000,000}{0.15}$$
$$= \$86,666,666,667$$

70. $\$9.4(10)^{11}$

71. The total number of people who will buy the product can be expressed as a geometric series.

$5,000,000(0.4) + 5,000,000(0.4)^2 + 5,000,000(0.4)^3 + 5,000,000(0.4)^4 + \ldots$

Note that $a_1 = 5,000,000(0.4)$ and $r = 0.4$.

$$S_\infty = \frac{a_1}{1 - r}$$
$$S_\infty = \frac{5,000,000(0.4)}{1 - 0.4} = \frac{2,000,000}{0.6} \approx 3,333,333, \text{ or } 3.3\overline{3} \times 10^6$$

3,333,333 represents what % of the population?

$$3,333,333 = x \cdot 5,000,000$$
$$\frac{3,333,333}{5,000,000} = x$$
$$0.666\overline{6} = x$$
$$66\tfrac{2}{3}\,\% = x$$

72. 3,000,000; 100%

73. See the answer section in the text.

74. a) $\frac{13}{3}, \frac{58}{3}$

b) $x = -\frac{11}{3}, a_4 = -\frac{50}{3}$ or $x = 5, a_4 = \frac{250}{3}$

75. $1 + x + x^2 + \ldots$

This is a geometric series with $a_1 = 1$ and $r = x$.

$$S_n = \frac{a_1(r^n - 1)}{r - 1} = \frac{1(x^n - 1)}{x - 1} = \frac{x^n - 1}{x - 1}, \text{ or } \frac{1 - x^n}{1 - x}$$

76. $\frac{x^2[(-x)^n - 1]}{-x - 1}$, or $\frac{x^2[1 - (-x)^n]}{x + 1}$

77. See the answer section in the text.

78.
$$\frac{a_n}{a_{n+1}} = r$$
$$\ln \frac{a_n}{a_{n+1}} = \ln r$$
$$\ln a_n - \ln a_{n+1} = \ln r$$

Thus, $\ln a_1, \ln a_2, \ln a_3, \ldots$ is an arithmetic sequence.

79. $a_{n+1} - a_n = d$ $\quad$ ($a_1, a_2, a_3, \ldots$ is arithmetic)

$$\frac{5^{a_{n+1}}}{5^{a_n}} = 5^{a_{n+1} - a_n} = 5^d, \text{ a constant}$$

Hence $5^{a_1}, 5^{a_2}, 5^{a_3}, \ldots$ is a geometric sequence.

80. 2.717

81. $a_{k+1} = \frac{1}{2}\left[a_k + \frac{2}{a_k}\right]$, $a_1 = 2$

$a_2 = \frac{1}{2}\left[2 + \frac{2}{2}\right] = 1.5$

$a_3 = \frac{1}{2}\left[1.5 + \frac{2}{1.5}\right] = 1.416667$

$a_4 = 1.414216$, $a_5 = 1.414214$, $a_6 = 1.414214$

The limit is $\sqrt{2}$.

82. 512 cm²

83. $s_1 = 2 = 2.$

$s_2 = 2 + \frac{1}{2} = 2.5$

$s_3 = 2 + \frac{1}{2} + \frac{1}{6} = 2.666667$

$s_4 = 2 + \frac{1}{2} + \frac{1}{6} + \frac{1}{24} = 2.708333$

$s_5 = 2 + \frac{1}{2} + \frac{1}{6} + \frac{1}{24} + \frac{1}{120} = 2.716667$

$s_6 = 2 + \frac{1}{2} + \frac{1}{6} + \frac{1}{24} + \frac{1}{120} + \frac{1}{720} = 2.718056$

$s_\infty = e \approx 2.7182818$

Exercise Set 12.4

1. $n^2 < n^3$

$1^2 < 1^3$, $2^2 < 2^3$, $3^2 < 3^3$, $4^2 < 4^3$, $5^2 < 5^3$

2. $1^2 - 1 + 41$ is prime, $2^2 - 2 + 41$ is prime, etc.

3. A polygon of n sides has $\frac{n(n-3)}{2}$ diagonals.

A polygon of 3 sides has $\frac{3(3-3)}{2}$ diagonals.

A polygon of 4 sides has $\frac{4(4-3)}{2}$ diagonals., etc.

4. The sum of the angles of a polygon of 3 sides is $(3 - 2)\cdot 180°$., etc.

5. See the answer section in the text.

6. S_n: $4 + 8 + 12 + \ldots + 4n = 2n(n + 1)$

S_1: $4 = 2\cdot 1\cdot (1 + 1)$

S_k: $4 + 8 + 12 + \ldots + 4k = 2k(k + 1)$

S_{k+1}: $4 + 8 + 12 + \ldots + 4k + 4(k + 1) = 2(k + 1)(k + 2)$

1) Basis step: S_1 is true by substitution.

6. (continued)

2) Induction step: Assume S_k. Deduce S_{k+1}. Starting with the left side of S_{k+1}, we have

$$\underbrace{4 + 8 + 12 + \ldots + 4k}_{} + 4(k + 1)$$

$$= 2k(k + 1) + 4(k + 1) \quad (\text{By } S_k)$$

$$= (k + 1)(2k + 4)$$

$$= 2(k + 1)(k + 2)$$

7. See the answer section in the text.

8. S_n: $3 + 6 + 9 + \ldots + 3n = \frac{3n(n + 1)}{2}$

S_1: $3 = \frac{3(1 + 1)}{2}$

S_k: $3 + 6 + 9 + \ldots + 3k = \frac{3k(k + 1)}{2}$

S_{k+1}: $3 + 6 + 9 + \ldots + 3k + 3(k + 1) = \frac{3(k + 1)(k + 2)}{2}$

1) Basis step: S_1 is true by substitution.

2) Induction step: Assume S_k. Deduce S_{k+1}. Starting with the left side of S_{k+1}, we have

$$\underbrace{3 + 6 + 9 + \ldots + 3k}_{} + 3(k + 1)$$

$$= \frac{3k(k + 1)}{2} + 3(k + 1) \quad (\text{By } S_k)$$

$$= \frac{3k(k + 1) + 6(k + 1)}{2}$$

$$= \frac{(k + 1)(3k + 6)}{2}$$

$$= \frac{3(k + 1)(k + 2)}{2}$$

9. See the answer section in the text for S_n, S_1, S_k, and S_{k+1}.

1) Basis step: S_1 is true by substitution.

2) Induction step: Assume S_k. Deduce S_{k+1}.

$$\frac{1}{1\cdot 2} + \frac{1}{2\cdot 3} + \ldots + \frac{1}{k(k + 1)} = \frac{k}{k + 1} \quad (S_k)$$

$$\frac{1}{1\cdot 2} + \frac{1}{2\cdot 3} + \ldots + \frac{1}{k(k+1)} + \frac{1}{(k+1)(k+2)} = \frac{k}{k+1} + \frac{1}{(k+1)(k+2)}$$

$\left[\text{Adding } \frac{1}{(k+1)(k+2)} \text{ on both sides}\right]$

$$= \frac{k(k + 2) + 1}{(k + 1)(k + 2)}$$

$$= \frac{k^2 + 2k + 1}{(k + 1)(k + 2)}$$

$$= \frac{(k + 1)(k + 1)}{(k + 1)(k + 2)}$$

$$= \frac{k + 1}{k + 2}$$

10. S_n: $2 + 4 + 8 + \dots + 2^n = 2(2^n - 1)$

S_1: $2 = 2(2 - 1)$

S_k: $2 + 4 + 8 + \dots + 2^k = 2(2^k - 1)$

S_{k+1}: $2 + 4 + 8 + \dots + 2^k + 2^{k+1} = 2(2^{k+1} - 1)$

1) Basis step: S_1 is true by substitution.

2) Induction step: Assume S_k. Deduce S_{k+1}. Starting with the left side of S_{k+1}, we have

$$\underbrace{2 + 4 + 8 + \dots + 2^k}_{} + 2^{k+1}$$

$= 2(2^k - 1) + 2^{k+1}$ (By S_k)

$= 2^{k+1} - 2 + 2^{k+1}$

$= 2 \cdot 2^{k+1} - 2$

$= 2(2^{k+1} - 1)$

11. S_n: $n < n + 1$

S_1: $1 < 1 + 1$

S_k: $k < k + 1$

S_{k+1}: $k + 1 < k + 2$

1) Basis step: S_1 is true by substitution.

2) Induction step: Assume S_k. Deduce S_{k+1}.

$k < k + 1$ (S_k)

$k + 1 < k + 1 + 1$ (Adding 1)

$k + 1 < k + 2$

12. S_n: $2 \leqslant 2^n$

S_1: $2 \leqslant 2^1$

S_k: $2 \leqslant 2^k$

S_{k+1}: $2 \leqslant 2^{k+1}$

1) Basis step: S_1 is true by substitution.

2) Induction step: Assume S_k. Deduce S_{k+1}.

$2 \leqslant 2^k$ (S_k)

$2 \cdot 2 \leqslant 2^k \cdot 2$ (Multiplying by 2)

$2 < 2 \cdot 2 \leqslant 2^{k+1}$ ($2 < 2 \cdot 2$)

$2 \leqslant 2^{k+1}$

13. S_n: $3^n < 3^{n+1}$

S_1: $3^1 < 3^{1+1}$

S_k: $3^k < 3^{k+1}$

S_{k+1}: $3^{k+1} < 3^{k+2}$

1) Basis step: S_1 is true by substitution.

2) Induction step: Assume S_k. Deduce S_{k+1}.

$3^k < 3^{k+1}$ (S_k)

$3^k \cdot 3 < 3^{k+1} \cdot 3$ (Multiplying by 3)

$3^{k+1} < 3^{k+1+1}$

$3^{k+1} < 3^{k+2}$

14. S_n: $2n \leqslant 2^n$

S_1: $2 \cdot 1 \leqslant 2^1$

S_k: $2k \leqslant 2^k$

S_{k+1}: $2(k + 1) \leqslant 2^{k+1}$

1) Basis step: S_1 is true by substitution.

2) Induction step: Assume S_k. Deduce S_{k+1}.

$2k \leqslant 2^k$ (S_k)

$2 \cdot 2k \leqslant 2 \cdot 2^k$ (Multiplying by 2)

$4k \leqslant 2^{k+1}$

Since $1 \leqslant k$, $k + 1 \leqslant k + k$, or $k + 1 \leqslant 2k$. Then $2(k + 1) \leqslant 4k$.

Thus $2(k + 1) \leqslant 4k \leqslant 2^{k+1}$, so $2(k + 1) \leqslant 2^{k+1}$.

15. S_n: $1^3 + 2^3 + 3^3 + \dots + n^3 = \dfrac{n^2(n + 1)^2}{4}$

S_1: $1^3 = \dfrac{1^2(1 + 1)^2}{4}$

S_k: $1^3 + 2^3 + 3^3 + \dots + k^3 = \dfrac{k^2(k + 1)^2}{4}$

S_{k+1}: $1^3 + 2^3 + 3^3 + \dots + k^3 + (k + 1)^3 = \dfrac{(k + 1)^2(k + 2)^2}{4}$

1) Basis step: S_1 is true by substitution.

2) Induction step: Assume S_k. Deduce S_{k+1}. Add $(k + 1)^3$ on both sides of S_k. Then simplify the right side.

$1^3 + 2^2 + 3^3 + \dots + k^3 + (k + 1)^3$

$= \dfrac{k^2(k + 1)^2}{4} + (k + 1)^3$

$= \dfrac{k^2(k + 1)^2 + 4(k + 1)^3}{4}$

$= \dfrac{(k + 1)^2[k^2 + 4(k + 1)]}{4}$

$= \dfrac{(k + 1)^2(k^2 + 4k + 4)}{4}$

$= \dfrac{(k + 1)^2(k + 2)^2}{4}$

16. S_n: $\dfrac{1}{1\cdot2\cdot3} + \dfrac{1}{2\cdot3\cdot4} + \dfrac{1}{3\cdot4\cdot5} + \dots + \dfrac{1}{n(n + 1)(n + 2)} = \dfrac{n(n + 3)}{4(n + 1)(n + 2)}$

S_1: $\dfrac{1}{1\cdot2\cdot3} = \dfrac{1(1 + 3)}{4\cdot2\cdot3}$

S_k: $\dfrac{1}{1\cdot2\cdot3} + \dfrac{1}{2\cdot3\cdot4} + \dots + \dfrac{1}{k(k + 1)(k + 2)} = \dfrac{k(k + 3)}{4(k + 1)(k + 2)}$

S_{k+1}: $\dfrac{1}{1\cdot2\cdot3} + \dfrac{1}{2\cdot3\cdot4} + \dots + \dfrac{1}{k(k + 1)(k + 2)} + \dfrac{1}{(k + 1)(k + 2)(k + 3)}$

$= \dfrac{(k + 1)(k + 1 + 3)}{4(k + 1 + 1)(k + 1 + 2)} = \dfrac{(k + 1)(k + 4)}{4(k + 2)(k + 3)}$

16. (continued)

1) Basis step: S_1 is true by substitution.

2) Induction step: Assume S_k. Deduce S_{k+1}.

Add $\frac{1}{(k + 1)(k + 2)(k + 3)}$ on both sides of S_k and simplify the right side.

Only the right side is shown here.

$$\frac{k(k + 3)}{4(k + 1)(k + 2)} + \frac{1}{(k + 1)(k + 2)(k + 3)}$$

$$= \frac{k(k + 3)(k + 3) + 4}{4(k + 1)(k + 2)(k + 3)}$$

$$= \frac{k^3 + 6k^2 + 9k + 4}{4(k + 1)(k + 2)(k + 3)}$$

$$= \frac{(k + 1)^2(k + 4)}{4(k + 1)(k + 2)(k + 3)}$$

$$= \frac{(k + 1)(k + 4)}{4(k + 2)(k + 3)}$$

17. See the answer section in the text.

18. S_n: $a_1 + (a_1 + d) + \ldots + [a_1 + (n - 1)d] = \frac{n}{2}[2a_1 + (n - 1)d]$

S_1: $a_1 = \frac{1}{2}[2a_1 + (1 - 1)d]$

S_k: $a_1 + (a_1 + d) + \ldots + [a_1 + (k - 1)d] = \frac{k}{2}[2a_1 + (k - 1)d]$

S_{k+1}: $a_1 + (a_1 + d) + \ldots + [a_1 + (k - 1)d] + [a_1 + kd] = \frac{k + 1}{2}[2a_1 + kd]$

1) Basis step: S_1 is true by substitution.

2) Induction step: Assume S_k. Deduce S_{k+1}. Starting with the left side of S_{k+1}, we have

$$a_1 + (a_1 + d) + \ldots + [a_1 + (k-1)d] + [a_1 + kd]$$

$$= \frac{k}{2}[2a_1 + (k - 1)d] + [a_1 + kd] \quad \text{(By } S_k\text{)}$$

$$= \frac{k[2a_1 + (k - 1)d]}{2} + \frac{2[a_1 + kd]}{2}$$

$$= \frac{2ka_1 + k(k - 1)d + 2a_1 + 2kd}{2}$$

$$= \frac{2a_1(k + 1) + k(k - 1)d + 2kd}{2}$$

$$= \frac{2a_1(k + 1) + (k - 1 + 2)kd}{2}$$

$$= \frac{2a_1(k + 1) + (k + 1)kd}{2}$$

$$= \frac{k + 1}{2}[2a_1 + kd]$$

19. See the answer section in the text.

20. S_n: $\cos n\pi = (-1)^n$

S_1: $\cos \pi = (-1)^1$

S_k: $\cos k\pi = (-1)^k$

S_{k+1}: $\cos [(k + 1)\pi] = (-1)^{k+1}$

1) Basis step: S_1 is true by substitution.

2) Induction step: Assume S_k. Deduce S_{k+1}. Starting with the left side of S_{k+1}, we have

$$\cos [(k + 1)\pi] = \cos (k\pi + \pi)$$
$$= \cos k\pi \cos \pi - \sin k\pi \sin \pi$$
$$= [\cos k\pi](-1) - [\sin k\pi]\cdot 0$$
$$= (-1) \cos k\pi$$
$$= (-1)(-1)^k \quad \text{(By } S_k\text{)}$$
$$= (-1)^{k+1}$$

21. See the answer section in the text.

22. S_n: $|\sin (nx)| \leq n|\sin x|$

S_1: $|\sin x| \leq |\sin x|$

S_k: $|\sin (kx)| \leq k|\sin x|$

S_{k+1}: $|\sin (k + 1)x| \leq (k + 1)|\sin x|$

1) Basis step: S_1 is true by substitution.

2) Induction step: Assume S_k. Deduce S_{k+1}. Starting with the left side of S_{k+1}, we have

$$|\sin (k + 1)x|$$
$$= |\sin (kx + x)|$$
$$= |\sin kx \cos x + \cos kx \sin x|$$
$$\leq |\sin kx \cos x| + |\cos kx \sin x|$$
$$= |\sin kx||\cos x| + |\cos kx||\sin x|$$
$$\leq k|\sin x||\cos x| + |\cos kx||\sin x| \quad \text{(By } S_k\text{)}$$
$$= |\sin x|(k|\cos x| + |\cos kx|)$$
$$\leq |\sin x|(k + 1),$$

since $|\cos x| \leq 1$ and $|\cos kx| \leq 1$

23. See the answer section in the text.

24. S_2: $\log_a (b_1b_2) = \log_a b_1 + \log_a b_2$

S_k: $\log_a (b_1b_2\ldots b_k) = \log_a b_1 + \log_a b_2 + \ldots + \log_a b_k$

$$\log_a (b_1b_2\ldots b_{k+1}) \quad \text{(Left side of } S_{k+1}\text{)}$$
$$= \log_a (b_1b_2\ldots b_k) + \log_a b_{k+1} \quad \text{(By } S_2\text{)}$$
$$= \log_a b_1 + \log_a b_2 + \ldots + \log_a b_k + \log_a b_{k+1}$$

25. See the answer section in the text.

26. $S_1 = \overline{z^1} = \overline{z}^1$

$S_k = \overline{z^k} = \overline{z}^k$

$\overline{z^k} \cdot \overline{z} = \overline{z}^k \cdot \overline{z}$ (Multiplying both sides of S_k by $\overline{z}$)

$\overline{z^k} \cdot \overline{z} = \overline{z}^{k+1}$

$\overline{z^{k+1}} = \overline{z}^{k+1}$

27. See the answer section in the text.

28. S_2: $\overline{z_1 z_2} = \overline{z_1} \cdot \overline{z_2}$

S_k: $\overline{z_1 z_2 \ldots z_k} = \overline{z_1}\,\overline{z_2} \ldots \overline{z_k}$

Starting with the left side of S_{k+1}, we have

$\overline{z_1 z_2 \ldots z_k z_{k+1}} = \overline{z_1 z_2 \ldots z_k} \cdot \overline{z_{k+1}}$ (By S_2)

$= \overline{z_1}\,\overline{z_2} \ldots \overline{z_k} \cdot \overline{z_{k+1}}$ (By S_k)

29. See the answer section in the text.

30. S_1: 3 is a factor of $1^3 + 2 \cdot 1$

S_k: 3 is a factor of $k^3 + 2k$, i.e. $k^3 + 2k = 3 \cdot m$

S_{k+1}: 3 is a factor of $(k + 1)^3 + 2(k + 1)$

Consider:

$(k + 1)^3 + 2(k + 1)$

$= k^3 + 3k^2 + 5k + 3$

$= (k^3 + 2k) + 3k^2 + 3k + 3$

$= 3m + 3(k^2 + k + 1)$ (A multiple of 3)

31. See the answer section in the text.

32. S_1: 5 is a factor of $1^5 - 1$.

S_k: 5 is a factor of $k^5 - k$, i.e., $k^5 - k = 5 \cdot m$

S_{k+1}: 5 is a factor of $(k + 1)^5 - (k + 1)$.

Consider:

$(k + 1)^5 - (k + 1)$.

$= (k^5 + 5k^4 + 10k^3 + 10k^2 + 5k + 1) - (k + 1)$

$= (k^5 - k) + 5(k^4 + 2k^3 + 2k^2 + k)$

$= 5 \cdot m + 5(k^4 + 2k^3 + 2k^2 + k)$ (A multiple of 5)

33. See the answer section in the text.

34. S_n: $\frac{1}{\sqrt{1}} + \frac{1}{\sqrt{2}} + \frac{1}{\sqrt{3}} + \cdots + \frac{1}{\sqrt{n}} > \sqrt{n}$

S_2: $\frac{1}{\sqrt{1}} + \frac{1}{\sqrt{2}} > \sqrt{2}$

S_k: $\frac{1}{\sqrt{1}} + \frac{1}{\sqrt{2}} + \frac{1}{\sqrt{3}} + \cdots + \frac{1}{\sqrt{k}} > \sqrt{k}$

S_{k+1}: $\frac{1}{\sqrt{1}} + \frac{1}{\sqrt{2}} + \frac{1}{\sqrt{3}} + \cdots + \frac{1}{\sqrt{k}} + \frac{1}{\sqrt{k + 1}}$

$> \sqrt{k + 1}$

1) Basis Step: $\frac{1}{\sqrt{1}} + \frac{1}{\sqrt{2}} = 1 + \frac{1}{\sqrt{2}} = \frac{\sqrt{2} + 1}{\sqrt{2}}$

$> \frac{2}{\sqrt{2}} = \sqrt{2}$

(Since $\sqrt{2} > 1$)

2) Induction Step: Assume S_k. Deduce S_{k+1}.

$\frac{1}{\sqrt{1}} + \frac{1}{\sqrt{2}} + \frac{1}{\sqrt{3}} + \cdots + \frac{1}{\sqrt{k}} > \sqrt{k}$, (by S_k).

Therefore, adding $\frac{1}{\sqrt{k + 1}}$,

$\frac{1}{\sqrt{1}} + \frac{1}{\sqrt{2}} + \frac{1}{\sqrt{3}} + \cdots + \frac{1}{\sqrt{k}} + \frac{1}{\sqrt{k + 1}}$

$> \sqrt{k} + \frac{1}{\sqrt{k + 1}}$

Then

$\sqrt{k} + \frac{1}{\sqrt{k + 1}} = \frac{\sqrt{k} \cdot \sqrt{k + 1} + 1}{\sqrt{k + 1}} > \frac{k + 1}{\sqrt{k + 1}}$

$= \sqrt{k + 1}$.

35. See the answer section in the text.

36. S_n: $n = n + 1$

S_1: $1 = 1 + 1$

S_k: $k = k + 1$

S_{k+1}: $k + 1 = (k + 1) + 1$

a) Basis step: $1 = 1 + 1$ cannot be proved.

b) Induction step: Assume S_k. Deduce S_{k+1}.

$k = k + 1$ (S_k)

$k + 1 = (k + 1) + 1$ (Adding 1)

The induction step can be proved.

c) The statement is not true.

37. See the answer section in the text.

38. a) Figure (2):

Number of sides: 12

Perimeter: 4a

Area: $\frac{a^2}{3}\sqrt{3} =$

$A_1\left[1 + \frac{1}{3}\right]$, where A_1 is the area of Figure (1)

Figure (3):

Number of sides: 48

Perimeter: $\frac{16a}{3}$

Area: $\frac{10a^2}{27}\sqrt{3} =$

$A_1\left[1 + \frac{1}{3} + \frac{1}{3}\cdot\frac{4}{9}\right]$, where A_1 is the area of Figure (1)

Figure (4):

Number of sides: 192

Perimeter: $\frac{64a}{9}$

Area: $\frac{94a^2}{243}\sqrt{3} = A_1\left[1 + \frac{1}{3} + \frac{1}{3}\cdot\frac{4}{9} + \frac{1}{3}\left(\frac{4}{9}\right)^2\right]$, where A_1 is the area of Figure (1)

b) Let N_n represent the number of sides of the nth figure.

Conjecture: $N_n = 3(4^{n-1})$

S_1: $N_1 = 3(4^{1-1}) = 3$

S_k: $N_k = 3(4^{k-1})$

S_{k+1}: $N_{k+1} = 3(4^k)$

1) Basis step: S_1 is true by part (a).

2) Induction step: Assume S_k. Deduce S_{k+1}.

$N_k = 3(4^{k-1})$

Adding another equilateral triangle to each side creates 4 sides where each side originally was. Then $N_{k+1} = 3(4^{k-1})(4) = 3(4^k)$.

c) Let P_n represent the perimeter of the nth figure.

Conjecture: $P_n = 3a\left(\frac{4}{3}\right)^{n-1}$ $\left(\text{or } \frac{4^{n-1}a}{3^{n-2}}\right)$

S_1: $P_1 = 3a\left(\frac{4}{3}\right)^{1-1} = 3a$

S_k: $P_k = 3a\left(\frac{4}{3}\right)^{k-1}$

S_{k+1}: $P_k = 3a\left(\frac{4}{3}\right)^{k}$

1) Basis step: S_1 is true by part (a).

38. (continued)

2) Induction step: Assume S_k. Deduce S_{k+1}.

$P_k = 3a\left(\frac{4}{3}\right)^{k-1}$

Adding another equilateral triangle to each side increases the length of each side by $\frac{1}{3}$, so the new length is $\frac{4}{3}$ of the former length. Then

$P_{k+1} = 3a\left(\frac{4}{3}\right)^{k-1}\left(\frac{4}{3}\right) = 3a\left(\frac{4}{3}\right)^k$.

d) Since Figure (2) has 12 sides, we observe that in going from Figure (2) to Figure (3) we gain 12 additional triangles of side $\frac{a}{3^2}$.

Hence, the additional area is $12\cdot\frac{\left(\frac{a}{9}\right)^2}{4}\sqrt{3}$, or $\frac{1}{3}r\cdot A_1$, where $r = \frac{4}{9}$. In going from Figure (3) to Figure (4) we gain

$48\cdot\frac{\left(\frac{a}{27}\right)^2}{4}\sqrt{3}$, or $\frac{1}{3}r^2\cdot A_1$, and so on.

Let A_n represent the area of the nth figure. Recall that N_n represents the number of sides of the nth figure.

Conjecture:

$A_n = A_1\left[1 + \left(\frac{1}{3} + \frac{1}{3}r + \cdots + \frac{1}{3}r^{n-2}\right)\right]$

$A_n = A_1\left[1 + \left(\frac{1}{3}\cdot\frac{1 - r^{n-1}}{1 - r}\right)\right]$

$A_n = \frac{a^2\sqrt{3}}{20}(8 - 3r^{n-1}),\ n \geq 1$

S_1: $A_1 = \frac{a^2\sqrt{3}}{20}(8 - 3\cdot 1) = \frac{a^2\sqrt{3}}{4}$

S_k: $A_k = \frac{a^2\sqrt{3}}{20}(8 - 3r^{k-1})$, or

$\frac{1}{5}A_1(8 - 3r^{k-1})$

S_{k+1}: $A_{k+1} = \frac{a^2\sqrt{3}}{20}(8 - 3r^k)$

1) Basis step: S_1 is true by part (a).

2) Induction step: Assume S_k. Deduce S_{k+1}.

Adding another equilateral triangle to each side of the kth figure adds the area of N_k equilateral triangles with side of length $\frac{a}{3^k}$. Thus

$$A_{k+1} = A_k + N_k\cdot\frac{\left(\frac{a}{3^k}\right)^2}{4}\sqrt{3}$$

$$= \frac{1}{5}A_1(8 - 3r^{k-1}) + N_k\cdot\frac{1}{9^k}\cdot A_1$$

$$= A_1\left(\frac{8}{5} - \frac{3}{5}r^{k-1} + 3\cdot 4^{k-1}\cdot\frac{1}{9^k}\right)$$

$$= A_1\left(\frac{8}{5} - \frac{3}{5}r^{k-1} + \frac{3}{4}r^k\right)$$

$$= \frac{a^2\sqrt{3}}{20}(8 - 3r^k)$$

38. (continued)

e) $\lim_{n\to\infty} A_n = \frac{2}{5} a^2 \sqrt{3}$, or $0.4\ a^2\sqrt{3}$

Exercise Set 12.5

1. ${}_4P_3 = 4\cdot3\cdot2 = 24$ (Theorem 8, Formula 1)

2. 2520

3. ${}_{10}P_7 = 10\cdot9\cdot8\cdot7\cdot6\cdot5\cdot4 = 604{,}800$ (Theorem 8, Formula 1)

4. 720

5. Without repetition: ${}_5P_5 = 5\cdot4\cdot3\cdot2\cdot1 = 120$ (Theorem 7)

With repetition: $5^5 = 3125$ (Theorem 11)

6. a) ${}_4P_4 = 24$

b) $4^4 = 256$

7. Line: ${}_5P_5 = 5\cdot4\cdot3\cdot2\cdot1 = 120$ (Theorem 7)

Circle: $(5 - 1)! = 4! = 4\cdot3\cdot2\cdot1 = 24$ (Theorem 10)

8. a) ${}_7P_7 = 7! = 5040$

b) $\frac{7!}{7} = 720$

9. DIGIT

There are 2 I's, 1 D, 1 G, and 1 T for a total of 5.

$P = \frac{5!}{2!\cdot1!\cdot1!\cdot1!}$ (Theorem 9)

$= \frac{5!}{2!} = \frac{5\cdot4\cdot3\cdot2!}{2!} = 5\cdot4\cdot3 = 60$

10. $\frac{6!}{2!} = 360$

11. There are only 9 choices for the first digit since 0 is excluded. There are also 9 choices for the second digit since 0 can be included and the first digit cannot be repeated. Because no digit is used more than once there are only 8 choices for the third digit, 7 for the fourth, 6 for the fifth, 5 for the sixth and 4 for the seventh. By the Fundamental Counting Principle the total number of permutations is

$9\cdot9\cdot8\cdot7\cdot6\cdot5\cdot4$, or 544,320.

Thus 544,320 7-digit phone numbers can be formed.

12. ${}_5P_5 \cdot {}_4P_4 = 2880$

13. $a^2b^3c^4 = a\cdot a\cdot b\cdot b\cdot b\cdot c\cdot c\cdot c\cdot c$

There are 2 a's, 3 b's, and 4 c's, for a total of 9.

Thus $P = \frac{9!}{2!\cdot3!\cdot4!}$ (Theorem 9)

$= \frac{9\cdot8\cdot7\cdot6\cdot5\cdot4!}{2\cdot1\cdot3\cdot2\cdot1\cdot4!} = \frac{9\cdot8\cdot7\cdot6\cdot5}{2\cdot3\cdot2} = 1260$

14. $\frac{6!}{3!2!} = 60$

15. The number of distinct circular arrangements of 13 objects (King Arthur plus 12 knights) is $(13 - 1)!$ (Theorem 10).

$(13 - 1)! = 12! = 12\cdot11\cdot10\cdot9\cdot8\cdot7\cdot6\cdot5\cdot4\cdot3\cdot2\cdot1$

$= 479{,}001{,}600$

16. $\frac{4!}{4} = 6$

17. a) ${}_5P_5 = 5!$ (Theorem 7)

$= 5\cdot4\cdot3\cdot2\cdot1 = 120$

b) There are 5 choices for the first coin and 2 possibilities (head or tail) for each choice. This results in a total of 10 choices for the first selection.

There are 4 choices (no coin can be used more than once) for the second coin and 2 possibilities (head or tail) for each choice. This results in a total of 8 choices for the second selection.

There are 3 choices for the third coin and 2 possibilities (head or tail) for each choice. This results in a total of 6 choices for the third selection.

Likewise there are 4 choices for the fourth selection and 2 choices for the fifth selection.

Using the Fundamental Counting Principle we know there are

$10\cdot8\cdot6\cdot4\cdot2$, or 3840

ways the coins can be lined up.

18. a) $4\cdot3\cdot2\cdot1 = 24$

b) $(4!)\cdot2^4 = 384$

19. ${}_{52}P_4 = 52\cdot51\cdot50\cdot49$ (Theorem 8, Formula 1)

$= 6{,}497{,}400$

20. ${}_{50}P_5 = \frac{50!}{45!} = 50\cdot49\cdot48\cdot47\cdot46 = 254{,}251{,}200$

21. $P = \frac{24!}{3!\cdot5!\cdot9!\cdot4!\cdot3!}$ (Theorem 9)

$= \frac{24\cdot23\cdot22\cdot21\cdot20\cdot19\cdot18\cdot17\cdot16\cdot15\cdot14\cdot13\cdot12\cdot11\cdot10\cdot9!}{3\cdot2\cdot1\cdot5\cdot4\cdot3\cdot2\cdot1\cdot4\cdot3\cdot2\cdot1\cdot3\cdot2\cdot1\cdot9!}$

$= 23\cdot11\cdot7\cdot19\cdot17\cdot8\cdot15\cdot14\cdot13\cdot12\cdot11\cdot10$ (Simplifying)

$= 16{,}491{,}024{,}950{,}400$

22. $\dfrac{20!}{2!\ 5!\ 8!\ 3!\ 2!} = 20{,}951{,}330{,}400$

23. MATH:

$_4P_4 = 4 \cdot 3 \cdot 2 \cdot 1 = 24$ (Theorem 7)

BUSINESS: 1 B, 1 U, 3 S's, 1 I, 1 N, 1 E, a total of 8.

$P = \dfrac{8!}{1! \cdot 1! \cdot 3! \cdot 1! \cdot 1! \cdot 1!}$ (Theorem 9)

$= \dfrac{8!}{3!} = \dfrac{8 \cdot 7 \cdot 6 \cdot 5 \cdot 4 \cdot 3!}{3!} = 8 \cdot 7 \cdot 6 \cdot 5 \cdot 4 = 6720$

PHILOSOPHICAL: 2 P's, 2 H's, 2 I's, 2 L's, 2 O's, 1 S, 1 C, and 1 A, a total of 13

$P = \dfrac{13!}{2! \cdot 2! \cdot 2! \cdot 2! \cdot 2! \cdot 1! \cdot 1! \cdot 1!}$ (Theorem 9)

$= \dfrac{13 \cdot 12 \cdot 11 \cdot 10 \cdot 9 \cdot 8 \cdot 7 \cdot 6 \cdot 5 \cdot 4 \cdot 3 \cdot 2 \cdot 1}{2 \cdot 1 \cdot 2 \cdot 1 \cdot 2 \cdot 1 \cdot 2 \cdot 1 \cdot 2 \cdot 1 \cdot 1 \cdot 1 \cdot 1}$

$= 13 \cdot 12 \cdot 11 \cdot 5 \cdot 9 \cdot 4 \cdot 7 \cdot 3 \cdot 5 \cdot 2 \cdot 3$ (Simplifying)

$= 194{,}594{,}400$

24. a) $6! = 720$

b) $\dfrac{7!}{2!} = 2520$

c) $\dfrac{11!}{2!\ 2!\ 2!} = 4{,}989{,}600$

25. There are 80 choices for the number of the county, 26 choices for the letter of the alphabet, and 9999 choices for the number that follows the letter. By the Fundamental Counting Principle we know there are $80 \cdot 26 \cdot 9999$, or 20,797,920 possible license plates.

26. a) $_5P_4 = 120$

b) $5 \cdot 5 \cdot 5 \cdot 5 = 625$

c) $1 \cdot {_4P_3} = 24$

d) $_3P_2 \cdot 1 = 6$

27. a) We want to find the number of permutations of 10 objects taken 5 at a time with repetition. This is 10^5, or 100,000.

b) Since there are 100,000 possible zip-codes, there could be 100,000 post offices.

28. a) 10^9, or 1,000,000,000

b) Yes

29. a) We want to find the number of permutations of 10 objects taken 9 at a time with repetition. This is 10^9, or 1,000,000,000.

b) Since there are more than 243 million social security numbers, each person can have one.

30. 41,437.7 yr (using 1 yr = 365.25 days)

31. $_nP_5 = 7 \cdot {_nP_4}$

$\dfrac{n!}{(n-5)!} = 7 \cdot \dfrac{n!}{(n-4)!}$ (Theorem 8)

$\dfrac{n!}{7(n-5)!} = \dfrac{n!}{(n-4)!}$ [Multiplying by $\frac{1}{7}$]

$7(n-5)! = (n-4)!$ (The denominators must be the same.)

$7(n-5)! = (n-4)(n-5)!$

$7 = n - 4$ [Multiplying by $\dfrac{1}{(n-5)!}$]

$11 = n$

32. 8

33. $_nP_5 = 9 \cdot {_{n-1}P_4}$

$\dfrac{n!}{(n-5)!} = 9 \cdot \dfrac{(n-1)!}{(n-5)!}$ (Theorem 8)

$n! = 9(n-1)!$ [Multiplying by $(n-5)!$]

$\dfrac{n!}{(n-1)!} = 9$ [Multiplying by $\dfrac{1}{(n-1)!}$]

$\dfrac{n(n-1)!}{(n-1)!} = 9$

$n = 9$

34. 11

35. There is one losing team per game. In order to leave one tournament winner there must be n - 1 losers produced in n - 1 games.

36. 2n - 1

37. Interpretation I: If one always looks at the same side of the key ring, there are, as at a round table, $(n - 1)!$ circular arrangements of n keys.

Interpretation II: Assuming the two sides of the key ring are indistinguishable (except for the order of the keys), then for each placement of the keys onto the ring, there would be, for each arrangement (say ABCD) on one side, the corresponding arrangement (DCBA) when viewing the same placement of keys from the other side. [See diagram.]

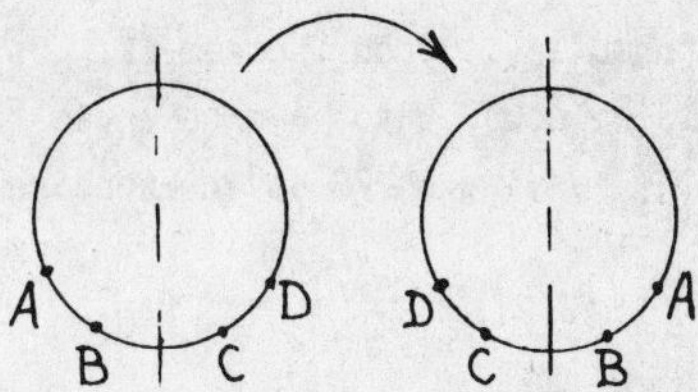

ABCD viewed from one side of ring.

Same placement of keys viewed from other side.

37. (continued)

In case of n = 4 there are (4 - 1)!, or 6, circular arrangements:
ABCD (same as BCDA or CDAB or DABC),
ABDC (same as BDCA or DCAB or CABD),
ACBD (same as CBDA or BDAC or DACB),
ACDB (same as CDBA or DBAC or BACD),
ADBC (same as DBCA or BCAD or CADB), and
ADCB (same as DCBA or CBAD or BADC).

We need only the first three for placing the keys onto the ring because, when viewed from the other side [see diagram],
ABCD becomes DCBA (or ADCB),
ABDC becomes CDBA (or ACDB), and
ACBD becomes DBCA (or ADBC).

Thus each side of the ring has $\frac{(n-1)!}{2}$ arrangements for a total of (n - 1)!.

Interpretation III: As suggested above, $\frac{(n-1)!}{2}$ placements of keys onto the ring produce all (n - 1)! circular arrangements when the ring is viewed from both sides. If, in the statement of the exercise, "arrangements on n keys on a key ring" is taken to mean "placements of n keys onto a key ring," then the answer would be $\frac{(n-1)!}{2}$.

Exercise Set 12.6

1. $_{13}C_2 = \frac{13\cdot 12}{2\cdot 1} = 78$

The 13 tells where to start.

The 2 tells how many factors there are in both numerator and denominator and where to start the denominator.

We used formula (2) from Theorem 12.

2. 84

3. $\binom{13}{11} = \frac{13!}{11!(13-11)!} = \frac{13!}{11!2!} = \frac{13\cdot 12\cdot 11!}{11!2!} =$
$\frac{13\cdot 12}{2\cdot 1} = 78$

We used formula (1) from Theorem 12. Using Theorem 13 we could also have observed that $\binom{13}{11} = \binom{13}{2}$, giving us 78 as in Exercise 1.

4. 84

5. $\binom{7}{1} = \frac{7}{1} = 7$ Formula (2)

6. 1

7. $\frac{_5P_3}{3!} = {_5C_3} = \frac{5\cdot 4\cdot 3}{3\cdot 2\cdot 1} = 10$ Formula (2)

8. 252

9. $\binom{6}{0} = \frac{6!}{0!(6-0)!} = \frac{6!}{1\cdot 6!} = 1$ Formula (1)

10. 6

11. $\binom{6}{2} = \frac{6\cdot 5}{2\cdot 1} = 15$ Formula (2)

12. 20

13. $_{12}C_{11} = {_{12}C_1}$ Theorem 13
$= \frac{12}{1}$ Formula (2)
$= 12$

14. 66

15. $_{12}C_9 = {_{12}C_3}$ Theorem 13
$= \frac{12\cdot 11\cdot 10}{3\cdot 2\cdot 1}$ Formula (2)
$= 220$

16. 495

17. $\binom{m}{2} = \frac{m(m-1)}{2\cdot 1}$ Formula (2)
$= \frac{m(m-1)}{2}$, or $\frac{m^2 - m}{2}$

Also $\binom{m}{2} = \frac{m!}{2!(m-2)!}$. Formula (1)

This is equivalent to the result using Formula (2).

18. $\frac{t!}{4!(t-4)!}$

19. $\binom{p}{3} = \frac{p(p-1)(p-2)}{3\cdot 2\cdot 1}$ Formula (2)
$= \frac{p(p-1)(p-2)}{6}$, or $\frac{p^3 - 3p^2 + 2p}{6}$

Also $\binom{p}{3} = \frac{p!}{3!(p-3)!}$. Formula (1)

This is equivalent to the result using Formula (2).

20. 1

21. By Theorem 13, $\binom{n}{r} = \binom{n}{n-r}$, so

$$\binom{7}{0} + \binom{7}{1} + \binom{7}{2} + \binom{7}{3} + \binom{7}{4} + \binom{7}{5} + \binom{7}{6} + \binom{7}{7}$$
$$= 2\left[\binom{7}{0} + \binom{7}{1} + \binom{7}{2} + \binom{7}{3}\right]$$
$$= 2\left[\frac{7!}{7!0!} + \frac{7!}{6!1!} + \frac{7!}{5!2!} + \frac{7!}{4!3!}\right]$$
$$= 2(1 + 7 + 21 + 35) = 2\cdot 64 = 128$$

22. 64

23. $${}_{100}C_0 + {}_{100}C_1 + \cdots + {}_{100}C_{100}$$
$$= 2({}_{100}C_0 + {}_{100}C_1 + \cdots + {}_{100}C_{49}) + {}_{100}C_{50}$$
$$= 2^{100}$$

(This computation is very tedious without the use of Theorem 17 in Section 12.7.)

24. 2^n

25. $${}_{23}C_4 = \binom{23}{4} = \frac{23!}{4!(23-4)!} \quad \text{(Theorem 12)}$$
$$= \frac{23!}{4!\cdot 19!} = \frac{23\cdot 22\cdot 21\cdot 20\cdot 19!}{4!\cdot 19!}$$
$$= \frac{23\cdot 22\cdot 21\cdot 20}{4\cdot 3\cdot 2\cdot 1} = 8855$$

26. a) ${}_9C_2 = 36$

b) $2 \cdot {}_9C_2 = 72$

27. $${}_{10}C_6 = \binom{10}{6} = \frac{10!}{6(10-6)!} \quad \text{(Theorem 12)}$$
$$= \frac{10!}{6!\cdot 4!} = \frac{10\cdot 9\cdot 8\cdot 7\cdot 6!}{6!\cdot 4!}$$
$$= \frac{10\cdot 9\cdot 8\cdot 7}{4\cdot 3\cdot 2\cdot 1} = 210$$

28. ${}_{11}C_7 = 330$

29. Since two points determine a line and no three of these 8 points are collinear, we need to find out the number of combinations of 8 points taken 2 at a time, ${}_8C_2$.

$${}_8C_2 = \binom{8}{2} = \frac{8!}{2!(8-2)!} \quad \text{(Theorem 12)}$$
$$= \frac{8\cdot 7\cdot 6!}{2\cdot 1\cdot 6!}$$
$$= \frac{8\cdot 7}{2} = 28$$

Thus 28 lines are determined.

29. (continued)

Since three noncollinear points determine a triangle and no four of these 8 points are coplanar, we need to find out the number of combinations of 8 points taken 3 at a time, ${}_8C_3$.

$${}_8C_3 = \binom{8}{3} = \frac{8!}{3!(8-3)!} \quad \text{(Theorem 12)}$$
$$= \frac{8\cdot 7\cdot 6\cdot 5!}{3\cdot 2\cdot 1\cdot 5!}$$
$$= \frac{8\cdot 7\cdot 6}{3\cdot 2} = 56$$

Thus 56 triangles are determined.

30. a) ${}_7C_2 = 21$

b) ${}_7C_3 = 35$

31. $${}_{10}C_7 \cdot {}_5C_3 = \binom{10}{7}\cdot\binom{5}{3} \quad \text{(Using the Fundamental Counting Principle)}$$
$$= \frac{10!}{7!(10-7)!} \cdot \frac{5!}{3!(5-3)!} \quad \text{(Theorem 12)}$$
$$= \frac{10\cdot 9\cdot 8\cdot 7!}{7!\cdot 3!} \cdot \frac{5\cdot 4\cdot 3!}{3!\cdot 2!}$$
$$= \frac{10\cdot 9\cdot 8}{3\cdot 2\cdot 1} \cdot \frac{5\cdot 4}{2\cdot 1} = 120\cdot 10 = 1200$$

32. ${}_8C_6 \cdot {}_4C_3 = 28\cdot 4 = 112$

33. $${}_{58}C_6 \cdot {}_{42}C_4 \quad \text{(Using the Fundamental Counting Principle)}$$
$$= \binom{58}{6}\cdot\binom{42}{4}$$
$$= \frac{58!}{6!(58-6)!} \cdot \frac{42!}{4!(42-4)!} \quad \text{(Theorem 12)}$$
$$= \frac{58\cdot 57\cdot 56\cdot 55\cdot 54\cdot 53\cdot 52!}{6!\cdot 52!} \cdot \frac{42\cdot 41\cdot 40\cdot 39\cdot 38!}{4!\cdot 38!}$$
$$= \frac{58\cdot 57\cdot 56\cdot 55\cdot 54\cdot 53}{6\cdot 5\cdot 4\cdot 3\cdot 2\cdot 1} \cdot \frac{42\cdot 41\cdot 40\cdot 39}{4\cdot 3\cdot 2\cdot 1}$$
$$= (29\cdot 19\cdot 14\cdot 11\cdot 9\cdot 53)\cdot(21\cdot 41\cdot 10\cdot 13)$$
$$= (40{,}475{,}358)(111{,}930)$$
$$= 4{,}530{,}406{,}820{,}940$$

34. $\binom{63}{8}\cdot\binom{37}{12}$

35. In a 52-card deck there are 4 aces and 48 cards that are not aces.

$${}_4C_3 \cdot {}_{48}C_2 = \binom{4}{3}\binom{48}{2} \quad \text{(Using the Fundamental Counting Principle)}$$
$$= \frac{4!}{3!(4-3)!} \cdot \frac{48!}{2!(48-2)!} \quad \text{(Theorem 12)}$$
$$= \frac{4!}{3!\cdot 1!} \cdot \frac{48!}{2!\cdot 46!}$$
$$= \frac{4\cdot 3!}{3!\cdot 1!} \cdot \frac{48\cdot 47\cdot 46!}{2!\cdot 46!}$$
$$= \frac{4}{1} \cdot \frac{48\cdot 47}{2}$$
$$= 4512$$

36. ${}_4C_2 \cdot {}_{48}C_3 = 103{,}776$

37. a) If order is considered and repetition is not allowed, the number of possible cones is

$${}_{33}P_3 = 33\cdot32\cdot31 = 32{,}736.$$

b) If order is considered and repetition is allowed, the number of possibilities is

$$33^3 = 35{,}937.$$

c) If order is not considered and there is no repetition, the number of possibilities is

$${}_{33}C_3 = \frac{33\cdot32\cdot31}{3\cdot2\cdot1} = 5456.$$

38. a) ${}_{31}P_2 = 930$

b) $31^2 = 961$

c) ${}_{31}C_2 = 465$

39. The number of plain pizzas (pizzas with no toppings) is $\binom{12}{0}$. The number of pizzas with 1 topping is $\binom{12}{1}$. The number of pizzas with 2 toppings is $\binom{12}{2}$, and so on. The number of pizzas with all 12 toppings is $\binom{12}{12}$. The total number of pizzas is $\binom{12}{0} + \binom{12}{1} + \binom{12}{2} + \cdots + \binom{12}{12} = 4096.$

40. $3\cdot2\cdot4096 = 24{,}576$

41. $${}_{52}C_5 = \binom{52}{5} = \frac{52!}{5!(52-5)!} \qquad \text{(Theorem 12)}$$
$$= \frac{52!}{5!\cdot47!}$$
$$= \frac{52\cdot51\cdot50\cdot49\cdot48\cdot47!}{5!\cdot47!}$$
$$= \frac{52\cdot51\cdot50\cdot49\cdot48}{5\cdot4\cdot3\cdot2\cdot1}$$
$$= 13\cdot17\cdot10\cdot49\cdot24$$
$$= 2{,}598{,}960$$

42. ${}_{52}C_{13} = 635{,}013{,}559{,}600$

43. $${}_8C_3 = \binom{8}{3} = \frac{8!}{3!(8-3)!} = \frac{8!}{3!\cdot5!}$$
$$= \frac{8\cdot7\cdot6\cdot5!}{3\cdot2\cdot1\cdot5!}$$
$$= 8\cdot7 = 56$$

44. $\binom{n}{4} = \frac{n(n-1)(n-2)(n-3)}{24}$

45. For <u>each</u> parallelogram 2 lines from each group of parallel lines are needed.

Two lines can be selected from the group of 5 parallel lines in ${}_5C_2$ ways and the other two lines can be selected from the group of 8 parallel lines in ${}_8C_2$ ways.

Using the Fundamental Counting Principle, it follows that the number of possible parallelograms is

$${}_5C_2 \cdot {}_8C_2 = \binom{5}{2}\binom{8}{2} \qquad \text{(Using the Fundamental Counting Principle)}$$
$$= \frac{5!}{2!(5-2)!}\cdot\frac{8!}{2!(8-2)!} \qquad \text{(Theorem 12)}$$
$$= \frac{5\cdot4\cdot3!}{2\cdot1\cdot3!}\cdot\frac{8\cdot7\cdot6!}{2\cdot1\cdot6!}$$
$$= \frac{5\cdot4}{2}\cdot\frac{8\cdot7}{2}$$
$$= 10\cdot28 = 280$$

Thus 280 parallelograms can be formed.

46. $$\binom{n}{r} = \frac{n!}{r!(n-r)!} = \frac{n!}{(n-r)!r!}$$
$$= \frac{n!}{(n-r)![n-(n-r)]!}$$
$$= \binom{n}{n-r}$$

47. If each team plays each other team once, we have $\binom{8}{2} = \frac{8!}{2!6!} = \frac{8\cdot7\cdot6!}{2\cdot1\cdot6!} = 28.$

If each team plays each other team twice, we have $2\binom{8}{2} = 2\cdot28 = 56.$

48. $$\binom{n}{2} = \frac{n!}{2!(n-2)!} = \frac{n(n-1)}{2};$$
$$2\binom{n}{2} = 2\left[\frac{n(n-1)}{2}\right] = n(n-1)$$

49. $$\binom{n+1}{3} = 2\cdot\binom{n}{2}$$
$$\frac{(n+1)!}{3!(n+1-3)!} = 2\cdot\frac{n!}{2!(n-2)!} \qquad \text{(Theorem 12)}$$
$$\frac{(n+1)(n)(n-1)(n-2)!}{3!(n-2)!} = \frac{n(n-1)(n-2)!}{(n-2)!}$$
$$\frac{(n+1)(n)(n-1)}{6} = n(n-1)$$
$$(n+1)(n)(n-1) = 6n(n-1)$$
$$n+1 = 6$$
$$n = 5$$

50. 4

51. $$\binom{n+2}{4} = 6\cdot\binom{n}{2}$$

$$\frac{(n+2)!}{4!(n+2-4)!} = 6\cdot\frac{n!}{2!(n-2)!} \quad \text{(Theorem 12)}$$

$$\frac{(n+2)!}{4!(n-2)!} = 6\cdot\frac{n!}{2!(n-2)!}$$

$$\frac{(n+2)!}{4!} = 6\cdot\frac{n!}{2!}$$

$$4!\cdot\frac{(n+2)!}{4!} = 4!\cdot 6\cdot\frac{n!}{2!}$$

$$(n+2)! = 72\cdot n!$$

$$(n+2)(n+1)n! = 72\cdot n!$$

$$(n+2)(n+1) = 72$$

$$n^2+3n+2 = 72$$

$$n^2+3n-70 = 0$$

$$(n+10)(n-7) = 0$$

$n + 10 = 0$ or $n - 7 = 0$

$n = -10$ or $n = 7$

The only solution is 7 since we cannot have a set of -10 objects.

52. 6

53. a) ${}_5C_2 = 10$

b) ${}_5C_2 - 5 = 5$

54. a) ${}_6C_2 = 15$

b) ${}_6C_2 - 6 = 9$

55. a) ${}_nC_2 = \frac{n(n-1)}{2}$

b) ${}_nC_2 - n = \frac{n(n-1)}{2} - \frac{2n}{2} = \frac{n(n-3)}{2}$, where $n = 4, 5, 6, \ldots$

c) Let D_n be the number of diagonals of an n-gon.

We must prove S_n (below) using mathematical induction.

We have

S_n: $D_n = \frac{n(n-3)}{2}$, for $n = 4, 5, 6, \ldots$

S_4: $D_4 = \frac{4\cdot 1}{2}$

S_k: $D_k = \frac{k(k-3)}{2}$

S_{k+1}: $D_{k+1} = \frac{(k+1)(k-2)}{2}$

1) Basis step: S_4 is true (a quadrilateral has 2 diagonals).

55. (continued)

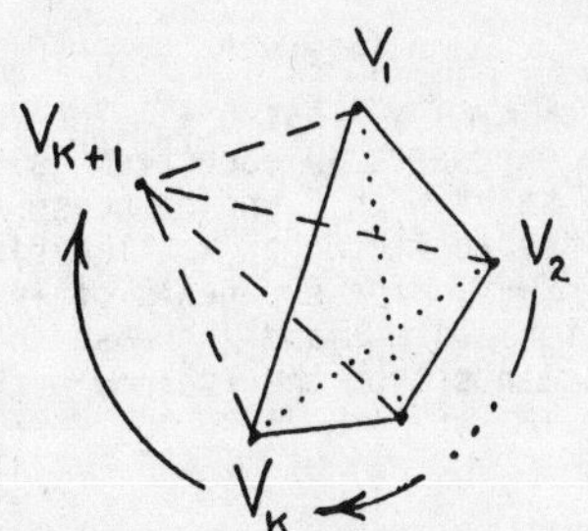

2) Induction step: Assume S_k. Observe that when an additional vertex V_{k+1} is added to the k-gon, we gain k segments, 2 of which are sides [of the (k + 1)-gon], and a former side $\overline{V_1V_k}$ becomes a diagonal.

Thus the additional number of diagonals is k - 2 + 1 or k - 1. Then the new total of diagonals is $D_k + (k - 1)$ or

$$D_{k+1} = D_k + (k-1)$$
$$= \frac{k(k-3)}{2} + (k-1) \quad (\text{by } S_k)$$
$$= \frac{(k+1)(k-2)}{2}.$$

56. $$\binom{n}{r-1} + \binom{n}{r}$$

$$= \frac{n!}{(r-1)!(n-r+1)!}\cdot\frac{r}{r} + \frac{n!}{r!(n-r)!}\cdot\frac{(n-r+1)}{(n-r+1)}$$

$$= \frac{n![r + (n-r+1)]}{r!\,(n-r+1)!}$$

$$= \frac{(n+1)!}{r!\,(n-r+1)!}$$

$$= \binom{n+1}{r}$$

Exercise Set 12.7

1. Pascal Method: Expand $(m + n)^5$.

The expansion of $(m + n)^5$ has 5 + 1, or 6, terms. The sum of the exponents in each term is 5. The exponents of m start with 5 and decrease to 0. The last term has no factor of m. The first term has no factor of n. The exponents of n start in the second term with 1 and increase to 5. We get the coefficients from the 6th row of Pascal's Triangle.

1
1 1
1 2 1
1 3 3 1
1 4 6 4 1
1 5 10 10 5 1

$$(m+n)^5 = 1\cdot m^5 + 5\cdot m^4n^1 + 10\cdot m^3\cdot n^2 + 10\cdot m^2\cdot n^3 + 5\cdot m\cdot n^4 + 1\cdot n^5$$
$$= m^5 + 5m^4n + 10m^3n^2 + 10m^2n^3 + 5mn^4 + n^5$$

2. $a^4 - 4a^3b + 6a^2b^2 - 4ab^3 + b^4$

3. Pascal Method: Expand $(x - y)^6$.

The expansion of $(x - y)^6$ has 6 + 1, or 7 terms. The sum of the exponents in each term is 6. The exponents of x start with 6 and decrease to 0. The last term has no factor of x. The first term has no factor of -y. The exponents of -y start in the second term with 1 and increase to 6. We get the coefficients from the 7th row of Pascal's Triangle.

1
1 1
1 2 1
1 3 3 1
1 4 6 4 1
1 5 10 10 5 1
1 6 15 20 15 6 1

$$\begin{aligned}(x - y)^6 &= 1\cdot x^6 + 6\cdot x^5\cdot(-y) + 15\cdot x^4\cdot(-y)^2 + \\ &\quad 20\cdot x^3\cdot(-y)^3 + 15\cdot x^2\cdot(-y)^4 + \\ &\quad 6\cdot x\cdot(-y)^5 + 1\cdot(-y)^6 \\ &= x^6 - 6x^5y + 15x^4y^2 - 20x^3y^3 + 15x^2y^4 - \\ &\quad 6xy^5 + y^6\end{aligned}$$

4. $p^7 + 7p^6q + 21p^5q^2 + 35p^4q^3 + 35p^3q^4 + 21p^2q^5 + 7pq^6 + q^7$

5. Pascal Method: Expand $(x^2 - 3y)^5$.

The expansion of $(x^2 - 3y)^5$ has 5 + 1, or 6, terms. The sum of the exponents in each term is 5. The exponents of x^2 start with 5 and decrease to 0. The last term has no factor of x^2. The first term has no factor of -3y. The exponents of -3y start in the second term with 1 and increase to 5. We get the coefficients from the 6th row of Pascal's Triangle.

1
1 1
1 2 1
1 3 3 1
1 4 6 4 1
1 5 10 10 5 1

$$\begin{aligned}(x^2 - 3y)^5 &= 1\cdot(x^2)^5 + 5\cdot(x^2)^4\cdot(-3y) + \\ &\quad 10\cdot(x^2)^3\cdot(-3y)^2 + 10\cdot(x^2)^2\cdot(-3y)^3 + \\ &\quad 5\cdot(x^2)\cdot(-3y)^4 + 1\cdot(-3y)^5 \\ &= x^{10} - 15x^8y + 90x^6y^2 - 270x^4y^3 + \\ &\quad 405x^2y^4 - 243y^5\end{aligned}$$

6. $2187c^7 - 5103c^6d + 5103c^5d^2 - 2835c^4d^3 + 945c^3d^4 - 189c^2d^5 + 21cd^6 - d^7$

7. Pascal Method: Expand $(3c - d)^6$.

The sum of the exponents in each term is 6. The exponents of 3c start with 6 and decrease to 0. The last term has no factor of 3c. The first term has no factor of -d. The exponents of -d start in the second term with 1 and increase to 6. We get the coefficients from the 7th row of Pascal's Triangle. The expansion has 6 + 1, or 7 terms.

1
1 1
1 2 1
1 3 3 1
1 4 6 4 1
1 5 10 10 5 1
1 6 15 20 15 6 1

$$\begin{aligned}(3c - d)^6 &= 1\cdot(3c)^6 + 6\cdot(3c)^5\cdot(-d) + \\ &\quad 15\cdot(3c)^4\cdot(-d)^2 + 20\cdot(3c)^3\cdot(-d)^3 + \\ &\quad 15\cdot(3c)^2\cdot(-d)^4 + 6\cdot(3c)\cdot(-d)^5 + \\ &\quad 1\cdot(-d)^6 \\ &= 3^6c^6 - 6\cdot 3^5c^5d + 15\cdot 3^4c^4d^2 - \\ &\quad 20\cdot 3^3c^3d^3 + 15\cdot 3^2c^2d^4 - 6\cdot 3cd^5 + d^6 \\ &= 729c^6 - 6\cdot 243c^5d + 15\cdot 81c^4d^2 - \\ &\quad 20\cdot 27c^3d^3 + 15\cdot 9c^2d^4 - 6\cdot 3cd^5 + d^6 \\ &= 729c^6 - 1458c^5d + 1215c^4d^2 - 540c^3d^3 + \\ &\quad 135c^2d^4 - 18cd^5 + d^6\end{aligned}$$

Preceding Coefficient Method: Expand $(3c - d)^6$.

We find each term in sequence. There are 7 terms in the expansion. When we get to the 4th coefficient, we can use symmetry to obtain the others.

The 1st term is $(3c)^6$.

The 2nd term is $\frac{1\cdot 6}{2-1}(3c)^5(-d)$, or $6(3c)^5(-d)$.

The 3rd term is $\frac{6\cdot 5}{3-1}(3c)^4(-d)^2$, or $15(3c)^4(-d)^2$.

The 4th term is $\frac{15\cdot 4}{4-1}(3c)^3(-d)^3$, or $20(3c)^3(-d)^3$.

The rest of the coefficients are 15, 6, and 1.

The complete expansion is

$$\begin{aligned}(3c - d)^6 &= (3c)^6 + 6(3c)^5(-d) + 15(3c)^4(-d)^2 + \\ &\quad 20(3c)^3(-d)^3 + 15(3c)^2(-d)^4 + \\ &\quad 6(3c)(-d)^5 + (-d)^6 \\ &= 729c^6 + 6(243c^5)(-d) + 15(81c^4)(d^2) + \\ &\quad 20(27c^3)(-d)^3 + 15(9c^2)(d^4) + \\ &\quad 6(3c)(-d)^5 + d^6 \\ &= 729c^6 - 1458c^5d + 1215c^4d^2 - 540c^3d^3 + \\ &\quad 135c^2d^4 - 18cd^5 + d^6\end{aligned}$$

8. $t^{-12} + 12t^{-10} + 60t^{-8} + 160t^{-6} + 240t^{-4} + 192t^{-2} + 64$

9. Pascal Method: Expand $(x - y)^3$.

The expansion of $(x - y)^3$ has 3 + 1, or 4 terms. The sum of the exponents in each term is 3. The exponents of x start with 3 and decrease to 0. The last term has no factor of x. The first term has no factor of -y. The exponents of -y start in the second term with 1 and increase to 3. We get the coefficients from the 4th row of Pascal's Triangle.

1

1 1

1 2 1

1 3 3 1

$$(x - y)^3 = 1\cdot x^3 + 3\cdot x^2\cdot(-y) + 3\cdot x\cdot(-y)^2 + 1\cdot(-y)^3$$
$$= x^3 - 3x^2y + 3xy^2 - y^3$$

Preceding Coefficient Method: Expand $(x - y)^3$.

We find each term in sequence. There are 4 terms in the expansion.

The 1st term is x^3.

The 2nd term is $\frac{1\cdot 3}{2-1}x^2(-y)$, or $3x^2(-y)$.

The 3rd term is $\frac{3\cdot 2}{3-1}x(-y)^2$, or $3x(-y)^2$.

The 4th term is $\frac{3\cdot 1}{4-1}(-y)^3$, or $(-y)^3$.

The complete expansion is
$(x - y)^3 = x^3 - 3x^2y + 3xy^2 - y^3$.

10. $x^5 - 5x^4y + 10x^3y^2 - 10x^2y^3 + 5xy^4 - y^5$

11. Preceding Coefficient Method: Expand $\left(\frac{1}{x} + y\right)^7$.

We find each term in sequence. There are 8 terms in the expansion. When we get to the 4th coefficient, we can use symmetry to obtain the others.

The 1st term is $\left(\frac{1}{x}\right)^7$.

The 2nd term is $\frac{1\cdot 7}{2-1}\left(\frac{1}{x}\right)^6 y$, or $7\left(\frac{1}{x}\right)^6 y$.

The 3rd term is $\frac{7\cdot 6}{3-1}\left(\frac{1}{x}\right)^5 y^2$, or $21\left(\frac{1}{x}\right)^5 y^2$.

The 4th term is $\frac{21\cdot 5}{4-1}\left(\frac{1}{x}\right)^4 y^3$, or $35\left(\frac{1}{x}\right)^4 y^3$.

The rest of the coefficients are 35, 21, 7 and 1.

The complete expansion is

$$\left(\frac{1}{x} + y\right)^7 = \left(\frac{1}{x}\right)^7 + 7\left(\frac{1}{x}\right)^6 y + 21\left(\frac{1}{x}\right)^5 y^2 + 35\left(\frac{1}{x}\right)^4 y^3 +$$
$$35\left(\frac{1}{x}\right)^3 y^4 + 21\left(\frac{1}{x}\right)^2 y^5 + 7\left(\frac{1}{x}\right) y^6 + y^7$$
$$= x^{-7} + 7x^{-6}y + 21x^{-5}y^2 + 35x^{-4}y^3 +$$
$$35x^{-3}y^4 + 21x^{-2}y^5 + 7x^{-1}y^6 + y^7$$

11. (continued)

Binomial Theorem Method: Expand $\left(\frac{1}{x} + y\right)^7$.

Note that $a = \frac{1}{x}$, $b = y$, and $n = 7$.

$$\left(\frac{1}{x} + y\right)^7 = \binom{7}{0}\left(\frac{1}{x}\right)^7 + \binom{7}{1}\left(\frac{1}{x}\right)^6 y + \binom{7}{2}\left(\frac{1}{x}\right)^5 y^2 +$$
$$\binom{7}{3}\left(\frac{1}{x}\right)^4 y^3 + \binom{7}{4}\left(\frac{1}{x}\right)^3 y^4 + \binom{7}{5}\left(\frac{1}{x}\right)^2 y^5 +$$
$$\binom{7}{6}\left(\frac{1}{x}\right) y^6 + \binom{7}{7} y^7$$
$$= \frac{7!}{7!0!}\left(\frac{1}{x}\right)^7 + \frac{7!}{6!1!}\left(\frac{1}{x}\right)^6 y + \frac{7!}{5!2!}\left(\frac{1}{x}\right)^5 y^2 +$$
$$\frac{7!}{4!3!}\left(\frac{1}{x}\right)^4 y^3 + \frac{7!}{3!4!}\left(\frac{1}{x}\right)^3 y^4 +$$
$$\frac{7!}{2!5!}\left(\frac{1}{x}\right)^2 y^5 + \frac{7!}{1!6!}\left(\frac{1}{x}\right) y^6 + \frac{7!}{0!7!} y^7$$
$$= x^{-7} + 7x^{-6}y + 21x^{-5}y^2 + 35x^{-4}y^3 +$$
$$35x^{-3}y^4 + 21x^{-2}y^5 + 7x^{-1}y^6 + y^7$$

12. $8s^3 - 36s^2t^2 + 54st^4 - 27t^6$

13. $\left(a - \frac{2}{a}\right)^9$

Let $a = a$, $b = -\frac{2}{a}$, and $n = 9$.

Expand using the Binomial Theorem Method.

$$\left(a - \frac{2}{a}\right)^9 = \binom{9}{0}a^9 + \binom{9}{1}a^8\left(-\frac{2}{a}\right) + \binom{9}{2}a^7\left(-\frac{2}{a}\right)^2 +$$
$$\binom{9}{3}a^6\left(-\frac{2}{a}\right)^3 + \binom{9}{4}a^5\left(-\frac{2}{a}\right)^4 +$$
$$\binom{9}{5}a^4\left(-\frac{2}{a}\right)^5 + \binom{9}{6}a^3\left(-\frac{2}{a}\right)^6 +$$
$$\binom{9}{7}a^2\left(-\frac{2}{a}\right)^7 + \binom{9}{8}a\left(-\frac{2}{a}\right)^8 +$$
$$\binom{9}{9}\left(-\frac{2}{a}\right)^9$$
$$= \frac{9!}{9!0!}a^9 + \frac{9!}{8!1!}a^8\left(-\frac{2}{a}\right) + \frac{9!}{7!2!}a^7\left(\frac{4}{a^2}\right) +$$
$$\frac{9!}{6!3!}a^6\left(-\frac{8}{a^3}\right) + \frac{9!}{5!4!}a^5\left(\frac{16}{a^4}\right) +$$
$$\frac{9!}{4!5!}a^4\left(-\frac{32}{a^5}\right) + \frac{9!}{3!6!}a^3\left(\frac{64}{a^6}\right) +$$
$$\frac{9!}{2!7!}a^2\left(-\frac{128}{a^7}\right) + \frac{9!}{1!8!}a\left(\frac{256}{a^8}\right) +$$
$$\frac{9!}{0!9!}\left(-\frac{512}{a^9}\right)$$
$$= a^9 - 9(2a^7) + 36(4a^5) - 84(8a^3) +$$
$$126(16a) - 126(32a^{-1}) + 84(64a^{-3}) -$$
$$36(128a^{-5}) + 9(256a^{-7}) - 512a^{-9}$$
$$= a^9 - 18a^7 + 144a^5 - 672a^3 + 2016a -$$
$$4032a^{-1} + 5376a^{-3} + 4608a^{-5} + 2304a^{-7} -$$
$$512a^{-9}$$

14. $512x^9 + 2304x^7 + 4608x^5 + 5376x^3 + 4032x + 2016x^{-1} + 672x^{-3} + 144x^{-5} + 18x^{-7} + x^{-9}$

15. $(1 - 1)^n$

Let a = 1 and b = -1.

Expand using the Binomial Theorem Method:

$$(1-1)^n = \binom{n}{0}1^n + \binom{n}{1}1^{n-1}(-1) + \binom{n}{2}1^{n-2}(-1)^2 + \binom{n}{3}1^{n-3}(-1)^3 + \ldots + \binom{n}{n}(-1)^n$$

$$= 1\cdot1 + n\cdot1\cdot(-1) + \binom{n}{2}\cdot1\cdot1 + \binom{n}{3}\cdot1\cdot(-1) + \ldots + \binom{n}{n}(-1)^n$$

$$= 1 - n + \binom{n}{2} - \binom{n}{3} + \ldots + \binom{n}{n}(-1)^n$$

16. $1 + 3n + \binom{n}{2}3^2 + \binom{n}{3}3^3 + \ldots + \binom{n}{n-2}3^{n-2} + \binom{n}{n-1}3^{n-1} + 3^n$

17. $(\sqrt{3} - t)^4$

Let $a = \sqrt{3}$, $b = -t$, and $n = 4$.

Expand using the Binomial Theorem Method.

$$(\sqrt{3} - t)^4 = \binom{4}{0}(\sqrt{3})^4 + \binom{4}{1}(\sqrt{3})^3(-t) + \binom{4}{2}(\sqrt{3})^2(-t)^2 + \binom{4}{3}(\sqrt{3})(-t)^3 + \binom{4}{4}(-t)^4$$

$$= 1\cdot9 + 4\cdot3\sqrt{3}\cdot(-t) + 6\cdot3\cdot t^2 + 4\cdot\sqrt{3}\cdot(-t^3) + 1\cdot t^4$$

$$= 9 - 12\sqrt{3}\,t + 18t^2 - 4\sqrt{3}\,t^3 + t^4$$

18. $125 + 150\sqrt{5}\,t + 375t^2 + 100\sqrt{5}\,t^3 + 75t^4 + 6\sqrt{5}\,t^5 + t^6$

19. $(\sqrt{2} + 1)^6 - (\sqrt{2} - 1)^6$

First, expand $(\sqrt{2} + 1)^6$.

$$(\sqrt{2}+1)^6 = \binom{6}{0}(\sqrt{2})^6 + \binom{6}{1}(\sqrt{2})^5(1) + \binom{6}{2}(\sqrt{2})^4(1)^2 + \binom{6}{3}(\sqrt{2})^3(1)^3 + \binom{6}{4}(\sqrt{2})^2(1)^4 + \binom{6}{5}(\sqrt{2})(1)^5 + \binom{6}{6}(1)^6$$

$$= \frac{6!}{6!0!}\cdot8 + \frac{6!}{5!1!}\cdot4\sqrt{2} + \frac{6!}{4!2!}\cdot4 + \frac{6!}{3!3!}\cdot2\sqrt{2} + \frac{6!}{2!4!}\cdot2 + \frac{6!}{1!5!}\cdot\sqrt{2} + \frac{6!}{0!6!}$$

$$= 8 + 24\sqrt{2} + 60 + 40\sqrt{2} + 30 + 6\sqrt{2} + 1$$

$$= 99 + 70\sqrt{2}$$

Next, expand $(\sqrt{2} - 1)^6$.

$$(\sqrt{2}-1)^6 = \binom{6}{0}(\sqrt{2})^6 + \binom{6}{1}(\sqrt{2})^5(-1) + \binom{6}{2}(\sqrt{2})^4(-1)^2 + \binom{6}{3}(\sqrt{2})^3(-1)^3 + \binom{6}{4}(\sqrt{2})^2(-1)^4 + \binom{6}{5}(\sqrt{2})(-1)^5 + \binom{6}{6}(-1)^6$$

$$= \frac{6!}{6!0!}\cdot8 - \frac{6!}{5!1!}\cdot4\sqrt{2} + \frac{6!}{4!2!}\cdot4 - \frac{6!}{3!3!}\cdot2\sqrt{2} + \frac{6!}{2!4!}\cdot2 - \frac{6!}{1!5!}\cdot\sqrt{2} + \frac{6!}{0!6!}$$

$$= 8 - 24\sqrt{2} + 60 - 40\sqrt{2} + 30 - 6\sqrt{2} + 1$$

$$= 99 - 70\sqrt{2}$$

$$(\sqrt{2}+1)^6 - (\sqrt{2}-1)^6 = (99 + 70\sqrt{2}) - (99 - 70\sqrt{2})$$

$$= 99 + 70\sqrt{2} - 99 + 70\sqrt{2}$$

$$= 140\sqrt{2}$$

20. 34

21. Expand $(x^{-2} + x^2)^4$ using the Binomial Theorem.

Note that $a = x^{-2}$, $b = x^2$, and $n = 4$.

$$(x^{-2} + x^2)^4 = \binom{4}{0}(x^{-2})^4 + \binom{4}{1}(x^{-2})^3(x^2) + \binom{4}{2}(x^{-2})^2(x^2)^2 + \binom{4}{3}(x^{-2})(x^2)^3 + \binom{4}{4}(x^2)^4$$

$$= \frac{4!}{4!0!}(x^{-8}) + \frac{4!}{3!1!}(x^{-6})(x^2) + \frac{4!}{2!2!}(x^{-4})(x^4) + \frac{4!}{1!3!}(x^{-2})(x^6) + \frac{4!}{0!4!}(x^8)$$

$$= x^{-8} + 4x^{-4} + 6x^0 + 4x^4 + x^8$$

$$= x^{-8} + 4x^{-4} + 6 + 4x^4 + x^8$$

22. $x^{-3} - 6x^{-2} + 15x^{-1} - 20 + 15x - 6x^2 + x^3$

23. Find the 3rd term of $(a + b)^6$.

First, we note that $3 = 2 + 1$, $a = a$, $b = b$, and $n = 6$. Then the 3rd term of the expansion of $(a + b)^6$ is

$\binom{6}{2}a^{6-2}b^2$, or $\frac{6!}{4!2!}a^4b^2$, or $15a^4b^2$.

24. $21x^2y^5$

25. Find the 12th term of $(a - 2)^{14}$.

First, we note that $12 = 11 + 1$, $a = a$, $b = -2$, and $n = 14$. Then the 12th term of the expansion of $(a - 2)^{14}$ is

$$\binom{14}{11} a^{14-11}\cdot(-2)^{11} = \frac{14!}{3!11!}a^3(-2048)$$

$$= 364a^3(-2048)$$

$$= -745{,}472a^3$$

26. $3{,}897{,}234x^2$

27. Find the 5th term of $(2x^3 - \sqrt{y})^8$.

First, we note that $5 = 4 + 1$, $a = 2x^3$, $b = -\sqrt{y}$, and $n = 8$. Then the 5th term of the expansion of $(2x^3 - \sqrt{y})^8$ is

$$\binom{8}{4}(2x^3)^{8-4}(-\sqrt{y})^4$$

$$= \frac{8!}{4!4!}(2x^3)^4(-\sqrt{y})^4$$

$$= 70(16x^{12})(y^2)$$

$$= 1120x^{12}y^2$$

28. $\frac{35}{27}b^{-5}$

29. The middle term of the expansion of $(2u - 3v^2)^{10}$ is the 6th term. Note that $6 = 5 + 1$, $a = 2u$, $b = -3v^2$, and $n = 10$. Then the 6th term of the expansion of $(2u - 3v^2)^{10}$ is

$$\binom{10}{5}(2u)^{10-5}(-3v^2)^5$$

$$= \frac{10!}{5!5!}(2u)^5(-3v^2)^5$$

$$= 252(32u^5)(-243v^{10})$$

$$= -1{,}959{,}552u^5v^{10}$$

30. $30x\sqrt{x}$, $30x\sqrt{3}$

31. The number of subsets is 2^7, or 128 (Theorem 17).

32. 2^6, or 64

33. The number of subsets is 2^{26}, or 67,108,864 (Theorem 17).

34. 2^{24}, or 16,777,216

35. We use the Binomial Theorem. Note that

$a = \sqrt{2}$, $b = -i$, and $n = 4$.

$$(\sqrt{2} - i)^4 = \binom{4}{0}(\sqrt{2})^4 + \binom{4}{1}(\sqrt{2})^3(-i) + \binom{4}{2}(\sqrt{2})^2(-i)^2 + \binom{4}{3}(\sqrt{2})(-i)^3 + \binom{4}{4}(-i)^4$$

$$= \frac{4!}{0!4!}(4) + \frac{4!}{1!3!}(2\sqrt{2})(-i) + \frac{4!}{2!2!}(2)(-1) + \frac{4!}{3!1!}(\sqrt{2})(i) + \frac{4!}{4!0!}(1)$$

$$= 4 - 8\sqrt{2}\,i - 12 + 4\sqrt{2}\,i + 1$$

$$= -7 - 4\sqrt{2}\,i$$

36. $-8i$

37. $$(\sin t - \csc t)^7 = \left(\sin t - \frac{1}{\sin t}\right)^7$$

$$= \binom{7}{0}\sin^7 t + \binom{7}{1}\sin^6 t\left(-\frac{1}{\sin t}\right) + \binom{7}{2}\sin^5 t\left(-\frac{1}{\sin t}\right)^2 + \binom{7}{3}\sin^4 t\left(-\frac{1}{\sin t}\right)^3 + \binom{7}{4}\sin^3 t\left(-\frac{1}{\sin t}\right)^4 + \binom{7}{5}\sin^2 t\left(-\frac{1}{\sin t}\right)^5 + \binom{7}{6}\sin t\left(-\frac{1}{\sin t}\right)^6 + \binom{7}{7}\left(-\frac{1}{\sin t}\right)^7$$

$$= \sin^7 t - 7\sin^5 t + 21\sin^3 t - 35\sin t + 35\left(\frac{1}{\sin t}\right) - 21\left(\frac{1}{\sin^3 t}\right) + 7\left(\frac{1}{\sin^5 t}\right) - \frac{1}{\sin^7 t}, \text{ or}$$

$$\sin^7 t - 7\sin^5 t + 21\sin^3 t - 35\sin t + 35\csc t - 21\csc^3 t + 7\csc^5 t - \csc^7 t$$

38. $\tan^{11}\theta + 11\tan^9\theta + 55\tan^7\theta + 165\tan^5\theta + 330\tan^3\theta + 462\tan\theta + 462\cot\theta + 330\cot^3\theta + 165\cot^5\theta + 55\cot^7\theta + 11\cot^9\theta + \cot^{11}\theta$

39. $(a-b)^n = \binom{n}{0}a^n + \binom{n}{1}a^{n-1}(-b) + \binom{n}{2}a^{n-2}(-b)^2 + \cdots + \binom{n}{n}(-b)^n$

$= \binom{n}{0}a^n + \binom{n}{1}(-1)a^{n-1}b + \binom{n}{2}(-1)^2a^{n-2}b^2 + \cdots + \binom{n}{n}(-1)^nb^n$

$= \sum_{r=0}^{n}\binom{n}{r}(-1)^ra^{n-r}b^r$

40. $\sum_{r=1}^{n}\binom{n}{r}x^{n-r}h^{r-1}$

41. $\sum_{r=0}^{8}\binom{8}{r}x^{8-r}3^r = 0$

The left side of the equation is sigma notation for the binomial series $(x+3)^8$, so we have:

$(x+3)^8 = 0$

$x + 3 = 0$ (Taking the 8th root on both sides)

$x = -3$

42. $-5 + 2\sqrt{2}$

43. $\sum_{r=0}^{5}\binom{5}{r}(-1)^rx^{5-r}3^r = \sum_{r=0}^{5}\binom{5}{r}x^{5-r}(-3)^r$, so the left side of the equation is sigma notation for the binomial series $(x-3)^5$. We have:

$(x-3)^5 = 32$

$x - 3 = 2$ (Taking the 5th root on both sides)

$x = 5$

44. $\frac{3\pi}{2} + 2k\pi$, k an integer

45. Find the third term of $(0.313 + 0.687)^5$:

$\binom{5}{2}(0.313)^{5-2}(0.687)^2 = \frac{5!}{3!2!}(0.313)^3(0.687)^2 \approx 0.14473$

46. $\binom{8}{5}(0.15)^3(0.85)^5 \approx 0.08386$

47. Find and add the 3rd through 6th terms of $(0.313 + 0.687)^5$:

$\binom{5}{2}(0.313)^3(0.687)^2 + \binom{5}{3}(0.313)^2(0.687)^3 + \binom{5}{4}(0.313)(0.687)^4 + \binom{5}{5}(0.687)^5 \approx 0.96403$

48. $\binom{8}{6}(0.15)^2(0.85)^6 + \binom{8}{7}(0.15)(0.85)^7 + \binom{8}{8}(0.85)^8 \approx 0.89479$

49. There are 11 terms in the expression of $(8u + 3v^2)^{10}$, so the 6th term is the middle term.

$\binom{10}{5}(8u)^5(3v^2)^5 = \frac{10!}{5!5!}(32{,}768u^5)(243v^{10}) = 2{,}006{,}581{,}248u^5v^{10}$

50. $30x\sqrt{x}$, $-30x\sqrt{3}$

51. The (r + 1)st term of $\left(\frac{3x^2}{2} - \frac{1}{3x}\right)^{12}$ is $\binom{12}{r}\left(\frac{3x^2}{2}\right)^{12-r}\left(-\frac{1}{3x}\right)^r$. In the term which does not contain x, the exponent of x in the numerator is equal to the exponent of x in the denominator.

$2(12 - r) = r$

$24 - 2r = r$

$24 = 3r$

$8 = r$

Find the (8 + 1)st, or 9th term:

$\binom{12}{8}\left(\frac{3x^2}{2}\right)^4\left(-\frac{1}{3x}\right)^8 = \frac{12!}{4!8!}\left(\frac{3^4x^8}{2^4}\right)\left(\frac{1}{3^8x^8}\right) = \frac{55}{144}$

52. $-4320x^6y^{9/2}$

53. $\dfrac{\binom{5}{3}(p^2)^2\left(-\frac{1}{2}p\sqrt[3]{q}\right)^3}{\binom{5}{2}(p^2)^3\left(-\frac{1}{2}p\sqrt[3]{q}\right)^2} = \dfrac{-\frac{1}{8}p^7q}{\frac{1}{4}p^8\sqrt[3]{q^2}} = -\dfrac{\sqrt[3]{q}}{2p}$

54. $-35x^{-1/6}$, or $-\frac{35}{x^{1/6}}$

55. The term of highest degree of $(x^5 + 3)^4$ is the first term, or

$\binom{4}{0}(x^5)^{4-0}3^0 = \frac{4!}{4!0!}x^{20} = x^{20}$.

Therefore, the degree of $(x^5 + 3)^4$ is 20.

56. 127

57. Familiarize: Let x, $x + 1$, $x + 2$, and $x + 3$ represent the integers.

Translate: We write an equation.

$x^3 + (x + 1)^3 + (x + 2)^3 = (x + 3)^3$

Carry out: We solve the equation.

$x^3 + x^3 + 3x^2 + 3x + 1 + x^3 + 6x^2 + 12x + 8 =$

$x^3 + 9x^2 + 27x + 27$

$2x^3 - 12x - 18 = 0$

$x^3 - 6x - 9 = 0$ $\quad$ [Multiplying by $\frac{1}{2}$]

Using the rational roots theorem and synthetic division, we see that 3 is a root of the equation.

$(x - 3)(x^2 + 3x + 3) = 0$

The quadratic factor does not have real number zeros, so the only real number solution is 3.

Check: If $x = 3$, then the integers are 3, 4, 5, and 6. It is true that $3^3 + 4^3 + 5^3 = 6^3$. The values check.

State: The integers are 3, 4, 5, and 6.

58. $[\log_a (xt)]^{23}$

59. $\ln \left[\sum_{r=0}^{15} \binom{15}{r}(\cos t)^{30-2r}(\sin t)^{2r}\right]$

$= \ln \left[\sum_{r=0}^{15} \binom{15}{r}(\cos^2 t)^{15-r}(\sin^2 t)^{r}\right]$

$= \ln (\cos^2 t + \sin^2 t)^{15} = \ln (1)^{15} = \ln 1 = 0$

60. Let S_k be the statement of the binomial theorem with n replaced by k. Multiply both sides of S_k by $(a + b)$ to obtain

$(a + b)^{k+1}$

$= \left[a^k + \cdots + \binom{k}{r-1}a^{k-(r-1)}\, b^{r-1} + \right.$

$\left.\binom{k}{r}a^{k-r}\, b^r + \cdots + b^k\right](a + b)$

$= a^{k+1} + \cdots + \left[\binom{k}{r-1} + \binom{k}{r}\right]a^{(k+1)-r}\, b^r +$

$\cdots + b^{k+1}$

$= a^{k+1} + \cdots + \binom{k+1}{r}a^{(k+1)-r}\, b^r + \cdots + b^{k+1}.$

This proves S_{k+1}, assuming S_k. S_1 is true. Hence S_n is true for $n = 1, 2, 3, \ldots$.

Exercise Set 12.8

1. 100 surveyed
   57 wore either glasses or contacts
   43 wore neither glasses nor contacts

   Using Principle P,

   P(wearing either glasses or contacts) = $\frac{57}{100}$, or 0.57

   P(wearing neither glasses nor contacts) = $\frac{43}{100}$, or 0.43

2. 0.18, 0.24, 0.23, 0.23, 0.12; Opinions might vary, but it would seem that people tend not to pick the first or last numbers.

3. 1044 Total letters in the three paragraphs
   78 A's occurred
   140 E's occurred
   60 I's occurred
   74 O's occurred
   31 U's occurred

   Using Principle P,

   P(the occurrence of letter A) = $\frac{78}{1044} \approx 0.075$

   P(the occurrence of letter E) = $\frac{140}{1044} \approx 0.134$

   P(the occurrence of letter I) = $\frac{60}{1044} \approx 0.057$

   P(the occurrence of letter O) = $\frac{74}{1044} \approx 0.071$

   P(the occurrence of letter U) = $\frac{31}{1044} \approx 0.030$

4. 0.367

5. 1044 Total letters
   383 Total vowels
   661 Total consonants

   Using Principle P,

   P(the occurrence of a consonant) = $\frac{661}{1044} \approx 0.633$

6. Z; 0.999

7. Let D represent the number of deer in the preserve. If there are D deer in the preserve and 318 of them are tagged, then the probability that a deer is tagged is $\frac{318}{D}$. Later, 168 deer are caught of which 56 were tagged. The ratio of deer tagged to deer caught is $\frac{56}{168}$. We assume the two ratios are the same. We solve the proportion for D.

$$\frac{318}{D} = \frac{56}{168}$$
$$168 \cdot 318 = D \cdot 56$$
$$53{,}424 = 56D$$
$$954 = D$$

Thus we estimate that there are 954 deer in the preserve.

8. 287

9. There are 52 equally likely outcomes.

10. $\frac{4}{52}$, or $\frac{1}{13}$

11. Since there are 52 equally likely outcomes and there are 13 ways to obtain a heart, by Principle P we have

P(drawing a heart) $= \frac{13}{52} = \frac{1}{4}$.

12. $\frac{13}{52}$, or $\frac{1}{4}$

13. Since there are 52 equally likely outcomes and there are 4 ways to obtain a 4, by Principle P we have

P(drawing a 4) $= \frac{4}{52} = \frac{1}{13}$.

14. $\frac{26}{52}$, or $\frac{1}{2}$

15. Since there are 52 equally likely outcomes and there are 26 ways to obtain a black card, by Principle P we have

P(drawing a black card) $= \frac{26}{52} = \frac{1}{2}$.

16. $\frac{2}{13}$

17. Since there are 52 equally likely outcomes and there are 8 ways to obtain a 9 or a king (four 9's and four kings), we have, by Principle P,

P(drawing a 9 or a king) $= \frac{8}{52} = \frac{2}{13}$.

18. $\frac{2}{7}$

19. Since there are 14 equally likely ways of selecting a marble from a bag containing 4 red marbles and 10 green marbles, we have, by Principle P,

P(selecting a green marble) $= \frac{10}{14} = \frac{5}{7}$.

20. 0

21. There are 14 equally likely ways of selecting any marble from a bag containing 4 red marbles and 10 green marbles. Since the bag does not contain any white marbles, there are 0 ways of selecting a white marble. By Principle P, we have

P(selecting a white marble) $= \frac{0}{14} = 0$.

22. $\frac{{}_{13}C_4}{{}_{52}C_4} = \frac{11}{4165}$

23. The number of ways of drawing 4 cards from a deck of 52 cards is ${}_{52}C_4$. Now 13 of the 52 cards are hearts, so the number of ways of drawing 4 hearts is ${}_{13}C_4$. Thus,

P(getting 4 hearts) $= \frac{{}_{13}C_4}{{}_{52}C_4} = \dfrac{\frac{13!}{4!\cdot 9!}}{\frac{52!}{4!\cdot 48!}}$

$$= \frac{13\cdot 12\cdot 11\cdot 10\cdot 9!}{4!\cdot 9!} \cdot \frac{4!\cdot 48!}{52\cdot 51\cdot 50\cdot 49\cdot 48!}$$

$$= \frac{13\cdot 12\cdot 11\cdot 10}{52\cdot 51\cdot 50\cdot 49} = \frac{11}{4165}$$

24. $\frac{{}_8C_2 \cdot {}_6C_2}{{}_{14}C_4} = \frac{60}{143}$

25. The number of ways of selecting 4 people from a group of 15 is ${}_{15}C_4$. Two men can be selected in ${}_8C_2$ ways, and 2 women can be selected in ${}_7C_2$ ways. By the Fundamental Counting Principle, the number of ways of selecting 2 men and 2 women is ${}_8C_2 \cdot {}_7C_2$. Thus,

P(2 men and 2 women are chosen) $= \frac{{}_8C_2 \cdot {}_7C_2}{{}_{15}C_4}$

$$= \dfrac{\frac{8!}{2!6!} \cdot \frac{7!}{2!5!}}{\frac{15!}{4!11!}} = \dfrac{\frac{8\cdot 7\cdot 6!}{2\cdot 1\cdot 6!} \cdot \frac{7\cdot 6\cdot 5!}{2\cdot 1\cdot 5!}}{\frac{15\cdot 14\cdot 13\cdot 12\cdot 11!}{4\cdot 3\cdot 2\cdot 1\cdot 11!}}$$

$$= \frac{28\cdot 21}{15\cdot 7\cdot 13} = \frac{28}{65}$$

26. $\frac{5}{36}$

27. On each die there are 6 possible outcomes. The outcomes are paired so there are 6·6, or 36 possible ways in which the two can fall. The pairs that total 3 are (1, 2) and (2, 1). Thus there are 2 possible ways of getting a total of 3, so the probability is $\frac{2}{36}$, or $\frac{1}{18}$.

28. $\frac{1}{36}$

29. On each die there are 6 possible outcomes. The outcomes are paired so there are 6·6, or 36 possible ways in which the two can fall. There is only 1 way of getting a total of 12, the pair (6, 6), so the probability is $\frac{1}{36}$.

30. $\frac{{}_5C_2 \cdot {}_8C_2 \cdot {}_7C_1}{{}_{20}C_5} = \frac{245}{1938}$

31. The number n of ways of getting 6 coins from a bag containing 20 coins is ${}_{20}C_6$. Three nickels can be selected in ${}_6C_3$ ways since the bag contains 6 nickels. Two dimes can be selected in ${}_{10}C_2$ ways since the bag contains 10 dimes. One quarter can be selected in ${}_4C_1$ ways since the bag contains 4 quarters. By the Fundamental Counting Principle the number of ways of selecting 3 nickels, 2 dimes, and 1 quarter is ${}_6C_3 \cdot {}_{10}C_2 \cdot {}_4C_1$. Thus,

P(getting 3 nickels, 2 dimes, and 1 quarter)

$$= \frac{{}_6C_3 \cdot {}_{10}C_2 \cdot {}_4C_1}{{}_{20}C_6} = \frac{\frac{6!}{3!3!} \cdot \frac{10!}{2!8!} \cdot \frac{4!}{1!3!}}{\frac{20!}{6 \cdot 14!}}$$

$$= \frac{\frac{6 \cdot 5 \cdot 4}{3 \cdot 2 \cdot 1} \cdot \frac{10 \cdot 9}{2 \cdot 1} \cdot \frac{4}{1}}{\frac{20 \cdot 19 \cdot 18 \cdot 17 \cdot 16 \cdot 15}{6 \cdot 5 \cdot 4 \cdot 3 \cdot 2 \cdot 1}} = \frac{20 \cdot 45 \cdot 4}{19 \cdot 17 \cdot 8 \cdot 15} = \frac{30}{323}$$

32. $\frac{18}{38}$, or $\frac{9}{19}$

33. The roulette wheel contains 38 equally likely slots. Eighteen of the 38 slots are colored red. Thus, by Principle P,

P(the ball falls in a red slot) $= \frac{18}{38} = \frac{9}{19}$

34. $\frac{36}{38}$, or $\frac{18}{19}$

35. The roulette wheel contains 38 equally likely slots. Only 1 slot is numbered 00. Then, by Principle P,

P(the ball falls in the 00 slot) $= \frac{1}{38}$.

36. $\frac{1}{38}$

37. The roulette wheel contains 38 equally likely slots, 2 of which are numbered 00 and 0. Thus, using Principle P,

$P\begin{bmatrix}\text{the ball falling in either}\\ \text{the 00 or the 0 slot}\end{bmatrix} = \frac{2}{38} = \frac{1}{19}$.

38. $\frac{18}{38}$, or $\frac{9}{19}$

39. ${}_{52}C_5 = \frac{52!}{5!47!} = \frac{52 \cdot 51 \cdot 50 \cdot 49 \cdot 48 \cdot 47!}{5 \cdot 4 \cdot 3 \cdot 2 \cdot 1 \cdot 47!}$

$= 26 \cdot 17 \cdot 10 \cdot 49 \cdot 12$

$= 2{,}598{,}960$

40. a) 4

b) $\frac{4}{{}_{52}C_5} = \frac{4}{2{,}598{,}960} \approx 1.54 \times 10^{-6}$

41. Consider a suit

A K Q J 10 9 8 7 6 5 4 3 2

A straight flush can be any of the following combinations in the same suit.

K Q J 10 9
Q J 10 9 8
J 10 9 8 7
10 9 8 7 6
9 8 7 6 5
8 7 6 5 4
7 6 5 4 3
6 5 4 3 2
5 4 3 2 A

Remember a straight flush does not include A K Q J 10 which is a royal flush.

a) Since there are 9 straight flushes per suit, there are 36 straight flushes in all 4 suits.

b) Since 2,598,960, or ${}_{52}C_5$, poker hands can be dealt from a standard 52-card deck and 36 of those hands are straight flushes, the probability of getting a straight flush is $\frac{36}{2{,}598{,}960}$, or 0.0000139.

42. a) 13·48 = 624

b) $\frac{13 \cdot 48}{{}_{52}C_5} = \frac{624}{2{,}598{,}960} \approx 2.4 \times 10^{-4}$

43. a) $[13 \cdot {}_4C_3] \cdot [12 \cdot {}_4C_2] = 3744$; that is, there are 13 ways to select a denomination. Then from that denomination there are ${}_4C_3$ ways to pick 3 of the 4 cards in that denomination. Now there are 12 ways to select any one of the remaining 12 denominations and ${}_4C_2$ ways to pick 2 cards from the 4 cards in that denomination.

b) $\frac{3744}{{}_{52}C_5} = \frac{3744}{2{,}598{,}960} \approx 0.00144$

44. a) $\left[13 \cdot \binom{4}{2}\right] \cdot \left[\binom{12}{3} \cdot \binom{4}{1}\binom{4}{1}\binom{4}{1}\right]$

$= 1,098,240$

b) $\frac{1,098,240}{_{52}C_5} = \frac{1,098,240}{2,598,960} \approx 0.423$

45. a) $\left[13 \cdot \binom{4}{3}\right] \cdot \left[\binom{48}{2}\right] - 3744 = 54,912.$

There are 13 ways to select a denomination and then $\binom{4}{3}$ ways to pick 3 cards from the 4 in this denomination. Now there are $\binom{48}{2}$ ways to pick 2 cards from the remaining 12 denominations (48 cards); but $13\binom{4}{3}\binom{48}{2}$ includes the 3744 hands (like QQQ-10-10) in a full house. Subtracting 3744, we have the above result.

b) $\frac{54,912}{_{52}C_5} = \frac{54,912}{2,598,960} \approx 0.0211$

46. a) $4 \cdot \binom{13}{5} - 4 - 36 = 5108$

b) $\frac{5108}{_{52}C_5} = \frac{5108}{2,598,960} \approx 0.00197$

47. a) $\binom{13}{2}\binom{4}{2}\binom{4}{2}\binom{44}{1} = 123,552$

There are $\binom{13}{2}$ ways to select 2 denominations from 13 denominations; then there are $\binom{4}{2}$ ways to pick 2 of the 4 cards in either of these 2 denominations, and $\binom{4}{2}$ ways to pick 2 of the 4 cards in the other of these 2 denominations. Finally, there are $\binom{44}{1}$ ways to select the fifth card from the remaining 44.

b) $\frac{123,552}{_{52}C_5} = \frac{123,552}{2,598,960} \approx 0.0475$

48. a) $\binom{10}{1}\binom{4}{1}\binom{4}{1}\binom{4}{1}\binom{4}{1}\binom{4}{1} - 4 - 36$

$= 10,200$

b) $\frac{10,200}{_{52}C_5} = \frac{10,200}{2,598,960} \approx 0.00392$

49. a) Area of n circles of radius r $= n \cdot \pi r^2$.
Area of n squares of side 2r $= n(2r)^2$.

Then $p = \frac{n \cdot \pi r^2}{n \cdot (2r)^2} = \frac{\pi}{4}$.

b) $1 - \frac{\pi}{4} = \frac{4 - \pi}{4}$

c) $\pi = 4p$

Since $p \approx \frac{78}{100}$, $\pi \approx 4(0.78) = 3.12$.